ENGINEERING CIRCUIT ANALYSIS

ENGINEERING CIRCUIT ANALYSIS

TENTH EDITION

William H. Hayt, Jr.
Purdue University

Jack E. Kemmerly
California State University, Fullerton

Jamie D. Phillips
University of Delaware

Steven M. Durbin
Western Michigan University

ENGINEERING CIRCUIT ANALYSIS

Published by McGraw Hill LLC, 1325 Avenue of the Americas, New York, NY 10019.

This book is printed on acid-free paper.

9 CPI 28 27 26 25

ISBN 978-1-266-26249-4
MHID 1-266-26249-0

Cover Image: *Timofeev Vladimir/Shutterstock*

To Sean and Kristi
The best part of every day.

ABOUT THE AUTHORS

WILLIAM H. HAYT, JR., received his B.S. and M.S. at Purdue University and his Ph.D. from the University of Illinois. After spending four years in industry, Professor Hayt joined the faculty of Purdue University, where he served as Professor and Head of the School of Electrical Engineering, and as Professor Emeritus after retiring in 1986. Besides *Engineering Circuit Analysis*, Professor Hayt authored three other texts, including *Engineering Electromagnetics*, now in its eighth edition with McGraw Hill. Professor Hayt's professional society memberships included Eta Kappa Nu, Tau Beta Pi, Sigma Xi, Sigma Delta Chi, Fellow of IEEE, ASEE, and NAEB. While at Purdue, he received numerous teaching awards, including the university's Best Teacher Award. He is also listed in Purdue's Book of Great Teachers, a permanent wall display in the Purdue Memorial Union, dedicated on April 23, 1999. The book bears the names of the inaugural group of 225 faculty members, past and present, who have devoted their lives to excellence in teaching and scholarship. They were chosen by their students and their peers as Purdue's finest educators.

JACK E. KEMMERLY received his B.S. magna cum laude from The Catholic University of America, M.S. from University of Denver, and Ph.D. from Purdue University. Professor Kemmerly first taught at Purdue University and later worked as principal engineer at the Aeronutronic Division of Ford Motor Company. He then joined California State University, Fullerton, where he served as Professor, Chairman of the Faculty of Electrical Engineering, Chairman of the Engineering Division, and Professor Emeritus. Professor Kemmerly's professional society memberships included Eta Kappa Nu, Tau Beta Pi, Sigma Xi, ASEE, and IEEE (Senior Member). His pursuits outside of academe included being an officer in the Little League and a scoutmaster in the Boy Scouts.

JAMIE PHILLIPS received his B.S., M.S., and Ph.D. degrees in Electrical Engineering from the University of Michigan, Ann Arbor, Michigan. He was a postdoctoral researcher at Sandia National Laboratories in Albuquerque, New Mexico, and a research scientist at the Rockwell Science Center in Thousand Oaks, California, before returning to the University of Michigan as a faculty member in the EECS Department in 2002. At the University of Michigan, Prof. Phillips taught and developed numerous courses in circuits and semiconductor devices spanning from first-year undergraduate courses to advanced graduate courses. There he received several teaching honors including the University Undergraduate Teaching Award and an Arthur F. Thurnau Professorship recognizing faculty for outstanding contributions to undergraduate education. In 2020, he joined the University of Delaware as a Professor and Chair of the Electrical and Computer Engineering Department. His research interests are on semiconductor optoelectronic devices with particular emphasis on infrared detectors and photovoltaics and engineering education. His professional memberships include IEEE (Senior Member), Eta Kappa Nu, Optica, Materials Research Society, Tau Beta Pi, and ASEE.

STEVEN M. DURBIN received the B.S., M.S. and Ph.D. degrees in Electrical Engineering from Purdue University, West Lafayette, Indiana. Subsequently, he was with the Department of Electrical Engineering at Florida State University and Florida A&M University before joining the University of Canterbury, New Zealand, in 2000. In 2010, he moved to the University at Buffalo, The State University of New York, where he held a joint tenured appointment between the Departments of Electrical Engineering and Physics. Since 2013, he has been with the Department of Electrical and Computer Engineering at Western Michigan University, where he is Professor of Electrical and Computer Engineering, and has served in a variety of administrative roles. His teaching interests include circuits, electronics, electromagnetics, solid-state electronics and nanotechnology. His research interests are primarily concerned with the development of novel electronic materials—in particular those based on oxide and nitride compounds—as well as studying disorder in both naturally occurring and synthetic systems as a route to physical understanding and engineering of properties. He is a founding principal investigator of the MacDiarmid Institute for Advanced Materials and Nanotechnology, a New Zealand National Centre of Research Excellence, and coauthor of over 100 technical publications, including several patents and patents pending. He is a senior member of the IEEE, and a member of Eta Kappa Nu, Tau Beta Pi, the Materials Research Society, the AVS (formerly the American Vacuum Society), the Society of Amateur Radio Astronomers, the American Physical Society, and the Royal Society of New Zealand. In his spare time, he enjoys building acoustic and electric guitars, and solid-top ukuleles.

BRIEF CONTENTS

CONTENTS

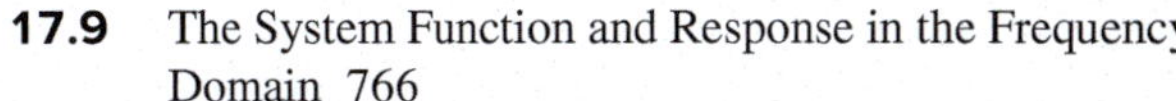

PREFACE

The target audience colors everything about a book, being a major factor in decisions big and small, particularly both the pace and the overall writing style. Consequently, it is important to note that the authors have made the conscious decision to write this book to the **student,** and not to the instructor. Our underlying philosophy is that reading the book should be enjoyable, despite the level of technical detail that it must incorporate. When we look back to the very first edition of *Engineering Circuit Analysis*, it's clear that it was developed specifically to be more of a conversation than a dry, dull discourse on a prescribed set of fundamental topics. To keep it conversational, we've had to work hard at updating the book so that it continues to speak to the increasingly diverse group of students using it all over the world.

Although in many engineering programs the introductory circuits course is preceded or accompanied by an introductory physics course in which electricity and magnetism are introduced (typically from a fields perspective), this is not required to use this book. After finishing the course, many students find themselves truly amazed that such a broad set of analytical tools have been derived from **only three simple scientific laws**—Ohm's law and Kirchhoff's voltage and current laws. The first six chapters assume only a familiarity with algebra and simultaneous equations; subsequent chapters assume a first course in calculus (derivatives and integrals) is being taken in tandem. Beyond that, we have tried to incorporate sufficient details to allow the book to be read on its own.

So, what key features have been designed into this book with the student in mind? First, individual chapters are organized into relatively short subsections, each having a single primary topic. The language has been updated to remain informal and to flow smoothly. Color is used to highlight important information as opposed to merely improve the aesthetics of the page layout, and white space is provided for jotting down short notes and questions. New terms are defined as they are introduced, and examples are placed strategically to demonstrate not only basic concepts, but problem-solving approaches as well. Practice problems relevant to the examples are placed in proximity so that students can try out the techniques for themselves before attempting the end-of-chapter exercises. The exercises represent a broad range of difficulties, generally ordered from simpler to more complex, and grouped according to the relevant section of each chapter.

Engineering is an intensive subject to study, and students often find themselves faced with deadlines and serious workloads. This does not mean that textbooks have to be dry and pompous, however, or that coursework should never contain any element of fun. In fact, successfully solving a problem often *is* fun, and learning how to do that can be fun as well. Determining how to best accomplish this within the context of a textbook is an ongoing process. The authors have always relied on the often very candid feedback received from our own students at Purdue University; the California State University, Fullerton; Fort Lewis College in Durango; the joint engineering program at Florida A&M University and Florida State University; the University of Canterbury (New Zealand); the University at Buffalo, and Western Michigan University. We also rely on comments, corrections, and suggestions from instructors and students worldwide.

The first edition of *Engineering Circuit Analysis* was written by Bill Hayt and Jack Kemmerly, two engineering professors who very much enjoyed teaching, interacting with their students, and training generations of future engineers. It was well received due to its compact structure, "to the point" informal writing style, and logical organization. There is no timidity when it comes to presenting the theory underlying a specific topic, or pulling punches when developing mathematical expressions. Everything, however, was carefully designed to assist students in their learning, present things in a straightforward fashion, and leave theory for theory's sake to other books. They clearly put a great deal of thought into writing the book, and their enthusiasm for the subject comes across to the reader.

KEY FEATURES OF THE TENTH EDITION

Great care has been taken to retain key features from the ninth edition which were clearly working well. These include the general layout and sequence of chapters, the basic style of both the text and line drawings, the use of four-color printing where appropriate, numerous worked examples and related practice problems, and grouping of end-of-chapter exercises according to section. Transformers continue to merit their own chapter, and complex frequency is briefly introduced through a student-friendly extension of the phasor technique, instead of indirectly by merely stating the Laplace transform integral. We also have retained the use of icons, an idea first introduced in the sixth edition:

Provides a heads-up to common mistakes and/or potential pitfalls;

Indicates a point that's worth noting;

Denotes a design problem to which there is no unique answer;

Indicates a problem which requires computer-aided analysis.

Indicates an Example that reinforces the flow chart illustrating a typical problem-solving methodology that is presented in Chapter 1.

Circuit analysis is a robust method for training engineering students to think analytically, step-by-step, and returning to check their answers. A flow chart illustrating a typical problem-solving methodology is presented in Chapter 1; these steps are explicitly included in one example in each of the subsequent chapters to reinforce the concept.

The introduction of engineering-oriented analysis and design software in the book has been done with the mind-set that it should assist, not replace, the learning process. Consequently, the computer icon denotes problems that are typically phrased such that the software is used to *verify* answers, and not simply provide them. Both MATLAB® and LTspice® are used in this context.

SPECIFIC CHANGES FOR THE TENTH EDITION INCLUDE:

- New and revised end-of-chapter exercises
- New figures and photos
- Updated screen captures and text descriptions of computer-aided analysis software, and continued use of LTspice as freeware software that is available natively on both Windows and Mac OS platforms
- Updated worked examples and practice problems
- Updates to the Practical Application feature, introduced to help students connect material in each chapter to broader concepts in engineering. Topics include distortion in amplifiers, circuits to measure an electrocardiogram, practical aspects of grounding, resistivity, and the memristor, sometimes called "the missing element"
- Streamlining of text, especially in the worked examples, to get to the point faster

DIGITAL RESOURCES

Proctorio
Remote Proctoring & Browser-Locking Capabilities

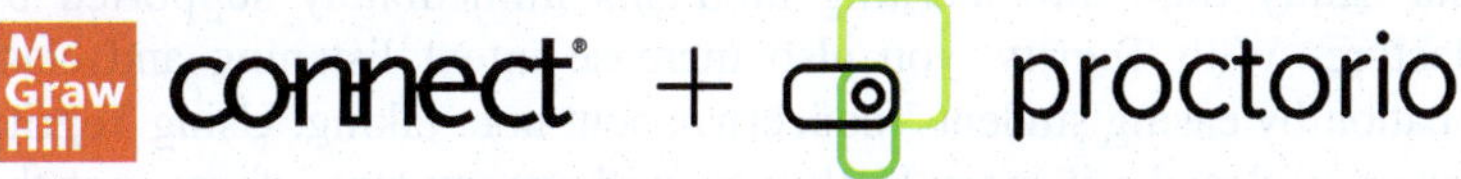

Remote proctoring and browser-locking capabilities, hosted by Proctorio within Connect, provide control of the assessment environment by enabling security options and verifying the identity of the student.

Seamlessly integrated within Connect, these services allow instructors to control the assessment experience by verifying identification, restricting browser activity, and monitoring student actions.

Instant and detailed reporting gives instructors an at-a-glance view of potential academic integrity concerns, thereby avoiding personal bias and supporting evidence-based claims.

ReadAnywhere®

Read or study when it's convenient for you with McGraw Hill's free ReadAnywhere® app. Available for iOS or Android smartphones or tablets, ReadAnywhere gives users access to McGraw Hill tools including the eBook and SmartBook® 2.0 or Adaptive Learning Assignments in Connect. Take notes, highlight, and complete assignments offline—all of your work will sync when you open the app with Wi-Fi access. Log in with your McGraw Hill Connect username and password to start learning—anytime, anywhere!

OLC-Aligned Courses Implementing High-Quality Instruction and Assessment through Preconfigured Courseware

In consultation with the Online Learning Consortium (OLC) and our certified Faculty Consultants, McGraw Hill has created pre-configured courseware using OLC's quality scorecard to align with best practices in online course delivery. This turnkey courseware contains a combination of formative assessments, summative assessments, homework, and application activities, and can easily be customized to meet an individual instructor's needs and desired course outcomes. For more information, visit https://www.mheducation.com/highered/olc.

Tegrity: Lectures 24/7

Tegrity in Connect is a tool that makes class time available 24/7 by automatically capturing every lecture. With a simple one-click start-and-stop process, you capture all computer screens and corresponding audio in a format that is easy to search, frame by frame. Students can replay any part of any class with easy-to-use, browser-based viewing on a PC, Mac, or mobile device.

Educators know that the more students can see, hear, and experience class resources, the better they learn. In fact, studies prove it. Tegrity's unique search feature helps students efficiently find what they need, when they need it, across an entire semester of class recordings. Help turn your students' study time into learning moments immediately supported by your lecture. With Tegrity, you also increase intent listening and class participation by easing students' concerns about note-taking. Using Tegrity in Connect will make it more likely you will see students' faces, not the tops of their heads.

Writing Assignment

Available within Connect and Connect Master, the Writing Assignment tool delivers a learning experience to help students improve their written communication skills and conceptual understanding. As an instructor, you can assign, monitor, grade, and provide feedback on writing more efficiently and effectively.

Create Your Book, Your Way

McGraw Hill's Content Collections Powered by Create® is a self-service website that enables instructors to create custom course materials—print and eBooks—by drawing upon McGraw Hill's comprehensive, cross-disciplinary content. Choose what you want from our high-quality textbooks, articles, and cases. Combine it with your content quickly and easily, and tap into other rights-secured, third-party content such as readings, cases, and articles. Content can be arranged in a way that makes the most sense for your course, and you can include the course name and information as well. Choose the best format for your course: color print, black-and-white print, or eBook. The eBook can be included in your Connect course and is available on the free ReadAnywhere® app for smartphone or tablet access as well. When you are finished customizing, you will receive a free digital copy to review in just minutes! Visit McGraw Hill Create®—www.mcgrawhillcreate.com—today and begin building!

Reflecting the Diverse World Around Us

McGraw Hill believes in unlocking the potential of every learner at every stage of life. To accomplish that, we are dedicated to creating products that reflect, and are accessible to, all the diverse, global customers we serve. Within McGraw Hill, we foster a culture of belonging, and we work with partners who share our commitment to equity, inclusion, and diversity in all forms. In McGraw Hill Higher Education, this includes, but is not limited to, the following:

- Refreshing and implementing inclusive content guidelines around topics including generalizations and stereotypes, gender, abilities/disabilities, race/ethnicity, sexual orientation, diversity of names, and age.
- Enhancing best practices in assessment creation to eliminate cultural, cognitive, and affective bias.
- Maintaining and continually updating a robust photo library of diverse images that reflect our student populations.
- Including more diverse voices in the development and review of our content.
- Strengthening art guidelines to improve accessibility by ensuring meaningful text and images are distinguishable and perceivable by users with limited color vision and moderately low vision.

ACKNOWLEDGMENTS

I joined the book as a co-author in 1999, and sadly never had the opportunity to speak to either Bill or Jack about the revision process. I count myself lucky to have taken a circuits course from Bill Hayt while I was a student at Purdue. He was, without a doubt, a truly excellent instructor.

I am very grateful to the many people at McGraw Hill who assisted with this project and helped make it a reality, especially Heather Ervolino, Product Developer extraordinaire, and an amazing person to work with.

I am also grateful to Jamie Phillips, now at the University of Delaware, who joined the author team for the ninth edition and who made many valuable contributions. I would also like to thank the following people for providing technical suggestions and/or photographs: Prof. Damon Miller of Western Michigan University, Prof. Masakazu Kobayashi of Waseda University, Dr. Wade Enright, Prof. Pat Bodger, Prof. Rick Millane, and Mr. Gary Turner of the University of Canterbury, Prof. Richard Blaikie of the University of Otago, and Profs. Reginald Perry and Jim Zheng of Florida A&M University and the Florida State University.

Additionally, I would like to thank the reviewers of the tenth edition text for their suggestions and valued feedback:

Andrew Affrunti
The University of Colorado

Mary C Baker
Texas Tech University

Mark M Budnik
Valparaiso University

MVS Chandrashekhar
University of South Carolina

Somnath Chattopadhyay
California State University Northridge

Yinchao Chen
University of South Carolina

Dorina Cornea-Hasegan
Portland Community College

Hoo Kim
LeTourneau University

Michael Maassel
North Dakota State University

Joseph Martel-Foley
Wentworth Institute of Technology

Katharine Polasek
Hope College

Emily Porter
University of Texas at Austin

Mohandas Raj
Cascadia College

Joseph Vincent Rispoli
Purdue University

Mostafa Shiva
California State University, Fullerton

Marcelo Godoy Simoes
Colorado School of Mines

Raman Menon Unnikrishnan
California State University Fullerton

Farah Vandrevala
University at Buffalo

Paul Villeneuve
University of Maine

Kun Yao
University of Georgia

Finally, I would like to briefly thank several other people who have contributed both directly and indirectly to the tenth edition: First and foremost, my wife, Kristi, and our son, Sean, for their patience, understanding, support, welcome distractions, and helpful advice. Throughout the day, it has always been a pleasure to talk to friends and colleagues about what should be taught, how it should be taught, and how to measure learning. In particular, Martin Allen, Richard Blaikie, Steve Carr, Peter Cottrell, Wade Enright, Jeff Gray, Mike Hayes, Bill Kennedy, Susan Lord, Bob

Makin, Philippa Martin, Chris McConville, Damon Miller, Reginald Perry, Joan Redwing, Roger Reeves, Dick Schwartz, Leonard Tung, Jim Zheng, and many others have provided me with many useful insights, as did my father, Jesse Durbin, an electrical engineering graduate of the Indiana Institute of Technology.

Steven M. Durbin, Kalamazoo, Michigan

Instructors

The Power of Connections

A complete course platform

Connect enables you to build deeper connections with your students through cohesive digital content and tools, creating engaging learning experiences. We are committed to providing you with the right resources and tools to support all your students along their personal learning journeys.

65%
Less Time Grading

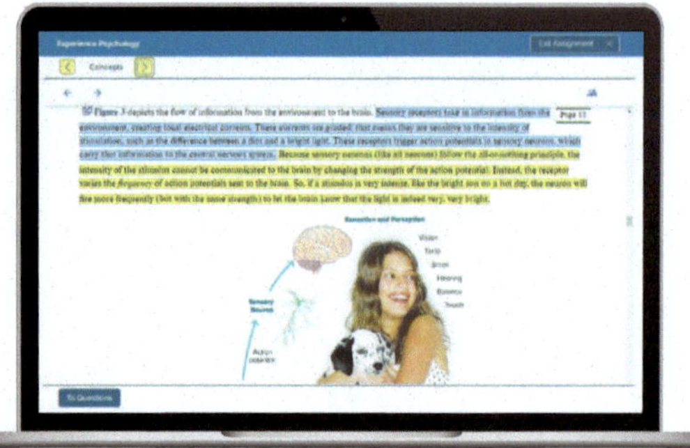

Laptop: Getty Images; Woman/dog: George Doyle/Getty Images

Every learner is unique

In Connect, instructors can assign an adaptive reading experience with SmartBook® 2.0. Rooted in advanced learning science principles, SmartBook 2.0 delivers each student a personalized experience, focusing students on their learning gaps, ensuring that the time they spend studying is time well-spent.
mheducation.com/highered/connect/smartbook

Affordable solutions, added value

Make technology work for you with LMS integration for single sign-on access, mobile access to the digital textbook, and reports to quickly show you how each of your students is doing. And with our Inclusive Access program, you can provide all these tools at the lowest available market price to your students. Ask your McGraw Hill representative for more information.

Solutions for your challenges

A product isn't a solution. Real solutions are affordable, reliable, and come with training and ongoing support when you need it and how you want it. Visit **supportateverystep.com** for videos and resources both you and your students can use throughout the term.

Students

Get Learning that Fits You

Effective tools for efficient studying

Connect is designed to help you be more productive with simple, flexible, intuitive tools that maximize your study time and meet your individual learning needs. Get learning that works for you with Connect.

Study anytime, anywhere

Download the free ReadAnywhere® app and access your online eBook, SmartBook® 2.0, or Adaptive Learning Assignments when it's convenient, even if you're offline. And since the app automatically syncs with your Connect account, all of your work is available every time you open it. Find out more at **mheducation.com/readanywhere**

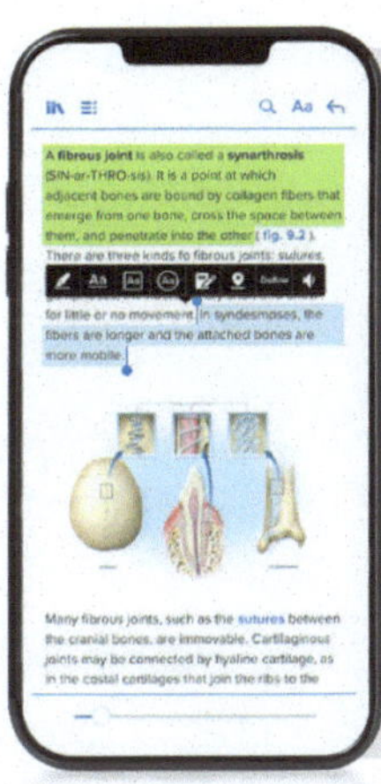

iPhone: Getty Images

"I really liked this app—it made it easy to study when you don't have your text-book in front of you."

- Jordan Cunningham, Eastern Washington University

Everything you need in one place

Your Connect course has everything you need—whether reading your digital eBook or completing assignments for class—Connect makes it easy to get your work done.

Learning for everyone

McGraw Hill works directly with Accessibility Services Departments and faculty to meet the learning needs of all students. Please contact your Accessibility Services Office and ask them to email accessibility@mheducation.com, or visit **mheducation.com/about/accessibility** for more information.

CHAPTER 1

Introduction

KEY CONCEPTS

Linear versus Nonlinear Circuits

Four Main Categories of Circuit Analysis:

- DC
- Transient
- Sinusoidal
- Frequency Domain

Circuit Analysis Beyond Circuits

Analysis and Design

Use of Engineering Software

A Problem-Solving Strategy

PREAMBLE

All engineers have a great deal in common, regardless of their specialty. This is particularly evident when it comes to problem-solving methodology. In fact, many practicing engineers find it is possible to work in a large variety of settings and even outside their traditional field, as their skill set is often transferable to other environments. Today's engineering graduates find themselves employed in a broad range of jobs, from design of individual components and systems, to solving socioeconomic problems such as air and water pollution, urban planning, communication, medical treatments, mass transportation, power generation and distribution, and efficient use and conservation of natural resources.

Circuit analysis has long been a traditional introduction to the **art of problem solving** from an engineering perspective, even for those whose interests lie outside electrical engineering. There are many reasons for this, but one of the best is that in today's world it's extremely unlikely for any engineer to encounter a system that does not in some way include electrical circuitry. As circuits shrink and require less power, and power sources become smaller and cheaper, embedded circuits are seemingly everywhere. Since most engineering situations require a team effort at some stage, a working knowledge of circuit analysis therefore helps to provide everyone on a project with the background needed for effective communication.

Consequently, this book is not just about "circuit analysis" from an engineering perspective it is also about developing basic problem-solving skills as they apply to situations an engineer is likely to encounter. Along the way, we also find that we're

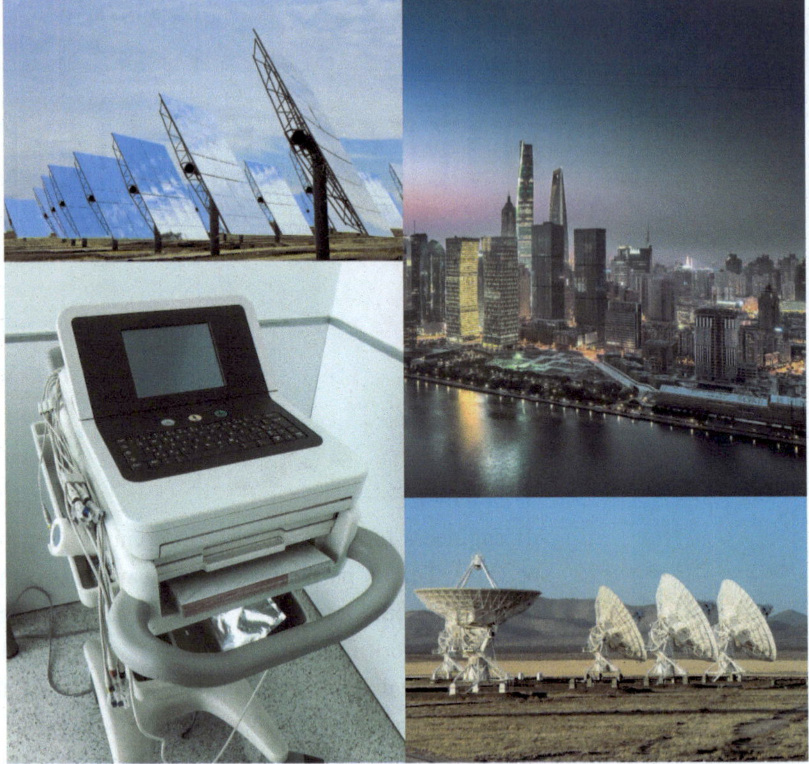

Not all electrical engineers routinely make use of circuit analysis, but they often bring to bear analytical and problem-solving skills learned early on in their careers. A circuit analysis course is one of the first exposures to such concepts.
Solar Mirrors: Darren Baker/Shutterstock; Skyline: Eugene Lu/Shutterstock; Electrocardiogram machine: David Prado Perucha/Shutterstock; Dish: Jonathan Larsen/iStock/Getty Images

developing an intuitive understanding at a general level, and often we can understand a complex system by its analogy to an electrical circuit. Before launching into all this, however, we should begin with a quick preview of the topics found in the remainder of the book, pausing briefly to ponder the difference between analysis and design, and the evolving role computer tools play in modern engineering.

1.1 OVERVIEW OF TEXT

The fundamental subject of this text is *linear circuit analysis,* which sometimes prompts a few readers to ask,

"Is there such a thing *nonlinear* circuit analysis?"

Sure! We encounter nonlinear circuits every day: they capture and decode signals for our TVs and radios, perform calculations hundreds of millions (even billions) of times a second inside microprocessors, convert speech into electrical signals for transmission over fiber-optic cables as well as cellular networks, and execute many other functions outside our field of view. In designing, testing, and implementing such nonlinear circuits, detailed analysis is unavoidable.

Flat panel displays include nonlinear circuits. Many of them, however, can be understood, designed, and analyzed with the assistance of linear models.
sdecoret/Shutterstock

"Then why study *linear* circuit analysis?"

you might ask. An excellent question. The simple fact of the matter is that no physical system (including electrical circuits) is ever perfectly linear. Fortunately for us, however, a great many systems behave in a reasonably linear fashion over a limited range—allowing us to model them as linear systems if we keep the range limitations in mind.

For example, consider the common function

$$f(x) = e^x$$

A linear approximation to this function is

$$f(x) \approx 1 + x$$

Let's test this out. Table 1.1 shows both the exact value and the approximate value of $f(x)$ for a range of x. Interestingly, the linear approximation is exceptionally accurate up to about $x = 0.1$, when the relative error is still less than 1%. Although many engineers are rather quick on a calculator, it's hard to argue that any approach is faster than just adding 1.

TABLE 1.1 Comparison of a Linear Model for e^x to Exact Value

x	$f(x)$*	$1 + x$	Relative Error**
0.0001	1.0001	1.0001	0.0000005%
0.001	1.0010	1.001	0.00005%
0.01	1.0101	1.01	0.005%
0.1	1.1052	1.1	0.5%
1.0	2.7183	2.0	26%

*Quoted to four significant figures.

**Relative error $\triangleq \left| 100 \times \frac{e^x - (1+x)}{e^x} \right|$

Linear problems are inherently more easily solved than their nonlinear counterparts. For this reason, we often seek reasonably accurate linear approximations (or *models*) to physical situations. Furthermore, linear models are easily manipulated and understood—making the design process more straightforward.

The circuits we will encounter in subsequent chapters all represent linear approximations to physical electric circuits. Where appropriate, brief discussions of potential inaccuracies or limitations to these models are provided, but generally speaking we find them to be suitably accurate for most applications. When greater accuracy is required in practice, nonlinear models are employed, but with a considerable increase in solution complexity. A detailed discussion of what constitutes a *linear electric circuit* can be found in Chap. 2.

Linear circuit analysis can be separated into four broad categories: (1) *dc analysis,* where the energy sources do not change with time; (2) *transient analysis,* where things often change quickly; (3) *sinusoidal analysis,* which applies to both ac power and signals; and (4) *frequency response,* which is the most general of the four categories, but typically assumes something is changing with time. We begin our journey with the topic of resistive circuits, such as an automotive rear window defogger.

Modern trains are powered by electric motors. Their electrical systems are best analyzed using ac or phasor analysis techniques.
Dr. Masakazu Kobayashi

Frequency-dependent circuits lie at the heart of many electronic devices, and they can be a great deal of fun to design.
prykhodov/123RF

This provides us with a perfect opportunity to learn a number of powerful engineering circuit analysis techniques, such as *nodal analysis, mesh analysis, superposition, source transformation, Thévenin's theorem, Norton's theorem,* and several methods for simplifying networks of components connected in series or parallel. The single most redeeming feature of resistive circuits is that the time dependence of any quantity of interest does not affect our analysis procedure. In other words, if asked for an electrical quantity of a resistive circuit at several specific instants in time, we do not need to analyze the circuit more than once. As a result, we will spend most of our initial effort considering dc circuits—those circuits whose electrical parameters do not vary with time.

Although dc circuits are undeniably important in everyday life, things are often much more interesting when something happens suddenly. In circuit analysis parlance, we refer to *transient analysis* as the suite of techniques used to study circuits that are suddenly energized or de-energized. To make such circuits interesting, we need to add elements that respond to the rate of change of electrical quantities, leading to circuit equations that include derivatives and integrals. Fortunately, we can obtain such equations using the simple techniques learned in the first part of our study.

Still, not all time-varying circuits are turned on and off suddenly. Air conditioners, fans, and lighting fixtures are only a few of the many examples we may see daily. In such situations, a calculus-based approach for every analysis can become tedious and time-consuming. Fortunately, there is a better alternative for situations where equipment has been allowed to run long enough for transient effects to die out, and this is commonly referred to as ac or sinusoidal analysis, or sometimes *phasor analysis*.

The final leg of our journey deals with a subject known as *frequency response*. Working directly with the differential equations obtained in time-domain analysis helps us develop an intuitive understanding of the operation of circuits containing energy storage elements (e.g., capacitors and inductors). As we shall see, however, circuits with even a relatively small number of components can be somewhat onerous to analyze, and much more straightforward methods have been developed. These methods, which include Laplace and Fourier analysis, allow us to transform differential equations into algebraic equations. Such methods also enable us to design circuits to respond in specific ways to particular frequencies. We make use of frequency-dependent circuits every day when we use a mobile phone, select our favorite radio station, or connect to the Internet.

1.2 RELATIONSHIP OF CIRCUIT ANALYSIS TO ENGINEERING

It is worth noting that there are several layers to the concepts under study in this text. Beyond the nuts and bolts of circuit analysis techniques lies the opportunity to develop a methodical approach to problem solving, the ability to determine the goal or goals of a particular problem, skill at collecting the information needed to effect a solution, and, perhaps equally importantly, opportunities for practice at verifying solution accuracy.

Students familiar with the study of other engineering topics such as fluid flow, automotive suspension systems, bridge design, supply chain management, or robotics will recognize the general form of many of the

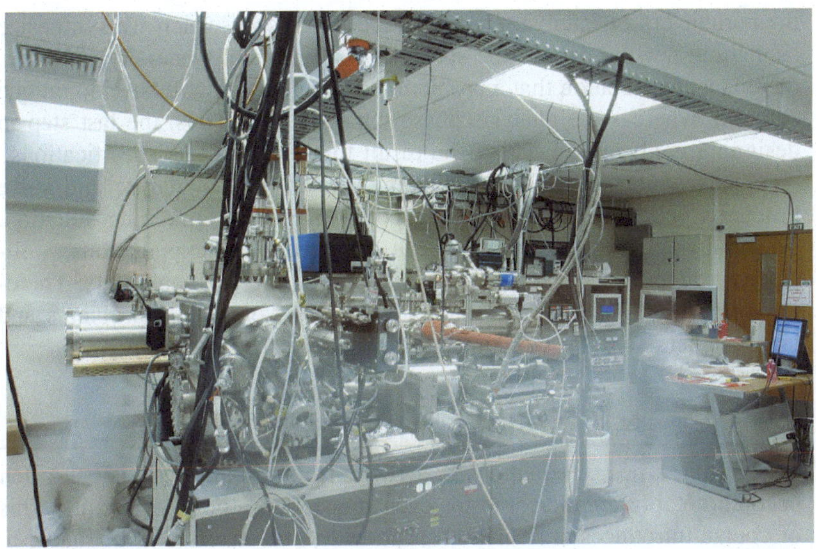

A molecular beam epitaxy crystal growth facility. The equations governing its operation closely resemble those used to describe simple linear circuits.
Steve Durbin/McGraw Hill

equations we develop to describe the behavior of various circuits. We simply need to learn how to "translate" the relevant variables (e.g., replacing *voltage* with *force, charge* with *distance, resistance* with *friction coefficient,* distance, or resistance with friction coefficient) to find that we already know how to work a new type of problem. Very often, if we have previous experience in solving a similar or related problem, our intuition can guide us through the solution of a totally new problem. An additional benefit to being able to recognize the form common to several problems in different fields is that engineers often cross traditional boundaries between disciplines by utilizing that knowledge and knowing how to apply it.

What we are about to learn regarding linear circuit analysis forms the basis for many subsequent electrical engineering courses. The study of electronics relies on the analysis of circuits with devices known as diodes and transistors, which are used to construct power supplies, amplifiers, and digital circuits. The skills which we will develop are typically applied in a rapid, methodical fashion by electronics engineers, who sometimes can analyze a complicated circuit without even reaching for a pencil. The time-domain and frequency-domain chapters of this text lead directly into discussions of signal processing, power transmission, control theory, and communications. We find that frequency-domain analysis in particular is an extremely powerful technique, easily applied to any physical system subjected to time-varying excitation, and particularly helpful in the design of filters.

An example of a robotic manipulator. The feedback control system can be modeled using linear circuit elements to determine situations in which the operation may become unstable.
THINK A/Shutterstock

1.3 ANALYSIS AND DESIGN

Engineers take a fundamental understanding of scientific principles, combine this with practical knowledge often expressed in mathematical terms, and (frequently with considerable creativity) arrive at a solution to a given problem. ***Analysis*** is the process through which we determine the scope of a problem, obtain the information required to understand it, and compute

Two proposed designs for a next-generation space shuttle. Although both contain similar elements, each is unique. NASA Dryden Flight Research Center

the parameters of interest. ***Design*** is the process by which we synthesize something new as part of the solution to a problem. Generally speaking, there is an expectation that a problem requiring design will have no unique solution, whereas the analysis phase typically will. Thus, the last step in designing is always analyzing the result to see if it meets specifications.

This text is focused on developing our ability to analyze and solve problems because it is the starting point in every engineering situation. The philosophy of this book is that we need clear explanations, well-placed examples, and plenty of practice to develop such an ability. Therefore, elements of design are integrated into end-of-chapter problems and later chapters so as to be enjoyable rather than distracting.

1.4 COMPUTER-AIDED ANALYSIS

Solving the types of equations that result from circuit analysis can often become notably cumbersome for even moderately complex circuits. This of course introduces an increased probability that errors will be made, in addition to considerable time in performing the calculations. The desire to find a tool to help with this process actually predates electronic computers, with purely mechanical computers such as the Analytical Engine designed by Charles Babbage in the 1880s proposed as possible solutions. Perhaps the earliest successful electronic computer designed for solution of differential equations was the 1940s-era ENIAC, whose vacuum tubes filled a large room. With the advent of low-cost desktop computers, however, computer-aided circuit analysis has developed into an invaluable everyday tool which has become an integral part of not only analysis but design as well. All of today's computer chips are first designed and analyzed using computer simulations based on a set of known physical rules, which are typically combined with empirical data to account for "real world" performance characteristics. Once the simulations show desired results, the design is then used to provide the information needed to fabricate the real circuit or system. Without computer-aided analysis and design, this process would be nearly impossible, as today's chips contain millions of devices in a single circuit!

One of the most powerful aspects of computer-aided design is the relatively recent integration of multiple programs in a fashion transparent to the user. This allows the circuit to be drawn schematically on the screen, reduced automatically to the format required by an analysis program (such as SPICE, introduced in Chap. 4), and the resulting output smoothly transferred to a third program capable of plotting various electrical quantities of interest that describe the operation of the circuit. Once the engineer is satisfied with the simulated performance of the design, the same software can generate the printed circuit board layout using geometrical parameters in the components library. This level of integration is continually increasing, to the point where it is now possible to draw a schematic, click a few buttons, and walk to the other side of the table to pick up a manufactured version of the circuit, ready to test!

The reader should be wary, however, of one thing. Circuit analysis software, although fun to use, is by no means a replacement for good old-fashioned paper-and-pencil analysis. We need to have a solid understanding of how circuits work in order to develop an ability to design

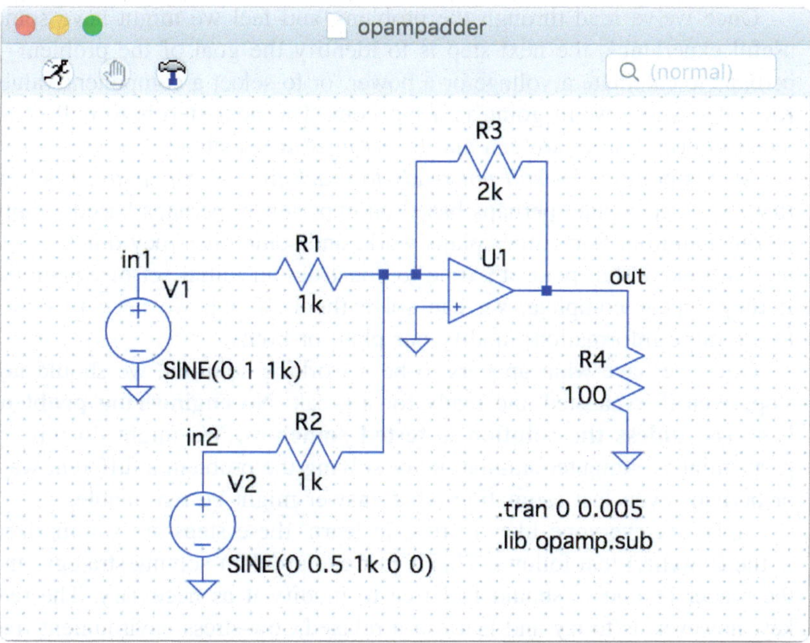

An amplifier circuit drawn using a commercial schematic capture software package.

them. Simply going through the motions of running a particular software package is a little like playing the lottery: with user-generated entry errors, hidden default parameters in the myriad of menu choices, and the occasional shortcoming of human-written code, there is no substitute for having at least an approximate idea of the expected behavior of a circuit. Then, if the simulation result does not agree with expectations, we can find the error early, rather than after it's too late.

Still, computer-aided analysis is a powerful tool. It allows us to vary parameter values and evaluate the change in circuit performance, and to consider several variations to a design in a straightforward manner. The result is a reduction of repetitive tasks, and more time to concentrate on engineering details.

1.5 SUCCESSFUL PROBLEM-SOLVING STRATEGIES

As the reader might have picked up, this book is just as much about problem solving as it is about circuit analysis. During your time as an engineering student, the expectation is that you are learning how to solve problems—just at this moment, those skills are not yet fully developed. As you proceed through your course of study, you will pick up techniques that work for you, and likely continue to do so as a practicing engineer.

By far the most common difficulty encountered by engineering students is *not knowing how to start* a problem. This improves with experience, but early on that's of no help. The best advice we can give is to adopt a methodical approach, beginning with reading the problem statement slowly and carefully (and more than once, if needed). Since experience usually gives us some type of insight into how to deal with a specific problem, worked examples appear throughout the book. Rather than just read them, however, it might be helpful to work through them with a pencil and a piece of paper.

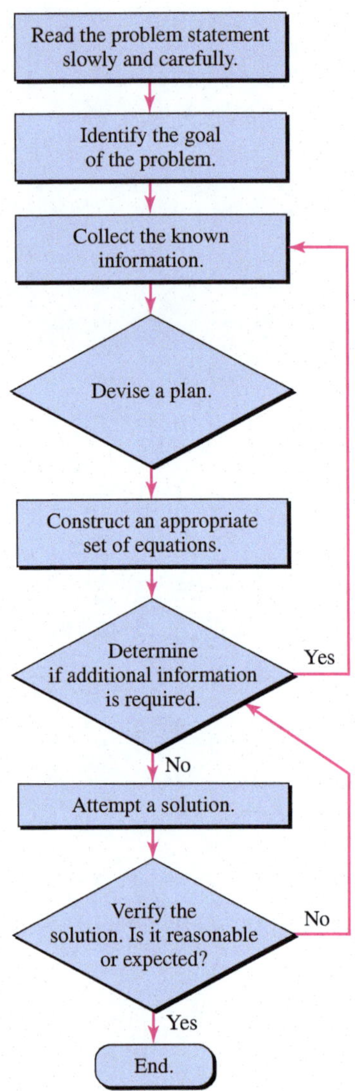

Once we've read through the problem, and feel we might have some useful experience, the next step is to identify the goal of the problem—perhaps to calculate a voltage or a power, or to select a component value. Knowing where we're going is a big help. The next step is to collect as much information as we can and to organize it somehow.

At this point *we're still not ready to reach for the calculator.* It's best first to devise a plan, perhaps based on experience, perhaps based simply on our intuition. Sometimes plans work, and sometimes they don't. Starting with our initial plan, it's time to construct an initial set of equations. If they appear complete, we can solve them. If not, we need to either locate more information, modify our plan, or both.

Once we have what appears to be a working solution, we should not stop, even if exhausted and ready for a break. **No engineering problem is solved unless the solution is tested somehow.** We might do this by performing a computer simulation, or solving the problem a different way, or perhaps even just estimating what answer might be reasonable.

Since not everyone likes to read to learn, these steps are summarized in the flowchart that follows. This is just one problem-solving strategy, and the reader of course should feel free to modify it as necessary. The real key, however, is to try and learn in a relaxed, low-stress environment free of distractions. Experience is the best teacher, and learning from our own mistakes will always be part of the process of becoming a skilled engineer.

READING FURTHER

This relatively inexpensive classic book teaches the reader how to develop winning strategies in the face of seemingly impossible problems:

G. Polya, *How to Solve It.* Princeton, N.J.: Princeton University Press, 1971.

EXERCISES

1. Consider the function $f(\theta) = 3\theta$ as an approximation to the function $g(\theta) = 3 \sin \theta$. Restricting your analysis to the range $0 \le \theta \le \pi/2$, determine the value of θ at which the relative error between f and g exceeds 1%. In this context, we define relative error as $100 \times \left| \frac{f(\theta) - g(\theta)}{g(\theta)} \right|$.

2. The function $\frac{1}{1-x}$ can be approximated with the linear function $g(x) = 1 + x$. For what value(s) of x does the linear approximation result in 1% relative error?

3. One common circuit element capable of storing energy is known as the capacitor. When discharging its stored energy, the voltage V_C across the device can be expressed as $V_C = V_0(1 - e^{-t/\tau})$, where V_0 is the voltage just prior to the start of the discharge process, t is the time, and τ is known as the "time constant." To avoid errors, t and τ should be measured in the same units (seconds, milliseconds, etc.).

   (*a*) Rearrange the expression to solve for t, resulting in an expression containing a natural logarithm.

   (*b*) Approximating the natural logarithm as $\ln x \approx x - 1$, determine the relative error of the approximation at $V_C/V_0 =$ (a) 0.10; (b) 0.50.

CHAPTER 2

Basic Components and Electric Circuits

KEY CONCEPTS

Basic Electrical Quantities and Associated Units: Charge, Current, Voltage, and Power

Current Direction and Voltage Polarity

The Passive Sign Convention for Calculating Power

Ideal Voltage and Current Sources

Dependent Sources

Resistance and Ohm's Law

INTRODUCTION

In conducting circuit analysis, we often find ourselves seeking specific *currents, voltages*, or *powers*, so here we begin with a brief description of these quantities. In terms of components that can be used to build electrical circuits, we have quite a few from which to choose. We initially focus on the *resistor*, a simple passive component, and a range of idealized active sources of voltage and current. As we move forward, new components will be added to the inventory to allow more complex (and useful) circuits to be considered.

A quick word of advice before we begin: Pay close attention to the role of "+" and "−" signs when labeling voltages, and the significance of the arrow in defining current: they often make the difference between wrong and right answers.

2.1 UNITS AND SCALES

In order to state the value of some measurable quantity, we must give both a *number* and a *unit*, such as "3 meters." Fortunately, we all use the same number system. This is not true for units, and a little time must be spent in becoming familiar with a suitable system. We must agree on a standard unit and be assured of its permanence and its general acceptability. The standard unit of length, for example, should not be defined in terms of the distance between two marks on a certain rubber band; this is not permanent, and furthermore everybody else is using another standard.

The most frequently used system of units is the International System of Units (abbreviated ***SI*** in all languages), adopted by the General Conference on Weights and Measures in 1960. Modified several times since, the SI is built upon seven basic units: the

meter, kilogram, second, ampere, kelvin, mole, and *candela* (see Table 2.1). This is a "metric system," some form of which is now in common use in most countries of the world, although it is not yet widely used in the United States, despite being adopted by the U.S. National Bureau of Standards in 1964, used by all major professional engineering societies, and the language in which textbook and professional publications are written. Units for other quantities such as volume, force, energy, etc., are derived from the seven base units.

There is some inconsistency regarding whether units named after a person should be capitalized. Here, we will adopt the most contemporary convention,[1,2] where such units are written out in lowercase (e.g., watt, joule), but abbreviated with an uppercase symbol (e.g., W, J).

(1) H. Barrell, *Nature* 220, 1968, p. 651.
(2) V. N. Krutikov, T. K. Kanishcheva, S. A. Kononogov, L. K. Isaev, and N. I. Khanov, *Measurement Techniques* 51, 2008, p. 1045.

TABLE 2.1 SI Base Units

Base Quantity	Name	Symbol
length	meter	m
mass	kilogram	kg
time	second	s
electric current	ampere	A
thermodynamic temperature	kelvin	K
amount of substance	mole	mol
luminous intensity	candela	cd

The "calorie" used with food, drink, and exercise is really a kilocalorie, 4.187 J.

The fundamental unit of work or energy is the ***joule*** (J). One joule (a kg m^2 s^{-2} in SI base units) is equivalent to 0.7376 foot pound-force (ft · lbf). Other energy units include the calorie (cal), equal to 4.187 J; the British thermal unit (Btu), which is 1055 J; and the kilowatthour (kWh), equal to 3.6×10^6 J. Power is defined as the *rate* at which work is done or energy is expended. The fundamental unit of power is the ***watt*** (W), defined as 1 J/s. One watt is equivalent to 0.7376 ft · lbf/s or, equivalently, 1/745.7 horsepower (hp).

The SI uses the decimal system to relate larger and smaller units to the basic unit, and it employs prefixes to signify the various powers of 10. A list of prefixes and their symbols is given in Table 2.2; the ones most commonly encountered in engineering are highlighted.

TABLE 2.2 SI Prefixes

Factor	Name	Symbol	Factor	Name	Symbol
10^{-24}	yocto	y	10^{24}	yotta	Y
10^{-21}	zepto	z	10^{21}	zetta	Z
10^{-18}	atto	a	10^{18}	exa	E
10^{-15}	femto	f	10^{15}	peta	P
10^{-12}	pico	p	10^{12}	tera	T
10^{-9}	nano	n	10^{9}	giga	G
10^{-6}	micro	μ	10^{6}	mega	M
10^{-3}	milli	m	10^{3}	kilo	k
10^{-2}	centi	c	10^{2}	hecto	h
10^{-1}	deci	d	10^{1}	deka	da

These prefixes are worth memorizing, for they will appear often both in this text and in other technical work. Combinations of several prefixes, such as the millimicrosecond, are unacceptable. It is worth noting that in terms of distance, it is common to see "micron (μm)" as opposed to "micrometer," and sometimes the angstrom (Å) is used for 10^{-10} meter. Also, in circuit analysis and engineering in general, it is fairly common to see numbers expressed in what are often termed "engineering units." In engineering notation, a quantity is represented by a number between 1 and 999 and an appropriate metric unit using a power divisible by 3. So, for example, it is preferable to express the quantity 0.048 W as 48 mW, instead of 4.8 cW, 4.8×10^{-2} W, or 48,000 μW.

PRACTICE

2.1 A krypton fluoride laser emits light at a wavelength of 248 nm. This is the same as: (*a*) 0.0248 mm; (*b*) 2.48 μm; (*c*) 0.248 μm; (*d*) 24,800 Å.

2.2 A single logic gate in a prototype integrated circuit is found to be capable of switching from the "on" state to the "off" state in 12 ps. This corresponds to: (*a*) 1.2 ns; (*b*) 120 ns; (*c*) 1200 ns; (*d*) 12,000 ns.

2.3 A typical solid-state light bulb runs at 8.5 W. If it is left on constantly, how much energy (J) is consumed per day, and what is the weekly cost if energy is charged at a rate of 12 cents per kilowatthour?

Ans: (*c*); (*d*); 7.3 kJ, $0.17.

2.2 CHARGE, CURRENT, VOLTAGE, POWER, AND ENERGY

Charge

One of the most fundamental concepts in electric circuit analysis is that of charge conservation. We know from basic physics that there are two types of charge: positive (corresponding to a proton) and negative (corresponding to an electron). For the most part, this text is concerned with circuits in which only electron flow is relevant. There are many devices (such as batteries, diodes, and transistors) in which positive charge motion is important to understanding internal operation, but external to the device we typically concentrate on the electrons which flow through the connecting wires. Although we continuously transfer charges between different parts of a circuit, we do nothing to change the total amount of charge. In other words, we neither create nor destroy electrons (or protons) when running electric circuits.[3] Charge in motion represents a *current*.

As seen in Table 2.1, the base units of the SI are not derived from fundamental physical quantities. Instead, they represent historically agreed-upon measurements, leading to definitions which occasionally seem backward. For example, it would make more sense physically to define the ampere based on electronic charge.

In the SI system, the fundamental unit of charge is the ***coulomb*** (C). It is defined in terms of the ***ampere*** by counting the total charge that passes through an arbitrary cross section of a wire during an interval of one second; one coulomb is measured each second for a wire carrying a current of 1 ampere (Fig. 2.1). In this system of units, a single electron has a charge of -1.602×10^{-19} C and a single proton has a charge of $+1.602 \times 10^{-19}$ C.

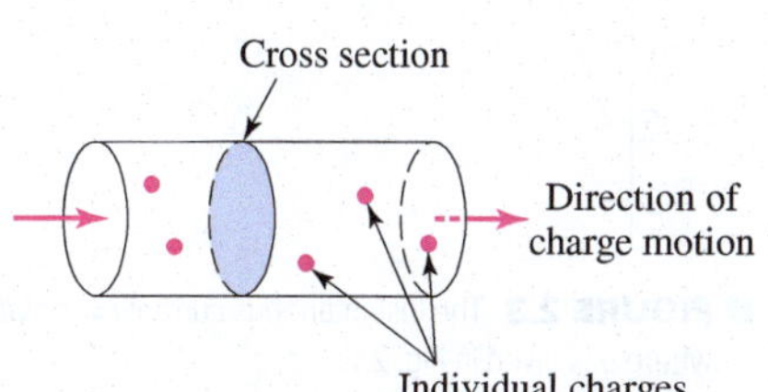

FIGURE 2.1 The definition of current illustrated using charge flowing through a wire; 1 ampere corresponds to 1 coulomb of charge passing through the arbitrarily chosen cross section in 1 second.

(3) Although the occasional appearance of smoke may seem to suggest otherwise…

A quantity of charge that does not change with time is typically represented by Q. The instantaneous amount of charge (which may or may not be time-invariant) is commonly represented by $q(t)$, or simply q. This convention is used throughout the remainder of the text: capital letters are reserved for constant (time-invariant) quantities, whereas lowercase letters represent the more general case. Thus, a constant charge may be represented by *either* Q or q, but an amount of charge that changes over time *must* be represented by the lowercase letter q.

Current

The idea of "transfer of charge" or "charge in motion" is of vital importance to us in studying electric circuits because, in moving a charge from place to place, we may also transfer energy from one point to another. The familiar cross-country power-transmission line is a practical example of a device that transfers energy. Of equal importance is the possibility of varying the rate at which the charge is transferred in order to communicate or transfer information. This process is the basis of modern communication systems, including satellite global positioning systems, 5G (and beyond), and WiFi Internet connections.

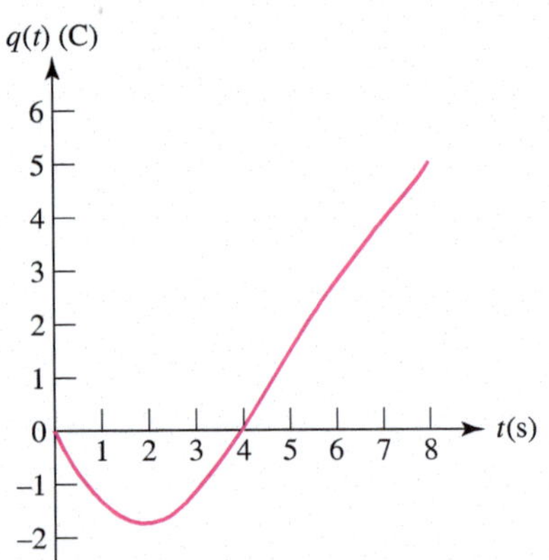

FIGURE 2.2 A graph of the instantaneous value of the total charge $q(t)$ that has passed a given reference point since $t = 0$.

The current flowing in a discrete path, such as a metallic wire, has both a *numerical value* and a *direction* associated with it; it is a measure of the rate at which charge is moving past a given reference point in a specified direction. Once we have specified a reference direction, we may then let $q(t)$ be the total charge that has passed the reference point since an arbitrary time $t = 0$, moving in the defined direction. A contribution to this total charge will be negative if negative charge is moving in the reference direction, or if positive charge is moving in the opposite direction. As an example, Fig. 2.2 shows a history of the total charge $q(t)$ that has passed a given reference point in a wire (such as the one shown in Fig. 2.1).

We define the current at a specific point and flowing in a specified direction as the instantaneous rate at which net positive charge is moving past that point in the specified direction. This, unfortunately, is the historical definition, which came into popular use before it was appreciated that current in wires is actually due to negative, not positive, charge motion. Current is symbolized by I or i, and so

$$i = \frac{dq}{dt} \qquad [1]$$

The unit of current is the ampere (A), named after A. M. Ampère, a French physicist. It is commonly abbreviated as an "amp," although this is unofficial and somewhat informal. One ampere equals 1 coulomb per second.

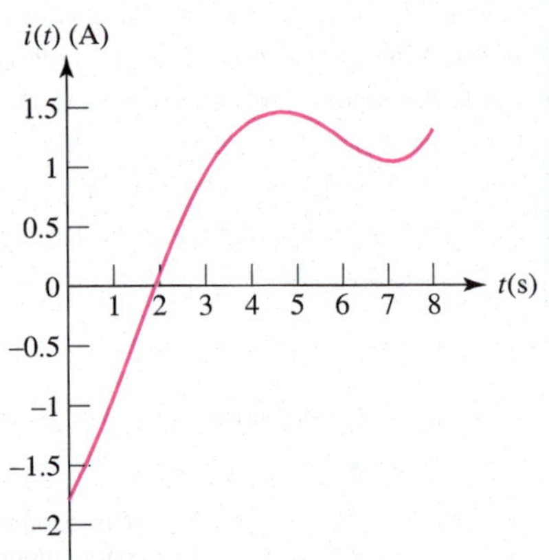

FIGURE 2.3 The instantaneous current $i = dq/dt$, where q is given in Fig. 2.2.

Using Eq. [1], we compute the instantaneous current corresponding to Fig. 2.2 and obtain Fig. 2.3. The use of the lowercase letter i is again to be associated with an instantaneous value; an uppercase I would denote a constant (i.e., time-invariant) quantity. The charge transferred between time t_0 and t may be expressed as a definite integral:

$$\int_{q(t_0)}^{q(t)} dq = \int_{t_0}^{t} i\,dt' \qquad [2]$$

The total charge transferred over all time is thus given by

$$q(t) = \int_{t_0}^{t} i\,dt' + q(t_0) \qquad [3]$$

Note that a current represented by i or $i(t)$ can be constant (dc) or time-varying, but currents represented by the symbol I must be non-time-varying.

Several different types of current are illustrated in Fig. 2.4. A current that is constant in time is termed a direct current, or simply dc, and is shown by Fig. 2.4*a*. We will find many practical examples of currents that vary sinusoidally with time (Fig. 2.4*b*); currents of this form are present in normal household circuits. Such a current is often referred to as alternating current, or ac. Exponential currents and damped sinusoidal currents (Fig. 2.4*c* and *d*) will also be encountered later.

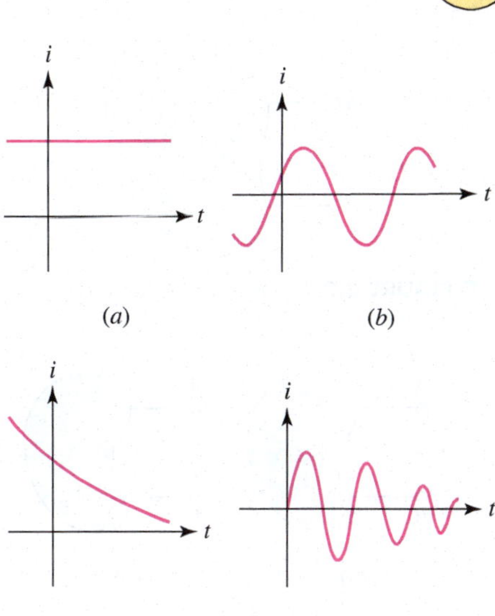

FIGURE 2.4 Several types of current: (*a*) Direct current (dc). (*b*) Sinusoidal current (ac). (*c*) Exponential current. (*d*) Damped sinusoidal current.

We create a graphical symbol for current by placing an arrow next to the conductor. Thus, in Fig. 2.5*a* the direction of the arrow and the value 3 A indicate either that a net positive charge of 3 C/s is moving to the right or that a net negative charge of −3 C/s is moving to the left each second. In Fig. 2.5*b* there are again two possibilities: either −3 A is flowing to the left or +3 A is flowing to the right. All four statements and both figures represent currents that are equivalent in their electrical effects, and we say that they are equal. A nonelectrical analogy that may be easier to visualize is to think in terms of a personal savings account: e.g., a deposit can be viewed as either a *negative* cash flow *out of* your account or a *positive* flow *into* your account.

It is convenient to think of current as the motion of positive charge, even though it is known that current flow in metallic conductors results from electron motion. In ionized gases, in electrolytic solutions, and in some semiconductor materials, however, positive charges in motion constitute part or all of the current. Thus, any definition of current can agree with the physical nature of conduction only part of the time. The definition and symbolism we have adopted are standard.

It is essential that we realize that the current arrow does not indicate the "actual" direction of current flow but is simply part of a convention that allows us to talk about "the current in the wire" in an unambiguous manner. The arrow is a fundamental part of the definition of a current! Thus, to talk about the value of a current $i_1(t)$ without specifying the arrow is to discuss an undefined entity. For example, Fig. 2.6*a* and *b* are meaningless representations of $i_1(t)$, whereas Fig. 2.6*c* is complete.

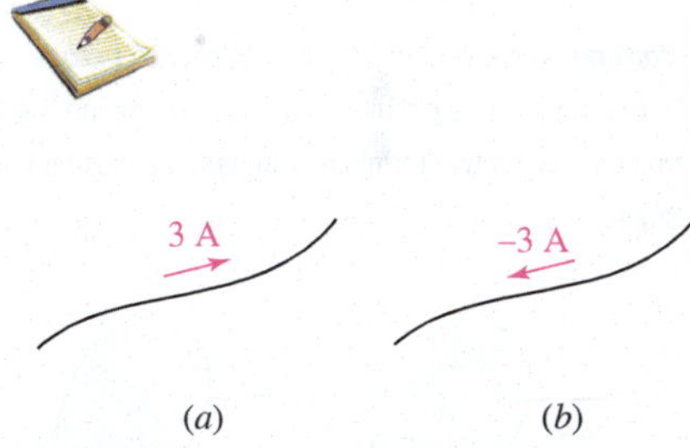

FIGURE 2.5 Two methods of representation for the exact same current.

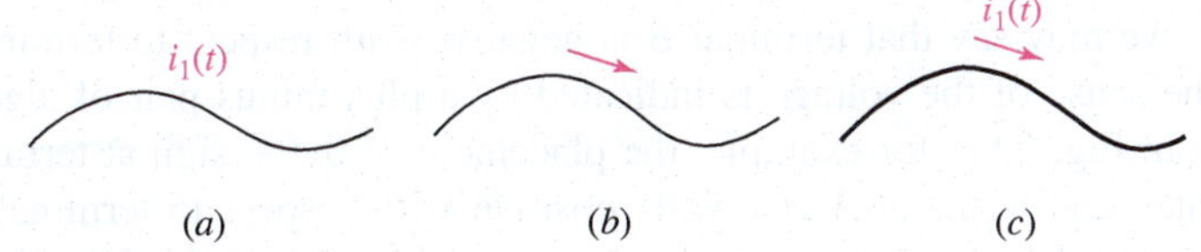

FIGURE 2.6 (*a*, *b*) Incomplete, improper, and incorrect definitions of a current. (*c*) The correct definition of $i_1(t)$.

Current is the flow of charge flowing ***through*** a wire or circuit component. We define the current path with an arrow, or flow of charge ***into*** or ***out of*** the wire or circuit component.

PRACTICE

2.4 In the wire of Fig. 2.7, electrons are moving *left* to *right* to create a current of 1 mA. Determine I_1 and I_2.

I_1

I_2

FIGURE 2.7

Ans: $I_1 = -1$ mA; $I_2 = +1$ mA.

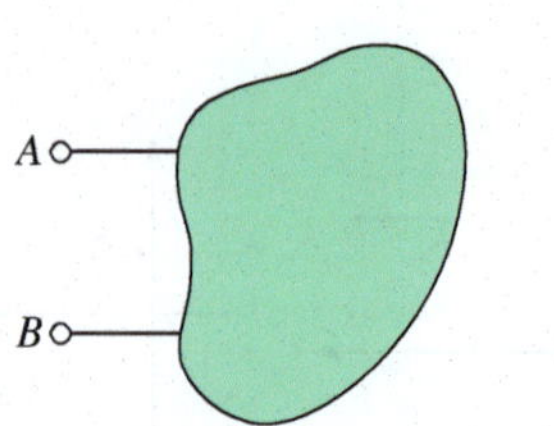

FIGURE 2.8 A general two-terminal circuit element.

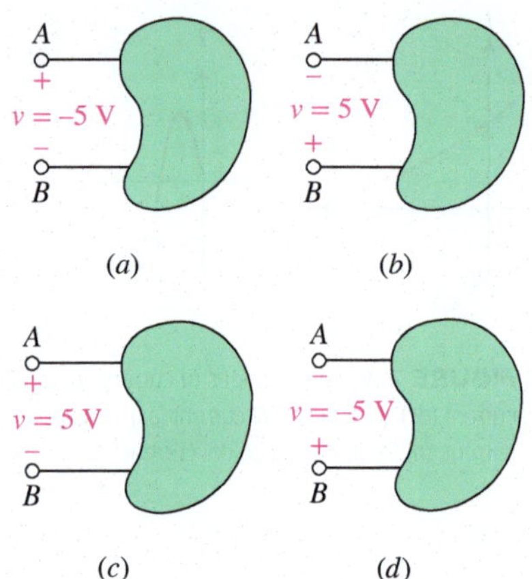

FIGURE 2.9 (*a*, *b*) Terminal *B* is 5 V positive with respect to terminal *A*; (*c*, *d*) terminal *A* is 5 V positive with respect to terminal *B*.

Voltage is the electric potential difference ***across*** two terminals of a circuit component. We define the voltage across two terminals with labeled plus-minus signs.

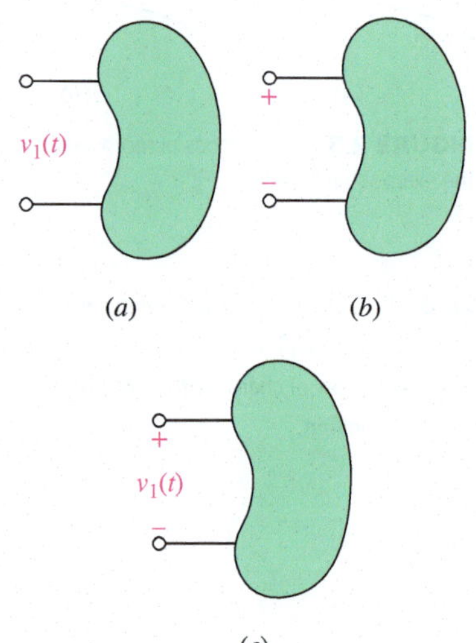

FIGURE 2.10 (*a*, *b*) These are inadequate definitions of a voltage. (*c*) A correct definition includes both a symbol for the variable and a plus-minus symbol pair.

Voltage

We must now begin to refer to a circuit element, something best defined in general terms to begin with. Such electrical devices as fuses, light bulbs, resistors, batteries, capacitors, generators, and spark coils can be represented by combinations of simple circuit elements. We begin by showing a very general circuit element as a shapeless object possessing two terminals at which connections to other elements may be made (Fig. 2.8).

There are two paths by which current may enter or leave the element. In subsequent discussions we will define particular circuit elements by describing the electrical characteristics that may be observed at their terminals.

In Fig. 2.8, let us suppose that a dc current is sent into terminal *A*, through the general element, and back out of terminal *B*. Let us also assume that pushing charge through the element requires an expenditure of energy. We then say that an electrical voltage (or a *potential difference*) exists between the two terminals, or that there is a voltage "across" the element. Thus, the voltage across a terminal pair is a measure of the work required to move charge through the element. The unit of voltage is the ***volt,***[4] and 1 volt is the same as 1 J/C. Voltage is represented by V or v.

A voltage can exist between a pair of electrical terminals whether a current is flowing or not. An automobile battery, for example, has a voltage of 12 V across its terminals even if nothing whatsoever is connected to the terminals.

According to the principle of conservation of energy, the energy that is expended in forcing charge through the element must appear somewhere else. When we later meet specific circuit elements, we will note whether that energy is stored in some form that is readily available as electric energy or whether it changes irreversibly into heat, light, or some other nonelectrical form of energy.

We must now establish a convention by which we can distinguish between energy supplied *to* an element and energy that is supplied *by* the element itself. We do this by our choice of sign for the voltage of terminal *A* with respect to terminal *B*. If a positive current is entering terminal *A* of the element and an external source must expend energy to establish this current, then terminal *A* is positive with respect to terminal *B*. (Alternatively, we may say that terminal *B* is negative with respect to terminal *A*.)

The sense of the voltage is indicated by a plus-minus pair of algebraic signs. In Fig. 2.9*a*, for example, the placement of the + sign at terminal *A* indicates that terminal *A* is v volts positive with respect to terminal *B*. If we later find that v happens to have a numerical value of −5 V, then we may say either that *A* is −5 V positive with respect to *B* or that *B* is 5 V positive with respect to *A*. Other cases are shown in Fig. 2.9*b*, *c*, and *d*.

Just as we noted in our definition of current, it is essential to realize that the plus-minus pair of algebraic signs does not indicate the "actual" polarity of the voltage but is simply part of a convention that enables us to talk unambiguously about "the voltage across the terminal pair." *The definition of any voltage must include a plus-minus sign pair*! Using a quantity $v_1(t)$ without specifying the location of the plus-minus sign pair is using an undefined term. Figure 2.10*a* and *b* do *not* serve as definitions of $v_1(t)$; Fig. 2.10*c* does.

(4) We are probably fortunate that the full name of the 18th-century Italian physicist, *Alessandro Giuseppe Antonio Anastasio Volta,* is not used for our unit of potential difference!

PRACTICE

2.5 For the element in Fig. 2.11, if $v_1 = 17$ V, determine v_2.

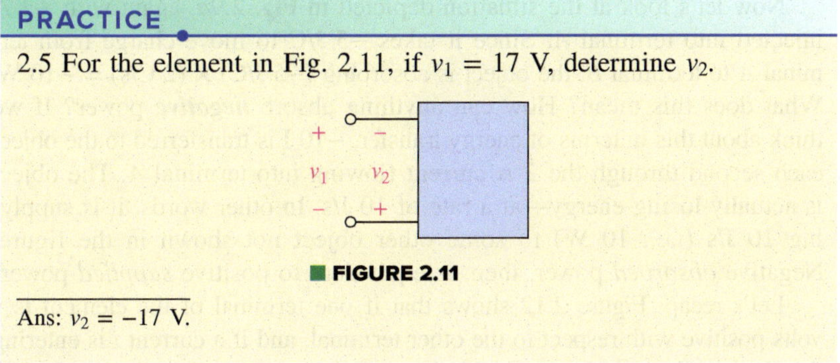

FIGURE 2.11

Ans: $v_2 = -17$ V.

Power

We have already defined power, and we will represent it by P or p. If one joule of energy is expended in transferring one coulomb of charge through the device in one second, then the rate of energy transfer is one watt. The absorbed power must be proportional both to the number of coulombs transferred per second (current) and to the energy needed to transfer one coulomb through the element (voltage). Thus,

$$p = vi \qquad [4]$$

Dimensionally, the right side of this equation is the product of joules per coulomb and coulombs per second, which produces the expected dimension of joules per second, or watts. The conventions for current, voltage, and power are shown in Fig. 2.12.

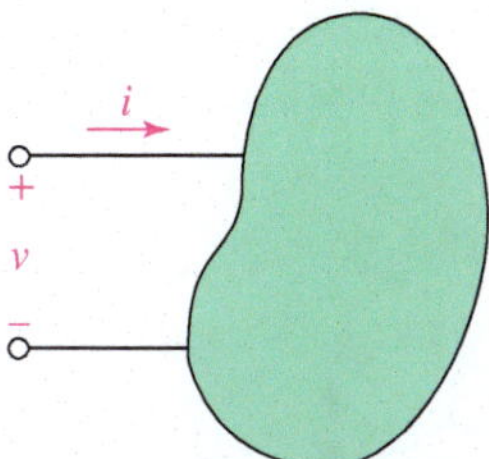

FIGURE 2.12 The power absorbed by the element is given by the product $p = vi$. Alternatively, we can say that the element generates or supplies a power $-vi$.

We now have an expression for the power being absorbed by a circuit element in terms of a voltage across it and current through it. Voltage was defined in terms of an energy expenditure, and power is the rate at which energy is expended. However, no statement can be made concerning energy transfer in any of the four cases shown in Fig. 2.9, for example, until the direction of the current is specified. Let us imagine that a current arrow is placed alongside each upper lead, directed to the right, and labeled "+2 A." First, consider the case shown in Fig. 2.9*c*. Terminal *A* is 5 V positive with respect to terminal *B*, which means that 5 J of energy is required to move each coulomb of positive charge into terminal *A*, through the object, and out terminal *B*. Since we are injecting +2 A (a current of 2 coulombs of positive charge per second) into terminal *A*, we are doing (5 J/C) × (2 C/s) = 10 J of work per second on the object. In other words, the object is absorbing 10 W of power from whatever is injecting the current.

We know from an earlier discussion that there is no difference between Fig. 2.9*c* and *d*, so we expect the object depicted in Fig. 2.9*d* to also be absorbing 10 W. We can check this easily enough: we are injecting +2 A into terminal *A* of the object, so +2 A flows out of terminal *B*. Another way of saying this is that we are injecting −2 A of current into terminal *B*. It takes −5 J/C to move charge from terminal *B* to terminal *A*, so the object is absorbing (−5 J/C) × (−2 C/s) = +10 W as expected. The only difficulty in describing this particular case is keeping the minus signs straight, but with a bit of care we see the correct answer can be obtained regardless of our choice of positive reference terminal (terminal *A* in Fig. 2.9*c*, and terminal *B* in Fig. 2.9*d*).

Now let's look at the situation depicted in Fig. 2.9*a*, again with +2 A injected into terminal *A*. Since it takes −5 J/C to move charge from terminal *A* to terminal *B*, the object is absorbing (−5 J/C) × (2 C/s) = −10 W. What does this mean? How can anything absorb ***negative*** power? If we think about this in terms of energy transfer, −10 J is transferred to the object each second through the 2 A current flowing into terminal *A*. The object is actually losing energy—at a rate of 10 J/s. In other words, it is supplying 10 J/s (i.e., 10 W) to some other object not shown in the figure. Negative *absorbed* power, then, is equivalent to positive *supplied* power.

If the current arrow is directed into the "+" marked terminal of an element, then $p = vi$ yields the ***absorbed*** power. A negative value indicates that power is actually being generated by the element.

Let's recap. Figure 2.12 shows that if one terminal of the element is v volts positive with respect to the other terminal, and if a current i is entering the element through that terminal, then a power $p = vi$ is being *absorbed* by the element; it is also correct to say that a power $p = vi$ is being *delivered* to the element. When the current arrow is directed into the element at the plus-marked terminal, we satisfy the ***passive sign convention.*** This convention should be studied carefully, understood, and memorized. In other words, it says that if the current arrow and the voltage polarity signs are placed such that the current enters the terminal on the element marked with the positive sign, then the power *absorbed* by the element can be expressed by the product of the specified current and voltage variables. If the numerical value of the product is negative, then we say that the element is absorbing negative power, or that it is actually generating power and delivering it to some external element. For example, in Fig. 2.12 with $v = 5$ V and $i = -4$ A, the element may be described as either absorbing −20 W or generating 20 W.

If the current arrow is directed out of the "+" terminal of an element, then $p = vi$ yields the ***supplied*** power. A negative value in this case indicates that power is being absorbed.

Conventions are only required when there is more than one way to do something, and confusion may result when two different groups try to communicate. For example, it is rather arbitrary to always place "North" at the top of a map; compass needles don't point "up," anyway. Still, if we were talking to people who had (unknown to us) chosen the opposite convention of placing "South" at the top of their maps, imagine the confusion that could result! In the same fashion, there is a general convention that always draws the current arrows pointing into the positive voltage terminal, regardless of whether the element supplies or absorbs power. This convention is not incorrect but sometimes results in counterintuitive currents labeled on circuit schematics. The reason for this is that it simply seems more natural to refer to positive current flowing out of a voltage or current source that is supplying positive power to one or more circuit elements.

EXAMPLE 2.1

Compute the power absorbed by each part in Fig. 2.13.

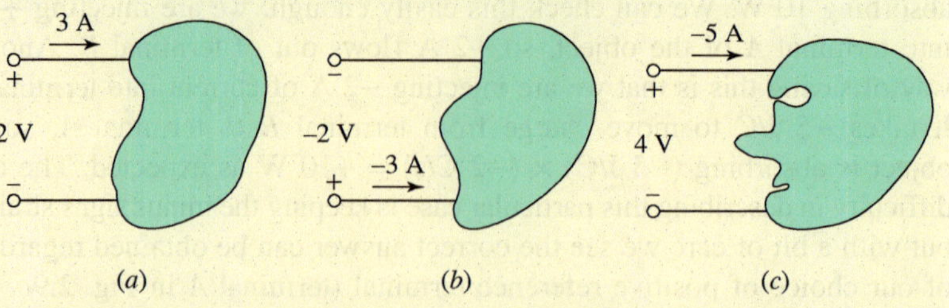

FIGURE 2.13 (*a*, *b*, *c*) Three examples of two-terminal elements.

In Fig. 2.13*a*, we see that the reference current is defined consistent with the passive sign convention, which assumes that the element is absorbing power. With +3 A flowing into the positive reference terminal, we compute

$$P = (2 \text{ V})(3 \text{ A}) = 6 \text{ W}$$

of power absorbed by the element.

Figure 2.13*b* shows a slightly different picture. Now, we have a current of −3 A flowing into the positive reference terminal. This gives us an absorbed power

$$P = (-2 \text{ V})(-3 \text{ A}) = 6 \text{ W}$$

Thus, we see that the two cases are actually equivalent: A current of +3 A flowing into the top terminal is the same as a current of +3 A flowing out of the bottom terminal, or, equivalently, a current of −3 A flowing into the bottom terminal.

Referring to Fig. 2.13*c*, we again apply the passive sign convention rules and compute an absorbed power

$$P = (4 \text{ V})(-5 \text{ A}) = -20 \text{ W}$$

Since we computed a negative *absorbed* power, this tells us that the element in Fig. 2.13*c* is actually *supplying* +20 W (i.e., it's a source of energy).

PRACTICE

2.6 Determine the power being absorbed by the circuit element in Fig. 2.14*a*.

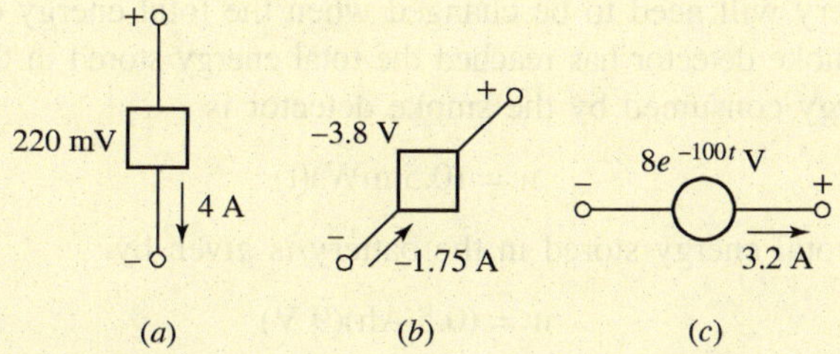

■ **FIGURE 2.14**

2.7 Determine the power being supplied by the circuit element in Fig. 2.14*b*.

2.8 Determine the power being delivered to the circuit element in Fig. 2.14*c* at $t = 5$ ms.

Ans: 880 mW; 6.65 W; −15.53 W.

Energy

In electrical circuits, attention is often devoted to power, but sometimes we would also like to know the total energy transferred for a given period of time. For example, energy usage determines how long the battery in your circuit will last, or what your electricity bill will be. Recalling that power is the rate of work, energy (w) is defined as

$$w(t) = \int_{t_0}^{t} p\, dt' = \int_{t_0}^{t} vi\, dt' \qquad [5]$$

The SI unit of energy is the *joule* (J). Noting that energy is the product of power and time (1 joule = 1 watt × 1 second), it is also convenient to define energy in terms of *watt hours* (Wh) or *kilowatt hours* (kWh). Electrical utilities typically charge electricity usage in units of kWh, and this unit is typically displayed on the dashboard or display of electric vehicles. Converting units yields the relations

$$1 \text{ Wh} = 3600 \text{ J} \quad [6]$$

$$1 \text{ kWh} = 3.6 \times 10^6 \text{ J} \quad [7]$$

Battery capacity (energy stored) can also be defined in terms of Wh. Since the voltage on a battery is constant, it becomes convenient to separate out the battery voltage and simply refer to the total charge storage on the battery (Q). Thus,

$$w = \int vi\, dt = V \int i\, dt = VQ \quad [8]$$

The total charge Q is given in units of *amp hours* (Ah) or *milliamp hours* (mAh)

$$1 \text{ Ah} = 3600 \text{ C} \quad [9]$$

$$1 \text{ mAh} = 3.6 \text{ C} \quad [10]$$

EXAMPLE 2.2

A battery-powered smoke detector has an average power consumption of 0.5 mW and runs off of a 9 V battery with a capacity of 500 mAh. How often do you expect to change the battery?

The battery will need to be changed when the total energy consumed by the smoke detector has reached the total energy stored in the battery. The energy consumed by the smoke detector is

$$w = (0.5 \text{ mW})(t)$$

and the total energy stored in the battery is given by

$$w = (0.5 \text{ Ah})(9 \text{ V})$$

Equating the two after converting units, and solving for t results in

$$t = \frac{(0.5 \text{ Ah})(9 \text{ V})}{(0.5 \times 10^{-3} \text{ W})} = 9000 \text{ h}$$

$$t = 9000 \text{ h} \times \frac{(1 \text{ day})}{(24 \text{ h})} \times \frac{(1 \text{ year})}{(365 \text{ days})} = 1.03 \text{ years}$$

(Note that every 6 months is a standard recommendation, for safety.)

PRACTICE

2.9 A certain rechargeable smartphone battery has a voltage of 3.8 V and capacity of 1800 mAh. You find that a single battery charge can provide 12 h of talk time, or 10 days of standby time. What is the average power consumption for (*a*) talk mode and (*b*) standby mode?

Ans: (*a*) 570 mW (*b*) 28.5 mW.

2.3 VOLTAGE AND CURRENT SOURCES

Using the concepts of current and voltage, it is now possible to be more specific in defining a ***circuit element.***

In so doing, it is important to differentiate between the physical device itself and the mathematical model that we will use to analyze its behavior in a circuit. The model is only an approximation.

We will use the expression *circuit element* to refer to the mathematical model. The choice of a particular model for any real device must be made on the basis of experimental data or experience; we will usually assume that this choice has already been made. For simplicity, we initially consider circuits with idealized components represented by simple models.

All of the simple circuit elements we will consider can be classified according to the relationship of the current through the element to the voltage across the element. For example, if the voltage across the element is linearly proportional to the current through it, we will call the element a resistor. Other types of simple circuit elements have terminal voltages which are proportional to the *derivative* of the current with respect to time (an inductor), or to the *integral* of the current with respect to time (a capacitor). There are also elements in which the voltage is completely independent of the current, or the current is completely independent of the voltage; these are termed *independent sources*. Furthermore, we will need to define special kinds of sources for which either the source voltage or current depends upon a current or voltage elsewhere in the circuit; such sources are referred to as *dependent sources*. Dependent sources are used a great deal in electronics to model both dc and ac behavior of transistors, especially in amplifier circuits.

By definition, a simple circuit element is the mathematical model of a two-terminal electrical device, and it can be completely characterized by its voltage–current relationship; it cannot be subdivided into other two-terminal devices.

Independent Voltage Sources

The first element we will consider is the ***independent voltage source.*** The circuit symbol is shown in Fig. 2.15*a*; the subscript *s* merely identifies the voltage as a "source" voltage, and is common but not required. *An independent voltage source is characterized by a terminal voltage which is completely independent of the current through it*. Thus, if we are given an independent voltage source and are notified that the terminal voltage is 12 V, then we always assume this voltage, regardless of the current flowing.

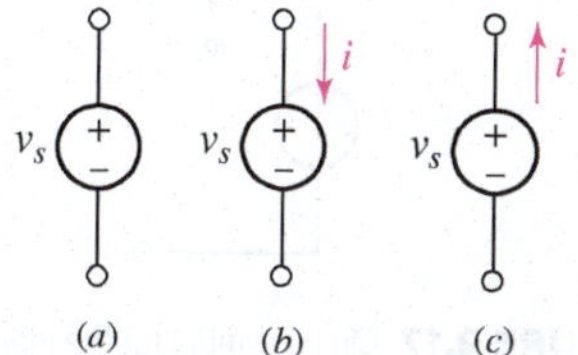

FIGURE 2.15 Circuit symbol of the independent voltage source.

The independent voltage source is an *ideal* source and does not represent exactly any real physical device, because the ideal source could theoretically deliver an infinite amount of energy from its terminals. This idealized voltage source does, however, furnish a reasonable approximation to several practical voltage sources. An internal combustion engine (ICE) automobile storage battery, for example, has a 12 V terminal voltage that remains essentially constant as long as the current through it does not exceed a few amperes. A small current may flow in either direction through the battery. If it is positive and flowing out of the positively marked terminal, then the battery is furnishing power to the headlights, for example; if the current is positive and flowing into the positive terminal, then the battery is charging by absorbing energy from the alternator.[5] An ordinary Canadian or U.S. household electrical outlet also approximates an independent voltage source, providing a voltage $v_s = 120\sqrt{2}\cos 2\pi 60t$ V; this representation is valid for currents less than 20 A or so.

If you've ever noticed the room lights dim when an air conditioner kicks on, it's because the sudden large current demand temporarily led to a (nonideal) voltage drop. After the motor starts moving, it takes less current to keep it in motion. At that point, the current demand is reduced, the voltage returns to its original value, and the wall outlet again provides a reasonable approximation of an ideal voltage source.

(5) Or the battery of a friend's car, if you accidentally left your headlights on...

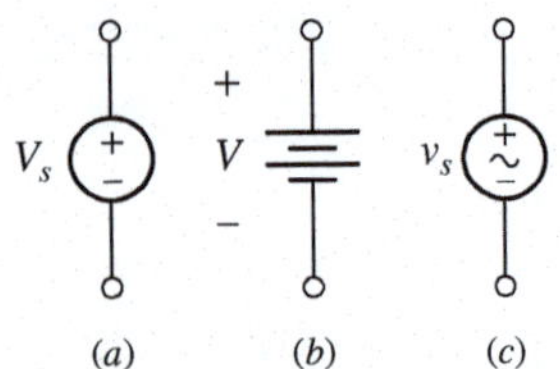

■ **FIGURE 2.16** (*a*) DC voltage source symbol; (*b*) battery symbol; (*c*) ac voltage source symbol.

A point worth repeating here is that the presence of the plus sign at the upper end of the symbol for the independent voltage source in Fig. 2.15*a* does not necessarily mean that the upper terminal is numerically positive with respect to the lower terminal. Instead, it means that the upper terminal is v_s volts positive with respect to the lower terminal. If at some instant v_s happens to be negative, then the upper terminal is actually negative with respect to the lower terminal at that instant.

Consider a current arrow labeled "i" placed adjacent to the upper conductor of the source as in Fig. 2.15*b*. The current i is entering the terminal at which the positive sign is located, the passive sign convention is satisfied, and the source thus *absorbs* power $p = v_s i$. More often than not, a source is expected to deliver power to a network and not to absorb it. Consequently, we might choose to direct the arrow as in Fig. 2.15*c* so that $v_s i$ will represent the power *delivered* by the source. Technically, either arrow direction may be chosen; whenever possible, we will adopt the convention of Fig. 2.15*c* in this text for voltage and current sources, which are not usually considered passive devices.

Terms like dc voltage source and dc current source are commonly used. Literally, they mean "direct-current voltage source" and "direct-current current source," respectively. Although these terms may seem a little odd or even redundant, the terminology is so widely used there's no point in fighting it.

An independent voltage source with a constant terminal voltage is often termed an independent dc voltage source and can be represented by either of the symbols shown in Fig. 2.16*a* and *b*. Note in Fig. 2.16*b* that when the physical plate structure of the battery is suggested, the longer plate is placed at the positive terminal; the plus and minus signs then represent redundant notation, but they are usually included anyway. For the sake of completeness, the symbol for an independent ac voltage source is shown in Fig. 2.16*c*.

Independent Current Sources

Another ideal source which we will need is the ***independent current source.*** Here, the current through the element is completely independent of the voltage across it. The symbol for an independent current source is shown in Fig. 2.17. If i_s is constant, we call the source an independent dc current source. An ac current source is often drawn with a tilde through the arrow, similar to the ac voltage source shown in Fig. 2.16*c*.

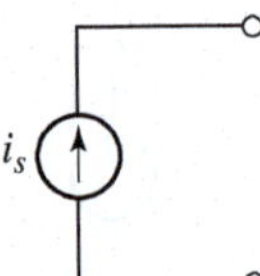

■ **FIGURE 2.17** Circuit symbol for the independent current source.

Like the independent voltage source, the independent current source is at best a reasonable approximation for a physical element. In theory it can deliver infinite power from its terminals because it produces the same finite current for any voltage across it, no matter how large that voltage may be. It is, however, a good approximation for many practical sources, particularly in electronic circuits.

Although most students seem happy enough with an independent voltage source providing a fixed voltage but essentially any current, *it is a common mistake* to view an independent current source as having zero voltage across its terminals while providing a fixed current. In fact, we do not know a priori what the voltage across a current source will be—it depends entirely on the circuit to which it is connected.

Dependent Sources

The two types of ideal sources that we have discussed up to now are called *independent* sources because the value of the source quantity is not affected in any way by activities in the remainder of the circuit. This is in contrast with yet another kind of ideal source, the *dependent*, or *controlled*, source, in which the source quantity is determined by a voltage or current existing at some other location in the system being analyzed. Sources such as these appear in the equivalent electrical models for many electronic devices, such as transistors, operational amplifiers, and integrated circuits. To distinguish between dependent and independent sources, we introduce the diamond symbols shown in Fig. 2.18. In Fig. 2.18*a* and *c*, K is a dimensionless scaling constant. In Fig. 2.18*b*, g is a scaling factor with units of A/V; in Fig. 2.18*d*, r is a scaling factor with units of V/A. The controlling current i_x and the controlling voltage v_x must be defined somewhere in the circuit.

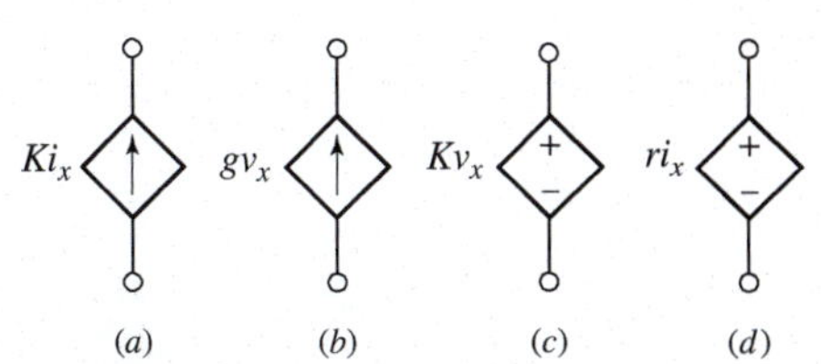

FIGURE 2.18 The four different types of dependent sources: (*a*) current-controlled current source; (*b*) voltage-controlled current source; (*c*) voltage-controlled voltage source; (*d*) current-controlled voltage source.

It does seem odd at first to have a current source whose value depends on a voltage, or a voltage source which is controlled by a current flowing through some other element. Even a voltage source depending on a remote voltage can appear strange. Such sources are invaluable for modeling complex systems, however, making the analysis algebraically straightforward. Examples include the drain current of a field effect transistor as a function of the gate voltage, or the output voltage of an analog integrated circuit as a function of differential input voltage. When encountered during circuit analysis, we write down the entire controlling expression for the dependent source just as we would if it was a numerical value attached to an independent source. This often results in the need for an additional equation to complete the analysis, unless the controlling voltage or current is already one of the specified unknowns in our system of equations.

EXAMPLE 2.3

In the circuit of Fig. 2.19*a*, determine v_L if v_2 is 3 V.

▶ ***Identify the goal of the problem.***

We need to determine the voltage across the far-right element using the pre-assigned polarity.

▶ ***Collect the known information.***

We have been provided with a partially labeled circuit diagram. We should add the additional information that $v_2 = 3$ V to the schematic (Fig. 2.19*b*).

▶ ***Devise a plan.***

We notice that the desired voltage is the same as the voltage across the dependent voltage source. If we find that voltage, we have our answer.

▶ ***Construct an appropriate set of equations.***

We write

$$v_L = 5v_2$$

and

$$v_2 = 3$$

Thus, we have two equations in two unknowns.

▶ ***Determine if additional information is required.***

We have two equations in two variables (v_L and v_2). This is sufficient to solve.

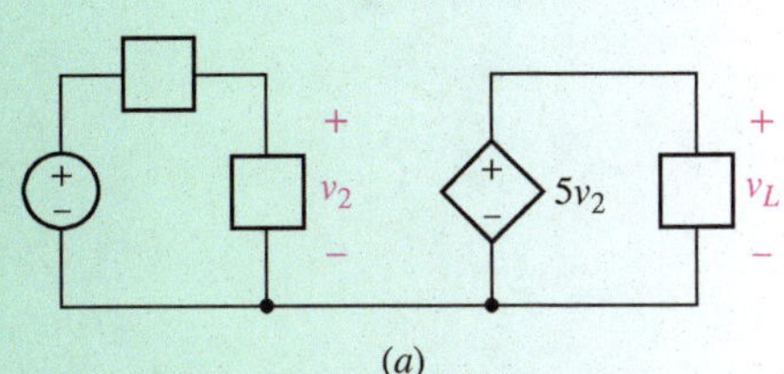

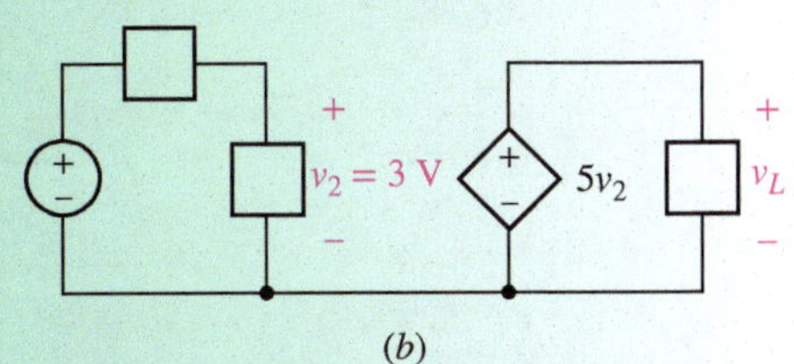

FIGURE 2.19 (*a*) An example circuit containing a voltage-controlled voltage source. (*b*) The additional information provided is included on the diagram.

(Continued on next page)

Attempt a solution.

This simple set of equations is easily solved to determine that $v_L = 15$ V. We see that the time it takes to completely label a circuit diagram is a good investment!

Verify the solution. Is it reasonable or expected?

We can (and should) go back and check our math. We have limited information about the circuit so our expectations might be equally limited. Still, we note a controlling voltage of 3 V and a voltage elsewhere of 15 V. These are within an order of magnitude of each other, which is a good indication (i.e., one of them is not 1,000,000 times the other).

PRACTICE

2.10 Find the power *absorbed* by each element in the circuit in Fig. 2.20.

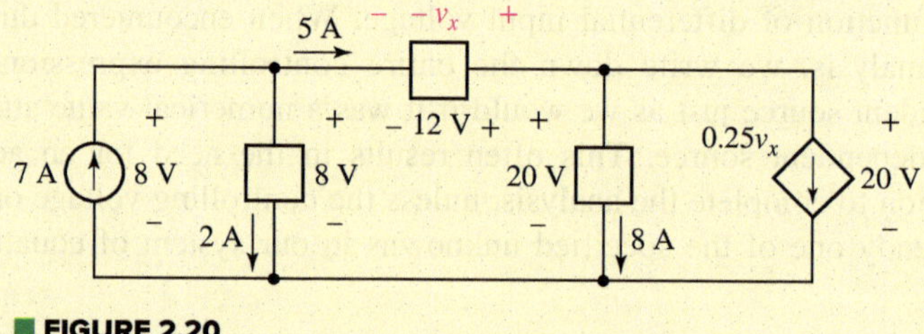

FIGURE 2.20

Ans: (left to right) −56 W; 16 W; −60 W; 160 W; −60 W.

Dependent and independent voltage and current sources are *active* elements; they are capable of delivering power to some external device. For the present we will think of a *passive* element as one which is capable only of receiving power. However, we will later see that several passive elements are able to store finite amounts of energy and then return that energy later to various external devices; since we still wish to call such elements passive, it will be necessary to improve upon our two definitions a little later.

Networks and Circuits

The interconnection of two or more simple circuit elements forms an electrical ***network.*** If the network contains at least one closed path, it is also an electric *circuit.* Note: Every circuit is a network, but not all networks are circuits (see Fig. 2.21)!

A network that contains at least one active element, such as an independent voltage or current source, is an active network. A network that does not contain any active elements is a passive network.

We have now defined what we mean by the term ***circuit element,*** and we have presented the definitions of several specific circuit elements, the independent and dependent voltage and current sources. Throughout the remainder of the book we will define only five additional circuit elements: the resistor, inductor, capacitor, transformer, and the ideal operational

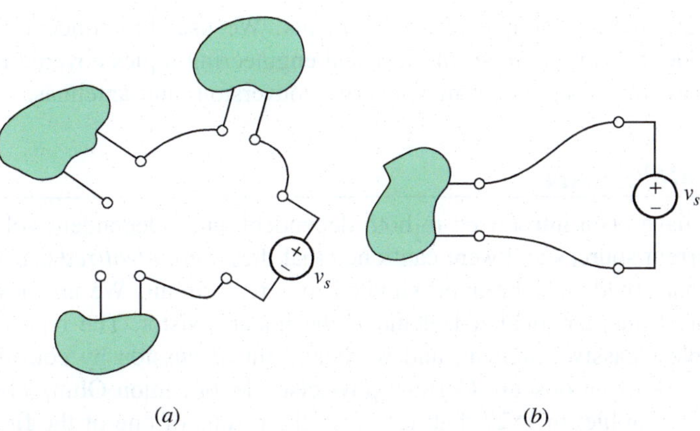

■ **FIGURE 2.21** (*a*) A network that is not a circuit. (*b*) A network that is a circuit.

amplifier ("op amp," for short). These are all ideal elements. They are important because we may combine them into networks and circuits that represent real devices as accurately as we require. Thus, the transistor shown in Fig. 2.22*a* and *b* may be modeled by the voltage terminals designated v_{gs} and the single dependent current source of Fig. 2.22*c*. Note that the dependent current source produces a current that depends on a voltage elsewhere in the circuit. The parameter g_m, commonly referred to as the transconductance, is calculated using transistor-specific details as well as the operating point determined by the circuit connected to the transistor. It is generally a small number, on the order of 10^{-2} to perhaps 10 A/V. This model works pretty well as long as the frequency of any sinusoidal source is neither very large nor very small; the model can be modified to account for frequency-dependent effects by including additional ideal circuit elements.

Similar (but much smaller) transistors typically constitute only one small part of an integrated circuit that may be less than 2 mm × 2 mm square and 200 μm thick and yet contain several thousand transistors plus various resistors and capacitors. Thus, we may have a physical device that is about the size of one letter on this page but requires a model composed

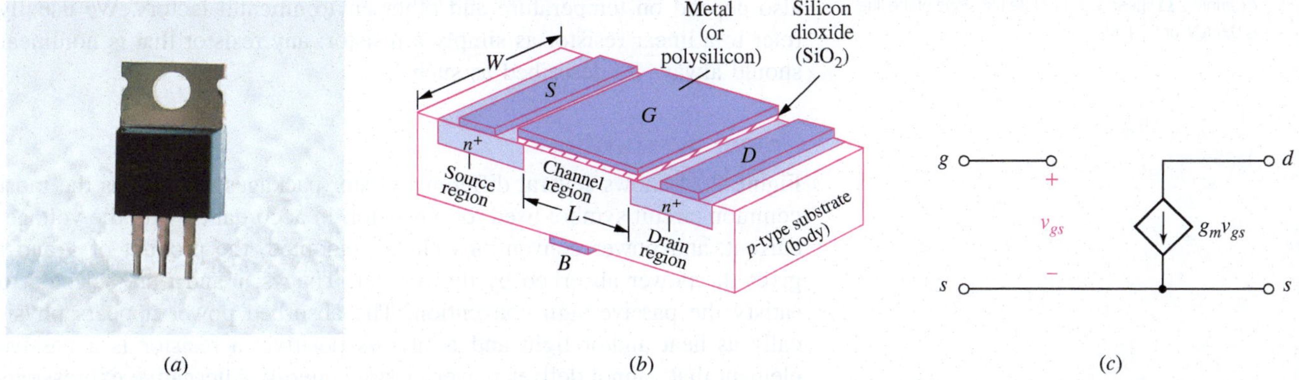

■ **FIGURE 2.22** The metal oxide semiconductor field effect transistor (MOSFET). (*a*) An IRF540 N-channel power MOSFET in a TO-220 package, rated at 100 V and 22 A; (*b*) cross-sectional view of a basic MOSFET; (*c*) equivalent circuit model for use in ac circuit analysis.

(*a*) Steve Durbin/McGraw Hill (*b*) R. Jaeger, *Microelectronic Circuit Design,* McGraw Hill, 1997 of ten thousand ideal simple circuit elements. We use this concept of "circuit modeling" in a number of electrical engineering topics covered in other courses, including electronics, energy conversion, and antennas.

2.4 OHM'S LAW

So far, we have been introduced to both dependent and independent voltage and current sources and were cautioned that they were *idealized* active elements that could only be approximated in a real circuit. We are now ready to meet another idealized element, the linear resistor. The resistor is the simplest passive element, and we begin our discussion by considering the work of an obscure German physicist, Georg Simon Ohm, who published a pamphlet in 1827 that described the results of one of the first efforts to measure currents and voltages, and to describe and relate them mathematically. One result was a statement of the fundamental relationship we now call ***Ohm's law,*** even though it has since been shown that this result was discovered 46 years earlier in England by Henry Cavendish, a brilliant semirecluse.

Ohm's law states that the voltage across conducting materials is directly proportional to the current flowing through the material, or

$$v = Ri \quad [11]$$

where the constant of proportionality R is called the ***resistance.*** The unit of resistance is the *ohm*, which is 1 V/A and customarily abbreviated by a capital omega, Ω.

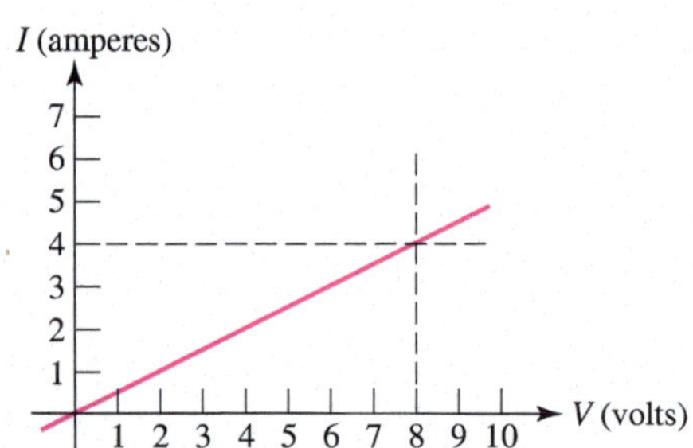

FIGURE 2.23 Current–voltage relationship for an example 2 Ω linear resistor. Note the slope of the line is 0.5 A/V, or 500 $m\Omega^{-1}$.

When this equation is plotted on i-versus-v axes, the graph is a straight line passing through the origin (Fig. 2.23). Equation [4] is a linear equation, and we will consider it to be the definition of a *linear resistor*. Resistance is normally considered to be a positive quantity, although negative resistances may be simulated with special circuitry.

Again, it must be emphasized that the linear resistor is an idealized circuit element; it is only a mathematical model of a real, physical device. "Resistors" may be easily purchased or manufactured, but it is soon found that the voltage–current ratios of these physical devices are reasonably constant only within certain ranges of current, voltage, or power, and they also depend on temperature and other environmental factors. We usually refer to a linear resistor as simply a resistor; any resistor that is nonlinear should always be described as such.

Power Absorption

Figure 2.24 shows several different resistor packages, as well as the most common circuit symbol used for a resistor. In accordance with the voltage, current, and power conventions already adopted, the product of v and i gives the power absorbed by the resistor. That is, v and i are selected to satisfy the passive sign convention. The absorbed power appears physically as heat and/or light and is always positive; a resistor is a passive element that cannot deliver power or store energy. Alternative expressions for the absorbed power are

$$p = vi = i^2R = v^2/R \quad [12]$$

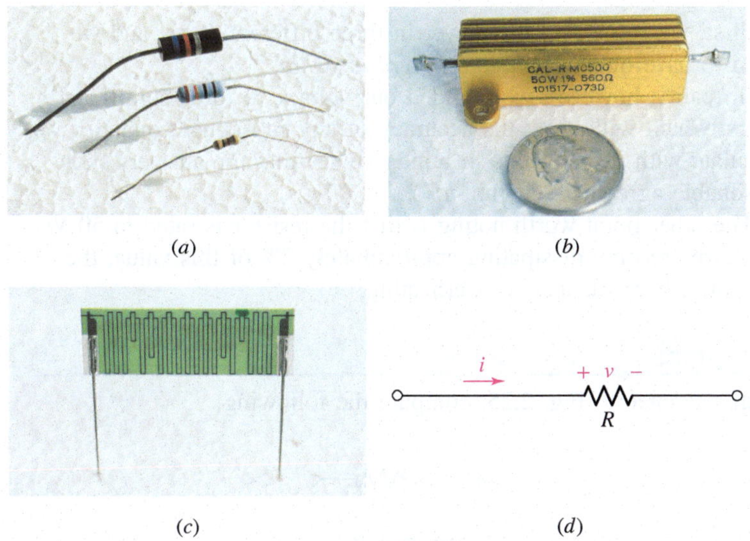

(*a*) (*b*)

i $+\ v\ -$ R

(*c*) (*d*)

FIGURE 2.24 (*a*) Several common resistor packages. (*b*) A 560 Ω power resistor rated at up to 50 W. (*c*) A 5% tolerance 10-teraohm (10,000,000,000,000 Ω) resistor manufactured by Ohmcraft. (*d*) Circuit symbol for the resistor, applicable to all of the devices in (*a*) through (*c*). (*a–c*) Steve Durbin/McGraw Hill

One of the authors (who shall remain anonymous) had the unfortunate experience of inadvertently connecting a 100 Ω, 2 W carbon resistor across a 110 V source. The ensuing flame, smoke, and fragmentation were rather disconcerting, demonstrating clearly that a practical resistor has definite limits to its ability to behave like the ideal linear model. In this case, the unfortunate resistor was called upon to absorb 121 W; since it was designed to handle only 2 W, its reaction was understandably violent.

EXAMPLE 2.4

The 560 Ω resistor shown in Fig. 2.24*b* is connected to a circuit which causes a current of 42.4 mA to flow through it. Calculate the voltage across the resistor and the power it is dissipating.

The voltage across the resistor is given by Ohm's law:

$$v = Ri = (560)(0.0424) = 23.7 \text{ V}$$

The dissipated power can be calculated in several different ways. For instance,

$$p = vi = (23.7)(0.0424) = 1.005 \text{ W}$$

Alternatively,

$$p = v^2/R = (23.7)^2/560 = 1.003 \text{ W}$$

or

$$p = i^2R = (0.0424)^2(560) = 1.007 \text{ W}$$

We note several things.

(*Continued on next page*)

First, we calculated the power in three different ways, and we seem to have obtained *three different answers*!

In reality, however, we rounded our voltage to three significant digits, which will affect the accuracy of any subsequent quantity we calculate with it. With this in mind, we see that the answers show reasonable agreement (within 1%).

The other point worth noting is that the resistor is rated to 50 W—since we are only dissipating approximately 2% of this value, the resistor is in no danger of overheating.

PRACTICE

With reference to Fig. 2.25, compute the following:

■ **FIGURE 2.25**

2.11 R if $i = -2\ \mu$A and $v = -44$ V.

2.12 The power absorbed by the resistor if $v = 1$ V and $R = 2$ kΩ.

2.13 The power absorbed by the resistor if $i = 3$ nA and $R = 4.7$ MΩ.

Ans: 22 MΩ; 500 μW; 42.3 pW.

PRACTICAL APPLICATION

Wire Gauge

Technically speaking, any material (except for a superconductor) will provide resistance to current flow. As in all introductory circuits texts, however, we tacitly assume that wires appearing in circuit diagrams have zero resistance. This implies that there is no potential difference between the ends of a wire, and hence no power absorbed or heat generated. Although usually not an unreasonable assumption, it does neglect practical considerations when choosing the appropriate wire diameter for a specific application.

Resistance is determined by (1) the inherent resistivity of a material and (2) the device geometry. ***Resistivity,*** represented by the symbol ρ, is a measure of the ease with which electrons can travel through a certain material. Since it is the ratio of the electric field (V/m) to the areal density of current flowing in the material (A/m^2), the general unit of ρ is an $\Omega \cdot$ m, although metric prefixes are often employed. Every material has a different inherent resistivity, which depends on temperature. Some examples are shown in Table 2.3; as can be seen, there is a small variation between different types of copper (less than 1%) but a very large difference between different metals. In particular, although physically stronger than copper, steel wire is several times more resistive. In some technical discussions, it is more common to see the conductivity (symbolized by σ) of a material quoted, which is simply the reciprocal of the resistivity.

The resistance of a particular object is obtained by multiplying the resistivity by the length ℓ of the resistor and dividing by the cross-sectional area (A) as in Eq. [6]; these parameters are illustrated in Fig. 2.26.

$$R = \rho \frac{\ell}{A} \qquad [13]$$

TABLE 2.3 Common Electrical Wire Materials and Resistivities*

ASTM Specification**	Temper and Shape	Resistivity at 20°C ($\mu\Omega \cdot$ cm)
B33	Copper, tinned soft, round	1.7654
B75	Copper, tube, soft, OF copper	1.7241
B188	Copper, hard bus tube, rectangular or square	1.7521
B189	Copper, lead-coated soft, round	1.7654
B230	Aluminum, hard, round	2.8625
B227	Copper-clad steel, hard, round, grade 40 HS	4.3971
B355	Copper, nickel-coated soft, round Class 10	1.9592
B415	Aluminum-clad steel, hard, round	8.4805

* C. B. Rawlins, "Conductor materials," *Standard Handbook for Electrical Engineering*, 13th ed., D. G. Fink and H. W. Beaty, eds. New York: McGraw Hill, 1993, pp. 4-4 to 4-8.

** American Society of Testing and Materials.

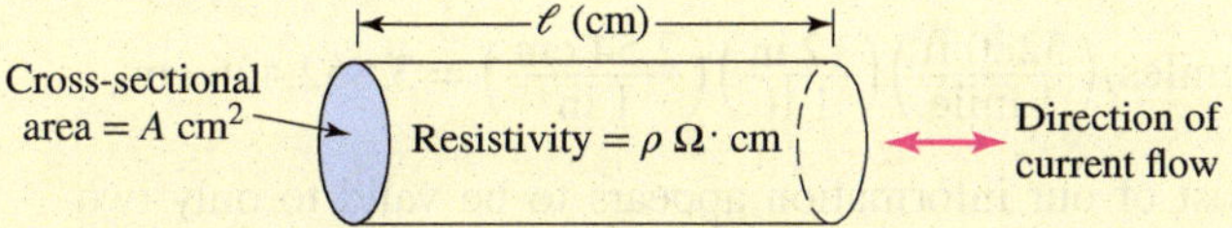

FIGURE 2.26 Definition of geometrical parameters used to compute the resistance of a wire. The resistivity of the material is assumed to be spatially uniform.

We determine the resistivity when we select the material from which to fabricate a wire and measure the temperature of the application environment. Since a finite amount of power is absorbed by the wire due to its resistance, current flow leads to the production of heat. Thicker wires have lower resistance and also dissipate heat more easily but are heavier, take up a larger volume, and are more expensive. Thus, we are motivated by practical considerations to choose the smallest wire that can safely do the job, rather than simply choosing the largest-diameter wire available in an effort to minimize resistive losses. The American Wire Gauge (AWG) is a standard system of specifying wire size. In selecting a wire gauge, smaller AWG corresponds to a larger wire diameter; an abbreviated table of common gauges is given in Table 2.4. Local fire and electrical safety codes typically dictate the required gauge for specific wiring applications, based on the maximum current expected as well as where the wires will be located.

TABLE 2.4 Some Common Wire Gauges and the Resistance of (Soft) Solid Copper Wire*

Conductor Size (AWG)	Cross-Sectional Area (mm^2)	Ohms per 1000 ft at 20°C
28	0.0804	65.3
24	0.205	25.7
22	0.324	16.2
18	0.823	6.39
14	2.08	2.52
12	3.31	1.59
6	13.3	0.3952
4	21.1	0.2485
2	33.6	0.1563

* C. B. Rawlins et al., *Standard Handbook for Electrical Engineering*, 13th ed., D. G. Fink and H. W. Beaty, eds. New York: McGraw Hill, 1993, p. 4–47.

EXAMPLE 2.5

A dc power link is to be made between two islands separated by a distance of 24 miles. The operating voltage is 500 kV and the system capacity is 600 MW. Calculate the maximum dc current flow, and estimate the resistivity of the cable, assuming a diameter of 2.5 cm and a solid (not stranded) wire.

Dividing the maximum power (600 MW, or 600×10^6 W) by the operating voltage (500 kV, or 500×10^3 V) yields a maximum current of

$$\frac{600 \times 10^6}{500 \times 10^3} = 1200 \text{ A}$$

The cable resistance is simply the ratio of the voltage to the current, or

$$R_{\text{cable}} = \frac{500 \times 10^3}{1200} = 417\ \Omega$$

We know the length:

$$\ell = (24 \text{ miles})\left(\frac{5280 \text{ ft}}{1 \text{ mile}}\right)\left(\frac{12 \text{ in}}{1 \text{ ft}}\right)\left(\frac{2.54 \text{ cm}}{1 \text{ in}}\right) = 3{,}862{,}426 \text{ cm}$$

Given that most of our information appears to be valid to only two significant figures, we round this to 3.9×10^6 cm. With the cable diameter specified as 2.5 cm, we know its cross-sectional area is 4.9 cm^2. Thus,

$$\rho_{\text{cable}} = R_{\text{cable}}\frac{A}{\ell} = 417\left(\frac{4.9}{3.9 \times 10^6}\right) = 520\ \mu\Omega \cdot \text{cm}$$

PRACTICE

2.14 A 500 ft long 24 AWG soft copper wire is carrying a current of 100 mA. What is the voltage dropped across the wire?

Ans: 3.26 V.

Fuses

The capacity for handling current flow is an important consideration when designing circuits. Electrical components and electrical leads should be capable of carrying the current flow that the circuit is designed for. For example, you would not want to use 1/8 watt resistors or hairline thin wiring for power handling in an electric car! Similarly, it is desirable to minimize hazards associated with unintentional short circuit conditions. Such unintentional short circuits could cause very large spikes in current that can damage electrical components, or more severely, cause a fire or electric shock. To protect against overcurrent conditions, fuses are often incorporated in a series connection to the circuit. A fuse is simply a resistor that is specially designed to (safely) fail at a particular current condition. At this level of current flow, the material in the fuse will melt and result in an open-circuit condition that protects the circuit from dangerous current levels. In addition to designing a fuse to fail at a particular current level, fuses also aim to achieve very low resistance to minimize power consumption. Blowing a fuse is very similar to the failure mechanism in burning out an incandescent light bulb, and it needs to be replaced following failure. Circuit breakers may also be used to provide overcurrent protection. The physical mechanism in a circuit breaker for preventing overcurrent is very different than a fuse, and it may also be reset and reused following a "trip" condition. The size and cost of circuit breakers are often much higher than fuses, and the choice of which device to use is dependent on the needs for the particular application.

Conductance

For a linear resistor, the ratio of current to voltage is also a constant

$$\frac{i}{v} = \frac{1}{R} = G \qquad [14]$$

where G is called the *conductance*. The SI unit of conductance is the siemens (S), or A/V. An older, unofficial unit for conductance is the mho, which was often abbreviated as ℧ and is still occasionally written as Ω^{-1}. You will occasionally see it used on some circuit diagrams, as well as in catalogs and texts. The same circuit symbol (Fig. 2.24*d*) is used to represent both resistance and conductance. The absorbed power is again necessarily positive and may be expressed in terms of the conductance by

$$p = vi = v^2G = \frac{i^2}{G} \qquad [15]$$

Thus a 2 Ω resistor has a conductance of $\frac{1}{2}$ S, and if a current of 5 A is flowing through it, then a voltage of 10 V is present across the terminals and a power of 50 W is being absorbed.

All the expressions given so far in this section were written in terms of instantaneous current, voltage, and power, such as $v = iR$ and $p = vi$. We should recall that this is a shorthand notation for $v(t) = Ri(t)$ and $p(t) = v(t)i(t)$. The current through and voltage across a resistor must both vary with time in the same manner. Thus, if $R = 10\ \Omega$ and $v = 2 \sin 100t$ V, then $i = 0.2 \sin 100t$ A. Note that the power is given by $0.4 \sin^2 100t$ W, and a simple sketch will illustrate the different nature of its variation with time. Although the current and voltage are each negative during certain time intervals, the absorbed power is *never* negative!

COMPUTER-AIDED ANALYSIS

There are a number of software packages designed to assist with engineering problems. Some are highly specialized, and others are rather general. One package that fits well within the latter category is MATLAB. Its syntax is similar to that of the C programming language, and commands can be executed in a line-by-line fashion or as a self-contained program. Comments can be included using the percent sign (%), and a semicolon at the end of a line suppresses "printing" of the output result for a line of code. Consider the calculations in Example 2.4. In MATLAB, we can write it line-by-line as shown in Fig. 2.27*a*.

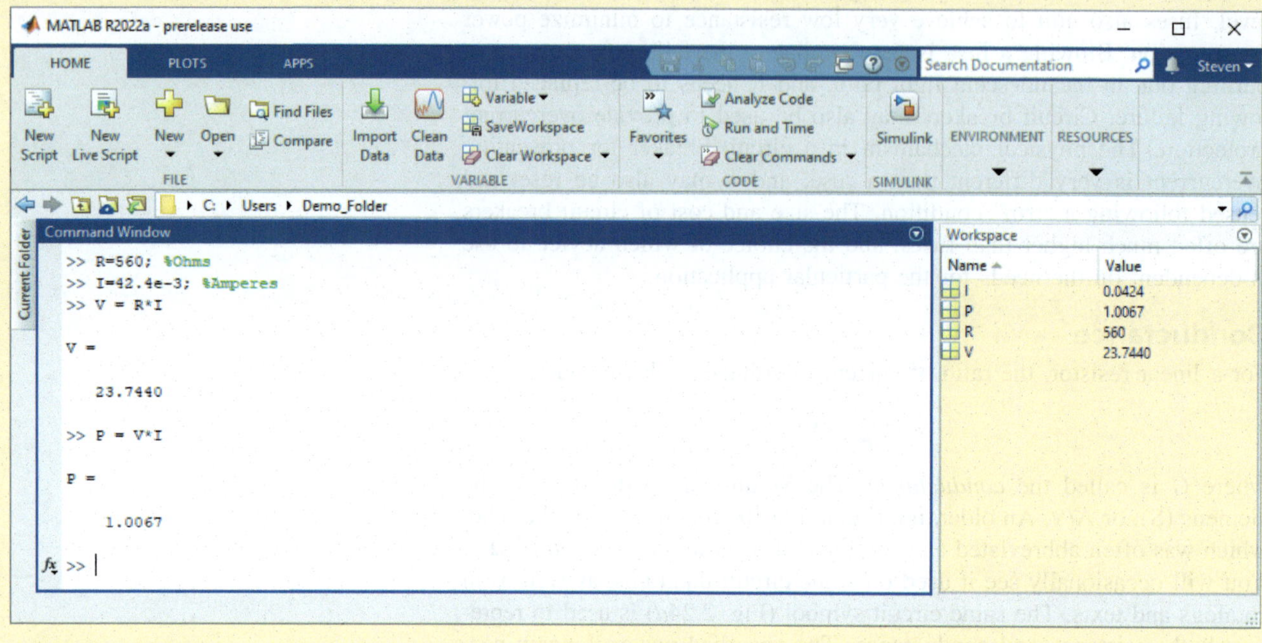

(*a*)

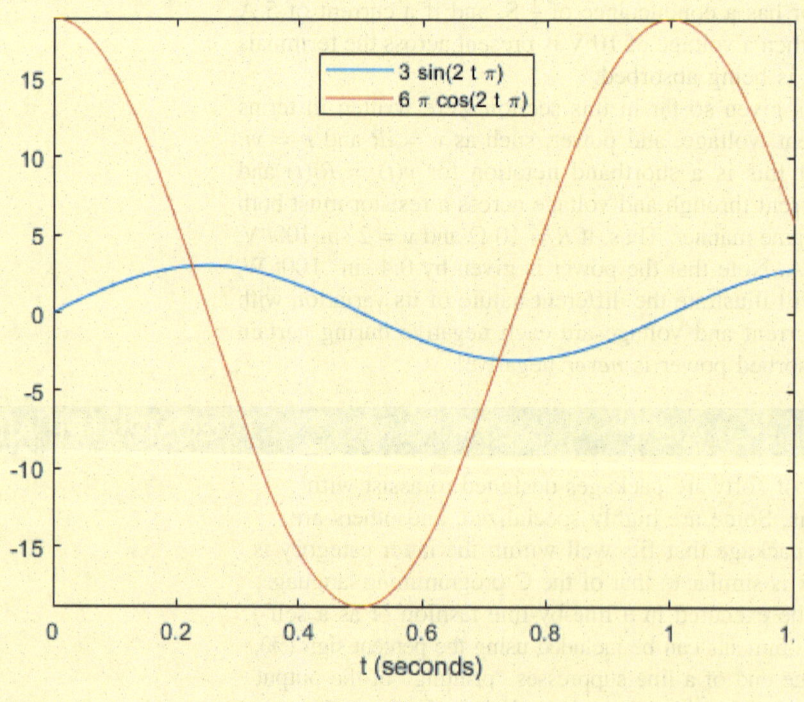

(*b*)

FIGURE 2.27 (*a*) Example 2.4 performed within MATLAB. The symbol >> is the cursor. (*b*) Example graphical output.

(*a*) MATLAB

MATLAB is also very helpful in plotting functions, and performing significantly more complex calculations. For example, consider the function $q(t) = 3*\sin(2\pi t)$. If $q(t)$ represents the charge in coulombs that crosses a cross-section of a wire, the current flowing through the wire is obtained by taking the first derivative of $q(t)$. This can be accomplished within MATLAB by using what is termed a symbolic variable:

```
>> syms t;
>> q = 3*sin(2*pi*t);
>> current = diff(q) % avoiding 'i' as a variable as reserved for
imaginary number

current =

6*pi*cos(2*pi*t)

>> fplot(q, [0 1.2]); % restricting range to 0 to 1.2 seconds
>> hold;             % keeps current plot so we can add to it
Current plot held
>> fplot(current, [0 1.2]);
>> xlabel ('t (seconds)'); % add axis label
>> legend            % add legend since we have two functions
plotted
```

The charge and current are both plotted in Fig. 2.27*b*.

Resistance may be used as the basis for defining two commonly used terms, *short circuit* and *open circuit*. We define a short circuit as a resistance of zero ohms; then, since $v = iR$, the voltage across a short circuit must be zero, although the current may have any value. In an analogous manner, we define an open circuit as an infinite resistance. It follows from Ohm's law that the current must be zero, regardless of the voltage across the open circuit. Although real wires have a small resistance associated with them, we always assume them to have zero resistance unless otherwise specified. Thus, in all of our circuit schematics, wires are taken to be perfect short circuits.

SUMMARY AND REVIEW

In this chapter, we introduced the topic of units—specifically those relevant to electrical circuits—and their relationship to fundamental (SI) units. We also discussed current and current sources, voltage and voltage sources, and the fact that the product of voltage and current yields power (the rate of energy consumption or generation). Since power can be either positive or negative depending on the current direction and voltage polarity, the passive sign convention was described to ensure we always know if an element is *absorbing* or *supplying* energy to the rest of the circuit. Four additional sources were introduced, forming a general class known as dependent sources. They are often used to model complex systems and electrical components, but the actual value of voltage or current supplied is typically unknown until the entire circuit is analyzed.

We concluded the chapter with the resistor—by far the most common circuit element—whose voltage and current are linearly related (described by Ohm's law). Whereas the *resistivity* of a material is one of its fundamental properties (measured in $\Omega \cdot$ cm), *resistance* describes a device property (measured in Ω) and hence depends not only on resistivity but on the device geometry (i.e., length and area) as well.

We wrap up with key points of this chapter to review, along with appropriate examples.

- The system of units most commonly used in electrical engineering is the SI.
- The direction in which positive charges are moving is the direction of positive current flow; alternatively, positive current flow is in the direction opposite that of moving electrons.
- To define a current, both a value and a direction must be given. Currents are typically denoted by the uppercase letter I for constant (dc) values, and either $i(t)$ or simply i otherwise.
- To define a voltage across an element, it is necessary to label the terminals with + and − signs as well as to provide a value (either an algebraic symbol or a numerical value).
- Any element is said to supply positive power if positive current flows out of the positive voltage terminal. Any element absorbs positive power if positive current flows into the positive voltage terminal. (Example 2.1 and Example 2.2)
- There are six types of sources: the independent voltage source, the independent current source, the current-controlled dependent current source, the voltage-controlled dependent current source, the voltage-controlled dependent voltage source, and the current-controlled dependent voltage source. (Example 2.3)
- Ohm's law states that the voltage across a linear resistor is directly proportional to the current flowing through it; i.e., $v = Ri$. (Example 2.4)
- The power dissipated by a resistor (which leads to the production of heat) is given by $p = vi = i^2R = v^2/R$. (Example 2.4)
- Wires are typically assumed to have zero resistance in circuit analysis. When selecting a wire gauge for a specific application, however, local electrical and fire codes must be consulted. (Example 2.5)

READING FURTHER

A good book that discusses the properties and manufacture of resistors in considerable depth:

F. Zandman, P.-R. Simon, and J. Szwarc, *Resistor Theory and Technology.* Raleigh, N.C.: SciTech Publishing, 2002.

A good all-purpose electrical engineering handbook:

D. G. Fink and H. Wayne Beaty, *Standard Handbook for Electrical Engineers*, 16th ed. New York: McGraw Hill, 2013.

A detailed reference for the SI is available on the Web from the National Institute of Standards:

A. Thompson and B. N. Taylor, *Guide for the Use of the International System of Units (SI)*, NIST Special Publication 811, 2008 edition, www.nist.gov.

EXERCISES

2.1 Units and Scales

1. Convert the following to engineering notation:
   (*a*) 0.045 W (*b*) 2000 pJ
   (*c*) 0.1 ns (*d*) 39,212 as
   (*e*) 3 Ω (*f*) 18,000 m
   (*g*) 2,500,000,000,000 bits (*h*) 10^{15} atoms/cm^3
2. Convert the following to engineering notation:
   (*a*) 1230 fs (*b*) 0.0001 decimeter
   (*c*) 1400 mK (*d*) 32 nm
   (*e*) 13,560 kHz (*f*) 2021 micromoles
   (*g*) 13 deciliters (*h*) 1 hectometer
3. Express the following in engineering units:
   (*a*) 1212 mV (*b*) 10^{11} pA
   (*c*) 1000 yoctoseconds (*d*) 33.9997 zeptoseconds
   (*e*) 13,100 attoseconds (*f*) 10^{-14} zettasecond
   (*g*) 10^{-5} second (*h*) 10^{-9} Gs
4. Expand the following distances in simple meters:
   (*a*) 1 Zm (*b*) 1 Em
   (*c*) 1 Pm (*d*) 1 Tm
   (*e*) 1 Gm (*f*) 1 Mm
5. Convert the following to SI units, taking care to employ proper engineering notation:
   (*a*) 212°F (*b*) 0°F
   (*c*) 0 K (*d*) 200 hp
   (*e*) 1 yard (*f*) 1 mile
6. Convert the following to SI units, taking care to employ proper engineering notation:
   (*a*) 100°C (*b*) 0°C
   (*c*) 4.2 K (*d*) 150 hp
   (*e*) 500 Btu (*f*) 100 J/s
7. It takes you approximately 2 hours to finish your homework on thermodynamics. Since it feels like it took forever, how many galactic years does this correspond to? (1 galactic year = 250 million years)
8. A certain krypton fluoride laser generates 15 ns long pulses, each of which contains 550 mJ of energy. (*a*) Calculate the peak instantaneous output power of the laser. (*b*) If up to 100 pulses can be generated per second, calculate the maximum average power output of the laser.
9. Your recommended daily food intake is 2500 food calories (kcal). If all of this energy is efficiently processed, what would your average power output be?
10. An electric vehicle is driven by a single motor rated at 40 hp. If the motor is run continuously for 3 h at maximum output, calculate the electrical energy consumed. Express your answer in SI units using engineering notation.
11. Under insolation conditions of 500 W/m^2 (direct sunlight), and 10% solar cell efficiency (defined as the ratio of electrical output power to incident solar power), calculate the area required for a photovoltaic (solar cell) array capable of running the vehicle in Exercise 10 at half power.

12. A certain metal oxide nanowire piezoelectricity generator is capable of producing 100 pW of usable electricity from the type of motion obtained from a person jogging at a moderate pace. (*a*) How many nanowire devices are required to operate a personal MP3 player that draws 1 W of power? (*b*) If the nanowires can be produced with a density of five devices per square micron directly onto a piece of fabric, what area is required, and would it be practical?
13. Assuming a global population of 9 billion people, each using approximately 100 W of power continuously throughout the day, calculate the total land area that would have to be set aside for photovoltaic power generation, assuming 800 W/m^2 of incident solar power and a conversion efficiency (sunlight to electricity) of 10%.

2.2 Charge, Current, Voltage, Power, and Energy

14. The total charge flowing out of one end of a small copper wire and into an unknown device is determined to follow the relationship $q(t) = 5e^{-t/2}$ C, where t is expressed in seconds. Calculate the current flowing into the device, taking note of the sign.
15. The current flowing into the collector lead of a certain bipolar junction transistor (BJT) is measured to be 1 nA. If no charge was transferred in or out of the collector lead prior to $t = 0$, and the current flows for 1 min, calculate the total charge which crosses into the collector.
16. The total charge stored on a 1 cm diameter insulating plate is -10^{13} C. (*a*) How many electrons are on the plate? (*b*) What is the areal density of electrons (number of electrons per square meter)? (*c*) If additional electrons are added to the plate from an external source at the rate of 10^6 electrons per second, what is the magnitude of the current flowing between the source and the plate?
17. A mysterious device found in a forgotten laboratory accumulates charge at a rate specified by the expression $q(t) = 9 - 10t$ C from the moment it is switched on. (*a*) Calculate the total charge contained in the device at $t = 0$. (*b*) Calculate the total charge contained at $t = 1$ s. (*c*) Determine the current flowing into the device at $t = 1$ s, 3 s, and 10 s.
18. A new type of device appears to accumulate charge according to the expression $q(t) = 10t^2 - 22t$ mC (t in s). (*a*) In the interval $0 \le t < 5$ s, at what time does the current flowing into the device equal zero? (*b*) Sketch $q(t)$ and $i(t)$ over the interval $0 \le t < 5$ s.
19. The current flowing through a tungsten-filament light bulb is determined to follow $i(t) = 114 \sin(100\pi t)$ A. (*a*) Over the interval defined by $t = 0$ and $t = 2$ s, how many times does the current equal zero amperes? (*b*) How much charge is transported through the light bulb in the first second?
20. The current waveform depicted in Fig. 2.28 is characterized by a period of 8 s. (*a*) What is the average value of the current over a single period? (*b*) If $q(0) = 0$, sketch $q(t)$, $0 < t < 20$ s.

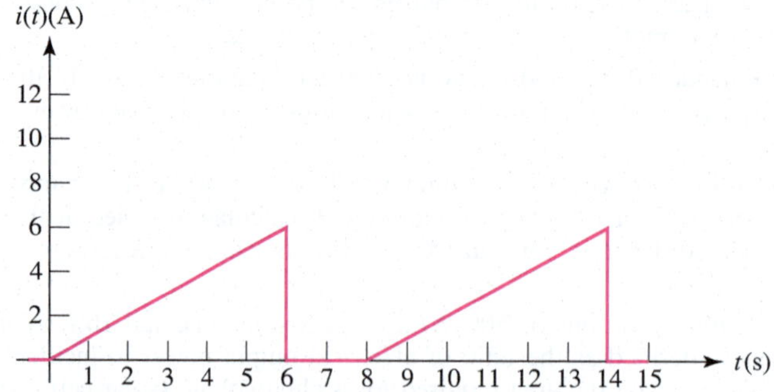

FIGURE 2.28 An example of a time-varying current.

21. The current waveform depicted in Fig. 2.29 is characterized by a period of 4 s. (*a*) What is the average value of the current over a single period? (*b*) Compute the average current over the interval $1 < t < 3$ s. (*c*) If $q(0) = 1\text{C}$, sketch $q(t)$, $0 < t < 4$ s.

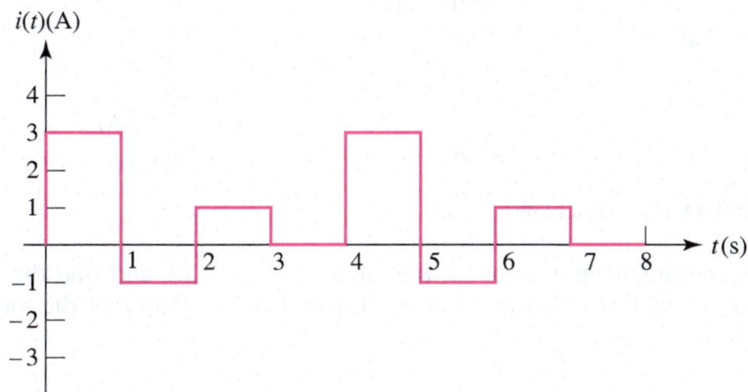

FIGURE 2.29 An example of a time-varying current.

22. A wind power system with increasing windspeed has the current waveform described by the equation below, delivered to an 80 Ω resistor. Plot the current, power, and energy waveform over a period of 60 s, and calculate the total energy collected over the 60 s time period.

$$i(t) = \frac{1}{2}t^2 \sin\left(\frac{\pi}{8}t\right)\cos\left(\frac{\pi}{4}t\right) A$$

23. Two metallic terminals protrude from a device. The terminal on the left is the positive reference for a voltage called v_x (the other terminal is the negative reference). The terminal on the right is the positive reference for a voltage called v_y (the other terminal being the negative reference). If it takes 1 mJ of energy to push a single electron into the left terminal, determine the voltages v_x and v_y.
24. The convention for voltmeters is to use a black wire for the negative reference terminal and a red wire for the positive reference terminal. (*a*) Explain why two wires are required to measure a voltage. (*b*) If it is dark and the wires into the voltmeter are swapped by accident, what will happen during the next measurement?
25. Determine the power absorbed by each of the elements in Fig. 2.30.

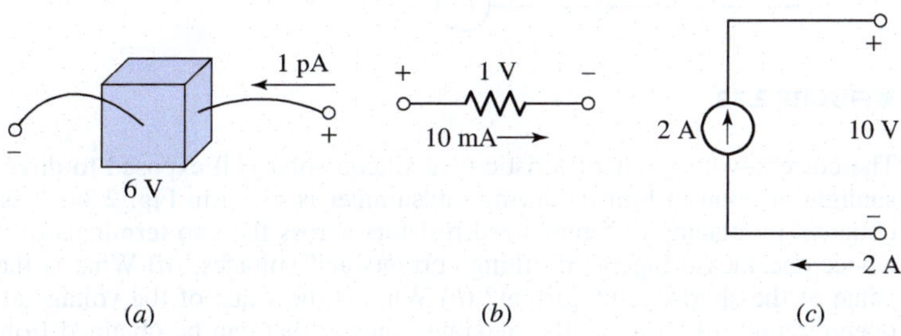

FIGURE 2.30 Elements for Exercise 25.

26. Determine the power absorbed by each of the elements in Fig. 2.31.

■ **FIGURE 2.31** Elements for Exercise 26.

27. Determine the unknown current for the circuit in Fig. 2.32, and find the power that is supplied or absorbed by each element. Confirm that the total power is zero.

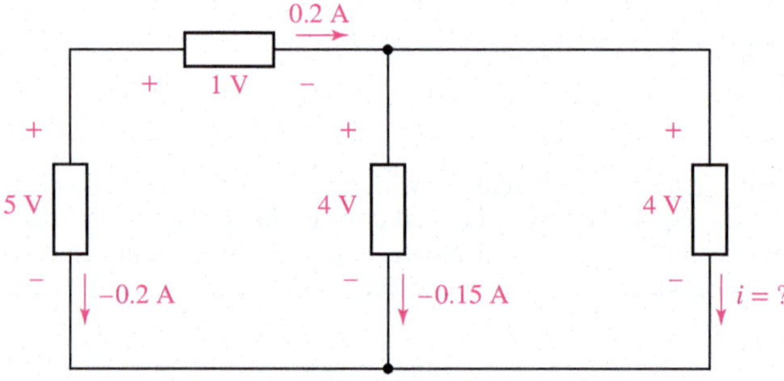

■ **FIGURE 2.32**

28. A constant current of 1 ampere is measured flowing into the positive reference terminal of a pair of leads whose voltage we'll call v_p. Calculate the absorbed power at $t = 1$ s if $v_p(t)$ equals (*a*) +1 V; (*b*) −1 V; (*c*) $2 + 5\cos(5t)$ V; (*d*) $4e^{-2t}$ V. (*e*) Explain the significance of a negative value for absorbed power.

29. Determine the power supplied by the leftmost element in the circuit of Fig. 2.33.

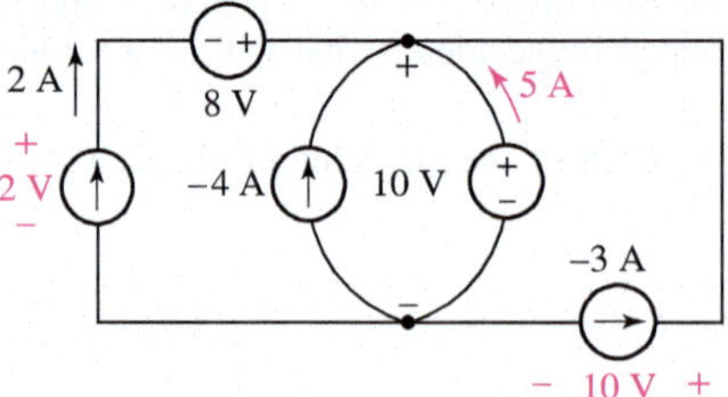

■ **FIGURE 2.33**

30. The current–voltage characteristic of a silicon solar cell exposed to direct sunlight at noon in Florida during midsummer is given in Fig. 2.34. It is obtained by placing different-sized resistors across the two terminals of the device and measuring the resulting currents and voltages. (*a*) What is the value of the short-circuit current? (*b*) What is the value of the voltage at open circuit? (*c*) Estimate the maximum power that can be obtained from the device.

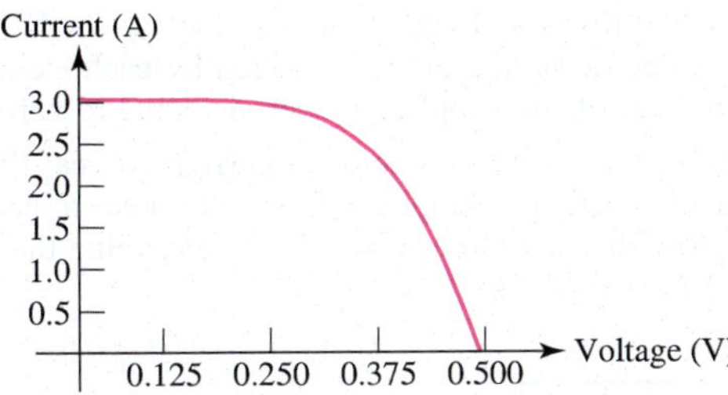

FIGURE 2.34

31. A particular electric utility charges customers different rates depending on their daily rate of energy consumption: \$0.05/kWh up to 20 kWh, and \$0.10/kWh for all energy usage above 20 kWh in any 24-hour period. (*a*) Calculate how many 100 W light bulbs can be run continuously for less than \$10 per week. (*b*) Calculate the daily energy cost if 2000 kW of power is used continuously.
32. The Tilting Windmill Electrical Cooperative LLC Inc. has instituted a differential pricing scheme aimed at encouraging customers to conserve electricity use during daylight hours, when local business demand is at its highest. If the price per kilowatthour is \$0.033 between the hours of 9 p.m. and 6 a.m., and \$0.057 for all other times, how much does it cost to run a 2.5 kW portable heater continuously for 30 days?
33. A laptop computer consumes an average power of 20 W. The rechargeable battery has a voltage of 12 V and capacity of 5800 mAh. How long can the laptop run on a single battery charge?
34. You have just installed a rooftop solar photovoltaic system that consists of 40 solar modules that each provide 180 W of power under peak sunlight conditions. Your location gets an average of 5 hours of peak sunlight per day. If electricity in your area is valued at 15 ¢/kWh, what is the annual value of the electricity generated by your installation?
35. A portable music player requiring 5 W is powered by a 3.7 V Li-ion battery with capacity of 4000 mAh. The battery can be charged by a charger providing a current of 2 A with an efficiency of 80%. (*a*) How long can the music player run on a full battery charge? (*b*) How long would it take to fully charge the battery?

2.3 Voltage and Current Sources

36. Some of the ideal sources in the circuit of Fig. 2.33 are supplying positive power, and others are absorbing positive power. Determine which are which, and show that the algebraic sum of the power absorbed by each element (taking care to preserve signs) is equal to zero.
37. You are comparing an old incandescent light bulb with a newer high-efficiency LED light bulb. You find that they both have the same output of 800 lumens, which corresponds to approximately 5 W of optical power. However, you find that the incandescent bulb is consuming 60 W of electrical power, and the LED bulb is consuming 12 W of electrical power. Why do the optical and electrical powers not agree? Doesn't conservation of energy require the two quantities to be equal?
38. Refer to the circuit represented in Fig. 2.35, while noting that the same current flows through each element. The voltage-controlled dependent source provides a current which is five times as large as the voltage V_x. (*a*) For $V_R = 10$ V and $V_x = 2$ V, determine the power absorbed by each element. (*b*) Is element *A* likely a passive or active source? Explain.
39. Refer to the circuit represented in Fig. 2.35, while noting that the same current flows through each element. The voltage-controlled dependent source

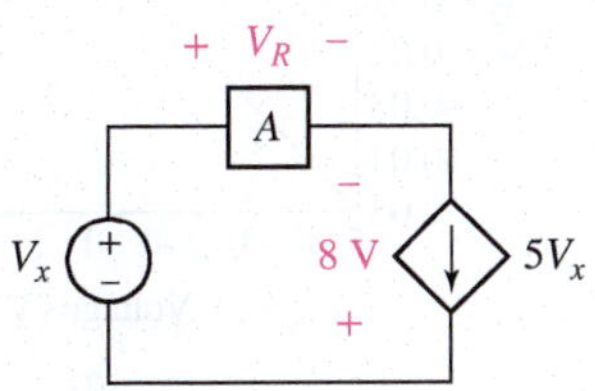

FIGURE 2.35

provides a current which is five times as large as the voltage V_x. (*a*) For $V_R = 100$ V and $V_x = 92$ V, determine the power supplied by each element. (*b*) Verify that the algebraic sum of the supplied powers is equal to zero.

40. The circuit depicted in Fig. 2.36 contains a dependent current source; the magnitude and direction of the current it supplies are directly determined by the voltage labeled v_1. Note that therefore $i_2 = -3\ v_1$. Determine the voltage v_1 if $v_2 = 33\ i_2$ and $i_2 = 100$ mA.

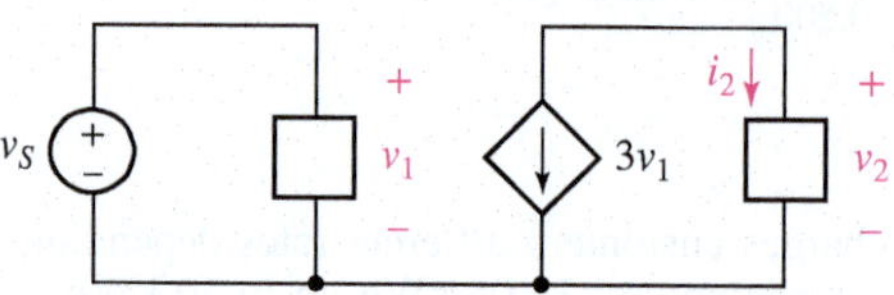

FIGURE 2.36

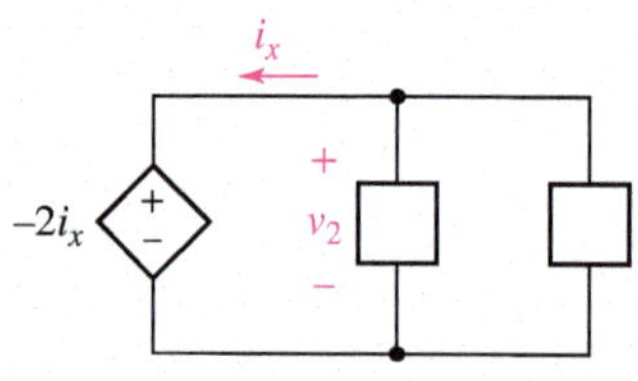

FIGURE 2.37

41. The dependent source in the circuit of Fig. 2.37 provides a voltage whose value depends on the current i_x. What value of i_x is required for the dependent source to be supplying 1 W?

2.4 Ohm's Law

42. Determine the magnitude of the current flowing through a 4.7 kΩ resistor if the voltage across it is (*a*) 1 mV; (*b*) 10 V; (*c*) $4e^{-t}$ V; (*d*) 100 cos(5*t*) V; (*e*) −7 V.

43. Real resistors can only be manufactured to a specific tolerance, so in effect the value of the resistance is uncertain. For example, a 1 Ω resistor specified as 5% tolerance could in practice be found to have a value anywhere in the range of 0.95 to 1.05 Ω. Calculate the voltage across a 2.2 kΩ 10% tolerance resistor if the current flowing through the element is (*a*) 1 mA; (*b*) 4 sin 44*t* mA.

44. (*a*) Sketch the current–voltage relationship (current on the *y* axis) of a 2 kΩ resistor over the voltage range of $-10\text{ V} \leq V_{\text{resistor}} \leq +10$ V. Be sure to label both axes appropriately. (*b*) What is the numerical value of the slope (express your answer in siemens)?

45. Sketch the voltage across a 33 Ω resistor over the range $0 < t < 2\pi$ s, if the current is given by 2.8 cos(*t*) A. Assume both the current and voltage are defined according to the passive sign convention.

46. Figure 2.38 depicts the current–voltage characteristic of three different resistive elements. Determine the resistance of each, assuming the voltage and current are defined in accordance with the passive sign convention.

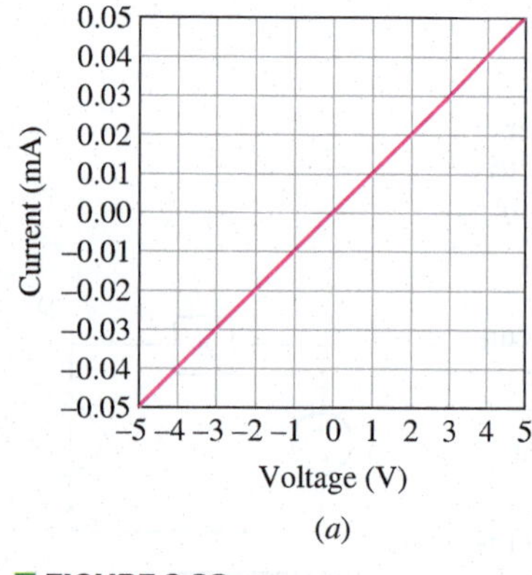

(*a*)

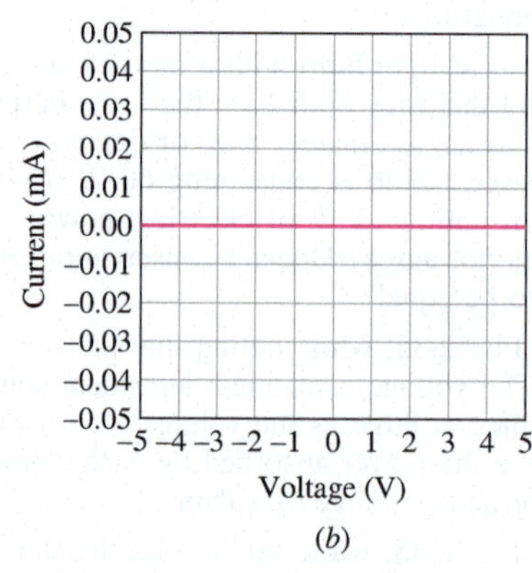

(*b*)

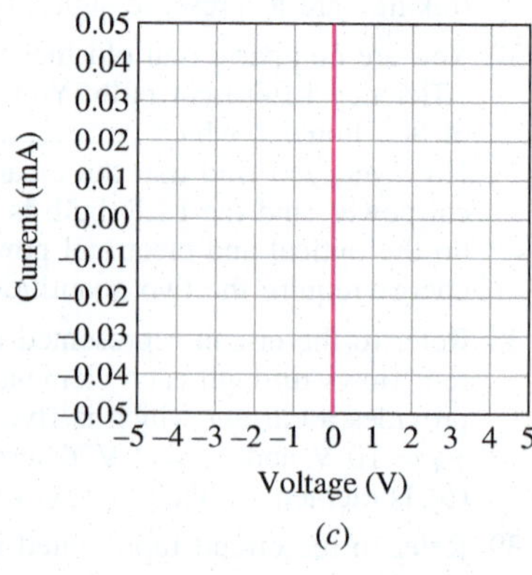

(*c*)

FIGURE 2.38

47. Examine the I-V characteristics in Fig. 2.38. Which would be the most desirable for a fuse? Explain.
48. Determine the conductance (in siemens) of the following: (*a*) 0 Ω; (*b*) 100 MΩ; (*c*) 200 mΩ.
49. Determine the magnitude of the current flowing through a 10 mS conductance if the voltage across it is (*a*) 2 mV; (*b*) −1 V; (*c*) $100e^{-2t}$ V; (*d*) 5 sin(5*t*) V; (*e*) 0 V.
50. A 1% tolerance 1 kΩ resistor may in reality have a value anywhere in the range of 990 to 1010 Ω. Assuming a voltage of 9 V is applied across it, determine (*a*) the corresponding range of current and (*b*) the corresponding range of absorbed power. (*c*) If the resistor is replaced with a 10% tolerance 1 kΩ resistor, repeat parts (*a*) and (*b*).
51. The following experimental data is acquired for an unmarked resistor, using a variable-voltage power supply and a current meter. The current meter readout is somewhat unstable, unfortunately, which introduces error into the measurement. (*a*) Plot the measured current-versus-voltage characteristic. (*b*) Using a best-fit line, estimate the value of the resistance.

Voltage (V)	Current (mA)
−2.0	−0.89
−1.2	−0.47
0.0	0.01
1.0	0.44
1.5	0.70

52. Utilize the fact that in the circuit of Fig. 2.39, the total power supplied by the voltage source must equal the total power absorbed by the two resistors to show that

$$V_{R_2} = V_S \frac{R_2}{R_1 + R_2}.$$

You may assume the same current flows through each element (a requirement of charge conservation).

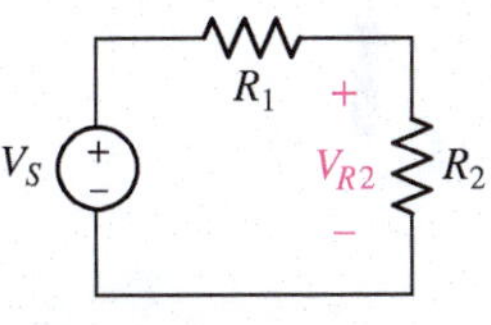

■ FIGURE 2.39

53. For the circuit in Fig. 2.39, suppose that the resistor R_2 represents a very sensitive and expensive piece of electronics. To ensure that the equipment is not damaged, R_1 is incorporated to represent a fuse, with a rating of 5 A and resistance of 0.1 Ω. If the voltage source is 12 V, what is the lowest resistance that could be encountered as a short circuit condition for R_2 before blowing the fuse?
54. For each of the circuits in Fig. 2.40, find the current *I* and compute the power absorbed by the resistor.
55. Sketch the power absorbed by a 100 Ω resistor as a function of voltage over the range $-2\text{ V} \le V_{\text{resistor}} \le +2\text{ V}$.
56. You built an android that has a subcircuit containing a power supply, a tactile sensor, and a fuse where safe operation should keep current below 250 mA. You measured that your sensor is dissipating 12 W, the power supply is providing 12.2 W, and the voltage drop across the fuse is 500 mV. Is your circuit properly protected?
57. Using the data in Table 2.4, calculate the resistance and conductance of 50 ft of wire with the following sizes: AWG 2, AWG 14, and AWG 28.

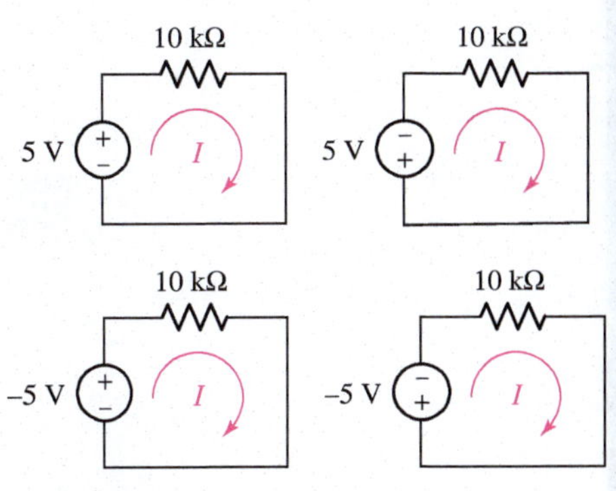

■ FIGURE 2.40

Chapter-Integrating Exercises

58. To protect an expensive circuit component from being delivered too much power, you decide to incorporate a fast-blowing fuse into the design. Knowing that the circuit component is connected to 12 V, its minimum power consumption is 12 W, and the maximum power it can safely dissipate is 100 W, which of the three available fuse ratings should you select: 1 A, 4 A, or 10 A? Explain your answer.

59. So-called n-type silicon has a resistivity given by $\rho = (-qN_D\mu_n)^{-1}$, where N_D is the volume density of phosphorus atoms (atoms/cm^3), μ_n is the electron mobility (cm^2/V · s), and $q = -1.602 \times 10^{-19}$ C is the charge of each electron. Conveniently, a relationship exists between mobility and N_D, as shown in Fig. 2.41. Assume an 8-inch-diameter silicon wafer (disk) having a thickness of 300 μm. Design a 10 Ω resistor by specifying a phosphorus concentration in the range of 2×10^{15} cm^{-3} $\leq N_D \leq 2 \times 10^{17}$ cm^{-3}, along with a suitable geometry (the wafer may be cut, but not thinned).

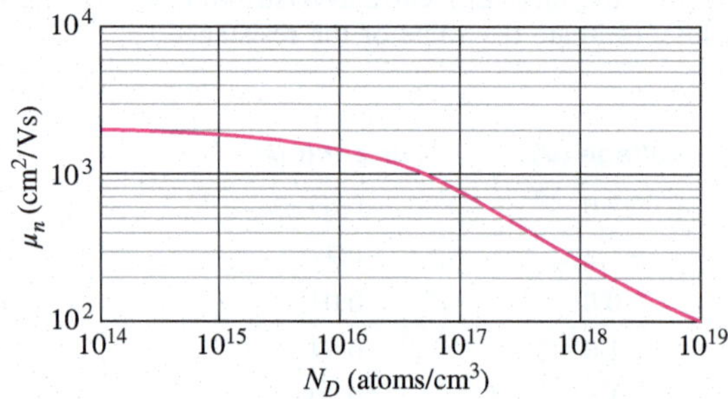

FIGURE 2.41

60. A 250 ft long span separates a dc power supply from a lamp which draws 25 A of current. If 14 AWG wire is used (note that two wires are needed for a total of 500 ft), calculate the amount of power wasted in the wire.

61. The resistance values in Table 2.4 are calibrated for operation at 20°C. They may be corrected for operation at other temperatures using the relationship[6]

$$\frac{R_2}{R_1} = \frac{234.5 + T_2}{234.5 + T_1}$$

where T_1 = reference temperature (20°C in present case)
T_2 = desired operating temperature
R_1 = resistance at T_1
R_2 = resistance at T_2

A piece of equipment relies on an external wire made of 28 AWG soft copper, which has a resistance of 50.0 Ω at 20°C. Unfortunately, the operating environment has changed, and it is now 110.5°F. (*a*) Calculate the length of the original wire. (*b*) Determine by how much the wire should be shortened so that it is once again 50.0 Ω.

62. Your favorite meter contains a precision (1% tolerance) 10 Ω resistor. Unfortunately, the last person who borrowed this meter somehow blew the resistor, and it needs to be replaced. Design a suitable replacement, assuming at least 1000 ft of each of the wire gauges listed in Table 2.4 is readily available to you.

63. If 1 mA of current is forced through a 1 mm diameter, 2.3-meter-long piece of hard, round, aluminum-clad steel (B415) wire, how much power is wasted

(6) D. G. Fink and H. W. Beaty, *Standard Handbook for Electrical Engineers,* 13th ed. New York: McGraw Hill, 1993, p. 2–9.

as a result of resistive losses? If instead wire of the same dimensions but conforming to B75 specifications is used, by how much will the power wasted due to resistive losses be reduced?

64. The network shown in Fig. 2.42 can be used to accurately model the behavior of a bipolar junction transistor provided that it is operating in the forward active mode. The parameter β is known as the current gain. If for this device $\beta = 100$, and I_B is determined to be 100 μA, calculate (*a*) I_C, the current flowing into the collector terminal, and (*b*) the power dissipated by the base-emitter region.

FIGURE 2.42 DC model for a bipolar junction transistor operating in forward active mode.

65. A 100 W tungsten filament light bulb functions by taking advantage of resistive losses in the filament, absorbing 100 joules each second of energy from the wall socket. How much *optical* energy per second do you expect it to produce, and does this violate the principle of energy conservation?

66. An LED operates at a current of 40 mA, with a forward voltage of 2.4 V. You construct the series circuit shown in Fig. 2.43 to power the LED using two 1.5 V batteries, each with a capacity of 2000 mAh. Determine the required value of the resistor and how long the circuit will operate before the batteries run out of energy.

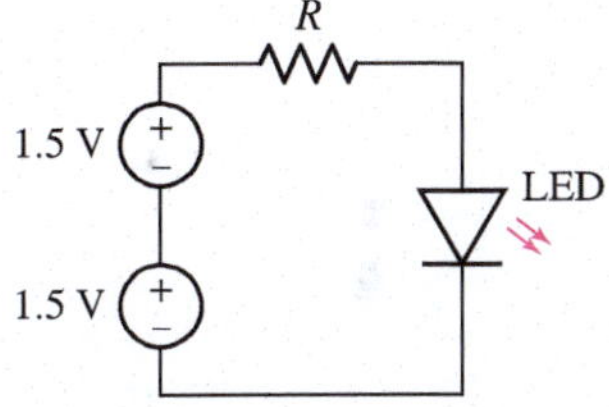

FIGURE 2.43

67. You have found a way to directly power your wall clock (consumes 0.5 mW of power) using a solar cell collecting ambient room light, rather than using an AA battery. The solar cell and battery each provide the required voltage of 1.5 V and the proper current for clock operation. Your solar cell has an efficiency of 15% and costs \$6, and each AA battery has a capacity of 1200 mAh and costs \$1. What is the payback time (point in time where the cost of solar cell would match the cost for supplying batteries) for using a solar cell instead of batteries?

68. The primary power system on NASA's Perseverance Mars rover utilizes what is known as a multimission radioisotope thermoelectric generator (MMRTG). The basic principle is to use a radioisotope (such as plutonium oxide) as a heat source, and solid-state thermocouples to convert heat directly to electricity. Its maximum power output degenerates slowly, with a value "at launch" of approximately 110 W. (*a*) If the MMRTG can provide 30 V, what is the maximum current it can deliver? (*b*) Not all of the power generated by the radioisotopes is converted to electricity. If the "beginning-of-life" (BOL) system efficiency is 6%, compute the thermal power output of the radioisotopes. (*c*) If the "end-of-design-life" (EODL) power is specified as 72 W, what is the corresponding end-of-design-life maximum current? (*d*) If the MMRTG employs 768 thermocouples, how much power does each thermocouple generate at BOL and EODL?

69. A certain electric vehicle is constructed with a multiple-cell battery pack having a total capacity of 82 kWh. (*a*) If the battery pack consists of 2170 cells, what is the capacity of each cell? (*b*) If the vehicle maximum range is 560 km, estimate the energy (in J) required to travel 1 km. (*c*) If the vehicle travels at a constant speed of 80 km/h, calculate the average power drawn

from the battery pack. (*d*) If recharging the battery pack from near-zero (not recommended) to 100% requires 7 hours using a Level 2 charger, how much power is drawn from the charging unit if we model the charging process as delivering energy at a constant rate regardless of the charge state of the battery pack? (*e*) If the Level 2 charger provides 240 V to the battery pack, what current does the pack draw while charging (on average)? (*f*) Generate and test a MATLAB script that allows the answers to this question to be calculated.

70. While estimates vary, Balasubramanian[7] notes that the human brain operates on a typical energy budget of approximately 20 W. (*a*) If 3 W is associated with cortical gray matter consumption, and 8 W powers the remainder of the brain functions, how much power is lost to heat? (*b*) If the power usage is 2 nW/neuron, what would be the power demand of an 86 billion neuron brain if all neurons fired simultaneously? (*c*) If the cortex of a certain brain contains 26 billion neurons, estimate the number that fire simultaneously.

[7]V. Balasubramanian, "Brain power," *Proceedings of the National Academy of Sciences* **118**(32), 2021, e2107022118, https://doi.org/10.1073/pnas.2107022118.

Voltage and Current Laws

KEY CONCEPTS

Circuit Terms: *Node*, *Path*, *Loop*, and *Branch*

Kirchhoff's Current Law (KCL)

Kirchhoff's Voltage Law (KVL)

Analysis of Basic Series and Parallel Circuits

Series and Parallel Connected Sources

Series and Parallel Resistor Combinations

Voltage and Current Division

Ground Connections

INTRODUCTION

In Chap. 2 we were introduced to independent voltage and current sources, dependent sources, and resistors. We discovered that *dependent* sources come in four varieties and are controlled by either a voltage or current which exists elsewhere in the same circuit. Once we know the voltage across a resistor, we know its current (and vice versa); this is *not* the case for sources, however. In general, circuits must be analyzed to determine a complete set of voltages and currents. This turns out to be reasonably straightforward, and only two simple laws are needed in addition to Ohm's law. These new laws are Kirchhoff's current law (KCL) and Kirchhoff's voltage law (KVL), and they are simply restatements of charge and energy conservation, respectively. They apply to any circuit we will ever encounter, although in later chapters we will learn additional techniques for specific types of situations.

3.1 NODES, PATHS, LOOPS, AND BRANCHES

Let's focus our attention on the current–voltage relationships in simple networks of two or more circuit elements. The elements will be connected by wires (sometimes referred to as "leads"), which have zero resistance. Since the network then appears as a number of simple elements and a set of connecting leads, it is called a ***lumped-parameter network.*** A more difficult analysis problem arises when we are faced with a ***distributed-parameter network,*** which contains an essentially infinite number of vanishingly small elements. We will concentrate on lumped-parameter networks in this text.

In circuits assembled in the real world, the wires will always have finite resistance. However, this resistance is typically so small compared to other resistances in the circuit that we can neglect it without introducing significant error. In our idealized circuits, we will therefore assume "zero resistance" wires from now on unless told otherwise.

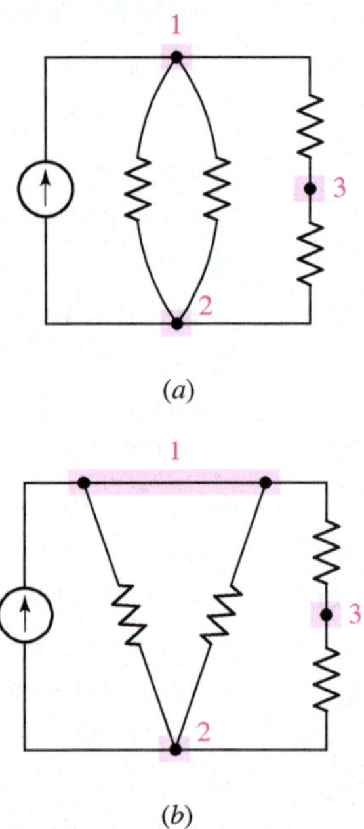

FIGURE 3.1 (*a*) A circuit containing three nodes and five branches. (*b*) Node 1 is redrawn to look like two nodes; it is still just one node, however.

A point at which two or more elements have a common connection is called a ***node.*** For example, Fig. 3.1*a* shows a circuit containing three nodes. Sometimes networks are drawn so as to trap an unwary student into believing that there are more nodes present than is actually the case. This occurs when a node, such as node 1 in Fig. 3.1*a*, is shown as two separate junctions connected by a (zero-resistance) conductor, as in Fig. 3.1*b*. However, all that has been done is to spread the common point out into a common zero-resistance line. Thus, we must necessarily consider all of the perfectly conducting leads or portions of leads attached to the node as part of the node. Note also that every element has a node at each of its ends.

Suppose that we start at one node in a network and move through a simple element to the node at the other end. We then continue from that node through a different element to the next node, and continue this movement until we have gone through as many elements as we wish. If no node was encountered more than once, then the set of nodes and elements that we have passed through is defined as a ***path.*** If the node at which we started is the same as the node on which we ended, then the path is, by definition, a *closed* path or a ***loop.***

For example, in Fig. 3.1*a*, if we move from node 2 through the current source to node 1, and then through the upper right resistor to node 3, we have established a path; since we have not continued on to node 2 again, we have not made a loop. If we proceeded from node 2 through the current source to node 1, down through the left resistor to node 2, and then up through the central resistor to node 1 again, we do not have a path, since a node (actually two nodes) was encountered more than once; we also do not have a loop, because a loop must be a path.

Another term whose use will prove convenient is ***branch.*** We define a branch as a single path in a network, composed of one simple element and the node at each end of that element. Thus, a path is a particular collection of branches. The circuit shown in Fig. 3.1*a* and *b* contains five branches.

3.2 KIRCHHOFF'S CURRENT LAW

We are now ready to consider the first of the two laws named for Gustav Robert Kirchhoff (two *h*'s and two *f*'s), a German university professor who was born about the time Ohm was doing his experimental work. This axiomatic law is called Kirchhoff's current law (abbreviated KCL), and it simply states that

> The algebraic sum of the currents entering any node is zero.

This law represents a mathematical statement of the fact that charge cannot accumulate at a node. *A node is not a circuit element,* and it certainly cannot store, destroy, or generate charge. Hence, the currents associated with the node must sum to zero. A hydraulic analogy might be useful here: for example, consider three water pipes joined in the shape of a Y. We define three water currents as flowing *into* each of the three pipes. If we insist that water is always flowing, then obviously we cannot have three positive water currents, or the joint would burst. This is a result of our defining currents independent of the direction that water is actually flowing. Therefore, the value of either one or two of the currents as defined must be negative.

Consider the node shown in Fig. 3.2. The algebraic sum of the four currents entering the node must be zero:

$$i_A + i_B + (-i_C) + (-i_D) = 0$$

However, the law could be equally well applied to the algebraic sum of the currents *leaving* the node:

$$(-i_A) + (-i_B) + i_C + i_D = 0$$

We might also decide to equate the sum of the currents directed into the node to the sum of those directed out of the node:

$$i_A + i_B = i_C + i_D$$

which simply states that the sum of the currents going in must equal the sum of the currents going out.

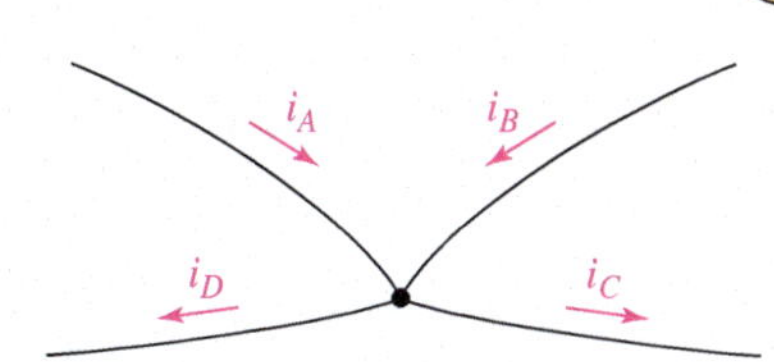

FIGURE 3.2 Example node to illustrate the application of Kirchhoff's current law.

EXAMPLE 3.1

For the circuit in Fig. 3.3*a*, compute the current through resistor R_3 if it is known that the voltage source supplies a current of 3 A.

▶ ***Identify the goal of the problem.***

The current through resistor R_3, labeled as i on the circuit diagram.

▶ ***Collect the known information.***

The node at the top of R_3 is connected to four branches.

Two of these currents are clearly labeled: 2 A flows out of the node into R_2, and 5 A flows into the node from the current source. We are told the current out of the 10 V source is 3 A.

▶ ***Devise a plan.***

If we label the current through R_1 (Fig. 3.3*b*), we may write a KCL equation at the top node of resistors R_2 and R_3.

▶ ***Construct an appropriate set of equations.***

Summing the currents flowing into the node:

$$i_{R_1} - 2 - i + 5 = 0$$

The currents flowing into this node are shown in the expanded diagram of Fig. 3.3*c* for clarity.

▶ ***Determine if additional information is required.***

We have one equation but two unknowns, which means we need to obtain an additional equation. At this point, the fact that we know the 10 V source is supplying 3 A comes in handy: KCL shows us that this is also the current i_{R_1}.

▶ ***Attempt a solution.***

Substituting, we find that $i = 3 - 2 + 5 = 6$ A.

▶ ***Verify the solution. Is it reasonable or expected?***

It is always worth the effort to recheck our work. Also, we can attempt to evaluate whether at least the magnitude of the solution is reasonable. In this case, we have two sources—one supplies 5 A, and the other supplies 3 A. There are no other sources, independent or dependent. Thus, we would not expect to find any current in the circuit in excess of 8 A.

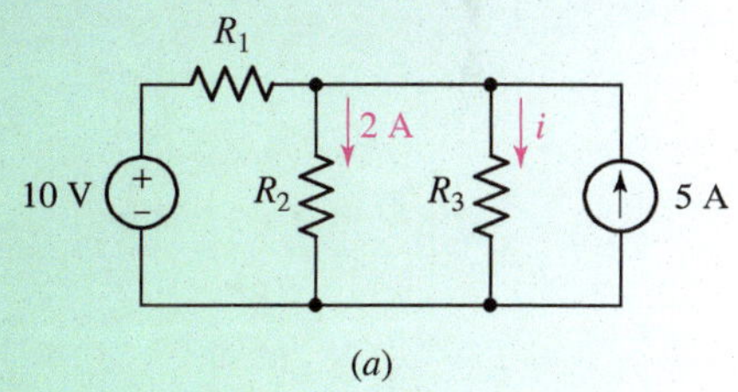

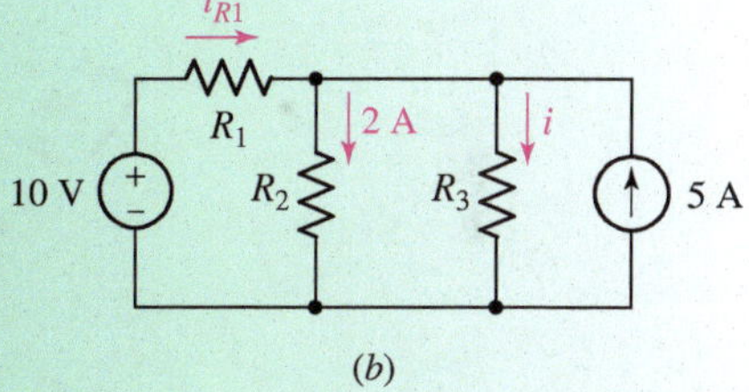

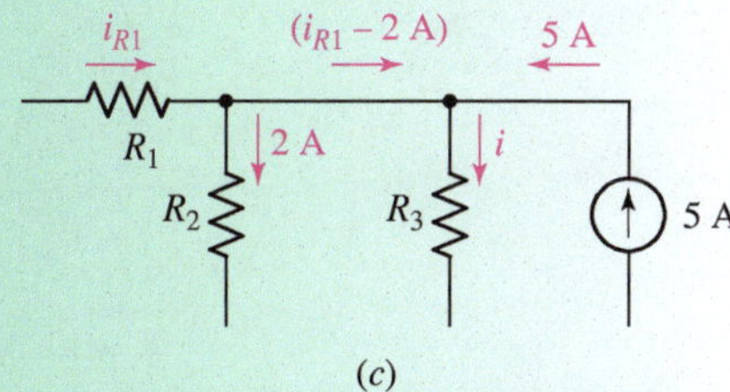

FIGURE 3.3 (*a*) Simple circuit for which the current through resistor R_3 is desired. Note that the connection between R_1, R_2, and R_3 is drawn to look like *two* nodes, but is really only *one* node. (*b*) The current through resistor R_1 is labeled so that a KCL equation can be written. (*c*) The currents into the top node of R_3 are redrawn for clarity.

(Continued on next page)

PRACTICE

3.1 (*a*) Count the number of branches and nodes in the circuit in Fig. 3.4. (*b*) If $i_x = 3$ A and the 18 V source delivers 8 A of current, what is the value of R_A? (*Hint:* You need Ohm's law as well as KCL.)

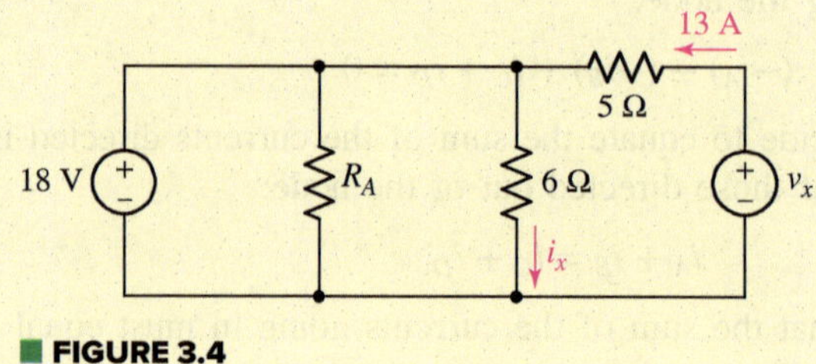

FIGURE 3.4

Ans: 5 branches, 3 nodes, 1 Ω.

A compact expression for Kirchhoff's current law is

$$\sum_{n=1}^{N} i_n = 0 \qquad [1]$$

which is just a shorthand statement for

$$i_1 + i_2 + i_3 + \cdots + i_N = 0 \qquad [2]$$

When Eq. [1] or Eq. [2] is used, it is understood that the N current arrows are either all directed *toward* the node in question or all directed *away* from it.

3.3 KIRCHHOFF'S VOLTAGE LAW

Current is related to the charge flowing *through* a circuit element, whereas voltage is a measure of potential energy difference *across* the element. These are often confused early on as a student learns circuit analysis, for some reason. There is a single unique value for any voltage in circuit theory. Thus, the energy required to move a charge from point A to point B in a circuit must have a value independent of the path chosen to get from A to B (there is often more than one such path). We may assert this fact through Kirchhoff's voltage law (abbreviated **KVL**):

The algebraic sum of the voltages around any closed path is zero.

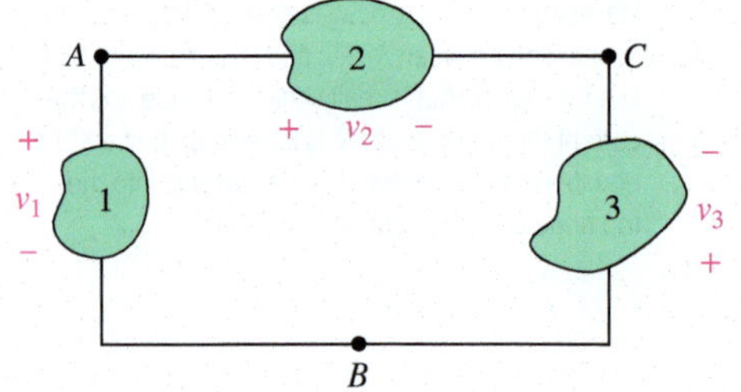

FIGURE 3.5 The potential difference between points A and B is independent of the path selected.

In Fig. 3.5, if we carry a charge of 1 C from A to B through element 1, the reference polarity signs for v_1 show that we do v_1 joules of work.[1] Now if, instead, we choose to the 1 C of charge from A to B via node C, then we expend $(v_2 - v_3)$ joules of energy. The work done, however, is

(1) Note that we chose a 1 C charge for the sake of numerical convenience: therefore, we performed (1 C)(v_1 J/C) = v_1 joules of work.

independent of the path in a circuit, and so any route must lead to the same value for the voltage. In other words,

$$v_1 = v_2 - v_3 \qquad [3]$$

We can apply KVL to a circuit in several different ways. One method that leads to fewer equation-writing errors than others consists of moving mentally around the closed path in a clockwise direction and writing down directly the voltage of each element whose (+) terminal is entered, and writing down the negative of every voltage first met at the (−) sign. Applying this to the single loop of Fig. 3.5, we have

$$-v_1 + v_2 - v_3 = 0$$

which agrees with our previous result, Eq. [3]. Now let's look at this more generally. Assuming we encounter the positive reference of each voltage as we trace out a closed path, it follows that the algebraic sum of the voltages across the individual elements around the closed path must be zero. Thus, we may write

$$v_1 + v_2 + v_3 + \cdots + v_N = 0$$

or, more compactly,

$$\sum_{n=1}^{N} v_n = 0 \qquad [4]$$

EXAMPLE 3.2

In the circuit of Fig. 3.6, determine v_x and i_x.

We know the voltage across two of the three elements in the circuit. Thus, KVL can be applied immediately to obtain v_x.

Beginning with the bottom node of the 5 V source, we apply KVL clockwise around the loop:

$$-5 - 7 + v_x = 0$$

so v_x = 12 V.

KCL can be applied to this circuit, but only tells us that the same current (i_x) flows through all three elements. We do know the voltage across the 100 Ω resistor now, however.

Invoking Ohm's law,

$$i_x = \frac{v_x}{100} = \frac{12}{100} \text{ A} = 120 \text{ mA}$$

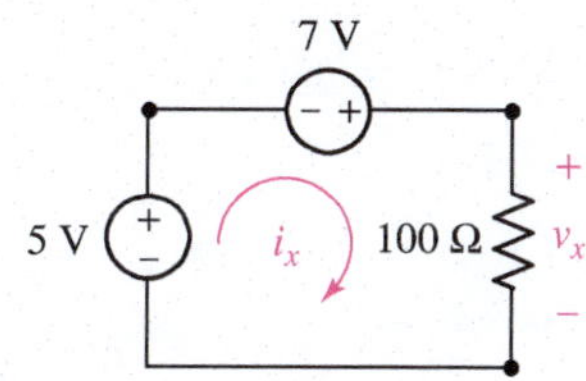

FIGURE 3.6 A simple circuit with two voltage sources and a single resistor.

PRACTICE

3.2 Determine i_x and v_x in the circuit of Fig. 3.7.

Ans: v_x = −4 V; i_x = −400 mA.

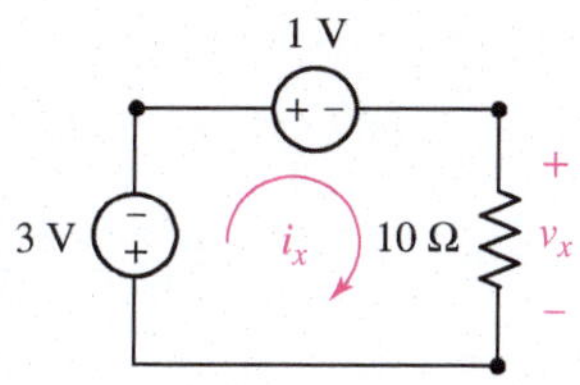

FIGURE 3.7

EXAMPLE 3.3

For the circuit of Fig. 3.8, calculate v_{R2} (the voltage across R_2) and the voltage labeled v_x.

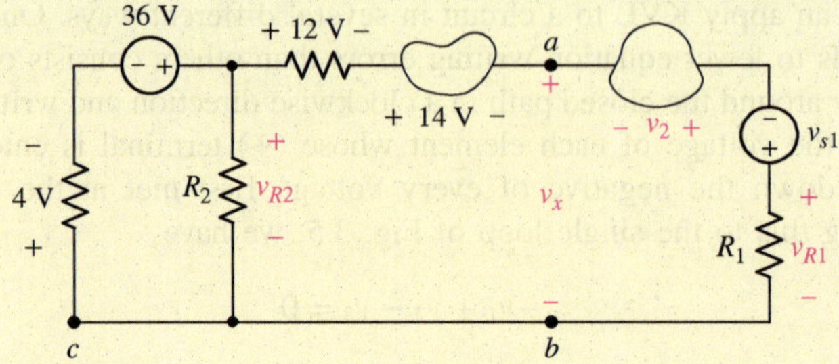

FIGURE 3.8 A circuit with eight elements for which we desire v_{R2} and v_x.

The best approach for finding v_{R2} is to look for a loop to which we can apply KVL. There are several options, but the leftmost loop offers a straightforward route; two of the voltages are clearly specified. Thus, we find v_{R2} by writing a KVL equation around the loop on the left, starting at point c:

$$4 - 36 + v_{R2} = 0$$

which leads to $v_{R2} = 32$ V.

Points b and c, as well as the wire between them, are all part of the same node.

To find v_x, we might think of this as the (algebraic) sum of the voltages across the three elements on the right. However, since we do not have values for these quantities, such an approach would not lead to a numerical answer. Instead, we apply KVL beginning at point c, moving up and across the top to a, through v_x to b, and through the conducting lead to the starting point:

$$+4 - 36 + 12 + 14 + v_x = 0$$

so that

$$v_x = 6 \text{ V}$$

An alternative approach: Knowing v_{R2}, we might have taken the shortcut through R_2:

$$-32 + 12 + 14 + v_x = 0$$

yielding $v_x = 6$ V once again.

PRACTICE

3.3 For the circuit of Fig. 3.9, if $v_{R1} = 1$ V, determine (*a*) v_{R2} and (*b*) v_2.

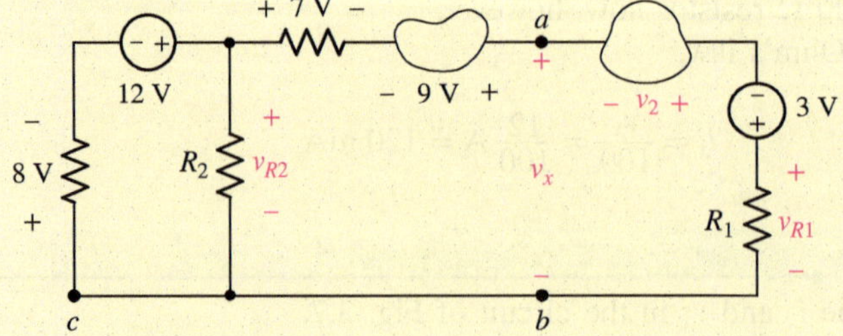

FIGURE 3.9

Ans: (*a*) 4 V; (*b*) −8 V.

As we have just seen, the key to correctly analyzing a circuit is to first methodically label all voltages and currents on the diagram. This way, carefully written KCL or KVL equations will yield correct relationships, and Ohm's law can be applied as necessary if more unknowns than equations are obtained initially.

EXAMPLE 3.4

Determine v_x in the circuit of Fig. 3.10*a*.

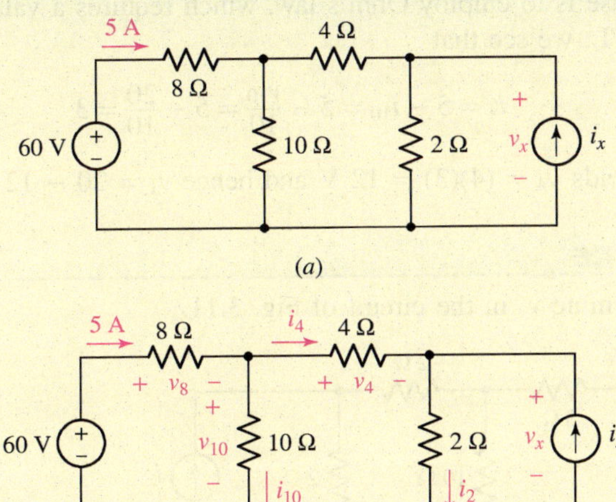

FIGURE 3.10 (*a*) A circuit for which v_x is to be determined using KVL. (*b*) Circuit with voltages and currents labeled.

We begin by labeling voltages and currents on the rest of the elements in the circuit (Fig. 3.10*b*). Note that v_x appears across the 2 Ω resistor and the current source i_x as well.

Plan *A*: If we can obtain the current through the 2 Ω resistor, Ohm's law will yield v_x. Writing the appropriate KCL equation, we see that

$$i_2 = i_4 + i_x$$

Unfortunately, we do not have values for any of these three quantities. Our solution has (temporarily) stalled. Fortunately we have a plan *B*.

Plan *B*: Since we were given the current flowing from the 60 V source, perhaps we should consider starting from that side of the circuit. Instead of finding v_x using i_2, it might be possible to find v_x directly using KVL. We can write the following KVL equations:

$$-60 + v_8 + v_{10} = 0$$

and

$$-v_{10} + v_4 + v_x = 0 \qquad [5]$$

(Continued on next page)

This is progress: we now have two equations in four unknowns, an improvement over one equation in which *all* terms were unknown. In fact, we know that $v_8 = 40$ V through Ohm's law, as we were told that 5 A flows through the 8 Ω resistor. Thus, $v_{10} = 0 + 60 - 40 = 20$ V, so Eq. [5] reduces to

$$v_x = 20 - v_4$$

If we can determine v_4, the problem is solved.

The best route to finding a numerical value for the voltage v_4 in this case is to employ Ohm's law, which requires a value for i_4. From KCL, we see that

$$i_4 = 5 - i_{10} = 5 - \frac{v_{10}}{10} = 5 - \frac{20}{10} = 3$$

which yields $v_4 = (4)(3) = 12$ V and hence $v_x = 20 - 12 = 8$ V.

PRACTICE

3.4 Determine v_x in the circuit of Fig. 3.11.

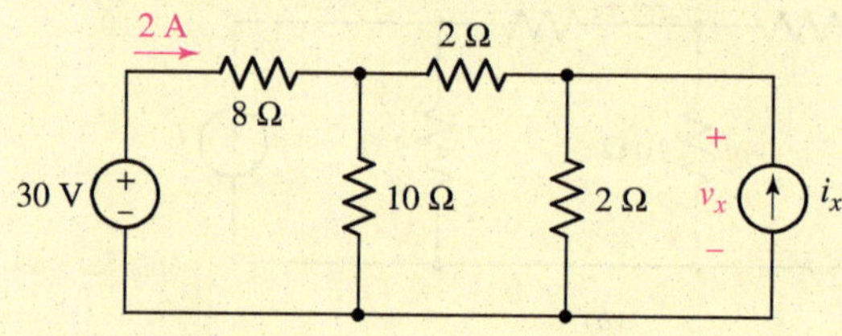

FIGURE 3.11

Ans: $v_x = 12.8$ V.

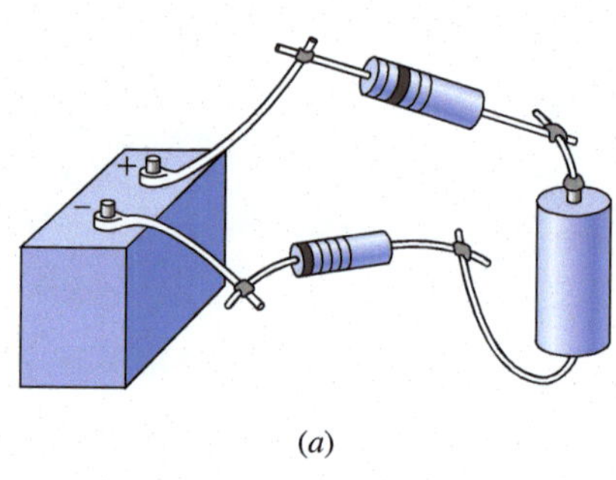

(*a*)

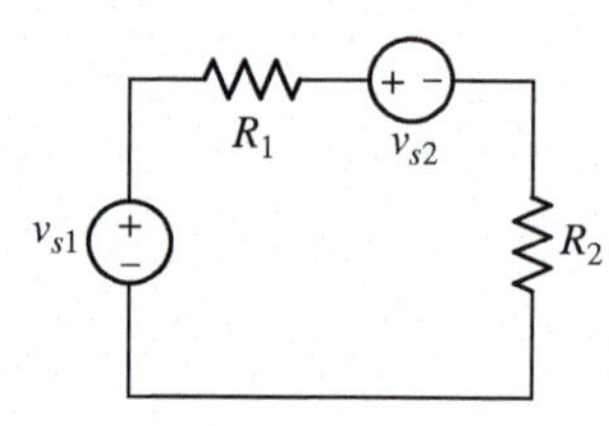

(*b*)

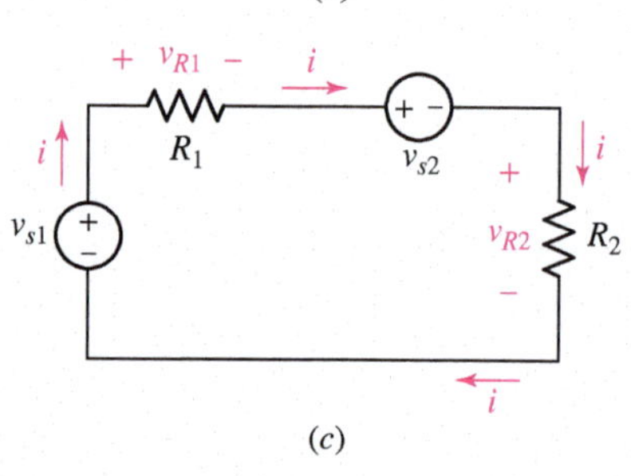

(*c*)

FIGURE 3.12 (*a*) A single-loop circuit with four elements. (*b*) The circuit model with source voltages and resistance values given. (*c*) Current and voltage reference signs have been added to the circuit.

3.4 THE SINGLE-LOOP CIRCUIT

We have seen that repeated use of KCL and KVL in conjunction with Ohm's law can be applied to nontrivial circuits containing several loops and a number of different elements. Before proceeding further, this is a good time to focus on the concept of series (and, in the next section, parallel) circuits, as they form the basis of any network we will encounter in the future.

All of the elements in a circuit that carry the same current are said to be connected in ***series.*** As an example, consider the circuit of Fig. 3.10. The 60 V source is in series with the 8 Ω resistor; they carry the same 5 A current. However, the 8 Ω resistor is not in series with the 4 Ω resistor; they carry different currents. Note that elements may carry equal currents and not be in series; two 100 W light bulbs in neighboring houses may very well carry equal currents, but they certainly do not carry the same current and are *not* connected in series.

Figure 3.12*a* shows a simple circuit consisting of two batteries and two resistors. Each terminal, connecting lead, and solder glob is assumed to have zero resistance; together they constitute an individual node of the circuit diagram in Fig. 3.12*b*. Both batteries are modeled by ideal voltage sources; any internal resistances they may have are assumed to be small enough to neglect. The two resistors are assumed to be ideal (linear) resistors.

We seek the current *through* each element, the voltage *across* each element, and the power *absorbed* by each element. Our first step in the analysis is the assumption of reference directions for the unknown currents. Arbitrarily, let us select a clockwise current i, which flows out of the upper terminal of the voltage source on the left. This choice is indicated by an arrow labeled i at that point in the circuit, as shown in Fig. 3.12*c*. A trivial application of Kirchhoff's current law assures us that this same current must also flow through every other element in the circuit; we emphasize this fact (just this once) by placing several other current symbols about the circuit.

Our second step in the analysis is a choice of the voltage reference for each of the two resistors. The passive sign convention requires that the resistor current and voltage variables be defined so that the current enters the terminal at which the positive voltage reference is located. Since we already (arbitrarily) selected the current direction, v_{R1} and v_{R2} are defined as in Fig. 3.12*c*.

The third step is the application of Kirchhoff's voltage law to the only closed path. Let us decide to move around the circuit in the clockwise direction, beginning at the lower left corner, and to write down directly every voltage first met at its positive reference, and to write down the negative of every voltage encountered at the negative terminal. Thus,

$$-v_{s1} + v_{R1} + v_{s2} + v_{R2} = 0 \qquad [6]$$

We then apply Ohm's law to the resistive elements:

$$v_{R1} = R_1 i \quad \text{and} \quad v_{R2} = R_2 i$$

Substituting into Eq. [6] yields

$$-v_{s1} + R_1 i + v_{s2} + R_2 i = 0$$

Since i is the only unknown, we find that

$$i = \frac{v_{s1} - v_{s2}}{R_1 + R_2}$$

The voltage or power associated with any element may now be obtained by applying $v = Ri$, $p = vi$, or $p = i^2R$.

PRACTICE

3.5 In the circuit of Fig. 3.12*b*, v_{s1} = 120 V, v_{s2} = 30 V, R_1 = 30 Ω, and R_2 = 15 Ω. Compute the power absorbed by each element. Do you notice anything by considering the full set of values?

Ans: $p_{120V} = -240$ W; $p_{30V} = +60$ W; $p_{30\Omega} = 120$ W; $p_{15\Omega} = 60$ W. Note these all sum to zero. (It's not a coincidence.)

EXAMPLE 3.5

Compute the power absorbed in each element for the circuit shown in Fig. 3.13*a*.

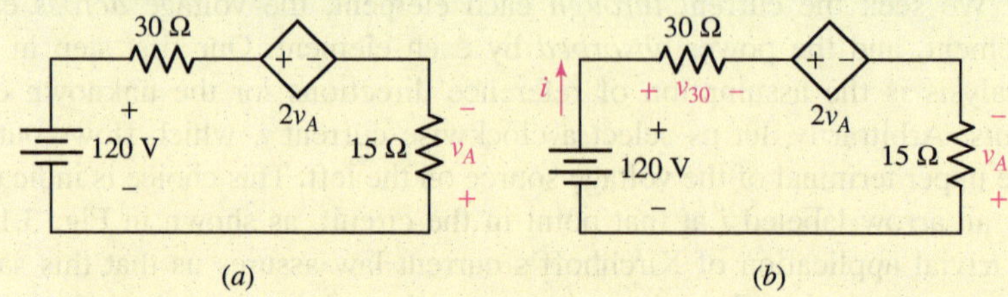

FIGURE 3.13 (*a*) A single-loop circuit containing a dependent source. (*b*) The current *i* and voltage v_{30} are assigned.

We first assign a reference direction for the current i and a reference polarity for the voltage v_{30} as shown in Fig. 3.13*b*.

This circuit contains a dependent voltage source, the value of which remains unknown until we determine v_A. However, its algebraic value $2v_A$ can be used in the same fashion as if a numerical value were available. Thus, applying KVL around the loop in a clockwise direction:

$$-120 + v_{30} + 2v_A - v_A = 0 \qquad [7]$$

Next we apply Ohm's law to introduce the known resistor values:

$$v_{30} = 30i \quad \text{and} \quad v_A = -15i$$

Note that the negative sign is required since i flows into the negative terminal of v_A.

Substituting into Eq. [7] yields

$$-120 + 30i - 30i + 15i = 0$$

and so we find that

$$i = 8 \text{ A}$$

Computing the power *absorbed* by each element:

$$\begin{aligned} p_{120\text{V}} &= (120)(-8) = -960 \text{ W} \\ p_{30\Omega} &= (8)^2(30) \quad = 1920 \text{ W} \\ p_{\text{dep}} &= (2v_A)(8) \quad = 2[(-15)(8)](8) \\ &\qquad\qquad\quad = -1920 \text{ W} \\ p_{15\Omega} &= (8)^2(15) \quad = 960 \text{ W} \end{aligned}$$

PRACTICE

3.6 In the circuit of Fig. 3.14, find the power absorbed by each of the five elements in the circuit.

Ans: (CW from left) 0.768 W; 1.92 W; 0.2048 W; 0.1792 W; −3.072 W.

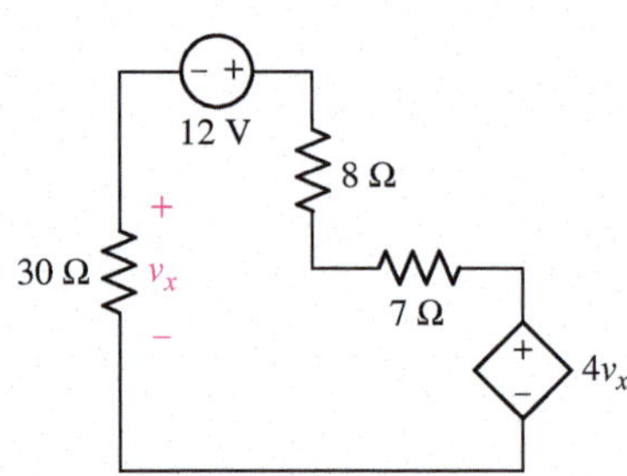

FIGURE 3.14 A simple loop circuit which includes a voltage-controlled dependent voltage source in addition to an independent voltage source.

In Example 3.5 and again in Practice Problem 3.6, we were asked to compute the power absorbed by each element of a circuit. It is difficult to think of a situation, however, in which *all* of the absorbed power quantities of a circuit would be positive, for the simple reason that the energy must come from somewhere. Thus, from simple conservation of energy, we expect that ***the sum of the absorbed power for each element of a circuit should be zero.*** In other words, at least one of the quantities should be negative (neglecting the trivial case where the circuit is not operating). Stated another way, the sum of the supplied power for each element should be zero. More pragmatically, ***the sum of the absorbed power equals the sum of the supplied power,*** which seems reasonable enough at face value.

Let's test this with the circuit of Fig. 3.13 from Example 3.5, which consists of two sources (one dependent and one independent) and two resistors. Adding the power absorbed by each element, we find

$$\sum_{\text{all elements}} p_{\text{absorbed}} = -960 + 1920 - 1920 + 960 = 0$$

In reality (noting the sign associated with its absorbed power) the 120 V source *supplies* +960 W, and the dependent source supplies +1920 W. Thus, the sources together supply a total of 960 + 1920 = 2880 W. The resistors are expected to absorb positive power, which in this case sums to a total of 1920 + 960 = 2880 W. Thus, if we take into account each element of the circuit,

$$\sum p_{\text{absorbed}} = \sum p_{\text{supplied}}$$

as we expect.

Turning our attention to Practice Problem 3.6, the solution to which the reader might want to verify, we see that the absorbed powers sum to 0.768 + 1.92 + 0.2048 + 0.1792 − 3.072 = 0. Interestingly enough, the 12 V independent voltage source is absorbing +1.92 W, which means it is *dissipating* power, not supplying it. Instead, the dependent voltage source appears to be supplying all the power in this particular circuit. Is such a thing possible? We usually expect an independent source to supply positive power, but since we are employing idealized sources in our circuits, it is in fact possible to have a net power flow into any source. Further, dependent sources are typically models of more complex circuits which often have their own power, but we do not see that from the simple model—only the overall behavior once it's connected to the rest of our circuit. If the circuit is changed in some way, the same source might then be found to supply positive power. The result is not known until a circuit analysis has been completed.

3.5 THE SINGLE-NODE-PAIR CIRCUIT

The companion of the single-loop circuit discussed in Sec. 3.4 is the single-node-pair circuit, in which any number of simple elements are connected between the same pair of nodes. An example of such a circuit is shown in Fig. 3.15*a*. KVL forces us to recognize that the voltage across each branch is the same as that across any other branch. *Elements in a circuit having a common voltage across them are said to be connected in* ***parallel.***

EXAMPLE 3.6

Find the voltage, current, and power associated with each element in the circuit of Fig. 3.15*a*.

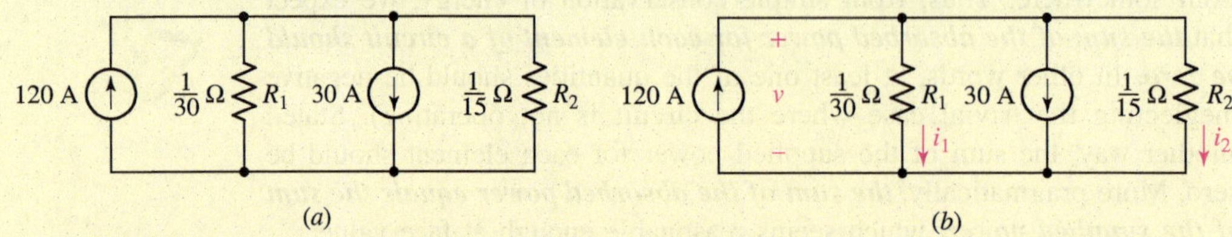

■ **FIGURE 3.15** (*a*) A single-node-pair circuit. (*b*) A voltage and two currents are assigned.

We first define a voltage v and arbitrarily select its polarity as shown in Fig. 3.15*b*. Two currents, flowing in the resistors, are selected in conformance with the passive sign convention, as shown in Fig. 3.15*b*.

Determining either current i_1 or i_2 will enable us to obtain a value for v. Thus, our next step is to apply KCL to either of the two nodes in the circuit. Equating the algebraic sum of the currents leaving the upper node to zero, we find:

$$-120 + i_1 + 30 + i_2 = 0$$

Next, writing both currents in terms of the voltage v using Ohm's law,

$$i_1 = 30v \quad \text{and} \quad i_2 = 15v$$

we obtain

$$-120 + 30v + 30 + 15v = 0$$

Solving this equation for v results in

$$v = 2\text{ V}$$

Invoking Ohm's law then yields

$$i_1 = 60\text{ A} \quad \text{and} \quad i_2 = 30\text{ A}$$

The absorbed power in each element can now be computed. In the two resistors,

$$p_{R1} = 30(2)^2 = 120\text{ W} \quad \text{and} \quad p_{R2} = 15(2)^2 = 60\text{ W}$$

and for the two sources,

$$p_{120A} = 120(-2) = -240\text{ W} \quad \text{and} \quad p_{30A} = 30(2) = 60\text{ W}$$

Since the 120 A source absorbs negative 240 W, it is actually *supplying* power to the other elements in the circuit. In a similar fashion, we find that the 30 A source is actually *absorbing* power rather than *supplying* it.

PRACTICE

3.7 Determine v in the circuit of Fig. 3.16.

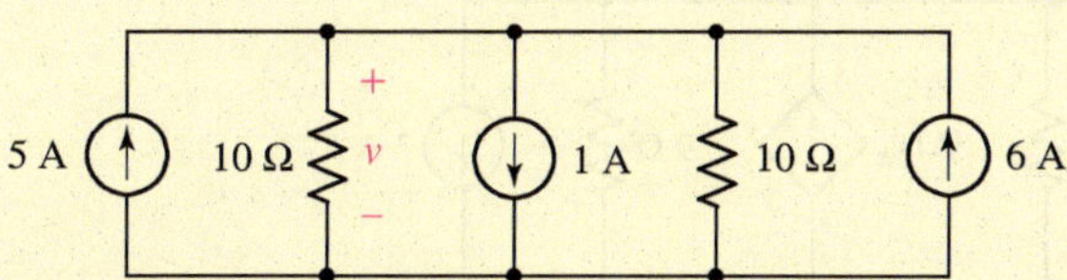

■ **FIGURE 3.16**

Ans: 50 V.

EXAMPLE 3.7

Determine the value of v and the power supplied by the independent current source in Fig. 3.17.

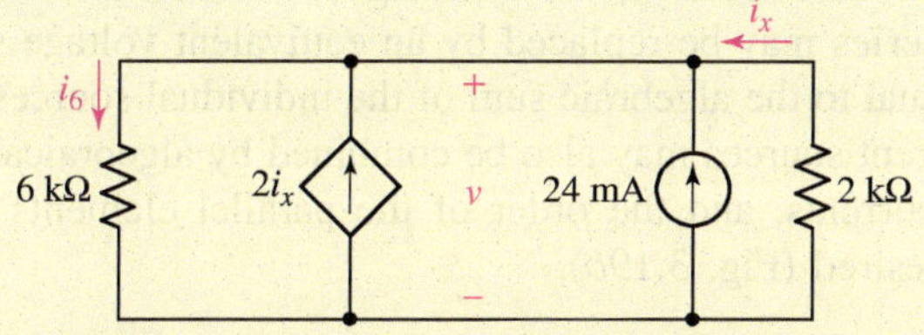

■ **FIGURE 3.17** A voltage v and a current i_6 are assigned in a single-node-pair circuit containing a dependent source.

By KCL, the sum of the currents leaving the upper node must be zero, so that

$$i_6 - 2i_x - 0.024 - i_x = 0$$

Again, note that the algebraic value of the dependent source ($2i_x$) is treated the same as a numerical value for the current would be, even though its exact value is not known until the circuit has been analyzed.

We next apply Ohm's law to each resistor:

$$i_6 = \frac{v}{6000} \quad \text{and} \quad i_x = \frac{-v}{2000}$$

Therefore,

$$\frac{v}{6000} - 2\left(\frac{-v}{2000}\right) - 0.024 - \left(\frac{-v}{2000}\right) = 0$$

and so $v = (600)(0.024) = 14.4$ V.

Any other information we may want to find for this circuit is now easily obtained, usually in a single step. For example, the power supplied by the independent source is $p_{24} = 14.4 \times 0.024 = 0.3456$ W (345.6 mW).

(Continued on next page)

PRACTICE

3.8 For the single-node-pair circuit of Fig. 3.18, compute i_A, i_B, and i_C.

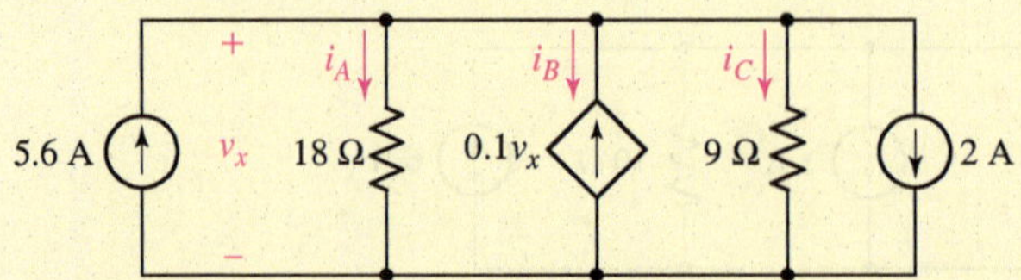

FIGURE 3.18

Ans: 3 A; −5.4 A; 6 A.

3.6 SERIES AND PARALLEL CONNECTED SOURCES

It turns out that some of the equation writing that we have been doing for series and parallel circuits can be avoided by combining sources. Note, however, that all the current, voltage, and power relationships in the remainder of the circuit will be unchanged. For example, several voltage sources in series may be replaced by an equivalent voltage source having a voltage equal to the algebraic sum of the individual sources (Fig. 3.19*a*). Parallel current sources may also be combined by algebraically adding the individual currents, and the order of the parallel elements may be rearranged as desired (Fig. 3.19*b*).

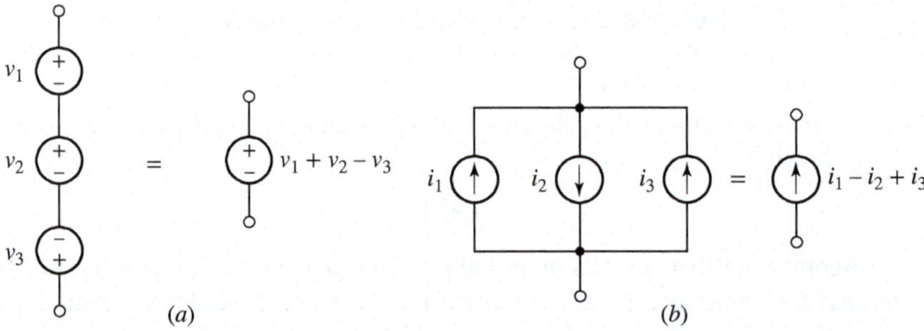

FIGURE 3.19 (*a*) Series-connected voltage sources can be replaced by a single source. (*b*) Parallel current sources can be replaced by a single source.

EXAMPLE **3.8**

Determine the current *i* in the circuit of Fig. 3.20*a* by first combining the sources into a single equivalent voltage source.

To be able to combine the voltage sources, they must be in series. Since the same current (i) flows through each, this condition is satisfied.

Starting from the bottom left-hand corner and proceeding clockwise,

$$-3 - 9 - 5 + 1 = -16 \text{ V}$$

so we may replace the four voltage sources with a single 16 V source having its negative reference as shown in Fig. 3.20*b*.

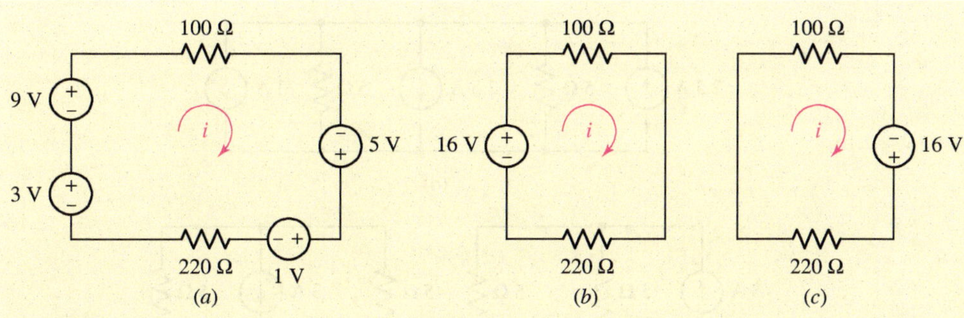

■ **FIGURE 3.20**

KVL combined with Ohm's law then yields

$$-16 + 100i + 220i = 0$$

or

$$i = \frac{16}{320} = 50 \text{ mA}$$

We should note that the circuit in Fig. 3.20*c* is also equivalent, a fact easily verified by computing *i*.

PRACTICE

3.9 Determine the current *i* in the circuit of Fig. 3.21 after first replacing the four sources with a single equivalent source.

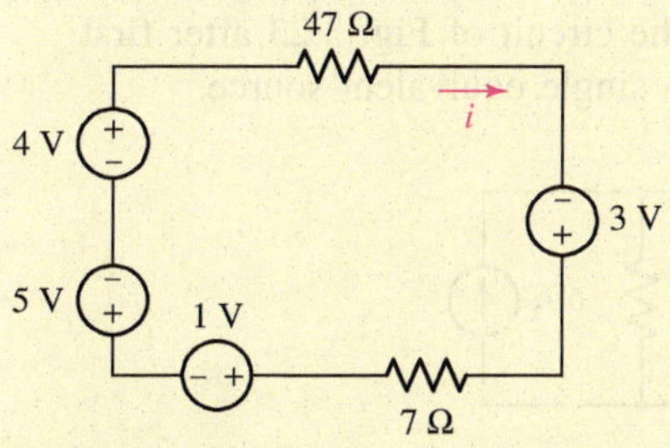

■ **FIGURE 3.21**

Ans: 18.5 mA.

EXAMPLE 3.9

Determine the voltage *v* in the circuit of Fig. 3.22*a* by first combining the sources into a single equivalent current source.

The sources may be combined since the same voltage appears across each one. Thus, we create a new source, arrow pointing upward into the top node, by adding the currents that flow into that node:

$$2.5 - 2.5 - 3 = -3 \text{ A}$$

(Continued on next page)

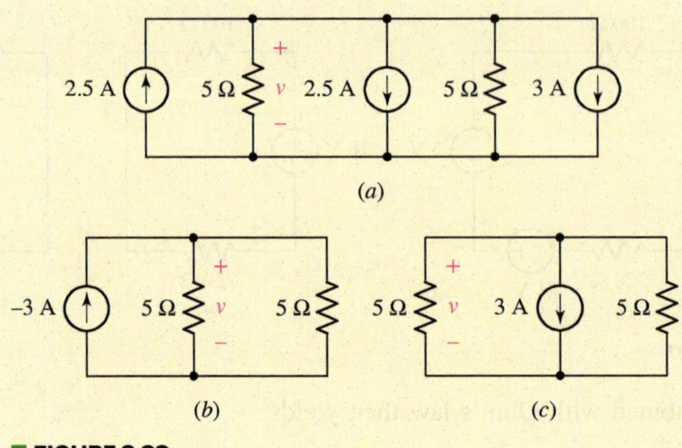

■ **FIGURE 3.22**

One equivalent circuit is shown in Fig. 3.22*b*.

KCL then allows us to write

$$-3 + \frac{v}{5} + \frac{v}{5} = 0$$

Solving, we find $v = 7.5$ V.

Another equivalent circuit is shown in Fig. 3.22*c*.

PRACTICE

3.10 Determine the voltage *v* in the circuit of Fig. 3.23 after first replacing the three sources with a single equivalent source.

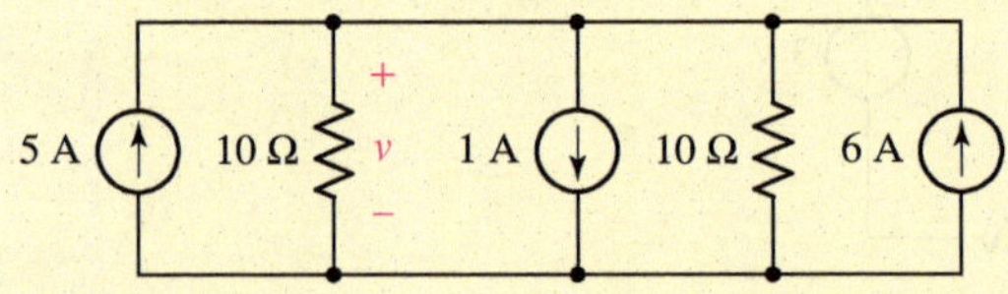

■ **FIGURE 3.23**

Ans: 50 V.

To conclude the discussion of parallel and series source combinations, we should consider the parallel combination of two voltage sources and the series combination of two current sources. For instance, what is the equivalent of a 5 V source in parallel with a 10 V source? By the definition of a voltage source, the voltage across the source cannot change; by Kirchhoff's voltage law, then, 5 equals 10 and we have hypothesized a physical impossibility. Thus, *ideal* voltage sources in parallel are permissible only when each has the same terminal voltage at every instant. In a similar way, two current sources may not be placed in series unless each has the same current, including sign, for every instant of time.

EXAMPLE 3.10

Determine which of the circuits of Fig. 3.24 are valid.

The circuit of Fig. 3.24*a* consists of two voltage sources in parallel. The value of each source is different, so this circuit violates KVL. For example, if a resistor is placed in parallel with the 5 V source, it is also in parallel with the 10 V source. The actual voltage across it is therefore ambiguous, and clearly the circuit cannot be constructed as indicated. If we attempt to build such a circuit in real life, we will find it impossible to locate "ideal" voltage sources—all real-world sources have an internal resistance. The presence of such resistance allows a voltage difference between the two *real* sources. Along these lines, the circuit of Fig. 3.24*b* is perfectly valid.

The circuit of Fig. 3.24*c* violates KCL: it is unclear what current actually flows through the resistor *R*.

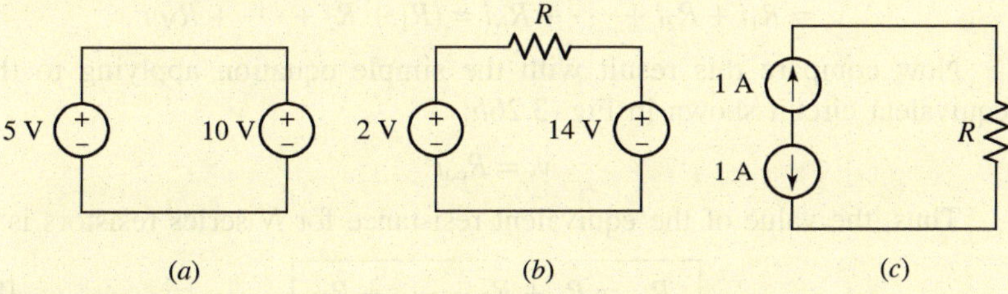

FIGURE 3.24 (*a*) to (*c*) Examples of circuits with multiple sources, some of which violate Kirchhoff's laws.

PRACTICE

3.11 Determine whether the circuit of Fig. 3.25 violates either of Kirchhoff's laws.

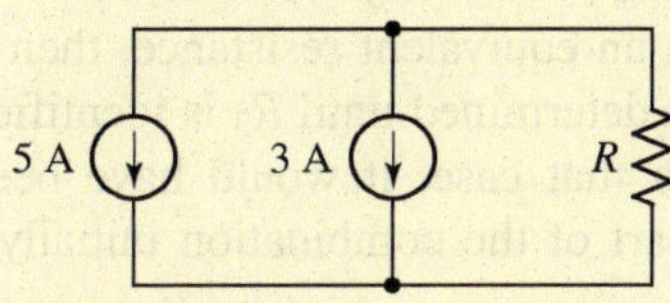

FIGURE 3.25

Ans: No. (If the resistor were removed, however, the resulting circuit would.)

3.7 RESISTORS IN SERIES AND PARALLEL

It is often possible to replace relatively complicated resistor combinations with a single equivalent resistor. This is useful when we are not specifically interested in the current, voltage, or power associated with any of the individual resistors in the combinations. *All the current, voltage, and power relationships in the remainder of the circuit will be unchanged.*

Consider the series combination of *N* resistors shown in Fig. 3.26*a*. We want to simplify the circuit by replacing the *N* resistors with a single resistor R_{eq} so that the remainder of the circuit, in this case only the voltage source,

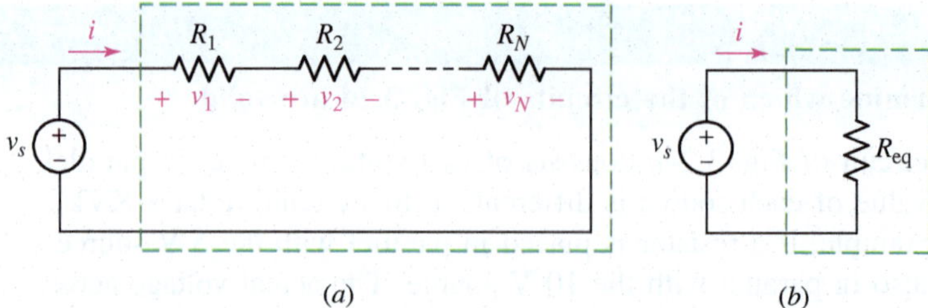

FIGURE 3.26 *(a)* Series combination of *N* resistors. *(b)* Electrically equivalent circuit.

does not realize that any change has been made. The current, voltage, and power of the source must be the same before and after the replacement.

First, apply KVL:

$$v_s = v_1 + v_2 + \cdots + v_N$$

and then Ohm's law:

$$v_s = R_1 i + R_2 i + \cdots + R_N i = (R_1 + R_2 + \cdots + R_N)i$$

Helpful Tip: Inspection of the KVL equation for any series circuit will show that *the order* in which elements are placed in such a circuit *makes no difference*.

Now compare this result with the simple equation applying to the equivalent circuit shown in Fig. 3.26*b*:

$$v_s = R_{eq} i$$

Thus, the value of the equivalent resistance for *N* series resistors is

$$\boxed{R_{eq} = R_1 + R_2 + \cdots + R_N} \qquad [8]$$

We are therefore able to replace a two-terminal network consisting of *N* series resistors with a single two-terminal element R_{eq} that has the same *v–i* relationship.

Of course, we might be interested in the current, voltage, or power of one of the original elements. For example, the voltage of a dependent voltage source may depend upon the voltage across R_3. Once R_3 is combined with several series resistors to form an equivalent resistance, then it is gone and the voltage across it cannot be determined until R_3 is identified by removing it from the combination. In that case, it would have been better to look ahead and not make R_3 a part of the combination initially.

PRACTICE

3.12 Determine a single-value equivalent resistance for the network shown in Fig. 3.27, as seen between the terminals marked *(a)* *a* and *b*, with *c* unconnected; *(b)* *a* and *c*, with *b* unconnected.

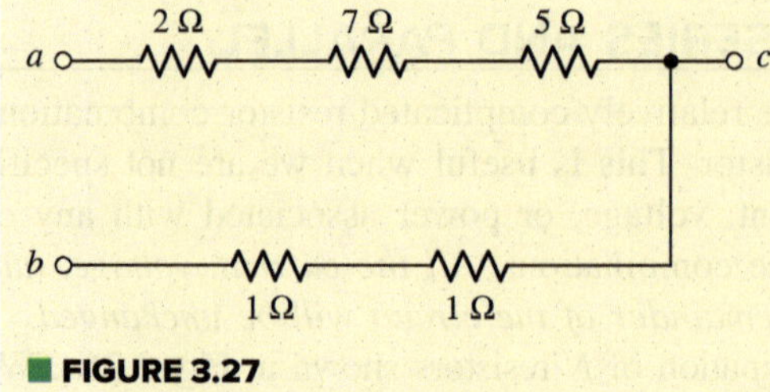

FIGURE 3.27

Ans: 16 Ω; 14 Ω.

EXAMPLE 3.11

Use resistance and source combinations to determine the current *i* in Fig. 3.28*a* and the power delivered by the 80 V source.

We first interchange the element positions in the circuit, being careful to preserve the proper sense of the sources, as shown in Fig. 3.28*b*. While not necessary, this step makes it a little easier to combine the three voltage sources into an equivalent 90 V source, and the four resistors into an equivalent 30 Ω resistance, as in Fig. 3.28*c*. Thus, instead of writing

$$-80 + 10i - 30 + 7i + 5i + 20 + 8i = 0$$

we have simply

$$-90 + 30i = 0$$

and so we find that

$$i = 3 \text{ A}$$

In order to calculate the power delivered to the circuit by the 80 V source appearing in the given circuit, it is necessary to return to Fig. 3.28*a* with the knowledge that the current is 3 A. The desired power is then 80 V × 3 A = 240 W.

It is interesting to note that no element of the original circuit remains in the equivalent circuit.

FIGURE 3.28 (*a*) A series circuit with several sources and resistors. (*b*) The elements are rearranged for the sake of clarity. (*c*) A simpler equivalent.

PRACTICE

3.13 Determine *i* in the circuit of Fig. 3.29.

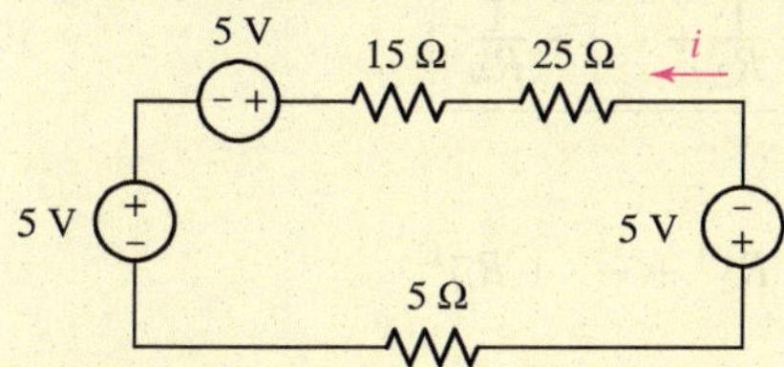

FIGURE 3.29

Ans: −333 mA.

COMPUTER-AIDED ANALYSIS

In Chap. 2 we were introduced to MATLAB, and performed calculations using the line-by-line editor. Another option, preferred in general, is to create text file with the commands, and then run it from within the MATLAB environment. Such files are saved or renamed with a ".m" suffix, and can be edited, updated, and combined with other MATLAB files. Here, we will write a short program to compute series resistance based on user input:

```
% Series Resistance Calculator
Rseries = 0;    %initialize value
prompt = 'Enter number of resistors in series string: ';
n = input(prompt)
for j = 1:n
    prompt = 'Enter resistance value (Ohms): ';
    R = input(prompt)
    Rseries = Rseries + R;
end;
Result = ['The equivalent series resistance is ',num2str(Rseries),' Ohms.'];
disp(Result)
%end of routine
```

After entering this in an editor such as Notepad and saving with a ".m" extension, we then locate the file in the directory from within MATLAB, identify the file, and click the **Run** icon. Note that the output appears in the Command Window. In order to print our result on one line, we had to create a series string and also convert the result (Rseries) to a string variable.

Similar simplifications can be applied to parallel circuits. A circuit containing N resistors in parallel, as in Fig. 3.30*a*, leads to the KCL equation

$$i_s = i_1 + i_2 + \cdots + i_N$$

or

$$i_s = \frac{v}{R_1} + \frac{v}{R_2} + \cdots + \frac{v}{R_N}$$

$$= \frac{v}{R_{\text{eq}}}$$

Thus,

$$\boxed{\frac{1}{R_{\text{eq}}} = \frac{1}{R_1} + \frac{1}{R_2} + \cdots + \frac{1}{R_N}} \qquad [9]$$

which can be written as

$$R_{\text{eq}}^{-1} = R_1^{-1} + R_2^{-1} + \cdots + R_N^{-1}$$

or, in terms of conductances, as

$$G_{\text{eq}} = G_1 + G_2 + \cdots + G_N$$

The simplified (equivalent) circuit is shown in Fig. 3.30*b*.

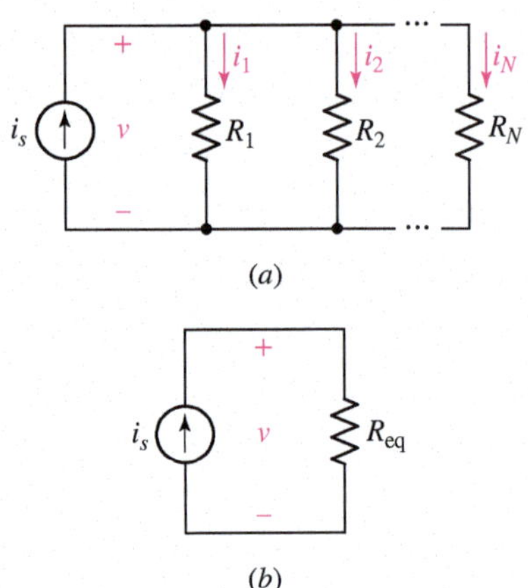

FIGURE 3.30 (*a*) A circuit with N resistors in parallel. (*b*) Equivalent circuit.

A parallel combination is routinely indicated by the following shorthand notation:

$$R_{eq} = R_1 \| R_2 \| R_3$$

The special case of only two parallel resistors is encountered fairly often and is given by

$$R_{eq} = R_1 \| R_2 = \frac{1}{\frac{1}{R_1} + \frac{1}{R_2}}$$

Or, more simply,

$$\boxed{R_{eq} = \frac{R_1 R_2}{R_1 + R_2}} \qquad [10]$$

The last form is worth memorizing, although it is a common error to attempt to generalize Eq. [10] to more than two resistors, e.g.,

$$R_{eq} \neq \frac{R_1 R_2 R_3}{R_1 + R_2 + R_3}$$

A quick look at the units of this equation will immediately show that the expression cannot possibly be correct.

PRACTICE

3.14 Determine v in the circuit of Fig. 3.31 by first combining the three current sources, and then the two 10 Ω resistors.

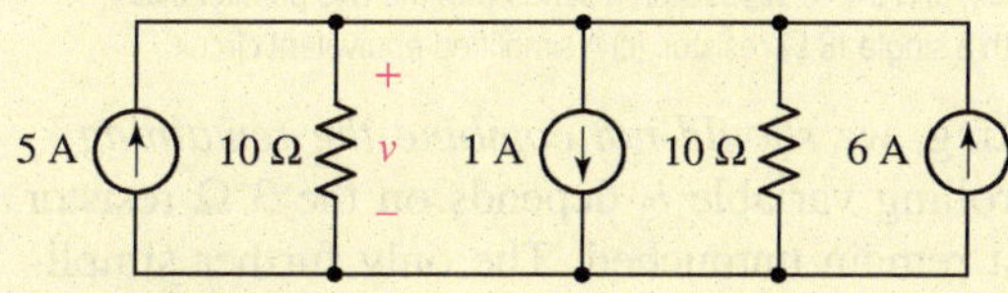

■ **FIGURE 3.31**

Ans: 50 V.

EXAMPLE 3.12

Calculate the power and voltage of the dependent source in Fig. 3.32*a*.

We will seek to simplify the circuit before analyzing it, but take care not to include the dependent source since its voltage and power characteristics are of interest.

Despite not being drawn adjacent to one another, the two independent current sources are in fact in parallel, so we replace them with a 2 A source, arrow upward.

The two 6 Ω resistors are in parallel and can be replaced with a single 3 Ω resistor in series with the 15 Ω resistor. Thus, the two 6 Ω resistors and the 15 Ω resistor are replaced by an 18 Ω resistor (Fig. 3.32*b*).

(Continued on next page)

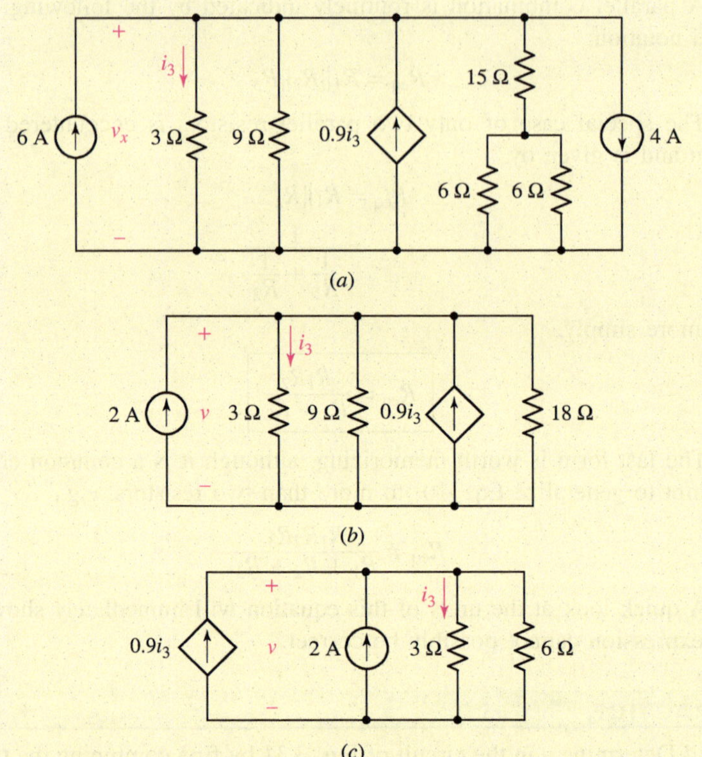

■ **FIGURE 3.32** (a) A multinode circuit. (b) The two independent current sources are combined into a 2 A source, and the 15 Ω resistor in series with the two parallel 6 Ω resistors are replaced with a single 18 Ω resistor. (c) A simplified equivalent circuit.

No matter how tempting, *we should not combine the remaining three resistors;* the controlling variable i_3 depends on the 3 Ω resistor, and so that resistor must remain untouched. The only further simplification, then, is 9 Ω‖18 Ω = 6 Ω, as shown in Fig. 3.32*c*.

Applying KCL at the top node of Fig. 3.32*c*, we have

$$-0.9i_3 - 2 + i_3 + \frac{v}{6} = 0$$

Employing Ohm's law,

$$v = 3\, i_3$$

which allows us to compute

$$i_3 = \frac{10}{3}\ \text{A}$$

Thus, the voltage across the dependent source (which is the same as the voltage across the 3 Ω resistor) is

$$v = 3i_3 = 10\ \text{V}$$

The dependent source therefore furnishes $v \times 0.9i_3 = 10(0.9)(10/3) =$ 30 W to the remainder of the circuit.

Now if we are later asked for the power dissipated in the 15 Ω resistor, we must return to the original circuit. This resistor is in series with an equivalent 3 Ω resistor; a voltage of 10 V is across the 18 Ω total; therefore, a current of 5/9 A flows through the 15 Ω resistor, and the power absorbed by this element is $(5/9)^2(15)$ or 4.63 W.

PRACTICE

3.15 For the circuit of Fig. 3.33, calculate the voltage v_x.

■ **FIGURE 3.33**

Ans: 2.819 V.

Three final comments on series and parallel combinations might be helpful. The first is illustrated by referring to Fig. 3.34*a* and asking, "*Are v_s and R in series or in parallel?*" The answer is "*Both.*" The two elements carry the same current and are therefore in series; they also enjoy the same voltage and consequently are in parallel.

The second comment is a word of caution. Circuits can be drawn in such a way as to make series or parallel combinations difficult to spot. In Fig. 3.34*b*, for example, the only two resistors in parallel are R_2 and R_3, while the only two in series are R_1 and R_8.

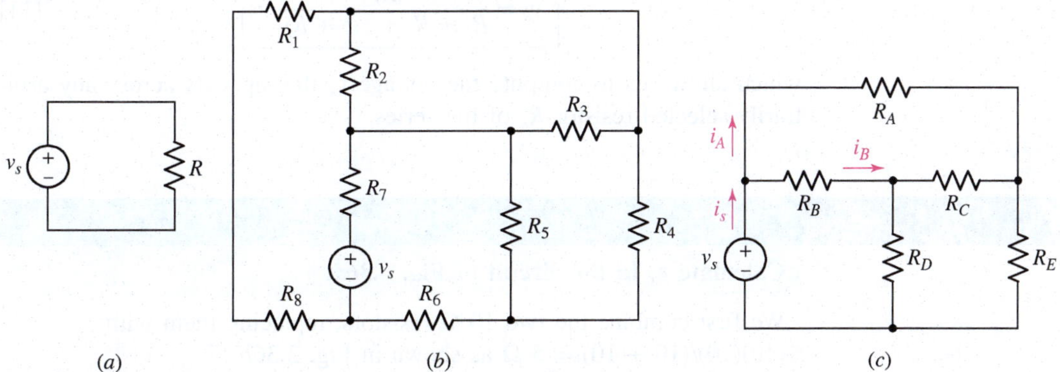

■ **FIGURE 3.34** (*a*) These two circuit elements are both in series and in parallel. (*b*) R_2 and R_3 are in parallel, and R_1 and R_8 are in series. (*c*) There are no circuit elements either in series or in parallel with one another.

The final comment is simply that a simple circuit element need not be in series or parallel with any other simple circuit element in a circuit. For example, R_4 and R_5 in Fig. 3.34*b* are not in series or parallel with any other simple circuit element, and there are no simple circuit elements in Fig. 3.34*c* that are in series or parallel with any other simple circuit element. In other words, we cannot simplify that circuit further using any of the techniques discussed in this chapter.

3.8 VOLTAGE AND CURRENT DIVISION

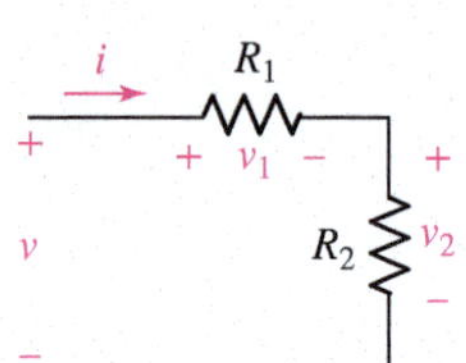

FIGURE 3.35 An illustration of voltage division.

By combining resistances and sources, we found one method of shortening the work of analyzing a circuit. Another useful shortcut is the application of the ideas of voltage and current division. Voltage division is used to express the voltage across one of several series resistors in terms of the voltage across the combination. In Fig. 3.35, the voltage across R_2 is found via KVL and Ohm's law:

$$v = v_1 + v_2 = iR_1 + iR_2 = i(R_1 + R_2)$$

so

$$i = \frac{v}{R_1 + R_2}$$

Thus,

$$v_2 = iR_2 = \left(\frac{v}{R_1 + R_2}\right)R_2$$

or

$$v_2 = \frac{R_2}{R_1 + R_2}v$$

and the voltage across R_1 is, similarly,

$$v_1 = \frac{R_1}{R_1 + R_2}v$$

If the network of Fig. 3.35 is generalized by removing R_2 and replacing it with the series combination of $R_2, R_3, \ldots, R_N$, then we have the general result for voltage division across a string of N series resistors

$$\boxed{v_k = \frac{R_k}{R_1 + R_2 + \cdots + R_N}v} \qquad [11]$$

which allows us to compute the voltage v_k that appears across any arbitrarily selected resistor R_k of the series.

EXAMPLE 3.13

Calculate v_x in the circuit of Fig. 3.36*a*.

We first combine the two 10 Ω resistors, replacing them with (10)(10)/(10 + 10) = 5 Ω as shown in Fig. 3.36*b*.

Note that if we allow ourselves to get too overzealous and combine the 2 Ω and 3 Ω resistors, we lose the voltage v_x we are trying to find.

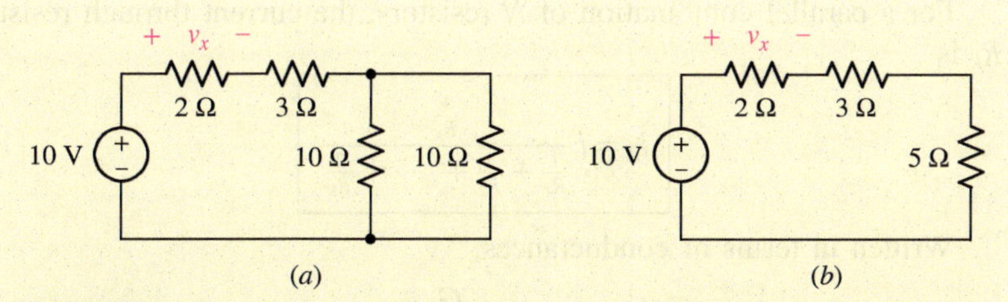

FIGURE 3.36 A numerical example illustrating resistance combination and voltage division. (*a*) Original circuit. (*b*) Simplified circuit.

Since v_x appears across the 2 Ω resistor, we proceed by simply applying voltage division to the circuit in Fig. 3.36*b*:

$$v_x = 10 \frac{2}{2 + 3 + 5} = 2 \text{ V}$$

PRACTICE

3.16 Use voltage division to determine v_x in the circuit of Fig. 3.37. (Don't let the sinusoidal source throw you.)

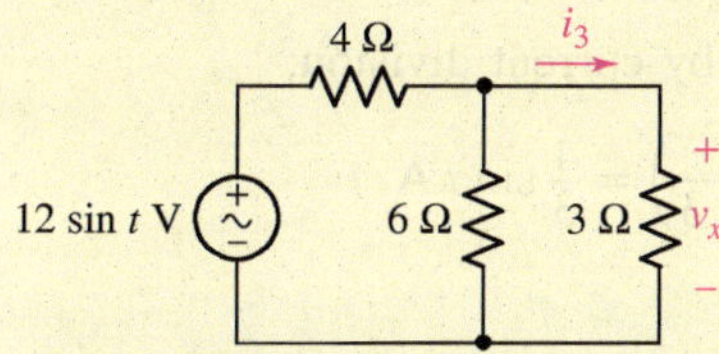

FIGURE 3.37

Ans: $v_x = 4 \sin t$ volts

The dual[2] of voltage division is current division. We are now given a total current supplied to several parallel resistors, as shown in the partial circuit of Fig. 3.38.

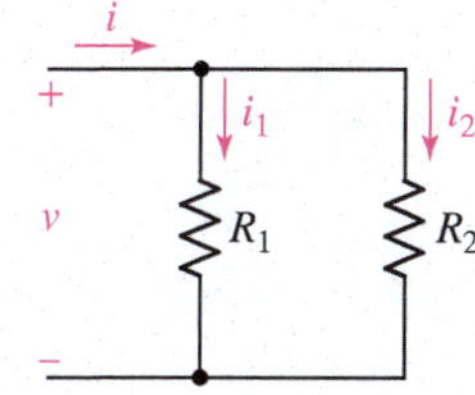

FIGURE 3.38 An illustration of current division.

The current flowing through R_2 is

$$i_2 = \frac{v}{R_2} = \frac{i(R_1 \| R_2)}{R_2} = \frac{i}{R_2} \frac{R_1 R_2}{R_1 + R_2}$$

or

$$i_2 = i \frac{R_1}{R_1 + R_2} \qquad [12]$$

and, similarly,

$$i_1 = i \frac{R_2}{R_1 + R_2} \qquad [13]$$

Nature has not smiled on us here, for these last two equations have a factor which differs subtly from the factor used with voltage division, and some effort is going to be needed to avoid errors. Often students look on the expression for voltage division as "obvious" and that for current division as being "different." It helps to realize that the larger of two parallel resistors always carries the smaller current.

(2) The principle of duality is encountered often in engineering. We will consider the topic briefly in Chap. 7 when we compare inductors and capacitors.

For a parallel combination of N resistors, the current through resistor R_k is

$$i_k = i \frac{\frac{1}{R_k}}{\frac{1}{R_1} + \frac{1}{R_2} + \cdots + \frac{1}{R_N}} \qquad [14]$$

Written in terms of conductances,

$$i_k = i \frac{G_k}{G_1 + G_2 + \cdots + G_N}$$

which strongly resembles Eq. [11] for voltage division.

EXAMPLE 3.14

Write an expression for the current through the 3 Ω resistor in the circuit of Fig. 3.39.

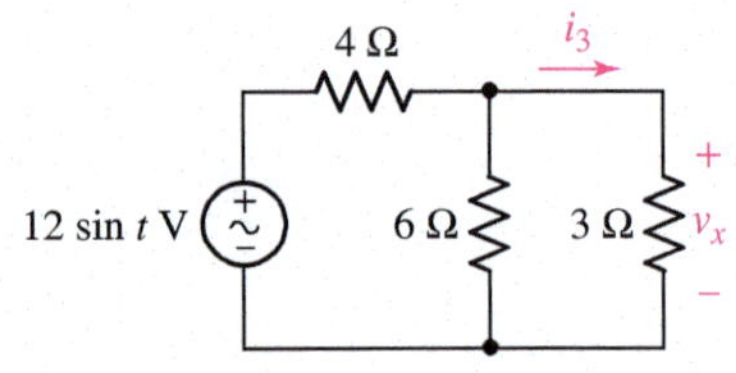

FIGURE 3.39 A circuit used as an example of current division. The wavy line in the voltage source symbol indicates a sinusoidal variation with time.

The total current flowing into the 3 Ω–6 Ω combination is

$$i(t) = \frac{12 \sin t}{4 + 3\|6} = \frac{12 \sin t}{4 + 2} = 2 \sin t \text{ A}$$

and thus the desired current is given by current division:

$$i_3(t) = (2 \sin t)\left(\frac{6}{6+3}\right) = \frac{4}{3} \sin t \text{ A}$$

PRACTICE

3.17 In the circuit of Fig. 3.40, use resistance combination methods and current division to find i_1, i_2, and v_3.

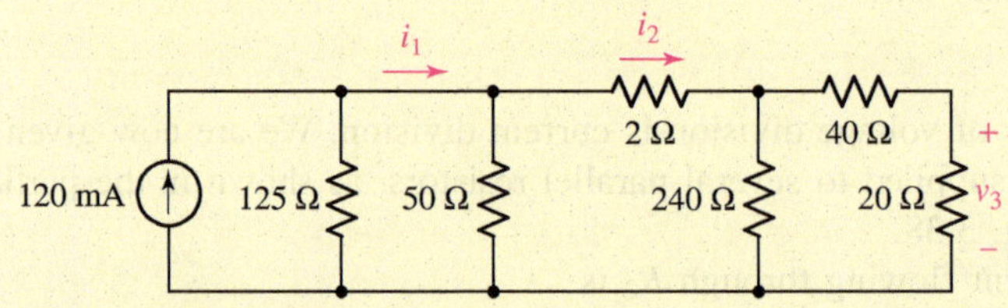

FIGURE 3.40

Ans: 100 mA; 50 mA; 0.8 V.

Unfortunately, current division is sometimes applied when it is not applicable! As one example, let us consider again the circuit shown in Fig. 3.34*c*, a circuit that we have already agreed contains no circuit elements that are in series or in parallel. Without parallel resistors, there is no way that current division can be applied. Even so, it can be tempting to take a quick look at resistors R_A and R_B and try to apply current division, writing an incorrect equation such as

$$i_A \neq i_S \frac{R_B}{R_A + R_B}$$

Remember, *parallel resistors must be branches between the same pair of nodes*.

PRACTICAL APPLICATION

Not the Earth Ground from Geology

Up to now, we have been drawing circuit schematics in a fashion similar to that of the one shown in Fig. 3.41, where voltages are defined across two clearly marked terminals. Special care was taken to emphasize the fact that voltage cannot be defined at a single point—it is by definition the *difference* in potential between *two* points. However, many schematics make use of the convention of taking the earth as zero volts, and all other voltages are implicitly referenced to this potential. The concept is often referred to as ***earth ground*** (as in the earth, assuming the circuit is connected to a distant power source), and is fundamentally tied to safety regulations designed to prevent fires, fatal electrical shocks, and related mayhem. The symbol for earth ground is shown in Fig. 3.42*a*.

Since earth ground is defined as zero volts, it is often convenient to use this as a common terminal in schematics. The circuit of Fig. 3.41 is shown redrawn in this fashion in Fig. 3.43, where the earth ground symbol represents a common node. It is important to note that the two circuits are equivalent in terms of our value for v_a (4.5 V in either case) but are no longer exactly the same. The circuit in Fig. 3.41 is said to be "floating" in that it could for all practical purposes be installed on a circuit board of a satellite in geosynchronous orbit (or on its way to Pluto). The circuit in Fig. 3.43, however, is somehow physically connected to the ground through a conducting path. For this reason, there are two other symbols that are occasionally used to denote a common terminal. Figure 3.42*b* shows what is commonly referred to as ***signal ground;*** there can be (and often is) a large voltage between earth ground and any terminal tied to signal ground.

FIGURE 3.41 A simple circuit with a voltage v_a defined between two terminals.

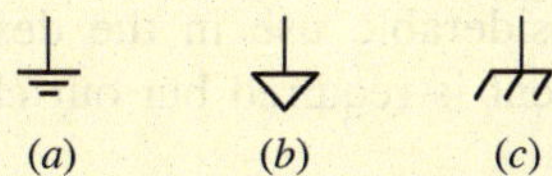

FIGURE 3.42 Three different symbols used to represent a ground or common terminal: (*a*) earth ground; (*b*) signal ground; (*c*) chassis ground.

The fact that the common terminal of a circuit may or may not be connected by some low-resistance pathway to earth ground can lead to potentially dangerous situations. Consider the diagram of Fig. 3.44*a*, which depicts an innocent bystander about to touch a piece of equipment powered by an ac outlet. The equipment has a conducting (i.e., metal) chassis. The common terminal of every circuit in the equipment has been tied together and electrically connected to the conducting equipment chassis; this terminal is often denoted using the ***chassis ground*** symbol of Fig. 3.42*c*. Unfortunately, a wiring fault exists here, due to either poor manufacturing or perhaps just wear and tear. At any rate, the chassis is not tied to earth ground, and so there is a very large (essentially infinite) resistance between chassis ground and earth ground; as a result, the equipment is not functioning. A pseudo-schematic (some liberty was taken with the person's equivalent resistance symbol) of the situation is shown in Fig. 3.44*b*. The resistance of the person, however, is not infinite, and it could be very small—especially if they are not wearing rubber-soled shoes. Once the person taps on the equipment to see why it isn't working properly . . . well, let's just say not all stories have happy endings.

The fact that "ground" is not always "earth ground" can cause a wide range of safety and electrical noise problems. One example is occasionally encountered in older buildings, where plumbing originally consisted of electrically conducting metal pipes. In such buildings, any water pipe was often treated as a low-resistance path to earth ground, and therefore was used in many electrical connections. However, when corroded pipes are replaced with more modern and cost-effective nonconducting

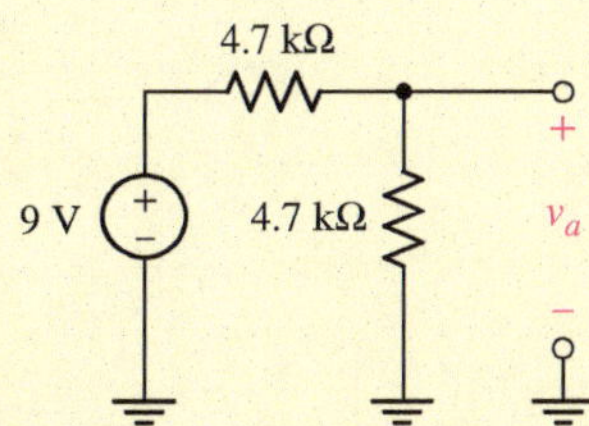

FIGURE 3.43 The circuit of Fig. 3.41, redrawn using the earth ground symbol. The rightmost ground symbol is redundant; it is only necessary to label the positive terminal of v_a; the negative reference is then implicitly ground, or zero volts.

(Continued on next page)

PVC piping, the low-resistance path to earth ground no longer exists. A related problem occurs when the composition of the earth varies greatly over a particular region. In such situations, it is possible to actually have two separated buildings in which the two "earth grounds" are not at the same electrical potential, and current can flow as a result.

Within this text, the earth ground symbol will be used exclusively. It is worth remembering, however, that not all grounds are created equal!

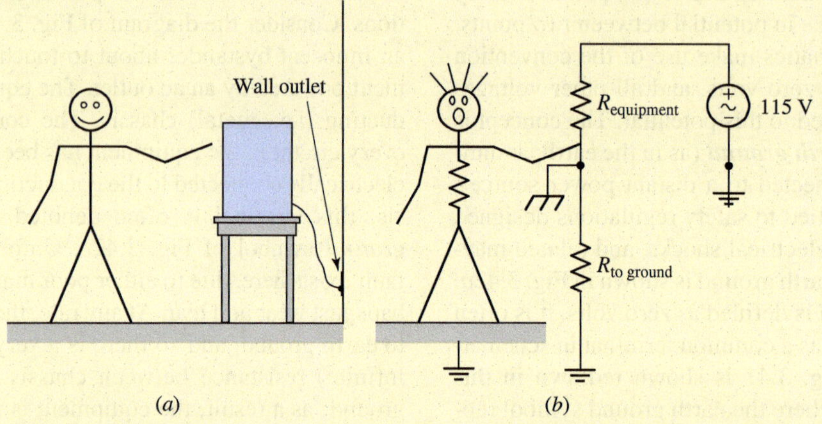

■ **FIGURE 3.44** (*a*) A sketch of an innocent person about to touch an improperly grounded piece of equipment. It's not going to be pretty. (*b*) A schematic of an equivalent circuit for the situation as it is about to unfold; the person has been represented by an equivalent resistance, as has the equipment. A resistor ($R_{\text{to ground}}$) has been used to represent the nonhuman path to ground, which in this example is so large it can be assumed to be essentially infinity. The human, with no insulated gloves or shoes, represents a low-resistance path to (earth) ground. $R_{\text{to ground}}$ represents a path to ground that exists because the chassis is not properly tied to ground (perhaps the connectors have been oxidized).

SUMMARY AND REVIEW

We began by discussing connections of circuit elements, and introduced the terms *node, path, loop,* and *branch*. The next two topics could be considered the two most important in the entire text, namely, Kirchhoff's current law (KCL) and Kirchhoff's voltage law. These two laws allow us to analyze *any* circuit, linear or otherwise, provided we have a way of relating the voltage and current associated with passive elements (e.g., Ohm's law for the resistor). In the case of a single-loop circuit, the elements are connected in *series* and hence each carries the same current. The single-node-pair circuit, in which elements are connected in *parallel* with one another, is characterized by a single voltage common to each element. Extending these concepts allowed us to develop expressions for series and parallel connected resistors. The final topic, that of voltage and current division, finds considerable use in the design of circuits where a specific voltage or current is required but our choice of source is limited.

Let's conclude with key points of this chapter to review, highlighting appropriate examples.

- Kirchhoff's current law (KCL) states that the algebraic sum of the currents entering any node is zero. (Examples 3.1, 3.4)
- Kirchhoff's voltage law (KVL) states that the algebraic sum of the voltages around any closed path in a circuit is zero. (Examples 3.2, 3.3)
- All elements in a circuit that carry the same current are said to be connected in series. (Example 3.5)
- Elements in a circuit having a common voltage across them are said to be connected in parallel. (Examples 3.6, 3.7)
- Voltage sources in series can be replaced by a single source, provided care is taken to note the individual polarity of each source. (Examples 3.8, 3.10)
- Current sources in parallel can be replaced by a single source, provided care is taken to note the direction of each current arrow. (Examples 3.9, 3.10)
- A series combination of N resistors can be replaced by a single resistor having the value $R_{eq} = R_1 + R_2 + \cdots + R_N$. (Example 3.11)
- A parallel combination of N resistors can be replaced by a single resistor having the value

$$\frac{1}{R_{eq}} = \frac{1}{R_1} + \frac{1}{R_2} + \cdots + \frac{1}{R_N}$$

(Example 3.12)
- Voltage division allows us to calculate what fraction of the total voltage across a series string of resistors is dropped across any one resistor (or group of resistors). (Example 3.13)
- Current division allows us to calculate what fraction of the total current into a parallel string of resistors flows through any one of the resistors. (Example 3.14)

READING FURTHER

A discussion of the principles of conservation of energy and conservation of charge, as well as Kirchhoff's laws, can be found in

R. Feynman, R. B. Leighton, and M. L. Sands, *The Feynman Lectures on Physics.* Reading, MA: Addison-Wesley, 1989, pp. 4-1, 4-7, and 25-9.

Detailed discussions of numerous aspects of grounding practices consistent with the 2017 National Electrical Code® can be found throughout

F. P. Hartwell, J. F. McPartland, B. McPartland, *McGraw Hill's National Electrical Code 2017 Handbook,* 29th ed. New York: McGraw Hill, 2017.

EXERCISES

3.1 Nodes, Paths, Loops, and Branches

1. Referring to the circuit depicted in Fig. 3.45, count the number of (*a*) nodes; (*b*) elements; (*c*) branches.

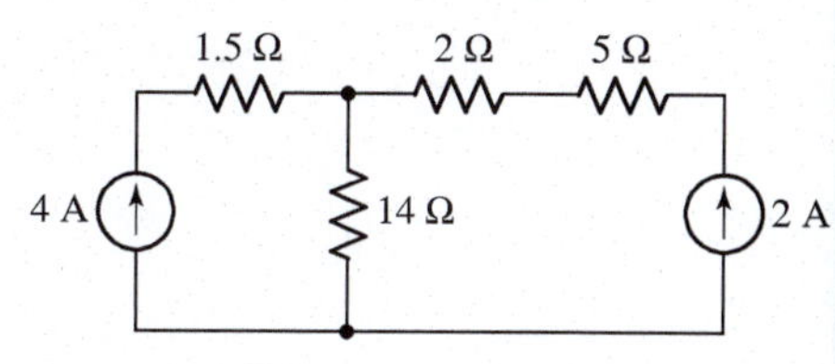

FIGURE 3.45

2. Referring to the circuit depicted in Fig. 3.46, count the number of (*a*) nodes; (*b*) elements; (*c*) branches.

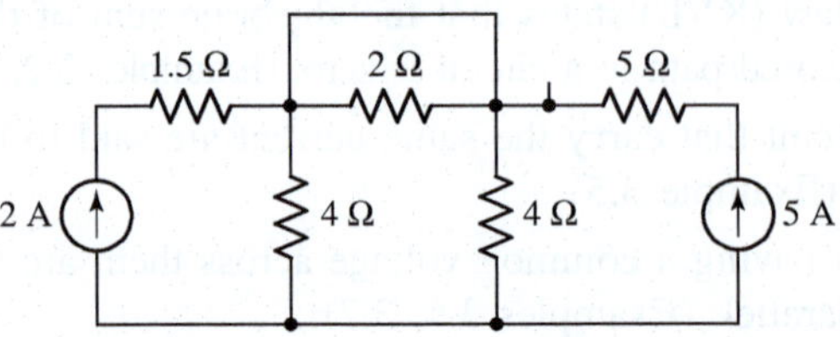

■ **FIGURE 3.46**

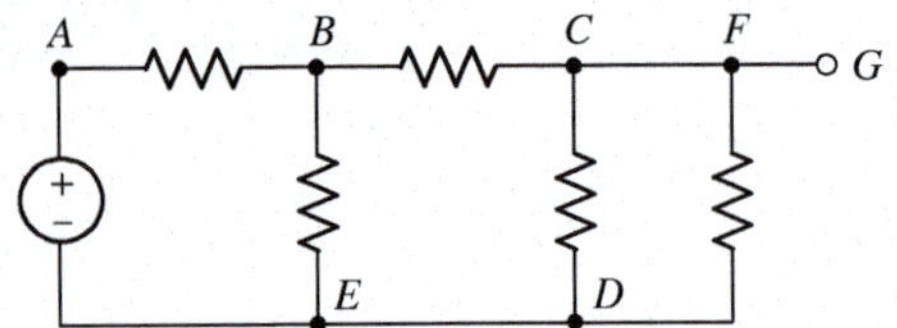

■ **FIGURE 3.47**

3. For the circuit of Fig. 3.47:
   (*a*) Count the number of nodes.
   (*b*) In moving from *A* to *B*, have we formed a path? Have we formed a loop?
   (*c*) In moving from *C* to *F* to *G*, have we formed a path? Have we formed a loop?
4. For the circuit of Fig. 3.47:
   (*a*) Count the number of circuit elements.
   (*b*) If we move from *B* to *C* to *D*, have we formed a path? Have we formed a loop?
   (*c*) If we move from *E* to *D* to *C* to *B*, have we formed a path? Have we formed a loop?

■ **FIGURE 3.48**

5. Refer to the circuit of Fig. 3.48, and answer the following:
   (*a*) How many distinct nodes are contained in the circuit?
   (*b*) How many elements are contained in the circuit?
   (*c*) How many branches does the circuit have?
   (*d*) Determine if each of the following represents a path, a loop, both, or neither:
      (i) *A* to *B*
      (ii) *B* to *D* to *C* to *E*
      (iii) *C* to *E* to *D* to *B* to *A* to *C*
      (iv) *C* to *D* to *B* to *A* to *C* to *E*

3.2 Kirchhoff's Current Law

6. A local restaurant has a neon sign constructed from 12 separate bulbs; when a bulb fails, it appears as an infinite resistance and cannot conduct current. In wiring the sign, the manufacturer offers two options (Fig. 3.49). From what you've learned about KCL, which one should the restaurant owner select? Explain.

■ **FIGURE 3.49**

7. Referring to the single-node diagram of Fig. 3.50, compute:
   (*a*) i_B, if $i_A = 1$ A, $i_D = -2$ A, $i_C = 3$ A, and $i_E = 0$;
   (*b*) i_E, if $i_A = -1$ A, $i_B = -1$ A, $i_C = -1$ A, and $i_D = -1$ A.

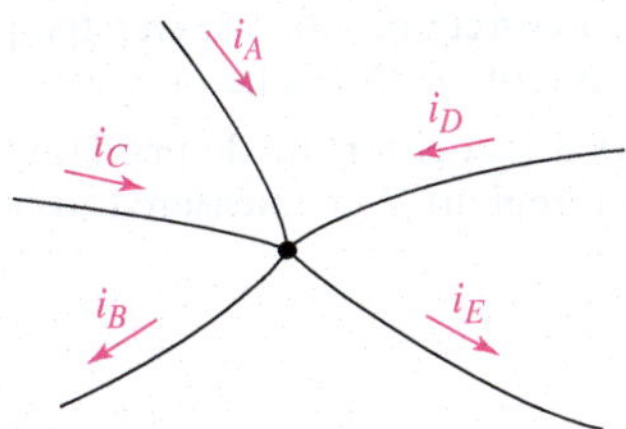

FIGURE 3.50

8. Determine the current labeled *I* in each of the circuits of Fig. 3.51.

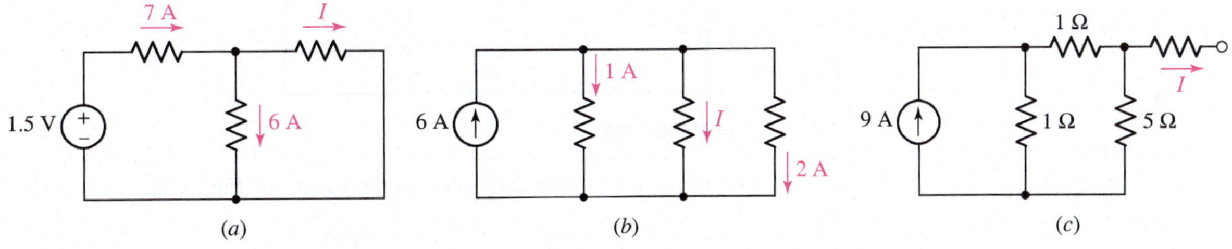

FIGURE 3.51

9. In the circuit shown in Fig. 3.52, the resistor values are unknown, but the 2 V source is known to be supplying a current of 7 A to the rest of the circuit. Calculate the current labeled i_2.

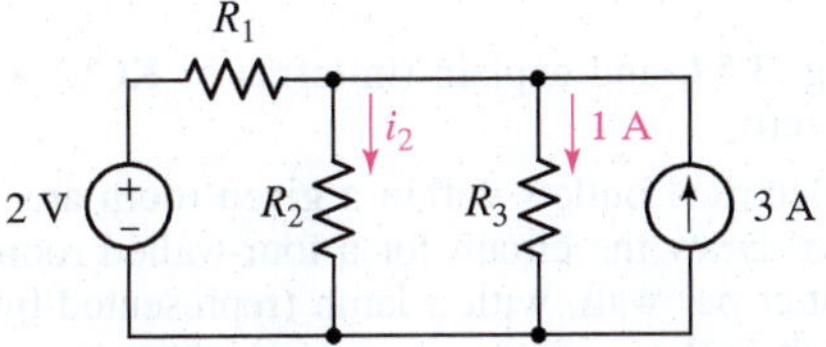

FIGURE 3.52

10. The circuit of Fig. 3.53 represents a system comprised of an LED sign powered by a combination of battery storage and three solar panels. The panels are not equally illuminated, so the current each supplies can vary, although

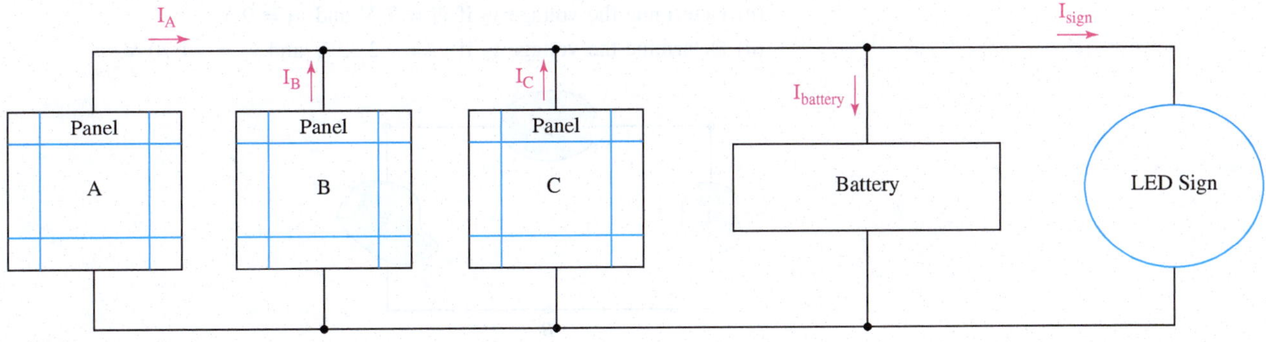

FIGURE 3.53

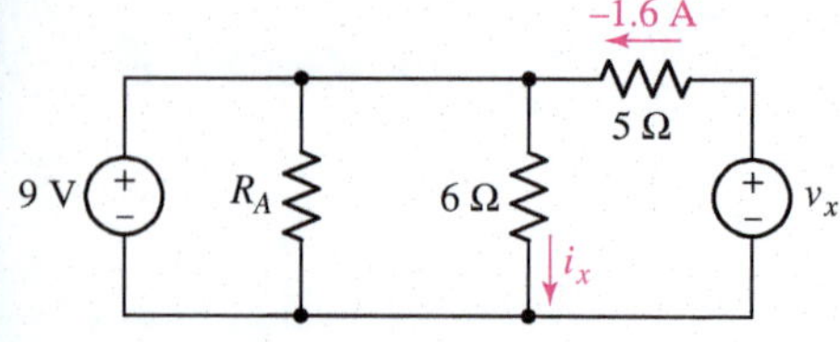

■ FIGURE 3.54

the voltage across each is forced to be the same. If I_A = 4.5 A, I_B = 4.3 A, and I_C = 4.6 A, calculate the current flowing into the battery if the LED sign draws 5.1 A.

11. In the circuit depicted in Fig. 3.54, i_x is determined to be 1.5 A, and the 9 V source supplies a current of 7.6 A (that is, a current of 7.6 A leaves the positive reference terminal of the 9 V source). Determine the value of resistor R_A.
12. For the circuit of Fig. 3.55 (which employs a model for the dc operation of a bipolar junction transistor biased in active region), I_C is measured to be 1.5 mA. Calculate I_B and I_E.

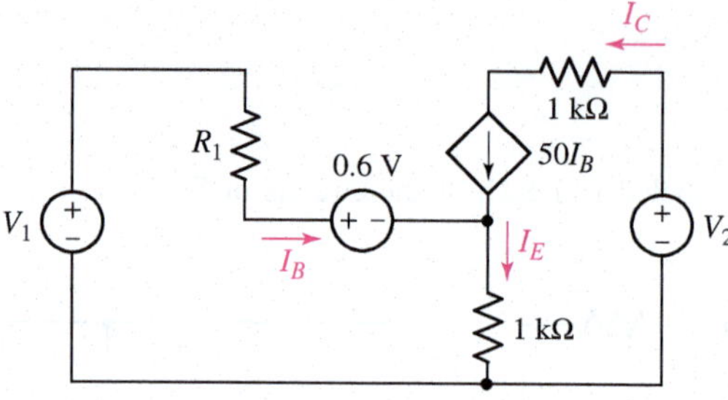

■ FIGURE 3.55

13. Determine the current labeled I_3 in the circuit of Fig. 3.56.

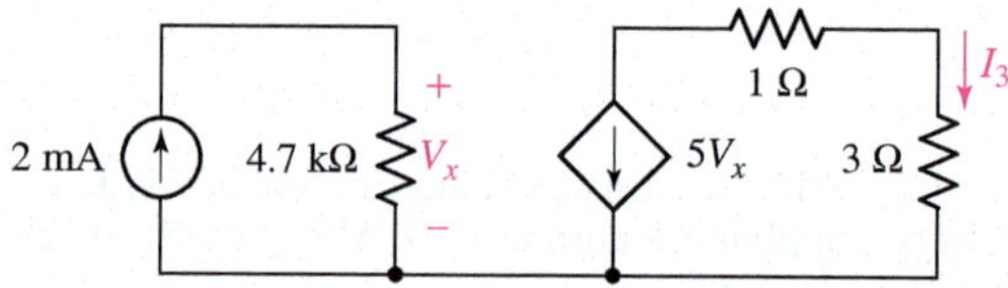

■ FIGURE 3.56

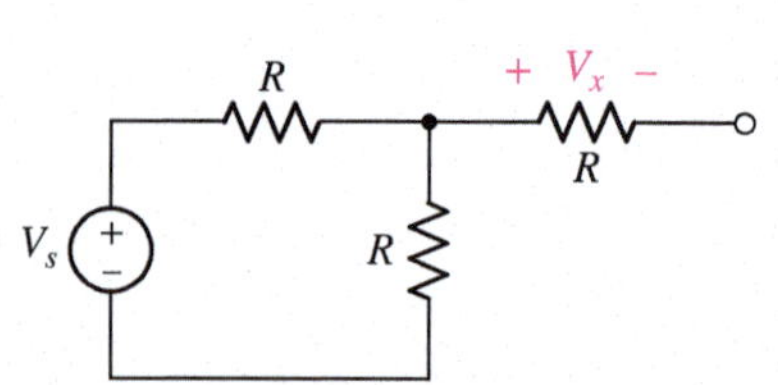

■ FIGURE 3.57

14. Study the circuit depicted in Fig. 3.57, and explain (in terms of KCL) why the voltage labeled V_x must be zero.
15. In many households, multiple electrical outlets within a given room are often all part of the same circuit. Draw the circuit for a four-walled room which has a single electrical outlet per wall, with a lamp (represented by a 400 Ω resistor) connected to each outlet.

3.3 Kirchhoff's Voltage Law

16. For the circuit of Fig. 3.58:
    (*a*) Determine the voltage v_1 if $v_2 = 0$ and $v_3 = -17$ V.
    (*b*) Determine the voltage v_1 if $v_2 = -2$ V and $v_3 = +2$ V.
    (*c*) Determine the voltage v_2 if $v_1 = 7$ V and $v_3 = 9$ V.
    (*d*) Determine the voltage v_3 if $v_1 = -2.33$ V and $v_2 = -1.70$ V.

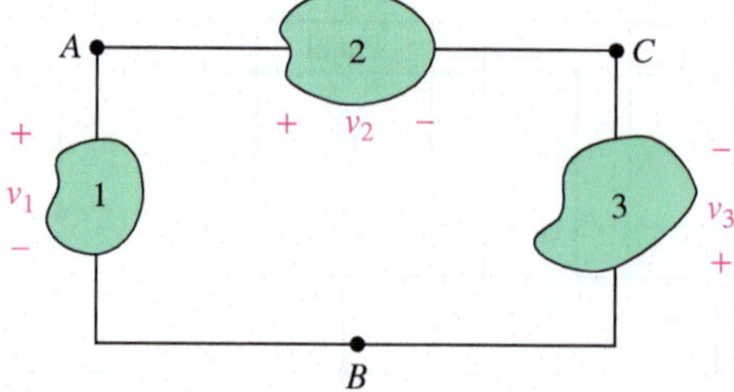

■ FIGURE 3.58

17. For each of the circuits in Fig. 3.59, determine the voltage v_x and the current i_x.

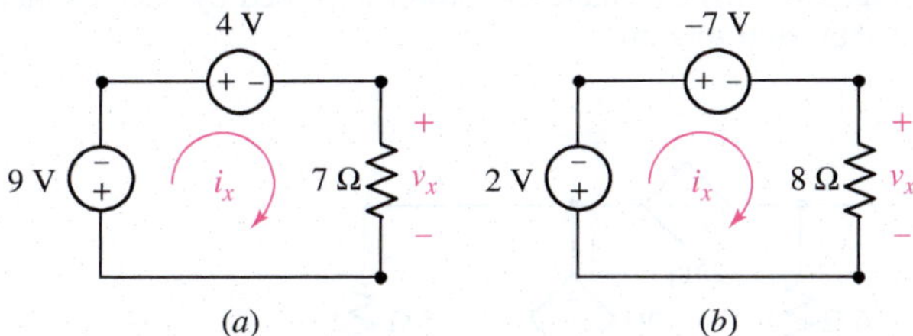

■ **FIGURE 3.59**

18. Use KVL to obtain a numerical value for the current labeled i in each circuit depicted in Fig. 3.60.

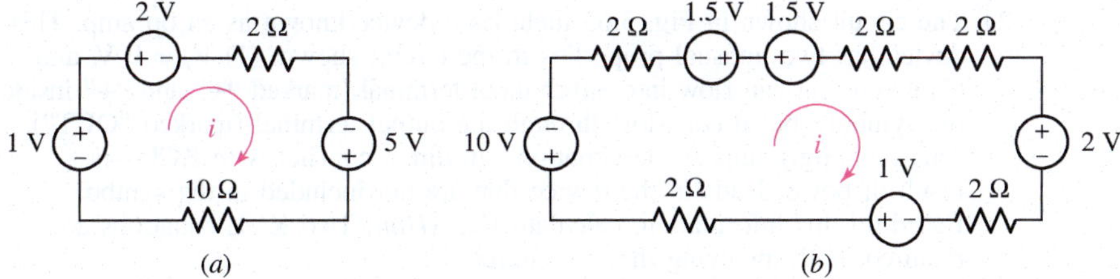

■ **FIGURE 3.60**

19. In the circuit of Fig. 3.61, it is determined that $v_1 = 3$ V and $v_3 = 1.5$ V. Calculate v_R and v_2.

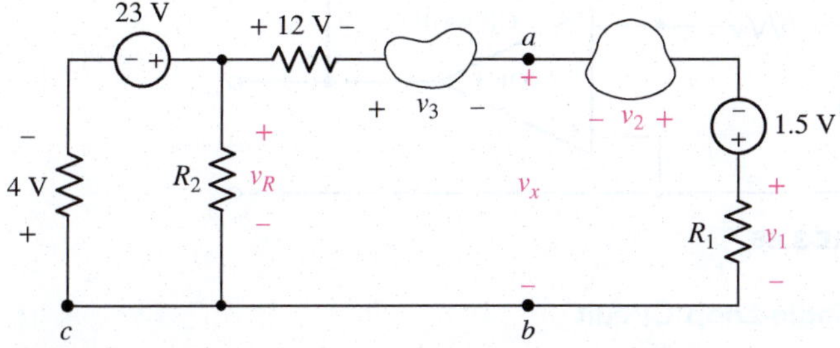

■ **FIGURE 3.61**

20. In the circuit of Fig. 3.55, calculate the voltage across the dependent source ("+" reference on top) if V_2 is 15 V, and I_B is 20 µA.
21. Determine the value of v_x as labeled in the circuit of Fig. 3.62.

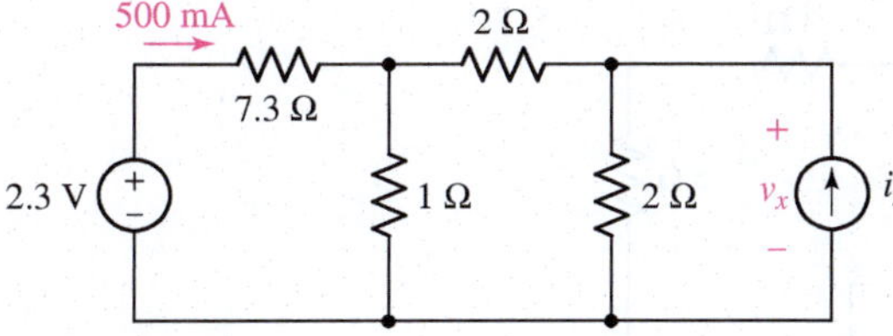

■ **FIGURE 3.62**

22. Consider the simple circuit shown in Fig. 3.63. (*a*) Using KVL, derive the expressions

$$v_1 = v_s \frac{R_1}{R_1 + R_2} \quad \text{and} \quad v_2 = v_s \frac{R_2}{R_1 + R_2}$$

(*b*) Under what conditions is it possible to find that $|v_2| < |v_s|$?

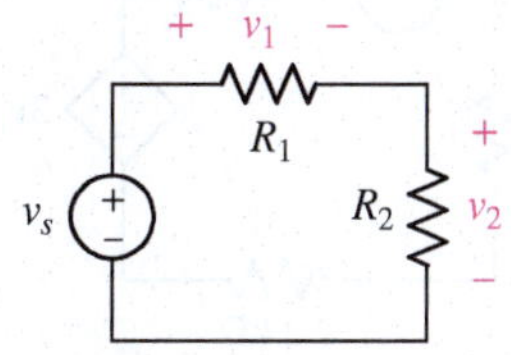

■ **FIGURE 3.63**

23. (*a*) Determine a numerical value for each current and voltage (i_1, v_1, etc.) in the circuit of Fig. 3.64. (*b*) Calculate the power absorbed by each element and verify that they sum to zero.

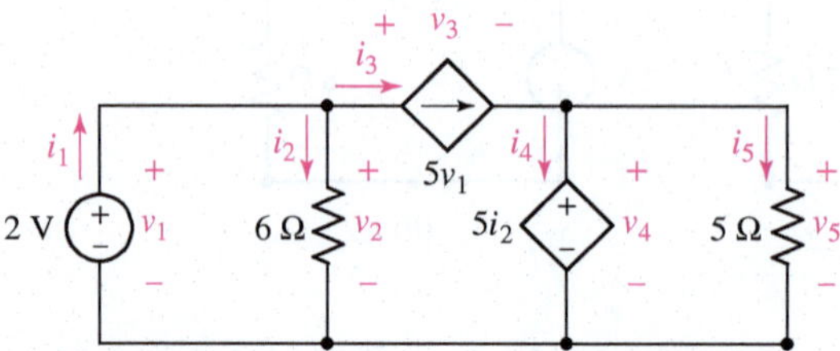

■ **FIGURE 3.64**

24. The circuit shown in Fig. 3.65 includes a device known as an op amp. This device has two unusual properties in the circuit shown: (1) $V_d = 0$ V, and (2) no current can flow into either input terminal (marked "−" and "+" inside the symbol), but it *can* flow through the output terminal (marked "OUT"). This seemingly impossible situation—in direct conflict with KCL—is a result of power leads to the device that are not included in the symbol. Based on this information, calculate V_{out}. (*Hint*: Two KVL equations are required, both involving the 5 V source.)

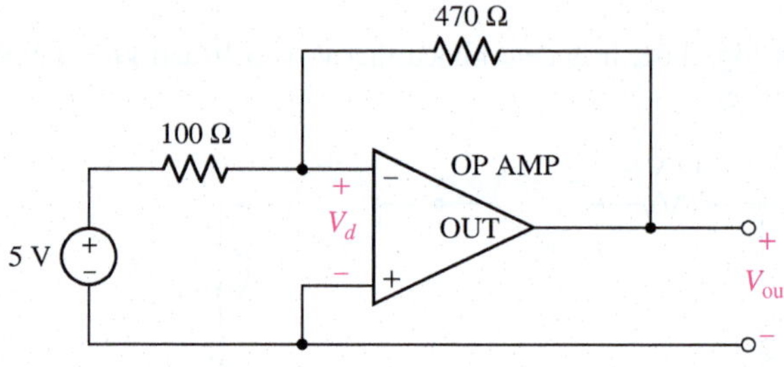

■ **FIGURE 3.65**

3.4 The Single-Loop Circuit

25. The circuit of Fig. 3.12*b* is constructed with the following: $v_{s1} = -8$ V, $R_1 = 1\ \Omega$, $v_{s2} = 16$ V, and $R_2 = 4.7\ \Omega$. Calculate the power absorbed by each element. Verify that the absorbed powers sum to zero.

26. Obtain a numerical value for the power absorbed by each element in the circuit shown in Fig. 3.66.

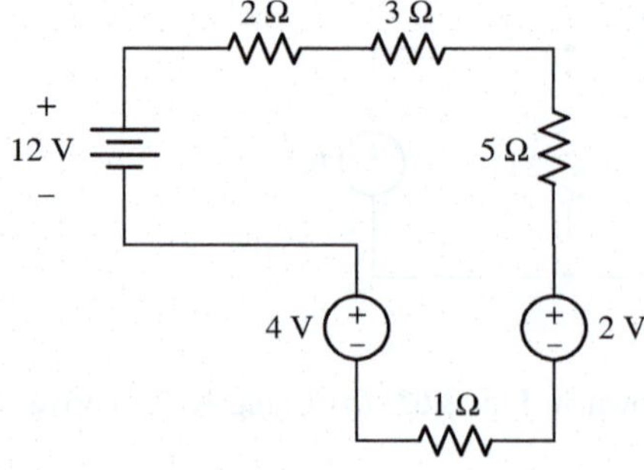

■ **FIGURE 3.66**

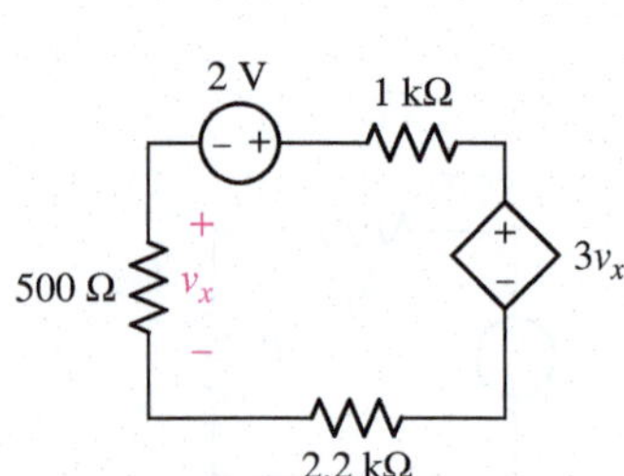

■ **FIGURE 3.67**

27. Compute the power absorbed by each element of the circuit of Fig. 3.67.

28. Compute the power absorbed by each element in the circuit of Fig. 3.68 if the mysterious element X is (*a*) a 13 Ω resistor; (*b*) a dependent voltage source labeled $4v_1$, "+" reference on top; (*c*) a dependent voltage source labeled $4i_x$, "+" reference on top.

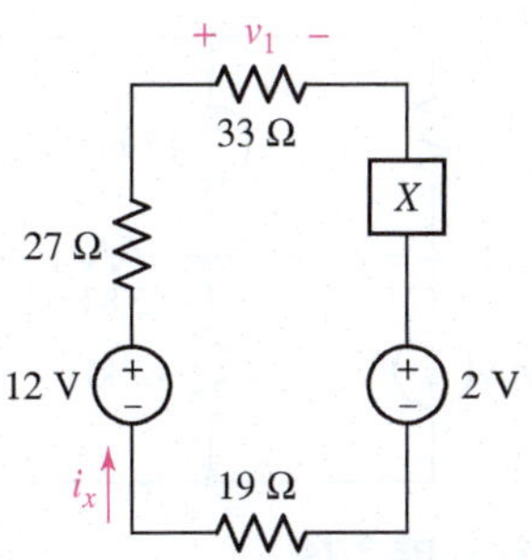

■ **FIGURE 3.68**

29. Kirchhoff's laws apply whether or not Ohm's law applies to a particular element. The *I–V* characteristic of a diode, for example, is given by

$$I_D = I_S(e^{V_D/V_T} - 1)$$

where $V_T = 27$ mV at room temperature and I_S can vary from 10^{-12} to 10^{-3} A. In the circuit of Fig. 3.69, use KVL/KCL to obtain I_D and V_D if $I_S = 45$ nA. (*Note: This problem results in a transcendental equation, requiring an iterative approach to obtaining a numerical solution. Most scientific calculators will perform such a function.*)

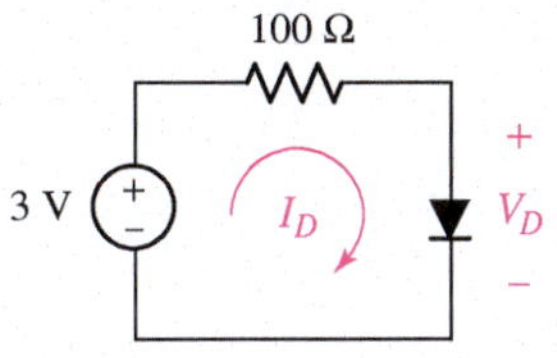

■ **FIGURE 3.69**

3.5 The Single-Node-Pair Circuit

30. Referring to the circuit of Fig. 3.70, (*a*) determine the two currents i_1 and i_2; (*b*) compute the power absorbed by each element.

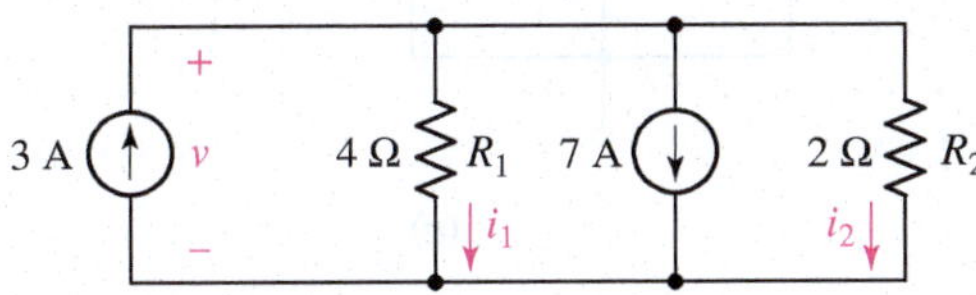

■ **FIGURE 3.70**

31. Determine a value for the voltage v as labeled in the circuit of Fig. 3.71, and compute the power supplied by the two current sources.

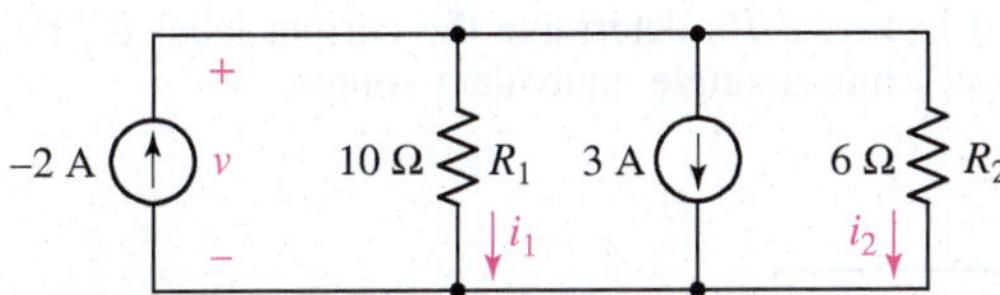

■ **FIGURE 3.71**

32. Referring to the circuit depicted in Fig. 3.72, determine the value of the voltage v if the element marked X is (*a*) a 2 A current source, arrow pointing down; (*b*) a 2 V voltage source, + reference on top; (*c*) a dependent current source, arrow pointing down, controlled by quantity $2v$.

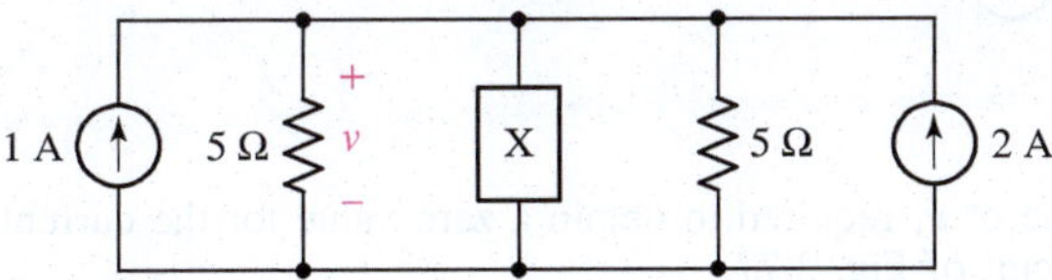

■ **FIGURE 3.72**

33. Determine the voltage v as labeled in Fig. 3.73, and calculate the power supplied by each current source.

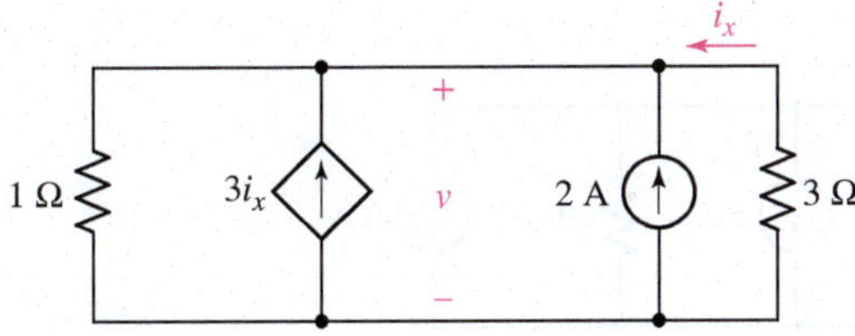

■ **FIGURE 3.73**

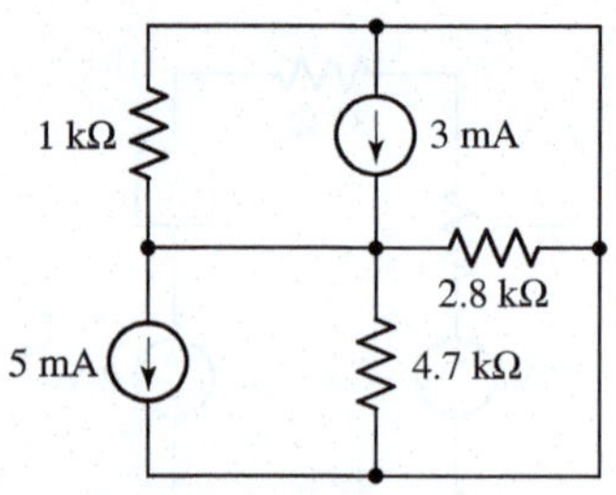

FIGURE 3.74

34. Although drawn so that it may not appear obvious at first glance, the circuit of Fig. 3.74 is in fact a single-node-pair circuit. (*a*) Determine the power absorbed by each resistor. (*b*) Determine the power supplied by each current source. (*c*) Show that the sum of the absorbed power calculated in (*a*) is equal to the sum of the supplied power calculated in (*b*).

3.6 Series and Parallel Connected Sources

35. Determine the numerical value for v_{eq} in Fig. 3.75*a*, if (*a*) $v_1 = 0$, $v_2 = -3$ V, and $v_3 = +3$ V; (*b*) $v_1 = v_2 = v_3 = 1$ V; (*c*) $v_1 = -9$ V, $v_2 = 4.5$ V, $v_3 = 1$ V.

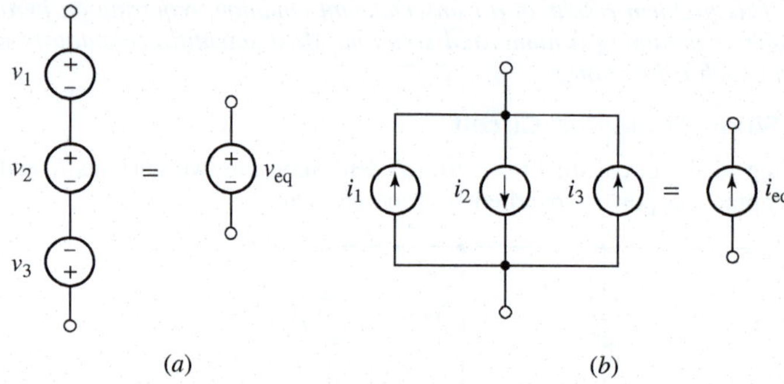

FIGURE 3.75

36. Determine the numerical value for i_{eq} in Fig. 3.75*b*, if (*a*) $i_1 = 0$, $i_2 = -3$ A, and $i_3 = +3$ A; (*b*) $i_1 = i_2 = i_3 = 1$ A; (*c*) $i_1 = -9$ A, $i_2 = 4.5$ A, $i_3 = 1$ A.

37. For the circuit presented in Fig. 3.76, determine the current labeled *i* by first combining the four sources into a single equivalent source.

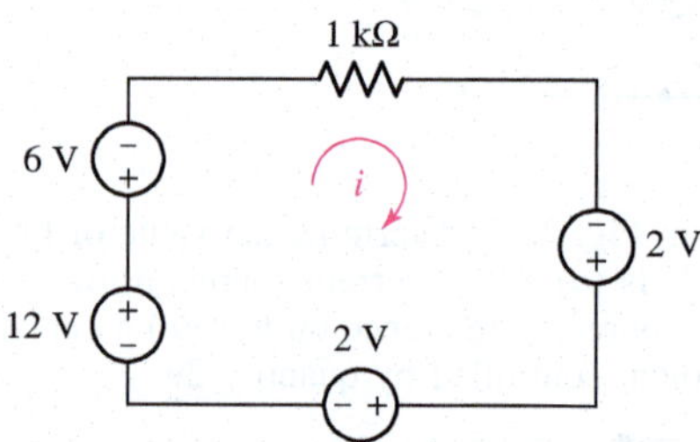

FIGURE 3.76

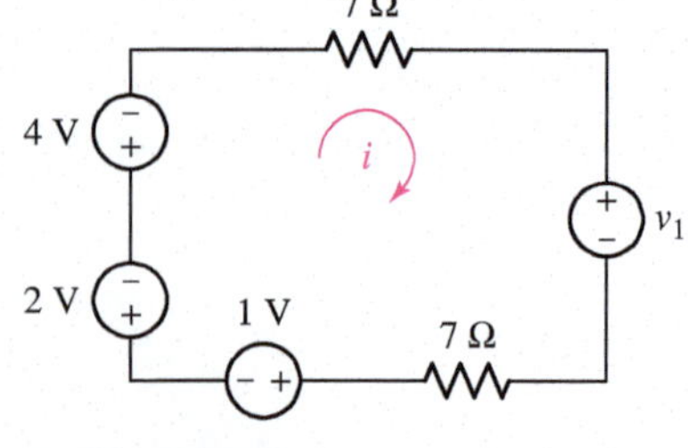

FIGURE 3.77

38. Determine the value of v_1 required to obtain a zero value for the current labeled *i* in the circuit of Fig. 3.77.

39. (*a*) For the circuit of Fig. 3.78, determine the value for the voltage labeled *v*, after first simplifying the circuit to a single current source in parallel with two resistors. (*b*) Verify that the power supplied by your equivalent source is equal to the sum of the supplied powers of the individual sources in the original circuit.

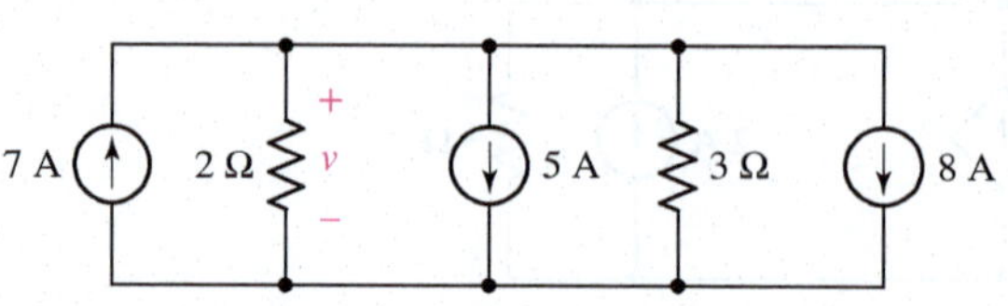

FIGURE 3.78

40. What value of I_S in the circuit of Fig. 3.79 will result in a zero voltage v?

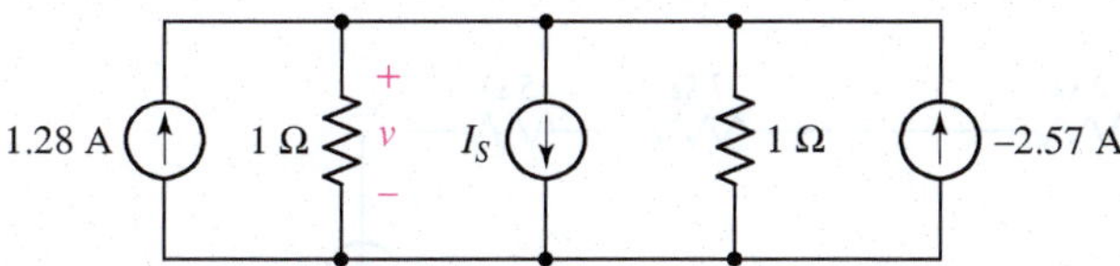

FIGURE 3.79

41. (*a*) Determine the values for I_X and V_Y in the circuit shown in Fig. 3.80. (*b*) Are those values necessarily unique for that circuit? Explain. (*c*) Simplify the circuit of Fig. 3.80 as much as possible and still maintain the values for v and i. (Your circuit must contain the 1 Ω resistor.)

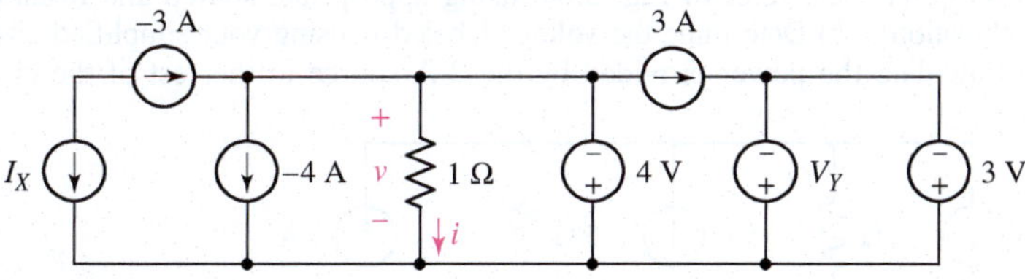

FIGURE 3.80

3.7 Resistors in Series and Parallel

42. Determine the equivalent resistance of each of the networks shown in Fig. 3.81.

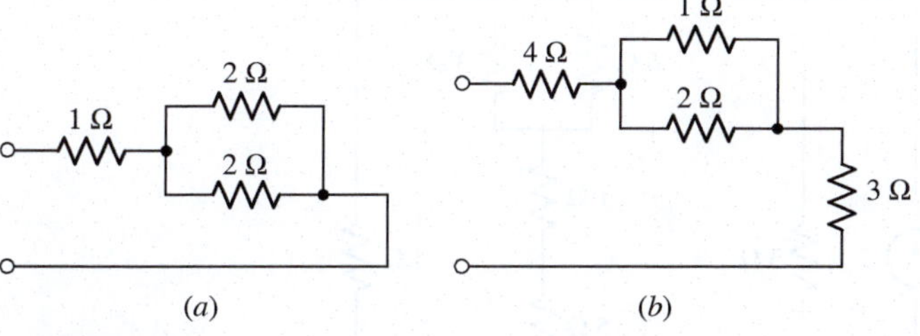

FIGURE 3.81

43. For each network depicted in Fig. 3.82, determine a single equivalent resistance if $R =$ (*a*) 2 Ω; (*b*) 4 Ω; (*c*) 0 Ω.

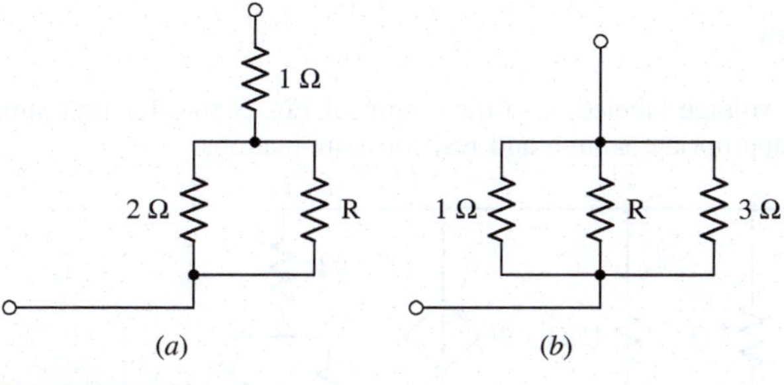

FIGURE 3.82

44. (*a*) Simplify the circuit of Fig. 3.83 as much as possible by using resistor combination techniques. (*b*) Calculate i, using your simplified circuit.

(*c*) To what voltage should the 1 V source be changed to reduce i to zero?
(*d*) Calculate the power absorbed by the 1 Ω resistor.

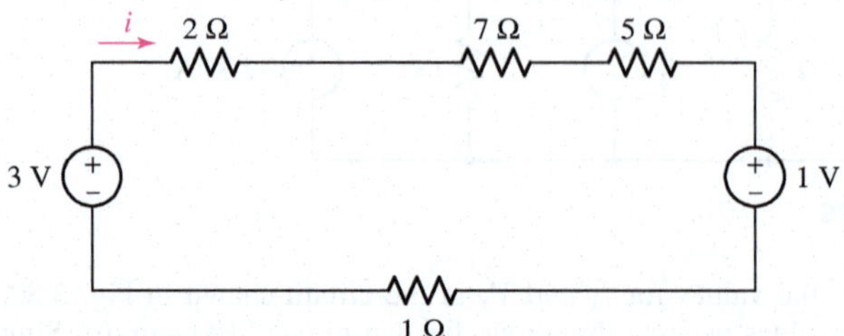

FIGURE 3.83

45. (*a*) Simplify the circuit of Fig. 3.84, using appropriate source and resistor combinations. (*b*) Determine the voltage labeled v, using your simplified circuit. (*c*) Calculate the power provided by the 2 A source to the rest of the circuit.

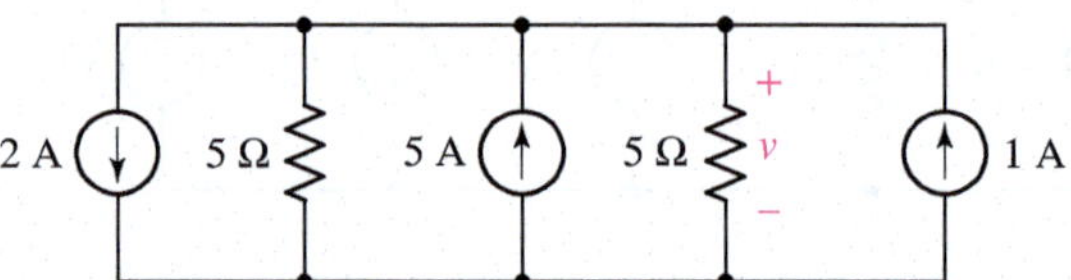

FIGURE 3.84

46. Making appropriate use of resistor combination techniques, calculate i_3 and v_x in the circuit of Fig. 3.85.

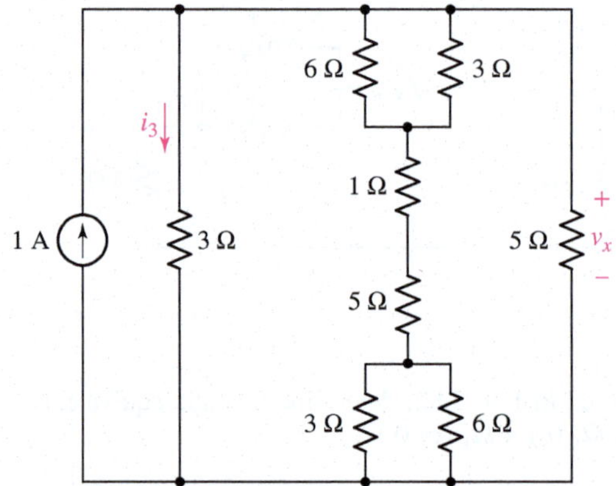

FIGURE 3.85

47. Calculate the voltage labeled v_x in the circuit of Fig. 3.86 after first simplifying, using appropriate source and resistor combinations.

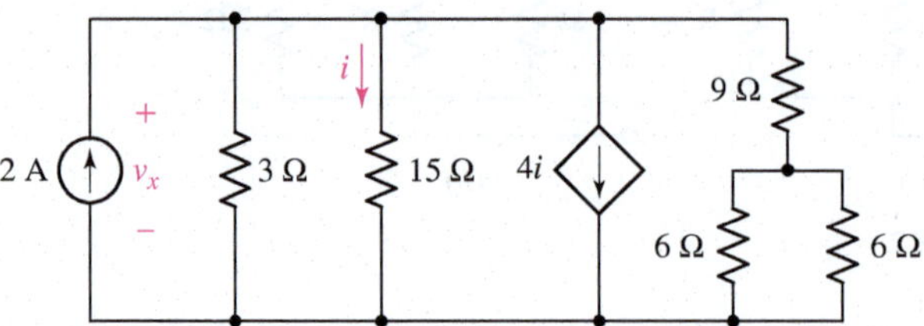

FIGURE 3.86

48. Determine the power absorbed by the 15 Ω resistor in the circuit of Fig. 3.87.

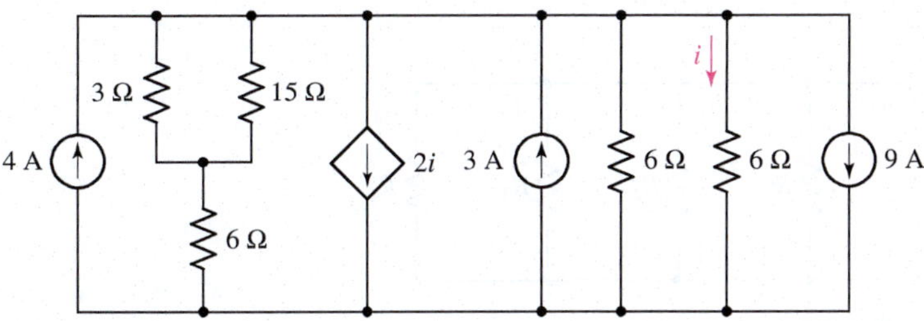

■ **FIGURE 3.87**

49. Calculate the equivalent resistance R_{eq} of the network shown in Fig. 3.88 if $R_1 = R_2 = \cdots = R_{11} = 10\ \Omega$.

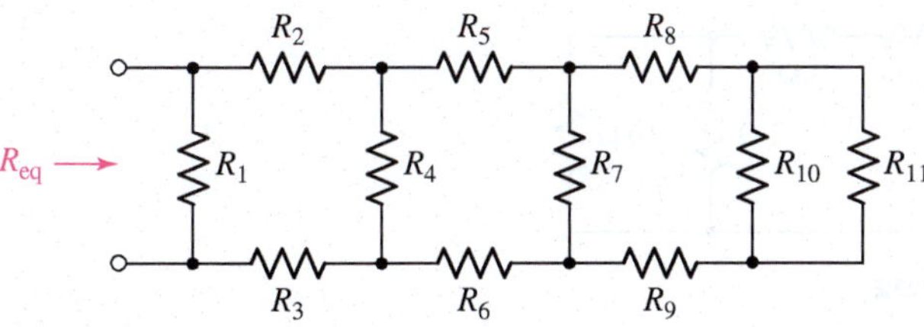

■ **FIGURE 3.88**

50. Show how to combine four 100 Ω resistors to obtain an equivalent resistance of (*a*) 25 Ω; (*b*) 60 Ω; (*c*) 40 Ω.

3.8 Voltage and Current Division

51. In the voltage divider network of Fig. 3.89, calculate (*a*) v_2 if $v = 9.2$ V and $v_1 = 3$ V; (*b*) v_1 if $v_2 = 1$ V and $v = 2$ V; (*c*) v if $v_1 = 3$ V and $v_2 = 6$ V; (*d*) R_1/R_2 if $v_1 = v_2$; (*e*) v_2 if $v = 3.5$ V and $R_1 = 2R_2$; (*f*) v_1 if $v = 1.8$ V, $R_1 = 1$ kΩ, and $R_2 = 4.7$ kΩ.

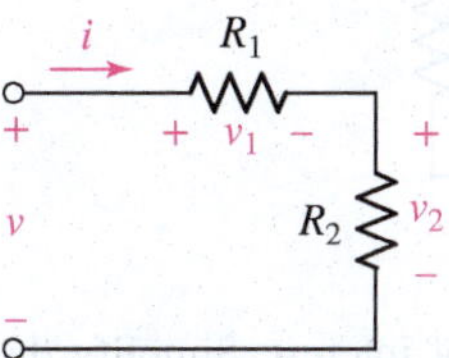

■ **FIGURE 3.89**

52. In the current divider network represented in Fig. 3.90, calculate (*a*) i_1 if $i = 8$ A and $i_2 = 1$ A; (*b*) v if $R_1 = 100$ kΩ, $R_2 = 100$ kΩ, and $i = 1$ mA; (*c*) i_2 if $i = 20$ mA, $R_1 = 1$ Ω, and $R_2 = 4$ Ω; (*d*) i_1 if $i = 10$ A, $R_1 = R_2 = 9$ Ω; (*e*) i_2 if $i = 10$ A, $R_1 = 100$ MΩ, and $R_2 = 1$ Ω.

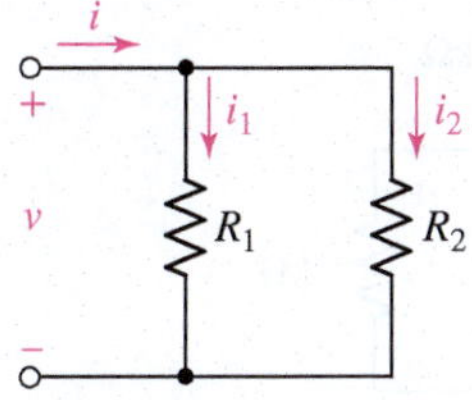

■ **FIGURE 3.90**

DP 53. Choose a voltage $v < 2.5$ V and values for the resistors R_1, R_2, R_3, and R_4 in the circuit of Fig. 3.91 so that $i_1 = 1$ A, $i_2 = 1.2$ A, $i_3 = 8$ A, and $i_4 = 3.1$ A.

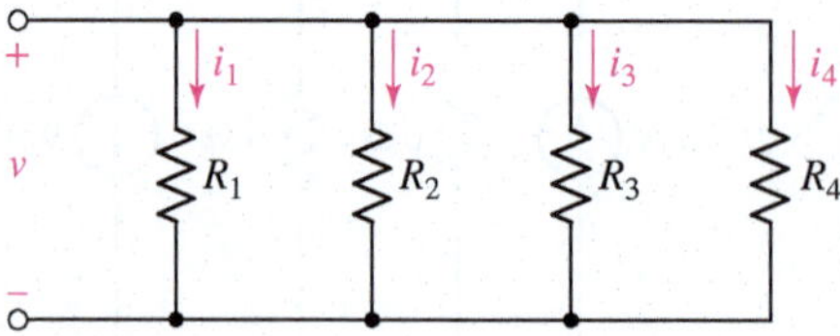

■ **FIGURE 3.91**

54. Employ voltage division to assist in the calculation of the voltage labeled v_x in the circuit of Fig. 3.92.

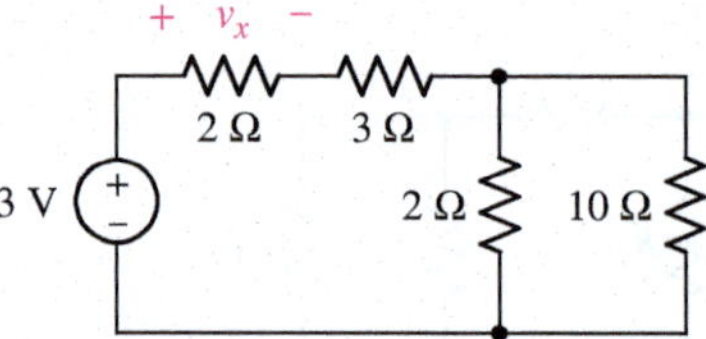

■ **FIGURE 3.92**

55. A network is constructed from a series connection of five resistors having values 1 Ω, 3 Ω, 5 Ω, 7 Ω, and 9 Ω. If 9 V is connected across the terminals of the network, employ voltage division to calculate the voltage across the 3 Ω resistor, and the voltage across the 7 Ω resistor.

56. Employing resistance combination and current division as appropriate, determine values for i_1, i_2, and v_3 in the circuit of Fig. 3.93.

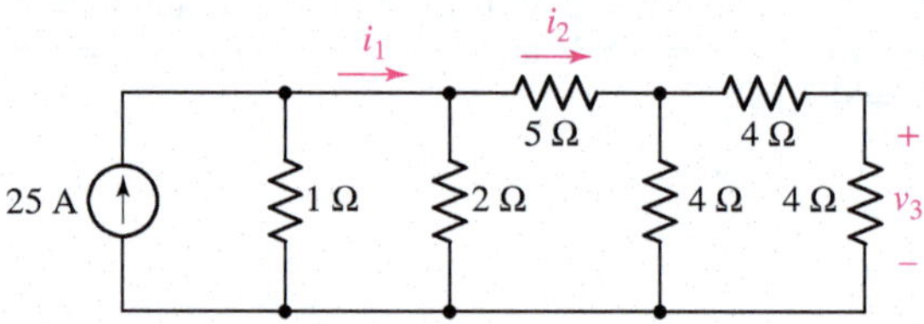

■ **FIGURE 3.93**

57. In the circuit of Fig. 3.94, only the voltage v_x is of interest. Simplify the circuit using appropriate resistor combinations and iteratively employ voltage division to determine v_x.

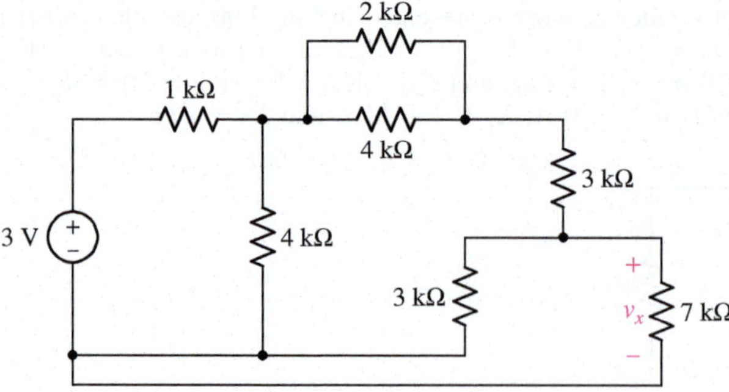

■ **FIGURE 3.94**

Chapter-Integrating Exercises

58. The circuit shown in Fig. 3.95 is a linear model of a bipolar junction transistor biased in the forward active region of operation. Explain why voltage division is not a valid approach for determining the voltage across the 47 kΩ resistor.

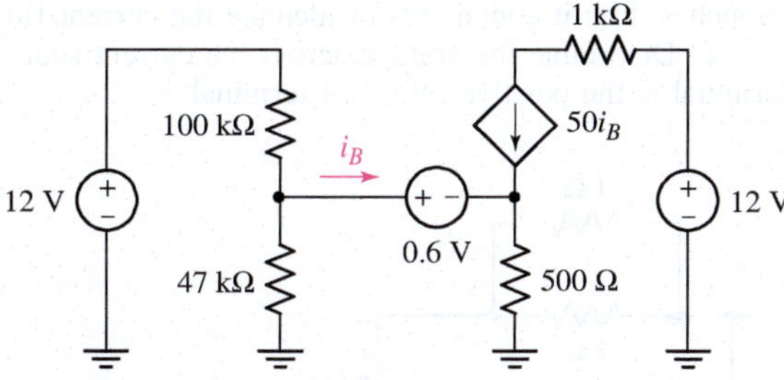

■ **FIGURE 3.95**

59. A common midfrequency model for a field effect–based amplifier circuit is shown in Fig. 3.96. If the controlling parameter g_m (known as the *transconductance*) is equal to 1.2 mS, (*a*) employ current division to obtain the current through the 1 kΩ resistor, and then (*b*) calculate the amplifier output voltage v_{out}. (*c*) Does the circuit "amplify" the signal (as measured at the sinusoidal source)? (*d*) If the input is instead assumed to be the voltage v_π, does the circuit amplify?

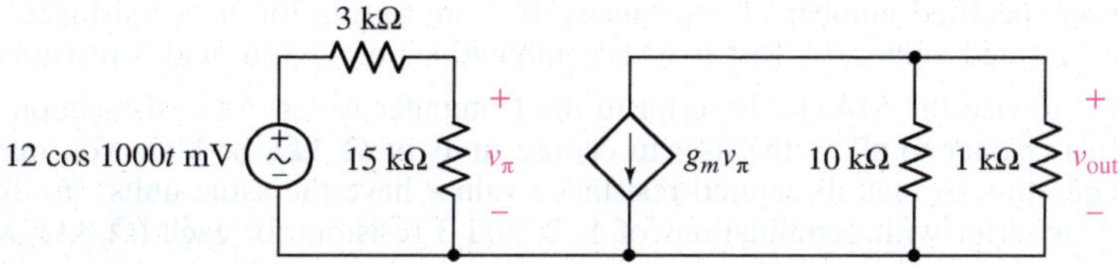

■ **FIGURE 3.96**

60. The circuit depicted in Fig. 3.97 is routinely employed to model the midfrequency operation of a bipolar junction transistor–based amplifier. Calculate the amplifier output v_{out} if the transconductance g_m is equal to 322 mS.

■ **FIGURE 3.97**

61. With regard to the circuit shown in Fig. 3.98, compute (*a*) the voltage across the two 10 Ω resistors, assuming the top terminal is the positive reference, and (*b*) the power dissipated by the 47 Ω resistor. (*c*) If the maximum rating of the 47 Ω resistor is 0.25 W, is it exceeded by this circuit? Explain.

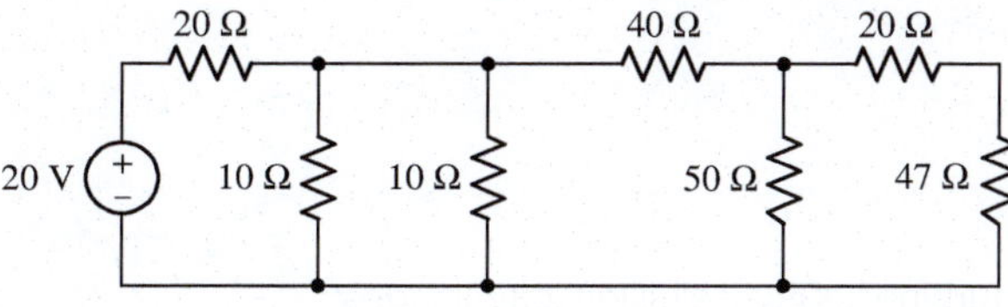

■ **FIGURE 3.98**

62. Delete the leftmost 10 Ω resistor in the circuit of Fig. 3.98, and compute (*a*) the current flowing into the left-hand terminal of the 40 Ω resistor; (*b*) the power supplied by the 20 V source; (*c*) the power dissipated by the 50 Ω resistor.
63. Consider the seven-element circuit depicted in Fig. 3.99. (*a*) How many nodes, loops, and branches does it contain? (*b*) Calculate the current flowing through each resistor. (*c*) Determine the voltage across the current source, assuming the top terminal is the positive reference terminal.

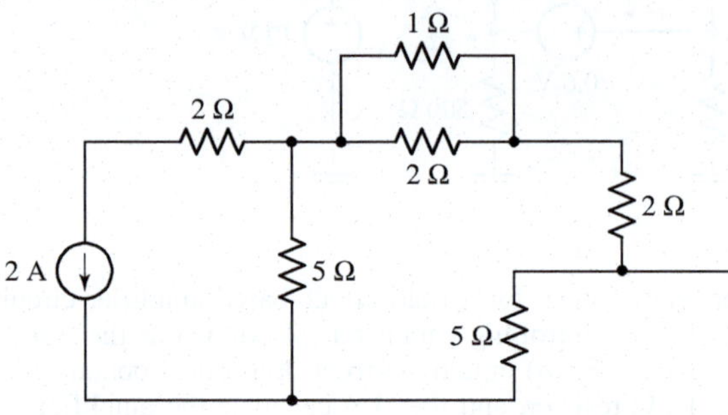

FIGURE 3.99

64. (*a*) Write a MATLAB script to compute the equivalent resistance of a user-specified number of resistances. Be sure to trap for zero resistance as an entered value. (*b*) Test it with combinations of 1, 3, 6, and 8 resistors.

65. (*a*) Revise the MATLAB script in the Computer-Aided Analysis section of this chapter to allow the user to choose units of Ω, kΩ, or MΩ. You may code this so that all entered resistance values have the same units. (*b*) Test your script with combinations of 1, 2, and 3 resistors for each (Ω, kΩ, MΩ).

66. (*a*) Write a MATLAB script to calculate the range of possible resistance values for a series-connected string composed of a user-supplied number of resistors. The user should specify the resistor tolerance of 1%, 5% or 10%; all resistors will have the same tolerance. (*b*) Verify your script using a combination of 0, 1, and 5 resistors for each of the specified tolerance options.

CHAPTER 4

Basic Nodal and Mesh Analysis

KEY CONCEPTS

Nodal Analysis

The Supernode Technique

Mesh Analysis

The Supermesh Technique

Choosing Between Nodal and Mesh Analysis

Computer-Aided Analysis, Including SPICE and MATLAB

INTRODUCTION

With our knowledge of Ohm's and Kirchhoff's laws, analyzing a simple linear circuit to obtain critical information such as the current, voltage, or power associated with a particular element is perhaps starting to seem a straightforward enough venture. Still, for the moment at least, every circuit seems unique, requiring (to some degree) a measure of creativity in approaching the analysis. In this chapter, we learn two basic circuit analysis techniques—***nodal analysis*** and ***mesh analysis***—both of which allow us to investigate many different circuits with a consistent, methodical approach. The result is a streamlined analysis process, a more uniform level of complexity in our equations, fewer errors and, perhaps most importantly, a reduced occurrence of "*I don't know how to even start!*"

Many of the circuits we have seen up to now have been rather simple and (to be honest) of questionable practical use. Such circuits are valuable, however, in helping us to learn to apply fundamental techniques. Although the more complex circuits appearing in this chapter may represent a variety of electrical systems, including control circuits, communication networks, motors, and integrated circuits, as well as electric circuit models of nonelectrical systems, we believe it best not to dwell on such specifics at this early stage. Rather, it is important to initially focus on the *methodology of problem solving* that we will continue to develop throughout the book.

4.1 NODAL ANALYSIS

We begin our study of general methods for methodical circuit analysis by considering a powerful method based on KCL, namely ***nodal analysis.*** In Chap. 3 we considered the analysis of a simple circuit containing only two nodes. We found that the major step of the analysis was obtaining a single equation in terms of a single unknown quantity—the voltage between the pair of nodes.

We will now let the number of nodes increase and correspondingly provide one additional unknown quantity and one additional equation for each added node. Thus, a three-node circuit should have two unknown voltages and two equations; a 10-node circuit will have nine unknown voltages and nine equations; an N-node circuit will need $(N - 1)$ voltages and $(N - 1)$ equations. Each equation is a KCL equation.

To illustrate the basic technique, consider the three-node circuit shown in Fig. 4.1*a*, redrawn in Fig. 4.1*b* to emphasize the fact that there are only three nodes, numbered for convenience. Our goal will be to determine the voltage across each element, and the next step in the analysis is critical. We designate one node as a ***reference node;*** it will be the negative terminal of our $N - 1 = 2$ nodal voltages, as shown in Fig. 4.1*c*.

A little simplification in the resultant equations is obtained if the node connected to the greatest number of branches is identified as the reference node. If there is a ground node, it is usually most convenient to select it as the reference node, although many people seem to prefer selecting the bottom node of a circuit as the reference, especially if no explicit ground is identified.

The voltage of node 1 *relative to the reference node* is named v_1, and v_2 is defined as the voltage of node 2 with respect to the reference node.

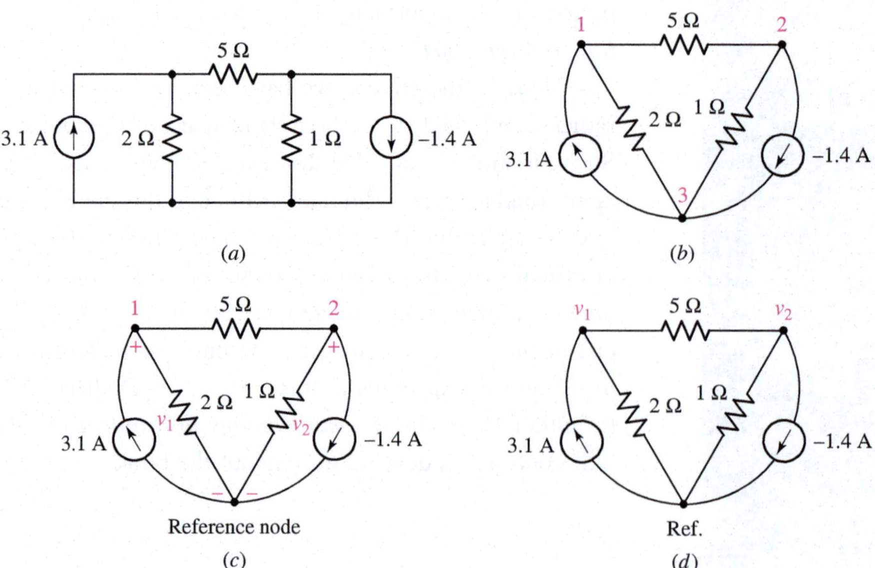

FIGURE 4.1 (*a*) A simple three-node circuit. (*b*) Circuit redrawn to emphasize nodes. (*c*) Reference node selected and voltages assigned. (*d*) Shorthand voltage references. If desired, an appropriate ground symbol may be substituted for "Ref."

These two voltages are all we need, as the voltage between any other pair of nodes may be found in terms of them. For example, the voltage of node 1 with respect to node 2 is $v_1 - v_2$. The voltages v_1 and v_2 and their reference signs are shown in Fig. 4.1*c*. It is common practice once a reference node has been labeled to omit the reference signs for the sake of clarity; the node labeled with the voltage is taken to be the positive terminal (Fig. 4.1*d*). This is understood to be a type of shorthand voltage notation.

The reference node in a schematic is implicitly defined as zero volts. However, it is important to remember that any terminal can be designated as the reference terminal. Thus, the reference node is at zero volts with respect to the other defined nodal voltages, and not necessarily with respect to *earth* ground.

We now apply KCL to nodes 1 and 2. We do this by equating the total current entering the node to the total current leaving the node through the several resistors. Thus,

$$3.1 = \frac{v_1}{2} + \frac{v_1 - v_2}{5} \qquad [1]$$

or

$$3.1 = 0.7v_1 - 0.2v_2 \qquad [2]$$

At node 2 we obtain

$$-(-1.4) = \frac{v_2}{1} + \frac{v_2 - v_1}{5} \qquad [3]$$

or

$$1.4 = -0.2v_1 + 1.2v_2 \qquad [4]$$

Equations [2] and [4] are the desired two equations in two unknowns, and they may be solved easily. The results are $v_1 = 5$ V and $v_2 = 2$ V. From this, it is straightforward to determine the voltage across the 5 Ω resistor: $v_{5\Omega} = v_1 - v_2 = 3$ V. The currents and absorbed powers may also be computed in one step.

We should note at this point that there is more than one way to write the KCL equations for nodal analysis. For example, the reader may prefer to sum all the currents entering a given node and set this quantity to zero. Thus, for node 1 we might have written

$$3.1 - \frac{v_1}{2} - \frac{v_1 - v_2}{5} = 0$$

or

$$3.1 + \frac{-v_1}{2} + \frac{v_2 - v_1}{5} = 0$$

either of which is equivalent to Eq. [1].

Is one way better than any other? Every instructor and every student develops a personal preference, and at the end of the day the most important thing is to be consistent. The authors prefer constructing KCL equations for nodal analysis in such a way as to end up with all resistor terms on one side and all current source terms on the other. Specifically,

$\sum$currents leaving the node through resistors
$= \sum$currents entering the node from current sources

There are several advantages to such an approach. First, there is never any confusion over whether a term should be "$v_1 - v_2$" or "$v_2 - v_1$;" the first voltage in every resistor current expression corresponds to the node for which a KCL equation is being written, as seen in Eqs. [1] and [3]. Second, it allows a quick check that a term has not been accidentally omitted. Simply count the current sources connected to a node and then the resistors; grouping them in the stated fashion makes the comparison a little easier, and helps us avoid sign errors.

EXAMPLE 4.1

Calculate the current flowing left to right through the 15 Ω resistor of Fig. 4.2*a*.

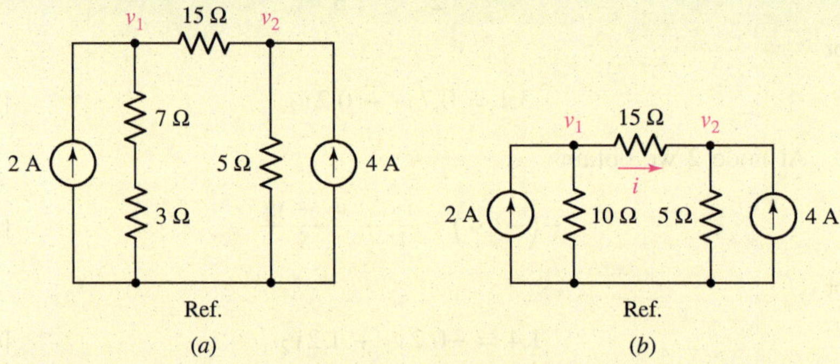

FIGURE 4.2 (*a*) A four-node circuit containing two independent current sources. (*b*) The two resistors in series are replaced with a single 10 Ω resistor, reducing the circuit to three nodes.

Nodal analysis will directly yield numerical values for the nodal voltages v_1 and v_2, and the desired current is given by $i = (v_1 - v_2)/15$.

Before launching into nodal analysis, however, we first note that no details regarding either the 7 Ω resistor or the 3 Ω resistor are of interest. Thus, we may replace their series combination with a 10 Ω resistor as in Fig. 4.2*b*. The result is a reduction in the number of equations to solve.

Writing an appropriate KCL equation for node 1,

$$\frac{v_1}{10} + \frac{v_1 - v_2}{15} = 2 \qquad [5]$$

and for node 2,

$$\frac{v_2}{5} + \frac{v_2 - v_1}{15} = 4 \qquad [6]$$

Rearranging, we obtain

$$5\,v_1 - 2\,v_2 = 60$$

and

$$-v_1 + 4\,v_2 = 60$$

Solving, we find that $v_1 = 20$ V and $v_2 = 20$ V so that $v_1 - v_2 = 0$. In other words, ***zero current*** is flowing through the 15 Ω resistor in this circuit!

PRACTICE

4.1 For the circuit of Fig. 4.3, determine the nodal voltages v_1 and v_2.

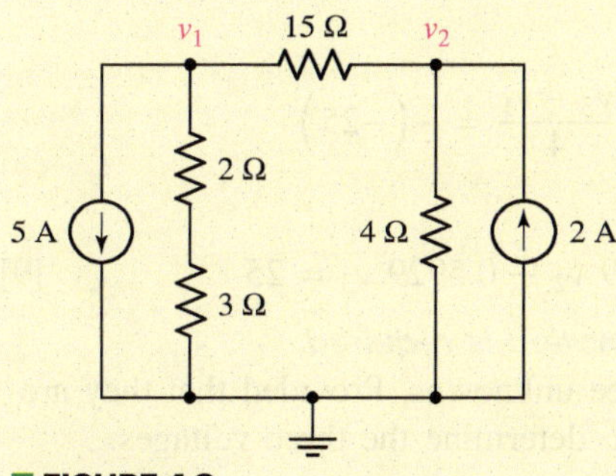

FIGURE 4.3

Ans: $v_1 = -145/8$ V; $v_2 = 5/2$ V.

Now let us increase the number of nodes so that we may use this technique to work a slightly more difficult problem.

EXAMPLE 4.2

Determine each of the nodal voltages for the circuit of Fig. 4.4*a*, as referenced to the bottom node.

Identify the goal of the problem.

There are four nodes in this circuit. With the bottom node as our reference, we label the other three nodes as shown in Fig. 4.4*b*. The circuit has been redrawn for clarity, taking care to identify the two relevant nodes for the 4 Ω resistor.

Collect the known information.

We have three unknown voltages, v_1, v_2, and v_3. All current sources and resistors have designated values, which are marked on the schematic.

Devise a plan.

This problem is well suited to nodal analysis, as three independent KCL equations may be written in terms of the current sources and the current through each resistor.

Construct an appropriate set of equations.

We begin by writing a KCL equation for node 1:

$$\frac{v_1 - v_2}{3} + \frac{v_1 - v_3}{4} = -8 - 3$$

or

$$0.5833\, v_1 - 0.3333\, v_2 - 0.25\, v_3 = -11 \qquad [7]$$

At node 2:

$$\frac{v_2 - v_1}{3} + \frac{v_2}{1} + \frac{v_2 - v_3}{7} = -(-3)$$

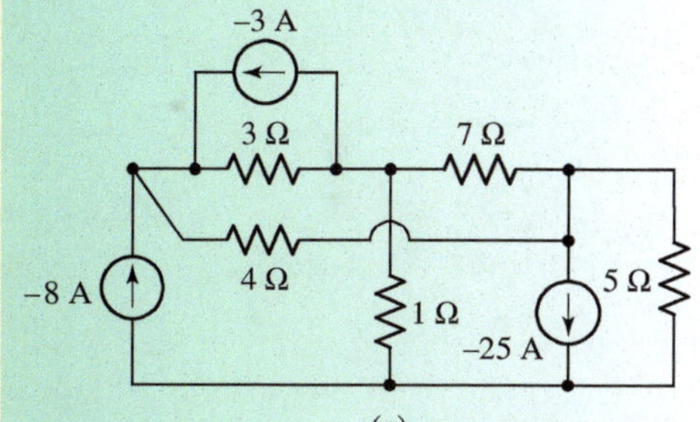

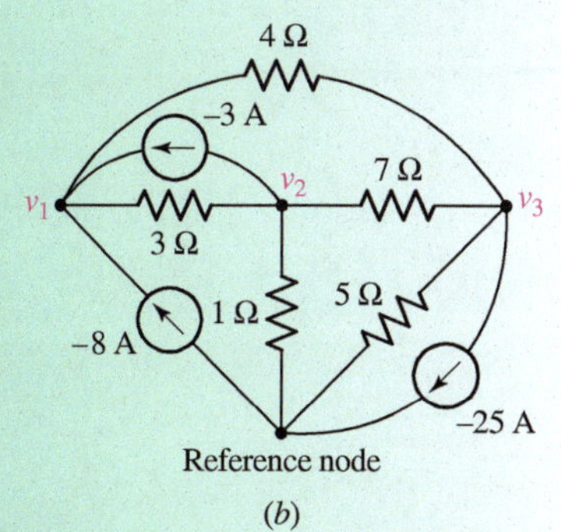

FIGURE 4.4 (*a*) A four-node circuit. (*b*) Redrawn circuit with reference node chosen and voltages labeled.

(Continued on next page)

or

$$-0.3333\ v_1 + 1.4762\ v_2 - 0.1429\ v_3 = 3 \qquad [8]$$

And, at node 3:

$$\frac{v_3}{5} + \frac{v_3 - v_2}{7} + \frac{v_3 - v_1}{4} = -(-25)$$

or, more simply,

$$-0.25\ v_1 - 0.1429\ v_2 + 0.5929\ v_3 = 25 \qquad [9]$$

▶ ***Determine if additional information is required.***

We have three equations in three unknowns. Provided that they are independent, this is sufficient to determine the three voltages.

▶ ***Attempt a solution.***

Equations [7] through [9] can be solved using a scientific calculator (Appendix 2), software packages such as MATLAB, or more traditional "plug-and-chug" techniques such as elimination of variables, matrix methods, or Cramer's rule. Using the method of matrix inversion with a calculator, described in Appendix 2, we can arrange our system of equations in the matrix format in the form $\mathbf{Av} = \mathbf{B}$:

$$\begin{bmatrix} 0.5833 & -0.3333 & -0.25 \\ -0.3333 & 1.4762 & -0.1429 \\ -0.25 & -0.1429 & 0.5929 \end{bmatrix} \begin{bmatrix} v_1 \\ v_2 \\ v_3 \end{bmatrix} = \begin{bmatrix} -11 \\ 3 \\ 25 \end{bmatrix}$$

where

$$\mathbf{A} = \begin{bmatrix} 0.5833 & -0.3333 & -0.25 \\ -0.3333 & 1.4762 & -0.1429 \\ -0.25 & -0.1429 & 0.5929 \end{bmatrix}$$

and

$$\mathbf{B} = \begin{bmatrix} -11 \\ 3 \\ 25 \end{bmatrix}$$

Entering the numbers for matrix **A** and vector **B** into a calculator and solving for $\mathbf{v} = \mathbf{A}^{-1}\mathbf{B}$ yields our final solution

$$\mathbf{v} = \begin{bmatrix} v_1 \\ v_2 \\ v_3 \end{bmatrix} = \begin{bmatrix} 5.4124 \\ 7.7375 \\ 46.313 \end{bmatrix} \text{V}$$

▶ ***Verify the solution. Is it reasonable or expected?***

Substituting the nodal voltages into any of our three nodal equations is sufficient to ensure we made no computational errors. Beyond that, is it possible to determine whether these voltages are "reasonable" values? We have a maximum possible current of $3 + 8 + 25 = 36$ amperes anywhere in the circuit. The largest resistor is 7 Ω, so we do not expect any voltage magnitude greater than $7 \times 36 = 252$ V.

There are, of course, numerous methods available for the solution of linear systems of equations, and we describe several in Appendix 2 in detail. Prior to the advent of the scientific calculator, Cramer's rule was very common in circuit analysis, although often tedious to implement. It is, however, straightforward to use on a simple four-function calculator, and so an awareness of the technique can be valuable. MATLAB, on the other hand, (although perhaps not likely to be available during an examination), is a powerful software package that can greatly simplify the solution process; a brief tutorial on getting started is provided in Appendix 6.

For the situation encountered in Example 4.2, there are several options available through MATLAB. First, we can represent Eqs. [7] to [9] in matrix form, as shown in the example.

In MATLAB, we write

```
>> A = [0.5833 −0.3333 −0.25; −0.3333 1.4762 −0.1429;
        −0.25 −0.1429 0.5929];
>> B = [−11; 3; 25];
>> v = A\B
v =
    5.4124
    7.7375
   46.3127
>>
```

where spaces separate elements along rows, and a semicolon separates rows. The matrix named **v**, which can also be referred to as a ***vector*** as it has only one column, is our solution. Thus, $v_1 = 5.412$ V, $v_2 = 7.738$ V, and $v_3 = 46.31$ V. Note the use of the backslash operator `v = A\B`, instead of `v = A^-1/B` or `v = inv(A)*B`, as recommended in MATLAB for solving systems of equations.

We could also use the KCL equations as we wrote them initially if we employ the symbolic processor of MATLAB.

```
>> syms v1 v2 v3
>> eqn1 = −8−3 == (v1 − v2)/3 + (v1 − v3)/4;
>> eqn2 = −(−3) == (v2 − v1)/3 + v2/1 + (v2 − v3)/7;
>> eqn3 = −(−25) == v3/5 + (v3 − v2)/7 + (v3 − v1)/4;
>> answer=solve(eqn1,eqn2,eqn3,[v1 v2 v3]);
>> answer.v1
ans =
720/133
>> answer.v2
ans =
147/19
>> answer.v3
ans =
880/19
>>
```

which results in exact answers, with no rounding. The `solve()` routine is invoked with the list of symbolic equations we named `eqn1`, `eqn2`, and `eqn3`, but the variables `v1`, `v2`, and `v3` must also be specified. If `solve()` is called with fewer variables than equations, an algebraic solution is returned. The form of the solution is worth a quick comment; it is returned in what is referred to in programming parlance as a *structure;* in this case, we called our structure answer. Each component of the structure is accessed separately by name as shown.

PRACTICE

4.2 For the circuit of Fig. 4.5, compute the voltage across each current source.

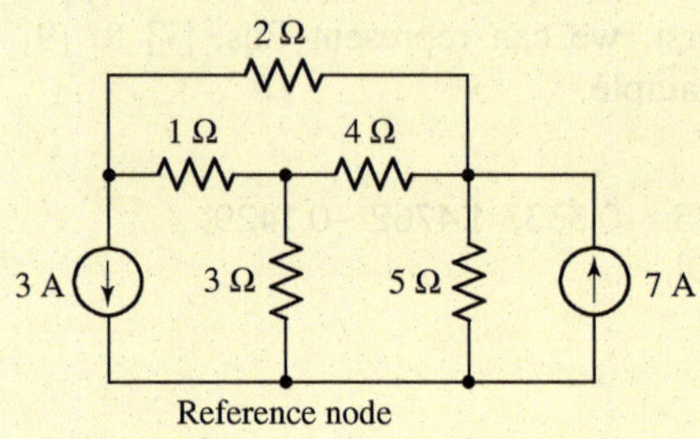

FIGURE 4.5

Ans: $v_{3A} = 5.235$ V; $v_{7A} = 11.47$ V.

The previous examples have demonstrated the basic approach to nodal analysis, but it is worth considering what happens if dependent sources are present as well.

EXAMPLE 4.3

Determine the power supplied by the dependent source of Fig. 4.6*a*.

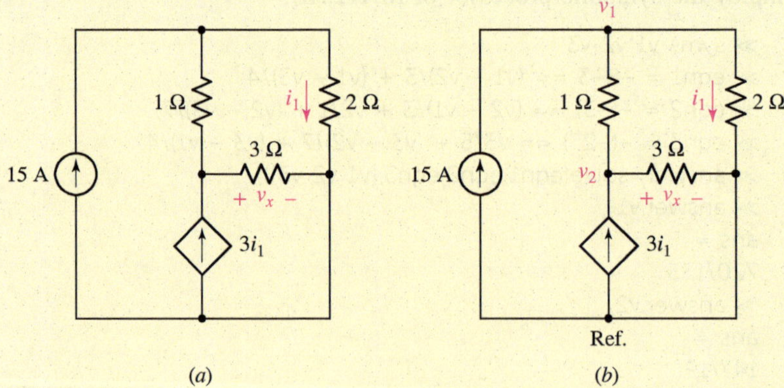

FIGURE 4.6 (*a*) A four-node circuit containing a dependent current source. (*b*) Circuit labeled for nodal analysis.

We choose the bottom node as our reference, since it has a large number of branch connections, and proceed to label the nodal voltages v_1 and v_2 as shown in Fig. 4.6*b*. The quantity labeled v_x is actually equal to v_2.

At node 1, we write

$$\frac{v_1 - v_2}{1} + \frac{v_1}{2} = 15 \quad [10]$$

and at node 2

$$\frac{v_2 - v_1}{1} + \frac{v_2}{3} = 3i_1 \qquad [11]$$

Unfortunately, we have only two equations but three unknowns; *this is a direct result of the presence of the dependent current source, since it is not controlled by a nodal voltage*. Thus, we need an additional equation that relates i_1 to one or more nodal voltages.

In this case, we can write

$$i_1 = \frac{v_1}{2} \qquad [12]$$

which upon substitution into Eq. [11] yields (with a little rearranging)

$$3v_1 - 2v_2 = 30 \qquad [13]$$

and Eq. [10] then simplifies to

$$-15v_1 + 8v_2 = 0 \qquad [14]$$

Solving, we find that $v_1 = -40$ V, $v_2 = -75$ V, and $i_1 = 0.5v_1 = -20$ A. Thus, the power supplied by the dependent source is equal to $(3i_1)(v_2) = (-60)(-75) = 4.5$ kW.

We see that the presence of a dependent source will create the need for an additional equation in our analysis if the controlling quantity is not a nodal voltage. Now let's look at the same circuit, but with the controlling variable of the dependent current source changed to a different quantity—the voltage across the 3 Ω resistor, which is in fact a nodal voltage. We will find that only *two* equations are required to complete the analysis.

EXAMPLE 4.4

Determine the power supplied by the dependent source of Fig. 4.7*a*.

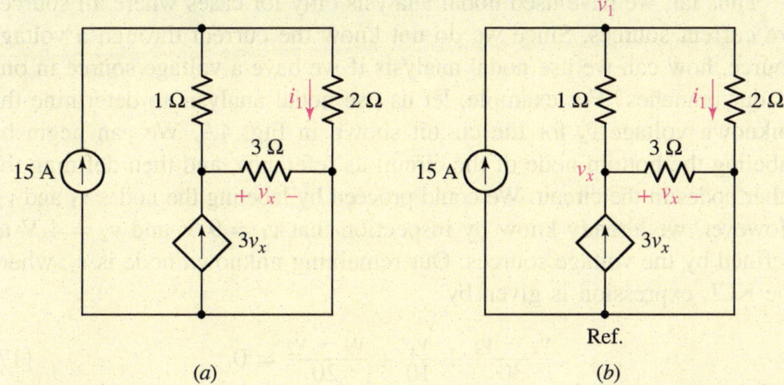

■ **FIGURE 4.7** (*a*) A four-node circuit containing a dependent current source. (*b*) Circuit labeled for nodal analysis.

(Continued on next page)

We select the bottom node as our reference and label the nodal voltages as shown in Fig. 4.7*b*. We have labeled the nodal voltage v_x explicitly for clarity. Note that our choice of reference node is important in this case; it led to the quantity v_x being a nodal voltage.

Our KCL equation for node 1 is

$$\frac{v_1 - v_x}{1} + \frac{v_1}{2} = 15 \quad [15]$$

and for node x is

$$\frac{v_x - v_1}{1} + \frac{v_2}{3} = 3v_x \quad [16]$$

Grouping terms and solving, we find that $v_1 = \frac{50}{7}$ V and $v_x = -\frac{30}{7}$ V. Thus, the dependent source in this circuit generates $(3v_x)(v_x) = 55.1$ W.

PRACTICE

4.3 For the circuit of Fig. 4.8, determine the nodal voltage v_1 if A is (*a*) $2i_1$; (*b*) $2v_1$.

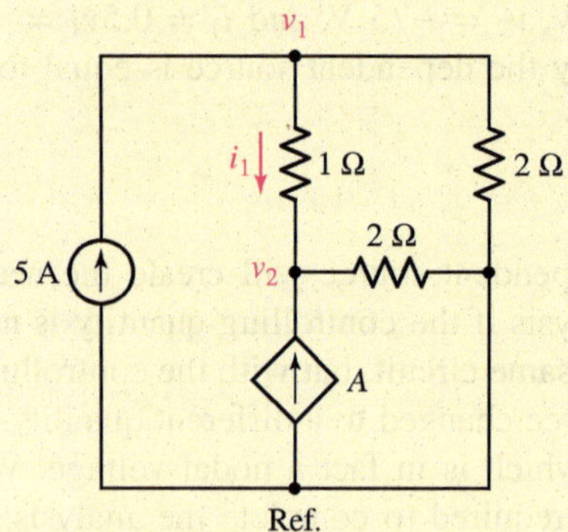

■ **FIGURE 4.8**

Ans: (*a*) $\frac{70}{9}$ V; (*b*) –10 V.

Thus far, we have used nodal analysis only for cases where all sources are current sources. Since we do not know the current through a voltage source, how can we use nodal analysis if we have a voltage source in one of our branches? For example, let us use nodal analysis to determine the unknown voltage v_x for the circuit shown in Fig. 4.9. We can begin by labeling the bottom node of the circuit as reference, and then defining the other nodes in the circuit. We could proceed by labeling the nodes v_1 and v_2. However, we already know by inspection that $v_1 = 9$ V and $v_2 = 4$ V as defined by the voltage sources. Our remaining unknown node is v_x, where the KCL expression is given by

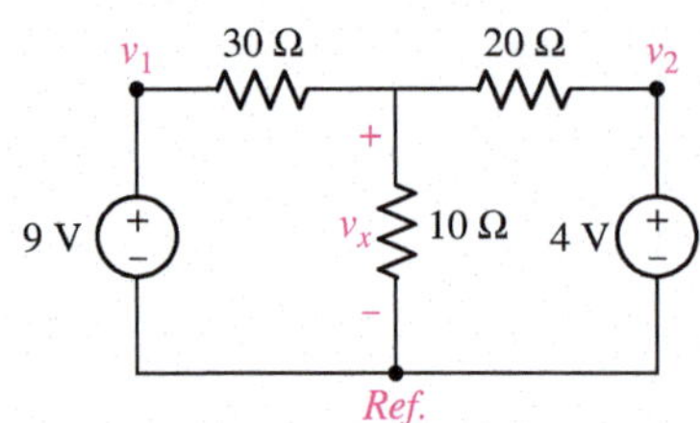

■ **FIGURE 4.9** Example of nodal analysis with voltage sources.

$$\frac{v_x - v_1}{30} + \frac{v_x}{10} + \frac{v_x - v_2}{20} = 0 \quad [17]$$

and substituting $v_1 = 9$ V and $v_2 = 4$ V,

$$\frac{v_x - 9}{30} + \frac{v_x}{10} + \frac{v_x - 4}{20} = 0 \quad [18]$$

Solving for v_x yields

$$v_x\left(\frac{1}{30} + \frac{1}{10} + \frac{1}{20}\right) = \frac{9}{30} + \frac{4}{20} \quad [19]$$

$$v_x = \frac{30}{11} = 2.7273 \text{ V} \quad [20]$$

Note that our KCL expression defined the current flowing through the 30 Ω and 20 Ω resistors, using proper definitions for voltage across them, $(v_x - 9)$ and $(v_x - 4)$, respectively. A common error in nodal analysis is to forget the voltage sources when writing the KCL expressions, for example, writing $(v_x)/30$ as the current through the 30 Ω resistor instead of $(v_x - 9)/30$.

Summary of Basic Nodal Analysis Procedure

1. **Select a reference node.** The number of terms in your nodal equations can be minimized by selecting the reference node as the one with the greatest number of branches connected to it.
2. **Count and label the voltage at each of the other nodes in the circuit,** relative to the reference node you have selected.
3. **Write a KCL equation for each of the nonreference nodes.** Sum the currents flowing *out of* the node through resistors on one side of the equation. On the other side, sum the currents flowing *into* a node from sources. Pay close attention to minus signs.
4. **Express any additional unknowns in terms of appropriate nodal voltages.** This situation can occur if voltage sources or dependent sources appear in our circuit.
5. **Organize the equations.** Group terms according to nodal voltages.
6. **Solve the system of equations for the nodal voltages.**

These basic steps will work on any circuit we ever encounter, although the presence of voltage sources directly connecting two nodes will require extra care. Such situations are discussed next.

4.2 THE SUPERNODE

In the previous example, we examined a case where nodal analysis can be used when a voltage source is present. However, this situation included a resistor directly in series with the voltage source, allowing us to define a current for our KCL equation. What if you have a voltage source, but no other means of defining current through it? As an example of how voltage sources are best handled when performing nodal analysis, consider the circuit shown in Fig. 4.10*a*. The original four-node circuit of Fig. 4.4 has

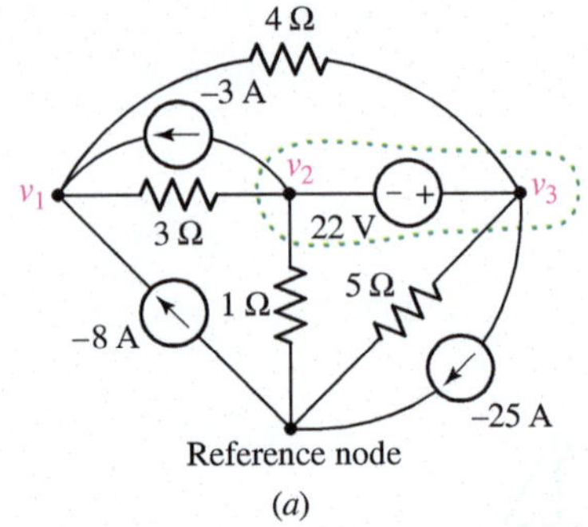

(*a*)

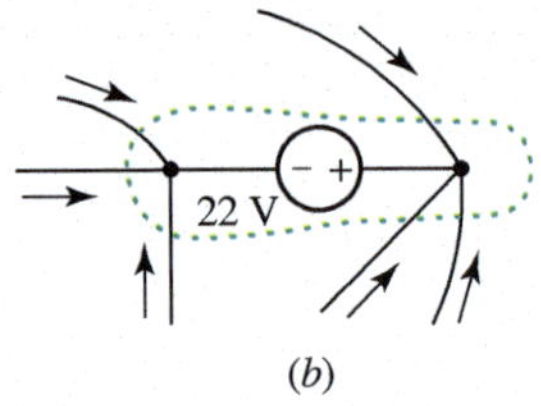

(*b*)

FIGURE 4.10 (*a*) The circuit of Example 4.2 with a 22 V source in place of the 7 Ω resistor. (*b*) Expanded view of the region defined as a supernode; KCL requires that all currents flowing into the region sum to zero, or we would pile up or run out of electrons.

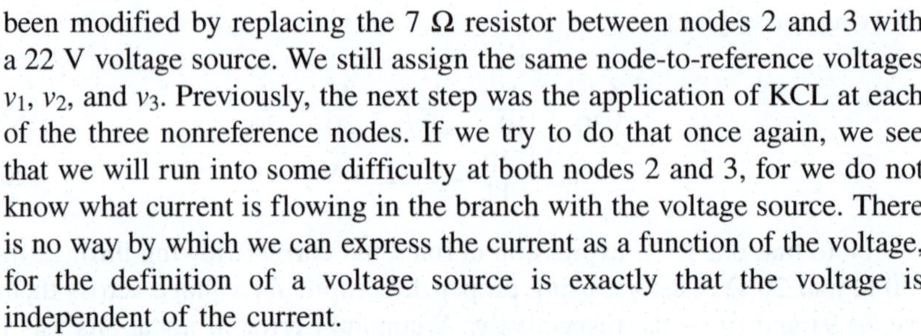

been modified by replacing the 7 Ω resistor between nodes 2 and 3 with a 22 V voltage source. We still assign the same node-to-reference voltages v_1, v_2, and v_3. Previously, the next step was the application of KCL at each of the three nonreference nodes. If we try to do that once again, we see that we will run into some difficulty at both nodes 2 and 3, for we do not know what current is flowing in the branch with the voltage source. There is no way by which we can express the current as a function of the voltage, for the definition of a voltage source is exactly that the voltage is independent of the current.

There are two ways out of this dilemma. The more tedious approach is to assign an unknown current to the branch that contains the voltage source, proceed to apply KCL three times, and then apply KVL ($v_3 - v_2 = 22$) once between nodes 2 and 3; the result is then four equations with four unknowns.

The easier method is to treat node 2, node 3, and the voltage source together as a ***supernode*** and apply KCL to both nodes at the same time; the supernode is indicated by the region enclosed by the broken line in Fig. 4.10*a*. This is okay because if the total current leaving node 2 is zero and the total current leaving node 3 is zero, then the total current leaving the combination of the two nodes is zero. Note that any current defined within the supernode will simply cancel in our KCL expressions; for example, the current leaving node 2 will be equal and opposite to the current leaving node 3. This concept is represented graphically in the expanded view of Fig. 4.10*b*.

The concept of the supernode uses a more general definition for KCL: *The algebraic sum of the currents entering any node* ***or closed surface*** *is zero.*

EXAMPLE 4.5

Determine the value of the unknown node voltage v_1 in the circuit of Fig. 4.10*a*.

The KCL equation at node 1 is unchanged from Example 4.2:

$$\frac{v_1 - v_2}{3} + \frac{v_1 - v_3}{4} = -8 - 3$$

or

$$0.5833v_1 - 0.3333v_2 - 0.2500v_3 = -11 \qquad [21]$$

Next we consider the 2–3 supernode. Two current sources are connected, and four resistors. Thus,

$$\frac{v_2 - v_1}{3} + \frac{v_3 - v_1}{4} + \frac{v_3}{5} + \frac{v_2}{1} = 3 + 25$$

or

$$-0.5833v_1 + 1.3333v_2 + 0.45v_3 = 28 \qquad [22]$$

Since we have three unknowns, we need one additional equation, and it must utilize the fact that there is a 22 V voltage source between nodes 2 and 3:

$$v_2 - v_3 = -22 \qquad [23]$$

Solving Eqs. [21] to [23], the solution for v_1 is 1.071 V.

PRACTICE

4.4 For the circuit of Fig. 4.11, compute the voltage across each current source.

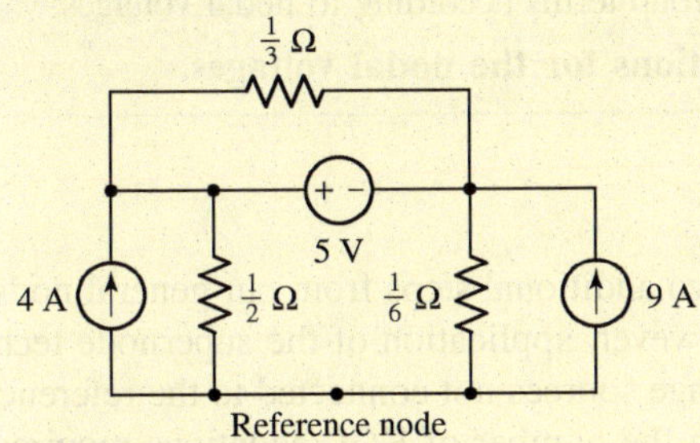

FIGURE 4.11

Ans: 5.375 V; 375 mV.

The presence of a voltage source thus reduces by 1 the number of nonreference nodes at which we must apply KCL, regardless of whether the voltage source extends between two nonreference nodes or is connected between a node and the reference. We should be careful in analyzing circuits such as that of Practice Problem 4.4. Since both ends of the resistor are part of the supernode, there must technically be *two* corresponding current terms in the KCL equation, but they cancel each other out. In fact, we can determine the current and/or power associated with that resistor without analyzing the rest of the circuit! We can summarize the supernode method as follows:

Summary of Supernode Analysis Procedure

1. **Select a reference node.** The number of terms in your nodal equations can be minimized by selecting the reference node as the one with the greatest number of branches connected to it.
2. **Count and label the voltage at each node in the circuit,** relative to the reference node you have selected.
3. **If the circuit contains voltage sources, form a supernode around each one.** This is done by enclosing the source, its two terminals, and any other elements connected between the two terminals within a broken-line enclosure.
4. **Write a KCL equation for each of the nonreference nodes and for each supernode *that does not contain the reference node*.** Sum the currents flowing *out of* the node/supernode

through resistors on one side of the equation. On the other side, sum the currents flowing *into* a node/supernode from sources. Pay close attention to minus signs.

5. **Relate the voltage across each voltage source to nodal voltages.** This is accomplished by simple application of KVL; one such equation is needed for each supernode defined.
6. **Express any additional unknowns in terms of appropriate nodal voltages.** This situation can occur if voltage sources or dependent sources appear in our circuit.
7. **Organize the equations.** Group terms according to nodal voltages.
8. **Solve the system of equations for the nodal voltages.**

We see that we have added two additional steps from our general nodal analysis procedure. In reality, however, application of the supernode technique to a circuit containing voltage sources not connected to the reference node will result in a reduction in the number of KCL equations required. With this in mind, let's consider the circuit of Fig. 4.12, which contains all four types of sources and has five nodes.

EXAMPLE 4.6

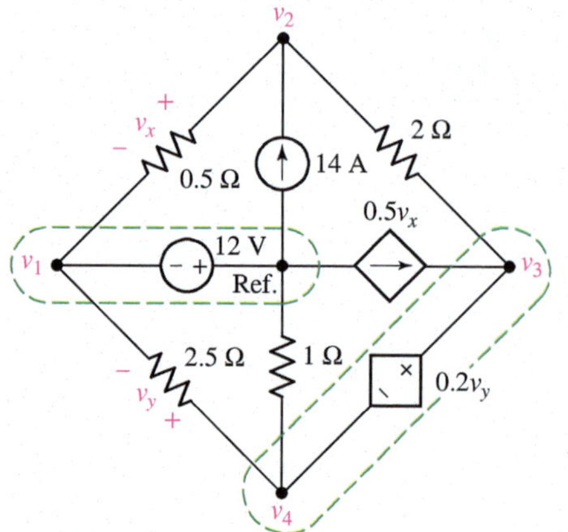

FIGURE 4.12 A five-node circuit with four different types of sources.

Determine the node-to-reference voltages in the circuit of Fig. 4.12.

After establishing a supernode around each *voltage* source, we see that we need to write KCL equations only at node 2 and at the supernode containing the dependent voltage source. By inspection, it is clear that $v_1 = -12$ V.

At node 2,

$$\frac{v_2 - v_1}{0.5} + \frac{v_2 - v_3}{2} = 14 \qquad [24]$$

while at the 3–4 supernode,

$$\frac{v_3 - v_2}{2} + \frac{v_4}{1} + \frac{v_4 - v_1}{2.5} = 0.5v_x \qquad [25]$$

We next relate the source voltages to the node voltages:

$$v_3 - v_4 = 0.2v_y \qquad [26]$$

and

$$0.2v_y = 0.2(v_4 - v_1) \qquad [27]$$

Finally, we express the dependent current source in terms of the assigned variables:

$$0.5v_x = 0.5(v_2 - v_1) \qquad [28]$$

Five nodes requires *four* KCL equations in general nodal analysis, but we have reduced this requirement to *only two,* as we formed two separate supernodes. Each supernode required a KVL equation (Eq. [26] and $v_1 = -12$, the latter written by inspection). Neither dependent source was controlled by a nodal voltage, so two additional equations were needed as a result.

With this done, we can now eliminate v_x and v_y to obtain a set of four equations in the four node voltages:

$$\begin{aligned} -2v_1 + 2.5v_2 - 0.5v_3 \qquad\qquad &= 14 \\ 0.1v_1 - v_2 + 0.5v_3 + 1.4v_4 &= 0 \\ v_1 \qquad\qquad\qquad\qquad\qquad &= -12 \\ 0.2v_1 \qquad\quad + v_3 - 1.2v_4 &= 0 \end{aligned}$$

Solving, $v_1 = -12$ V, $v_2 = -4$ V, $v_3 = 0$ V, and $v_4 = -2$ V.

PRACTICE

4.5 Determine the nodal voltages in the circuit of Fig. 4.13.

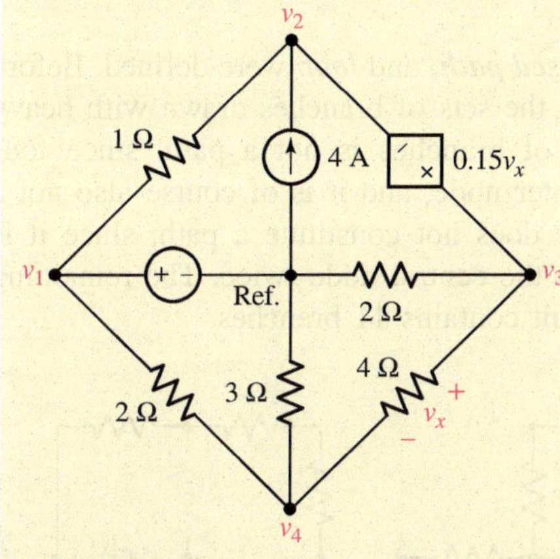

FIGURE 4.13

Ans: $v_1 = 3$ V; $v_2 = -2.33$ V; $v_3 = -1.91$ V; $v_4 = 0.945$ V.

4.3 MESH ANALYSIS

As we have seen, nodal analysis is a straightforward analysis technique when only current sources are present, and voltage sources are easily accommodated with the supernode concept. Still, nodal analysis is based on KCL, and the reader might at some point wonder if there isn't a similar approach based on KVL. There is—it's known as ***mesh analysis***—and although only strictly speaking applicable to what we will shortly define as a planar circuit[1], it can in many cases prove simpler to apply than nodal analysis.

If it is possible to draw the diagram of a circuit on a surface in such a way that no branch passes over or under any other branch, then that

(1) For an example of how mesh analysis can be generalized to nonplanar circuits, see, for example, B. Skromme and W. Barnard, "Turning mesh analysis inside out," 2020 American Society of Engineering Education Conference. DOI:10.18260/1-2-35403.

circuit is said to be a ***planar circuit.*** Thus, Fig. 4.14*a* shows a planar network, Fig. 4.14*b* shows a nonplanar network, and Fig. 4.14*c* also shows a planar network, although it is drawn in such a way as to make it appear nonplanar at first glance.

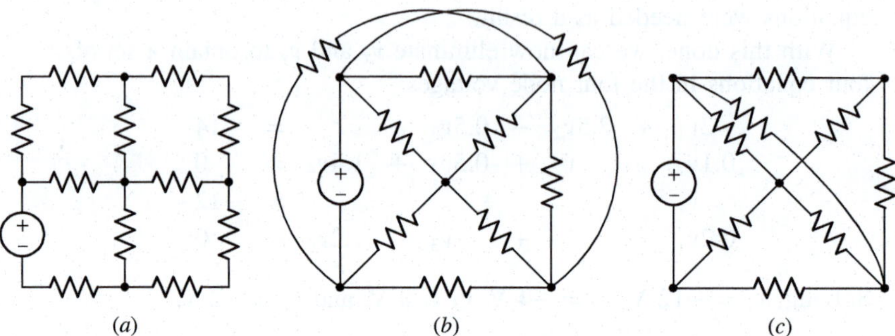

FIGURE 4.14 Examples of planar and nonplanar networks; crossed wires without a solid dot are not in physical contact with each other.

In Sec. 3.1, the terms ***path, closed path,*** and ***loop*** were defined. Before we define a mesh, let us consider the sets of branches drawn with heavy lines in Fig. 4.15. The first set of branches is not a path, since four branches are connected to the center node, and it is of course also not a loop. The second set of branches does not constitute a path, since it is traversed only by passing through the central node twice. The remaining four paths are all loops. The circuit contains 11 branches.

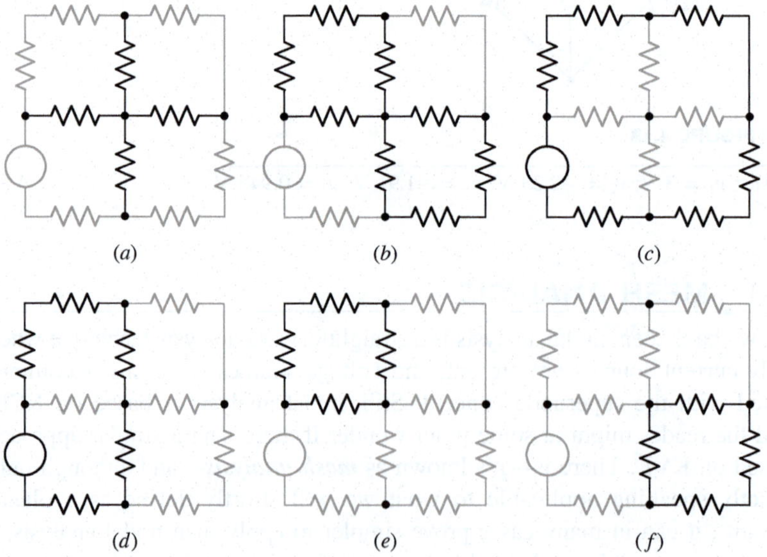

FIGURE 4.15 (*a*) The set of branches identified by the heavy lines is neither a path nor a loop. (*b*) The set of branches here is not a path, since it can be traversed only by passing through the central node twice. (*c*) This path is a loop but not a mesh, since it encloses other loops. (*d*) This path is also a loop but not a mesh. (*e, f*) Each of these paths is both a loop and a mesh.

The mesh is a property of a planar circuit and is undefined for a nonplanar circuit. We define a ***mesh*** as a loop that does not contain any other loops within it. Thus, the loops indicated in Fig. 4.15*c* and *d* are not meshes, whereas those of parts *e* and *f* are meshes. Once a circuit has been drawn neatly in planar form, it often has the appearance of a multipaned window; the boundary of each pane in the window may be considered to be a mesh.

If a network is planar, mesh analysis can be used to accomplish the analysis. This technique involves the concept of a ***mesh current,*** which we introduce by considering the analysis of the two-mesh circuit of Fig. 4.16*a*.

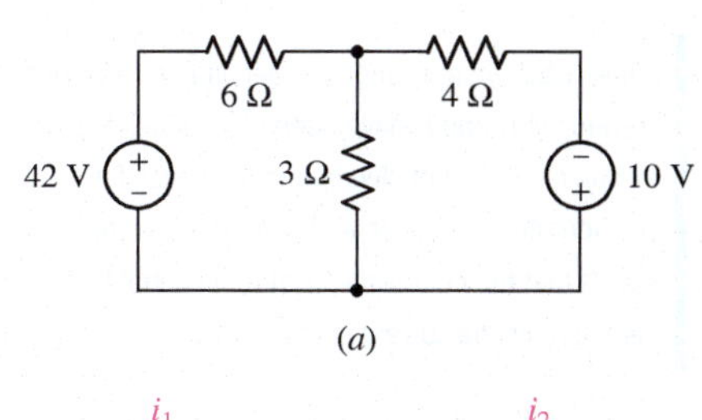

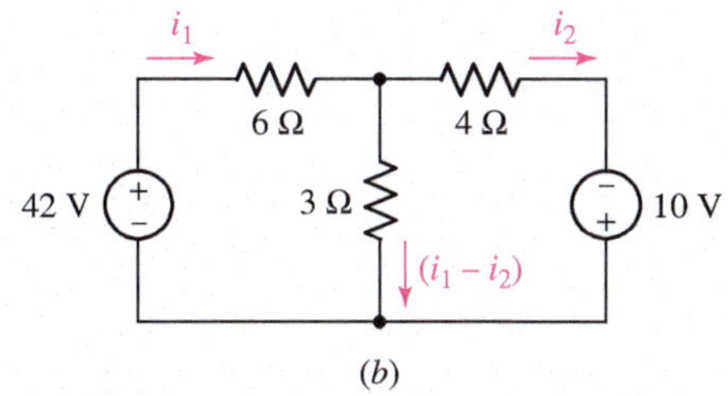

FIGURE 4.16 (*a, b*) A simple circuit for which currents are required.

As we did in the single-loop circuit, we will begin by defining a current through one of the branches. Let us call the current flowing to the right through the 6 Ω resistor i_1. We will apply KVL around each of the two meshes, and the two resulting equations are sufficient to determine two unknown currents. We next define a second current i_2 flowing to the right in the 4 Ω resistor. We might also choose to call the current flowing downward through the central branch i_3, but from KCL we know that i_3 may be expressed in terms of the two previously assumed currents as $(i_1 - i_2)$. The assumed currents are shown in Fig. 4.16*b*.

Following the method of solution for the single-loop circuit, we now apply KVL to the left-hand mesh,

$$-42 + 6i_1 + 3(i_1 - i_2) = 0$$

or

$$9i_1 - 3i_2 = 42 \qquad [29]$$

Applying KVL to the right-hand mesh,

$$-3(i_1 - i_2) + 4i_2 - 10 = 0$$

or

$$-3i_1 + 7i_2 = 10 \qquad [30]$$

Equations [29] and [30] are independent equations; one cannot be derived from the other. With two equations and two unknowns, the solution is easily obtained:

$$i_1 = 6\,\text{A} \qquad i_2 = 4\,\text{A} \qquad \text{and} \qquad (i_1 - i_2) = 2\,\text{A}$$

If our circuit contains *M* meshes, then we expect to have *M* mesh currents and therefore will be required to write *M* independent equations.

Now let us consider this same problem in a slightly different manner by using mesh currents. We define a ***mesh current*** as a current that flows only around the perimeter of a mesh. If we call the left-hand mesh of our problem mesh 1, then we may establish a mesh current i_1 flowing in a clockwise direction about this mesh. A mesh current is indicated by a curved arrow that almost closes on itself and is drawn inside the appropriate mesh, as shown in Fig. 4.17. The mesh current i_2 is established in the remaining mesh, again in a clockwise direction. Although the directions are arbitrary, we will always choose clockwise mesh currents because a certain error-minimizing symmetry then results in the equations.

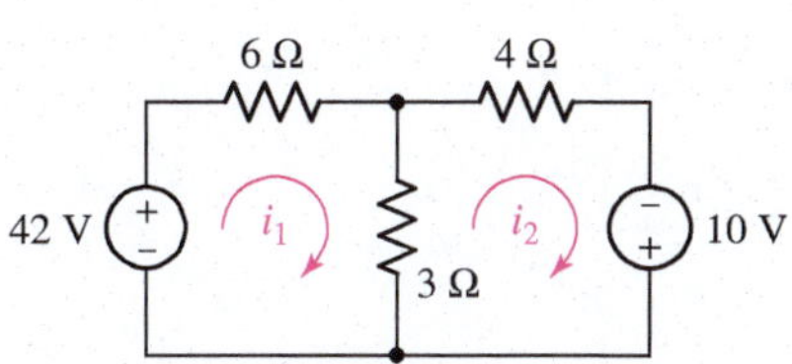

FIGURE 4.17 The same circuit considered in Fig. 4.16*b*, but viewed a slightly different way.

A mesh current may often be identified as a branch current, as i_1 and i_2 have been identified in this example. This is not always true, however, for consideration of a square nine-mesh network soon shows that the central mesh current cannot be identified as the current in any branch.

We no longer have a current or current arrow shown directly on each branch in the circuit. The current through any branch must be determined by considering the mesh currents flowing in every mesh in which that branch appears. This is not difficult, because no branch can appear in more than two meshes. For example, the 3 Ω resistor appears in both meshes, and the current flowing downward through it is $i_1 - i_2$. The 6 Ω resistor appears only in mesh 1, and the current flowing to the right in that branch is equal to the mesh current i_1.

For the left-hand mesh,

$$-42 + 6i_1 + 3(i_1 - i_2) = 0 \qquad [31]$$

while for the right-hand mesh,

$$3(i_2 - i_1) + 4i_2 - 10 = 0 \qquad [32]$$

and these two equations are equivalent to Eqs. [29] and [30].

EXAMPLE 4.7

Determine the power supplied by the 2 V source of Fig. 4.18*a*.

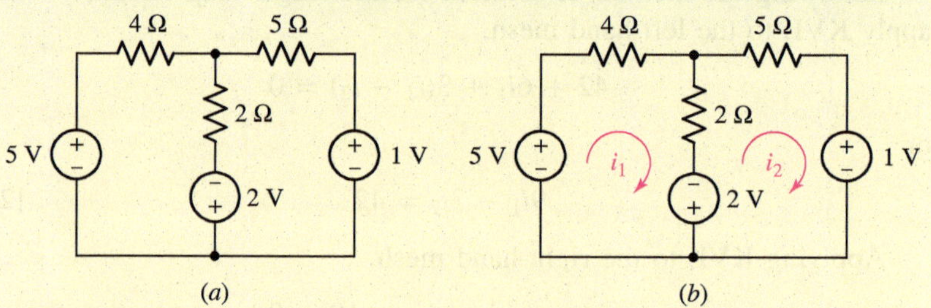

FIGURE 4.18 (*a*) A two-mesh circuit containing three sources. (*b*) Circuit labeled for mesh analysis.

We first define two clockwise mesh currents as shown in Fig. 4.18*b*.

Beginning at the bottom left node of mesh 1, we write the following KVL equation as we proceed clockwise through the branches:

$$-5 + 4i_1 + 2(i_1 - i_2) - 2 = 0$$

Doing the same for mesh 2, we write

$$+2 + 2(i_2 - i_1) + 5i_2 + 1 = 0$$

Rearranging and grouping terms,

$$6i_1 - 2i_2 = 7$$

and

$$-2i_1 + 7i_2 = -3$$

Solving, $i_1 = \frac{43}{38} = 1.132$ A and $i_2 = -\frac{2}{19} = -0.1053$ A.

The current flowing out of the positive reference terminal of the 2 V source is $i_1 - i_2$. Thus, the 2 V source supplies (2)(1.237) = 2.474 W.

PRACTICE

4.6 Determine i_1 and i_2 in the circuit in Fig. 4.19.

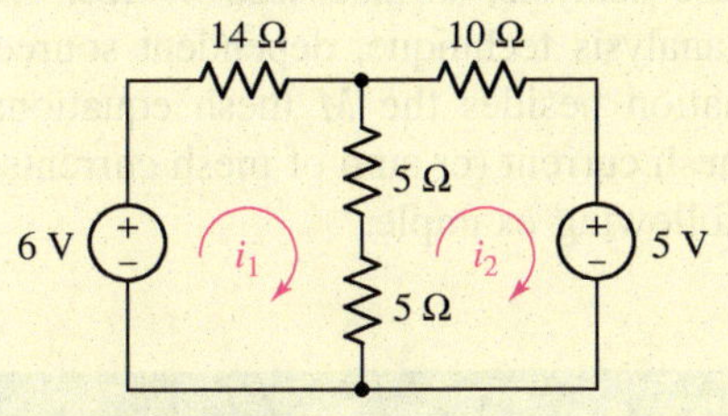

FIGURE 4.19

Ans: +184.2 mA; −157.9 mA.

Let us next consider the five-node, seven-branch, three-mesh circuit shown in Fig. 4.20. With three meshes, we expect three equations in three unknowns, ultimately (the three mesh currents being the three unknowns).

EXAMPLE 4.8

Use mesh analysis to determine the three mesh currents in the circuit of Fig. 4.20.

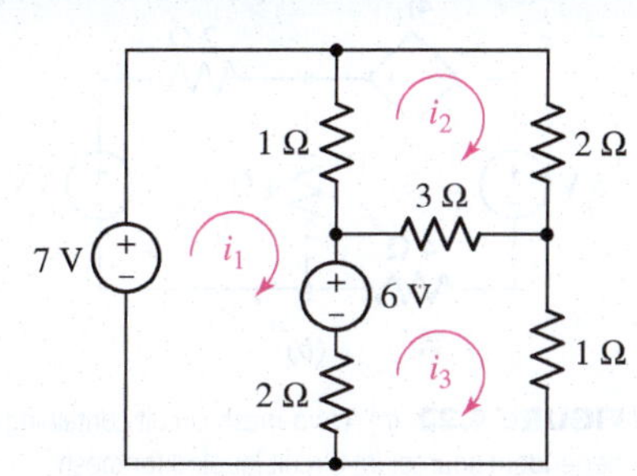

FIGURE 4.20 A five-node, seven-branch, three-mesh circuit.

The three required mesh currents are assigned as indicated in Fig. 4.20, and we methodically apply KVL about each mesh:

$$-7 + 1(i_1 - i_2) + 6 + 2(i_1 - i_3) = 0$$
$$1(i_2 - i_1) + 2i_2 + 3(i_2 - i_3) = 0$$
$$2(i_3 - i_1) - 6 + 3(i_3 - i_2) + 1i_3 = 0$$

Simplifying,

$$3i_1 - i_2 - 2i_3 = 1$$
$$-i_1 + 6i_2 - 3i_3 = 0$$
$$-2i_1 - 3i_2 + 6i_3 = 6$$

and solving, we obtain $i_1 = 3$ A, $i_2 = 2$ A, and $i_3 = 3$ A.

PRACTICE

4.7 Obtain values for i_1 and i_2 in the circuit of Fig 4.21.

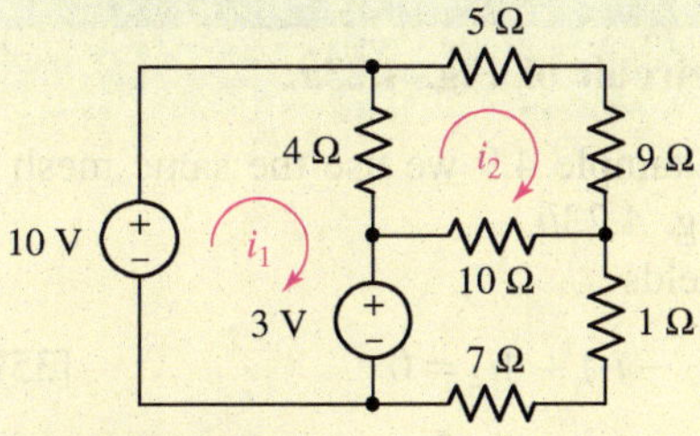

FIGURE 4.21

Ans: 2.220 A; 470.0 mA.

The previous examples dealt with circuits powered exclusively by independent voltage sources. If a current source is included in the circuit, it may either simplify or complicate the analysis, as discussed in Sec. 4.4. As seen in our study of the nodal analysis technique, dependent sources generally require an additional equation besides the M mesh equations, unless the controlling variable is a mesh current (or sum of mesh currents). We explore such a situation in the following example.

EXAMPLE 4.9

Determine the current i_1 in the circuit of Fig. 4.22*a*.

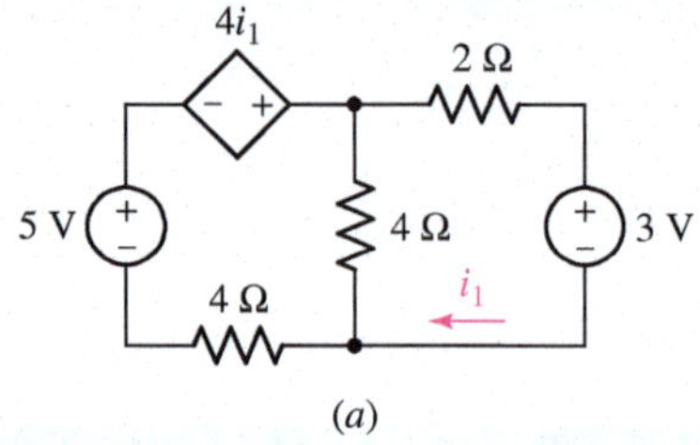

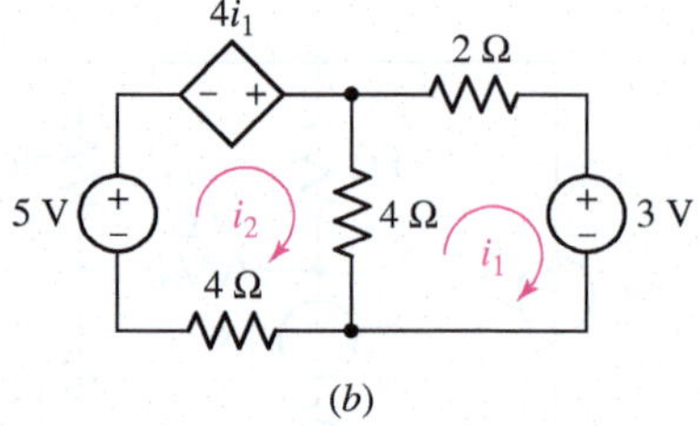

FIGURE 4.22 (*a*) A two-mesh circuit containing a dependent source. (*b*) Circuit labeled for mesh analysis.

The current i_1 is actually a mesh current, so rather than redefine it we label the rightmost mesh current i_1 and define a clockwise mesh current i_2 for the left mesh, as shown in Fig. 4.22*b*.

For the left mesh, KVL yields

$$-5 - 4i_1 + 4(i_2 - i_1) + 4i_2 = 0 \qquad [33]$$

and for the right mesh we find

$$4(i_1 - i_2) + 2i_1 + 3 = 0 \qquad [34]$$

Grouping terms, these equations may be written more compactly as

$$-8i_1 + 8i_2 = 5$$

and

$$6i_1 - 4i_2 = -3$$

Solving, $i_1 = -250$ mA.

Since the dependent source of Fig. 4.22 is controlled by a mesh current (i_1), only two equations—Eqs. [33] and [34]—were required to analyze the two-mesh circuit. In the following example, we explore the situation that arises if the controlling variable is *not* a mesh current.

EXAMPLE 4.10

Determine the current i_1 in the circuit of Fig. 4.23*a*.

In order to draw comparisons to Example 4.9 we use the same mesh current definitions, as shown in Fig. 4.23*b*.

For the left mesh, KVL now yields

$$-5 - 2v_x + 4(i_2 - i_1) + 4i_2 = 0 \qquad [35]$$

and for the right mesh we find the same as before, namely,

$$4(i_1 - i_2) + 2i_1 + 3 = 0 \qquad [36]$$

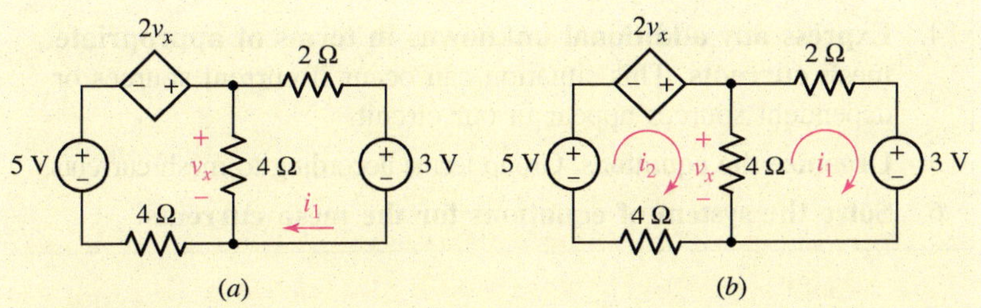

■ **FIGURE 4.23** (*a*) A circuit with a dependent source controlled by a voltage. (*b*) Circuit labeled for mesh analysis.

Since the dependent source is controlled by the unknown voltage v_x, we are faced with *two* equations in *three* unknowns. The way out of our dilemma is to construct an equation for v_x in terms of mesh currents, such as

$$v_x = 4(i_2 - i_1) \qquad [37]$$

We simplify our system of equations by substituting Eq. [37] into Eq. [35], resulting in

$$4i_1 = 5$$

Solving, we find that $i_1 = 1.25$ A. In this particular instance, Eq. [36] is not needed unless a value for i_2 is desired.

PRACTICE

4.8 Determine i_1 in the circuit of Fig. 4.24 if the controlling quantity A is equal to (*a*) $2i_2$; (*b*) $2v_x$.

Ans: (*a*) 1.35 A; (*b*) 546 mA.

■ **FIGURE 4.24**

The mesh analysis procedure can be summarized by the seven basic steps that follow. It will work on any *planar* circuit we ever encounter, although the presence of current sources will require extra care. Such situations are discussed in Sec. 4.4.

Summary of Basic Mesh Analysis Procedure

1. **Determine if the circuit is a planar circuit.** If not, perform nodal analysis instead.
2. **Count and label each mesh current in the circuit.** Redraw the circuit if necessary. Generally, defining all mesh currents to flow clockwise results in a simpler analysis.
3. **Write a KVL equation around each mesh.** Begin with a convenient node and proceed in the direction of the mesh current. Pay close attention to minus signs. If a current source lies on the periphery of a mesh, no KVL equation is needed since the mesh current is already defined!

4. **Express any additional unknowns in terms of appropriate mesh currents.** This situation can occur if current sources or dependent sources appear in our circuit.
5. **Organize the equations.** Group terms according to mesh currents.
6. **Solve the system of equations for the mesh currents.**

4.4 THE SUPERMESH

How must we modify this straightforward procedure when a current source is present in the network? Taking our lead from nodal analysis, there are two possible methods. First, we could assign an unknown voltage across the current source, apply KVL around each mesh as before, and then relate the source current to the assigned mesh currents. This is generally the more tedious approach.

A better technique is one that is quite similar to the supernode approach in nodal analysis. There we formed a supernode, completely enclosing the voltage source inside the supernode and reducing the number of nonreference nodes by 1 for each voltage source. Now we create a ***"supermesh"*** from two meshes that have a current source as a common element; the current source is in the interior of the supermesh. We thus reduce the number of meshes by 1 for each current source present. If the current source lies on the *perimeter* of the circuit, then the single mesh in which it is found is ignored. Kirchhoff's voltage law is thus applied only to those meshes or supermeshes in the reinterpreted network.

EXAMPLE **4.11**

Determine the three mesh currents in Fig. 4.25*a*.

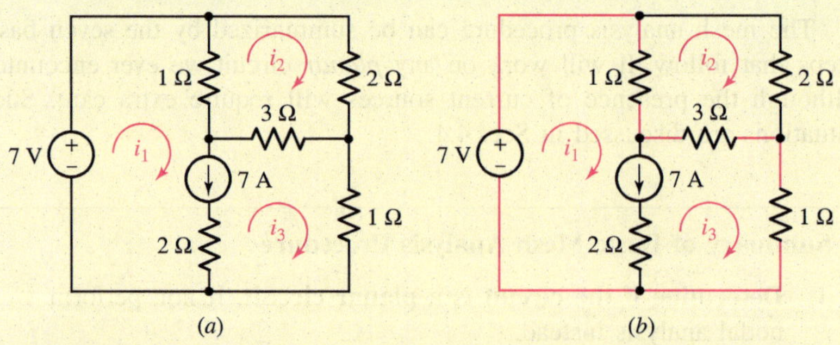

■ **FIGURE 4.25** (*a*) A three-mesh circuit with an independent current source. (*b*) A supermesh is defined by the colored line.

We note that a 7 A independent current source is in the common boundary of two meshes, which leads us to create a supermesh whose interior is that of meshes 1 and 3 as shown in Fig. 4.25*b*. Applying KVL around this loop,

$$-7 + 1(i_1 - i_2) + 3(i_3 - i_2) + 1i_3 = 0$$

or

$$i_1 - 4i_2 + 4i_3 = 7 \qquad [38]$$

and around mesh 2,

$$1(i_2 - i_1) + 2i_2 + 3(i_2 - i_3) = 0$$

or

$$-i_1 + 6i_2 - 3i_3 = 0 \qquad [39]$$

Finally, the independent source current is related to the mesh currents,

$$i_1 - i_3 = 7 \qquad [40]$$

Solving Eqs. [38] through [40], we find $i_1 = 9$ A, $i_2 = 2.5$ A, and $i_3 = 2$ A.

PRACTICE

4.9 Determine the current i_1 in the circuit of Fig. 4.26.

Ans: −1.93 A.

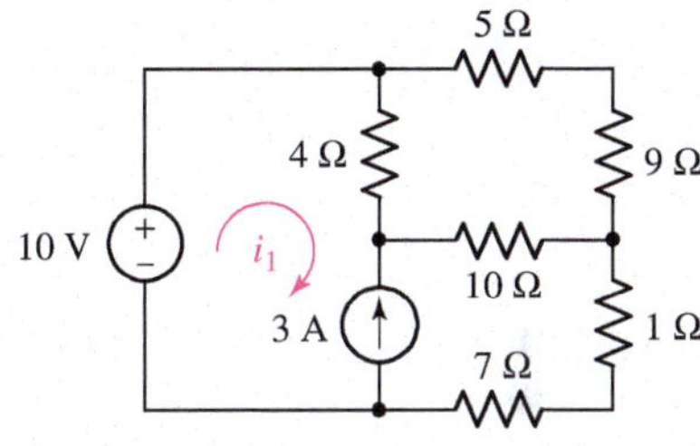

FIGURE 4.26

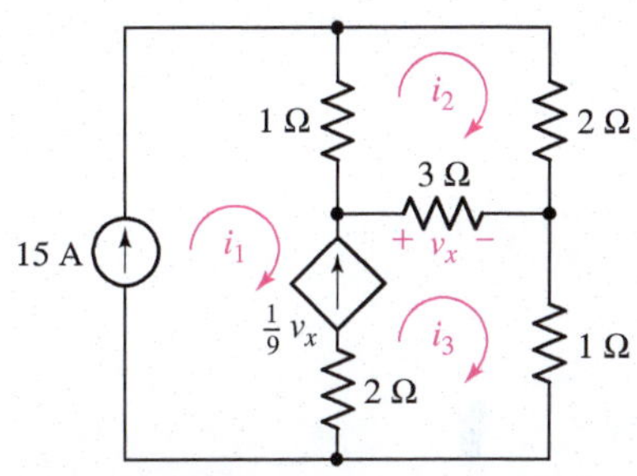

FIGURE 4.27 A three-mesh circuit with one dependent and one independent current source.

The presence of one or more dependent sources merely requires each of these source quantities and the variable on which it depends to be expressed in terms of the assigned mesh currents. In Fig. 4.27, for example, we note that both a dependent and an independent current source are included in the network. Let's see how their presence affects the analysis of the circuit and actually simplifies it.

EXAMPLE 4.12

Evaluate the three unknown currents in the circuit of Fig. 4.27.

The current sources appear in meshes 1 and 3. Since the 15 A source is located on the perimeter of the circuit, we may eliminate mesh 1 from consideration—it is clear, "by inspection," that $i_1 = 15$ A.

We find that because we now know one of the two mesh currents relevant to the dependent current source, there is no need to write a supermesh equation around meshes 1 and 3. Instead, we simply relate i_1 and i_3 to the current from the dependent source using KCL:

$$\frac{v_x}{9} = i_3 - i_1 = \frac{3(i_3 - i_2)}{9}$$

which can be written more compactly as

$$-i_1 + \frac{1}{3}i_2 + \frac{2}{3}i_3 = 0 \quad \text{or} \quad \frac{1}{3}i_2 + \frac{2}{3}i_3 = 15 \qquad [41]$$

(Continued on next page)

With one equation in two unknowns, all that remains is to write a KVL equation around mesh 2:

$$1(i_2 - i_1) + 2i_2 + 3(i_2 - i_3) = 0$$

or

$$6i_2 - 3i_3 = 15 \qquad [42]$$

Solving Eqs. [41] and [42], we find that $i_2 = 11$ A and $i_3 = 17$ A; we already determined that $i_1 = 15$ A by inspection.

PRACTICE

4.10 Determine v_3 in the circuit of Fig. 4.28.

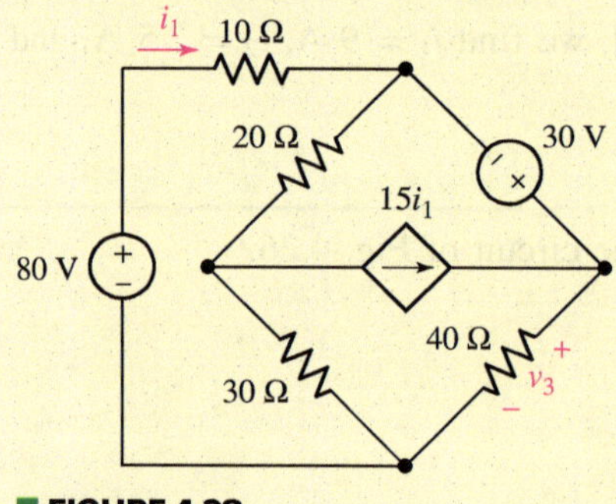

FIGURE 4.28

Ans: 104.2 V.

We can now summarize the general approach to writing mesh equations, whether or not dependent sources, voltage sources, and/or current sources are present, provided that the circuit can be drawn as a planar circuit:

Summary of Supermesh Analysis Procedure

1. **Determine if the circuit is a planar circuit.** If not, perform nodal analysis instead.
2. **Count and label each mesh current in the circuit.** Redraw the circuit if necessary. Generally, defining all mesh currents to flow clockwise results in a simpler analysis.
3. **If the circuit contains current sources shared by two meshes, form a supermesh to enclose both meshes.** A highlighted enclosure helps when writing KVL equations.
4. **Write a KVL equation around each mesh/supermesh.** Begin with a convenient node and proceed in the direction of the mesh current. Pay close attention to minus signs. If a current source lies on the periphery of a mesh, no KVL equation is needed since the mesh current is already defined!

5. **Relate the current flowing from each current source to mesh currents.** This is accomplished by simple application of KCL; one such equation is needed for each supermesh defined.
6. **Express any additional unknowns in terms of appropriate mesh currents.** This situation can occur if current sources or dependent sources appear in our circuit.
7. **Organize the equations.** Group terms according to mesh currents.
8. **Solve the system of equations for the mesh currents.**

4.5 NODAL VS. MESH ANALYSIS: A COMPARISON

Now that we have examined two distinctly different approaches to circuit analysis, it seems logical to ask if there is ever any advantage to using one over the other. If the circuit is nonplanar, then there is really no choice: only nodal analysis may be applied.

Provided that we are indeed considering the analysis of a *planar* circuit, however, there are situations where one technique has a small advantage over the other. If we plan to use nodal analysis, then a circuit with N nodes will lead to at most $(N - 1)$ KCL equations. Each supernode defined will further reduce this number by 1. If the same circuit has M distinct meshes, then we will obtain at most M KVL equations; each supermesh will reduce this number by 1. Based on these facts, we should select the approach that will result in the smaller number of simultaneous equations.

If one or more dependent sources are included in the circuit, then each controlling quantity may influence our choice of nodal or mesh analysis. For example, a dependent voltage source controlled by a nodal voltage does not require an additional equation when we perform nodal analysis. Likewise, a dependent current source controlled by a mesh current does not require an additional equation when we perform mesh analysis. *What about the situation where a dependent voltage source is controlled by a current? Or the converse, where a dependent current source is controlled by a voltage?* Provided that the controlling quantity can be easily related to mesh currents, we might expect mesh analysis to be the more straightforward option. Likewise, if the controlling quantity can be easily related to nodal voltages, nodal analysis may be preferable. One final point in this regard is to keep in mind the *location* of the source; current sources which lie on the periphery of a mesh, whether dependent or independent, are easily treated in mesh analysis; voltage sources connected to the reference terminal are easily treated in nodal analysis.

When either method results in essentially the same number of equations, it may be worthwhile to also consider what quantities are being sought. Nodal analysis results in direct calculation of nodal voltages, whereas mesh analysis provides currents. If we are asked to find currents through a set of resistors, for example, after performing nodal analysis, we must still invoke Ohm's law at each resistor to determine the current.

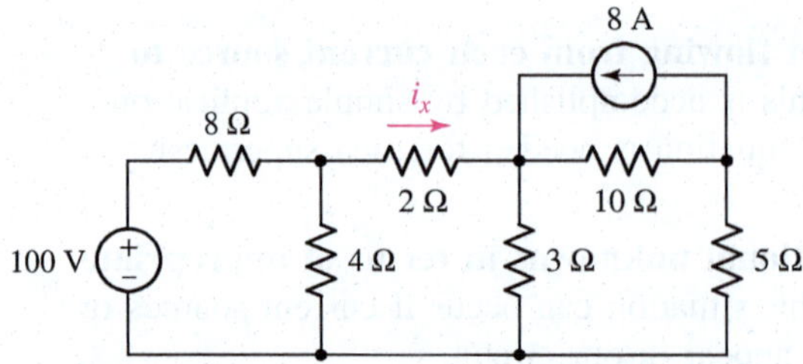

FIGURE 4.29 A planar circuit with five nodes and four meshes.

As an example, consider the circuit in Fig. 4.29. We wish to determine the current i_x.

We choose the bottom node as the reference node, and note that there are four nonreference nodes. Although this means that we can write four distinct equations, there is no need to label the node between the 100 V source and the 8 Ω resistor, since that node voltage is clearly 100 V. Thus, we label the remaining node voltages v_1, v_2, and v_3 as in Fig. 4.30.

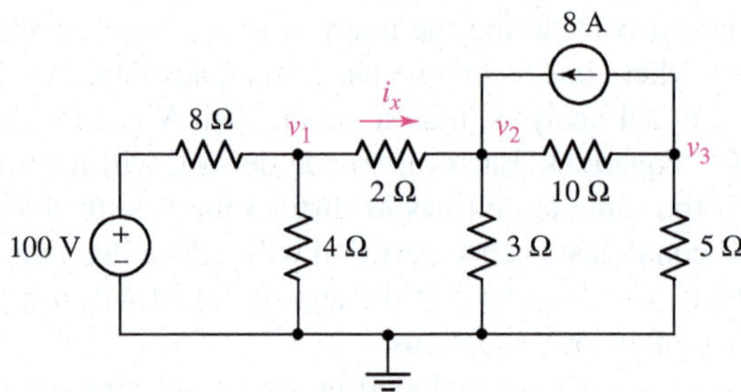

FIGURE 4.30 The circuit of Fig. 4.29 with node voltages labeled. Note that an earth ground symbol was chosen to designate the reference terminal.

We write the following three equations:

$$\frac{v_1 - 100}{8} + \frac{v_1}{4} + \frac{v_1 - v_2}{2} = 0 \quad \text{or} \quad 0.875v_1 - 0.5v_2 = 12.5 \qquad [43]$$

$$\frac{v_2 - v_1}{2} + \frac{v_2}{3} + \frac{v_2 - v_3}{10} - 8 = 0 \quad \text{or} \quad -0.5v_1 - 0.9333v_2 - 0.1v_3 = 8 \qquad [44]$$

$$\frac{v_3 - v_2}{10} + \frac{v_3}{5} + 8 = 0 \quad \text{or} \quad -0.1v_2 + 0.3v_3 = -8 \qquad [45]$$

Solving, we find that $v_1 = 25.89$ V and $v_2 = 20.31$ V. We determine the current i_x by application of Ohm's law:

$$i_x = \frac{v_1 - v_2}{2} = 2.79 \text{ A} \qquad [46]$$

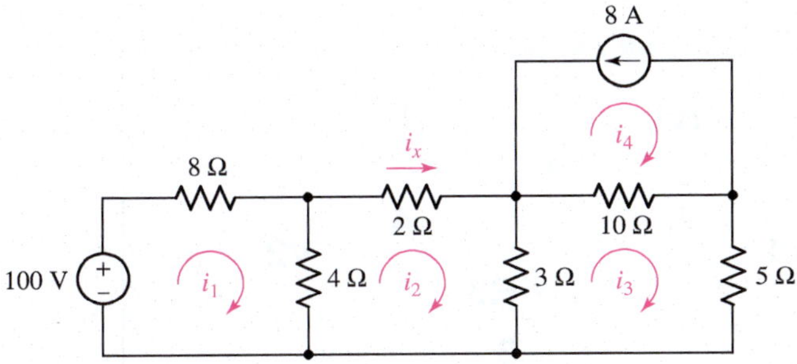

FIGURE 4.31 The circuit of Fig. 4.29 with mesh currents labeled.

Next, we consider the same circuit using mesh analysis. We see in Fig. 4.31 that we have four distinct meshes, although by inspection we see that $i_4 = -8$ A; we therefore need to write three distinct equations.

Writing a KVL equation for meshes 1, 2, and 3:

$$-100 + 8i_1 + 4(i_1 - i_2) = 0 \quad \text{or} \quad 12i_1 - 4i_2 = 100 \qquad [47]$$

$$4(i_2 - i_1) + 2i_2 + 3(i_2 - i_3) = 0 \quad \text{or} \quad -4i_1 + 9i_2 - 3i_3 = 0 \qquad [48]$$

$$3(i_3 - i_2) + 10(i_3 + 8) + 5i_3 = 0 \quad \text{or} \quad -3i_2 + 18i_3 = -80 \qquad [49]$$

Solving, we find that i_2 (= i_x) = 2.79 A. For this particular problem, mesh analysis proved to be simpler. Since either method is valid, however, working the same problem both ways can also serve as a means to check our answers.

4.6 COMPUTER-AIDED CIRCUIT ANALYSIS

We have seen that it does not take many components at all to create a circuit of respectable complexity. As we continue to examine even more complex circuits, it will become obvious rather quickly that it is easy to make errors during the analysis, and verifying solutions by hand can be time-consuming. Computer-aided design (CAD) software is commonly employed for rapid analysis of circuits, often including schematic capture tools that can be integrated with either printed circuit board or circuit layout tools. A general-purpose, open-source circuit simulator known as SPICE (*Simulation Program with Integrated Circuit Emphasis*), originally developed in the early 1970s at the University of California at Berkeley, is an industry standard. There are many software packages now available that have built intuitive graphical interfaces around the core SPICE program, each with their own strengths and limitations. In this book, we employ LTspice, a freeware package from Linear Technology (now part of Analog Devices) that is widely used and runs on both Windows and Mac OS X.

Although computer-aided analysis is a relatively quick means of determining voltages and currents in a circuit, we should be careful not to allow simulation packages to completely replace traditional "paper and pencil" analysis. There are several reasons for this. First, in order to design we must be able to analyze. Overreliance on software tools can inhibit the development of necessary analytical skills, similar to introducing calculators too early in school. Second, it is virtually impossible to use a complicated

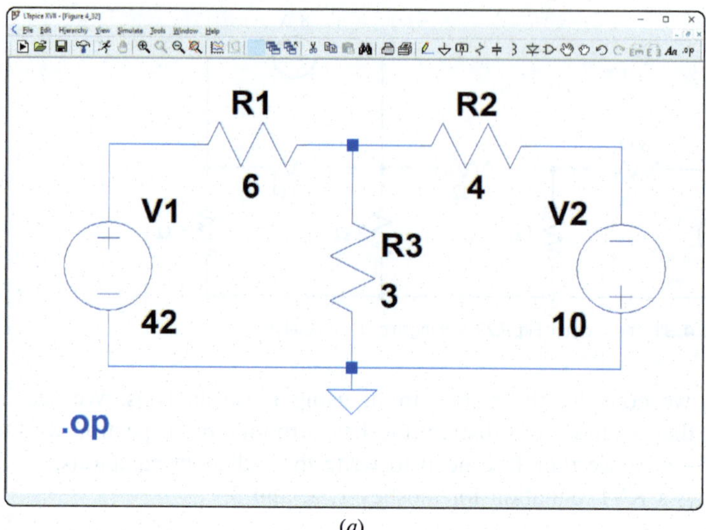

(*a*)

* C:\Users\Steve\Dropbox\HKD10E\Revised Chapters\Chapter 4\Figure ...

```
        --- Operating Point ---

V(n001):         42              voltage
V(n003):         -10             voltage
V(n002):         6               voltage
I(R3):           2               device_current
I(R2):           -4              device_current
I(R1):           -6              device_current
I(V2):           -4              device_current
I(V1):           -6              device_current
```

(*b*)

FIGURE 4.32 (*a*) Circuit of Fig. 4.16*a* drawn using LTspice. (*b*) Output log after simulation, showing the voltage at each node and current through each component. Note all currents are defined using the passive sign convention, i.e., into the positive reference terminal. This is the terminal about which resistors can be rotated in the schematic.
(*a–b*) Analog Devices

software package over a long period of time without making some type of data-entry error. If we have no basic intuition as to what type of answer to expect from a simulation, then there is no way to assess whether or not it is valid. Thus, the generic name really is a fairly accurate description: computer-*aided* analysis. Human brains are not obsolete. Not yet, anyway.

As an example, consider the circuit of Fig. 4.16*b*, which includes two dc voltage sources and three resistors. We wish to simulate this circuit using SPICE so that we may determine the currents i_1 and i_2. Figure 4.32*a* shows the circuit as drawn using LTspice.

In order to determine the mesh currents, we only need to run a bias point simulation. Under Simulate, we select Run. This opens the Edit Simulation Command; under the DC op pnt tab (far right) we see a text box with .op (for DC operating point). Selecting OK, the circuit analysis is performed, with the results appearing as shown in Fig. 4.32*b*.

We see that the two currents i_1 and i_2 are 6 A and 4 A, respectively, as we found previously. Be sure to note the assigned current direction! LTspice and other packages typically do not show the current direction assigned. For LTspice, after the simulation, you can move your cursor over an element in the schematic to see the current direction defined (a current probe icon with an arrow defining the direction should appear). LTspice assumes the passive sign convention, though you need to know the polarity assigned to the component.

As a further example, consider the circuit shown in Fig. 4.33*a*. It contains a dc voltage source, a dc current source, and a voltage-controlled current source. We are interested in the three nodal voltages, which from either nodal or mesh analysis are found to be 82.91 V, 69.9 V, and 59.9 V, respectively, as we move from left to right across the top of the circuit. Figure 4.33*b* shows this circuit after the simulation was performed. The three nodal voltages are indicated directly on the schematic, which was accomplished by clicking on wires connected to nodes of interest. Note that the dependent source has four terminals: two for the current it delivers as a source, and two which enable the controlling voltage to be defined; in LTspice, crossed wires do not constitute an electrical connection unless a node is explicitly shown.

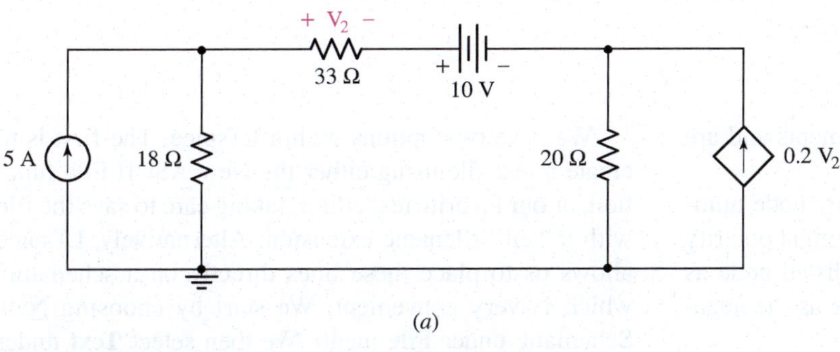

(*a*)

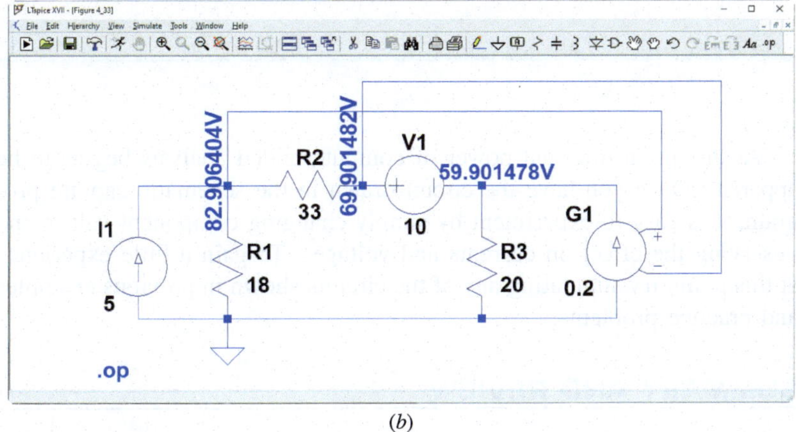

(*b*)

FIGURE 4.33 (*a*) Circuit with dependent current source. (*b*) Circuit drawn using a schematic capture tool, with simulation results presented directly on the schematic.
(*b*) Analog Devices

PRACTICAL APPLICATION

Node-Based Circuit Definition

The most common method of describing a circuit in conjunction with computer-aided circuit analysis is with some type of graphical schematic drawing package, an example output of which was shown in Fig. 4.33. SPICE, however, was written before the advent of such software, and as such requires circuits to be described in a specific text-based format. The format has its roots in the syntax used for punch cards, which gives it a somewhat distinct appearance. The basis for circuit description is the definition of elements, each terminal of which is assigned a node number. So, although we have just studied two different generalized circuit analysis methods—the nodal and mesh techniques—it is interesting that SPICE and was written using a clearly defined nodal analysis approach.

Even though graphics-oriented interactive software is convenient, the ability to read the text-based "input deck" generated by the schematic capture tool can be invaluable in tracking down specific problems. The easiest way to develop such an ability is to learn how to run SPICE directly from a user-written input deck. Proficient users sometimes also find that a text-based definition of the circuit can be faster and more convenient than using a graphical interface!

Consider, for example, the voltage divider circuit shown in Fig. 4.34*a*. The corresponding input deck is:

```
* Example SPICE input deck for simple voltage divider circuit.
.op                 (Requests dc operating point)
R1 1 2 1k           (Locates R1 between nodes 1 and 2; value is 1 kΩ)
R2 2 0 1k           (Locates R2 between nodes 2 and 0; also 1 kΩ)
V1 1 0 5            (Locates 5 V source between nodes 1 and 0)
.end                (End of input deck)
```

(lines beginning with an asterisk are comments and are skipped by SPICE).

Each component is defined, followed by node number, followed by component value. The terminal polarity for the component is defined as the first listed node as the positive terminal, and the second node as the negative terminal.

We have two options within LTspice. The first is to create a text file using either the New ASCII File function, or our favorite text editor, taking care to save the file with a ".cir" filename extension. Alternatively, LTspice allows us to place these lines directly on a schematic, which is very convenient. We start by choosing **New** Schematic under **File** menu. We then select **Text** under

At this point, the real power of computer-aided analysis begins to be apparent: Once you have the circuit drawn in the schematic capture program, it is easy to experiment by simply changing component values and observing the effect on currents and voltages. To gain a little experience at this point, try simulating any of the circuits shown in previous examples and practice problems.

SUMMARY AND REVIEW

Although Chap. 3 introduced KCL and KVL, both of which are sufficient to enable us to analyze any circuit, a more methodical approach proves helpful in everyday situations. Thus, in this chapter we developed the **Edit** menu. We next enter the text shown in blue here, taking care to select the SPICE directive radio button in the popup menu, as shown in Fig. 4.34*b*.

We run the simulation by clicking the **Run** command (running person icon on top left) under **Simulate.** The results are shown in a window such as depicted in Fig. 4.34*c*.

Here it is worth noting that the output provides the expected nodal voltages (5 V at node 1, 2.5 V across resistor R2), and the current quoted using the passive sign convention (i.e., currents into the positive reference terminal of each element, taken as the first node listed in the input deck).

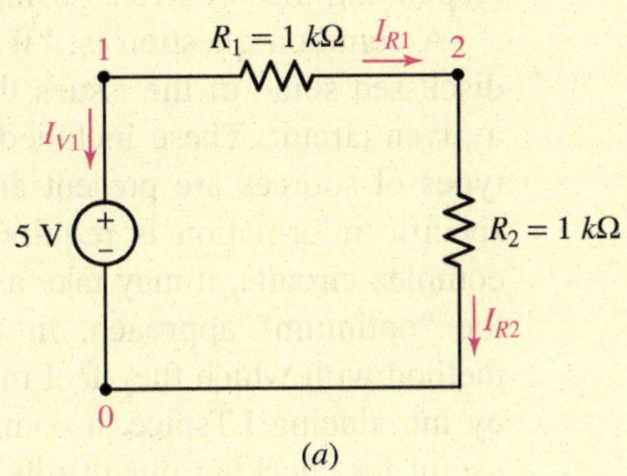

(*a*)

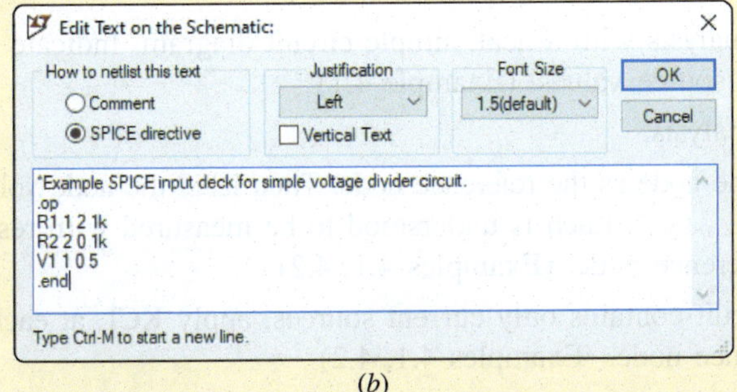

(*b*)

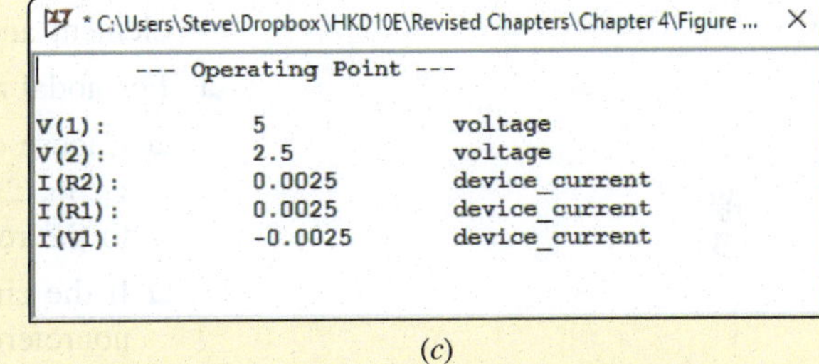

(*c*)

■ **FIGURE 4.34** (*a*) Schematic diagram of voltage divider, (*b*) LTspice window showing the input deck describing the voltage divider, and (*c*) output log file, showing nodal voltages and current through each component. Note that the voltage across R1 requires post-simulation subtraction. (*b–c*) Analog Devices

the nodal analysis technique based on KCL, which results in a voltage at each node (with respect to some designated "reference" node). We generally need to solve a system of simultaneous equations, unless voltage sources are connected so that they automatically provide nodal voltages. The controlling quantity of a *dependent* source is written down just as we would write down the numerical value of an "independent" source. Typically an additional equation is then required, unless the dependent source is controlled by a nodal voltage. When a voltage source bridges two nodes, the basic technique can be extended by creating a *supernode*; KCL dictates that the sum of the currents flowing into a group of connections so defined is equal to the sum of the currents flowing out.

As an alternative to nodal analysis, the mesh analysis technique was developed through application of KVL; it yields the complete set of *mesh* currents, which do not always represent the *net* current flowing through any particular element (for example, if an element is shared by two meshes). The presence of a current source will simplify the analysis if it lies on the periphery of a mesh; if the source is shared, then the *supermesh* technique is best. In that case, we write a KVL equation around a path that avoids the shared current source, then algebraically link the two corresponding mesh currents using the source.

A common question is: "*Which analysis technique should I use?*" We discussed some of the issues that might go into choosing a technique for a given circuit. These included whether or not the circuit is planar, what types of sources are present and how they are connected, and also what specific information is required (i.e., a voltage, current, or power). For complex circuits, it may take a greater effort than it is worth to determine the "optimum" approach, in which case most people will opt for the method with which they feel most comfortable. We concluded the chapter by introducing LTspice, a common circuit simulation tool, which is very useful for checking our results.

At this point we wrap up by identifying key points of this chapter to review, along with relevant example(s).

- Start each analysis with a neat, simple circuit diagram. Indicate all element and source values. (Example 4.1)
- For nodal analysis,
  - Choose one node as the reference node. Then label the node voltages $v_1, v_2, \ldots, v_{N-1}$. Each is understood to be measured with respect to the reference node. (Examples 4.1, 4.2)
  - If the circuit contains only current sources, apply KCL at each nonreference node. (Examples 4.1, 4.2)
  - If the circuit contains voltage sources, form a supernode about each one, and then apply KCL at all nonreference nodes and supernodes. (Examples 4.5, 4.6)
- For mesh analysis, first make certain that the network is a planar network.
  - Assign a clockwise mesh current in each mesh: $i_1, i_2, \ldots, i_M$. (Example 4.7)
  - If the circuit contains only voltage sources, apply KVL around each mesh. (Examples 4.7, 4.8, 4.9)
  - If the circuit contains current sources, create a supermesh for each one that is common to two meshes, and then apply KVL around each mesh and supermesh. (Examples 4.11, 4.12)
- Dependent sources will add an additional equation to nodal analysis if the controlling variable is a current, but not if the controlling variable is a nodal voltage. (Conversely, a dependent source will add an additional equation to mesh analysis if the controlling variable is a voltage,

but not if the controlling variable is a mesh current.) (Examples 4.3, 4.4, 4.6, 4.9, 4.10, 4.12)

- In deciding whether to use nodal or mesh analysis for a planar circuit, a circuit with fewer nodes/supernodes than meshes/supermeshes will result in fewer equations using nodal analysis.
- Computer-aided analysis is useful for checking results and analyzing circuits with large numbers of elements. However, common sense must be used to check simulation results.

READING FURTHER

A detailed treatment of nodal and mesh analysis can be found in:

R. A. DeCarlo and P. M. Lin, *Linear Circuit Analysis,* 2nd ed. New York: Oxford University Press, 2001.

A solid guide to SPICE is

P. Tuinenga, *SPICE: A Guide to Circuit Simulation and Analysis Using PSPICE,* 3rd ed. Upper Saddle River, N.J.: Prentice-Hall, 1995.

EXERCISES

4.1 Nodal Analysis

1. Solve the following systems of equations:

   (*a*) $2v_2 - 4v_1 = 9$ and $v_1 - 5v_2 = -4$;

   (*b*) $-v_1 + 2v_3 = 8$; $2v_1 + v_2 - 5v_3 = -7$; $4v_1 + 5v_2 + 8v_3 = 6$.

2. (*a*) Solve the following system of equations:

$$3 = \frac{v_1}{5} - \frac{v_2 - v_1}{22} + \frac{v_1 - v_3}{3}$$

$$2 - 1 = \frac{v_2 - v_1}{22} + \frac{v_2 - v_3}{14}$$

$$0 = \frac{v_3}{10} + \frac{v_3 - v_1}{3} + \frac{v_3 - v_2}{14}$$

   (*b*) Verify your solution using MATLAB.

3. (*a*) Solve the following system of equations:

$$7 = \frac{v_1}{2} - \frac{v_2 - v_1}{12} + \frac{v_1 - v_3}{19}$$

$$15 = \frac{v_2 - v_1}{12} + \frac{v_2 - v_3}{2}$$

$$4 = \frac{v_3}{7} + \frac{v_3 - v_1}{19} + \frac{v_3 - v_2}{2}$$

   (*b*) Verify your solution using MATLAB.

4. Correct (and verify by running) the following MATLAB code:

```
syms e1 e2 e3
e1 = 3 = v1/7 – (v2 – v1)/2 + (v1 – v3)/3;
e2 = 2 == (v2 – v1)/2 + (v2 – v3)/14;
e = 0 == v3/10 + (v3 – v1)/3 + (v3 – v2)/14;
a = sove(e e2 e3,[v1 v2 v3]);
```

5. In the circuit of Fig. 4.35, determine the current labeled i with the assistance of nodal analysis techniques.

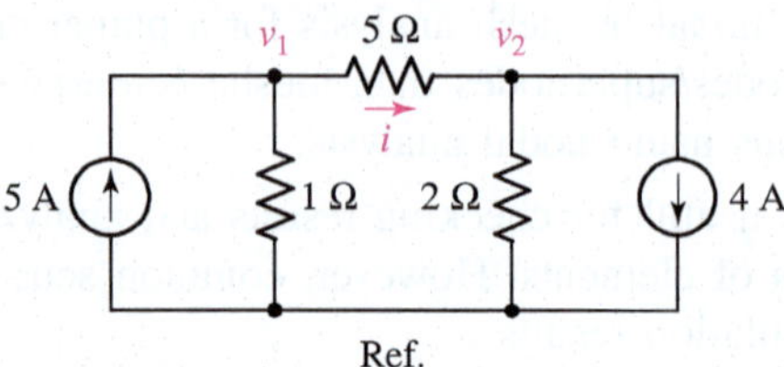

■ **FIGURE 4.35**

6. Calculate the power dissipated in the 1 Ω resistor of Fig. 4.36.

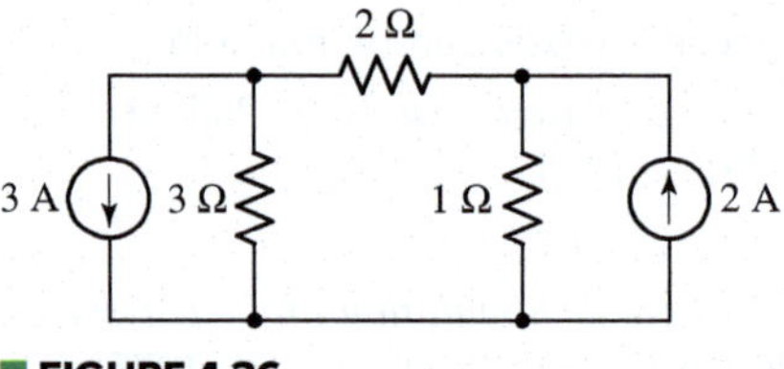

■ **FIGURE 4.36**

7. For the circuit in Fig. 4.37, determine the value of the current i_x.

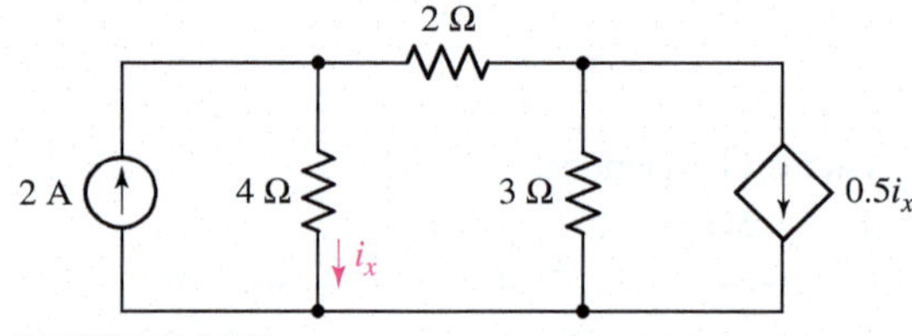

■ **FIGURE 4.37**

8. With the assistance of nodal analysis, determine $v_1 - v_2$ in the circuit shown in Fig. 4.38.

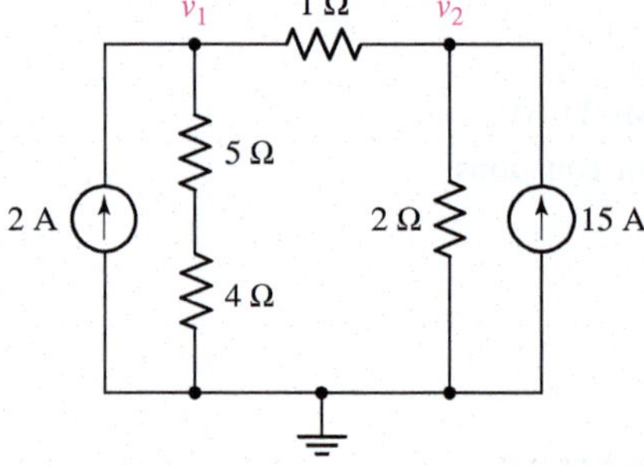

■ **FIGURE 4.38**

9. For the circuit of Fig. 4.39, determine the value of the voltage labeled v_x.

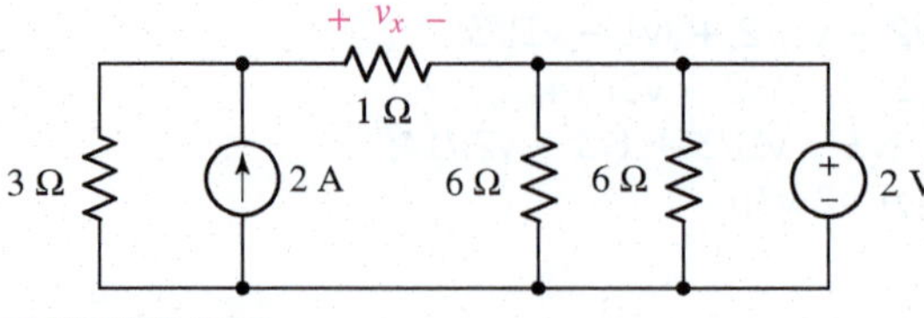

■ **FIGURE 4.39**

10. For the circuit of Fig. 4.40, determine the value of the voltage labeled v_o.

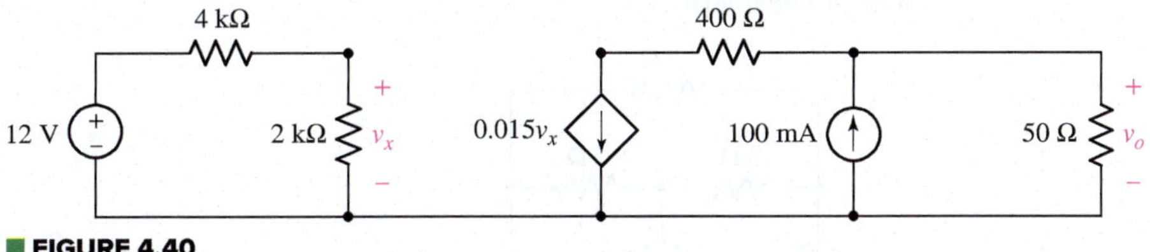

FIGURE 4.40

11. Use nodal analysis to find v_P in the circuit shown in Fig. 4.41.

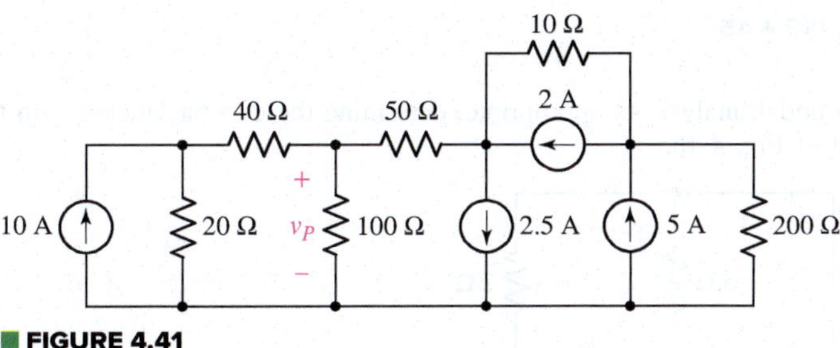

FIGURE 4.41

12. Using the bottom node as reference, determine the voltage across the 5 Ω resistor in the circuit of Fig. 4.42, and calculate the power dissipated by the 7 Ω resistor.

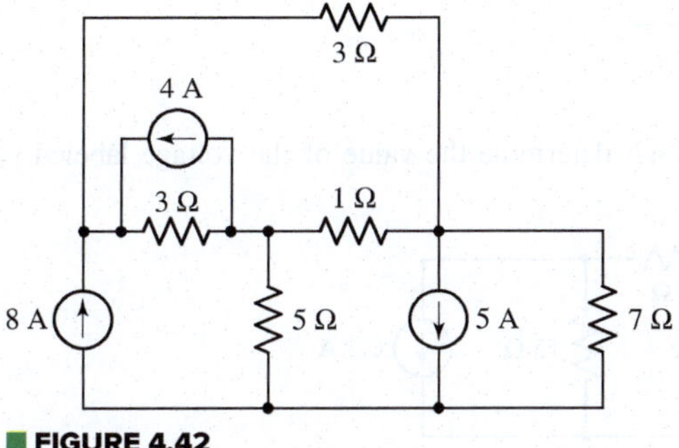

FIGURE 4.42

13. For the circuit of Fig. 4.43, use nodal analysis to determine the current i_5.

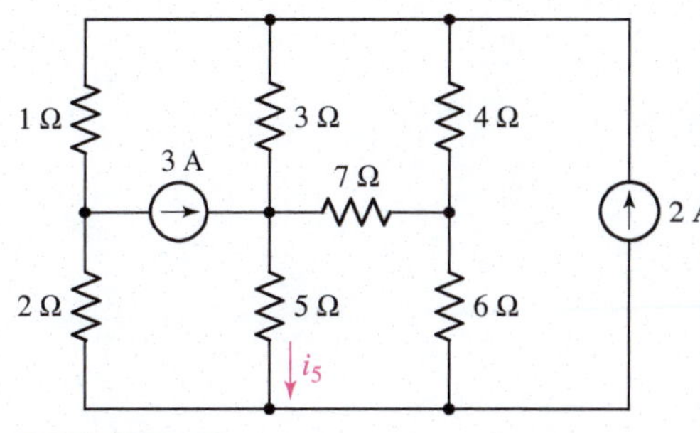

FIGURE 4.43

14. Determine a numerical value for each nodal voltage in the circuit of Fig. 4.44.

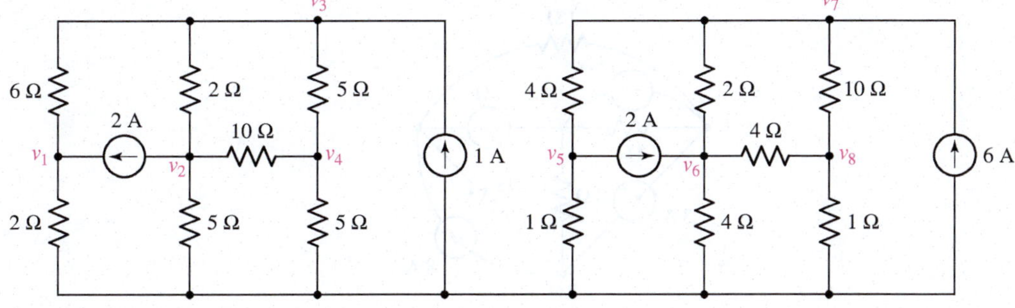

FIGURE 4.44

15. Determine the current i_2 as labeled in the circuit of Fig. 4.45, with the assistance of nodal analysis.

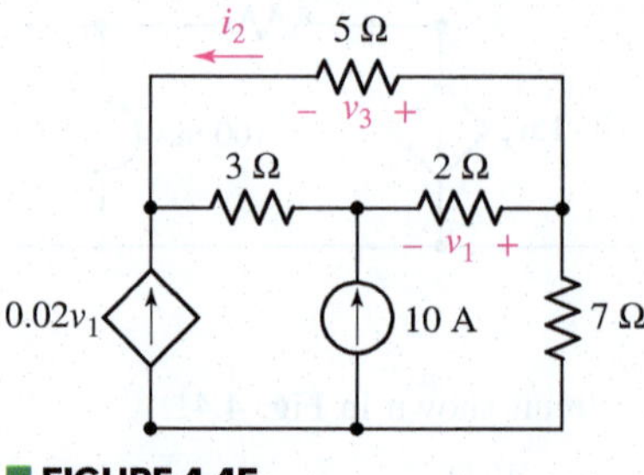

FIGURE 4.45

16. Using nodal analysis as appropriate, determine the current labeled i_1 in the circuit of Fig. 4.46.

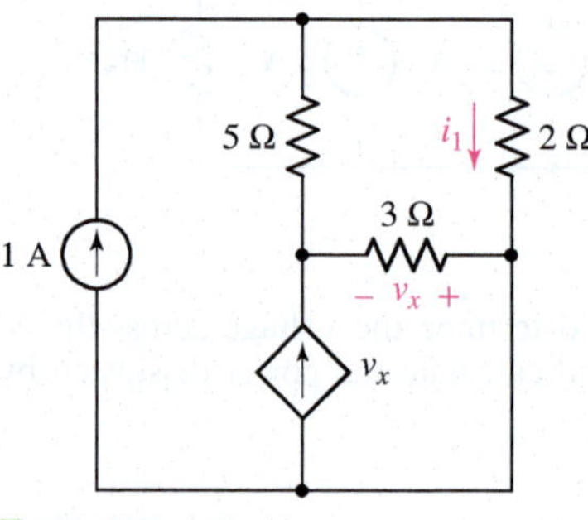

FIGURE 4.46

17. For the circuit of Fig. 4.47, determine the value of the voltage labeled v_x.

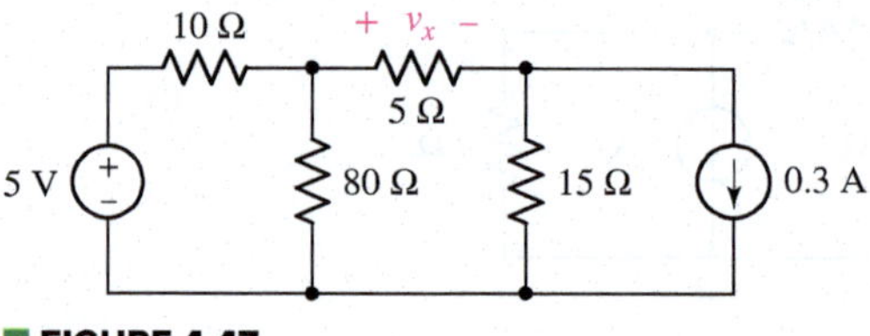

FIGURE 4.47

4.2 The Supernode

18. Determine the nodal voltages as labeled in Fig. 4.48, making use of the supernode technique as appropriate.

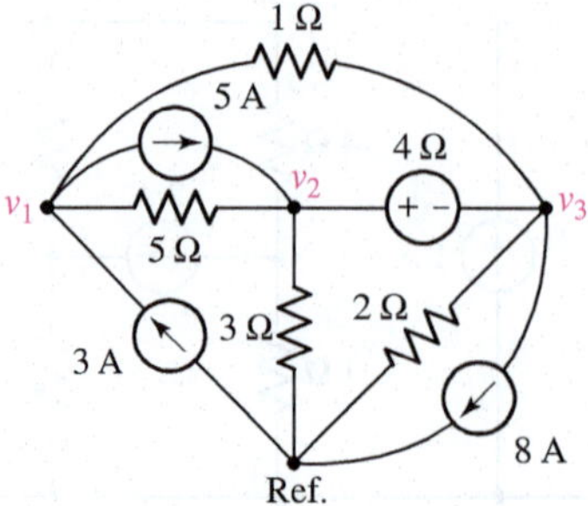

FIGURE 4.48

19. For the circuit shown in Fig. 4.49, determine a numerical value for the voltage labeled v_1.

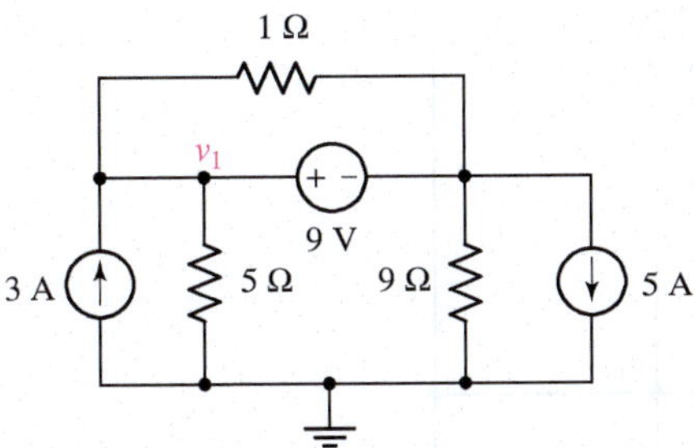

FIGURE 4.49

20. For the circuit of Fig. 4.50, determine all four nodal voltages.

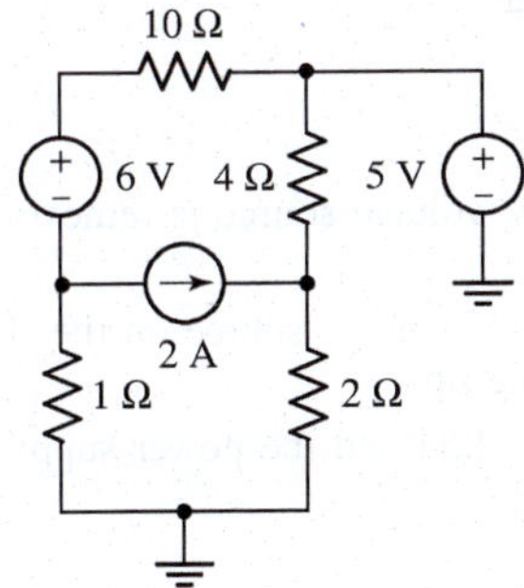

FIGURE 4.50

21. Employing supernode/nodal analysis techniques as appropriate, determine the power dissipated by the 1 Ω resistor in the circuit of Fig. 4.51.

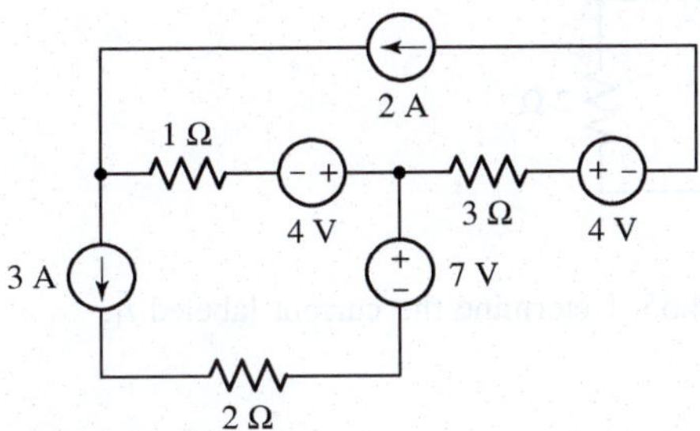

FIGURE 4.51

22. Referring to the circuit of Fig. 4.52, obtain a numerical value for the power supplied by the 1 V source.

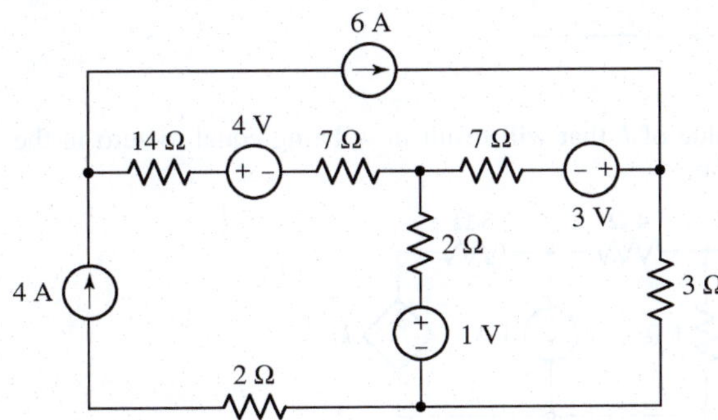

FIGURE 4.52

23. For the circuit in Fig. 4.53, determine the node voltages v_1 and v_2.

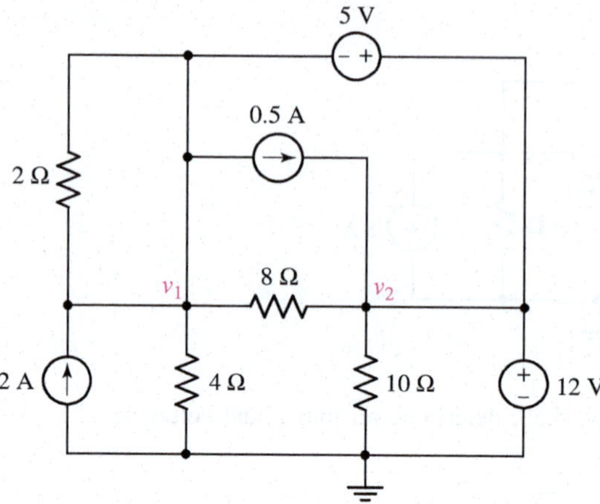

■ **FIGURE 4.53**

24. Repeat Exercise 23 for the case if the top 5 V voltage source is removed (open circuit).
25. Repeat Exercise 23 for the case where the 12 V voltage source on the right is replaced by a current source of 1 A pointing up.
26. Determine the voltage v_x in the circuit of Fig. 4.54 and the power supplied by the 1 A source.

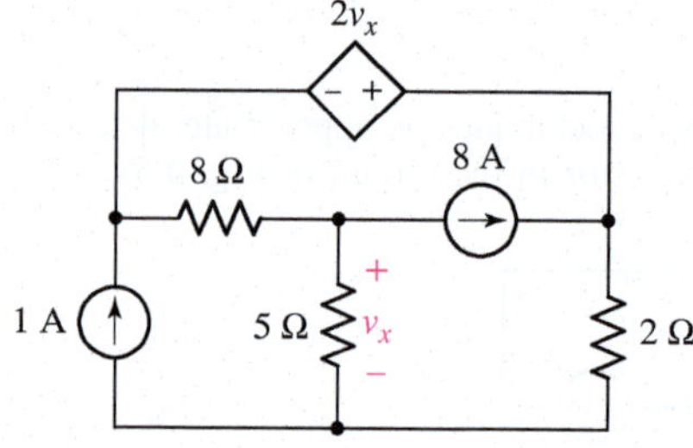

■ **FIGURE 4.54**

27. Consider the circuit of Fig. 4.55. Determine the current labeled i_1.

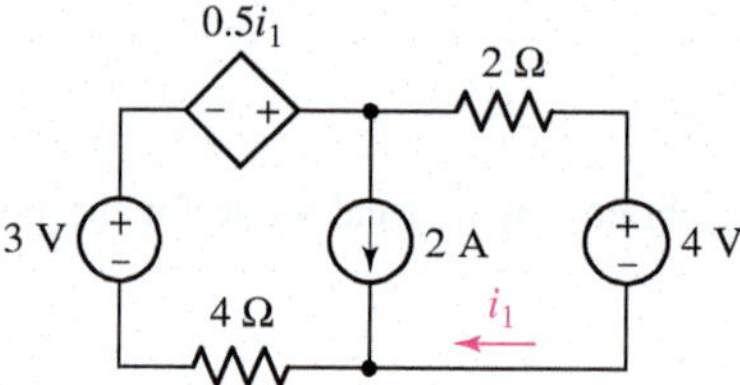

■ **FIGURE 4.55**

28. Determine the value of k that will result in v_x being equal to zero in the circuit of Fig. 4.56.

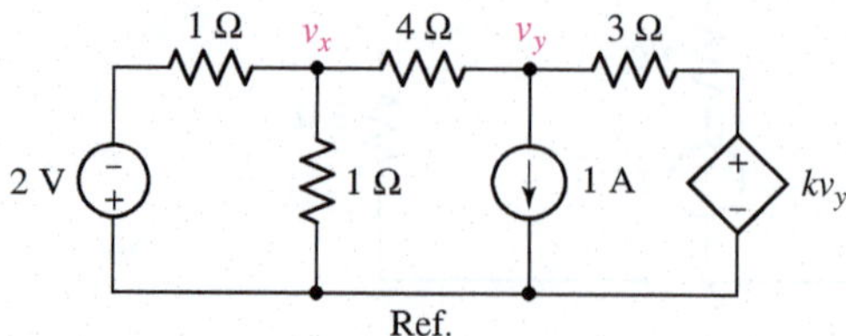

■ **FIGURE 4.56**

29. For the circuit depicted in Fig. 4.57, determine the voltage labeled v_1 across the 3 Ω resistor.

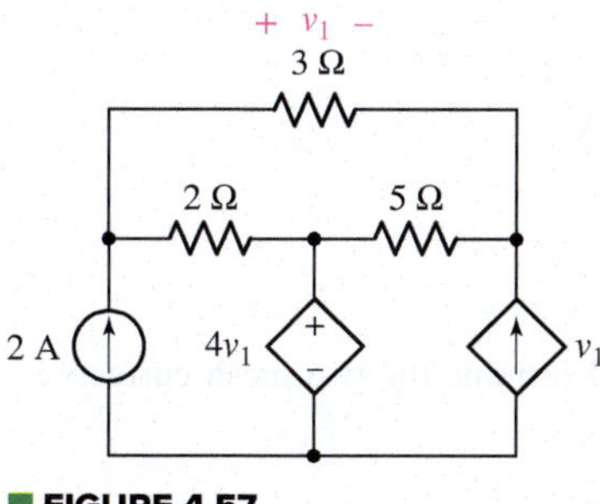

FIGURE 4.57

30. For the circuit of Fig. 4.58, determine all four nodal voltages.

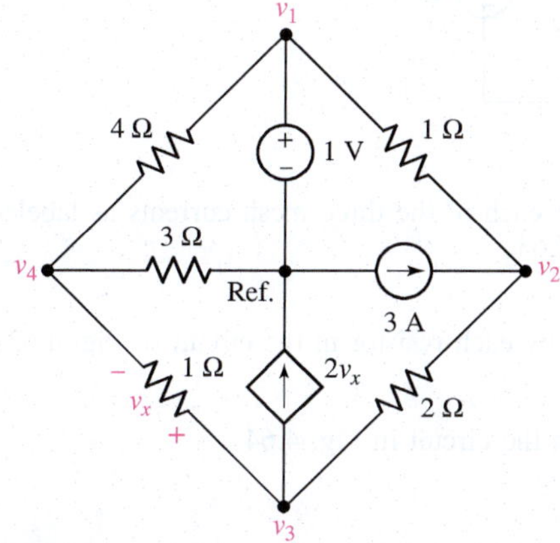

FIGURE 4.58

31. For the circuit of Fig. 4.59, determine the unknown node voltages v_1, v_2, v_3, and v_4.

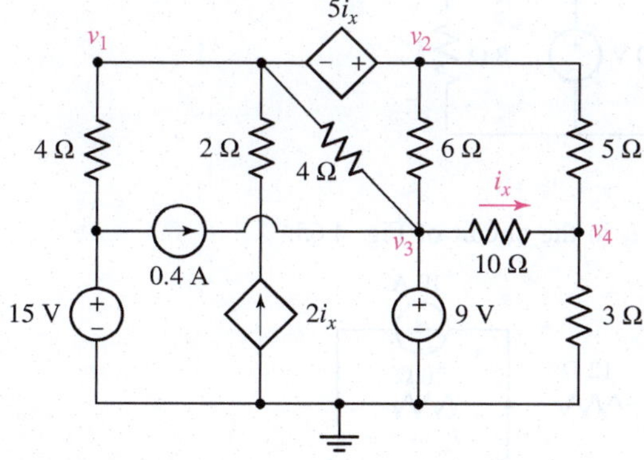

FIGURE 4.59

4.3 Mesh Analysis

32. Determine the currents flowing out of the positive terminal of each voltage source in the circuit of Fig. 4.60.

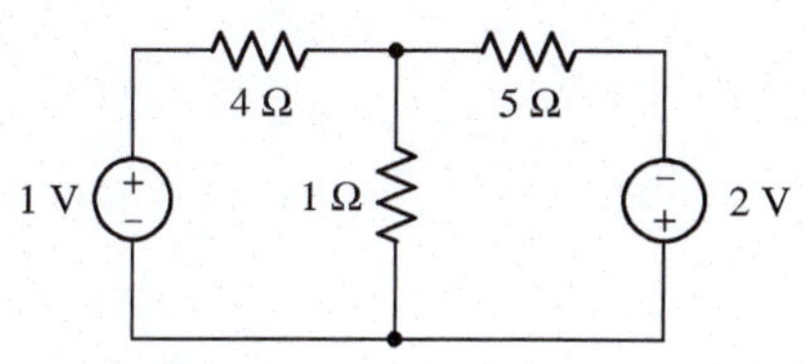

FIGURE 4.60

33. Obtain numerical values for the two mesh currents i_1 and i_2 in the circuit shown in Fig. 4.61.

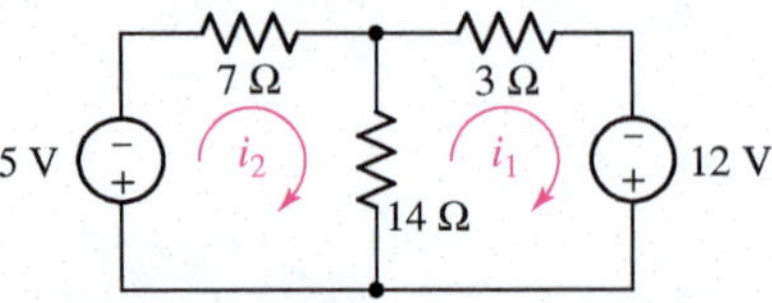

FIGURE 4.61

34. Use mesh analysis as appropriate to determine the two mesh currents labeled in Fig. 4.62.

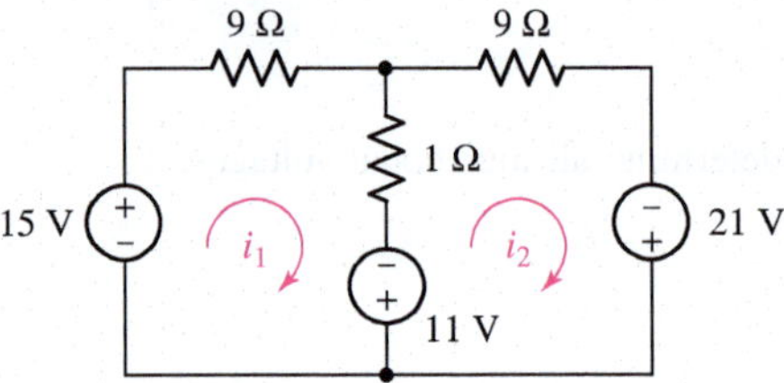

FIGURE 4.62

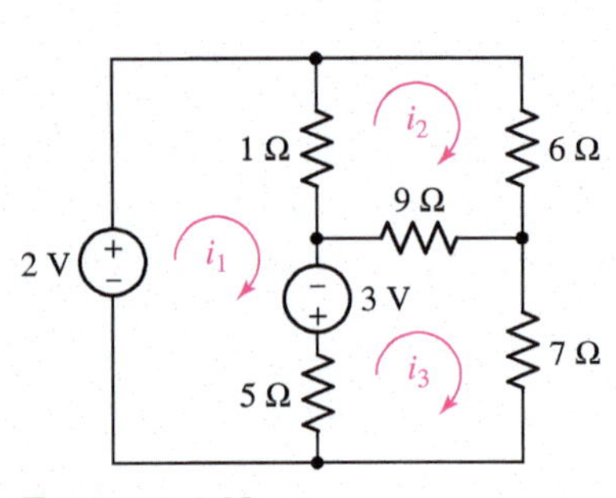

FIGURE 4.63

35. Determine numerical values for each of the three mesh currents as labeled in the circuit diagram of Fig. 4.63.

36. Calculate the power dissipated by each resistor in the circuit of Fig. 4.63.

37. Find the unknown voltage v_x in the circuit in Fig. 4.64.

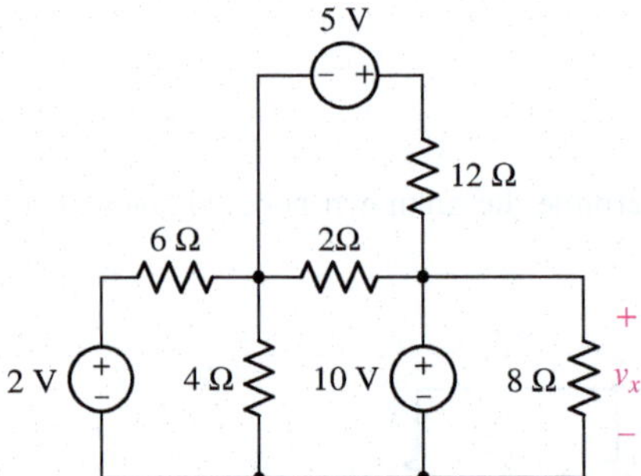

FIGURE 4.64

38. Calculate the current i_x in the circuit of Fig. 4.65.

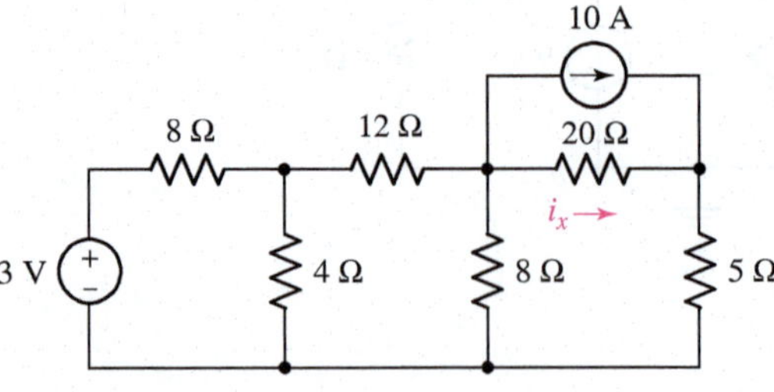

FIGURE 4.65

39. Employing mesh analysis procedures, obtain a value for the current labeled i in the circuit represented by Fig. 4.66.

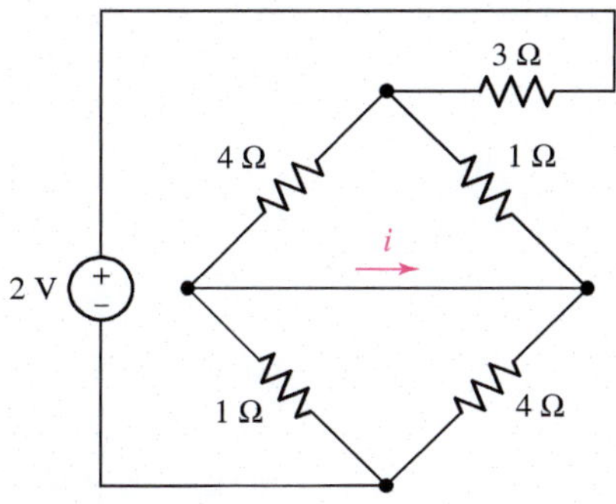

FIGURE 4.66

40. Determine the power dissipated in the 4 Ω resistor of the circuit shown in Fig. 4.67.

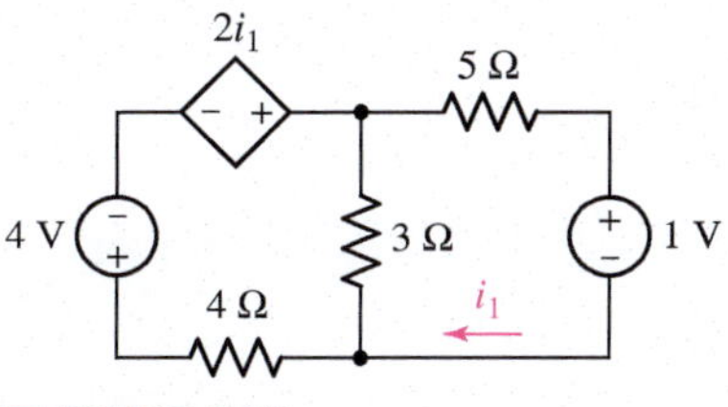

FIGURE 4.67

41. (*a*) Employ mesh analysis to determine the power dissipated by the 1 Ω resistor in the circuit represented schematically by Fig. 4.68. (*b*) Check your answer using nodal analysis.

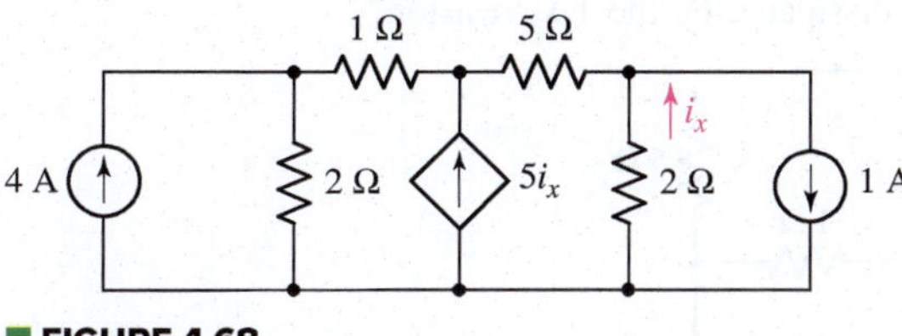

FIGURE 4.68

42. Define three clockwise mesh currents for the circuit of Fig. 4.69, and employ mesh analysis to obtain a value for each.

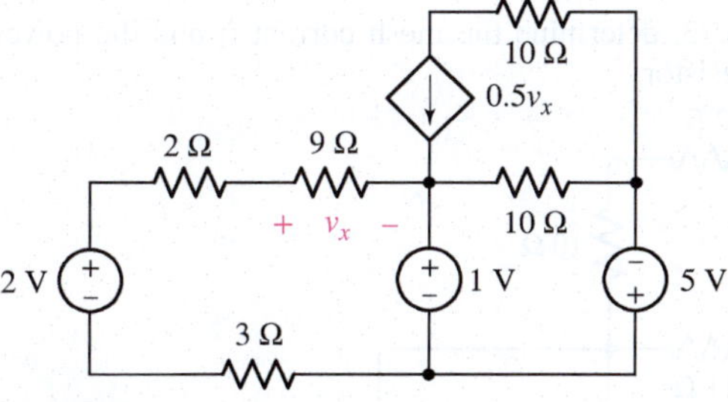

FIGURE 4.69

43. Employ mesh analysis to obtain values for i_x and v_a in the circuit of Fig. 4.70.

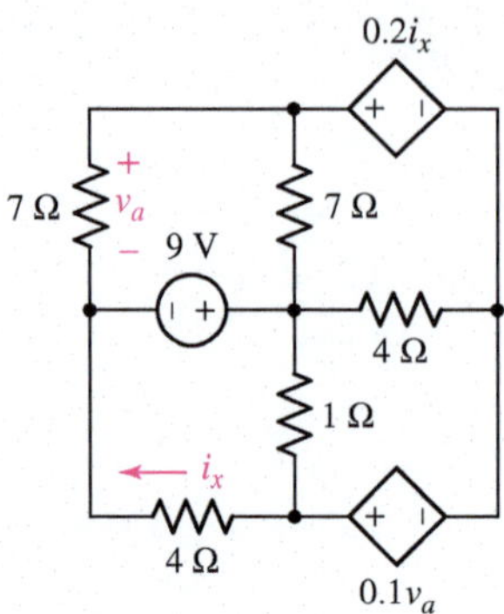

■ **FIGURE 4.70**

4.4 The Supermesh

44. Determine values for the three mesh currents of Fig. 4.71.

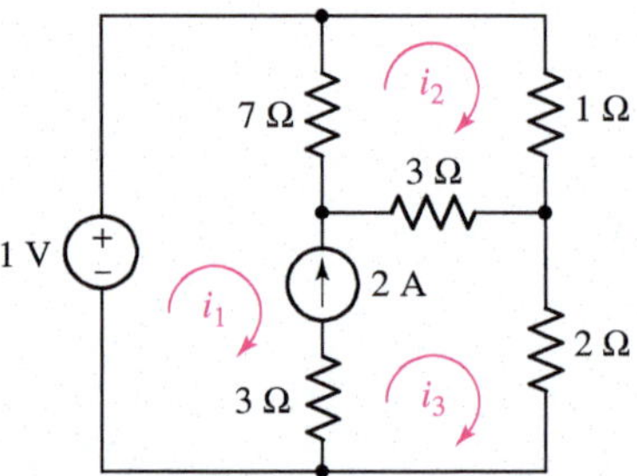

■ **FIGURE 4.71**

45. Through appropriate application of the supermesh technique, obtain a numerical value for the mesh current i_3 in the circuit of Fig. 4.72, and calculate the power dissipated by the 1 Ω resistor.

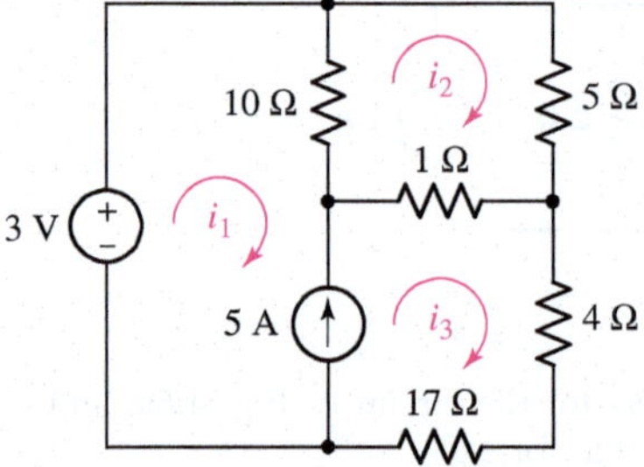

■ **FIGURE 4.72**

46. For the circuit of Fig. 4.73, determine the mesh current i_1 and the power dissipated by the 1 Ω resistor.

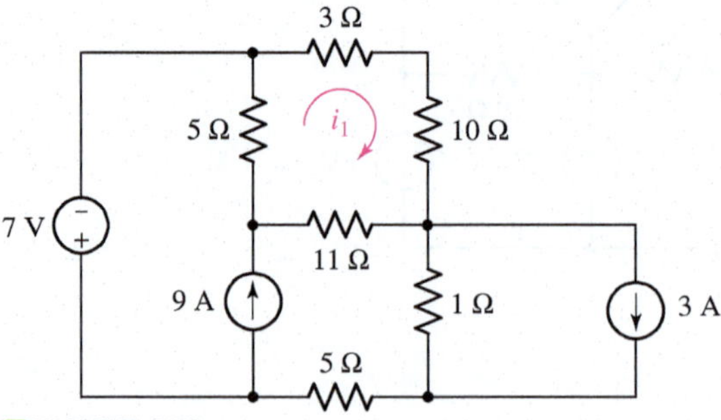

■ **FIGURE 4.73**

47. Calculate the three mesh currents labeled in the circuit diagram of Fig. 4.74.

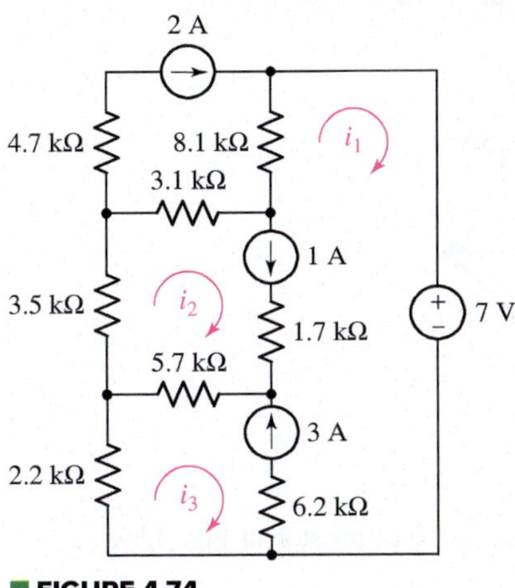

■ **FIGURE 4.74**

48. Use mesh analysis to find the current i_x in the circuit of Fig. 4.75.

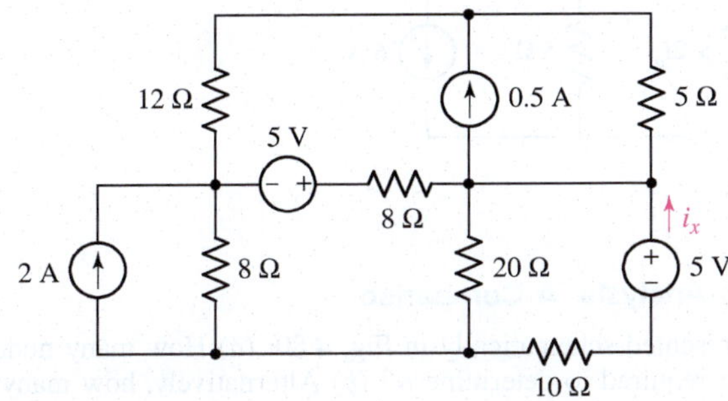

■ **FIGURE 4.75**

49. Through careful application of the supermesh technique, obtain values for all three mesh currents as labeled in Fig. 4.76.

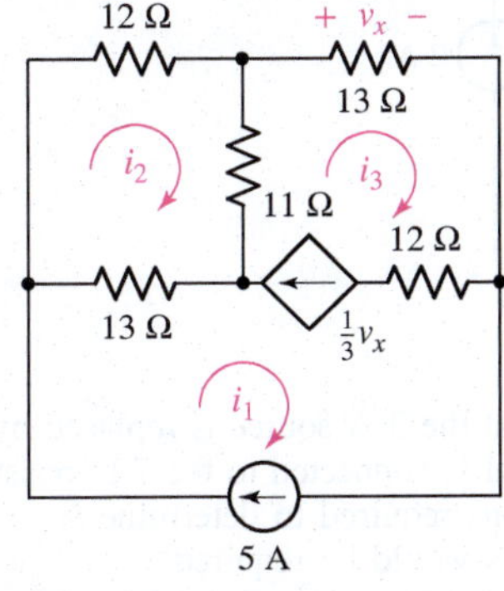

■ **FIGURE 4.76**

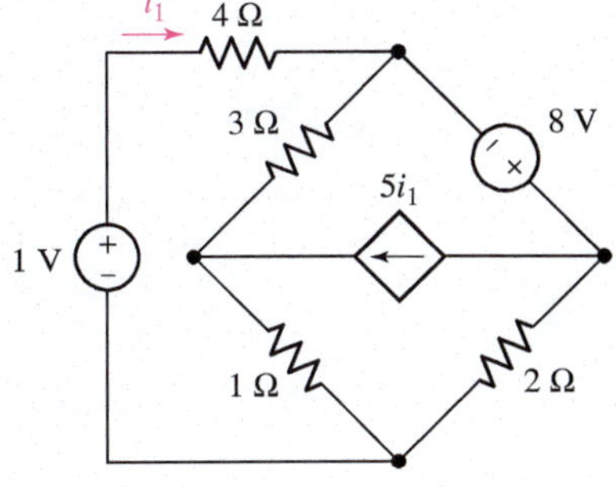

■ **FIGURE 4.77**

50. Determine the power supplied by the 1 V source in Fig. 4.77.

51. Define three clockwise mesh currents for the circuit of Fig. 4.78, and employ the supermesh technique to obtain a value for v_3.

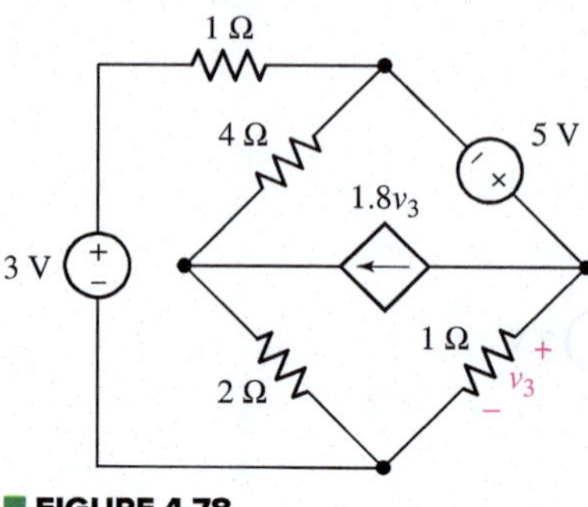

■ **FIGURE 4.78**

52. Determine the power absorbed by the 10 Ω resistor in Fig. 4.79.

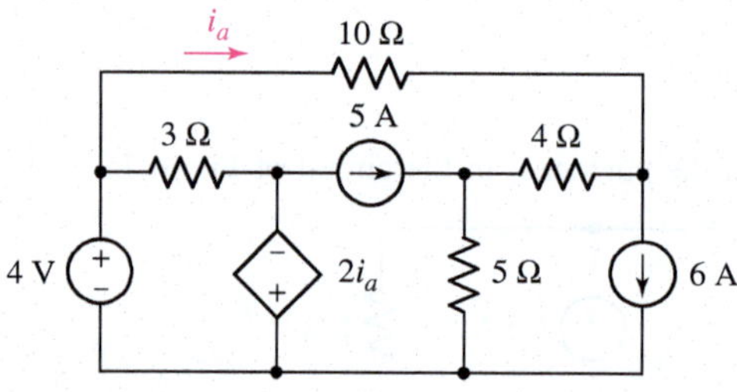

■ **FIGURE 4.79**

4.5 Nodal vs. Mesh Analysis: A Comparison

53. For the circuit represented schematically in Fig. 4.80: (*a*) How many nodal equations would be required to determine i_5? (*b*) Alternatively, how many mesh equations would be required? (*c*) Would your preferred analysis method change if only the voltage across the 7 Ω resistor were needed? *Explain.*

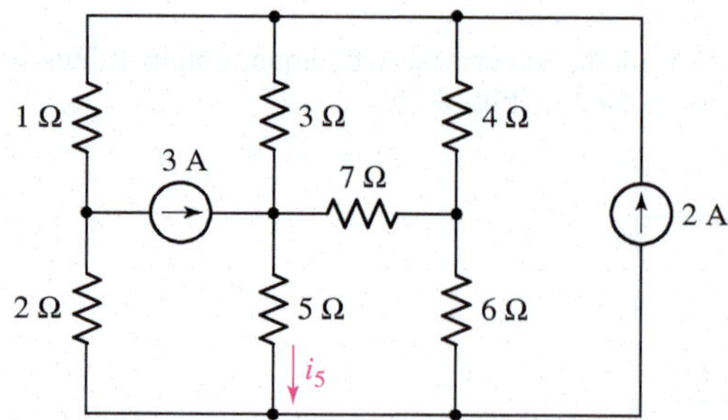

■ **FIGURE 4.80**

54. The circuit of Fig. 4.80 is modified such that the 3 A source is replaced by a 3 V source whose positive reference terminal is connected to the 7 Ω resistor. (*a*) Determine the number of nodal equations required to determine i_5. (*b*) Alternatively, how many mesh equations would be required? (*c*) Would your preferred analysis method change if only the voltage across the 7 Ω resistor were needed? *Explain.*

55. The circuit of Fig. 4.81 contains three sources. (*a*) As it is now drawn, would nodal or mesh analysis result in fewer equations to determine the voltages v_1 and v_2? *Explain.* (*b*) If the voltage source were replaced with current sources, and the current source replaced with a voltage source, would your answer to part (*a*) change? *Explain.*

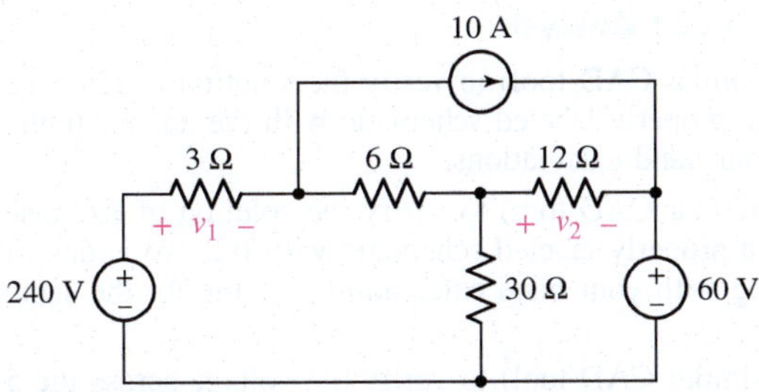

FIGURE 4.81

56. Solve for the voltage v_x as labeled in the circuit of Fig. 4.82 using (*a*) mesh analysis. (*b*) Repeat using nodal analysis. (*c*) Which approach was easier, and why?

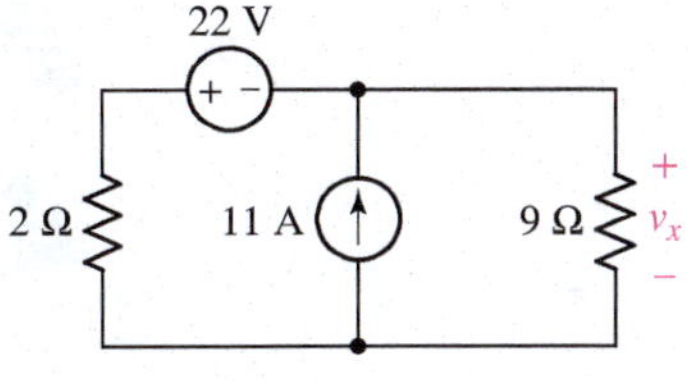

FIGURE 4.82

57. Consider the five-source circuit of Fig. 4.83. Determine the total number of simultaneous equations that must be solved in order to determine v_1 using (*a*) nodal analysis; (*b*) mesh analysis. (*c*) Which method is preferred, and does it depend on which side of the 40 Ω resistor is chosen as the reference node? *Explain your answer.*

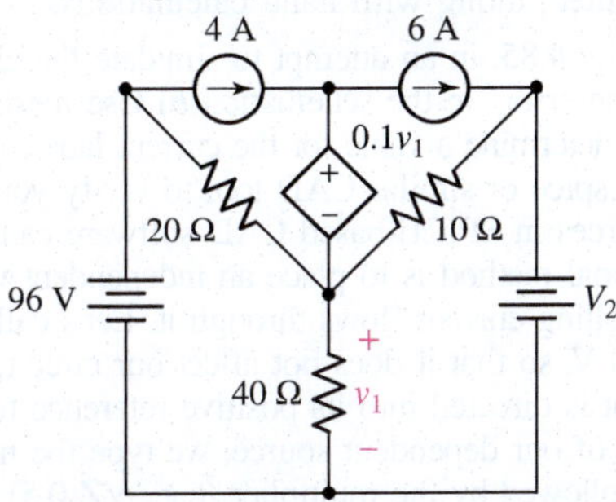

FIGURE 4.83

58. Replace the dependent voltage source in the circuit of Fig. 4.83 with a dependent current source oriented such that the arrow points upward. The controlling expression 0.1 v_1 remains unchanged. The value V_2 is zero. (*a*) Determine the total number of simultaneous equations required to obtain the power dissipated by the 40 Ω resistor if nodal analysis is employed. (*b*) Is mesh analysis preferred instead? *Explain.*

59. After studying the circuit of Fig. 4.84, determine the total number of simultaneous equations that must be solved to determine voltages v_1 and v_3 using (*a*) nodal analysis; (*b*) mesh analysis.

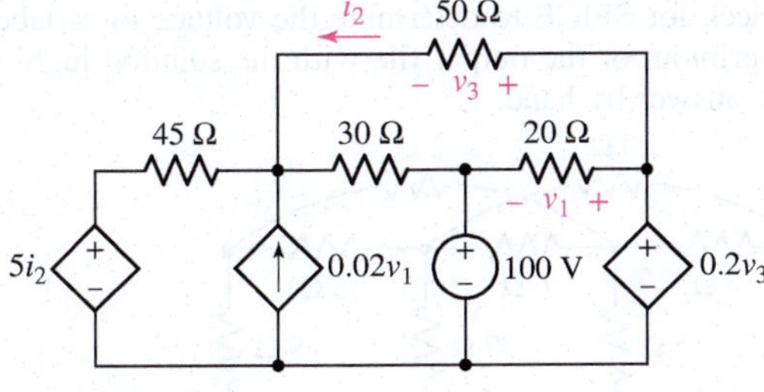

FIGURE 4.84

60. From the perspective of determining voltages and currents associated with all components, (*a*) design a five-node, four-mesh circuit that is analyzed more easily using nodal techniques. (*b*) Modify your circuit by replacing only one component such that it is now more easily analyzed using mesh techniques.

4.6 Computer-Aided Circuit Analysis

61. Employ LTspice (or similar CAD tool) to verify the solution of Exercise 5. Submit a printout of a properly labeled schematic with the answer highlighted, along with your hand calculations.

62. Employ LTspice (or similar CAD tool) to verify the solution of Exercise 8. Submit a printout of a properly labeled schematic with the two nodal voltages highlighted, along with your hand calculations solving for the same quantities.

63. Employ LTspice (or similar CAD tool) to verify the voltage across the 5 Ω resistor in the circuit of Exercise 12. Submit a printout of a properly labeled schematic with the answer highlighted, along with your hand calculations.

64. Verify numerical values for each nodal voltage in Exercise 14 by employing LTspice or a similar CAD tool. Submit a printout of an appropriately labeled schematic with the nodal voltages highlighted, along with your hand calculations.

65. Verify the numerical values for i_1 and v_x as indicated in Fig. 4.46, using LTspice or a similar CAD tool. Submit a printout of a properly labeled schematic with the answers highlighted, along with hand calculations.

66. An LTspice schematic is shown in Fig. 4.85, in an attempt to simulate the circuit in Fig. 4.55. (*a*) Specify all of the errors in the schematic. (*b*) Use mesh analysis on the circuit of Fig. 4.55 to determine a value for the current labeled i_1. (*c*) Correct the schematic and use LTspice or similar CAD tool to verify your analysis. *Hint*: current-controlled sources in SPICE-based CAD software can be handled in several ways. One traditional method is to place an independent voltage source inline such that the controlling current flows through it. Let's call that source VZ. We set its value to 0 V, so that it does not affect our circuit, and place it so the controlling current is directed into its positive reference terminal. Then, for the parameter value of our dependent source, we type the name of the source, followed by a space, followed by the multiplier (e.g., VZ 0.5).

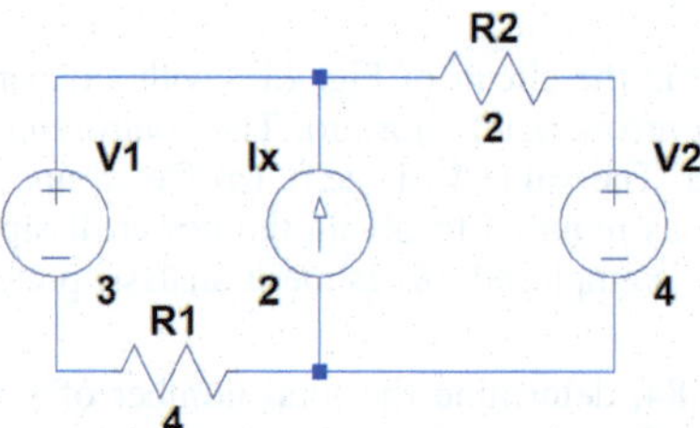

FIGURE 4.85

67. (*a*) Generate an input deck for SPICE to determine the voltage v_9 as labeled in Fig. 4.86. Submit a printout of the output file with the solution highlighted. (*b*) Verify your answer by hand.

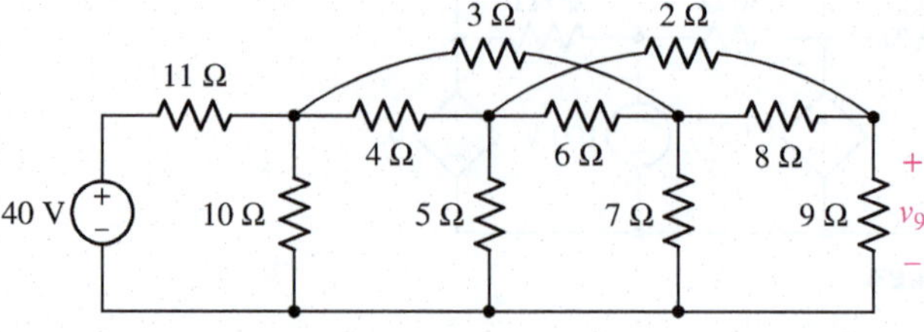

FIGURE 4.86

Chapter-Integrating Exercises

68. A decorative string of multicolored outdoor lights is installed on a home in a quiet residential area. After plugging the 12 V ac adapter into the electrical socket, the homeowner immediately notes that two bulbs are burned out. (*a*) Are the individual lights connected in series or parallel? *Explain.* (*b*) Simulate the string by writing a SPICE input deck, assuming 44 lights, 12 V dc power supply, 24 AWG soft solid copper wire, and individual bulbs rated at 10 mW each. Submit a printout of the output file, with the power supplied by the 12 V supply highlighted. (*c*) Verify your simulation with hand calculations.

69. Consider the circuit depicted in Fig. 4.87. Employ either nodal or mesh analysis as a design tool to obtain a value of 200 mA for i_1, if elements *A*, *B*, *C*, *D*, *E*, and *F* must be either current or voltage sources with nonzero values.

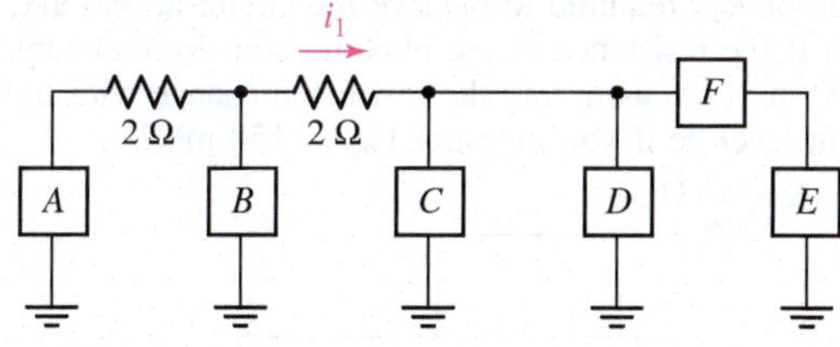

■ **FIGURE 4.87**

70. (*a*) Under what circumstances does the presence of an independent voltage source greatly simplify nodal analysis? *Explain.* (*b*) Under what circumstances does the presence of an independent current source significantly simplify mesh analysis? *Explain.* (*c*) On which fundamental physical principle do we base nodal analysis? (*d*) On which fundamental physical principle do we base mesh analysis?

71. Referring to Fig. 4.88, (*a*) determine whether nodal or mesh analysis is more appropriate in determining i_2 if element *A* is replaced with a short circuit, then carry out the analysis. (*b*) Verify your answer with an appropriate LTspice simulation. Submit a properly labeled schematic along with the answer highlighted.

72. Consider the LED circuit containing a red, green, and blue LED as shown in Fig. 4.89. The LEDs behave much like a voltage source resulting in the circuit in Fig. 4.89, where the light output from each LED will be proportional to the current flowing through the LED. (*a*) Calculate the current flowing through each LED (I_{Red}, I_{Green}, and I_{Blue}) if $R_1 = R_2 = R_3 = 100\ \Omega$. (*b*) Determine the resistor values R_1, R_2, and R_3 needed to ensure that the LEDs each have a current of 4 mA flowing through them.

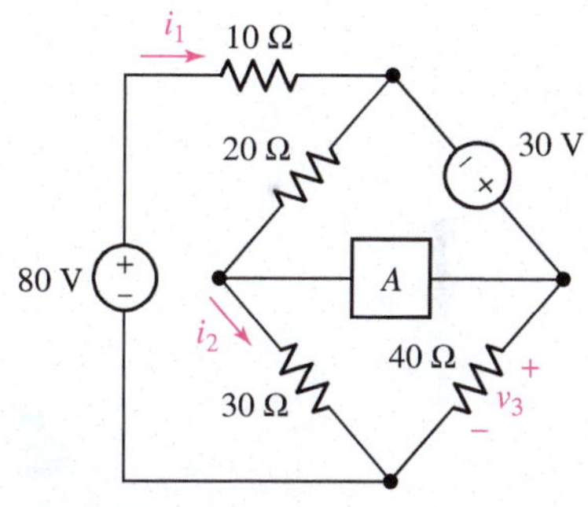

■ **FIGURE 4.88**

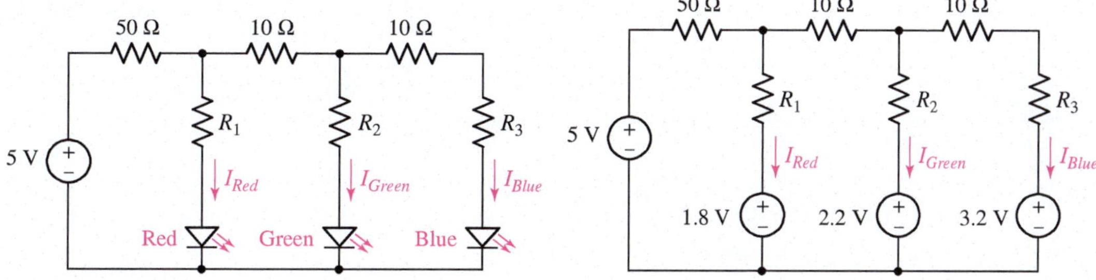

■ **FIGURE 4.89**

73. The LED circuit in Fig. 4.89 is used to mix colors to achieve any desired color in the RGB color palette. Use LTspice and the circuit model representing LEDs as voltage sources to see how changing the resistance R_1 from 100 to 1 kΩ

affects the color, with the other resistors fixed at $R_2 = R_3 = 100\ \Omega$. You can do this using a parameter sweep statement by defining a variable such as **{rvariable}** (including the curly brackets) in the value for R_1. Then include a SPICE directive such as **.step param rvariable 100 1k 20** to step the variable from 100 to 1000 in steps of 20. (*a*) Plot the current of all three LEDs as a function of R_1 and explain the result. (*b*) Find an RGB color chart and describe how the color changes with R_1, increasing from 100 to 1 kΩ. (*c*) Find a value of R_1 that could be used to achieve a khaki color approximating RGB hex code C2BD23, RGB (194,189,35).

74. A light-sensing circuit is in Fig. 4.90, including a resistor that changes value under illumination (photoresistor R_{light}) and a variable resistor (potentiometer R_{pot}). The circuit is in the Wheatstone bridge configuration such that a "balanced" condition results in $V_{\text{out}} = 0$ for a defined value of incident light and a corresponding value for R_{light}. (*a*) Derive an algebraic expression for V_{out} in terms of R_S, R_1, R_2, R_{light}, and R_{pot}. (*b*) Using the numerical values given in the circuit, calculate the value of R_{pot} required to balance the circuit at 500 lux, where $R_{\text{light}} = 200\ \Omega$. (*c*) If the resistance of the photoresistor decreases by 2% for a light increase to 600 lux (and assuming the resistance change with light is linear), what will the light level be if you measure $V_{\text{out}} = 150$ mV?

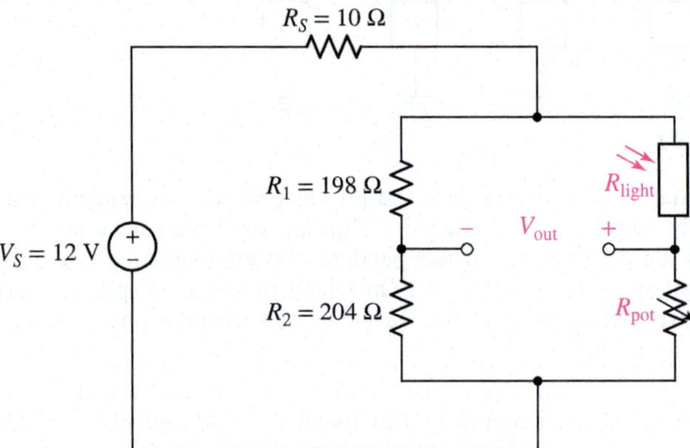

FIGURE 4.90

75. Use SPICE to analyze the circuit in Exercise 74 by doing the following. (*a*) Simulate the circuit for varying values of R_{pot} to balance the circuit at 500 lux, where $R_{\text{light}} = 200\ \Omega$. It is helpful to use a parameter sweep by defining a variable such as **{potentiometer}** (including the curly brackets) in the value for R_{pot}, and a SPICE directive such as **.step param potentiometer 150 250 2** to step the variable from 150 to 250 in steps of 2. (*b*) If the resistance of the photoresistor decreases by 2% for a light increase to 600 lux, use SPICE to find the resulting output voltage V_{out}.

CHAPTER 5

Handy Circuit Analysis Techniques

KEY CONCEPTS

Superposition: Determining the *Individual Contributions* of Different Sources

Source Transformation for Simplifying Circuits

Thévenin Equivalents

Norton Equivalents

Thévenin and Norton Equivalent Networks

Maximum Power Transfer

$\Delta \leftrightarrow$ Y Transformations for Resistive Networks

Choosing a Specific Analysis Technique

DC Sweep Simulations

INTRODUCTION

The techniques of nodal and mesh analysis described in Chap. 4 are reliable and extremely powerful methods. However, both require that we develop a complete set of equations to describe a particular circuit as a general rule, even if only one current, voltage, or power quantity is of interest. In this chapter, we investigate a variety of different techniques for isolating specific parts of a circuit in order to simplify the analysis. After examining each of these techniques, we focus on how one might go about selecting one method over another.

5.1 LINEARITY AND SUPERPOSITION

All of the circuits we plan to analyze can be classified as *linear circuits,* so this is a good time to be more specific in defining exactly what we mean by that. Having done this, we can then consider the most important consequence of linearity: the principle of ***superposition.*** This concept will appear repeatedly in our study of linear circuit analysis, as well as in many other fields. As a matter of fact, the nonapplicability of superposition to nonlinear circuits is the very reason they can be so challenging to analyze.

The principle of superposition states that the *response* (a desired current or voltage) in a linear circuit having more than one independent source can be obtained by adding the responses caused by the separate independent sources *acting alone.*

Linear Elements and Linear Circuits

We define a ***linear element*** as a passive element that has a linear voltage–current relationship. By a "linear voltage–current relationship" we mean that multiplication of the current through the element by a constant K results in the multiplication of the voltage across the element by the same constant K. So far, we have encountered only one passive element (the resistor), and its voltage–current relationship

$$v(t) = Ri(t)$$

is clearly linear. As a matter of fact, if $v(t)$ is plotted as a function of $i(t)$, the result is a straight line.

The dependent voltage source $v_s = 0.6i_1 - 14v_2$ is linear, but $v_s = 0.6i_1^2$ and $v_s = 0.6i_1v_2$ are not.

We define a ***linear dependent source*** as a dependent current or voltage source whose output current or voltage is proportional only to the first power of a specified current or voltage variable in the circuit (or to the *sum* of such quantities).

We now define a ***linear circuit*** as a circuit composed entirely of independent sources, linear dependent sources, and linear elements. From this definition, it is possible to show[1] that "the response is proportional to the source," or that multiplication of *all* independent source voltages and currents by a constant K increases *all* the current and voltage responses by the same factor K (including the voltage or current output of any dependent sources).

The Superposition Principle

The most important consequence of linearity is ***superposition.***

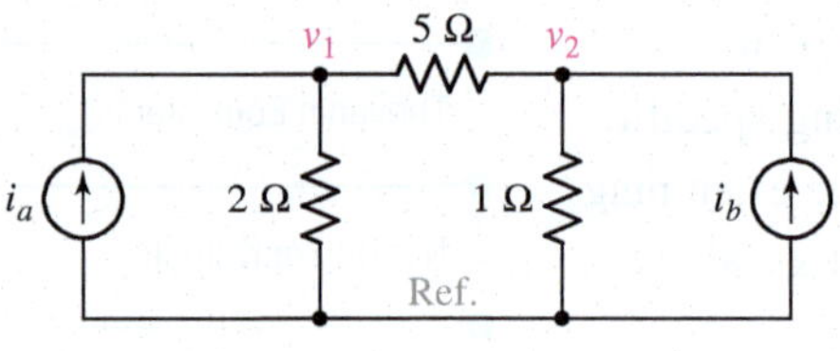

FIGURE 5.1 A circuit with two independent current sources.

Let us explore the superposition principle by considering first the circuit of Fig. 5.1, which contains two independent sources, the current generators that force the currents i_a and i_b into the circuit. Sources are often called *forcing functions* for this reason, and the nodal voltages that they produce can be termed *response functions,* or simply *responses.* Both the forcing functions and the responses may be functions of time. The two nodal equations for this circuit are

$$0.7v_1 - 0.2v_2 = i_a \qquad [1]$$
$$-0.2v_1 + 1.2v_2 = i_b \qquad [2]$$

Now let us perform an experiment. We change the two forcing functions to i_{ax} and i_{bx}; the two unknown voltages will now be different, so we will call them v_{1x} and v_{2x}. Thus,

$$0.7v_{1x} - 0.2v_{2x} = i_{ax} \qquad [3]$$
$$-0.2v_{1x} + 1.2v_{2x} = i_{bx} \qquad [4]$$

If we next perform another experiment by changing the source currents to i_{ay} and i_{by} and measure the responses v_{1y} and v_{2y}, we obtain:

$$0.7v_{1y} - 0.2v_{2y} = i_{ay} \qquad [5]$$
$$-0.2v_{1y} + 1.2v_{2y} = i_{by} \qquad [6]$$

These three sets of equations describe the same circuit with three different sets of source currents.

(1) The proof involves first showing that the use of nodal analysis on the linear circuit can produce only linear equations of the form

$$a_1v_1 + a_2v_2 + \cdots + a_Nv_N = b$$

where the a_i are constants (combinations of resistance or conductance values, constants appearing in dependent source expressions, 0, or ±1), the v_i are the unknown node voltages (responses), and b is an independent source value or a sum of independent source values. Given a set of such equations, if we multiply all the b's by K, then it is evident that the solution of this new set of equations will be the node voltages $Kv_1, Kv_2, \ldots, Kv_N$.

Let us *add* or "*superpose*" the last two sets of equations. Adding Eqs. [3] and [5],

$$\begin{array}{ccccccc} (0.7v_{1x} + 0.7v_{1y}) & - & (0.2v_{2x} + 0.2v_{2y}) & = & i_{ax} + i_{ay} & & [7] \\ 0.7v_1 & - & 0.2v_2 & = & i_a & & [1] \end{array}$$

and adding Eqs. [4] and [6],

$$\begin{array}{ccccccc} -(0.2v_{1x} + 0.2v_{1y}) & + & (1.2v_{2x} + 1.2v_{2y}) & = & i_{bx} + i_{by} & & [8] \\ -0.2v_1 & + & 1.2v_2 & = & i_b & & [2] \end{array}$$

where Eq. [1] has been written immediately below Eq. [7] and Eq. [2] below Eq. [8] for easy comparison.

The linearity of all these equations allows us to draw an interesting conclusion. If we select i_{ax} and i_{ay} such that their sum is i_a and select i_{bx} and i_{by} such that their sum is i_b, then the desired responses v_1 and v_2 may be found by adding v_{1x} to v_{1y} and v_{2x} to v_{2y}, respectively. In other words, we can perform the first experiment and note the responses, perform the next experiment and note the responses, and finally add the two sets of responses. This leads to the fundamental concept involved in the superposition principle: to look at each independent source (and the response it generates) one at a time with the other independent sources "turned off" or "zeroed out."

If we reduce a voltage source to zero volts, we have effectively made it into a short circuit (Fig. 5.2*a*). If we reduce a current source to zero amperes, we have effectively created an open circuit (Fig. 5.2*b*). Thus, the ***superposition theorem*** can be stated as:

> In any linear resistive network, the voltage across or the current through any resistor or source may be calculated by adding algebraically all the individual voltages or currents caused by the separate **independent** sources acting alone, with all other independent voltage sources replaced by short circuits and all other independent current sources replaced by open circuits.

Thus, if there are N independent sources, we must perform N experiments, each having only one of the independent sources active and the others inactive/turned off/zeroed out. Note that *dependent* sources are in general active in every experiment.

There is also no reason that an independent source must assume only its given value or a zero value in the several experiments; it is necessary

■ **FIGURE 5.2** (*a*) A voltage source set to zero acts like a short circuit. (*b*) A current source set to zero acts like an open circuit. In each case, the equivalent representation on the right can be substituted into a circuit schematic to aid in analysis.

only for the sum of the several values to be equal to the original value. (An inactive source almost always leads to the simplest circuit, however.)

The circuit we have just used as an example should indicate that a much stronger theorem might be written; a *group* of independent sources may be made active and inactive collectively, if we wish. For example, suppose there are three independent sources. The theorem states that we may find a given response by considering each of the three sources acting alone and adding the three results. Alternatively, we may find the response due to the first and second sources operating with the third inactive, and then add to this the response caused by the third source acting alone.

EXAMPLE 5.1

For the circuit of Fig. 5.3*a*, use superposition to determine the unknown branch current i_x.

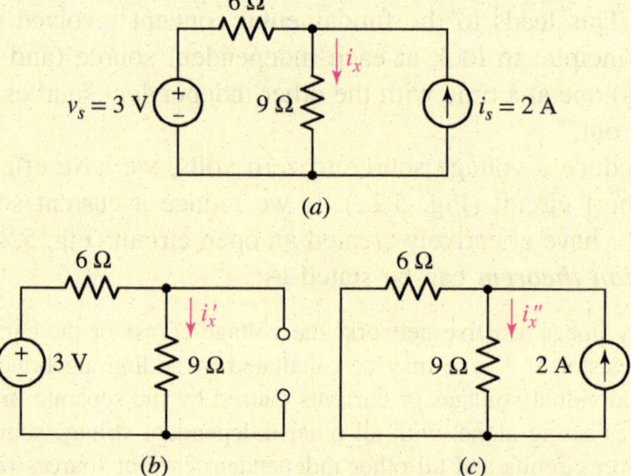

FIGURE 5.3 (*a*) An example circuit with two independent sources for which the branch current i_x is desired; (*b*) same circuit with current source open-circuited; (*c*) original circuit with voltage source short-circuited.

We first set the current source equal to zero and redraw the circuit as shown in Fig. 5.3*b*, where the deactivated current source is represented by an open circuit. The portion of i_x due to the voltage source has been designated i'_x to avoid confusion and is easily found to be 0.2 A.

Next we set the voltage source in Fig. 5.3*a* to zero and again redraw the circuit, as shown in Fig. 5.3*c*. We have replaced the deactivated voltage source with a short circuit in the schematic. Current division lets us determine that i''_x (the portion of i_x due to the 2 A current source) is 0.8 A.

Now we can compute the total current i_x by adding the two individual components:

$$i_x = i_{x|3\text{ V}} + i_{x|2\text{ A}} = i'_x + i''_x$$

or

$$i_x = \frac{3}{6+9} + 2\left(\frac{6}{6+9}\right) = 0.2 + 0.8 = 1.0 \text{ A}$$

PRACTICE

5.1 For the circuit of Fig. 5.4, use superposition to compute the current i_x.

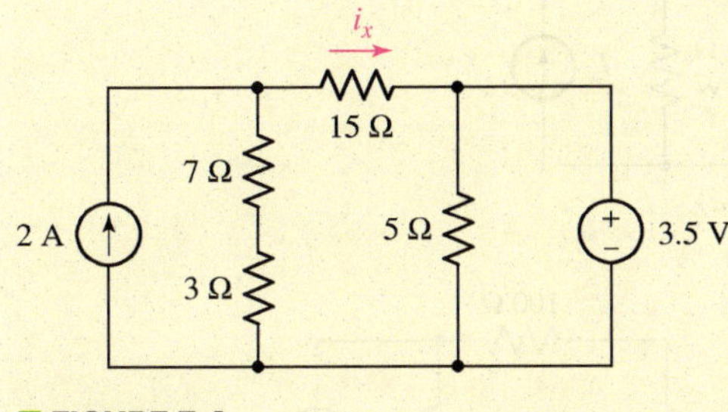

FIGURE 5.4

Ans: 660 mA.

Another way of looking at Example 5.1 is that the 3 V source and the 2 A source are each performing work on the circuit, resulting in a total current i_x flowing through the 9 Ω resistor. *However, the contribution of the 3 V source to i_x does not depend on the contribution of the 2 A source, and vice versa.* For example, if we double the output of the 2 A source to 4 A, it will now contribute 1.6 A to the total current i_x flowing through the 9 Ω resistor. However, the 3 V source will still contribute only 0.2 A to i_x, for a new total current of $0.2 + 1.6 = 1.8$ A.

As we will see, superposition does not generally reduce our workload when considering a particular circuit, since it leads to the analysis of several new circuits to obtain the desired response. However, it is particularly useful in identifying the significance of various parts of a more complex circuit, and in separating dc sources and their effects from those of time-varying sources—something often done in transistor circuit analysis. It also forms the basis of phasor analysis, which is introduced in Chap. 10.

EXAMPLE 5.2

Referring to the circuit of Fig. 5.5*a*, determine the maximum *positive* current to which the source I_x can be set before any resistor exceeds its power rating and overheats.

▶ ***Identify the goal of the problem.***

Each resistor is rated to a maximum of 250 mW. If the circuit allows this value to be exceeded (by forcing too much current through either resistor), excessive heating will occur—possibly leading to an accident. The 6 V source cannot be changed, so we are looking for an equation involving I_x and the maximum current through each resistor.

▶ ***Collect the known information.***

Based on its 250 mW power rating, the maximum current the 100 Ω resistor can tolerate is

$$\sqrt{\frac{P_{\text{max}}}{R}} = \sqrt{\frac{0.250}{100}} = 50 \text{ mA}$$

and, similarly, the current through the 64 Ω resistor must be less than 62.5 mA.

(Continued on next page)

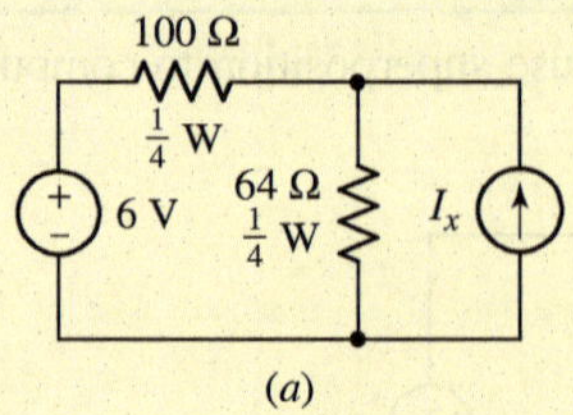

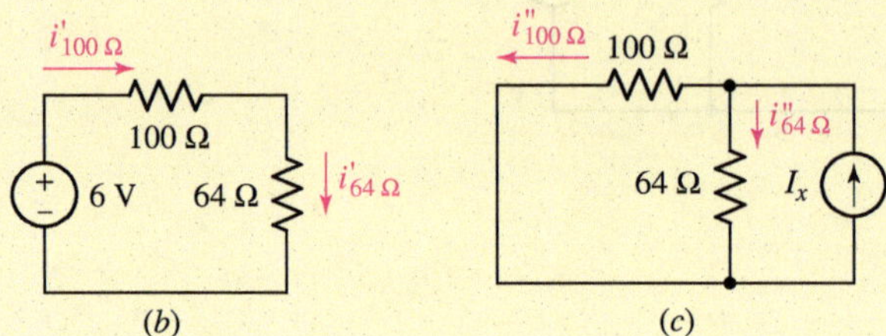

■ **FIGURE 5.5** (*a*) A circuit with two resistors each rated at $\frac{1}{4}$ W. (*b*) Circuit with only the 6 V source active. (*c*) Circuit with the source I_x active.

▶ ***Devise a plan.***

Either nodal or mesh analysis may be applied to the solution of this problem, but superposition may give us a slight edge, since we are primarily interested in the effect of the current source.

▶ ***Construct an appropriate set of equations.***

Using superposition, we redraw the circuit as in Fig. 5.5*b* and find that the 6 V source contributes a current

$$i'_{100\,\Omega} = \frac{6}{100 + 64} = 36.59 \text{ mA}$$

to the 100 Ω resistor and, since the 64 Ω resistor is in series here, $i'_{64\,\Omega} = 36.59$ mA as well.

Recognizing the current divider in Fig. 5.5*c*, we note that $i''_{64\,\Omega}$ will *add* to $i'_{64\,\Omega}$, but $i''_{100\,\Omega}$ is *opposite* in direction to $i'_{100\,\Omega}$. Therefore, I_X can safely contribute 62.5 − 36.59 = 25.91 mA to the 64 Ω resistor current, and 50 − (−36.59) = 86.59 mA to the 100 Ω resistor current.

The 100 Ω resistor therefore places the following constraint on I_x:

$$I_x < (86.59 \times 10^{-3})\left(\frac{100 + 64}{64}\right)$$

and the 64 Ω resistor requires that

$$I_x < (25.91 \times 10^{-3})\left(\frac{100 + 64}{100}\right)$$

▶ ***Attempt a solution.***

Considering the 100 Ω resistor first, we see that I_x is limited to $I_x < 221.9$ mA. The 64 Ω resistor limits I_x such that $I_x < 42.49$ mA. In order to satisfy both constraints, I_x must be less than 42.49 mA. If the value is increased, the 64 Ω resistor will overheat before the 100 Ω resistor does.

▶ ***Verify the solution. Is it reasonable or expected?***

One particularly useful way to evaluate our solution is to perform a dc sweep analysis in LTspice as described after the next example. An interesting question, however, is whether we would have expected the 64 Ω resistor to overheat first.

Originally we found that the 100 Ω resistor has a smaller maximum current, so it might be reasonable to expect it to limit I_x. However, because I_x *opposes* the current sent by the 6 V source through the 100 Ω resistor but *adds* to the 6 V source's contribution to the current through the 64 Ω resistor, it turns out to work the other way—it's the 64 Ω resistor that sets the limit on I_x.

EXAMPLE 5.3

In the circuit of Fig. 5.6*a*, employ superposition to determine the value of i_x.

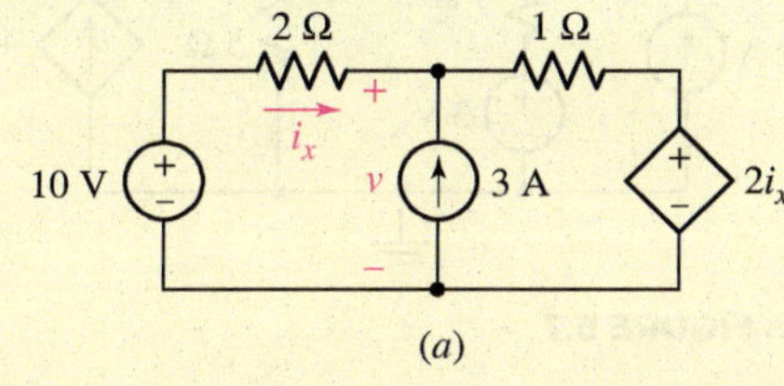

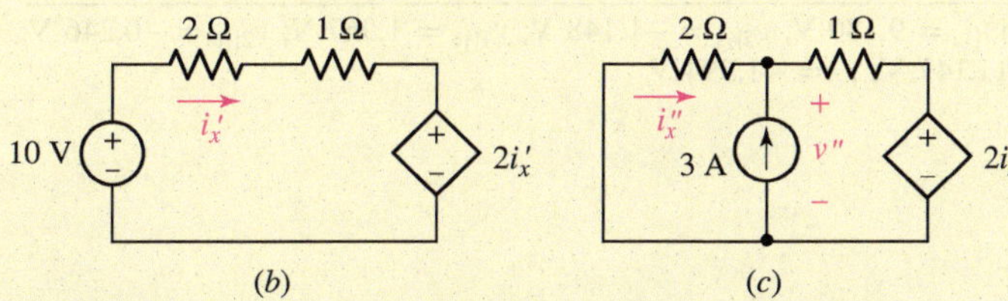

■ **FIGURE 5.6** (*a*) An example circuit with two independent sources and one dependent source for which the branch current i_x is desired. (*b*) Circuit with the 3 A source open-circuited. (*c*) Original circuit with the 10 V source short-circuited.

First open-circuit the 3 A source (Fig. 5.6*b*). The single mesh equation is

$$-10 + 2i'_x + i'_x + 2i'_x = 0$$

so that

$$i'_x = 2 \text{ A}$$

Next, short-circuit the 10 V source (Fig. 5.6*c*) and write the single-node equation

$$\frac{v''}{2} + \frac{v'' - 2i''_x}{1} = 3$$

and relate the dependent-source-controlling quantity to v'':

$$v'' = 2(-i''_x)$$

(Continued on next page)

Solving, we find

$$i_x'' = -0.6 \text{ A}$$

and, thus,

$$i_x = i_x' + i_x'' = 2 + (-0.6) = 1.4 \text{ A}$$

Note that in redrawing each subcircuit, we are always careful to use some type of notation to indicate that we are not working with the original variables. This prevents the possibility of rather disastrous errors when we add the individual results.

PRACTICE

5.2 For the circuit of Fig. 5.7, use superposition to obtain the voltage across each current source.

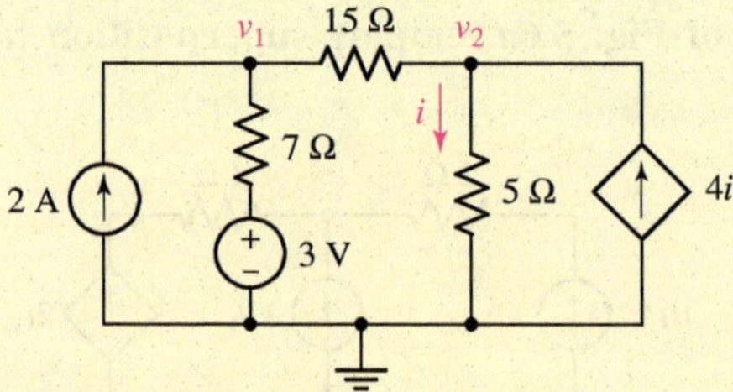

■ **FIGURE 5.7**

Ans: $v_{1|_{2A}} = 9.180$ V; $v_{2|_{2A}} = -1.148$ V; $v_{1|_{3V}} = 1.967$ V; $v_{2|_{3V}} = -0.246$ V; $v_1 = 11.147$ V; $v_2 = -1.394$ V.

Summary of Basic Superposition Procedure

1. **Select one of the independent sources. Set all other independent sources to zero.** This means voltage sources are replaced with short circuits and current sources are replaced with open circuits. Leave dependent sources in the circuit.
2. **Relabel voltages and currents using suitable notation** (e.g., v', i_2''). Be sure to also relabel controlling variables of dependent sources to avoid confusion.
3. **Analyze the simplified circuit to find the desired currents and/or voltages.**
4. **Repeat steps 1 through 3 until each independent source has been considered.**
5. **Add the partial currents and/or voltages obtained from the separate analyses.** Pay careful attention to voltage signs and current directions when summing.
6. **Do not add power quantities.** Calculate power quantities only after partial voltages and/or currents have been summed.

Note that step 1 may be altered in several ways. First, independent sources can be considered in groups as opposed to individually if it simplifies the analysis, as long as no independent source is included in more than one subcircuit. Second, it is technically not necessary to set sources to zero, although this is almost always the best route. For example, a 3 V source may appear in two subcircuits as a 1.5 V source, since 1.5 + 1.5 = 3 V just as 0 + 3 = 3 V. Because it is unlikely to simplify our analysis, however, there may be little benefit to such an exercise.

COMPUTER-AIDED ANALYSIS

LTspice is extremely useful in verifying that we have analyzed a complete circuit correctly, but it can also assist us in determining the contribution of each source to a particular response. To do this, we employ what is known as a *dc parameter sweep*.

Consider the circuit presented in Example 5.2, when we were asked to determine the maximum positive current that could be obtained from the current source without exceeding the power rating of either resistor in the circuit. The circuit is shown redrawn in Fig. 5.8 using the LTspice schematic capture tool within a Windows environment. Note that no value has been assigned to the current source.

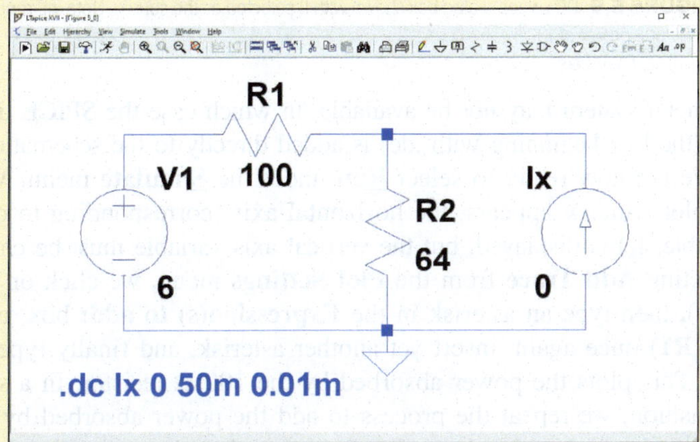

FIGURE 5.8 The circuit from Example 5.2.
Analog Devices

After the schematic has been entered and saved, the next step is to specify the dc sweep parameters. This option allows us to specify a range of values for a voltage or current source (in the present case, the current source I_x), rather than a specific value. Selecting **SPICE Analysis** under **Edit,** we are provided with the dialog box shown in Fig. 5.9.

Next, we select the **DC sweep** tab, choose the **1st Source** tab, and then type I_x in the **Name of 1st source to sweep:** box. There are several options under **Type of sweep: Linear, Octave, Decade,** and **List.** The last option allows us to specify each value to assign to I_x. In order to generate a smooth plot, however, we choose to perform a **Linear** sweep, with a **Start value** of 0 mA, a **Stop Value** of 50 mA, and a value of 0.01 mA for the **Increment.** Note that in a non-Windows environment,

(Continued on next page)

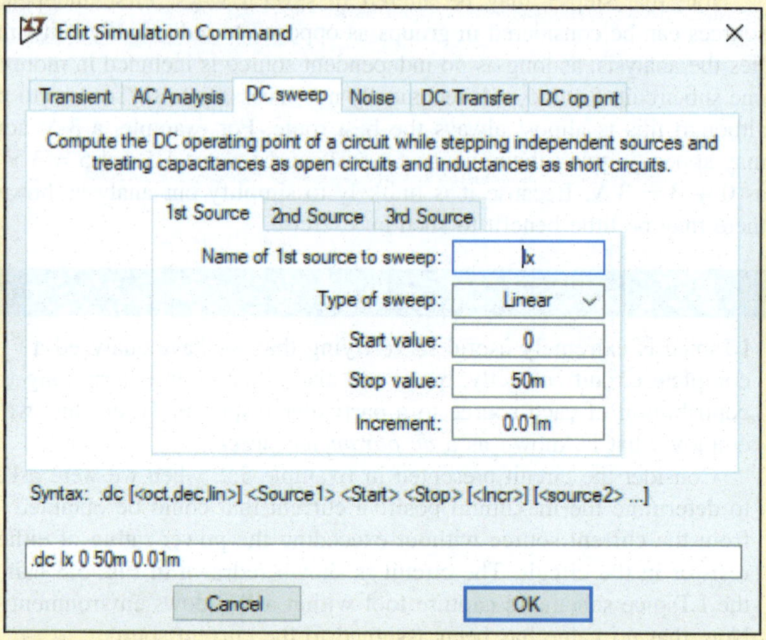

FIGURE 5.9 DC sweep dialog box shown with I_x selected as the source. Note 'm' represents 'milli' or a power of 10^{-3}.
Analog Devices

the menu system may not be available, in which case the SPICE directive (the line beginning with .dc) is added directly to the schematic.

We are now ready to select **Run** under the **Simulate** menu. When the plot window appears, the horizontal axis (corresponding to our variable, I_x) is displayed, but the vertical axis variable must be chosen. Selecting **Add Trace** from the **Plot Settings** menu, we click on **I(R1),** then type an asterisk in the **Expression(s) to add:** box, click on **I(R1)** once again, insert yet another asterisk, and finally type in 100. This plots the power absorbed by the 100 Ω resistor. In a similar fashion, we repeat the process to add the power absorbed by the 64 Ω resistor, resulting in a plot similar to that shown in Fig. 5.10*a*. A horizontal reference line at 250 mW was also added to the plot by typing 0.25 in the **Expression(s) to add:** box after selecting **Add Trace** from the **Plot Settings** menu a third time. We should note that there are two *y* axes. The one on the left corresponds to our 250 mW baseline, and we had to manually adjust the limits for it to correspond to the scale of the right-hand axis, which defaulted to units of mA^2 since two currents were multiplied in the expression.

We see from the plot that the 64 Ω resistor *does* exceed its 250 mW power rating in the vicinity of $I_x = 43$ mA. In contrast, however, we also see that regardless of the value of the current source I_x (provided that it is between 0 and 50 mA), the 100 Ω resistor will never dissipate 250 mW; in fact, the absorbed power *decreases* with increasing current from the current source. If we want a more precise answer, we can make use of the cursor tool, which is invoked by selecting the expression

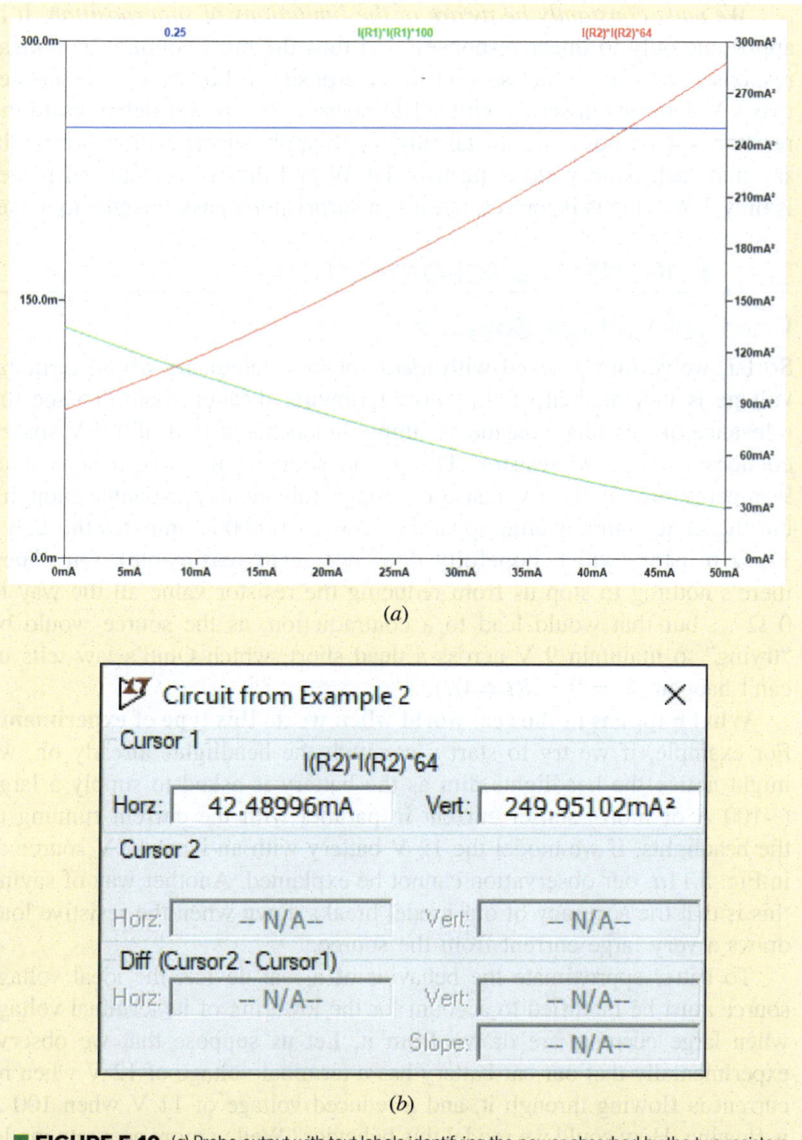

(*a*)

(*b*)

■ **FIGURE 5.10** (*a*) Probe output with text labels identifying the power absorbed by the two resistors individually. A horizontal line indicating 250 mW has also been included, as well as text labels to improve clarity. (*b*) Cursor dialog box.
(*b*) Analog Devices

of interest from the top of the plot window. Figure 5.10*b* shows the result of dragging the cursor to where the 64 Ω resistor is operating at its maximum rated power of 250 mW. Increased precision can be obtained by decreasing the increment value used in the dc sweep.

Unfortunately, it usually turns out that little if any time is saved in analyzing a circuit containing one or more dependent sources by use of the superposition principle, for there must always be at least two sources in operation: one independent source and all the dependent sources.

We must constantly be aware of the limitations of superposition. It is applicable only to linear responses, and thus the most common nonlinear response—power—is not subject to superposition. For example, consider two 1 V batteries in series with a 1 Ω resistor. The power delivered to the resistor is 4 W, but if we mistakenly try to apply superposition, we might say that each battery alone furnished 1 W and thus the calculated power is only 2 W. This is incorrect, but it is a surprisingly easy mistake to make.

5.2 SOURCE TRANSFORMATIONS

Practical Voltage Sources

So far, we've only worked with *ideal* sources—elements whose terminal voltage is independent of the current flowing through them. To see the relevance of this fact, consider a simple independent ("ideal") 9 V source connected to a 1 Ω resistor. The 9 volt source will force a current of 9 amperes through the 1 Ω resistor (perhaps this seems reasonable enough), but the same source would apparently force 9,000,000 amperes through a 1 $\mu\Omega$ resistor (which hopefully does not seem reasonable). On paper, there's nothing to stop us from reducing the resistor value all the way to 0 Ω … but that would lead to a contradiction, as the source would be "trying" to maintain 9 V across a dead short, which Ohm's law tells us can't happen ($V = 9 = RI = 0$?).

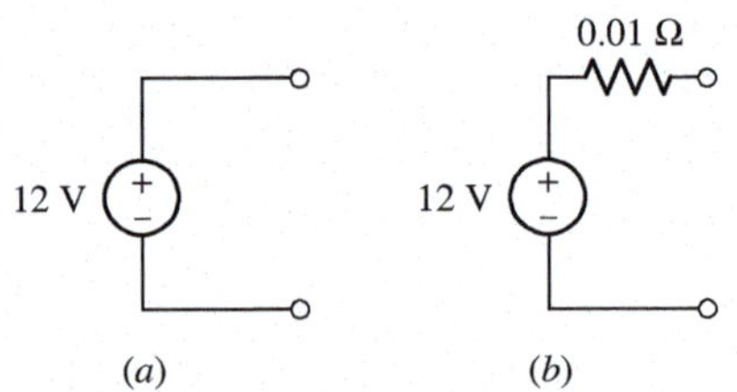

FIGURE 5.11 (*a*) An ideal 12 V dc voltage source used to model a lead-acid car battery. (*b*) A more accurate model that accounts for the observed reduction in terminal voltage at large currents. The series resistance of a "dead" battery increases significantly, allowing us to measure a reasonable voltage at open circuit, but a small voltage when "loaded" by the starter circuit.

What happens in the real world when we do this type of experiment? For example, if we try to start a car with the headlights already on, we might notice the headlights dim as the battery is asked to supply a large (~100 A or more) starter current in parallel with the current running to the headlights. If we model the 12 V battery with an ideal 12 V source as in Fig. 5.11*a*, our observation cannot be explained. Another way of saying this is that the accuracy of our model breaks down when the resistive load draws a very large current from the source.

To better approximate the behavior of a real device, the ideal voltage source must be modified to account for the lowering of its terminal voltage when large currents are drawn from it. Let us suppose that we observe experimentally that our car battery has a terminal voltage of 12 V when no current is flowing through it, and a reduced voltage of 11 V when 100 A is flowing. How could we model this behavior? Well, a more accurate model might be an ideal voltage source of 12 V in series with a resistor across which 1 V appears when 100 A flows through it. A quick calculation shows that the resistor must be 1 V/100 A = 0.01 Ω, and the ideal voltage source and this series resistor constitute a ***practical voltage source*** (Fig. 5.11*b*). Thus, we are using the series combination of two ideal circuit elements, an independent voltage source and a resistor, to model a real device.

We do not expect to find such an arrangement of ideal elements inside our car battery, of course. Any real device is characterized by a certain current–voltage relationship at its terminals, and our problem is to develop some combination of ideal elements that can furnish a similar current–voltage characteristic, at least over some useful range of current, voltage, or power.

In Fig. 5.12*a*, we show our two-piece practical model of the car battery now connected to some load resistor R_L. The terminal voltage of the practical source is the same as the voltage across R_L and is marked[2] V_L. Figure 5.12*b* shows a plot of load voltage V_L as a function of the load current I_L for this practical source. The KVL equation for the circuit of Fig. 5.12*a* may be written in terms of I_L and V_L:

$$12 = 0.01\ I_L + V_L$$

and thus

$$V_L = -0.01\ I_L + 12$$

This is a linear equation in I_L and V_L, and the plot in Fig. 5.12*b* is a straight line. Each point on the line corresponds to a different value of R_L. For example, the midpoint of the straight line is obtained when the load resistance is equal to the internal resistance of the practical source, or $R_L = 0.01\ \Omega$. Here, the load voltage is exactly one-half the ideal source voltage.

When $R_L = \infty$ and no current whatsoever is being drawn by the load, the practical source is open-circuited and the terminal voltage, or open-circuit voltage, is $V_{Loc} = 12$ V. If, on the other hand, $R_L = 0$, thereby short-circuiting the load terminals, then a load current or short-circuit current, $I_{Lsc} = 1200$ A, would flow. (*In practice, such an experiment would probably result in the destruction of the short circuit, the battery, and any measuring instruments incorporated in the circuit!*)

Since the plot of V_L versus I_L is a straight line for our practical voltage source, we should note that the values of V_{Loc} and I_{Lsc} uniquely determine its entire V_L–I_L curve.

The horizontal broken line of Fig. 5.12*b* represents the V_L–I_L plot for an *ideal* voltage source; the terminal voltage remains constant for any value of load current. For the practical voltage source, the terminal voltage has a value near that of the ideal source only when the load current is relatively small.

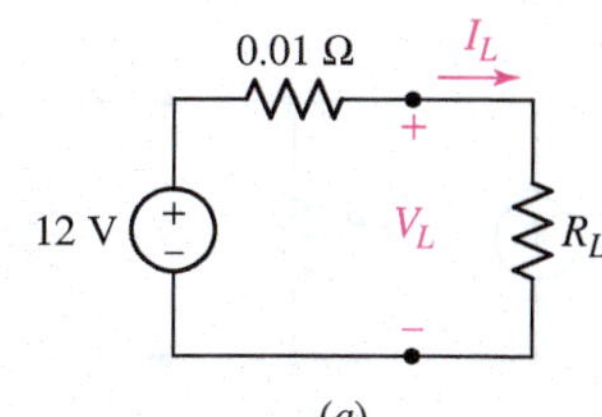

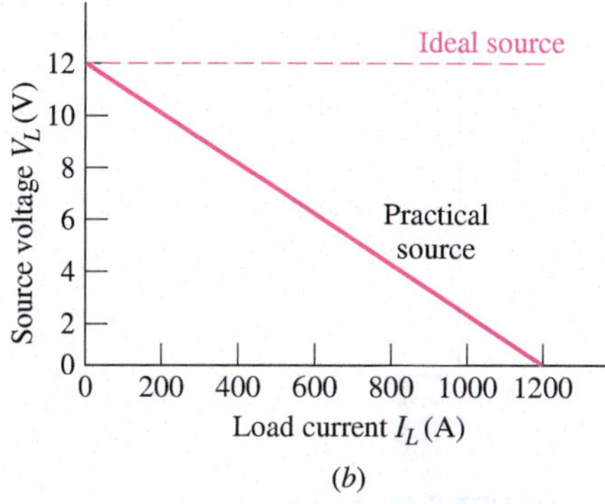

FIGURE 5.12 (*a*) A practical source, which approximates the behavior of a certain 12 V automobile battery, is shown connected to a load resistor R_L. (*b*) The relationship between I_L and V_L is linear.

Let us now consider a *general* practical voltage source, as shown in Fig. 5.13*a*. The voltage of the ideal source is v_s, and a resistance R_s, often called an *internal resistance* or *output resistance,* is placed in series with it. Again, we must note that the resistor is not really present as a separate component but merely serves to account for a terminal voltage that decreases as the load current increases. Its presence enables us to model the behavior of a physical voltage source more closely.

The linear relationship between v_L and i_L is

$$v_L = v_s - R_s i_L \qquad [9]$$

and this is plotted in Fig. 5.13*b*. The open-circuit voltage ($R_L = \infty$, so $i_L = 0$) is

$$v_{Loc} = v_s \qquad [10]$$

and the short-circuit current ($R_L = 0$, so $v_L = 0$) is

$$i_{Lsc} = \frac{v_s}{R_s} \qquad [11]$$

Once again, these values are the intercepts for the straight line in Fig. 5.13*b*, and they serve to define it completely.

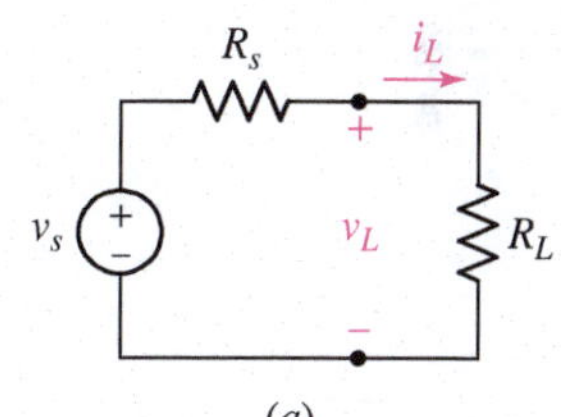

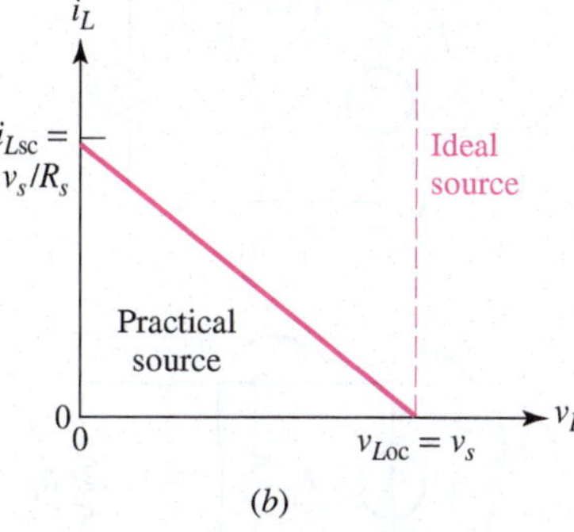

FIGURE 5.13 (*a*) A general practical voltage source connected to a load resistor R_L. (*b*) The terminal voltage of a practical voltage source decreases as i_L increases and $R_L = v_L/i_L$ decreases. The terminal voltage of an ideal voltage source (also plotted) remains the same for any current delivered to a load.

(2) From this point on we will try to adhere to the standard convention of referring to strictly dc quantities using capital letters, whereas lowercase letters denote a quantity that we know to have some time-varying component. However, in describing general theorems that apply to either dc or ac, we will continue to use lowercase to emphasize the general nature of the concept.

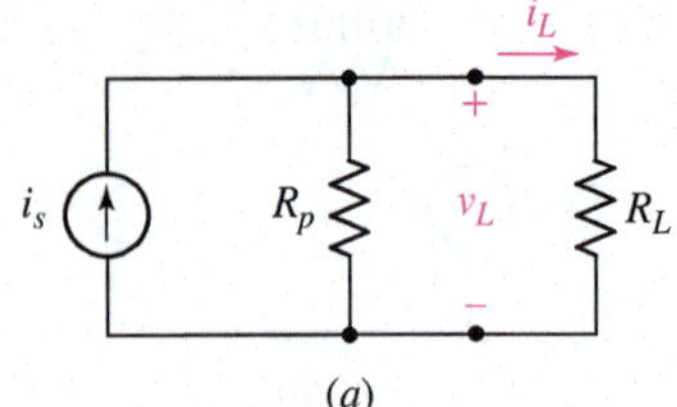

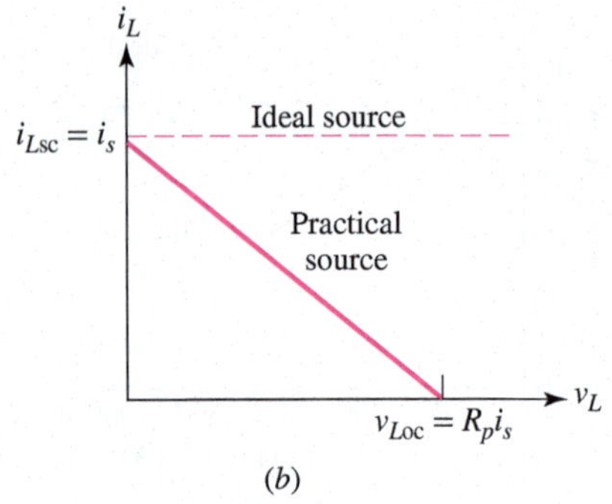

FIGURE 5.14 *(a)* A general practical current source connected to a load resistor R_L. *(b)* The load current provided by the practical current source is shown as a function of the load voltage.

Practical Current Sources

An ideal current source is also nonexistent in the real world; there is no physical device that will deliver a constant current regardless of the load resistance to which it is connected or the voltage across its terminals. Certain transistor circuits will deliver a constant current to a wide range of load resistances, but the load resistance can always be made sufficiently large that the current through it becomes very small. Infinite power is simply never available (unfortunately).

A practical current source is defined as an ideal current source in parallel with an internal resistance R_p. Such a source is shown in Fig. 5.14*a*, and the current i_L and voltage v_L associated with a load resistance R_L are indicated. Application of KCL yields

$$i_L = i_s - \frac{v_L}{R_p} \qquad [12]$$

which is again a linear relationship. The open-circuit voltage and the short-circuit current are

$$v_{Loc} = R_p i_s \qquad [13]$$

and

$$i_{Lsc} = i_s \qquad [14]$$

The variation of load current with changing load voltage may be investigated by changing the value of R_L as shown in Fig. 5.14*b*. The straight line is traversed from the short-circuit, or "northwest," end to the open-circuit termination at the "southeast" end by increasing R_L from zero to infinite ohms. The midpoint occurs for $R_L = R_p$. The load current i_L and the ideal source current are approximately equal only for small values of load voltage, which are obtained with values of R_L that are small compared to R_p.

Equivalent Practical Sources

It may be no surprise that we can improve upon models to increase their accuracy; at this point we now have a practical voltage source model and also a practical current source model. Before we proceed, however, let's take a moment to compare Fig. 5.13*b* and Fig. 5.14*b*. One is for a circuit with a voltage source and the other, with a current source, *but the graphs are indistinguishable!*

It turns out that this is no coincidence. In fact, we are about to show that a practical voltage source *can be* electrically equivalent to a practical current source—meaning that a load resistor R_L connected to either will have the same v_L and i_L. This means we can replace one practical source with the other and the rest of the circuit will not know the difference.

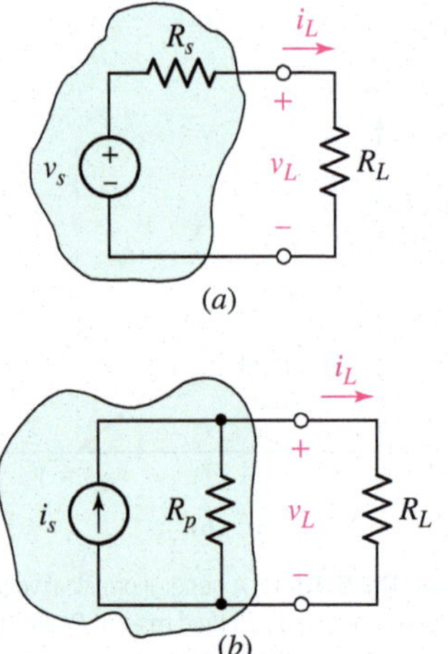

FIGURE 5.15 *(a)* A given practical voltage source connected to a load R_L. *(b)* The equivalent practical current source connected to the same load.

Consider the practical voltage source and resistor R_L shown in Fig. 5.15*a*, and the circuit composed of a practical current source and resistor R_L shown in Fig. 5.15*b*. A simple calculation shows that the voltage across the load R_L of Fig. 5.15*a* is

$$v_L = v_s \frac{R_L}{R_s + R_L} \qquad [15]$$

A similar calculation shows that the voltage across the load R_L in Fig. 5.15*b* is

$$v_L = \left(i_s \frac{R_p}{R_p + R_L} \right) \cdot R_L$$

The two practical sources are electrically equivalent, then, if

$$R_s = R_p \quad [16]$$

and

$$v_s = R_p i_s = R_s i_s \quad [17]$$

(From this point forward, we let R_s represent the internal resistance of either practical source, which is the conventional notation.)

Let's try this with the practical current source shown in Fig. 5.16*a*. Since its internal resistance is 2 Ω, the internal resistance of the equivalent practical voltage source is also 2 Ω; the voltage of the ideal voltage source contained within the practical voltage source is (2)(3) = 6 V. The equivalent practical voltage source is shown in Fig. 5.16*b*.

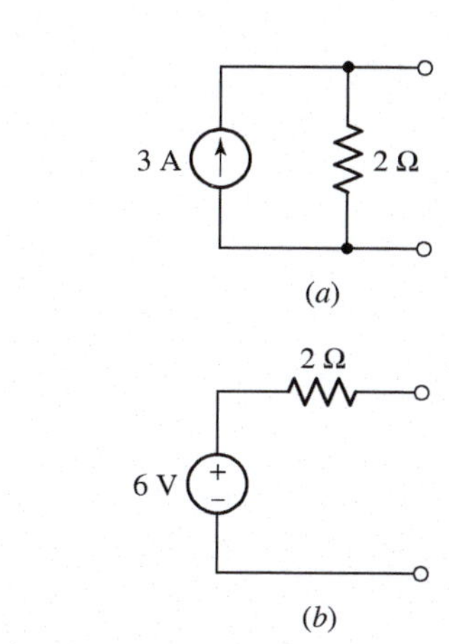

FIGURE 5.16 (*a*) A given practical current source. (*b*) The equivalent practical voltage source.

To check the equivalence, let us visualize a 4 Ω resistor connected to each source. In both cases a current of 1 A, a voltage of 4 V, and a power of 4 W are associated with the 4 Ω load. However, we should note very carefully that the ideal current source is delivering a total power of 12 W, while the ideal voltage source is delivering only 6 W. Furthermore, the internal resistance of the practical current source is absorbing 8 W, whereas the internal resistance of the practical voltage source is absorbing only 2 W. Thus we see that the two practical sources are equivalent only with respect to what transpires at the load terminals; they are *not* equivalent internally!

EXAMPLE **5.4**

Compute the current through the 4.7 kΩ resistor in Fig. 5.17*a* after first transforming the 9 mA source into an equivalent voltage source.

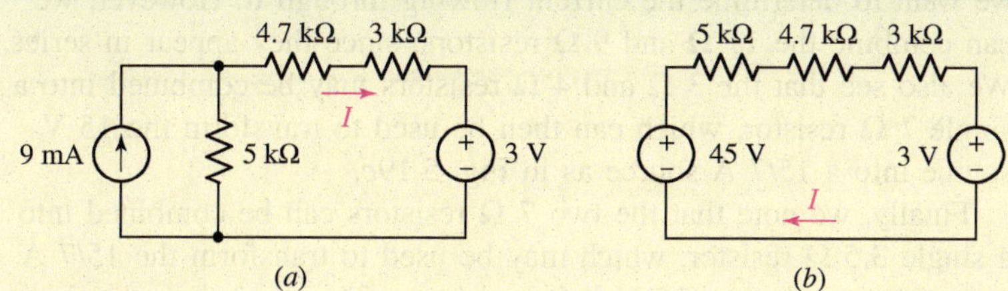

FIGURE 5.17 (*a*) A circuit with both a voltage source and a current source. (*b*) The circuit after the 9 mA source is transformed into an equivalent voltage source.

It's not just the 9 mA source at issue, but also the resistance in parallel with it (5 kΩ). We remove these components, leaving two terminals "dangling." We then replace them with a voltage source in series with a 5 kΩ resistor, as shown in Fig. 5.17*b*. The value of the voltage source must be (0.09)(5000) = 45 V.

(Continued on next page)

We can now write a simple KVL equation

$$-45 + 5000I + 4700I + 3000I + 3 = 0$$

which is easily solved to yield $I = 3.307$ mA.

Our answer can be verified by analyzing the circuit of Fig. 5.17*a* using either nodal or mesh techniques.

PRACTICE

5.3 For the circuit of Fig. 5.18, compute the current I_X through the 47 kΩ resistor after performing a source transformation on the voltage source.

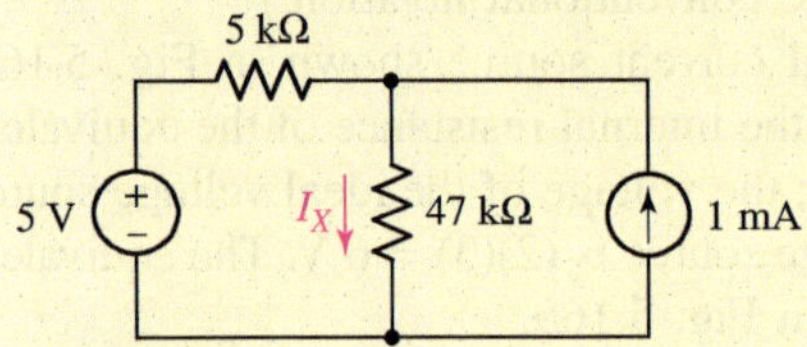

FIGURE 5.18

Ans: 192 μA.

EXAMPLE 5.5

Calculate the current through the 2 Ω resistor in Fig. 5.19*a* by making use of source transformations to first simplify the circuit.

We begin by transforming each current source into a voltage source (Fig. 5.19*b*), the strategy being to convert the circuit into a simple loop.

We must be careful to retain the 2 Ω resistor for two reasons: first, the dependent source controlling variable appears across it, and second, we want to determine the current flowing through it. However, we can combine the 17 Ω and 9 Ω resistors, since they appear in series. We also see that the 3 Ω and 4 Ω resistors may be combined into a single 7 Ω resistor, which can then be used to transform the 15 V source into a 15/7 A source as in Fig. 5.19*c*.

Finally, we note that the two 7 Ω resistors can be combined into a single 3.5 Ω resistor, which may be used to transform the 15/7 A current source into a 7.5 V voltage source. The result is a simple loop circuit, shown in Fig. 5.19*d*.

The current I can now be found using KVL:

$$-7.5 + 3.5I - 51V_x + 28I + 9 = 0$$

where

$$V_x = 2I$$

Thus,

$$I = 21.28 \text{ mA}$$

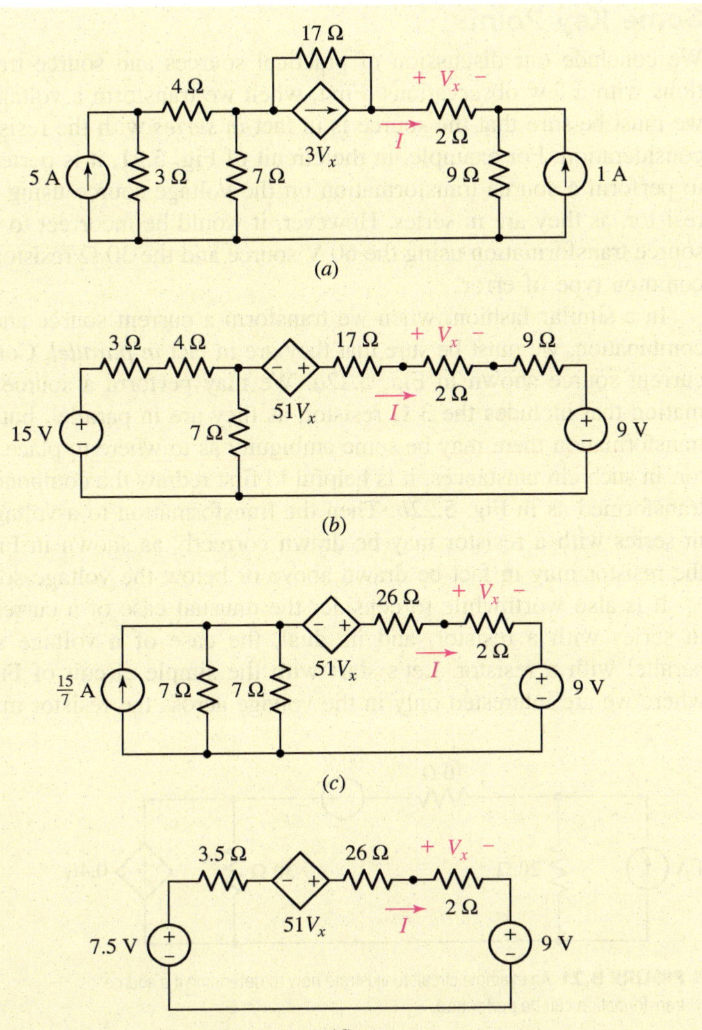

FIGURE 5.19 (*a*) A circuit with two independent current sources and one dependent source. (*b*) The circuit after each source is transformed into a voltage source. (*c*) The circuit after further combinations. (*d*) The final single-loop circuit.

PRACTICE

5.4 For the circuit of Fig. 5.20, compute the voltage V across the 1 MΩ resistor using repeated source transformations and resistor combinations to first simplify the circuit as much as possible.

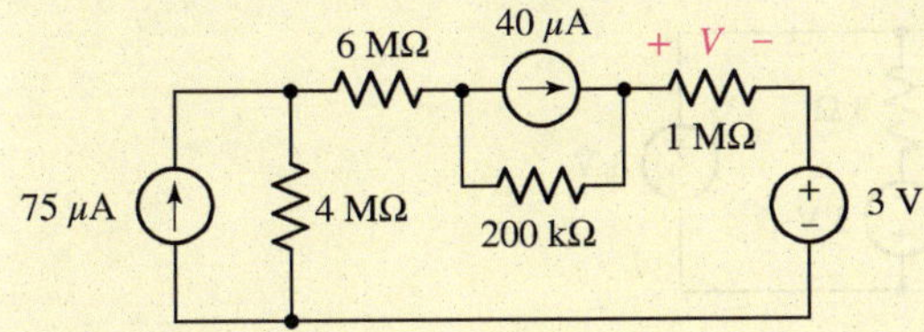

FIGURE 5.20

Ans: 27.2 V.

Some Key Points

We conclude our discussion of practical sources and source transformations with a few observations. First, when we transform a voltage source, we must be sure that the source is in fact *in series* with the resistor under consideration. For example, in the circuit of Fig. 5.21, it is perfectly valid to perform a source transformation on the voltage source using the 10 Ω resistor, as they are in series. However, it would be incorrect to attempt a source transformation using the 60 V source and the 30 Ω resistor—a very common type of error.

In a similar fashion, when we transform a current source and resistor combination, we must be sure that they are in fact *in parallel*. Consider the current source shown in Fig. 5.22*a*. We may perform a source transformation that includes the 3 Ω resistor, as they are in parallel, but after the transformation there may be some ambiguity as to where to place the resistor. In such circumstances, it is helpful to first redraw the components to be transformed as in Fig. 5.22*b*. Then the transformation to a voltage source in series with a resistor may be drawn correctly as shown in Fig. 5.22*c*; the resistor may in fact be drawn above or below the voltage source.

It is also worthwhile to consider the unusual case of a current source in series with a resistor, and its dual, the case of a voltage source in parallel with a resistor. Let's start with the simple circuit of Fig. 5.23*a*, where we are interested only in the voltage across the resistor marked R_2.

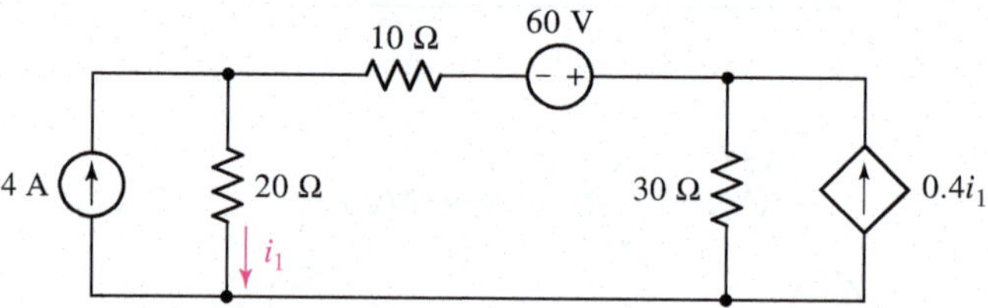

FIGURE 5.21 An example circuit to illustrate how to determine if a source transformation can be performed.

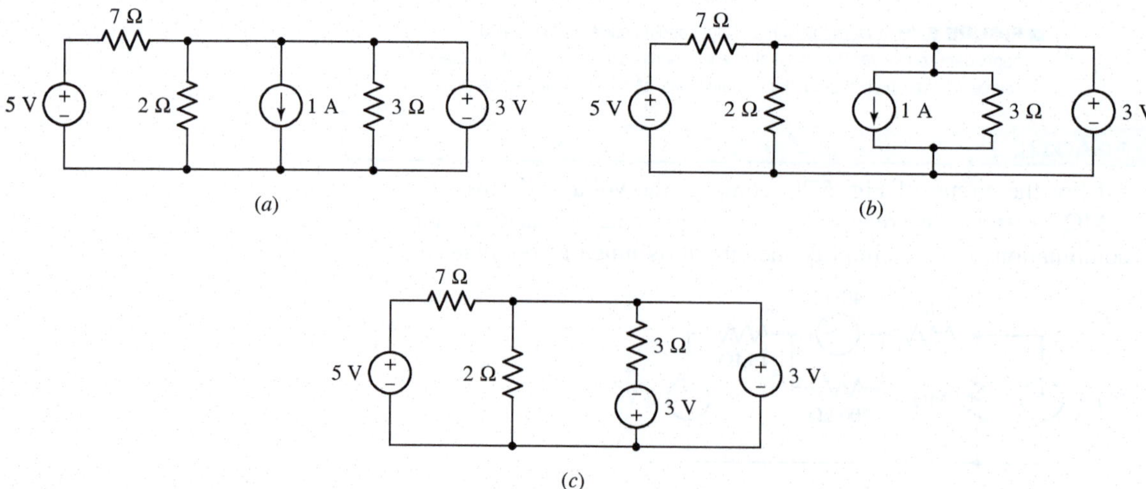

FIGURE 5.22 (*a*) A circuit with a current source to be transformed to a voltage source. (*b*) Circuit redrawn so as to avoid errors. (*c*) Transformed source/resistor combination.

We note that regardless of the value of resistor R_1, $V_{R2} = I_x R_2$. Although we might be tempted to perform an inappropriate source transformation on such a circuit, in fact *we may simply omit resistor* R_1 (provided that it is of no interest to us itself). A similar situation arises with a voltage source in parallel with a resistor, as depicted in Fig. 5.23*b*. Again, if we are only interested in some quantity regarding resistor R_2, we may find ourselves tempted to perform some strange (and incorrect) source transformation on the voltage source and resistor R_1. In reality, we may omit resistor R_1 from our circuit as far as resistor R_2 is concerned—its presence does not alter the voltage across, the current through, or the power dissipated by resistor R_2.

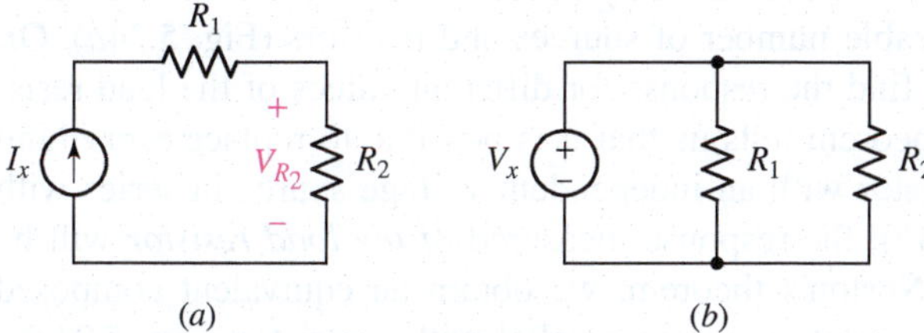

FIGURE 5.23 (*a*) Circuit with a resistor R_1 in series with a current source. (*b*) A voltage source in parallel with two resistors.

Summary of Source Transformation

1. **A common goal in source transformation is to end up with either all current sources or all voltage sources in the circuit.** This is especially true if it makes nodal or mesh analysis easier.
2. **Repeated source transformations can be used to simplify a circuit by allowing resistors and sources to eventually be combined.**
3. **The resistor *value* does not change during a source transformation, but it is *not* the same resistor.** This means that currents or voltages associated with the original resistor are irretrievably lost when we perform a source transformation.
4. **If the voltage or current associated with a particular resistor is used as a controlling variable for a dependent source, it should not be included in any source transformation.** The original resistor must be retained in the final circuit, untouched.
5. **If the voltage or current associated with a particular element is of interest, that element should not be included in any source transformation.** The original element must be retained in the final circuit, untouched.
6. **In a source transformation, the head of the current source arrow corresponds to the "+" terminal of the voltage source.**
7. **A source transformation on a current source and resistor requires that the two elements be in parallel.**
8. **A source transformation on a voltage source and resistor requires that the two elements be in series.**

5.3 THÉVENIN AND NORTON EQUIVALENT CIRCUITS

Now that we have been introduced to source transformations and the superposition principle, it is possible to develop two more techniques that will greatly simplify the analysis of many linear circuits. The first of these theorems is named after L. C. Thévenin, a French engineer working in telegraphy who published the theorem in 1883; the second may be considered a corollary of the first and is credited to E. L. Norton, a scientist with the Bell Telephone Laboratories.

Let us suppose that we need to make only a partial analysis of a circuit. For example, perhaps we need to determine the current, voltage, and power delivered to a single "load" resistor by the remainder of the circuit, which may consist of a sizable number of sources and resistors (Fig. 5.24*a*). Or, perhaps we wish to find the response for different values of the load resistance. Thévenin's theorem tells us that it is possible to replace everything except the load resistor with an independent voltage source in series with a resistor (Fig. 5.24*b*); the response measured *at the load resistor* will be unchanged. Using Norton's theorem, we obtain an equivalent composed of an independent current source in parallel with a resistor (Fig. 5.24*c*).

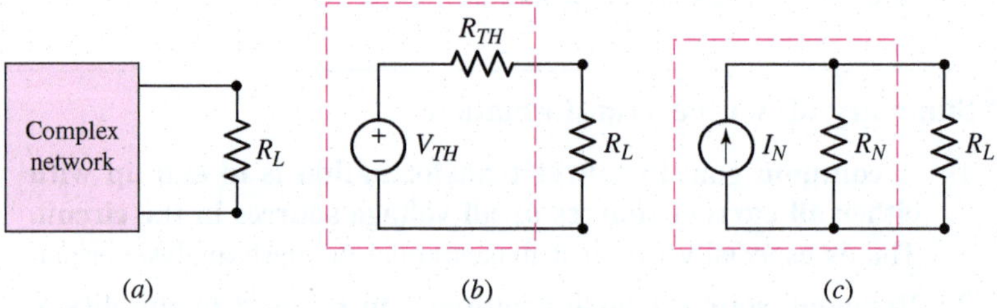

FIGURE 5.24 (*a*) A complex network including a load resistor R_L. (*b*) A Thévenin equivalent network connected to the load resistor R_L. (*c*) A Norton equivalent network connected to the load resistor R_L.

Thus, one of the main uses of Thévenin's and Norton's theorems is the replacement of a large part of a circuit, often a complicated and uninteresting part, with a very simple equivalent. The new, simpler circuit enables us to make rapid calculations of the voltage, current, and power that the original circuit can deliver to a load. It also helps us to choose the best value of this load resistance. In a transistor power amplifier, for example, the Thévenin or Norton equivalent enables us to determine the maximum power that can be taken from the amplifier and delivered to the speakers.

EXAMPLE 5.6

Consider the circuit shown in Fig. 5.25*a*. Determine the Thévenin equivalent of network *A*, and compute the power delivered to the load resistor R_L.

The dashed regions separate the circuit into networks *A* and *B*; our main interest is in network *B*, which consists only of the load resistor R_L. Network *A* may be simplified by making repeated source transformations.

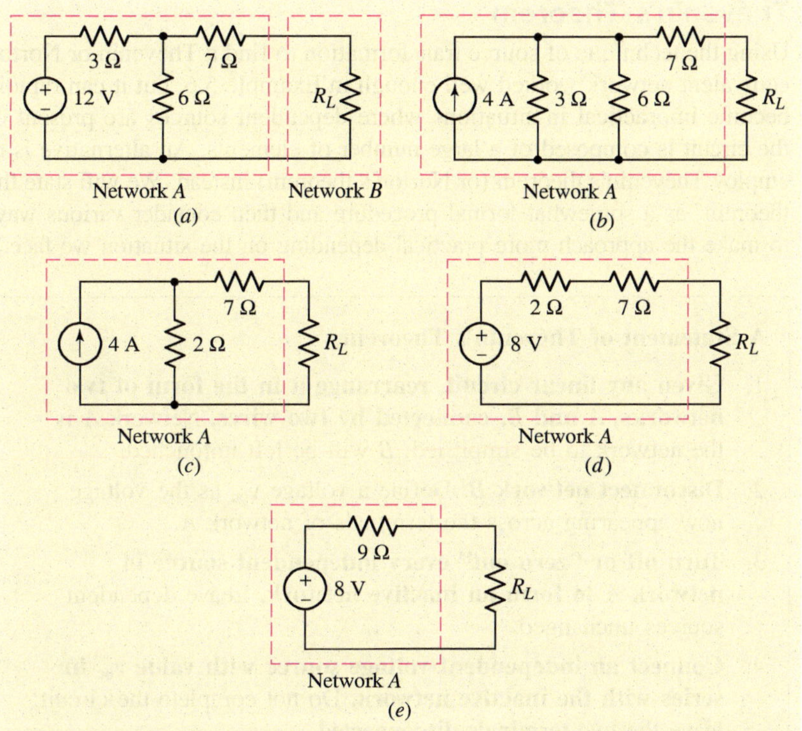

FIGURE 5.25 *(a)* A circuit separated into two networks. *(b)*–*(d)* Intermediate steps to simplifying network *A*. *(e)* The Thévenin equivalent circuit.

We first treat the 12 V source and the 3 Ω resistor as a practical voltage source and replace it with a practical current source consisting of a 4 A source in parallel with 3 Ω (Fig. 5.25*b*). The parallel resistances are then combined into 2 Ω (Fig. 5.25*c*), and the practical current source that results is transformed back into a practical voltage source (Fig. 5.25*d*). The final result is shown in Fig. 5.25*e*.

From the viewpoint of the load resistor R_L, this network *A* (the Thévenin equivalent) is equivalent to the original network *A*; from our viewpoint, the circuit is much simpler, and we can now easily compute the power delivered to the load:

$$P_L = \left(\frac{8}{9 + R_L}\right)^2 R_L$$

Furthermore, we can see from the equivalent circuit that the maximum voltage that can be obtained across R_L is 8 V and that it corresponds to $R_L = \infty$. A quick transformation of network *A* to a practical current source (the Norton equivalent) indicates that the maximum current that may be delivered to the load is 8/9 A, which occurs when $R_L = 0$. Neither of these facts is readily apparent from the original circuit.

PRACTICE

5.5 Using repeated source transformations, determine the Norton equivalent of the highlighted network in the circuit of Fig. 5.26.

Ans: 1 A; 5 Ω.

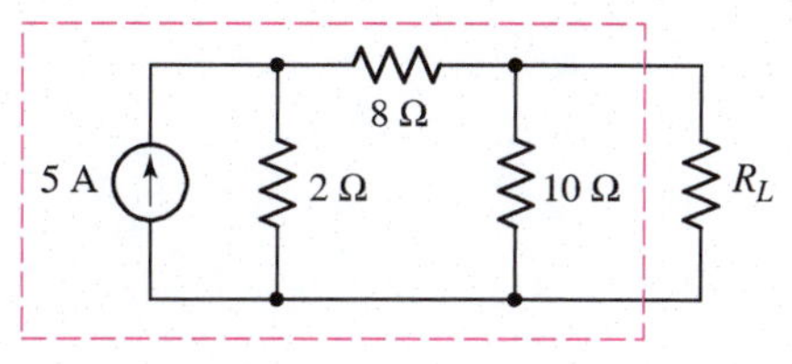

FIGURE 5.26

Thévenin's Theorem

Using the technique of source transformation to find a Thévenin or Norton equivalent network worked well enough in Example 5.6, but it can rapidly become impractical in situations where dependent sources are present or the circuit is composed of a large number of elements. An alternative is to employ Thévenin's theorem (or Norton's theorem) instead. We will state the theorem[3] as a somewhat formal procedure and then consider various ways to make the approach more practical depending on the situation we face.

A Statement of Thévenin's Theorem

1. **Given any linear circuit, rearrange it in the form of two networks, *A* and *B*, connected by two wires.** Network *A* is the network to be simplified; *B* will be left untouched.
2. **Disconnect network *B*.** Define a voltage v_{oc} as the voltage now appearing across the terminals of network *A*.
3. **Turn off or "zero out" every independent source in network *A* to form an inactive network.** Leave dependent sources unchanged.
4. **Connect an independent voltage source with value v_{oc} in series with the inactive network.** Do not complete the circuit; leave the two terminals disconnected.
5. **Connect network *B* to the terminals of the new network *A*.** All currents and voltages in *B* will remain unchanged.

Note that if either network contains a dependent source, *its control variable must be in the same network.*

Let us see if we can apply Thévenin's theorem successfully to the circuit we considered in Fig. 5.25. We have already found the Thévenin equivalent of the circuit to the left of R_L in Example 5.6, but we want to see if there is an easier way to obtain the same result.

EXAMPLE 5.7

Use Thévenin's theorem to determine the Thévenin equivalent for that part of the circuit in Fig. 5.25*a* to the left of R_L.

We begin by disconnecting R_L, and note that no current flows through the 7 Ω resistor in the resulting partial circuit shown in Fig. 5.27*a*. Thus, V_{oc} appears across the 6 Ω resistor (with no current through the 7 Ω resistor there is no voltage drop across it), and voltage division enables us to determine that

$$V_{oc} = 12\left(\frac{6}{3+6}\right) = 8 \text{ V}$$

(3) A proof of Thévenin's theorem in the form in which we have stated it is rather lengthy, and therefore it has been placed in Appendix 3, where the curious may peruse it.

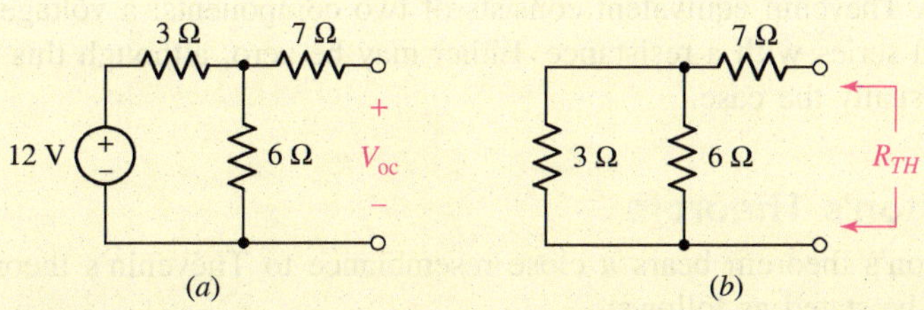

FIGURE 5.27 (*a*) The circuit of Fig. 5.25*a* with network *B* (the resistor R_L) disconnected and the voltage across the connecting terminals labeled as V_{oc}. (*b*) The independent source in Fig. 5.25*a* has been zeroed out, and we look into the terminals where network *B* was connected to determine the effective resistance of network *A*.

Turning off network *A* (i.e., replacing the 12 V source with a short circuit) and looking back into the dead network, we see a 7 Ω resistor connected in series with the parallel combination of 6 Ω and 3 Ω (Fig. 5.27*b*).

Thus, the inactive network can be represented here by a 9 Ω resistor, referred to as the *Thévenin equivalent resistance* of network *A*. The Thévenin equivalent then is V_{oc} in series with a 9 Ω resistor, which agrees with our previous result.

PRACTICE

5.6 Use Thévenin's theorem to find the current through the 2 Ω resistor in the circuit of Fig. 5.28. (*Hint:* Designate the 2 Ω resistor as network *B*.)

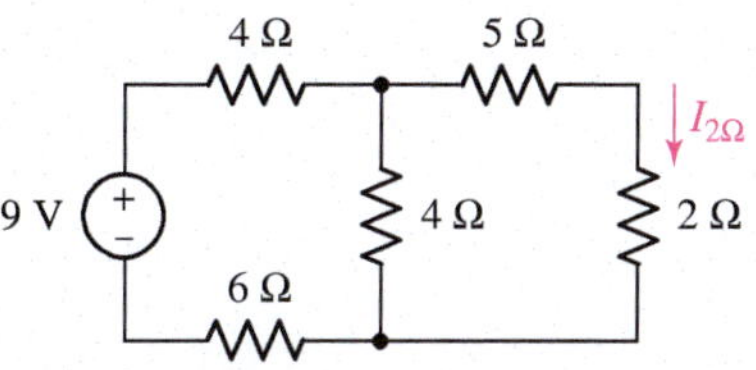

FIGURE 5.28

Ans: V_{TH} = 2.571 V; R_{TH} = 7.857 Ω; $I_{2\Omega}$ = 260.8 mA.

A Few Key Points

The equivalent circuit we have learned how to obtain is completely independent of network *B:* we have been instructed to first remove network *B* and then measure the open-circuit voltage produced by network *A*, an operation that certainly does not depend on network *B* in any way. The *B* network is mentioned only to indicate that an equivalent for *A* may be obtained no matter what arrangement of elements is connected to the *A* network; the *B* network represents this general network.

Several points about the theorem deserve emphasis.

- The only restriction that we must impose on *A* or *B* is that all *dependent* sources in *A* have their control variables in *A*, and similarly for *B*.
- No restrictions are imposed on the complexity of *A* or *B;* either one may contain any combination of independent voltage or current sources, linear dependent voltage or current sources, resistors, or any other circuit elements which are linear.
- The deactivated network *A* can be represented by a single equivalent resistance R_{TH}, which we will call the Thévenin equivalent resistance. This holds true whether or not dependent sources exist in the inactive *A* network, an idea we will explore shortly.

- A Thévenin equivalent consists of two components: a voltage source in series with a resistance. Either may be zero, although this is not usually the case.

Norton's Theorem

Norton's theorem bears a close resemblance to Thévenin's theorem and may be stated as follows:

A Statement of Norton's Theorem

1. **Given any linear circuit, rearrange it in the form of two networks, *A* and *B*, connected by two wires.** Network *A* is the network to be simplified; *B* will be left untouched. As before, if either network contains a dependent source, *its controlling variable must be in the same network.*
2. **Disconnect network *B*, and short the terminals of *A*.** Define a current i_{sc} as the current now flowing through the shorted terminals of network *A*.
3. **Turn off or "zero out" every independent source in network *A* to form an inactive network.** Leave dependent sources unchanged.
4. **Connect an independent current source with value i_{sc} in parallel with the inactive network.** Do not complete the circuit; leave the two terminals disconnected.
5. **Connect network *B* to the terminals of the new network *A*.** All currents and voltages in *B* will remain unchanged.

The Norton equivalent of a linear network is the Norton current source i_{sc} in parallel with the Thévenin resistance R_{TH}. Thus, we see that in fact it is possible to obtain the Norton equivalent of a network by performing a source transformation on the Thévenin equivalent. This results in a direct relationship between v_{oc}, i_{sc}, and R_{TH}:

$$v_{oc} = R_{TH} i_{sc} \quad [18]$$

In circuits containing dependent sources, we will often find it more convenient to determine either the Thévenin or Norton equivalent by finding both the open-circuit voltage and the short-circuit current and then determining the value of R_{TH} as their quotient. It is therefore advisable to become adept at finding both open-circuit voltages and short-circuit currents, even in the simple problems that follow. If the Thévenin and Norton equivalents are determined independently, Eq. [18] can serve as a useful check.

Let's consider three different examples of the determination of a Thévenin or Norton equivalent circuit.

EXAMPLE 5.8

Find the Thévenin and Norton equivalent circuits for the network faced by the 1 kΩ resistor in Fig. 5.29*a*.

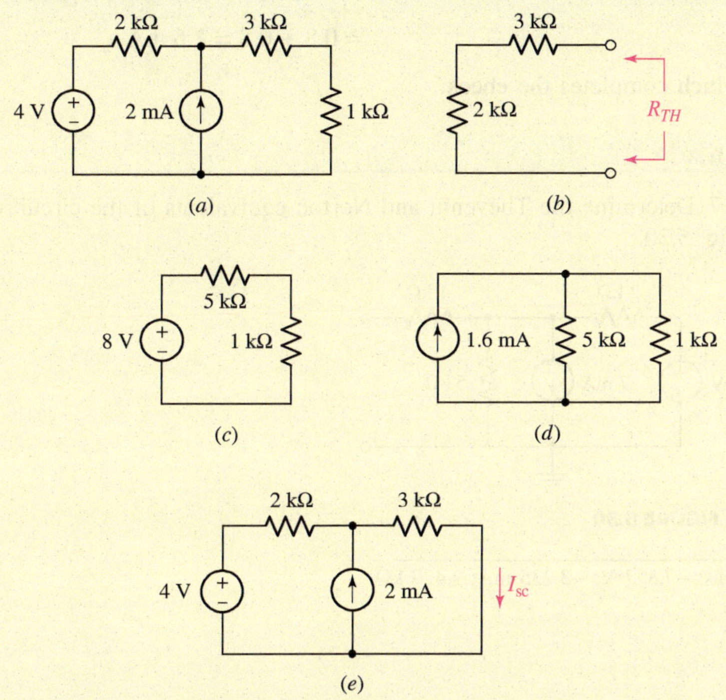

■ **FIGURE 5.29** (*a*) A given circuit in which the 1 kΩ resistor is identified as network *B*. (*b*) Network *A* with all independent sources killed. (*c*) The Thévenin equivalent is shown for network *A*. (*d*) The Norton equivalent is shown for network *A*. (*e*) Circuit for determining I_{SC}.

From the wording of the problem statement, network *B* is the 1 kΩ resistor, so network *A* is everything else.

Choosing to find the Thévenin equivalent of network *A* first, we apply superposition, noting that no current flows through the 3 kΩ resistor once network *B* is disconnected. With the current source set to zero, $V_{oc|4V} = 4$ V. With the voltage source set to zero,

$$V_{oc|_{2\,mA}} = (0.002)(2000) = 4 \text{ V. Thus, } V_{oc} = 4 + 4 = 8 \text{ V.}$$

To find R_{TH}, set both sources to zero as in Fig. 5.29*b*. By inspection, $R_{TH} = 2$ kΩ + 3 kΩ = 5 kΩ. The complete Thévenin equivalent, with network *B* reconnected, is shown in Fig. 5.29*c*.

The Norton equivalent is found by a simple source transformation of the Thévenin equivalent, resulting in a current source of 8/5000 = 1.6 mA in parallel with a 5 kΩ resistor (Fig. 5.29*d*).

Check: Find the Norton equivalent directly from Fig. 5.29*a*. Removing the 1 kΩ resistor and shorting the terminals of network *A*,

(Continued on next page)

we find I_{sc} as shown in Fig. 5.29*e* by superposition and current division:

$$I_{sc} = I_{sc|_{4\,V}} + I_{sc|_{2\,mA}} = \frac{4}{2+3} + (2)\frac{2}{2+3}$$

$$= 0.8 + 0.8 = 1.6 \text{ mA}$$

which completes the check.

PRACTICE

5.7 Determine the Thévenin and Norton equivalents of the circuit of Fig. 5.30.

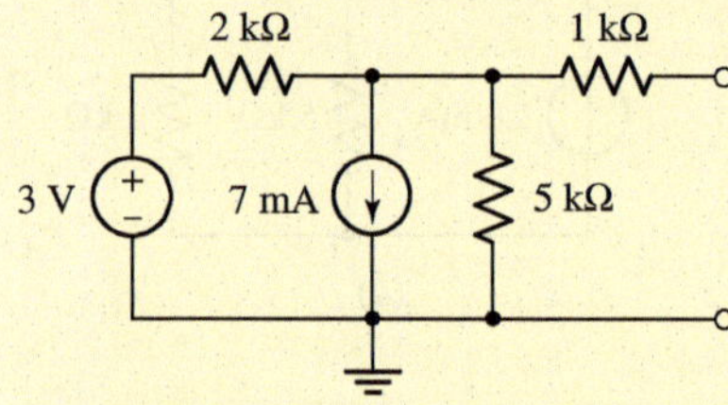

FIGURE 5.30

Ans: −7.857 V; −3.235 mA; 2.429 kΩ.

When Dependent Sources Are Present

Technically speaking, there does not always have to be a "network *B*" for us to invoke either Thévenin's theorem or Norton's theorem; we could instead be asked to find the equivalent of a network with two terminals not yet connected to another network. If there *is* a network *B* that we do not want to involve in the simplification procedure, however, we must use a little caution if it contains dependent sources. In such situations, the controlling variable and the associated element(s) must be included in network *B* and excluded from network *A*. Otherwise, there will be no way to analyze the final circuit because the controlling quantity will be lost.

If network *A* contains a dependent source, then again we must ensure that the controlling variable and its associated element(s) cannot be in network *B*. Up to now, we have only considered circuits with resistors and independent sources. Although technically speaking it is correct to leave a dependent source in the "inactive" network when creating a Thévenin or Norton equivalent, in practice this does not result in any kind of simplification. What we really want is an independent voltage source in series with a single resistor, or an independent current source in parallel with a single resistor—in other words, a two-component equivalent. In the following examples, we consider various means of reducing networks with dependent sources and resistors into a single resistance.

EXAMPLE 5.9

Determine the Thévenin equivalent of the circuit in Fig. 5.31*a*.

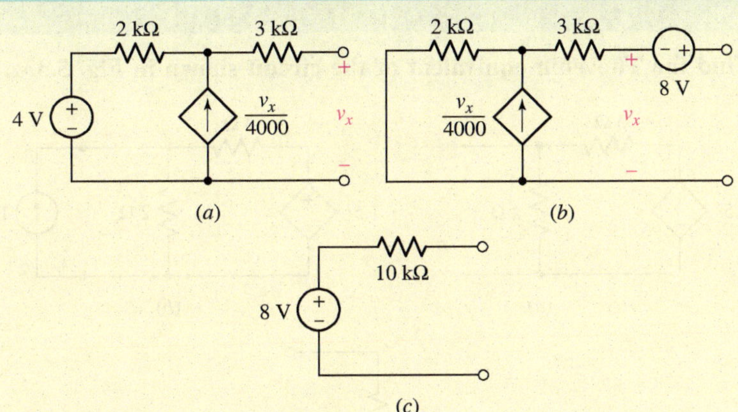

■ **FIGURE 5.31** (*a*) A given network whose Thévenin equivalent is desired. (*b*) A possible, but rather useless, form of the Thévenin equivalent. (*c*) The best form of the Thévenin equivalent for this linear resistive network.

To find V_{oc} we note that $v_x = V_{oc}$ and that the dependent source current must pass through the 2 kΩ resistor, since no current can flow through the 3 kΩ resistor. Using KVL around the outer loop:

$$-4 + 2 \times 10^3\left(-\frac{v_x}{4000}\right) + 3 \times 10^3(0) + v_x = 0$$

and

$$v_x = 8 \text{ V} = V_{oc}$$

By Thévenin's theorem, then, the equivalent circuit could be formed with the inactive *A* network in series with an 8 V source, as shown in Fig. 5.31*b*. This is correct, but not very simple and not very helpful; in the case of linear resistive networks, we really want a simpler equivalent for the inactive *A* network, namely, R_{TH}.

The dependent source prevents us from determining R_{TH} directly for the inactive network through resistance combination; we therefore seek I_{sc}. Upon short-circuiting the output terminals in Fig. 5.31*a*, it is apparent that $V_x = 0$ and the dependent current source is not active. Hence, $I_{sc} = 4/(5 \times 10^3) = 0.8$ mA. Thus,

$$R_{TH} = \frac{V_{oc}}{I_{sc}} = \frac{8}{0.8 \times 10^{-3}} = 10 \text{ k}\Omega$$

and the acceptable Thévenin equivalent of Fig. 5.31*c* is obtained.

PRACTICE

5.8 Find the Thévenin equivalent for the network of Fig. 5.32. (*Hint:* a quick source transformation on the dependent source might help.)

Ans: −502.5 mV; −100.5 Ω.

Note: A negative resistance might seem strange—and it is! Such a thing is physically possible only if, for example, we do a bit of clever electronic circuit design to create something that behaves like the dependent current source we represented in Fig. 5.32.

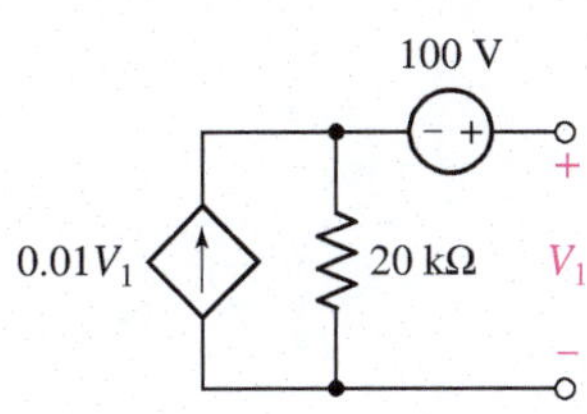

■ **FIGURE 5.32**

As another example, let us consider a network having a dependent source but no independent source.

EXAMPLE **5.10**

Find the Thévenin equivalent of the circuit shown in Fig. 5.33*a*.

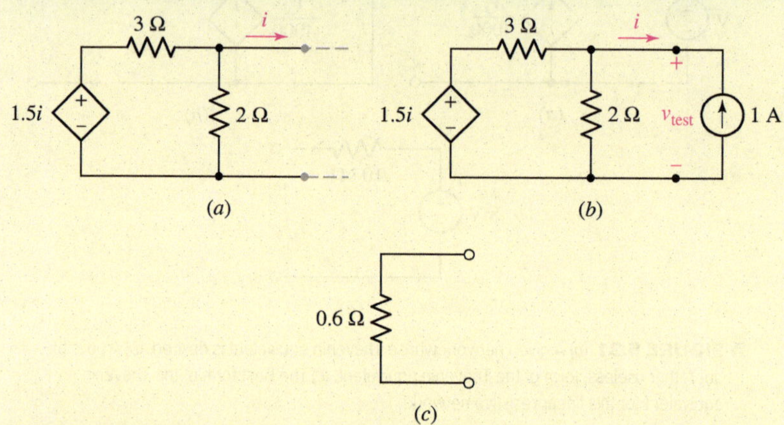

FIGURE 5.33 (*a*) A network with no independent sources. (*b*) A hypothetical measurement to obtain R_{TH}. (*c*) The Thévenin equivalent to the original circuit.

The rightmost terminals are already open-circuited, hence $i = 0$. Consequently, the dependent source is inactive, so $v_{oc} = 0$.

We next seek the value of R_{TH} represented by this two-terminal network. However, we cannot find v_{oc} and i_{sc} and take their quotient, for there is no independent source in the network, and both v_{oc} and i_{sc} are zero. Let us, therefore, be a little tricky.

We apply a 1 A source externally, measure the voltage v_{test} that results, and then set $R_{TH} = v_{test}/1$. Referring to Fig. 5.33*b*, we see that $i = -1$ A. Applying nodal analysis,

$$\frac{v_{test} - 1.5(-1)}{3} + \frac{v_{test}}{2} = 1$$

so that

$$v_{test} = 0.6 \text{ V}$$

and thus

$$R_{TH} = 0.6\ \Omega$$

The Thévenin equivalent is shown in Fig. 5.33*c*.

A Quick Recap of Procedures

We have now looked at three examples in which we determined a Thévenin or Norton equivalent circuit. The first example (Fig. 5.29) contained only independent sources and resistors, and several different methods could have been applied to it. One would involve calculating R_{TH} for the inactive network and then V_{oc} for the live network. We could also have found R_{TH} and I_{sc}, or V_{oc} and I_{sc}.

PRACTICAL APPLICATION

The Digital Multimeter

One of the most common pieces of electrical test equipment is the DMM, or digital multimeter (Fig. 5.34), which is designed to measure voltage, current, and resistance values.

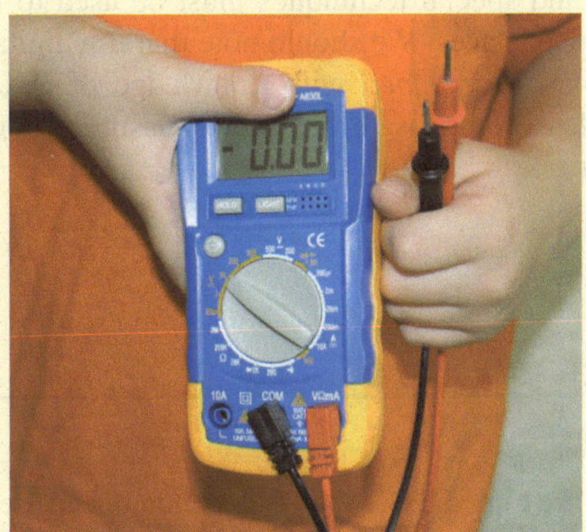

■ **FIGURE 5.34** A handheld digital multimeter.
Steve Durbin/McGraw Hill

In a voltage measurement, two leads from the DMM are connected across the appropriate circuit element, as depicted in Fig. 5.35. The positive reference terminal of the meter is typically marked "V/Ω," and the negative reference terminal—often referred to as the *common terminal*—is typically designated by "COM." The convention is to use a red lead for the positive reference terminal and a black lead for the common terminal.

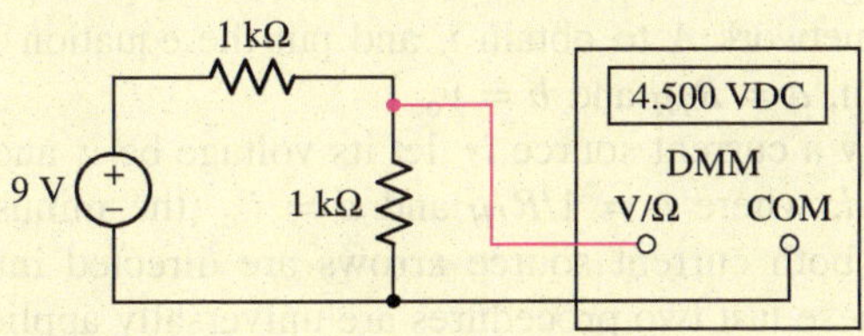

■ **FIGURE 5.35** A DMM connected to measure voltage.

From our discussion of Thévenin and Norton equivalents, it may now be apparent that the DMM has its own Thévenin equivalent resistance. This Thévenin equivalent resistance will appear in parallel with our circuit, and its value can affect the measurement (Fig. 5.36). The DMM does not supply power to the circuit to measure voltage, so its Thévenin equivalent consists of only a resistance, which we will name R_{DMM}.

The input resistance of a good DMM is typically 10 MΩ or more. The measured voltage V thus appears across 1 kΩ||10 MΩ = 999.9 Ω. Using voltage division, we find that $V = 4.4998$ volts, slightly less than the expected value of 4.5 volts. Thus, the finite input resistance of the voltmeter introduces a small error in the measured value.

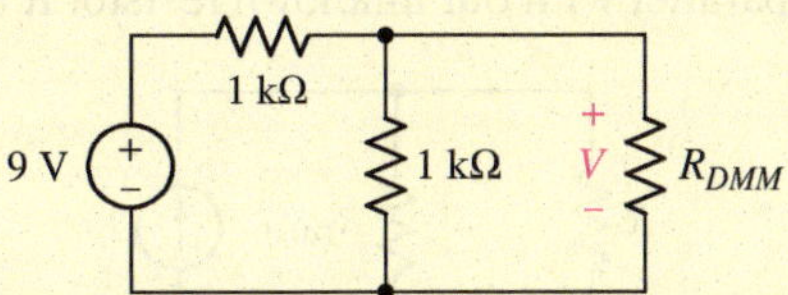

■ **FIGURE 5.36** DMM in Fig. 5.35 shown as its Thévenin equivalent resistance, R_{DMM}.

To measure current, the DMM must be placed in series with a circuit element, generally requiring that we cut a wire (Fig. 5.37). One DMM lead is connected to the common terminal of the meter, and the other lead is placed in a connector usually marked "A" to signify current measurement. Again, the DMM does not supply power to the circuit in this type of measurement.

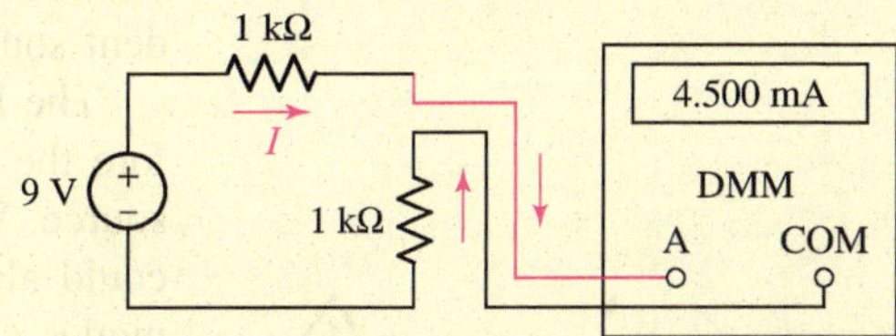

■ **FIGURE 5.37** A DMM connected to measure current.

We see that the Thévenin equivalent resistance (R_{DMM}) of the DMM is in series with our circuit, so its value can affect the measurement. Writing a simple KVL equation around the loop,

$$-9 + 1000I + R_{\mathrm{DMM}}I + 1000I = 0$$

Note that since we have reconfigured the meter to perform a current measurement, the Thévenin equivalent resistance is not the same as when the meter is configured to measure voltages. In fact, we would ideally like R_{DMM} to be 0 Ω for current measurements, and ∞ for voltage measurements. If R_{DMM} is now 0.1 Ω, we see that the measured current I is 4.4998 mA, which is only slightly different from the expected value of 4.5 mA. Depending on the number of digits that can be displayed by the meter, we may not even notice the effect of nonzero DMM resistance on our measurement.

The same meter can be used to determine resistance, provided no independent sources are active during the measurement. Internally, a known current is passed

(Continued on next page)

through the resistor being measured, and the voltmeter circuitry is used to measure the resulting voltage. Replacing the DMM with its Norton equivalent (which now includes an active independent current source to generate the predetermined current), we see that R_{DMM} appears in parallel with our unknown resistor R (Fig. 5.38).

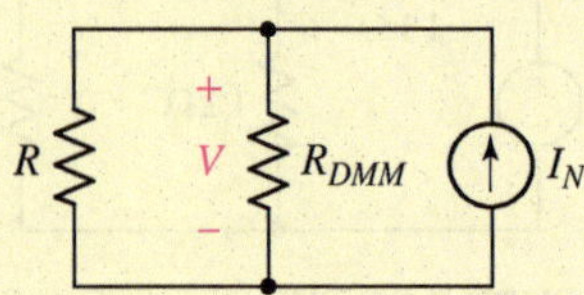

FIGURE 5.38 DMM in resistance measurement configuration replaced by its Norton equivalent, showing R_{DMM} in parallel with the unknown resistor R to be measured.

As a result, the DMM actually measures $R \| R_{DMM}$. If $R_{DMM} = 10\ \text{M}\Omega$ and $R = 10\ \Omega$, $R_{measured} = 9.99999\ \Omega$, which is more than accurate enough for most purposes. However, if $R = 10\ \text{M}\Omega$, $R_{measured} = 5\ \text{M}\Omega$. The input resistance of a DMM therefore places a practical upper limit on the values of resistance that can be measured, and special techniques must be used to measure larger resistances. We should note that if a digital multimeter is *programmed* with knowledge of R_{DMM}, it is possible to compensate and allow measurement of larger resistances.

In the second example (Fig. 5.31), both independent and dependent sources were present, and the method we used required us to find V_{oc} and I_{sc}. We could not easily find R_{TH} for the inactive network because the dependent source could not be made inactive.

The last example did not contain any independent sources, and therefore the Thévenin and Norton equivalents do not contain an independent source. We found R_{TH} by applying 1 A and finding $v_{test} = 1 \times R_{TH}$. We could also apply 1 V and determine $i = 1/R_{TH}$. These two related techniques can be applied to any circuit with dependent sources, *as long as all independent sources are set to zero first.*

Two other methods have a certain appeal because they can be used for any of the three types of networks considered. In the first, simply replace network B with a voltage source vs, define the current leaving its positive terminal as i, analyze network A to obtain i, and put the equation in the form $vs = ai + b$. Then, $a = R_{TH}$ and $b = v_{oc}$.

We could also apply a current source is, let its voltage be v, and then determine $is = cv - d$, where $c = 1/R_{TH}$ and $d = i_{sc}$ (the minus sign arises from assuming both current source arrows are directed into the same node). Both of these last two procedures are universally applicable, but some other method can usually be found that is easier and more rapid.

Although we are devoting our attention almost entirely to the analysis of linear circuits, it is good to know that Thévenin's and Norton's theorems are both valid if network B is nonlinear; only network A must be linear.

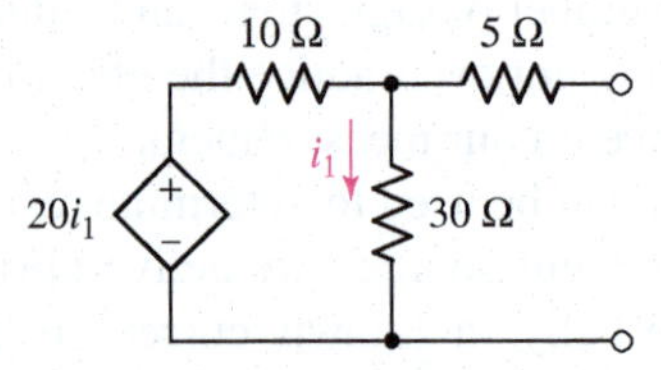

FIGURE 5.39 See Practice Problem 5.9.

PRACTICE

5.9 Find the Thévenin equivalent for the network of Fig. 5.39. (*Hint:* Try a 1 V test source.)

Ans: $I_{test} = 50$ mA so $R_{TH} = 20\ \Omega$.

5.4 MAXIMUM POWER TRANSFER

A very useful power theorem may be developed with reference to a practical voltage or current source. For the practical voltage source (Fig. 5.40), the power delivered to the load R_L is

$$p_L = i_L^2 R_L = \frac{v_s^2 R_L}{(R_s + R_L)^2} \quad [19]$$

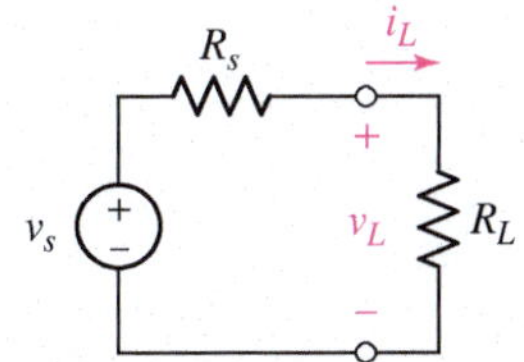

■ **FIGURE 5.40** A practical voltage source connected to a load resistor R_L.

We can find the value of R_L that will absorb maximum power from the given practical source if we differentiate with respect to R_L

$$\frac{dp_L}{dR_L} = \frac{(R_s + R_L)^2 v_s^2 - v_s^2 R_L(2)(R_s + R_L)}{(R_s + R_L)^4}$$

and equate this to zero, obtaining

$$2R_L(R_s + R_L) = (R_s + R_L)^2$$

which leads us to

$$R_s = R_L$$

Since the values $R_L = 0$ and $R_L = \infty$ both give a minimum ($p_L = 0$), and since we have already developed the equivalence between practical voltage and current sources, we have therefore proved the following ***maximum power transfer theorem:***

> An independent voltage source in series with a resistance R_s (or an independent current source in parallel with a resistance R_s) delivers maximum power to a load resistance R_L such that $R_L = R_s$.

An alternative way to view the maximum power theorem is possible in terms of the Thévenin equivalent resistance of a network:

> A network delivers maximum power to a load resistance R_L when R_L is equal to the Thévenin equivalent resistance of the network.

Thus, the maximum power transfer theorem tells us that a 2 Ω resistor draws the greatest power (4.5 W) from either practical source of Fig. 5.16, whereas a resistance of 0.01 Ω receives the maximum power (3.6 kW) in Fig. 5.11.

There is a distinct difference between *drawing* maximum power from a *source* and *delivering* maximum power to a *load.* If the load is sized such that its Thévenin resistance is equal to the Thévenin resistance of the network to which it is connected, it will receive maximum power from that network. *Any change to the load resistance will reduce the power delivered to the load.* However, consider just the Thévenin equivalent of the network itself. We draw the maximum possible power from the voltage source by drawing the maximum possible current—which is achieved by shorting the network terminals! However, in this extreme example we deliver *zero power* to the "load"—a short circuit in this case—as $p = i^2R$, and we just set $R = 0$ by shorting the network terminals.

A minor amount of algebra applied to Eq. [19], coupled with the maximum power transfer requirement that $R_L = R_s = R_{TH}$ yields

$$p_{\text{max|delivered to load}} = \frac{v_s^2}{4\,R_s} = \frac{v_{TH}^2}{4\,R_{TH}}$$

where v_{TH} and R_{TH} recognize that the practical voltage source of Fig. 5.40 can also be viewed as a Thévenin equivalent of some specific source.

We should pause here and mention that it is not uncommon for the maximum power theorem to be misinterpreted. It is designed to help us select an optimum load in order to maximize power absorption. If the load resistance is already specified, however, the maximum power theorem is of *no assistance*. If for some reason we can affect the size of the Thévenin equivalent resistance of the network connected to our load, setting it equal to the load does not guarantee maximum power transfer to our predetermined load. A quick consideration of the power lost in the Thévenin resistance will clarify this point.

EXAMPLE 5.11

The circuit shown in Fig. 5.41 is a model for the common-emitter bipolar junction transistor amplifier. Choose a resistance R_L so that maximum power is transferred to the load from the amplifier, and calculate the actual power absorbed.

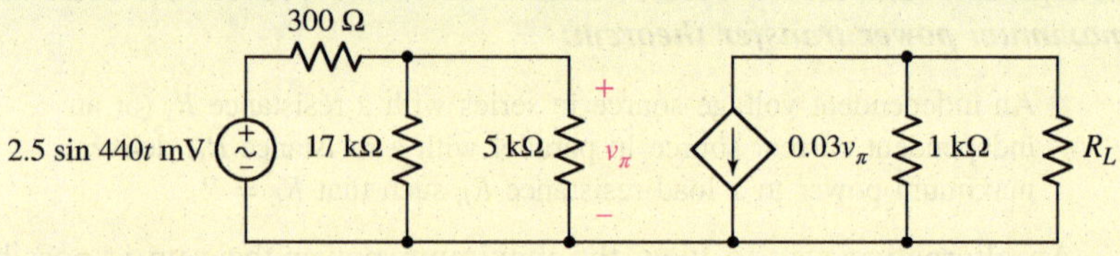

FIGURE 5.41 A small-signal model of the common-emitter amplifier, with the load resistance unspecified.

Since it is the load resistance we are asked to determine, the maximum power theorem applies. The first step is to find the Thévenin equivalent of the rest of the circuit.

We first determine the Thévenin equivalent resistance, which requires that we remove R_L and short-circuit the independent source as in Fig. 5.42*a*.

Since $v_\pi = 0$, the dependent current source is an open circuit, and $R_{TH} = 1\ \text{k}\Omega$. This can be verified by connecting an independent 1A current source across the 1 kΩ resistor; v_π will still be zero, so the dependent source remains inactive and hence contributes nothing to R_{TH}.

Thus, in order to obtain maximum power delivered into the load, R_L should be set to $R_{TH} = 1\ \text{k}\Omega$.

To find v_{TH} we consider the circuit shown in Fig. 5.42*b*, which is Fig. 5.41 with R_L removed. We may write

$$v_{oc} = -0.03\ v_\pi(1000) = -30\ v_\pi$$

where the voltage v_π may be found from simple voltage division:

$$v_\pi = (2.5 \times 10^{-3} \sin 440t)\left(\frac{3864}{300 + 3864}\right)$$

so that our Thévenin equivalent is a voltage −69.6 sin 440*t* mV in series with 1 kΩ.

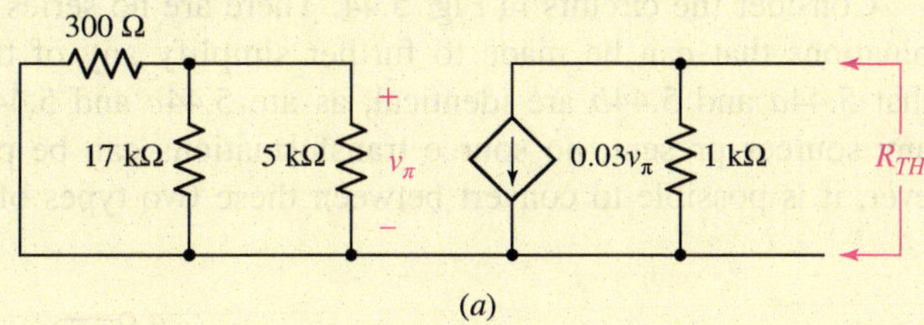

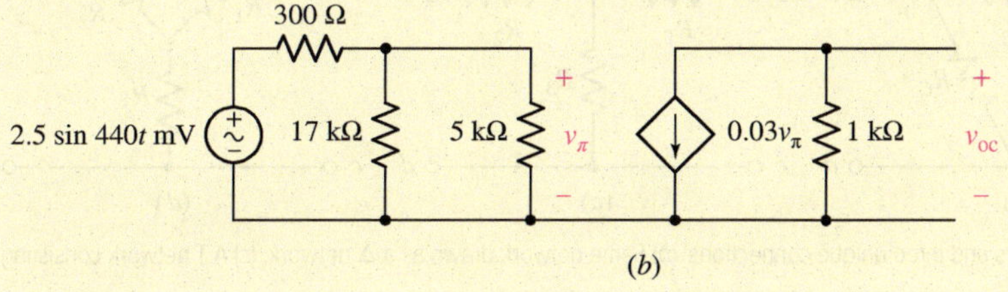

■ **FIGURE 5.42** (*a*) Circuit with R_L removed and independent source short-circuited. (*b*) Circuit for determining v_{TH}.

The maximum power is therefore

$$p_{max} = \frac{v_{TH}^2}{4\,R_{TH}} = 1.211 \sin^2 440t\ \mu\text{W}$$

PRACTICE

5.10 Consider the circuit of Fig. 5.43.

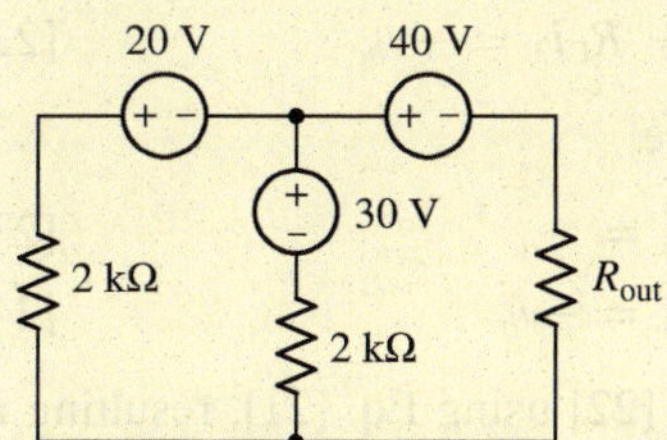

■ **FIGURE 5.43**

(*a*) Compute the maximum power that can be delivered to R_{out}.

(*b*) If R_{out} = 3 kΩ, determine the power delivered to it.

(*c*) What two different values of R_{out} will have exactly 20 mW delivered to them?

Ans: 306 mW; 230 mW; 59.2 kΩ and 16.9 Ω.

5.5 DELTA-WYE CONVERSION

We saw previously that identifying parallel and series combinations of resistors can often lead to a significant reduction in the complexity of a circuit. In situations where such combinations do not exist, we can often make use of source transformations to enable such simplifications. There is another useful technique, called ***Δ–Y (delta–wye)*** conversion, which arises out of network theory.

Consider the circuits in Fig. 5.44. There are no series or parallel combinations that can be made to further simplify any of the circuits (note that 5.44*a* and 5.44*b* are identical, as are 5.44*c* and 5.44*d*), and without any sources present, no source transformations can be performed. However, it is possible to convert between these two types of networks.

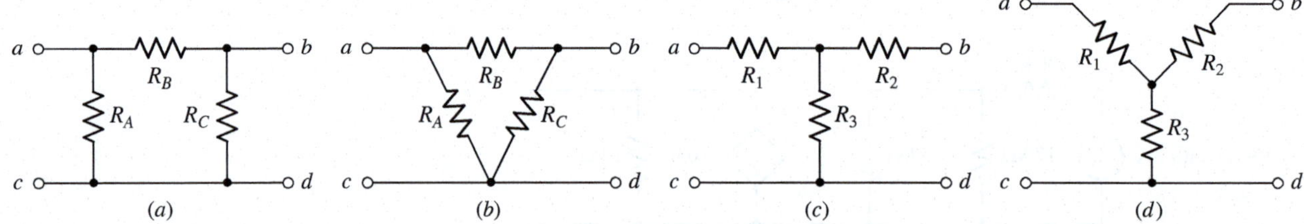

FIGURE 5.44 (*a*) Π network consisting of three resistors and three unique connections. (*b*) Same network drawn as a Δ network. (*c*) A T network consisting of three resistors. (*d*) Same network drawn as a Y network.

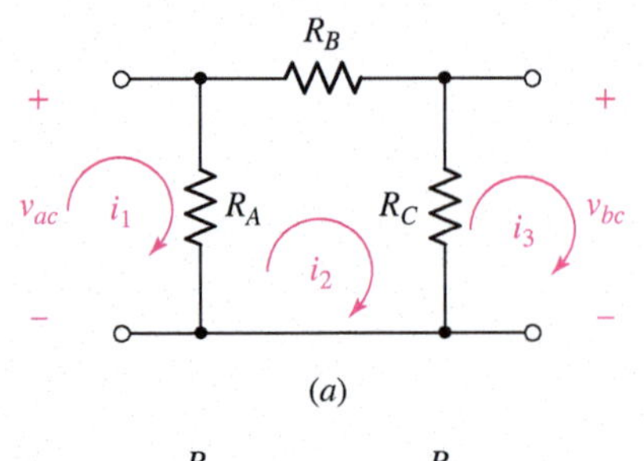

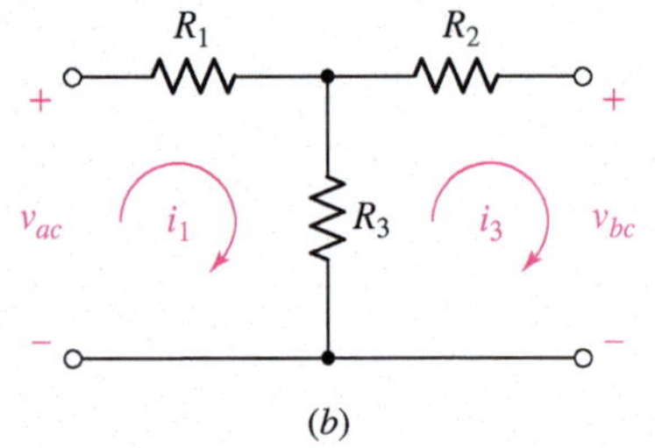

FIGURE 5.45 (*a*) Labeled Π network; (*b*) labeled T network.

We first define two voltages v_{ac} and v_{bc}, and three currents i_1, i_2, and i_3 as depicted in Fig. 5.45. If the two networks are equivalent, then the terminal voltages and currents must be equal (there is no current i_2 in the T-connected network). A set of relationships between R_A, R_B, R_C and R_1, R_2, and R_3 can now be defined simply by performing mesh analysis. For example, for the network of Fig. 5.45*a* we may write

$$R_A i_1 - R_A i_2 = v_{ac} \quad [20]$$

$$-R_A i_1 + (R_A + R_B + R_C) i_2 - R_C i_3 \quad [21]$$

$$-R_C i_2 + R_C i_3 = -v_{bc} \quad [22]$$

and for the network of Fig. 5.45*b* we have

$$(R_1 + R_3) i_1 - R_3 i_3 = v_{ac} \quad [23]$$

$$-R_3 i_1 + (R_2 + R_3) i_3 = -v_{bc} \quad [24]$$

We next remove i_2 from Eqs. [20] and [22] using Eq. [21], resulting in

$$\left(R_A - \frac{R_A^2}{R_A + R_B + R_C}\right) i_1 - \frac{R_A R_C}{R_A + R_B + R_C} i_3 = v_{ac} \quad [25]$$

and

$$-\frac{R_A R_C}{R_A + R_B + R_C} i_1 + \left(\frac{R_C - R_C^2}{R_A + R_B + R_C}\right) i_3 = -v_{bc} \quad [26]$$

Comparing terms between Eq. [23] and Eq. [25], we see that

$$R_3 = \frac{R_A R_C}{R_A + R_B + R_C}$$

In a similar fashion, we may find expressions for R_1 and R_2 in terms of R_A, R_B, and R_C, as well as expressions for R_A, R_B, and R_C in terms of R_1, R_2, and R_3; we leave the remainder of the derivations as an exercise for the reader. Thus, to convert from a Y network to a Δ network, the new resistor values are calculated using

$$R_A = \frac{R_1R_2 + R_2R_3 + R_3R_1}{R_2}$$

$$R_B = \frac{R_1R_2 + R_2R_3 + R_3R_1}{R_3}$$

$$R_C = \frac{R_1R_2 + R_2R_3 + R_3R_1}{R_1}$$

and to convert from a Δ network to a Y network,

$$R_1 = \frac{R_AR_B}{R_A + R_B + R_C}$$

$$R_2 = \frac{R_BR_C}{R_A + R_B + R_C}$$

$$R_3 = \frac{R_CR_A}{R_A + R_B + R_C}$$

Application of these equations is straightforward, although identifying the actual networks sometimes requires a little concentration.

EXAMPLE 5.12

Use the technique of Δ–Y conversion to find the Thévenin equivalent resistance of the circuit in Fig. 5.46*a*.

We see that the network in Fig. 5.46*a* is composed of two Δ-connected networks that share the 3 Ω resistor. We must be careful at this point not to be too eager, attempting to convert both Δ-connected networks to two Y-connected networks. The reason for this may be more obvious after we convert the top network consisting of the 1 Ω, 4 Ω, and 3 Ω resistors into a Y-connected network (Fig. 5.46*b*).

Note that in converting the upper network to a Y-connected network, we have removed the 3 Ω resistor. As a result, there is no way to convert the original Δ-connected network consisting of the 2 Ω, 5 Ω, and 3 Ω resistors into a Y-connected network.

We proceed by combining the $\frac{3}{8}$ Ω and 2 Ω resistors and the $\frac{3}{2}$ Ω and 5 Ω resistors (Fig. 5.46*c*). We now have a $\frac{19}{8}$ Ω resistor in parallel with a $\frac{13}{2}$ Ω resistor, and this parallel combination is in series with the $\frac{1}{2}$ Ω resistor. Thus, we can replace the original network of Fig. 5.46*a* with a single $\frac{159}{71}$ Ω resistor (Fig. 5.46*d*).

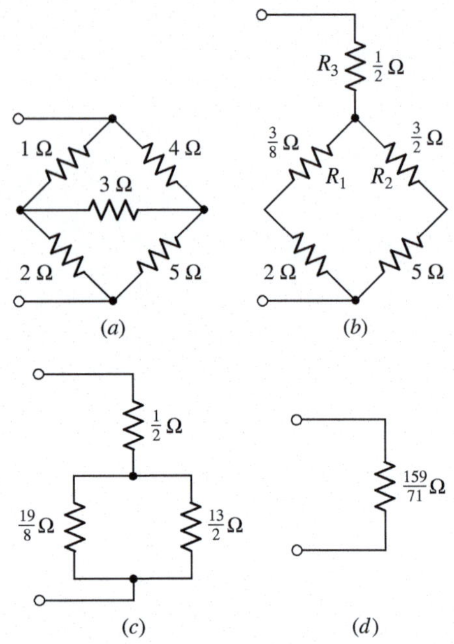

FIGURE 5.46 (*a*) A given resistive network whose input resistance is desired. (*b*) The upper Δ network is replaced by an equivalent Y network. (*c, d*) Series and parallel combinations result in a single resistance value.

PRACTICE

5.11 Use the technique of Y–Δ conversion to find the Thévenin equivalent resistance of the circuit of Fig. 5.47.

R_{in}

Each R is 10 Ω

FIGURE 5.47

Ans: 11.43 Ω.

5.6 SELECTING AN APPROACH: A SUMMARY OF VARIOUS TECHNIQUES

In Chap. 3, we were introduced to Kirchhoff's current law (KCL) and Kirchhoff's voltage law (KVL). These two laws apply to *any* circuit we will ever encounter, provided that we take care to consider the entire system that the circuits represent. The reason for this is that KCL and KVL enforce charge and energy conservation, respectively, which are fundamental principles. Based on KCL, we developed the very powerful method of nodal analysis. A similar technique based on KVL (unfortunately only applicable to planar circuits) is known as mesh analysis and is also a useful circuit analysis approach.

For the most part, this text is concerned with developing analytical skills that apply to *linear* circuits. If we know a circuit is constructed of only linear components (in other words, all voltages and currents are related by linear functions), then we can often simplify circuits before employing either mesh or nodal analysis. Perhaps the most important result that comes from the knowledge that we are dealing with a completely linear system is that the *principle of superposition applies*: given a number of independent sources acting on our circuit, we can add the contribution of each source independently of the other sources. This technique is pervasive throughout the field of engineering, and we will encounter it often. In many real situations, we will find that although several "sources" are acting simultaneously on our "system," typically one of them dominates the system response. Superposition allows us to quickly identify that source, provided that we have a reasonably accurate linear model of the system.

However, from a circuit analysis standpoint, unless we are asked to find which independent source contributes the most to a particular response, we find that rolling up our sleeves and launching straight into either nodal or mesh analysis is often a more straightforward tactic. The reason for this is that applying superposition to a circuit with 12 independent sources will require us to redraw the original circuit 12 times, and often we will have to apply nodal or mesh analysis to each partial circuit, anyway.

The technique of source transformations, on the other hand, is often a very useful tool in circuit analysis. Performing source transformations can allow us to consolidate resistors or sources that are not in series or parallel in the original circuit. Source transformations may also allow us to convert all or at least most of the sources in the original circuit to the same type (either all voltage sources or all current sources), so nodal or mesh analysis is more straightforward.

Thévenin's theorem is extremely important for a number of reasons. In working with electronic circuits, we are always aware of the Thévenin equivalent resistance of different parts of our circuit, especially the input and output resistances of amplifier stages. The reason for this is that matching of resistances is often the best route to optimizing the performance of a given circuit. We have seen a small preview of this in our discussion of maximum power transfer, where the load resistance should be chosen to match the Thévenin equivalent resistance of the network to which the load is connected. In terms of day-to-day circuit analysis, however, we find that converting part of a circuit to its Thévenin or Norton equivalent is almost as much work as analyzing the complete circuit. Therefore, as in the case of superposition, Thévenin's and Norton's theorems are typically applied when we need specialized information about part of our circuit.

SUMMARY AND REVIEW

Although we asserted in Chap. 4 that nodal analysis and mesh analysis are sufficient to analyze any circuit we might encounter (provided we have the means to relate voltage and current for any passive element, such as Ohm's law for resistors), the simple truth is that often we do not really need *all* voltages, or *all* currents. Sometimes, it is simply *one* element, or a *small portion* of a larger circuit, that has our attention. In such cases, we can exploit the fact that at the moment we have confined ourselves to *linear* circuits. This allows the development of other tools: *superposition,* where individual contributions of sources can be identified; *source transformations,* where a voltage source in series with a resistor can be replaced with a current source in parallel with a resistor; and the most powerful of all—*Thévenin* (and *Norton*) *equivalents.*

An interesting offshoot of these topics is the idea of *maximum power transfer.* Assuming we can represent our (arbitrarily complex) circuit by two networks, one passive and one active, maximum power transfer to the passive network is achieved when its Thévenin resistance is equal to the Thévenin resistance of the active network. Finally, we introduced the concept of delta–wye conversion, a process that allows us to simplify some resistive networks that at face value are not reducible using standard series–parallel combination techniques.

We are still faced with the perpetual question, *"Which tool should I use to analyze this circuit?"* The answer typically lies in the type of information required about our circuit. Experience will eventually guide us a bit, but it is not always true that there is one "best" approach. Certainly one issue to focus on is whether one or more components might be changed—this can suggest whether superposition, a Thévenin equivalent, or a partial simplification such as can be achieved with source or delta–wye transformation is the most practical route.

We conclude this chapter by reviewing key points, along with identifying relevant example(s).

- The principle of superposition states that the *response* in a linear circuit can be obtained by adding the individual responses caused by the separate *independent* sources *acting alone.* (Examples 5.1, 5.2, 5.3)
- Superposition is most often used when it is necessary to determine the individual contribution of each source to a particular response. (Examples 5.2, 5.3)
- A practical model for a real voltage source is a resistor in series with an independent voltage source. A practical model for a real current source is a resistor in parallel with an independent current source.
- Source transformations allow us to convert a practical voltage source into a practical current source, and vice versa. (Example 5.4)
- Repeated source transformations can greatly simplify analysis of a circuit by providing the means to combine resistors and sources. (Example 5.5)
- The Thévenin equivalent of a network is a resistor in series with an independent voltage source. The Norton equivalent is the same resistor in parallel with an independent current source. (Example 5.6)
- There are several ways to obtain the Thévenin equivalent resistance, depending on whether or not dependent sources are present in the network. (Examples 5.7, 5.8, 5.9, 5.10)
- Maximum power transfer occurs when the load resistor matches the Thévenin equivalent resistance of the network to which it is connected. (Example 5.11)
- When faced with a Δ-connected resistor network, it is straightforward to convert it to a Y-connected network. This can be useful in simplifying the network prior to analysis. Conversely, a Y-connected resistor network can be converted to a Δ-connected network to assist in simplification of the network. (Example 5.12)

READING FURTHER

A book about battery technology, including characteristics of built-in resistance:

T. B. Reddy, ed., *Linden's Handbook of Batteries,* 4th ed. New York: McGraw Hill Education, 2010.

An excellent discussion of pathological cases and various circuit analysis theorems can be found in:

R. A. DeCarlo and P. M. Lin, *Linear Circuits,* 3rd ed. Dubuque, IA: Kendall Hunt Publishing, 2009.

EXERCISES

5.1 Linearity and Superposition

1. Linear systems are so easy to work with that engineers often construct linear models of real (nonlinear) systems to assist in analysis and design. Such models are often surprisingly accurate over a limited range. For example, consider the simple exponential function e^x. The Taylor series representation of this function is

$$e^x \approx 1 + x + \frac{x^2}{2} + \frac{x^3}{6} + \cdots$$

(*a*) Construct a linear model for this function by truncating the Taylor series expansion after the linear (first-order) term. (*b*) Evaluate your model function at $x = 0.000005$, 0.0005, 0.05, 0.5, and 5.0. (*c*) For which values of x does your model yield a "reasonable" approximation to e^x? *Explain your reasoning*.

2. Construct a linear approximation to the function $y(t) = 4 \sin 2t$. (*a*) Evaluate your approximation at $t = 0$, 0.001, 0.01, 0.1, and 1.0. (*b*) For which values of t does your model provide a "reasonable" approximation to the actual (nonlinear) function $y(t)$? *Explain your reasoning*.

3. Considering the circuit of Fig. 5.48, employ superposition to determine the two components of i_8 arising from the action of the two independent sources, respectively.

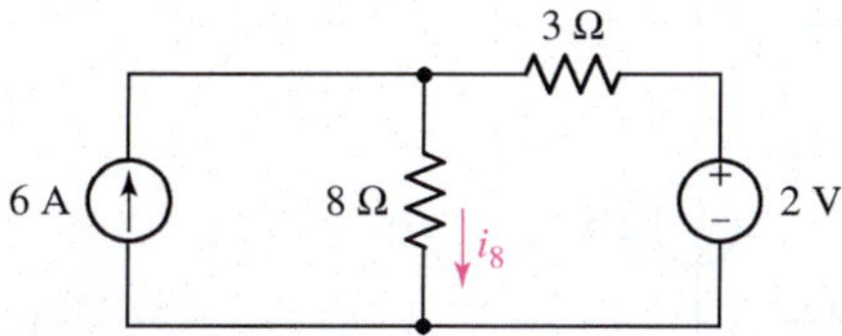

FIGURE 5.48

4. (*a*) Employ superposition to determine the current labeled i in the circuit of Fig. 5.49. (*b*) Express the contribution the 1 V source makes to the total current i in terms of a percentage. (*c*) Changing only the value of the 10 A source, adjust the circuit of Fig. 5.49 so that the two sources contribute equally to the current i.

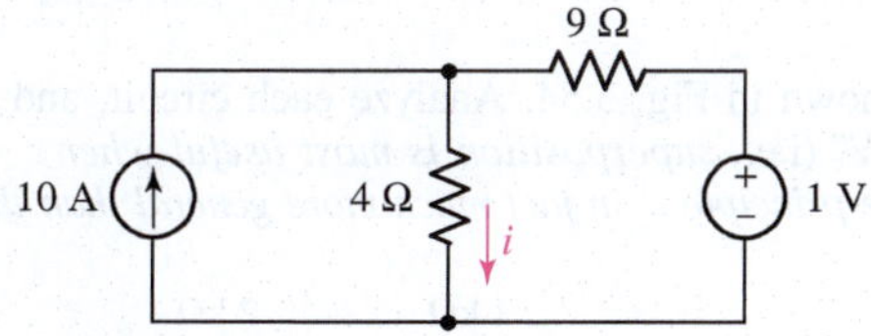

FIGURE 5.49

5. Regarding the circuit of Fig. 5.50, (*a*) Using superposition to consider each source one at a time, compute i_x. (*b*) Determine the percentage of i_x arising from each source. (*c*) Adjusting only the current source, alter the circuit to double i_x.

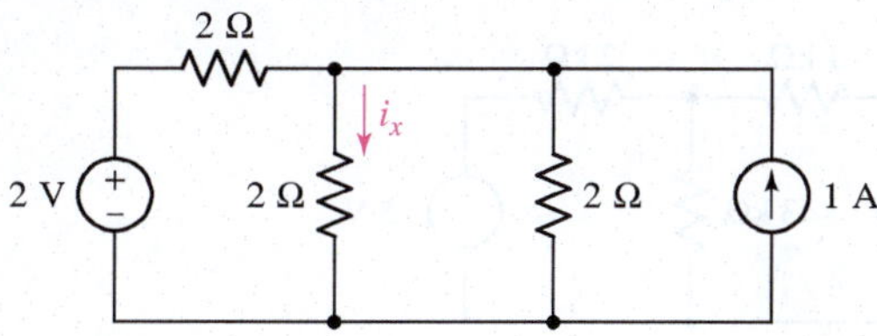

FIGURE 5.50

6. (*a*) Determine the individual contributions of each of the two current sources in the circuit of Fig. 5.51 to the nodal voltage v_1. (*b*) Determine the power dissipated by the 1 Ω resistor.

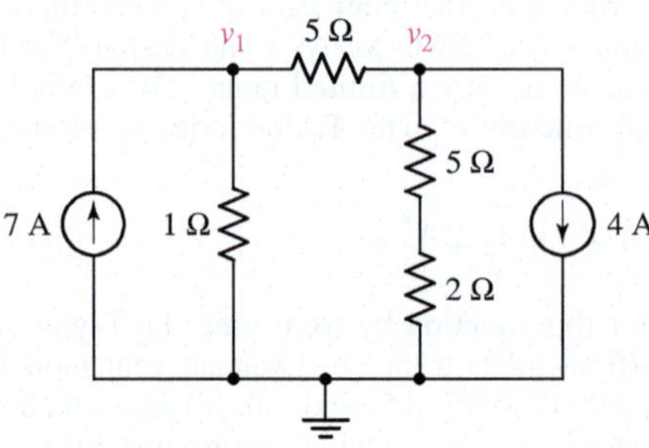

■ **FIGURE 5.51**

7. (*a*) Determine the individual contributions of each of the two current sources shown in Fig. 5.52 to the nodal voltage labeled v_2. (*b*) Instead of performing two separate simulations, verify your answer by using a single dc sweep. Submit a labeled schematic, relevant graphical output, and a short description of the results.

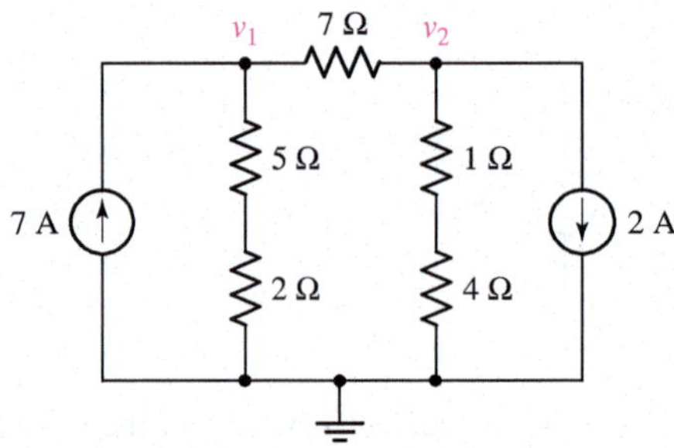

■ **FIGURE 5.52**

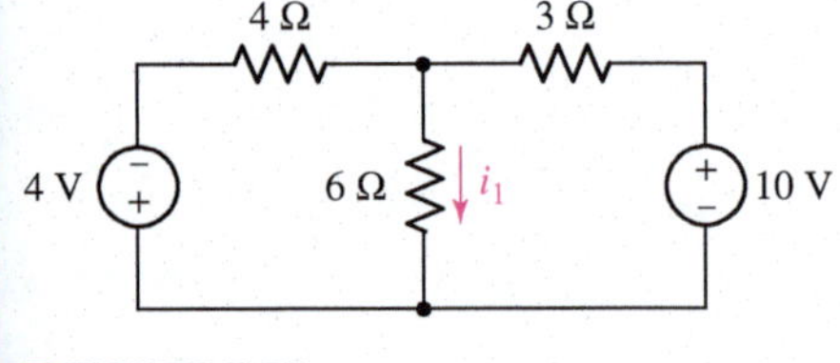

■ **FIGURE 5.53**

8. After studying the circuit of Fig. 5.53, change both voltage source values such that (*a*) i_1 doubles; (*b*) the direction of i_1 reverses, but its magnitude is unchanged.

9. Consider the three circuits shown in Fig. 5.54. Analyze each circuit, and demonstrate that $V_x = V_x' + V_x''$ (i.e., *superposition is most useful when sources are set to zero, but the principle is in fact much more general than that*).

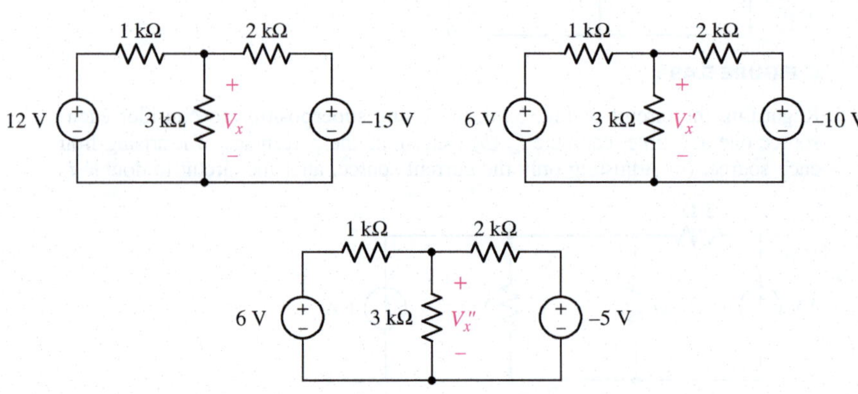

■ **FIGURE 5.54**

10. (*a*) Using superposition, determine the voltage labeled v_x in the circuit represented in Fig. 5.55. (*b*) To what value should the 2 A source be changed to reduce v_x by 10%? (*c*) Verify your answers by performing three dc sweep simulations (one for each source). Submit a labeled schematic, relevant graphical output, and a short description of the results.

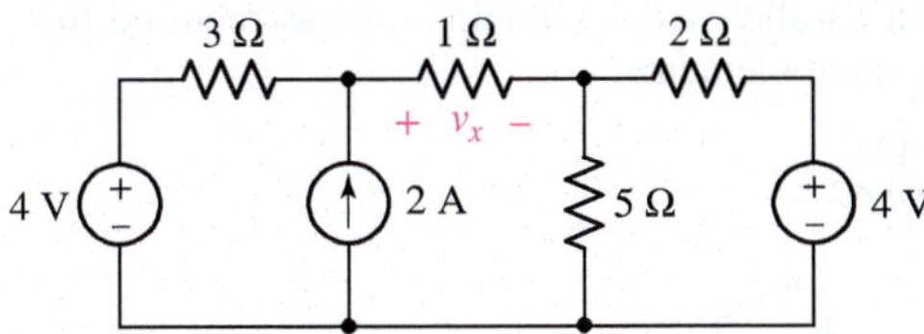

FIGURE 5.55

11. Employ superposition principles to obtain a value for the current I_x as labeled in Fig. 5.56.

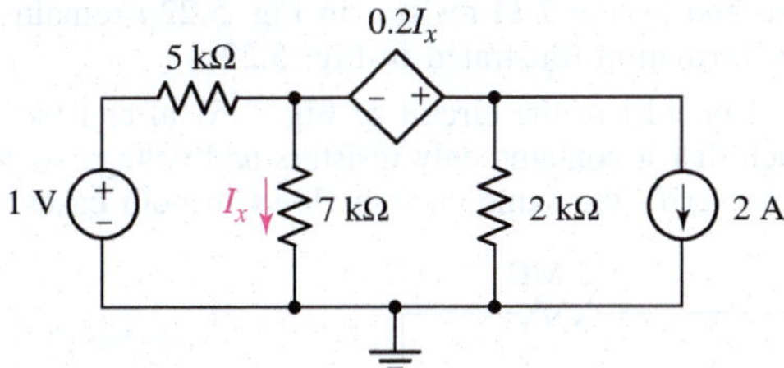

FIGURE 5.56

12. (*a*) Employ superposition to determine the individual contribution from each independent source to the current i_x as labeled in the circuit shown in Fig. 5.57. (*b*) Compute the power absorbed by each 1 Ω resistor.

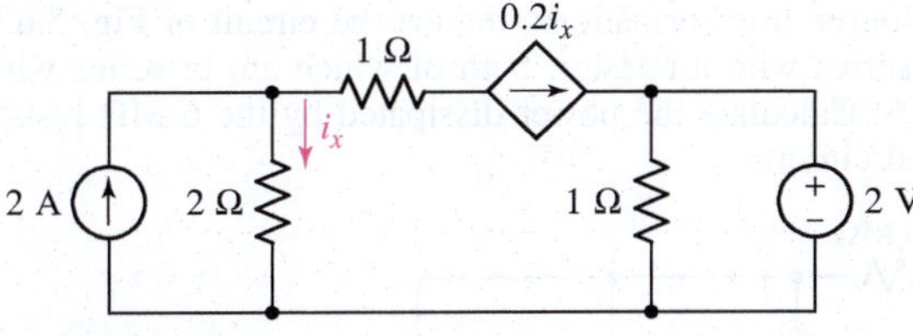

FIGURE 5.57

5.2 Source Transformations

13. Perform an appropriate source transformation on each of the circuits depicted in Fig. 5.58, taking care to retain the 4 Ω resistor in each final circuit.

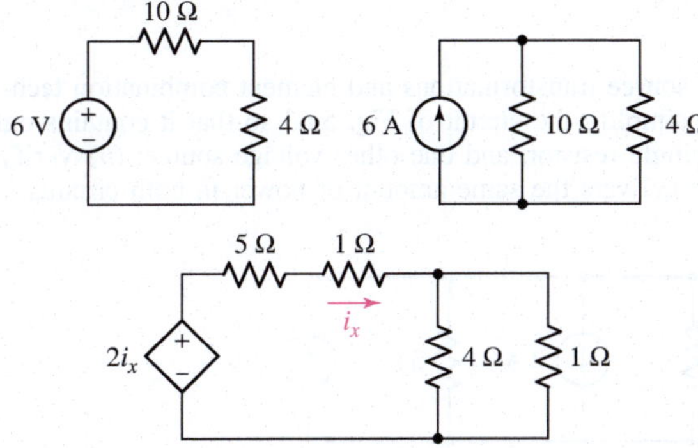

FIGURE 5.58

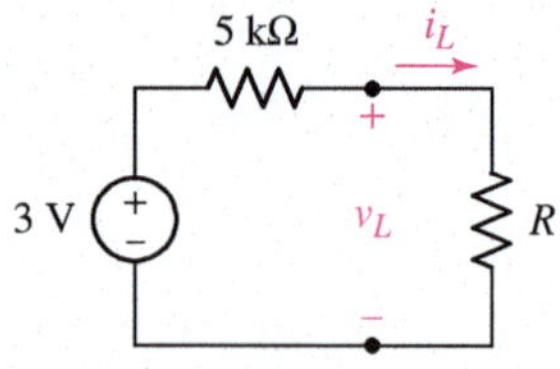

■ **FIGURE 5.59**

14. (*a*) For the circuit of Fig. 5.59, plot i_L versus v_L corresponding to the range of $0 \leq R \leq \infty$. (*b*) Plot the power delivered by the network to *R*, using the same range of resistance values as in part (*a*). (*c*) Repeat (*b*) after first performing a source transformation.

15. Determine the current labeled *I* in the circuit of Fig. 5.60 by first performing source transformations and parallel–series combinations as required to reduce the circuit to only three elements.

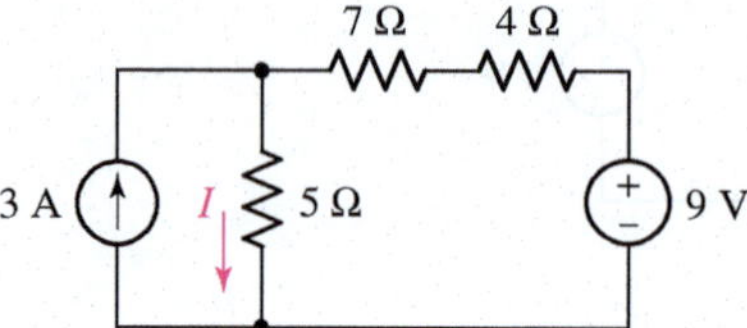

■ **FIGURE 5.60**

16. Verify that the power absorbed by the 7 Ω resistor in Fig. 5.22*a* remains the same after the source transformation illustrated in Fig. 5.22*c*.

17. (*a*) Determine the current labeled *i* in the circuit of Fig. 5.61 after first transforming the circuit such that it contains only resistors and voltage sources. (*b*) Simulate each circuit to verify the same current flows in both cases.

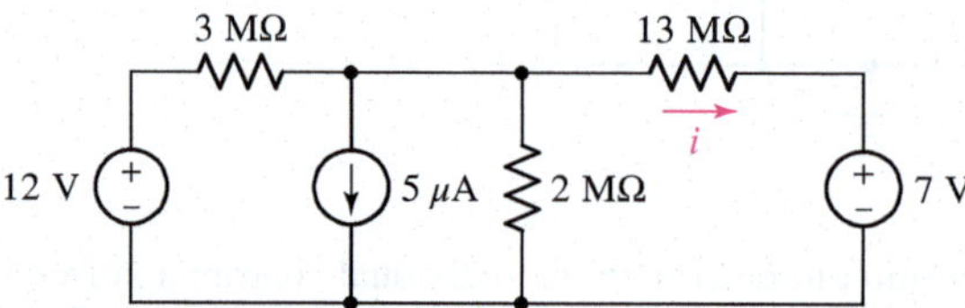

■ **FIGURE 5.61**

18. (*a*) Using repeated source transformations, reduce the circuit of Fig. 5.62 to a voltage source in series with a resistor, both of which are in series with the 6 MΩ resistor. (*b*) Calculate the power dissipated by the 6 MΩ resistor using your simplified circuit.

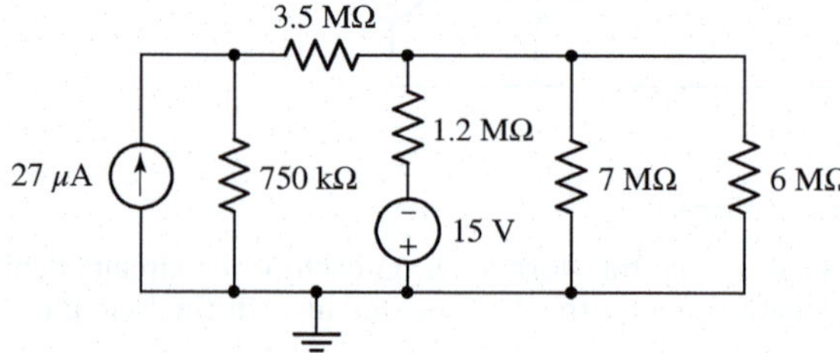

■ **FIGURE 5.62**

19. (*a*) Using as many source transformations and element combination techniques as required, simplify the circuit of Fig. 5.63 so that it contains only the 7 V source, a single resistor, and one other voltage source. (*b*) Verify that the 7 V source delivers the same amount of power in both circuits.

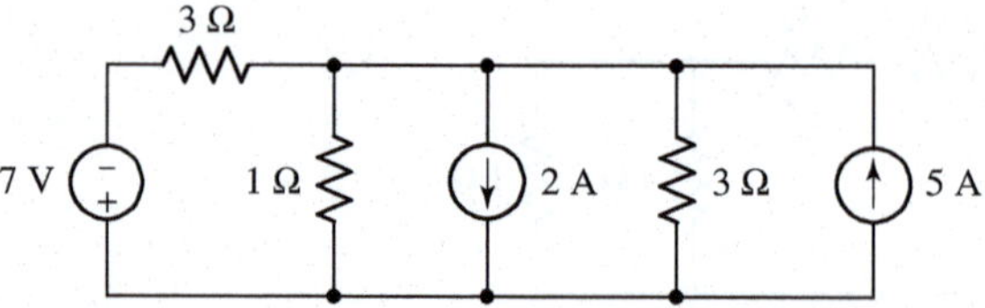

■ **FIGURE 5.63**

20. (*a*) Making use of repeated source transformations, reduce the circuit of Fig. 5.64 such that it contains a single voltage source, the 17 Ω resistor, and one other resistor. (*b*) Calculate the power dissipated by the 17 Ω resistor. (*c*) Verify your results by simulating both circuits.

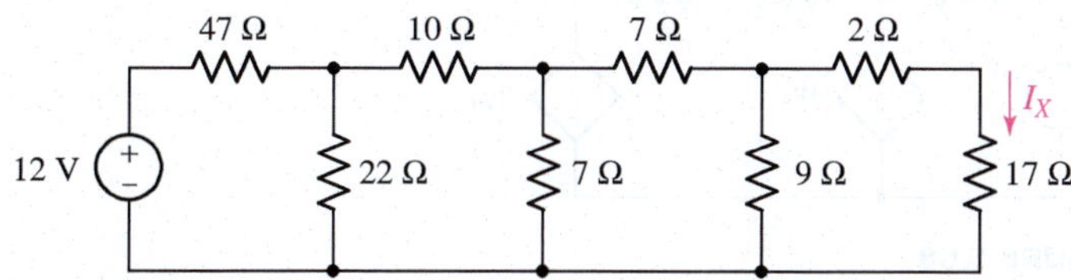

FIGURE 5.64

21. Make use of source transformations to first convert all three sources in Fig. 5.65 to voltage sources, then simplify the circuit as much as possible and calculate the voltage V_x which appears across the 4 Ω resistor. Be sure to draw and label your simplified circuit.

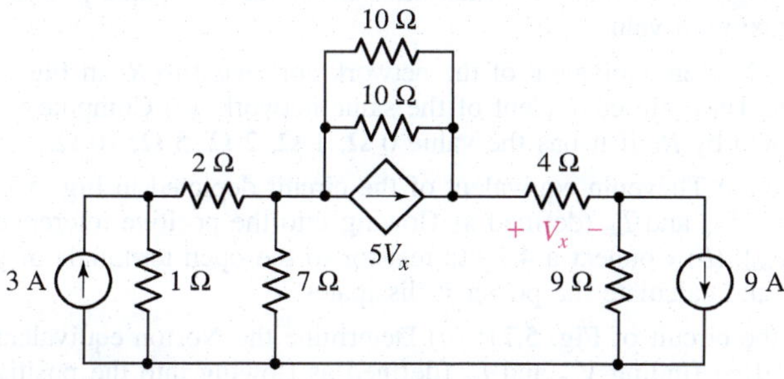

FIGURE 5.65

22. (*a*) With the assistance of source transformations, transform the two voltage sources in Fig. 5.66 into appropriate current sources. (*b*) Using your new circuit, calculate the power dissipated in the 7 Ω resistor. (*c*) Verify your solution by simulating both circuits.

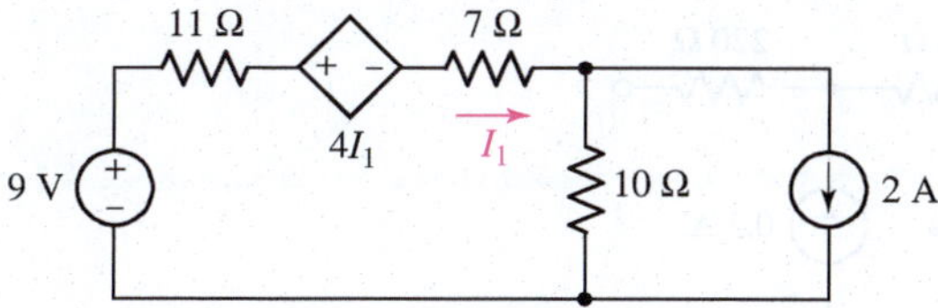

FIGURE 5.66

23. For the circuit in Fig. 5.67 transform all independent sources to current sources, then obtain an expression for I_B.

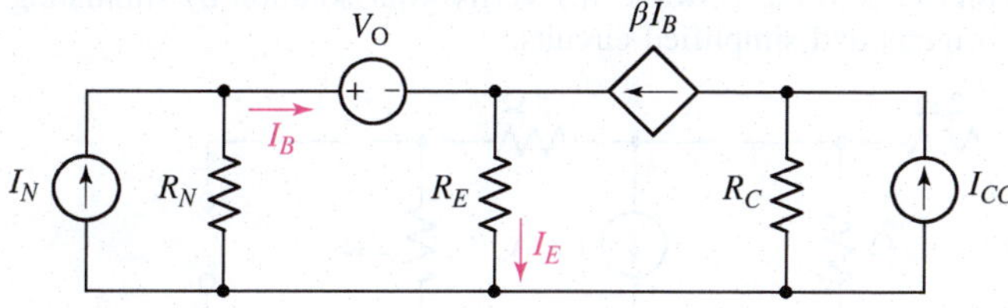

FIGURE 5.67

24. With regard to the circuit represented in Fig. 5.68, first transform both voltage sources to current sources, reduce the number of elements as much as possible, and determine the voltage v_3.

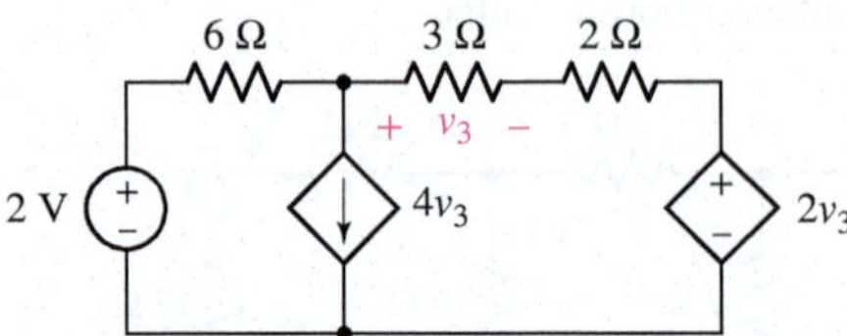

■ FIGURE 5.68

5.3 Thévenin and Norton Equivalent Circuits

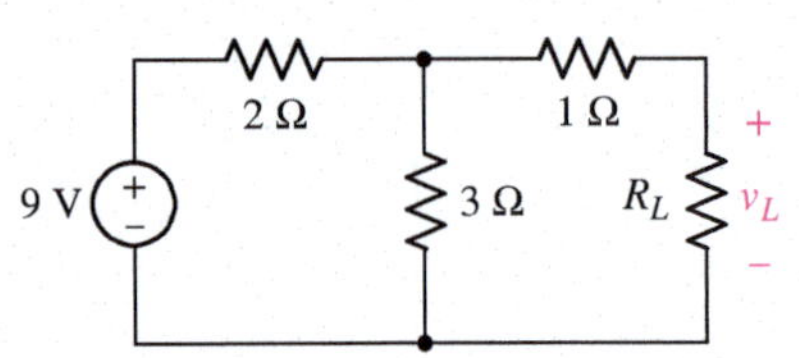

■ FIGURE 5.69

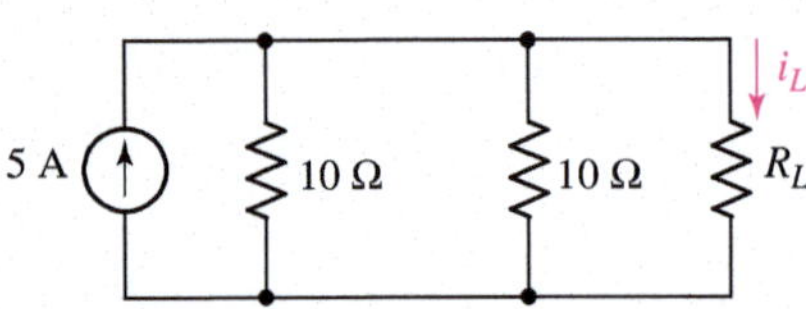

■ FIGURE 5.70

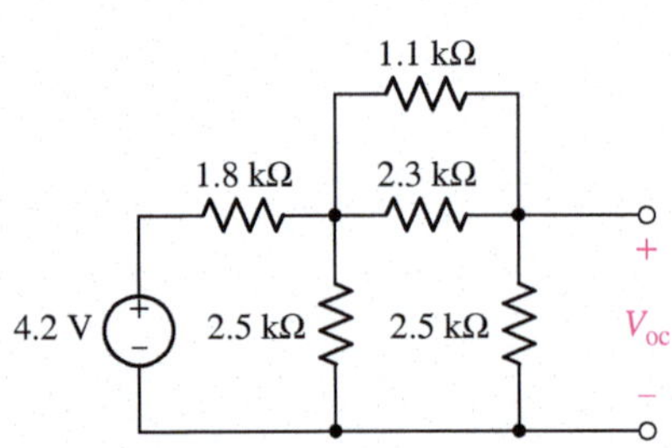

■ FIGURE 5.71

25. (*a*) Referring to Fig. 5.69, determine the Thévenin equivalent of the network connected to R_L. (*b*) Determine v_L for R_L = 1 Ω, 3.5 Ω, 6.257 Ω, and 9.8 Ω.

26. (*a*) With respect to the circuit depicted in Fig. 5.69, obtain the Norton equivalent of the network connected to R_L. (*b*) Plot the power dissipated in resistor R_L as a function of i_L corresponding to the range of $0 < R_L < 5\ \Omega$. (*c*) Using your graph, estimate at what value of R_L the dissipated power reaches its maximum value.

27. (*a*) Obtain the Norton equivalent of the network connected to R_L in Fig. 5.70. (*b*) Obtain the Thévenin equivalent of the same network. (*c*) Compute the power dissipated by R_L if it has the value 0 Ω, 1 Ω, 2 Ω, 5 Ω, 10 Ω.

28. (*a*) Determine the Thévenin equivalent of the circuit depicted in Fig. 5.71 by first finding V_{oc} and I_{sc} (defined as flowing into the positive reference terminal of V_{oc}). (*b*) Connect a 4.7 kΩ resistor to the open terminals of your new network and calculate the power it dissipates.

29. Referring to the circuit of Fig. 5.71: (*a*) Determine the Norton equivalent of the circuit by first finding V_{oc} and I_{sc} (defined as flowing into the positive reference terminal of V_{oc}). (*b*) Connect a 1.7 kΩ resistor to the open terminals of your new network and calculate the power supplied to that resistor.

30. (*a*) Employ Thévenin's theorem to obtain a simple two-component equivalent of the circuit shown in Fig. 5.72. (*b*) Use your equivalent circuit to determine the power delivered to a 100 Ω resistor connected to the open terminals. (*c*) Verify your solution by analyzing the original circuit with the same 100 Ω resistor connected across the open terminals.

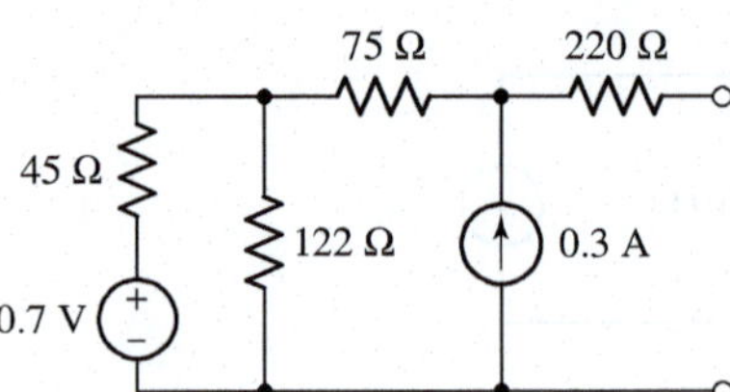

■ FIGURE 5.72

31. (*a*) Employ Thévenin's theorem to obtain a two-component equivalent for the network shown in Fig. 5.73. (*b*) Determine the power supplied to a 1 Ω resistor connected to the network. (*c*) Verify your solution by simulating both the original and simplified circuits.

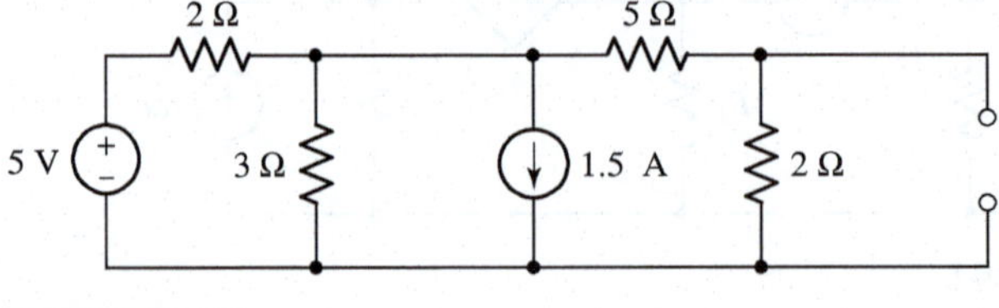

■ FIGURE 5.73

32. Determine the Thévenin equivalent of the network shown in Fig. 5.74 as seen looking into the two open terminals.

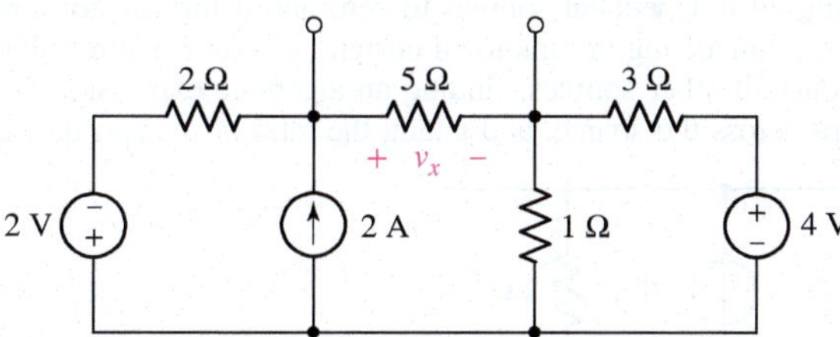

FIGURE 5.74

33. (*a*) Determine the Norton equivalent of the circuit depicted in Fig. 5.74 as seen looking into the two open terminals. (*b*) Compute power dissipated in a 5 Ω resistor connected in parallel with the existing 5 Ω resistor. (*c*) Compute the current flowing through a short circuit connecting the two terminals.

34. For the circuit of Fig. 5.75: (*a*) Employ Norton's theorem to reduce the network connected to R_L to only two components. (*b*) Calculate the downward-directed current flowing through R_L if it is a 3.3 kΩ resistor. (*c*) Verify your answer by simulating both circuits.

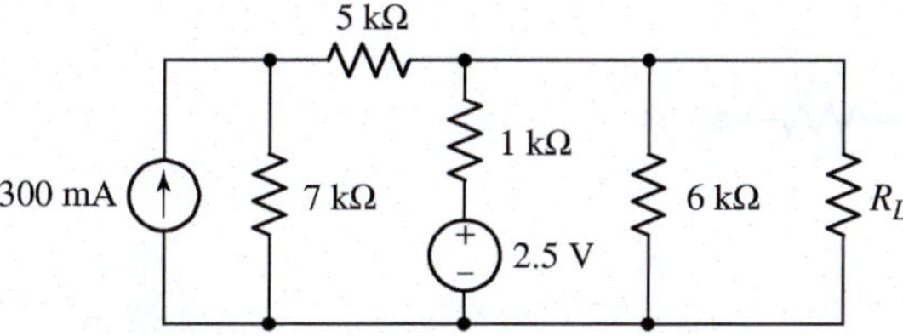

FIGURE 5.75

35. (*a*) Obtain a value for the Thévenin equivalent resistance seen looking into the open terminals of the circuit in Fig. 5.76 by first finding V_{oc} and I_{sc}. (*b*) Connect a 1 A test source to the open terminals of the original circuit after shorting the voltage source, and use this to obtain R_{TH}. (*c*) Connect a 1 V test source to the open terminals of the original circuit after again zeroing the 2 V source, and use this now to obtain R_{TH}.

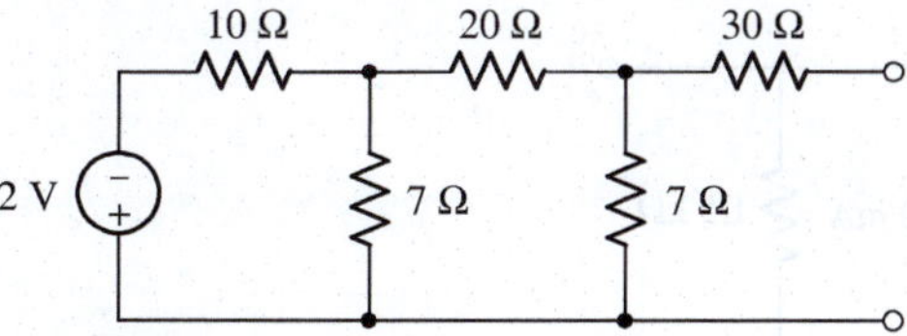

FIGURE 5.76

36. Refer to the circuit depicted in Fig. 5.77. (*a*) Obtain a value for the Thévenin equivalent resistance seen looking into the open terminals by first finding V_{oc} and I_{sc}. (*b*) Connect a 1 A test source to the open terminals of the original circuit after deactivating the other current source, and use this to obtain R_{TH}. (*c*) Connect a 1 V test source to the open terminals of the original circuit, once again zeroing out the original source, and use this now to obtain R_{TH}.

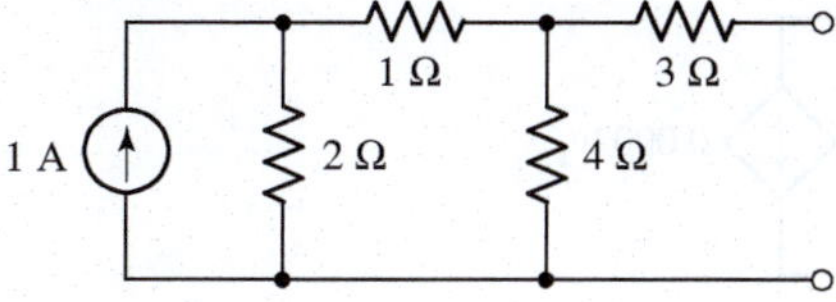

FIGURE 5.77

37. Obtain a value for the Thévenin equivalent resistance seen looking into the open terminals of the circuit in Fig. 5.78 by (*a*) finding V_{oc} and I_{sc}, and then taking their ratio; (*b*) setting all independent sources to zero and using resistor combination techniques; (*c*) connecting an unknown current source to the terminals, deactivating (zero out) all other sources, finding an algebraic expression for the voltage that develops across the source, and taking the ratio of the two quantities.

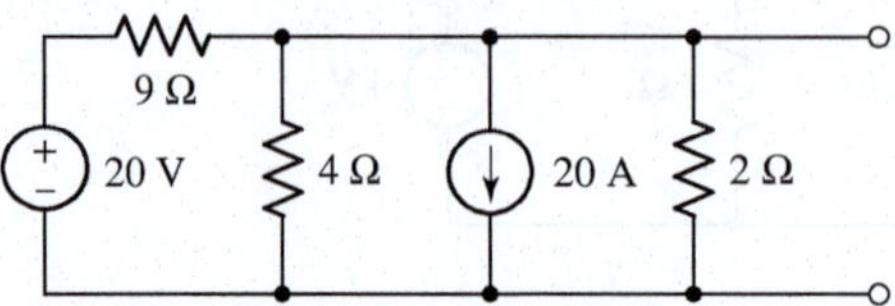

FIGURE 5.78

38. With regard to the network depicted in Fig. 5.79, determine the Thévenin equivalent as seen by an element connected to terminals (*a*) *a* and *b*; (*b*) *a* and *c*; (*c*) *b* and *c*. (*d*) Verify your answers using an appropriate circuit simulation. (*Hint: Connect a test source to the terminals of interest.*)

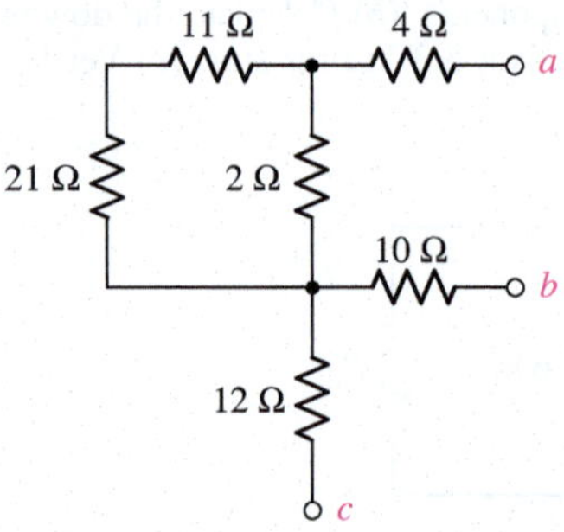

FIGURE 5.79

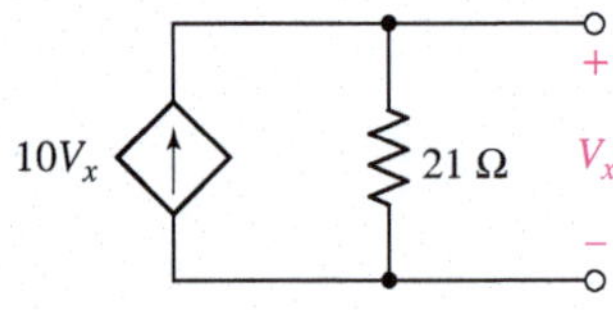

FIGURE 5.80

39. Determine the Thévenin and Norton equivalents of the circuit represented in Fig. 5.80 from the perspective of the open terminals. (There should be no dependent sources in your answer.)

40. Determine the Norton equivalent of the circuit drawn in Fig. 5.81 as seen by terminals *a* and *b*. (There should be no dependent sources in your answer.)

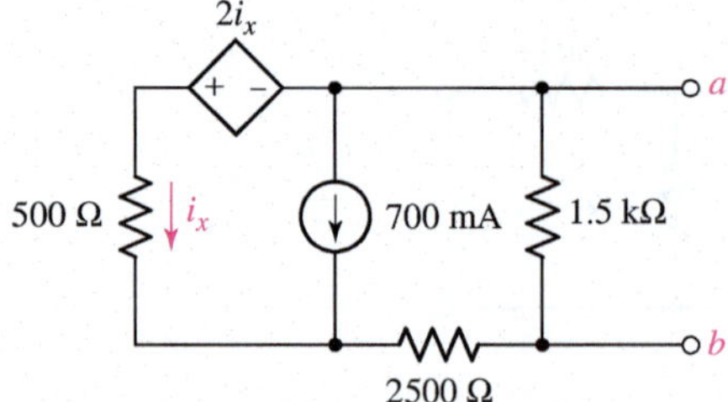

FIGURE 5.81

41. With regard to the circuit of Fig. 5.82, determine the power dissipated by (*a*) a 1 kΩ resistor connected between *a* and *b*; (*b*) a 4.7 kΩ resistor connected between *a* and *b*; (*c*) a 10.54 kΩ resistor connected between *a* and *b*.

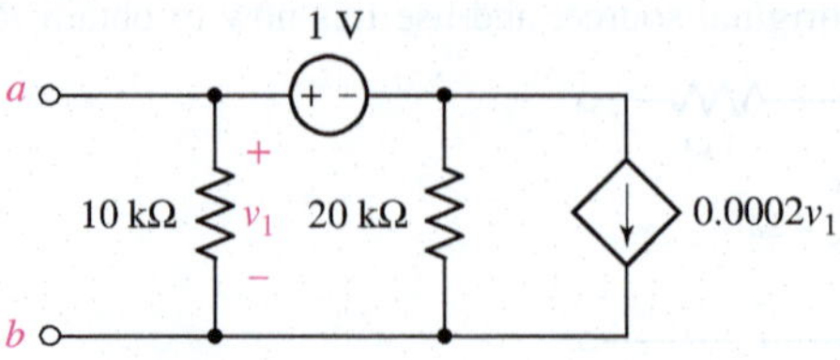

FIGURE 5.82

42. Determine the Thévenin and Norton equivalents of the circuit shown in Fig. 5.83, as seen by an unspecified element connected between terminals a and b.

43. Referring to the circuit of Fig. 5.84, determine the Thévenin equivalent resistance of the circuit to the right of the dashed line. This circuit is a common-source transistor amplifier, and you are calculating its input resistance.

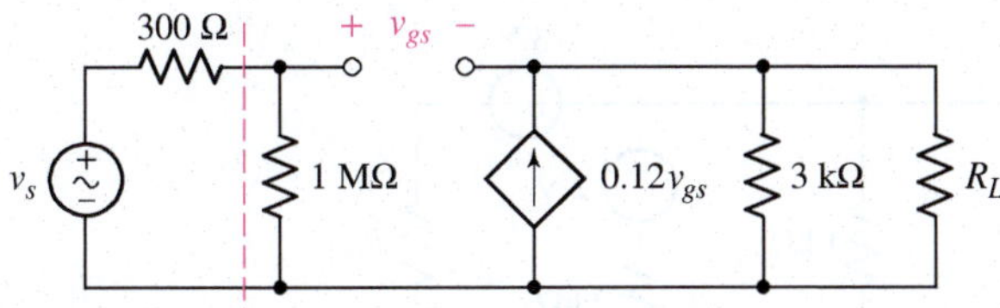

■ **FIGURE 5.84**

44. Referring to the circuit of Fig. 5.85, determine the Thévenin equivalent resistance of the circuit to the right of the dashed line. This circuit is a common-collector transistor amplifier, and you are calculating its input resistance.

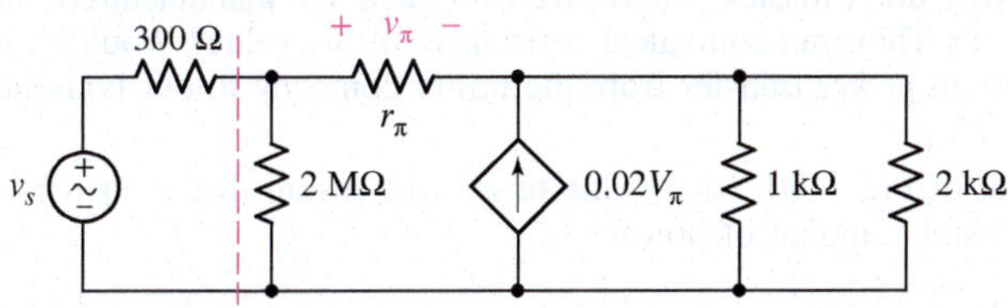

■ **FIGURE 5.85**

45. The circuit shown in Fig. 5.86 is a reasonably accurate model of an operational amplifier. In cases where R_i and A are very large and $R_o \sim 0$, a resistive load (such as a speaker) connected between ground and the terminal labeled v_{out} will see a voltage $-R_f/R_1$ times larger than the input signal v_{in}. Find the Thévenin equivalent of the circuit, taking care to label v_{out}.

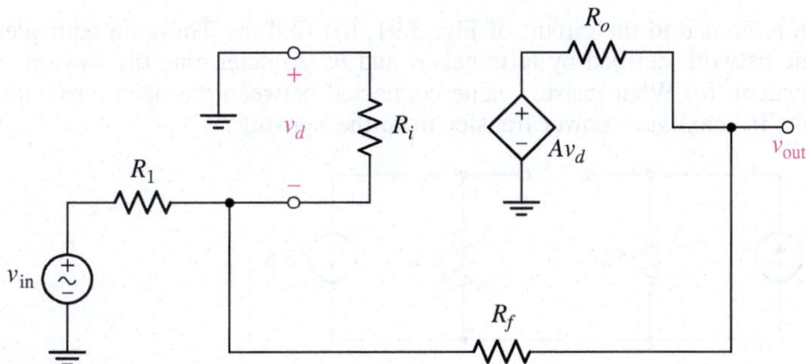

■ **FIGURE 5.86**

5.4 Maximum Power Transfer

46. (*a*) For the simple circuit of Fig. 5.87, find the Thévenin equivalent connected to resistor R_L. (*b*) Plot the power delivered to R_L (as a function of R_L) if its value is constrained by $0 \leq R_L \leq 10$ kΩ. (*c*) What value of R_L results in maximum power transferred from the network? (*d*) What value of R_L results in 50% of the power in part (*c*)?

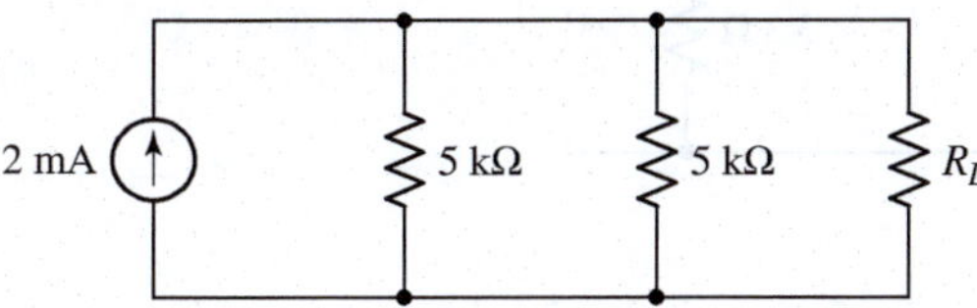

■ **FIGURE 5.87**

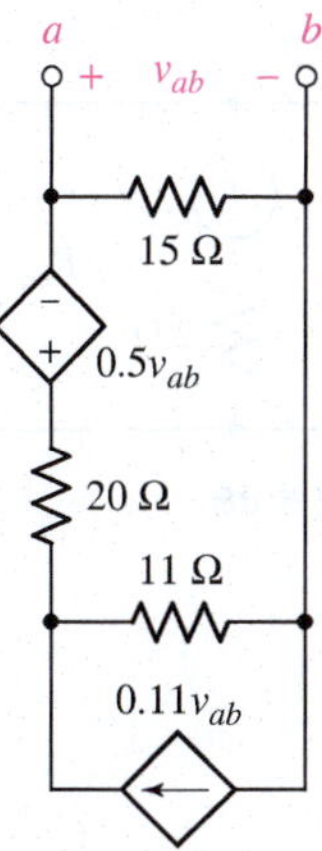

■ **FIGURE 5.83**

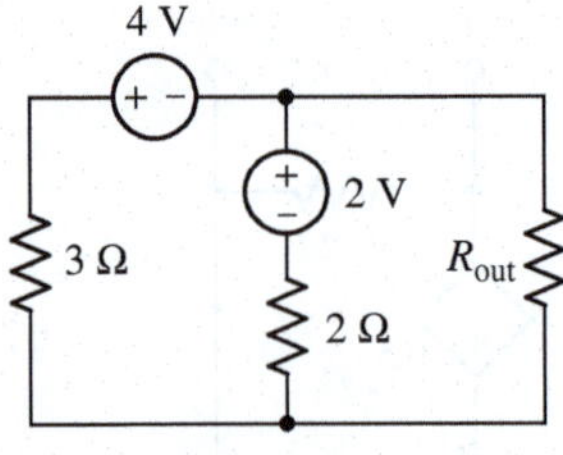

■ **FIGURE 5.88**

47. For the circuit drawn in Fig. 5.88, (*a*) determine the Thévenin equivalent connected to R_{out}. (*b*) Choose R_{out} such that maximum power is delivered to it.
48. Study the circuit of Fig. 5.89. (*a*) Determine the Norton equivalent connected to resistor R_{out}. (*b*) Select a value for R_{out} such that maximum power will be delivered to it.

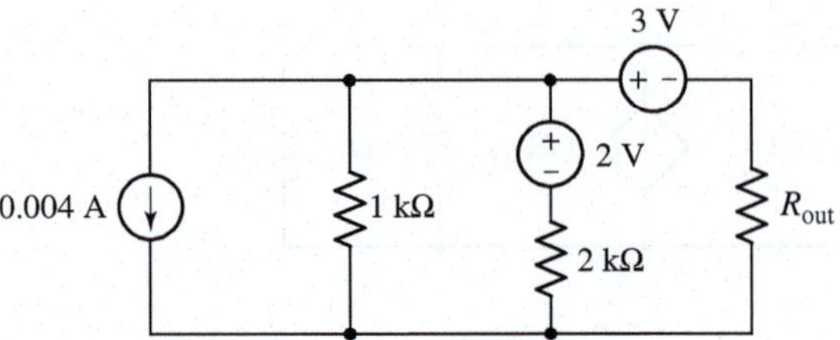

■ **FIGURE 5.89**

49. Assuming that we can determine the Thévenin equivalent resistance of our wall socket, why don't toaster, microwave oven, and TV manufacturers match each appliance's Thévenin equivalent resistance to this value? Wouldn't it permit maximum power transfer from the utility company to our household appliances?
50. For the circuit of Fig. 5.90, what value of R_L will ensure that it absorbs the maximum possible amount of power?

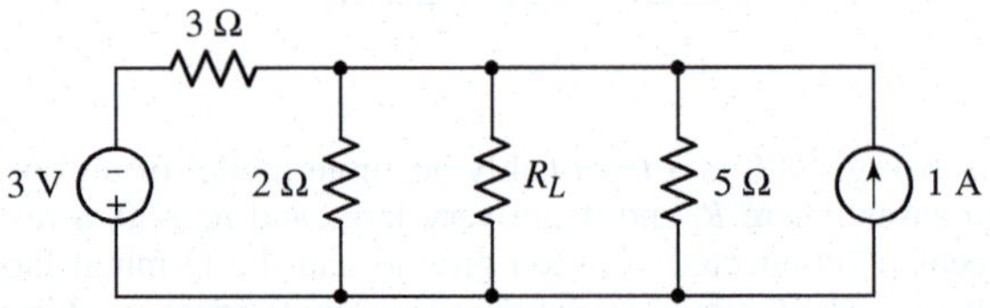

■ **FIGURE 5.90**

51. With reference to the circuit of Fig. 5.91, (*a*) find the Thévenin equivalent of the network defined by terminals *a* and *b*; (*b*) determine the Norton equivalent. (*c*) What resistor value connected between the open terminals results in maximum power transfer from the network?

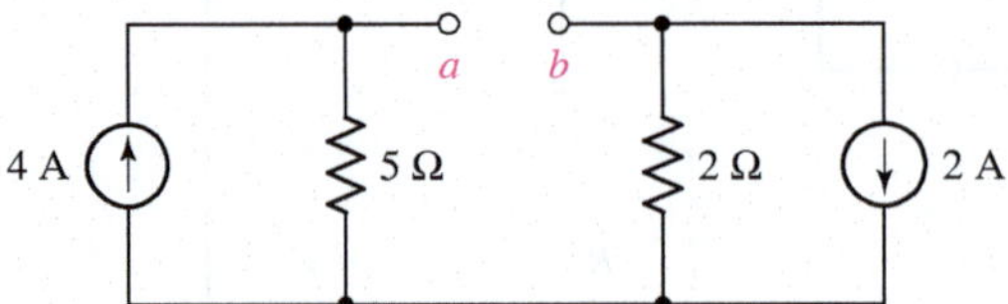

■ **FIGURE 5.91**

52. Referring to the circuit of Fig. 5.92, (*a*) determine the power absorbed by the 3.3 Ω resistor; (*b*) replace the 3.3 Ω resistor with another resistor such that it absorbs maximum power from the rest of the circuit.

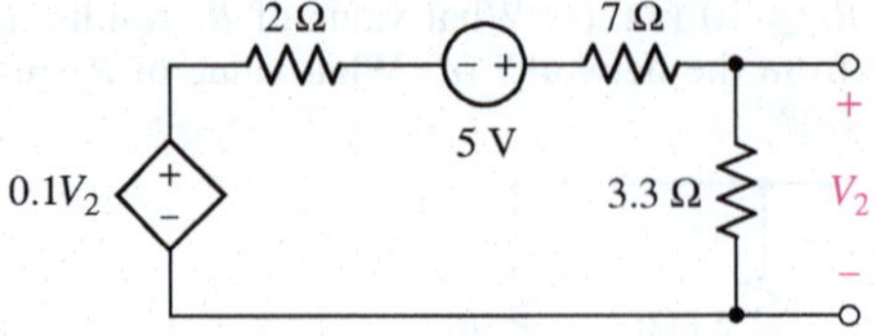

■ **FIGURE 5.92**

53. Select a value for R_L in Fig. 5.93 such that it will absorb maximum power from the circuit.

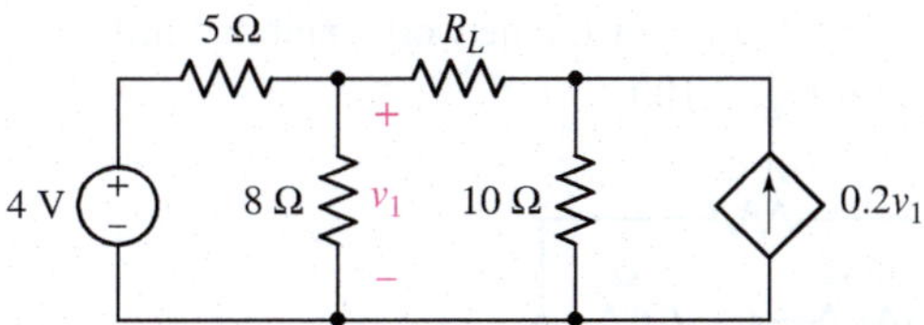

FIGURE 5.93

54. Determine what value of resistance would absorb maximum power from the circuit of Fig. 5.94 when connected across terminals a and b.

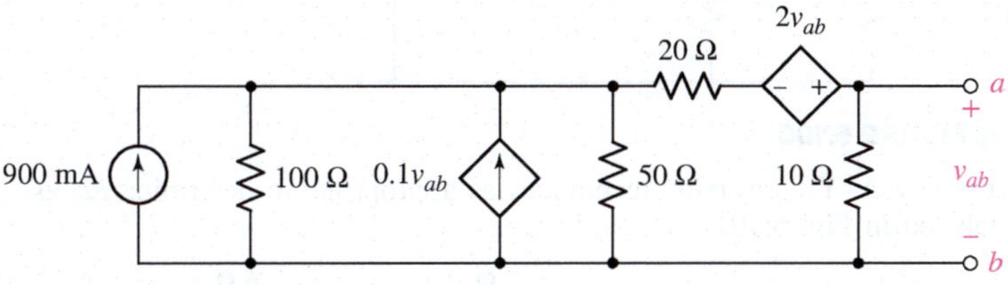

FIGURE 5.94

5.5 Delta–Wye Conversion

55. Derive the equations required to convert from a Y-connected network to a Δ-connected network.
56. Convert the Δ- (or "Π-") connected networks in Fig. 5.95 to Y-connected networks.

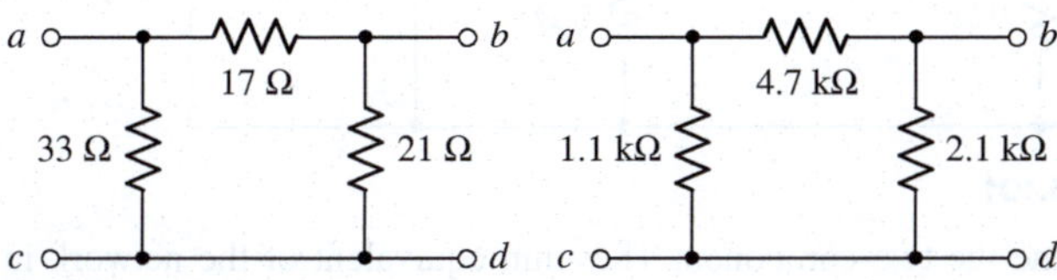

FIGURE 5.95

57. Convert the Y- (or "T-") connected networks in Fig. 5.96 to Δ-connected networks.

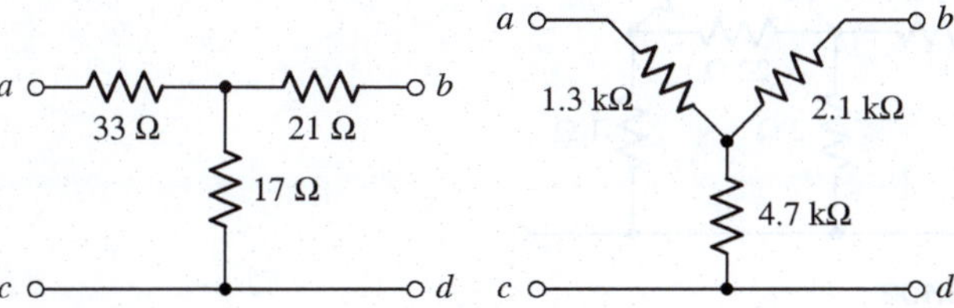

FIGURE 5.96

58. For the network of Fig. 5.97, select a value of R such that the network has an equivalent resistance of 9 Ω. Round your answer to two significant figures.
59. For the network of Fig. 5.98, select a value of R such that the network has an equivalent resistance of 70.6 Ω.

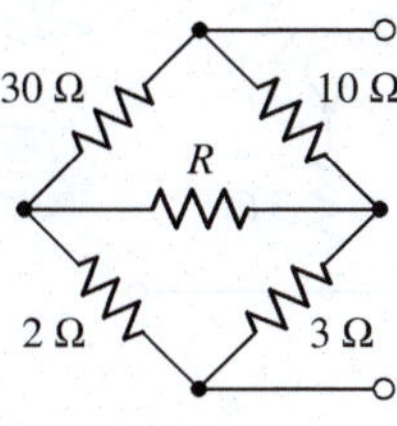

FIGURE 5.97

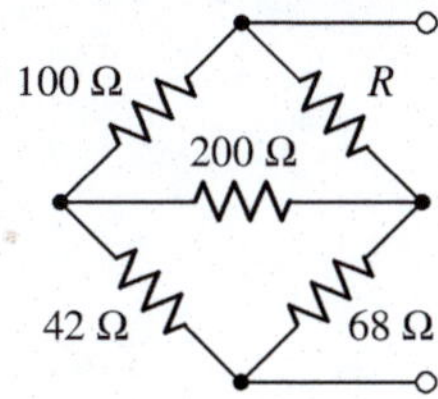

FIGURE 5.98

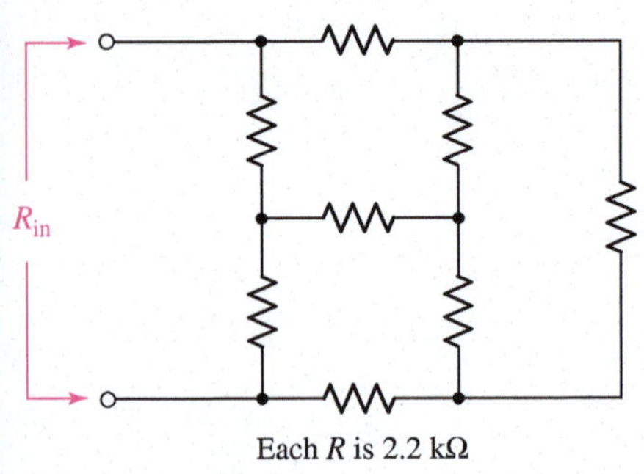

FIGURE 5.99

60. Determine the effective resistance R_{in} of the network exhibited in Fig. 5.99.
61. Calculate R_{in} as indicated in Fig. 5.100.

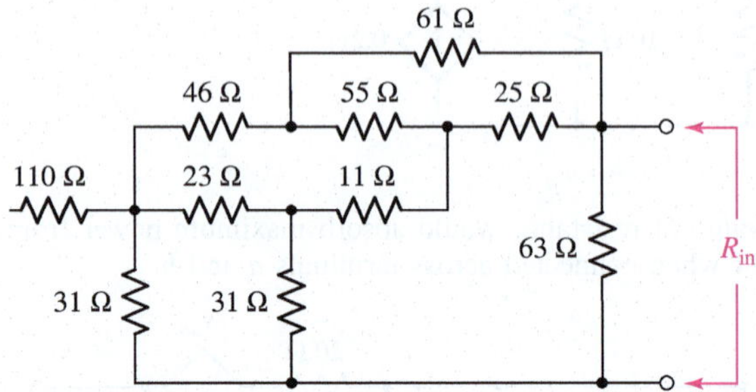

FIGURE 5.100

62. Employ Δ–Y conversion techniques as appropriate to determine R_{in} as labeled in Fig. 5.101.

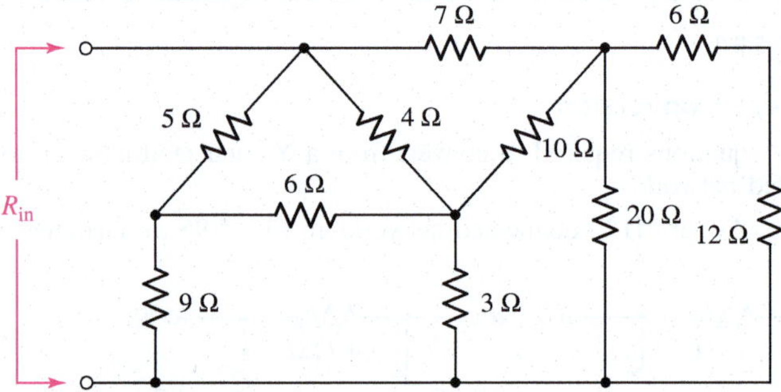

FIGURE 5.101

63. (*a*) Determine the two-component Thévenin equivalent of the network in Fig. 5.102. (*b*) Calculate the power dissipated by a 1 Ω resistor connected between the open terminals.

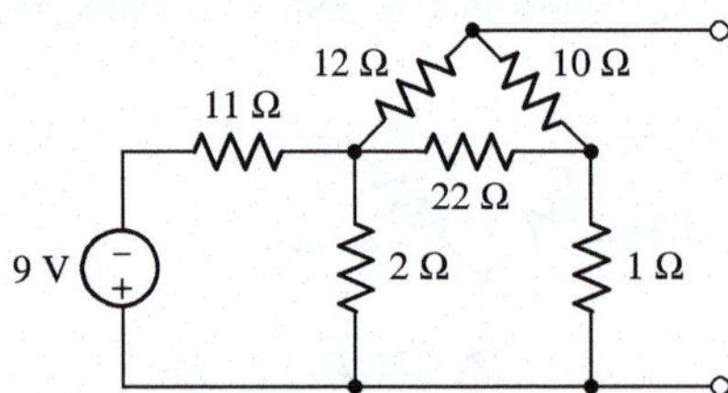

FIGURE 5.102

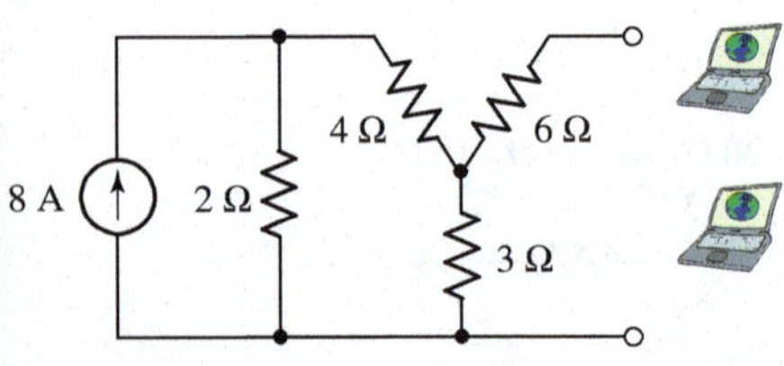

FIGURE 5.103

64. (*a*) Use appropriate techniques to obtain both the Thévenin and Norton equivalents of the network drawn in Fig. 5.103. (*b*) Verify your answers by simulating each of the three circuits connected to a 1 Ω resistor.
65. (*a*) For the network in Fig. 5.104, replace the leftmost Δ network with an equivalent T network. (*b*) Perform a computer simulation to verify that your answer is in fact equivalent. (*Hint:* Try adding a load resistor.)

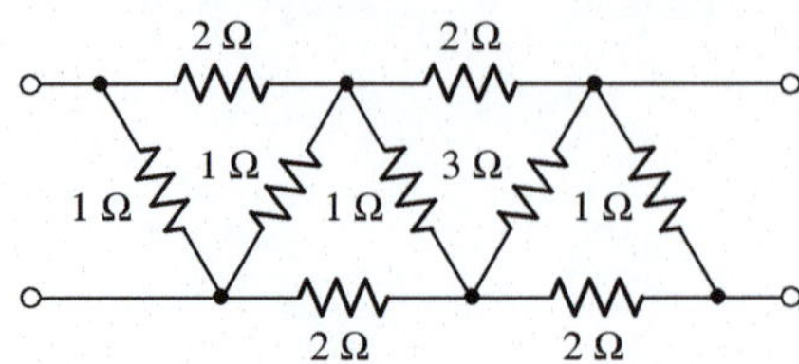

FIGURE 5.104

5.6 Selecting an Approach: A Summary of Various Techniques

66. Determine the power absorbed by a resistor connected between the open terminal of the circuit shown in Fig. 5.105 if it has a value of (*a*) 1 Ω; (*b*) 100 Ω; (*c*) 2.65 kΩ; (*d*) 1.13 MΩ.

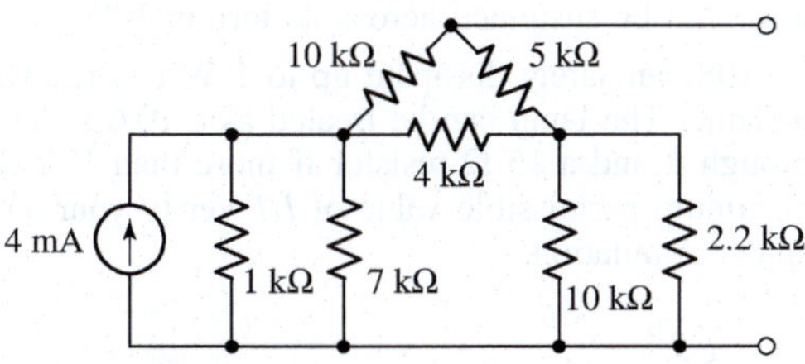

FIGURE 5.105

67. It is known that a load resistor of some type will be connected between terminals *a* and *b* of the network of Fig. 5.106. (*a*) Change the value of the 25 V source such that both voltage sources contribute equally to the current delivered to the load resistor, assuming its value is chosen such that it absorbs maximum power. (*b*) Calculate the value of the load resistor.

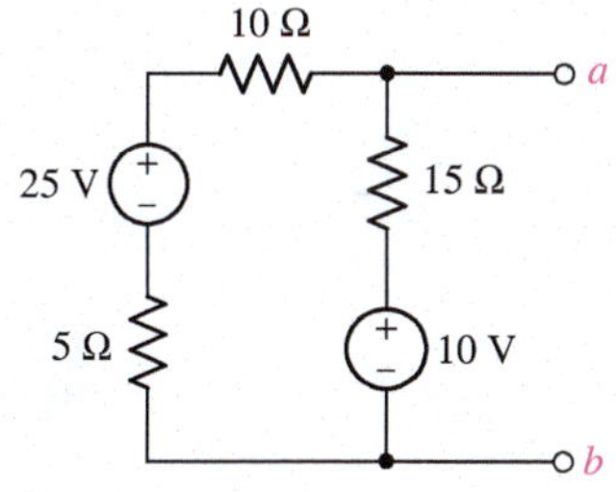

FIGURE 5.106

68. A 2.57 Ω load is connected between terminals *a* and *b* of the network drawn in Fig. 5.106. Unfortunately, the power delivered to the load is only 50% of the required amount. Altering only voltage sources, modify the circuit so that the required power is delivered.

69. A load resistor is connected across the open terminals of the circuit shown in Fig. 5.107, and its value was chosen carefully to ensure maximum power transfer from the rest of the circuit. (*a*) What is the value of the resistor? (*b*) If the power absorbed by the load resistor is three times as large as required, modify the circuit so that it performs as desired, without losing the maximum power transfer condition already enjoyed.

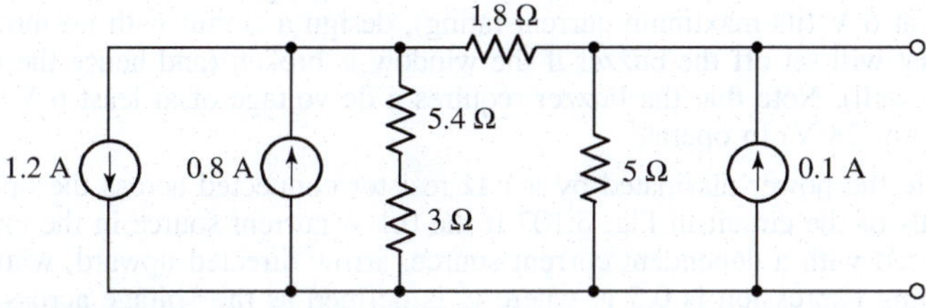

FIGURE 5.107

70. A backup is required for the circuit depicted in Fig. 5.107. It is unknown what will be connected to the open terminals, or whether it will be purely linear. If a simple battery is to be used, what no-load ("open circuit") voltage should it have, and what is the desired internal resistance?

Chapter-Integrating Exercises

71. Three 45 W light bulbs originally wired in a Y network configuration with a 120 V ac source connected across each port are rewired as a Δ network. The neutral, or center, connection is not used. If the intensity of each light is proportional to the power it draws, design a new 120 V ac power circuit so that the three lights have the same intensity in the Δ configuration as they did when connected in a Y configuration. Verify your design using LTspice by comparing the power drawn by each light in your circuit (modeled as an appropriately chosen resistor value) with the power each would draw in the original Y-connected circuit.

72. (*a*) Explain in general terms how source transformation can be used to simplify a circuit prior to analysis. (*b*) Even if source transformations can greatly simplify a particular circuit, when might it not be worth the effort?

(*c*) Multiplying all the independent sources in a circuit by the same scaling factor results in all other voltages and currents being scaled by the same amount. Explain why we don't scale the dependent sources as well. (*d*) In a general circuit, if we set an independent voltage source to zero, what current can flow through it? (*e*) In a general circuit, if we set an independent current source to zero, what voltage can be sustained across its terminals?

73. The load resistor in Fig. 5.108 can safely dissipate up to 1 W before overheating and bursting into flame. The lamp can be treated as a 10.6 Ω resistor if less than 1 A flows through it and a 15 Ω resistor if more than 1 A flows through it. What is the maximum permissible value of I_s? Verify your answer with an appropriate computer simulation.

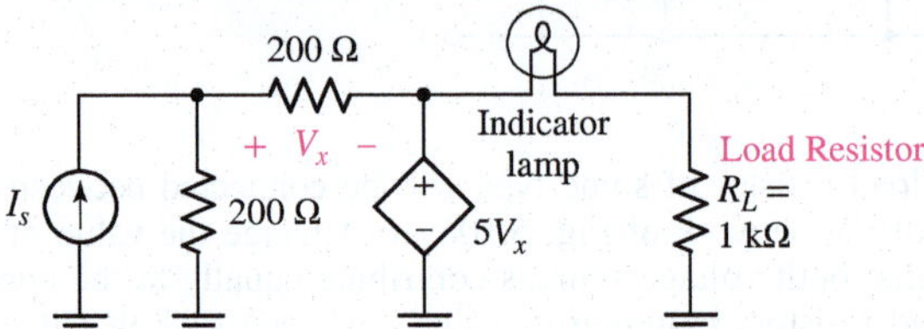

■ **FIGURE 5.108**

74. A certain red LED has a maximum current rating of 35 mA, and if this value is exceeded, overheating and catastrophic failure will result. The resistance of the LED is a nonlinear function of its current, but the manufacturer warrants a minimum resistance of 47 Ω and a maximum resistance of 117 Ω. Only 9 V batteries are available to power the LED. Design a suitable circuit to deliver the maximum power possible to the LED without damaging it. Use only combinations of the standard resistor values given in the inside front cover.

75. As part of a security system, a very thin 100 Ω wire is attached to a window using nonconducting epoxy. Given only a box of 12 rechargeable 1.5 V AAA batteries, one thousand 1 Ω resistors, and a 2.9 kHz piezo buzzer that draws 15 mA at 6 V (its maximum current rating), design a circuit with no moving parts that will set off the buzzer if the window is broken (and hence the thin wire as well). Note that the buzzer requires a dc voltage of at least 6 V (maximum 28 V) to operate.

76. Compute the power dissipated by a 1 Ω resistor connected across the open terminals of the circuit in Fig. 5.107 if the 0.1 A current source in the circuit is replaced with a dependent current source, arrow directed upward, whose controlling expression is 0.5 v_x where v_x is defined as the voltage across the 1.8 Ω resistor, "+" reference on the left. Provide your rationale for the analysis approach you chose (there is no one "right" approach—explain *why* you took the route you did, not *what* route you chose).

77. Design an experiment to determine if the circuit contained within an opaque box is linear over the range of ±5 V. Only two wires protrude from the box. The only power source is an external 9 V battery which can be connected or disconnected at will. Include a table specifying measurements to be performed, and provide each step of the required analysis as well as a guide as to how one should interpret the results to make a determination.

78. For the circuit of Fig. 5.51, the 5 Ω resistor between the two labeled nodes is to be replaced by a voltage source which will precisely balance the two halves of the circuit such that no current flows between the labeled nodes. Determine the necessary voltage of this source, and specify its polarity.

CHAPTER 6

The Operational Amplifier

KEY CONCEPTS

- Characteristics of Ideal Op Amps
- Inverting and Noninverting Amplifiers
- Summing and Difference Amplifier Circuits
- Cascaded Op Amp Stages
- Voltage Gain and Feedback
- Nonideal Characteristics of Op Amps
- Comparators and Schmitt Triggers
- Instrumentation Amplifiers

INTRODUCTION

At this point we have a good set of circuit analysis tools at our disposal, but we have focused on somewhat general circuits composed of only sources and resistors, or *passive* devices where electron flow cannot be directly controlled. Electronics requires the ability to control electron flow using an electrical input, such as the input/output relationships in computing or amplification of an audio signal. *Active* devices that can control electron flow are typically nonlinear, but they can often be treated effectively with linear models. In this chapter, we introduce one of these elements, known as the *operational amplifier* or *op amp* for short. Op amps are among the most versatile devices that find daily usage in applications such as sensor circuits, control, signal processing, and of course, amplifiers!

6.1 BACKGROUND

The origins of the operational amplifier date to the 1940s, when basic circuits were constructed using vacuum tubes to perform mathematical operations such as addition, subtraction, multiplication, division, differentiation, and integration. This enabled the construction of analog (as opposed to digital) computers tasked with the solution of complex differential equations. The first commercially available op amp *device* is generally considered to be the K2-W, manufactured by Philbrick Researches, Inc. of Boston from about 1952 through the early 1970s (Fig. 6.1). These early vacuum tube devices weighed 3 oz (85 g), measured $1\frac{33}{64}$ in × $2\frac{9}{64}$ in × $4\frac{7}{64}$ in (3.8 cm × 5.4 cm × 10.4 cm), and sold for about US$22. In contrast, integrated circuit (IC) op amps such as the TLV9061 weigh approximately 22.5 mg,

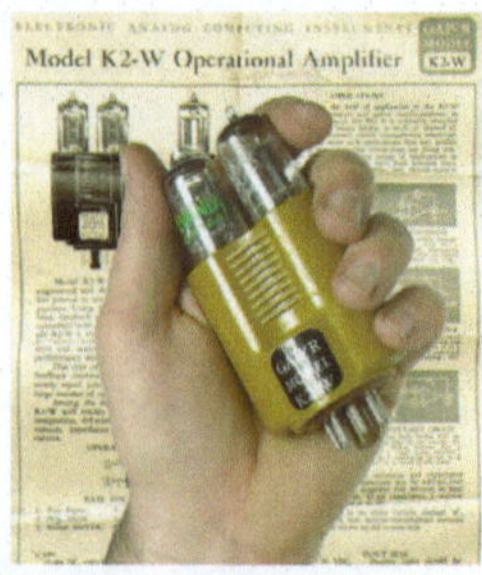

FIGURE 6.1 A Philbrick K2-W op amp, based on a matched pair of 12AX7A vacuum tubes.
Steve Durbin/McGraw Hill

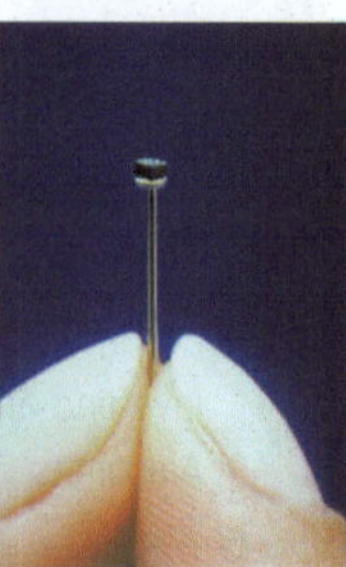

FIGURE 6.2 LMC6035 operational amplifier, which packs 114 transistors into a package so small that it fits on the head of a pin.
Courtesy of Texas Instruments

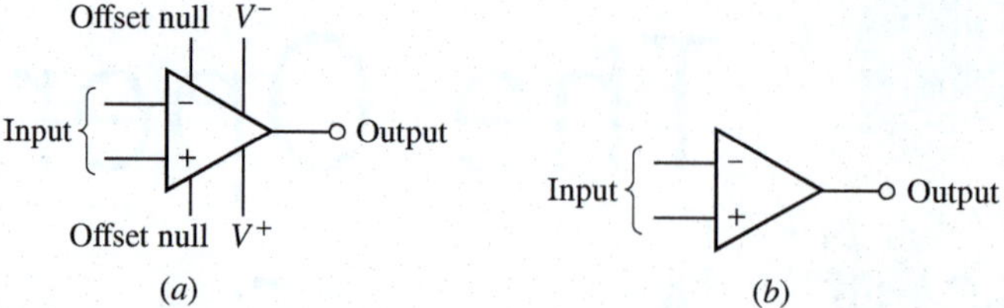

FIGURE 6.3 (*a*) Electrical symbol for the op amp. (*b*) Minimum required connections to be shown on a circuit schematic.

measure 3.05 mm × 3.0 mm × 1.45 mm, and sell for approximately US$0.79. In contrast to op amps based on vacuum tubes, modern integrated circuit (IC) op amps are constructed using perhaps 25 or more transistors all on the same silicon "chip," as well as resistors and capacitors needed to obtain the desired performance characteristics. As a result, they run at much lower dc supply voltages (±18 V or lower, for example, as opposed to ±300 V for the K2-W), are more reliable, and are considerably smaller (Fig. 6.2). In some cases, the IC may contain several op amps.

The op amp is a voltage amplifier with two inputs and one output. Two input terminals are denoted by a "+" for the ***noninverting input*** and by a "−" for the ***inverting input.*** In addition to the output pin and the two inputs, other pins enable power to be supplied to run the transistors in the IC and to make external adjustments to balance and compensate the op amp (not always available for all op amps). The symbol commonly used for an op amp is shown in Fig. 6.3*a*. At this point, we are not concerned with the internal circuitry of the op amp or the IC, but only with the voltage and current relationships that exist between the input and output terminals. Thus, for the time being we will use a simpler electrical symbol, shown in Fig. 6.3*b*.

6.2 THE IDEAL OP AMP

In practice, we find that most op amps perform so well that we can often make the assumption that we are dealing with an "ideal" op amp. The characteristics of an ***ideal op amp*** form the basis for two fundamental rules for analyzing circuits with ideal op amps:

Ideal Op Amp Rules

1. No current ever flows into either input terminal. (Current can flow at the output terminal!)
2. There is no voltage difference between the two input terminals.

In a real op amp, a very small leakage current will flow into the input (sometimes as low as 40 femtoamperes). It is also possible to obtain a very small voltage across the two input terminals. However, compared to other voltages and currents in most circuits, such values are so small that including them in the analysis does not typically affect our calculations.

When analyzing op amp circuits, we should keep one other point in mind. As opposed to the circuits that we have studied so far, an op amp circuit always has an *output* that depends on some type of *input*. Therefore, we will analyze op amp circuits with the goal of obtaining an expression for the output in terms of the input quantities. *We will find that it is usually a good idea to begin the analysis of an op amp circuit at the input, and proceed from there.*

The circuit shown in Fig. 6.4 is known as an ***inverting amplifier.*** We choose to analyze this circuit using KVL, beginning with the input voltage source, with the goal of determining the output v_{out} in terms of the input v_{in} and circuit resistor values. The current labeled i flows only through the two resistors R_1 and R_f, since ideal op amp rule 1 states that no current flows into the input terminal. Thus, we can write

$$-v_{in} + R_1 i + R_f i + v_{out} = 0$$

which can be rearranged to obtain an equation that relates the output to the input:

$$v_{out} = v_{in} - (R_1 + R_f)i \qquad [1]$$

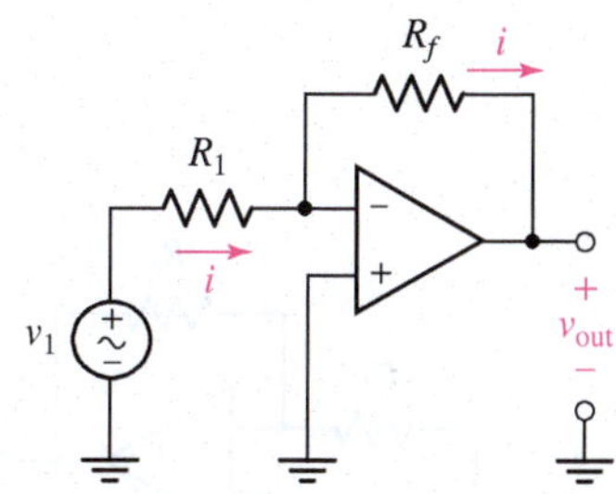

FIGURE 6.4 An op amp used to construct an inverting amplifier circuit. The current i flows to ground through the output pin of the op amp. The resistor R_f connected between the output pin and the inverting input pin provides for a negative feedback path; this serves to limit the output voltage range.

This is a good time to mention that we have not yet made use of ideal op amp rule 2. Since the noninverting input is grounded, it is at zero volts. By ideal op amp rule 2, the inverting input is therefore also at zero volts! *This does not mean that the two inputs are physically shorted together, and we should be careful not to make such an assumption.* Rather, the two input voltages simply track each other: if we try to change the voltage at one pin, the other pin will be driven by internal circuitry to the same value. Thus, we can write one more KVL equation:

$$-v_{in} + R_1 i + 0 = 0$$

or

$$i = \frac{v_{in}}{R_1} \qquad [2]$$

The fact that the inverting input terminal finds itself at zero volts in this type of circuit configuration leads to what is often referred to as a "virtual ground." This does not mean that the pin is actually grounded, which is sometimes a source of confusion for students. The op amp makes whatever internal adjustments are necessary to prevent a voltage difference between the input terminals. The input terminals are not shorted together.

Combining Eq. [2] with Eq. [1], we obtain an expression for v_{out} in terms of v_{in}:

$$v_{out} = -\frac{R_f}{R_1} v_{in} \qquad [3]$$

The resulting answer shows that the output v_{out} is proportional to the input v_{in} by the factor $(-R_f/R_1)$, or it amplifies the input by a negative constant defined by resistor values. Given values of $v_{in} = 5 \sin 3t$ mV, $R_1 = 4.7$ kΩ, and $R_f = 47$ kΩ for the circuit, we get

$$v_{out} = -50 \sin 3t \text{ mV}$$

Since $R_f > R_1$, this circuit amplifies the input voltage signal v_{in}. If we choose $R_f < R_1$, the signal will be attenuated instead. We also note that the output voltage has the opposite sign of the input voltage,[1] hence the name "inverting amplifier." The output is sketched in Fig. 6.5, along with the input waveform for comparison.

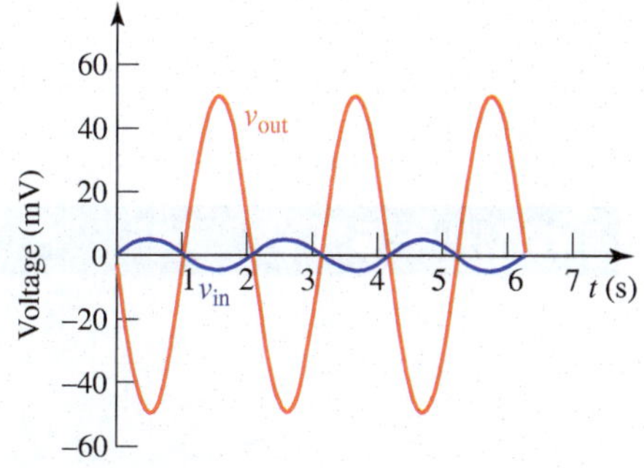

FIGURE 6.5 Input and output waveforms of the inverting amplifier circuit.

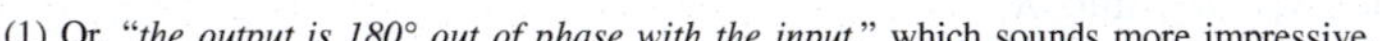

(1) Or, "*the output is 180° out of phase with the input,*" which sounds more impressive.

At this point, it is worth mentioning that the ideal op amp seems to be violating KCL. Specifically, in the preceding circuit, no current flows into or out of either input terminal, but somehow current is able to flow into the output pin! This would imply that the op amp is somehow able to either create electrons out of nowhere or store them forever (depending on the direction of current flow). Obviously, this is not possible. The conflict arises because we have been treating the op amp the same way we treated passive elements such as the resistor. In reality, however, the op amp cannot function unless it is connected to external power sources. It is through those power sources that we can direct current flow into or out of the output terminal.

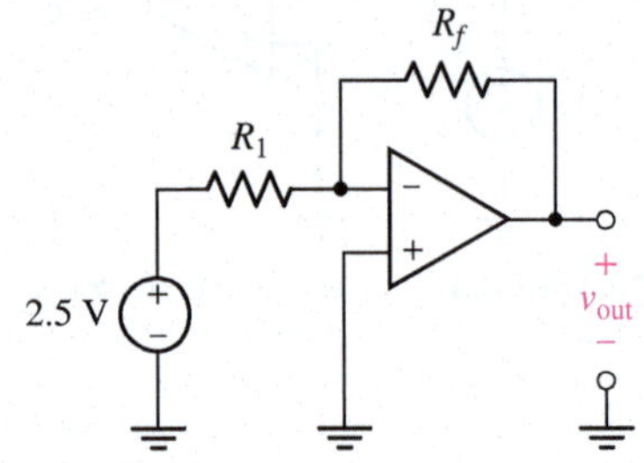

■ **FIGURE 6.6** An inverting amplifier circuit with a 2.5 V input.

Although we have shown that the inverting amplifier circuit of Fig. 6.4 can amplify an ac signal (a sine wave in this case having a frequency of 3 rad/s and an amplitude of 5 mV), it works just as well with dc inputs. We consider this type of situation in Fig. 6.6, where values for R_1 and R_f are to be selected to obtain an output voltage of -10 V. This is the same circuit as shown in Fig. 6.4, but with a 2.5 V dc input. Since no other change has been made, the expression we presented as Eq. [3] is valid for this circuit as well. To obtain the desired output, we seek a ratio of R_f to R_1 of 10/2.5, or 4. Since it is only the ratio that is important here, we simply need to pick a convenient value for one resistor, and the other resistor value is then fixed at the same time. For example, we could choose $R_1 = 100\ \Omega$ (so $R_f = 400\ \Omega$), or even $R_f = 8\ \text{M}\Omega$ (so $R_1 = 2\ \text{M}\Omega$). In practice, other constraints (such as bias current) may limit our choices.

This circuit configuration therefore acts as a convenient type of voltage amplifier (or ***attenuator,*** if the ratio of R_f to R_1 is less than 1), but it does have the sometimes inconvenient property of inverting the sign of the input. There is an alternative, however, which is analyzed just as easily—the noninverting amplifier shown in Fig. 6.7. We examine such a circuit in the following example.

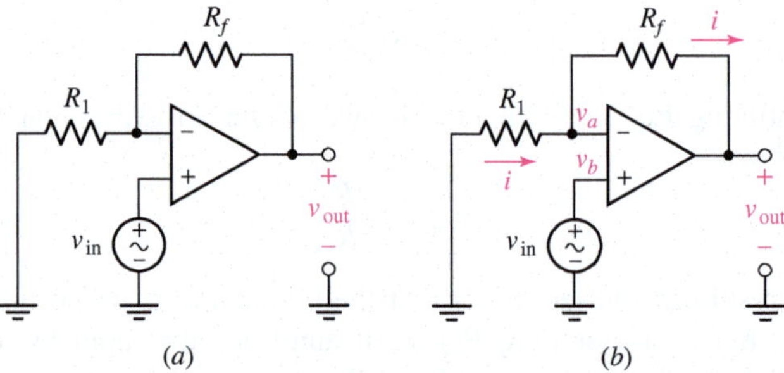

■ **FIGURE 6.7** (*a*) An op amp used to construct a noninverting amplifier circuit. (*b*) Circuit with the current through R_1 and R_f defined, as well as both input voltages labeled.

EXAMPLE 6.1

Sketch the output waveform of the noninverting amplifier circuit in Fig. 6.7*a*. Use $v_{in} = 5 \sin 3t$ mV, $R_1 = 4.7\ \text{k}\Omega$, and $R_f = 47\ \text{k}\Omega$.

▶ ***Identify the goal of the problem.***

We require an expression for v_{out} that only depends on the known quantities v_{in}, R_1, and R_f.

▶ ***Collect the known information.***

Since values have been specified for the resistors and the input waveform, we begin by labeling the current i and the two input voltages as shown in Fig. 6.7*b*. We will assume that the op amp is an ideal op amp.

▶ ***Devise a plan.***

Although mesh analysis is a favorite technique of students, it turns out to be more practical in most op amp circuits to apply nodal analysis, since there is no direct way to determine the current flowing out of the op amp output.

▶ ***Construct an appropriate set of equations.***

Note that we are using ideal op amp rule 1 implicitly by defining the same current through both resistors: no current flows into the inverting input terminal. Employing nodal analysis to obtain our expression for v_{out} in terms of v_{in}, we thus find that

at node a:

$$0 = \frac{v_a}{R_1} + \frac{v_a - v_{out}}{R_f} \qquad [4]$$

and at node b:

$$v_b = v_{in} \qquad [5]$$

▶ ***Determine if additional information is required.***

Our goal is to obtain a single expression that relates the input and output voltages, although neither Eq. [4] nor Eq. [5] appears to do so. However, we have not yet employed ideal op amp rule 2, and we will find that in almost every op amp circuit *both* rules need to be invoked in order to complete our analysis.

Thus, we recognize that $v_a = v_b = v_{in}$, and Eq. [4] becomes

$$0 = \frac{v_{in}}{R_1} + \frac{v_{in} - v_{out}}{R_f}$$

▶ ***Attempt a solution.***

Rearranging, we obtain an expression for the output voltage in terms of the input voltage v_{in}:

$$v_{out} = \left(1 + \frac{R_f}{R_1}\right) v_{in} = 11\, v_{in} = 55 \sin 3t \text{ mV}$$

▶ ***Verify the solution. Is it reasonable or expected?***

The output waveform is sketched in Fig. 6.8, along with the input waveform for comparison. In contrast to the output waveform of the inverting amplifier circuit, we note that the input and output are in phase for the noninverting amplifier. This should not be entirely unexpected: it is implicit in the name "noninverting amplifier."

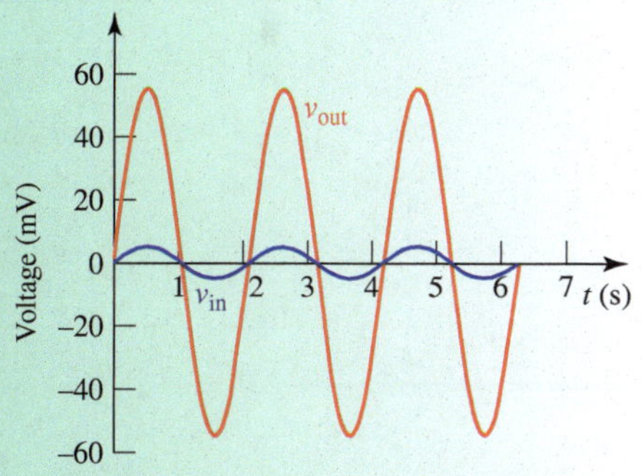

FIGURE 6.8 Input and output waveforms for the noninverting amplifier circuit.

PRACTICE

6.1 Derive an expression for v_{out} in terms of v_{in} for the circuit shown in Fig. 6.9.

Ans: $v_{out} = v_{in}$. The circuit is known as a "*voltage follower,*" since the output voltage tracks or "follows" the input voltage.

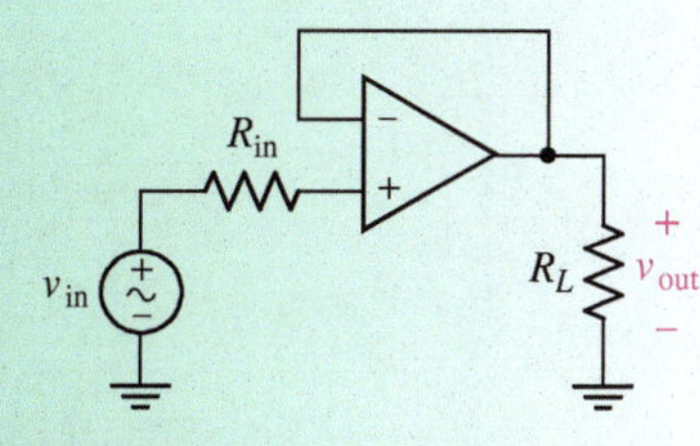

FIGURE 6.9

Just like the inverting amplifier, the noninverting amplifier works with dc as well as ac inputs, but it has a voltage gain of $v_{out}/v_{in} = 1 + (R_f/R_1)$. Thus, if we set $R_f = 9\ \Omega$ and $R_1 = 1\ \Omega$, we obtain an output v_{out} which is 10 times larger than the input voltage v_{in}. In contrast to the inverting amplifier, the output and input of the noninverting amplifier always have the same sign, and the output voltage cannot be less than the input; the minimum gain is 1. Which amplifier we choose depends on the application we are considering.

In the special case of the voltage follower circuit shown in Fig. 6.9, which represents a noninverting amplifier with R_1 set to ∞ and R_f set to zero, the output is identical to the input in both sign *and* magnitude. This may seem rather pointless as a general type of circuit, but we should keep in mind that *the voltage follower draws no current from the input* (in the ideal case)—it therefore can act as a ***buffer*** between the voltage v_{in} and some resistive load R_L connected to the output of the op amp.

We mentioned earlier that the name "operational amplifier" originates from using such devices to perform arithmetical operations on analog (i.e., nondigitized, real-time, real-world) signals. As we see in the following two circuits, this includes both addition and subtraction of input voltage signals.

EXAMPLE 6.2

Obtain an expression for v_{out} in terms of v_1, v_2, and v_3 for the op amp circuit in Fig. 6.10, also known as a *summing amplifier*.

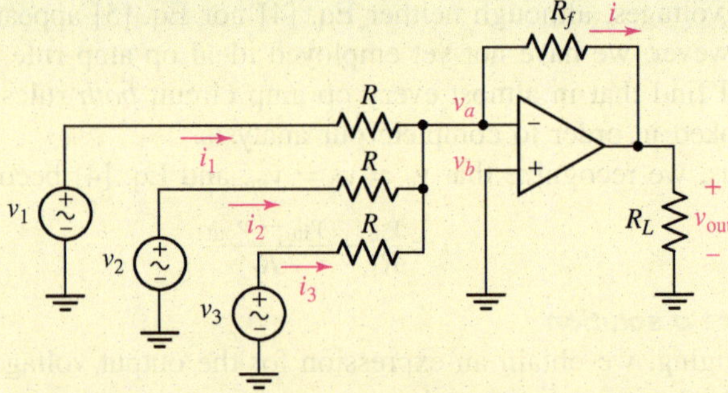

FIGURE 6.10 Basic summing amplifier circuit with three inputs.

We first note that this circuit is similar to the inverting amplifier circuit of Fig. 6.4. Again, the goal is to obtain an expression for v_{out} (which in this case appears across a load resistor R_L) in terms of the inputs (v_1, v_2, and v_3). Since no current can flow into the inverting input terminal, we can write

$$i_1 + i_2 + i_3 = i$$

Therefore, we can write the following equation at the node labeled v_a:

$$\frac{v_1 - v_a}{R} + \frac{v_2 - v_a}{R} + \frac{v_3 - v_a}{R} = \frac{v_a - v_{out}}{R_f}$$

This equation contains both v_{out} and the input voltages, but unfortunately it also contains the nodal voltage v_a. To remove this unknown quantity from our expression, we need to write an additional equation that relates v_a to v_{out}, the input voltages, R_f, and/or R. At this point, we remember that we have not yet used ideal op amp rule 2, and that we will almost certainly require the use of both rules when analyzing an op amp circuit. Thus, since $v_a = v_b = 0$, we can write the following:

$$\frac{v_1}{R} + \frac{v_2}{R} + \frac{v_3}{R} = -\frac{v_{out}}{R_f}$$

Rearranging, we obtain the following expression for v_{out}:

$$v_{out} = -\frac{R_f}{R}(v_1 + v_2 + v_3) \qquad [6]$$

In the special case where $v_2 = v_3 = 0$, we see that our result agrees with Eq. [3], which was derived for essentially the same circuit.

PRACTICE

6.2 Derive an expression for v_{out} in terms of v_1 and v_2 for the circuit shown in Fig. 6.11, also known as a ***difference amplifier.***

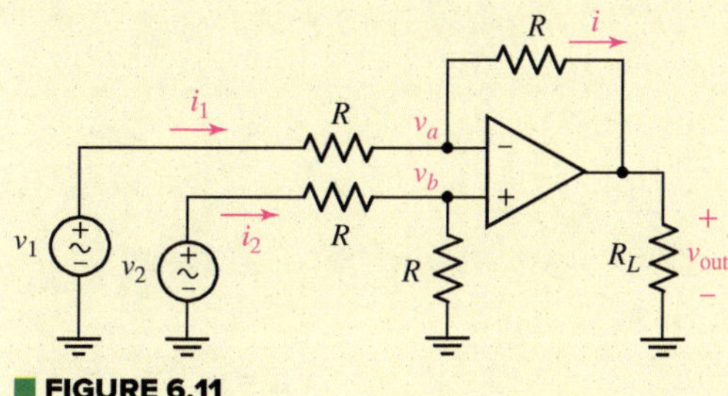

FIGURE 6.11

Ans: $v_{out} = v_2 - v_1$. *Hint:* Use voltage division to obtain v_b.

There are several interesting features about the result we have just derived. First, if we select R_f so that it is equal to R, then the output is the (negative of the) sum of the three input signals v_1, v_2, and v_3. Further, we can select the ratio of R_f to R to multiply this sum by a fixed constant.

Also, we notice that R_L did not appear in our final expression. As long as its value is not too low such that the output is shorted, the operation of the circuit will not be affected; at present, we have not considered a detailed enough model of an op amp to predict such an occurrence. This resistor represents the Thévenin equivalent of whatever we use to monitor the amplifier output. If our output device is a simple voltmeter, then R_L represents the Thévenin equivalent resistance seen looking into the voltmeter terminals (typically 10 MΩ or more). Or, our output device might be a speaker (typically 8 Ω), in which case we hear the sum of the three separate sources of sound; v_1, v_2, and v_3 might represent microphones in that case.

One word of caution: It is often tempting to assume that the current labeled i in Fig. 6.10 flows not only through R_f but through R_L also. Not true! It is very possible that current is flowing through the output terminal of the op amp as well, so that *the currents through the two resistors are not the same*. It is for this reason that we almost universally avoid writing KCL equations at the output pin of an op amp, which leads to the preference of nodal over mesh analysis when working with most op amp circuits.

For convenience, we summarize the most common op amp circuits in Table 6.1.

TABLE 6.1 Summary of Basic Op Amp Circuits

Name	Circuit Schematic	Input-Output Relation
Inverting Amplifier		$v_{out} = -\frac{R_f}{R_1} v_{in}$
Noninverting Amplifier		$v_{out} = \left(1 + \frac{R_f}{R_1}\right) v_{in}$
Voltage Follower (also known as a Unity Gain Amplifier)		$v_{out} = v_{in}$
Summing Amplifier		$v_{out} = -\frac{R_f}{R}(v_1 + v_2 + v_3)$
Difference Amplifier		$v_{out} = v_2 - v_1$

PRACTICAL APPLICATION

A Fiber Optic Intercom

A point-to-point intercom system can be constructed using a number of different approaches, depending on the intended application environment. Low-power radio frequency (RF) systems work very well and are generally cost-effective, but are subject to interference from other RF sources and are also prone to eavesdropping. Use of a simple wire to connect the two intercom systems instead can eliminate a great deal of the RF interference as well as increase privacy. However, wires are subject to corrosion and short circuits when the plastic insulation wears, and their weight can be a concern in aircraft and related applications (Fig. 6.12).

■ **FIGURE 6.12** The application environment often dictates design constraints.
Lucian Coman/EyeEm/Getty Images

An alternative design would be to convert the electrical signal from the microphone to an optical signal, which could then be transmitted through a thin (~50 μm diameter) optical fiber. The optical signal is then converted back to an electrical signal, which is amplified and delivered to a speaker. A schematic diagram of such a system is shown in Fig. 6.13; two such systems would be needed for two-way communication.

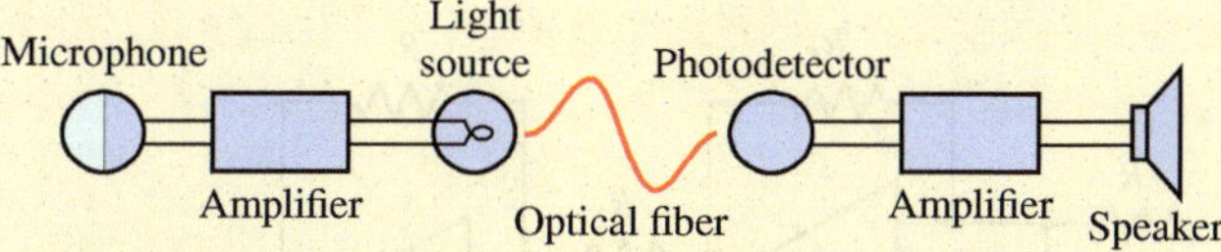

■ **FIGURE 6.13** Schematic diagram of one-half of a simple fiber optic intercom.

We can consider the design of the transmission and reception circuits separately, since the two circuits are in fact electrically independent. Figure 6.14 shows a simple

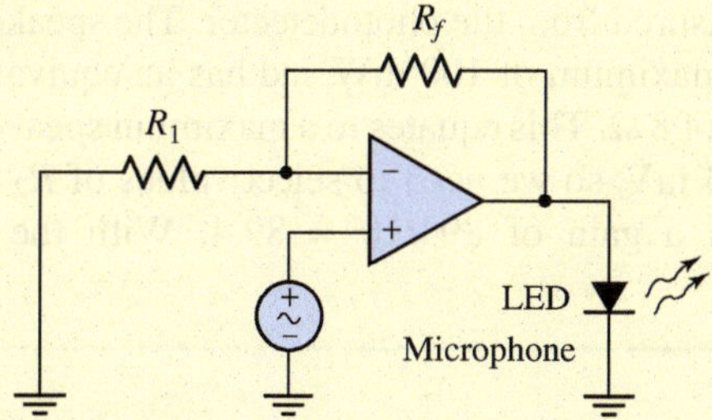

■ **FIGURE 6.14** Circuit used to convert the electrical microphone signal into an optical signal for transmission through a fiber.

signal generation circuit consisting of a microphone, a light-emitting diode (LED), and an op amp used in a noninverting amplifier circuit to drive the LED; not shown are the power connections required for the op amp itself. The light output of the LED is roughly proportional to its current, although less so for very small and very large values of current.

We know the gain of the amplifier is given by

$$\frac{v_{\text{out}}}{v_{\text{in}}} = 1 + \frac{R_f}{R_1}$$

which is independent of the resistance of the LED. In order to select values for R_f and R_1, we need to know the input voltage from the microphone and the necessary output voltage to power the LED. A quick measurement indicates that the typical voltage output of the microphone peaks at 40 mV when someone is using a normal speaking voice. The LED manufacturer recommends operating at approximately 1.6 V, so we design for a gain of 1.6/0.04 = 40. Arbitrarily choosing $R_1 = 1$ kΩ leads to a required value of 39 kΩ for R_f.

The circuit of Fig. 6.15 is the receiver part of our one-way intercom system. It converts the optical signal from the fiber into an electrical signal, amplifying it so that an audible sound emanates from the speaker.

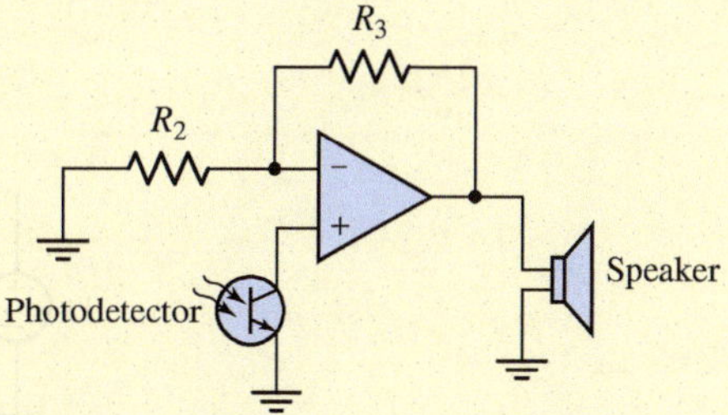

■ **FIGURE 6.15** Receiver circuit used to convert the optical signal into an audio signal.

(Continued on next page)

After coupling the LED output of the transmitting circuit to the optical fiber, a signal of approximately 10 mV is measured from the photodetector. The speaker is rated for a maximum of 100 mW and has an equivalent resistance of 8 Ω. This equates to a maximum speaker voltage of 894 mV, so we need to select values of R_2 and R_3 to obtain a gain of 894/10 = 89.4. With the arbitrary selection of $R_2 = 10$ kΩ, we find that a value of 884 kΩ completes our design.

This circuit will work in practice, although the nonlinear characteristics of the LED lead to a noticeable distortion of the audio signal. We leave improved designs for more advanced texts.

6.3 CASCADED STAGES

Although the op amp is an extremely versatile device, there are many applications in which a single op amp will not suffice. In such cases, it is often possible to meet application requirements by cascading several individual op amp circuits together into a larger circuit. An example of this is shown in Fig. 6.16, which consists of the summing amplifier circuit of Fig. 6.10 with only two input sources, and the output fed into a simple inverting amplifier. The result is a two-stage op amp circuit.

We have already analyzed each of these op amp circuits separately. Based on our previous experience, if the two op amp circuits were disconnected, we would expect

$$v_x = -\frac{R_f}{R}(v_1 + v_2) \qquad [7]$$

and

$$v_{\text{out}} = -\frac{R_2}{R_1} v_x \qquad [8]$$

In fact, since the two circuits are connected at a single point and the voltage v_x is not influenced by the connection, we can combine Eqs. [7] and [8] to obtain

$$v_{\text{out}} = \frac{R_2}{R_1}\frac{R_f}{R}(v_1 + v_2) \qquad [9]$$

which describes the input/output characteristics of the circuit shown in Fig. 6.16. We may not always be able to reduce such a circuit to familiar stages, however, so it is worth seeing how the two-stage circuit of Fig. 6.16 can be analyzed as a whole.

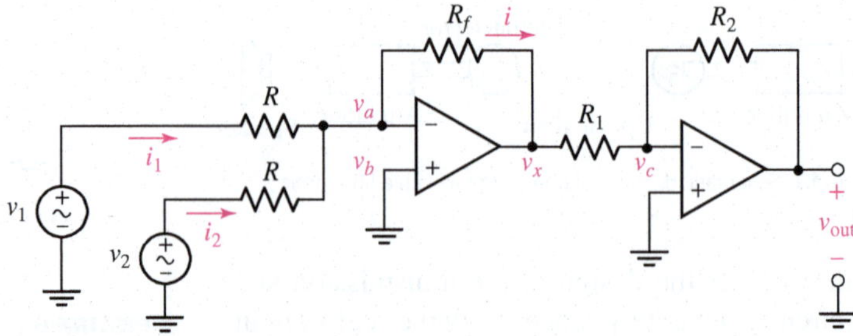

FIGURE 6.16 A two-stage op amp circuit consisting of a summing amplifier cascaded with an inverting amplifier circuit.

When analyzing cascaded circuits, it is sometimes helpful to begin with the last stage and work backward toward the input stage. Referring to ideal op amp rule 1, the same current flows through R_1 and R_2. Writing the appropriate nodal equation at the node labeled v_c yields

$$\frac{v_c - v_x}{R_1} + \frac{v_c - v_{\text{out}}}{R_2} = 0 \qquad [10]$$

Applying ideal op amp rule 2, we can set $v_c = 0$ in Eq. [10], resulting in

$$\frac{v_x}{R_1} + \frac{v_{\text{out}}}{R_2} = 0 \qquad [11]$$

Since our goal is an expression for v_{out} in terms of v_1 and v_2, we proceed to the first op amp in order to obtain an expression for v_x in terms of the two input quantities.

Applying ideal op amp rule 1 at the inverting input of the first op amp,

$$\frac{v_a - v_x}{R_f} + \frac{v_a - v_1}{R} + \frac{v_a - v_2}{R} = 0 \qquad [12]$$

Ideal op amp rule 2 allows us to replace v_a in Eq. [12] with zero, since $v_a = v_b = 0$. Thus, Eq. [12] becomes

$$\frac{v_x}{R_f} + \frac{v_1}{R} + \frac{v_2}{R} = 0 \qquad [13]$$

We now have an equation for v_{out} in terms of v_x (Eq. [11]), and an equation for v_x in terms of v_1 and v_2 (Eq. [13]). These equations are identical to Eqs. [7] and [8], respectively, which means that cascading the two separate circuits as in Fig. 6.16 did not affect the input/output relationship of either stage. Combining Eqs. [11] and [13], we find that the input/output relationship for the cascaded op amp circuit is

$$v_{\text{out}} = \frac{R_2}{R_1}\frac{R_f}{R}(v_1 + v_2) \qquad [14]$$

which is identical to Eq. [9].

Thus, the cascaded circuit acts as a summing amplifier, but without inverting the sign between the input and output. By choosing the resistor values carefully, we can either amplify or attenuate the sum of the two input voltages. If we select $R_2 = R_1$ and $R_f = R$, we can also obtain an amplifier circuit where $v_{\text{out}} = v_1 + v_2$, if desired.

EXAMPLE **6.3**

A multiple-tank gas propellant fuel system is installed in a small lunar orbiter, such as the one in Fig. 6.17. The amount of fuel in any tank is monitored by measuring the tank pressure (in psia).[2] Technical details for tank capacity as well as sensor pressure and voltage range are given in Table 6.2. Design a circuit which provides a positive dc voltage signal proportional to the total fuel remaining, such that 1 V = 100 percent.

(2) Pounds per square inch, absolute. This is a differential pressure measurement relative to a vacuum reference.

(Continued on next page)

FIGURE 6.17 An example lunar orbiter.
ullstein bild/Getty Images

TABLE 6.2 Technical Data for Tank Pressure Monitoring System

Tank 1 Capacity	10,000 psia
Tank 2 Capacity	10,000 psia
Tank 3 Capacity	2000 psia
Sensor Pressure Range	0 to 12,500 psia
Sensor Voltage Output	0 to 5 Vdc

We see from Table 6.2 that the system has three separate gas tanks, requiring three separate sensors. Each sensor is rated up to 12,500 psia, with a corresponding output of 5 V. Thus, when tank 1 is full, its sensor will provide a voltage signal of $5 \times (10{,}000/12{,}500) = 4$ V; the same is true for the sensor monitoring tank 2. The sensor connected to tank 3, however, will only provide a maximum voltage signal of $5 \times (2000/12{,}500) = 800$ mV.

One possible solution is the circuit shown in Fig. 6.18*a*, which employs a summing amplifier stage with v_1, v_2, and v_3 representing the three sensor outputs, followed by an inverting amplifier to adjust the voltage sign and magnitude. Since we are not told the output resistance of the sensor, we employ a buffer for each one as shown in Fig. 6.18*b*; the result is (in the ideal case) no current flow from the sensor.

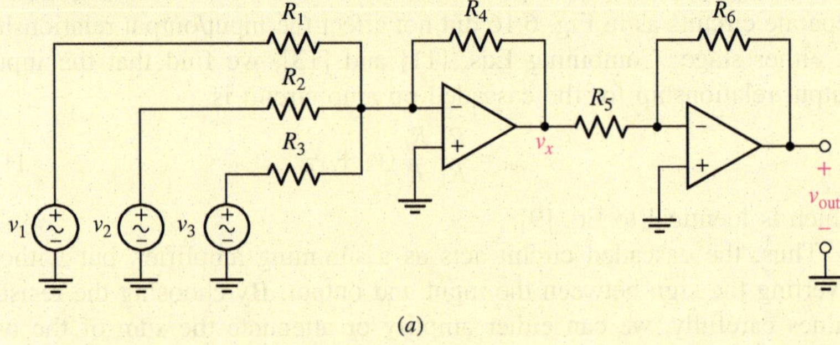

(*a*)

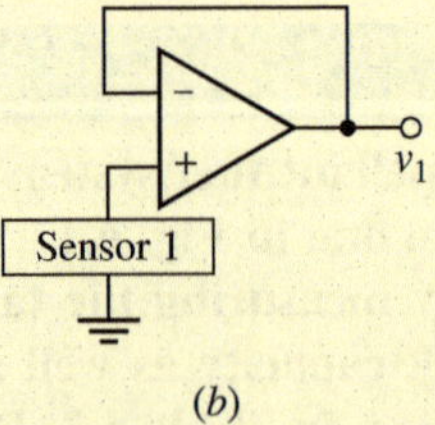

(*b*)

FIGURE 6.18 (*a*) A proposed circuit to provide a total fuel remaining readout. (*b*) Buffer design to avoid errors associated with the internal resistance of the sensor and limitations on its ability to provide current. One such buffer is used for each sensor, providing the inputs v_1, v_2, and v_3 to the summing amplifier stage.

To keep the design as simple as possible, we begin by choosing R_1, R_2, R_3, and R_4 to be 1 kΩ; any value will do as long as all four resistors are equal. Thus, the output of the summing stage is

$$v_x = -(v_1 + v_2 + v_3)$$

The final stage must invert this voltage and scale it such that the output voltage is 1 V when all three tanks are full. The full condition results in $v_x = -(4 + 4 + 0.8) = -8.8$ V. Thus, the final stage needs a voltage ratio of $R_6/R_5 = 1/8.8$. Arbitrarily choosing $R_6 = 1$ kΩ, we find that a value of 8.8 kΩ for R_5 completes the design.

PRACTICE

6.3 An historic bridge is showing signs of deterioration. Until renovations can be performed, it is decided that only cars weighing less than 1600 kg will be allowed across. To monitor this, a four-pad weighing system is designed. There are four independent voltage signals, one from each wheel pad, with 1 mV = 1 kg. Design a circuit to provide a positive voltage signal to be displayed on a DMM (digital multimeter) that represents the total weight of a vehicle, such that 1 mV = 1 kg. You may assume there is no need to buffer the wheel pad voltage signals.

Ans: See Fig. 6.19.

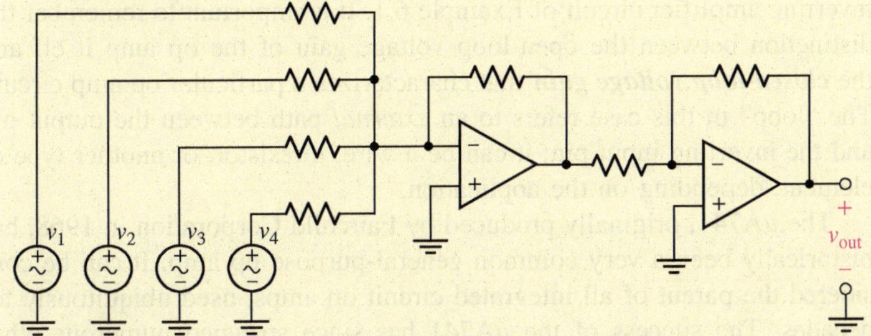

FIGURE 6.19 One possible solution to Practice Problem 6.3: all resistors are 10 kΩ (although any value will do as long as they are all equal). Input voltages v_1, v_2, v_3, and v_4 represent the voltage signals from the four wheel pad sensors, and v_{out} is the output signal to be connected to the positive input terminal of the DMM. All five voltages are referenced to ground, and the common terminal of the DMM should be connected to ground as well.

6.4 PRACTICAL CONSIDERATIONS

A More Detailed Op Amp Model

Reduced to its essentials, the op amp can be thought of as a voltage-controlled dependent voltage source. The dependent voltage source provides the output of the op amp, and the voltage on which it depends is applied to the input terminals. A schematic diagram of a reasonable model for a practical op amp is shown in Fig. 6.20; it includes a dependent voltage

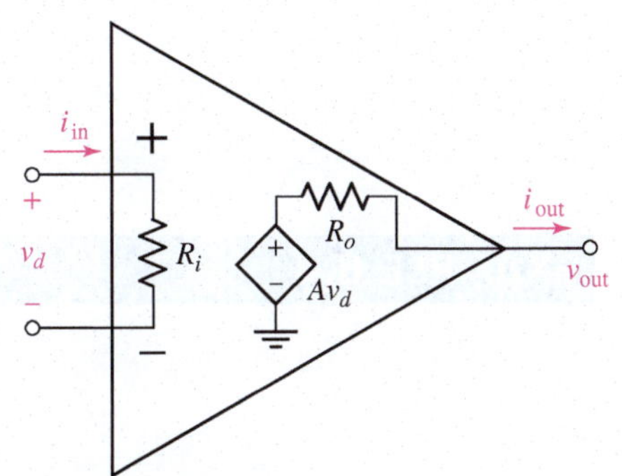

FIGURE 6.20 A more detailed model for the op amp.

TABLE 6.3 Typical Parameter Values for Several Types of Op Amps

Part Number	μA741	LM324	LT1001	OPA690	AD549K
Description	General purpose	Low-power quad	Precision	Wideband video frequency op amp	Ultralow input bias current
Open-loop gain A	2×10^5 V/V	10^5 V/V	8×10^5 V/V	2800 V/V	10^6 V/V
Input resistance	2 MΩ	*	100 MΩ	190 kΩ	10 TΩ
Output resistance	75 Ω	*	*	*	~15 Ω
Input bias current	80 nA	45 nA	0.5 nA	3 μA	75 fA
Input offset voltage	1.0 mV	2.0 mV	7 μV	±1.0 mV	0.150 mV
CMRR	90 dB	85 dB	110 dB	65 dB	100 dB
Slew Rate	0.5 V/μs	*	0.25 V/μs	1800 V/μs	3 V/μs

* Not provided by manufacturer.

source with voltage gain A, an output resistance R_o, and an input resistance R_i. Table 6.3 gives typical values for these parameters for several types of commercially available op amps.

The parameter A is referred to as the ***open-loop voltage gain*** of the op amp, and is typically in the range of 10^5 to 10^6. We notice that all of the op amps listed in Table 6.3 have extremely large open-loop voltage gain, especially compared to the voltage gain of 11 that characterized the noninverting amplifier circuit of Example 6.1. It is important to remember the distinction between the open-loop voltage gain of the op amp itself and the ***closed-loop voltage gain*** that characterizes a particular op amp circuit. The "loop" in this case refers to an *external* path between the output pin and the inverting input pin; it can be a wire, a resistor, or another type of element, depending on the application.

The μA741, originally produced by Fairchild Corporation in 1968, has historically been a very common general-purpose op amp. It can be considered the parent of all integrated circuit op amps, used ubiquitously for decades. The success of the μA741 has since spawned numerous other devices, which offer higher performance for specific applications, such as those shown in Table 6.3. The μA741 is characterized by an open-loop voltage gain of 200,000, an input resistance of 2 MΩ, and an output resistance of 75 Ω. In order to evaluate how well the ideal op amp model approximates the behavior of this device, let's revisit the inverting amplifier circuit of Fig. 6.4.

EXAMPLE 6.4

Using the appropriate values for the μA741 op amp in the model of Fig. 6.20, reanalyze the inverting amplifier circuit of Fig. 6.4.

We begin by replacing the ideal op amp symbol of Fig. 6.4 with the detailed model, resulting in the circuit shown in Fig. 6.21.

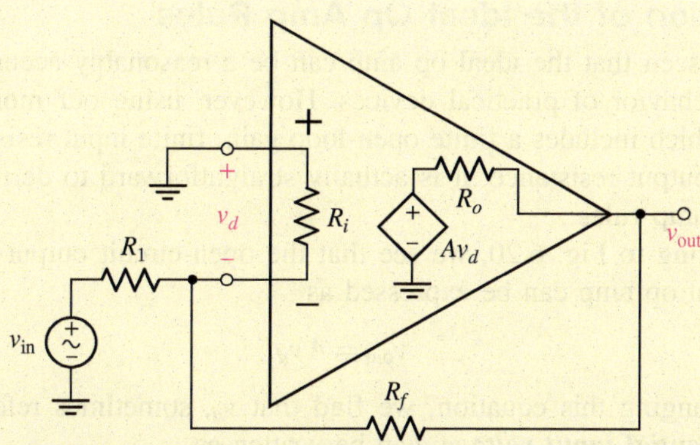

■ **FIGURE 6.21** Inverting amplifier circuit drawn using detailed op amp model.

Note that we can no longer invoke the ideal op amp rules, since we are not using the ideal op amp model. Thus, we write two nodal equations:

$$\frac{-v_d - v_{in}}{R_1} + \frac{-v_d - v_{out}}{R_f} + \frac{-v_d}{R_i} = 0$$

$$\frac{v_{out} + v_d}{R_f} + \frac{v_{out} - Av_d}{R_o} = 0$$

Performing some straightforward but rather lengthy algebra, we eliminate v_d and combine these two equations to obtain the following expression for v_{out} in terms of v_{in}:

$$v_{out} = \left[\frac{R_o + R_f}{R_o - AR_f}\left(\frac{1}{R_1} + \frac{1}{R_f} + \frac{1}{R_i}\right) - \frac{1}{R_f}\right]^{-1} \frac{v_{in}}{R_1} \qquad [15]$$

Substituting $v_{in} = 5 \sin 3t$ mV, $R_1 = 4.7$ kΩ, $R_f = 47$ kΩ, $R_o = 75$ Ω, $R_i = 2$ MΩ, and $A = 2 \times 10^5$, we obtain

$$v_{out} = -9.999448\ v_{in} = -49.99724 \sin 3t \text{ mV}$$

Upon comparing this to the expression that was found assuming an ideal op amp ($v_{out} = -10v_{in} = -50 \sin 3t$ mV), we see that the ideal op amp is indeed a reasonably accurate model. Further, assuming an ideal op amp leads to a significant reduction in the algebra required to perform the circuit analysis. Note that if we allow $A \to \infty$, $R_o \to 0$, and $R_i \to \infty$, Eq. [15] reduces to

$$v_{out} = -\frac{R_f}{R_1} v_{in}$$

which is what we derived earlier for the inverting amplifier when assuming the op amp was ideal.

PRACTICE

6.4 Assuming a finite open-loop gain (A), a finite input resistance (R_i), and zero output resistance (R_o), derive an expression for v_{out} in terms of v_{in} for the op amp circuit of Fig. 6.4.

Ans: $v_{out}/v_{in} = -AR_f R_i/[(1 + A)R_1R_i + R_1R_f + R_f R_i]$.

Derivation of the Ideal Op Amp Rules

We have seen that the ideal op amp can be a reasonably accurate model for the behavior of practical devices. However, using our more detailed model, which includes a finite open-loop gain, finite input resistance, and nonzero output resistance, it is actually straightforward to derive the two ideal op amp rules.

Referring to Fig. 6.20, we see that the open-circuit output voltage of a practical op amp can be expressed as

$$v_{\text{out}} = A\, v_d \qquad [16]$$

Rearranging this equation, we find that v_d, sometimes referred to as the ***differential input voltage,*** can be written as

$$v_d = \frac{v_{\text{out}}}{A} \qquad [17]$$

As we might expect, there are practical limits to the output voltage v_{out} that can be obtained from a real op amp. As we will consider momentarily, op amp saturation occurs near the values of the power supply, typically in the range of 5 to 24 V. If we divide 24 V by the open-loop gain of the μA741 (2×10^5), we obtain $v_d = 120\ \mu\text{V}$. Although this is not the same as zero volts, such a small value compared to the output voltage of 24 V is *practically* zero. An ideal op amp would have infinite open-loop gain, resulting in $v_d = 0$ regardless of v_{out}; this leads to ideal op amp rule 2.

Ideal op amp rule 1 states that "*No current ever flows into either input terminal.*" Referring to Fig. 6.20, the input current of an op amp is simply

$$i_{\text{in}} = \frac{v_d}{R_i}$$

We have just determined that v_d is typically a very small voltage. As we can see from Table 6.3, the input resistance of an op amp is very large, ranging from the megaohms to the teraohms! Using the value of $v_d = 120\ \mu\text{V}$ from above and $R_i = 2\ \text{M}\Omega$, we compute an input current of 60 pA. This is an extremely small current, and we would require a specialized ammeter (known as a picoammeter) to measure it. We see from Table 6.3 that the typical input current (more accurately termed the ***input bias current***) of a μA741 is 80 nA, three orders of magnitude larger than our estimate. This is a shortcoming of the op amp model we are using, which is not designed to provide accurate values for input bias current. Compared to the other currents flowing in a typical op amp circuit, however, either value is essentially zero. More modern op amps (such as the AD549) have even lower input bias currents. Thus, we conclude that ideal op amp rule 1 is a fairly reasonable assumption.

From our discussion, it is clear that an ideal op amp has infinite open-loop voltage gain, and infinite input resistance. However, we have not yet considered the output resistance of the op amp and its possible effects on our circuit. Referring to Fig. 6.20, we see that

$$v_{\text{out}} = A\, v_d - R_o i_{\text{out}}$$

where i_{out} flows from the output pin of the op amp. Thus, a nonzero value of R_o acts to reduce the output voltage, an effect which becomes more pronounced as the output current increases. For this reason, an *ideal* op amp has an output resistance of zero ohms. The μA741 has a maximum output resistance of 75 Ω, and more modern devices such as the AD549 have even lower output resistance.

Common-Mode Rejection

The op amp is occasionally referred to as a *difference amplifier,* since the output is proportional to the voltage difference between the two input terminals. This means that if we apply identical voltages to both input terminals, we expect the output voltage to be zero. This ability of the op amp is one of its most attractive qualities, and is known as ***common-mode rejection.*** The circuit shown in Fig. 6.22 is connected to provide an output voltage

$$v_{\text{out}} = v_2 - v_1$$

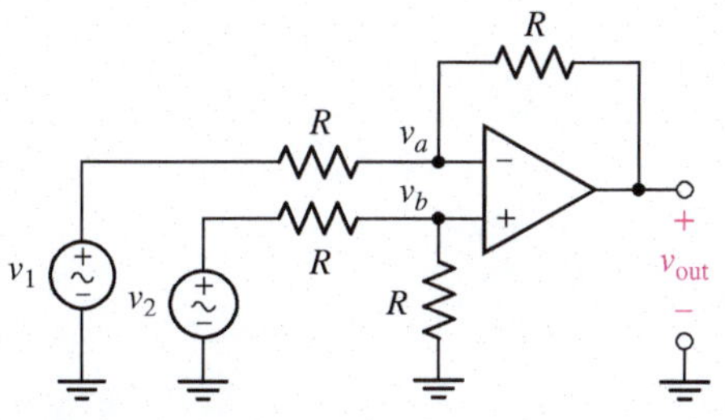

■ **FIGURE 6.22** An op amp connected as a difference amplifier.

If $v_1 = 2 + 3 \sin 3t$ volts and $v_2 = 2$ volts, we would expect the output to be $-3 \sin 3t$ volts; the 2 V component common to v_1 and v_2 would not be amplified, nor does it appear in the output.

For practical op amps, we do in fact find a small contribution to the output in response to common-mode signals. In order to compare one op amp type to another, it is often helpful to express the ability of an op amp to reject common-mode signals through a parameter known as the common-mode rejection ratio, or **CMRR.** Defining v_{oCM} as the output obtained when both inputs are equal ($v_1 = v_2 = v_{CM}$), we can determine A_{CM}, the common-mode gain of the op amp

$$A_{CM} = \left|\frac{v_{oCM}}{v_{CM}}\right|$$

We then define CMRR in terms of the ratio of differential-mode gain A to the common-mode gain A_{CM}, or

$$\text{CMRR} \equiv \left|\frac{A}{A_{CM}}\right| \qquad [18]$$

although this is often expressed in decibels (dB), a logarithmic scale:

$$\text{CMRR}_{(\text{dB})} \equiv 20 \log_{10} \left|\frac{A}{A_{CM}}\right| \qquad [19]$$

Typical values for several different op amps are provided in Table 6.3; a value of 100 dB corresponds to an absolute ratio of 10^5 for A to A_{CM}.

Saturation

So far, we have treated the op amp as a purely linear device, assuming that its characteristics are independent of the way in which it is connected in a circuit. In reality, it is necessary to supply power to an op amp in order to run the internal circuitry, as shown in Fig. 6.23. A positive supply, typically in the range of 5 to 24 V dc, is connected to the terminal marked V^+, and a negative supply of equal magnitude is connected to the terminal marked V^-. There are also a number of applications where a single voltage supply is acceptable, as well as situations where the two voltage magnitudes may be unequal. The op amp manufacturer will usually specify a maximum power supply voltage, beyond which damage to the internal transistors will occur.

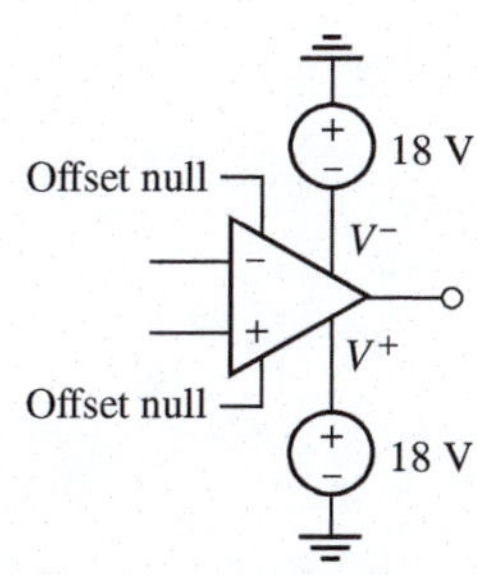

■ **FIGURE 6.23** Op amp with positive and negative voltage supplies connected. Two 18 V supplies are used as an example; note the polarity of each source.

COMPUTER-AIDED ANALYSIS

Up to now, we have performed the analysis of op amp circuits completely ignoring the practical aspect that they require power. Such circuits frequently make use of a dual power supply configuration—one positive voltage supply and one negative voltage supply—connected to the corresponding pins on the op amp. An example of how to construct such a circuit is provided in the schematic of Fig. 6.24.

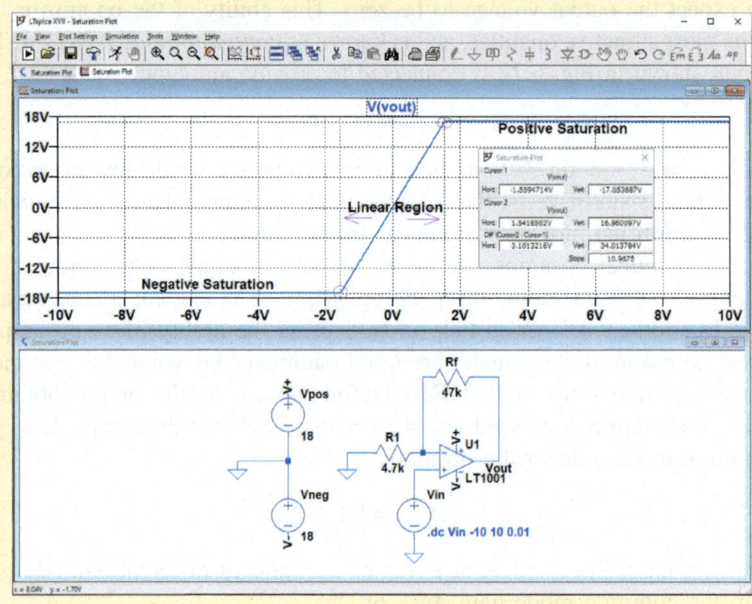

■ **FIGURE 6.24** Simulated input/output characteristics of an LT 1001 precision op amp connected as a noninverting amplifier with a gain of 10, and powered by ±18 V supplies. *Inset*: Cursor window used to determine linear region of operation.
Analog Devices

The power supply voltages are a critical choice when designing an op amp circuit because they represent the maximum possible output voltage of the op amp.[3] For example, consider the op amp circuit shown in Fig. 6.23, now connected as a noninverting amplifier having a gain of 10. As shown in the SPICE simulation in Fig. 6.24, we do in fact observe linear behavior from the op amp, but only in the range of ±1.55 V for the input voltage. Outside of this range, the output voltage is no longer proportional to the input, reaching a peak magnitude of 17.05 V. This important nonlinear effect is known as ***saturation,*** which refers to the fact that further increases in the input voltage do not result in a change in the output voltage. This phenomenon refers to the fact that the output of a real op amp cannot exceed its supply voltages. For example, if we choose to run the op amp with a +9 V supply and a −5 V supply, then our output voltage will be limited to

(3) In practice, we find the maximum output voltage is slightly less than the supply voltage by as much as a volt or so.

the range of −5 to +9 V. The output of the op amp is a linear response bounded by the positive and negative saturation regions, and as a general rule, we try to design our op amp circuits so that we do not accidentally enter the saturation region. This requires us to select the operating voltage carefully based on the closed-loop gain and maximum expected input voltage.

PRACTICE

6.5 Use SPICE to simulate a voltage follower using an LT1001 op amp with an input of 2 V and power supplies of ±12 V. How close is the value of the output voltage to the ideal value of 2 V?

Ans: 1.9999965 V, difference of 3.5 microvolt

Input Offset Voltage

As we are discovering, there are a number of practical considerations to keep in mind when working with op amps. One particular nonideality worth mentioning is the tendency for real op amps to have a nonzero output even when the two input terminals are shorted together. The value of the output under such conditions is known as the offset voltage, and the input voltage required to reduce the output to zero is referred to as the ***input offset voltage.*** Referring to Table 6.3, we see that typical values for the input offset voltage are on the order of a few millivolts or less.

Most op amps are provided with two pins marked either "offset null" or "balance." These terminals can be used to adjust the output voltage by connecting them to a variable resistor. A variable resistor is a three-terminal device commonly used for such applications as volume controls on radios. The device comes with a knob that can be rotated to select the actual value of resistance, and it has three terminals. Measured between the two extreme terminals, its resistance is fixed regardless of the position of the knob. Using the middle terminal and one of the end terminals creates a resistor whose value depends on the knob position. Figure 6.25 shows a typical circuit used to adjust the output voltage of an op amp; the manufacturer's data sheet may suggest alternative circuitry for a particular device.

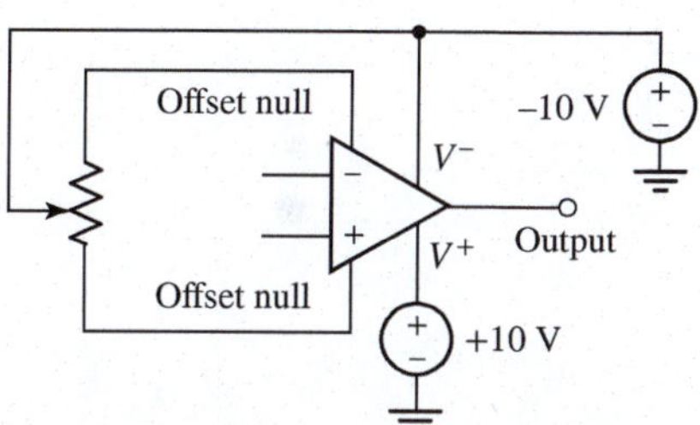

FIGURE 6.25 Suggested external circuitry for obtaining a zero output voltage. The ±10 V supplies are shown as an example; the actual supply voltages used in the final circuit would be chosen in practice.

Slew Rate

Up to now, we have tacitly assumed that the op amp will respond equally well to signals of any frequency, no matter how extreme. Experience, however, leads us to suspect that perhaps we should not be surprised to find that in practice there is some type of limitation in this regard. Since we know that op amp circuits work well at dc, which is essentially zero frequency, it is the performance as the signal frequency is increased that we must consider. One measure of the frequency performance of an op amp is its ***slew rate,*** which is the rate at which the output voltage can track (or respond to) changes in the input; it is most often expressed in V/μs. The typical slew rate specification for several commercially available devices is provided in Table 6.3, showing values on the order of a few volts per microsecond. One notable exception is the OPA690, which is designed as a high-speed op amp for video applications requiring operation at several hundred MHz. As can be seen, a respectable slew rate of 1800 V/μs is not unrealistic for this device, although its other parameters, particularly input bias current and CMRR, suffer somewhat as a result. The LTspice simulations shown in Fig. 6.26 illustrate the degradation in performance of an

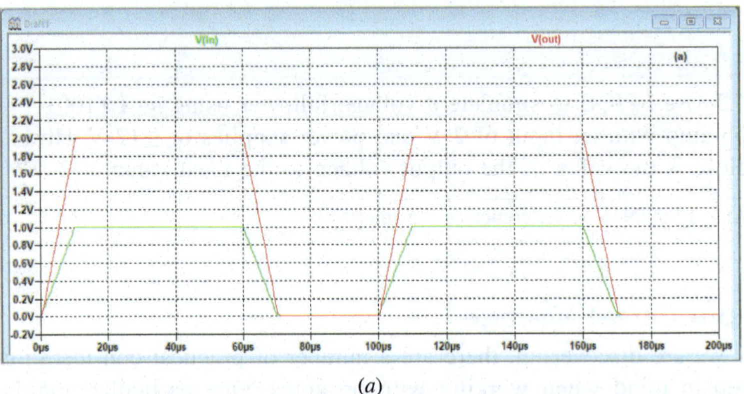

(*a*)

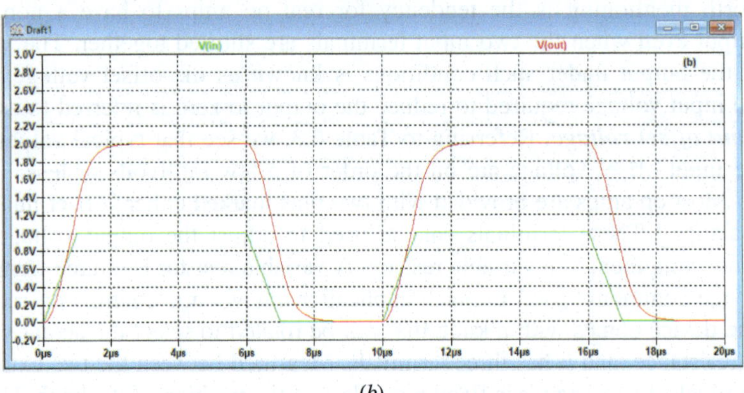

(*b*)

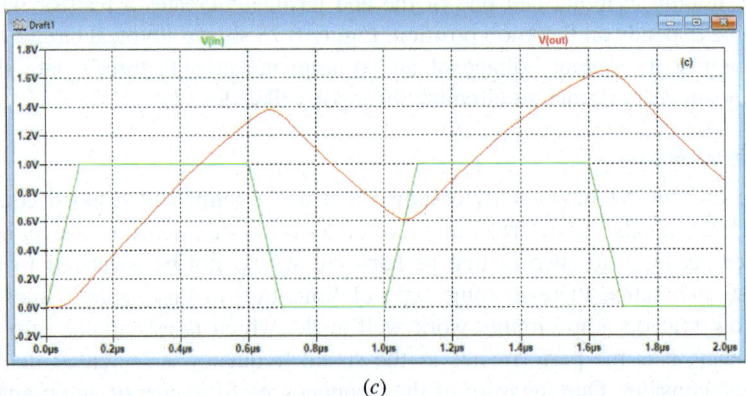

(*c*)

FIGURE 6.26 Simulated performance of an AD549 op amp connected as a noninverting amplifier having a gain of 2, with ±15 V supplies and a pulsed input waveform. (*a*) 10 μs rise and fall times, pulse width 50 μs; (*b*) 1 μs rise and fall times, pulse width 5 μs; (*c*) 100 ns rise and fall times, pulse width 500 ns. In this last simulation, we see the additional effects of what is termed the initial transient response of the op amp; we discuss this concept in Chap. 7.
(*a–c*) Analog Devices

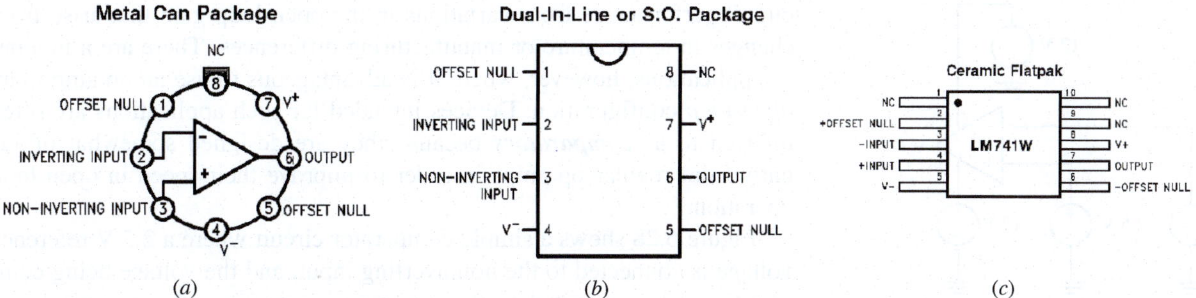

FIGURE 6.27 Several different package styles for the LM741 op amp: (*a*) metal can; (*b*) dual-in-line package; (*c*) ceramic flatpak. (Copyright © 2011 National Semiconductor Corporation (www.national.com). All rights reserved. Used with permission.)

op amp due to slew rate limitations. The circuit simulated is an AD549 configured as a noninverting amplifier with a gain of 2 and powered by ±15 V supplies. The input waveform is shown in green, and has a peak voltage of 1 V; the output voltage is shown in red. The simulation of Fig. 6.26*a* corresponds to a rise and fall time of 10 μs which, although a short time span for humans, is easily coped with by the AD549. As the rise and fall times are decreased by a factor of 10 to 1 μs (Fig. 6.26*b*), we begin to see that the AD549 is having a small difficulty in tracking the input. In the case of a 100 ns rise and fall time (Fig. 6.26*c*), we see that not only is there a significant delay between the output and the input, but the waveform is noticeably distorted as well—not a good feature of an amplifier. This observed behavior is consistent with the typical slew rate of 3 V/μs specified in Table 6.3, which indicates that the output might be expected to require something on the order of at least 670 ns to change from 0 to 2 V (or 2 V to 0 V).

Packaging

Modern op amps are available in a number of different types of packages. Some styles are better suited to high temperatures, and there are a variety of different ways to mount ICs on printed-circuit boards. Figure 6.27 shows several different styles of the LM741, manufactured by National Semiconductor. The label "NC" next to a pin means "no connection." The package styles shown in the figure are standard configurations and are used for a large number of different integrated circuits; occasionally there are more pins available on a package than required.

6.5 COMPARATORS AND THE INSTRUMENTATION AMPLIFIER

The Comparator

Every one of the op amp circuits we have considered up to this point has featured an electrical connection between the output pin and the inverting input pin. This is known as *closed-loop* operation, and is used to provide negative feedback as discussed previously. Closed loop is the preferred method of using an op amp as an amplifier, as it serves to isolate the

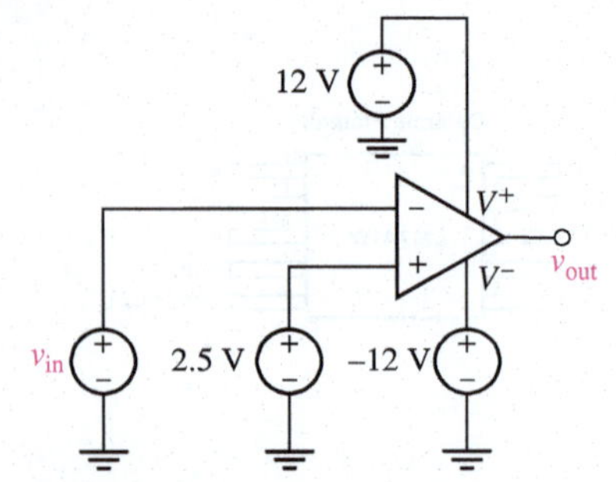

■ **FIGURE 6.28** An example comparator circuit with a 2.5 V reference voltage.

circuit performance from variations in the open-loop gain that arise from changes in temperature or manufacturing differences. There are a number of applications, however, where it is advantageous to use an op amp in an *open-loop* configuration. Devices intended for such applications are often referred to as ***comparators*** because they are designed somewhat differently from regular op amps in order to improve their speed in open-loop operation.

Figure 6.28 shows a simple comparator circuit where a 2.5 V reference voltage is connected to the noninverting input, and the voltage being compared (v_{in}) is connected to the inverting input. Since the op amp has a very large open-loop gain A, it does not take a large voltage difference between the input terminals to drive it into saturation. Thus, a positive 12 V output from the comparator indicates that the input voltage is *less than* the reference voltage, and a negative 12 V output indicates an input voltage *greater than* the reference (opposite behavior is obtained if we connect the reference voltage to the inverting input instead).

$$v_{out} = \begin{cases} 12\ V, & v_{in} < 2.5\ V \\ -12\ V, & v_{in} > 2.5\ V \end{cases}$$

EXAMPLE 6.5

Design a circuit that provides a "logic 1" 5 V output if a certain voltage signal drops below 3 V, and zero volts otherwise.

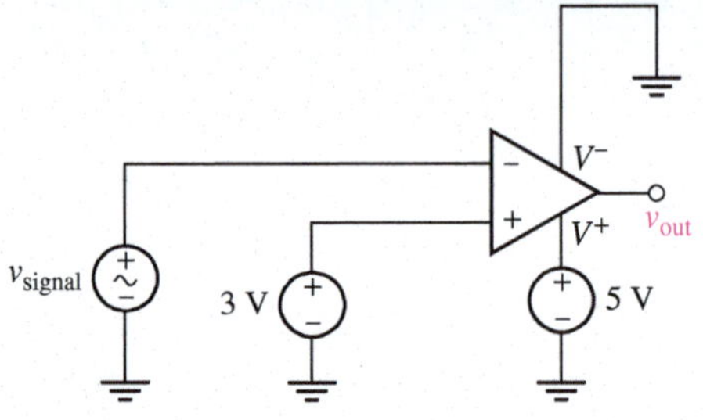

■ **FIGURE 6.29** One possible design for the required circuit.

Since we want the output of our comparator to swing between 0 and 5 V, we will use an op amp with a single-ended +5 V supply, connected as shown in Fig. 6.29. We connect a +3 V reference voltage to the noninverting input, which may be provided by two 1.5 V batteries in series. The input voltage signal (designated v_{signal}) is then connected to the inverting input. In reality, the saturation voltage range of a comparator circuit will be slightly less than that of the supply voltages, so some adjustment may be required in conjunction with simulation or testing.

PRACTICE

6.6 Design a circuit that provides a 12 V output if a certain voltage (v_{signal}) exceeds 0 V, and a −2 V output otherwise.

Ans: One possible solution is shown in Fig. 6.30.

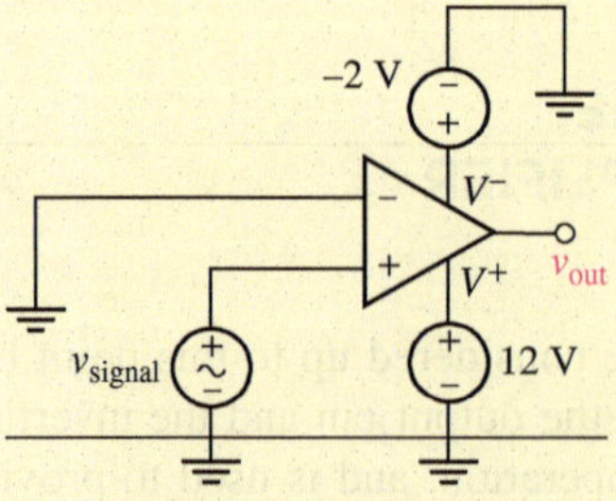

■ **FIGURE 6.30** One possible solution to Practice Problem 6.6.

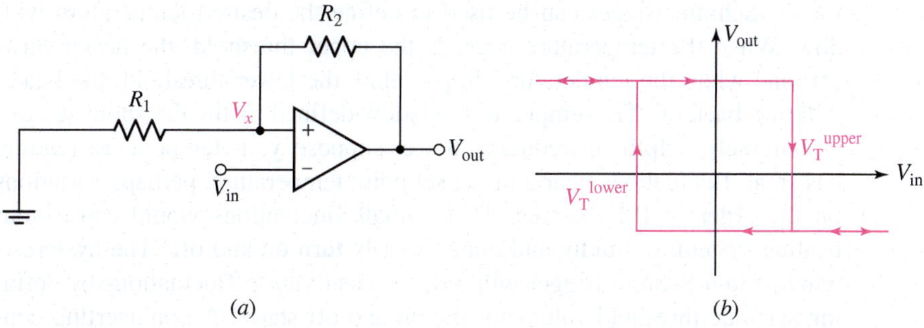

FIGURE 6.31 (*a*) Inverting comparator Schmitt trigger circuit and (*b*) voltage output versus voltage input showing hysteresis.

Comparator circuits can also be used in a *positive feedback* configuration, such as that shown in Fig. 6.31*a*. In this circuit, the output voltage V_{out} will be forced to one of the supply voltages (V^+ or V^-), depending on the sign of the voltage difference between V_x and V_{in}. The voltage at the noninverting input V_x is given by the voltage divider

$$V_x = \frac{R_1}{R_1 + R_2} V_{out}$$

Assuming that we begin with an output voltage in the high state ($V_{out} = V^+$), the output will remain in this state until V_{in} exceeds V_x, which corresponds to a negative voltage difference at the op amp terminals ($V_x - V_{in}$). The negative voltage difference at the op amp input forces the output to the low state ($V_{out} = V^-$). The output will therefore switch at the two threshold input voltages $V_T{}^{upper}$ and $V_T{}^{lower}$, given by

$$V_{in} = V_T{}^{upper} = \frac{R_1}{R_1 + R_2} V^+$$

and

$$V_{in} = V_T{}^{lower} = \frac{R_1}{R_1 + R_2} V^-$$

The circuit behavior depends on the present state of V_{out} and therefore has ***memory***. The output response is shown in Fig. 6.31*b*. The behavior of two thresholds for switching that depend on the present state is referred to as ***hysteresis***. A comparator circuit with hysteresis is called a ***Schmitt trigger***, and it is commonly used in applications for control and signal conditioning. A major advantage of the hysteresis "window" over a singular comparator voltage is that it can remove or reduce sensitivity to noisy signals.

The output of the circuit in Fig. 6.31*a* will remain high until V_{in} exceeds the upper threshold voltage $V_T{}^{upper}$, at which point the circuit will trigger a change to the low state. Once the output is low, it will remain low until V_{in} is decreased below the lower threshold $V_T{}^{lower}$. The behavior of a Schmitt trigger is very similar to thermostats used to control temperature. For example, consider the output voltage as a signal to a heater, while the input voltage indicates the temperature. The lower and upper thresholds for the Schmitt trigger can be used to define the desired temperature window. When the temperature exceeds the upper threshold, the heater turns off; and when the temperature drops below the lower threshold, the heater will turn back on. The temperature window defined by the threshold settings is extremely helpful in reducing noise: suppose your temperature reading has small fluctuations near a target set point temperature, perhaps variations on the order of 0.1 degrees. These small fluctuations would cause your heating system to rapidly and unnecessarily turn on and off! The hysteresis window in a Schmitt trigger will reduce sensitivity to fluctuations by defining separate threshold values for the on and off states. A noninverting configuration is also shown in the following example, which could similarly be used for temperature control for a cooling unit such as an air conditioner. While temperature control is an example we can all relate to, Schmitt triggers and related circuits with hysteresis are very valuable in many applications, including signal conditioning for digital circuits.

EXAMPLE 6.6

The circuit in Fig. 6.32 is a noninverting configuration of a comparator Schmitt trigger. Determine the threshold voltages for triggering the circuit output and plot the V_{out} versus V_{in} for the case where $R_1 = 400\ \Omega$, $R_2 = 1.2\ k\Omega$, $V^+ = 15$ V, and $V^- = -12$ V.

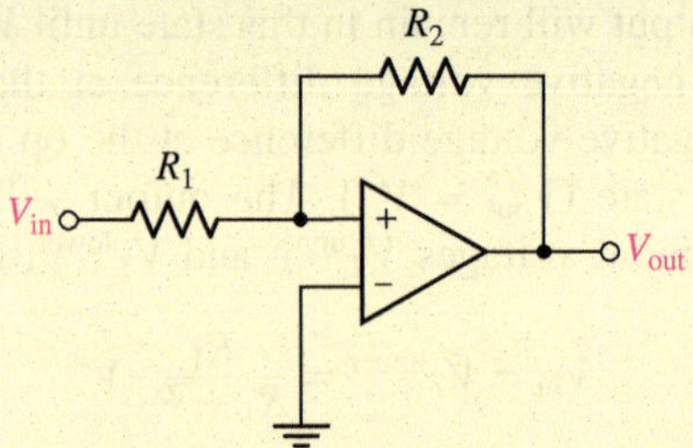

FIGURE 6.32 Noninverting comparator Schmitt trigger circuit.

We begin by relating the input and output by writing a KCL expression at the noninverting terminal of the op amp (let us again assign the node voltage V_x).

$$\frac{(V_x - V_{in})}{R_1} + \frac{(V_x - V_{out})}{R_2} = 0$$

Solving for V_x and rearranging terms,

$$V_x = \frac{R_2}{R_1 + R_2} V_{in} + \frac{R_1}{R_1 + R_2} V_{out}$$

The threshold values for switching/triggering will occur when the voltage difference between the inverting and noninverting terminals of the op amp is zero, i.e. $V_x = 0$. Assigning $V_x = 0$ and solving for V_{in} yields

$$V_{in} = -\left[\left(\frac{R_1}{R_1 + R_2}\right)\Big/\left(\frac{R_2}{R_1 + R_2}\right)\right] V_{out} = -\frac{R_1}{R_2} V_{out}$$

The two output voltages will be determined by the power supplies V^+ and V^-, where the lower and upper threshold voltages for triggering are

$$V_{in} = V_T^{lower} = -\frac{R_1}{R_2} V^+$$

and

$$V_{in} = V_T^{upper} = -\frac{R_1}{R_2} V^-$$

When the circuit starts in the low state where $V_{out} = V^-$, the output will remain low until the noninverting terminal of the op amp becomes positive (providing a positive voltage difference between the noninverting and inverting op amp terminals), occurring when $V_{in} > V_T^{upper}$. This will trigger an output voltage $V_{out} = V^+$. The high state will be maintained until $V_{in} < V_T^{lower}$, triggering an output voltage $V_{out} = V^-$. Inserting the numerical values of the circuit, we obtain

$$V_{out} = \begin{cases} -12 \text{ V}, & \textit{low state} \\ 15 \text{ V}, & \textit{high state} \end{cases}$$

with triggering values

$$V_T^{lower} = -\frac{400}{1{,}200}\, 15 = -5 \text{ V}$$

and

$$V_T^{upper} = -\frac{400}{1{,}200}\,(-12) = 4 \text{ V}$$

The circuit response characteristic is plotted in Fig. 6.33.

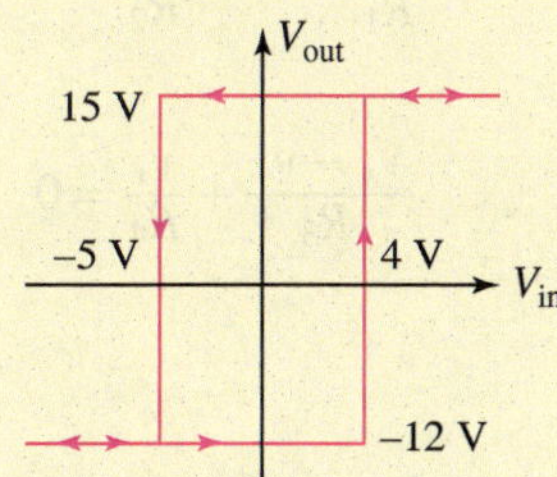

FIGURE 6.33 Voltage output of noninverting comparator Schmitt trigger circuit.

PRACTICE

6.7 Design a noninverting Schmitt trigger that will output ±5 V with threshold voltages of ±2 V.

Ans: Using noninverting configuration from Fig. 6.32, power supplies with $V^+ = 5$ V, $V^- = -5$ V; the ratio of $R_1/R_2 = 2/5$.

The Instrumentation Amplifier

The basic comparator circuit acts on the voltage difference between the two input terminals to the device, although it does not technically amplify signals as the output is not proportional to the input. The difference amplifier of Fig. 6.11 also acts on the voltage difference between the inverting and noninverting inputs, and as long as care is taken to avoid saturation, it *does* provide an output directly proportional to this difference. When dealing with a very small input voltage, however, a better alternative is a device known as an ***instrumentation amplifier,*** which is actually three op amp devices in a single package.

An example of the common instrumentation amplifier configuration is shown in Fig. 6.34*a*, and its symbol is shown in Fig. 6.33*b*. Each input is fed directly into a voltage follower stage, and the output of both voltage followers is fed into a difference amplifier stage. It is particularly well suited to applications where the input voltage signal is very small (for example, on the order of millivolts), such as that produced by thermocouples or strain gauges, and where a significant common-mode noise signal (a common signal on both inputs) of several volts may be present.

If components of the instrumentation amplifier are fabricated all on the same silicon chip, then it is possible to obtain well-matched device characteristics and to achieve precise ratios for the two sets of resistors. In order to maximize the rejection of the common-mode signal of the instrumentation amplifier, we expect $R_4/R_3 = R_2/R_1$, so that equal amplification of common-mode components of the input signals is obtained. To explore this further, we identify the voltage at the output of the top voltage follower as "v_-," and the voltage at the output of the bottom voltage follower as "v_+." Assuming all three op amps are ideal and naming the voltage at either input of the difference stage v_x, we may write the following nodal equations:

$$\frac{v_x - v_-}{R_1} + \frac{v_x - v_{\text{out}}}{R_2} = 0 \qquad [20]$$

and

$$\frac{v_x - v_+}{R_3} + \frac{v_x}{R_4} = 0 \qquad [21]$$

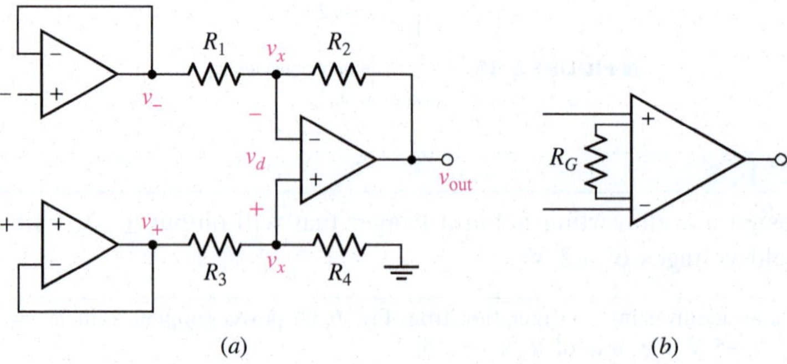

FIGURE 6.34 (*a*) The basic instrumentation amplifier. (*b*) Commonly used symbol.

Solving Eq. [21] for v_x, we find that

$$v_x = \frac{v_+}{1 + R_3/R_4} \quad [22]$$

and upon substituting into Eq. [20], we obtain an expression for v_{out} in terms of the input:

$$v_{\text{out}} = \frac{R_4}{R_3}\left(\frac{1 + R_2/R_1}{1 + R_4/R_3}\right) v_+ - \frac{R_2}{R_1} v_- \quad [23]$$

From Eq. [23] it is clear that the general case allows amplification of common-mode components to the two inputs. In the specific case where $R_4/R_3 = R_2/R_1 = K$, however, Eq. [23] reduces to $K(v_+ - v_-) = Kv_d$, so that (assuming ideal op amps) only the difference is amplified, and the gain is set by the resistor ratio. Since these resistors are internal to the instrumentation amplifier and not accessible to the user, devices such as the AD622 allow the gain to be set anywhere in the range of 1 to 1000 by connecting an external resistor between two pins (shown as R_G in Fig. 6.34*b*).

SUMMARY AND REVIEW

In this chapter we introduced a new circuit element—a three-terminal device—called the operational amplifier (or more commonly, the *op amp*). In many circuit analysis situations it is approximated as an ideal device, which leads to two rules that are applied. We studied several op amp circuits in detail, including the *inverting amplifier* with gain $-R_f/R_1$, the *noninverting amplifer* with gain $1 + R_f/R_1$, and the *summing amplifier*. We were also introduced to the *voltage follower* and the *difference amplifier*, although the analysis of these two circuits was left for the reader. The concept of cascaded stages was found to be particularly useful, as it allows a design to be broken down into distinct units, each of which has a specific function.

The inverting and noninverting op amp circuits, and mathematical operations such as summing and difference amplifiers, all used *negative feedback* and circuit designs with resistor values to define a desired circuit gain/amplification. *Positive feedback* can result in unstable circuit operation, and output in *saturation*. Operation in saturation can provide useful circuits, including *comparators* that convert signals to high and low output values according to the input, and *Schmitt triggers* that are comparator circuits that exhibit *memory* and a *hysteresis* characteristic. A special case of an op amp circuit that compares the voltage difference at two input terminals is the *instrumentation amplifier*, which is routinely used to amplify very small voltages.

Modern op amps have nearly ideal characteristics, as we found when we opted for a more detailed model based on a dependent source. Still, nonidealities are encountered occasionally, so we considered the role *of negative feedback* in reducing the effect of temperature and manufacturing-related variations in various parameters, *common-mode rejection*, *saturation* and *slew rate*.

This is a good point to pause, take a breath, and recap some of the key points. At the same time, we will highlight relevant examples as an aid to the reader.

- ❑ There are two fundamental rules that must be applied when analyzing *ideal* op amp circuits (Example 6.1):
  1. No current ever flows into either input terminal.
  2. No voltage ever exists between the input terminals.
- ❑ Op amp circuits are usually analyzed for an output voltage in terms of some input quantity or quantities. (Examples 6.1, 6.2)
- ❑ Nodal analysis is typically the best choice in analyzing op amp circuits, and it is usually better to begin at the input and work toward the output. (Examples 6.1, 6.2)
- ❑ The output current of an op amp cannot be assumed; it must be found after the output voltage has been determined independently. (Example 6.2)
- ❑ Cascaded stages may be analyzed one stage at a time to relate the output to the input. (Example 6.3)
- ❑ The ideal op amp model is based on the approximation of infinite open-loop gain A, infinite input resistance R_i, and zero output resistance R_o. (Example 6.4)
- ❑ Comparators are op amps designed to be driven into saturation. These circuits operate in open loop, and hence have no external feedback resistor. (Examples 6.5, 6.6)
- ❑ In practice, the output voltage range of an op amp is limited by the supply voltages used to power the device.

READING FURTHER

Two very readable books that deal with a variety of op amp applications are:

R. Mancini (ed.), *Op Amps Are for Everyone,* 2nd ed. Amsterdam: Newnes, 2003. Also available on the Texas Instruments website (www.ti.com).

W. G. Jung, *Op Amp Cookbook,* 3rd ed. Upper Saddle River, N.J.: Prentice-Hall, 1997.

One of the first reports of the implementation of an "operational amplifier" can be found in

J. R. Ragazzini, R. M. Randall, and F. A. Russell, "Analysis of problems in dynamics by electronic circuits," *Proceedings of the IRE* **35**(5), 1947, pp. 444–452.

and an early applications guide for the op amp can be found on the Analog Devices, Inc. website (www.analog.com):

G. A. Philbrick Researches, Inc., *Applications Manual for Computing Amplifiers for Modelling, Measuring, Manipulating & Much Else.* Norwood, Mass.: Analog Devices, 1998.

EXERCISES

6.2 The Ideal Op Amp

1. For the op amp circuit shown in Fig. 6.35, calculate v_{out} if (*a*) $R_1 = R_2 = 100\ \Omega$ and $v_{in} = 5$ V; (*b*) $R_2 = 200R_1$ and $v_{in} = 1$ V; (*c*) $R_1 = 4.7\ \text{k}\Omega$, $R_2 = 47\ \text{k}\Omega$, and $v_{in} = 20 \sin 5t$ V.

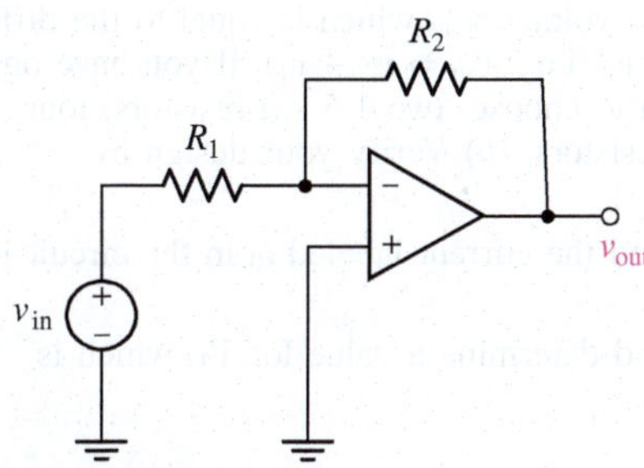

■ **FIGURE 6.35**

2. Determine the power dissipated by a 100 Ω resistor connected between ground and the output pin of the op amp of Fig. 6.35 if $v_{in} = 4$ V and (*a*) $R_1 = 2R_2$; (*b*) $R_1 = 1\ \text{k}\Omega$ and $R_2 = 22\ \text{k}\Omega$; (*c*) $R_1 = 100\ \Omega$ and $R_2 = 101\ \Omega$.

3. For the circuit of Fig. 6.36, calculate v_{out} if (*a*) $R_1 = R_2 = 100\ \text{k}\Omega$, $R_L = 100\ \Omega$, and $v_{in} = 5$ V; (*b*) $R_1 = 0.1R_2$, $R_L = \infty$, and $v_{in} = 2$ V; (*c*) $R_1 = 1\ \text{k}\Omega$, $R_2 = 0$, $R_L = 1\ \Omega$, and $v_{in} = 43.5$ V.

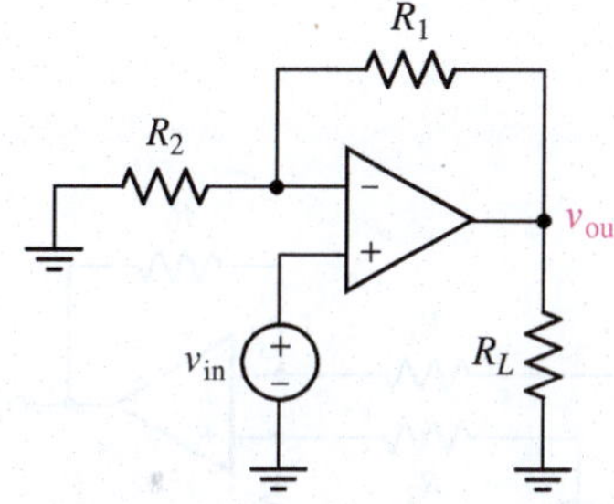

■ **FIGURE 6.36**

4. For the circuit in Fig. 6.36, find the values of current at all terminals of the op amp for the case where $R_l = 500\ \Omega$, $R_2 = 100\ \Omega$, and $R_L = 50\ \Omega$.

5. (*a*) Design a circuit which converts a voltage $v_1(t) = 9 \cos 5t$ V into $-4 \cos 5t$ V. (*b*) Verify your design by analyzing the final circuit.

6. A certain load resistor requires a constant 5 V dc supply. Unfortunately, its resistance value changes with temperature. Design a circuit which supplies the requisite voltage if only 9 V batteries and standard 10% tolerance resistor values are available.

7. For the circuit of Fig. 6.36, $R_1 = R_L = 50\ \Omega$. Calculate the value of R_2 required to deliver 5 W to R_L if V_{in} equals (*a*) 5 V; (*b*) 1.5 V. (*c*) Repeat parts (*a*) and (*b*) if R_L is reduced to 22 Ω.

8. Calculate v_{out} as labeled in the schematic of Fig. 6.37 if (*a*) $i_{in} = 1$ mA, $R_p = 2.2\ \text{k}\Omega$, and $R_3 = 1\ \text{k}\Omega$; (*b*) $i_{in} = 2$ A, $R_p = 1.1\ \Omega$, and $R_3 = 8.5\ \Omega$. (*c*) For each case, state whether the circuit is wired as a noninverting or an inverting amplifier. *Explain your reasoning.*

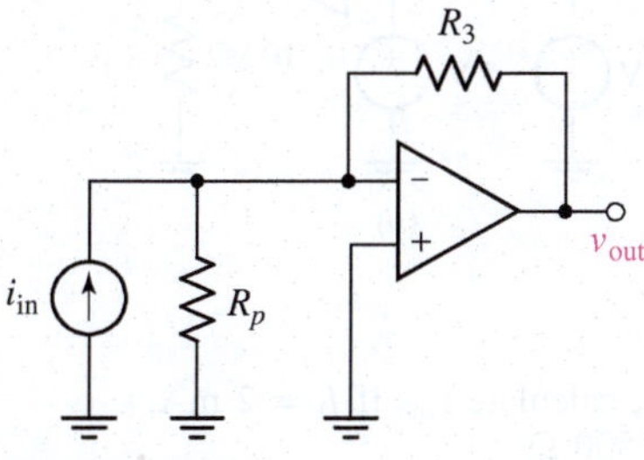

■ **FIGURE 6.37**

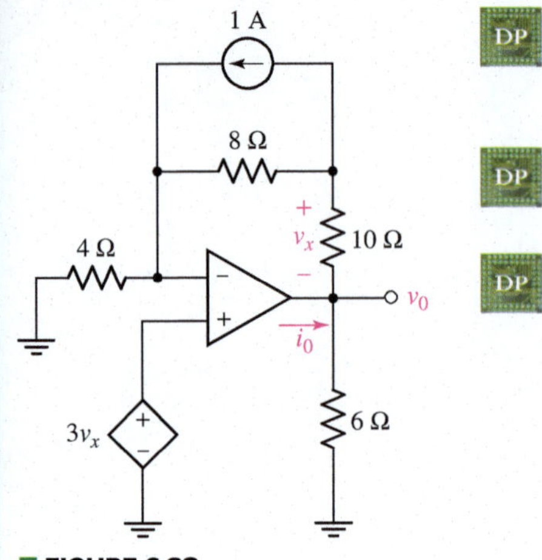

■ **FIGURE 6.38**

DP 9. (*a*) Design a circuit using only a single op amp which adds two voltages v_1 and v_2 and provides an output voltage twice their sum (negative values acceptable, i.e., $|v_{out}| = 2v_1 + 2v_2$). (*b*) Verify your design by analyzing the final circuit.

DP 10. (*a*) Design a circuit that provides a current *i* which is equal in magnitude to the sum of three input voltages v_1, v_2, and v_3. (Compare volts to amperes.) (*b*) Verify your design by analyzing the final circuit.

DP 11. (*a*) Design a circuit that provides a voltage v_{out} which is equal to the difference between two voltages v_2 and v_1 (i.e., $v_{out} = v_2 - v_1$), if you have only the following resistors from which to choose: two 1.5 kΩ resistors, four 6 kΩ resistors, and three 500 Ω resistors. (*b*) Verify your design by analyzing the final circuit.

12. Determine the output voltage v_0 and the current labeled i_0 in the circuit in Fig. 6.38.

13. Analyze the circuit of Fig. 6.39 and determine a value for V_1, which is referenced to ground.

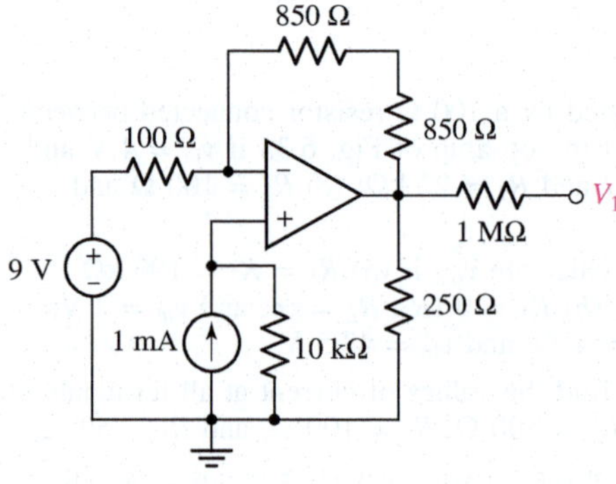

■ **FIGURE 6.39**

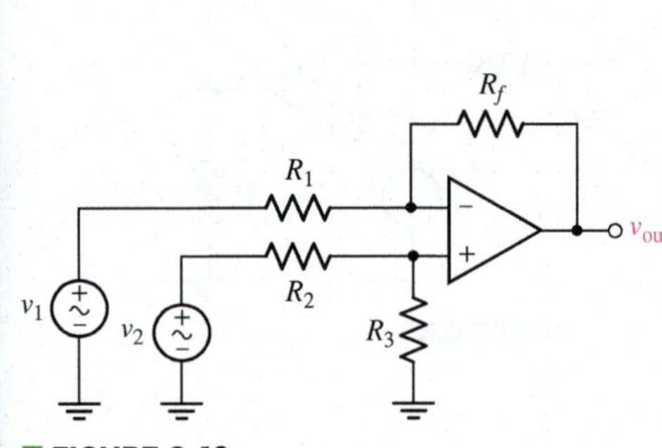

■ **FIGURE 6.40**

14. Derive an expression for v_{out} as a function of v_1 and v_2 for the circuit represented in Fig. 6.40.

15. Explain what is wrong with each diagram in Fig. 6.41 if the two op amps are known to be *perfectly ideal.*

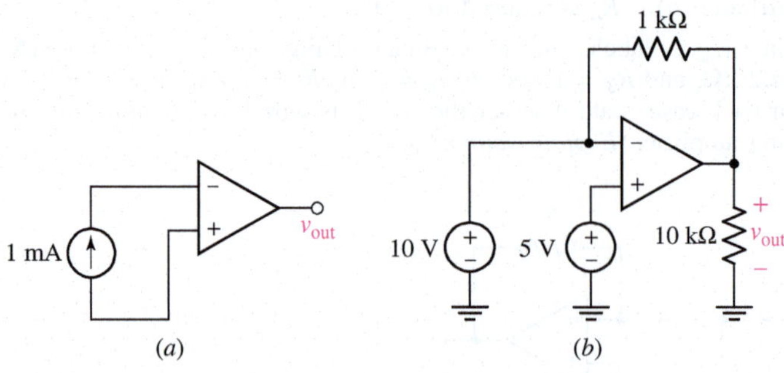

■ **FIGURE 6.41**

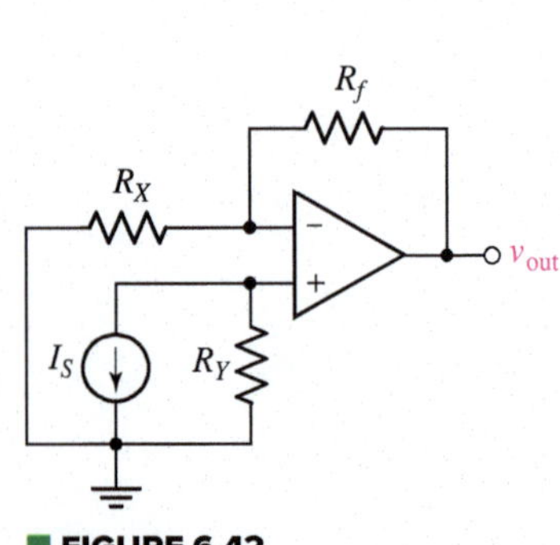

■ **FIGURE 6.42**

16. For the circuit depicted in Fig. 6.42, calculate v_{out} if $I_s = 2$ mA, $R_Y = 4.7$ kΩ, $R_X = 1$ kΩ, and $R_f = 500$ Ω.

17. Consider the amplifier circuit shown in Fig. 6.42. What value of R_f will yield $v_{out} = 2$ V when $I_s = -5/3$ mA and $R_Y = 2R_X = 500$ Ω?

18. With respect to the circuit shown in Fig. 6.43, calculate v_{out} if v_s equals (*a*) 2 cos 100*t* mV; (*b*) 2 sin (4*t* + 19°) V.

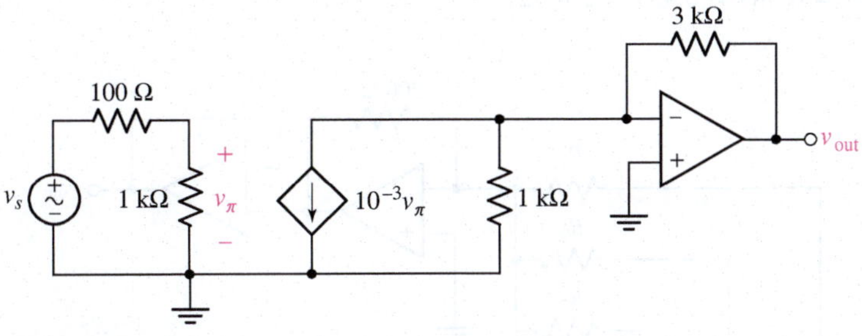

■ **FIGURE 6.43**

6.3 Cascaded Stages

19. Calculate v_{out} as labeled in the circuit of Fig. 6.44 if R_x = 1 kΩ.

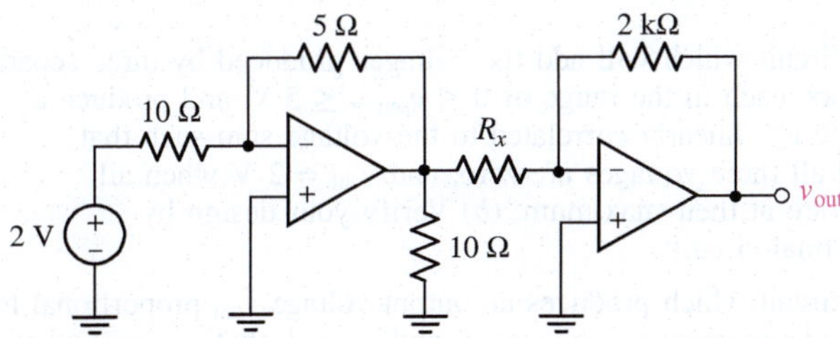

■ **FIGURE 6.44**

20. For the circuit of Fig. 6.44, determine the value of R_x that will result in a value of v_{out} = 10 V.
21. Referring to Fig. 6.45, sketch v_{out} as a function of (*a*) v_{in} over the range of $-2\text{ V} \le v_{in} \le +2\text{ V}$, if R_4 = 2 kΩ; (*b*) R_4 over the range of $1\text{ k}\Omega \le R_4 \le 10\text{ k}\Omega$, if v_{in} = 300 mV.

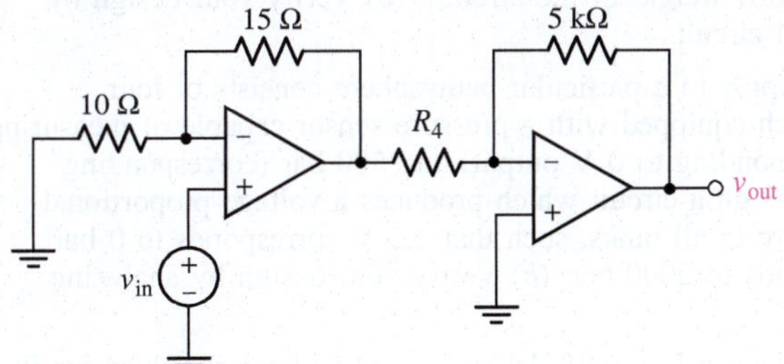

■ **FIGURE 6.45**

22. Repeat Exercise 21 using a parameter sweep in SPICE.
23. Obtain an expression for v_{out} as labeled in the circuit of Fig. 6.46 if v_1 equals (*a*) 0 V; (*b*) 1 V; (*c*) −5 V; (*d*) 2 sin 100*t* V.
24. The 1.5 V source of Fig. 6.46 is disconnected, and the output of the circuit shown in Fig. 6.45 is connected to the left-hand terminal of the 500 Ω resistor instead. Calculate v_{out} if R_4 = 2 kΩ and (*a*) v_{in} = 2 V, v_1 = 1 V; (*b*) v_{in} = 1 V, v_1 = 0; (*c*) v_{in} = 1 V, v_1 = −1 V.

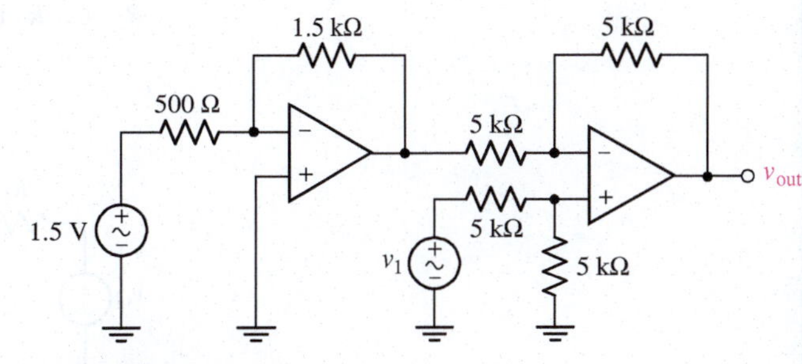

■ **FIGURE 6.46**

25. For the circuit shown in Fig. 6.47, compute v_{out} if (*a*) $v_1 = 2v_2 = 0.5v_3 = 2.2$ V and $R_1 = R_2 = R_3 = 50$ kΩ; (*b*) $v_1 = 0$, $v_2 = -8$ V, $v_3 = 9$ V, and $R_1 = 0.5R_2 = 0.4R_3 = 100$ kΩ.

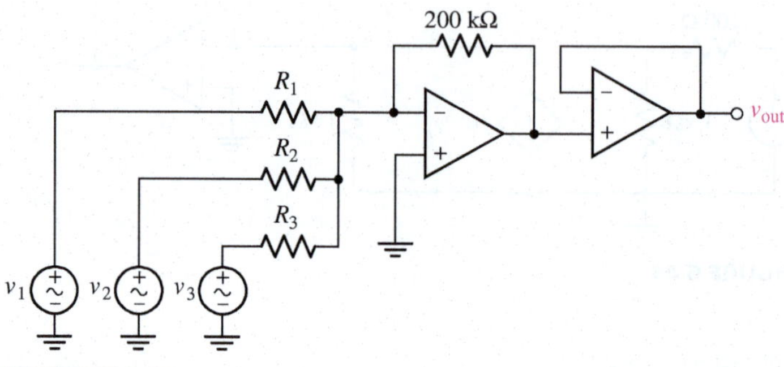

FIGURE 6.47

DP 26. (*a*) Design a circuit which will add the voltages produced by three separate pressure sensors, each in the range of $0 \le v_{sensor} \le 5$ V, and produce a positive voltage v_{out} linearly correlated to the voltage sum such that $v_{out} = 0$ when all three voltages are zero, and $v_{out} = 2$ V when all three voltages are at their maximum. (*b*) Verify your design by analyzing the final circuit.

DP 27. (*a*) Design a circuit which produces an output voltage v_{out} proportional to the difference of two positive voltages v_1 and v_2 such that $v_{out} = 0$ when both voltages are equal, and $v_{out} = 10$ V when $v_1 - v_2 = 1$ V. (*b*) Verify your design by analyzing the final circuit.

DP 28. (*a*) Three pressure-sensitive sensors are used to double-check the weight readings obtained from the suspension systems of a long-range jet airplane. Each sensor is calibrated such that 10 μV corresponds to 1 kg. Design a circuit which adds the three voltage signals to produce an output voltage calibrated such that 10 V corresponds to 400,000 kg, the maximum takeoff weight of the aircraft. (*b*) Verify your design by analyzing the final circuit.

DP 29. (*a*) The oxygen supply to a particular bathysphere consists of four separate tanks, each equipped with a pressure sensor capable of measuring between 0 (corresponding to 0 V output) and 500 bar (corresponding to 5 V output). Design a circuit which produces a voltage proportional to the total pressure in all tanks, such that 1.5 V corresponds to 0 bar and 3 V corresponds to 2000 bar. (*b*) Verify your design by analyzing the final circuit.

30. For the circuit shown in Fig. 6.48, let $v_{in} = 8$ V, and select values for R_1, R_2, and R_3 to ensure an output voltage $v_{out} = 4$ V.

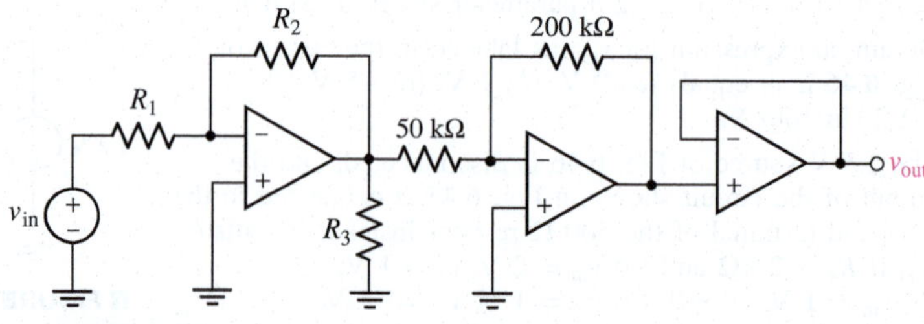

FIGURE 6.48

31. For the circuit of Fig. 6.49, derive an expression for v_{out} in terms of v_{in}.

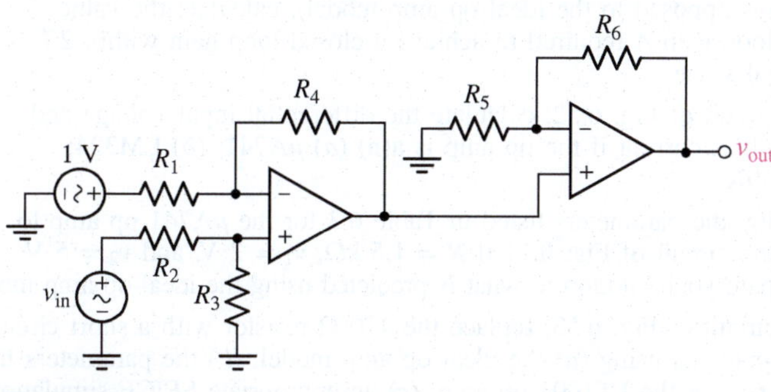

■ FIGURE 6.49

32. Determine the value of V_{out} for the circuit in Fig. 6.50.

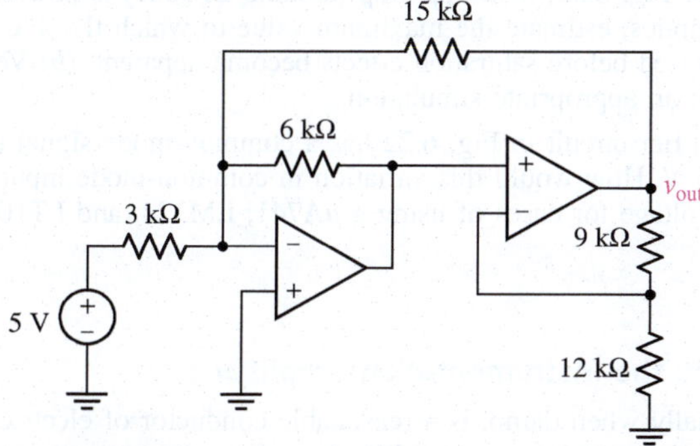

■ FIGURE 6.50

33. Calculate V_0 for the circuit in Fig. 6.51.

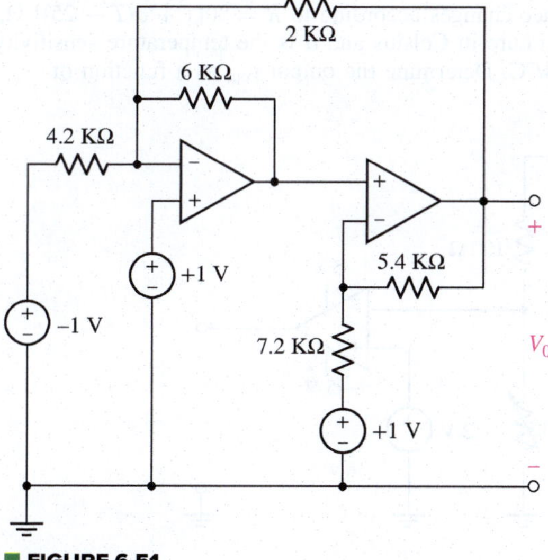

■ FIGURE 6.51

6.4 Practical Considerations

34. (*a*) Employ the parameters listed in Table 6.3 for the μA741 op amp to analyze the circuit of Fig. 6.52 and compute a value for v_{out}. (*b*) Compare your result to what is predicted using the ideal op amp model.

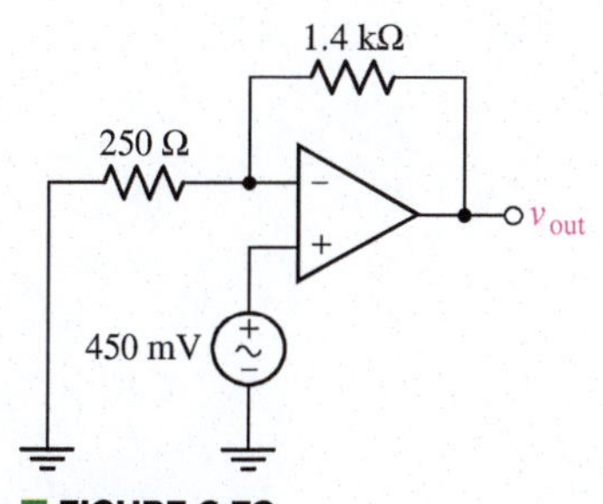

■ FIGURE 6.52

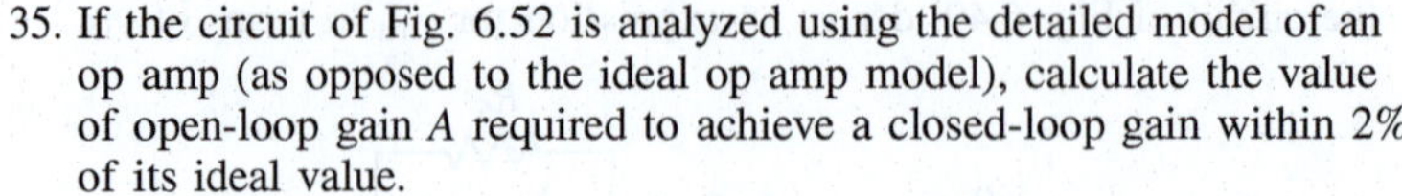

35. If the circuit of Fig. 6.52 is analyzed using the detailed model of an op amp (as opposed to the ideal op amp model), calculate the value of open-loop gain A required to achieve a closed-loop gain within 2% of its ideal value.

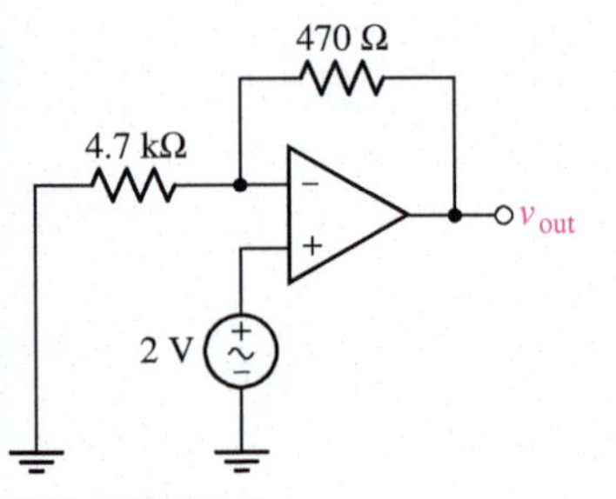

■ FIGURE 6.53

36. For the circuit of Fig. 6.52, calculate the differential input voltage and the input bias current if the op amp is a(n) (*a*) μA741; (*b*) LM324; (*c*) AD549K.

37. (*a*) Employ the parameters listed in Table 6.3 for the μA741 op amp to analyze the circuit of Fig. 6.11 if $R = 1.5$ kΩ, $v_1 = 2$ V, and $v_2 = 5$ V. (*b*) Compare your solution to what is predicted using the ideal op amp model.

38. For the circuit of Fig. 6.53, replace the 470 Ω resistor with a short circuit, and compute v_{out} using (*a*) the ideal op amp model; (*b*) the parameters listed in Table 6.3 for the LT1001 op amp; (*c*) an appropriate SPICE simulation. (*d*) Compare the values obtained in parts (*a*) to (*c*) and comment on the possible origin of any discrepancies.

39. (*a*) For the circuit of Fig. 6.53, if the op amp (assume LT1001) is powered by matched 9 V supplies, estimate the maximum value to which the 470 Ω resistor can be increased before saturation effects become apparent. (*b*) Verify your prediction with an appropriate simulation.

40. The difference amplifier circuit in Fig. 6.22 has a common-mode signal that can vary by up to 5 V. How would this variation in common-mode input change the output voltage for cases of using a μA741, LM324, and LT1001 op amp?

6.5 Comparators and the Instrumentation Amplifier

41. Human skin, especially when damp, is a reasonable conductor of electricity. If we assume a resistance of less than 10 MΩ for a fingertip pressed across two terminals, design a circuit which provides a +1 V output if this nonmechanical switch is "closed" and −1 V if it is "open."

42. The temperature alarm circuit in Fig. 6.54 utilizes a temperature sensor whose resistance changes according to $R = 80[1 + \alpha(T - 25)]$ Ω, where T is the temperature in Celsius and α is the temperature sensitivity with a value of 0.004/°C. Determine the output v_{out} as a function of temperature.

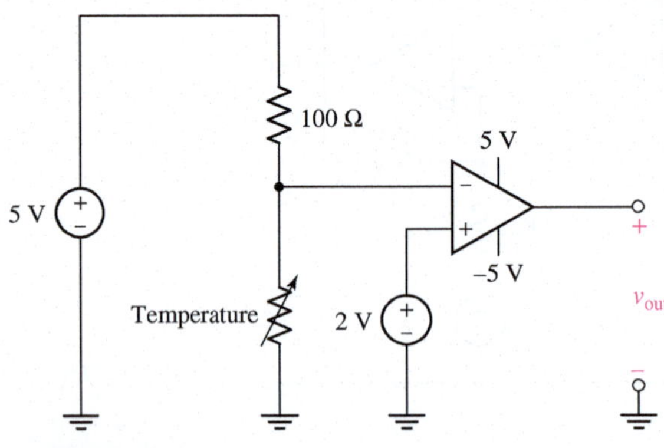

■ FIGURE 6.54

43. Design a circuit which provides an output voltage v_{out} based on the behavior of another voltage v_{in}, such that

$$v_{out} \begin{cases} 2.5 \text{ V} & v_{in} > 1 \text{ V} \\ 1.2 \text{ V} & \text{otherwise} \end{cases}$$

44. For the circuit depicted in Fig. 6.55, sketch the expected output voltage v_{out} as a function of v_{active} for $-5 \text{ V} \leq v_{active} \leq +5 \text{ V}$, if v_{ref} is equal to (*a*) −3 V; (*b*) +3 V.

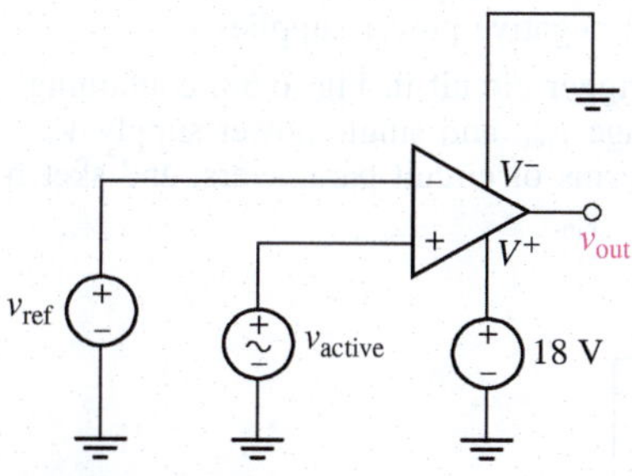

FIGURE 6.55

45. For the circuit depicted in Fig. 6.56, (*a*) sketch the expected output voltage v_{out} as a function of v_1 for $-5 \text{ V} \leq v_1 \leq +5 \text{ V}$, if $v_2 = +2$ V; (*b*) sketch the expected output voltage v_{out} as a function of v_2 for $-5 \text{ V} \leq v_2 \leq +5 \text{ V}$, if $v_1 = +2$ V.

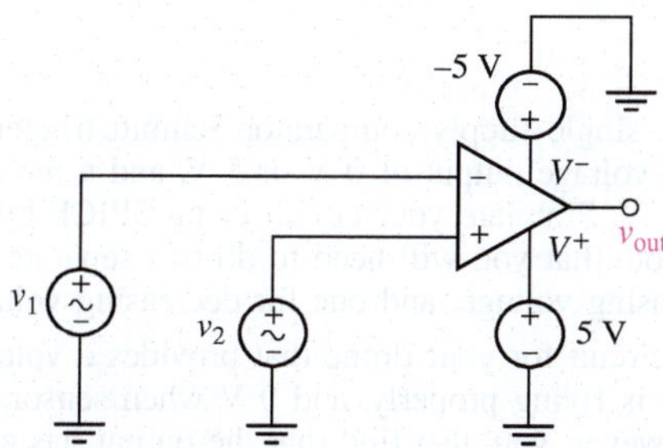

FIGURE 6.56

46. For the circuit depicted in Fig. 6.57, sketch the expected output voltage v_{out} as a function of v_{active}, if $-2 \text{ V} \leq v_{active} \leq +2 \text{ V}$. Verify your solution in SPICE using an op amp model of your choosing (be sure the model includes power supply terminals). Submit a properly labeled schematic with your results.

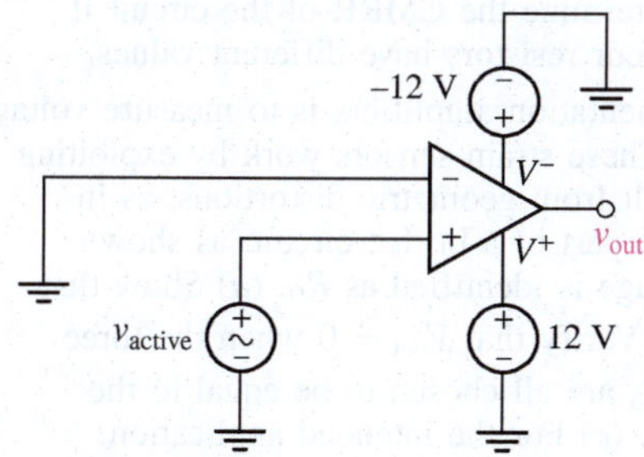

FIGURE 6.57

47. In digital logic applications, a +5 V signal represents a logic "1" state, and a 0 V signal represents a logic "0" state. In order to process real-world information using a digital computer, some type of interface is required, which typically includes an analog-to-digital (A/D) converter—a device that converts analog signals into digital signals. Design a circuit that acts as a simple 1-bit A/D, with any signal less than 1.5 V resulting in a logic "0" and any signal greater than 1.5 V resulting in a logic "1."

48. Using the temperature sensor in the circuit in Prob. 42, design a temperature alarm circuit that outputs a voltage of +5 V when the temperature exceeds 100°C and a voltage of −5 V when the temperature goes below 10°C. (*Hint:* It may be a good idea to place the temperature sensor in a resistor network that uses both positive and negative power supplies.)

49. Examine the comparator Schmitt trigger circuit in Fig. 6.58, containing an input voltage v_{in}, reference voltage v_{ref}, and single power supply V_s. Determine the trigger voltages in terms of circuit parameters, and sketch the output characteristics v_{out} versus v_{in}.

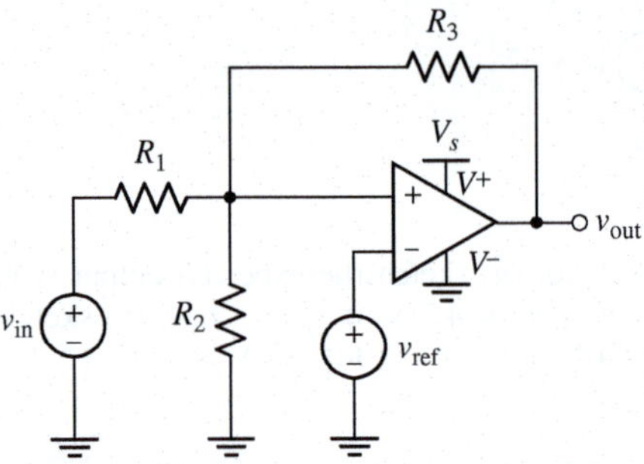

■ **FIGURE 6.58**

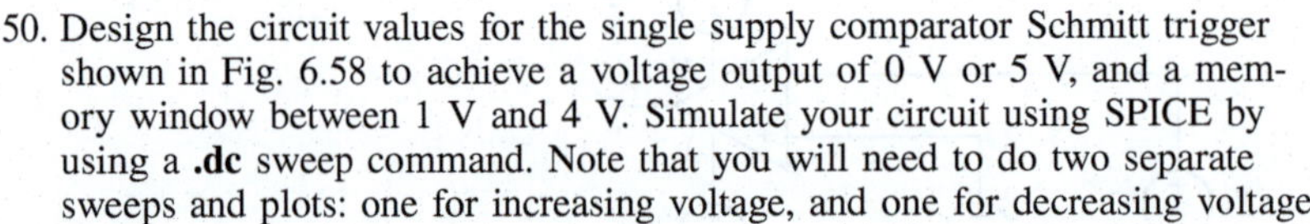

50. Design the circuit values for the single supply comparator Schmitt trigger shown in Fig. 6.58 to achieve a voltage output of 0 V or 5 V, and a memory window between 1 V and 4 V. Simulate your circuit using SPICE by using a **.dc** sweep command. Note that you will need to do two separate sweeps and plots: one for increasing voltage, and one for decreasing voltage.

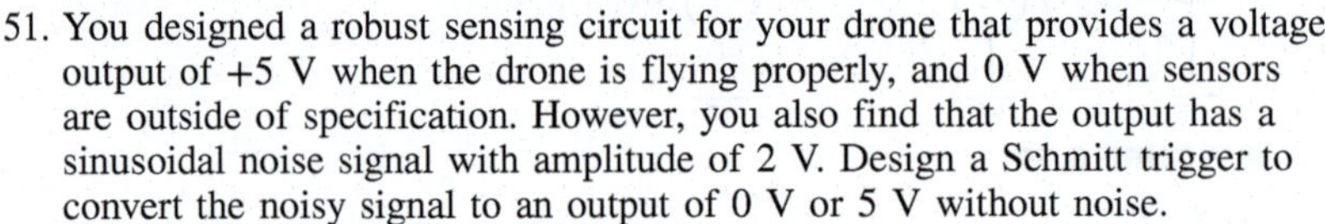

51. You designed a robust sensing circuit for your drone that provides a voltage output of +5 V when the drone is flying properly, and 0 V when sensors are outside of specification. However, you also find that the output has a sinusoidal noise signal with amplitude of 2 V. Design a Schmitt trigger to convert the noisy signal to an output of 0 V or 5 V without noise.

52. Use an appropriate SPICE simulation to verify a Schmitt trigger circuit that accomplishes the goals of Exercise 51. It may be helpful to use a behavioral source such as the **bv** component in LTspice to define a source with dc and ac components.

53. For the instrumentation amplifier shown in Fig. 6.34*a*, assume that the three internal op amps are ideal, and determine the CMRR of the circuit if (*a*) $R_1 = R_3$ and $R_2 = R_4$; (*b*) all four resistors have different values.

54. A common application for instrumentation amplifiers is to measure voltages in resistive strain gauge circuits. These strain sensors work by exploiting the changes in resistance that result from geometric distortions, as in Eq. [6] of Chap. 2. They are often part of a bridge circuit, as shown in Fig. 6.59*a*, where the strain gauge is identified as R_G. (*a*) Show that $V_{out} = V_{in}\left[\frac{R_2}{R_1 + R_2} - \frac{R_3}{R_3 + R_{Gauge}}\right]$. (*b*) Verify that $V_{out} = 0$ when the three fixed-value resistors R_1, R_2, and R_3 are all chosen to be equal to the unstrained gauge resistance R_{Gauge}. (*c*) For the intended application, the gauge selected has an unstrained resistance of 5 kΩ, and a maximum resistance increase of 50 mΩ is expected. Only ±12 V supplies are available. Using the instrumentation amplifier of Fig. 6.59*b*, design a circuit that will provide a voltage signal of +1 V when the strain gauge is at its maximum loading.

AD622 Specifications

Amplifier gain G can be varied from 2 to 1000 by connecting a resistor between pins 1 and 8 with a value calculated by $R = \frac{50.5}{G-1}$ kΩ.

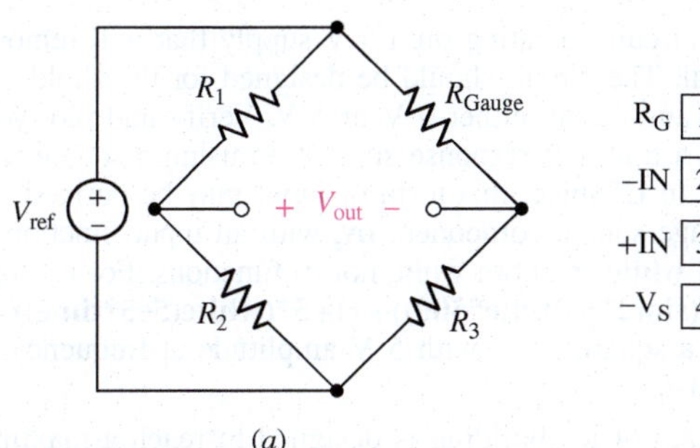

FIGURE 6.59

Chapter-Integrating Exercises

55. The circuit depicted in Fig. 6.60 is known as a Howland current source. Derive expressions for v_{out} and I_L, respectively, as a function of V_1 and V_2.

56. For the circuit depicted in Fig. 6.60, known as a Howland current source, set $V_2 = 0$, $R_1 = R_3$, and $R_2 = R_4$; then solve for the current I_L when $R_1 = 2R_2 = 1$ kΩ and $R_L = 100$ Ω.

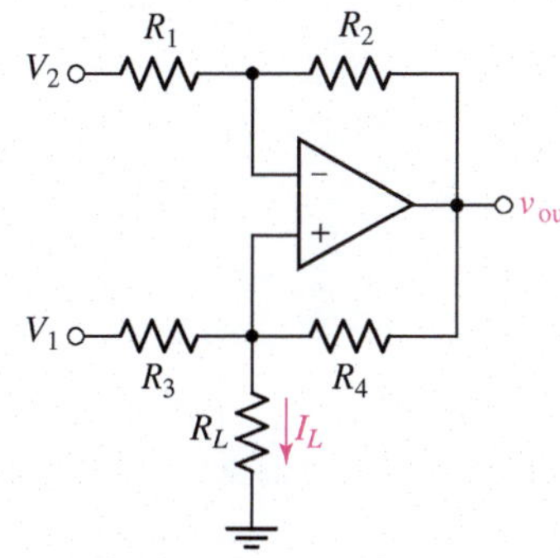

FIGURE 6.60

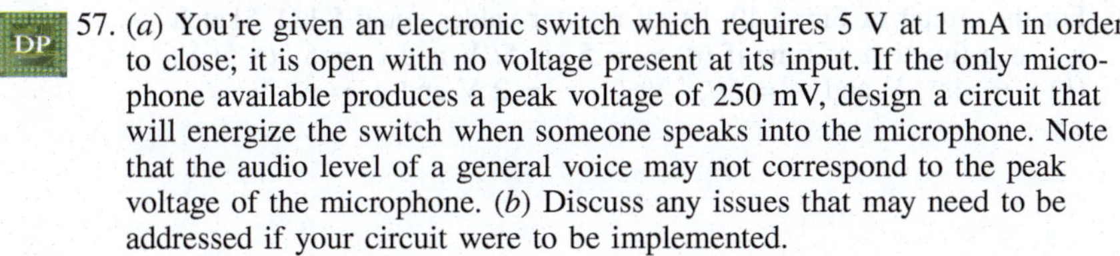

DP 57. (a) You're given an electronic switch which requires 5 V at 1 mA in order to close; it is open with no voltage present at its input. If the only microphone available produces a peak voltage of 250 mV, design a circuit that will energize the switch when someone speaks into the microphone. Note that the audio level of a general voice may not correspond to the peak voltage of the microphone. (b) Discuss any issues that may need to be addressed if your circuit were to be implemented.

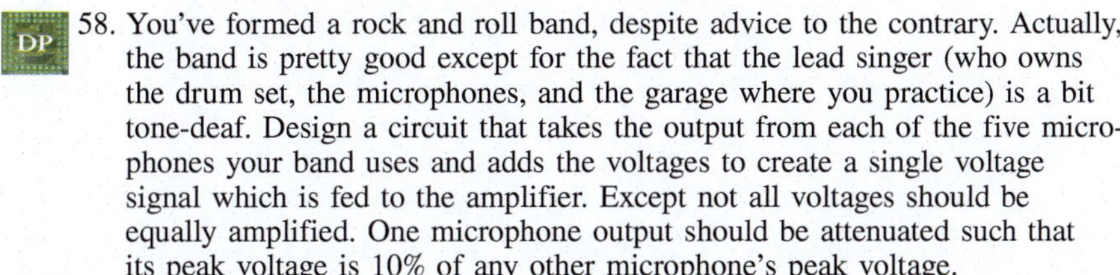

DP 58. You've formed a rock and roll band, despite advice to the contrary. Actually, the band is pretty good except for the fact that the lead singer (who owns the drum set, the microphones, and the garage where you practice) is a bit tone-deaf. Design a circuit that takes the output from each of the five microphones your band uses and adds the voltages to create a single voltage signal which is fed to the amplifier. Except not all voltages should be equally amplified. One microphone output should be attenuated such that its peak voltage is 10% of any other microphone's peak voltage.

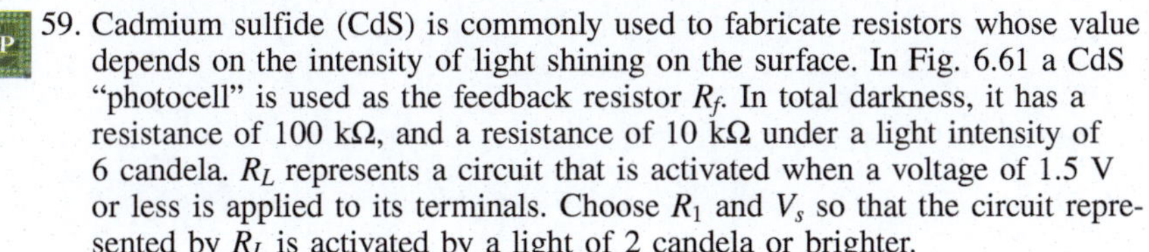

DP 59. Cadmium sulfide (CdS) is commonly used to fabricate resistors whose value depends on the intensity of light shining on the surface. In Fig. 6.61 a CdS "photocell" is used as the feedback resistor R_f. In total darkness, it has a resistance of 100 kΩ, and a resistance of 10 kΩ under a light intensity of 6 candela. R_L represents a circuit that is activated when a voltage of 1.5 V or less is applied to its terminals. Choose R_1 and V_s so that the circuit represented by R_L is activated by a light of 2 candela or brighter.

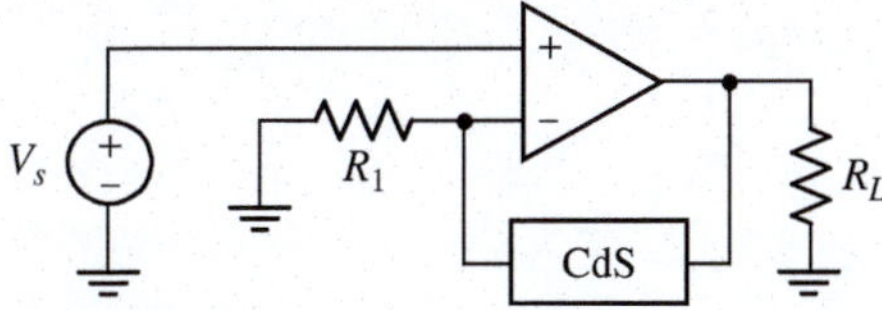

FIGURE 6.61

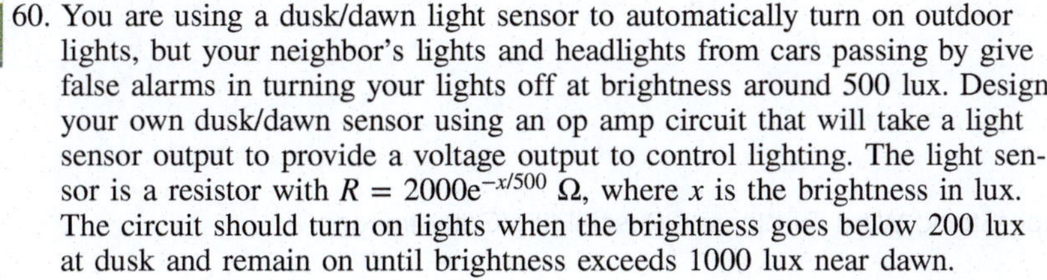

DP 60. You are using a dusk/dawn light sensor to automatically turn on outdoor lights, but your neighbor's lights and headlights from cars passing by give false alarms in turning your lights off at brightness around 500 lux. Design your own dusk/dawn sensor using an op amp circuit that will take a light sensor output to provide a voltage output to control lighting. The light sensor is a resistor with $R = 2000e^{-x/500}$ Ω, where x is the brightness in lux. The circuit should turn on lights when the brightness goes below 200 lux at dusk and remain on until brightness exceeds 1000 lux near dawn.

61. Design a Schmitt trigger circuit operating on a 5 V supply that will remove noise from a digital signal. The circuit should be designed for threshold values of 1 V and 3.5 V, and output either 0 V or 5 V. Verify and plot your results in SPICE through a transient response simulation using a square wave input that includes noise. In LTSpice, this form of input may be defined using the behavioral voltage source component **bv,** with an input function using **sgn** (sign), **sin,** and **white** (random white noise) functions. For example, the function **V=2.5*(sgn(sin(2*pi*time*500))+1)+5*(white(5e5*time))** can be used to represent a square wave with 5 V amplitude at frequency of 500 Hz with white noise.

62. A fountain outside a certain office building is designed to reach a maximum height of 5 meters at a flow rate of 100 l/s. A variable position valve in line with the water supply to the fountain can be controlled electrically, such that 0 V applied results in the valve being fully open, and 5 V results in the valve being closed. In adverse wind conditions the maximum height of the fountain needs to be adjusted; if the wind velocity exceeds 50 km/h, the height cannot exceed 2 meters. A wind velocity sensor is available which provides a voltage calibrated such that 1 V corresponds to a wind velocity of 25 km/h. Design a circuit which uses the velocity sensor to control the fountain according to specifications.

63. For the circuit of Fig. 6.40, let all resistor values equal 5 kΩ. Sketch v_{out} as a function of time if (*a*) $v_1 = 5 \sin 5t$ V and $v_2 = 5 \cos 5t$ V; (*b*) $v_1 = 4e^{-t}$ V and $v_2 = 5e^{-2t}$ V; (*c*) $v_1 = 2$ V and $v_2 = e^{-t}$ V.

CHAPTER 7

Capacitors and Inductors

KEY CONCEPTS

Voltage–Current Relationship of an Ideal Capacitor

Current–Voltage Relationship of an Ideal Inductor

Calculating Energy Stored in Capacitors and Inductors

Response of Capacitors and Inductors to Time-Varying Waveforms

Series and Parallel Combinations

Op Amp Circuits with Capacitors

Computer Modeling of Energy Storage Elements

INTRODUCTION

In this chapter we introduce two new passive circuit elements: the *capacitor* and the *inductor*. Each element can both store and deliver *finite* amounts of energy, which is a significant difference compared to the resistor of Chap. 2. Although they are classed as linear elements, the current–voltage relationships for these new elements are time-dependent, leading to many interesting circuits. The range of capacitance and inductance values we might encounter can be huge, so at times they may dominate circuit behavior and at other times they may be essentially insignificant. Such issues continue to be relevant in modern circuit applications, particularly as computer and communication systems move to increasingly higher operating frequencies and component densities.

7.1 THE CAPACITOR

Ideal Capacitor Model

Previously, we referred to independent and dependent sources as *active* elements, and the linear resistor as a *passive* element, although our definitions of active and passive are still slightly fuzzy and need to be brought into sharper focus. We now define an ***active element*** as an element that is capable of furnishing an average power greater than zero to some external device, where the average is taken over an infinite time interval. Ideal sources are active elements, and the operational amplifier is also an active device. A ***passive element***, however, is defined as an element that cannot supply an average power that is greater than zero over an infinite time interval. The resistor falls into this category; the energy it receives is usually transformed into heat (sometimes smoke), and it never supplies energy.

We now introduce a new passive circuit element, the *capacitor*. We define capacitance C by the voltage–current relationship

$$\boxed{i = C\frac{dv}{dt}} \qquad [1]$$

FIGURE 7.1 Electrical symbol and current–voltage conventions for a capacitor.

where v and i satisfy the conventions for a passive element, as shown in Fig. 7.1. We should bear in mind that v and i are functions of time; if needed, we can emphasize this fact by writing $v(t)$ and $i(t)$ instead. From Eq. [1], we may determine the unit of capacitance as an ampere-second per volt, or coulomb per volt. We will now define the ***farad***[1] (F) as one coulomb per volt, and will use this as our unit of capacitance.

The ideal capacitor defined by Eq. [1] is only a mathematical model of a real device. A capacitor consists of two conducting surfaces on which charge may be stored, separated by a thin insulating layer that has a very large resistance. If we assume that this resistance is sufficiently large that it may be considered infinite, then equal and opposite charges placed on the capacitor "plates" can never meet, at least by any path within the element. The construction of the physical device is suggested by the circuit symbol shown in Fig. 7.1.

Let's visualize some external device connected to this capacitor and causing a positive current to flow into one plate of the capacitor and out of the other plate. Equal currents are entering and leaving the two terminals, and this is no more than we expect for any circuit element. Now let us examine the interior of the capacitor. The positive current entering one plate represents positive charge moving toward that plate through its terminal lead; this charge cannot pass through the interior of the capacitor, and it therefore accumulates on the plate. As a matter of fact, the current and the increasing charge are related by the familiar equation

$$i = \frac{dq}{dt}$$

Now let us consider this plate as an overgrown node and apply Kirchhoff's current law. It apparently does not hold; current is approaching the plate from the external circuit, but it is not flowing out of the plate into the "internal circuit." This dilemma bothered a famous Scottish scientist, James Clerk Maxwell, more than a century ago. The unified electromagnetic theory that he subsequently developed hypothesizes a "displacement current" that is present wherever an electric field or a voltage is varying with time. The displacement current flowing internally between the capacitor plates is exactly equal to the conduction current flowing in the capacitor leads; Kirchhoff's current law is therefore satisfied if we include both conduction and displacement currents. However, circuit analysis is not concerned with this internal displacement current, and since it is fortunately equal to the conduction current, we may consider Maxwell's hypothesis as relating the conduction current to the changing voltage across the capacitor.

A capacitor constructed of two parallel conducting plates of area A, separated by a distance d, has a capacitance $C = \varepsilon A/d$, where ε is the permittivity, a property of the insulating material between the plates; this assumes the linear dimensions of the conducting plates are all much greater than d.

(1) Named in honor of Michael Faraday, a 19th-century English scientist.

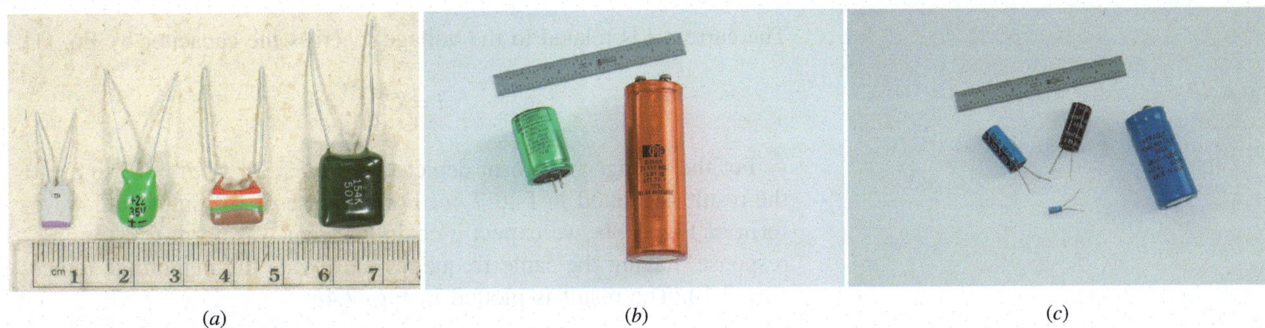

(a) (b) (c)

■ **FIGURE 7.2** Several examples of commercially available capacitors. (*a*) Left to right: 270 pF ceramic, 20 μF tantalum, 15 nF polyester, 150 nF polyester. (*b*) Left: 2000 μF 40 VDC rated electrolytic, 25,000 μF 35 VDC rated electrolytic. (*c*) Clockwise from smallest: 100 μF 63 VDC rated electrolytic, 2200 μF 50 VDC rated electrolytic, 55 F 2.5 VDC rated electrolytic, and 4800 μF 50 VDC rated electrolytic. Note that, generally speaking, larger capacitance values require larger packages, with one notable exception here. What was the trade-off in that case? Note the scale for (*b*) and (*c*) is shown in inches.
(*a–c*) Steve Durbin/McGraw Hill

For air or vacuum, $\varepsilon = \varepsilon_0 = 8.854$ pF/m. Most capacitors employ a thin dielectric layer with a larger permittivity than air in order to minimize the device size. Examples of various types of commercially available capacitors are shown in Fig. 7.2, although we should remember that any two conducting surfaces not in direct contact with each other may be characterized by a nonzero (although probably small) capacitance. We should also note that a capacitance of several hundred *microfarads* (μF) is considered "large."

Several important characteristics of our new mathematical model can be discovered from the defining equation, Eq. [1]. A constant voltage across a capacitor results in zero current passing through it; a capacitor is thus an "*open circuit to dc*." This fact is pictorially represented by the capacitor symbol. It is also apparent that a sudden jump in the voltage requires an infinite current. Since this is physically impossible, we will therefore prohibit the voltage across a capacitor to change in zero time.

EXAMPLE 7.1

Determine the current i flowing through the capacitor of Fig. 7.1 for the two voltage waveforms of Fig. 7.3 if C = 2 F.

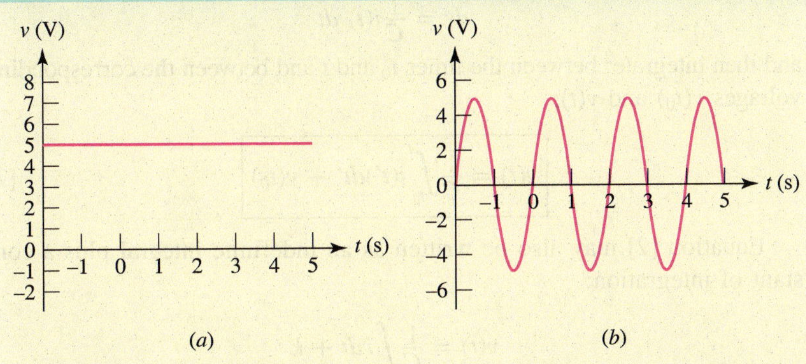

■ **FIGURE 7.3** (*a*) A dc voltage applied to the terminals of the capacitor. (*b*) A sinusoidal voltage waveform applied to the capacitor terminals.

(Continued on next page)

The current i is related to the voltage v across the capacitor by Eq. [1]:

$$i = C\frac{dv}{dt}$$

For the voltage waveform depicted in Fig. 7.3*a*, $dv/dt = 0$, so $i = 0$; the result is plotted in Fig. 7.4*a*. For the case of the sinusoidal waveform of Fig. 7.3*b*, we expect a cosine current waveform to flow in response, having the same frequency and twice the magnitude (since $C = 2$ F). The result is plotted in Fig. 7.4*b*.

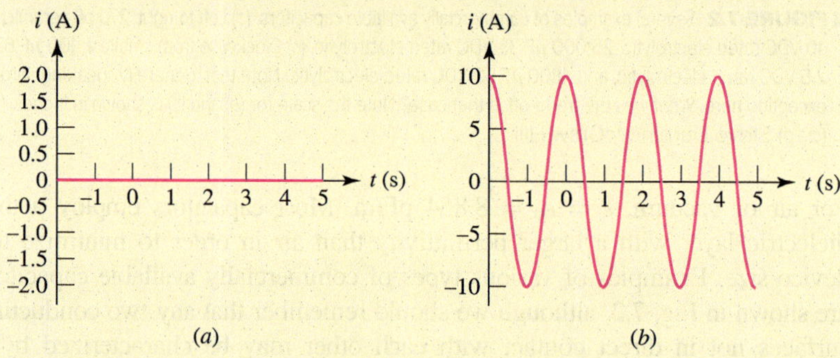

■ **FIGURE 7.4** (*a*) $i = 0$ since the voltage applied is dc. (*b*) The current has a cosine form in response to an applied sine wave voltage.

PRACTICE

7.1 Determine the current flowing through a 5 mF capacitor in response to a voltage v equal to: (*a*) −20 V; (*b*) $2e^{-5t}$ V.

Ans: 0 A; $-50e^{-5t}$ mA.

Integral Voltage–Current Relationships

The capacitor voltage may be expressed in terms of the current by integrating Eq. [1]. We first obtain

$$dv = \frac{1}{C} i(t)\, dt$$

and then integrate[2] between the times t_0 and t and between the corresponding voltages $v(t_0)$ and $v(t)$:

$$v(t) = \frac{1}{C}\int_{t_0}^{t} i(t')dt' + v(t_0) \qquad [2]$$

Equation [2] may also be written as an indefinite integral plus a constant of integration:

$$v(t) = \frac{1}{C}\int i\, dt + k$$

(2) Note that we are employing the mathematically correct procedure of defining a *dummy variable* t' in situations where the integration variable t is also a limit.

Finally, in many situations we will find that $v(t_0)$, the voltage initially across the capacitor, cannot be discerned. In such cases it is mathematically convenient to set $t_0 = -\infty$ and $v(-\infty) = 0$, so that

$$v(t) = \frac{1}{C}\int_{-\infty}^{t} i\, dt'$$

Since the integral of the current over any time interval is the corresponding charge accumulated on the capacitor plate into which the current is flowing, we may also define capacitance as

$$q(t) = Cv(t)$$

where $q(t)$ and $v(t)$ represent instantaneous values of the charge on either plate and the voltage between the plates, respectively.

EXAMPLE 7.2

Find the capacitor voltage that is associated with the current shown graphically in Fig. 7.5*a*. The value of the capacitance is 5 μF.

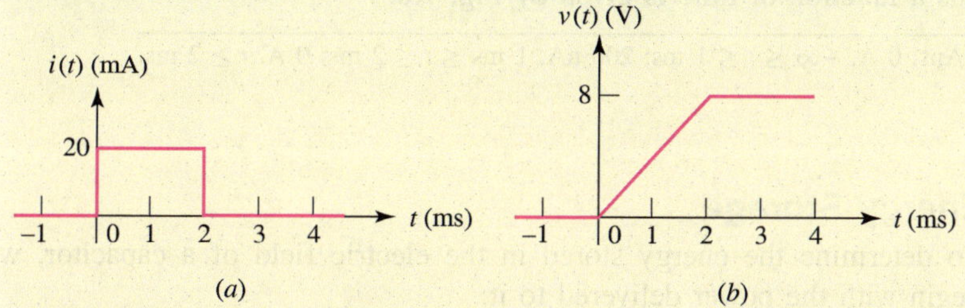

FIGURE 7.5 (*a*) The current waveform applied to a 5 μF capacitor. (*b*) The resultant voltage waveform obtained by graphical integration.

Equation [2] is the appropriate expression here:

$$v(t) = \frac{1}{C}\int_{t_0}^{t} i(t')dt' + v(t_0)$$

but now it needs to be interpreted graphically. To do this, we note that the difference in voltage between times t and t_0 is proportional to the area under the current curve defined by the same two times. The constant of proportionality is $1/C$.

From Fig. 7.5*a*, we see three separate intervals: $t \le 0$, $0 \le t \le 2$ ms, and $t \ge 2$ ms. Defining the first interval more specifically as between $-\infty$ and 0, so that $t_0 = -\infty$, we note two things, both a consequence of the fact that the current has always been zero up to $t = 0$: First,

$$v(t_0) = v(-\infty) = 0$$

Second, the integral of the current between $t_0 = -\infty$ and 0 is simply zero, since $i = 0$ in that interval. Thus,

$$v(t) = 0 + v(-\infty) \qquad -\infty \le t \le 0$$

or

$$v(t) = 0 \qquad t \le 0$$

(Continued on next page)

If we now consider the time interval represented by the rectangular pulse, we obtain

$$v(t) = \frac{1}{5 \times 10^{-6}} \int_0^t 20 \times 10^{-3}\, dt' + v(0)$$

Since $v(0) = 0$,

$$v(t) = 4000t \qquad 0 \le t \le 2 \text{ ms}$$

For the semi-infinite interval following the pulse, the integral of $i(t)$ is once again zero, so that

$$v(t) = 8 \qquad t \ge 2 \text{ ms}$$

The results are shown graphically in Fig. 7.5*b*.

PRACTICE

7.2 Determine the current through a 100 pF capacitor if its voltage as a function of time is given by Fig. 7.6.

Ans: 0 A, $-\infty \le t \le 1$ ms; 200 nA, 1 ms $\le t \le 2$ ms; 0 A, $t \ge 2$ ms.

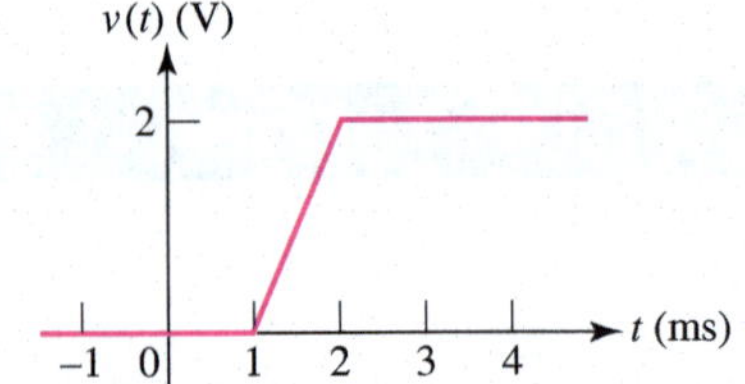

FIGURE 7.6

Energy Storage

To determine the energy stored in the electric field of a capacitor, we begin with the power delivered to it:

$$p = vi = Cv\frac{dv}{dt}$$

and simply integrate over the time interval of interest:

$$\int_{t_0}^{t} p\, dt' = C\int_{t_0}^{t} v\frac{dv}{dt'}\,dt' = C\int_{v(t_0)}^{v(t)} v'dv' = \frac{1}{2}C\left\{[v(t)]^2 - [v(t_0)]^2\right\}$$

Thus,

$$w_C(t) - w_C(t_0) = \frac{1}{2}C\left\{[v(t)]^2 - [v(t_0)]^2\right\} \qquad [3]$$

where the stored energy is w_C, measured in joules (J), and the voltage at t_0 is $v(t_0)$. If we select a zero-energy reference at t_0, implying that the capacitor voltage is also zero at that instant, then

$$\boxed{w_C(t) = \frac{1}{2}Cv^2} \qquad [4]$$

Let us consider a numerical example. As sketched in Fig. 7.7, a sinusoidal voltage source is in parallel with a 1 MΩ resistor and a 20 μF capacitor. The parallel resistor may be assumed to represent the finite resistance of the dielectric between the plates of the physical capacitor (an *ideal* capacitor has infinite resistance associated with it).

EXAMPLE 7.3

Find the maximum energy stored in the capacitor of Fig. 7.7 and the energy dissipated in the resistor over the interval $0 < t < 0.5$ s.

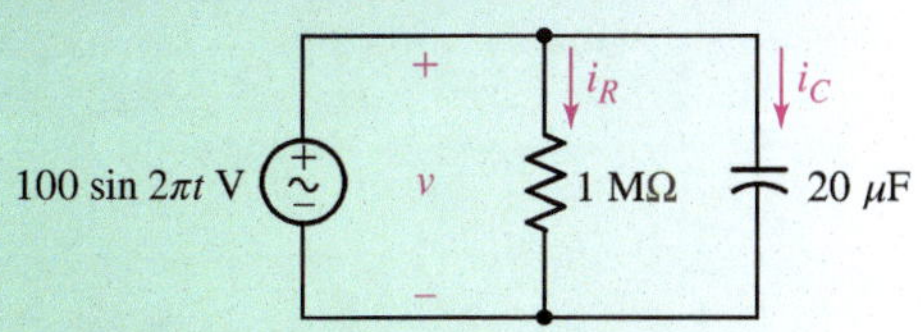

FIGURE 7.7 A sinusoidal voltage source is applied to a parallel *RC* network. The 1 MΩ resistor might represent the finite resistance of the "real" capacitor's dielectric layer.

▶ ***Identify the goal of the problem.***

The energy stored in the capacitor varies with time; we are asked for the *maximum* value over a specific time interval. We are also asked to find the *total* amount of energy dissipated by the resistor over this interval. These are two completely different questions.

▶ ***Collect the known information.***

The only source of energy in the circuit is the independent voltage source, which has a value of 100 sin $2\pi t$ V. We are only interested in the time interval of $0 < t < 0.5$ s. The circuit is properly labeled.

▶ ***Devise a plan.***

Determine the energy in the capacitor by evaluating the voltage and using Eq. [4]. To find the energy dissipated in the resistor during the same time interval, integrate the dissipated *power*, $p_R = i_R^2 \cdot R$.

▶ ***Construct an appropriate set of equations.***

The energy stored in the capacitor is

$$w_C(t) = \frac{1}{2}Cv^2 = 0.1 \sin^2 2\pi t \text{ J}$$

We obtain an expression for the power dissipated by the resistor in terms of the current i_R:

$$i_R = \frac{v}{R} = 10^{-4} \sin 2\pi t \text{ A}$$

so

$$p_R = i_R^2 R = (10^{-4})(10^6) \sin^2 2\pi t$$

and the energy dissipated in the resistor between 0 and 0.5 s is

$$w_R = \int_0^{0.5} p_R dt = \int_0^{0.5} 10^{-2} \sin^2 2\pi t \, dt \text{ J}$$

▶ ***Determine if additional information is required.***

We have an expression for the energy stored in the capacitor; a sketch is shown in Fig. 7.8. The expression derived for the energy dissipated by the resistor does not involve any unknown quantities, and so it may also be readily evaluated.

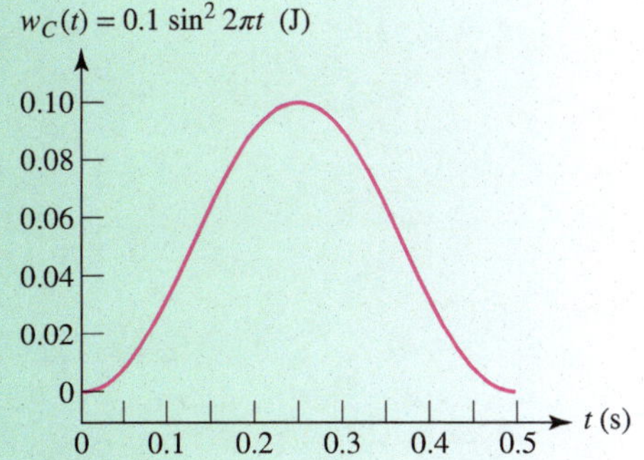

FIGURE 7.8 A sketch of the energy stored in the capacitor as a function of time.

▶ ***Attempt a solution.***

From our sketch of the expression for the energy stored in the capacitor, we see that it increases from zero at $t = 0$ to a maximum of 100 mJ at $t = \frac{1}{4}$s, and then decreases to zero in another $\frac{1}{4}$s. Thus, $w_{C_{max}} = 100$ mJ. Evaluating our integral expression for the energy dissipated in the resistor, we find that $w_R = 2.5$ mJ.

(Continued on next page)

▶ ***Verify the solution. Is it reasonable or expected?***

We do not expect to calculate a *negative* stored energy, which is borne out in our sketch. Further, since the maximum value of sin $2\pi t$ is 1, the maximum energy expected anywhere would be $(1/2)(20 \times 10^{-6})(100)^2 = 100$ mJ.

The resistor dissipated 2.5 mJ in the period of 0 to 500 ms, although the capacitor stored a maximum of 100 mJ at one point during that interval. What happened to the "other" 97.5 mJ? To answer this, we compute the capacitor current

$$i_C = 20 \times 10^{-6}\frac{dv}{dt} = 0.004\pi \cos 2\pi t$$

and the current i_s defined as flowing *into* the voltage source

$$i_s = -i_C - i_R$$

both of which are plotted in Fig. 7.9. We observe that the current flowing through the resistor is a small fraction of the source current—not entirely surprising as 1 MΩ is a relatively large resistance value. As current flows from the source, a small amount is diverted to the resistor, with the rest flowing into the capacitor as it charges. After $t = 250$ ms, the source current is seen to change sign; current is now flowing from the capacitor back into the source. Most of the energy stored in the capacitor is being returned to the ideal voltage source, except for the small fraction dissipated in the resistor.

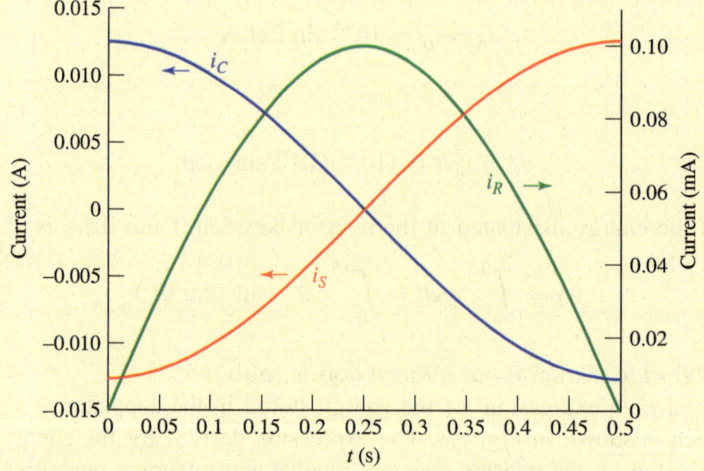

FIGURE 7.9 Plot of the resistor, capacitor, and source currents during the interval of 0 to 500 ms. Note that i_s is defined as flowing into the positive terminal of the source.

PRACTICE

7.3 Calculate the energy stored in a 1000 μF capacitor at $t = 50$ μs if the voltage across it is given by 1.5 cos 10^5 t volts.

Ans: 90.52 μJ.

Important Characteristics of an Ideal Capacitor

1. There is no current through a capacitor unless the voltage across it is changing with time. A capacitor is therefore an *open circuit to dc*.
2. A finite amount of energy can be stored in a capacitor even if the current through the capacitor is zero, such as when the voltage across it is constant.
3. It is impossible to change the voltage across a capacitor by a finite amount in zero time, as this requires an infinite current through the capacitor. (A capacitor resists an abrupt change in the voltage across it in a manner analogous to the way a spring resists an abrupt change in its displacement.)
4. An ideal capacitor never dissipates energy, but only stores it. Although this is true for the *mathematical model,* it is not true for a *physical* capacitor due to finite resistances associated with the dielectric as well as the packaging. Thus, a real capacitor will eventually discharge once disconnected from a power source.

7.2 THE INDUCTOR

Ideal Inductor Model

In the early 1800s Danish scientist Hans Christian Ørsted showed that a current-carrying conductor produced a magnetic field (i.e. compass needles were affected in the presence of a wire when current was flowing). Shortly thereafter, Ampère made some careful measurements which demonstrated that this magnetic field was *linearly* related to the current which produced it. The next step occurred some 20 years later when English experimentalist Michael Faraday and American inventor Joseph Henry discovered almost simultaneously[3] that a changing magnetic field could induce a voltage in a neighboring circuit. They showed that this voltage was proportional to the time rate of change of the current producing the magnetic field. The constant of proportionality is what we now call the ***inductance,*** symbolized by L, and therefore

$$\boxed{v = L\frac{di}{dt}} \qquad [5]$$

where we must realize that v and i are both functions of time. When we wish to emphasize this, we may do so by using the symbols $v(t)$ and $i(t)$.

The circuit symbol for the inductor is shown in Fig. 7.10, and it should be noted that the passive sign convention is used, just as it was with the resistor and the capacitor. The unit in which inductance is measured is the ***henry*** (H), and the defining equation shows that the henry is just a shorter expression for a volt-second per ampere.

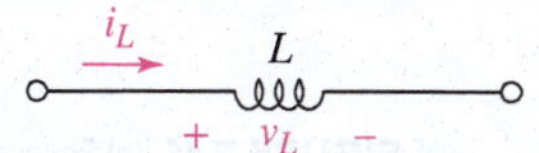

FIGURE 7.10 Electrical symbol and current–voltage conventions for an inductor.

(3) Faraday won.

The inductor whose inductance is defined by Eq. [5] is a mathematical model; it is an *ideal* element which we may use to approximate the behavior of a *real* device. A physical inductor may be constructed by winding a length of wire into a coil. This serves effectively to increase the current that is causing the magnetic field and also to increase the "number" of neighboring circuits into which Faraday's voltage may be induced. The result of this twofold effect is that the inductance of a coil is approximately proportional to the square of the number of complete turns made by the conductor out of which it is formed. For example, an inductor or "coil" that has the form of a long helix of very small pitch is found to have an inductance of $\mu N^2 A/s$, where A is the cross-sectional area, s is the axial length of the helix, N is the number of complete turns of wire, and μ (mu) is a constant of the material inside the helix, called the permeability. For free space (and very closely for air), $\mu = \mu_0 = 4\pi \times 10^{-7}$ H/m $= 4\pi$ nH/cm. Several examples of commercially available inductors are shown in Fig. 7.11.

Let us now scrutinize Eq. [5] to determine some of the electrical characteristics of the mathematical model. This equation shows that the voltage across an inductor is proportional to the time rate of change of the current through it. In particular, it shows that there is no voltage across an inductor carrying a constant current, regardless of the magnitude of this current. Accordingly, we may view an inductor as a *short circuit to dc*.

Another fact that can be obtained from Eq. [5] is that a sudden or discontinuous change in the current must be associated with an infinite voltage across the inductor. In other words, if we wish to produce an abrupt change in an inductor current, we must apply an infinite voltage. Although an infinite-voltage forcing function might be amusing theoretically,

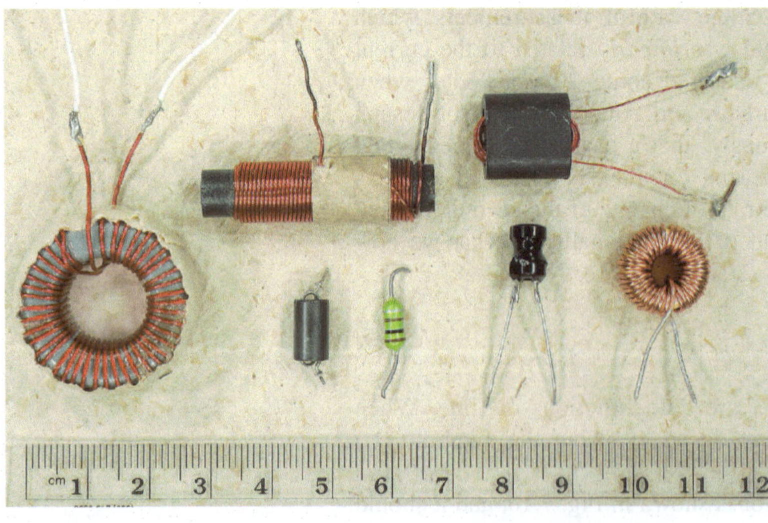

(*a*)

(*b*)

FIGURE 7.11 (*a*) Several different types of commercially available inductors, sometimes also referred to as "chokes." Clockwise, starting from far left: 287 μH ferrite core toroidal inductor, 266 μH ferrite core cylindrical inductor, 215 μH ferrite core inductor designed for VHF frequencies, 85 μH iron powder core toroidal inductor, 10 μH bobbin-style inductor, 100 μH axial lead inductor, and 7 μH lossy-core inductor used for RF suppression. (*b*) An 11 H inductor, measuring 10 cm (tall) × 8 cm (wide) × 8 cm (deep).
(*a–b*) Steve Durbin/McGraw Hill

it can never be a part of the phenomena displayed by a real physical device. As we shall see shortly, an abrupt change in the inductor current also requires an abrupt change in the energy stored in the inductor, and this sudden change in energy requires infinite power at that instant; infinite power is again not a part of the real physical world. In order to avoid infinite voltage and infinite power, an inductor current must not be allowed to jump *instantaneously* from one value to another.

If an attempt is made to open-circuit a physical inductor through which a finite current is flowing, an arc may appear across the switch. This is useful in the ignition system of some automobiles, where the current through the spark coil is interrupted by the distributor and the arc appears across the spark plug. Although this does not occur instantaneously, it happens in a very short timespan, leading to the creation of a large voltage. The presence of a large voltage across a short distance equates to a very large electric field; the stored energy is dissipated in ionizing the air in the path of the arc.

Equation [5] may also be interpreted (and solved, if necessary) by graphical methods, as seen in Example 7.4.

EXAMPLE 7.4

Given the waveform of the current in a 3 H inductor as shown in Fig. 7.12*a*, determine the inductor voltage and sketch it.

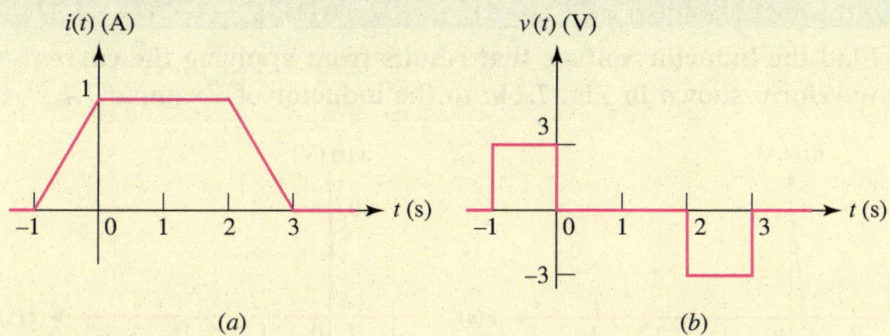

FIGURE 7.12 (*a*) The current waveform in a 3 H inductor. (*b*) The corresponding voltage waveform, $v = 3\,di/dt$.

Defining the voltage v and the current i to satisfy the passive sign convention, we may obtain v from Fig. 7.12*a* using Eq. [5]:

$$v = 3\frac{di}{dt}$$

Since the current is zero for $t < -1$ s, the voltage is zero in this interval. The current then begins to increase at the linear rate of 1 A/s, and thus a constant voltage of $L\,di/dt = 3$ V is produced. During the following 2 s interval, the current is constant and the voltage is therefore zero. The final decrease of the current results in $di/dt = -1$ A/s, yielding $v = -3$ V. For $t > 3$ s, $i(t)$ is a constant (zero), so $v(t) = 0$ for that interval. The complete voltage waveform is sketched in Fig. 7.12*b*.

PRACTICE

7.4 The current through a 200 mH inductor is shown in Fig. 7.13. Assume the passive sign convention, and find v_L at t equal to (*a*) 0; (*b*) 2 ms; (*c*) 6 ms.

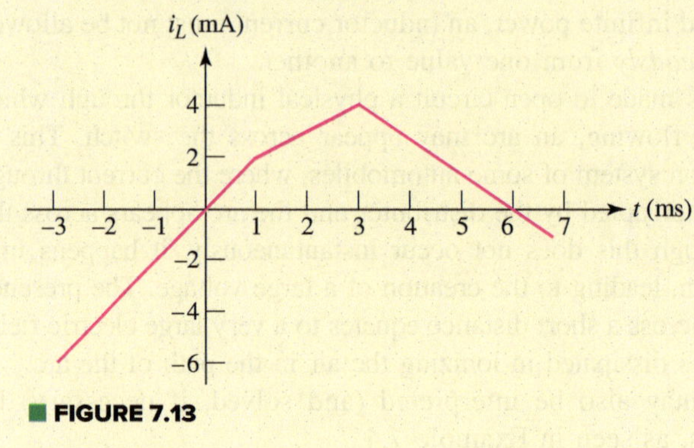

FIGURE 7.13

Ans: 0.4 V; 0.2 V; −0.267 V.

Let us now investigate the effect of a more rapid rise and decay of the current between the 0 and 1 A values.

EXAMPLE 7.5

Find the inductor voltage that results from applying the current waveform shown in Fig. 7.14*a* to the inductor of Example 7.4.

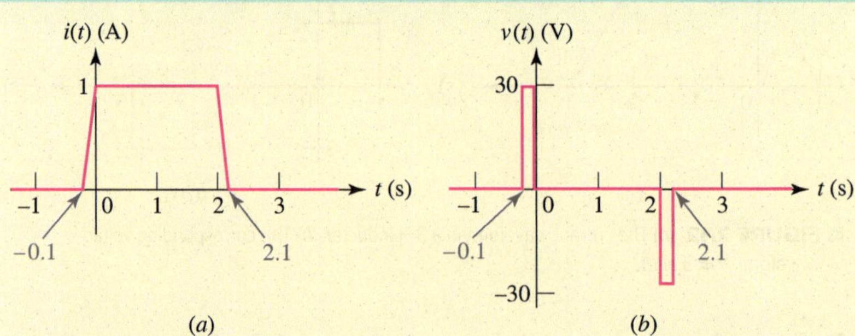

FIGURE 7.14 (*a*) The time required for the current of Fig. 7.12*a* to change from 0 to 1 and from 1 to 0 is decreased by a factor of 10. (*b*) The resultant voltage waveform. The pulse widths are exaggerated for clarity.

Note that the intervals for the rise and fall have decreased to 0.1 s. Thus, the magnitude of each derivative will be 10 times larger; this condition is shown in the current and voltage sketches of Fig. 7.14*a* and *b*. In the voltage waveforms of Fig. 7.12*b* and 7.14*b*, it is interesting to note that the area under each voltage pulse is 3 V·s.

Just for curiosity's sake, let's continue in the same vein for a moment. A further decrease in the rise and fall times of the current waveform will produce a proportionally larger voltage magnitude, but only within the interval in which the current is increasing or decreasing. An abrupt change in the current will cause the infinite voltage "spikes" (each having an area of 3 V · s) as suggested by the waveforms of Fig. 7.15*a* and *b*; or, from the equally valid but opposite point of view, infinite voltage spikes are required to produce abrupt changes in the current.

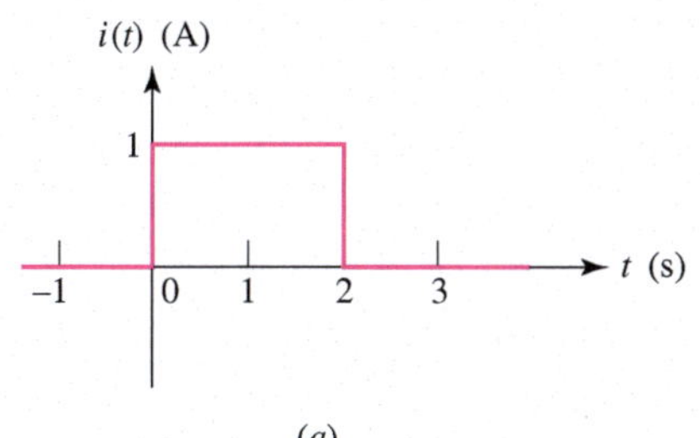

(*a*)

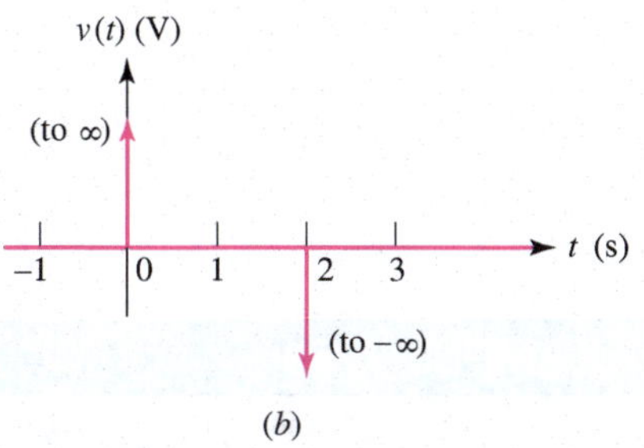

(*b*)

FIGURE 7.15 (*a*) The time required for the current of Fig. 7.14*a* to change from 0 to 1 and from 1 to 0 is decreased to zero; the rise and fall are abrupt. (*b*) The resultant voltage across the 3 H inductor consists of a positive and a negative infinite spike.

PRACTICE

7.5 The current waveform of Fig. 7.14*a* has equal rise and fall times of duration 0.1 s (100 ms). Calculate the maximum positive and negative voltages across the same inductor if the rise and fall times, respectively, are changed to (*a*) 1 ms, 1 ms; (*b*) 12 μs, 64 μs; (*c*) 1 s, 1 ns.

Ans: 3 kV, –3 kV; 250 kV, –46.88 kV; 3 V, –3 GV.

Integral Voltage–Current Relationships

We have defined inductance by a simple differential equation,

$$v = L\frac{di}{dt}$$

and we have been able to draw several conclusions about the characteristics of an inductor from this relationship. For example, we have found that we may consider an inductor to be a short circuit to direct current, and we have agreed that we cannot permit an inductor current to change abruptly from one value to another, because this would require that an infinite voltage and power be associated with the inductor. The simple defining equation for inductance contains still more information, however. Rewritten in a slightly different form,

$$di = \frac{1}{L}v\,dt$$

it invites integration. Let us first consider the limits to be placed on the two integrals. We desire the current i at time t, and this pair of quantities therefore provides the upper limits on the integrals appearing on the left and right sides of the equation, respectively; the lower limits may also be kept general by merely assuming that the current is $i(t_0)$ at time t_0. Thus,

$$\int_{i(t_0)}^{i(t)} di' = \frac{1}{L}\int_{t_0}^{t} v(t')\,dt'$$

which leads to the equation

$$i(t) - i(t_0) = \frac{1}{L}\int_{t_0}^{t} v\,dt'$$

or

$$\boxed{i(t) = \frac{1}{L}\int_{t_0}^{t} v\,dt' + i(t_0)} \qquad [6]$$

Equation [5] expresses the inductor voltage in terms of the current, whereas Eq. [6] gives the current in terms of the voltage. Other forms are also possible for the latter equation. We may write the integral as an indefinite integral and include a constant of integration k:

$$\boxed{i(t) = \frac{1}{L}\int v\,dt + k} \qquad [7]$$

We also may assume that we are solving a realistic problem in which the selection of t_0 as $-\infty$ ensures no current or energy in the inductor. Thus, if $i(t_0) = i(-\infty) = 0$, then

$$\boxed{i(t) = \frac{1}{L}\int_{-\infty}^{t} v\,dt'} \qquad [8]$$

Let us investigate the use of these several integrals by working a simple example where the voltage across an inductor is specified.

EXAMPLE 7.6

The voltage across a 2 H inductor is known to be 6 cos 5*t* V. Determine the resulting inductor current if $i(t = -\pi/2) = 1$ A.

From Eq. [6],

$$i(t) = \frac{1}{2}\int_{t_0}^{t} 6\cos 5t'\,dt' + i(t_0)$$

or

$$\begin{aligned} i(t) &= \frac{1}{2}\left(\frac{6}{5}\right)\sin 5t - \frac{1}{2}\left(\frac{6}{5}\right)\sin 5t_0 + i(t_0) \\ &= 0.6\sin 5t - 0.6\sin 5t_0 + i(t_0) \end{aligned}$$

The first term indicates that the inductor current varies sinusoidally; the second and third terms together represent a constant which becomes known when the current is numerically specified at some instant of time. Using the fact that the current is 1 A at $t = -\pi/2$ s, we identify t_0 as $-\pi/2$ with $i(t_0) = 1$, and find that

$$i(t) = 0.6\sin 5t - 0.6\sin(-2.5\pi) + 1$$

or

$$i(t) = 0.6\sin 5t + 1.6$$

Alternatively, from Eq. [6],

$$i(t) = 0.6\sin 5t + k$$

and we establish the numerical value of k by forcing the current to be 1 A at $t = -\pi/2$:

$$1 = 0.6\sin(-2.5\pi) + k$$

or

$$k = 1 + 0.6 = 1.6$$

and so, as before,

$$i(t) = 0.6\sin 5t + 1.6$$

Equation [8] is going to cause trouble with this particular voltage. We based the equation on the assumption that the current was zero when $t = -\infty$. To be sure, this must be true in the real, physical world, but we are working in the land of the mathematical model; our elements and forcing functions are all idealized. The difficulty arises after we integrate, obtaining

$$i(t) = 0.6 \sin 5t' \big|_{-\infty}^{t}$$

and attempt to evaluate the integral at the lower limit:

$$i(t) = 0.6 \sin 5t - 0.6 \sin(-\infty)$$

The sine of $\pm\infty$ is indeterminate, and therefore we cannot evaluate our expression. Equation [8] is only useful if we are evaluating functions which approach zero as $t \to -\infty$.

PRACTICE

7.6 A 100 mH inductor has voltage $v_L = 2e^{-3t}$ V across its terminals. Determine the resulting inductor current if $i_L(-0.5) = 1$ A.

Ans: $-\frac{20}{3}e^{-3t} + 30.9$ A

We should not make any snap judgments as to which single form of Eqs. [6], [7], and [8] we are going to use forever after; each has its advantages, depending on the problem and the application. Equation [6] represents a long, general method, but it shows clearly that the constant of integration is a current. Equation [7] is a somewhat more concise expression of Eq. [6], but the nature of the integration constant is suppressed. Finally, Eq. [8] is an excellent expression, since no constant is necessary; however, it applies only when the current is zero at $t = -\infty$ and when the analytical expression for the current is not indeterminate there.

Energy Storage

Let us now turn our attention to power and energy. The absorbed power is given by the current–voltage product

$$p = vi = Li\frac{di}{dt}$$

The energy w_L accepted by the inductor is stored in the magnetic field around the coil. The change in this energy is expressed by the integral of the power over the desired time interval:

$$\int_{t_0}^{t} p\, dt' = L\int_{t_0}^{t} i\frac{di}{dt'}dt' = L\int_{i(t_0)}^{i(t)} i'\,di'$$
$$= \frac{1}{2}L\{[i(t)]^2 - [i(t_0)]^2\}$$

Thus,

$$w_L(t) - w_L(t_0) = \frac{1}{2}L\{[i(t)]^2 - [i(t_0)]^2\} \qquad [9]$$

where we have again assumed that the current is $i(t_0)$ at time t_0. In using the energy expression, it is customary to assume that a value of t_0 is selected at which the current is zero; it is also customary to assume that the energy is zero at this time. We then have simply

$$\boxed{w_L(t) = \frac{1}{2}Li^2} \qquad [10]$$

where we now understand that our reference for zero energy is any time at which the inductor current is zero. At any subsequent time at which the current is zero, we also find no energy stored in the coil. Whenever the current is not zero, and regardless of its direction or sign, energy is stored in the inductor. It follows, therefore, that energy may be delivered to the inductor for a part of the time and recovered from the inductor later. All of the stored energy may be recovered from an ideal inductor; there are no storage charges or agent's commissions in the mathematical model. A physical coil, however, must be constructed out of real wire and thus will always have a resistance associated with it. Energy can no longer be stored and recovered without loss.

These ideas may be illustrated by a simple example. In Fig. 7.16, a 3 H inductor is shown in series with a 0.1 Ω resistor and a sinusoidal current source, $i_s = 12 \sin\frac{\pi t}{6}$ A. The resistor should be interpreted as the resistance of the wire which must be associated with the physical coil.

EXAMPLE 7.7

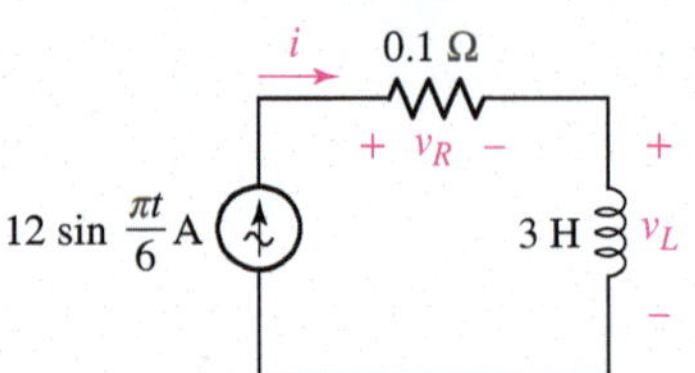

FIGURE 7.16 A sinusoidal current is applied as a forcing function to a series *RL* circuit. The 0.1 Ω represents the inherent resistance of the wire from which the inductor is fabricated.

Find the maximum energy stored in the inductor of Fig. 7.16, and calculate how much energy is dissipated in the resistor in the time during which the energy is being stored in, and then recovered from, the inductor.

The energy stored in the inductor is

$$w_L = \frac{1}{2}Li^2 = 216 \sin^2\frac{\pi t}{6}\ \text{J}$$

and this energy increases from zero at $t = 0$ to 216 J at $t = 3$ s. Thus, the maximum energy stored in the inductor is 216 J.

After reaching its peak value at $t = 3$ s, the energy has completely left the inductor 3 s later. Let us see what price we have paid in this coil for the privilege of storing and removing 216 J in these 6 seconds. The power dissipated in the resistor is easily found as

$$p_R = i^2R = 14.4 \sin^2\frac{\pi t}{6}\ \text{W}$$

and the energy converted into heat in the resistor within this 6 s interval is therefore

$$w_R = \int_0^6 p_R dt = \int_0^6 14.4 \sin^2\frac{\pi}{6}t\, dt$$

or

$$w_R = \int_0^6 14.4\left(\frac{1}{2}\right)\left(1 - \cos\frac{\pi}{3}t\right)dt = 43.2 \text{ J}$$

Thus, we have expended 43.2 J in the process of storing and then recovering 216 J in a 6 s interval. This represents 20% of the maximum stored energy, but it is a reasonable value for many coils having this large an inductance. For coils having an inductance of about 100 μH, we might expect a figure closer to 2% or 3%.

PRACTICE

7.7 Let $L = 25$ mH for the inductor of Fig. 7.10. (*a*) Find v_L at $t = 12$ ms if $i_L = 10te^{-100t}$ A. (*b*) Find i_L at $t = 0.1$ s if $v_L = 6e^{-12t}$ V and $i_L(0) = 10$ A. If $i_L = 8(1 - e^{-40t})$ mA, find (*c*) the power being delivered to the inductor at $t = 50$ ms and (*d*) the energy stored in the inductor at $t = 40$ ms.

Ans: −15.06 mV; 24.0 A; 7.49 μW; 0.510 μJ.

We summarize by listing four key characteristics of an inductor that result from its defining equation $v = L\, di/dt$:

Important Characteristics of an Ideal Inductor

1. There is no voltage across an inductor unless the current through it is changing with time. An inductor is therefore a *short circuit to dc.*
2. A finite amount of energy can be stored in an inductor even if the voltage across the inductor is zero, such as when the current through it is constant.
3. It is impossible to change the current through an inductor by a finite amount in zero time, for this requires an infinite voltage across the inductor. (An inductor resists an abrupt change in the current through it in a manner analogous to the way a mass resists an abrupt change in its velocity.)
4. The inductor never dissipates energy, but only stores it. Although this is true for the *mathematical* model, it is not true for a *physical* inductor due to series resistances. An interesting exception is created when a superconducting wire is used to build the inductor.

It is interesting to anticipate our discussion of ***duality*** in Sec. 7.6 by rereading the previous four statements with certain words replaced by their "duals." If *capacitor* and *inductor, capacitance* and *inductance, voltage* and *current, across* and *through, open circuit* and *short circuit, spring* and *mass*, and *displacement* and *velocity* are interchanged (in either direction), the four statements previously given for capacitors are obtained.

PRACTICAL APPLICATION

In Search of the Missing Element

We have been introduced to three different two-terminal passive elements: the *resistor*, the *capacitor*, and the *inductor*. Each has been defined in terms of its current–voltage relationship ($v = Ri$, $i = C\,dv/dt$, and $v = L\,di/dt$, respectively). From a more fundamental perspective, however, we can view these three elements as part of a larger picture relating four basic quantities, namely, charge q, current i, voltage v, and flux linkage φ. Charge, current, and voltage are discussed in Chap. 2. Flux linkage is the product of magnetic flux and the number of turns of conducting wire linked by the flux, and it can be expressed in terms of the voltage v across the coil as $\varphi = \int v\,dt$ or $v = d\varphi/dt$.

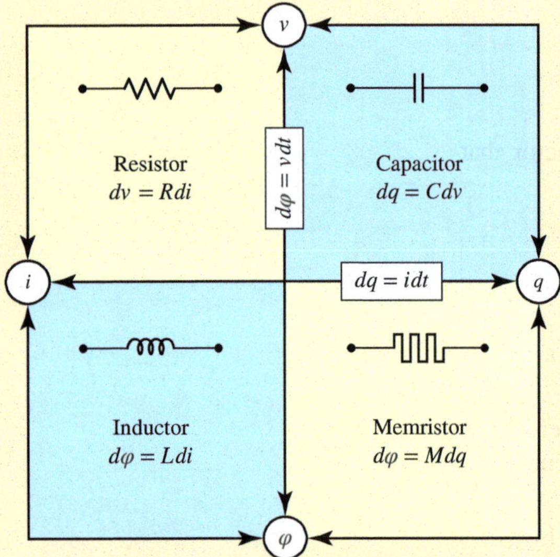

■ **FIGURE 7.17** A graphical representation of the four basic two-terminal passive elements (resistor, capacitor, inductor, and memristor) and their interrelationships. Note that flux linkage is more commonly represented by the Greek letter λ to distinguish it from flux: then $\lambda = N\varphi$ where N is the number of turns and φ is the flux. (Reprinted by permission from Macmillan Publishers Ltd. Nature Publishing Group, "Electronics: The Fourth Element," Vol. 453, pg. 42, 2008.)

Figure 7.17 graphically represents how these four quantities are interrelated. First, apart from any circuit elements and their characteristics, we have $dq = i\,dt$ (Chap. 2) and now $d\varphi = v\,dt$. Charge is related to voltage in the context of a capacitor, since $C = dq/dv$ or $dq = C\,dv$. The element we call a resistor provides a direct relationship between voltage and current, which for consistency can be expressed as $dv = R\,di$. Continuing our counterclockwise journey around the perimeter of Fig. 7.17, we note that our original expression connecting the voltage and current associated with an inductor can be written in terms of current i and flux linkage φ, since rearranging yields $v\,dt = L\,di$, and we know $d\varphi = v\,dt$. Thus, for the inductor, we can write $d\varphi = L\,di$.

So far, we have traveled from q to v with the aid of a capacitor, v to i using the resistor, and i to ϕ using the inductor. However, we have not yet used any element to connect φ and q, although symmetry suggests it should be possible. In the early 1970s, Leon Chua found himself thinking along these lines, and he postulated a new device—a "missing" two-terminal circuit element—and named it the ***memristor***.[1] He went on to show that the electrical characteristics of a memristor should be nonlinear and depend on its history—in other words, a memristor might be characterized by having a memory (hence its name). Independent of his work, others had proposed a similar device, not so much for its practical use in real circuits, but for its potential in device modeling and signal processing.

Not a great deal was heard of this hypothetical element afterward, at least until Dmitri Strukov and coworkers at HP Labs in Palo Alto published a short paper in 2008 claiming to have "found" the memristor.[2] They offer several reasons why it took almost four decades to realize the general type of device Chua hypothesized in 1971, but one of the most interesting has to do with size. In making their prototype memristor, *nanotechnology* (the art of fabricating devices with at least one dimension less than 1000 nm, which is approximately 1% of the diameter of human hair) played a key role. A 5 nm thick oxide layer sandwiched between platinum electrodes comprises the entire device. The nonlinear electrical characteristics of the prototype immediately generated considerable excitement, most notably for its potential applications in integrated circuits, where devices are already approaching their smallest realistic size; many believe new types of devices will be required to further extend integrated circuit density and functionality. Whether the memristor is the circuit element that will allow this remains to be seen—despite many reports of a variety of device geometries and current examples of commercial products, there remains much work to be done before the memristor technology becomes widespread.

(1) L. O. Chua, "Memristor—The missing circuit element," *IEEE Transactions on Circuit Theory* **CT–18 (5),** 1971, p. 507.
(2) D. B. Strukov, G. S. Snider, D. R. Stewart, and R. S. Williams, "The missing memristor found," *Nature* **453**, 2008, p. 80.

7.3 INDUCTANCE AND CAPACITANCE COMBINATIONS

Now that we have added the inductor and capacitor to our list of passive circuit elements, we need to decide whether or not the methods we have developed for resistive circuit analysis are still valid. It will also be convenient to learn how to replace series and parallel combinations of either of these elements with simpler equivalents, just as we did with resistors in Chap. 3.

We look first at Kirchhoff's two laws, both of which are axiomatic. However, when we hypothesized these two laws, we did so with no restrictions as to the types of elements constituting the network. Both, therefore, remain valid.

Inductors in Series

We first consider an ideal voltage source applied to the series combination of N inductors, as shown in Fig. 7.18*a*. We desire a single equivalent inductor, with inductance L_{eq}, which may replace the series combination so that the source current $i(t)$ is unchanged. The equivalent circuit is sketched in Fig. 7.18*b*. Applying KVL to the original circuit,

$$\begin{aligned} v_s &= v_1 + v_2 + \cdots + v_N \\ &= L_1\frac{di}{dt} + L_2\frac{di}{dt} + \cdots + L_N\frac{di}{dt} \\ &= (L_1 + L_2 + \cdots + L_N)\frac{di}{dt} \end{aligned}$$

or, written more concisely,

$$v_s = \sum_{n=1}^{N} v_n = \sum_{n=1}^{N} L_n\frac{di}{dt} = \frac{di}{dt}\sum_{n=1}^{N} L_n$$

But for the equivalent circuit we have

$$v_s = L_{eq}\frac{di}{dt}$$

and thus the equivalent inductance is

$$L_{eq} = L_1 + L_2 + \cdots + L_N$$

or

$$L_{eq} = \sum_{n=1}^{N} L_n \qquad [11]$$

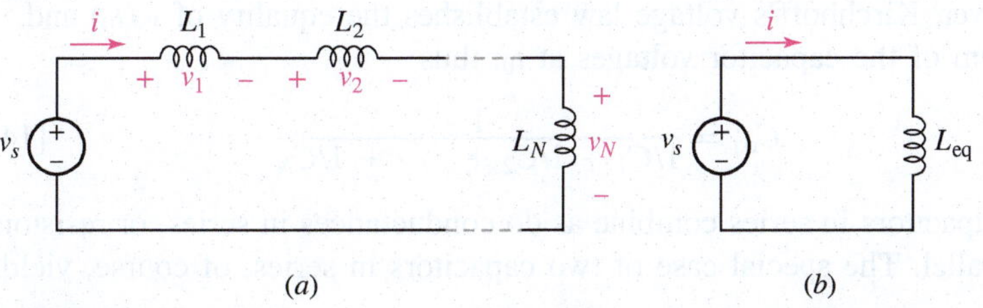

FIGURE 7.18 (*a*) A circuit containing N inductors in series. (*b*) The desired equivalent circuit, in which $L_{eq} = L_1 + L_2 + \cdots + L_N$.

The inductor which is equivalent to several inductors connected in series is one whose inductance is the sum of the inductances in the original circuit. *This is exactly the same result we obtained for resistors in series.*

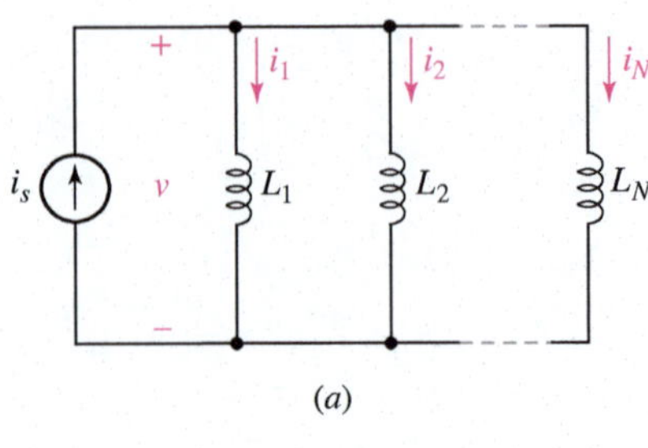

(*a*)

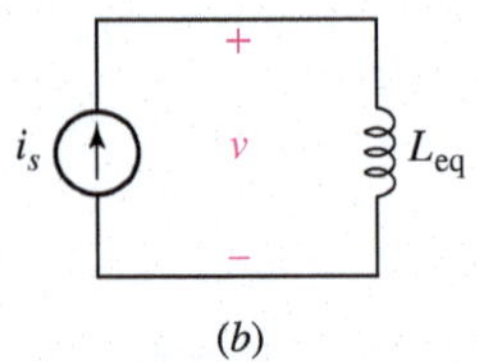

(*b*)

FIGURE 7.19 (*a*) The parallel combination of *N* inductors. (*b*) The equivalent circuit, where $L_{eq} = [1/L_1 + 1/L_2 + \cdots + 1/L_N]^{-1}$.

Inductors in Parallel

The combination of a number of parallel inductors is accomplished by writing the single nodal equation for the original circuit, shown in Fig. 7.19*a*,

$$i_s = \sum_{n=1}^{N} i_n = \sum_{n=1}^{N} \left[\frac{1}{L_n}\int_{t_0}^{t} v\,dt' + i_n(t_0)\right]$$

$$= \left(\sum_{n=1}^{N} \frac{1}{L_n}\right)\int_{t_0}^{t} v\,dt' + \sum_{n=1}^{N} i_n(t_0)$$

and comparing it with the result for the equivalent circuit of Fig. 7.19*b*,

$$i_s = \frac{1}{L_{eq}}\int_{t_0}^{t} v\,dt' + i_s(t_0)$$

Since Kirchhoff's current law demands that $i_s(t_0)$ be equal to the sum of the branch currents at t_0, the two integral terms must also be equal; hence,

$$L_{eq} = \frac{1}{1/L_1 + 1/L_2 + \cdots + 1/L_N} \qquad [12]$$

For the special case of two inductors in parallel,

$$L_{eq} = \frac{L_1 L_2}{L_1 + L_2} \qquad [13]$$

and we note that inductors in parallel combine exactly as do resistors in parallel.

Capacitors in Series

In order to find a capacitor that is equivalent to *N* capacitors in series, we use the circuit of Fig. 7.20*a* and its equivalent in Fig. 7.20*b* to write

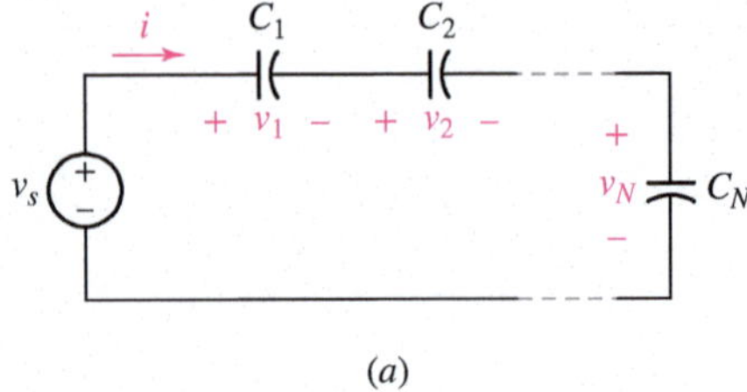

(*a*)

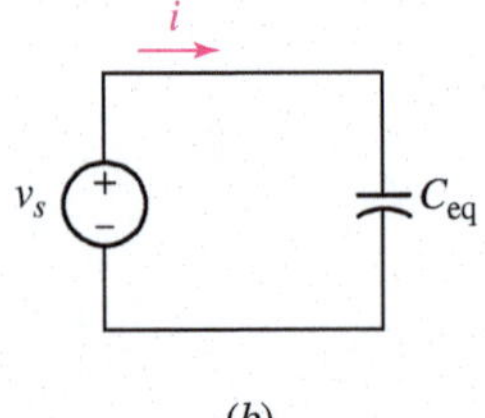

(*b*)

FIGURE 7.20 (*a*) A circuit containing *N* capacitors in series. (*b*) The desired equivalent circuit, where $C_{eq} = [1/C_1 + 1/C_2 + \cdots + 1/C_N]^{-1}$.

$$v_s = \sum_{n=1}^{N} v_n = \sum_{n=1}^{N} \left[\frac{1}{C_n}\int_{t_0}^{t} i\,dt' + v_n(t_0)\right]$$

$$= \left(\sum_{n=1}^{N} \frac{1}{C_n}\right)\int_{t_0}^{t} i\,dt' + \sum_{n=1}^{N} v_n(t_0)$$

and

$$v_s = \frac{1}{C_{eq}}\int_{t_0}^{t} i\,dt' + v_s(t_0)$$

However, Kirchhoff's voltage law establishes the equality of $v_s(t_0)$ and the sum of the capacitor voltages at t_0; thus

$$C_{eq} = \frac{1}{1/C_1 + 1/C_2 + \cdots + 1/C_N} \qquad [14]$$

and capacitors in series combine as do conductances in series, or resistors in parallel. The special case of two capacitors in series, of course, yields

$$C_{eq} = \frac{C_1 C_2}{C_1 + C_2} \qquad [15]$$

Capacitors in Parallel

Finally, the circuits of Fig. 7.21 enable us to establish the value of the capacitor which is equivalent to N parallel capacitors as

$$C_{eq} = C_1 + C_2 + \cdots + C_N \qquad [16]$$

and it is no great source of amazement to note that capacitors in parallel combine in the same manner in which we combine resistors in series, that is, by simply adding all the individual capacitances.

These formulas are well worth memorizing. The formulas applying to series and parallel combinations of inductors are identical to those for resistors, so they typically seem "obvious." Care should be exercised, however, in the case of the corresponding expressions for series and parallel combinations of capacitors, as they are opposite those of resistors and inductors, often leading to errors when calculations are made too hastily.

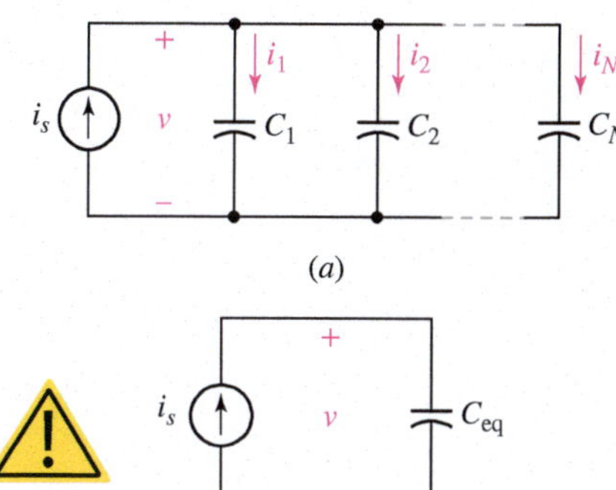

■ **FIGURE 7.21** (a) The parallel combination of N capacitors. (b) The equivalent circuit, where $C_{eq} = C_1 + C_2 + \cdots + C_N$.

EXAMPLE 7.8

Simplify the network of Fig. 7.22*a* using series–parallel combinations.

The 6 μF and 3 μF series capacitors are first combined into a 2 μF equivalent, and this capacitor is then combined with the 1 μF element with which it is in parallel to yield an equivalent capacitance of 3 μF. In addition, the 3 H and 2 H inductors are replaced by an equivalent 1.2 H inductor, which is then added to the 0.8 H element to give a total equivalent inductance of 2 H. The much simpler (and probably less expensive) equivalent network is shown in Fig. 7.22*b*.

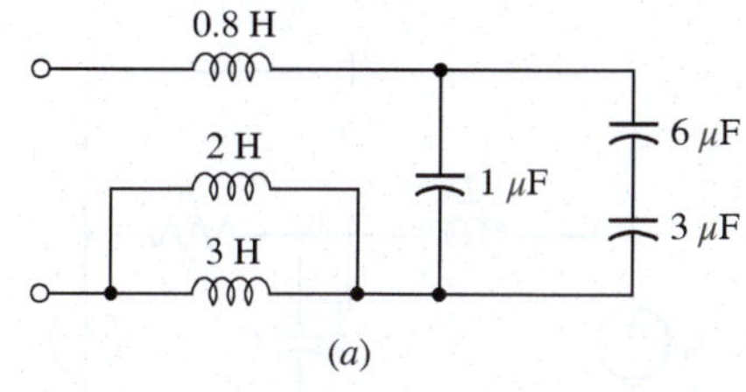

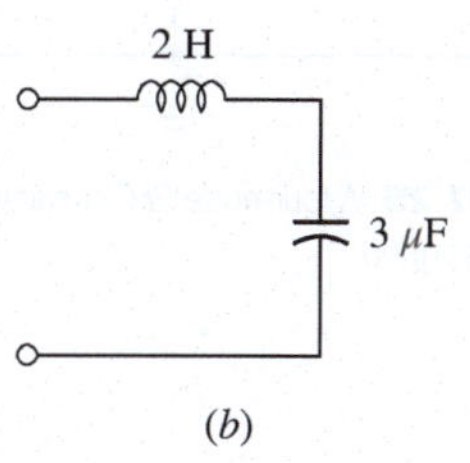

■ **FIGURE 7.22** (a) A given LC network. (b) A simpler equivalent circuit.

PRACTICE

7.8 Find C_{eq} for the network of Fig. 7.23.

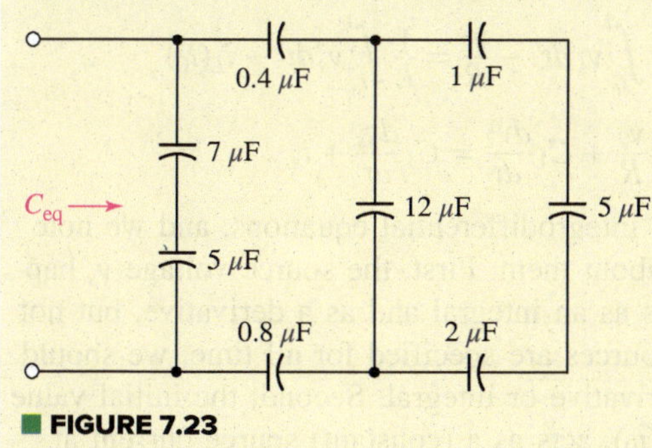

■ **FIGURE 7.23**

Ans: 3.18 μF.

The network shown in Fig. 7.24 contains three inductors and three capacitors, but no series or parallel combinations of either the inductors or the capacitors can be achieved. Simplification of this network cannot be accomplished using the techniques presented here.

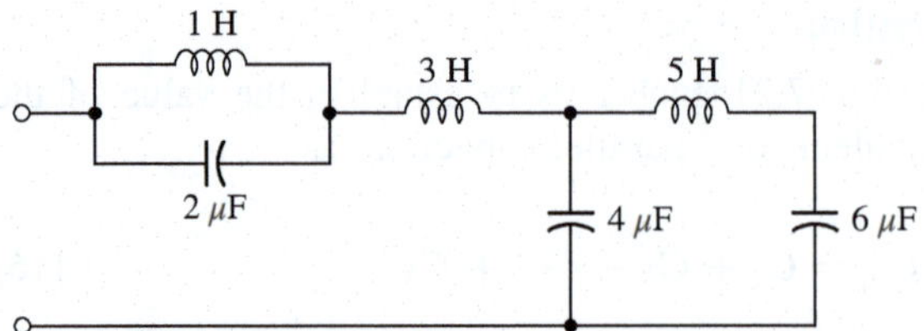

FIGURE 7.24 An *LC* network in which no series or parallel combinations of either the inductors or the capacitors can be made.

7.4 LINEARITY AND ITS CONSEQUENCES

Next let us turn to nodal and mesh analysis. Since we already know that we may safely apply Kirchhoff's laws, we can apply them in writing a set of equations that are both sufficient and independent. They will be constant-coefficient linear "integrodifferential" equations, however, which are hard enough to pronounce, let alone solve. Consequently, we shall write them now to gain familiarity with the use of Kirchhoff's laws in *RLC* circuits and discuss the solution of the simpler cases in subsequent chapters.

EXAMPLE 7.9

Write appropriate nodal equations for the circuit of Fig. 7.25.

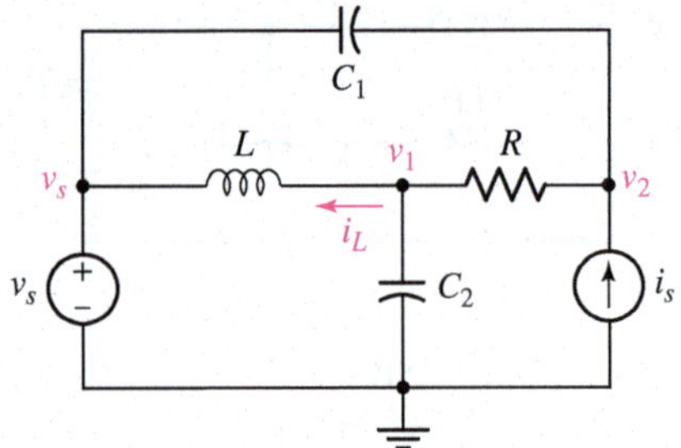

FIGURE 7.25 A four-node *RLC* circuit with node voltages assigned.

Node voltages are already chosen, so we sum currents leaving the central node:

$$\frac{1}{L}\int_{t_0}^{t}(v_1 - v_s)\,dt' + i_L(t_0) + \frac{v_1 - v_2}{R} + C_2\frac{dv_1}{dt} = 0$$

where $i_L(t_0)$ is the value of the inductor current at the time the integration begins. At the right-hand node,

$$C_1\frac{d(v_2 - v_s)}{dt} + \frac{v_2 - v_1}{R} - i_s = 0$$

Rewriting these two equations, we have

$$\frac{v_1}{R} + C_2\frac{dv_1}{dt} + \frac{1}{L}\int_{t_0}^{t} v_1\,dt' - \frac{v_2}{R} = \frac{1}{L}\int_{t_0}^{t} v_s\,dt' - i_L(t_0)$$

$$-\frac{v_1}{R} + \frac{v_2}{R} + C_1\frac{dv_2}{dt} = C_1\frac{dv_s}{dt} + i_s$$

These are the promised integrodifferential equations, and we note several interesting points about them. First, the source voltage v_s happens to enter the equations as an integral and as a derivative, but not simply as v_s. Since both sources are specified for all time, we should be able to evaluate the derivative or integral. Second, the initial value of the inductor current, $i_L(t_0)$, acts as a (constant) source current at the center node.

PRACTICE

7.9 If $v_C(t) = 4\cos 10^5 t$ V in the circuit in Fig. 7.26, find $v_s(t)$.

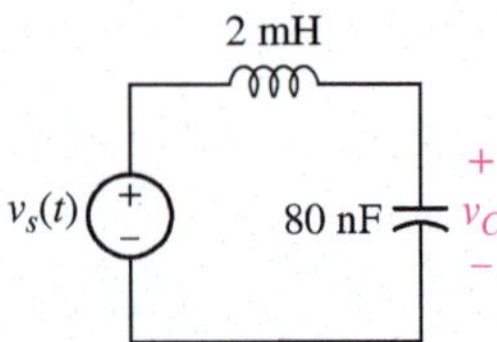

FIGURE 7.26

Ans: $-2.4\cos 10^5 t$ V

We will not attempt the solution of integrodifferential equations here. It is worthwhile pointing out, however, that when the voltage forcing functions are sinusoidal functions of time, it will be possible to define a voltage–current ratio (called *impedance*) or a current–voltage ratio (called *admittance*) for each of the three passive elements. The factors operating on the two node voltages in the preceding equations will then become simple multiplying factors, and the equations will be linear algebraic equations once again. These we may solve by determinants or a simple elimination of variables as before.

We may also show that the benefits of linearity apply to *RLC* circuits as well. In accordance with our previous definition of a linear circuit, these circuits are also linear because the voltage–current relationships for the inductor and capacitor are linear relationships. For the inductor, we have

$$v = L\frac{di}{dt}$$

and multiplication of the current by some constant K leads to a voltage that is also greater by a factor K. In the integral formulation,

$$i(t) = \frac{1}{L}\int_{t_0}^{t} v\,dt' + i(t_0)$$

it can be seen that, if each term is to increase by a factor of K, then the initial value of the current must also increase by this same factor.

A corresponding investigation of the capacitor shows that it, too, is linear. Thus, a circuit composed of independent sources, linear dependent sources, and linear resistors, inductors, and capacitors is a linear circuit.

In this linear circuit the response is again proportional to the forcing function. The proof of this statement is accomplished by first writing a general system of integrodifferential equations. Let us place all the terms having the form of Ri, $L\,di/dt$, and $1/C \int i\,dt$ on the left side of each equation, and keep the independent source voltages on the right side. As a simple example, one of the equations might have the form

$$Ri + L\frac{di}{dt} + \frac{1}{C}\int_{t_0}^{t} i\,dt' + v_C(t_0) = v_s$$

If every independent source is now increased by a factor K, then the right side of each equation is greater by the factor K. Now each term on the left side is either a linear term involving some loop current or an initial capacitor voltage. In order to cause all the responses (loop currents) to increase by a factor K, it is apparent that we must also increase the initial capacitor voltages by a factor K. That is, we must treat the initial capacitor voltage as an independent source voltage and increase it also by a factor K. In a similar manner, initial inductor currents appear as independent source currents in nodal analysis.

The principle of proportionality between source and response can thus be extended to the general *RLC* circuit, and it follows that the principle of superposition also applies. It should be emphasized that initial inductor currents and capacitor voltages must be treated as independent sources in applying the superposition principle; each initial value must take its turn in being rendered inactive. In Chap. 5 we learned that the principle of superposition is a natural consequence of the linear nature of resistive circuits. The resistive circuits are linear because the voltage–current relationship for the resistor is linear and Kirchhoff's laws are linear.

Before we can apply the superposition principle to *RLC* circuits, however, it is first necessary to develop methods of solving the equations describing these circuits when only one independent source is present. At this time we should understand that a linear circuit will have a response whose amplitude is proportional to the amplitude of the source. We should be prepared to apply superposition later, considering an inductor current or capacitor voltage specified at $t = t_0$ as a source that must be deactivated when its turn comes.

Thévenin's and Norton's theorems are based on the linearity of the initial circuit, the applicability of Kirchhoff's laws, and the superposition principle. The general *RLC* circuit conforms perfectly to these requirements, and it follows, therefore, that all linear circuits that contain any combinations of independent voltage and current sources, linear dependent voltage and current sources, and linear resistors, inductors, and capacitors may be analyzed with the use of these two theorems, if we wish.

7.5 SIMPLE OP AMP CIRCUITS WITH CAPACITORS

In Chap. 6 we were introduced to several different types of amplifier circuits based on the ideal op amp. In almost every case, we found that the output was related to the input voltage through some combination of resistance ratios. If we replace one or more of these resistors with a capacitor, it is possible to obtain some interesting circuits in which the output is proportional to either the derivative or the integral of the input voltage. Such circuits find widespread use in practice. For example, a velocity sensor can be connected to a differentiating op amp circuit to provide a signal proportional to the acceleration, or an output signal can be obtained that represents the total charge incident on a semiconductor surface during a specific period of time by simply integrating the measured current.

To create an integrator using an ideal op amp, we ground the noninverting input, install an ideal capacitor as a feedback element from the output back to the inverting input, and connect a signal source v_s to the inverting input through an ideal resistor as shown in Fig. 7.27.

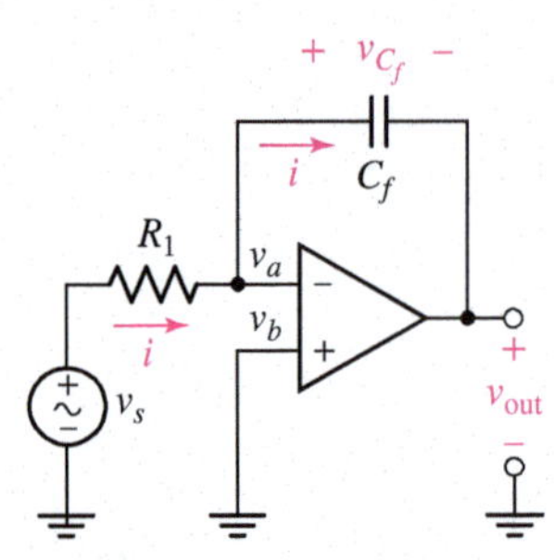

FIGURE 7.27 An ideal op amp connected as an integrator.

Performing nodal analysis at the inverting input,

$$0 = \frac{v_a - v_s}{R_1} + i$$

We can relate the current i to the voltage across the capacitor,

$$i = C_f \frac{dv_{C_f}}{dt}$$

resulting in

$$0 = \frac{v_a - v_s}{R_1} + C_f \frac{dv_{C_f}}{dt}$$

Invoking ideal op amp rule 2, we know that $v_a = v_b = 0$, so

$$0 = \frac{-v_s}{R_1} + C_f \frac{dv_{C_f}}{dt}$$

Integrating and solving for v_{out}, we obtain

$$v_{C_f} = v_a - v_{\text{out}} = 0 - v_{\text{out}} = \frac{1}{R_1 C_f}\int_0^t v_s\,dt' + v_{C_f}(0)$$

or

$$v_{\text{out}} = -\frac{1}{R_1 C_f}\int_0^t v_s\,dt' - v_{C_f}(0) \qquad [17]$$

We therefore have combined a resistor, a capacitor, and an op amp to form an integrator. Note that the first term of the output is $1/RC$ times the negative of the integral of the input from $t' = 0$ to t, and the second term is the negative of the initial value of v_{C_f}. The value of $(RC)^{-1}$ can be made equal to unity, if we wish, by choosing $R = 1$ MΩ and $C = 1$ μF, for example; other selections may be made that will increase or decrease the output voltage.

Before we leave the integrator circuit, we might anticipate a question from the reader: "*Could we use an inductor in place of the capacitor and obtain a differentiator?*" Indeed we could, but circuit designers usually avoid the use of inductors whenever possible because of their size, weight, cost, and associated resistance and capacitance. Instead, it is possible to interchange the positions of the resistor and capacitor in Fig. 7.27 and obtain a differentiator.

EXAMPLE 7.10

Derive an expression for the output voltage of the op amp circuit shown in Fig. 7.28.

We begin by writing a nodal equation at the inverting input pin, with $v_{C_1} \triangleq v_a - v_s$:

$$0 = C_1\frac{dv_{C_1}}{dt} + \frac{v_a - v_{\text{out}}}{R_f}$$

Invoking ideal op amp rule 2, $v_a = v_b = 0$. Thus,

$$C_1\frac{dv_{C_1}}{dt} = \frac{v_{\text{out}}}{R_f}$$

Solving for v_{out},

$$v_{\text{out}} = R_f C_1\frac{dv_{C_1}}{dt}$$

Since $v_{C_1} = v_a - v_s = -v_s$,

$$v_{\text{out}} = -R_f C_1\frac{dv_s}{dt}$$

So, simply by swapping the resistor and capacitor in the circuit of Fig. 7.27, we obtain a differentiator instead of an integrator.

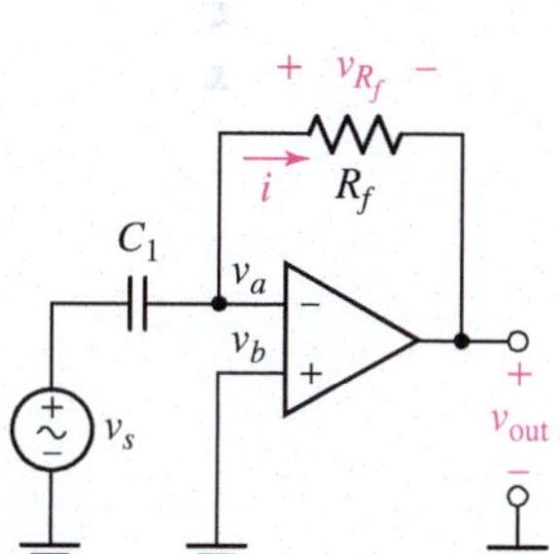

■ **FIGURE 7.28** An ideal op amp connected as a differentiator.

PRACTICE

7.10 Derive an expression for v_{out} in terms of v_s for the circuit shown in Fig. 7.29.

Ans: $v_{\text{out}} = -(L_f/R_1)\,dv_s/dt$.

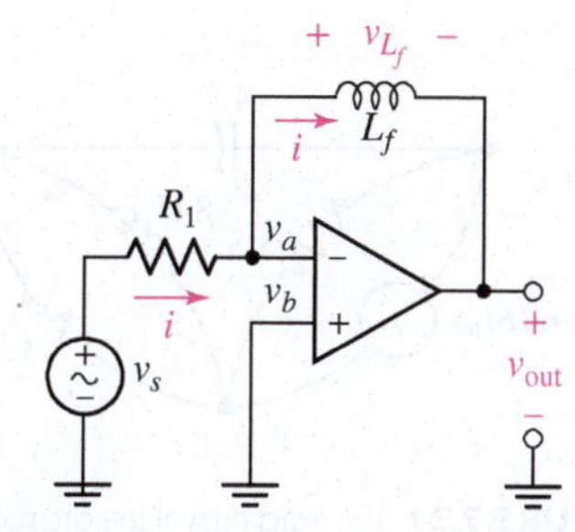

■ **FIGURE 7.29**

7.6 DUALITY

The concept of ***duality*** applies to many fundamental engineering concepts. In this section, we shall define duality in terms of circuit equations. Two circuits are "duals" if the mesh equations that characterize one of them have the *same mathematical form* as the nodal equations that characterize the other. They are said to be exact duals if each mesh equation of one circuit is numerically identical with the corresponding nodal equation of the other; the current and voltage variables themselves cannot be identical, of course. Duality itself merely refers to any of the properties exhibited by dual circuits.

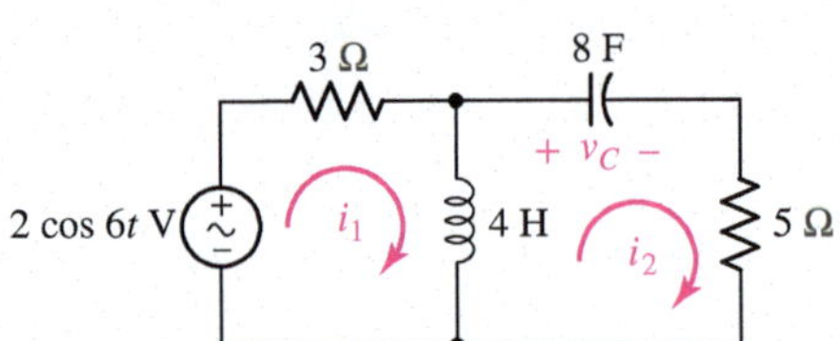

FIGURE 7.30 A given circuit to which the definition of duality may be applied to determine the dual circuit. Note that $v_C(0) = 10$ V.

Let us use the definition to construct an exact dual circuit by writing the two mesh equations for the circuit shown in Fig. 7.30. Two mesh currents i_1 and i_2 are assigned, and the mesh equations are

$$3i_1 + 4\frac{di_1}{dt} - 4\frac{di_2}{dt} = 2\cos 6t \qquad [18]$$

$$-4\frac{di_1}{dt} + 4\frac{di_2}{dt} + \frac{1}{8}\int_0^t i_2 dt' + 5i_2 = -10 \qquad [19]$$

We may now construct the two equations that describe the exact dual of our circuit. We wish these to be nodal equations, and thus we begin by replacing the mesh currents i_1 and i_2 in Eqs. [18] and [19] with the two nodal voltages v_1 and v_2 respectively. We obtain

$$3v_1 + 4\frac{dv_1}{dt} - 4\frac{dv_2}{dt} = 2\cos 6t \qquad [20]$$

$$-4\frac{dv_1}{dt} + 4\frac{dv_2}{dt} + \frac{1}{8}\int_0^t v_2 dt' + 5v_2 = -10 \qquad [21]$$

and we now seek the circuit represented by these two nodal equations.

Let us first draw a line to represent the reference node, and then we may establish two nodes at which the positive references for v_1 and v_2 are located. Equation [20] indicates that a current source of 2 cos 6t A is connected between node 1 and the reference node, oriented to provide a current entering node 1. This equation also shows that a 3 S conductance appears between node 1 and the reference node. Turning to Eq. [21], we first consider the nonmutual terms, i.e., those terms which do not appear in Eq. [20], and they instruct us to connect an 8 H inductor and a 5 S conductance (in parallel) between node 2 and the reference. The two similar terms in Eqs. [20] and [21] represent a 4 F capacitor present mutually at nodes 1 and 2; the circuit is completed by connecting this capacitor between the two nodes. The constant term on the right side of Eq. [21] is the value of the inductor current at $t = 0$; in other words, $i_L(0) = 10$ A. The dual circuit is shown in Fig. 7.31; since the two sets of equations are numerically identical, the circuits are exact duals.

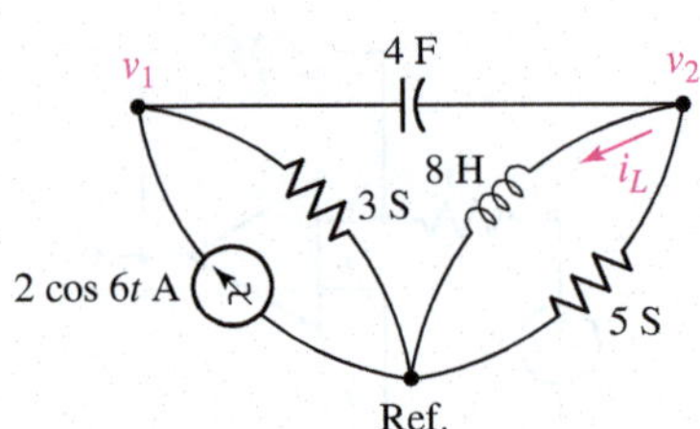

FIGURE 7.31 The exact dual of the circuit of Fig. 7.30.

Dual circuits may be obtained more readily than by this method, for the equations need not be written. In order to construct the dual of a given circuit, we think of the circuit in terms of its mesh equations. With each mesh we must associate a nonreference node, and, in addition, we must supply the reference node. On a diagram of the given circuit we therefore place a node in the center of each mesh and supply the reference node as a line near the diagram or a loop enclosing the diagram. Each element that appears jointly in two meshes is a mutual element and gives rise to identical terms, except for sign, in the two corresponding mesh equations. It must be replaced by an element that supplies the dual term in the two corresponding nodal equations. This dual element must therefore be connected directly between the two nonreference nodes that are within the meshes in which the given mutual element appears.

The nature of the dual element itself is easily determined; the mathematical form of the equations will be the same only if inductance is replaced by capacitance, capacitance by inductance, conductance by resistance, and resistance by conductance. Thus, the 4 H inductor which is common to meshes 1 and 2 in the circuit of Fig. 7.30 appears as a 4 F capacitor connected directly between nodes 1 and 2 in the dual circuit.

Elements that appear only in one mesh must have duals that appear between the corresponding node and the reference node. Referring again to Fig. 7.30, the voltage source 2 cos 6*t* V appears only in mesh 1; its *dual* is a current source 2 cos 6*t* A, which is connected only to node 1 and the reference node. Since the voltage source is clockwise-sensed, the current source must be into-the-nonreference-node-sensed. Finally, provision must be made for the dual of the initial voltage present across the 8 F capacitor in the given circuit. The equations have shown us that the dual of this initial voltage across the capacitor is an initial current through the inductor in the dual circuit; the numerical values are the same, and the correct sign of the initial current may be determined most readily by considering both the initial voltage in the given circuit and the initial current in the dual circuit as sources. Thus, if v_C in the given circuit is treated as a source, it would appear as $-v_C$ on the right side of the mesh equation; in the dual circuit, treating the current i_L as a source would yield a term $-i_L$ on the right side of the nodal equation. Since each has the same sign when treated as a source, then, if $v_C(0) = 10$ V, $i_L(0)$ must be 10 A.

The circuit of Fig. 7.30 is repeated in Fig. 7.32, and its exact dual is constructed on the circuit diagram itself by merely drawing the dual of each given element between the two nodes that are inside the two meshes that are common to the given element. A reference node that surrounds the given circuit may be helpful. After the dual circuit is redrawn in more standard form, it appears as shown in Fig. 7.31.

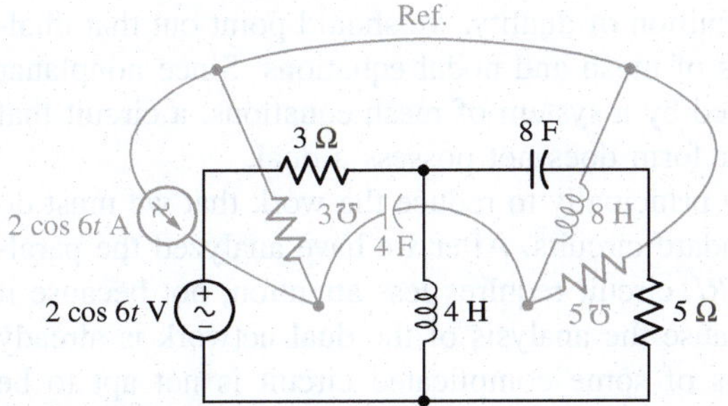

FIGURE 7.32 The dual of the circuit of Fig. 7.30 is constructed directly from the circuit diagram.

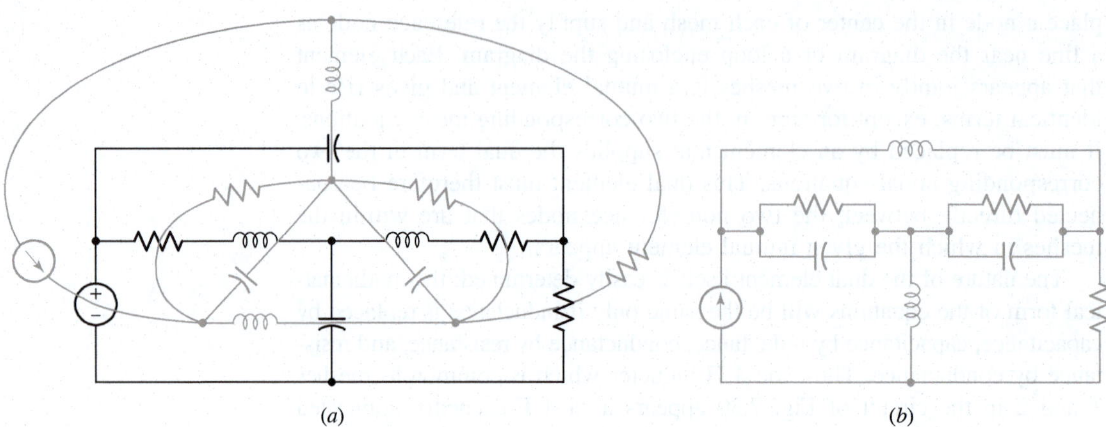

FIGURE 7.33 (*a*) The dual (in gray) of a given circuit (in black) is constructed on the given circuit. (*b*) The dual circuit is drawn in more conventional form for comparison to the original.

An additional example of the construction of a dual circuit is shown in Fig. 7.33*a* and *b*. Since no particular element values are specified, these two circuits are duals, but not necessarily exact duals. The original circuit may be recovered from the dual by placing a node in the center of each of the five meshes of Fig. 7.33*b* and proceeding as before.

The concept of duality may also be carried over into the language by which we describe circuit analysis or operation. For example, if we are given a voltage source in series with a capacitor, we might wish to make the important statement, "*The voltage source causes a current to flow through the capacitor.*" The dual statement is, "*The current source causes a voltage to exist across the inductor.*" The dual of a less carefully worded statement, such as "*The current goes round and round the series circuit*," may require a little inventiveness.[4]

Practice in using dual language can be obtained by reading Thévenin's theorem in this sense; Norton's theorem should result.

We have spoken of dual elements, dual language, and dual circuits. What about a dual network? Consider a resistor R and an inductor L in series. The dual of this two-terminal network exists and is most readily obtained by connecting some ideal source to the given network. The dual circuit is then obtained as the dual source in parallel with a conductance G with the same magnitude as R, and a capacitance C having the same magnitude as L. We consider the dual network as the two-terminal network that is connected to the dual source; it is thus a pair of terminals between which G and C are connected in parallel.

Before leaving the definition of duality, we should point out that duality is defined on the basis of mesh and nodal equations. Since nonplanar circuits cannot be described by a system of mesh equations, a circuit that cannot be drawn in planar form does not possess a dual.

We plan to use duality principally to reduce the work that we must do to analyze the simple standard circuits. After we have analyzed the parallel RC circuit, the series RL circuit requires less attention, not because it is less important, but because the analysis of the dual network is already known. Since the analysis of some complicated circuit is not apt to be well known, duality will usually not provide us with any quick solution.

(4) Someone suggested, "The voltage is across all over the parallel circuit."

PRACTICE

7.11 Write the single nodal equation for the circuit of Fig. 7.34*a*, and show, by direct substitution, that $v = -80e^{-10^6 t}$ mV is a solution. Knowing this, find (*a*) v_1; (*b*) v_2; and (*c*) i for the circuit of Fig. 7.34*b*.

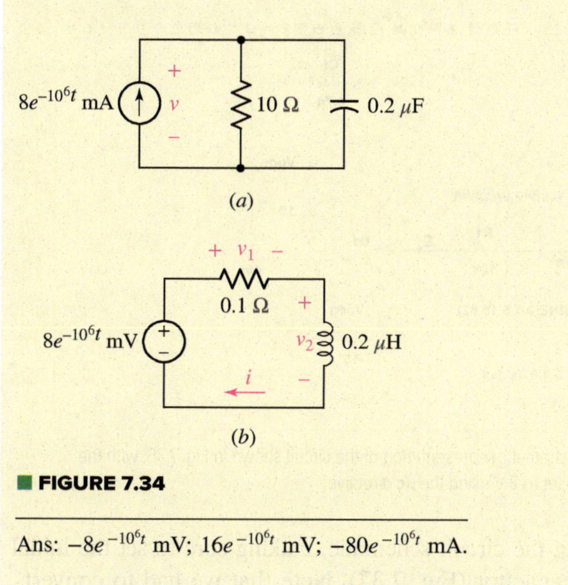

FIGURE 7.34

Ans: $-8e^{-10^6 t}$ mV; $16e^{-10^6 t}$ mV; $-80e^{-10^6 t}$ mA.

7.7 COMPUTER MODELING OF CIRCUITS WITH CAPACITORS AND INDUCTORS

When using software to analyze circuits containing capacitors and inductors, we might find it necessary to be able to specify the initial condition of each element [i.e., $v_C(0)$ and $i_L(0)$]. There are several approaches currently in use to accomplish this, depending on the specific software package. Within LTspice, this is achieved by specifying voltages at the nodes to which the capacitor is attached, or the current through the inductor. The directive named **.ic**, created using the **SPICE Directive** command (found under the **Edit** menu), allows us to do this, as illustrated in Fig. 7.35. Upon closing the dialog box, the text may be placed anywhere on the schematic.

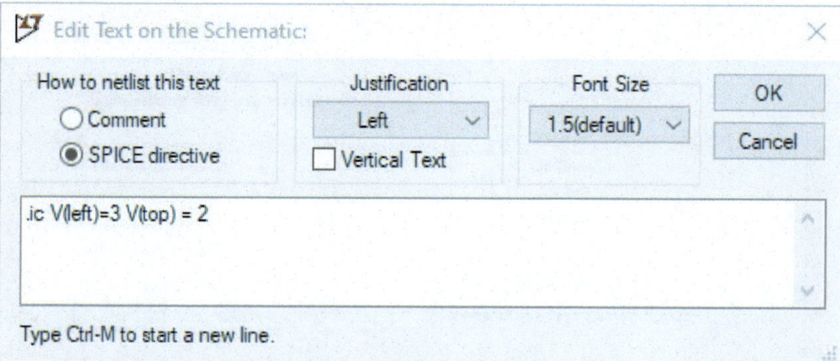

FIGURE 7.35 Edit Text dialog box used to enter initial condition for a capacitor connected between a node named *left* and a node named *top*. The initial voltage across the capacitor is 1 V, with the positive reference on the node named *left*.
Analog Devices

EXAMPLE 7.11

Simulate the output voltage waveform of the circuit in Fig. 7.36 if $v_s = 1.5 \sin 100t$ V, $R_1 = 10$ kΩ, $C_f = 4.7$ μF, and $v_C(0) = 2$ V.

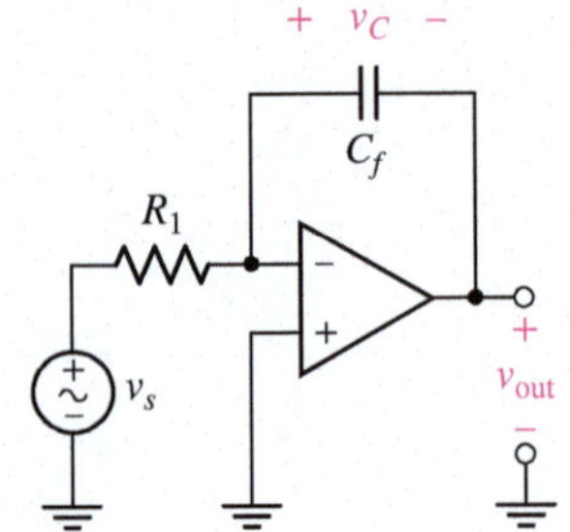

■ **FIGURE 7.36** An integrating op amp circuit.

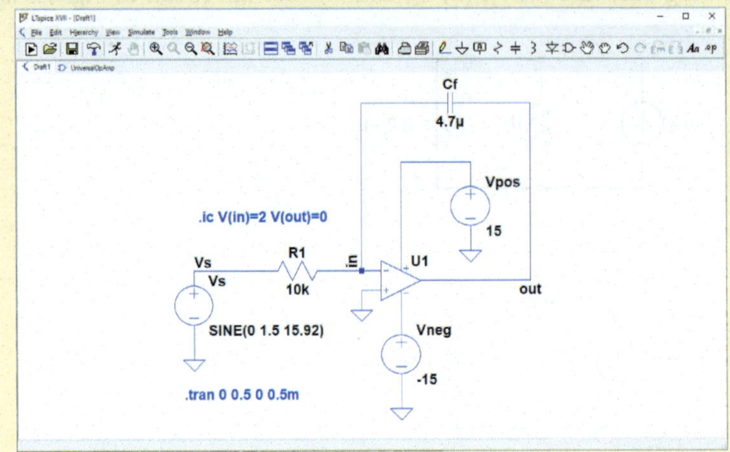

■ **FIGURE 7.37** The schematic representation of the circuit shown in Fig. 7.36, with the initial capacitor voltage set to 2 V using the **.ic** directive.
Analog Devices

We begin by drawing the circuit schematic, making sure to set the initial voltage across the capacitor (Fig. 7.37). Note that we had to convert the frequency from 100 rad/s to $100/2\pi = 15.92$ Hz. For simplicity, we also named three nodes (Vs, In, Out) using the **Label Net** function under **Edit**. The ac parameters were set for source Vs by right-clicking on the source, then selecting **Advanced** to access the menu in Fig. 7.38. The UniversalOpAmp was chosen, and ±15 V supplies selected.

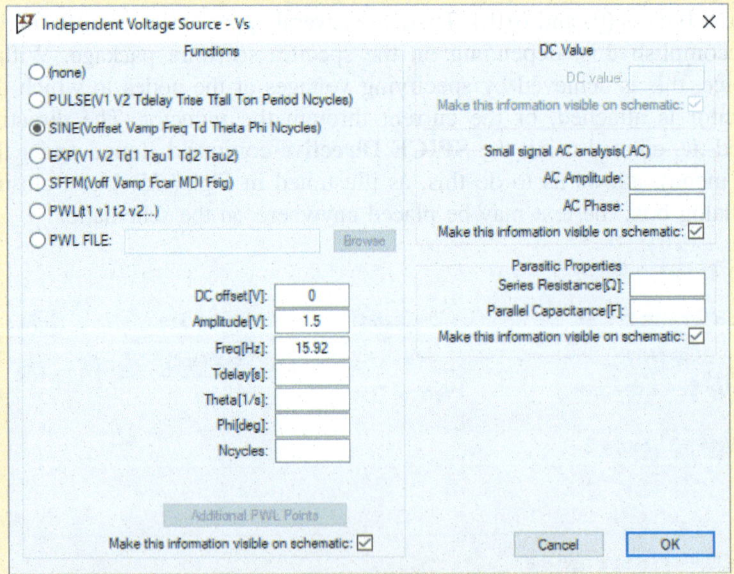

■ **FIGURE 7.38** Dialog box for setting parameters of the sinusoidal source.
Analog Devices

In order to obtain time-varying voltages and currents, we need to perform what is referred to as a *transient analysis*. Under the **Edit** menu, we select **SPICE Analysis,** which leads to the dialog box re-created in Fig. 7.39. **Stop time** represents the time at which the simulation is terminated; LTspice will select its own discrete times at which to calculate the various voltages and currents. Occasionally we obtain an error message stating that the transient solution could not converge, or the output waveform does not appear as smooth as we would like. In such situations, it is useful to set a value for **Maximum Timestep,** which has been set to 0.5 ms in this example. Alternatively, the **SPICE directive** can be directly entered on the schematic using the text **.tran 0 0.5 0 0.5m.**

Edit Simulation Command
Transient | AC Analysis | DC sweep | Noise | DC Transfer | DC op pnt
Perform a non-linear, time-domain simulation.
Stop time: 0.5
Time to start saving data: 0
Maximum Timestep: 0.5m
Start external DC supply voltages at 0V:
Stop simulating if steady state is detected:
Don't reset T=0 when steady state is detected:
Step the load current source:
Skip initial operating point solution:
Syntax: .tran <Tprint> <Tstop> [<Tstart> [<Tmaxstep>]] [<option> [<option>] ...]
.tran 0 0.5 0 0.5m
Cancel OK

FIGURE 7.39 Dialog box for setting up a transient analysis. We choose a final time of 0.5 s to obtain several periods of the output waveform (1/15.92 ≈ 0.06 s).
Analog Devices

From our earlier analysis and Eq. [17], we expect the output to be proportional to the negative integral of the input waveform, that is, $v_{\text{out}} = 0.319 \cos 100t - 2.319$ V, as shown in Fig. 7.40. The initial condition of 2 V across the capacitor has combined with a constant term from the integration to result in *a nonzero average value* for the output, unlike the input which has an average value of zero.

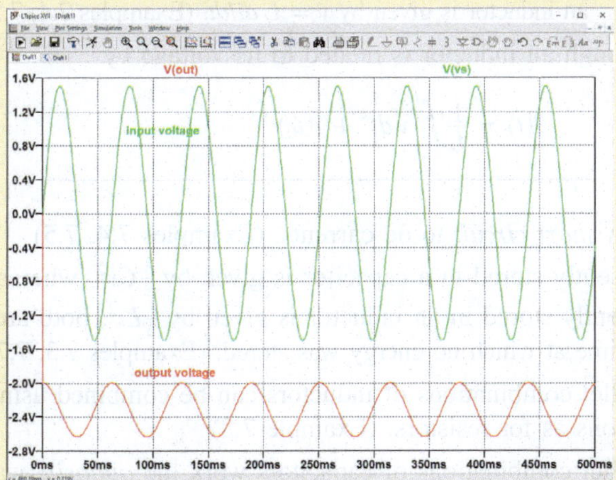

FIGURE 7.40 Output for the simulated integrator circuit along with the input waveform for comparison.
Analog Devices

SUMMARY AND REVIEW

A large number of practical circuits can be effectively modeled using only resistors and voltage/current sources. However, most interesting everyday occurrences somehow involve something changing with time, and in such cases intrinsic capacitances and/or inductances can become important. We employ such energy storage elements consciously as well, for example, in the design of frequency-selective filters, capacitor banks, and electric vehicle motors.

An *ideal* capacitor is modeled as having infinite shunt resistance and a current which depends on the time rate of change of the terminal voltage. Capacitance is measured in units of *farads* (F). Conversely, an *ideal* inductor is modeled as having zero series resistance and a terminal voltage which depends on the time rate of change of the current. Inductance is measured in units of *henrys* (H). Either element can *store energy;* the amount of energy present in a capacitor (stored in its electric field) is proportional to the *square* of the terminal voltage, and the amount of energy present in an inductor (stored in its magnetic field) is proportional to the *square* of its current.

As we found for resistors, we can simplify some connections of capacitors (or inductors) using series/parallel combinations. The validity of such equivalents arises from KCL and KVL. Once we have simplified a circuit as much as possible (taking care not to "combine away" a component which is used to define a current or voltage of interest to us), nodal and mesh analysis can be applied to circuits with capacitors and inductors.

As an additional review aid, here we list some key points from the chapter and identify relevant example(s).

- The current through a capacitor is given by $i = C\,dv/dt$. (Example 7.1)
- The voltage across a capacitor is related to its current by

  $$v(t) = \frac{1}{C}\int_{t_0}^{t} i(t')dt' + v(t_0)$$

  (Example 7.2)
- A capacitor is an *open circuit* to dc voltages. (Example 7.1)
- The voltage across an inductor is given by $v = L\,di/dt$. (Examples 7.4, 7.5)
- The current through an inductor is related to its voltage by

  $$i(t) = \frac{1}{L}\int_{t_0}^{t} v\,dt' + i(t_0)$$

  (Example 7.6)
- An inductor is a *short circuit* to dc currents. (Examples 7.4, 7.5)
- The energy presently stored in a capacitor is given by $\frac{1}{2}Cv^2$, whereas the energy presently stored in an inductor is given by $\frac{1}{2}Li^2$; both are referenced to a time at which no energy was stored. (Examples 7.3, 7.7)
- Series and parallel combinations of inductors can be combined using the same equations as for resistors. (Example 7.8)
- Series and parallel combinations of capacitors work the *opposite* way from the way they do for resistors. (Example 7.8)

- Since capacitors and inductors are linear elements, KVL, KCL, superposition, Thévenin's and Norton's theorems, and nodal and mesh analysis apply to their circuits as well. (Example 7.9)
- A capacitor as the feedback element in an inverting op amp leads to an output voltage proportional to the *integral* of the input voltage. Swapping the input resistor and the feedback capacitor leads to an output voltage proportional to the *derivative* of the input voltage. (Example 7.10)
- LTspice allows us to set the initial voltage across a capacitor and the initial current through an inductor. A transient analysis provides details of the time-dependent response of circuits containing these types of elements. (Example 7.11)

READING FURTHER

A detailed guide to characteristics and selection of various capacitor and inductor types can be found in:

H. B. Drexler, *Passive Electronic Component Handbook*, 2nd ed., C. A. Harper, ed. New York: McGraw Hill, 2003, pp. 69–203.

C. J. Kaiser, *The Inductor Handbook*. Olathe, Kans.: C.J. Publishing, 1996.

Two books that describe capacitor-based op amp circuits are:

B. Carter, *Op Amps Are for Everyone*, 4th ed. Boston: Newnes, 2013.

W. G. Jung, *IC Op Amp Cookbook*, 3rd ed. Upper Saddle River, N.J.: Prentice-Hall, 1997.

There are multiple resources now available for learning more about memristor-based technology. For example,

R. Tetzlaff, ed., *Memristors and Memristive Systems*. Heidelberg, Germany: Springer, 2016.

EXERCISES

7.1 The Capacitor

1. Making use of the passive sign convention, determine the current flowing through a 100 pF capacitor for $t \geq 0$ if its voltage $v_C(t)$ is given by (*a*) 5 V; (*b*) $10e^{-t}$ V; (*c*) 2 sin 0.01t V; (*d*) −5 + 2 sin 0.01t V.
2. Sketch the current flowing through a 10 nF capacitor for $t \geq 0$ as a result of the waveforms shown in Fig. 7.41. Assume the passive sign convention.

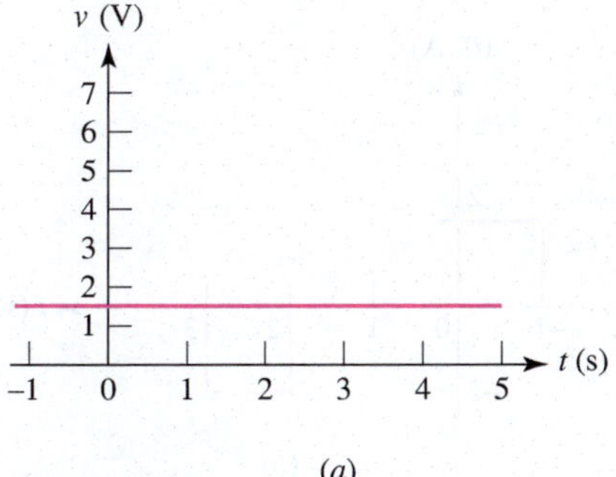

(*a*)

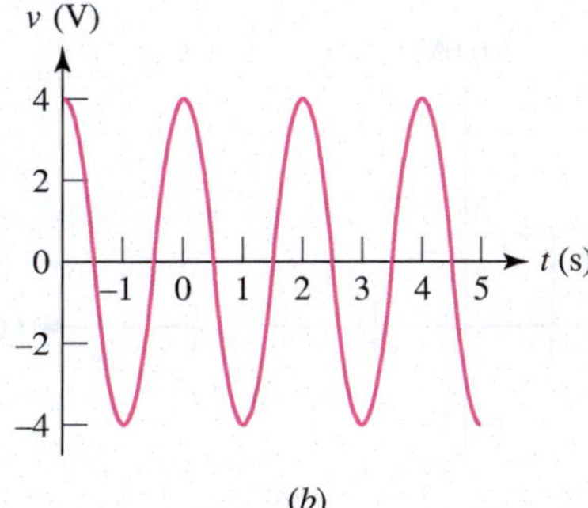

(*b*)

FIGURE 7.41

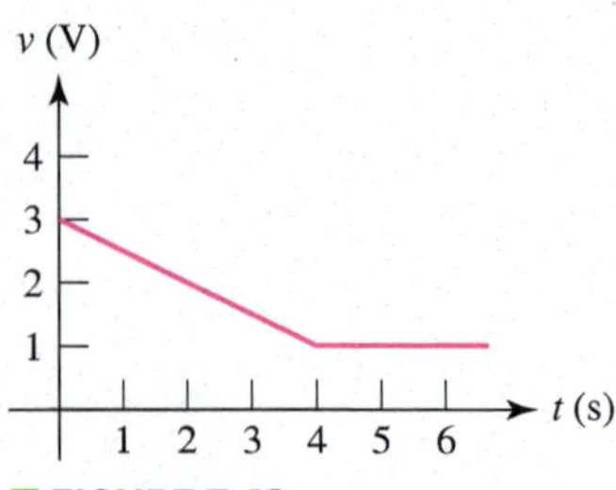

■ **FIGURE 7.42**

3. (*a*) If the voltage waveform depicted in Fig. 7.42 is applied across the terminals of a 1 μF electrolytic capacitor, graph the resulting current for $t > 0$, assuming the passive sign convention. (*b*) Repeat part (*a*) if the capacitor is replaced with a 20 nF capacitor.
4. A capacitor is constructed from two brass plates, each measuring 2.5 mm × 2.5 mm and 300 μm thick. The two plates are placed such that they face each other and are separated by a 25 μm gap. Calculate the resulting capacitance if (*a*) the intervening dielectric has a permittivity of $15\varepsilon_0$; (*b*) the intervening dielectric has a permittivity of $1.5\varepsilon_0$; (*c*) the plate separation is doubled and the gap is filled with air; (*d*) the plate area is doubled and the gap is filled with air.
5. Two conducting metal discs, each having diameter 25 mm, are placed facing each other with a uniform gap of 0.1 mm. Compute the capacitance if the gap is filled with (*a*) air; (*b*) mylar; (*c*) SiO_2 (silicon dioxide).

6. Design a 100 nF capacitor constructed from 1 μm thick gold foil, and which fits entirely within a volume equal to that of a standard AAA battery, if the only dielectric available has a permittivity of $3.1\varepsilon_0$.

7. Design a capacitor whose capacitance can be varied mechanically between 100 pF and 200 pF with a simple linear motion of 10 mm.

8. Design a capacitor whose capacitance can be varied between 250 pF and 500 pF mechanically by squeezing the plates. No more than 0.1 mm motion is allowable.
9. A silicon *pn* junction diode is characterized by a junction capacitance C_j defined as

$$C_j = \frac{K_s\varepsilon_0 A}{W}$$

where $K_s = 11.8$ for silicon, ε_0 is the vacuum permittivity, A = the cross-sectional area of the junction, and W is known as the depletion width of the junction. Width W depends not only on how the diode is fabricated, but also on the voltage applied to its two terminals. It can be computed using

$$W = \sqrt{\frac{2K_s\varepsilon_0}{qN}(V_{\text{bi}} - V_A)}$$

Thus, diodes are often used in electronic circuits since they can be viewed in this context as voltage-controlled capacitors. Assuming parameter values of $N = 5.0 \times 10^{18}\ \text{cm}^{-3}$, $V_{\text{bi}} = 0.62$ V, and using $q = 1.6 \times 10^{-19}$ C, calculate the capacitance of a diode with cross-sectional area $A = 2\ \mu\text{m}^2$ at applied voltages of $V_A = -1$, -3, and -10 V.
10. Assuming the passive sign convention, sketch the voltage which develops across the terminals of a 2 F capacitor in response to the current waveforms shown in Fig. 7.43.

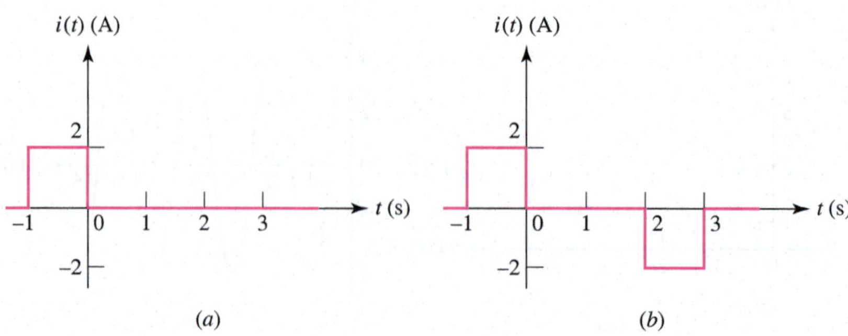

■ **FIGURE 7.43**

11. The current flowing through a 1 mF capacitor is shown graphically in Fig. 7.44. (*a*) Assuming the passive sign convention, sketch the resulting voltage waveform across the device. (*b*) Compute the voltage at 200 ms, 600 ms, and 1.2 s.

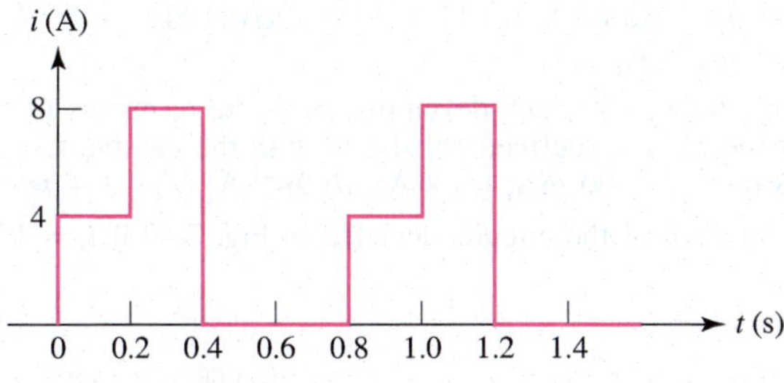

■ FIGURE 7.44

12. Calculate the energy stored in a capacitor at time $t = 1$ s if (*a*) $C = 1.4$ F and $v_C = 8$ V, $t > 0$; (*b*) $C = 22$ pF and $v_C = 0.8$ V, $t > 0$; (*c*) $C = 18$ nF, $v_C(1) = 12$ V, $v_C(0) = 2$ V, and $w_C(0) = 295$ nJ.

13. A 150 pF capacitor is connected to a voltage source such that $v_C(t) = 12e^{-2t}$ V, $t \geq 0$ and $v_C(t) = 12$ V, $t < 0$. Calculate the energy stored in the capacitor at t equal to (*a*) 0; (*b*) 200 ms; (*c*) 500 ms; (*d*) 1 s.

14. Calculate the power dissipated in the 40 Ω resistor and the voltage labeled v_C in each of the circuits depicted in Fig. 7.45.

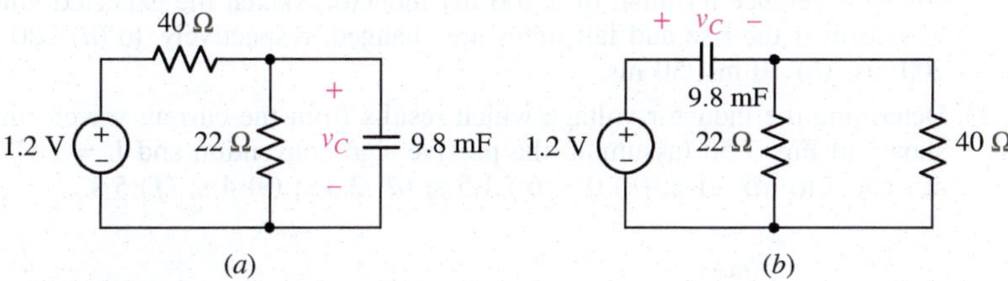

■ FIGURE 7.45

15. For each circuit shown in Fig. 7.46, calculate the voltage labeled v_C.

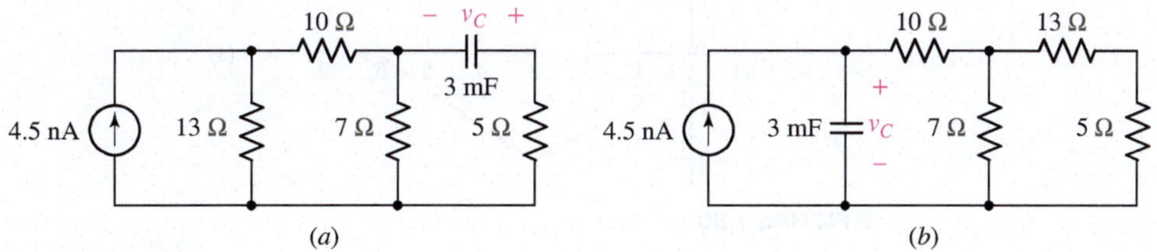

■ FIGURE 7.46

7.2 The Inductor

16. Design a 30 nH inductor using 28 AWG solid soft copper wire. Include a sketch of your design, and label geometrical parameters as necessary for clarity. Assume the coil is filled with air only.

17. If the current flowing through a 75 mH inductor has the waveform shown in Fig. 7.47, (*a*) sketch the voltage which develops across the inductor terminals for $t \geq 0$, assuming the passive sign convention; and (*b*) calculate the voltage at $t = 1$ s, 2.9 s, and 3.1 s.

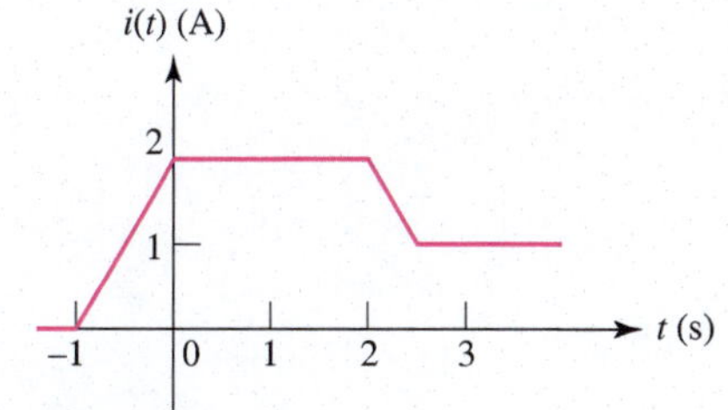

■ FIGURE 7.47

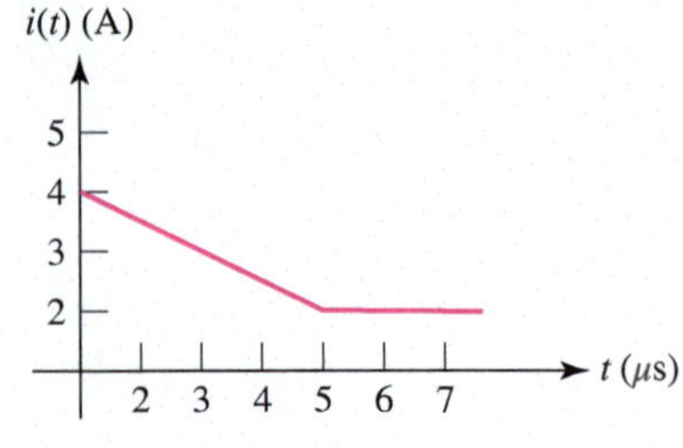

■ FIGURE 7.48

18. The current through a 17 nH aluminum inductor is shown in Fig. 7.48. Sketch the resulting voltage waveform for $t \geq 0$, assuming the passive sign convention.

19. Determine the voltage for $t \geq 0$ which develops across the terminals of a 4 mH inductor if the current (defined consistent with the passive sign convention) is (*a*) -10 mA; (*b*) $3 \sin 6t$ A; (*c*) $11 + 115\sqrt{2}\cos(100\pi t - 9°)$ A; (*d*) $15e^{-t}$ nA; (*e*) $3 + te^{-10t}$ A.

20. Determine the voltage for $t \geq 0$ which develops across the terminals of an 8 pH inductor if the current (defined consistent with the passive sign convention) is (*a*) 8 mA; (*b*) 800 mA; (*c*) 8 A; (*d*) $4e^{-t}$ A; (*e*) $-3 + te^{-t}$ A.

21. Calculate v_L and i_L for each of the circuits depicted in Fig. 7.49 if $i_s = 1$ mA and $v_s = 2$ V.

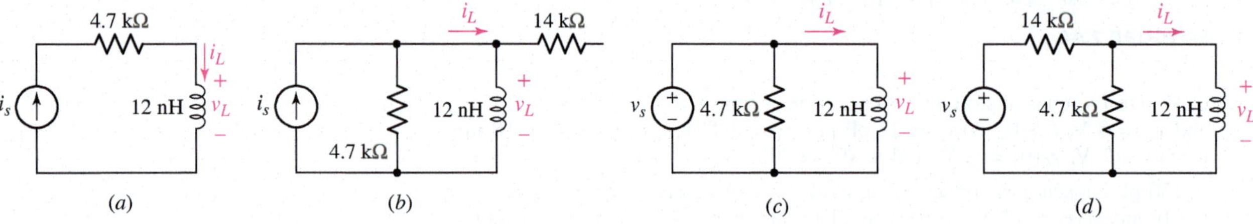

■ FIGURE 7.49

22. The current waveform shown in Fig. 7.14 has a rise time of 0.1 s (100 ms) and a fall time of the same duration. If the current is applied to the "+" voltage reference terminal of a 200 nH inductor, sketch the expected voltage waveform if the rise and fall times are changed, respectively, to (*a*) 200 ms, 200 ms; (*b*) 10 ms, 50 ms.

23. Determine the inductor voltage which results from the current waveform shown in Fig. 7.50 (assuming the passive sign convention and $L = 1$ H) at t equal to (*a*) -1 s; (*b*) 0 s; (*c*) 1.5 s; (*d*) 2.5 s; (*e*) 4 s; (*f*) 5 s.

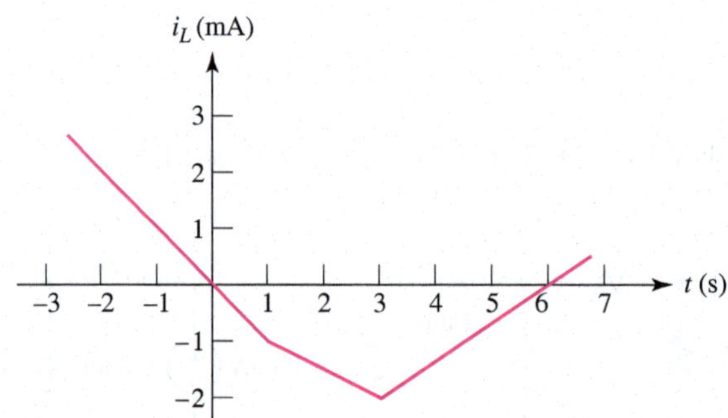

■ FIGURE 7.50

24. Determine the current flowing through a 6 mH inductor if the voltage (defined to be consistent with the passive sign convention) is given by (*a*) 5 V; (*b*) $100 \sin 120\pi t$, $t \geq 0$ and 0, $t < 0$.

25. The voltage across a 2 H inductor is given by $v_L = 4t$. With the knowledge that $i_L(-0.1) = 100\ \mu$A, calculate the current (assuming it is defined consistent with the passive sign convention) at t equal to (*a*) 0; (*b*) 1.5 ms; (*c*) 45 ms.

26. Calculate the energy stored in a 1 nH inductor if the current flowing through it is (*a*) 0 mA; (*b*) 1 mA; (*c*) 20 A; (*d*) $5 \sin 6t$ mA, $t > 0$.

27. Determine the amount of energy stored in a 33 mH inductor at $t = 1$ ms as a result of a current i_L given by (*a*) 7 A; (*b*) $3 - 9e^{-103t}$ mA.

28. Making the assumption that the circuits in Fig. 7.51 have been connected for a very long time, determine the value for each current labeled i_x.

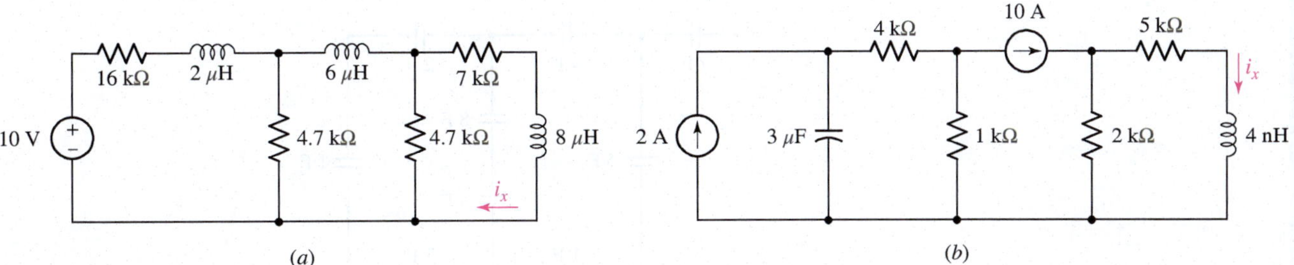

■ **FIGURE 7.51**

29. Calculate the voltage labeled v_x in Fig. 7.52, assuming the circuit has been running a very long time, if (*a*) a 10 Ω resistor is connected between terminals *x* and *y*; (*b*) a 1 H inductor is connected between terminals *x* and *y*; (*c*) a 1 F capacitor is connected between terminals *x* and *y*; (*d*) a 4 H inductor in parallel with a 1 Ω resistor is connected between terminals *x* and *y*.

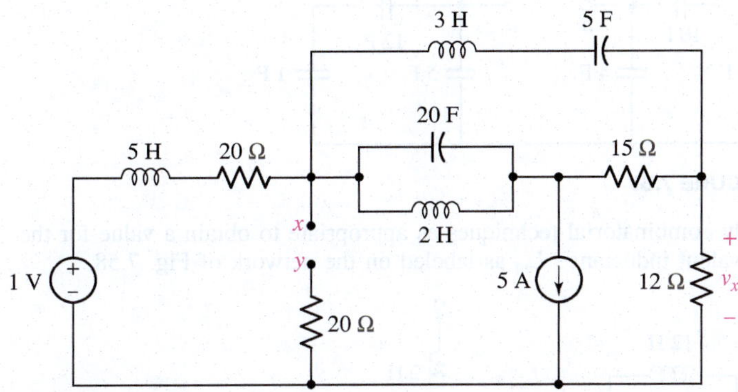

■ **FIGURE 7.52**

30. For the circuit shown in Fig. 7.53, (*a*) compute the Thévenin equivalent seen by the inductor; (*b*) determine the power being dissipated by both resistors; (*c*) calculate the energy stored in the inductor.

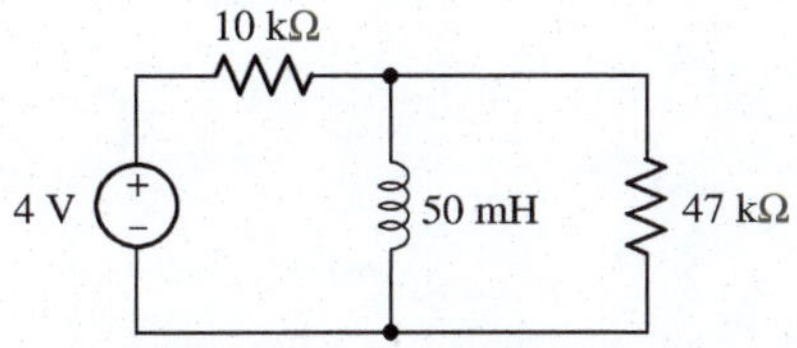

■ **FIGURE 7.53**

7.3 Inductance and Capacitance Combinations

31. If each capacitor has a value of 1 F, determine the equivalent capacitance of the network shown in Fig. 7.54.

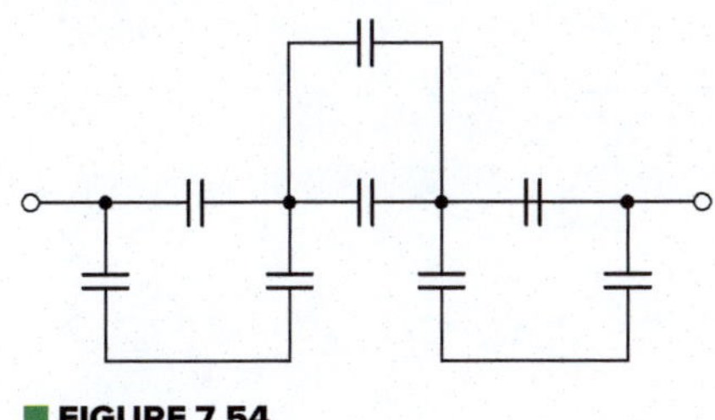

■ **FIGURE 7.54**

32. Determine an equivalent inductance for the network shown in Fig. 7.55 if each inductor has value *L*.

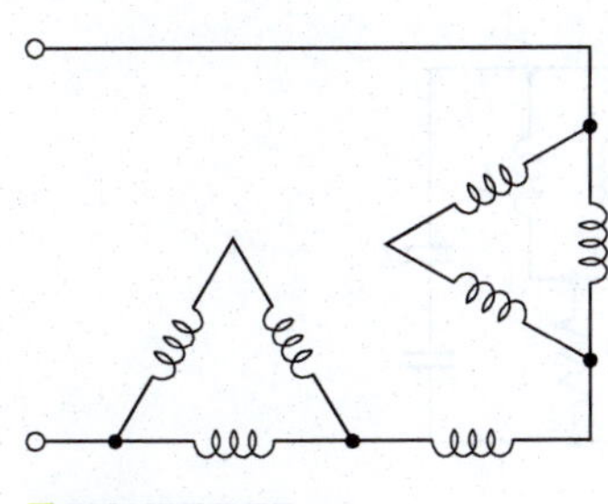

■ **FIGURE 7.55**

33. Using as many 1 nH inductors as you like, design two networks, each of which has an equivalent inductance of 1.25 nH.

34. Compute the equivalent capacitance C_{eq} as labeled in Fig. 7.56.

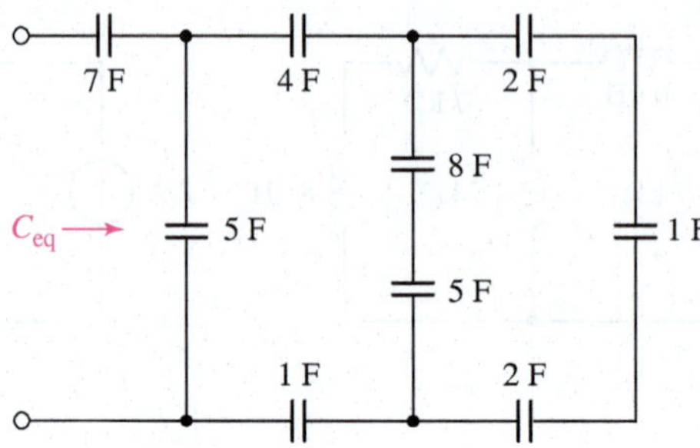

FIGURE 7.56

35. Determine the equivalent capacitance C_{eq} of the network shown in Fig. 7.57.

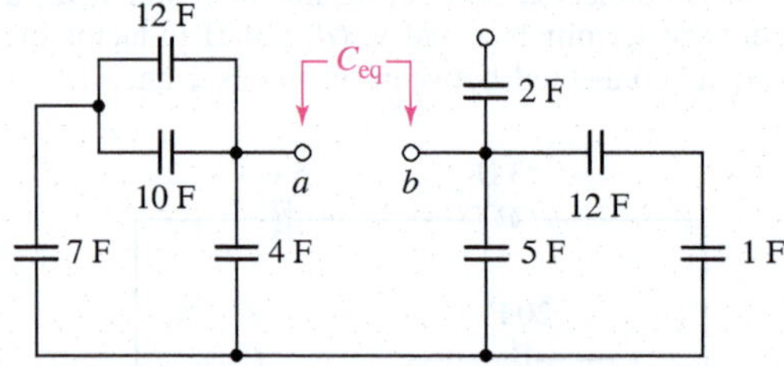

FIGURE 7.57

36. Apply combinatorial techniques as appropriate to obtain a value for the equivalent inductance L_{eq} as labeled on the network of Fig. 7.58.

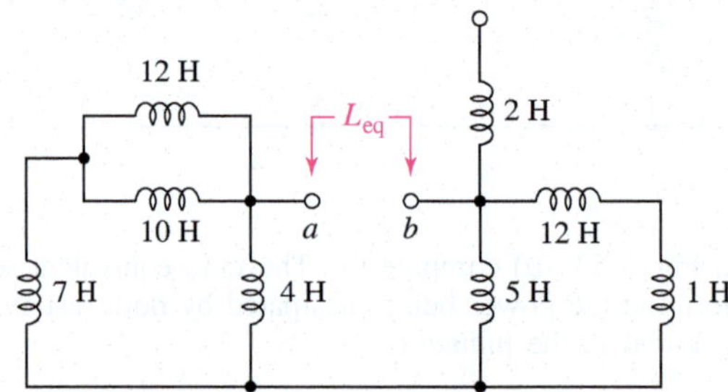

FIGURE 7.58

37. Reduce the circuit depicted in Fig. 7.59 to as few components as possible, noting that the voltage v_x is important.

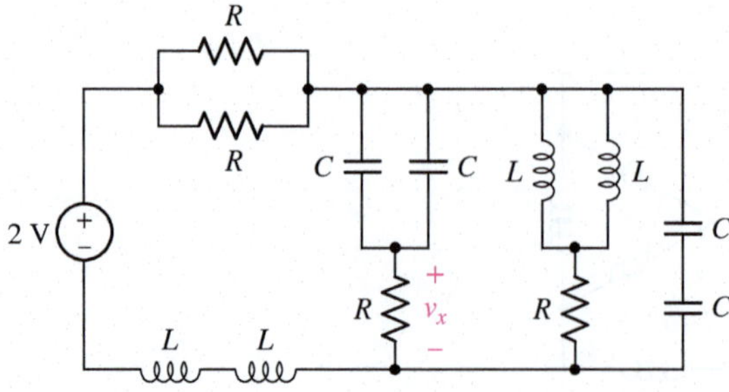

FIGURE 7.59

38. Refer to the network shown in Fig. 7.60 and find (*a*) R_{eq} if each element is a 10 Ω resistor; (*b*) L_{eq} if each element is a 10 H inductor; and (*c*) C_{eq} if each element is a 10 F capacitor.

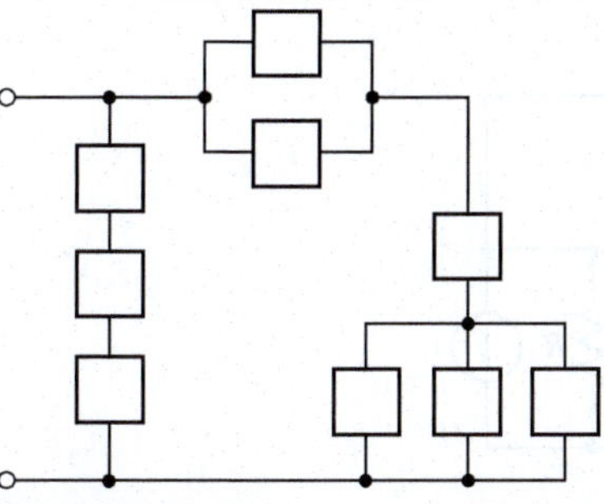

FIGURE 7.60

39. Determine the equivalent inductance seen looking into the terminals marked *a* and *b* of the network represented in Fig. 7.61.

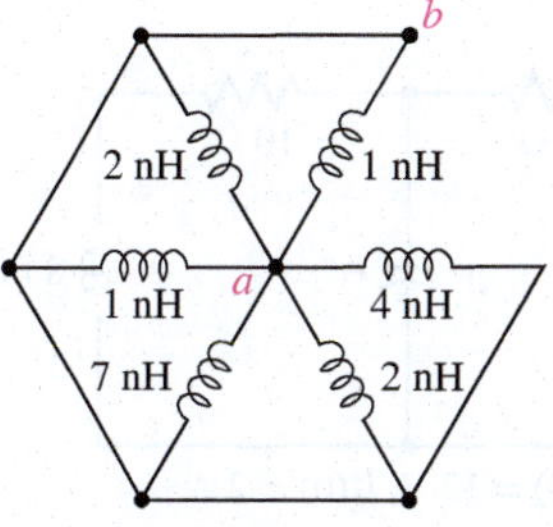

FIGURE 7.61

40. Reduce the circuit represented in Fig. 7.62 to the smallest possible number of components.

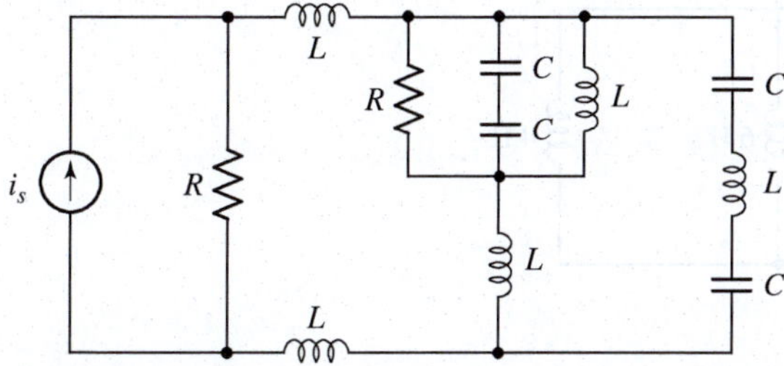

FIGURE 7.62

41. Reduce the network of Fig. 7.63 to the smallest possible number of components if each inductor is 2 nH and each capacitor is 2 nF.

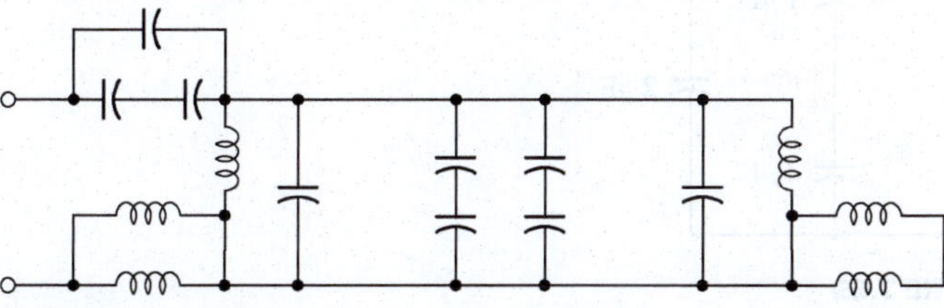

FIGURE 7.63

42. For the network of Fig. 7.64, $L_1 = 1$ H, $L_2 = L_3 = 2$ H, $L_4 = L_5 = L_6 = 3$ H. (*a*) Find the equivalent inductance. (*b*) Derive an expression for a general network of this type having *N* stages, assuming stage *N* is composed of *N* inductors, each having inductance *L* henrys.

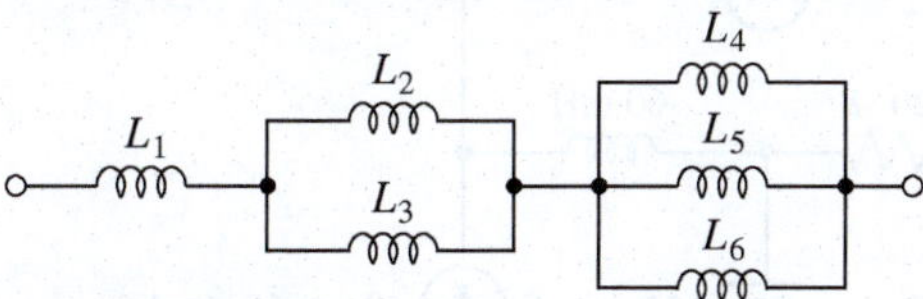

FIGURE 7.64

43. Simplify the network of Fig. 7.65 if each element is a 10 F capacitor.

44. Simplify the network of Fig. 7.65 if each element is a 10 H inductor.

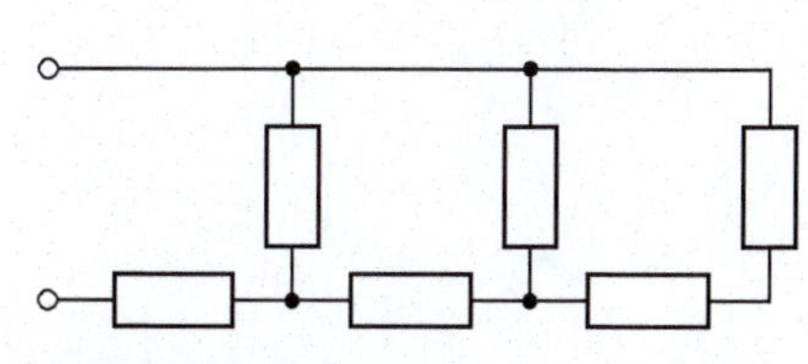

FIGURE 7.65

7.4 Linearity and Its Consequences

45. With regard to the circuit represented in Fig. 7.66, (*a*) write a complete set of nodal equations and (*b*) write a complete set of mesh equations.

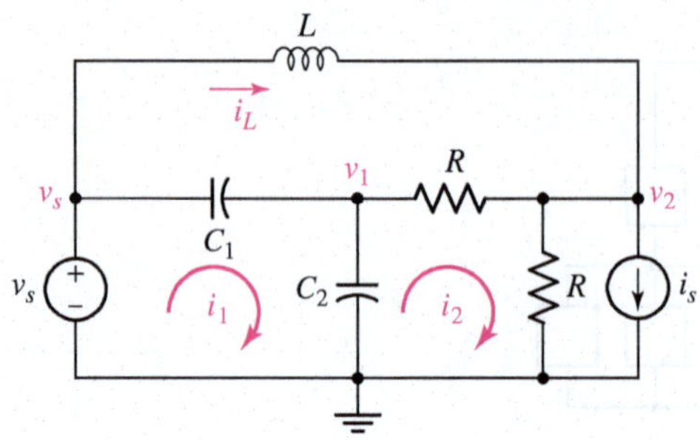

FIGURE 7.66

46. Write mesh equations for the circuit of Fig. 7.67.

$v_C(0) = 12$ V, $i_L(0) = 2$ A

FIGURE 7.67

47. In the circuit shown in Fig. 7.68, let $i_s = 60e^{-200t}$ mA with $i_1(0) = 20$ mA. (*a*) Find $v(t)$ for all t. (*b*) Find $i_1(t)$ for $t \geq 0$. (*c*) Find $i_2(t)$ for $t \geq 0$.

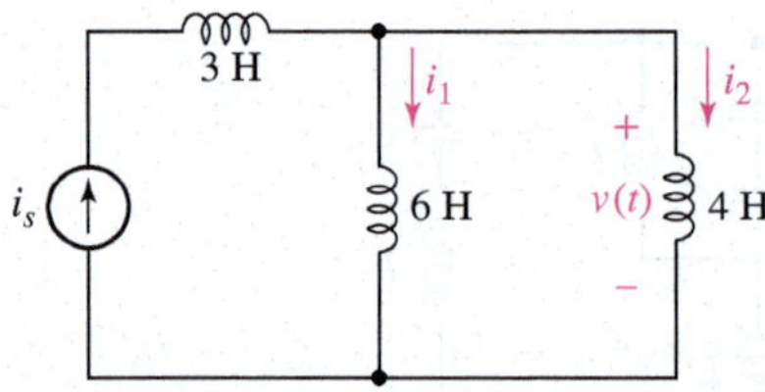

FIGURE 7.68

48. Let $v_s = 100e^{-80t}$ V with no initial energy stored in the circuit of Fig. 7.69. (*a*) Find $i(t)$ for all t. (*b*) Find $v_1(t)$ for $t \geq 0$. (*c*) Find $v_2(t)$ for $t \geq 0$.

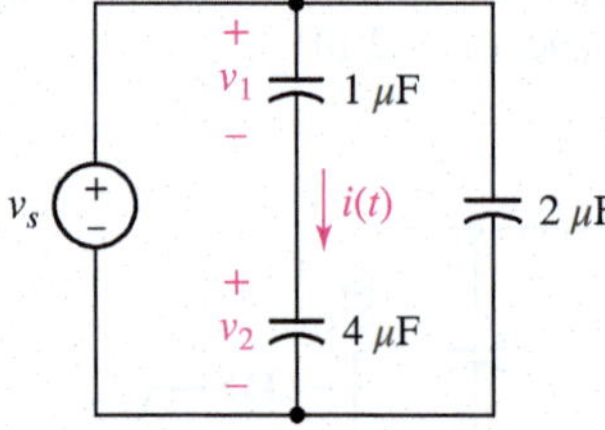

FIGURE 7.69

49. If it is assumed that all the sources in the circuit of Fig. 7.70 have been connected and operating for a very long time, use the superposition principle to find $v_C(t)$ and $v_L(t)$.

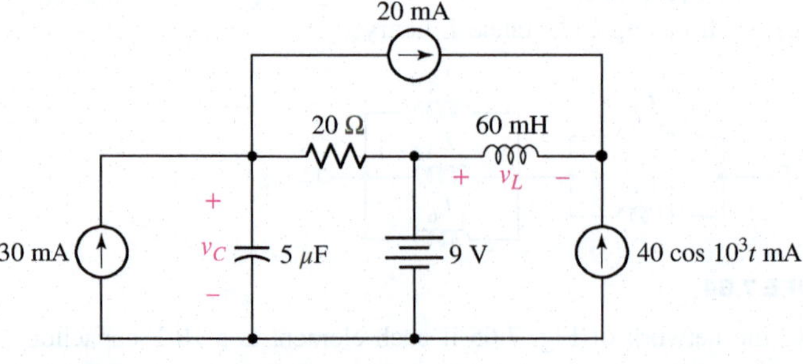

FIGURE 7.70

50. For the circuit of Fig. 7.71, assume no energy is stored at $t = 0$, and write a complete set of mesh equations.

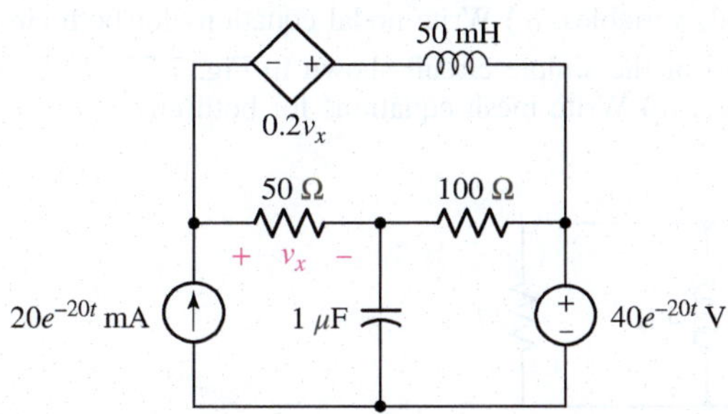

■ **FIGURE 7.71**

7.5 Simple Op Amp Circuits with Capacitors

51. Interchange the location of R_1 and C_f in the circuit of Fig. 7.27, and assume that $R_i = \infty$, $R_o = 0$, and $A = \infty$ for the op amp. Find $v_{\text{out}}(t)$ as a function of $v_s(t)$.

52. For the integrating amplifier circuit of Fig. 7.27, $R_1 = 100$ kΩ, $C_f = 500$ μF, and $v_s = 20 \sin 540t$ mV. Calculate v_{out}.

53. Derive an expression for v_{out} in terms of v_s for the amplifier circuit shown in Fig. 7.72.

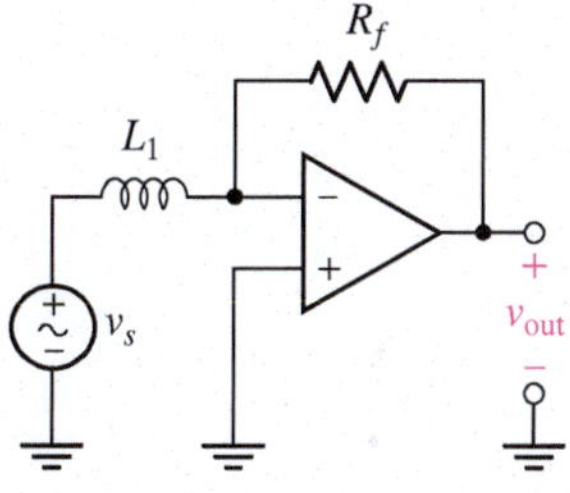

■ **FIGURE 7.72**

54. For the circuit shown in Fig. 7.73, assume no energy is initially stored in the capacitor, and determine v_{out} if v_s is given by (*a*) 5 sin 20*t* mV; (*b*) $2e^{-t}$ V.

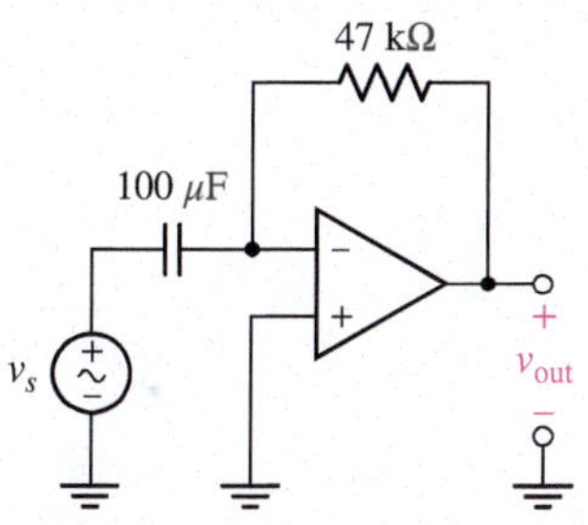

■ **FIGURE 7.73**

DP 55. A new piece of equipment designed to make crystals from molten constituents is experiencing too many failures (cracked products). The production manager wants to monitor the cooling rate to see if this is related to the problem. The system has two output terminals available, where the voltage across them is linearly proportional to the crucible temperature such that 30 mV corresponds to 30°C and 1 V corresponds to 1000°C. Design a circuit whose voltage output represents the cooling rate, calibrated such that 1 V = 1°C/s.

DP 56. An altitude sensor on a weather balloon provides a voltage calibrated to 1 mV = 1 meter (above sea level). Design a circuit to provide a voltage signal proportional to the ascent rate (positive) or descent rate (negative) such that 1 mV corresponds to a 1 m/s ascent rate. The maximum altitude is 1000 m.

DP 57. One problem satellites face is exposure to high-energy particles, which can cause damage to sensitive electronics as well as solar arrays used to provide power. A new communications satellite is equipped with a high-energy proton detector measuring 1 cm × 1 cm. It provides a current directly equal to the number of protons impinging the surface per second. Design a circuit whose output voltage provides a running total of the number of proton hits, calibrated such that 1 V = 1 million hits.

DP 58. The output of a velocity sensor attached to a sensitive piece of mobile equipment is calibrated to provide a signal such that 10 mV corresponds to linear motion at 1 m/s. If the equipment is subjected to sudden shock, it can be damaged. Since force = mass × acceleration, monitoring of the rate of change of velocity can be used to determine if the equipment is transported improperly. (*a*) Design a circuit to provide a voltage proportional to the linear acceleration such that 10 mV = 1 m/s^2. (*b*) How many sensor-circuit combinations does this application require?

DP 59. A floating sensor in a certain fuel tank is connected to a variable resistor (often called a potentiometer) such that a full tank (100 liters) corresponds to a resistance of 10 Ω, and an empty tank corresponds to a resistance of 0 Ω. (*a*) Design a circuit that provides an output voltage proportional to the fuel remaining, such that a full tank yields a voltage of 5 V, whereas an empty tank yields 0 V. (*b*) Design a circuit to indicate the rate of fuel consumption by providing a voltage output calibrated such that 1 V = 1 liter per second.

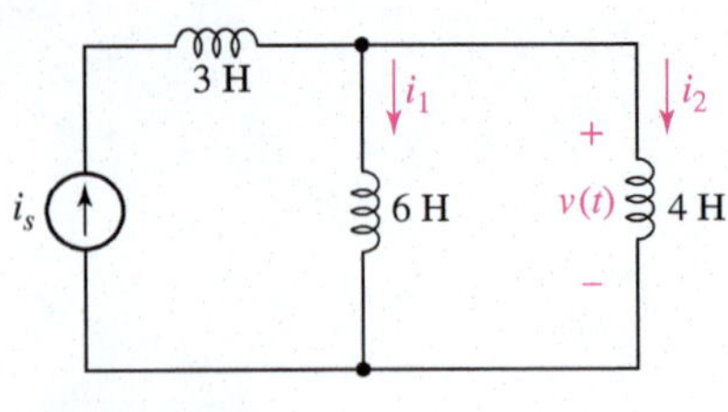

■ **FIGURE 7.74**

7.6 Duality

60. (*a*) If $I_s = 3 \sin t$ A, draw the exact dual of the circuit depicted in Fig. 7.74. (*b*) Label the new (dual) variables. (*c*) Write nodal equations for both circuits.

61. (*a*) Draw the exact dual of the simple circuit shown in Fig. 7.75. (*b*) Label the new (dual) variables. (*c*) Write mesh equations for both circuits.

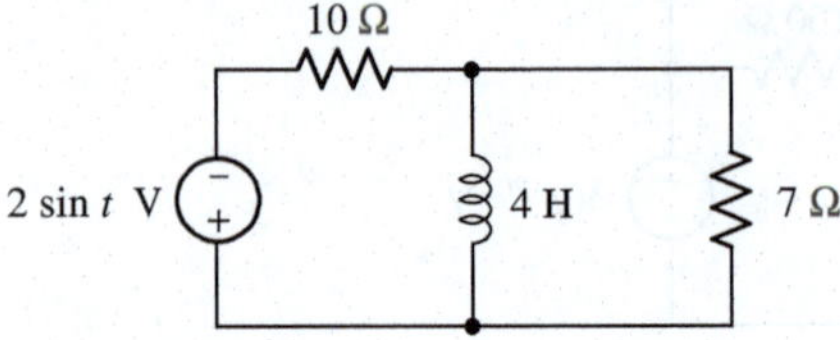

■ **FIGURE 7.75**

62. (*a*) Draw the exact dual of the simple circuit shown in Fig. 7.76 and label the new (dual) variables if $v_s = 2 \sin t$ V. (*b*) Write a nodal equation for the original circuit and a mesh equation for the dual circuit.

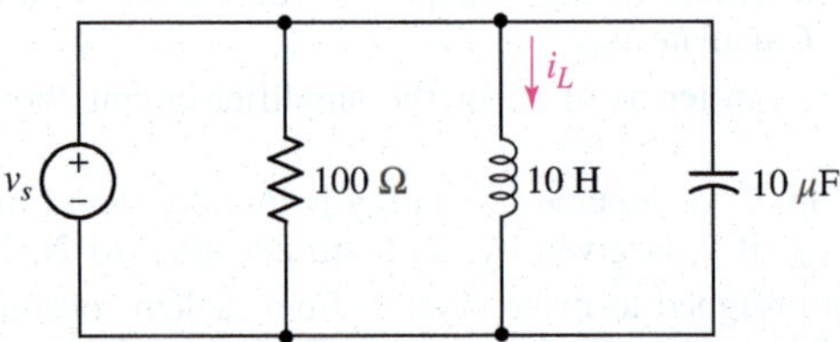

■ **FIGURE 7.76**

63. (*a*) Draw the exact dual of the simple circuit shown in Fig. 7.77. (*b*) Label the new (dual) variables.

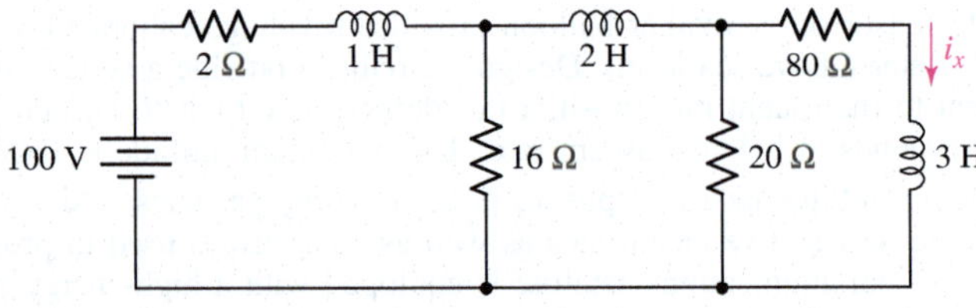

■ **FIGURE 7.77**

64. Draw the exact dual of the circuit shown in Fig. 7.78.

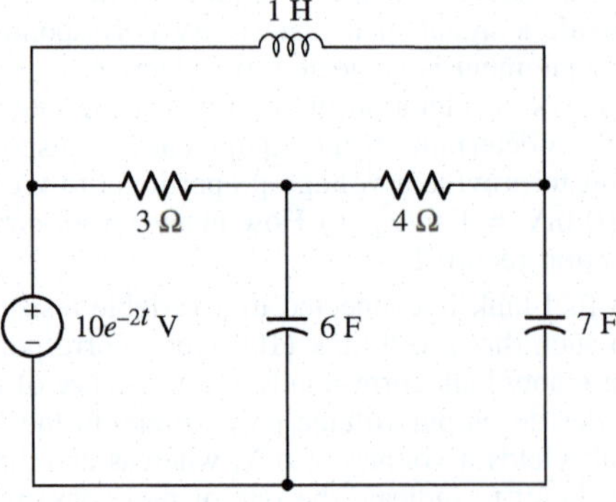

■ **FIGURE 7.78**

7.7 Computer Modeling of Circuits with Capacitors and Inductors

65. Taking the bottom node in the circuit of Fig. 7.79 as the reference terminal, calculate (*a*) the current through the inductor and (*b*) the power dissipated by the 46 kΩ resistor. (*c*) Verify your answers with an appropriate simulation.

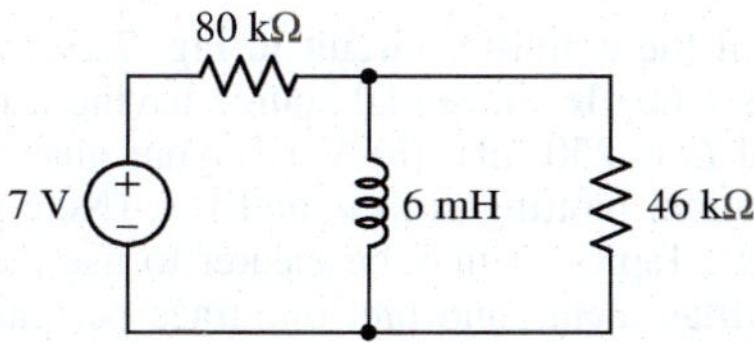

■ **FIGURE 7.79**

66. For the four-element circuit shown in Fig. 7.80, (*a*) calculate the power absorbed in each resistor, (*b*) determine the voltage across the capacitor, (*c*) compute the energy stored in the capacitor, and (*d*) verify your answers with an appropriate simulation. (Recall that calculations can be performed by selecting multiple traces and an operator.)

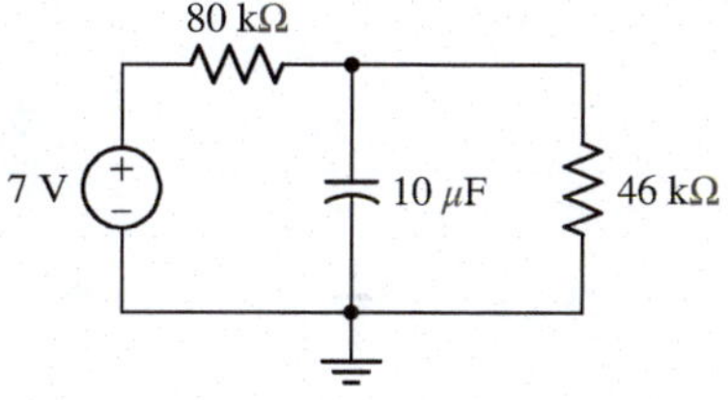

■ **FIGURE 7.80**

67. (*a*) Compute i_L and v_x as indicated in the circuit of Fig. 7.81. (*b*) Determine the energy stored in the inductor and in the capacitor. (*c*) Verify your answers with an appropriate simulation.

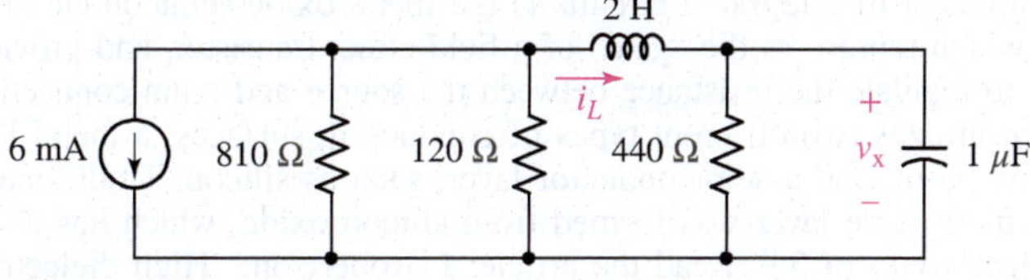

■ **FIGURE 7.81**

68. For the circuit depicted in Fig. 7.82, the value of $i_L(0) = 1$ mA. (*a*) Compute the energy stored in the element at $t = 0$. (*b*) Perform a transient simulation of the circuit over the range of $0 \leq t \leq 500$ ns. Determine the value of i_L at $t = 0$, 130 ns, 260 ns, and 500 ns. (*c*) What fraction of the initial energy remains in the inductor at $t = 130$ ns? At $t = 500$ ns?

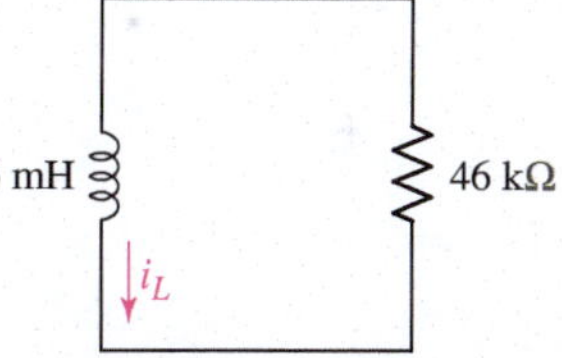

■ **FIGURE 7.82**

69. Assume an initial voltage of 9 V across the 10 μF capacitor shown in Fig. 7.83 (i.e., $v(0) = 9$ V). (*a*) Compute the initial energy stored in the capacitor. (*b*) For $t > 0$, do you expect the energy to remain in the capacitor? *Explain.* (*c*) Perform a transient simulation of the circuit over the range of $0 \leq t \leq 2.5$ s and determine $v(t)$ at $t = 460$ ms, 920 ms, and 2.3 s. (*d*) What fraction of the initial energy remains stored in the capacitor at $t = 460$ ms? At $t = 2.3$ s?

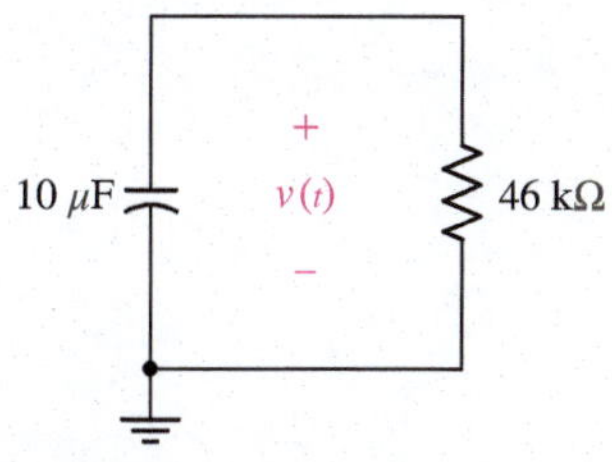

■ **FIGURE 7.83**

70. Referring to the circuit of Fig. 7.84, (*a*) calculate the energy stored in each energy storage element; (*b*) verify your answers with an appropriate simulation.

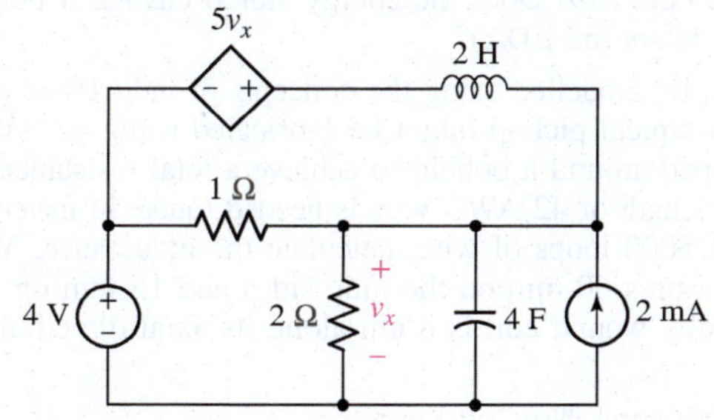

■ **FIGURE 7.84**

Chapter-Integrating Exercises

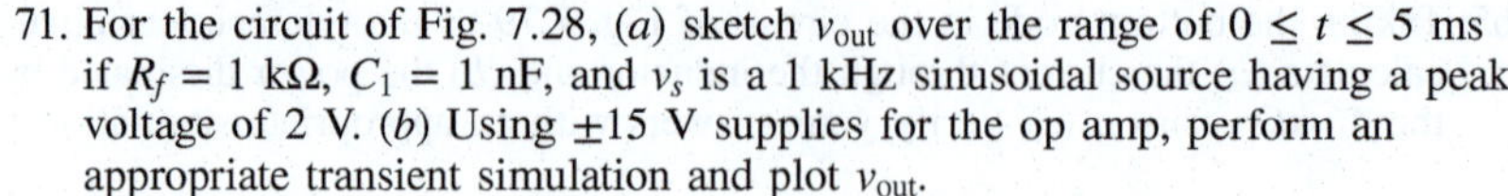

71. For the circuit of Fig. 7.28, (*a*) sketch v_{out} over the range of $0 \le t \le 5$ ms if $R_f = 1$ kΩ, $C_1 = 1$ nF, and v_s is a 1 kHz sinusoidal source having a peak voltage of 2 V. (*b*) Using ±15 V supplies for the op amp, perform an appropriate transient simulation and plot v_{out}.

72. (*a*) Sketch the output function v_{out} of the amplifier circuit in Fig. 7.29 over the range of $0 \le t \le 100$ ms if v_s is a 60 Hz sinusoidal source having a peak voltage of 400 mV, R_1 is 1 kΩ, and L_f is 250 mH. (*b*) Verify your answer with an appropriate transient simulation, plotting both v_s and v_{out}. (Note the scales are very different, so if using LTspice, it may be clearer to use the **Add Plot Pane** under the **Plot Settings** menu, and plot one trace per pane.)

73. For the circuit of Fig. 7.72, (*a*) sketch v_{out} over the range of $0 \le t \le 2.5$ ms if $R_f = 47$ Ω, $L_1 = 100$ mH, and v_s is a 2 kHz sinusoidal source having a peak voltage of 2 V. (*b*) Verify your answer with an appropriate transient simulation, plotting both v_s and v_{out}.

74. (*a*) Reconstruct the plot of Fig. 7.9 using MATLAB or a similar software resource. (*b*) Add a fourth curve representing the sum of the three currents, verifying zero total value within the specified time range of the plot. (*c*) Repeat the analysis of Example 7.3 with the resistor replaced with two 100 kΩ resistors in parallel, and the capacitor value increased to 2000 μF, to determine the three currents i_s, i_C, and i_R. (*d*) Plot the results of part (*c*), and compare to the plot you created for part (*a*).

75. A key component of integrated circuits is the metal-oxide-semiconductor capacitor, which represents the "gate" of a field-effect transistor, and providing the means to regulate the resistance between the source and drain connections. This device utilizes two different types of conducting surfaces: a metal layer, such as aluminum, and a semiconductor layer, such as silicon. Traditionally, the intervening oxide layer was formed from silicon oxide, which has a relative permittivity of 3.9. Read the article: J. Robertson, "High dielectric constant oxides," *The European Physical Journal Applied Physics* **28,** 2004, pp. 256–291. DOI: 10.1051/epjap:2004206. (*a*) If the oxide layer is 1.6 nm thick, construct a table of the gate capacitance (per unit area) for each of the high-K oxides in table 2 of the article. (*b*) If the gate metal is 1 μm in diameter, calculate the capacitance obtained using an HfO_2 layer. (*c*) What problem is intended to be solved by considering high-K oxides as an alternative to SiO_2?

76. Human skin is conducting! The conductance (or resistance) is variable—this phenomenon is known sometimes as galvanic skin response, or more conventionally as electrodermal activity (EDA). Regardless, this means that if two people are sitting next to each other, without touching, they constitute a capacitor! (*a*) Estimate how much energy might be stored if, for some reason, they each touch a different terminal of a 9 V battery. For simplicity, let's assume that both people are far apart except for their elbows—which they point at one another at a spacing of 1 mm (one right elbow pointed at the other person's left elbow, just to be pedantic)—and that each elbow has a cross-sectional area of 25 cm^2. (*b*) Does the energy stored change if both elbows are sweaty (which alters the EDA)?

77. Electric guitar pickups can be modelled using the concepts of inductance and resistance. For example, a typical pickup might be fabricated using 42 AWG copper magnet wire, wrapped around a bobbin to achieve a total resistance of 4.7 kΩ at 20 °C. (*a*) What length of 42 AWG wire is needed (quote in meters)? (*b*) If the bobbin is full at 5000 loops of wire, calculate the inductance. You can assume each loop measures 60 mm on the long sides and 15 mm on the two short sides, and the fully wound coil is 8 nm along its axial direction.

CHAPTER 8

Basic *RC* and *RL* Circuits

KEY CONCEPTS

RC and *RL* Time Constants

Natural and Forced Response

Calculating the Time-Dependent Response to DC Excitation

How to Determine Initial Conditions and Their Effect on the Circuit Response

Analyzing Circuits with Step Function Input and with Switches

Construction of Pulse Waveforms Using Unit-Step Functions

Response of Sequentially Switched Circuits

INTRODUCTION

In Chap. 7 we wrote equations for the response of several circuits containing both inductance and capacitance, but we did not solve any of them. Now we are ready to proceed with the solution of the simpler circuits, namely, those which contain only resistors and inductors, or only resistors and capacitors.

Although the circuits we are about to consider have a very elementary appearance, they are also of practical importance. Networks of this form find use in electronic amplifiers, automatic control systems, operational amplifiers, communications equipment, and many other applications. Familiarity with these simple circuits will enable us to predict the accuracy with which the output of an amplifier can follow an input that is changing rapidly with time, or to predict how quickly the speed of a motor will change in response to a change in its field current. Our understanding of simple *RC* and *RL* circuits will also enable us to suggest modifications to the amplifier or motor in order to obtain a more desirable response, as well as develop an intuition regarding other engineering systems.

8.1 THE SOURCE-FREE *RC* CIRCUIT

Capacitors and inductors are energy storage elements, and their current–voltage relations are described by differential equations (namely $i = C\, dv/dt$ and $v = L\, di/dt$). Circuit analysis with these elements will require the solution of a differential equation to determine the instantaneous time dependence of current and voltage. The circuit response can be divided into two components: the

steady-state response (does not change with time) and the ***transient*** response (changes with time). Alternatively, the circuit response can be divided into the ***natural*** response and the ***forced*** response. The natural response describes the behavior in the absence of all external sources, depending on the "nature" of the circuit (the types of elements, their sizes, the interconnection of elements). The forced response describes the additional response or "contribution" from external sources. The complete response will be the addition of the steady-state and transient, or the natural and forced response. For example, the complete solution for a voltage at a node in a circuit can be described by

$$v_{\text{complete}} = v_{\text{steady-state}} + v_{\text{transient}}$$

or

$$v_{\text{complete}} = v_{\text{natural}} + v_{\text{forced}}$$

It may seem unusual to discuss a time-varying voltage in a circuit with no sources! Keep in mind that we only know the voltage at the time specified as $t = 0$; we don't know the voltage prior to that time. In the same vein, we don't know what the circuit looked like prior to $t = 0$, either. A nonzero voltage implies energy storage in the capacitor and that a source was present at some point. Fortunately, we do not need to know the details prior to $t = 0$ in order to analyze the circuit.

Let us begin by considering an *RC* circuit as shown in Fig. 8.1, subject to the initial condition $v(0) = V_0$. Note that this circuit does not have an external source or forcing function. The solution to this ***source-free*** circuit will therefore be the ***natural*** response of the circuit. An equation for the voltage v can be described by writing a KCL equation at the top node of the circuit.

$$C\frac{dv}{dt} + \frac{v}{R} = 0 \qquad [1]$$

Our goal is an expression for $v(t)$ which satisfies this equation *and* also has the value V_0 at $t = 0$. The solution may be obtained by several different methods.

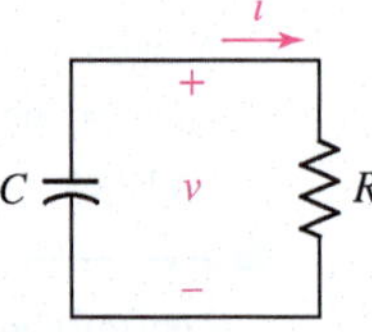

FIGURE 8.1 A parallel *RC* circuit for which $v(t)$ is to be determined, subject to the initial condition that $v(0) = V_0$.

Solution by Direct Integration

The solution can be solved directly by separating voltage and time variables and then integrating. Combining voltage variables on the left and time and circuit constants on the right gives

$$\frac{dv}{v} = -\frac{1}{RC}dt$$

Integrating both sides over voltage and time, respectively, results in

$$\int_{V_0}^{v(t)} \frac{dv'}{v'} = \int_0^t -\frac{1}{RC}dt'$$

Following integration and evaluating for $t > 0$ and the initial condition V_0 at $t = 0$,

$$\ln v'\Big|_{V_0}^{v(t)} = -\frac{1}{RC}dt'\Big|_0^t$$

and so,

$$\ln v(t) - \ln V_0 = -\frac{1}{RC}(t - 0)$$

Taking the exponential of both sides and solving for $v(t)$ gives the final solution.

$$v(t) = V_0 e^{-t/RC}$$

EXAMPLE 8.1

For the circuit in Fig. 8.2, calculate the time where the voltage v decreases to half of the voltage at time $t = 0$ ($v(0) = V_0$).

The time-dependent voltage is given by $v(t) = V_0 e^{-t/RC}$, where we are asked to find the time where $v(t) = V_0/2$. Thus,

$$v(t) = V_0 e^{-t/RC} = V_0/2$$

Solving for t (noting that the initial value V_0 cancels in the equation),

$$\begin{aligned} t &= -RC\ln\left(\frac{1}{2}\right) \\ &= -(2 \times 10^3)(6 \times 10^{-6})\ln\left(\frac{1}{2}\right) \\ &= 8.3178 \times 10^{-3}\ \text{s} = 8.3178\ \text{ms} \end{aligned}$$

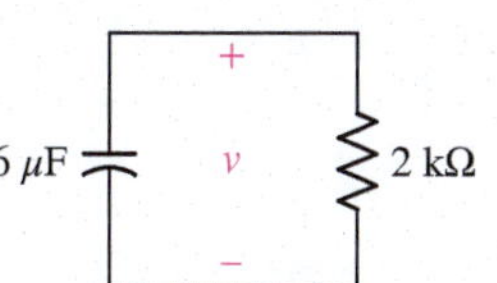

FIGURE 8.2

PRACTICE

8.1 For the circuit in Fig. 8.2, what value of capacitance would be required to ensure that $v(t) < V_0/10$ for $t > 10$ ms?

Ans: $C < 2.17\ \mu\text{F}$

A More General Solution Approach

Direct integration works well for the source-free RC circuit, but it is limited to cases where variables can be separated. An alternative method is to "guess" or assume a form for the solution and then to test our assumptions. Let us examine this approach, outlined in Fig. 8.3, as it will help us to analyze more complex circuits. Many of the differential equations encountered in circuit analysis have a solution that may be represented by the exponential function, or by the sum of exponential functions. Let's assume a solution of Eq. [1] as an exponential

$$v(t) = Ae^{st}$$

where A and s are unknown constants. Substituting the assumed solution into Eq. [1] we have

$$C\frac{d(Ae^{st})}{dt} + \frac{Ae^{st}}{R} = 0$$

$$sCAe^{st} + \frac{(Ae^{st})}{R} = 0$$

$$\left(sC + \frac{1}{R}\right)Ae^{st} = 0$$

In order to satisfy this equation for all values of time, it is necessary that $A = 0$, $s = -\infty$, or $(sC + 1/R) = 0$. The cases $A = 0$ *or* $s = -\infty$ would not offer a solution to our problem. Therefore, we choose

Assume a general solution with appropriate constants.

Substitute the trial solution into the differential equation and simplify the result.

Determine the value for one constant that does not result in a trivial solution.

Invoke the initial condition(s) to determine values for the remaining constant(s).

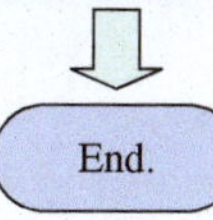

FIGURE 8.3 Flowchart for the general approach to the solution of first-order differential equations where, based on experience or general knowledge, we can guess the form of the solution.

$$\left(sC + \frac{1}{R}\right) = 0$$

$$s = -\frac{1}{RC}$$

Substituting into our assumed solution yields

$$v(t) = Ae^{-t/RC}$$

The remaining unknown constant A can be evaluated by applying the initial condition at $t = 0$, $v(0) = V_0$, resulting in

$$v(t) = V_0 e^{-t/RC} \quad [2]$$

EXAMPLE **8.2**

For the circuit of Fig. 8.4*a*, find the voltage labeled *v* at $t = 200\ \mu$s.

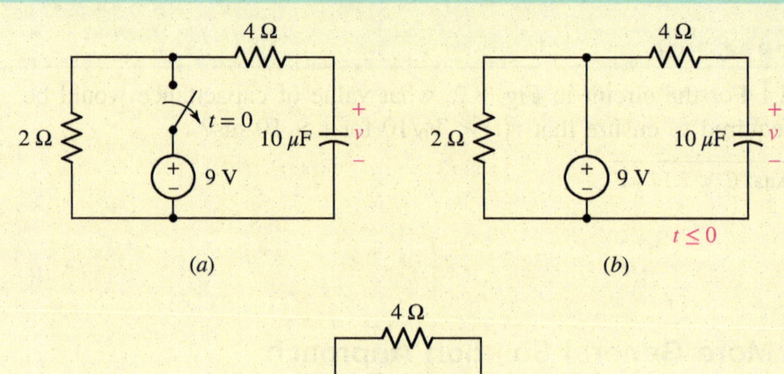

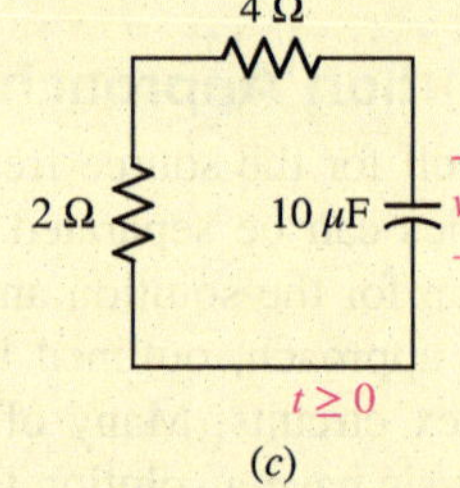

■ **FIGURE 8.4** (*a*) A simple *RC* circuit with a switch thrown at time $t = 0$. (*b*) The circuit as it exists prior to $t = 0$. (*c*) The circuit after the switch is thrown, and the 9 V source is removed.

To find the requested voltage, we will need to draw and analyze two separate circuits: one corresponding to before the switch is thrown (Fig. 8.4*b*), and one corresponding to after the switch is thrown (Fig. 8.4*c*).

The sole purpose of analyzing the circuit of Fig. 8.4*b* is to obtain an initial capacitor voltage; we assume any transients in that circuit died out long ago, leaving a purely dc circuit. With no current through either the capacitor or the 4 Ω resistor, then,

$$v(0) = 9\ \text{V} \quad [3]$$

We next turn our attention to the circuit of Fig. 8.4*c*, recognizing that

$$\tau = RC = (2 + 4)(10 \times 10^{-6}) = 60 \times 10^{-6}\ \text{s}$$

Thus, from Eq. [2],

$$v(t) = v(0)e^{-t/RC} = v(0)e^{-t/60\times 10^{-6}} \quad [4]$$

The capacitor voltage must be the same in both circuits at $t = 0$; no such restriction is placed on any other voltage or current. Substituting Eq. [3] into Eq. [4],

$$v(t) = 9e^{-t/60\times10^{-6}}\ \text{V}$$

so that $v(200 \times 10^{-6}) = 321.1$ mV (less than 4% of its maximum value).

PRACTICE

8.2 Noting carefully how the circuit changes once the switch in the circuit of Fig. 8.5 is thrown, determine $v(t)$ at $t = 0$ and at $t = 160\ \mu$s.

Ans: 50 V, 18.39 V.

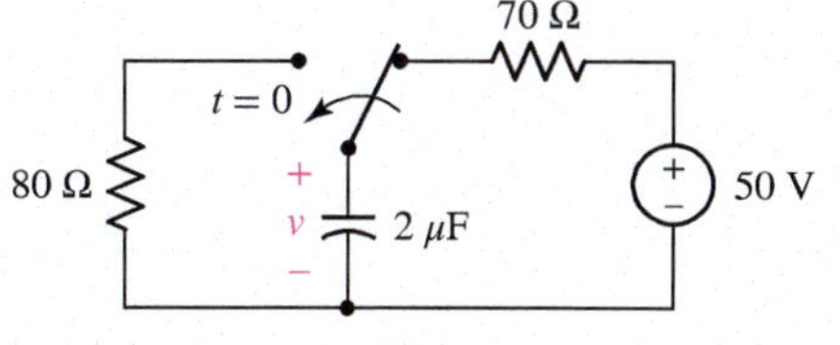

■ **FIGURE 8.5**

Accounting for the Energy

Before we turn our attention to the interpretation of the response, let us return to the circuit of Fig. 8.1 and check the power and energy relationships. The power being dissipated in the resistor is

$$p_R = \frac{v^2}{R} = \frac{V_0^2}{R}e^{-2t/RC}$$

and the total energy turned into heat in the resistor is found by integrating the instantaneous power from zero time to infinite time:

$$w_R = \int_0^{\infty} p_R dt = \frac{(V_0)^2}{R}\int_0^{\infty} e^{-2t/RC}dt$$

$$= \frac{(V_0)^2}{R}\left(\frac{-RC}{2}\right)e^{-2t/RC}\Big|_0^{\infty} = \frac{1}{2}C(V_0)^2$$

This is the result we expect because the total energy stored initially in the capacitor is $\frac{1}{2}C(V_0)^2$, and there is no longer any energy stored in the capacitor at infinite time since its voltage eventually drops to zero. All the initial energy therefore is accounted for by dissipation in the resistor.

8.2 PROPERTIES OF THE EXPONENTIAL RESPONSE

Now consider the nature of the response in the series RC circuit. We have found that the inductor current is represented by

$$v(t) = V_0 e^{-t/RC}$$

At $t = 0$, the voltage has value V_0, but as time increases, the voltage decreases and approaches zero. The shape of this decaying exponential is seen by the plot of $v(t)/V_0$ versus t shown in Fig. 8.6. Since the argument of the exponent is unitless, the product of RC has a unit of time. This effective unit of time is called the *time constant*, and it reflects the rate of decay of the voltage response of the RC circuit. The time constant is denoted by the Greek letter τ

$$\tau = RC$$

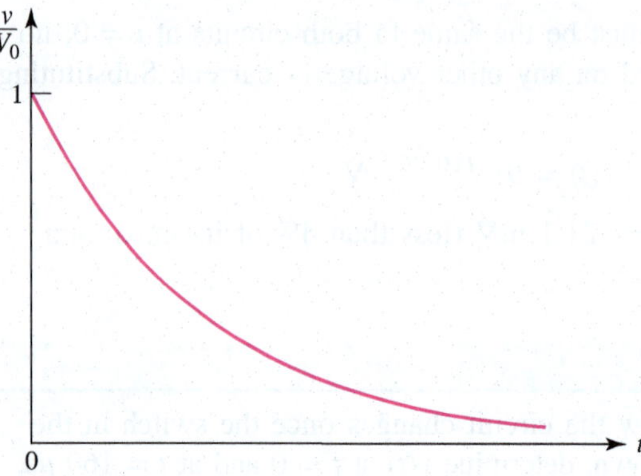

FIGURE 8.6 A plot of $e^{-t/RC}$ versus *t*.

Since the function we are plotting is $e^{-t/RC}$, the curve will not change if *RC* remains unchanged. Thus, the same curve must be obtained for every series *RC* circuit having the same *RC* product and time constant.

The time constant represents the approximate time decay from the initial value. Mathematically, the time constant is also the time that would be required for the voltage to drop to zero *if it continued to drop at its initial rate.* The rate of decrease (slope) of the voltage response is given by

$$\left.\frac{d}{dt}\frac{v}{V_0}\right|_{t=0} = -RC\,e^{-\frac{t}{RC}}\Big|_{t=0} = -RC$$

Continuing this rate of decay would lead to an intercept on the time axis given by $t = RC = \tau$.

The time constant of a series RC circuit is shown graphically in Fig. 8.7; it is necessary only to draw the tangent to the curve at $t = 0$ and determine the intercept of this tangent line with the time axis. This is often a convenient way of approximating the time constant from the display on an oscilloscope.

An equally important interpretation of the time constant τ is obtained by determining the value of $v(t)/V_0$ at $t = \tau$. We have

$$\frac{v(\tau)}{V_0} = e^{-1} = 0.3679 \qquad \text{or} \qquad v(\tau) = 0.3679\,V_0$$

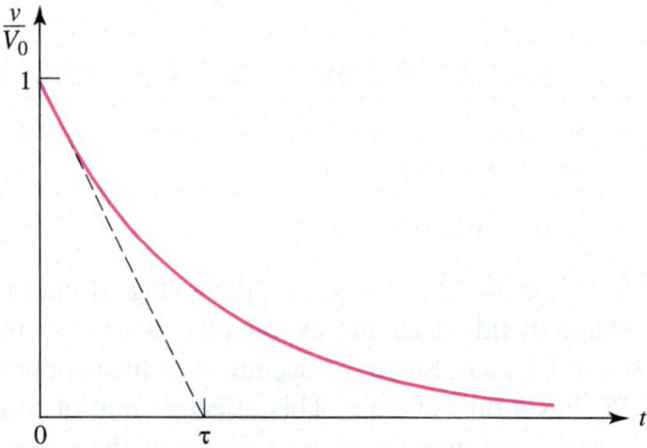

FIGURE 8.7 The time constant τ is *RC* for a series *RC* circuit. It is the time required for the response curve to drop to zero if it decays at a constant rate equal to its initial rate of decay.

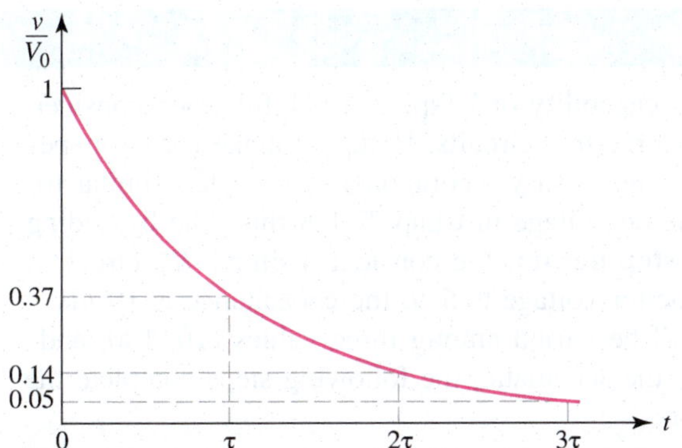

FIGURE 8.8 The voltage in a series *RC* circuit is reduced to 37% of its initial value at $t = \tau$, 14% at $t = 2\tau$, and 5% at $t = 3\tau$.

Thus, in one time constant the response has dropped to 36.8% of its initial value; the value of τ may also be determined graphically from this fact, as indicated by Fig. 8.8. It is convenient to measure the decay of the voltage at intervals of one time constant, and recourse to a calculator shows that $v(t)/V_0$ is 0.3679 at $t = \tau$, 0.1353 at $t = 2\tau$, 0.04979 at $t = 3\tau$, 0.01832 at $t = 4\tau$, and 0.006738 at $t = 5\tau$. After approximately three to five time constants, most of us would agree that the voltage is a negligible fraction of its initial value. Thus, if we are asked, "*How long does it take for the voltage to decay to zero?*" our answer might be, "*About five time constants.*" At that point, the voltage is less than 1% of its original value!

EXAMPLE 8.3

The time constant of a series RC circuit has a time constant τ_0. If the time constant is now increased by a factor of four, how would the voltage change at a time $t = 2\tau_0$?

$$v(t) = V_0 e^{-t/\tau}$$

The ratio of voltages for the two different time constants is given by

$$\frac{V_0 e^{-2\tau_0/\tau_0}}{V_0 e^{-2\tau_0/4\tau_0}} = \frac{e^{-2}}{e^{-1/2}} = 0.2231$$

Increasing the time constant by a factor of four will result in a much faster decay in the voltage response, and the voltage at time $t = 2\tau_0$ will be decreased to approximately 22.3% of what it would have been with the original time constant.

PRACTICE

8.3 In a source-free series *RC* circuit, find the numerical value of the ratio: *(a)* $v(2\tau)/v(\tau)$; *(b)* $v(0.5\tau)/v(0)$; *(c)* t/τ if $v(t)/v(0) = 0.2$; *(d)* t/τ if $v(0) - v(t) = v(0) \ln 2$.

Ans: 0.368; 0.607; 1.609; 1.181.

COMPUTER-AIDED ANALYSIS

The transient analysis capability of LTspice is helpful when considering the response of source-free circuits. Here, we make use of a special feature that allows us to vary a component parameter, similar to the way we varied the dc voltage in Chap. 5, but this time by adding the SPICE directive **.step param.** We consider a simple *RC* circuit where the initial capacitor voltage is 5 V, the capacitance is 10 μF, and the resistance will be varied among three values (10, 100, and 1 kΩ). After drawing the schematic, the following steps complete the simulation preparation:

1. **Set the initial condition for the circuit.** First, label the node for the capacitor using **Net Name** (Cap_voltage for this example). Set the initial condition of this node to 5 V using the SPICE Directive *.ic V(Cap_voltage)=5.*
2. **Define the resistor values.** For the value of resistance for the resistor component, enter the text *{Resistance}*, which will replace a numerical value with a variable. Define the values you wish to simulate using the SPICE Directive *.step param Resistance list 10 100 1k.* To help interpret, this directive is stepping through the parameter *Resistance* according to the *list* of numbers provided.
3. **Define the simulation.** Since we intend to examine the transient response, use the SPICE Directive *.tran* <*Tstop*>, where <*Tstop*> is the time where you wish to end the simulation. In this case, let us choose 5 ms.

The resulting circuit schematic, including the labeled SPICE directives, is shown in Fig. 8.9.

After running the simulation, the Waveform Data window will open. To plot the transient response, use the Add Traces command to select the *Cap_voltage* node, or directly click on the node on the circuit

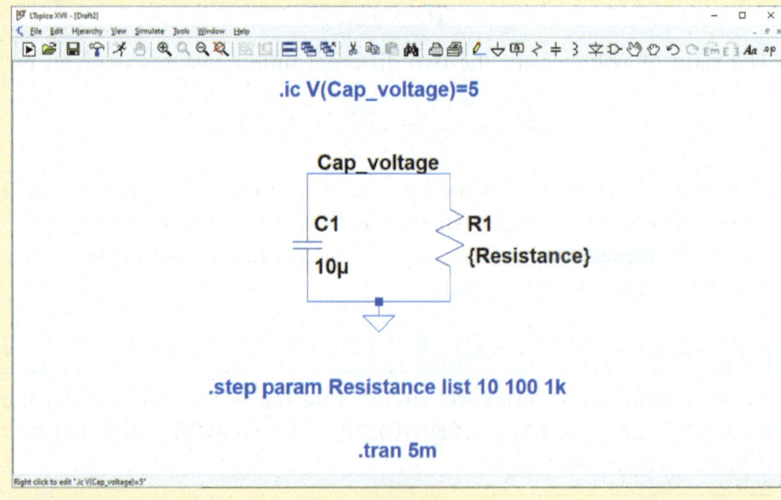

FIGURE 8.9 Simple *RC* circuit drawn in LTspice.
Analog Devices

schematic. The resulting response will then appear in the Waveform Data window as shown in Fig. 8.10 (note data labels were added, and the Color Palette Editor was used (Color Preferences under Tools).

Why does a larger value of resistance produce a response curve that decays more slowly? In terms of the time constant τ, the response of the series *RC* circuit may be written simply as

$$v(t) = V_0 e^{-t/\tau}$$

The time constant (*RC*) increases with increasing resistance, and increasing capacitance. In this example, we see that the energy stored in the capacitor takes longer to dissipate for larger resistor values; this is because the current (and hence power) flowing into the resistor is lower.

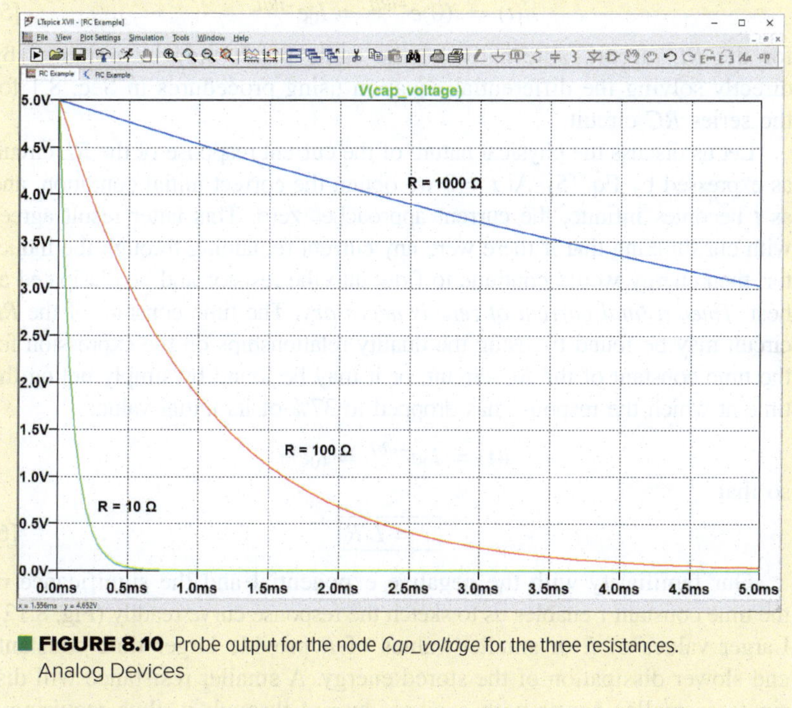

FIGURE 8.10 Probe output for the node *Cap_voltage* for the three resistances.
Analog Devices

8.3 THE SOURCE-FREE *RL* CIRCUIT

Inductors also store energy, similar to capacitors, and would also be expected to show similar time-dependent response in a circuit. Let us see how closely the analysis of the parallel (or is it series?) *RL* circuit shown in Fig. 8.11 corresponds to that of the *RC* circuit.

Let us designate the time-varying current as $i(t)$; with initial value at $t = 0$ as $i(0) = I_0$. Applying KVL to the circuit, we get an equation in terms of the unknown current

$$Ri + v_L = Ri + L\frac{di}{dt} = 0$$

or

$$\frac{di}{dt} + \frac{R}{L}i = 0$$

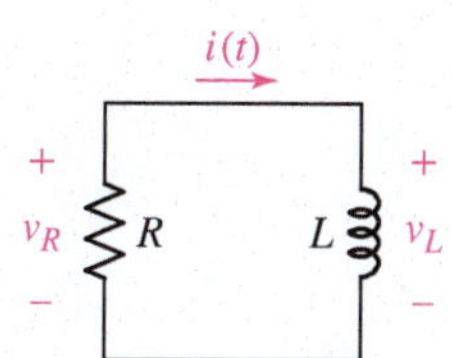

FIGURE 8.11 A series *RL* circuit for which $i(t)$ is to be determined, subject to the initial condition that $i(0) = I_0$.

The differential equation has a familiar form; comparison with

$$\frac{dv}{dt} + \frac{v}{RC} = 0$$

shows that the replacement of v by i and RC by L/R produces the identical equation we considered previously. It should, since the RL circuit we are now analyzing is the dual of the RC circuit we considered first. This *duality* forces $v(t)$ for the RC circuit and $i(t)$ for the RL circuit to have identical expressions if the resistance of one circuit is equal to the reciprocal of the resistance of the other circuit, and if L is numerically equal to C. Thus, the response of the RC circuit

$$v(t) = v(0)e^{-t/RC} = V_0 e^{-t/RC}$$

enables us to immediately write

$$i(t) = i(0)e^{-tR/L} = I_0 e^{-tR/L} \quad [5]$$

for the RL circuit. Alternatively, the same result may be obtained by directly solving the differential equation using procedures in Sec. 8.1 for the series RC circuit.

Let us discuss the physical nature of the current response of the RL circuit as expressed by Eq. [5]. At $t = 0$ we obtain the correct initial condition, and as t becomes infinite, the current approaches zero. This latter result agrees with our thinking that if there were any current remaining through the inductor, then energy would continue to flow into the resistor and be dissipated as heat. *Thus, a final current of zero is necessary.* The time constant of the RL circuit may be found by using the duality relationships on the expression for the time constant of the RC circuit, or it may be found by simply noting the time at which the response has dropped to 37% of its initial value:

$$i(t) = I_0 e^{-tR/L} = I_0 e^{-t/\tau}$$

so that

$$\boxed{\tau = L/R} \quad [6]$$

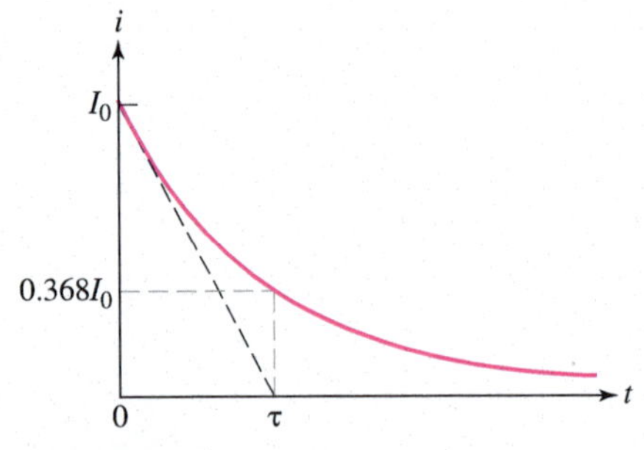

FIGURE 8.12 The inductor current $i(t)$ in the parallel RL circuit is plotted as a function of time. The initial value of $i(t)$ is I_0.

Our familiarity with the negative exponential and the significance of the time constant τ enables us to sketch the response curve readily (Fig. 8.12). Larger values of L or smaller values of R provide larger time constants and slower dissipation of the stored energy. A smaller resistance will dissipate a smaller power with a given current through it, thus requiring a greater time to convert the stored energy into heat; a larger inductance stores a larger energy with a given current through it, again requiring a greater time to lose this initial energy.

EXAMPLE 8.4

If the inductor of Fig. 8.13 has a current $i_L = 2$ A at $t = 0$, find an expression for $i_L(t)$ valid for $t > 0$, and its value at $t = 200$ μs.

This is the identical type of circuit just considered, so we expect an inductor current of the form

$$i_L = I_0 e^{-Rt/L}$$

where $R = 200\ \Omega$, $L = 50$ mH, and I_0 is the initial current flowing through the inductor at $t = 0$. Thus,

$$i_L(t) = 2e^{-4000t}$$

Substituting $t = 200 \times 10^{-6}$ s, we find that $i_L(t) = 898.7$ mA, less than half the initial value.

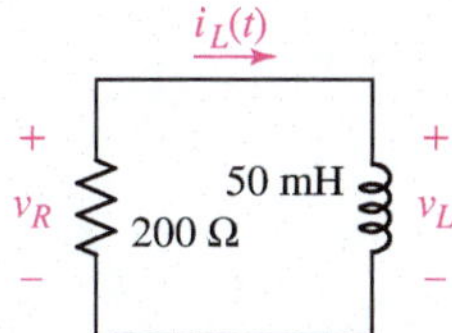

FIGURE 8.13 A simple *RL* circuit in which energy is stored in the inductor at $t = 0$.

PRACTICE

8.4 Determine the current i_R through the resistor of Fig. 8.14 at $t = 2$ ns if $i_R(0) = 5$ mA.

Ans: 91 μA.

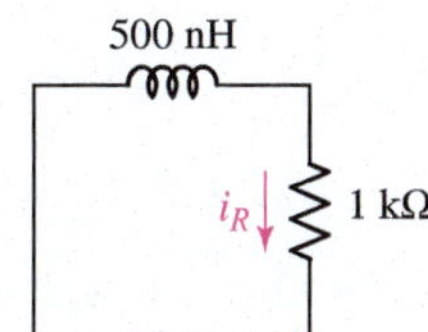

FIGURE 8.14 Circuit for Practice Problem 8.4.

EXAMPLE 8.5

For the circuit of Fig. 8.15*a*, find the voltage labeled *v* at $t = 200$ ms.

▶ ***Identify the goal of the problem.***

The schematic of Fig. 8.15*a* actually represents *two different* circuits: one with the switch closed (Fig. 8.15*b*) and one with the switch open (Fig. 8.15*c*). We are asked to find $v(0.2)$ for the circuit shown in Fig. 8.15*c*.

▶ ***Collect the known information.***

Both new circuits are drawn and labeled correctly. We next make the assumption that the circuit in Fig. 8.15*b* has been connected for a long time, so any transients have dissipated. We may make such an assumption as a general rule unless instructed otherwise. This circuit determines $i_L(0)$.

▶ ***Devise a plan.***

The circuit of Fig. 8.15*c* may be analyzed by writing a KVL equation. Ultimately we want a differential equation with only v and t as variables; we will then solve the differential equation for $v(t)$.

▶ ***Construct an appropriate set of equations.***

Referring to Fig. 8.15*c*, we write

$$-v + 10i_L + 5\frac{di_L}{dt} = 0$$

Substituting $i_L = -v/40$ we find that

$$\frac{5}{40}\frac{dv}{dt} + \left(\frac{10}{40} + 1\right)v = 0$$

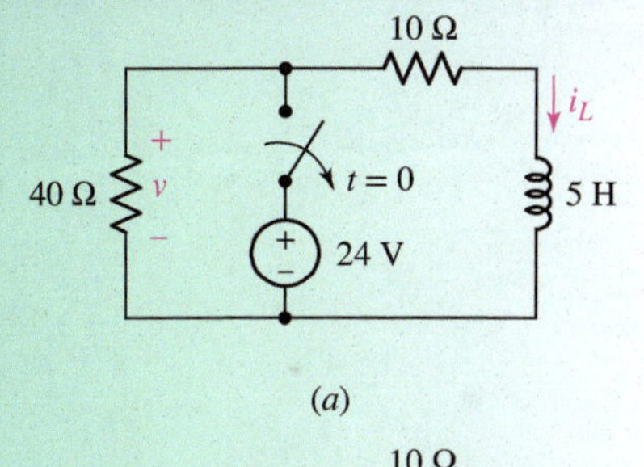

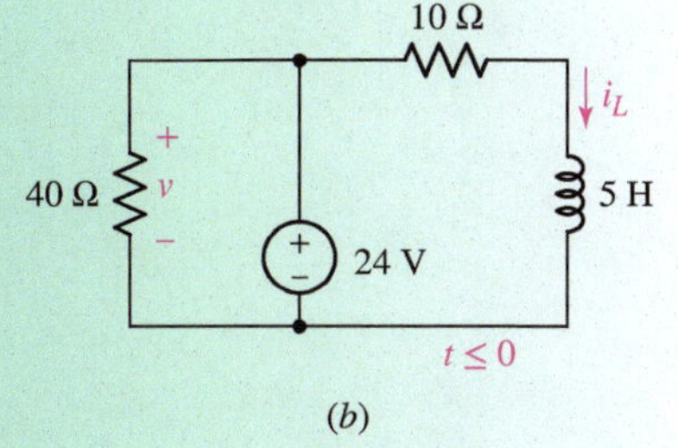

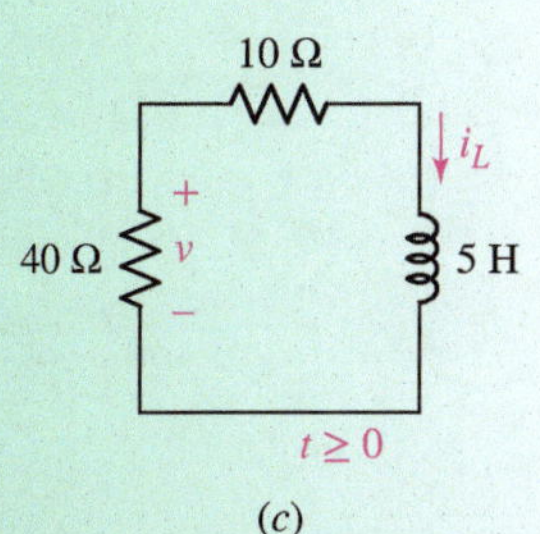

FIGURE 8.15 (*a*) A simple *RL* circuit with a switch thrown at time $t = 0$. (*b*) The circuit as it exists prior to $t = 0$. (*c*) The circuit after the switch is thrown and the 24 V source is removed.

(Continued on next page)

or, more simply,

$$\frac{dv}{dt} + 10v = 0 \qquad [7]$$

▶ *Determine if additional information is required.*

From previous experience, we know that a complete expression for v will require knowledge of v at a specific instant of time, with $t = 0$ being the most convenient. We might be tempted to look at Fig. 8.15*b* and write $v(0) = 24$ V, but this is only true *just before the switch opens.* The resistor voltage can change to any value in the instant that the switch is thrown; only the inductor current must remain unchanged.

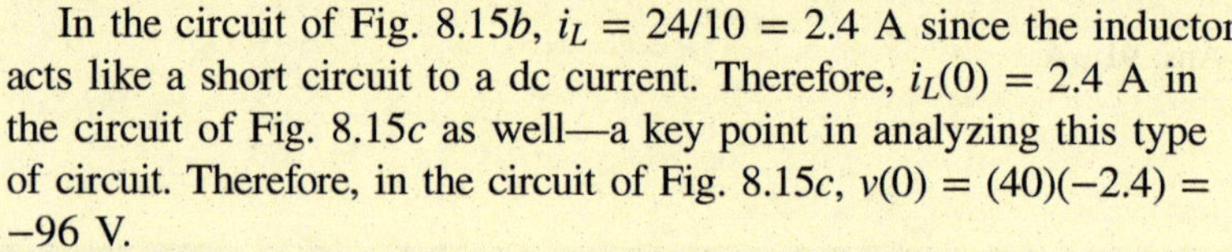

In the circuit of Fig. 8.15*b*, $i_L = 24/10 = 2.4$ A since the inductor acts like a short circuit to a dc current. Therefore, $i_L(0) = 2.4$ A in the circuit of Fig. 8.15*c* as well—a key point in analyzing this type of circuit. Therefore, in the circuit of Fig. 8.15*c*, $v(0) = (40)(-2.4) = -96$ V.

▶ *Attempt a solution.*

Any of the three basic solution techniques can be brought to bear; let's start by writing the characteristic equation corresponding to Eq. [7]:

$$s + 10 = 0$$

Solving, we find that $s = -10$, so

$$v(t) = Ae^{-10t} \qquad [8]$$

(which, upon substitution into the left-hand side of Eq. [7], results in

$$-10Ae^{-10t} + 10Ae^{-10t} = 0$$

as expected).

We find A by setting $t = 0$ in Eq. [8] and employing the fact that $v(0) = -96$ V. Thus,

$$v(t) = -96e^{-10t} \qquad [9]$$

and so $v(0.2) = -12.99$ V, down from a maximum of -96 V.

▶ *Verify the solution. Is it reasonable or expected?*

Instead of writing a differential equation in v, we could have written our differential equation in terms of i_L:

$$40i_L + 10i_L + 5\frac{di_L}{dt} = 0$$

or

$$\frac{di_L}{dt} + 10i_L = 0$$

which has the solution $i_L = Be^{-10t}$. With $i_L(0) = 2.4$, we find that $i_L(t) = 2.4e^{-10t}$. Since $v = -40i_L$, we once again obtain Eq. [9]. We should note: it is **no coincidence** that the inductor current and the resistor voltage have the same exponential dependence!

PRACTICE

8.5 Determine the inductor voltage v in the circuit of Fig. 8.16 for $t > 0$.

Ans: $-25e^{-2t}$ V.

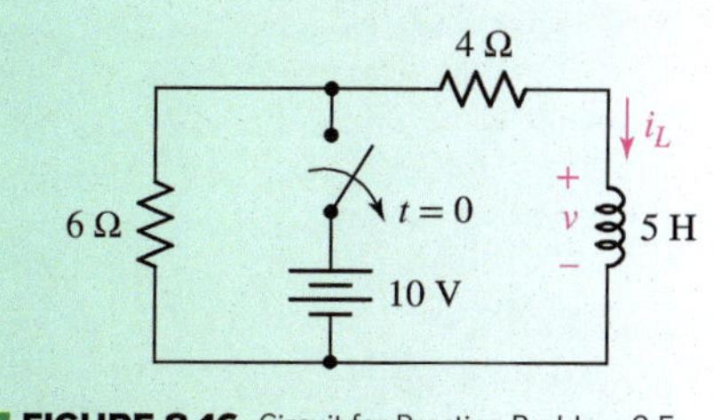

FIGURE 8.16 Circuit for Practice Problem 8.5.

8.4 A MORE GENERAL PERSPECTIVE

As seen indirectly from Examples 8.2 and 8.5, regardless of how many resistors we have in the circuit, we obtain a single time constant (either $\tau = RC$ or $\tau = L/R$) when only one energy storage element is present. We can formalize this by realizing that the value needed for R is in fact the Thévenin equivalent resistance seen by our energy storage element. (*Strange as it may seem, it is even possible to compute a time constant for a circuit containing dependent sources!*)

Many of the *RC* or *RL* circuits for which we would like to find the natural response contain more than a single resistor and capacitor/inductor. Let us suppose that we are faced with a circuit containing a single capacitor or inductor, but any number of resistors. It is possible to replace the two-terminal resistive network connected across the capacitor or inductor terminals with an equivalent resistor, and we may then write down the expression for the effective time constant given by

$$\tau = R_{eq}C$$

or

$$\tau = L/R_{eq}$$

for *RC* and *RL* equivalent circuits, respectively. If there are several capacitors (or several inductors) present in a circuit, and they can be combined using series and/or parallel combination, then the circuit can be further generalized as C_{eq} (or L_{eq}) with a single time constant. However, be aware that this only works if you can combine capacitor or inductor elements in a singular equivalent C_{eq} or L_{eq}; otherwise your circuit will contain multiple time constants!

Based on the exponential time dependence of the capacitor voltage or inductor current, *every other voltage and current throughout the circuit must follow the same functional behavior.* This is made clear by considering the capacitor as a voltage source applied to a resistive network. Every current and voltage in the resistive network will have the same time dependence as the source, as the resistor network will respond instantaneously to any changes in the source according to Ohm's law $v = iR$.

Using this time dependence $e^{-t/\tau}$, we can therefore use a general procedure to solve for any current or voltage in a circuit with a single energy storage element. This technique can be applied to any circuit with one inductor and any number of resistors, as well as to those special circuits containing two or more inductors and also two or more resistors that may be simplified by resistance or inductance combination to one inductor and one resistor.

Source-Free *RC* and *RL* Circuit with Single Energy Storage Element

1. Determine the time constant using a Thévenin equivalent circuit.
2. Find the initial condition of the variable of interest.
3. Arrive at the solution.

$$v(t) = v(0^+)e^{-t/\tau}$$

or

$$i(t) = i(0^+)e^{-t/\tau}$$

Slicing Thinly: The Difference Between $t = 0^+$ and $t = 0^-$

Finding the initial conditions can perhaps be the trickiest part of solving general *RC* and *RL* circuits, and the most likely source of error. In the previous procedure, why did we use the notation (0^+) rather than simply (0)? In the case of capacitors, we know that the voltage across a capacitor cannot change instantaneously, but the current through the capacitor *can* change instantaneously! For example, when a capacitor is charged to a given voltage and holding at steady state, the current is zero. If this capacitor is then switched at time $t = 0$ to a resistive network, the current will instantly change from $i = 0$ to a nonzero current to discharge through the network. In other words, the voltage just before and after switching are equal: $v(0^-) = v(0^+)$; the current, however, will not be the same; that is, $i(0^-) \neq i(0^+)$. Conversely, the current through an inductor will not change instantaneously, but the voltage across the inductor may change instantaneously such that $i(0^-) = i(0^+)$ and $v(0^-) \neq v(0^+)$.

Therefore, to find our initial condition for an *RC* or *RL* circuit, we need to analyze it just before and just after the event that triggers the time response (e.g., a switch). As an example, let us find the current i_2 in the circuit shown in Fig. 8.17, assuming that some finite amount of energy is stored in the inductor at $t = 0$ so that $i_L(0) = I_0$.

The equivalent resistance the inductor faces is

$$R_{\text{eq}} = R_3 + R_4 + \frac{R_1 R_2}{R_1 + R_2}$$

and the time constant is therefore

$$\tau = \frac{L}{R_{\text{eq}}}$$

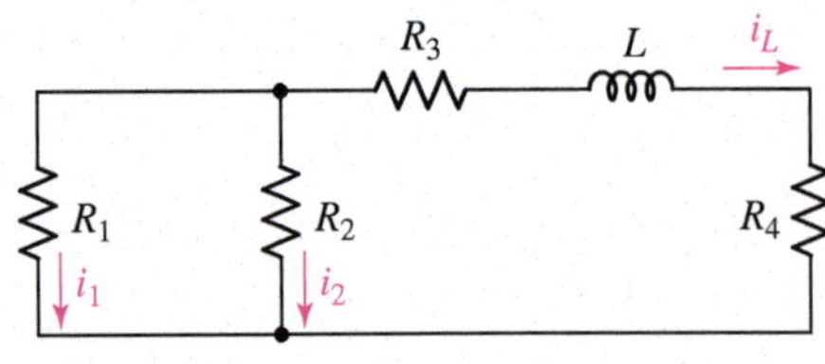

■ **FIGURE 8.17** A source-free circuit containing one inductor and several resistors is analyzed by determining the time constant $\tau = L/R_{\text{eq}}$.

We now need the initial condition, which will be related to our given value $i_L(0^+) = I_0$. Using current division,

$$i_2(0^+) = -\frac{R_1}{R_1 + R_2} I_0$$

Finally, we arrive at the solution

$$i_2(t) = i_2(0^+)e^{-t/\tau} = -\frac{R_1}{R_1 + R_2} I_0 e^{-t/\tau}$$

EXAMPLE **8.6**

Find $v(0^+)$ and $i_1(0^+)$ for the circuit shown in Fig. 8.18*a* if $v(0^-) = V_0$.

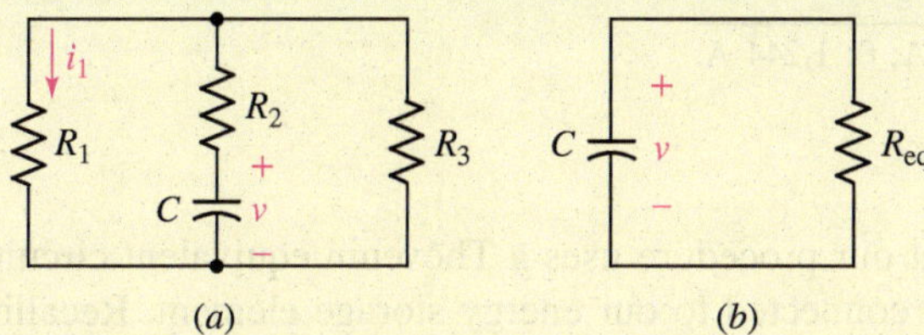

FIGURE 8.18 (*a*) A given circuit containing one capacitor and several resistors. (*b*) The resistors have been replaced by a single equivalent resistor; the time constant is simply $\tau = R_{eq}C$.

We first simplify the circuit of Fig. 8.18*a* to that of Fig. 8.18*b*, enabling us to write

$$v = V_0 e^{-t/R_{eq}C}$$

where

$$v(0^+) = v(0^-) = V_0 \qquad \text{and} \qquad R_{eq} = R_2 + \frac{R_1 R_3}{R_1 + R_3}$$

Every current and voltage in the resistive portion of the network must have the form $Ae^{-t/R_{eq}C}$, where A is the initial value of that current or voltage. Thus, the current in R_1, for example, may be expressed as

$$i_1 = i_1(0^+)e^{-t/\tau}$$

where

$$\tau = \left(R_2 + \frac{R_1 R_3}{R_1 + R_3}\right)C$$

and $i_1(0^+)$ remains to be determined from the initial condition. Any current flowing in the circuit at $t = 0^+$ must come from the capacitor. Therefore, since v cannot change instantaneously, $v(0^+) = v(0^-) = V_0$ and

$$i_1(0^+) = \frac{V_0}{R_2 + R_1R_3/(R_1 + R_3)} \frac{R_3}{R_1 + R_3}$$

PRACTICE

8.6 Find values of v_C and v_o in the circuit of Fig. 8.19 at t equal to (*a*) 0^-; (*b*) 0^+; (*c*) 1.3 ms.

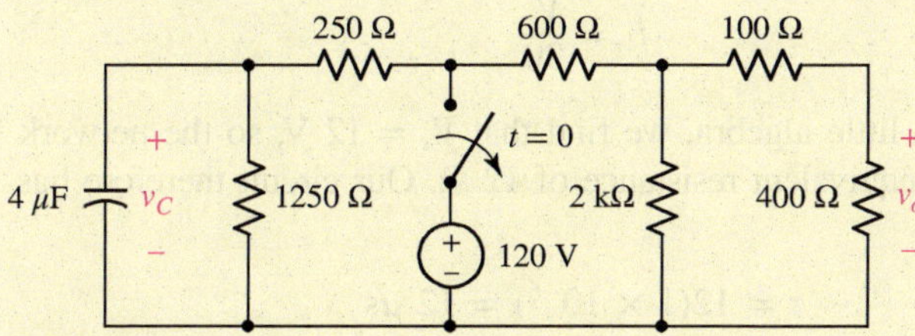

FIGURE 8.19

Ans: 100 V, 38.4 V; 100 V, 25.6 V; 59.5 V, 15.22 V.

(Continued on next page)

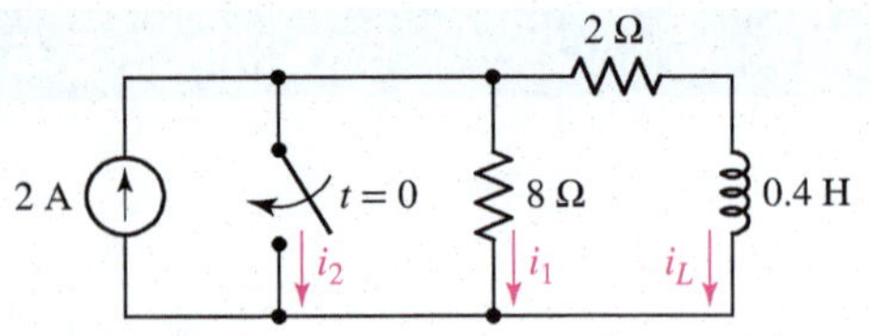

FIGURE 8.20

PRACTICE

8.7 At $t = 0.15$ s in the circuit of Fig. 8.20, find the value of (*a*) i_L; (*b*) i_1; (*c*) i_2.

Ans: 0.756 A; 0; 1.244 A.

Note that our procedure uses a Thévenin equivalent circuit to describe the network connected to our energy storage element. Recalling our prior discussions of Thévenin equivalent circuits, they can also contain *dependent sources*. Our procedure for examining *RC* and *RL* circuits similarly applies to any network that can be described by a Thévenin equivalent, including those with dependent sources. Let us examine cases in the following example and practice problem.

EXAMPLE 8.7

For the circuit of Fig. 8.21*a*, find the voltage labeled v_C for $t > 0$ if $v_C(0^-) = 2$ V.

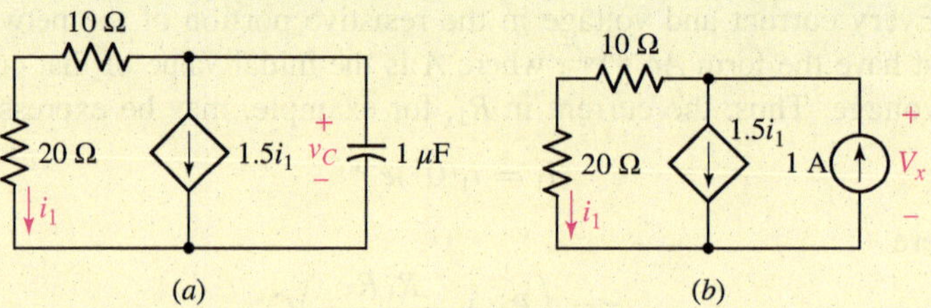

FIGURE 8.21 (*a*) A simple *RC* circuit containing a dependent source not controlled by a capacitor voltage or current. (*b*) Circuit for finding the Thévenin equivalent of the network connected to the capacitor.

The dependent source is not controlled by a capacitor voltage or current, so we can start by finding the Thévenin equivalent of the network to the left of the capacitor. Connecting a 1 A test source as in Fig. 8.21*b*,

$$\frac{V_x}{30} + 1.5i_1 = 1$$

where

$$i_1 = \frac{V_x}{30}$$

Performing a little algebra, we find that $V_x = 12$ V, so the network has a Thévenin equivalent resistance of 12 Ω. Our circuit therefore has a time constant

$$\tau = 12(1 \times 10^{-6}) = 12\ \mu\text{s}$$

The initial condition $v_C(0^+) = v_C(0^-) = 2$ V. Thus,

$$v_C(t) = 2e^{-t/12\times10^{-6}}\ \text{V} \qquad [10]$$

PRACTICE

8.8 (*a*) Regarding the circuit of Fig. 8.22, determine the voltage $v_c(t)$ for $t > 0$ if $v_c(0^-) = 11$ V. (*b*) Is the circuit "stable"?

Ans: $v_C(t) = 11e^{-2\times10^3 t/3}$ V, $t > 0$; Yes, it decays (exponentially) rather than grows with time.

FIGURE 8.22 Circuit for Practice Problem 8.8.

Some circuits containing a number of both resistors and capacitors may be replaced by an equivalent circuit containing only one resistor and one capacitor; it is necessary that the original circuit be one which can be broken into two parts, one containing all resistors and the other containing all capacitors, such that the two parts are connected by only two ideal conductors. Otherwise, multiple time constants and multiple exponential terms will be required to describe the behavior of the circuit (one time constant for each energy storage element remaining in the circuit after it is reduced as much as possible).

EXAMPLE 8.8

Determine both i_1 and i_L in the circuit shown in Fig. 8.23*a* for $t > 0$.

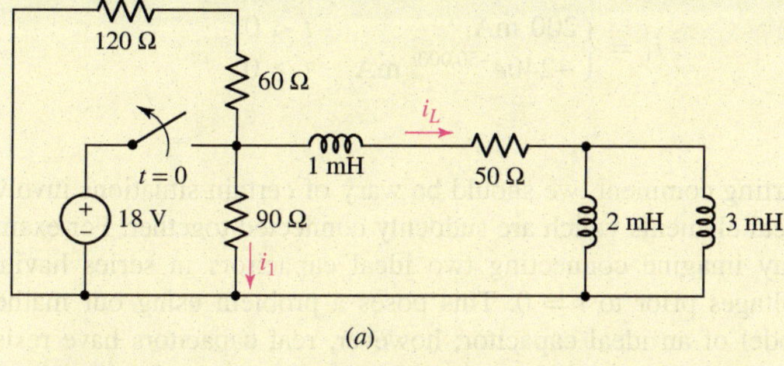

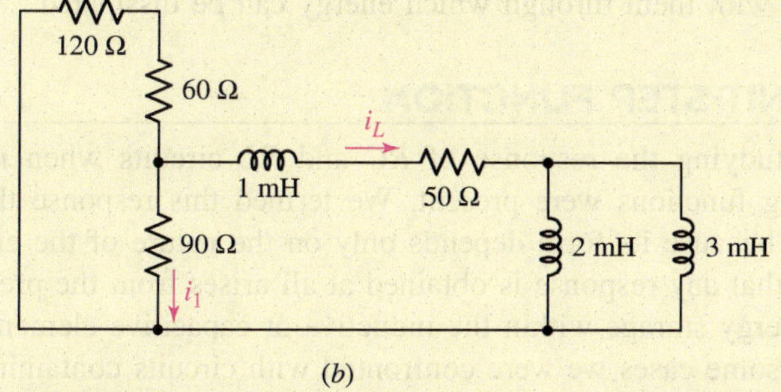

FIGURE 8.23 (*a*) A circuit with multiple resistors and inductors. (*b*) After $t = 0$, the circuit simplifies to an equivalent resistance of 110 Ω in series with $L_{eq} = 2.2$ mH.

After $t = 0$, when the voltage source is disconnected as shown in Fig. 8.23*b*, we easily calculate an equivalent inductance,

$$L_{eq} = \frac{2 \times 3}{2 + 3} + 1 = 2.2 \text{ mH}$$

(Continued on next page)

an equivalent resistance, in series with the equivalent inductance,

$$R_{\text{eq}} = \frac{90(60 + 120)}{90 + 180} + 50 = 110\ \Omega$$

and the time constant,

$$\tau = \frac{L_{\text{eq}}}{R_{\text{eq}}} = \frac{2.2 \times 10^{-3}}{110} = 20\ \mu\text{s}$$

Thus, the form of the natural response is $Ke^{-50{,}000t}$, where K is an unknown constant. Considering the circuit just prior to the switch opening ($t = 0^-$), $i_L = 18/50$ A. Since $i_L(0^+) = i_L(0^-)$, we know that $i_L = 18/50$ A or 360 mA at $t = 0^+$, and so

$$i_L = \begin{cases} 360 \text{ mA} & t < 0 \\ 360e^{-50{,}000t} \text{ mA} & t \geq 0 \end{cases}$$

There is no restriction on i_1 changing instantaneously at $t = 0$, so its value at $t = 0^-$(18/90 A or 200 mA) is not relevant to finding i_1 for $t > 0$. Instead, we must find $i_1(0^+)$ through our knowledge of $i_L(0^+)$. Using current division,

$$i_1(0^+) = -i_L(0^+)\frac{120 + 60}{120 + 60 + 90} = -240 \text{ mA}$$

Hence,

$$i_1 = \begin{cases} 200 \text{ mA} & t < 0 \\ -240e^{-50{,}000t} \text{ mA} & t \geq 0 \end{cases}$$

As a parting comment, we should be wary of certain situations involving only ideal elements which are suddenly connected together. For example, we may imagine connecting two ideal capacitors in series having unequal voltages prior to $t = 0$. This poses a problem using our mathematical model of an ideal capacitor; however, real capacitors have resistances associated with them through which energy can be dissipated.

8.5 THE UNIT-STEP FUNCTION

We have been studying the response of *RC* and *RL* circuits when no sources or forcing functions were present. We termed this response the *natural response* because its form depends only on the nature of the circuit. The reason that any response is obtained at all arises from the presence of initial energy storage within the inductive or capacitive elements in the circuit. In some cases we were confronted with circuits containing sources and switches; we were informed that certain switching operations were performed at $t = 0$ in order to remove sources from the circuit, while leaving known amounts of energy stored here and there. In other words, we have been solving problems in which energy sources are suddenly *removed* from the circuit; now we must consider that type of response which results when energy sources are suddenly *applied* to a circuit.

We will focus on the response that occurs when the energy sources suddenly applied are dc sources. Since every electrical device is intended to be energized at least once, and since most devices are turned on and off many times in the course of their lifetimes, our study applies to many practical cases. Even though we are now restricting ourselves to dc sources, there are still many cases in which these simpler examples correspond to the operation of physical devices. For example, the first circuit we will analyze could represent the buildup of the current when a dc motor is started. The generation and use of the rectangular voltage pulses needed to represent a number or a command in a microprocessor provide many examples in the field of electronic or transistor circuitry. Similar circuits are found in many other instances, including communications systems using pulse modulation, and in radar equipment.

We have been speaking of the "sudden application" of an energy source, and by this phrase we imply its application in zero time.[1] The operation of a switch in series with a battery is thus equivalent to a forcing function which is zero up to the instant that the switch is closed and is equal to the battery voltage thereafter. The forcing function has a break, or discontinuity, at the instant the switch is closed. Certain special forcing functions which are discontinuous or have discontinuous derivatives are called ***singularity functions,*** the two most important of these singularity functions being the ***unit-step function*** and the ***unit-impulse function.***

We define the unit-step forcing function as a function of time which is zero for all values of its argument less than zero and which is unity for all positive values of its argument. If we let $(t - t_0)$ be the argument and represent the unit-step function with u, then $u(t - t_0)$ must be zero for all values of t less than t_0, and it must be unity for all values of t greater than t_0. At $t = t_0$, $u(t - t_0)$ changes *abruptly* from 0 to 1. Its value at $t = t_0$ is not defined, but its value is known for all instants of time that are arbitrarily close to $t = t_0$. We often indicate this by writing $u(t_0^-) = 0$ and $u(t_0^+) = 1$. The concise mathematical definition of the unit-step forcing function is

$$u(t - t_0) = \begin{cases} 0 & t < t_0 \\ 1 & t > t_0 \end{cases}$$

and the function is shown graphically in Fig. 8.24. Note that a vertical line of unit length is shown at $t = t_0$. Although this "riser" is not strictly a part of the definition of the unit step, it is usually shown in each drawing.

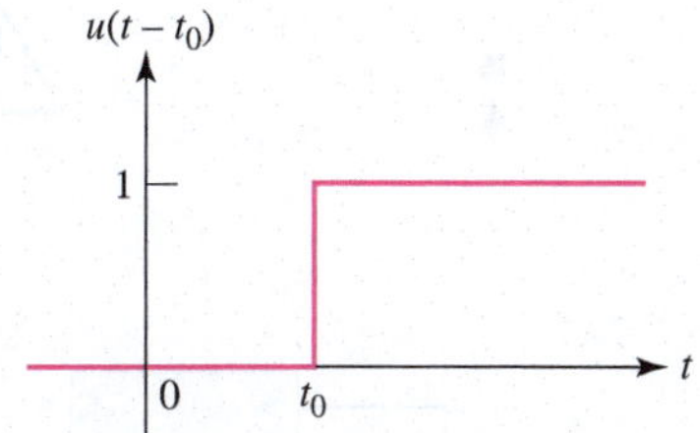

FIGURE 8.24 The unit-step forcing function, $u(t - t_0)$.

We also note that the unit step need not be a time function. For example, $u(x - x_0)$ could be used to denote a unit-step function where x might be a distance in meters, for example, or a frequency.

Very often in circuit analysis a discontinuity or a switching action takes place at an instant that is defined as $t = 0$. In that case $t_0 = 0$, and we then represent the corresponding unit-step forcing function with $u(t - 0)$, or more simply $u(t)$. This is shown in Fig. 8.25. Thus

$$u(t) = \begin{cases} 0 & t < 0 \\ 1 & t > 0 \end{cases}$$

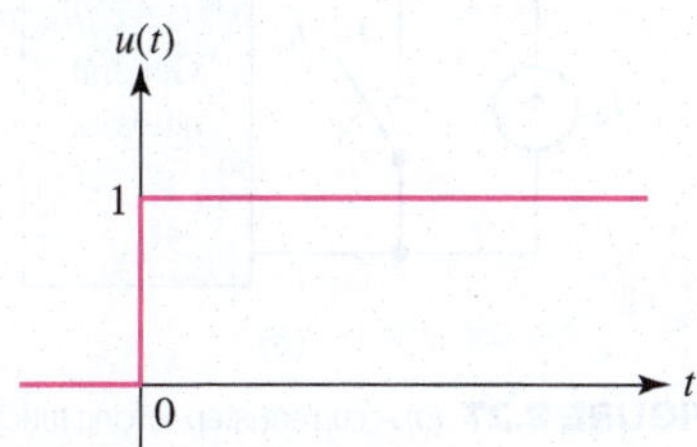

FIGURE 8.25 The unit-step forcing function $u(t)$ is shown as a function of t.

The unit-step forcing function is in itself dimensionless. If we wish it to represent a voltage, it is necessary to multiply $u(t - t_0)$ by some constant voltage, such as 5 V. Thus, $v(t) = 5u(t - 0.2)$ V is an ideal voltage source which is zero before $t = 0.2$ s and a constant 5 V after $t = 0.2$ s. This forcing function is shown connected to a general network in Fig. 8.26*a*.

(1) Of course, this is not physically possible. However, if the time scale over which such an event occurs is very short compared to all other relevant time scales that describe the operation of a circuit, this is approximately true and mathematically convenient.

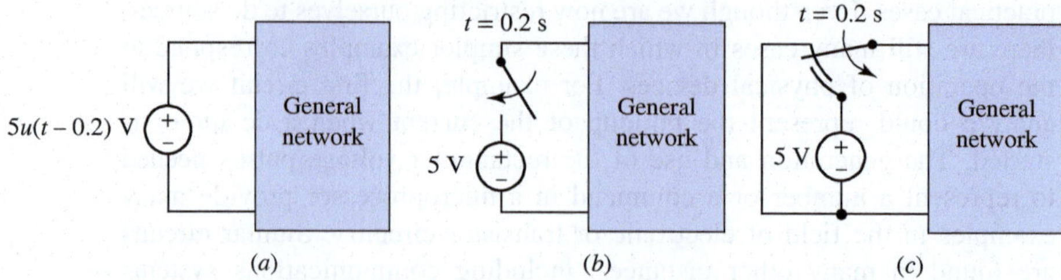

FIGURE 8.26 (*a*) A voltage-step forcing function is shown as the source driving a general network. (*b*) A simple circuit which, although not the exact equivalent of part (*a*), may be used as its equivalent in many cases. (*c*) An exact equivalent of part (*a*), using what is often called a "make before break" switch.

Physical Sources and the Unit-Step Function

Perhaps we should ask what physical source is the equivalent of this discontinuous forcing function. By equivalent, we mean simply that the voltage–current characteristics of the two networks are identical. For the step-voltage source of Fig. 8.26*a*, the voltage–current characteristic is simple: the voltage is zero prior to $t = 0.2$ s, it is 5 V after $t = 0.2$ s, and the current may be any (finite) value in either time interval. Our first thoughts might produce the attempt at an equivalent shown in Fig. 8.26*b*, a 5 V dc source in series with a switch which closes at $t = 0.2$ s. This network is not equivalent for $t < 0.2$ s, however, because the voltage across the battery and switch is completely unspecified in this time interval. The "equivalent" source is an open circuit, and the voltage across it *may be anything*. After $t = 0.2$ s, the networks are equivalent, and if this is the only time interval in which we are interested, and if the initial currents which flow from the two networks are identical at $t = 0.2$ s, then Fig. 8.26*b* becomes a useful equivalent of Fig. 8.26*a*.

In order to obtain an exact equivalent for the voltage-step forcing function, we may provide a single-pole double-throw switch. Before $t = 0.2$ s, the switch serves to ensure zero voltage across the input terminals of the general network. After $t = 0.2$ s, the switch is thrown to provide a constant input voltage of 5 V. At $t = 0.2$ s, the voltage is indeterminate (as is the step forcing function), and the battery is momentarily short-circuited (it is fortunate that we are dealing with mathematical models!). This exact equivalent of Fig. 8.26*a* is shown in Fig. 8.26*c*.

Figure 8.27*a* shows a current-step forcing function driving a general network. If we try to replace this circuit with a dc source in parallel with a switch (which opens at $t = t_0$), we must realize that the circuits are equivalent after $t = t_0$ but that the responses after $t = t_0$ are alike only if the initial conditions are the same. The circuit in Fig. 8.27*b* implies that no voltage exists across the current source terminals for $t < t_0$. This is not the case for the circuit of Fig. 8.27*a*. However, we may often use the circuits of Fig. 8.27*a* and *b* interchangeably. The exact equivalent of Fig. 8.27*a* is, in fact, the dual of the circuit of Fig. 8.26*c*; the exact equivalent of Fig. 8.27*b* cannot be constructed with current- and voltage-step forcing functions alone.[2]

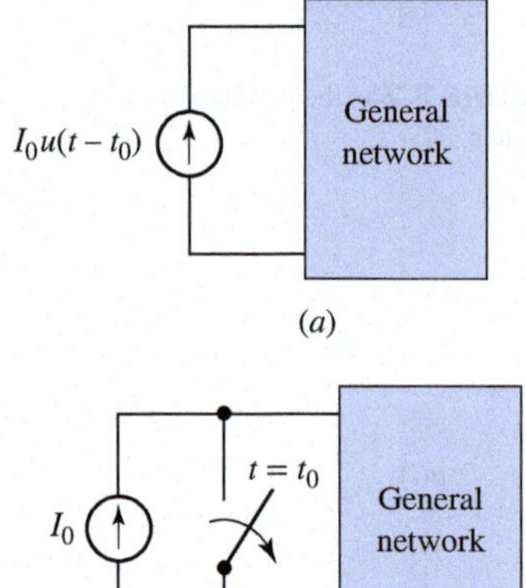

FIGURE 8.27 (*a*) A current-step forcing function is applied to a general network. (*b*) A simple circuit which, although not the exact equivalent of part (*a*), may be used as its equivalent in many cases.

(2) The equivalent can be drawn if the current through the switch prior to $t = t_0$ is known.

The Rectangular Pulse Function

Some very useful forcing functions may be obtained by manipulating the unit-step forcing function. Let us define a rectangular voltage pulse by the following conditions:

$$u(t) = \begin{cases} 0 & t < t_0 \\ V_0 & t_0 < t < t_1 \\ 0 & t > t_1 \end{cases}$$

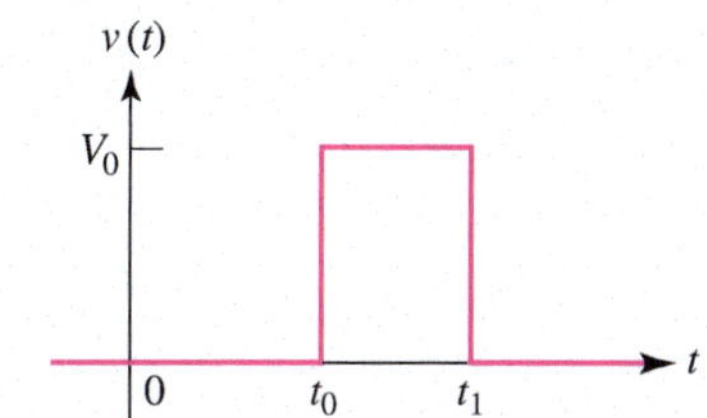

FIGURE 8.28 A useful forcing function, the rectangular voltage pulse.

The pulse is drawn in Fig. 8.28. Can this pulse be represented in terms of the unit-step forcing function? Let us consider the difference of the two unit steps, $u(t - t_0) - u(t - t_1)$. The two step functions are shown in Fig. 8.29*a*, and their difference is a rectangular pulse. The source $V_0 u(t - t_0) - V_0 u(t - t_1)$ which provides us with the desired voltage is indicated in Fig. 8.29*b*.

If we have a sinusoidal voltage source $V_m \sin \omega t$ which is suddenly connected to a network at $t = t_0$, then an appropriate voltage forcing function would be $v(t) = V_m u(t - t_0) \sin \omega t$. If we wish to represent one burst of energy from the transmitter for a radio-controlled car operating at 47 MHz (295 Mrad/s), we may turn the sinusoidal source off 70 ns later by a second unit-step forcing function.[3] The voltage pulse is thus

$$v(t) = V_m[u(t - t_0) - u(t - t_0 - 7 \times 10^{-8})] \sin(295 \times 10^6 t)$$

which is sketched in Fig. 8.30.

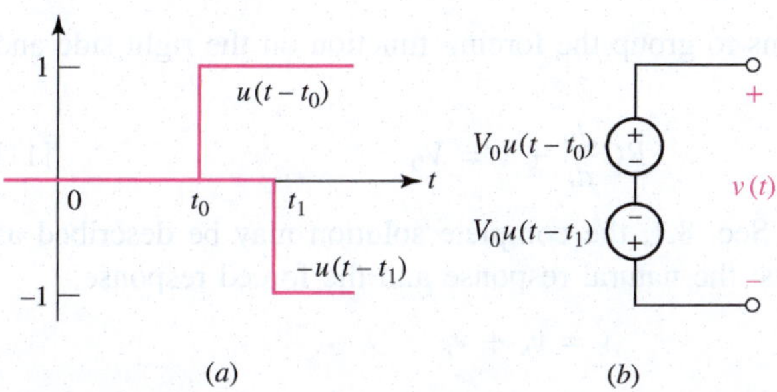

FIGURE 8.29 (*a*) The unit steps $u(t - t_0)$ and $-u(t - t_1)$. (*b*) A source which yields the rectangular voltage pulse of Fig. 8.28.

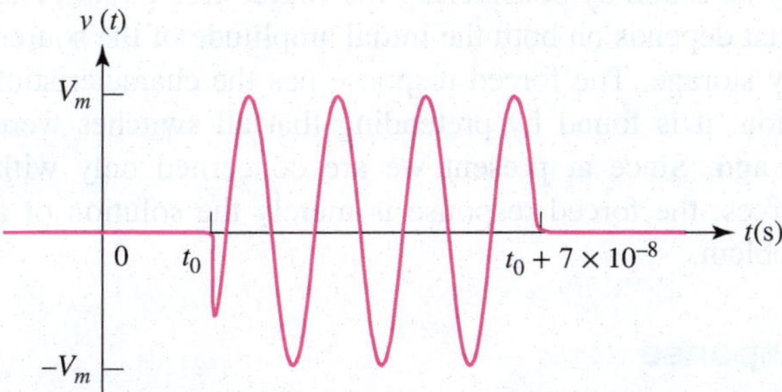

FIGURE 8.30 A 47 MHz radio-frequency pulse, described by $v(t) = V_m[u(t - t_0) - u(t - t_0 - 7 \times 10^{-8})] \sin(259 \times 10^6 t)$.

(3) Apparently, we're pretty good at the controls of this car. A reaction time of 70 ns?

PRACTICE

8.9 Evaluate each of the following at $t = 0.8$: (*a*) $3u(t) - 2u(-t) + 0.8u(1 - t)$; (*b*) $[4u(t)]u(-t)$; (*c*) $2u(t)\sin \pi t$.

Ans: 3.8; 0; 1.176.

8.6 DRIVEN *RC* CIRCUITS

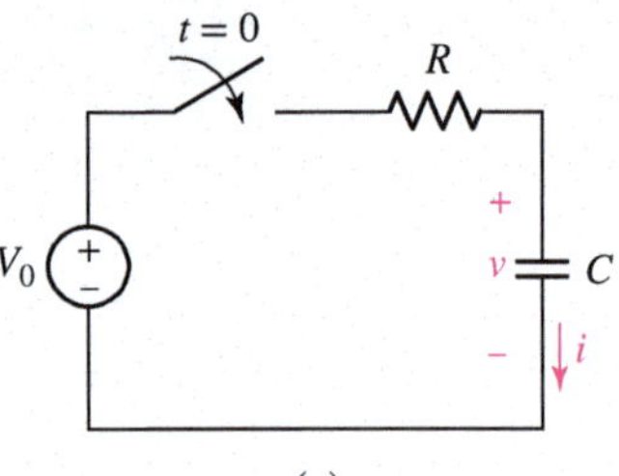

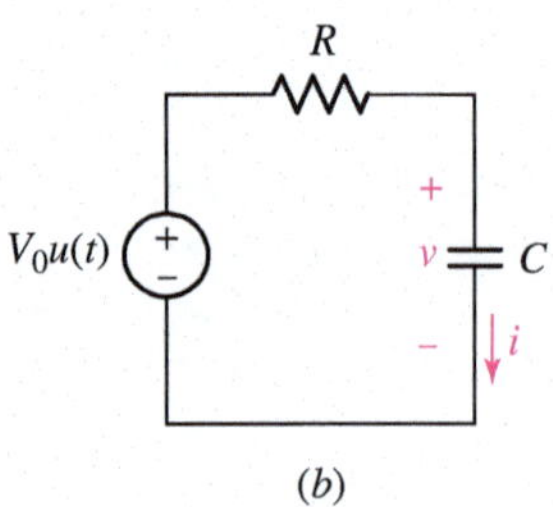

■ **FIGURE 8.31** (*a*) The given circuit. (*b*) A functionally equivalent circuit, possessing the same response $v(t)$ in the case of an initially uncharged capacitor. Note that although both circuits yield zero current for $t < 0$, technically they are not identical, since the circuit in (*a*) has an open circuit for $t < 0$, whereas for the situation in (*b*), the resistor and capacitor are connected in series for $t < 0$.

We are now ready to analyze a circuit with the sudden application of a dc source. The circuit consists of a battery whose voltage is V_0 in series with a switch, a resistor R, and a capacitor C. The switch is closed at $t = 0$, as indicated on the circuit diagram of Fig 8.31. It is evident that the current $i(t)$ is zero before $t = 0$, and assuming that the capacitor is not charged prior to $t = 0$, $v(t)$ is also zero before $t = 0$. We can replace the battery and switch by a voltage-step forcing function $V_0u(t)$, which also produces no response prior to $t = 0$. After $t = 0$, the two circuits are clearly identical. Hence, we seek the voltage $v(t)$ either in the given circuit of Fig. 8.31*a* or in the equivalent circuit of Fig. 8.31*b*.

We will find $v(t)$ by writing the appropriate circuit equation and then by solving the differential equation. Applying KCL to the resistor and capacitor currents we get

$$C\frac{dv}{dt} + \frac{v - V_0}{R} = 0 \text{ (for } t > 0)$$

Rearranging terms to group the forcing function on the right side and multiplying by R,

$$RC\frac{dv}{dt} + v = V_0 \qquad [11]$$

As described in Sec. 8.1, the complete solution may be described as the sum of two parts: the natural response and the forced response,

$$v = v_n + v_f$$

where v is the complete response, v_n is the natural response, and v_f is the forced response.

The natural response is a characteristic of the circuit and not of the sources. Its form may be found by considering the source-free circuit, and it has an amplitude that depends on both the initial amplitude of the source and the initial energy storage. The forced response has the characteristics of the forcing function; it is found by pretending that all switches were thrown a long time ago. Since at present we are concerned only with switches and dc sources, the forced response is merely the solution of a simple dc circuit problem.

The Natural Response

The natural response is the solution for the case where there is no source or forcing function. Removing the source term from Eq. [11] gives

$$RC\frac{dv}{dt} + v = 0$$

Our analyses of RC (and RL) circuits thus far have all been source-free, where we have already found the natural response

$$v_n = Ae^{-t/RC}$$

Since the complete solution depends on both the natural response and the forced response, the amplitude A may not be the same as the initial value. We should only determine the unknown constant A after finding the complete solution and applying the initial conditions of the circuit.

The Forced Response

We can use a similar approach to the general solution procedure in Sec. 8.1 to find the forced response. In this approach we assume or "guess" a solution and apply it to the differential equation. In this case, we have a forcing function that is a constant, so let us assume that the forced solution will also be a constant.

$$v_f = K$$

Substituting the forced solution into the differential equation (Eq. [11]) gives

$$RC\frac{d(K)}{dt} + K = V_0$$

Since the constant K is not time dependent, we arrive at

$$v_f = K = V_0$$

We should see that the forced response might have been obtained without evaluating the differential equation because it must be the complete response at infinite time. The forced response is thus obtained by inspection of the final circuit after the natural response has died out. However, the technique outlined may be applied to more complicated forcing functions, which may also have a time dependence.

Determination of the Complete Response

The complete solution can be obtained by adding the natural and forced solutions, and then applying initial conditions to determine unknown constants. The complete solution is given by

$$v = v_n + v_f = Ae^{-t/RC} + V_0 \qquad [12]$$

For $t < 0$, the voltage source was zero, and the corresponding capacitor voltage is $v(0^-) = 0$. Since the capacitor voltage cannot change instantaneously, we also know that $v(0^+) = v(0^-) = 0$. Substituting in our complete response for $t > 0$

$$v(0^+) = Ae^{-(0)/RC} + V_0 = 0$$

$$A = -V_0$$

Substituting the value for A into Eq. [12]

$$v = -V_0e^{-t/RC} + V_0$$

or rearranging,

$$v = V_0(1 - e^{-t/RC})$$

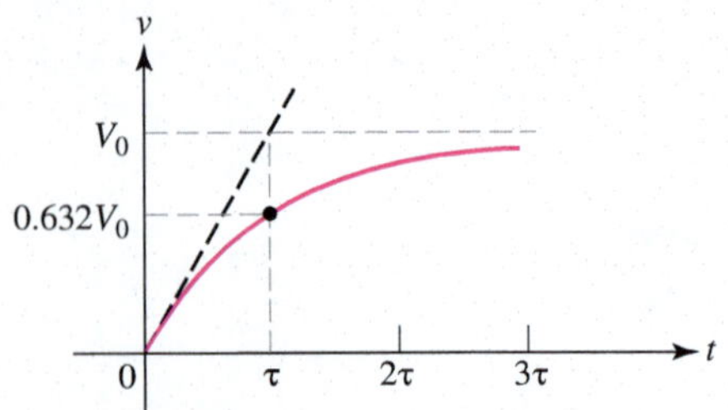

FIGURE 8.32 The current flowing through the inductor of Fig. 8.31 is shown graphically. A line extending the initial slope meets the constant forced response at $t = \tau$.

A plot of the result is shown in Fig. 8.32, where we see that the capacitor voltage builds up from its initial value of zero to its final value of V_0, with a transient response that has a time constant $\tau = RC$. The transition is effectively accomplished in a time 3τ. In one time constant, the voltage has attained 63.2% of its final value.

Now suppose that the capacitor in the preceding circuit has an initial value that is nonzero, and then it switches at time $t = 0$ to a larger value, as shown in Fig. 8.33. In this case, the capacitor is switching between two different voltage sources. Mathematically, the circuit could also be described by a single voltage source of $2 + 3u(t)$ V. Physical intuition tells us that the capacitor would begin with an initial value $v_C(0^+) = v_C(0^-) = 2$ V and charge to a final value of $v_C(\infty) = 5$ V according to a time constant. Let us confirm this mathematically.

We can find the solution using the same procedure as above, but writing variables in more general terms of $v_C(0^+)$ and $v_C(\infty)$. The natural solution is the same as before,

$$v_{Cn} = Ae^{-t/RC}$$

while the forced solution may be written more generally as

$$v_{Cf} = v_C(\infty)$$

to describe the final voltage after the natural response has died out.

The complete solution is given by

$$v_C = v_{Cn} + v_{Cf} = Ae^{-t/RC} + v_C(\infty)$$

where we again need to find the unknown constant A according to the initial conditions. Evaluating at $t = 0^+$,

$$v_C(0^+) = Ae^{-(0)/RC} + v_C(\infty) = A + v_C(\infty)$$

Thus,

$$A = v_C(0^+) - v_C(\infty)$$

and

$$v_C = [v_C(0^+) - v_C(\infty)]e^{-t/RC} + v_C(\infty)$$

Substituting numerically, we get

$$v_C = [2 - 5]e^{-t/(120\times5\times10^{-3})} + 5 = 5 - 3e^{-t/(0.6)} \text{ V}$$

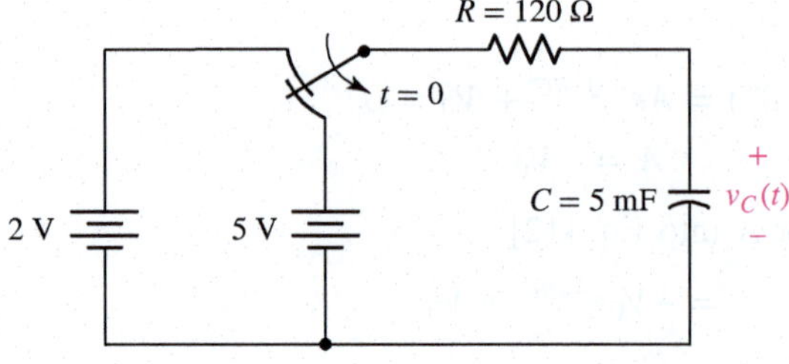

FIGURE 8.33 Capacitor circuit switching between two different voltage supplies.

The voltage response for this circuit is plotted in Fig 8.34. Analyzing the solution, we see that the capacitor voltage increases from 2 V at $t = 0$ to a final value of 5 V according to an exponential relation with time constant $\tau = RC = 0.6$ s.

Recall that a circuit with a single capacitor (or equivalent capacitance C_{eq}) will require that *every voltage and current throughout the circuit must follow the same functional behavior*. We can therefore use a general procedure to find the current and voltage anywhere for the step response of an *RC* circuit—without repeating the solution to the differential equation!

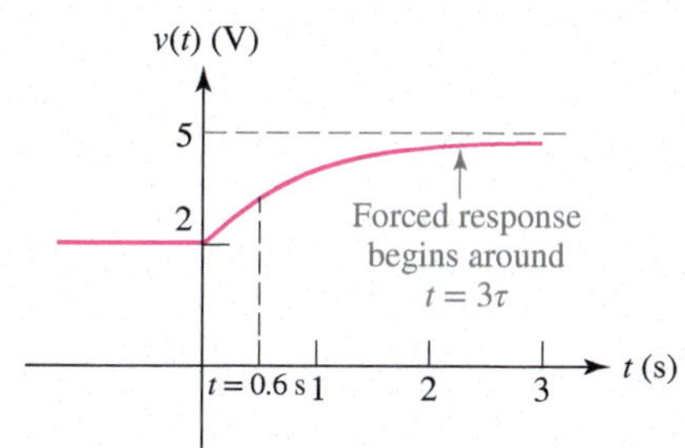

FIGURE 8.34 Voltage response for the circuit in Fig. 8.33.

Procedure for Step Response of *RC* Circuit

1. With all independent sources zeroed out, simplify the circuit to determine R_{eq} and C_{eq}, and the time constant $\tau_{eq} = R_{eq}C_{eq}$.
2. Determine the initial condition $v(0^+)$ or $i(0^+)$ [recall the requirement that for any capacitor voltage $v_c(0^-) = v_c(0^+)$].
3. Determine the final condition $v(\infty)$ or $i(\infty)$.
4. The final response is given by

$$v = v(\infty) + [v(0^+) - v(\infty)]e^{-t/\tau}$$

or

$$i = i(\infty) + [i(0^+) - i(\infty)]e^{-t/\tau}$$

EXAMPLE **8.9**

Find the capacitor voltage $v_C(t)$ and the current $i(t)$ in the 200 Ω resistor of Fig. 8.35*a* for all time.

We begin by examining the circuit for $t > 0$ to find R_{eq}, C_{eq} and the time constant. We only have a single capacitor, given by $C = 50$ mF. Replacing the 50 V source with a short circuit and evaluating the equivalent resistance (the Thévenin equivalent resistance "seen" by the capacitor) yields

$$R_{eq} = \frac{1}{\frac{1}{50} + \frac{1}{200} + \frac{1}{60}} = 24\ \Omega$$

and thus

$$\tau = R_{eq}C = (24)(0.050) = 1.2\text{ s}$$

To determine the initial conditions, we consider the state of the circuit at $t < 0$, corresponding to the switch at position *a* as represented in Fig. 8.35*b*. As usual, we assume no transients are present, so only a forced response due to the 120 V source is relevant to finding $v_c(0^-)$. Simple voltage division then gives us the initial voltage,

$$v_C(0) = \frac{50}{50 + 10}(120) = 100\text{ V}$$

(Continued on next page)

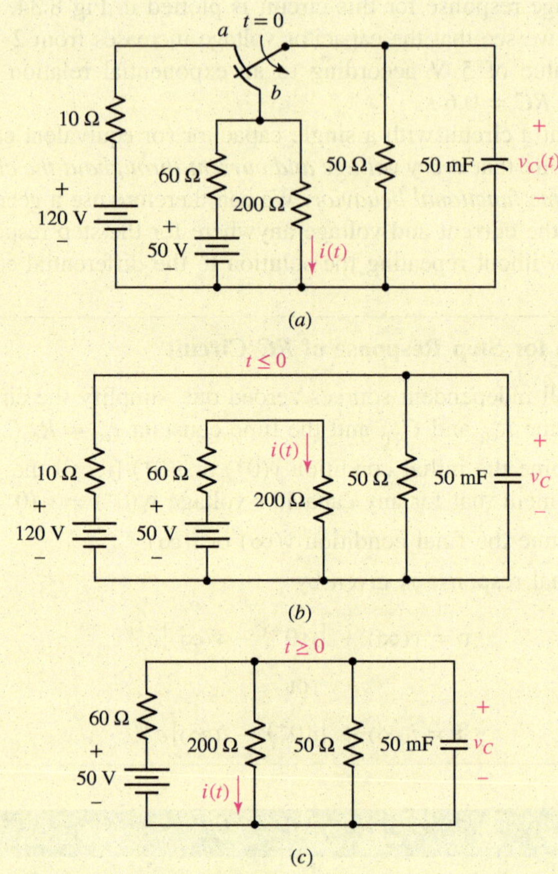

FIGURE 8.35 (*a*) An *RC* circuit in which the complete responses v_C and i are obtained by adding a forced response and a natural response. (*b*) Circuit for $t \leq 0$. (*c*) Circuit for $t \geq 0$.

Since the capacitor voltage cannot change instantaneously, this voltage is equally valid at $t = 0^-$ and $t = 0^+$.

The corresponding circuit for $t > 0$ has been redrawn in Fig. 8.35*c* for convenience. In order to evaluate the forced response with the switch at *b*, we wait until all the voltages and currents have stopped changing—thus treating the capacitor as an open circuit—and use voltage division once more:

$$v_C(\infty) = 50\left(\frac{200 \parallel 50}{60 + 200 \parallel 50}\right)$$

$$= 50\left(\frac{(50)(200)/250}{60 + (50)(200)/(250)}\right) = 20 \text{ V}$$

Consequently,

$$v_C = v_C(\infty) + [v_C(0^+) - v_C(\infty)]e^{-t/\tau}$$

$$v_C = v_C = 20 + (100 - 20)e^{-t/1.2} \text{ V}$$

or

$$v_C = 20 + 80e^{-t/1.2} \text{ V} \qquad t \geq 0$$

and

$$v_C = 100 \text{ V} \qquad t < 0$$

This response is sketched in Fig. 8.36*a*; again the natural response is seen to form a transition from the initial to the final response.

Next we consider $i(t)$. This specific response need not remain constant during the instant of switching. With the contact at a, it is evident that $i(0^-) = 50/260 = 192.3$ mA. We also need to know $i(0^+)$ when the contact switches to position b, which may be found by fixing our attention on the energy-storage element (the capacitor). The fact that v_c must remain 100 V during the switching interval is the governing condition which establishes the other currents and voltages at $t = 0^+$. Since $v_c(0^+) = 100$ V, and since the capacitor is in parallel with the 200 Ω resistor, we find $i(0^+) = 0.5$ A. When the switch moves to position b, the forced response for this current becomes

$$i(\infty) = \frac{50}{60 + (50)(200)/(50 + 200)}\left(\frac{50}{50 + 200}\right) = 0.1 \text{ A}$$

Combining the forced and natural responses, we obtain

$$i = i(\infty) + [i(0^+) - i(\infty)]e^{-t/\tau} \text{ A}$$

$$i = 0.1 + [0.5 - 0.1]e^{-t/1.2} \text{ A}$$

and thus

$$i(t) = 0.1923 \text{ A} \qquad t < 0$$

$$i(t) = 0.1 + 0.4e^{-t/1.2} \text{ A} \qquad t > 0$$

or

$$i(t) = 0.1923 + (-0.0923 + 0.4e^{-t/1.2})u(t) \text{ A}$$

where the last expression is correct for all t.

The complete response for all t may also be written concisely by using $u(-t)$, which is unity for $t < 0$ and 0 for $t > 0$. Thus,

$$i(t) = 0.1923u(-t) + (0.1 + 0.4e^{-t/1.2})u(t) \text{ A}$$

This response is sketched in Fig. 8.36*b*. Note that only four numbers are needed to write the functional form of the response for this single-energy-storage-element circuit, or to prepare the sketch: the constant value prior to switching (0.1923 A), the instantaneous value just after switching (0.5 A), the constant forced response (0.1 A), and the time constant (1.2 s). The appropriate negative exponential function is then easily written or drawn.

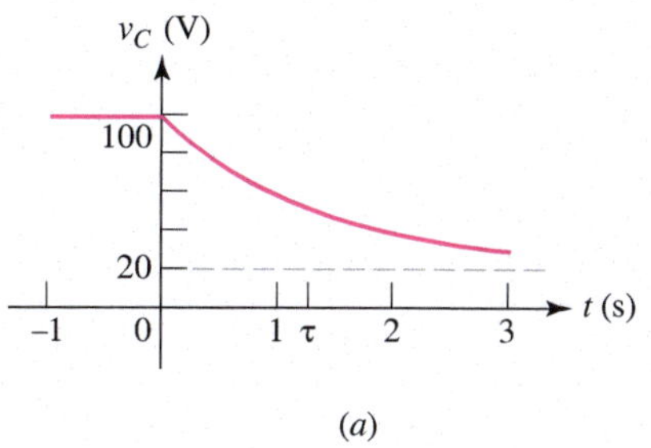

(*a*)

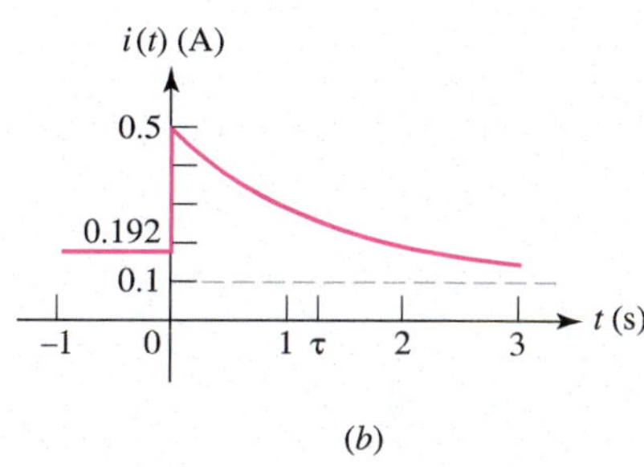

(*b*)

FIGURE 8.36 The responses (*a*) v_C and (*b*) i are plotted as functions of time for the circuit of Fig. 8.35.

PRACTICE

8.10 For the circuit of Fig. 8.37, find $v_C(t)$ at t equal to (*a*) 0^-; (*b*) 0^+; (*c*) ∞; (*d*) 0.08 s.

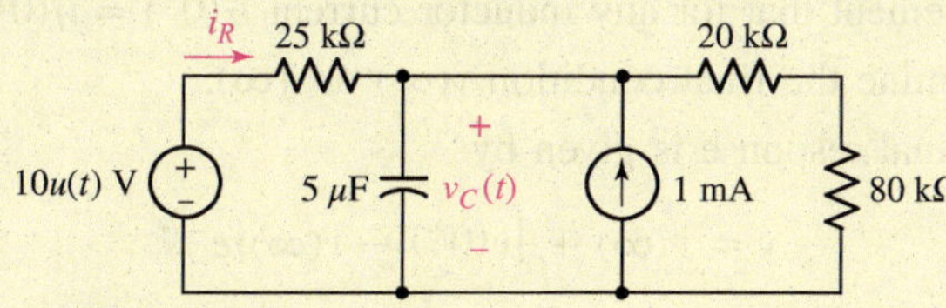

FIGURE 8.37

Ans: 20 V; 20 V; 28 V; 24.4 V.

Developing an Intuitive Understanding

The reason for the two responses, forced and natural, may be seen from physical arguments. We know that our circuit will eventually assume the forced response. However, at the instant the switches are thrown, the initial capacitor voltages (or, in *RL* circuits, the currents through the inductors) will have values that depend only on the energy stored in these elements. These currents or voltages cannot be expected to be the same as the currents and voltages demanded by the forced response. Hence, there must be a transient period during which the currents and voltages change from their given initial values to their required final values. The portion of the response that provides the transition from initial to final values is the natural response (often called the *transient* response, as we found earlier). If we describe the response of the simple source-free *RC* circuit in these terms, then we should say that the forced response is zero and that the natural response serves to connect the initial response dictated by the stored energy with the zero value of the forced response.

This description is appropriate only for those circuits in which the natural response eventually dies out. This always occurs in physical circuits where some resistance is associated with every element, but there are a number of "pathologic" circuits in which the natural response is nonvanishing as time becomes infinite. Those circuits in which trapped currents circulate around inductive loops, or voltages are trapped in series strings of capacitors, are examples. We should note that in theory, at least, a persistent dc current can be induced in a superconductor loop. There may be a dual/analog of such a situation for capacitors—an interesting avenue of research!

8.7 DRIVEN *RL* CIRCUITS

The complete response of any *RL* circuit may also be obtained using a virtually identical procedure to what we have already discussed in detail for *RC* circuits. The primary distinction, as in the case of the unforced solution for *RC* and *RL* circuits, is the difference in time constant and the condition for inductors that $i_L(0^-) = i_L(0^+)$. The general procedure can be outlined as

Procedure for Step Response of *RL* Circuit

1. With all independent sources zeroed out, simplify the circuit to determine R_{eq} and L_{eq}, and the time constant $\tau_{eq} = L_{eq}/R_{eq}$.
2. Determine the initial condition $v(0^+)$ or $i(0^+)$ [recall the requirement that for any inductor current $i_L(0^-) = i_L(0^+)$].
3. Determine the final condition $v(\infty)$ or $i(\infty)$.
4. The final response is given by

$$v = v(\infty) + [v(0^+) - v(\infty)]e^{-t/\tau}$$

or

$$i = i(\infty) + [i(0^+) - i(\infty)]e^{-t/\tau}$$

EXAMPLE 8.10

Determine $i(t)$ for all values of time in the circuit of Fig. 8.38.

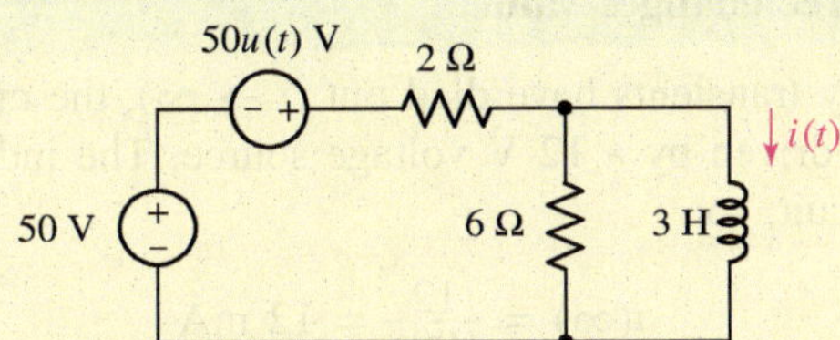

FIGURE 8.38 The circuit of Example 8.10.

The circuit contains a dc voltage source as well as a step-voltage source. We might choose to replace everything to the left of the inductor with a Thévenin equivalent, but instead let us merely recognize the form of that equivalent as a resistor in series with some voltage source. The circuit contains only one energy storage element, the inductor. We first note that

$$\tau = \frac{L}{R_{eq}} = \frac{3}{1.5} = 2 \text{ s}$$

Prior to $t = 0$, the 50 V source is dropped across the 2 Ω resistor (note that the 6 Ω resistor is short-circuited by the inductor). The resulting current through the inductor is $i(0) = 25$ A, and it cannot change instantaneously. Similarly, the forced response after a long time will have a total voltage drop of 100 V across the 2 Ω resistor, resulting in $i(\infty) = 50$ A. Thus,

$$i = i(\infty) + [i(0^+) - i(\infty)]e^{-t/\tau}$$

$$i = 50 + (25 - 50)e^{-0.5t} \text{ A} \qquad t > 0$$

or

$$i = 50 - 25e^{-0.5t} \text{ A} \qquad t > 0$$

We complete the solution by also stating

$$i = 25 \text{ A} \qquad t < 0$$

or by writing a single expression valid for all t,

$$i = 25 + 25(1 - e^{-0.5t})u(t) \text{ A}$$

The complete response is sketched in Fig. 8.39. Note how the natural response serves to connect the response for $t < 0$ with the constant forced response.

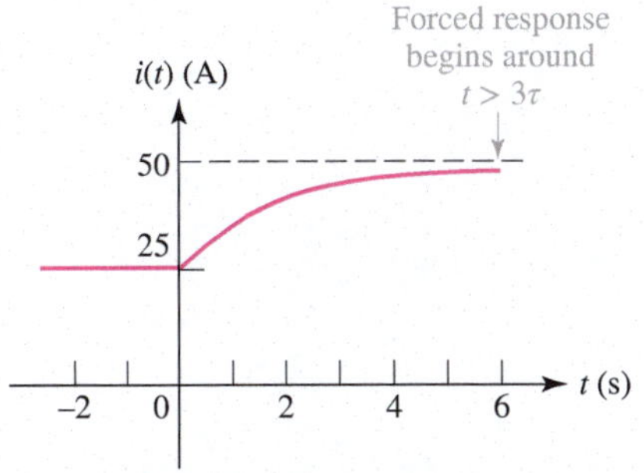

FIGURE 8.39 The response $i(t)$ of the circuit shown in Fig. 8.38 is sketched for values of times less and greater than zero.

PRACTICE

8.11 A voltage source, $v_s = 20u(t)$ V, is in series with a 200 Ω resistor and a 4 H inductor. Find the magnitude of the inductor current at t equal to (*a*) 0^-; (*b*) 0^+; (*c*) 8 ms; (*d*) 15 ms.

Ans: 0; 0; 32.97 mA; 52.76 mA

EXAMPLE 8.11

For the circuit of Fig. 8.40, find $i(t)$ for $t = \infty$, 3^-, 3^+, and 100 μs after the source changes value.

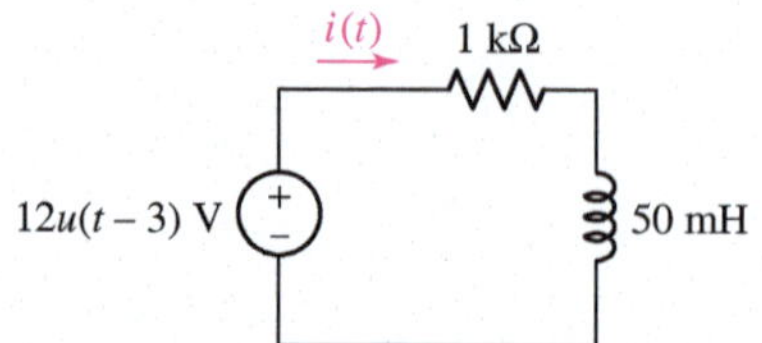

■ **FIGURE 8.40** A simple *RL* circuit driven by a voltage-step forcing function.

Long after any transients have died out ($t \to \infty$), the circuit is a simple dc circuit driven by a 12 V voltage source. The inductor appears as a short circuit, so

$$i(\infty) = \frac{12}{1000} = 12 \text{ mA}$$

What is meant by $i(3^-)$? This is simply a notational convenience to indicate the instant before the voltage source changes value. For $t < 3$, $u(t - 3) = 0$. Thus, $i(3^-) = 0$ as well.

At $t = 3^+$, the forcing function $12u(t - 3) = 12$ V. However, since the inductor current cannot change in zero time, $i(3^+) = i(3^-) = 0$.

The most straightforward approach to analyzing the circuit for $t > 3$ s is to write our solution as

$$i(t') = \left(\frac{V_0}{R} - \frac{V_0}{R}e^{-Rt'/L}\right) u(t')$$

where the variable t' represents a shift in the time axis such that

$$t' = t - 3$$

Therefore, with $V_0/R = 12$ mA and $R/L = 20{,}000 \text{ s}^{-1}$,

$$i(t - 3) = (12 - 12e^{-20{,}000(t-3)})\, u(t - 3) \text{ mA} \qquad [13]$$

which can be written more simply as

$$i(t) = (12 - 12e^{-20{,}000(t-3)})\, u(t - 3) \text{ mA} \qquad [14]$$

since the unit-step function forces a zero value for $t < 3$, as required. Substituting $t = 3.0001$ s into Eq. [13] or [14], we find that $i = 10.38$ mA at a time 100 μs after the source changes value.

PRACTICE

8.12 The voltage source $60 - 40u(t)$ V is in series with a 10 Ω resistor and a 50 mH inductor. Find the magnitudes of the inductor current and voltage at t equal to (*a*) 0^-; (*b*) 0^+; (*c*) ∞; (*d*) 3 ms.

Ans: 6 A, 0 V; 6 A, 40 V; 2 A, 0 V; 4.20 A, 22.0 V.

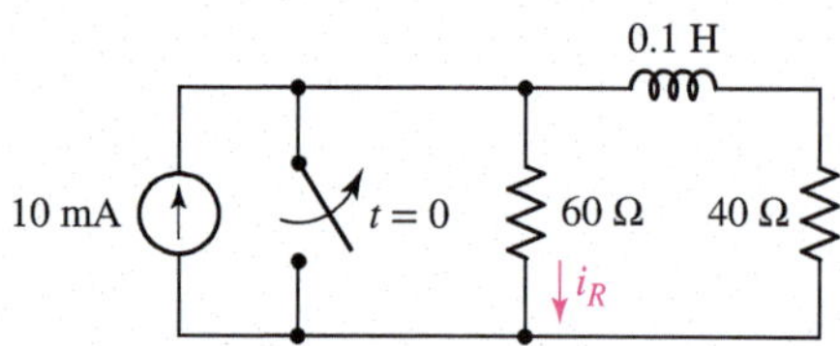

■ **FIGURE 8.41**

PRACTICE

8.13 The circuit shown in Fig. 8.41 has been in the form shown for a very long time. The switch opens at $t = 0$. Find i_R at t equal to (*a*) 0^-; (*b*) 0^+; (*c*) ∞; (*d*) 1.5 ms.

Ans: 0; 10 mA; 4 mA; 5.34 mA.

8.8 PREDICTING THE RESPONSE OF SEQUENTIALLY SWITCHED CIRCUITS

In Example 8.11 we briefly considered the response of an *RL* circuit to a pulse waveform, in which a source was effectively switched into and subsequently switched out of the circuit. This type of situation is common in practice, as few circuits are designed to be energized only once (passenger vehicle airbag triggering circuits, are one exception). In predicting the response of simple *RL* and *RC* circuits subjected to pulses and series of pulses—sometimes referred to as ***sequentially switched circuits***—the key is the relative size of the circuit time constant to the various times that define the pulse sequence. The underlying principle behind the analysis will be whether the energy storage element has time to fully charge before the pulse ends, and whether it has time to fully discharge before the next pulse begins.

Consider the circuit shown in Fig. 8.42*a*, which is connected to a pulsed voltage source described by seven separate parameters defined in Fig. 8.42*b*. The waveform is bounded by two values, **V1** and **V2.** The time

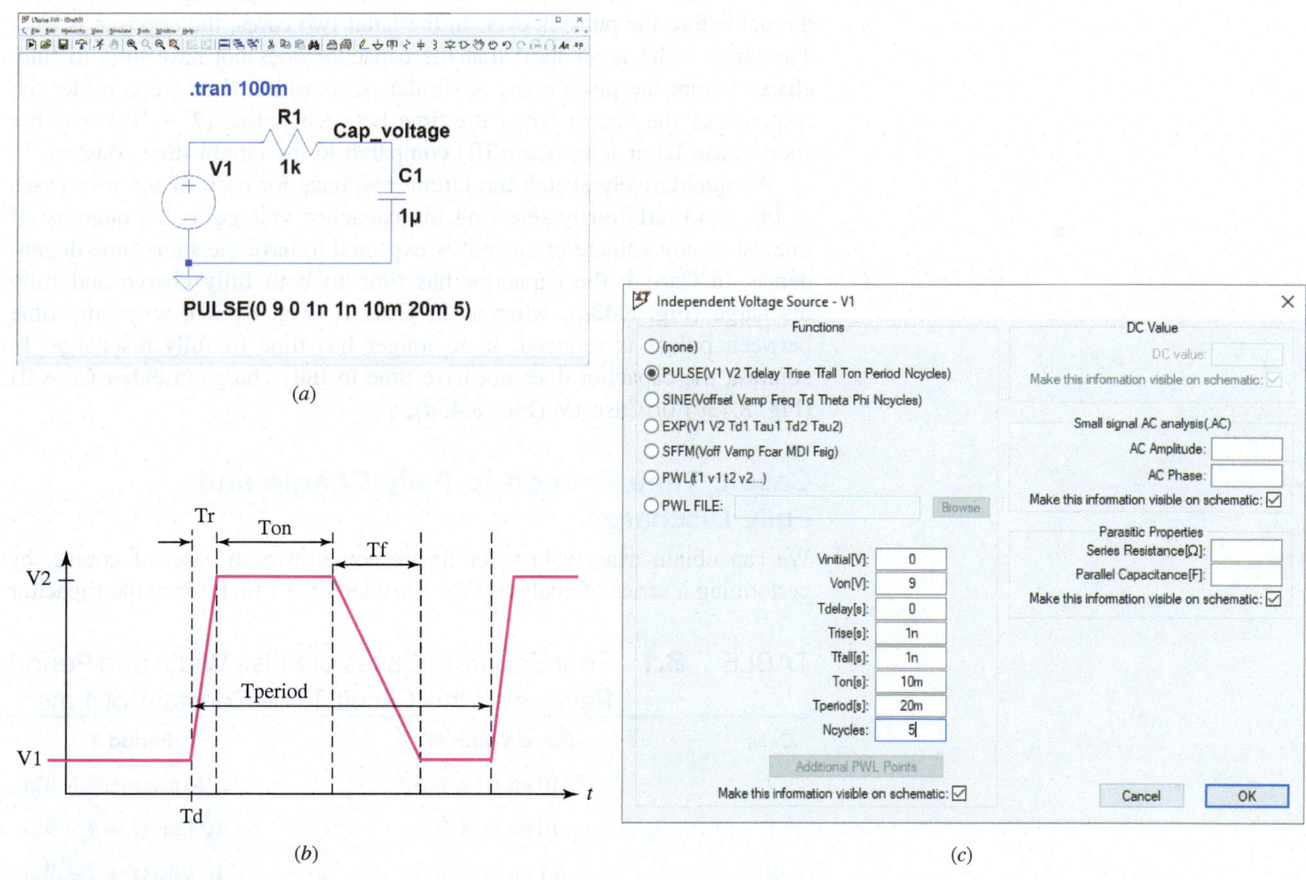

FIGURE 8.42 (*a*) Schematic of a simple *RC* circuit connected to a pulsed voltage waveform. (*b*) Diagram of the SPICE PULSE parameter definitions, and (*c*) definitions for PULSE voltage source in LTspice.
(*a* and *c*) Analog Devices

t_r required to change from **V1** to **V2** is called the ***rise time*** **(Tr),** and the time t_f required to change from **V2** to **V1** is called the ***fall time*** **(Tf).** The duration W_P of the pulse is referred to as the ***pulse width*** **(Ton),** and the ***period*** **(Tperiod)** of the waveform is the time it takes for the pulse to repeat. Note also that SPICE allows a time delay **(Td)** before the pulse train begins, which can be useful in allowing initial transient responses to decay for some circuit configurations. The number of cycles can also be specified by **Ncycles.** To edit these parameters on the voltage source, right-click on the source, click on the **Advanced** button, and choose the style **PULSE** under Functions. The window for specifying the voltage source is shown in Fig. 8.42*c*.

For the purposes of this discussion, we set a zero time delay, V1 = 0, and V2 = 9 V. The circuit time constant is $\tau = RC = 1$ ms, so we set the rise and fall times to be 1 ns. Although SPICE will not allow a voltage to change in zero time since it solves the differential equations using discrete time intervals, relative to our circuit time constant, 1 ns is a reasonable approximation to "instantaneous."

We will consider four basic cases, summarized in Table 8.1. In the first two cases, the pulse width W_p is much longer than the circuit time constant τ, so we expect the transients resulting from the beginning of the pulse to die out before the pulse is over. In the latter two cases, the opposite is true: The pulse width is so short that the capacitor does not have time to fully charge before the pulse ends. A similar issue arises when we consider the response of the circuit when the time between pulses $(T - W_p)$ is either short (Case II) or long (Case III) compared to the circuit time constant.

We qualitatively sketch the circuit response for each of the four cases in Fig. 8.43, arbitrarily selecting the capacitor voltage as the quantity of interest as any voltage or current is expected to have the same time dependence. In Case I, the capacitor has time to both fully charge and fully discharge (Fig. 8.43*a*), whereas in Case II (Fig. 8.43*b*), when the time between pulses is reduced, it no longer has time to fully discharge. In contrast, the capacitor does not have time to fully charge in either Case III (Fig. 8.43*c*) or Case IV (Fig. 8.43*d*).

Case I: Time Enough to Fully Charge and Fully Discharge

We can obtain exact values for the response in each case, of course, by performing a series of analyses. We consider Case I first. Since the capacitor

TABLE **8.1** Four Separate Cases of Pulse Width and Period Relative to the Circuit Time Constant of 1 ms

Case	Pulse Width W_p	Period T
I	10 ms ($\tau \ll W_p$)	20 ms ($\tau \ll T - W_p$)
II	10 ms ($\tau \ll W_p$)	10.1 ms ($\tau \gg T - W_p$)
III	0.1 ms ($\tau \gg W_p$)	10.1 ms ($\tau \ll T - W_p$)
IV	0.1 ms ($\tau \gg W_p$)	0.2 ms ($\tau \gg T - W_p$)

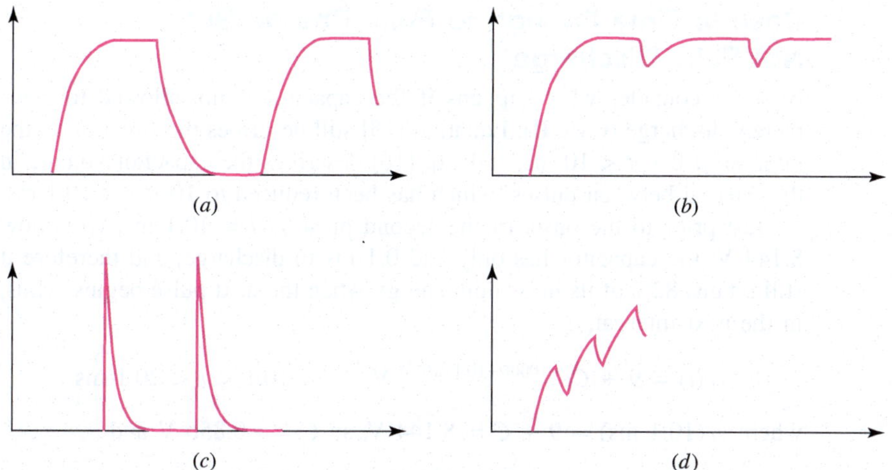

FIGURE 8.43 Capacitor voltage for the *RC* circuit, with pulse width and period as in (*a*) Case I; (*b*) Case II; (*c*) Case III; and (*d*) Case IV.

has time to fully charge, the forced response will correspond to the 9 V dc driving voltage. The complete response to the first pulse is therefore

$$v_C(t) = 9 + Ae^{-1000t} \text{ V}$$

With $v_C(0) = 0$, $A = -9$ V and so

$$v_C(t) = 9(1 - e^{-1000t}) \text{ V} \qquad [15]$$

in the interval of $0 < t < 10$ ms. At $t = 10$ ms, the source drops suddenly to 0 V, and the capacitor begins to discharge through the resistor. In this time interval we are faced with a simple "source-free" *RC* circuit, and we can write the response as

$$v_C(t) = Be^{-1000(t-0.01)} \qquad 10 < t < 20 \text{ ms} \qquad [16]$$

where $B = 8.99959$ V is found by substituting $t = 10$ ms in Eq. [15]; we will be pragmatic here and round this to 9 V, noting that the value calculated is consistent with our assumption that the initial transient dissipates before the pulse ends.

At $t = 20$ ms, the voltage source jumps immediately back to 9 V. The capacitor voltage just prior to this event is given by substituting $t = 20$ ms in Eq. [16], leading to $v_C(20 \text{ ms}) = 408.6\ \mu$V, essentially zero compared to the peak value of 9 V.

If we keep to our convention of rounding to four significant digits, the capacitor voltage at the beginning of the second pulse is zero, which is the same as our starting point. Thus, Eqs. [15] and [16] form the basis of the response for all subsequent pulses, and we may write

$$v_C(t) = \begin{cases} 9(1 - e^{-1000t}) \text{ V} & 0 \leq t \leq 10 \text{ ms} \\ 9e^{-1000(t-0.01)} \text{ V} & 10 < t \leq 20 \text{ ms} \\ 9(1 - e^{-1000(t-0.02)}) \text{ V} & 20 < t \leq 30 \text{ ms} \\ 9e^{-1000(t-0.03)} \text{ V} & 30 < t \leq 40 \text{ ms} \end{cases}$$

and so on.

Case II: Time Enough to Fully Charge But Not Fully Discharge

Next we consider what happens if the capacitor is not allowed to completely discharge (Case II). Equation [15] still describes the situation in the interval of $0 < t < 10$ ms, and Eq. [16] describes the capacitor voltage in the interval between pulses, which has been reduced to $10 < t < 10.1$ ms.

Just prior to the onset of the second pulse at $t = 10.1$ ms, v_C is now 8.144 V; the capacitor has only had 0.1 ms to discharge, and therefore it still retains 82% of its maximum energy when the next pulse begins. Thus, in the next interval,

$$v_C(t) = 9 + Ce^{-1000(t-10.1\times10^{-3})}\ \text{V} \qquad 10.1 < t < 20.1\ \text{ms}$$

where $v_C(10.1\ \text{ms}) = 9 + C = 8.144$ V, so $C = -0.856$ V and

$$v_C(t) = 9 - 0.856e^{-1000(t-10.1\times10^{-3})}\ \text{V} \qquad 10.1 < t < 20.1\ \text{ms}$$

which reaches the peak value of 9 V much more quickly than for the previous pulse.

Case III: No Time to Fully Charge But Time to Fully Discharge

What if it isn't clear that the transient will dissipate before the end of the voltage pulse? In fact, this situation arises in Case III. Just as we wrote for Case I,

$$v_C(t) = 9 + Ae^{-1000t}\ \text{V} \qquad [17]$$

still applies to this situation, but now only in the interval $0 < t < 0.1$ ms. Our initial condition has not changed, so $A = -9$ V as before. Now, however, just before this first pulse ends at $t = 0.1$ ms, we find that $v_C =$ 0.8565 V. This is a far cry from the maximum of 9 V possible if we allow the capacitor time to fully charge, and it is a direct result of the pulse lasting only one-tenth of the circuit time constant.

The capacitor now begins to discharge, so that

$$v_C(t) = Be^{-1000(t-1\times10^{-4})}\ \text{V} \qquad 0.1 < t < 10.1\ \text{ms} \qquad [18]$$

We have already determined that $v_C(0.1^-\ \text{ms}) = 0.8565$ V, so $v_C(0.1^+\ \text{ms}) = 0.8565$ V and substitution into Eq. [18] yields $B = 0.8565$ V. Just prior to the onset of the second pulse at $t = 10.1$ ms, the capacitor voltage has decayed to essentially 0 V; this is the initial condition at the start of the second pulse, and so Eq. [17] can be rewritten as

$$v_C(t) = 9 - 9e^{-1000(t-10.1\times10^{-3})}\ \text{V} \qquad 10.1 < t < 10.2\ \text{ms} \qquad [19]$$

to describe the corresponding response.

Case IV: No Time to Fully Charge or Even Fully Discharge

In the last case, we consider the situation where the pulse width and period are so short that the capacitor can neither fully charge nor fully discharge in any one period. Based on experience, we can write

$$v_C(t) = 9 - 9e^{-1000t} \text{ V} \qquad 0 < t < 0.1 \text{ ms} \qquad [20]$$

$$v_C(t) = 0.8565e^{-1000(t-1\times10^{-4})} \text{ V} \qquad 0.1 < t < 0.2 \text{ ms} \qquad [21]$$

$$v_C(t) = 9 + C_e^{-1000(t-2\times10^{-4})} \text{ V} \qquad 0.2 < t < 0.3 \text{ ms} \qquad [22]$$

$$v_C(t) = D_e^{-1000(t-3\times10^{-4})} \text{ V} \qquad 0.3 < t < 0.4 \text{ ms} \qquad [23]$$

Just prior to the onset of the second pulse at $t = 0.2$ ms, the capacitor voltage has decayed to $v_C = 0.7750$ V; with insufficient time to fully discharge, it retains a large fraction of the little energy it had time to store initially. For the interval of $0.2 < t < 0.3$ ms, substitution of $v_C(0.2^+) = v_C(0.2^-) = 0.7750$ V into Eq. [22] yields $C = -8.225$ V. Continuing, we evaluate Eq. [22] at $t = 0.3$ ms and calculate $v_C = 1.558$ V just prior to the end of the second pulse. Thus, $D = 1.558$ V, and our capacitor is slowly charging to ever increasing voltage levels over several pulses. At this stage it might be useful if we plot the detailed responses, so we show the LTspice simulation results of Cases I through IV in Fig. 8.44. Note in

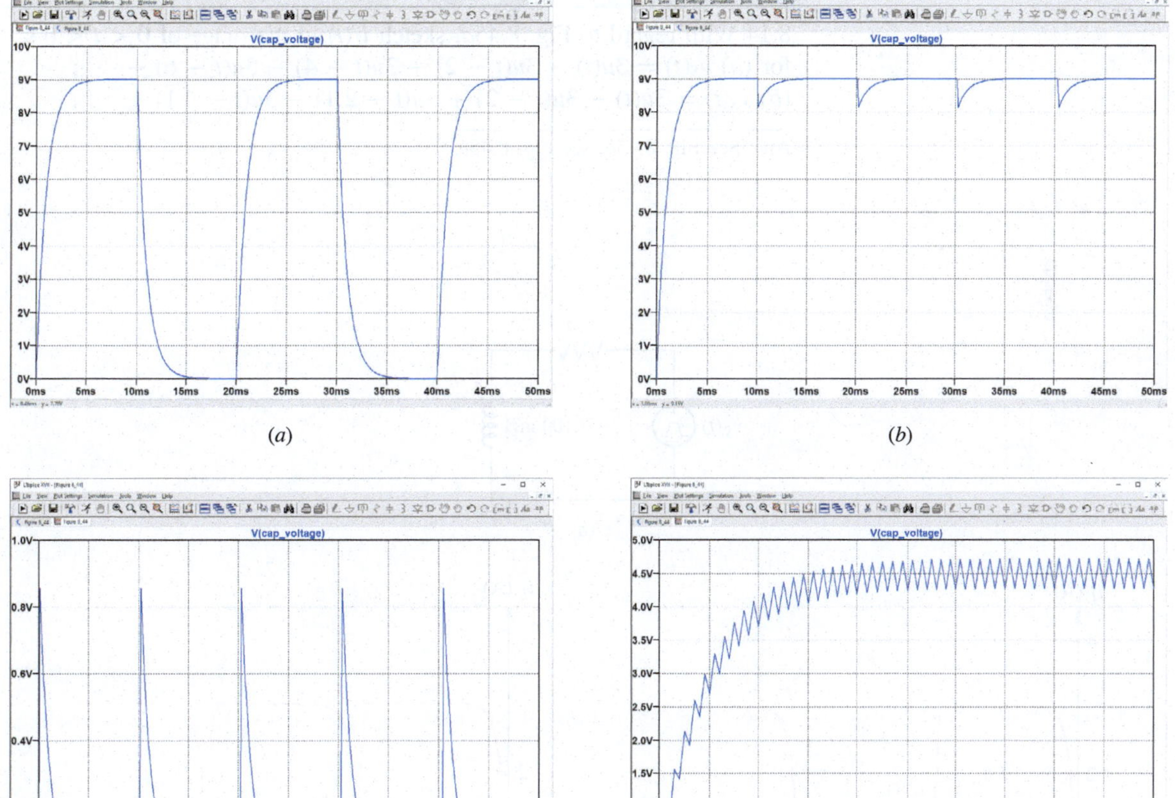

(a) (b)

(c) (d)

FIGURE 8.44 LTspice simulation results corresponding to (*a*) Case I; (*b*) Case II; (*c*) Case III; (*d*) Case IV.
(*a–d*) Analog Devices

particular that in Fig. 8.44*d*, the small charge/discharge transient response similar in shape to that shown in Fig. 8.44*a–c* is superimposed on a charging-type response of the form $(1 - e^{-t/\tau})$. Thus, it takes about 3 to 5 circuit time constants for the capacitor to charge to its maximum value in situations where a single period does not permit it to fully charge or discharge!

What we have not yet done is predict the behavior of the response for $t \gg 5\tau$, although we would be interested in doing so, especially if it was not necessary to consider a very long sequence of pulses one at a time. We note that the response of Fig. 8.44*d* has an *average* value of 4.50 V from about 4 ms onward. This is exactly half the value we would expect if the voltage source pulse width allowed the capacitor to fully charge. In fact, this long-term average value can be computed by multiplying the dc capacitor voltage by the ratio of the pulse width to the period.

PRACTICE

8.14 With regard to Fig. 8.45*a*, sketch $i_L(t)$ in the range of $0 < t < 6$ s for (*a*) $v_S(t) = 3u(t) - 3u(t-2) + 3u(t-4) - 3u(t-6) + \cdots$; (*b*) $v_S(t) = 3u(t) - 3u(t-2) + 3u(t-2.1) - 3u(t-4.1) + \cdots$.

Ans: See Fig. 8.45*b*; see Fig. 8.45*c*.

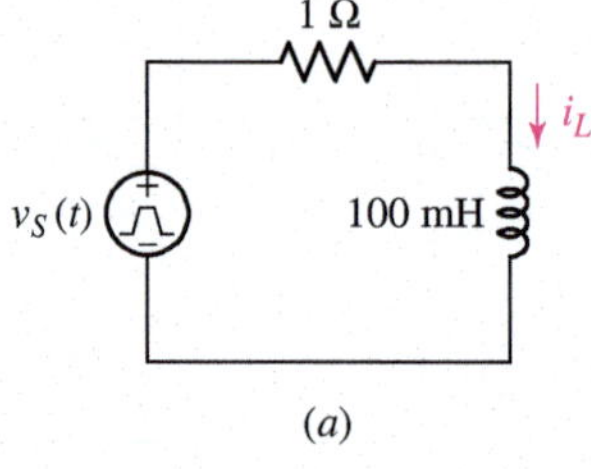

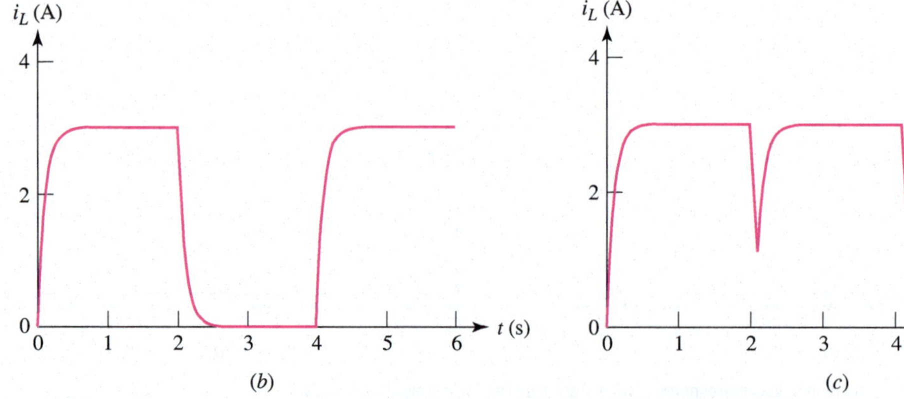

FIGURE 8.45 (*a*) Circuit for Practice Problem 8.14; (*b*) solution to part (*a*); (*c*) solution to part (*b*).

PRACTICAL APPLICATION

Frequency Limits in Digital Integrated Circuits

Modern digital integrated circuits such as programmable array logic (PALs) and microprocessors (Fig. 8.46) are composed of interconnected transistor circuits known as *gates*.

Digital signals are represented symbolically by combinations of ones and zeros and can be either data or instructions (such as "add" or "subtract"). Electrically, we represent a logic "1" by a "high" voltage, and a logic "0" by a "low" voltage. In practice, there is a range of voltages that correspond to each; for example, in the 7400 series of TTL logic integrated circuits, any voltage between 2 and 5 V will be interpreted as a logic 1, and any voltage between 0 and 0.8 V will be interpreted as a logic 0. Voltages between 0.8 and 2 V do not correspond to either logic state, as shown in Fig. 8.47.

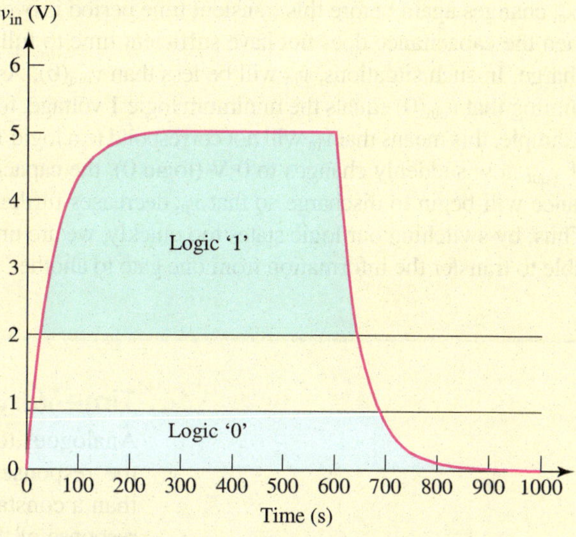

FIGURE 8.47 Charge/discharge characteristic of a pathway capacitance identifying the TTL voltage ranges for logic 1 and logic 0, respectively.

A key parameter in digital circuits is the speed at which we can effectively use them. In this sense, "speed" refers to how quickly we can switch a gate from one logic state to another (either logic 0 to logic 1 or vice versa), and the time delay required to convey the output of one gate to the input of the next gate. Although transistors contain "built-in" capacitances that affect their switching speed, it is the *interconnect pathways* that currently limit the speed of the fastest digital integrated circuits. We can model the interconnect pathway between two logic gates using a simple *RC* circuit (although as feature sizes continue to decrease in modern designs, more detailed models are required to accurately predict circuit performance). For example, consider a 2000 μm long pathway 2 μm wide. We can model this pathway in a typical silicon-based integrated circuit as having a capacitance of 0.5 pF and a resistance of 100 Ω, shown schematically in Fig. 8.48.

FIGURE 8.46 A silicon integrated circuit die with dimensions smaller than a US 1 cent coin.
(Lucidio Studio Inc/Photographer's Choice/Getty Images)

Let's assume the voltage v_{out} represents the output voltage of a gate that is changing from a logic 0 state to a logic 1 state. The voltage v_{in} appears across the input of a second gate, and we are interested in how long it takes v_{in} to reach the same value as v_{out}.

Assuming the 0.5 pF capacitance that characterizes the interconnect pathway is initially discharged [i.e., $v_{in}(0) = 0$], calculating the *RC* time constant for our pathway as $\tau = RC = 50$ ps, and defining $t = 0$ as the time when v_{out} changes, we obtain the expression

$$v_{in}(t) = Ae^{-t/\tau} + v_{out}(0)$$

Setting $v_{in}(0) = 0$, we find that $A = -v_{out}(0)$ so that

$$v_{in}(t) = v_{out}(0)[1 - e^{-t/\tau}]$$

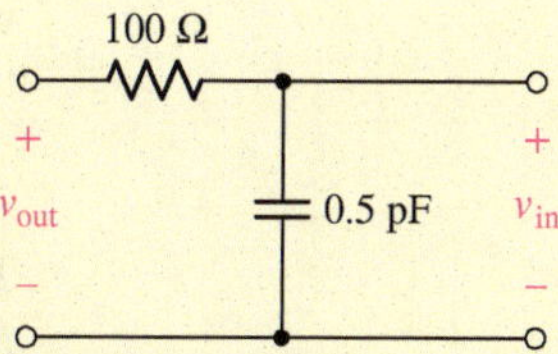

FIGURE 8.48 Circuit model for an integrated circuit pathway.

(Continued on next page)

Upon examining this equation, we see that v_{in} will reach the value $v_{out}(0)$ after $\sim 5\tau$ or 250 ps. If the voltage v_{out} changes again before this transient time period is over, then the capacitance does not have sufficient time to fully charge. In such situations, v_{in} will be less than $v_{out}(0)$. Assuming that $v_{out}(0)$ equals the minimum logic 1 voltage, for example, this means that v_{in} will not correspond to a logic 1. If v_{out} now suddenly changes to 0 V (logic 0), the capacitance will begin to discharge so that v_{in} decreases further. Thus, by switching our logic states too quickly, we are unable to transfer the information from one gate to another.

The fastest speed at which we can change logic states is therefore $(5\tau)^{-1}$. This can be expressed in terms of the maximum operating frequency:

$$f_{max} = \frac{1}{2(5\tau)} = 2 \text{ GHz}$$

where the factor of 2 represents a charge/discharge period. If we wish to operate our integrated circuit at a higher frequency so that calculations can be performed more quickly, we need to reduce the interconnect capacitance and/or the interconnect resistance.

Time-Varying Forced Response

Analogous to the switching response just described, how can we determine the response of an *RC* or *RL* circuit subject to a forcing function other than a constant? We can use a similar procedure in determining the forced response of the circuit, but now the forced response will have a time-varying behavior.

EXAMPLE 8.12

Determine an expression for $v(t)$ in the circuit of Fig. 8.49 valid for $t > 0$.

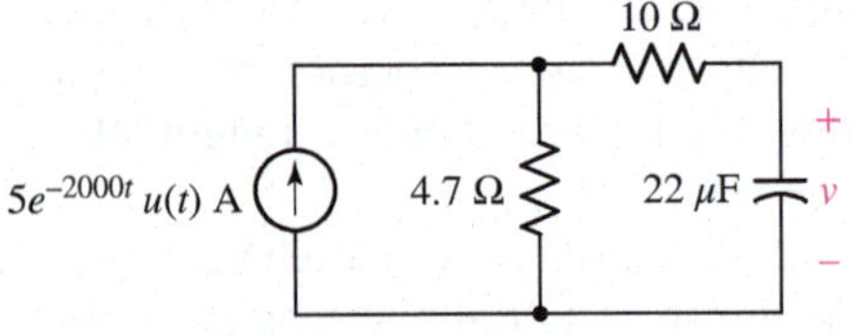

■ **FIGURE 8.49** A simple *RC* circuit driven by an exponentially decaying forcing function.

Based on experience, we expect a complete response of the form

$$v(t) = v_f + v_n$$

where v_f will likely resemble our forcing function and v_n will have the form $Ae^{-t/\tau}$.

There are several ways to proceed, although perhaps the most straightforward is to perform a source transformation, resulting in a voltage source $23.5e^{-2000t}\,u(t)$ V in series with 14.7 Ω and 22 μF.

Writing a simple KVL equation for $t > 0$, we find that

$$23.5e^{-2000t} = (14.7)(22 \times 10^{-6})\frac{dv}{dt} + v$$

A little rearranging results in

$$\frac{dv}{dt} + 3.092 \times 10^3 v = 72.67 \times 10^3\, e^{-2000t} \quad [24]$$

where the natural response will be given by an exponential with a time constant τ

$$v_n(t) = Ae^{-t/\tau} = Ae^{-3092t}$$

The forced solution will follow the exponential time response $v_f(t) = Ke^{-2000t}$. Substituting the forced solution into Eq. [24] and solving for K,

$$v_f(t) = 66.55e^{-2000t}$$

Combining natural and forced solutions,

$$v(t) = 66.55e^{-2000t} + Ae^{-3092t} \text{ V} \qquad [25]$$

Our only source is controlled by a step function with zero value for $t < 0$, so we know that $v(0^-) = 0$. Since v is a capacitor voltage, $v(0^+) = v(0^-)$, and we therefore find our initial condition $v(0) = 0$ easily enough. Substituting this into Eq. [25], we find $A = -66.55$ V, and so

$$v(t) = 66.55(e^{-2000t} - e^{-3092t}) \text{ V} \qquad t > 0$$

PRACTICE

8.15 Determine the capacitor voltage v in the circuit of Fig. 8.50 for $t > 0$.

Ans: $23.5 \cos 3t + 22.8 \times 10^{-3} \sin 3t - 23.5e^{-3092t}$ V.

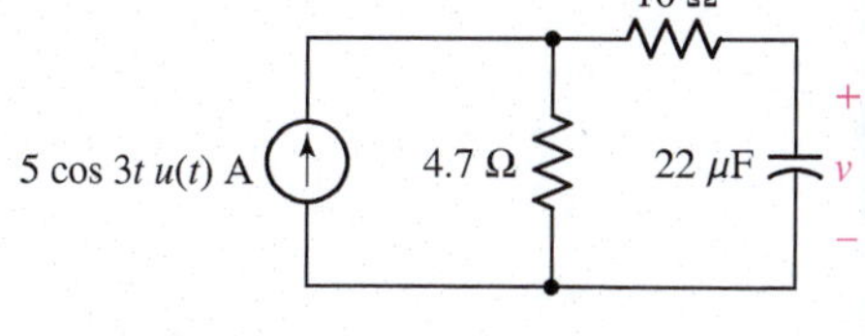

FIGURE 8.50 A simple *RC* circuit driven by a sinusoidal forcing function.

SUMMARY AND REVIEW

In this chapter we learned that circuits containing a single energy storage element (either a capacitor or an inductor) can be described by a characteristic time scale, namely, the *circuit time constant* ($\tau = RC$ or $\tau = L/R$, respectively). If we try to change the amount of energy stored in the element (either charging or discharging), *every* voltage and current in the circuit will include an exponential term of the form $e^{-t/\tau}$. After approximately *five time constants* from the moment we tried to change the amount of stored energy, the *transient* response has *essentially disappeared* and we are left simply with a *forced* response which arises from the independent sources driving the circuit at times $t > 0$. When determining the forced response in a purely dc circuit, we may treat inductors as short circuits and capacitors as open circuits.

We started our analysis with so-called source-free circuits to introduce the idea of time constants without unnecessary distractions; such circuits have zero forced response and a transient response derived entirely from the energy stored at $t = 0$. We reasoned that a capacitor cannot change its voltage in zero time (or an infinite current results), and we indicated this by introducing the notation $v_C(0^+) = v_C(0^-)$. Similarly, the current through an inductor cannot change in zero time, or $i_L(0^+) = i_L(0^-)$. The *complete* response is always the sum of the transient response and the forced response. Applying the initial condition to the complete response allows us to determine the unknown constant which multiplies the transient term.

We spent a little time discussing modeling switches, both analytically and within the context of CAD. A common mathematical representation makes use of the unit-step function $u(t - t_0)$, which has zero value for $t < t_0$, unity value for $t > t_0$, and is indeterminate for $t = t_0$. Unit-step functions can "activate" a circuit (connecting sources so current can flow) for values of t preceding a specific time as well as after. Combinations of step functions can be used to create pulses and more complex waveforms. In the case of sequentially switched circuits, where sources are connected and disconnected repeatedly, we found the behavior of the circuits to depend strongly on both period and pulse width as they compare to the circuit time constant.

This is a good time to highlight some key points worth reviewing, along with relevant example(s).

- The response of a circuit having sources suddenly switched in or out of a circuit containing capacitors and inductors will always be composed of two parts: a *natural* response and a *forced* response.
- The form of the natural response (also referred to as the *transient response*) depends only on the component values and the way they are wired together. (Examples 8.1, 8.2, 8.4)
- A circuit reduced to a single equivalent capacitance C and a single equivalent resistance R will have a natural response given by $v(t) = V_0 e^{-t/\tau}$, where $\tau = RC$ is the circuit time constant. (Examples 8.1, 8.2, 8.3)
- A circuit reduced to a single equivalent inductance L and a single equivalent resistance R will have a natural response given by $i(t) = I_0 e^{-t/\tau}$, where $\tau = L/R$ is the circuit time constant. (Examples 8.4 and 8.5)
- Circuits with dependent sources can be represented by a resistance using Thévenin procedures. (Examples 8.6, 8.7, 8.8)
- The unit-step function is a useful way to model the closing or opening of a switch, provided we are careful to keep an eye on the initial conditions. (Example 8.10)
- The form of the forced response mirrors the form of the forcing function. Therefore, a dc forcing function always leads to a constant forced response. (Examples 8.9 and 8.10)
- The *complete* response of an *RC* or *RL* circuit excited by a dc source will have the form $f(t) = f(\infty) + [f(0^+) - f(\infty)]e^{-t/\tau}$, or total response = final value + (initial value − final value)$e^{-t/\tau}$. (Examples 8.9, 8.10, 8.11)
- The complete response for an *RC* or *RL* circuit may also be determined by writing a single differential equation for the quantity of interest and solving. (Example 8.12)
- When dealing with sequentially switched circuits, or circuits connected to pulsed waveforms, the relevant issue is whether the energy storage element has sufficient time to fully charge and to fully discharge, as measured relative to the circuit time constant.

READING FURTHER

A guide to solution techniques for differential equations can be found in:

W. E. Boyce and R. C. DiPrima, *Elementary Differential Equations and Boundary Value Problems,* 7th ed. New York: Wiley, 2002.

A detailed description of transients in electric circuits is given in:

E. Weber, *Linear Transient Analysis Volume I.* New York: Wiley, 1954. (Out of print, but in many university libraries.)

EXERCISES

8.1 The Source-Free *RC* Circuit

1. A source-free RC circuit has $R = 4\ \text{k}\Omega$ and $C = 22\ \mu\text{F}$, and with the knowledge that $v(0) = 5$ V, (*a*) write an expression for $v(t)$ valid for $t > 0$; (*b*) compute $v(t)$ at $t = 0$, $t = 50$ ms, and $t = 500$ ms; and (*c*) calculate the energy stored in the capacitor at $t = 0$, $t = 50$ ms, and $t = 500$ ms.
2. A source-free RC circuit has $v(0) = 12$ V and $R = 100\ \Omega$. (*a*) Select C such that $v(250\ \mu\text{s}) = 5.215$ V; (*b*) compute the energy stored in the capacitor at $t = 0$, $t = 250\ \mu\text{s}$, $t = 500\ \mu\text{s}$, and $t = 1$ ms.
3. The resistor in the circuit of Fig. 8.51 has been included to model the dielectric layer separating the plates of the 3.1 nF capacitor, and it has a value of 55 MΩ. The capacitor is storing 200 mJ of energy just prior to $t = 0$. (*a*) Write an expression for $v(t)$ valid for $t \geq 0$. (*b*) Compute the energy remaining in the capacitor at $t = 170$ ms. (*c*) Graph $v(t)$ over the range of $0 < t < 850$ ms, and identify the value of $v(t)$ when $t = 2\tau$.
4. The resistor in the circuit of Fig. 8.51 has a value of 1 Ω and is connected to a 22 mF capacitor. The capacitor dielectric has infinite resistance, and the device is storing 891 mJ of energy just prior to $t = 0$. (*a*) Write an expression for $v(t)$ valid for $t \geq 0$. (*b*) Compute the energy remaining in the capacitor at $t = 11$ ms and 33 ms. (*c*) If it is determined that the capacitor dielectric is much leakier than expected, having a resistance as low as 100 kΩ, repeat parts (*a*) and (*b*).

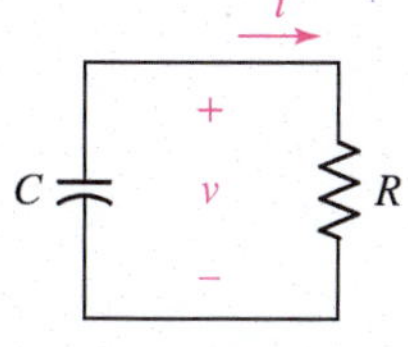

FIGURE 8.51

5. Calculate the time constant of the circuit depicted in Fig. 8.51 if $C = 10$ mF and R is equal to (*a*) 1 Ω; (*b*) 10 Ω; (*c*) 100 Ω. (*d*) Verify your answers with an appropriate parameter sweep simulation. (*Hint:* The cursor tool might come in handy, and the time constant does not depend on the initial voltage across the capacitor.)
6. It is safe to assume that the switch drawn in the circuit of Fig. 8.52 has been closed such a long time that any transients which might have arisen from first connecting the voltage source have disappeared. (*a*) Determine the circuit time constant. (*b*) Calculate the voltage $v(t)$ at $t = \tau$, 2τ, and 5τ.

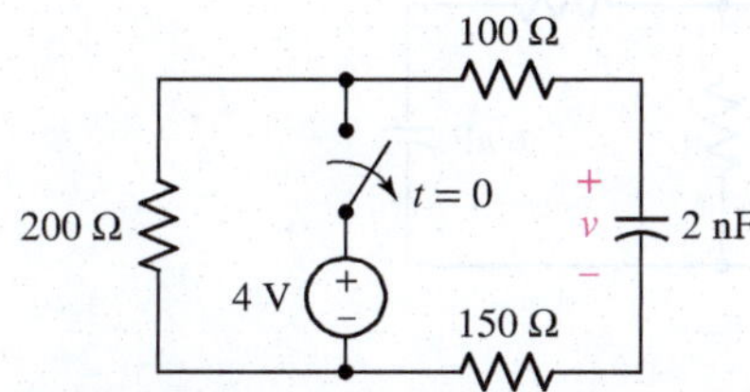

FIGURE 8.52

7. We can safely assume the switch in the circuit of Fig. 8.53 was closed a very long time prior to being thrown open at $t = 0$. (*a*) Determine the circuit time constant. (*b*) Obtain an expression for $i_1(t)$ which is valid for $t > 0$. (*c*) Determine the power dissipated by the 12 Ω resistor at $t = 500$ ms.

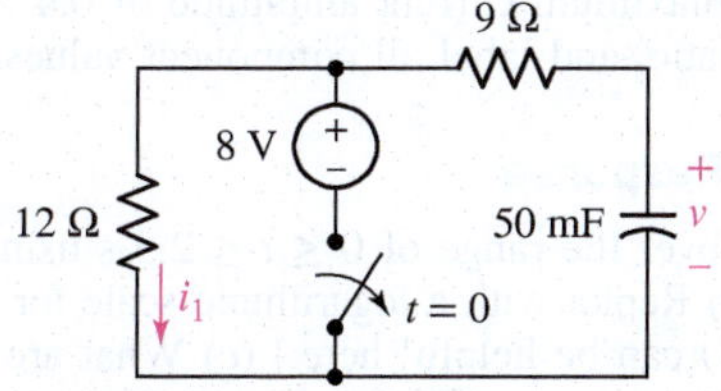

FIGURE 8.53

8. The switch above the 12 V source in the circuit of Fig. 8.54 has been closed since just after the wheel was invented. It is finally thrown open at $t = 0$. (*a*) Compute the circuit time constant. (*b*) Obtain an expression for $v(t)$ valid for $t > 0$. (*c*) Calculate the energy stored in the capacitor 170 ms after the switch is opened.

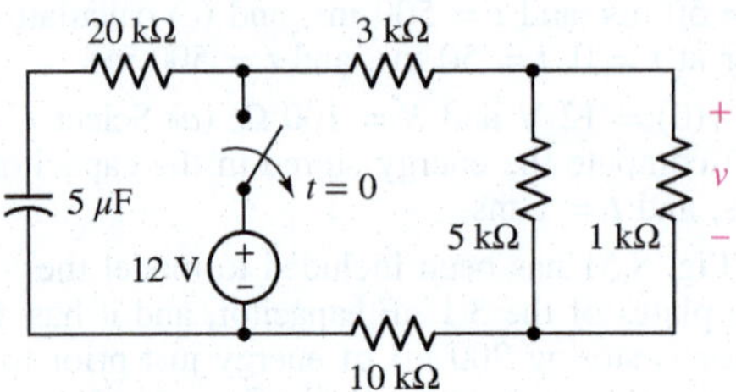

FIGURE 8.54

9. For the circuit represented schematically in Fig. 8.55, (*a*) calculate $v(t)$ at $t = 0$, $t = 984$ s, and $t = 1236$ s; (*b*) determine the energy still stored in the capacitor at $t = 100$ s.

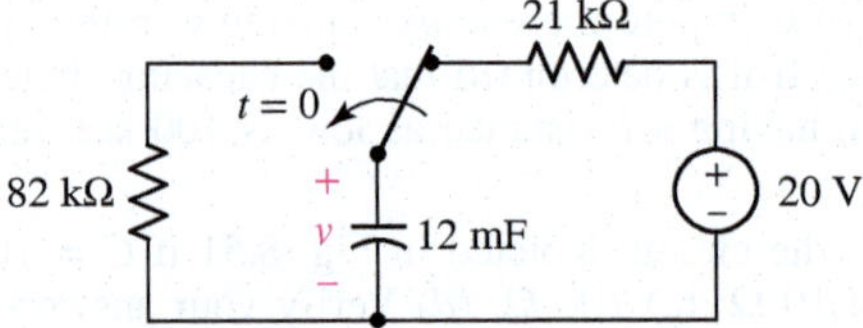

FIGURE 8.55

10. The switch in Fig. 8.56 has been closed for a long time, before being opened at $t = 0$. (*a*) Find an expression for $v(t)$ and $i(t)$ in the figure for $t > 0$, and (*b*) evaluate v and i at $t = 1$ ms.

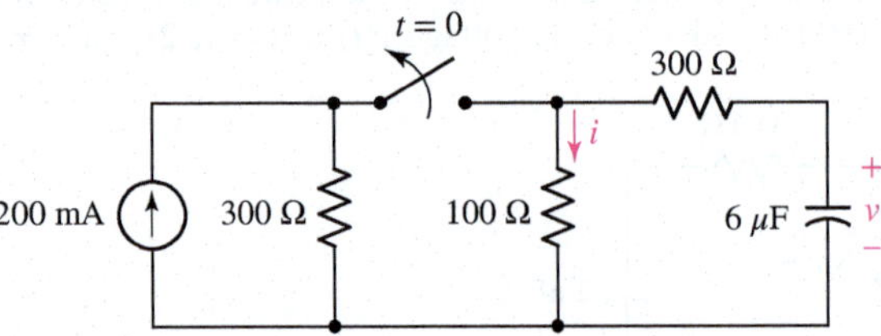

FIGURE 8.56

11. For the circuit in Fig. 8.56, find (*a*) the total energy stored in the capacitor before the switch is opened, (*b*) find an expression for the energy stored in the capacitor and power supplied by the capacitor for $t > 0$, and (*c*) plot the resulting expressions for $w(t)$ and $p(t)$ using an appropriate range of time.

12. Design a capacitor-based circuit that can achieve the following specifications simultaneously: (*a*) an initial voltage of 9 V at $t = 0$, (*b*) a voltage that decays to 1.2 V at $t = 2$ ms, (*c*) a maximum current amplitude (absolute value) of 1 mA for $t > 0$, and (*d*) a maximum current amplitude of 0.4 mA for $t > 100$ ns. Draw a circuit schematic, and label all component values.

8.2 Properties of the Exponential Response

13. (*a*) Graph the function $f(t) = 10e^{-2t}$ over the range of $0 \leq t \leq 2.5$ s using linear scales for both y and x axes. (*b*) Replot with a logarithmic scale for the y axis. [*Hint:* The function *semilogy()* can be helpful here.] (*c*) What are the units of the 2 in the argument of the exponential? (*d*) At what time does the function reach a value of 9? 8? 1?

14. The current $i(t)$ flowing through a 1 kΩ resistor is given by $i(t) = 5e^{-10t}$ mA, $t \geq 0$. (*a*) Determine the values of t for which the resistor voltage magnitude is equal to 5 V, 2.5 V, 0.5 V, and 5 mV. (*b*) Graph the function over the range of $0 \leq t \leq 1$ s using linear scales for both axes. (*c*) Draw a tangent to your curve at $t = 0$, and determine where the tangent intersects the time axis.

15. Radiocarbon dating has a similar exponential time relation to our circuits. A living object contains the same proportion of radioactive ^{14}C as its surroundings, but after the object dies, it no longer acquires ^{14}C, for which the radioactive isotopes decay into ^{12}C. The relationship is given by

$$N = N_0 e^{-\lambda t}$$

where N_0 is the concentration of ^{14}C at time of death, N is the concentration at a given time after death, and λ is a constant. Given that the half-life of ^{14}C is 5700 years (where $N/N_0 = 1/2$), determine (*a*) the value (and units!) of the constant λ; (*b*) the approximate age of an extraterrestrial alien fossil discovered that has 16 g of ^{14}C, and expected initial concentration of 42 g.

16. For the circuit of Fig. 8.4, compute the time constant if the 4 Ω resistor is replaced with (*a*) a short circuit; (*b*) a 1 Ω resistor; (*c*) a series connection of two 5 Ω resistors; (*d*) a 100 Ω resistor. (*e*) Verify your answers with a suitable parameter sweep simulation in SPICE.

17. Design a circuit which will produce a current of 1 mA at some initial time and a current of 368 μA at a time 5 s later. You may specify an initial capacitor voltage without showing how it arises.

8.3 The Source-Free *RL* Circuit

18. Setting $R = 1$ kΩ and $L = 1$ nH for the circuit represented in Fig. 8.11, and with the knowledge that $i(0) = -3$ mA, (*a*) write an expression for $i(t)$ valid for all $t \geq 0$; (*b*) compute $i(t)$ at $t = 0$, $t = 1$ ps, 2 ps, and 5 ps; and (*c*) calculate the energy stored in the inductor at $t = 0$, $t = 1$ ps, and $t = 5$ ps.

19. If $i(0) = 1$ A and $R = 100$ Ω for the circuit of Fig. 8.11, (*a*) select L such that $i(50 \text{ ms}) = 368$ mA; (*b*) compute the energy stored in the inductor at $t = 0$, 50 ms, 100 ms, and 150 ms.

20. Referring to the circuit shown in Fig. 8.11, select values for both elements such that $L/R = 1$, $i(0) = -5$ A, and $v_R(0) = 10$ V. (*a*) Calculate $v_R(t)$ at $t = 0$, 1, 2, 3, 4, and 5 s; (*b*) compute the power dissipated in the resistor at $t = 0$, 1 s, and 5 s. (*c*) At $t = 5$ s, what is the percentage of the initial energy still stored in the inductor?

21. The circuit depicted in Fig. 8.11 is constructed from components whose value is unknown. If a current $i(0)$ of 6 μA initially flows through the inductor, and it is determined that $i(1 \text{ ms}) = 2.207$ μA, calculate the ratio of R to L.

22. With the assumption that the switch in the circuit of Fig. 8.57 has been closed a long, long, long time, calculate $i_L(t)$ at (*a*) the instant just before the switch opens; (*b*) the instant just after the switch opens; (*c*) $t = 15.8$ μs; (*d*) $t = 31.5$ μs; (*e*) $t = 78.8$ μs.

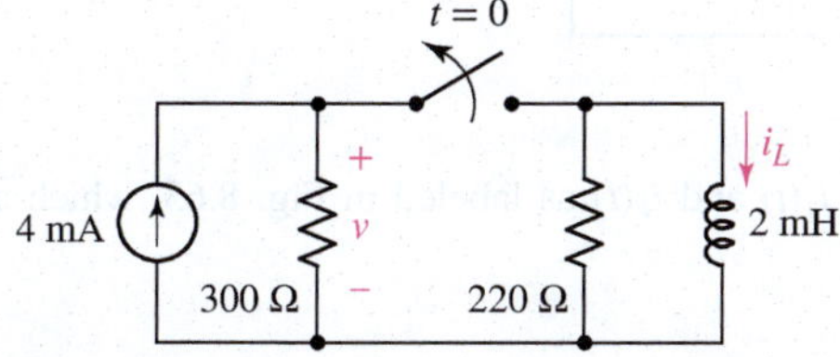

FIGURE 8.57

23. The switch in Fig. 8.57 has been closed since Catfish Hunter last pitched for the New York Yankees. Calculate the voltage labeled v as well as the energy stored in the inductor at (*a*) the instant just prior to the switch being thrown open; (*b*) the instant just after the switch is opened; (*c*) $t = 8$ μs; (*d*) $t = 80$ μs.

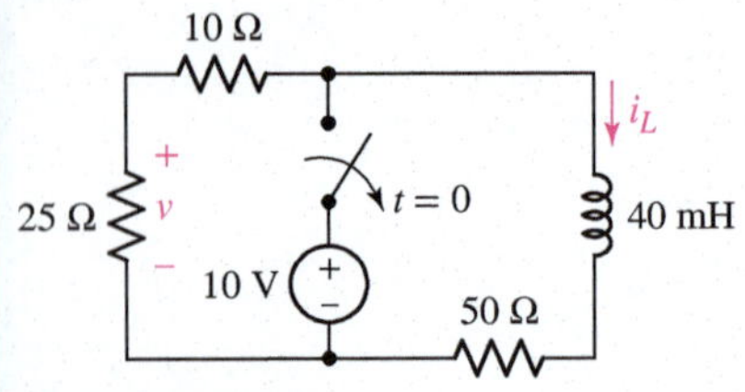

■ **FIGURE 8.58**

24. The switch in the circuit of Fig. 8.58 has been closed a ridiculously long time before suddenly being thrown open at $t = 0$. (*a*) Obtain expressions for i_L and v in the circuit of Fig. 8.58, which are valid for all $t \geq 0$. (*b*) Calculate $i_L(t)$ and $v(t)$ at the instant just prior to the switch opening, at the instant just after the switch opening, and at $t = 470\ \mu s$.

25. Assuming the switch initially has been open for a really, really long time, (*a*) obtain an expression for i_W in the circuit of Fig. 8.59, which is valid for all $t \geq 0$; (*b*) calculate i_W at $t = 0$ and $t = 1.3$ ns.

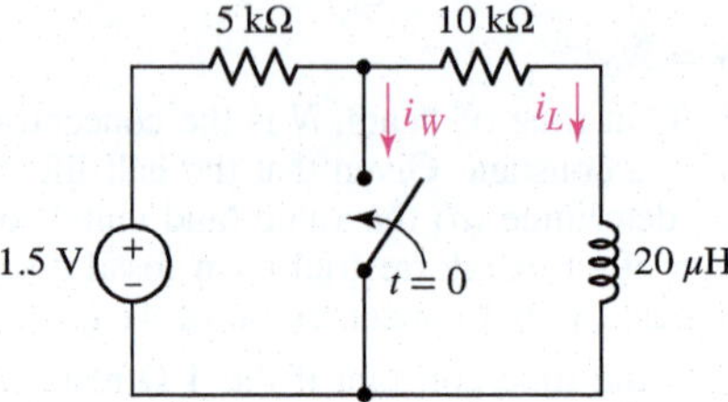

■ **FIGURE 8.59**

8.4 A More General Perspective

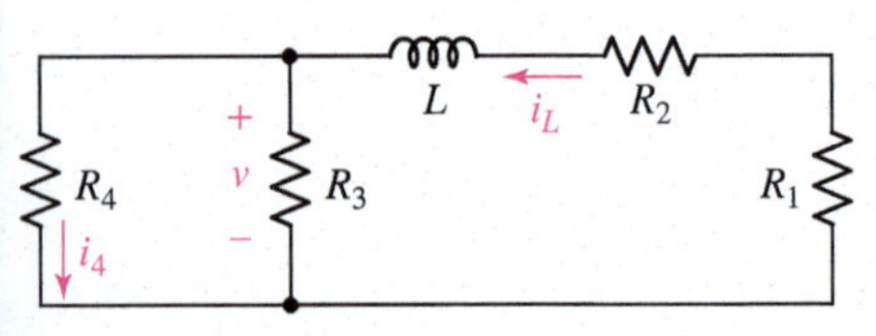

■ **FIGURE 8.60**

26. (*a*) Obtain an expression for $v(t)$, the voltage which appears across resistor R_3 in the circuit of Fig. 8.60, which is valid for $t > 0$. (*b*) If $R_1 = 2R_2 = 3R_3 = 4R_4 = 1.2$ kΩ, $L = 1$ mH, and $i_L(0^-) = 3$ mA, calculate $v(t = 500$ ns).

27. For the circuit of Fig. 8.61, determine i_x, i_L, and v_L at t equal to (*a*) 0^-; (*b*) 0^+.

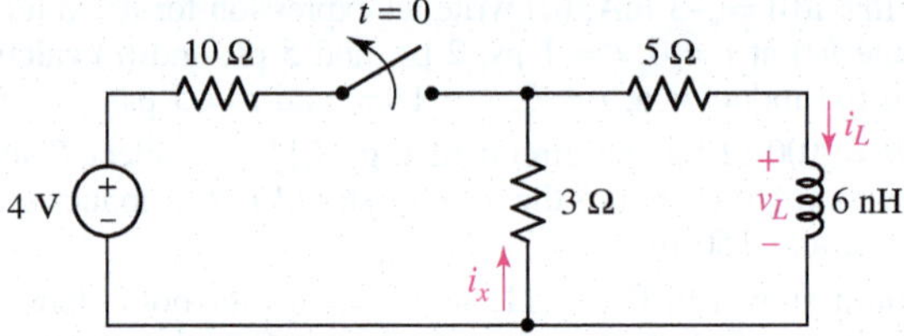

■ **FIGURE 8.61**

28. The switch shown in Fig. 8.62 had been closed for 6 years prior to being flipped open at $t = 0$. Determine i_L, v_L, and v_R at t equal to (*a*) 0^-; (*b*) 0^+; (*c*) 1 μs; (*d*) 10 μs.

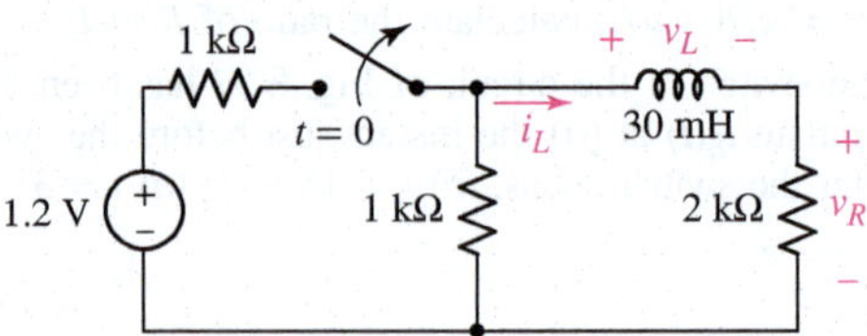

■ **FIGURE 8.62**

29. Obtain expressions for both $i_1(t)$ and $i_L(t)$ as labeled in Fig. 8.63, which are valid for $t > 0$.

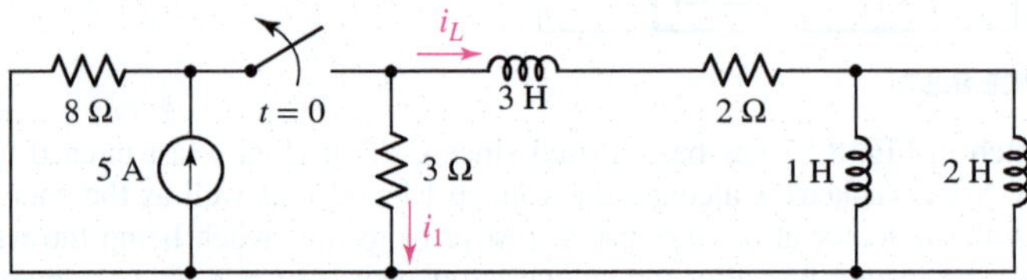

■ **FIGURE 8.63**

30. The voltage across the resistor in a simple source-free *RL* circuit is given by $5e^{-90t}$ V, $t > 0$. The inductor value is not known. (*a*) At what time will the inductor voltage be exactly one-half of its maximum value? (*b*) At what time will the inductor current reach 10% of its maximum value?

31. Referring to Fig. 8.64, calculate the currents i_1 and i_2 at t equal to (*a*) 1 ms; (*b*) 3 ms.

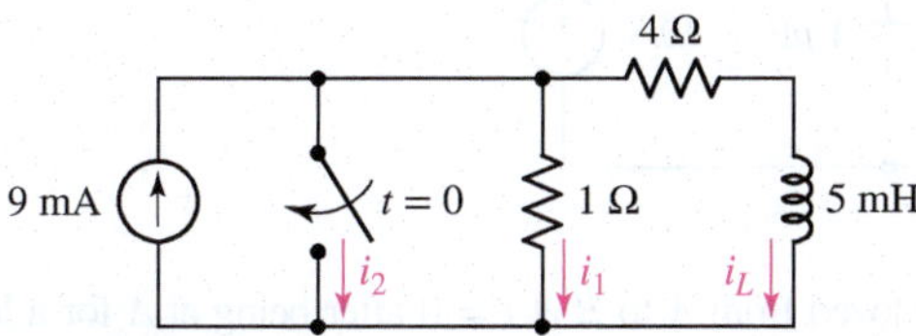

FIGURE 8.64

32. (*a*) Obtain an expression for v_x as labeled in the circuit of Fig. 8.65. (*b*) Evaluate v_x at $t = 5$ ms. (*c*) Verify your answer with an appropriate SPICE simulation. (*Hint:* Define the circuit for $t > 0$ and define an initial value using the .ic SPICE directive.)

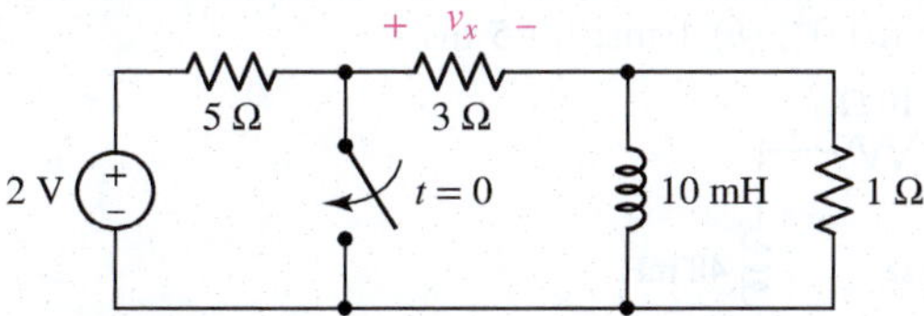

FIGURE 8.65

33. Design a complete circuit which provides a voltage v_{ab} across two terminals labeled *a* and *b*, respectively, such that $v_{ab} = 5$ V at $t = 0^-$, 2 V at $t = 1$ s, and less than 60 mV at $t = 5$. Verify the operation of your circuit using an appropriate SPICE simulation. (*Hint:* Define the circuit for $t > 0$ and define an initial value using the .ic SPICE directive.)

34. Select values for the resistors R_0 and R_1 in the circuit of Fig. 8.66 such that $v_C(0.65) = 5.22$ V and $v_C(2.21) = 1$ V.

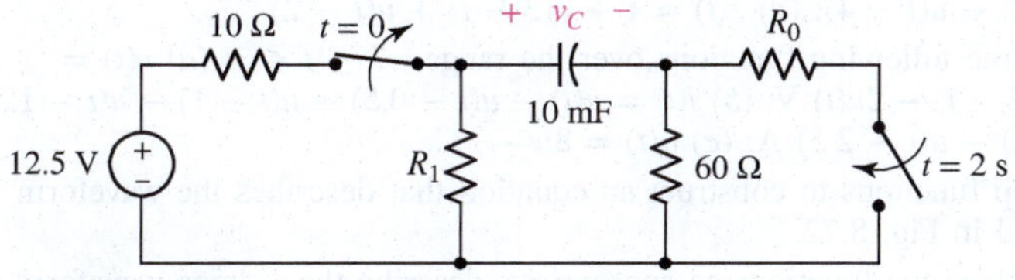

FIGURE 8.66

35. A quick measurement determines that the capacitor voltage v_C in the circuit of Fig. 8.67 is 2.5 V at $t = 0^-$. (*a*) Determine $v_C(0^+)$, $i_1(0^+)$, and $v(0^+)$. (*b*) Select a value of *C* so that the circuit time constant is equal to 14 s.

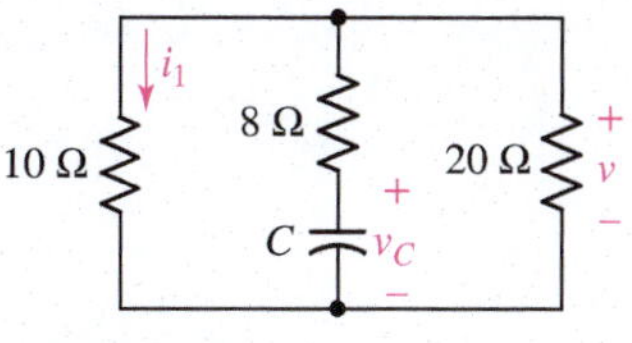

FIGURE 8.67

36. Determine $v_C(t)$ and $v_o(t)$ as labeled in the circuit represented by Fig. 8.68 for t equal to (*a*) 0^-; (*b*) 0^+; (*c*) 10 ms; (*d*) 12 ms.

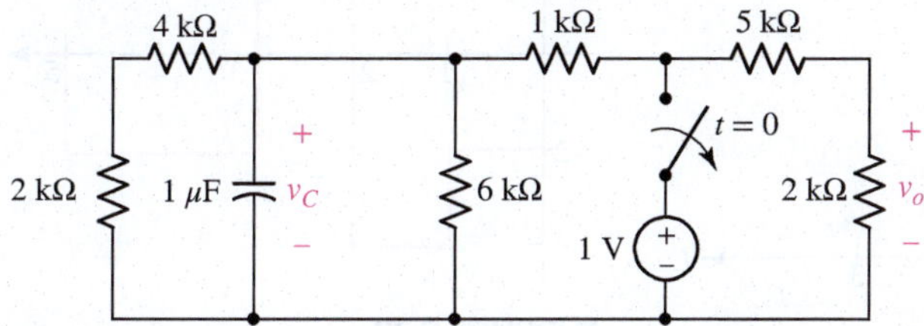

FIGURE 8.68

37. For the circuit shown in Fig. 8.69, determine (*a*) $v_C(0^-)$; (*b*) $v_C(0^+)$; (*c*) the circuit time constant; (*d*) $v_C(3\text{ ms})$.

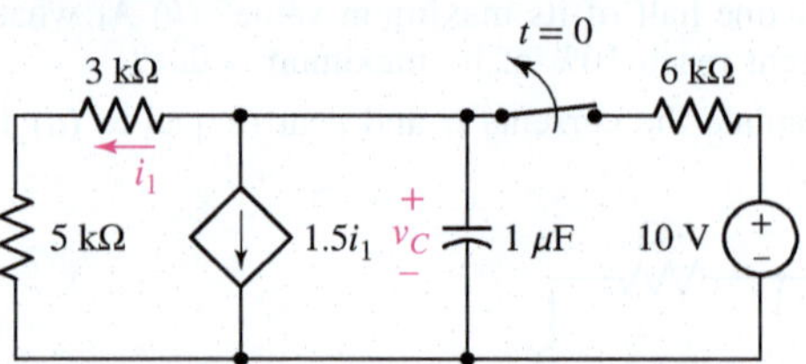

FIGURE 8.69

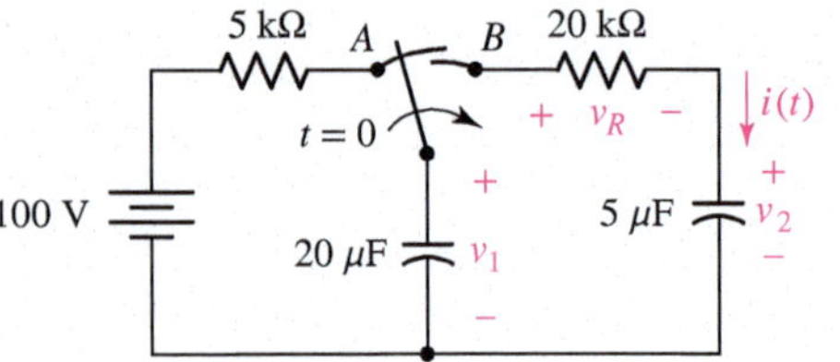

FIGURE 8.70

38. The switch in Fig. 8.70 is moved from *A* to *B* at $t = 0$ after being at *A* for a long time. This places the two capacitors in series, thus allowing equal and opposite dc voltages to be trapped on the capacitors. (*a*) Determine $v_1(0^-)$, $v_2(0^-)$, and $v_R(0^-)$. (*b*) Find $v_1(0^+)$, $v_2(0^+)$, and $v_R(0^+)$. (*c*) Determine the time constant of $v_R(t)$. (*d*) Find $v_R(t)$, $t > 0$. (*e*) Find $i(t)$. (*f*) Find $v_1(t)$ and $v_2(t)$ from $i(t)$ and the initial values. (*g*) Show that the stored energy at $t = \infty$ plus the total energy dissipated in the 20 kΩ resistor is equal to the energy stored in the capacitors at $t = 0$.

39. The inductor in Fig. 8.71 is storing 54 nJ at $t = 0^-$. Compute the energy remaining at t equal to (*a*) 0^+; (*b*) 1 ms; (*c*) 5 ms.

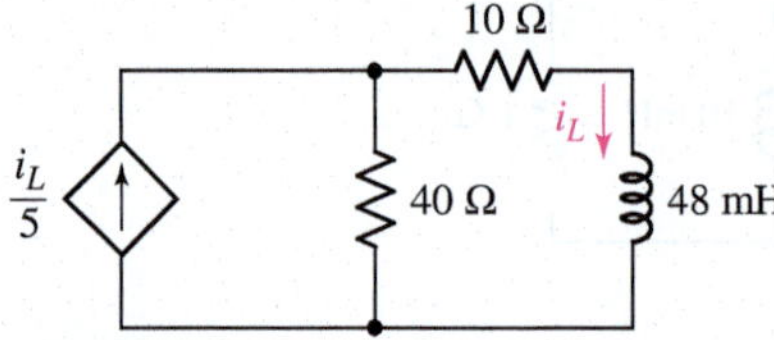

FIGURE 8.71

8.5 The Unit-Step Function

40. Evaluate the following functions at $t = -2$, 0^+, and $+2$: (*a*) $f(t) = 3u(t)$; (*b*) $g(t) = 5u(-t) + 3$; (*c*) $h(t) = 5u(t - 3)$; (*d*) $z(t) = 7u(1 - t) + 4u(t + 3)$.

41. Evaluate the following functions at $t = -1$, 0, and $+3$ (assume $u(0) = 1$): (*a*) $f(t) = tu(1 - t)$; (*b*) $g(t) = 8 + 2u(2 - t)$; (*c*) $h(t) = u(t + 1) - u(t - 1) + u(t + 2) - u(t - 4)$; (*d*) $z(t) = 1 + u(3 - t) + u(t - 2)$.

42. Sketch the following functions over the range $-3 \le t \le 3$: (*a*) $v(t) = 3 - u(2 - t) - 2u(t)$ V; (*b*) $i(t) = u(t) - u(t - 0.5) + u(t - 1) - u(t - 1.5) + u(t - 2) - u(t - 2.5)$ A; (*c*) $q(t) = 8u(-t)$ C.

43. Use step functions to construct an equation that describes the waveform sketched in Fig. 8.72.

44. Employing step functions as appropriate, describe the voltage waveform graphed in Fig. 8.73.

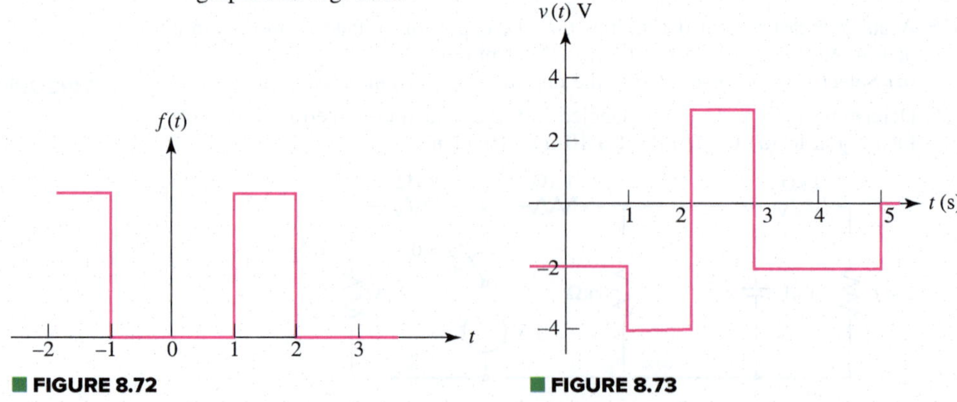

FIGURE 8.72

FIGURE 8.73

45. You can use MATLAB to represent the unit-step function using the function *heaviside(x)*. Use MATLAB to plot the function shown in Fig. 8.30.

8.6 Driven *RC* Circuits

46. With reference to the circuit depicted in Fig. 8.74, compute $v(t)$ for (*a*) $t = 0^-$; (*b*) $t = 0^+$; (*c*) $t = 2$ ms; (*d*) $t = 5$ ms.

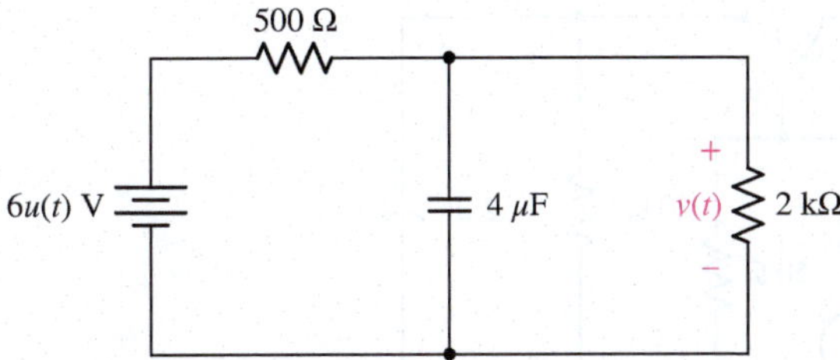

FIGURE 8.74

47. For the circuit given in Fig. 8.75, (*a*) determine $v_C(0^-)$, $v_C(0^+)$, $i_C(0^-)$, and $i_C(0^+)$; (*b*) calculate $v_C(20 \text{ ms})$ and $i_C(20 \text{ ms})$. (*c*) Verify your answer to part (*b*) with an appropriate SPICE simulation.

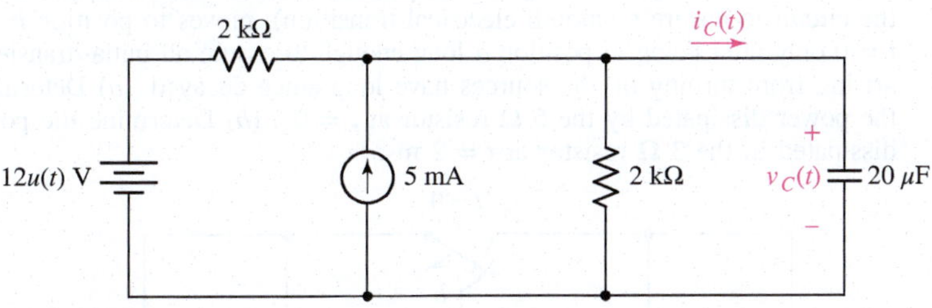

FIGURE 8.75

48. (*a*) Obtain an expression for v_C in the circuit of Fig. 8.76 valid for all values of t. (*b*) Sketch $v_C(t)$ over the range $0 \le t \le 4$ μs.

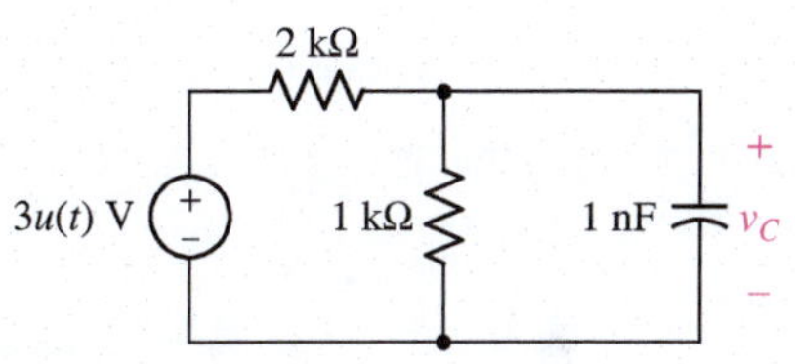

FIGURE 8.76

49. Obtain an equation which describes the behavior of i_A as labeled in Fig. 8.77 over the range of $-1 \text{ ms} \le t \le 5$ ms.

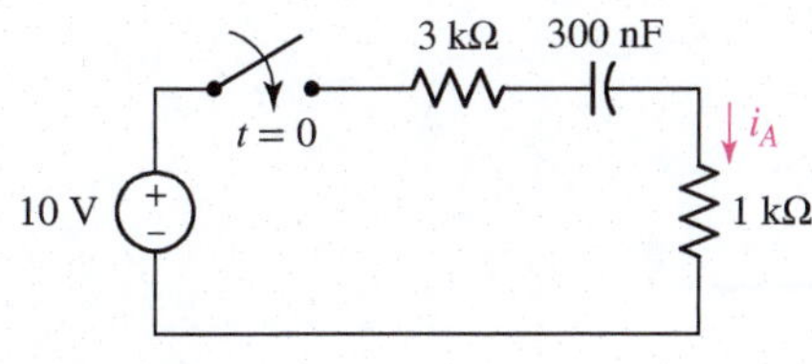

FIGURE 8.77

50. You build a portable solar charging circuit consisting of a supercapacitor and a solar cell that provides 100 mA and 3 V. If the series resistance is 10 Ω and the supercapacitor is 50 F, determine (*a*) the total energy storage that can be achieved by the capacitor, and (*b*) the time it takes to charge to 95% of full capacity.

51. The switch in the circuit of Fig. 8.78 has been closed an incredibly long time, before being thrown open at $t = 0$. (*a*) Evaluate the current labeled i_x at $t = 70$ ms. (*b*) Verify your answer with an appropriate SPICE simulation.

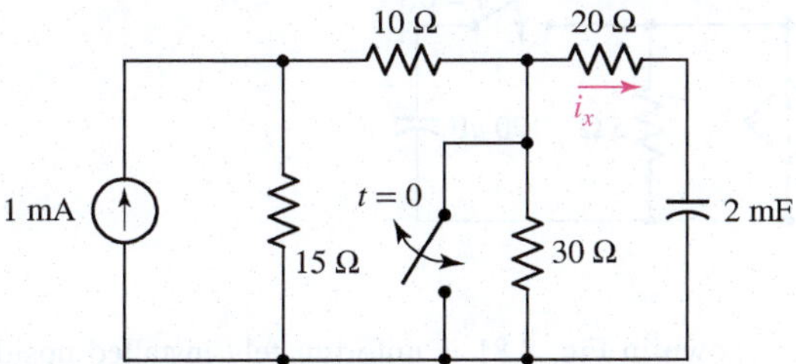

FIGURE 8.78

52. The switch in the circuit of Fig. 8.78 has been open a really, really, incredibly long time, before being closed without further fanfare at $t = 0$. (*a*) Evaluate the current labeled i_x at $t = 70$ ms. (*b*) Verify your answer with an appropriate SPICE simulation.

53. The "make-before-break" switch shown in Fig. 8.79 has been in position a since the first episode of *Jonny Quest* aired on television. It is moved to position b, finally, at time $t = 0$. (*a*) Obtain expressions for $i(t)$ and $v_C(t)$ valid for all values of t. (*b*) Determine the energy remaining in the capacitor at $t = 33\ \mu s$.

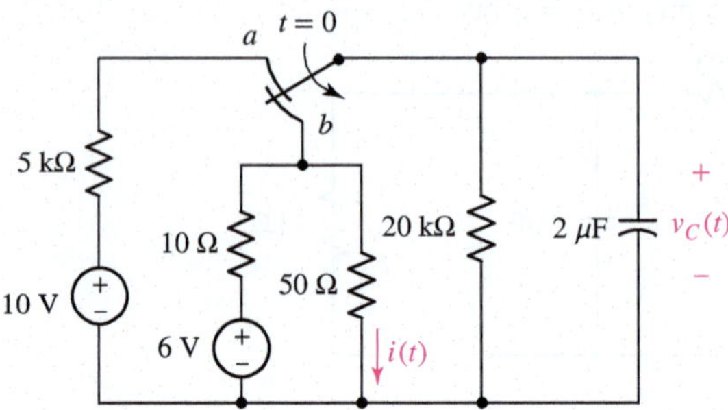

■ **FIGURE 8.79**

54. The switch in the circuit of Fig. 8.80, often called a *make-before-break* switch (since during switching it briefly makes contact with both parts of the circuit to ensure a smooth electrical transition), moves to position b at $t = 0$ only after being in position a long enough to ensure all initial transients arising from turning on the sources have long since decayed. (*a*) Determine the power dissipated by the 5 Ω resistor at $t = 0^-$. (*b*) Determine the power dissipated in the 3 Ω resistor at $t = 2$ ms.

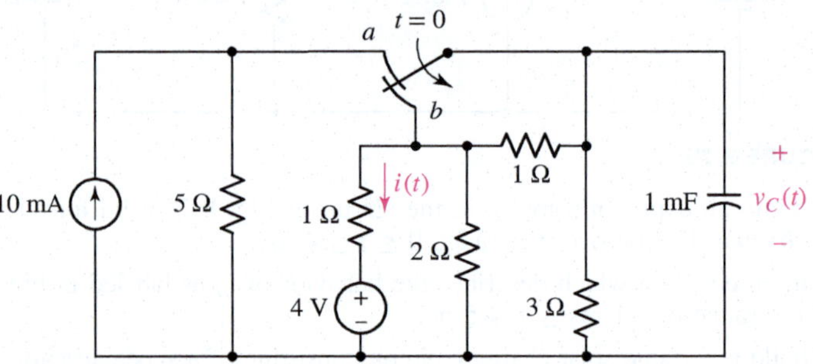

■ **FIGURE 8.80**

55. Referring to the circuit represented in Fig. 8.81, (*a*) obtain an equation which describes v_C valid for all values of t; (*b*) determine the energy remaining in the capacitor at $t = 0^+$, $t = 25\ \mu s$, and $t = 150\ \mu s$.

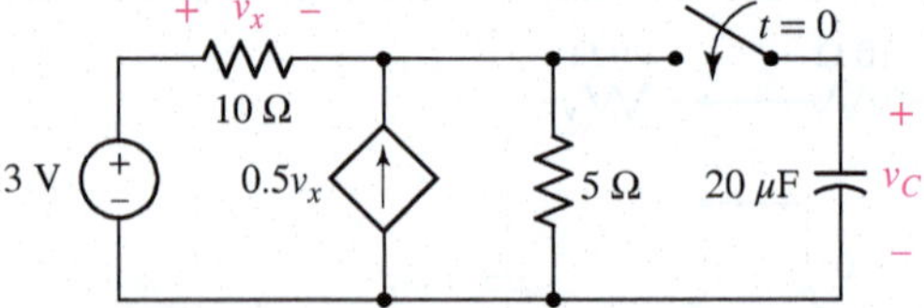

■ **FIGURE 8.81**

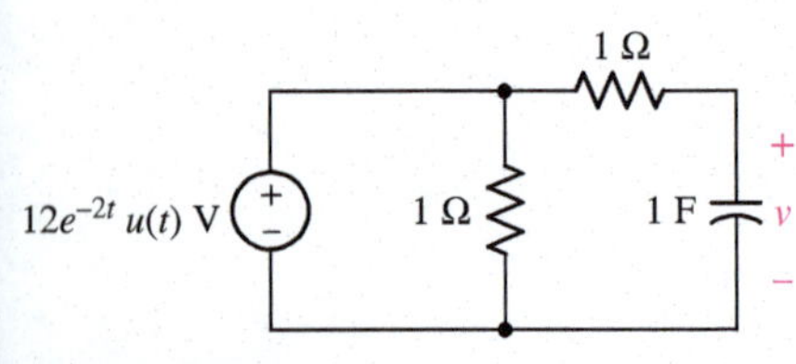

■ **FIGURE 8.82**

56. The dependent source shown in Fig. 8.81 is unfortunately installed upside down during manufacturing, so that the current arrow is pointing downward. This is not detected by the quality assurance team, so the unit ships out wired improperly. The capacitor is initially discharged. If the 5 Ω resistor is only rated to 2 W, after what time t is the circuit likely to fail?

57. For the circuit represented in Fig. 8.82, (*a*) obtain an expression for v which is valid for all values of t; (*b*) sketch your result for $0 \le t \le 3$ s.

58. Obtain an expression for the voltage v_x as labeled in the op amp circuit of Fig. 8.83.

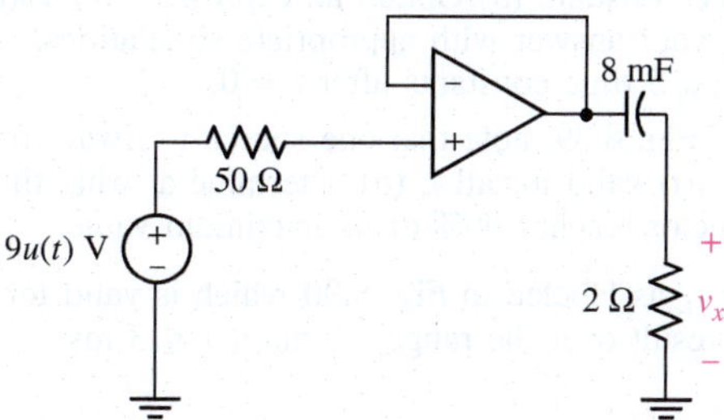

FIGURE 8.83

8.7 Driven *RL* Circuits

59. With reference to the simple circuit depicted in Fig. 8.84, compute $i(t)$ for (*a*) $t = 0^-$; (*b*) $t = 0^+$; (*c*) $t = 1^-$; (*d*) $t = 1^+$; (*e*) $t = 2$ ms.

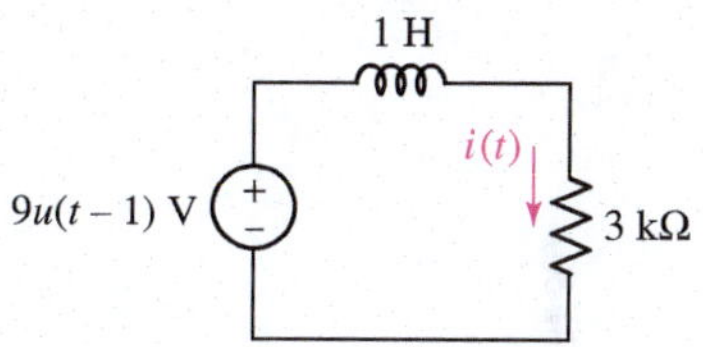

FIGURE 8.84

60. For the circuit given in Fig. 8.85, (*a*) determine $v_L(0^-)$, $v_L(0^+)$, $i_L(0^-)$, and $i_L(0^+)$; (*b*) calculate $i_L(150 \text{ ns})$. (*c*) Verify your answer to part (*b*) with an appropriate SPICE simulation.

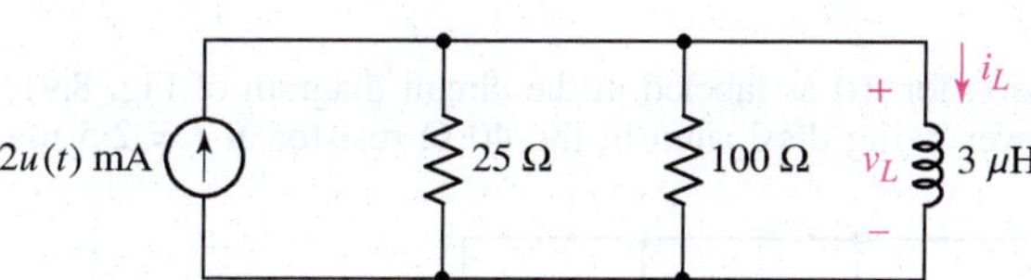

FIGURE 8.85

61. The circuit depicted in Fig. 8.86 contains two independent sources, one of which is only active for $t > 0$. (*a*) Obtain an expression for $i_L(t)$ valid for all t; (*b*) calculate $i_L(t)$ at $t = 10\ \mu$s, $20\ \mu$s, and $50\ \mu$s.

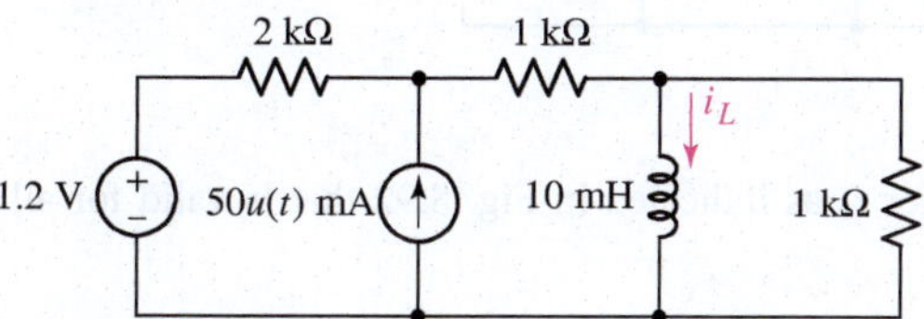

FIGURE 8.86

62. The circuit shown in Fig. 8.87 is powered by a source which is inactive for $t < 0$. (*a*) Obtain an expression for $i(t)$ valid for all t. (*b*) Graph your answer over the range of $-1 \text{ ms} \leq t \leq 10$ ms.

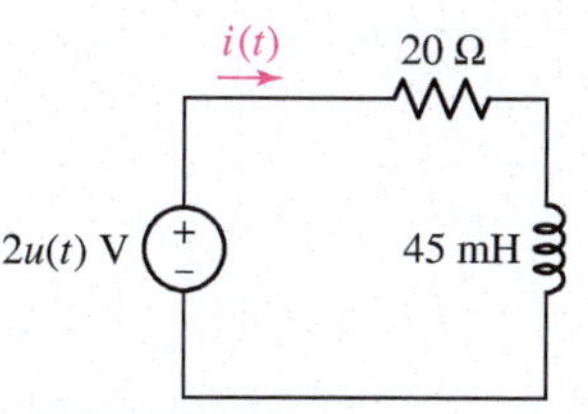

FIGURE 8.87

63. For the circuit shown in Fig. 8.88, (*a*) obtain an expression for $i(t)$ valid for all time; (*b*) obtain an expression for $v_R(t)$ valid for all time; and (*c*) graph both $i(t)$ and $v_R(t)$ over the range of $-1 \text{ s} \leq t \leq 6$ s.

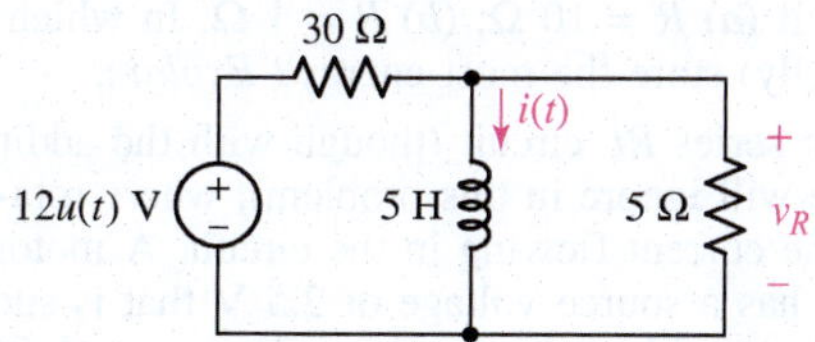

FIGURE 8.88

64. The circuit of Fig. 8.88 is modified such that the 5 H inductor is replaced with a 5 mH inductor, and the 30 resistor is replaced with a 5 Ω resistor. (*a*) Determine the circuit time constant. (*b*) Obtain an expression for $v_R(t)$ valid for all time. (*c*) Verify your answer with appropriate simulations, taking care to simulate at least 5 time constants after $t = 0$.

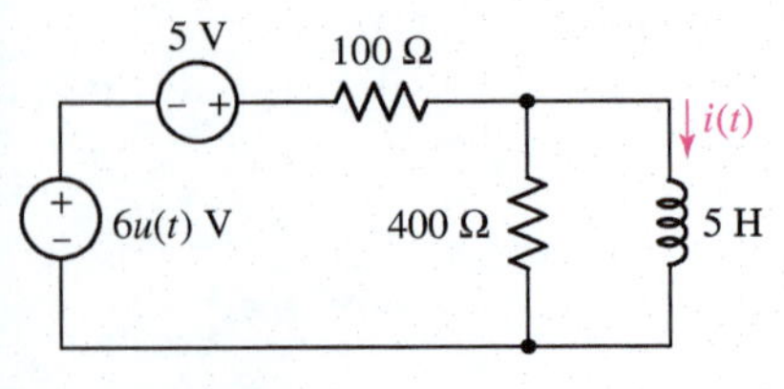

■ **FIGURE 8.89**

65. For the two-source circuit of Fig. 8.89, note that one source is always on. (*a*) Obtain an expression for $i(t)$ valid for all t; (*b*) determine at what time the energy stored in the inductor reaches 99% of its maximum value.

66. (*a*) Obtain an expression for i_L as labeled in Fig. 8.90 which is valid for all values of t. (*b*) Sketch your result over the range $-1 \text{ ms} \leq t \leq 3 \text{ ms}$.

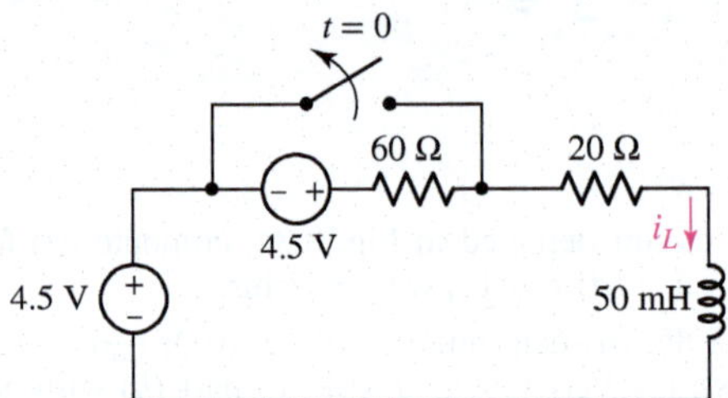

■ **FIGURE 8.90**

67. Obtain an expression for $i(t)$ as labeled in the circuit diagram of Fig. 8.91, and determine the power being dissipated in the 40 Ω resistor at $t = 2.5$ ms.

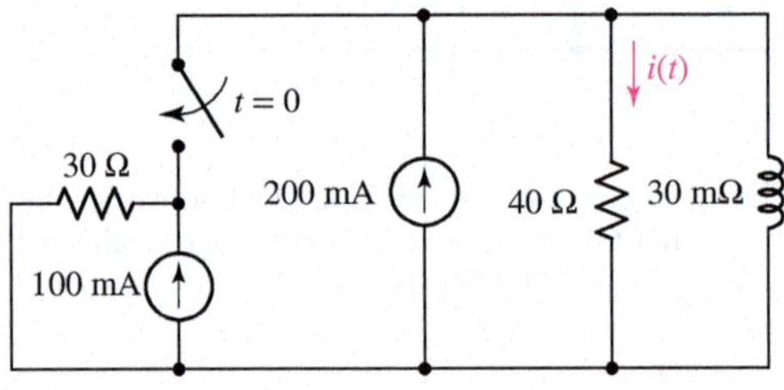

■ **FIGURE 8.91**

68. Obtain an expression for i_1 as indicated in Fig. 8.92 that is valid for all values of t.

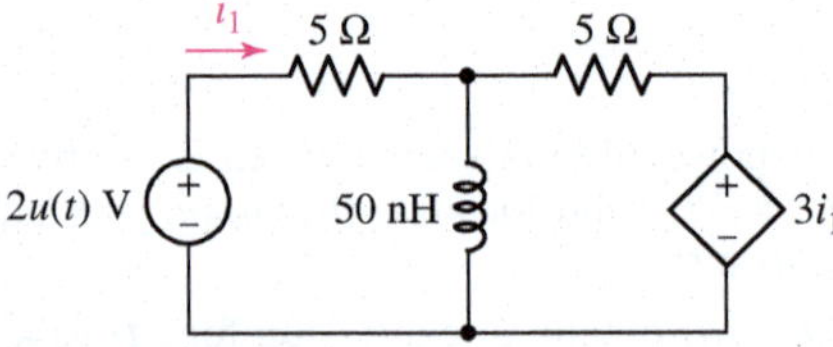

■ **FIGURE 8.92**

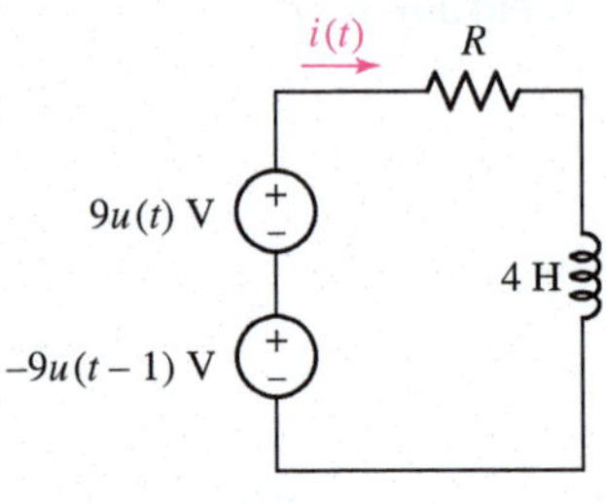

■ **FIGURE 8.93**

69. Plot the current $i(t)$ in Fig. 8.93 if (*a*) $R = 10$ Ω; (*b*) $R = 1$ Ω. In which case does the inductor (temporarily) store the most energy? *Explain.*

70. A dc motor can be modeled as a series *RL* circuit (though with the addition of a back emf voltage, which we will ignore in this problem), where rotational speed is proportional to the current flowing in the circuit. A motor with $R = 10$ Ω and $L = 20$ mH has a source voltage of 2.5 V that is suddenly increased to 5 V. How long would it take for the motor to reach 95% of full speed?

8.8 Predicting the Response of Sequentially Switched Circuits

71. Sketch the current i_L of the circuit in Fig. 8.45*a* if the 100 mH inductor is replaced by a 1 nH inductor and is subjected to the waveform $v_s(t)$ equal to (*a*) $5u(t) - 5u(t - 10^{-9}) + 5u(t - 2 \times 10^{-9})$ V, $0 \le t \le 4$ ns; (*b*) $9u(t) - 5u(t - 10^{-8}) + 5u(t - 2 \times 10^{-8})$ V, $0 \le t \le 40$ ns.

72. The 100 mH inductor in the circuit of Fig. 8.45*a* is replaced with a 1 H inductor. Sketch the inductor current i_L if the source $v_s(t)$ is equal to (*a*) $5u(t) - 5u(t - 0.01) + 5u(t - 0.02)$ V, $0 \le t \le 40$ ms; (*b*) $5u(t) - 5u(t - 10) + 5u(t - 10.1)$ V, $0 \le t \le 11$ s.

73. Sketch the voltage v_C across the capacitor of Fig. 8.94 for at least three periods if $R = 1\ \Omega$, $C = 1$ F, and $v_s(t)$ is a pulsed waveform having (*a*) minimum of 0 V, maximum of 2 V, rise and fall times of 1 ms, pulse width of 10 s, and period of 20 s; (*b*) minimum of 0 V, maximum of 2 V, rise and fall times of 1 ms, pulse width of 10 ms, and period of 20 ms. (*c*) Verify your answers with appropriate SPICE simulations.

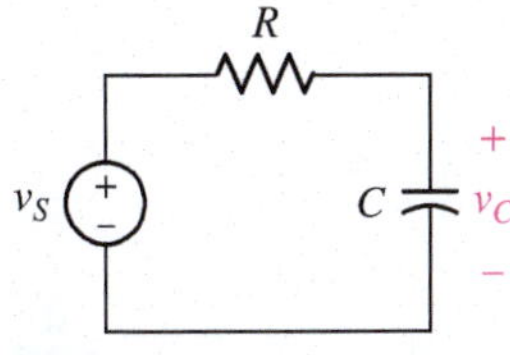

FIGURE 8.94

74. Sketch the voltage v_C across the capacitor of Fig. 8.94 for at least three periods if $R = 1\ \Omega$, $C = 1$ F, and $v_s(t)$ is a pulsed waveform having (*a*) minimum of 0 V, maximum of 2 V, rise and fall times of 1 ms, pulse width of 10 s, and period of 10.01 s; (*b*) minimum of 0 V, maximum of 2 V, rise and fall times of 1 ms, pulse width of 10 ms, and period of 10 s. (*c*) Verify your answers with appropriate SPICE simulations.

75. A series *RC* sequentially switched circuit has $R = 200\ \Omega$ and $C = 50\ \mu$F. The input is a 5 V pulsed voltage source of 50 ms pulse width with a period of 55 ms (source is zero volts outside of pulse), and it has zero rise and fall time. Calculate the capacitor voltage for three full periods and plot using MATLAB.

Chapter-Integrating Exercises

76. Refer to the circuit of Fig. 8.95, which contains a voltage-controlled dependent voltage source in addition to two resistors. (*a*) Compute the circuit time constant. (*b*) Obtain an expression for v_x valid for all *t*. (*c*) Plot the power dissipated in the 4 Ω resistor over the range of six time constants. (*d*) Repeat parts (*a*) to (*c*) if the dependent source is installed in the circuit upside down. (*e*) Are both circuit configurations "stable"? *Explain.*

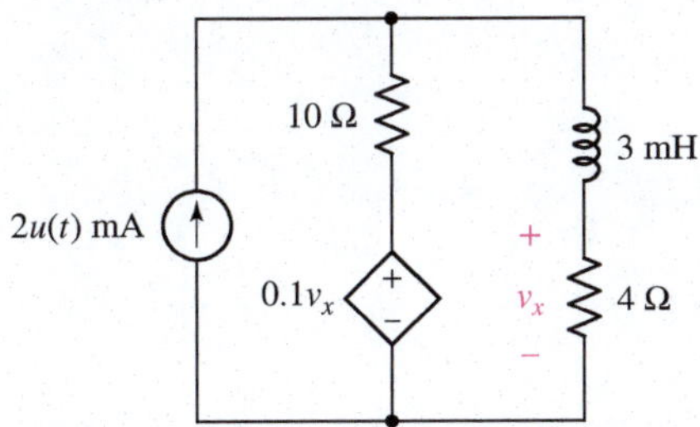

FIGURE 8.95

77. In the circuit of Fig. 8.95, a 3 mF capacitor is accidentally installed instead of the inductor. Unfortunately, that's not the end of the problems, as it's later determined that the real capacitor is not really well modeled by an ideal capacitor, and the dielectric has a resistance of 10 kΩ (which should be viewed as connected in parallel to the ideal capacitor). (*a*) Compute the circuit time constant with and without taking the dielectric resistance into account. By how much does the dielectric change your answer? (*b*) Calculate v_x at $t = 200$ ms. Does the dielectric resistance affect your answer significantly? *Explain.*

78. For the circuit of Fig. 8.96, assuming an ideal op amp, derive an expression for $v_o(t)$ if v_s is equal to (a) $4u(t)$ V; (b) $4e^{-130{,}000t}u(t)$ V.

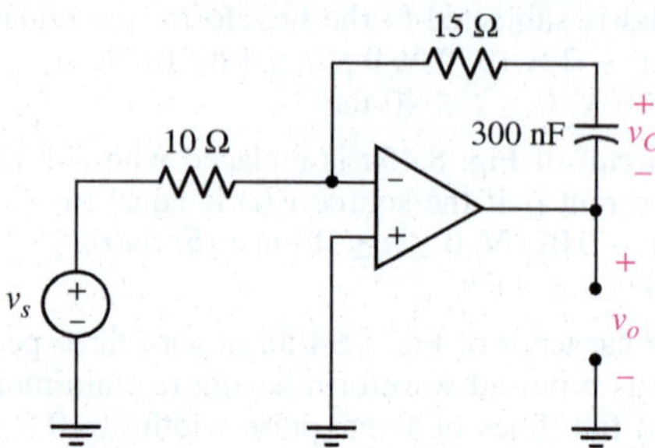

■ **FIGURE 8.96**

79. Consider the circuit of Fig. 8.42a. (a) With a 9 V peak voltage, select pulse width and period, as well as capacitor and resistor values such that the capacitor voltage (with respect to ground) approaches an average value of 5 V in 10 ms. Use standard 5% ("E24") tolerance resistor values and 10% ("E12") tolerance electrolytic capacitor values. (b) Use an appropriate simulation to verify the range of the voltage in the interval of 9 to 10 ms, including the effects of the component tolerances.

80. The circuit of Fig. 8.97 contains two switches that always operate together perfectly except for one quirk: when one switch opens, the other closes (and vice versa). Assume at $t = 0^-$, that switch A is open while switch B is closed. Starting at $t = 0$, they change positions every 30 ms. Calculate the capacitor voltage (with respect to the bottom node, which we take as the reference node) at $t =$ (a) 0^-; (b) 0^+; (c) 10 ms; (d) 30^- ms; (d) 30^+ ms; and (e) 40 ms.

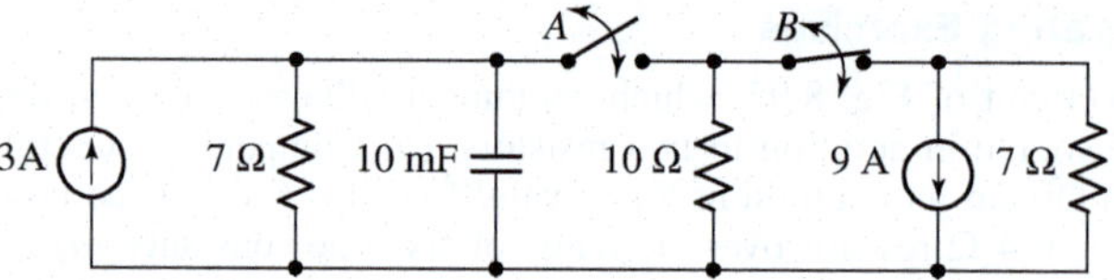

■ **FIGURE 8.97**

CHAPTER 9

The *RLC* Circuit

KEY CONCEPTS

Resonant Frequency and Damping Factor of Series and Parallel *RLC* Circuits

Overdamped Response

Critically Damped Response

Underdamped Response

Making Use of Two Initial Conditions

Complete (Natural + Forced) Response of *RLC* Circuits

Representing Differential Equations Using Op Amp Circuits

INTRODUCTION

In Chap. 8 we studied circuits which contained only **one** energy storage element, combined with a passive network which partly determined how long it took either the capacitor or the inductor to charge/discharge. The differential equations which resulted from analysis were always first-order. In this chapter, we consider more complex circuits which contain **both** an inductor **and** a capacitor. The result is a **second-order** differential equation for any voltage or current of interest. What we learned in Chap. 8 is easily extended to the study of these so-called *RLC* circuits, although now we need **two** initial conditions to solve each differential equation. Such circuits occur routinely in a wide variety of applications, including oscillators and frequency filters. They are also very useful in modeling a number of practical situations, such as automobile suspension systems, temperature controllers, and even the response of an airplane to changes in elevator and aileron positions.

9.1 THE SOURCE-FREE PARALLEL CIRCUIT

There are two basic types of *RLC* circuits: *parallel connected* and *series connected*. We could start with either, but we somewhat arbitrarily choose to begin by considering parallel *RLC* circuits. This combination of ideal elements is a reasonable model for portions of many communication networks. It represents, for example, an important part of certain electronic amplifiers found in radios, enabling the amplifiers to produce a large voltage amplification over a narrow band of signal frequencies (with almost zero amplification outside this band).

Frequency selectivity of this kind enables us to listen to the transmission of one station while rejecting the transmission of any

other station. Other applications include the use of parallel *RLC* circuits in frequency multiplexing and harmonic-suppression filters. However, even a simple discussion of these principles requires an understanding of such terms as *resonance, frequency response,* and *impedance,* which we have not yet discussed. Let it suffice to say, therefore, that an understanding of the natural behavior of the parallel *RLC* circuit is fundamentally important to future studies of communications networks and filter design, as well as many other applications.

When a *physical* capacitor is connected in parallel with an inductor and the capacitor has associated with it a finite resistance, the resulting network can be shown to have an equivalent circuit model like that shown in Fig. 9.1. The presence of this resistance can be used to model energy loss in the capacitor; over time, all real capacitors will eventually discharge, even if disconnected from a circuit. Energy losses in the physical inductor can also be taken into account by adding an ideal resistor in series with the ideal inductor. For simplicity, however, we restrict our discussion to the case of an essentially ideal inductor in parallel with a "leaky" capacitor.

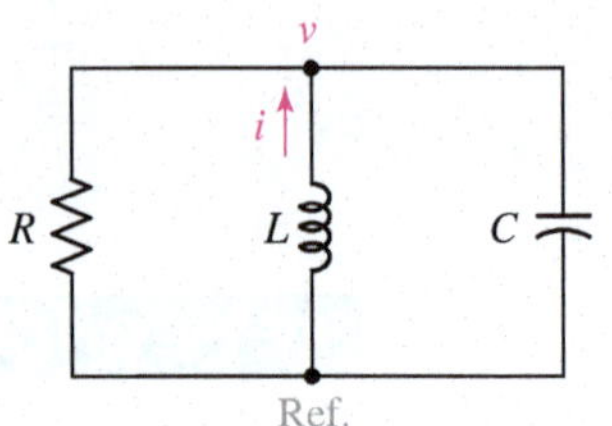

FIGURE 9.1 The source-free parallel *RLC* circuit.

Just as we did with *RL* and *RC* circuits, we first consider the natural response of a parallel *RLC* circuit, where one or both of the energy storage elements have some nonzero initial energy (the origin of which for now is unimportant). This is represented by the inductor current and the capacitor voltage, both specified at $t = 0^+$. Once we're comfortable with this part of *RLC* circuit analysis, we can easily include dc sources, switches, or step sources in the circuit. We then find the total response, which will be the sum of the natural response and the forced response.

Starting with Physical Intuition

Before we solve any equations, let us think about how the circuit might behave. Suppose we begin with a charged capacitor and zero current flowing through the inductor. Our intuition from Chap. 8 suggests that the capacitor will begin to discharge, with current flowing through both the resistor and the inductor. As current flows through the inductor, it will store energy and try to maintain current flow. Energy will continuously transfer back and forth between the capacitor and the inductor, resulting in ***oscillation.*** The resistor will dissipate some fraction of the energy, or ***dampen*** the oscillation. The behavior is similar to a mass on a spring, where the spring produces oscillations while outside forces such as friction will provide damping. In the absence of damping, oscillations will continue indefinitely, while large damping effects can stifle oscillations altogether.

Examining our circuit more carefully, let us follow the progression of energy transfer, as illustrated in Fig. 9.2 where the capacitor is initially storing energy while the inductor is not:

1. Capacitor discharges, energy stored in inductor, positive v, negative i
2. Inductor releases energy, energy stored in capacitor, negative v, negative i
3. Capacitor charges, energy stored in inductor, negative v, positive i
4. Inductor releases, energy stored in capacitor, positive v, positive i

Throughout the energy transfer, energy is dissipated in the resistor. If the energy is dissipated before transfer between the capacitor and inductor

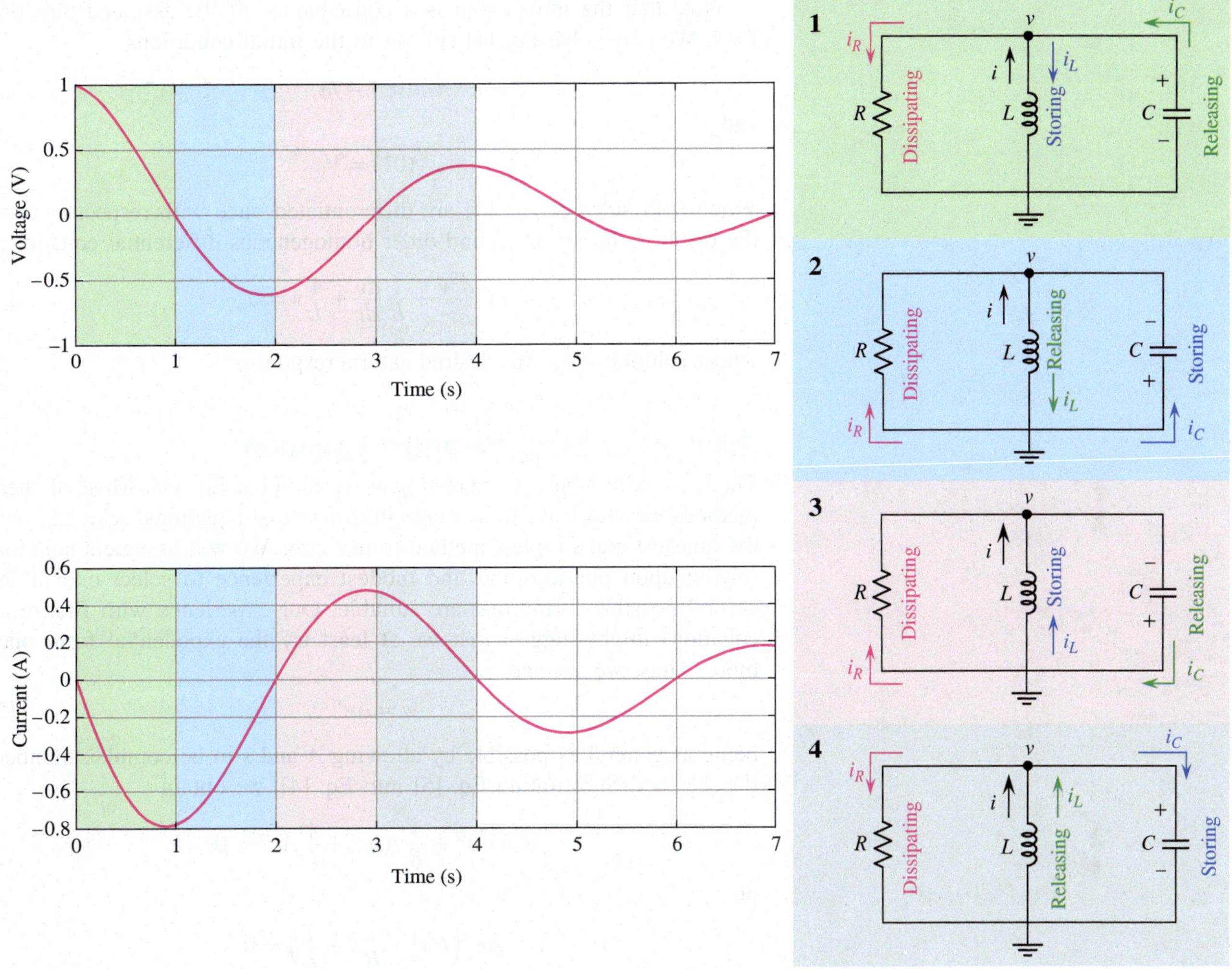

■ **FIGURE 9.2** Illustration of general energy transfer, oscillation, and damping characteristics of a parallel RLC circuit.

is completed, there will be no oscillation! This qualitative description provides a general intuition for energy transfer in *RLC* circuits (and holds strictly true for the case of no damping), though we will find that there are time segments in the oscillations for damped circuits where the inductor and capacitor may both be releasing energy in the circuit. As a result, you will not always have a direct correlation of nodes/antinodes in the oscillating voltage and current waveforms.

Obtaining the Differential Equation for a Parallel *RLC* Circuit

In the following analysis we will assume that energy may be stored initially in both the inductor and the capacitor; in other words, nonzero initial values of both inductor current and capacitor voltage may be present. With reference to the circuit of Fig. 9.1, we may then write the single nodal equation

$$\frac{v}{R} + \frac{1}{L}\int_{t_0}^{t} v\,dt' - i(t_0) + C\frac{dv}{dt} = 0 \qquad [1]$$

Note that the minus sign is a consequence of the assumed direction for i. We may solve Eq. [1] subject to the initial conditions

$$i(0^+) = I_0 \tag{2}$$

and

$$v(0^+) = V_0 \tag{3}$$

When both sides of Eq. [1] are differentiated once with respect to time, the result is the linear second-order homogeneous differential equation

$$C\frac{d^2v}{dt^2} + \frac{1}{R}\frac{dv}{dt} + \frac{1}{L}v = 0 \tag{4}$$

whose solution $v(t)$ is the desired natural response.

Solution of the Differential Equation

There are a number of interesting ways to solve Eq. [4]. Most of these methods we will leave to a course in differential equations, selecting only the quickest and simplest method to use now. We will assume a solution, relying upon our intuition and modest experience to select one of the several possible forms that are suitable. Our experience with first-order equations might suggest that we at least try the exponential form once more. Thus, we *assume*

$$v = Ae^{st} \tag{5}$$

being as general as possible by allowing A and s to be complex numbers if necessary. Substituting Eq. [5] into Eq. [4], we obtain

$$CAs^2e^{st} + \frac{1}{R}Ase^{st} + \frac{1}{L}Ae^{st} = 0$$

or

$$Ae^{st}\left(Cs^2 + \frac{1}{R}s + \frac{1}{L}\right) = 0$$

In order for this equation to be satisfied for all time, at least one of the three factors (A, e^{st}, or the expression in parenthesis) must be zero. If either of the first two factors is set equal to zero, then $v(t) = 0$. This is a trivial solution of the differential equation which cannot satisfy our given initial conditions. We therefore equate the remaining factor to zero:

$$Cs^2 + \frac{1}{R}s + \frac{1}{L} = 0 \tag{6}$$

This equation is usually called the *auxiliary equation* or the ***characteristic equation,*** as we discussed in Sec. 8.1. If it can be satisfied, then our assumed solution is correct. Since Eq. [6] is a quadratic equation, there are two solutions, identified as s_1 and s_2:

$$s_1 = -\frac{1}{2RC} + \sqrt{\left(\frac{1}{2RC}\right)^2 - \frac{1}{LC}} \tag{7}$$

and

$$s_2 = -\frac{1}{2RC} - \sqrt{\left(\frac{1}{2RC}\right)^2 - \frac{1}{LC}} \tag{8}$$

If *either* of these two values is used for s in the assumed solution, then that solution satisfies the given differential equation; it thus becomes a valid solution of the differential equation.

Let us assume that we replace s with s_1 in Eq. [5], obtaining

$$v_1 = A_1 e^{s_1 t}$$

and, similarly,

$$v_2 = A_2 e^{s_2 t}$$

The former satisfies the differential equation

$$C\frac{d^2 v_1}{dt^2} + \frac{1}{R}\frac{dv_1}{dt} + \frac{1}{L}v_1 = 0$$

and the latter satisfies

$$C\frac{d^2 v_2}{dt^2} + \frac{1}{R}\frac{dv_2}{dt} + \frac{1}{L}v_2 = 0$$

Adding these two differential equations and combining similar terms, we have

$$C\frac{d^2(v_1 + v_2)}{dt^2} + \frac{1}{R}\frac{d(v_1 + v_2)}{dt} + \frac{1}{L}(v_1 + v_2) = 0$$

Linearity triumphs, and it is seen that the *sum* of the two solutions is also a solution. We thus have the general form of the natural response

$$v(t) = A_1 e^{s_1 t} + A_2 e^{s_2 t} \qquad [9]$$

where s_1 and s_2 are given by Eqs. [7] and [8]; A_1 and A_2 are two arbitrary constants which are to be selected to satisfy the two specified initial conditions.

Definition of Frequency Terms

The form of the natural response as given in Eq. [9] offers little insight into the nature of the curve we might obtain if $v(t)$ were plotted as a function of time. The relative amplitudes of A_1 and A_2, for example, will certainly be important in determining the shape of the response curve. Furthermore, the constants s_1 and s_2 can be real numbers or conjugate complex numbers, depending upon the values of R, L, and C in the given network. These two cases will produce fundamentally different response forms. Therefore, it will be helpful to make some simplifying substitutions in Eq. [9].

Since the exponents $s_1 t$ and $s_2 t$ must be dimensionless, s_1 and s_2 must have the unit of some dimensionless quantity "per second." From Eqs. [7] and [8] we therefore see that the units of $1/2RC$ and $1/\sqrt{LC}$ must also be s^{-1} (i.e., seconds^{-1}). Units of this type are called ***frequencies.***

Let us define a new term, ω_0 (omega-sub-zero, or just omega-zero):

$$\omega_0 = \frac{1}{\sqrt{LC}} \qquad [10]$$

and reserve the term ***resonant frequency*** for it. On the other hand, we will call $1/2RC$ the ***neper frequency,*** or the ***exponential damping coefficient,*** and represent it by the symbol α (alpha):

$$\alpha = \frac{1}{2RC} \qquad [11]$$

This latter descriptive expression is used because α is a measure of how rapidly the natural response decays or damps out to its steady, final value (usually zero). Finally, s, s_1, and s_2, which are quantities that will form the basis for some of our later work, are called ***complex frequencies.***

We should note that s_1, s_2, α, and ω_0 are merely symbols used to simplify the discussion of *RLC* circuits; they are not mysterious new properties of any kind. It is easier, for example, to say "*alpha*" than it is to say "*the reciprocal of 2RC*."

Let us collect these results. The natural response of the parallel *RLC* circuit is

$$v(t) = A_1 e^{s_1 t} + A_2 e^{s_2 t} \qquad [9]$$

where

$$s_1 = -\alpha + \sqrt{\alpha^2 - \omega_0^2} \qquad [10]$$

$$s_2 = -\alpha - \sqrt{\alpha^2 - \omega_0^2} \qquad [11]$$

The ratio of α to ω_0 is called the *damping ratio* by control system engineers and is designated by ζ (zeta).

$$\alpha = \frac{1}{2RC} \qquad [12]$$

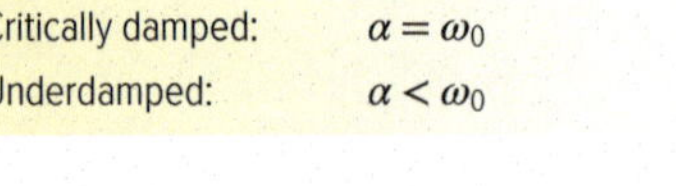

Overdamped:	$\alpha > \omega_0$
Critically damped:	$\alpha = \omega_0$
Underdamped:	$\alpha < \omega_0$

$$\omega_0 = \frac{1}{\sqrt{LC}} \qquad [13]$$

and A_1 and A_2 must be found by applying the given initial conditions.

We note two basic scenarios possible with Eqs. [10] and [11] depending on the relative sizes of α and ω_0 (dictated by the values of R, L, and C). If $\alpha > \omega_0$, s_1 and s_2 will both be real numbers, leading to what is referred to as an ***overdamped response.*** In the opposite case, where $\alpha < \omega_0$, both s_1 and s_2 will have nonzero imaginary components, leading to what is known as an ***underdamped response.*** Both of these situations are considered separately in the following sections, along with the special case of $\alpha = \omega_0$, which leads to what is called a ***critically damped response.*** We should also note that the general response comprised by Eqs. [9] through [13] describes not only the voltage but all three branch currents in the parallel *RLC* circuit; the constants A_1 and A_2 will be different for each, of course.

EXAMPLE 9.1

Consider a parallel *RLC* circuit having an inductance of 10 mH and a capacitance of 100 μF. Determine the resistor values that would lead to overdamped and underdamped responses.

We first calculate the resonant frequency of the circuit:

$$\omega_0 = \sqrt{\frac{1}{LC}} = \sqrt{\frac{1}{(10 \times 10^{-3})(100 \times 10^{-6})}} = 10^3 \text{ rad/s}$$

An *overdamped* response will result if $\alpha > \omega_0$; an *underdamped* response will result if $\alpha < \omega_0$. Thus, for the overdamped case,

$$\frac{1}{2RC} > 10^3$$

and so

$$R < \frac{1}{(2000)(100 \times 10^{-6})}$$

or

$R < 5\ \Omega$ leads to an overdamped response; $R > 5\ \Omega$ leads to an underdamped response.

PRACTICE

9.1 A parallel *RLC* circuit contains a 100 Ω resistor and has the parameter values $\alpha = 1000\ \text{s}^{-1}$ and $\omega_0 = 800$ rad/s. Find (*a*) *C*; (*b*) *L*; (*c*) s_1; (*d*) s_2.

Ans: 5 μF; 312.5 mH; $-400\ \text{s}^{-1}$; $-1600\ \text{s}^{-1}$.

9.2 THE OVERDAMPED PARALLEL *RLC* CIRCUIT

A comparison of Eqs. [12] and [13] shows that α will be greater than ω_0 if $LC > 4R^2C^2$. In this case the radical used in calculating s_1 and s_2 will be real, and both s_1 and s_2 will be real. Moreover, the following inequalities

$$\alpha > \sqrt{\alpha^2 - \omega_0^2}$$

$$\left(-\alpha - \sqrt{\alpha^2 - \omega_0^2}\right) < \left(-\alpha + \sqrt{\alpha^2 - \omega_0^2}\right) < 0$$

may be applied to Eqs. [10] and [11] to show that both s_1 and s_2 are *negative* real numbers. Thus, the response $v(t)$ can be expressed as the (algebraic) sum of two decreasing exponential terms, each of which approaches zero as time increases. In fact, since the absolute value of s_2 is larger than that of s_1, the term containing s_2 has the more rapid rate of decrease, and, for large values of time, we may write the limiting expression

$$v(t) \rightarrow A_1 e^{s_1 t} \rightarrow 0 \qquad \text{as } t \rightarrow \infty$$

The next step is to determine the arbitrary constants A_1 and A_2 in conformance with the initial conditions. We select a parallel *RLC* circuit with $R = 6\ \Omega$, $L = 7$ H, and, for ease of computation, $C = \frac{1}{42}$ F. The initial energy storage is specified by choosing an initial voltage across the circuit $v(0) = 0$ and an initial inductor current $i(0) = 10$ A, where v and i are defined in Fig. 9.3.

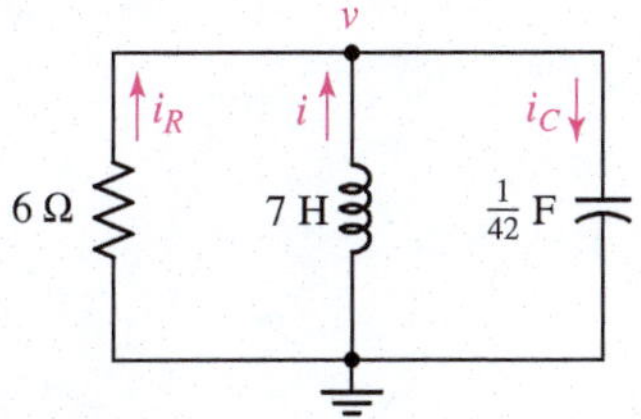

■ **FIGURE 9.3** A parallel *RLC* circuit used as a numerical example. The circuit is overdamped. The capacitor stores zero initial energy, whereas the initial current of the inductor is 10 A.

We may easily determine the values of the several parameters

$$\alpha = 3.5 \qquad \omega_0 = \sqrt{6} \qquad (\text{all s}^{-1})$$
$$s_1 = -1 \qquad s_2 = -6$$

and immediately write the general form of the natural response

$$v(t) = A_1 e^{-t} + A_2 e^{-6t} \qquad [14]$$

Finding Values for A_1 and A_2

Only the evaluation of the two constants A_1 and A_2 remains. If we knew the response $v(t)$ at two different values of time, these two values could be substituted in Eq. [14] and A_1 and A_2 easily found. However, we know only one instantaneous value of $v(t)$:

$$v(0) = 0$$

and, therefore,

$$0 = A_1 + A_2 \qquad [15]$$

We can obtain a second equation relating A_1 and A_2 by taking the derivative of $v(t)$ with respect to time in Eq. [14], determining the initial value of this derivative through the use of the remaining initial condition $i(0) = 10$, and equating the results. So, taking the derivative of both sides of Eq. [14],

$$\frac{dv}{dt} = -A_1 e^{-t} - 6A_2 e^{-6t}$$

and evaluating the derivative at $t = 0$,

$$\left.\frac{dv}{dt}\right|_{t=0} = -A_1 - 6A_2$$

For the remainder of our discussions concerning *RLC* circuits, we will always require two initial conditions in order to completely specify the response. One condition will usually be very easy to apply—either a voltage or current at $t = 0$. It is the second condition that usually requires a little effort. Although we will often have both an initial current and an initial voltage at our disposal, one of these will need to be applied indirectly through the derivative of our assumed solution.

we obtain a second equation. Although this may appear to be helpful, we do not have a numerical value for the initial value of the derivative, so we do not yet have two equations in two unknowns . . . or do we? The expression dv/dt suggests a capacitor current, since

$$i_C = C\frac{dv}{dt}$$

Kirchhoff's current law must hold at any instant in time, as it is based on conservation of electrons. Thus, we may write

$$-i_C(0) + i(0) + i_R(0) = 0$$

Substituting our expression for capacitor current and dividing by C,

$$\left.\frac{dv}{dt}\right|_{t=0} = \frac{i_C(0)}{C} = \frac{i(0) + i_R(0)}{C} = \frac{i(0)}{C} = 420 \text{ V/s}$$

since zero initial voltage across the resistor requires zero initial current through it. We thus have our second equation,

$$420 = -A_1 - 6A_2 \qquad [16]$$

and simultaneous solution of Eqs. [15] and [16] provides the two amplitudes $A_1 = 84$ and $A_2 = -84$. Therefore, the final numerical solution for the natural response of this circuit is

$$v(t) = 84(e^{-t} - e^{-6t}) \text{ V} \qquad [17]$$

EXAMPLE 9.2

Find an expression for $v_C(t)$ valid for $t > 0$ in the circuit of Fig. 9.4*a*.

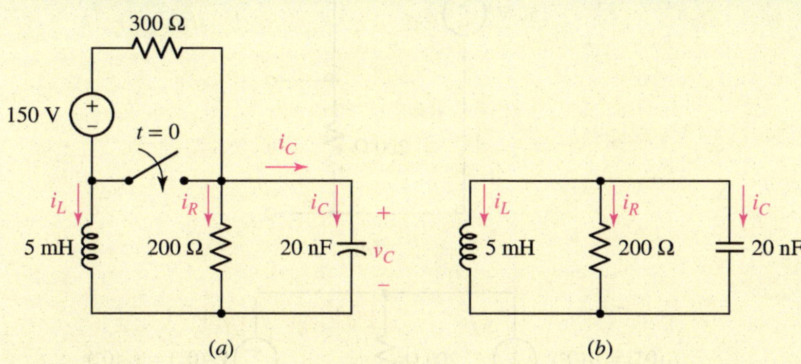

■ **FIGURE 9.4** (*a*) An *RLC* circuit that becomes source-free at $t = 0$. (*b*) The circuit for $t > 0$, in which the 150 V source and the 300 Ω resistor have been shorted out by the switch, and so are of no further relevance to v_C.

▶ ***Identify the goal of the problem.***

We are asked to find the capacitor voltage after the switch is thrown. This action leads to no sources remaining connected to either the inductor or the capacitor.

▶ ***Collect the known information.***

After the switch is thrown, the capacitor is left in parallel with a 200 Ω resistor and a 5 mH inductor (Fig. 9.4*b*). Thus, $\alpha = 1/2RC = 125{,}000\ \text{s}^{-1}$, $\omega_0 = 1/\sqrt{LC} = 100{,}000$ rad/s, $s_1 = -\alpha + \sqrt{\alpha^2 - \omega_0^2} = -50{,}000\ \text{s}^{-1}$ and $s_2 = -\alpha - \sqrt{\alpha^2 - \omega_0^2} = -200{,}000\ \text{s}^{-1}$.

▶ ***Devise a plan.***

Since $\alpha > \omega_0$, the circuit is overdamped, and so we expect a capacitor voltage of the form

$$v_C(t) = A_1 e^{s_1 t} + A_2 e^{s_2 t}$$

We know s_1 and s_2; we need to obtain and invoke two initial conditions to determine A_1 and A_2. To do this, we will analyze the circuit at $t = 0^-$ (Fig. 9.5*a*) to find $i_L(0^-)$ and $v_C(0^-)$. We will then analyze the circuit at $t = 0^+$ with the assumption that neither value changes.

▶ ***Construct an appropriate set of equations.***

From Fig. 9.5*a*, in which the inductor has been replaced with a short circuit and the capacitor with an open circuit, we see that

$$i_L(0^-) = -\frac{150}{200 + 300} = -300\ \text{mA}$$

and

$$v_C(0^-) = 150\frac{200}{200 + 300} = 60\ \text{V}$$

In Fig. 9.5*b*, we draw the circuit at $t = 0^+$, representing the inductor current and capacitor voltage by ideal sources for simplicity. Since neither can change in zero time, we know that $v_C(0^+) = 60$ V.

(Continued on next page)

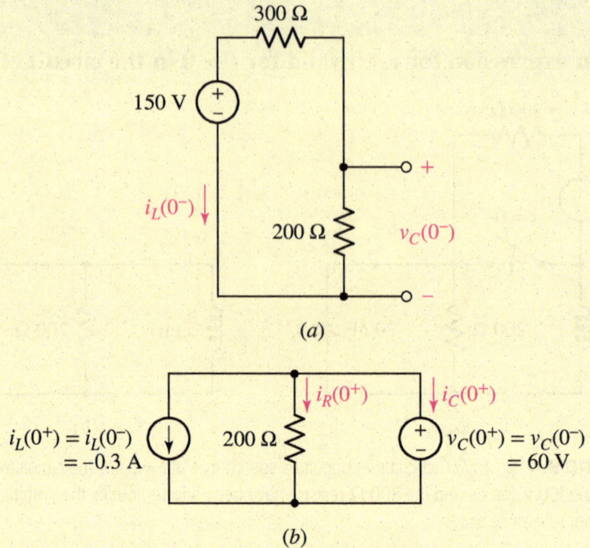

■ **FIGURE 9.5** (*a*) The equivalent circuit at $t = 0^-$; (*b*) equivalent circuit at $t = 0^+$, drawn using ideal sources to represent the initial inductor current and initial capacitor voltage.

▶ ***Determine if additional information is required.***

We have an equation for the capacitor voltage: $v_C(t) = A_1e^{-50{,}000t} + A_2e^{-200{,}000t}$. We now know $v_C(0) = 60$ V, but a third equation is still required. Differentiating our capacitor voltage equation, we find

$$\frac{dv_C}{dt} = -50{,}000A_1e^{-50{,}000t} - 200{,}000A_2e^{-200{,}000t}$$

which can be related to the capacitor current as $i_C = C(dv_C/dt)$. Returning to Fig. 9.5*b*, KCL yields

$$i_C(0^+) = -i_L(0^+) - i_R(0^+) = 0.3 - [v_C(0^+)/200] = 0$$

▶ ***Attempt a solution.***

Application of our first initial condition yields

$$v_C(0) = A_1 + A_2 = 60$$

and application of our second initial condition yields

$$i_C(0) = -20 \times 10^{-9}(50{,}000A_1 + 200{,}000A_2) = 0$$

Solving, $A_1 = 80$ V and $A_2 = -20$ V, so that

$$v_C(t) = 80e^{-50{,}000t} - 20e^{-200{,}000t} \text{ V}, \qquad t > 0$$

▶ ***Verify the solution. Is it reasonable or expected?***

At the very least, we can check our solution at $t = 0$, verifying that $v_C(0) = 60$ V. Differentiating and multiplying by 20×10^{-9}, we can also verify that $i_C(0) = 0$. Also, since we have a source-free circuit for $t > 0$, we expect that $v_C(t)$ must eventually decay to zero as t approaches ∞, which our solution does.

PRACTICE

9.2 After being open for a long time, the switch in Fig. 9.6 closes at $t = 0$. Find (*a*) $i_L(0^-)$; (*b*) $v_C(0^-)$; (*c*) $i_R(0^+)$; (*d*) $i_C(0^+)$; (*e*) $v_C(0.2)$.

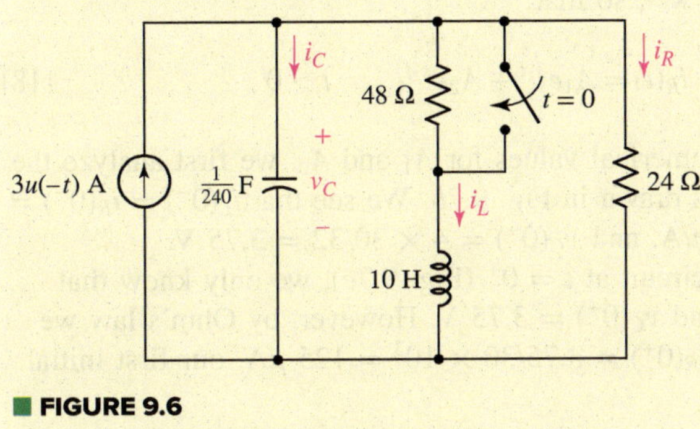

FIGURE 9.6

Ans: 1 A; 48 V; 2 A; −3 A; −17.54 V.

As noted previously, the form of the overdamped response applies to any voltage or current quantity, as we explore in the following example.

EXAMPLE **9.3**

The circuit of Fig. 9.7*a* reduces to a simple parallel *RLC* circuit after $t = 0$. Determine an expression for the resistor current i_R valid for all time.

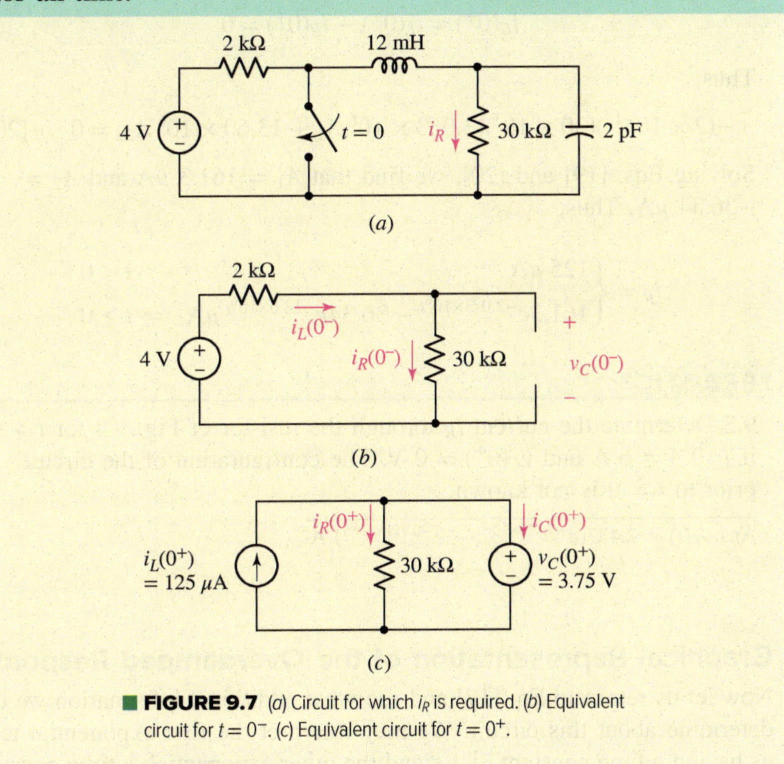

FIGURE 9.7 (*a*) Circuit for which i_R is required. (*b*) Equivalent circuit for $t = 0^-$. (*c*) Equivalent circuit for $t = 0^+$.

(*Continued on next page*)

For $t > 0$, we have a parallel *RLC* circuit with $R = 30\ \text{k}\Omega$, $L = 12$ mH, and $C = 2$ pF. Thus, $\alpha = 8.333 \times 10^6\ \text{s}^{-1}$ and $\omega_0 = 6.455 \times 10^6$ rad/s. We expect an overdamped response, with $s_1 = -3.063 \times 10^6\ \text{s}^{-1}$ and $s_2 = -13.60 \times 10^6\ \text{s}^{-1}$, so that

$$i_R(t) = A_1 e^{s_1 t} + A_2 e^{s_2 t}, \qquad t > 0 \qquad [18]$$

To determine numerical values for A_1 and A_2, we first analyze the circuit at $t = 0^-$, as drawn in Fig. 9.7*b*. We see that $i_L(0^-) = i_R(0^-) = 4/32 \times 10^3 = 125\ \mu\text{A}$, and $v_C(0^-) = 4 \times 30/32 = 3.75$ V.

In drawing the circuit at $t = 0^+$ (Fig. 9.7*c*), we only know that $i_L(0^+) = 125\ \mu\text{A}$ and $v_C(0^+) = 3.75$ V. However, by Ohm's law we can calculate that $i_R(0^+) = 3.75/30 \times 10^3 = 125\ \mu\text{A}$, our first initial condition. Thus,

$$i_R(0) = A_1 + A_2 = 125 \times 10^{-6}\ \text{A} \qquad [19]$$

How do we obtain a *second* initial condition? If we multiply Eq. [18] by 30×10^3, we obtain an expression for $v_C(t)$. Taking the derivative and multiplying by 2 pF yields an expression for $i_C(t)$:

$$i_C = C\frac{dv_C}{dt} = (2 \times 10^{-12})(30 \times 10^3)(A_1 s_1 e^{s_1 t} + A_2 s_2 e^{s_2 t})$$

By KCL,

$$i_C(0^+) = i_L(0^+) - i_R(0^+) = 0$$

Thus,

$$-(2 \times 10^{-12})(30 \times 10^3)(3.063 \times 10^6 A_1 + 13.60 \times 10^6 A_2) = 0 \qquad [20]$$

Solving Eqs. [19] and [20], we find that $A_1 = 161.3\ \mu\text{A}$ and $A_2 = -36.34\ \mu\text{A}$. Thus,

$$i_R = \begin{cases} 125\ \mu\text{A} & t < 0 \\ 161.3e^{-3.0^63\times 10^6 t} - 36.34e^{-13.6\times 10^6 t}\ \mu\text{A} & t \geq 0 \end{cases}$$

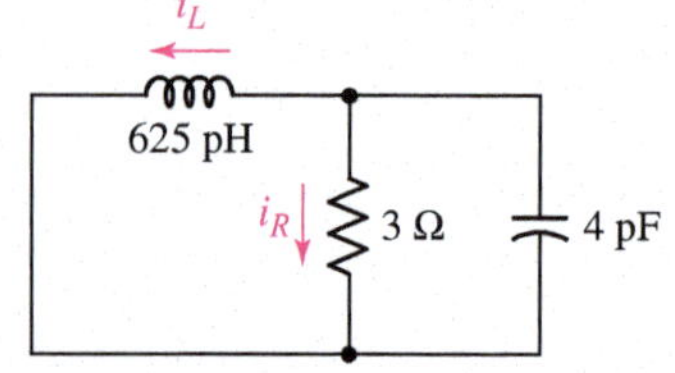

FIGURE 9.8 Circuit for Practice Problem 9.3.

PRACTICE

9.3 Determine the current i_R through the resistor of Fig. 9.8 for $t > 0$ if $i_L(0^-) = 6$ A and $v_C(0^+) = 0$ V. The configuration of the circuit prior to $t = 0$ is not known.

Ans: $i_R(t) = 2.437(e^{-7.823\times 10^{10} t} - e^{-0.511\times 10^{10} t})$ A.

Graphical Representation of the Overdamped Response

Now let us return to Eq. [17] and see what additional information we can determine about this circuit. We may interpret the first exponential term as having a time constant of 1 s and the other exponential, a time constant of $\frac{1}{6}$ s. Each starts with unity amplitude, but the latter decays more rapidly; $v(t)$ is never negative. As time becomes infinite, each term approaches zero, and the response itself dies out as it should. We therefore have a response curve which is zero at $t = 0$, is zero at $t = \infty$, and is never negative; since it is not everywhere zero, it must possess at least one maximum, and this is not a difficult point to determine exactly. We differentiate the response

$$\frac{dv}{dt} = 84(-e^{-t} + 6e^{-6t})$$

set the derivative equal to zero to determine the time t_m at which the voltage becomes maximum,

$$0 = -e^{-t_m} + 6e^{-6t_m}$$

manipulate once,

$$e^{5t_m} = 6$$

and obtain

$$t_m = 0.358 \text{ s}$$

and

$$v(t_m) = 48.9 \text{ V}$$

A reasonable sketch of the response may be made by plotting the two exponential terms $84e^{-t}$ and $84e^{-6t}$ and then taking their difference. This technique is illustrated by the curves of Fig. 9.9; the two exponentials are shown lightly, and their difference, the total response $v(t)$, is drawn as a colored line. The curves also verify our previous prediction that the functional behavior of $v(t)$ for very large t is $84e^{-t}$, the exponential term containing the smaller magnitude of s_1 and s_2.

A frequently asked question is the length of time it actually takes for the transient part of the response to disappear (or "*damp out*"). In practice, it is often desirable to have this transient response approach zero as rapidly as possible, that is, to minimize the ***settling time*** t_s**.** Theoretically, of course, t_s is infinite, because $v(t)$ never settles to zero in a finite time.

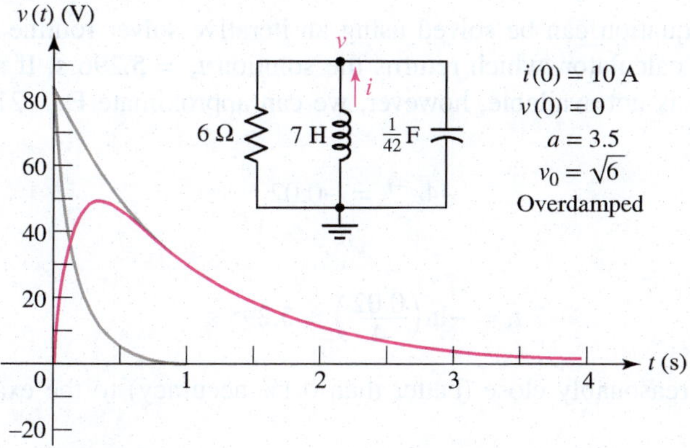

FIGURE 9.9 The response $v(t) = 84(e^{-t} - e^{-6t})$ of the network shown in Fig. 9.3.

However, a negligible response is present after the magnitude of $v(t)$ has settled to values that remain less than 1 percent of its maximum absolute value $|v_m|$. The time that is required for this to occur we define as the settling time. Since $|v_m| = v_m = 48.9$ V for our example, the settling time is the time required for the response to drop to 0.489 V. Substituting this value for $v(t)$ in Eq. [17] and neglecting the second exponential term, known to be negligible here, the settling time is found to be 5.15 s.

EXAMPLE 9.4

For $t > 0$, the capacitor current of a certain source-free parallel *RLC* circuit is given by $i_C(t) = 2e^{-2t} - 4e^{-t}$ A. Sketch the current in the range $0 < t < 5$ s, and determine the settling time.

We first sketch the two terms as shown in Fig. 9.10, then subtract them to find $i_C(t)$. The maximum value is clearly $|-2| = 2$ A. We therefore need to find the time at which $|i_C|$ has decreased to 20 mA, or

$$2e^{-2t_s} - 4e^{-t_s} = -0.02 \quad [21]$$

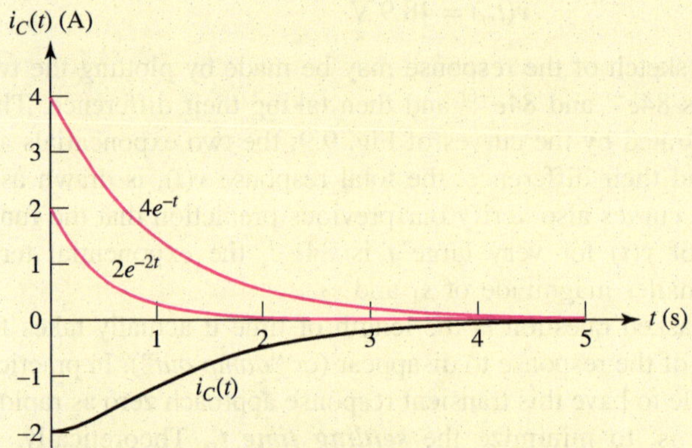

■ **FIGURE 9.10** The current response $i_C(t) = 2e^{-2t} - 4e^{-t}$ A, sketched alongside its two components.

This equation can be solved using an iterative solver routine on a scientific calculator, which returns the solution $t_s = 5.296$ s. If such an option is not available, however, we can approximate Eq. [21] for $t \geq t_s$ as

$$-4e^{-t_s} = -0.02 \quad [22]$$

Solving,

$$t_s = -\ln\left(\frac{0.02}{4}\right) = 5.298 \text{ s} \quad [23]$$

which is reasonably close (better than 0.1% accuracy) to the exact solution.

PRACTICE

9.4 (*a*) Sketch the voltage $v_R(t) = 2e^{-t} - 4e^{-3t}$ V in the range $0 < t < 5$ s. (*b*) Estimate the settling time. (*c*) Calculate the maximum positive value and the time at which it occurs.

Ans: See Fig. 9.11; 5.9 s; 544 mV, 896 ms.

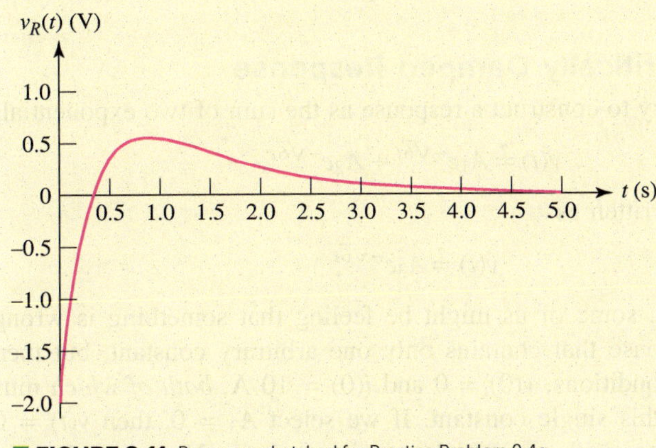

FIGURE 9.11 Response sketched for Practice Problem 9.4*a*.

9.3 CRITICAL DAMPING

The overdamped case is characterized by

$$\alpha > \omega_0$$

or

$$LC > 4R^2C^2$$

and leads to negative real values for s_1 and s_2 and to a response expressed as the algebraic sum of two negative exponentials.

Now let us adjust the element values until α and ω_0 are equal. This is a very special case which is termed ***critical damping.*** If we were to try to construct a parallel *RLC* circuit that is critically damped, we would be attempting an essentially impossible task, for we could unlikely ever make α exactly equal to ω_0. For completeness, however, we will discuss the critically damped circuit here because it shows an interesting transition between overdamping and underdamping.

"Impossible" is a pretty strong term. We make this statement because in practice it is unusual to obtain components that are closer than 1 percent of their specified values. Thus, obtaining L *precisely* equal to $4R^2C$ is theoretically possible, but not very likely, even if we're willing to measure a drawer full of components until we find the right ones.

Critical damping is achieved when

or

$$\left.\begin{aligned} \alpha &= \omega_0 \\ LC &= 4R^2C^2 \\ L &= 4R^2C \end{aligned}\right\} \text{ critical damping}$$

We can produce critical damping by changing the value of any of the three elements in the numerical example discussed at the end of Sec. 9.1.

We will select R, increasing its value until critical damping is obtained, and thus leave ω_0 unchanged. The necessary value of R is $7\sqrt{6}/2\ \Omega$; L is still 7 H, and C remains $\frac{1}{42}$ F. We thus find

$$\alpha = \omega_0 = \sqrt{6}\ \mathrm{s}^{-1}$$
$$s_1 = s_2 = -\sqrt{6}\ \mathrm{s}^{-1}$$

and recall the initial conditions that were specified, $v(0) = 0$ and $i(0) = 10$ A.

Form of a Critically Damped Response

We proceed to try to construct a response as the sum of two exponentials,

$$v(t) \stackrel{?}{=} A_1 e^{-\sqrt{6}t} + A_2 e^{-\sqrt{6}t}$$

which may be written as

$$v(t) \stackrel{?}{=} A_3 e^{-\sqrt{6}t}$$

At this point, some of us might be feeling that something is wrong. We have a response that contains only one arbitrary constant, but there are two initial conditions, $v(0) = 0$ and $i(0) = 10$ A, *both of which* must be satisfied by this single constant. If we select $A_3 = 0$, then $v(t) = 0$, which is consistent with our initial capacitor voltage. However, although there is no energy stored in the capacitor at $t = 0^+$, we have 350 J of energy initially stored in the inductor. This energy will lead to a transient current flowing out of the inductor, giving rise to a nonzero voltage across all three elements. This seems to be in direct conflict with our proposed solution.

If a mistake has not led to our difficulties, we must have begun with an incorrect assumption, and only one assumption has been made. We originally hypothesized that the differential equation could be solved by assuming an exponential solution, and this turns out to be incorrect for this single special case of critical damping. When $\alpha = \omega_0$, the differential equation, Eq. [4], becomes

$$\frac{d^2v}{dt^2} + 2\alpha\frac{dv}{dt} + \alpha^2 v = 0$$

The solution of this equation is not a difficult process, but we will avoid developing it here, since the equation is a standard type found in the usual differential equation texts. The solution is

$$v = e^{-\alpha t}(A_1 t + A_2) \qquad [24]$$

It should be noted that the solution is still expressed as the sum of two terms, where one term is the familiar negative exponential and the second is t times a negative exponential. We should also note that the solution contains the *two* expected arbitrary constants.

Finding Values for A_1 and A_2

Let us now complete our numerical example. After we substitute the known value of α in Eq. [24], obtaining

$$v = A_1 t e^{-\sqrt{6}t} + A_2 e^{-\sqrt{6}t}$$

we establish the values of A_1 and A_2 by first imposing the initial condition on $v(t)$ itself, $v(0) = 0$. Thus, $A_2 = 0$. This simple result occurs because the initial value of the response $v(t)$ was selected as zero; the more general case will require the solution of two equations simultaneously. The second initial condition must be applied to the derivative dv/dt just as in the overdamped case. We therefore differentiate, remembering that $A_2 = 0$:

$$\frac{dv}{dt} = A_1 t(-\sqrt{6})e^{-\sqrt{6}t} + A_1 e^{-\sqrt{6}t}$$

evaluate at $t = 0$:

$$\left.\frac{dv}{dt}\right|_{t=0} = A_1$$

and express the derivative in terms of the initial capacitor current:

$$\left.\frac{dv}{dt}\right|_{t=0} = \frac{i_C(0)}{C} = \frac{i_R(0)}{C} + \frac{i(0)}{C}$$

where reference directions for i_C, i_R, and i are defined in Fig. 9.3. Thus,

$$A_1 = 420 \text{ V}$$

The response is, therefore,

$$v(t) = 420te^{-2.45t} \text{ V} \qquad [25]$$

Graphical Representation of the Critically Damped Response

Before plotting this response in detail, let us again try to anticipate its form by qualitative reasoning. The specified initial value is zero, and Eq. [25] concurs. It is not immediately apparent that the response also approaches zero as t becomes infinitely large, because $te^{-2.45t}$ is an indeterminate form. However, this obstacle is easily overcome by use of L'Hôspital's rule, which yields

$$\lim_{t\to\infty} v(t) = 420 \lim_{t\to\infty} \frac{t}{e^{2.45t}} = 420 \lim_{t\to\infty} \frac{1}{2.45e^{2.45t}} = 0$$

and once again we have a response that begins and ends at zero and has positive values at all other times. A maximum value v_m occurs at time t_m; for our example,

$$t_m = 0.408 \text{ s} \qquad \text{and} \qquad v_m = 63.1 \text{ V}$$

This maximum is larger than that obtained in the overdamped case and is a result of the smaller losses that occur in the larger resistor; the time of the maximum response is slightly later than it was with overdamping. The settling time may also be determined by solving

$$\frac{v_m}{100} = 420t_s e^{-2.45t_s}$$

for t_s (by trial-and-error methods or a calculator's SOLVE routine):

$$t_s = 3.12 \text{ s}$$

which is a considerably smaller value than that which arose in the overdamped case (5.15 s). As a matter of fact, it can be shown that, for given

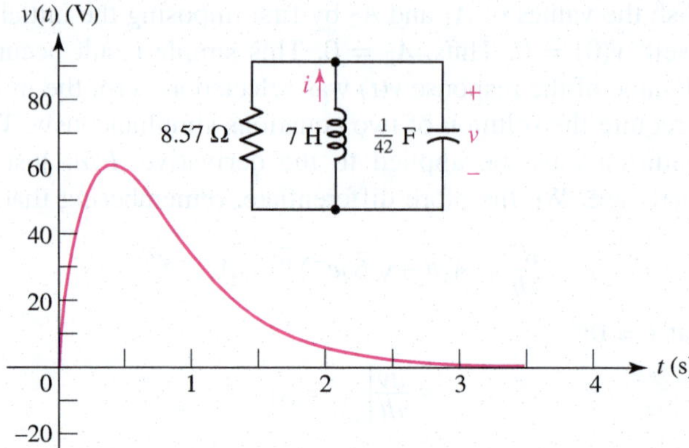

■ **FIGURE 9.12** The response $v(t) = 420te^{-2.45t}$ of the network shown in Fig. 9.3 with R changed to provide critical damping.

values of L and C, the selection of that value of R which provides critical damping will always give a shorter settling time than any choice of R that produces an overdamped response. However, a slight improvement (reduction) in settling time may be obtained by a further slight increase in resistance; a slightly underdamped response that will undershoot the zero axis before it dies out will yield the shortest settling time.

The response curve for critical damping is drawn in Fig. 9.12; it may be compared with the overdamped (and underdamped) case by reference to Fig. 9.17.

EXAMPLE 9.5

Select a value for R_1 such that the circuit of Fig. 9.13 will be characterized by a critically damped response for $t > 0$, and select a value for R_2 such that $v(0) = 2$ V.

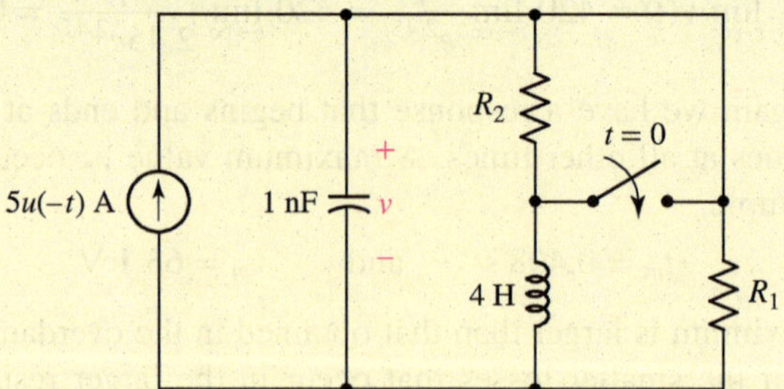

■ **FIGURE 9.13** A circuit that reduces to a parallel *RLC* circuit after the switch is thrown.

We note that at $t = 0^-$, the current source is on, and the inductor can be treated as a short circuit. Thus, $v(0^-)$ appears across R_2, and is given by

$$v(0^-) = 5R_2$$

and a value of 400 mΩ should be selected for R_2 to obtain $v(0) = 2$ V.

After the switch is thrown, the current source has turned itself off and R_2 is shorted. We are left with a parallel *RLC* circuit comprised of R_1, a 4 H inductor, and a 1 nF capacitor.

We may now calculate (for $t > 0$)

$$\alpha = \frac{1}{2RC} = \frac{1}{2 \times 10^{-9} R_1}$$

and

$$\omega_0 = \frac{1}{\sqrt{LC}} = \frac{1}{\sqrt{4 \times 10^{-9}}} = 15{,}810 \text{ rad/s}$$

Therefore, to establish a critically damped response in the circuit for $t > 0$, we need to set $R_1 = 31.63$ kΩ. (*Note: Since we have rounded to four significant figures, the pedantic can rightly argue that this is still not exactly a critically damped response—a difficult situation to create.*)

PRACTICE

9.5 (*a*) Choose R_1 in the circuit of Fig. 9.14 so that the response after $t = 0$ will be critically damped. (*b*) Now select R_2 to obtain $v(0) = 100$ V. (*c*) Find $v(t)$ at $t = 1$ ms.

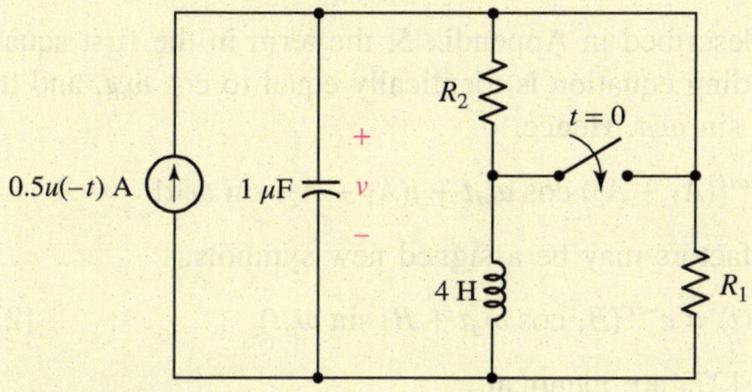

FIGURE 9.14

Ans: 1 kΩ; 250 Ω; −212 V.

9.4 THE UNDERDAMPED PARALLEL *RLC* CIRCUIT

Let us continue the process begun in Sec. 9.3 by increasing R once more to obtain what we will refer to as an ***underdamped*** response. Thus, the damping coefficient α decreases while ω_0 remains constant, α^2 becomes smaller than ω_0^2, and the radicand appearing in the expressions for s_1 and s_2 becomes negative. This causes the response to take on a much different character, but it is fortunately not necessary to return to the basic differential equation again. By using complex numbers, the exponential response turns into a *damped sinusoidal response;* this response is composed entirely of real quantities, the complex quantities being necessary only for the derivation.[1]

The Form of the Underdamped Response

We begin with the exponential form

$$v(t) = A_1 e^{s_1 t} + A_2 e^{s_2 t}$$

where

$$s_{1,2} = -\alpha \pm \sqrt{\alpha^2 - \omega_0^2}$$

and then let

$$\sqrt{\alpha^2 - \omega_0^2} = \sqrt{-1}\sqrt{\omega_0^2 - \alpha^2} = j\sqrt{\omega_0^2 - \alpha^2}$$

where $j \equiv \sqrt{-1}$.

Electrical engineers use "*j*" instead of "*i*" to represent $\sqrt{-1}$ to avoid confusion with currents.

We now take the new radical, which is real for the underdamped case, and call it ω_d, the ***natural resonant frequency:***

$$\omega_d = \sqrt{\omega_0^2 - \alpha^2}$$

The response may now be written as

$$v(t) = e^{-\alpha t}(A_1 e^{j\omega_d t} + A_2 e^{-j\omega_d t}) \qquad [26]$$

or, in the longer but equivalent form,

$$v(t) = e^{-\alpha t}\left\{(A_1 + A_2)\left[\frac{e^{j\omega_d t} + e^{-j\omega_d t}}{2}\right] + j(A_1 - A_2)\left[\frac{e^{j\omega_d t} - e^{-j\omega_d t}}{j2}\right]\right\}$$

Applying identities described in Appendix 5, the term in the first square brackets in the preceding equation is identically equal to $\cos \omega_d t$, and the second is identically $\sin \omega_d t$. Hence,

$$v(t) = e^{-\alpha t}[(A_1 + A_2)\cos \omega_d t + j(A_1 - A_2)\sin \omega_d t]$$

and the multiplying factors may be assigned new symbols:

$$v(t) = e^{-\alpha t}(B_1 \cos \omega_d t + B_2 \sin \omega_d t) \qquad [27]$$

where Eqs. [26] and [27] are identical.

It may seem a little odd that our expression originally appeared to have a complex component and now is purely real. However, we should remember that we originally allowed for A_1 and A_2 to be complex as well as s_1 and s_2. In any event, if we are dealing with the underdamped case, we have now left complex numbers behind. This must be true since α, ω_d, and t are real quantities, so $v(t)$ itself must be a real quantity (which might be presented on an oscilloscope, a voltmeter, or a sheet of graph paper). Equation [27] is the desired functional form for the underdamped response, and its validity may be checked by direct substitution in the original differential equation; this exercise is left to the doubters. The two real constants B_1 and B_2 are again selected to fit the given initial conditions.

(1) A review of complex numbers is presented in Appendix 5.

We return to our simple parallel *RLC* circuit of Fig. 9.3 with $R = 6\ \Omega$, $C = 1/42$ F, and $L = 7$ H, but we now increase the resistance further to 10.5 Ω. Thus,

$$\alpha = \frac{1}{2RC} = \quad 2\ \text{s}^{-1}$$

$$\omega_0 = \frac{1}{RLC} = \sqrt{6}\ \text{s}^{-1}$$

and

$$\omega_d = \sqrt{\omega_0^2 - \alpha^2} = \sqrt{2}\ \text{rad/s}$$

Except for the evaluation of the arbitrary constants, the response is now known:

$$v(t) = e^{-2t}(B_1 \cos \sqrt{2}t + B_2 \sin \sqrt{2}t)$$

Finding Values for B_1 and B_2

The determination of the two constants proceeds as before. If we still assume that $v(0) = 0$ and $i(0) = 10$, then B_1 must be zero. Hence

$$v(t) = B_2 e^{-2t} \sin \sqrt{2}t$$

The derivative is

$$\frac{dv}{dt} = \sqrt{2}B_2 e^{-2t} \cos \sqrt{2}t - 2B_2 e^{-2t} \sin \sqrt{2}t$$

and at $t = 0$ it becomes

$$\left.\frac{dv}{dt}\right|_{t=0} = \sqrt{2}B_2 = \frac{i_C(0)}{C} = 420$$

where i_C is defined in Fig. 9.3. Therefore,

$$v(t) = 210\sqrt{2}e^{-2t} \sin \sqrt{2}t$$

Graphical Representation of the Underdamped Response

We should note that, as before, this response function has an initial value of zero because of the initial voltage condition we imposed, and a final value of zero because the exponential term vanishes for large values of t. As t increases from zero through small positive values, $v(t)$ increases as $210\sqrt{2} \sin \sqrt{2}t$, because the exponential term remains essentially equal to unity. But, at some time t_m, the exponential function begins to decrease more rapidly than $\sin \sqrt{2}t$ is increasing; thus $v(t)$ reaches a maximum v_m and begins to decrease. The time t_m is not the value of t for which $\sin \sqrt{2}t$ is a maximum but must occur somewhat before then.

When $t = \pi/\sqrt{2}$, $v(t)$ is zero. Thus, in the interval $\pi/\sqrt{2} < t < \sqrt{2}\pi$, the response is negative, becoming zero again at $t = \sqrt{2}\pi$. Hence, $v(t)$ is an *oscillatory* function of time and crosses the time axis an infinite number of times at $t = n\pi/\sqrt{2}$, where n is any positive integer. In our example, however, the response is only slightly underdamped, and the exponential term causes the function to die out so rapidly that most of the zero crossings will not be evident in a sketch.

The oscillatory nature of the response becomes more noticeable as α decreases. If α is zero, which corresponds to an infinitely large resistance, then $v(t)$ is an undamped sinusoid that oscillates with constant amplitude. There is never a time at which $v(t)$ drops and stays below 1 percent of its maximum value; the settling time is therefore infinite. This is not perpetual motion; we have merely assumed an initial energy in the circuit and have not provided any means to dissipate this energy. It is transferred from its initial location in the inductor to the capacitor, then returns to the inductor, and so on, forever. This is possible with ideal (lossless) energy storage elements, at least.

The Role of Finite Resistance

A finite R in the parallel *RLC* circuit acts as a kind of electrical transfer agent. Every time energy is transferred from L to C or from C to L, the agent exacts a commission. Before long, the agent has taken all the energy, dissipating every last joule. The L and C are left without a joule of their own, without voltage and without current. Actual parallel *RLC* circuits can be made to have effective values of R so large that a natural undamped sinusoidal response can be maintained for years without supplying any additional energy.

Returning to our specific numerical problem, differentiation locates the first maximum of $v(t)$,

$$v_{m_1} = 71.8 \text{ V} \qquad \text{at} \qquad t_{m_1} = 0.435 \text{ s}$$

the succeeding minimum,

$$v_{m_2} = -0.845 \text{ V} \qquad \text{at} \qquad t_{m_2} = 2.66 \text{ s}$$

and so on. The response curve is shown in Fig. 9.15. Additional response curves for increasingly more underdamped circuits are shown in Fig. 9.16.

The settling time may be obtained by a trial-and-error solution, and for $R = 10.5\ \Omega$, it turns out to be 2.92 s, somewhat smaller than for critical damping. Note that t_s is *greater* than t_{m_2} because the magnitude of v_{m_2} is greater than 1 percent of the magnitude of v_{m_1}. This suggests that a slight decrease in R would reduce the magnitude of the undershoot and permit t_s to be less than t_{m_2}.

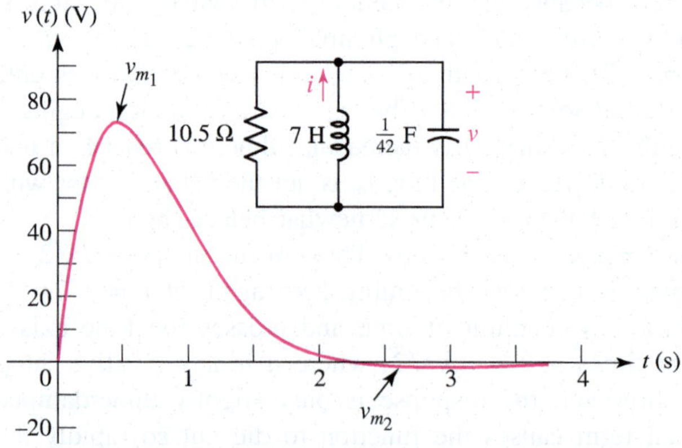

FIGURE 9.15 The response $v(t) = 210\sqrt{2}e^{-2t} \sin \sqrt{2}t$ of the network shown in Fig. 9.3 with R increased to produce an underdamped response.

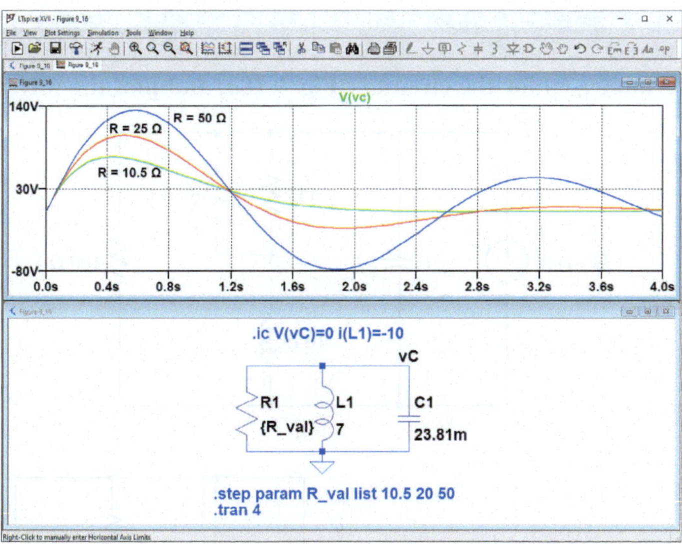

FIGURE 9.16 Simulated underdamped voltage response of the parallel *RLC* network for three different resistance values, showing an increase in the oscillatory behavior as *R* is increased. The simulated circuit is also shown, including initial conditions.
Analog Devices

The overdamped, critically damped, and underdamped responses for our network as simulated by LTspice are shown on the same graph in Fig. 9.17. A comparison of the three curves makes the following general conclusions plausible:

- When the damping is changed by increasing the size of the parallel resistance, the maximum magnitude of the response is greater and the amount of damping is smaller.
- The response becomes oscillatory when underdamping is present, and the minimum settling time is obtained for slight underdamping.

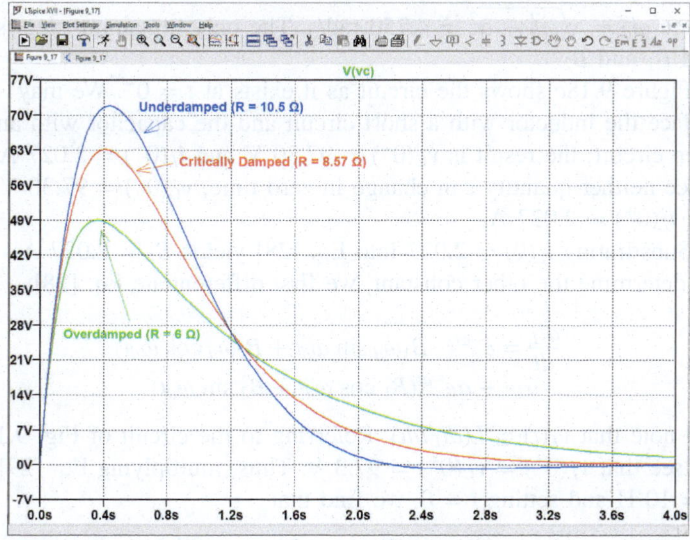

FIGURE 9.17 Simulated overdamped, critically damped, and underdamped voltage response for the example network, obtained by varying the value of the parallel resistance *R*.
Analog Devices

EXAMPLE 9.6

Determine $i_L(t)$ for the circuit of Fig. 9.18*a*, and plot the waveform.

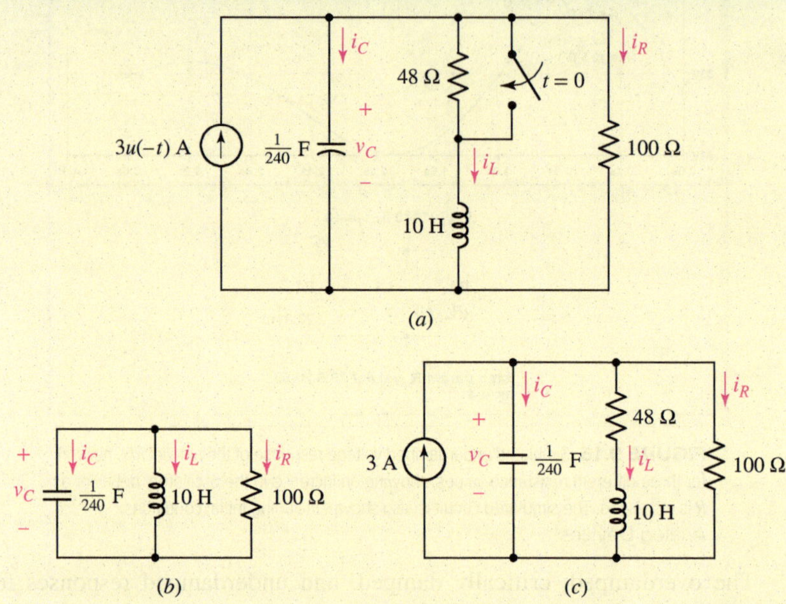

■ **FIGURE 9.18** (*a*) A parallel *RLC* circuit for which the current $i_L(t)$ is desired. (*b*) Circuit for $t \geq 0$. (*c*) Circuit for determining the initial conditions.

At $t = 0$, both the 3 A source and the 48 Ω resistor are removed, leaving the circuit shown in Fig. 9.18*b*. Thus, $\alpha = 1.2\ \text{s}^{-1}$ and $\omega_0 = 4.899$ rad/s. Since $\alpha < \omega_0$, the circuit is *underdamped*, and we therefore expect a response of the form

$$i_L(t) = e^{-\alpha t}(B_1 \cos \omega_d t + B_2 \sin \omega_d t) \qquad [28]$$

where $\omega_d = \sqrt{\omega_0^2 - \alpha^2} = 4.750$ rad/s. The only remaining step is to find B_1 and B_2.

Figure 9.18*c* shows the circuit as it exists at $t = 0^-$. We may replace the inductor with a short circuit and the capacitor with an open circuit; the result is $v_C(0^-) = 97.30$ V and $i_L(0^-) = 2.027$ A. Since neither quantity can change in zero time, $v_C(0^+) = 97.30$ V and $i_L(0^+) = 2.027$ A.

Substituting $i_L(0) = 2.027$ into Eq. [28] yields $B_1 = 2.027$ A. To determine the other constant, we first differentiate Eq. [28]:

$$\begin{aligned}\frac{di_L}{dt} &= e^{-\alpha t}(-B_1\omega_d \sin \omega_d t + B_2\omega_d \cos \omega_d t)\\ &\quad - \alpha e^{-\alpha t}(B_1 \cos \omega_d t + B_2 \sin \omega_d t)\end{aligned} \qquad [29]$$

and note that $v_L(t) = L(di_L/dt)$. Referring to the circuit of Fig. 9.18*b*, we see that $v_L(0^+) = v_C(0^+) = 97.3$ V. Thus, multiplying Eq. [29] by $L = 10$ H and setting $t = 0$, we find that

$$v_L(0) = 10(B_2\omega_d) - 10\alpha B_1 = 97.3$$

Solving, $B_2 = 2.561$ A, so that

$$i_L = e^{-1.2t}(2.027 \cos 4.75t + 2.561 \sin 4.75t) \text{ A}$$

which we have plotted in Fig. 9.19.

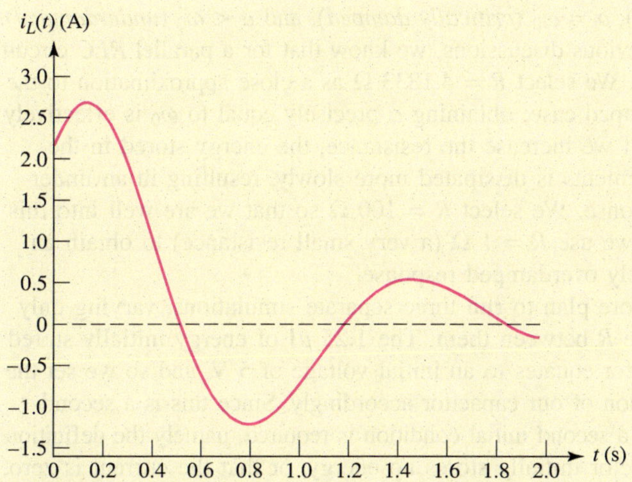

FIGURE 9.19 Plot of $i_L(t)$, showing unambiguous signs of being an underdamped response.

PRACTICE

9.6 The switch in the circuit of Fig. 9.20 has been in the left position for a long time; it is moved to the right at $t = 0$. Find (*a*) dv/dt at $t = 0^+$; (*b*) v at $t = 1$ ms; (*c*) t_0, the first value of t greater than zero at which $v = 0$.

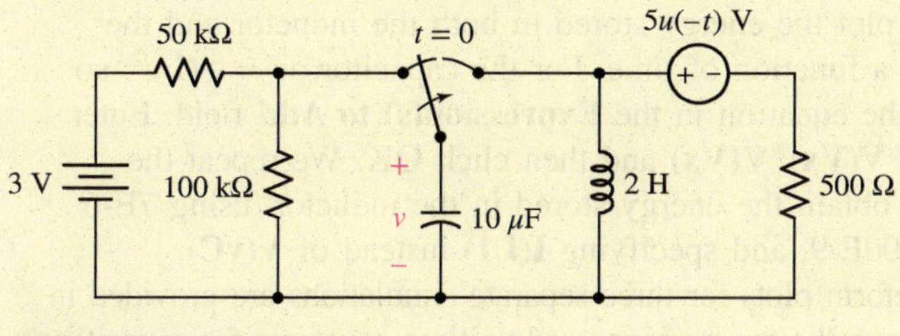

FIGURE 9.20

Ans: −1400 V/s; 0.695 V; 1.609 ms.

COMPUTER-AIDED ANALYSIS

One useful feature in SPICE is the ability to perform mathematical operations on the voltages and currents that result from a simulation. Here, we will make use of that ability to show the transfer of energy in a parallel *RLC* circuit from a capacitor that initially stores a specific amount of energy (1.25 μJ) to an inductor that initially stores no energy.

(Continued on next page)

We choose a 100 nF capacitor and a 7 μH inductor, which immediately enables us to calculate $\omega_0 = 1.195 \times 10^6\ s^{-1}$. In order to consider overdamped, critically damped, and underdamped cases, we need to select the parallel resistance in such a way as to obtain $\alpha > \omega_0$ (*overdamped*), $\alpha = \omega_0$ (*critically damped*), and $\alpha < \omega_0$ (*underdamped*). From our previous discussions, we know that for a parallel *RLC* circuit $\alpha = (2RC)^{-1}$. We select $R = 4.1833\ \Omega$ as a close approximation to the critically damped case; obtaining α precisely equal to ω_0 is effectively impossible. If we increase the resistance, the energy stored in the other two elements is dissipated more slowly, resulting in an underdamped response. We select $R = 100\ \Omega$ so that we are well into this regime, and we use $R = 1\ \Omega$ (a very small resistance) to obtain an unambiguously overdamped response.

We therefore plan to run three separate simulations, varying only the resistance R between them. The 1.25 μJ of energy initially stored in the capacitor equates to an initial voltage of 5 V, and so we set the initial condition of our capacitor accordingly. Since this is a second-order circuit, a second initial condition is required, namely the definition that the inductor initially stores no energy, or that the current is zero.

The schematic diagram of the circuit drawn in LTspice is shown in Fig. 9.21*a*. The initial conditions for the voltage on the capacitor and current through the inductor are defined using the SPICE directive **.ic V(vC)=5 I(L1)=0,** where **vC** was defined as the net name for the node voltage on the capacitor. A transient simulation with an end time of 3 microseconds is defined using the SPICE directive **.tran 3u** (the underdamped case will be simulated out to 30 microseconds to illustrate oscillations).

Following simulation, we click on **Add Trace** under **Plot Settings.** We wish to plot the energy stored in both the inductor and the capacitor as a function of time. For the capacitor, $w = \frac{1}{2}C\,v^2$, so we type in the equation in the **Expression(s) to Add** field. Enter **0.5*100E-9*V(Vx)*V(Vx)** and then click **OK**. We repeat the sequence to obtain the energy stored in the inductor, using 7E-6 instead of 100E-9, and specifying **I(L1)** instead of **V(vC)**.

The waveform plots for three separate simulations are provided in Fig. 9.21. Since the expressions used unitless constants for capacitance and inductance, rather than units of farads and henries, note that the output units are shown in V^2 and A^2 rather than joules. In Fig. 9.21*b*, we see that the energy remaining in the circuit is continuously transferred back and forth between the capacitor and the inductor until it is (eventually) completely dissipated by the resistor. Decreasing the resistance to 4.1833 Ω yields a critically damped circuit, resulting in the energy plot of Fig. 9.21*c*. The oscillatory energy transfer between the capacitor and the inductor has been dramatically reduced. We see that the energy transferred to the inductor peaks at approximately 0.8 μs and then drops to zero. The overdamped response is plotted in Fig. 9.21*d*. We note that the energy is dissipated much more quickly in the case of the overdamped response and that very little energy is transferred to the inductor, since most of it is now quickly dissipated in the resistor.

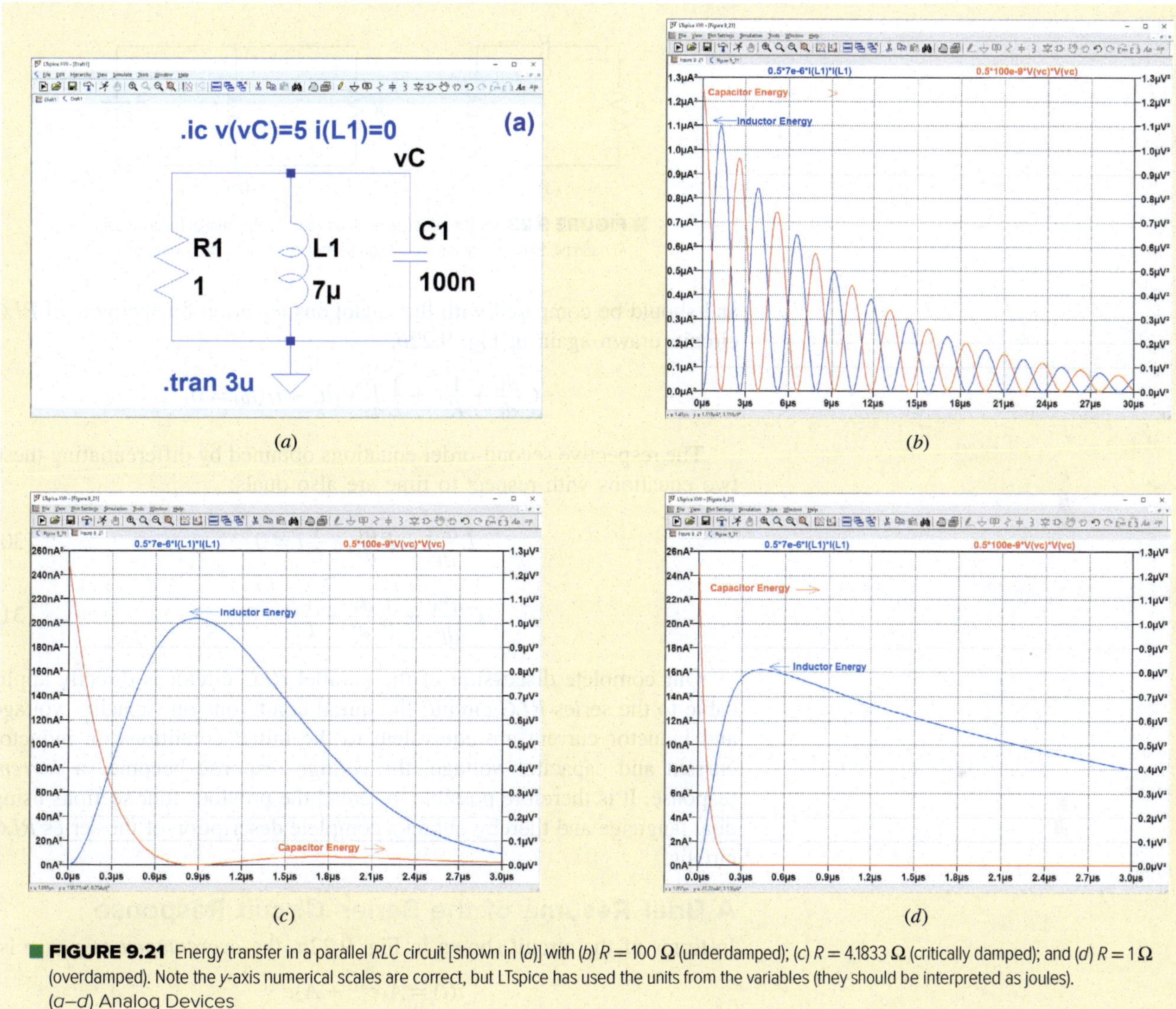

■ **FIGURE 9.21** Energy transfer in a parallel *RLC* circuit [shown in (*a*)] with (*b*) $R = 100\ \Omega$ (underdamped); (*c*) $R = 4.1833\ \Omega$ (critically damped); and (*d*) $R = 1\ \Omega$ (overdamped). Note the *y*-axis numerical scales are correct, but LTspice has used the units from the variables (they should be interpreted as joules).
(*a–d*) Analog Devices

9.5 THE SOURCE-FREE SERIES *RLC* CIRCUIT

We now wish to determine the natural response of a circuit model composed of an ideal resistor, an ideal inductor, and an ideal capacitor connected in *series*. The ideal resistor may represent a physical resistor connected into a series *LC* or *RLC* circuit; it may represent the ohmic losses and the losses in the ferromagnetic core of the inductor; or it may be used to represent all these and other energy-absorbing devices.

The series *RLC* circuit is the *dual* of the parallel *RLC* circuit, and this single fact is sufficient to make its analysis a trivial undertaking. Figure 9.22*a* shows the series circuit. The fundamental integrodifferential equation is

$$L\frac{di}{dt} + Ri + \frac{1}{C}\int_{t_0}^{t} i\, dt' - v_C(t_0) = 0$$

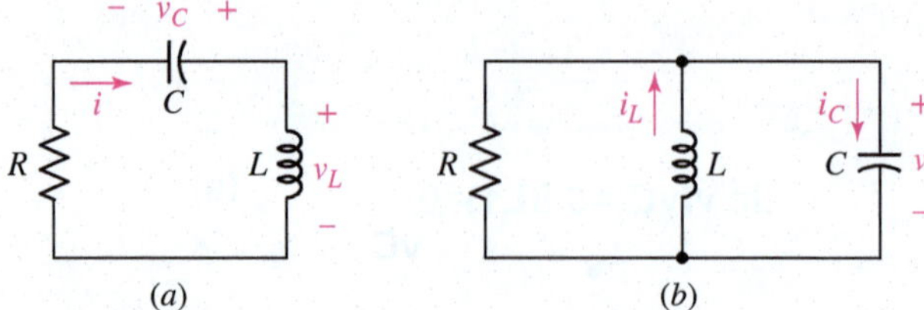

FIGURE 9.22 (*a*) The series *RLC* circuit which is the dual of (*b*), a parallel *RLC* circuit. Element values are, of course, not identical in the two circuits.

and should be compared with the analogous equation for the parallel *RLC* circuit, drawn again in Fig. 9.22*b*,

$$C\frac{dv}{dt} + \frac{1}{R}v + \frac{1}{L}\int_{t_0}^{t} v\, dt' - i_L(t_0) = 0$$

The respective second-order equations obtained by differentiating these two equations with respect to time are also duals:

$$L\frac{d^2i}{dt^2} + R\frac{di}{dt} + \frac{1}{C}i = 0 \qquad [30]$$

$$C\frac{d^2v}{dt^2} + \frac{1}{R}\frac{dv}{dt} + \frac{1}{L}v = 0 \qquad [31]$$

Our complete discussion of the parallel *RLC* circuit is directly applicable to the series *RLC* circuit; the initial conditions on capacitor voltage and inductor current are equivalent to the initial conditions on inductor current and capacitor voltage; the *voltage* response becomes a *current* response. It is therefore possible to reread the previous four sections using dual language and thereby obtain a complete description of the series *RLC* circuit.

A Brief Résumé of the Series Circuit Response

In terms of the circuit shown in Fig. 9.22*a*, the *overdamped response* is

$$i(t) = A_1 e^{s_1 t} + A_2 e^{s_2 t}$$

where

$$s_{1,2} = -\frac{R}{2L} \pm \sqrt{\left(\frac{R}{2L}\right)^2 - \frac{1}{LC}} = -\alpha \pm \sqrt{\alpha^2 - \omega_0^2}$$

and thus

$$\alpha = \frac{R}{2L}$$

$$\omega_0 = \frac{1}{\sqrt{LC}}$$

The form of the *critically damped response* is

$$i(t) = e^{-\alpha t}(A_1 t + A_2)$$

and the *underdamped response* may be written

$$i(t) = e^{-\alpha t}(B_1 \cos \omega_d t + B_2 \sin \omega_d t)$$

TABLE 9.1 Summary of Relevant Equations for Source-Free *RLC* Circuits

Condition	Criteria	α	ω_0	Response
Overdamped	$\alpha > \omega_0$	$\frac{1}{2RC}$ (parallel) $\frac{R}{2L}$ (series)	$\frac{1}{\sqrt{LC}}$	$A_1 e^{s_1 t} + A_2 e^{s_2 t}$, where $s_{1,2} = -\alpha \pm \sqrt{\alpha^2 - \omega_0^2}$
Critically damped	$\alpha = \omega_0$	$\frac{1}{2RC}$ (parallel) $\frac{R}{2L}$ (series)	$\frac{1}{\sqrt{LC}}$	$e^{-\alpha t}(A_1 t + A_2)$
Underdamped	$\alpha < \omega_0$	$\frac{1}{2RC}$ (parallel) $\frac{R}{2L}$ (series)	$\frac{1}{\sqrt{LC}}$	$e^{-\alpha t}(B_1 \cos \omega_d t + B_2 \sin \omega_d t)$, where $\omega_d = \sqrt{\omega_0^2 - \alpha^2}$

where

$$\omega_d = \sqrt{\omega_0^2 - \alpha^2}$$

If we work in terms of the parameters α, ω_0, and ω_d, the mathematical forms of the responses for the dual situations are identical. An increase in α in either the series or the parallel circuit, while keeping ω_0 constant, tends toward an overdamped response. We must only exercise caution in the computation of α, which is $1/2RC$ for the parallel circuit and $R/2L$ for the series circuit; thus, α is increased by increasing the series resistance or decreasing the parallel resistance. The key equations for parallel and series *RLC* circuits are summarized in Table 9.1 for convenience.

EXAMPLE 9.7

Given the series *RLC* circuit of Fig. 9.23 in which $L = 1$ H, $R = 2$ kΩ, $C = 1/401$ μF, $i(0) = 2$ mA, and $v_C(0) = 2$ V, find and sketch $i(t)$, $t > 0$.

We find that $\alpha = R/2L = 1000\ \text{s}^{-1}$ and $\omega_0 = 1/\sqrt{LC} = 20{,}025$ rad/s. This indicates an *underdamped* response; we therefore calculate the value of ω_d and obtain 20,000 rad/s. Except for the evaluation of the two arbitrary constants, the response is now known:

$$i(t) = e^{-1000t}(B_1 \cos 20{,}000t + B_2 \sin 20{,}000t)$$

Since $i(0) = 2$ mA, we may substitute this value into our equation for $i(t)$ to obtain

$$B_1 = 0.002 \text{ A}$$

and thus

$$i(t) = e^{-1000t}(0.002 \cos 20{,}000t + B_2 \sin 20{,}000t) \text{ A}$$

(Continued on next page)

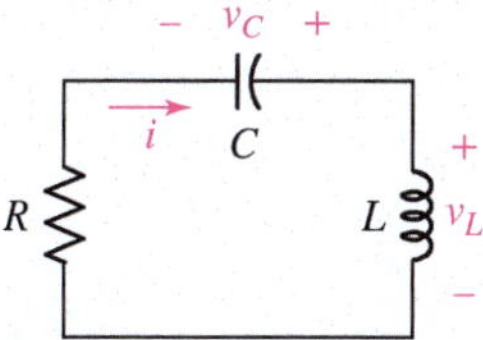

FIGURE 9.23 A simple source-free *RLC* circuit with energy stored in both the inductor and the capacitor at $t = 0$.

The remaining initial condition must be applied to the derivative; thus,

$$\frac{di}{dt} = e^{-1000t}(-40 \sin 20{,}000t + 20{,}000\, B_2 \cos 20{,}000t$$
$$- 2 \cos 20{,}000t - 1000\, B_2 \sin 20{,}000t)$$

and

$$\left.\frac{di}{dt}\right|_{t=0} = 20{,}000\, B_2 - 2 = \frac{v_L(0)}{L}$$
$$= \frac{v_C(0) - Ri(0)}{L}$$
$$= \frac{2 - 2000(0.002)}{1} = -2 \text{ A/s}$$

so that

$$B_2 = 0$$

The desired response is therefore

$$i(t) = 2e^{-1000t} \cos 20{,}000t \text{ mA}$$

A good sketch may be made by first drawing in the two portions of the exponential *envelope,* $2e^{-1000t}$ and $-2e^{-1000t}$ mA, as shown by the dashed lines in Fig. 9.24. The location of the quarter-cycle points of the sinusoidal wave at $20{,}000t = 0, \pi/2, \pi$, etc., or $t = 0.07854k$ ms, $k = 0, 1, 2, \ldots$, by light marks on the time axis then permits the oscillatory curve to be sketched in quickly.

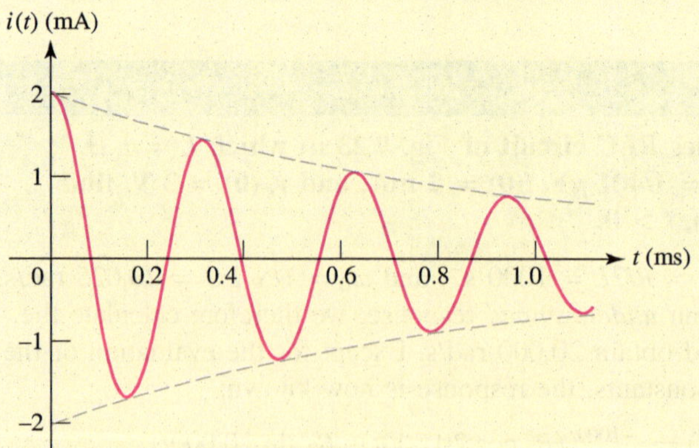

FIGURE 9.24 The current response in an underdamped series *RLC* circuit for which = 1000 s^{-1}, ω_0 = 20,000 s^{-1}, $i(0)$ = 2 mA, and $v_C(0)$ = 2 V. The graphical construction is simplified by drawing in the envelope, shown as a pair of dashed lines.

The settling time can be determined easily here by using the upper portion of the envelope. That is, we set $2e^{-1000t_s}$ mA equal to 1 percent of its maximum value, 2 mA. Thus, $e^{-1000t_s} = 0.01$, and $t_s = 4.61$ ms.

PRACTICE

9.7 With reference to the circuit shown in Fig. 9.25, find (*a*) α; (*b*) ω_0; (*c*) $i(0^+)$; (*d*) $di/dt|_{t=0^+}$; (*e*) $i(12\text{ ms})$.

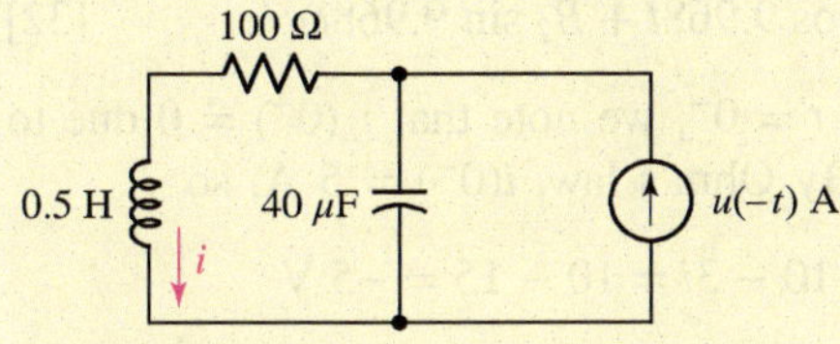

■ **FIGURE 9.25**

Ans: 100 s^{-1}; 224 rad/s; 1 A; 0; −0.1204 A.

As a final example, we pause to consider situations where the circuit includes a dependent source. If no controlling current or voltage associated with the dependent source is of interest, we may simply find the Thévenin equivalent connected to the inductor and capacitor. Otherwise, we are likely faced with having to write an appropriate integrodifferential equation, take the indicated derivative, and solve the resulting differential equation as best we can.

EXAMPLE 9.8

Find an expression for $v_C(t)$ in the circuit of Fig. 9.26*a*, valid for $t > 0$.

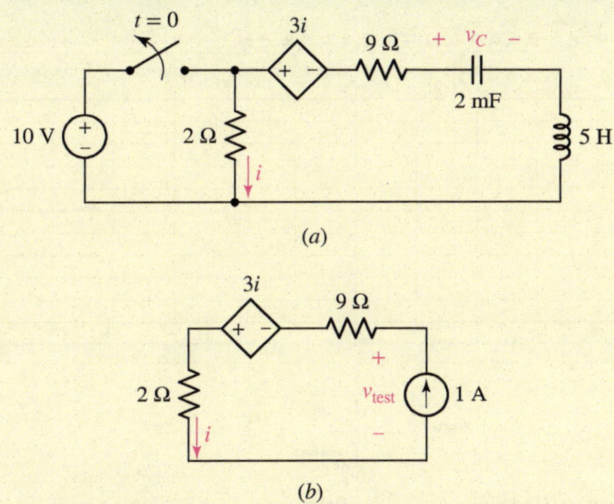

■ **FIGURE 9.26** (*a*) An *RLC* circuit containing a dependent source. (*b*) Circuit for finding R_{eq}.

As we are interested only in $v_C(t)$, it is perfectly acceptable to begin by finding the Thévenin equivalent resistance connected in series with the inductor and capacitor at $t = 0^+$. We do this by connecting a 1 A source as shown in Fig. 9.26*b*, from which we deduce that

$$v_{\text{test}} = 11i - 3i = 8i = 8(1) = 8\text{ V}$$

(*Continued on next page*)

Thus, $R_{eq} = 8\ \Omega$, so $\alpha = R/2L = 0.8\ \text{s}^{-1}$ and $\omega_0 = 1/\sqrt{LC} = 10$ rad/s, meaning that we expect an underdamped response with $\omega_d = 9.968$ rad/s and the form

$$v_C(t) = e^{-0.8t}(B_1 \cos 9.968t + B_2 \sin 9.968t) \qquad [32]$$

In considering the circuit at $t = 0^-$, we note that $i_L(0^-) = 0$ due to the presence of the capacitor. By Ohm's law, $i(0^-) = 5$ A, so

$$v_C(0^+) = v_C(0^-) = 10 - 3i = 10 - 15 = -5\ \text{V}$$

This last condition substituted into Eq. [32] yields $B_1 = -5$ V. Taking the derivative of Eq. [32] and evaluating at $t = 0$ yield

$$\left.\frac{dv_C}{dt}\right|_{t=0} = -0.8B_1 + 9.968B_2 = 4 + 9.968B_2 \qquad [33]$$

We see from Fig. 9.26*a* that

$$i = -C\frac{dv_C}{dt}$$

Thus, making use of the fact that $i(0^+) = i_L(0^-) = 0$ in Eq. [33] yields $B_2 = -0.4013$ V, and we may write

$$v_C(t) = -e^{-0.8t}(5 \cos 9.968t + 0.4013 \sin 9.968t)\ \text{V} \quad t > 0$$

The SPICE simulation of this circuit, shown in Fig. 9.27, confirms our analysis.

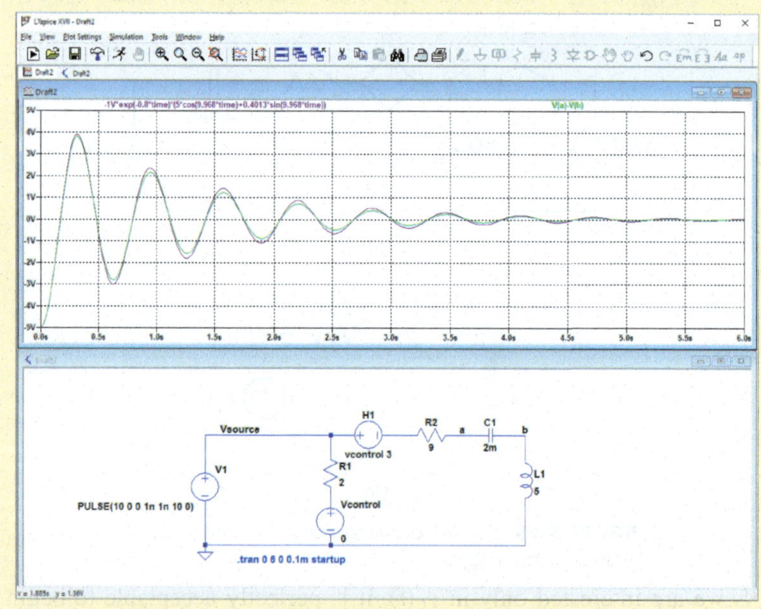

FIGURE 9.27 SPICE simulation of the circuit of Fig. 9.26*a* (schematic shown), plotting the voltage across the capacitor [defined by **V(a)–V(b)** in the simulation] and comparison to the analytical result. Note that in LTspice you may need to check box **"Use radian measure in waveform expression"** in the Control Panel settings for waveforms.
Analog Devices

PRACTICE

9.8 Find an expression for $i_L(t)$ in the circuit of Fig. 9.28, valid for $t > 0$, if $v_C(0^-) = 10$ V and $i_L(0^-) = 0$. *Note that although it is not helpful to apply Thévenin techniques in this instance, the action of the dependent source links v_C and v_L such that a first-order linear differential equation results.*

FIGURE 9.28 Circuit for Practice Problem 9.8.

Ans: $i_L(t) = -30e^{-300t}$ A, $t > 0$.

9.6 THE COMPLETE RESPONSE OF THE *RLC* CIRCUIT

We now consider those *RLC* circuits in which dc sources are switched into the network and produce forced responses that do not necessarily vanish as time becomes infinite.

The general solution is obtained by the same procedure that was followed for *RL* and *RC* circuits. The basic steps are (not necessarily in this order) as follows:

Summary of Procedure for Solving *RLC* Circuits

1. Determine the initial conditions.
2. Obtain a numerical value for the forced response.
3. Write the appropriate form of the natural response with the necessary number of unknown constants. Calculate α, ω_0, and cases of underdamped, critically damped, or overdamped.
4. Add the forced response and natural response to form the complete response.
5. Evaluate the response and its derivative at $t = 0$, and employ the initial conditions to solve for unknown constants.

We note that *it is generally this last step that causes the most trouble for students*, as the circuit must be carefully evaluated at $t = 0$ to make full use of the initial conditions. Consequently, although the determination of the initial conditions is basically no different for a circuit containing dc sources from what it is for the source-free circuits that we have already covered in some detail, this topic will receive particular emphasis in the examples that follow.

Most of the confusion in determining and applying the initial conditions arises for the simple reason that we do not have a rigorous set of rules laid down for us to follow. At some point in each analysis, a situation usually arises in which some thinking is involved that is more or less unique to that particular problem. This is almost always the source of the difficulty.

The Easy Part

The *complete* response (arbitrarily assumed to be a voltage response) of a second-order system consists of a *forced* response,

$$v_f(t) = V_f$$

which is a constant for dc excitation, and a *natural* response,

$$v_n(t) = Ae^{s_1 t} + Be^{s_2 t}$$

Thus,

$$v(t) = V_f + Ae^{s_1 t} + Be^{s_2 t}$$

We assume that s_1, s_2, and V_f have already been determined from the circuit and the given forcing functions; A and B remain to be found. The last equation shows the functional interdependence of A, B, v, and t; substitution of the known value of v at $t = 0^+$ thus provides us with a single equation relating A and B, $v(0^+) = V_f + A + B$. *This is the easy part.*

The Other Part

Another relationship between A and B is necessary, unfortunately, and this is normally obtained by taking the derivative of the response,

$$\frac{dv}{dt} = 0 + s_1 Ae^{s_1 t} + s_2 Be^{s_2 t}$$

and inserting the known value of dv/dt at $t = 0^+$. We thus have two equations relating A and B, and these may be solved simultaneously to evaluate the two constants.

The only remaining problem is that of determining the values of v and dv/dt at $t = 0^+$. Let us suppose that v is a capacitor voltage, v_C. Since $i_C = C\, dv_C/dt$, we should recognize the relationship between the initial value of dv/dt and the initial value of some capacitor current. If we can establish a value for this initial capacitor current, then we will automatically establish the value of dv/dt. Students can usually get $v(0^+)$ very easily, but they are inclined to stumble a bit in finding the initial value of dv/dt. If we had selected an inductor current i_L as our response, then the initial value of di_L/dt would be intimately related to the initial value of some inductor voltage. Variables other than capacitor voltages and inductor currents are determined by expressing their initial values and the initial values of their derivatives in terms of the corresponding values for v_C and i_L.

We will illustrate the procedure and find all these values by the careful analysis of the circuit shown in Fig. 9.29. To simplify the analysis, an unusual value of capacitance is used again.

EXAMPLE 9.9

There are three passive elements in the circuit shown in Fig. 9.29*a*, and a voltage and a current are defined for each. Find the values of these six quantities at both $t = 0^-$ and $t = 0^+$.

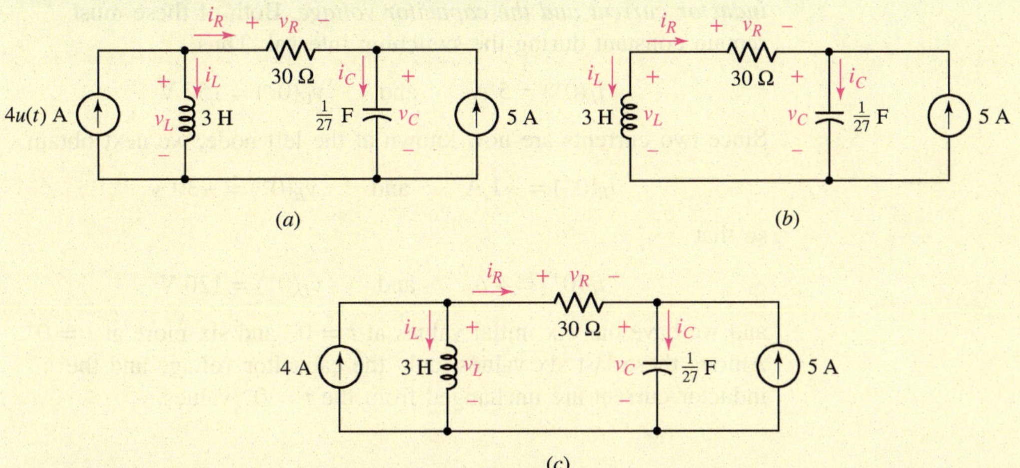

■ **FIGURE 9.29** (*a*) An *RLC* circuit that is used to illustrate several procedures by which the initial conditions may be obtained. The desired response is nominally taken to be $v_C(t)$. (*b*) $t = 0^-$. (*c*) $t > 0$.

Our object is to find the value of each current and voltage at both $t = 0^-$ and $t = 0^+$. Once these quantities are known, the initial values of the derivatives may be found easily.

1. $t = 0^-$ At $t = 0^-$, only the right-hand current source is active as depicted in Fig. 9.29*b*. The circuit is assumed to have been in this state forever, so all currents and voltages are constant. Thus, a dc current through the inductor requires zero voltage across it:

$$v_L(0^-) = 0$$

and a dc voltage across the capacitor ($-v_R$) requires zero current through it:

$$i_C(0^-) = 0$$

We next apply Kirchhoff's current law to the right-hand node to obtain

$$i_R(0^-) = -5 \text{ A}$$

which also yields

$$v_R(0^-) = -150 \text{ V}$$

We may now use Kirchhoff's voltage law around the left-hand mesh, finding

$$v_C(0^-) = 150 \text{ V}$$

while KCL enables us to find the inductor current,

$$i_L(0^-) = 5 \text{ A}$$

(Continued on next page)

2. $t = 0^+$ During the interval from $t = 0^-$ to $t = 0^+$, the left-hand current source becomes active and many of the voltage and current values at $t = 0^-$ will change abruptly. The corresponding circuit is shown in Fig. 9.29*c*. However, we should *begin by focusing our attention on those quantities which cannot change, namely, the inductor current and the capacitor voltage*. Both of these must remain constant during the switching interval. Thus,

$$i_L(0^+) = 5\text{ A} \qquad \text{and} \qquad v_C(0^+) = 150\text{ V}$$

Since two currents are now known at the left node, we next obtain

$$i_R(0^+) = -1\text{ A} \qquad \text{and} \qquad v_R(0^+) = -30\text{ V}$$

so that

$$i_C(0^+) = 4\text{ A} \qquad \text{and} \qquad v_L(0^+) = 120\text{ V}$$

and we have our six initial values at $t = 0^-$ and six more at $t = 0^+$. Among these last six values, only the capacitor voltage and the inductor current are unchanged from the $t = 0^-$ values.

We could have employed a slightly different method to evaluate these currents and voltages at $t = 0^-$ and $t = 0^+$. Prior to the switching operation, only direct currents and voltages exist in the circuit. The inductor may therefore be replaced by a short circuit, its dc equivalent, while the capacitor is replaced by an open circuit. Redrawn in this manner, the circuit of Fig. 9.29*a* appears as shown in Fig. 9.30*a*. Only the current source at the right is active, and its 5 A flow through the resistor and the inductor. We therefore have $i_R(0^-) = -5$ A and $v_R(0^-) = -150$ V, $i_L(0^-) = 5$ A and $v_L(0^-) = 0$, and $i_C(0^-) = 0$ and $v_C(0^-) = 150$ V, as before.

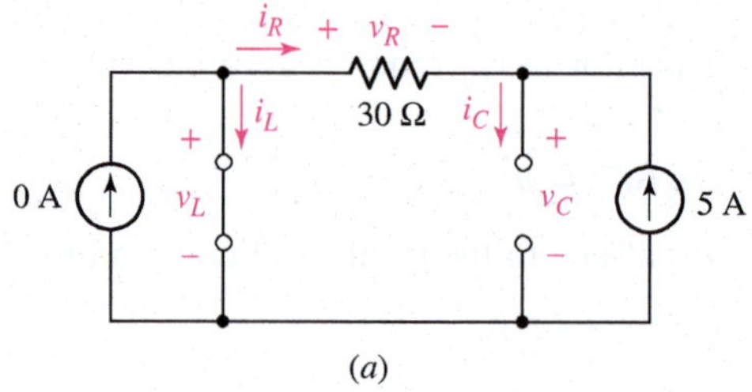

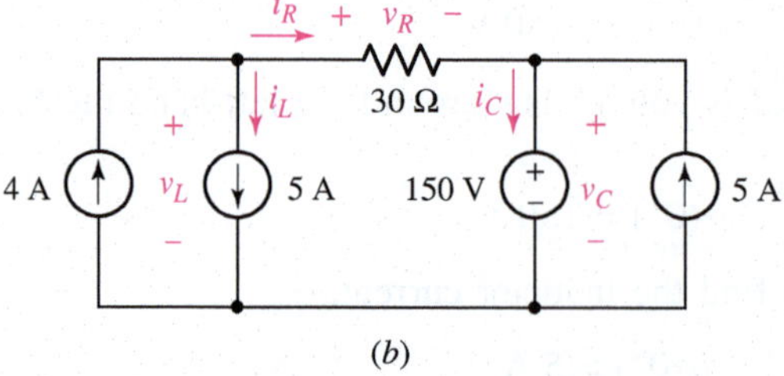

FIGURE 9.30 (*a*) A simple circuit equivalent to the circuit of Fig. 9.29*a* for $t = 0^-$. (*b*) Equivalent circuit with labeled voltages and currents valid at the instant defined by $t = 0^+$.

We now turn to the problem of drawing an equivalent circuit that will assist us in determining the several voltages and currents at $t = 0^+$. *Each capacitor voltage and each inductor current must remain constant during the switching interval.* These conditions are ensured by replacing the inductor with a current source and the capacitor with a voltage source. Each source serves to maintain a constant response during the discontinuity. The equivalent circuit of Fig. 9.30*b* results. It should be noted that the circuit shown in Fig. 9.30*b* is valid *only for the interval between* 0^- *and* 0^+.

The voltages and currents at $t = 0^+$ are obtained by analyzing this dc circuit. The solution is not difficult, but the relatively large number of sources present in the network does produce a somewhat strange sight. However, problems of this type were solved in Chap. 3, and nothing new is involved. Attacking the currents first, we begin at the upper left node and see that $i_R(0^+) = 4 - 5 = -1$ A. Moving to the upper right node, we find that $i_C(0^+) = -1 + 5 = 4$ A. And, of course, $i_L(0^+) = 5$ A.

Next we consider the voltages. Using Ohm's law, we see that $v_R(0^+) = 30(-1) = -30$ V. For the inductor, KVL gives us $v_L(0^+) = -30 + 150 = 120$ V. Finally, including $v_C(0^+) = 150$ V, we have all the values at $t = 0^+$.

PRACTICE

9.9 Let $i_s = 10u(-t) - 20u(t)$ A in Fig. 9.31. Find (*a*) $i_L(0^-)$; (*b*) $v_C(0^+)$; (*c*) $v_R(0^+)$; (*d*) $i_L(\infty)$; (*e*) $i_L(0.1\text{ ms})$.

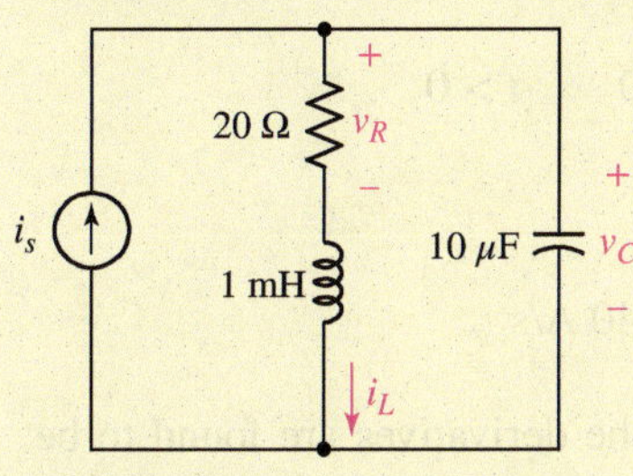

FIGURE 9.31

Ans: 10 A; 200 V; 200 V; −20 A; 2.07 A.

EXAMPLE 9.10

Complete the determination of the initial conditions in the circuit of Fig. 9.29, repeated in Fig. 9.32, by finding values at $t = 0^+$ for the first derivatives of the three voltage and three current variables defined on the circuit diagram.

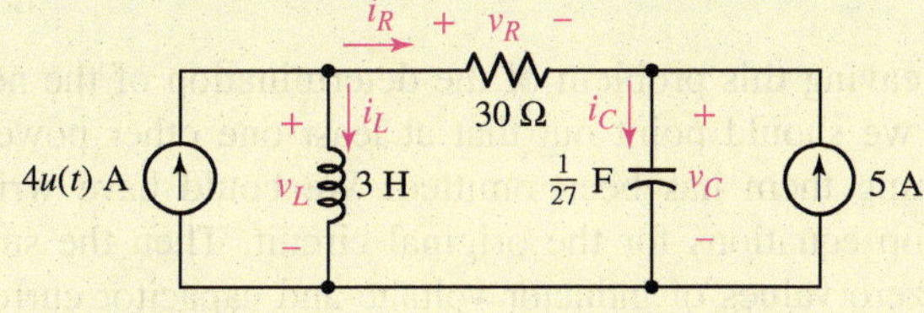

FIGURE 9.32 Circuit of Fig. 9.29, repeated for Example 9.10.

(Continued on next page)

We begin with the two energy storage elements. For the inductor,

$$v_L = L\frac{di_L}{dt}$$

and, specifically,

$$v_L(0^+) = L\left.\frac{di_L}{dt}\right|_{t=0^+}$$

Thus,

$$\left.\frac{di_L}{dt}\right|_{t=0^+} = \frac{v_L(0^+)}{L} = \frac{120}{3} = 40 \text{ A/s}$$

Similarly,

$$\left.\frac{dv_C}{dt}\right|_{t=0^+} = \frac{i_C(0^+)}{C} = \frac{4}{1/27} = 108 \text{ V/s}$$

The other four derivatives may be determined by realizing that KCL and KVL are both satisfied by the derivatives also. For example, at the left-hand node in Fig. 9.32,

$$4 - i_L - i_R = 0 \qquad t > 0$$

and thus,

$$0 - \frac{di_L}{dt} - \frac{di_R}{dt} = 0 \qquad t > 0$$

and therefore,

$$\left.\frac{di_R}{dt}\right|_{t=0^+} = -40 \text{ A/s}$$

The three remaining initial values of the derivatives are found to be

$$\left.\frac{dv_R}{dt}\right|_{t=0^+} = -1200 \text{ V/s}$$

$$\left.\frac{dv_L}{dt}\right|_{t=0^+} = -1092 \text{ V/s}$$

and

$$\left.\frac{di_C}{dt}\right|_{t=0^+} = -40 \text{ A/s}$$

Before leaving this problem of the determination of the necessary initial values, we should point out that at least one other powerful method of determining them has been omitted: We could have written general nodal or loop equations for the original circuit. Then the substitution of the known zero values of inductor voltage and capacitor current at $t = 0^-$ would uncover several other response values at $t = 0^-$ and enable the remainder to be found easily. A similar analysis at $t = 0^+$ must then be made. This is an important method, and it becomes a necessary one in more complicated circuits which cannot be analyzed by our simpler step-by-step procedures.

Now let us briefly complete the determination of the response $v_C(t)$ for the original circuit of Fig. 9.32. With both sources inactive, the circuit appears as a series *RLC* circuit, and s_1 and s_2 are easily found to be -1 and -9, respectively. The forced response may be found by inspection or, if necessary, by drawing the dc equivalent, which is similar to Fig. 9.30*a*, with the addition of a 4 A current source. The forced response is 150 V. Thus,

$$v_C(t) = 150 + Ae^{-t} + Be^{-9t}$$

and

$$v_C(0^+) = 150 = 150 + A + B$$

or

$$A + B = 0$$

Then,

$$\frac{dv_C}{dt} = -Ae^{-t} - 9Be^{-9t}$$

and

$$\left.\frac{dv_C}{dt}\right|_{t=0^+} = 108 = -A - 9B$$

Finally,

$$A = 13.5 \qquad B = -13.5$$

and

$$v_C(t) = 150 + 13.5(e^{-t} - e^{-9t}) \text{ V}$$

A Quick Summary of the Solution Process

In summary, then, whenever we wish to determine the transient behavior of a simple three-element *RLC* circuit, we must first decide whether we are confronted with a series or a parallel circuit, so that we may use the correct relationship for α. The two equations are

$$\alpha = \frac{1}{2RC} \qquad \text{(parallel } RLC\text{)}$$

$$\alpha = \frac{R}{2L} \qquad \text{(series } RLC\text{)}$$

Our second decision is made after comparing α with ω_0, which is given for either circuit by

$$\omega_0 = \frac{1}{\sqrt{LC}}$$

If $\alpha > \omega_0$, the circuit is *overdamped,* and the natural response has the form

$$f_n(t) = A_1e^{s_1t} + A_2e^{s_2t}$$

where

$$s_{1,2} = -\alpha \pm \sqrt{\alpha^2 - \omega_0^2}$$

If $\alpha = \omega_0$, then the circuit is *critically damped* and

$$f_n(t) = e^{-\alpha t}(A_1 t + A_2)$$

And finally, if $\alpha < \omega_0$, then we are faced with the *underdamped* response,

$$f_n(t) = e^{-\alpha t}(A_1 \cos \omega_d t + A_2 \sin \omega_d t)$$

where

$$\omega_d = \sqrt{\omega_0^2 - \alpha^2}$$

PRACTICAL APPLICATION

Automated External Defibrillators

Sudden cardiac arrest, a situation where the heart abruptly stops beating, is a life-threatening emergency that often results in death if not treated in minutes. One of the primary causes of sudden cardiac arrest is ventricular fibrillation, where the lower chambers of the heart cease to pump blood, and the electrical rhythm of the heart is disrupted. Application of an electrical shock can be a life-saving intervention to restore a normal heart rhythm. Many public locations now offer automated external defibrillators (AEDs) for emergency response. The AED includes a device to check heart rhythm and to apply an electric shock if necessary. The electric shock delivers a dose of electricity to the heart to "reset" the heart's electrical system and cardiac muscle.

In a typical AED, a power supply or battery is first used to charge a capacitor to an appropriate level. A voltage of approximately 5 kV is routinely used. An electrical shock is then delivered to the patient through a discharge from the capacitor via paddles in contact with the chest. The electrical current should be sustained on the order of milliseconds to be effective. In order to sustain this current flow, an inductance in the circuit is typically incorporated. The electrical current flows through the patient, where the tissue and contact to the paddles represents a resistance, completing a series *RLC* circuit. Examples of typical values for circuit elements are $C = 50\ \mu\text{F}$, $L = 50$ mH, and $R = 50\ \Omega$. Under a charging voltage of 5 kV, this would result in an electrical shock delivered for approximately 6 ms.

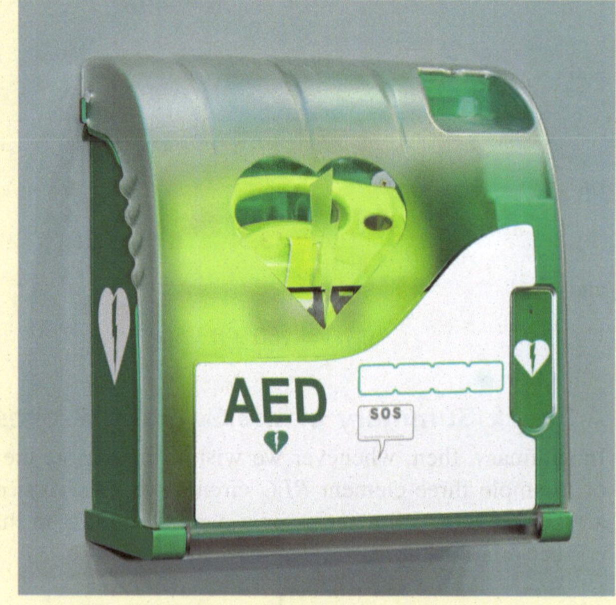

(*a*)

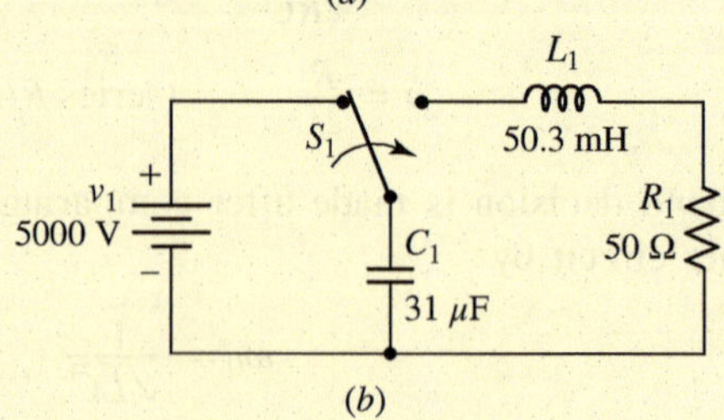

(*b*)

■ **FIGURE 9.33** (*a*) Illustration of an AED device often located in buildings, used to restore normal heart rhythm. (*b*) Example circuit schematic for an AED unit. (*a*) Baloncici/123RF

Our last decision depends on the independent sources. If there are none acting in the circuit after the switching or discontinuity is completed, then the circuit is source-free, and the natural response accounts for the complete response. If independent sources are still present, then the circuit is driven, and a forced response must be determined. The complete response is then the sum

$$f(t) = f_f(t) + f_n(t)$$

This is applicable to any current or voltage in the circuit. Our final step is to solve for unknown constants given the initial conditions.

PRACTICE

9.10 Let $v_s = 10 + 20u(t)$ V in the circuit of Fig. 9.34. Find (*a*) $i_L(0)$; (*b*) $v_C(0)$; (*c*) $i_{L,f}$; (*d*) $i_L(0.1 \text{ s})$.

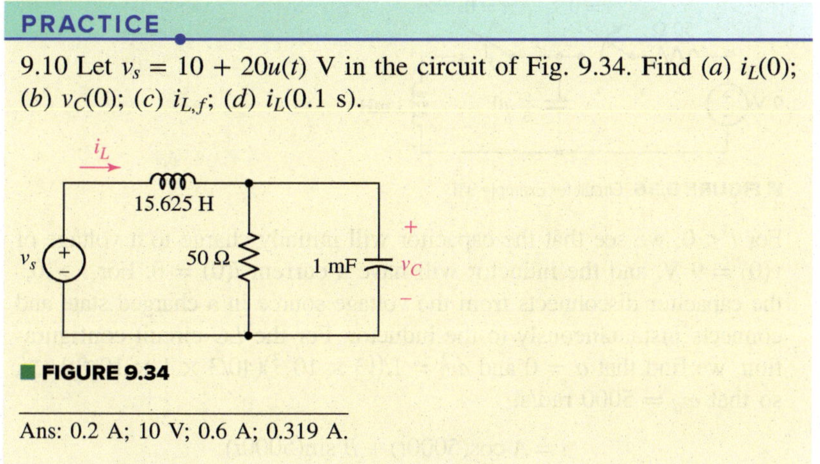

FIGURE 9.34

Ans: 0.2 A; 10 V; 0.6 A; 0.319 A.

9.7 THE LOSSLESS *LC* CIRCUIT

When we considered the source-free *RLC* circuit, it became apparent that the resistor served to dissipate any initial energy stored in the circuit. At some point it might occur to us to ask: what would happen if we could remove the resistor? If the value of the resistance in a parallel *RLC* circuit becomes infinite, or zero in the case of a series *RLC* circuit, we have a simple *LC* loop in which an oscillatory response can be maintained forever! Let us look briefly at an example of such a circuit, and then discuss another means of obtaining an identical response without the need of supplying any inductance.

Consider the source-free circuit of Fig. 9.35, in which the large values $L = 4$ H and $C = \frac{1}{36}$ F are used so that the calculations will be simple. We let $i(0) = -\frac{1}{6}$ A and $v(0) = 0$. We find that $\alpha = 0$ and $\omega_0^2 = 9 \text{ s}^{-2}$, so that $\omega_d = 3$ rad/s. In the absence of exponential damping, the voltage v is simply

$$v = A \cos 3t + B \sin 3t$$

Since $v(0) = 0$, we see that $A = 0$. Next,

$$\left.\frac{dv}{dt}\right|_{t=0} = 3B = -\frac{i(0)}{1/36}$$

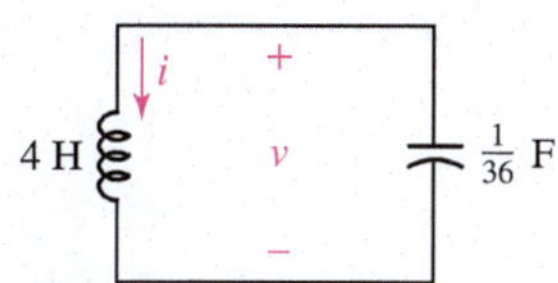

FIGURE 9.35 This circuit is lossless, and it provides the undamped response $v = 2 \sin 3t$ V, if $v(0) = 0$ and $i(0) = -\frac{1}{6}$ A.

But $i(0) = -\frac{1}{6}$ A, and therefore $dv/dt = 6$ V/s at $t = 0$. We must have $B = 2$ V and so

$$v = 2 \sin 3t \text{ V}$$

which is an undamped sinusoidal response; in other words, our voltage response does not decay.

EXAMPLE 9.11

Determine $i(t)$ for $t > 0$ for the circuit shown in Fig 9.36.

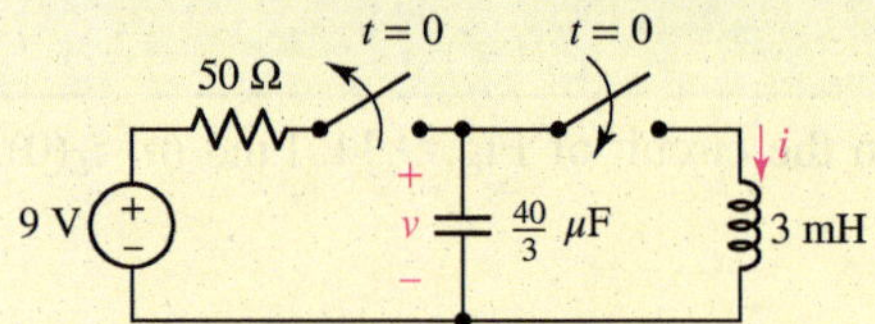

FIGURE 9.36 Circuit for Example 9.11.

For $t < 0$, we see that the capacitor will initially charge to a voltage of $v(0) = 9$ V, and the inductor will have a current $i(0) = 0$. For $t > 0$, the capacitor disconnects from the voltage source in a charged state and connects instantaneously to the inductor. For the *LC* circuit configuration, we find that $\alpha = 0$ and $\omega_0^2 = 1/((3 \times 10^{-3})(40/3 \times 1 \times 10^{-6}))$ s^{-2}, so that $\omega_d = 5000$ rad/s.

$$i = A \cos(5000t) + B \sin(5000t)$$

Since $i(0) = 0$, $A = 0$. Applying initial conditions to the derivative of current at $t = 0^+$,

$$\left.\frac{di}{dt}\right|_{t=0^+} = 5000B = \frac{v(0^+)}{L}$$

We know $v(0^+) = 9$ V, thus $di/dt = 3000$ A/s at $t = 0^+$. B is therefore 0.6 A. The resulting current for $t > 0$ is

$$i = 0.6 \sin(5000t) \text{ A}$$

PRACTICE

9.11 Alter the capacitor value and voltage source in Fig. 9.36 to oscillate at a frequency of 1 kHz with total energy of 0.96 mJ.

Ans: 8.44 μF; 15.1 V

Now let us see how we might obtain this voltage without using an *LC* circuit. Our intentions are to write the differential equation that v satisfies and then to develop a configuration of op amps that will yield the solution to the equation. Although we are working with a specific example, the technique is a general one that can be used to solve any linear homogeneous differential equation.

For the *LC* circuit of Fig. 9.35, we select v as our variable and set the sum of the downward inductor and capacitor currents equal to zero:

$$i + C\frac{dv}{dt} = \frac{1}{L}\int v\,dt + C\frac{dv}{dt} = 0$$

Differentiating once, we have

$$\frac{1}{L}v + C\frac{d^2v}{dt^2} = 0$$

or

$$\frac{d^2v}{dt^2} = -\frac{1}{LC}v$$

Substituting circuit values yields

$$\frac{d^2v}{dt^2} = -9v$$

In order to solve this differential equation, note that the solution can be obtained by integrating twice. Suppose we would like to obtain the same oscillator circuit result but eliminate the use of the inductor (which are often difficult to obtain, expensive, and not easy to find for integrated circuits). Alternatively, we learned that the operational amplifier can behave as an integrator, where two cascaded integrators could obtain the same result!

We assume that the highest-order derivative appearing in the differential equation here, d^2v/dt^2, is available in our configuration of op amps at an arbitrary point *A*. We now make use of the integrator, with $RC = 1$, as discussed in Sec. 7.5. The input is d^2v/dt^2, and the output must be $-dv/dt$, where the sign change results from using an inverting op amp configuration for the integrator. The negative of the first derivative now forms the input to a second integrator, where the output is $v(t)$. To complete the design, we need to multiply v by -9 to obtain the second derivative we assumed at point *A*. This is amplification by 9 with a sign change, and it is easily accomplished by using the op amp as an inverting amplifier.

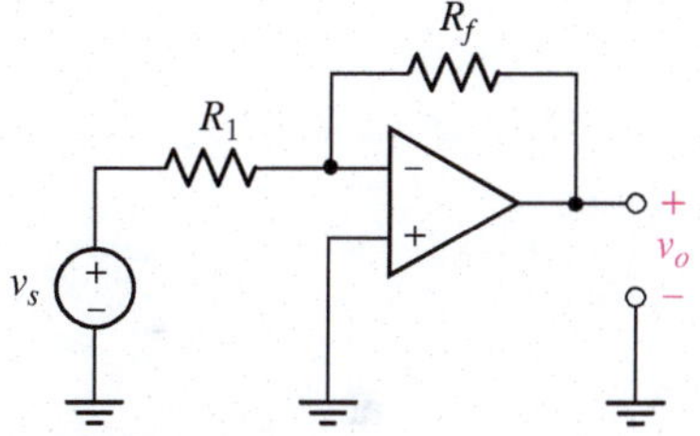

■ **FIGURE 9.37** The inverting operational amplifier provides a gain $v_o/v_s = -R_f/R_1$, assuming an ideal op amp.

Figure 9.37 shows the circuit of an inverting amplifier. For an ideal op amp, both the input current and the input voltage are zero. Thus, the current going "east" through R_1 is v_s/R_1, while that traveling west through R_f is v_o/R_f. Since their sum is zero, we have

$$\frac{v_o}{v_s} = -\frac{R_f}{R_1}$$

Thus, we can design for a gain of -9 by setting $R_f = 90$ kΩ and $R_1 = 10$ kΩ, for example.

If we let R be 1 MΩ and C be 1 μF in each of the integrators, then

$$v_o = -\int_0^t v_s\,dt' + v_o(0)$$

in each case. The output of the inverting amplifier now forms the assumed input at point *A*, leading to the configuration of op amps shown in Fig. 9.38. The initial conditions will be set by storage in the capacitors and will need to be matched to the case of the *LC* circuit. The *LC* circuit defined zero energy storage in the capacitor with $v = 0$, and energy storage in the inductor according to $i(0) = -\frac{1}{6}$ A. We can similarly define an initial voltage for the capacitors, drawn in the circuit schematically by a voltage source or short circuit connected by switches that open at $t = 0$. Analysis

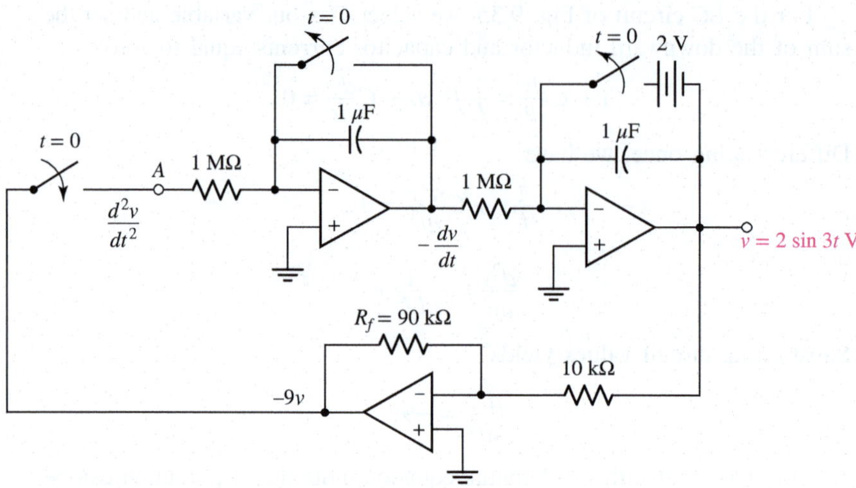

■ **FIGURE 9.38** Two integrators and an inverting amplifier are connected to provide the solution of the differential equation $d^2v/dt^2 = -9v$.

of our circuit at $t = 0$ indicates that a voltage of $v = 2$ V is required at the output, and similarly for the voltage across the capacitor on the right (and a short circuit across the capacitor on the left). If the left switch is closed at $t = 0$ while the two initial-condition switches are opened at the same time, the output of the second integrator will be the undamped sine wave $v = 2 \sin 3t$ V.

Note that both the *LC* circuit of Fig. 9.35 and the op amp circuit of Fig. 9.38 have the same output, but the op circuit does not contain a single inductor. It simply *acts* as though it contained an inductor, providing the appropriate sinusoidal voltage between its output terminal and ground. This can be a considerable practical or economic advantage in circuit design, as inductors are typically bulky, more costly than capacitors, and have more losses associated with them (and therefore are not as well approximated by the "ideal" model).

PRACTICE

9.12 Give new values for R_f and the two initial voltages in the circuit of Fig. 9.38 to provide a circuit that would provide an equivalent output to the voltage $v(t)$ in the circuit of Fig. 9.39.

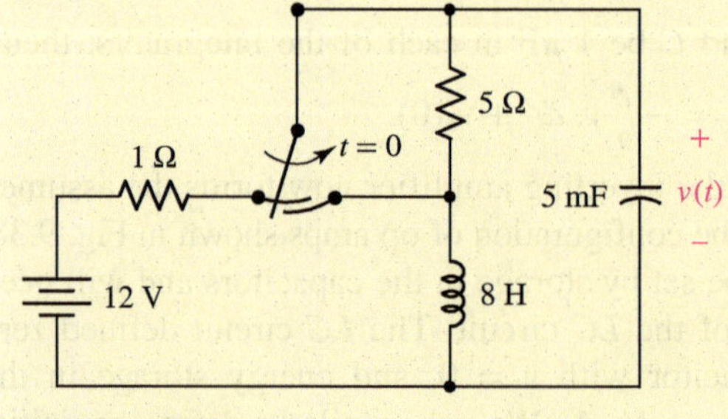

■ **FIGURE 9.39**

Ans: 250 kΩ; 400 V; 10 V.

SUMMARY AND REVIEW

The simple *RL* and *RC* circuits examined in Chap. 8 essentially did one of two things as the result of throwing a switch: *charge* or *discharge*. Which one happened was determined by the initial charge state of the energy storage element. In this chapter, we considered circuits that had two energy storage elements (a capacitor and an inductor) and found that things could get pretty interesting. There are two basic configurations of such *RLC* circuits: *parallel* connected and *series* connected. Analysis of such a circuit yields a *second-order* partial differential equation, consistent with the number of distinct energy storage elements (if we construct a circuit using only resistors and capacitors such that the capacitors cannot be combined using series/parallel techniques, we also obtain—eventually—a second-order partial differential equation).

Depending on the value of the resistance connected to our energy storage elements, we found the transient response of an *RLC* circuit could be either *overdamped* (decaying exponentially) or *underdamped* (decaying, but oscillatory), with a "special case" of *critically damped* which is difficult to achieve in practice. Oscillations can be useful (for example, in transmitting information over a wireless network) and not so useful (for example, in accidental feedback situations between an amplifier and a microphone at a concert). Although the oscillations are not sustained in the circuits we examined, we have at least seen one way to create them at will, and we can design for a specific frequency of operation if so desired. We didn't end up spending a great deal of time with the series-connected *RLC* circuit because with the exception of α, the equations are the same; we need only a minor adjustment in how we employ initial conditions to find the two unknown constants characterizing the transient response. Along those lines, there were two "tricks," if you will, that we encountered. One is that to employ the second initial condition, we need to take the derivative of our response equation. The second is that whether we're employing KCL or KVL to make use of that initial condition, we're doing so at the instant that $t = 0$; appreciating this fact can simplify equations dramatically by setting $t = 0$ early.

We wrapped up the chapter by considering the *complete response*, and our approach to this did not differ significantly from what we did in Chap. 8. We closed with a brief section on a topic that might have occurred to us at some point—what happens when we remove the resistive losses completely (by setting parallel resistance to ∞, or series resistance to 0)? We end up with an *LC* circuit, and we saw that we can approximate such an animal with an op amp circuit.

By now the reader is likely ready to finish reviewing key concepts of the chapter, so we'll stop here and list them, along with corresponding examples in the text.

- ❑ Circuits that contain two energy storage devices that cannot be combined using series-parallel combination techniques are described by a second-order differential equation.
- ❑ Series and parallel *RLC* circuits fall into one of three categories, depending on the relative values of *R*, *L*, and *C*: (Example 9.1)

Overdamped	$\alpha > \omega_0$
Critically damped	$\alpha = \omega_0$
Underdamped	$\alpha < \omega_0$

- For parallel *RLC* circuits, $\alpha = 1/2RC$ and $\omega_0 = 1\sqrt{LC}$. (Example 9.1)
- For series *RLC* circuits, $\alpha = R/2L$ and $\omega_0 = 1\sqrt{LC}$. (Example 9.7)
- The typical form of an overdamped response is the sum of two exponential terms, one of which decays more quickly than the other: e.g., $A_1e^{-t} + A_2e^{-6t}$. (Examples 9.2, 9.3, 9.4)
- The typical form of a critically damped response is $e^{-\alpha t}(A_1t + A_2)$. (Example 9.5)
- The typical form of an underdamped response is an exponentially damped sinusoid: $e^{-\alpha t}(B_1 \cos \omega_d t + B_2 \sin \omega_d t)$. (Examples 9.6, 9.7, 9.8)
- During the transient response of an *RLC* circuit, energy is transferred between energy storage elements to the extent allowed by the resistive component of the circuit, which acts to dissipate the energy initially stored. (See Computer-Aided Analysis section.)
- The complete response is the sum of the forced and natural responses. In this case the total response must be determined before solving for the constants. (Examples 9.9, 9.10)
- An RLC circuit without damping can lead to an oscillatory response that could be maintained forever. This is possible for an LC circuit (though finite resistance always occurs in practice) or a properly designed op amp circuit involving capacitors. (Example 9.11)

READING FURTHER

Many detailed descriptions of analogous networks can be found in Chap. 3 of

E. Weber, *Linear Transient Analysis Volume I.* New York: Wiley, 1954. (Out of print, but in many university libraries.)

EXERCISES

9.1 The Source-Free Parallel Circuit

1. For a certain source-free parallel *RLC* circuit, $R = 1$ kΩ, $C = 3$ μF, and L is such that the circuit response is overdamped. (*a*) Determine a suitable value of L. (*b*) Write the equation for the voltage v across the resistor if it is known that $v(0^-) = 9$ V and $dv/dt|_{t=0^+} = 2$ V/s.
2. Element values of 10 mF and 2 nH are employed in the construction of a simple source-free parallel *RLC* circuit. (*a*) Select R so that the circuit is just barely overdamped. (*b*) Write the equation for the resistor current if its initial value is $i_R(0^+) = 13$ pA and $di_R/dt|_{t=0^+} = 1$ nA/s.
3. If a parallel *RLC* circuit is constructed from component values $C = 16$ mF and $L = 1$ mH, choose R such that the circuit is (*a*) just barely overdamped; (*b*) just barely underdamped; (*c*) critically damped. (*d*) Does your answer for part (*a*) change if the resistor tolerance is 1%? 10%? (*e*) Increase the exponential damping coefficient for part (*c*) by 20%. Is the circuit now underdamped, overdamped, or still critically damped? *Explain.*
4. Calculate α, ω_0, s_1, and s_2 for a source-free parallel *RLC* circuit if (*a*) $R = 4$ Ω, $L = 2.22$ H, and $C = 12.5$ mF; (*b*) $L = 1$ nH, $C = 1$ pF, and R is 1% of the value required to make the circuit underdamped. (*c*) Calculate the damping ratio for the circuits of parts (*a*) and (*b*).

5. You go to construct the circuit in Exercise 1, only to find no 1 kΩ resistors. In fact, all you can find in addition to the capacitor and inductor is a 1-meter-long piece of 24 AWG soft solid copper wire. Connecting it in parallel to the two components you did find, compute the value of α, ω_0, s_1, and s_2, and verify that the circuit is still overdamped.
6. A parallel *RLC* circuit has inductance 2 mH and resistance 50 Ω. For capacitance values ranging from 10 nF to 10 μF, (*a*) plot α and ω_0 versus capacitance on a log-log plot (and indicate regions for underdamped and overdamped response), (*b*) extract values for C, α, and ω_0 for the critical damping case.

7. A source-free parallel *RLC* circuit is constructed using a 1 μF capacitor and a 1.00 kΩ resistor. (*a*) Compute the inductance required to achieve a critically-damped circuit. (*b*) If the value of R can be any 2% tolerance resistor value in the range of 1 to 2.05 kΩ, write the lines of code required to compute the corresponding inductor values required for critical damping. (*c*) Other than the case when R = 1.00 kΩ, do any of the other resistor values lead to integer inductor values to ensure critical damping? (*d*) How does the damping ratio change if the resistor value is doubled? *Explain*.

9.2 The Overdamped Parallel *RLC* Circuit

8. The circuit of Fig. 9.48 is constructed with a 500 Ω resistor instead of the specified 500 mΩ value. (*a*) Compute α, ω_d, and ς. (*b*) Obtain an expression for $v_C(t)$ valid for $t > 0$. (*c*) Calculate the time required for the energy stored in the capacitor to reach 10% of its maximum value. (*d*) Verify your answers to part (*c*) using an appropriate SPICE simulation.
9. The voltage across a capacitor is found to be given by $v_C(t) = 10e^{-10t} - 5e^{-4t}$ V.

   (*a*) Plot each of the two components over the range of $0 \le t \le 1.5$ s.

   (*b*) Plot the capacitor voltage over the same time range.
10. The current flowing through a certain inductor is found to be given by $i_L(t) = 0.20e^{-2t} - 0.6e^{-3t}$ V. (*a*) Plot each of the two components over the range of $0 \le t \le 1.5$ s. (*b*) Plot the inductor current over the same time range. (*c*) Plot the energy remaining in the inductor (assuming inductance of 1 H) over $0 \le t \le 1.5$ s.
11. The current flowing through a 5 Ω resistor in a source-free parallel *RLC* circuit is determined to be $i_R(t) = 2e^{-t} - 3e^{-8t}$ A, $t > 0$. Determine (*a*) the maximum current and the time at which it occurs; (*b*) the settling time; (*c*) the time t corresponding to the resistor absorbing 2.5 W of power.
12. For the circuit of Fig. 9.40, obtain an expression for $v_C(t)$ valid for all $t > 0$.
13. Consider the circuit depicted in Fig. 9.40. (*a*) Obtain an expression for $i_L(t)$ valid for all $t > 0$. (*b*) Obtain an expression for $i_R(t)$ valid for all $t > 0$. (*c*) Determine the settling time for both i_L and i_R.
14. With regard to the circuit represented in Fig. 9.41, determine (*a*) $i_C(0^-)$; (*b*) $i_L(0^-)$; (*c*) $i_R(0^-)$; (*d*) $v_C(0^-)$; (*e*) $i_C(0^+)$; (*f*) $i_L(0^+)$; (*g*) $i_R(0^+)$; (*h*) $v_C(0^+)$.

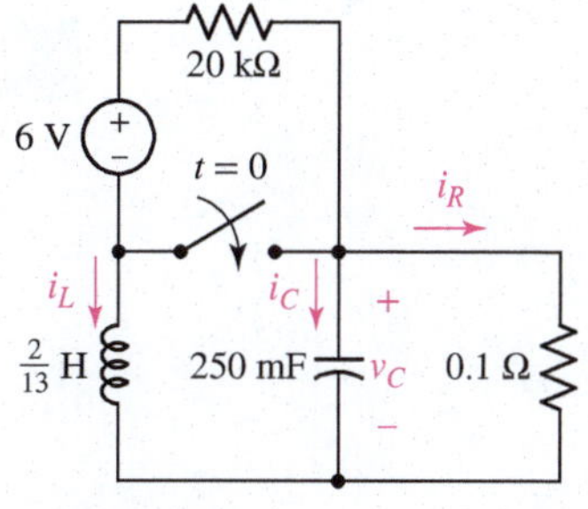

FIGURE 9.40

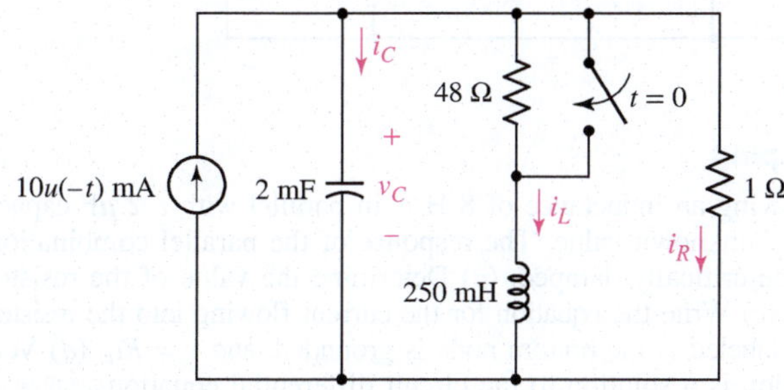

FIGURE 9.41

15. (*a*) Assuming the passive sign convention, obtain an expression for the voltage across the 1 Ω resistor in the circuit of Fig. 9.41 which is valid for all $t > 0$. (*b*) Determine the settling time of the resistor voltage.

16. With regard to the circuit presented in Fig. 9.42, (*a*) obtain an expression for $v(t)$ which is valid for all $t > 0$; (*b*) calculate the maximum inductor current and identify the time at which it occurs; (*c*) determine the settling time.

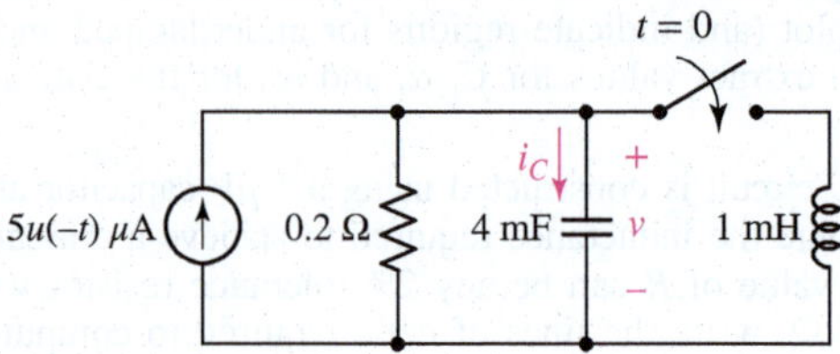

FIGURE 9.42

17. Obtain expressions for the current $i(t)$ and voltage $v(t)$ as labeled in the circuit of Fig. 9.43 which are valid for all $t > 0$.

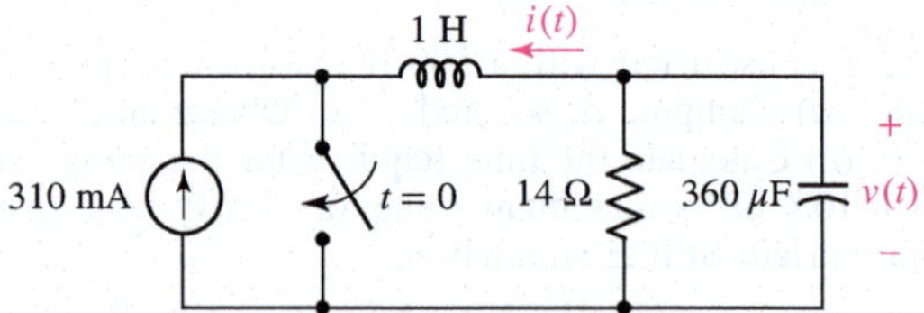

FIGURE 9.43

18. Replace the 14 Ω resistor in the circuit of Fig. 9.43 with a 1 Ω resistor. (*a*) Obtain an expression for the energy stored in the capacitor as a function of time, valid for $t > 0$. (*b*) Determine the time at which the energy in the capacitor has been reduced to one-half its maximum value. (*c*) Verify your answer with an appropriate SPICE simulation.

19. Design a complete source-free parallel *RLC* circuit which exhibits an overdamped response, has a settling time of 1 s, and has a damping ratio of 15.

20. For the circuit represented by Fig. 9.44, the two resistor values are $R_1 = 0.752\ \Omega$ and $R_2 = 1.268\ \Omega$, respectively. (*a*) Obtain an expression for the energy stored in the capacitor, valid for all $t > 0$; (*b*) determine the settling time of the current labeled i_A.

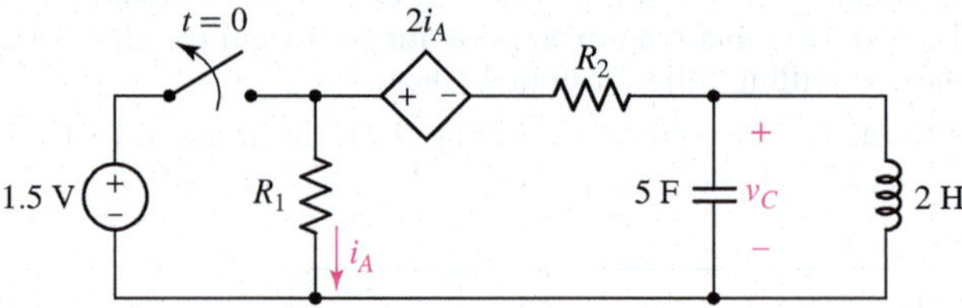

FIGURE 9.44

9.3 Critical Damping

21. A motor coil having an inductance of 8 H is in parallel with a 2 μF capacitor and a resistor of unknown value. The response of the parallel combination is determined to be critically damped. (*a*) Determine the value of the resistor. (*b*) Compute α. (*c*) Write the equation for the current flowing into the resistor if the top node is labeled v, the bottom node is grounded, and $v = Ri_r$. (*d*) Verify that your equation is a solution to the circuit differential equation,

$$\frac{di_r}{dt} + 2\alpha\frac{di_r}{dt} + \alpha^2 i_r = 0$$

22. The condition for critical damping in an *RLC* circuit is that the resonant frequency ω_0 and the exponential damping factor α are equal. This leads to the relationship $L = 4R^2C$, which implies that $1\ \text{H} = 1\ \Omega^2 \cdot \text{F}$. Verify this equivalence by breaking down each of the three units to fundamental SI units (see Chap. 2).

23. A critically damped parallel *RLC* circuit is constructed from component values 40 Ω, 8 nF, and 51.2 μH, respectively. (*a*) Verify that the circuit is indeed critically damped. (*b*) Explain why, in practice, the circuit once fabricated is unlikely to be truly critically damped. (*c*) The inductor initially stores 1 mJ of energy while the capacitor is initially discharged. Determine the magnitude of the capacitor voltage at $t = 500$ ns, the maximum absolute capacitor voltage, and the settling time.

24. A source-free parallel *RLC* circuit has an initial capacitor voltage of 9 V and inductor current of zero. Design a circuit that is critically damped that ensures that voltage oscillations have decayed below 100 mV for time great than 20 μs. The resistance can range from 10 Ω to 1 kΩ.

25. A critically damped parallel *RLC* circuit is constructed from component values 40 Ω and 2 pF. (*a*) Determine the value of *L*, taking care not to over round. (*b*) Explain why, in practice, the circuit once fabricated is unlikely to be truly critically damped. (*c*) The inductor initially stores no energy while the capacitor is initially storing 10 pJ. Determine the power absorbed by the resistor at $t = 2$ ns.

26. For the circuit of Fig. 9.45, $i_s(t) = 30u(-t)$ mA. (*a*) Select R_1 so that $v(0^+) = 6$ V. (*b*) Compute $v(2\ \text{ms})$. (*c*) Determine the settling time of the capacitor voltage. (*d*) Is the inductor current settling time the same as your answer to part (*c*)?

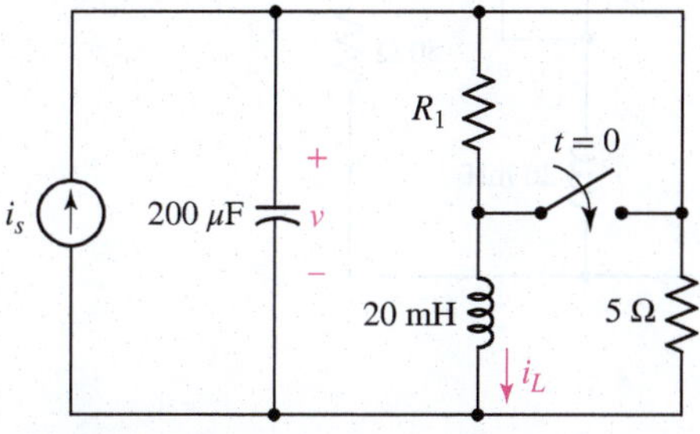

FIGURE 9.45

27. The inductor in the circuit of Fig. 9.43 is changed such that the circuit response is now critically damped. (*a*) Determine the new inductor value. (*b*) Calculate the energy stored in both the inductor and the capacitor at $t = 10$ ms.

28. The circuit of Fig. 9.44 is rebuilt such that the quantity controlling the dependent source is now $-60i_A$, a 2 μF capacitor is used instead, and $R_1 = R_2 = 10\ \Omega$. (*a*) Calculate the inductor value required to obtain a critically damped response. (*b*) Determine the power being absorbed by R_2 at $t = 300$ μs.

9.4 The Underdamped Parallel *RLC* Circuit

29. (*a*) With respect to the parallel *RLC* circuit, derive an expression for *R* in terms of *C* and *L* to ensure that the response is underdamped. (*b*) If $C = 1$ nF and $L = 10$ mH, select *R* such that an underdamped response is (just barely) achieved. (*c*) If the damping ratio is increased, does the circuit become more or less underdamped? *Explain.* (*d*) Compute α and ω_d for the value of *R* you selected in part (*b*).

30. The circuit of Fig. 9.1 is constructed using component values 10 kΩ, 72 μH, and 18 pF. (*a*) Compute α, ω_d, and ω_0. Is the circuit overdamped, critically damped, or underdamped? (*b*) Write the form of the natural capacitor voltage response $v(t)$. (*c*) If the capacitor initially stores 1 nJ of energy, compute *v* at $t = 300$ ns.

31. The source-free circuit depicted in Fig. 9.1 is constructed using a 10 mH inductor, a 1 mF capacitor, and a 1.5 kΩ resistor. (*a*) Calculate α, ω_d, and ω_0. (*b*) Write the equation which describes the current i for $t > 0$. (*c*) Determine the maximum value of i, and the time at which it occurs, if the inductor initially stores no energy and $v(0^-) = 9$ V.

32. (*a*) Graph the current i for the circuit described in Exercise 31 for resistor values 1.5 kΩ, 15 kΩ, and 150 kΩ. Make three separate graphs, and be sure to extend the corresponding time axis to observe the settling time in each case. (*b*) Determine the corresponding settling times.

33. Analyze the circuit described in Exercise 31 to find $v(t)$, $t > 0$, if R is equal to (*a*) 2 kΩ; (*b*) 2 Ω. (*c*) Graph both responses over the range of $0 \le t \le 60$ ms. (*d*) Verify your answers with appropriate SPICE simulations.

34. A source-free parallel *RLC* circuit is constructed using two 10 mF capacitors soldered together such that they are connected in series with one another and a single inductor having a component value of 10 mH. The capacitor has an initial voltage of 2 V, and the inductor current is initially zero. (*a*) Compute α, ω_0, ω_d, and ς, if $R = 50$ Ω. (*b*) Simulate the circuit with R varying between 50 and 200 Ω in 10 Ω increments to determine the resistance range which leads to voltage oscillations below ±200 kV for $t > 300$ ms.

35. For the circuit of Fig. 9.46, determine (*a*) $i_C(0^-)$; (*b*) $i_L(0^-)$; (*c*) $i_R(0^-)$; (*d*) $v_C(0^-)$; (*e*) $i_C(0^+)$; (*f*) $i_L(0^+)$; (*g*) $i_R(0^+)$; (*h*) $v_C(0^+)$.

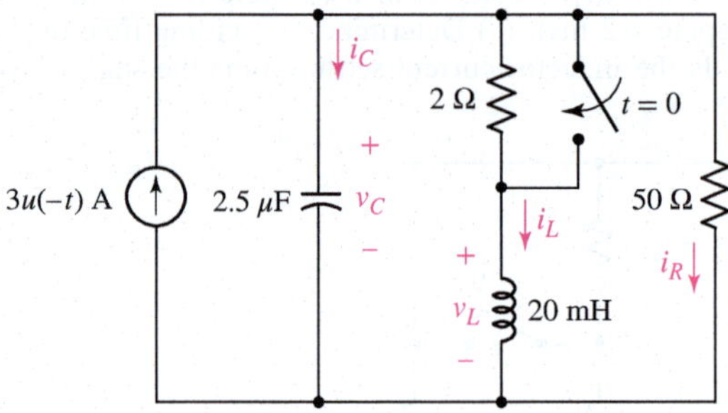

FIGURE 9.46

36. Obtain an expression for $v_L(t)$, $t > 0$, for the circuit shown in Fig. 9.46. Plot the waveform, ensuring that you observe the settling time.

37. For the circuit of Fig. 9.47, determine (*a*) the first time $t > 0$ when $v(t) = 0$; (*b*) the settling time.

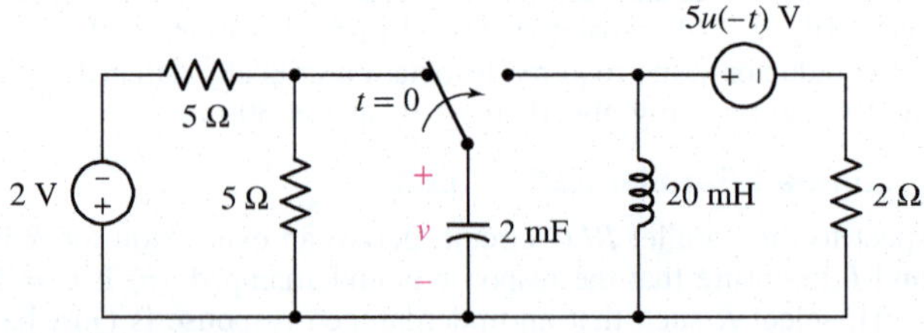

FIGURE 9.47

38. (*a*) Design a parallel *RLC* circuit that provides a capacitor voltage which oscillates with a frequency of 100 rad/s, with a maximum value of 10 V occurring at $t = 0$, and the second and third maxima both in excess of 6 V. (*b*) Verify your design with an appropriate SPICE simulation.

39. The circuit depicted in Fig. 9.48 is just barely underdamped. (*a*) Compute α and ω_d. (*b*) Obtain an expression for $i_L(t)$ valid for $t > 0$. (*c*) Determine how much energy is stored in the capacitor, and in the inductor, at $t = 200$ ms.

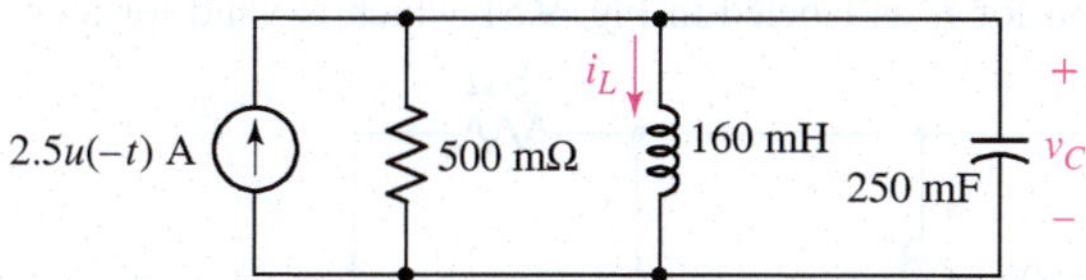

FIGURE 9.48

40. When constructing the circuit of Fig. 9.48, you inadvertently install a 500 MΩ resistor by mistake. (*a*) Compute α and ω_d. (*b*) Obtain an expression for $i_L(t)$ valid for $t > 0$. (*c*) Determine how long it takes for the energy stored in the inductor to reach 10% of its maximum value.

9.5 The Source-Free Series *RLC* Circuit

41. The circuit of Fig. 9.22*a* is constructed with a 160 mF capacitor and a 250 mH inductor. Determine the value of R needed to obtain (*a*) a critically damped response; (*b*) a "just barely" underdamped response. (*c*) Compare your answers to parts (*a*) and (*b*) if the circuit is a parallel *RLC* circuit.

42. Component values of $R = 2\ \Omega$, $C = 1$ mF, and $L = 2$ mH are used to construct the circuit represented in Fig. 9.22*a*. If $v_C(0^-) = 1$ V and no current initially flows through the inductor, calculate $i(t)$ at $t = 1$ ms, 2 ms, and 3 ms.

43. A source-free series *RLC* circuit has $R = 15\ \Omega$, $L = 25$ mH, and $C = 50\ \mu$F. If the current flow is initially 300 mA, and the capacitor is initially discharged, find an expression for the current dependence of the circuit and evaluate at time $t = 6$ ms.

44. The simple three-element series *RLC* circuit of Exercise 42 is constructed with the same component values, but the initial capacitor voltage $v_C(0^-) = 2$ V and the initial inductor current $i(0^-) = 1$ mA. (*a*) Obtain an expression for $i(t)$ valid for all $t > 0$. (*b*) Verify your solution with an appropriate SPICE simulation.

45. The series *RLC* circuit of Fig. 9.23 is constructed using $R = 1$ kΩ, $C = 2$ mF, and $L = 1$ mH. The initial capacitor voltage v_C is -4 V at $t = 0^-$. There is no current initially flowing through the inductor. (*a*) Obtain an expression for $v_C(t)$ valid for $t > 0$. (*b*) Graph over $0 \le t \le 6\ \mu$s.

46. With reference to the circuit depicted in Fig. 9.49, calculate (*a*) α; (*b*) ω_0; (*c*) $i(0^+)$; (*d*) $di/dt|_{0^+}$; (*e*) $i(t)$ at $t = 6$ s.

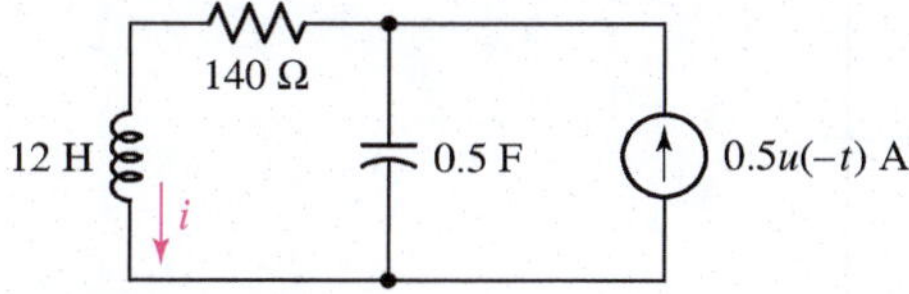

FIGURE 9.49

47. Obtain an equation for v_C as labeled in the circuit of Fig. 9.50 valid for all $t > 0$.

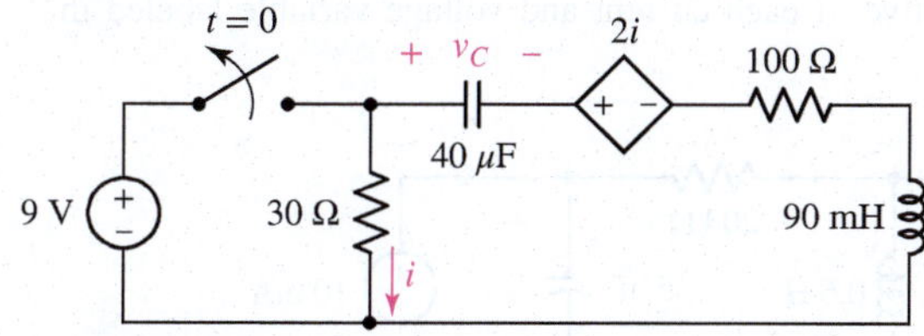

FIGURE 9.50

48. With reference to the series *RLC* circuit of Fig. 9.50, (*a*) obtain an expression for i, valid for $t > 0$; (*b*) calculate $i(0.8\text{ ms})$ and $i(4\text{ ms})$; (*c*) verify your answers to part (*b*) with an appropriate SPICE simulation.
49. Obtain an expression for i_1 as labeled in Fig. 9.51 which is valid for all $t > 0$.

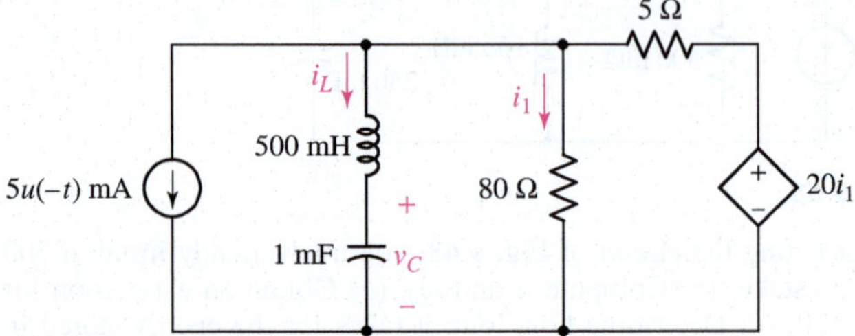

■ **FIGURE 9.51**

9.6 The Complete Response of the *RLC* Circuit

50. The circuit in Fig. 9.52 has the switch in position *a* for a long time, with the capacitor discharged. At time $t = 0$, the switch is moved to position *b*. Determine the initial and final conditions for the capacitor and inductor (both current and voltage for each element).

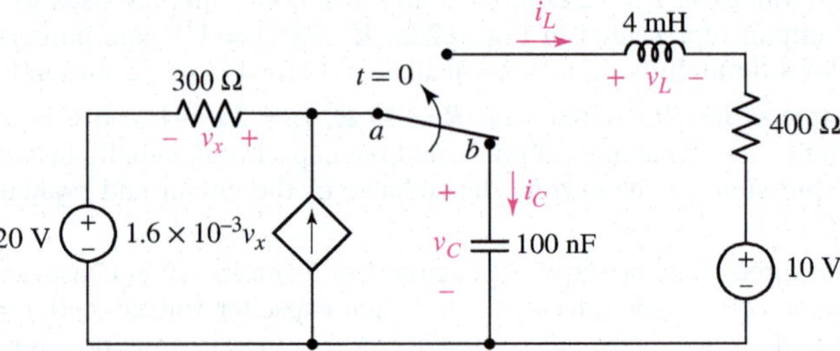

■ **FIGURE 9.52**

51. For the circuit in Fig. 9.52, determine the value for the capacitor voltage $v_C(t)$ for time $t > 0$.
52. In the series circuit of Fig. 9.53, set $R = 1\ \Omega$. (*a*) Compute α and ω_0. (*b*) If $i_s = 3u(-t) + 2u(t)$ mA, determine $v_R(0^-)$, $i_L(0^-)$, $v_C(0^-)$, $v_R(0^+)$, $i_L(0^+)$, $v_C(0^+)$, $i_L(\infty)$, and $v_C(\infty)$.

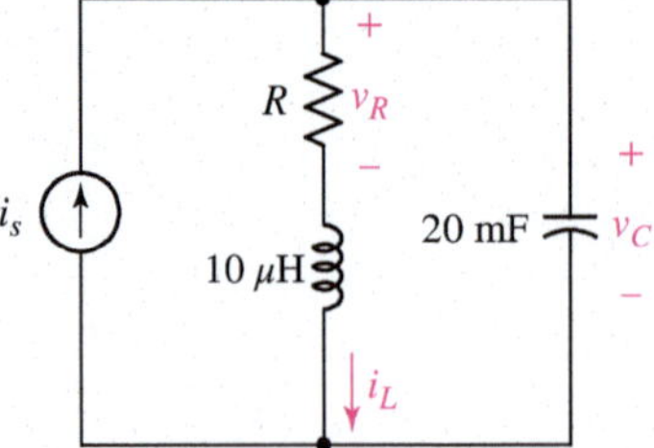

■ **FIGURE 9.53**

53. Evaluate the derivative of each current and voltage variable labeled in Fig. 9.54 at $t = 0^+$.

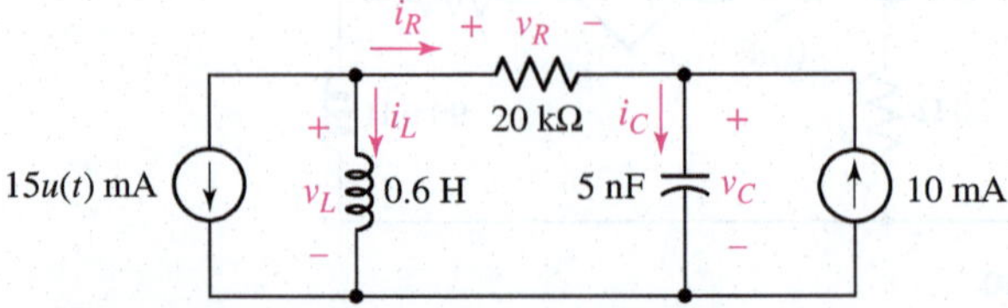

■ **FIGURE 9.54**

54. Consider the circuit depicted in Fig. 9.55. If $v_s(t) = -8 + 2u(t)$ V, determine (*a*) $v_C(0^+)$; (*b*) $i_L(0^+)$; (*c*) $v_C(\infty)$; (*d*) $v_C(t = 150 \text{ ms})$.

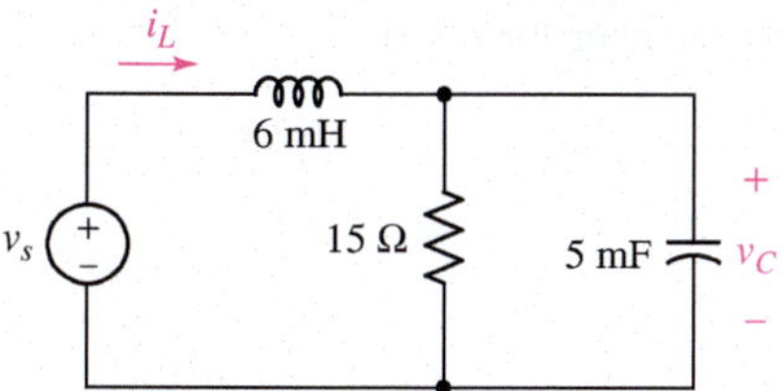

■ **FIGURE 9.55**

55. The 15 Ω resistor in the circuit of Fig. 9.55 is replaced with a 500 mΩ alternative. If the source voltage is given by $v_s = 1 - 2u(t)$ V, determine (*a*) $i_L(0^+)$; (*b*) $v_C(0^+)$; (*c*) $i_L(\infty)$; (*d*) $v_C(4 \text{ ms})$.

56. In the circuit shown in Fig. 9.56, (*a*) obtain an expression for i_L valid for $t > 0$ if $i_1 = 8 - 10u(t)$ mA, (*b*) graph the result for $0 \le t \le 2$ ms.

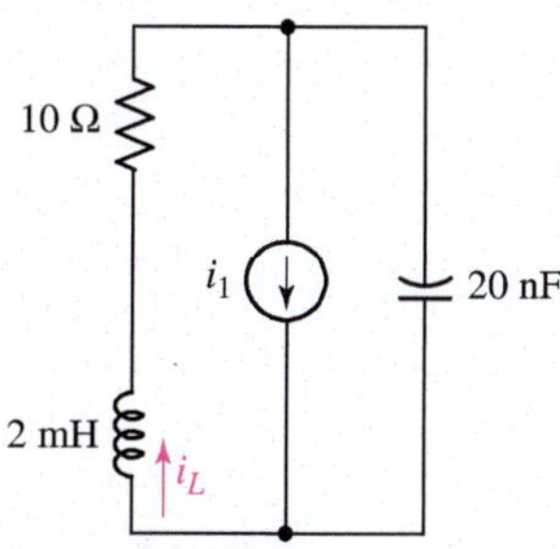

■ **FIGURE 9.56**

57. The 10 Ω resistor in the series *RLC* circuit of Fig. 9.56 is replaced with a 1 kΩ resistor. The source $i_1 = 5u(t) - 4$ mA. (*a*) Obtain an expression for i_L valid for all $t > 0$. (*b*) Graph the result for $0 \le t \le 200$ μs.

58. For the circuit represented in Fig. 9.57, (*a*) obtain an expression for $v_C(t)$ valid for all $t > 0$. (*b*) Determine v_C at $t = 10$ ms and $t = 600$ ms. (*c*) Verify your answers to part (*b*) with an appropriate SPICE simulation.

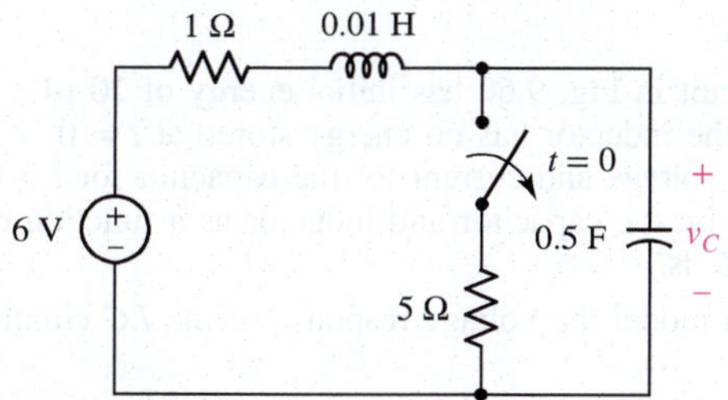

■ **FIGURE 9.57**

59. Replace the 1 Ω resistor in Fig. 9.57 with a 100 mΩ resistor, and the 5 Ω resistor with a 200 mΩ resistor. Assuming the passive sign convention, (*a*) obtain an expression for the capacitor current which is valid for $t > 0$, (*b*) graph the result for $0 \le t \le 2$ s.

60. A circuit has an inductive load of 2 μH, a capacitance of 500 nF, and a load resistance of 50 Ω. Using strictly a series or parallel configuration with these values results in either a circuit with undesirable "ringing" (oscillations) or a transient that is too fast. Design a circuit that responds to a voltage pulse which reaches half of the final value at a time of 50 μs. You must use the given components, but you may add additional resistors.

61. (*a*) Adjust the value of the 3 Ω resistor in the circuit represented in Fig. 9.58 to obtain a "just barely" overdamped response. Using the new resistor value, (*b*) determine expressions for $v_C(t)$ and $i_L(t)$ for $t > 0$, and (*c*) graph the energy stored in the capacitor and inductor for $t > 0$.

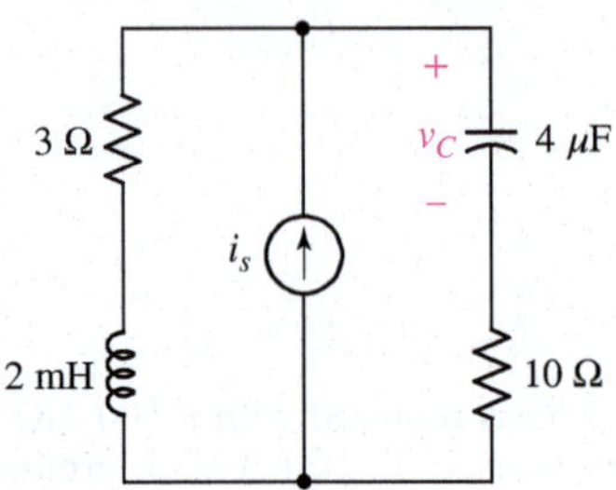

■ **FIGURE 9.58**

62. Determine expressions for $v_C(t)$ and $i_L(t)$ in Fig. 9.59 for the time windows (*a*) $0 < t < 2$ μs and (*b*) $t > 2$ μs.

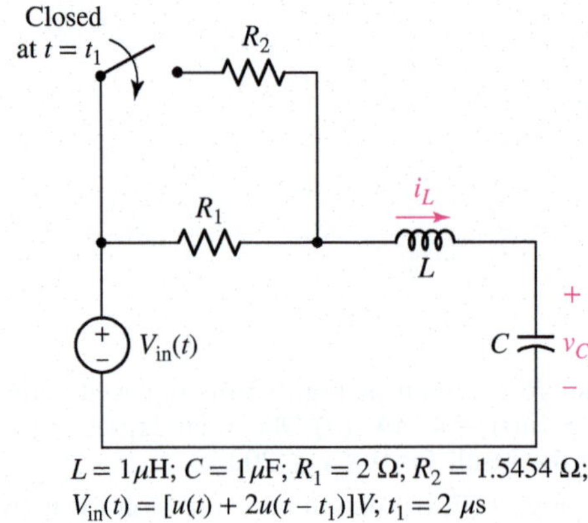

■ **FIGURE 9.59**

9.7 The Lossless *LC* Circuit

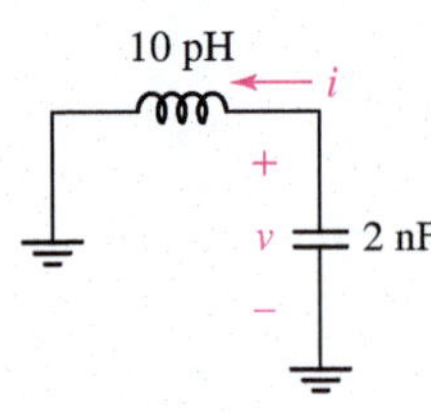

■ **FIGURE 9.60**

63. The capacitor in the *LC* circuit in Fig. 9.60 has initial energy of 20 pJ stored at time $t = 0$, while the inductor has no energy stored at $t = 0$. (*a*) Determine the capacitor voltage and current for the capacitor for $t > 0$. (*b*) Graph the energy stored on the capacitor and inductor as a function of time for the range $0 < t < 5$ ns.

DP 64. Design an op amp circuit to model the voltage response of the *LC* circuit shown in Fig. 9.60.

DP 65. Refer to Fig. 9.61, and design an op amp circuit whose output will be $i(t)$ for $t > 0$.

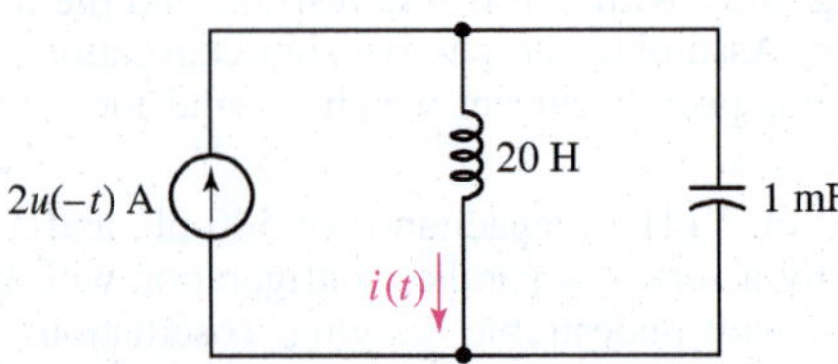

■ **FIGURE 9.61**

66. Suppose that the switch in the circuit in Fig. 9.62 is closed for a long time. The switch is then opened at time $t = 0$, and again closed at time $t = t_1 = 2\pi$ s. (*a*) Determine an expression for $v(t)$ for $0 < t < t_1$ and $t > t_1$. (*b*) Graph the results.

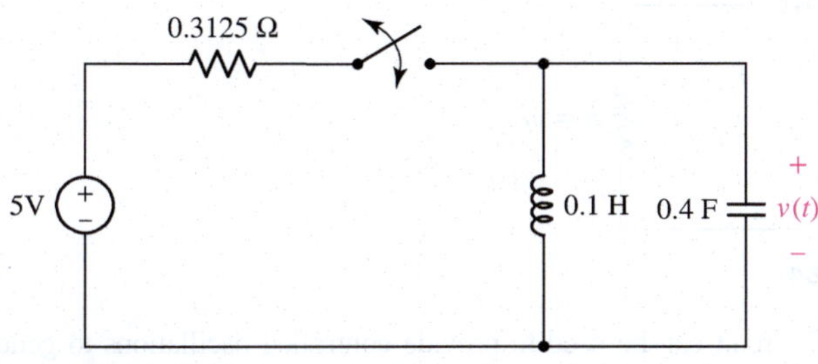

■ **FIGURE 9.62**

Chapter-Integrating Exercises

67. The capacitor in the circuit of Fig. 9.63 is set to 1 F. Determine $v_C(t)$ at (*a*) $t = -1$ s; (*b*) $t = 0^+$; (*c*) $t = 20$ s.

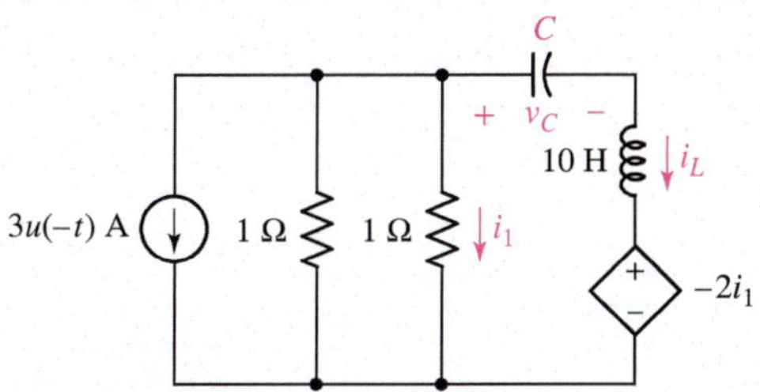

■ **FIGURE 9.63**

68. A particular robotic arm requires a current pulse to start motion. Design an *RLC* circuit that provides a current pulse to a 50 Ω load with amplitude of 500 mA at 1 ms and maintains a current of 250 mA or higher for a duration of at least 1 ms. The current pulse should decay to below 50 mA for $t > 15$ ms. Plot and analyze your result using MATLAB or SPICE simulation to verify your design.

69. A spark generator requires a short voltage pulse at 10 kV. Design an *RLC* circuit that provides a voltage pulse to a 1 kΩ load with amplitude of 10 kV at 5 μs and maintains a voltage of 5 kV or higher for a duration of at least 5 μs. The voltage pulse should decay to below 500 V for $t > 50$ μs. Plot and analyze your result using MATLAB or SPICE simulation to verify your design. (Before you build such a circuit, of course, be sure that your circuit components can operate at high voltage!)

70. The physical behavior of automotive suspension systems is similar to an *RLC* circuit. The differential equation is defined by

$$m\frac{d^2p(t)}{dt^2} + \mu_f\frac{dp(t)}{dt} + Kp(t) = 0$$

where $p(t)$ is the position variable of a piston in the cylinder of a shock absorber, m is the mass of the wheel, μ_f is the coefficient of friction, and K is the spring constant. The equivalent circuit representation is shown in Fig. 9.64. Suppose that the suspension is in its initial position at $t = 0$ ($p(0) = 0$), but it experiences a bump such that dp/dt at $t = 0$ is $\frac{1}{15}$ m/s. Use the equivalent circuit analogy and either MATLAB or SPICE to plot $p(t)$ for the cases where $\mu_f = 300$, 750, and 1500 N · s/m. The system has a spring constant of $K = 22$ kN/m and wheel mass of $m = 30$ kg.

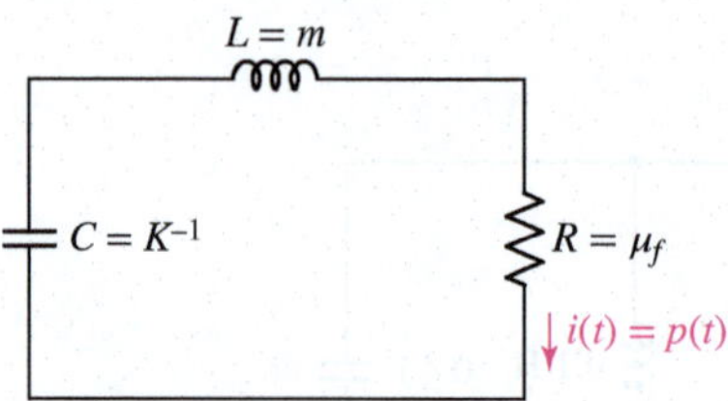

FIGURE 9.64

71. A lossless *LC* circuit can be used to provide controlled oscillations to generate a controlled frequency for wireless communications. (*a*) Design an *LC* circuit with amplitude of 5 V and frequency of 400 kHz, where the largest possible inductor available is 400 nH. Now suppose that you have an undesired resistance of 0.2 mΩ in series with the *LC* oscillator. (*b*) Determine if, and how much, the frequency changes as a result of the resistance. (*c*) Determine the maximum time that the oscillator can run before the voltage amplitude decays to 4.8 V, and (*d*) determine the energy dissipation during this time period (it will be very useful to use software such as MATLAB for calculating the energy dissipation!).

CHAPTER 10

Sinusoidal Steady-State Analysis

KEY CONCEPTS

Characteristics of Sinusoidal Functions

Phasor Representation of Sinusoids

Converting Between the Time and Frequency Domains

Impedance and Admittance

Reactance and Susceptance

Parallel and Series Combinations in the Frequency Domain

Determination of Forced Response Using Phasors

Application of Circuit Analysis Techniques in the Frequency Domain

INTRODUCTION

The complete response of a linear electric circuit is composed of two parts, the *natural* response and the *forced* response. The natural response is the short-lived transient response of a circuit to a sudden change in its condition. The forced response is the long-term steady-state response of a circuit to any independent sources present. Up to this point, the only forced response we have considered is that due to dc sources. Another very common forcing function is the sinusoidal waveform—this function describes the voltage available at household electrical sockets as well as the voltage of power lines connected to residential and industrial areas.

In this chapter, we assume that the transient response is of little interest, and the steady-state response of a circuit (a phone charger, a toaster, or a power distribution network) to a sinusoidal voltage or current is needed. We will analyze such circuits using a powerful technique that transforms integrodifferential equations into algebraic equations. Before we see how that works, it's useful to quickly review a few important attributes of general sinusoids, which will describe pretty much all currents and voltages throughout the chapter.

10.1 CHARACTERISTICS OF SINUSOIDS

Consider a sinusoidally varying voltage

$$v(t) = V_m \sin \omega t$$

shown graphically in Figs. 10.1*a* and *b*. The *amplitude* of the sine wave is V_m, and the *argument* is ωt. The *radian frequency,* or *angular frequency,* is ω. In Fig. 10.1*a*, $V_m \sin \omega t$ is plotted as a function of the argument ωt, and the periodic nature of the sine wave is evident. The

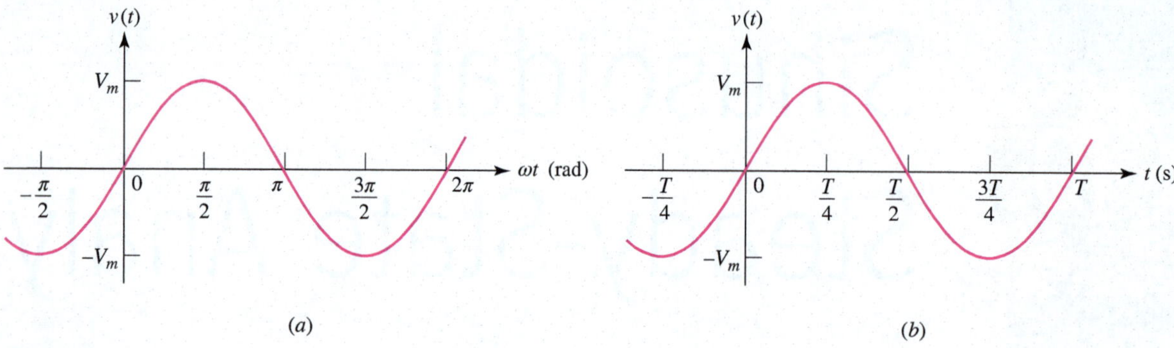

FIGURE 10.1 The sinusoidal function $v(t) = V_m \sin \omega t$ is plotted (*a*) versus ωt and (*b*) versus t.

function repeats itself every 2π radians, and its ***period*** is therefore 2π radians. In Fig. 10.1*b*, $V_m \sin \omega t$ is plotted as a function of t and the *period* is now T. A sine wave having a period T must execute $1/T$ periods each second; its ***frequency*** f is $1/T$ hertz, abbreviated Hz. Thus,

$$f = \frac{1}{T}$$

and since

$$\omega T = 2\pi$$

we obtain the common relationship between frequency and radian frequency,

$$\omega = 2\pi f$$

Lagging and Leading

A more general form of the sinusoid,

$$v(t) = V_m \sin(\omega t + \theta) \qquad [1]$$

includes a *phase angle* θ in its argument. Equation [1] is plotted in Fig. 10.2 as a function of ωt, and the phase angle appears as the number of radians by which the original sine wave (shown in green color in the sketch) is shifted to the left, or earlier in time. Since corresponding points on the sinusoid $V_m \sin(\omega t + \theta)$ occur θ rad, or θ/ω seconds, earlier, we say that $V_m \sin(\omega t + \theta)$ *leads* $V_m \sin \omega t$ by θ rad. Therefore, it is correct to describe $\sin \omega t$ as ***lagging*** $\sin(\omega t + \theta)$ by θ rad, or as ***leading*** $\sin(\omega t - \theta)$ by θ rad.

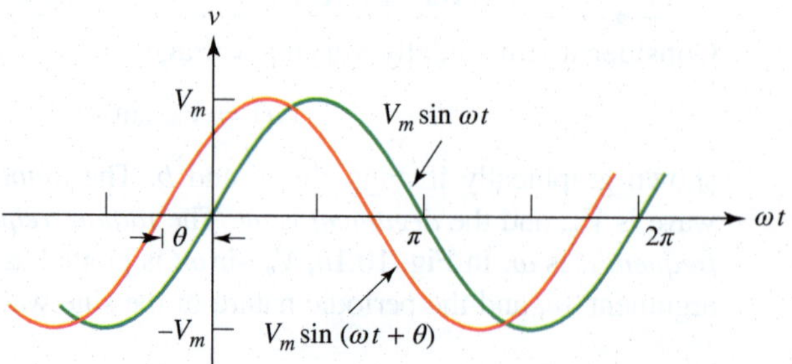

FIGURE 10.2 The sine wave $V_m \sin(\omega t + \theta)$ leads $V_m \sin \omega t$ by θ rad.

Recall that to convert radians to degrees, we simply multiply the angle by $180/\pi$.

In either case (leading or lagging) we say that the sinusoids are *out of phase*. If the phase angles are equal, the sinusoids are said to be *in phase*.

In electrical engineering, the phase angle is commonly given in degrees, rather than radians; to avoid confusion we should be sure to always use the degree symbol. Thus, instead of writing

$$v = 100 \sin\left(2\pi 1000t - \frac{\pi}{6}\right)$$

we customarily use

$$v = 100 \sin(2\pi 1000t - 30^\circ)$$

In evaluating this expression at a specific instant of time, e.g., $t = 10^{-4}$ s, $2\pi 1000t$ becomes 0.2π *radian,* which should be expressed as 36° before 30° is subtracted from it. (Don't confuse your apples with your oranges.)

Two sinusoidal waves whose phases are to be compared must:

1. Both be written as sine waves, or both as cosine waves.
2. Both be written with positive amplitudes.
3. Each have the same frequency.

Converting Sines to Cosines

Note that:

$-\sin \omega t = \sin(\omega t \pm 180^\circ)$

$-\cos \omega t = \cos(\omega t \pm 180^\circ)$

$\mp\sin \omega t = \cos(\omega t \pm 90^\circ)$

$\pm\cos \omega t = \sin(\omega t \pm 90^\circ)$

The sine and cosine are essentially the same function, but with a 90° phase difference. Thus, $\sin \omega t = \cos(\omega t - 90^\circ)$. Multiples of 360° may be added to or subtracted from the argument of any sinusoidal function without changing the value of the function. Hence, we may say that

$$\begin{aligned} v_1 &= V_{m_1} \cos(5t + 10^\circ) \\ &= V_{m_1} \sin(5t + 10^\circ + 90^\circ) \\ &= V_{m_1} \sin(5t + 100^\circ) \end{aligned}$$

leads

$$v_2 = V_{m_2} \sin(5t - 30^\circ)$$

by 130°. It is also correct to say that v_1 *lags* v_2 by 230°, since v_1 may be written as

$$v_1 = V_{m_1} \sin(5t - 260^\circ)$$

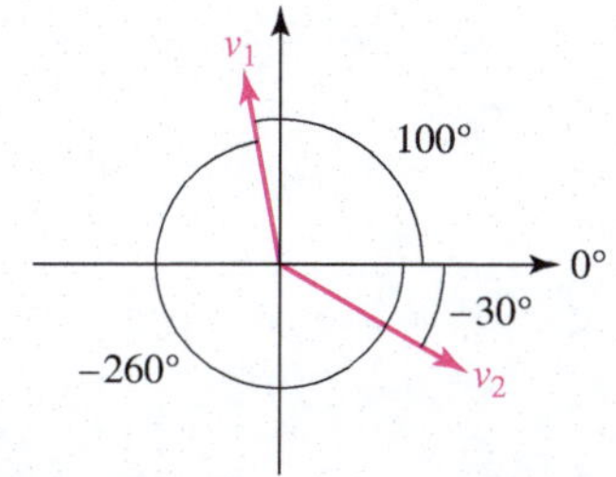

FIGURE 10.3 A graphical representation of the two sinusoids v_1 and v_2. The magnitude of each sine function is represented by the length of the corresponding arrow and the phase angle by the orientation with respect to the positive *x* axis. In this diagram, v_1 leads v_2 by $100^\circ + 30^\circ = 130^\circ$, although it could also be argued that v_2 leads v_1 by 230°. It is customary, however, to express the phase difference by an angle less than or equal to 180° in magnitude.

We assume that V_{m_1} and V_{m_2} are both positive quantities. A graphical representation is provided in Fig. 10.3; note that the frequency of both sinusoids (5 rad/s in this case) must be the same, or the comparison is meaningless. Normally, the difference in phase between two sinusoids is expressed by that angle which is less than or equal to 180° in magnitude.

The concept of a leading or lagging relationship between two sinusoids will be used extensively, and the relationship is recognizable both mathematically and graphically.

PRACTICE

10.1 Find the angle by which i_1 lags v_1 if $v_1 = 120\cos(120\pi t - 40°)$ V and i_1 equals (*a*) $2.5\cos(120\pi t + 20°)$ A; (*b*) $1.4\sin(120\pi t - 70°)$ A; (*c*) $-0.8\cos(120\pi t - 110°)$ A.

10.2 Determine values for *A*, *B*, *C*, and θ if $40\cos(100t - 40°) - 20\sin(100t + 170°) = A\cos 100t + B\sin 100t = C\cos(100t + \theta)$.

Ans: 10.1: −60°; 120°; −110°. 10.2: 27.2; 45.4; 52.9; −59.1°.

10.2 FORCED RESPONSE TO SINUSOIDAL FUNCTIONS

Now that we are familiar with the mathematical characteristics of sinusoids, we are ready to apply a sinusoidal forcing function to a simple circuit and obtain the forced response. We will first write the differential equation that applies to the given circuit. The complete solution of this equation is composed of two parts, the complementary solution (which we call the *natural response*) and the particular integral (or *forced response*). The methods we plan to develop in this chapter assume that we are not interested in the short-lived transient or natural response of our circuit, but only in the long-term or "steady-state" response.

The Steady-State Response

The term *steady-state response* is used synonymously with *forced response,* and the circuits we are about to analyze are commonly said to be in the "sinusoidal steady state." Unfortunately, *steady state* carries the connotation of "not changing with time" in the minds of many students. This is true for dc forcing functions, but the sinusoidal steady-state response is definitely changing with time. The steady state simply refers to the condition that is reached after the transient or natural response has died out.

The forced response has the mathematical form of the forcing function, plus all its derivatives and its first integral. With this knowledge, one of the methods by which the forced response may be found is to assume a solution composed of a sum of such functions, where each function has an unknown amplitude to be determined by direct substitution in the differential equation. As we are about to see, this can be a lengthy process, so we will be sufficiently motivated to seek out a simpler alternative!

Consider the series *RL* circuit shown in Fig. 10.4. The sinusoidal source voltage $v_s = V_m \cos\omega t$ has been switched into the circuit at some remote time in the past, and the natural response has died out completely. We seek the forced (or "steady-state") response, which must satisfy the differential equation

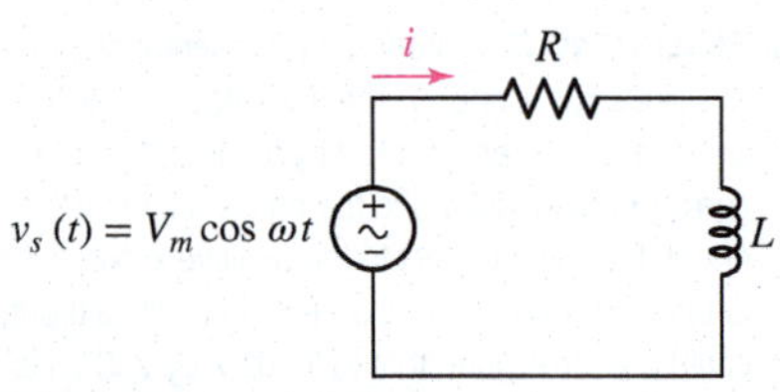

■ **FIGURE 10.4** A series *RL* circuit for which the forced response is desired.

$$L\frac{di}{dt} + Ri = V_m \cos\omega t$$

obtained by applying KVL around the loop. At any instant where the derivative is equal to zero, we see that the current must have the form $i \propto \cos\omega t$. Similarly, at an instant where the current is equal to zero, the *derivative* must be proportional to $\cos\omega t$, implying a current of the form $\sin\omega t$. We might expect, therefore, that the forced response will have the general form

$$i(t) = I_1 \cos\omega t + I_2 \sin\omega t$$

where I_1 and I_2 are real constants whose values depend upon V_m, R, L, and ω. No constant or exponential function can be present. Substituting the assumed form for the solution in the differential equation yields

$$L(-I_1\omega \sin\omega t + I_2\omega \cos\omega t) + R(I_1 \cos\omega t + I_2 \sin\omega t) = V_m \cos\omega t$$

If we collect the cosine and sine terms, we obtain

$$(-LI_1\omega + RI_2) \sin\omega t + (LI_2\omega + RI_1 - V_m) \cos\omega t = 0$$

This equation must be true for all values of t, which can be achieved only if the factors multiplying $\cos\omega t$ and $\sin\omega t$ are each zero. Thus,

$$-\omega LI_1 + RI_2 = 0 \qquad \text{and} \qquad \omega LI_2 + RI_1 - V_m = 0$$

and simultaneous solution for I_1 and I_2 leads to

$$I_1 = \frac{RV_m}{R^2 + \omega^2L^2} \qquad \text{and} \qquad I_2 = \frac{\omega LV_m}{R^2 + \omega^2L^2}$$

Thus, the forced response is obtained:

$$i(t) = \frac{RV_m}{R^2 + \omega^2L^2} \cos\omega t + \frac{\omega LV_m}{R^2 + \omega^2L^2} \sin\omega t \qquad [2]$$

A More Compact and User-Friendly Form

Although accurate, this expression is slightly cumbersome; a clearer picture of the response can be obtained by expressing it as a single sinusoid or cosinusoid with a phase angle. We choose to express the response as a cosine function,

$$i(t) = A \cos(\omega t - \theta) \qquad [3]$$

At least two methods of obtaining the values of A and θ suggest themselves. We might substitute Eq. [3] directly in the original differential equation, or we could simply equate the two solutions, Eqs. [2] and [3]. Selecting the latter method, and expanding the function $\cos(\omega t - \theta)$:

Several useful trigonometric identities are provided in the back of the book.

$$A \cos\theta \cos\omega t + A \sin\theta \sin\omega t = \frac{RV_m}{R^2 + \omega^2L^2} \cos\omega t + \frac{\omega LV_m}{R^2 + \omega^2L^2} \sin\omega t$$

All that remains is to collect terms and perform a bit of algebra. The result is

$$\theta = \tan^{-1}\frac{\omega L}{R}$$

and

$$A = \frac{V_m}{\sqrt{R^2 + \omega^2L^2}}$$

and so the *alternative form of* the forced response therefore becomes

$$i(t) = \frac{V_m}{\sqrt{R^2 + \omega^2L^2}} \cos\left(\omega t - \tan^{-1}\frac{\omega L}{R}\right) \qquad [4]$$

Once upon a time, the symbol *E* (for electromotive force) was used to designate voltages. Then every student learned the phase "ELI the ICE man" as a reminder that *voltage* leads *current* in an *inductive* circuit, while *current* leads *voltage* in a *capacitive* circuit. Now that we use *V* instead, it just isn't the same.

With this form, it is easy to see that the amplitude of the *response* is proportional to the amplitude of the *forcing function;* if it were not, the linearity concept would have to be discarded. The current is seen to lag the applied voltage by $\tan^{-1}(\omega L/R)$, an angle between 0 and 90°. When $\omega = 0$ or $L = 0$, the current must be in phase with the voltage; since the former situation is direct current and the latter provides a resistive circuit, the result agrees with our previous experience. If $R = 0$, the current lags the voltage by 90°. In an inductor, then, if the passive sign convention is satisfied, the current lags the voltage by exactly 90°. In a similar manner we can show that the current through a capacitor *leads* the voltage across it by 90°.

The phase difference between the current and voltage depends upon the ratio of the quantity ωL to R. We call ωL the *inductive reactance* of the inductor; it is measured in ohms, and it is a measure of the opposition that is offered by the inductor to the passage of a sinusoidal current.

EXAMPLE 10.1

Find the current i_L in the circuit shown in Fig. 10.5*a*, if the transients have already died out.

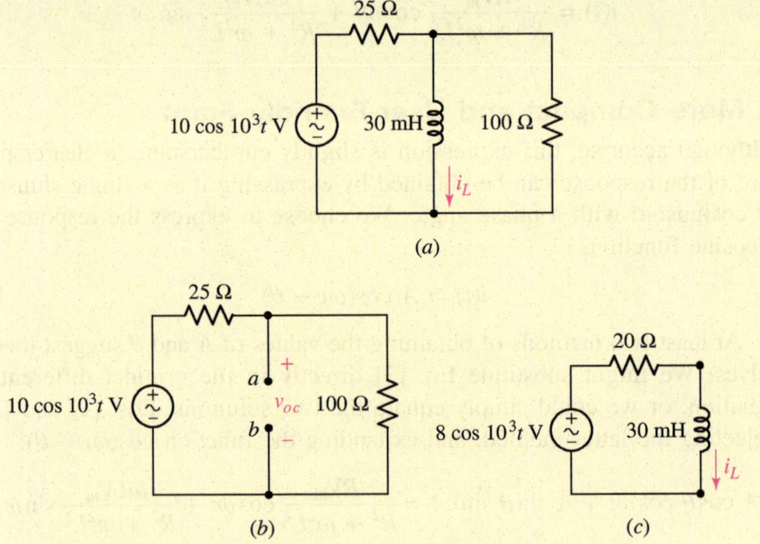

FIGURE 10.5 (*a*) The circuit for Example 10.1, in which the current i_L is desired. (*b*) The Thévenin equivalent is desired at terminals *a* and *b*. (*c*) The simplified circuit.

Although this circuit has a sinusoidal source and a single inductor, it contains two resistors and is not a single loop. In order to apply the results of the preceding analysis, we seek the Thévenin equivalent as viewed from terminals *a* and *b* in Fig. 10.5*b*.

The open-circuit voltage v_{oc} is

$$v_{oc} = (10 \cos 10^3 t)\frac{100}{100 + 25} = 8 \cos 10^3 t \text{ V}$$

Since there are no dependent sources in sight, we find R_{TH} by shorting out the independent source and calculating the resistance of the passive network, so $R_{TH} = (25 \times 100)/(25 + 100) = 20\ \Omega$.

Now we do have a series RL circuit, with $L = 30$ mH, $R_{TH} = 20\ \Omega$, and a source voltage of $8 \cos 10^3 t$ V, as shown in Fig. 10.5c. Thus, applying Eq. [4], which was derived for a general RL series circuit,

$$i_L = \frac{8}{\sqrt{20^2 + (10^3 \times 30 \times 10^{-3})^2}} \cos\left(10^3 t - \tan^{-1}\frac{30}{20}\right)$$

$$= 222 \cos(10^3 t - 56.3°)\ \text{mA}$$

The voltage and current waveforms are plotted in Fig. 10.6.

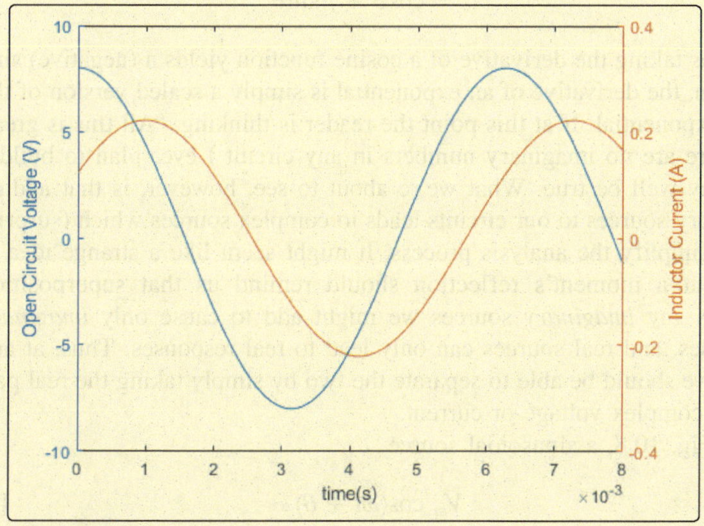

FIGURE 10.6 Voltage and current waveforms on a dual axis plot, generated using MATLAB:

```
≫ t = linspace(0,8e – 3,1000);
≫ v = 8*cos(1000*t);
≫ i = 0.222*cos(1000*t – 56.3*pi/180);
≫ axis_id = plotyy(t,v,t,i); xlabel('time (s)');
≫ ylabel(axis_id(1),'Open Circuit Voltage (V)');
≫ ylabel(axis_id(2),'Inductor Current (A)');
```

Note that there is not a 90° phase difference between the current and voltage waveforms of the plot. This is because we are not plotting the inductor voltage, which is left as an exercise for the reader.

PRACTICE

10.3 Let $v_s = 40 \cos 8000t$ V in the circuit of Fig. 10.7. Use Thévenin's theorem where it will do the most good, and find the value at $t = 0$ for (*a*) i_L; (*b*) v_L; (*c*) i_R; (*d*) i_s.

Ans: 18.71 mA; 15.97 V; 5.32 mA; 24.0 mA.

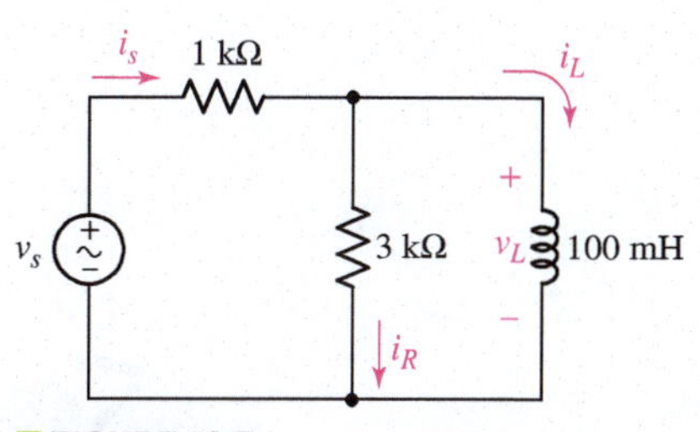

FIGURE 10.7

10.3 THE COMPLEX FORCING FUNCTION

The method we just employed works—the correct answer is obtained in a straightforward manner. However, it isn't particularly graceful, and after being applied to a few circuits, it remains as clunky and cumbersome as the first time we use it. The real problem isn't the time-varying source—it's the inductor (or capacitor), since a purely resistive circuit is no more difficult to analyze with sinusoidal sources than with dc sources, as only algebraic equations result. It turns out that there is an alternative approach for obtaining the sinusoidal steady-state response of any linear circuit, using simple algebraic expressions.

Appendix 5 defines the complex number and related terms, reviews complex arithmetic, and develops Euler's identity and the relationship between exponential and polar forms.

The basic idea is that sinusoids and exponentials are related through complex numbers. Euler's identity, for example, tells us that

$$e^{j\theta} = \cos\theta + j\sin\theta$$

Whereas taking the derivative of a cosine function yields a (negative) sine function, the derivative of an exponential is simply a scaled version of the same exponential. If at this point the reader is thinking, "All this is great, but there are no imaginary numbers in any circuit I ever plan to build!" that may well be true. What we're about to see, however, is that adding imaginary sources to our circuits leads to complex sources which (surprisingly) simplify the analysis process. It might seem like a strange idea at first, but a moment's reflection should remind us that superposition requires any *imaginary* sources we might add to cause only *imaginary* responses, and real sources can only lead to real responses. Thus, at any point, we should be able to separate the two by simply taking the real part of any complex voltage or current.

In Fig. 10.8, a sinusoidal source

$$V_m \cos(\omega t + \theta) \qquad [5]$$

is connected to a general network, which we will assume to contain only passive elements (i.e., no independent sources) in order to keep things simple. A current response in some other branch of the network is to be determined, and the parameters appearing in Eq. [5] are all real quantities.

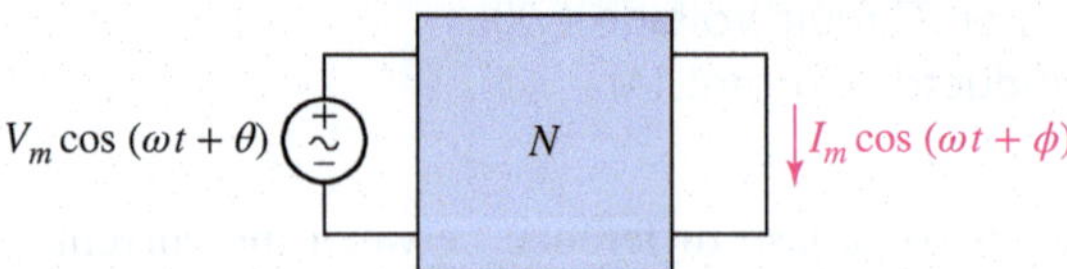

FIGURE 10.8 The sinusoidal forcing function $V_m \cos(\omega t + \theta)$ produces the steady-state sinusoidal response $I_m \cos(\omega t + \phi)$.

We have shown that we may represent the response by the general cosine function

$$I_m \cos(\omega t + \phi) \qquad [6]$$

A sinusoidal forcing function always produces a sinusoidal forced response of the same frequency in a linear circuit.

Now let us change our time reference by shifting the phase of the forcing function by 90°, or changing the instant that we call $t = 0$. Thus, the forcing function

$$V_m \cos(\omega t + \theta - 90°) = V_m \sin(\omega t + \phi) \qquad [7]$$

when applied to the same network will produce a corresponding response

$$I_m \cos(\omega t + \phi - 90°) = I_m \sin(\omega t + \phi) \qquad [8]$$

We next depart from physical reality by applying an imaginary forcing function, one that cannot be applied in the laboratory but can be applied mathematically. It's worth the effort, as we're about to see.

Imaginary Sources Lead to . . . Imaginary Responses

We construct an imaginary source very simply; it is only necessary to multiply Eq. [7] by j, the imaginary operator. We thus apply

$$jV_m \sin(\omega t + \theta) \qquad [9]$$

What is the response? If we had doubled the source, then the principle of linearity would require that we double the response; multiplication of the forcing function by a constant k would result in the multiplication of the response by the same constant k. The fact that our constant is $\sqrt{-1}$ does not destroy this relationship. The response to the imaginary source of Eq. [9] is thus

$$jI_m \sin(\omega t + \phi) \qquad [10]$$

The imaginary source and response are indicated in Fig. 10.9.

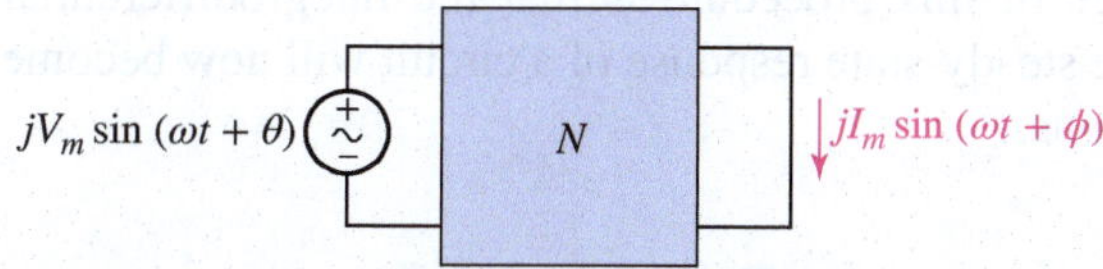

FIGURE 10.9 The imaginary sinusoidal forcing function $jV_m \sin(\omega t + \theta)$ produces the imaginary sinusoidal response $jI_m \cos(\omega t + \phi)$ in the network of Fig. 10.8.

Applying a Complex Forcing Function

We have applied a *real source* and obtained a *real* response; we have also applied an *imaginary* source and obtained an *imaginary* response. Since we are dealing with a *linear* circuit, we may use the superposition theorem to find the response to a complex forcing function which is the sum of the real and imaginary forcing functions. Thus, the sum of the forcing functions of Eqs. [5] and [9],

$$V_m \cos(\omega t + \theta) + jV_m \sin(\omega t + \theta) \qquad [11]$$

must produce a response that is the sum of Eqs. [6] and [10],

$$I_m \cos(\omega t + \phi) + jI_m \sin(\omega t + \phi) \qquad [12]$$

The complex source and response may be represented more simply by applying Euler's identity, i.e., $\cos(\omega t + \theta) + j\sin(\omega t + \theta) = e^{j(\omega t+\theta)}$. Thus, the source of Eq. [11] may be written as

$$V_m e^{j(\omega t+\theta)} \qquad [13]$$

and the response of Eq. [12] is

$$I_m e^{j(\omega t+\phi)} \qquad [14]$$

The complex source and response are illustrated in Fig. 10.10.

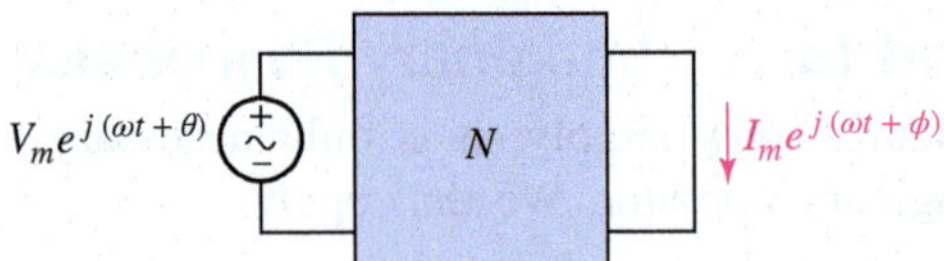

FIGURE 10.10 The complex forcing function $V_m e^{j(\omega t+\theta)}$ produces the complex response $I_m e^{j(\omega t+\theta)}$ in the network of Fig. 10.8.

Again, linearity assures us that the *real* part of the complex response is produced by the *real* part of the complex forcing function, while the *imaginary* part of the response is caused by the *imaginary* part of the complex forcing function. Our plan is that instead of applying a *real* forcing function to obtain the desired real response, we will substitute a *complex* forcing function whose real part is the given real forcing function; we expect to obtain a complex response whose real part is the desired real response. The advantage of this procedure is that the integrodifferential equations describing the steady-state response of a circuit will now become simple algebraic equations.

An Algebraic Alternative to Differential Equations

Let's try out this idea on the simple *RL* series circuit shown in Fig. 10.11. The real source $V_m \cos \omega t$ is applied; the real response $i(t)$ is desired. Since

$$V_m \cos \omega t = \text{Re}\{V_m \cos \omega t + j\, V_m \sin \omega t\} = \text{Re}\{V_m e^{j\omega t}\}$$

the necessary complex source is

$$V_m e^{j\omega t}$$

We express the complex response that results in terms of an unknown amplitude I_m and an unknown phase angle ϕ:

$$I_m e^{j(\omega t+\phi)}$$

Writing the differential equation for this particular circuit,

$$Ri + L\frac{di}{dt} = v_s$$

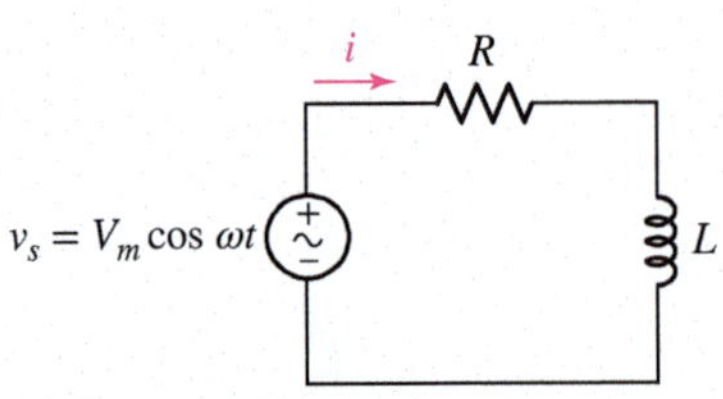

FIGURE 10.11 A simple circuit in the sinusoidal steady state is to be analyzed by the application of a complex forcing function.

we insert our complex expressions for v_s and i:

$$RI_m e^{j(\omega t+\phi)} + L\frac{d}{dt}(I_m e^{j(\omega t+\phi)}) = V_m e^{j\omega t}$$

take the indicated derivative:

$$RI_m e^{j(\omega t+\phi)} + j\omega LI_m e^{j(\omega t+\phi)} = V_m e^{j\omega t}$$

and obtain an *algebraic* equation. In order to determine the values of I_m and ϕ, we divide throughout by the common factor $e^{j\omega t}$:

$$RI_m e^{j\phi} + j\omega LI_m e^{j\phi} = V_m$$

factor the left side:

$$I_m e^{j\phi}(R + j\omega L) = V_m$$

rearrange:

$$I_m e^{j\phi} = \frac{V_m}{R + j\omega L}$$

and identify I_m and ϕ by expressing the right side of the equation in exponential or polar form:

$$I_m e^{j\phi} = \frac{V_m}{\sqrt{R^2 + \omega^2 L^2}} e^{j[-\tan^{-1}(\omega L/R)]} \qquad [15]$$

Thus,

$$I_m = \frac{V_m}{\sqrt{R^2 + \omega^2 L^2}}$$

and

$$\phi = -\tan^{-1}\frac{\omega L}{R}$$

In polar notation, this may be written as

$$I_m \underline{/\phi}$$

or

$$V_m/\sqrt{R^2 + \omega^2 L^2} \underline{/-\tan^{-1}(\omega L/R)}$$

The complex response is given by Eq. [15]. Since I_m and ϕ are readily identified, we can write the expression for $i(t)$ immediately. However, if we feel like using a more rigorous approach, we may obtain the real response $i(t)$ by reinserting the $e^{j\omega t}$ factor on both sides of Eq. [15] and taking the real part. Either way, we find that

$$i(t) = I_m \cos(\omega t + \phi) = \frac{V_m}{\sqrt{R^2 + \omega^2 L^2}} \cos\left(\omega t - \tan^{-1}\frac{\omega L}{R}\right)$$

which agrees with the response obtained in Eq. [4] for the same circuit.

EXAMPLE 10.2

For the simple *RC* circuit of Fig. 10.12*a*, substitute an appropriate complex source and use it to solve for the steady-state capacitor voltage.

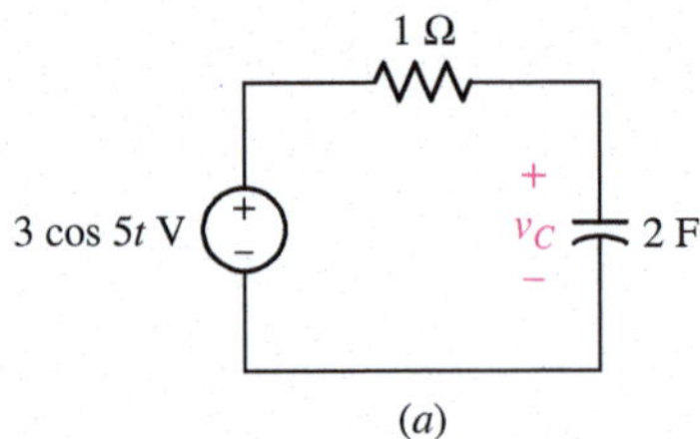

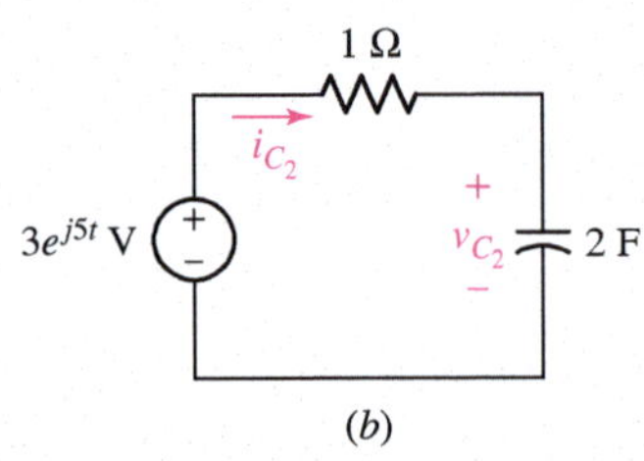

FIGURE 10.12 (*a*) An *RC* circuit for which the sinusoidal steady-state capacitor voltage is required. (*b*) Modified circuit, with the real source replaced with a complex source.

Since the real source is $3\cos 5t$ V, we "replace" it with a complex source $3e^{j5t}$ V. We'll call the new capacitor voltage v_{C_2} and define a capacitor current i_{C_2} consistent with the passive sign convention (Fig. 10.12*b*).

The differential equation can be now obtained by simple application of KVL,

$$-3\,e^{j5t} + 1i_{C_2} + v_{C_2} = 0$$

or

$$-3\,e^{j5t} + 2\frac{dv_{C_2}}{dt} + v_{C_2} = 0$$

We anticipate a steady-state response of the same form as our source; in other words,

$$v_{C_2} = V_m e^{j5t}$$

Substituting this into our differential equation and rearranging terms yields

$$j\,10V_m e^{j5t} + V_m e^{j5t} = 3\,e^{j5t}$$

Canceling the exponential term, we find that

$$V_m = \frac{3}{1+j10} = \frac{3}{\sqrt{1+10^2}}\underline{/-\tan^{-1}(10/1)}\text{ V}$$

and our steady-state capacitor voltage is given by

$$\text{Re}\{v_{C_2}\} = \text{Re}\{29.85e^{-j84.3^\circ}e^{j5t}\text{ mV}\} = 29.85\cos(5t - 84.3^\circ)\text{ mV}$$

PRACTICE

(If you have trouble working this practice problem, check out to Appendix 5.)

10.4 Evaluate and express the result in rectangular form:

(*a*) $[(2\underline{/30^\circ})(5\underline{/-110^\circ})](1 + j2)$; (*b*) $(5\underline{/-200^\circ}) + 4\underline{/20^\circ}$. Evaluate and express the result in polar form: (*c*) $(2 - j7)/(3 - j)$; (*d*) $8 - j4 + [(5\underline{/80^\circ})/(2\underline{/20^\circ})]$.

10.5 If the use of the passive sign convention is specified, find the (*a*) complex voltage that results when the complex current $4e^{j800t}$ A is applied to the series combination of a 1 mF capacitor and a 2 Ω resistor; (*b*) complex current that results when the complex voltage $100e^{j2000t}$ V is applied to the parallel combination of a 10 mH inductor and a 50 Ω resistor.

Ans: 10.4: $21.4 - j6.38$; $-0.940 + j3.08$; $2.30\underline{/-55.6^\circ}$; $9.43\underline{/-11.22^\circ}$.
10.5: $9.43e^{j(800t-32.0^\circ)}$ V; $5.39e^{j(2000t-68.2^\circ)}$ A.

10.4 THE PHASOR

In the last section, we saw that the addition of an imaginary sinusoidal source led to algebraic equations which describe the sinusoidal steady-state response of a circuit. An intermediate step of our analysis was the "canceling" of the complex exponential term—once its derivative was taken, we apparently had no further use for it until the real form of the response was desired. Even then, it was possible to read the magnitude and phase angle directly from our analysis, and hence skip the step where we overtly take the real part. Another way of looking at this is that every voltage and current in our circuit contains the same factor $e^{j\omega t}$, and the frequency, although relevant to our analysis, *does not change* as we move through the circuit. Dragging it around, then, is a bit of a waste of time.

Looking back at Example 10.2, then, we could have represented our source as

$$3e^{j0°}\text{ V} \qquad \text{(or even just 3 V)}$$

$e^{j0} = \cos 0 + j \sin 0 = 1$

and our capacitor voltage as $V_m e^{j\phi}$, which we ultimately found was $0.02985e^{-j84.3°}$ V. Knowledge of the source frequency is implicit here; without it, we cannot reconstruct any voltage or current.

These complex quantities are usually written in polar form rather than exponential form in order to achieve a slight additional saving of time and effort. For example, a source voltage

$$v(t) = V_m \cos \omega t = V_m \cos(\omega t + 0°)$$

we now represent in complex form as

$$V_m \underline{/0°}$$

and its current response

$$i(t) = I_m \cos(\omega t + \phi)$$

becomes

$$I_m \underline{/\phi}$$

This abbreviated complex representation is called a ***phasor.***[1]

Remember that none of the steady-state circuits we are considering will respond at a frequency other than that of the excitation source, so that the value of ω is always known.

Let us review the steps by which a real sinusoidal voltage or current is transformed into a phasor, and then we will be able to define a phasor more meaningfully and to assign a symbol to represent it.

A real sinusoidal current

$$i(t) = I_m \cos(\omega t + \phi)$$

is expressed as the real part of a complex quantity by invoking Euler's identity

$$i(t) = \text{Re}\{I_m e^{j(\omega t + \phi)}\}$$

We then represent the current as a complex quantity by dropping the instruction Re{ }, thus adding an imaginary component to the current without affecting the real component; further simplification is achieved by suppressing the factor $e^{j\omega t}$:

$$\mathbf{I} = I_m e^{j\phi}$$

and writing the result in polar form:

$$\mathbf{I} = I_m \underline{/\phi}$$

(1) Not to be confused with the *phaser,* an interesting device featured in a life-changing, ground-breaking, truly amazing television series....

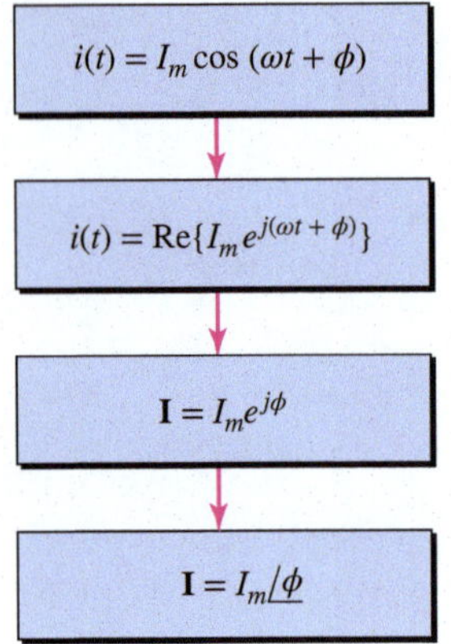

The process by which we change $i(t)$ into $\mathbf{I}$ is called a *phasor transformation* from the time domain to the frequency domain.

This abbreviated complex representation is the *phasor representation;* phasors are complex quantities and hence are printed in boldface type. Capital letters are used for the phasor representation of an electrical quantity because the phasor is not an instantaneous function of time; it contains only amplitude and phase information. We recognize this difference in viewpoint by referring to $i(t)$ as a *time-domain representation* and terming the phasor $\mathbf{I}$ a *frequency-domain representation.* It should be noted that the frequency-domain expression of a current or voltage does not explicitly include the frequency. The process of returning to the time domain from the frequency domain is exactly the reverse of the previous sequence. Thus, given the phasor voltage

$$\mathbf{V} = 115\angle{-45^\circ} \text{ volts}$$

and the knowledge that $\omega = 500$ rad/s, we can write the time-domain equivalent directly:

$$v(t) = 115 \cos(500t - 45^\circ) \text{ volts}$$

If desired as a sine wave, $v(t)$ could also be written

$$v(t) = 115 \sin(500t + 45^\circ) \text{ volts}$$

PRACTICE

10.6 Let $\omega = 2000$ rad/s and $t = 1$ ms. Determine the instantaneous value of each of the currents given here in phasor form: (*a*) $j10$ A; (*b*) $20 + j10$ A; (*c*) $20 + j(10\angle 20^\circ)$ A.

Ans: −9.09 A; −17.42 A; −15.44 A.

EXAMPLE 10.3

Transform the time-domain voltage $v(t) = 100 \cos(400t - 30^\circ)$ volts into the frequency domain.

The time-domain expression is already in the form of a cosine wave with a phase angle. Thus, suppressing $\omega = 400$ rad/s,

$$\mathbf{V} = 100\angle{-30^\circ} \text{ volts}$$

Note that we skipped several steps in writing this representation directly. Occasionally, this is a source of confusion for students, as they may forget that the phasor representation is *not* equal to the time-domain voltage $v(t)$. Rather, it is a simplified form of a complex function formed by adding an imaginary component to the real function $v(t)$.

Several useful trigonometric identities are provided in the back of the book for convenience.

PRACTICE

10.7 Transform each of the following functions of time into phasor form: (*a*) $-5 \sin(580t - 110^\circ)$; (*b*) $3 \cos 600t - 5 \sin(600t + 110^\circ)$; (*c*) $8 \cos(4t - 30^\circ) + 4 \sin(4t - 100^\circ)$. *Hint:* First convert each into a single cosine function with a positive magnitude.

Ans: $5\angle{-20^\circ}$; $2.41\angle{-134.8^\circ}$; $4.46\angle{-47.9^\circ}$.

The real power of the phasor-based analysis technique lies in the fact that it is possible to define *algebraic* relationships between the voltage and current for inductors and capacitors, just as we have always been able to do in the case of resistors. Now that we can transform into and out of the frequency domain, we proceed to our simplification of sinusoidal steady-state analysis by establishing the relationship between the phasor voltage and phasor current for each of the three passive elements.

The Resistor

The resistor provides the simplest case. In the time domain, as indicated by Fig. 10.13*a*, the defining equation is

$$v(t) = Ri(t)$$

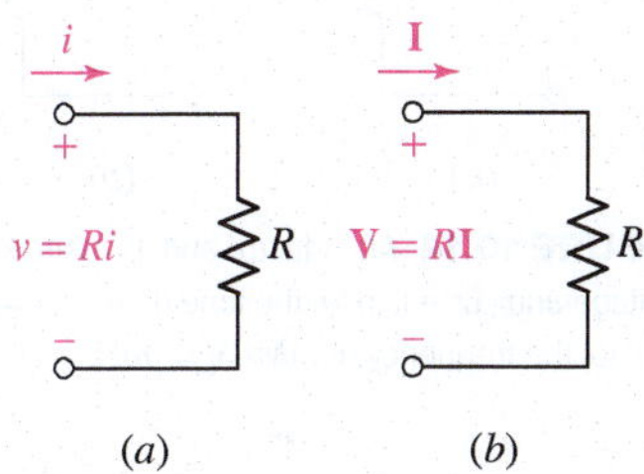

FIGURE 10.13 A resistor and its associated voltage and current in (*a*) the time domain, $v = Ri$; and (*b*) the frequency domain, $\mathbf{V} = R\mathbf{I}$.

Now let us apply the complex voltage

$$v(t) = V_m e^{j(\omega t+\theta)} = V_m \cos(\omega t + \theta) + jV_m \sin(\omega t + \theta) \qquad [16]$$

and assume the complex current response

$$i(t) = I_m e^{j(\omega t+\phi)} = I_m \cos(\omega t + \phi) + jI_m \sin(\omega t + \phi) \qquad [17]$$

so that

$$V_m e^{j(\omega t+\theta)} = Ri(t) = RI_m e^{j(\omega t+\phi)}$$

Dividing throughout by $e^{j\omega t}$, we find

$$V_m e^{j\theta} = RI_m e^{j\phi}$$

or, in polar form,

$$V_m \underline{/\theta} = RI_m \underline{/\phi}$$

However, $V_m \underline{/\theta}$ and $I_m \underline{/\phi}$ merely represent the general voltage and current phasors **V** and **I**. Thus,

$$\mathbf{V} = R\mathbf{I} \qquad [18]$$

Ohm's law holds true both in the time domain and in the frequency domain. In other words, the voltage across a resistor is always given by the resistance times the current flowing through the element.

The voltage–current relationship in phasor form for a resistor has the same form as the relationship between the time-domain voltage and current. The defining equation in phasor form is illustrated in Fig. 10.13*b*. The angles θ and ϕ are equal, so the current and voltage are always in phase.

As an example of the use of both the time-domain and frequency-domain relationships, let us assume that a voltage of 8 cos(100*t* − 50°) V is across a 4 Ω resistor. Working in the time domain, we find that the current must be

$$i(t) = \frac{v(t)}{R} = 2 \cos(100t - 50°) \text{ A}$$

The phasor form of the same voltage is $8\underline{/-50°}$ V, and therefore

$$\mathbf{I} = \frac{\mathbf{V}}{R} = 2\underline{/-50°} \text{ A}$$

If we transform this answer back to the time domain, it is evident that the same expression for the current is obtained. We conclude that there is no saving in time or effort when a *resistive* circuit is analyzed in the frequency domain.

The Inductor

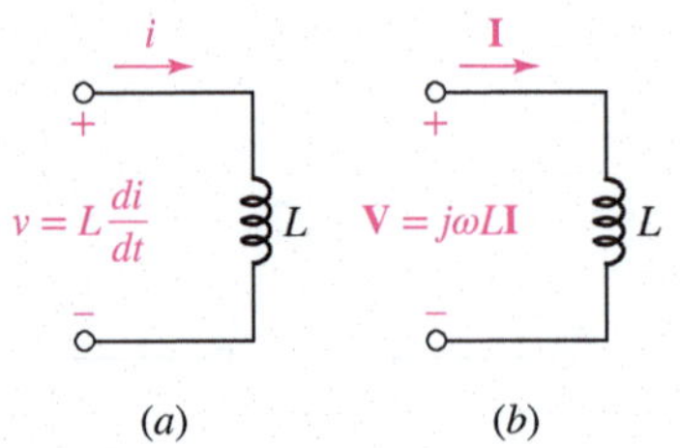

FIGURE 10.14 An inductor and its associated voltage and current in (*a*) the time domain, $v = L\, di/dt$; and (*b*) the frequency domain, $\mathbf{V} = j\omega L\mathbf{I}$.

Let us now turn to the inductor. The time-domain representation is shown in Fig. 10.14*a*, and the defining equation, a time-domain expression, is

$$v(t) = L\frac{di(t)}{dt} \qquad [19]$$

After substituting the complex voltage equation [16] and complex current equation [17] into Eq. [19], we have

$$V_m e^{j(\omega t+\theta)} = L\frac{d}{dt} I_m e^{j(\omega t+\phi)}$$

Taking the indicated derivative:

$$V_m e^{j(\omega t+\theta)} = j\omega L I_m e^{j(\omega t+\phi)} e^{j(\omega t+\phi)}$$

and dividing through by $e^{j\omega t}$:

$$V_m e^{j\theta} = j\omega L I_m e^{j\phi} e^{j\phi}$$

we obtain the desired phasor relationship

$$\mathbf{V} = j\omega L\mathbf{I}\mathbf{I} \qquad [20]$$

The time-domain differential equation [19] has become the algebraic equation [20] in the frequency domain. The phasor relationship is indicated in Fig. 10.14*b*. Note that the angle of the factor $j\omega L$ is exactly +90° and that **I** must therefore lag **V** by 90° in an inductor.

EXAMPLE 10.4

Apply the voltage $8\angle{-50°}$ V at a frequency $\omega = 100$ rad/s to a 4 H inductor, and determine the phasor current and the time-domain current.

We make use of the expression we just obtained for the inductor,

$$\mathbf{I} = \frac{\mathbf{V}}{j\omega L} = \frac{8\angle{-50°}}{j100(4)} = -j0.02\angle{-50°} = (1\angle{-90°})(0.02 - 50°)$$

or

$$\mathbf{I} = 0.02\angle{-140°}\text{ A}$$

If we express this current in the time domain, it becomes

$$i(t) = 0.02\cos(100t - 140°)\text{ A} = 20\cos(100t - 140°)\text{ mA}$$

The Capacitor

We next consider the capacitor. The time-domain current–voltage relationship is

$$i(t) = C\frac{dv(t)}{dt}$$

The equivalent expression in the frequency domain is obtained once more by letting $v(t)$ and $i(t)$ be the complex quantities of Eqs. [16] and [17], taking the indicated derivative, suppressing $e^{j\omega t}$, and recognizing the phasors **V** and **I**. Doing this, we find

$$\mathbf{I} = j\omega C\mathbf{V} \qquad [21]$$

Thus, **I** leads **V** by 90° in a capacitor. This, of course, does not mean that a current response is present one-quarter of a period earlier than the voltage that caused it! We are studying steady-state response, and we find that the current maximum is caused by the increasing voltage that occurs 90° earlier than the voltage maximum.

The time-domain and frequency-domain representations are compared in Fig. 10.15*a* and *b*. We have now obtained the **V**–**I** relationships for the three passive elements. These results are summarized in Table 10.1, where the time-domain *v*–*i* expressions and the frequency-domain **V**–**I** relationships are shown in adjacent columns for the three circuit elements. All the phasor equations are algebraic. Each is also linear, and the equations relating to inductance and capacitance bear a great similarity to Ohm's law. In fact, we will indeed *use* them as we use Ohm's law.

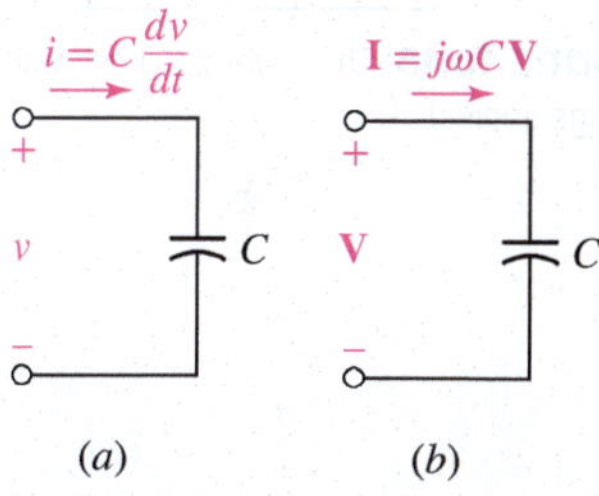

FIGURE 10.15 (*a*) The time-domain and (*b*) the frequency-domain relationships between capacitor current and voltage.

TABLE 10.1 Comparison of Time-Domain and Frequency-Domain Voltage–Current Expressions

Time Domain		Frequency Domain	
i → R, + v −	$v = Ri$	$\mathbf{V} = R\mathbf{I}$	**I** → R, + **V** −
i → L, + v −	$v = L\frac{di}{dt}$	$\mathbf{V} = j\omega L\mathbf{I}$	**I** → $j\omega L$, + **V** −
i → C, + v −	$v = \frac{1}{C}\int i\,dt$	$\mathbf{V} = \frac{1}{j\omega C}\mathbf{I}$	**I** → $1/j\omega C$, + **V** −

Kirchhoff's Laws Using Phasors

Kirchhoff's voltage law in the time domain is

$$v_1(t) + v_2(t) + \cdots + v_N(t) = 0$$

We now use Euler's identity to replace each real voltage v_i with a complex voltage having the same real part, suppress $e^{j\omega t}$ throughout, and obtain

$$\mathbf{V}_1 + \mathbf{V}_2 + \cdots + \mathbf{V}_N = 0$$

Thus, we see that Kirchhoff's voltage law applies to phasor voltages just as it did in the time domain. Kirchhoff's current law can be shown to hold for phasor currents by a similar argument.

Now let us look briefly at the series RL circuit that we have considered several times before. The circuit is shown in Fig. 10.16, and a phasor current and several phasor voltages are indicated. We may obtain the desired response, a time-domain current, by first finding the phasor current. From Kirchhoff's voltage law,

$$\mathbf{V}_R + \mathbf{V}_L = \mathbf{V}_s$$

and using the recently obtained **V–I** relationships for the elements, we have

$$R\mathbf{I} + j\omega L\mathbf{I} = \mathbf{V}_s$$

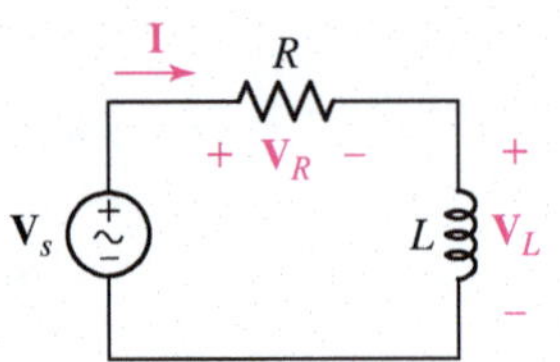

FIGURE 10.16 The series RL circuit with a phasor voltage applied.

The phasor current is then found in terms of the source voltage $\mathbf{V}_s$:

$$\mathbf{I} = \frac{\mathbf{V}_s}{R + j\omega L}$$

Let us select a source-voltage amplitude of V_m and a phase angle of 0°. Thus,

$$\mathbf{I} = \frac{V_m \underline{/0^\circ}}{R + j\omega L}$$

The current may be transformed to the time domain by first writing it in polar form:

$$\mathbf{I} = \frac{V_m}{\sqrt{R^2 + \omega^2 L^2}} \underline{/-\tan^{-1}(\omega L/R)}$$

and then following the familiar sequence of steps to obtain in a very simple manner the same result we obtained the "hard way" earlier in this chapter.

EXAMPLE 10.5

For the RLC circuit of Fig. 10.17, determine $\mathbf{I}_s$ and $i_s(t)$ if both sources operate at $\omega = 2$ rad/s, and $\mathbf{I}_C = 2\underline{/28^\circ}$ A.

The fact that we are given $\mathbf{I}_C$ and asked for $\mathbf{I}_s$ is all the prompting we need to consider applying KCL. If we label the capacitor voltage $\mathbf{V}_C$ consistent with the passive sign convention, then

$$\mathbf{V}_C = \frac{1}{j\omega C}\mathbf{I}_C = \frac{-j}{2}\mathbf{I}_C = \frac{-j}{2}(2\underline{/28^\circ}) = (0.5\underline{/-90^\circ})(2\underline{/28^\circ}) = 1\underline{/-62^\circ}\text{ V}$$

This voltage also appears across the 2 Ω resistor, so the current $\mathbf{I}_{R_2}$ flowing downward through that branch is

$$\mathbf{I}_{R_2} = \frac{1}{2}\mathbf{V}_C = \frac{1}{2}\underline{/-62^\circ}\text{ A}$$

KCL then yields

$$\mathbf{I}_s = \mathbf{I}_{R_2} + \mathbf{I}_C = \frac{1}{2}\underline{/-62^\circ} + 2\underline{/-28^\circ} = 2.06\underline{/14^\circ}\text{ A}$$

Thus $\mathbf{I}_s$ and a knowledge of ω permit us to write $i_s(t)$ directly:

$$i_s(t) = 2.06\cos(2t + 14^\circ)\text{ A}$$

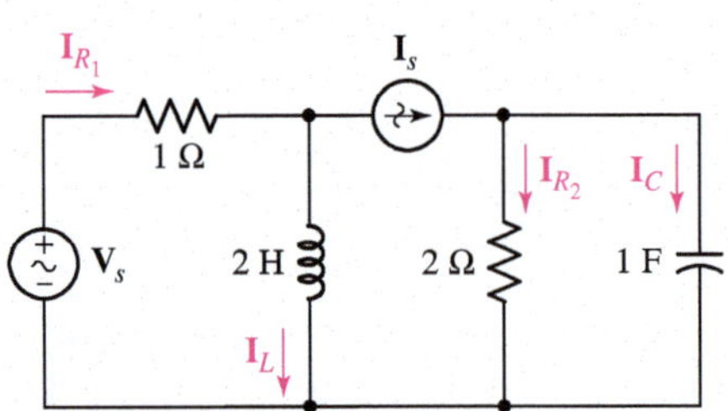

FIGURE 10.17 A three-mesh circuit. Each source operates at the same frequency ω.

PRACTICE

10.8 In the circuit of Fig. 10.17, both sources operate at $\omega = 1$ rad/s. If $\mathbf{I}_C = 2\underline{/28^\circ}$ A and $\mathbf{I}_L = 3\underline{/53^\circ}$ A, calculate (a) $\mathbf{I}_s$; (b) $\mathbf{V}_s$; (c) $i_{R_1}(t)$.

Ans: $3\underline{/-62^\circ}$ A; $3.71\underline{/-4.5^\circ}$ V; $3.22\cos(t - 4.5^\circ)$ A.

10.5 IMPEDANCE AND ADMITTANCE

The current–voltage relationships for the three passive elements in the frequency domain are (assuming that the passive sign convention is satisfied)

$$\mathbf{V} = R\mathbf{I} \qquad \mathbf{V} = j\omega L\mathbf{I} \qquad \mathbf{V} = \frac{\mathbf{I}}{j\omega C}$$

If these equations are written as phasor voltage/phasor current ratios

$$\frac{\mathbf{V}}{\mathbf{I}} = R \qquad \frac{\mathbf{V}}{\mathbf{I}} = j\omega L \qquad \frac{\mathbf{V}}{\mathbf{I}} = \frac{1}{j\omega C}$$

we find that these ratios are simple quantities that depend on element values (and frequency also, in the case of inductance and capacitance). We treat these ratios in the same manner that we treat resistances, provided we remember that they are complex quantities.

Let us define the ratio of the phasor voltage to the phasor current as ***impedance,*** symbolized by the letter **Z**. The impedance is a complex quantity having the dimensions of ohms. Impedance is not a phasor and cannot be transformed to the time domain by multiplying by $e^{j\omega t}$ and taking the real part. Instead, we think of an inductor as being represented in the time domain by its inductance L and in the frequency domain by its impedance $j\omega L$. A capacitor in the time domain has a capacitance C; in the frequency domain, it has an impedance $1/j\omega C$. Impedance is a part of the frequency domain and not a concept that is a part of the time domain.

$\mathbf{Z}_R = R$
$\mathbf{Z}_L = j\omega L$
$\mathbf{Z}_C = \frac{1}{j\omega C}$

Series Impedance Combinations

The validity of Kirchhoff's two laws in the frequency domain leads to the fact that impedances may be combined in series and parallel by the same rules we established for resistances. For example, at $\omega = 10 \times 10^3$ rad/s, a 5 mH inductor in series with a 100 μF capacitor may be replaced by the sum of the individual impedances. The impedance of the inductor is

$$\mathbf{Z}_L = j\omega L = j50\ \Omega$$

and the impedance of the capacitor is

$$\mathbf{Z}_C = \frac{1}{j\omega C} = \frac{-j}{\omega C} = -j\ \Omega$$

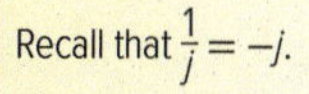

The impedance of the series combination is therefore

$$\mathbf{Z}_{\text{eq}} = \mathbf{Z}_L + \mathbf{Z}_C = j50 - j = j49\ \Omega$$

The impedance of inductors and capacitors is a function of frequency, and this equivalent impedance is thus applicable only at the single frequency at which it was calculated, $\omega = 10{,}000$ rad/s. If we change the frequency to $\omega = 5000$ rad/s, for example, $\mathbf{Z}_{\text{eq}} = j23\ \Omega$.

Parallel Impedance Combinations

The *parallel* combination of the 5 mH inductor and the 100 μF capacitor at $\omega = 10{,}000$ rad/s is calculated in exactly the same fashion in which we calculated parallel resistances:

$$\mathbf{Z}_{\text{eq}} = \frac{(j50)(-j)}{j50 - j} = \frac{50}{j49} = -j1.020\ \Omega$$

At $\omega = 5000$ rad/s, the parallel equivalent is $-j2.17\ \Omega$.

Reactance

Of course, we may choose to express impedance in either *rectangular* ($\mathbf{Z} = R + jX$) or *polar* ($\mathbf{Z} = |\mathbf{Z}|\underline{/\theta}$) form. In rectangular form, we can see clearly the real part which arises only from real resistances, and an imaginary component, termed the ***reactance,*** which arises from the energy storage elements. Both resistance and reactance have units of ohms, but reactance will always depend upon frequency. An ideal resistor has zero reactance; an ideal inductor or capacitor is purely reactive (i.e., characterized by zero resistance). Can a series or parallel combination include *both* a capacitor and an inductor, and yet have *zero reactance?* Sure! Consider the series connection of a 1 Ω resistor, a 1 F capacitor, and a 1 H inductor driven at $\omega = 1$ rad/s. $\mathbf{Z}_{\text{eq}} = 1 - j(1)(1) + j(1)(1) = 1\ \Omega$. At that frequency, the equivalent is a simple 1 Ω resistor. However, even small deviations from $\omega = 1$ rad/s lead to nonzero reactance.

EXAMPLE 10.6

Determine the equivalent impedance of the network shown in Fig. 10.18*a*, given an operating frequency of 5 rad/s.

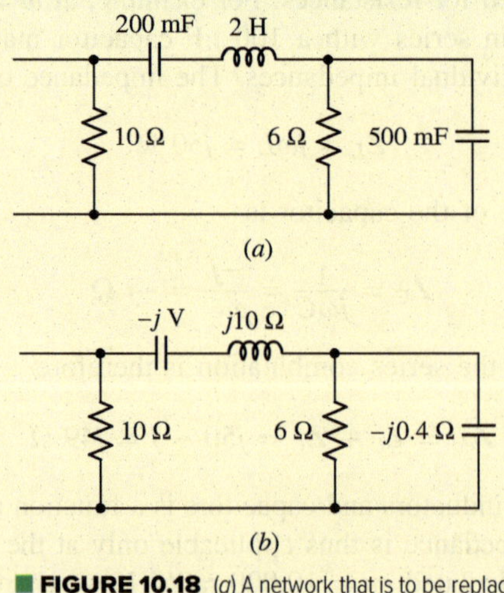

FIGURE 10.18 (*a*) A network that is to be replaced by a single equivalent impedance. (*b*) The elements are replaced by their impedances at $\omega = 5$ rad/s.

We begin by converting the resistors, capacitors, and inductor into the corresponding impedances as shown in Fig. 10.18*b*.

Upon examining the resulting network, we observe that the $6\ \Omega$ impedance is in parallel with $-j0.4\ \Omega$. This combination is equivalent to

$$\frac{(6)(-j0.4)}{6-j0.4} = 0.02655 - j0.3982\ \Omega$$

which is in series with both the $-j\ \Omega$ and $j10\ \Omega$ impedances, so that we have

$$0.0265 - j0.3982 - j + j10 = 0.02655 + j8.602\ \Omega$$

This new impedance is in parallel with $10\ \Omega$, so the equivalent impedance of the network is

$$10 \parallel (0.02655 + j8.602) = \frac{10(0.02655 + j8.602)}{10 + 0.02655 + j8.602}$$
$$= 4.255 + j4.929\ \Omega$$

Alternatively, we can express the impedance in polar form as $6.511\underline{/49.20^\circ}\ \Omega$.

PRACTICE

10.9 With reference to the network shown in Fig. 10.19, find the input impedance $\mathbf{Z}_{\text{in}}$ that would be measured between terminals: (*a*) *a* and *g*; (*b*) *b* and *g*; (*c*) *a* and *b*.

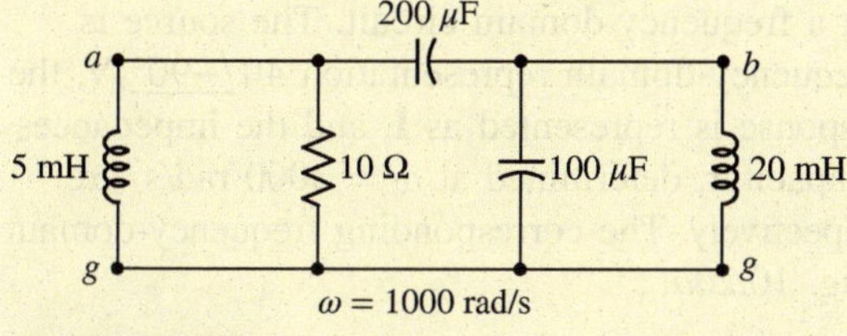

FIGURE 10.19

Ans: $2.81 + j4.49\ \Omega$; $1.798 - j1.124\ \Omega$; $0.1124 - j3.82\ \Omega$

It is important to note that the resistive component of the impedance is not necessarily equal to the resistance of the resistor that is present in the network. For example, a $10\ \Omega$ resistor and a 5 H inductor in series at $\omega = 4$ rad/s have an equivalent impedance $\mathbf{Z} = 10 + j20\ \Omega$, or, in polar form, $22.4\underline{/63.4^\circ}\ \Omega$. In this case, the resistive component of the impedance is equal to the resistance of the series resistor because the network is a simple series network. However, if these same two elements are placed in parallel, the equivalent impedance is $10(j20)/(10 + j20)\ \Omega$, or $8 + j4\ \Omega$. The resistive component of the impedance is now $8\ \Omega$.

EXAMPLE 10.7

Find the current $i(t)$ in the circuit shown in Fig. 10.20*a*.

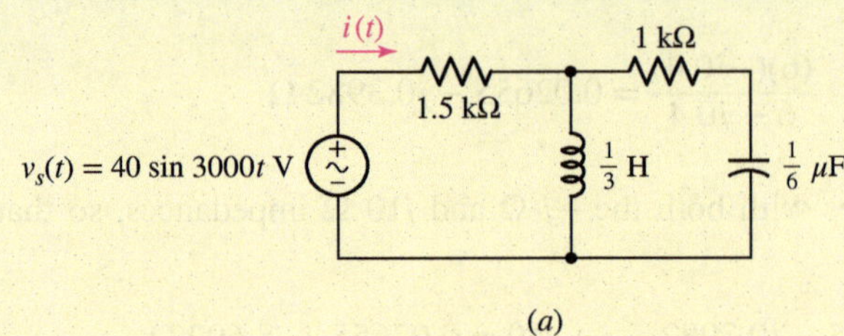

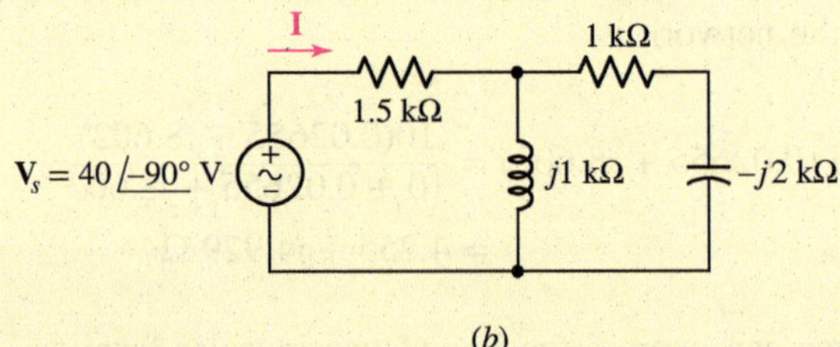

FIGURE 10.20 (*a*) An *RLC* circuit for which the sinusoidal forced response $i(t)$ is desired. (*b*) The frequency-domain equivalent of the given circuit at $\omega = 3000$ rad/s.

Identify the goal of the problem.

We need to find the sinusoidal steady-state current flowing through the 1.5 kΩ resistor due to the 3000 rad/s voltage source.

Collect the known information.

We begin by drawing a frequency-domain circuit. The source is transformed to the frequency-domain representation $40\underline{/-90^\circ}$ V, the frequency domain response is represented as **I**, and the impedances of the inductor and capacitor, determined at $\omega = 3000$ rad/s, are j kΩ and $-j2$ kΩ, respectively. The corresponding frequency-domain circuit is shown in Fig. 10.20*b*.

Devise a plan.

We will analyze the circuit of Fig. 10.20*b* to obtain **I**; combining impedances and invoking Ohm's law is one possible approach. We will then make use of the fact that we know $\omega = 3000$ rad/s to convert **I** into a time-domain expression.

Construct an appropriate set of equations.

$$\mathbf{Z}_{\text{eq}} = 1.5 + \frac{(j)(1 - 2j)}{j + 1 - 2j} = 1.5 + \frac{2 + j}{1 - j}$$

$$= 1.5 + \frac{2 + j}{1 - j}\,\frac{1 + j}{1 + j} = 1.5 + \frac{1 + j3}{2}$$

$$= 2 + j1.5 = 2.5\underline{/36.87^\circ} \text{ k}\Omega$$

The phasor current is then simply

$$\mathbf{I} = \frac{\mathbf{V}_s}{\mathbf{Z}_{eq}}$$

▶ ***Determine if additional information is required.***
Substituting known values, we find that

$$\mathbf{I} = \frac{40\underline{/-90^\circ}}{2.5\underline{/36.87^\circ}} \text{ mA}$$

which, along with the knowledge that $\omega = 3000$ rad/s, is sufficient to solve for $i(t)$.

▶ ***Attempt a solution.***
This complex expression is easily simplified to a single complex number in polar form:

$$\mathbf{I} = \frac{40}{2.5}\underline{/-90^\circ - 36.87^\circ} \text{ mA} = 16.00\underline{/-126.9^\circ} \text{ mA}$$

Upon transforming the current to the time domain, the desired response is obtained:

$$i(t) = 16\cos(3000t - 126.9^\circ) \text{ mA}$$

▶ ***Verify the solution. Is it reasonable or expected?***
The effective impedance connected to the source has an angle of +36.87°, indicating that it has a net inductive character, or that the current will lag the voltage. Since the voltage source has a phase angle of −90° (once converted to a cosine source), we see that our answer is consistent.

PRACTICE

10.10 In the frequency-domain circuit of Fig. 10.21, find (*a*) $\mathbf{I}_1$; (*b*) $\mathbf{I}_2$; (*c*) $\mathbf{I}_3$.

Ans: $28.3\underline{/45^\circ}$ A; $20\underline{/90^\circ}$ A; $20\underline{/0^\circ}$ A.

■ **FIGURE 10.21**

Before we begin to write great numbers of equations in the time domain or in the frequency domain, it is very important that we shun the construction of equations that are partly in the time domain, partly in the frequency domain, and wholly incorrect. One clue that a faux pas of this type has been committed is the sight of both a complex number and a t in the same equation, except in the factor $e^{j\omega t}$. And, since $e^{j\omega t}$ plays a much bigger role in derivations than in applications, it is pretty safe to say that students who find they have just created an equation containing j and t, or $\angle$ and t, have created a rather scary monster that the world would be better off without.

For example, a few equations back we saw

$$\mathbf{I} = \frac{\mathbf{V}_s}{\mathbf{Z}_{\text{eq}}} = \frac{40\underline{/-90^\circ}}{2.5\underline{/36.9^\circ}} = 16\underline{/-126.9^\circ} \text{ mA}$$

Please do not try anything like the following:

$$i(t) \not= \frac{40 \sin 3000t}{2.5\underline{/36.9^\circ}} \quad \text{or} \quad i(t) \not= \frac{40 \sin 3000t}{2 + j1.5}$$

Admittance

There is a general (unitless) term for both impedance and admittance—***immitance***—which is sometimes used, but not very often.

Although the concept of impedance is very useful, and familiar in a way based on our experience with resistors, the reciprocal is often just as valuable. We define this quantity as the ***admittance*** **Y** of a circuit element or passive network, and it is simply the ratio of current to voltage.

The real part of the admittance is the ***conductance*** G, and the imaginary part of the admittance is the ***susceptance*** B. Thus,

$$\mathbf{Y}_R = \frac{1}{R}$$
$$\mathbf{Y}_L = \frac{1}{j\omega L}$$
$$\mathbf{Y}_C = j\omega C$$

$$\mathbf{Y} = G + jB = \frac{1}{\mathbf{Z}} = \frac{1}{R + jX} \qquad [22]$$

The real part of the admittance is the ***conductance*** G, and the imaginary part is the ***susceptance*** B. All three quantities (**Y**, G, and B) are measured in siemens.

Equation [22] should be scrutinized carefully; it does *not* state that the real part of the admittance is equal to the reciprocal of the real part of the impedance, or that the imaginary part of the admittance is equal to the reciprocal of the imaginary part of the impedance!

PRACTICE

10.11 Determine the admittance (in rectangular form) of (*a*) an impedance $\mathbf{Z} = 1000 + j400\ \Omega$; (*b*) a network consisting of the parallel combination of an 800 Ω resistor, a 1 mH inductor, and a 2 nF capacitor, if $\omega = 1$ Mrad/s; (*c*) a network consisting of the series combination of an 800 Ω resistor, a 1 mH inductor, and a 2 nF capacitor, if $\omega = 1$ Mrad/s.

Ans: $0.862 - j0.345$ mS; $1.25 + j1$ mS; $0.899 - j0.562$ mS.

10.6 NODAL AND MESH ANALYSIS

We previously achieved a great deal with nodal and mesh analysis techniques, and it's reasonable to ask if a similar procedure might be valid in terms of phasors and impedances for the sinusoidal steady state. We already know that both of Kirchhoff's laws are valid for phasors; also, we have an Ohm-like law for the passive elements $\mathbf{V} = \mathbf{ZI}$. We may therefore analyze circuits by nodal techniques in the sinusoidal steady state. Using similar arguments, we can establish that mesh analysis methods are valid (and often useful) as well.

EXAMPLE 10.8

Find the time-domain node voltages $v_1(t)$ and $v_2(t)$ in the circuit shown in Fig. 10.22.

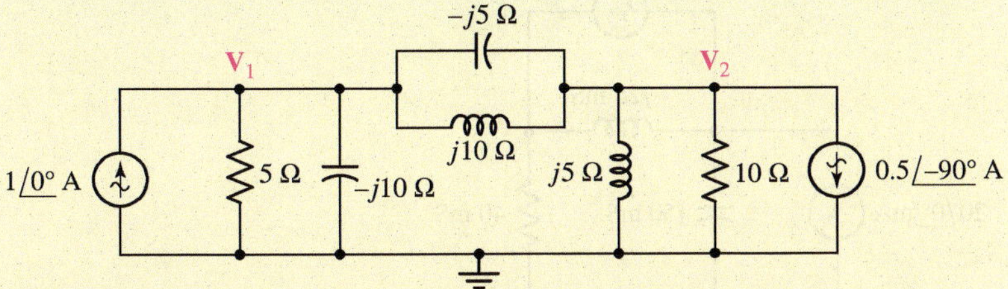

■ **FIGURE 10.22** A frequency-domain circuit for which node voltages $\mathbf{V}_1$ and $\mathbf{V}_2$ are identified.

Two current sources are given as phasors, and phasor node voltages $\mathbf{V}_1$ and $\mathbf{V}_2$ are indicated. At the left node we apply KCL, yielding:

$$\frac{\mathbf{V}_1}{5} + \frac{\mathbf{V}_1}{-j10} + \frac{\mathbf{V}_1 - \mathbf{V}_2}{-j5} + \frac{\mathbf{V}_1 - \mathbf{V}_2}{j10} = 1\,\underline{/0°} = 1 + j0$$

At the right node,

$$\frac{\mathbf{V}_2 - \mathbf{V}_1}{-j5} + \frac{\mathbf{V}_2 - \mathbf{V}_1}{j10} + \frac{\mathbf{V}_2}{j5} + \frac{\mathbf{V}_2}{10} = -(0.5\,\underline{/-90°}) = j0.5$$

Combining terms, we have

$$(0.2 + j0.2)\mathbf{V}_1 - j0.1\,\mathbf{V}_2 = 1$$

and

$$-j0.1\,\mathbf{V}_1 + (0.1 - j0.1)\mathbf{V}_2 = j0.5$$

These equations are easily solved on most scientific calculators, resulting in $\mathbf{V}_1 = 1 - j2$ V and $\mathbf{V}_2 = -2 + j4$ V.

The time-domain solutions are obtained by expressing $\mathbf{V}_1$ and $\mathbf{V}_2$ in polar form:

$$\mathbf{V}_1 = 2.24\,\underline{/-63.4°}$$
$$\mathbf{V}_2 = 4.47\,\underline{/116.6°}$$

and passing to the time domain:

$$v_1(t) = 2.24\cos(\omega t - 63.4°)\text{ V}$$
$$v_2(t) = 4.47\cos(\omega t + 116.6°)\text{ V}$$

Note that the value of ω would have to be known in order to compute the impedance values given on the circuit diagram. Also, *both sources must be operating at the same frequency.*

(Continued on next page)

PRACTICE

10.12 Use nodal analysis on the circuit of Fig. 10.23 to find $\mathbf{V}_1$ and $\mathbf{V}_2$.

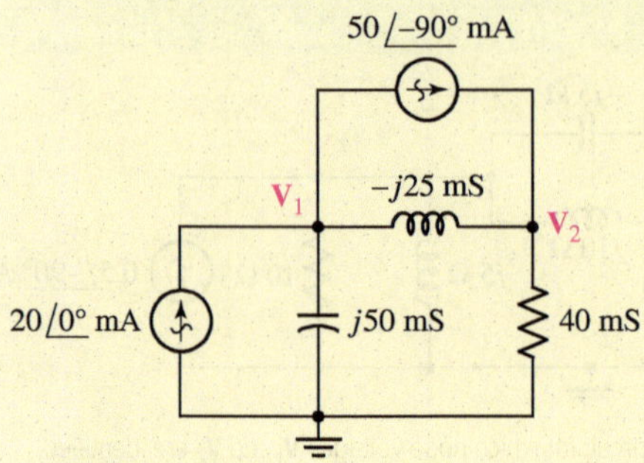

■ **FIGURE 10.23**

Ans: $1.062\angle 23.3°$ V; $1.593\angle -50.0°$ V.

Now let us look at an example of mesh analysis, keeping in mind again that all sources must be operating at the same frequency. Otherwise, it is impossible to define a numerical value for any reactance in the circuit. As we see in the next section, the only way out of such a dilemma is to apply superposition.

EXAMPLE 10.9

Obtain expressions for the time-domain currents i_1 and i_2 in the circuit given as Fig. 10.24*a*.

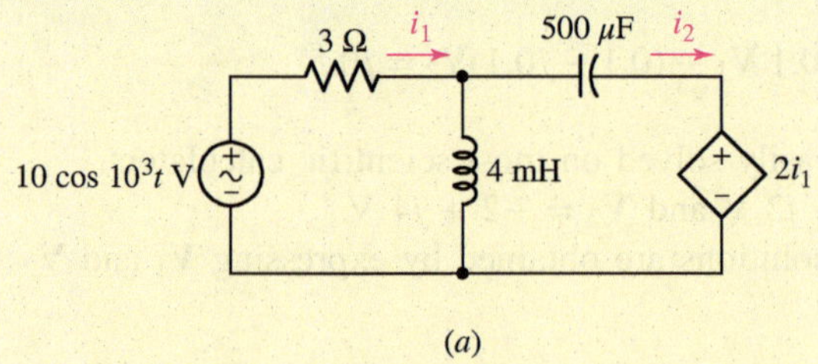

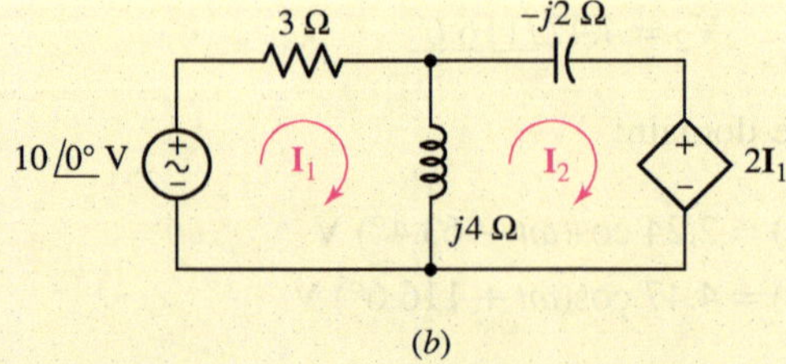

■ **FIGURE 10.24** (*a*) A time-domain circuit containing a dependent source. (*b*) The corresponding frequency-domain circuit.

Noting from the left source that $\omega = 10^3$ rad/s, we draw the frequency-domain circuit of Fig. 10.24*b* and assign mesh currents $\mathbf{I}_1$ and $\mathbf{I}_2$. Around mesh 1,

$$3\mathbf{I}_1 + j4(\mathbf{I}_1 - \mathbf{I}_2) = 10\underline{/0^\circ}$$

or

$$(3 + j4)\mathbf{I}_1 - j4\mathbf{I}_2 = 10$$

while mesh 2 leads to

$$j4(\mathbf{I}_2 - \mathbf{I}_1) - j2\mathbf{I}_2 + 2\mathbf{I}_1 = 0$$

or

$$(2 - j4)\mathbf{I}_1 + j2\mathbf{I}_2 = 0$$

Solving,

$$\mathbf{I}_1 = \frac{14 + j8}{13} = 1.24\underline{/29.7^\circ}\text{ A}$$

$$\mathbf{I}_2 = \frac{20 + j30}{13} = 2.77\underline{/56.3^\circ}\text{ A}$$

Hence,

$$\mathbf{I}_1(t) = 1.24\cos(10^3 t + 29.7^\circ)\text{ A}$$

$$\mathbf{I}_2(t) = 2.77\cos(10^3 t + 56.3^\circ)\text{ A}$$

PRACTICE

10.13 Use mesh analysis on the circuit of Fig. 10.25 to find $\mathbf{I}_1$ and $\mathbf{I}_2$.

Ans: $4.87\underline{/-164.6^\circ}$ A; $7.17\underline{/-144.9^\circ}$ A.

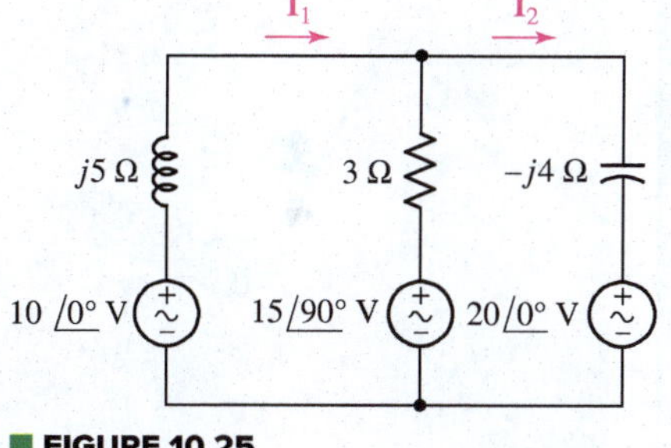

FIGURE 10.25

10.7 SUPERPOSITION, SOURCE TRANSFORMATIONS, AND THÉVENIN'S THEOREM

After inductors and capacitors were introduced in Chap. 7, we found that circuits containing these linear elements were still linear, and hence that the benefits of linearity were again available. Included among these were the superposition principle, Thévenin's and Norton's theorems, and source transformations. Thus, we know that these methods may be used on the circuits we are now considering; the fact that we happen to be applying sinusoidal sources and are seeking only the forced response is immaterial. The fact that we are analyzing the circuits in terms of phasors is also immaterial; they are still linear circuits. We might also remember that linearity and superposition were invoked when we combined real and imaginary sources to obtain a complex source.

PRACTICAL APPLICATION

Cutoff Frequency of a Transistor Amplifier

Transistor-based amplifier circuits are an integral part of many modern electronic instruments. One common application is in mobile telephones (Fig. 10.26), where audio signals are superimposed on high-frequency carrier waves. Unfortunately, transistors have built-in capacitances that lead to limitations in the frequencies at which they can be used, and this fact must be considered when choosing a transistor for a particular application.

Figure 10.27*a* shows what is commonly referred to as a *high-frequency hybrid-π model* for a bipolar junction transistor. In practice, although transistors are *nonlinear* devices, we find that this simple *linear* circuit does a reasonably accurate job of modeling the actual device behavior. The two capacitors C_π and C_μ are used to represent internal capacitances that characterize the particular transistor being used; additional capacitors as well as resistors can be added to increase the accuracy of the model as needed. Figure 10.27*b* shows the transistor model inserted into an amplifier circuit known as a common emitter amplifier.

Assuming a sinusoidal steady-state signal represented by its Thévenin equivalent $\mathbf{V}_s$ and R_s, we are interested in the ratio of the output voltage $\mathbf{V}_{\text{out}}$ to the input voltage $\mathbf{V}_{\text{in}}$. The presence of the internal transistor capacitances leads to a reduction in amplification as the frequency of $\mathbf{V}_s$ is increased; this ultimately limits the frequencies at which the circuit will operate properly. Writing a single nodal equation at the output yields

$$-g_m \mathbf{V}_\pi = \frac{\mathbf{V}_{\text{out}} - \mathbf{V}_{\text{in}}}{1/j\omega C_\mu} + \frac{\mathbf{V}_{\text{out}}}{R_C \parallel R_L}$$

Solving for $\mathbf{V}_{\text{out}}$ in terms of $\mathbf{V}_{\text{in}}$, and noting that $\mathbf{V}_\pi = \mathbf{V}_{\text{in}}$, we obtain an expression for the amplifier gain

$$\frac{\mathbf{V}_{\text{out}}}{\mathbf{V}_{\text{in}}} = \frac{-g_m(R_C \parallel R_L)(1/j\omega C_\mu) + (R_C \parallel R_L)}{(R_C \parallel R_L) + (1/j\omega C_\mu)}$$

$$= \frac{-g_m(R_C \parallel R_L) + j\omega(R_C \parallel R_L)C_\mu}{1 + j\omega(R_C \parallel R_L)C_\mu}$$

Given the typical values $g_m = 30$ mS, $R_C = R_L = 2$ kΩ, and $C_\mu = 5$ pF, we can plot the magnitude of the gain as a function of frequency (recalling that $\omega = 2\pi f$). The semilogarithmic plot is shown in Fig. 10.28*a*, and the MATLAB script used to generate the figure is given in

FIGURE 10.26 Transistor amplifiers are used in many devices, including mobile phones. Linear circuit models are often used to analyze their performance as a function of frequency.
pim pic/Shutterstock

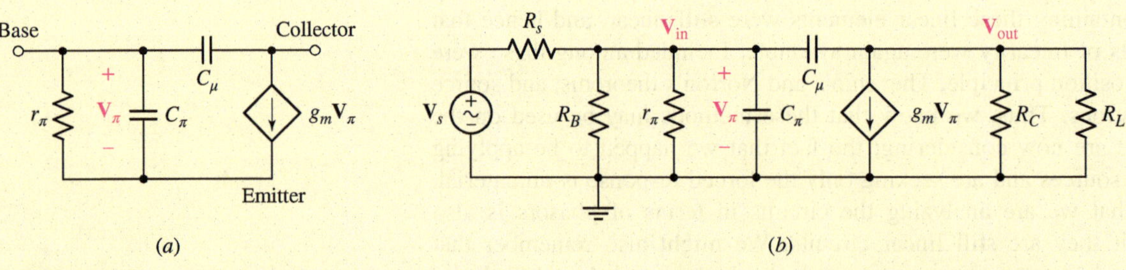

FIGURE 10.27 (*a*) High-frequency hybrid-π transistor model. (*b*) Common-emitter amplifier circuit using the hybrid-π transistor model.

Fig. 10.28*b*. It is interesting, but maybe not totally surprising, to see that a characteristic such as the amplifier gain is dependent on frequency. In fact, we might be able to contemplate using such a circuit as a means of filtering out frequencies we aren't interested in. However, at least for relatively low frequencies, we see that the gain is essentially independent of the frequency of our input source.

When characterizing amplifiers, it is common to reference the frequency at which the voltage gain is reduced to $1/\sqrt{2}$ times its maximum value. From Fig. 10.28*a*, we see that the maximum gain magnitude is 30, and the gain magnitude is reduced to $30/\sqrt{2} = 21$ at a frequency of approximately 30 MHz. This frequency is often called the *cutoff* or *corner* frequency of the amplifier. If operation at a higher frequency is required, either the internal capacitances must be reduced (i.e., a different transistor must be used) or the circuit must be redesigned in some way.

We should note at this point that defining the gain relative to $\mathbf{V}_{\text{in}}$ does not present a complete picture of the frequency-dependent behavior of the amplifier. This may be apparent if we briefly consider the capacitance $\omega \to \infty$, $\mathbf{Z}_{C_\pi} \to 0$, so $\mathbf{V}_{\text{in}} \to 0$. This effect does not manifest itself in the simple equation we derived. A more comprehensive approach is to develop an equation for $\mathbf{V}_{\text{out}}$ in terms of $\mathbf{V}_s$, in which case both capacitances will appear in the expression; this requires a little bit more algebra.

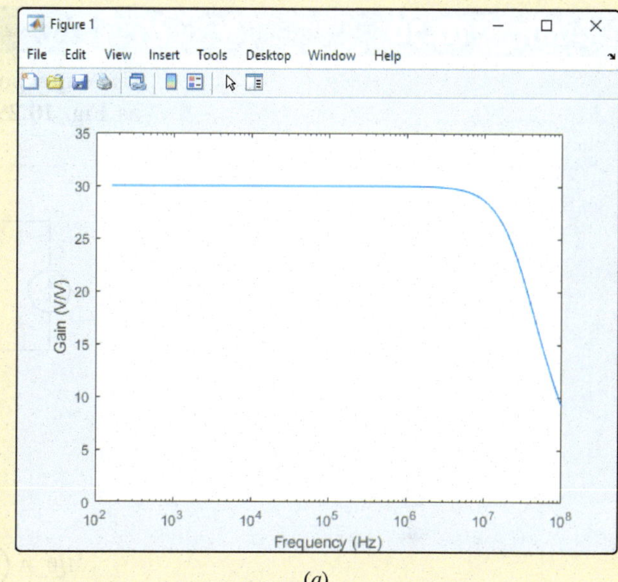

(*a*)

```
>> freq = logspace(3,9,100); %angular frequency (rad/s)
>> num = (30e-3)*(1e3) + i*freq*1000*5e-12;
>> den = 1 + i*freq*1000*5e-12;
>> gain = abs(num./den);
>> semilogx(freq/2/pi,gain);
>> axis([100 1e8 0 35]);
>> xlabel('Frequency (Hz)');
>> ylabel('Gain (V/V)');
```

(*b*)

FIGURE 10.28 (*a*) Amplifier gain as a function of frequency. We see that at high frequencies, it is no longer amplifying effectively. (*b*) MATLAB script used to create plot. Note that '*i*' is used for the imaginary number, not '*j*'. (*a–b*) Hayt/McGraw Hill

One final comment is in order. Up to this point, we have restricted ourselves to considering either single-source circuits or multiple-source circuits in which *every source operates at the exact same frequency*. This is necessary in order to define specific impedance values for inductive and capacitive elements. However, the concept of phasor analysis can be easily extended to circuits with multiple sources operating at different frequencies. In such instances, we simply employ superposition to determine the voltages and currents due to each source, and then add the results *in the time domain*. If several sources are operating at the same frequency, superposition will also allow us to consider those sources at the same time and to add the resulting response to the response(s) of any other source(s) operating at a different frequency.

EXAMPLE 10.10

Use superposition to find v_1 for the circuit of Fig. 10.22, repeated as Fig. 10.29a for convenience.

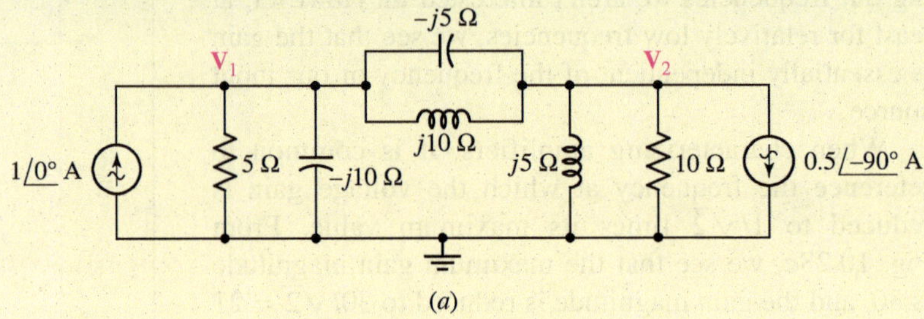

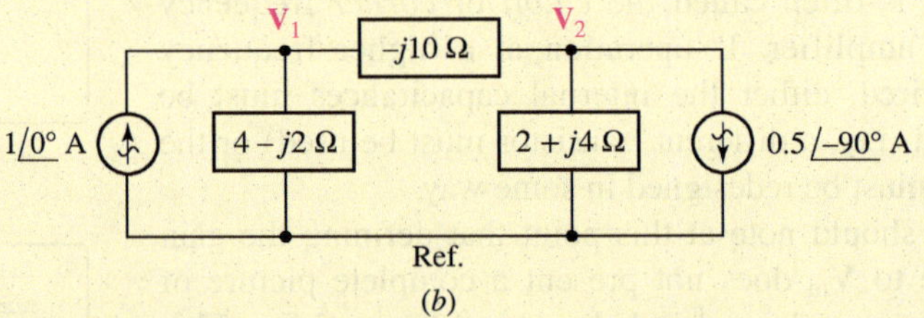

■ **FIGURE 10.29** (a) Circuit of Fig. 10.22 for which $\mathbf{V}_1$ is desired, (b) $\mathbf{V}_1$ may be found by using superposition of the separate phasor responses.

First we redraw the circuit as Fig. 10.29*b*, where each pair of parallel impedances is replaced by a single equivalent impedance. That is, $5 \parallel -j10\ \Omega$ is $4 - j2\ \Omega$; $j10 \parallel -j5\ \Omega$ is $-j10\ \Omega$; and $10 \parallel j5$ is equal to $2 + j4\ \Omega$. To find $\mathbf{V}_1$, we first activate only the left source and find the partial response, $\mathbf{V}_{1L}$. The $1\underline{/0°}$ source is in parallel with an impedance of

$$(4 - j2) \parallel (-j10 + 2 + j4)$$

so that

$$\mathbf{V}_{1L} = 1\underline{/0°}\,\frac{(4 - j2)(-j10 + 2 + j4)}{4 - j2 - j10 + 2 + j4}$$

$$= \frac{-4 - j28}{6 - j8} = 2 - j2 \text{ V}$$

With only the right source active, current division and Ohm's law yield

$$\mathbf{V}_{1R} = (-0.5\underline{/-90°})\left(\frac{2 + j4}{4 - j2 - j10 + 2 + j4}\right)(4 - j2) = -1 \text{ V}$$

Summing, then,

$$\mathbf{V}_1 = \mathbf{V}_{1L} + \mathbf{V}_{1R} = 2 - j2 - 1 = 1 - j2 \text{ V}$$

which agrees with our previous result from Example 10.8.

As we will see, superposition is also extremely useful when dealing with a circuit in which not all sources operate at the same frequency.

PRACTICE

10.14 If superposition is used on the circuit of Fig. 10.30, find $\mathbf{V}_1$ with (*a*) only the $20\underline{/0°}$ mA source operating; (*b*) only the $50\underline{/-90°}$ mA source operating.

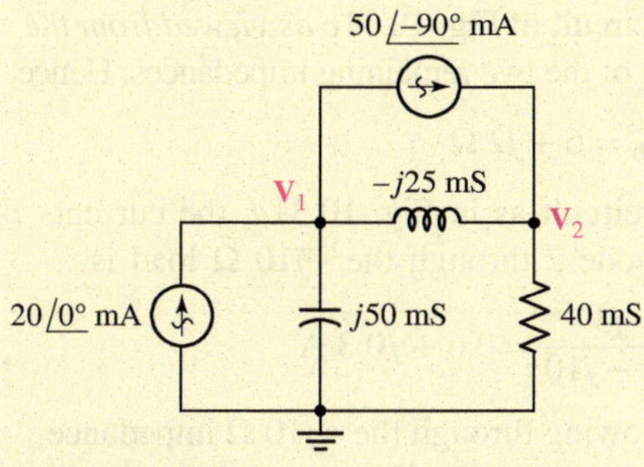

■ **FIGURE 10.30**

Ans: 0.1951 – *j*0.556 V; 0.780 + *j*0.976 V.

EXAMPLE **10.11**

Determine the Thévenin equivalent seen by the –*j*10 Ω impedance of Fig. 10.31*a*, and use this to compute $\mathbf{V}_1$.

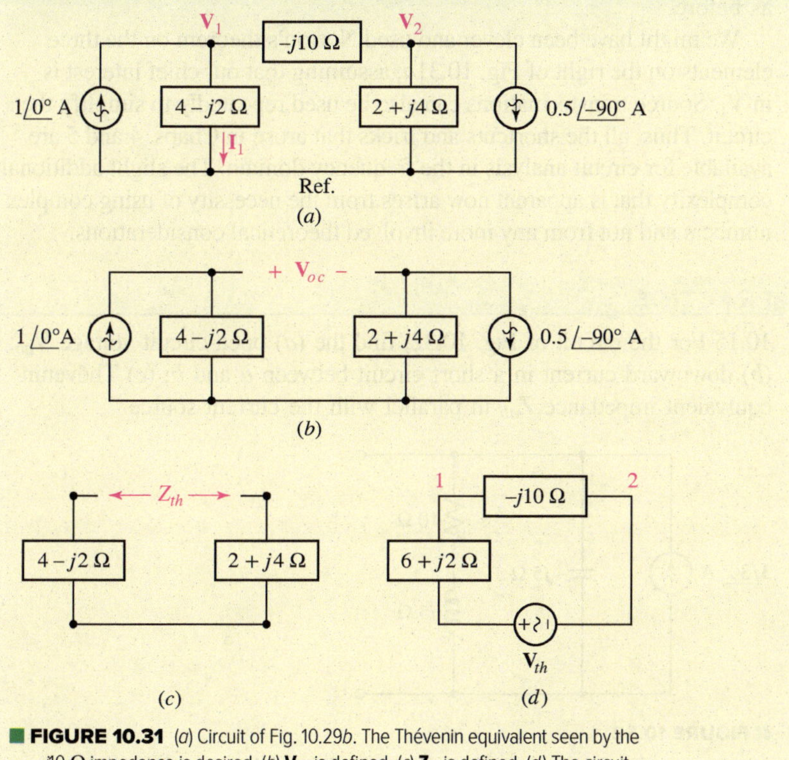

■ **FIGURE 10.31** (*a*) Circuit of Fig. 10.29*b*. The Thévenin equivalent seen by the –*j*10 Ω impedance is desired. (*b*) $\mathbf{V}_{oc}$ is defined. (*c*) $\mathbf{Z}_{th}$ is defined. (*d*) The circuit is redrawn using the Thévenin equivalent.

(*Continued on next page*)

The open-circuit voltage, defined in Fig. 10.31*b*, is

$$\mathbf{V}_{oc} = 1(\underline{/0^\circ})(4 - j2) - (-0.5\underline{/-90^\circ})(2 + j4)$$
$$= 4 - j2 + 2 - j1 = 6 - j3 \text{ V}$$

The impedance of the inactive circuit of Fig. 10.31*c as viewed from the load terminals* is simply the sum of the two remaining impedances. Hence,

$$\mathbf{Z}_{th} = 6 + j2\ \Omega$$

Thus, when we reconnect the circuit as in Fig. 10.31*d*, the current directed from node 1 toward node 2 through the $-j10\ \Omega$ load is

$$\mathbf{I}_{12} = \frac{6 - j3}{6 + j2 - j10} = 0.6 + j0.3 \text{ A}$$

We now know the current flowing through the $-j10\ \Omega$ impedance of Fig. 10.31*a*. *Note that we cannot compute* $\mathbf{V}_1$ *using the circuit of* Fig. 10.31*d as the reference node no longer exists*. Returning to the original circuit, then, and subtracting the $0.6 + j0.3$ A current from the left source current, the downward current through the $(4 - j2)\ \Omega$ branch is found:

$$\mathbf{I}_1 = 1 - 0.6 - j0.3 = 0.4 - j0.3 \text{ A}$$

and, thus,

$$\mathbf{I}_1 = (0.4 - j0.3)(4 - j2) = 1 - j2 \text{ V}$$

as before.

We might have been clever and used Norton's theorem on the three elements on the right of Fig. 10.31*a*, assuming that our chief interest is in $\mathbf{V}_1$. Source transformations can also be used repeatedly to simplify the circuit. Thus, all the shortcuts and tricks that arose in Chaps. 4 and 5 are available for circuit analysis in the frequency domain. The slight additional complexity that is apparent now arises from the necessity of using complex numbers and not from any more involved theoretical considerations.

PRACTICE

10.15 For the circuit of Fig. 10.32, find the (*a*) open-circuit voltage $\mathbf{V}_{ab}$; (*b*) downward current in a short circuit between *a* and *b*; (*c*) Thévenin equivalent impedance $\mathbf{Z}_{ab}$ in parallel with the current source.

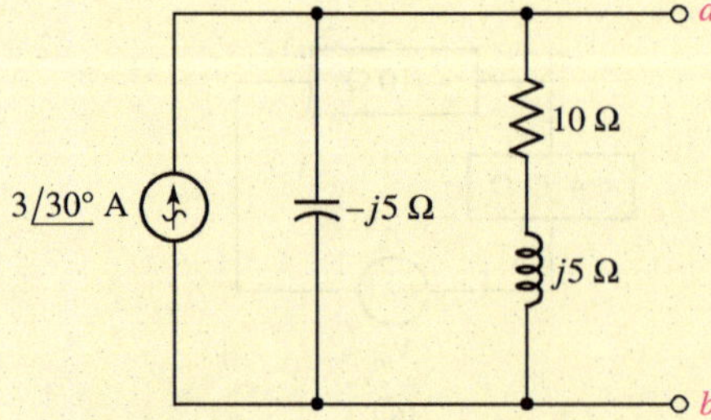

FIGURE 10.32

Ans: $16.77\underline{/-33.4^\circ}$ V; $2.60 + j1.500$ A; $2.5 - j5\ \Omega$.

EXAMPLE 10.12

Determine the power dissipated by the 10 Ω resistor in the circuit of Fig. 10.33*a*, which utilizes two sources each operating at a different frequency.

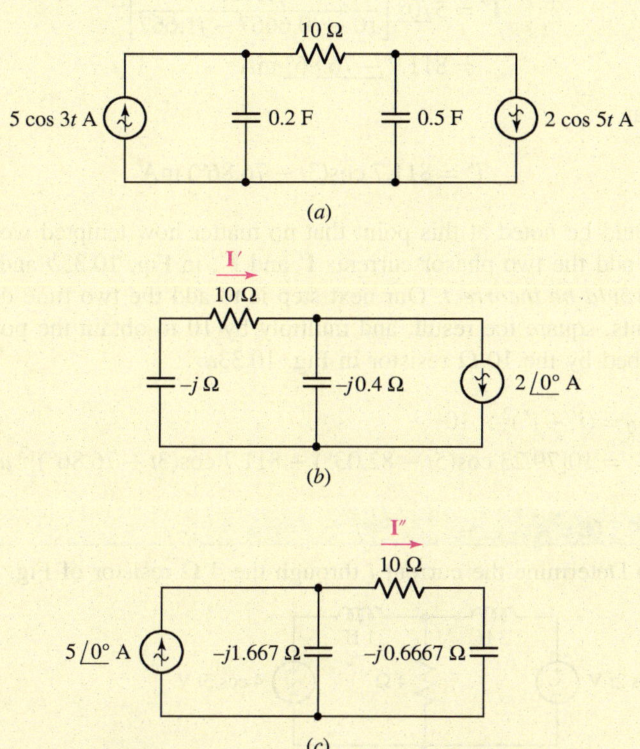

FIGURE 10.33 (*a*) A simple circuit having sources operating at different frequencies. (*b*) Circuit with the left source killed. (*c*) Circuit with the right source killed.

After glancing at the circuit, we might be tempted to write two quick nodal equations, or perhaps perform two sets of source transformations and launch immediately into finding the voltage across the 10 Ω resistor.

Unfortunately, this is impossible, since we have *two* sources operating at *different* frequencies. In such a situation, there is no way to compute the impedance of any capacitor or inductor in the circuit—which ω would we use?

The only way out of this dilemma is to employ superposition, grouping all sources with the same frequency in the same subcircuit, as shown in Fig. 10.33*b* and *c*.

In the subcircuit of Fig. 10.33*b*, we quickly compute the current $\mathbf{I}'$ using current division:

$$\mathbf{I}' = 2\underline{/0^\circ}\left[\frac{-j0.4}{10 - j - j0.4}\right]$$

$$= 79.23\underline{/-82.03^\circ}\ \text{mA}$$

In future studies of signal processing, we will also be introduced to the method of Jean-Baptiste Joseph Fourier, a French mathematician who developed a technique for representing almost any arbitrary function by a combination of sinusoids. When working with linear circuits, once we know the response of a particular circuit to a general sinusoidal forcing function, we can easily predict the response of the circuit to an arbitrary waveform represented by a Fourier series function, simply by using superposition.

(Continued on next page)

so that

$$i' = 79.23 \cos(5t - 82.03°) \text{ mA}$$

Likewise, we find that

$$\mathbf{I}'' = 5\underline{/0°}\left[\frac{-j1.667}{10 - j0.6667 - j1.667}\right]$$
$$= 811.7\underline{/-76.86°} \text{ mA}$$

so that

$$i'' = 811.7 \cos(3t - 76.86°) \text{ mA}$$

It should be noted at this point that no matter how tempted we might be to add the two phasor currents $\mathbf{I}'$ and $\mathbf{I}''$, in Fig. 10.33*b* and *c*, this *would be incorrect*. Our next step is to add the two time-domain currents, square the result, and multiply by 10 to obtain the power absorbed by the 10 Ω resistor in Fig. 10.33*a*:

$$p_{10} = (i' + i'')^2 \times 10$$
$$= 10[79.23 \cos(5t - 82.03°) + 811.7 \cos(3t - 76.86°)]^2 \ \mu\text{W}$$

PRACTICE

10.16 Determine the current i through the 4 Ω resistor of Fig. 10.34.

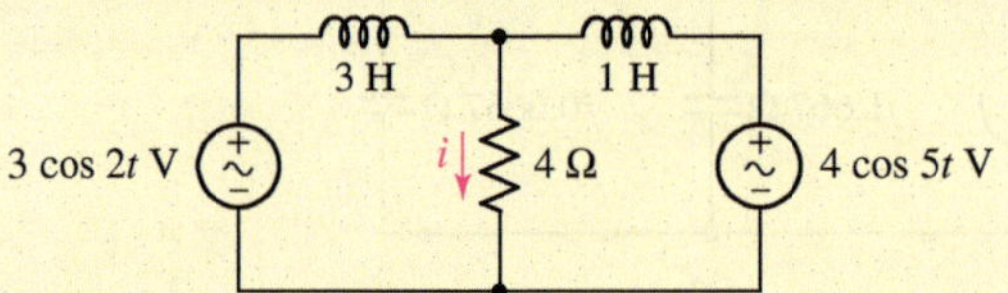

■ **FIGURE 10.34**

Ans: $i = 175.6 \cos(2t - 20.55°) + 547.1 \cos(5t - 43.16°)$ mA.

COMPUTER-AIDED ANALYSIS

Sinusoidal steady state circuit analysis can be accomplished in LTspice by defining AC parameters of magnitude and phase for current and voltage sources.

Let's simulate the circuit of Fig. 10.20*a*, shown redrawn in Fig. 10.35.

The parameters for the source are accessed by right-clicking on the voltage source, then selecting **Advanced** (Fig. 10.36). The frequency of the source for this simulation is actually defined through the **SPICE Analysis** dialog box, which creates the .ac command on the schematic. We select a **List** sweep and set 1st frequency to 477.5.

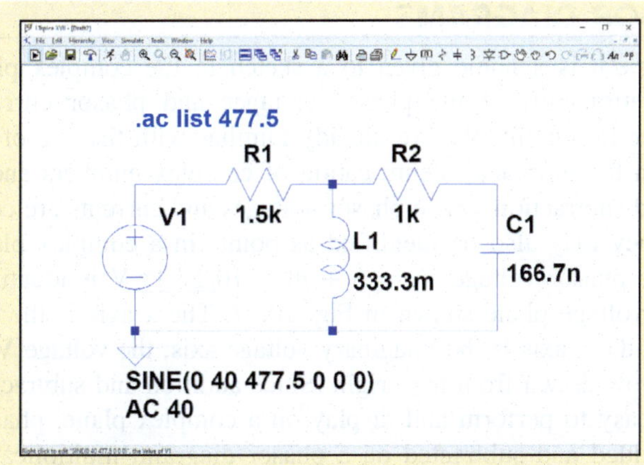

FIGURE 10.35 The circuit of Fig. 10.20*a*, operating at $\omega = 3000$ rad/s, or approximately 477.5 Hz. The current through the 1.5 kΩ resistor is desired. Note that if the resistor is rotated during placement, a 180° phase shift is introduced in the simulation result for I(R1), since the direction of current flow for the component is defined as being into the first node in its netlist description.
Analog Devices

Independent Voltage Source - V1

Functions
(none)
PULSE(V1 V2 Tdelay Trise Tfall Ton Period Ncycles)
SINE(Voffset Vamp Freq Td Theta Phi Ncycles)
EXP(V1 V2 Td1 Tau1 Td2 Tau2)
SFFM(Voff Vamp Fcar MDI Fsig)
PWL(t1 v1 t2 v2...)
PWL FILE: Browse

DC offset[V]: 0
Amplitude[V]: 40
Freq[Hz]: 477.5
Tdelay[s]: 0
Theta[1/s]: 0
Phi[deg]: 0
Ncycles:

Additional PWL Points
Make this information visible on schematic: ☑

DC Value
DC value:
Make this information visible on schematic: ☑

Small signal AC analysis(.AC)
AC Amplitude: 40
AC Phase:
Make this information visible on schematic: ☑

Parasitic Properties
Series Resistance[Ω]:
Parallel Capacitance[F]:
Make this information visible on schematic: ☑

Cancel OK

FIGURE 10.36 Dialog box for setting source parameters.
Analog Devices

The simulation results appear after we select **Run** under **Simulate.** We find that for resistor R1 we have a magnitude of 0.015999 A, or 16 mA, at a phase angle of −36.8657°. Thus, the current through R1 is

$$\mathbf{I} = 16 \sin(3000t - 36.9°) \text{ mA}$$

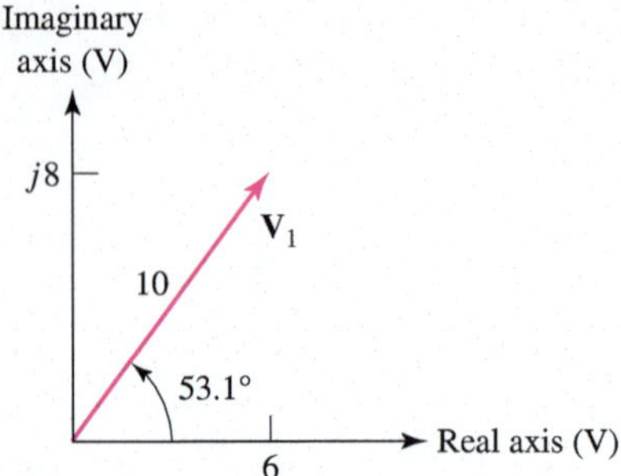

■ **FIGURE 10.37** A simple phasor diagram shows the single voltage phasor $\mathbf{V}_1 = 6 + j8 = 10\angle 53.1°$ V.

10.8 PHASOR DIAGRAMS

The *phasor diagram* is a name given to a sketch in the complex plane showing the relationships of the phasor voltages and phasor currents throughout a specific circuit. We are already familiar with the use of the complex plane in the graphical identification of complex numbers and in their addition and subtraction. Since phasor voltages and currents are complex numbers, they may also be identified as points in a complex plane. For example, the phasor voltage $\mathbf{V}_1 = 6 + j8 = 10\angle 53.1°$ V is identified on the complex voltage plane shown in Fig. 10.37. The x axis is the real voltage axis, and the y axis is the imaginary voltage axis; the voltage $\mathbf{V}_1$ is located by an arrow drawn from the origin. Since addition and subtraction are particularly easy to perform and display on a complex plane, phasors may be easily added and subtracted on a phasor diagram; multiplication and division result in the addition and subtraction of angles and a change of amplitude. Figure 10.38*a* shows the sum of $\mathbf{V}_1$ and a second phasor voltage $\mathbf{V}_2 = 3 - j4 = 5\angle -53.1°$ V, and Fig. 10.38*b* shows the current $\mathbf{I}_1$, which is the product of $\mathbf{V}_1$ and the admittance $\mathbf{Y} = 1 + j$ S.

This last phasor diagram shows both current and voltage phasors on the same complex plane; it is understood that each will have its own amplitude scale, but a common angle scale. For example, a phasor voltage 1 cm long might represent 100 V, while a phasor current 1 cm long could indicate 3 mA. Plotting both phasors on the same diagram enables us to easily determine which waveform is leading and which is lagging.

The phasor diagram also offers an interesting interpretation of the time-domain to frequency-domain transformation, since the diagram may be interpreted from either the time- or the frequency-domain viewpoint. Up to this point, we have been using the frequency-domain interpretation, as we have been showing phasors directly on the phasor diagram. However, let us proceed to a time-domain viewpoint by first showing the phasor voltage $\mathbf{V} = V_m\angle\alpha$ as sketched in Fig. 10.39*a*. In order to transform $\mathbf{V}$ to the time domain, the next necessary step is the multiplication of the phasor by $e^{j\omega t}$; thus we now have the complex voltage $V_m e^{j\alpha} e^{j\omega t} = V_m\angle \omega t + \alpha$. This voltage may also be interpreted as a phasor-like quantity one which possesses a phase angle that increases linearly with time. On a phasor diagram it therefore represents a rotating line segment, the instantaneous position being ωt radians ahead (counterclockwise) of $V_m\angle\alpha$. Both $V_m\angle\alpha$ and $V_m\angle\omega t + \alpha$ are shown on the phasor diagram of Fig. 10.39*b*. The passage to the time domain is now completed by taking the real part of

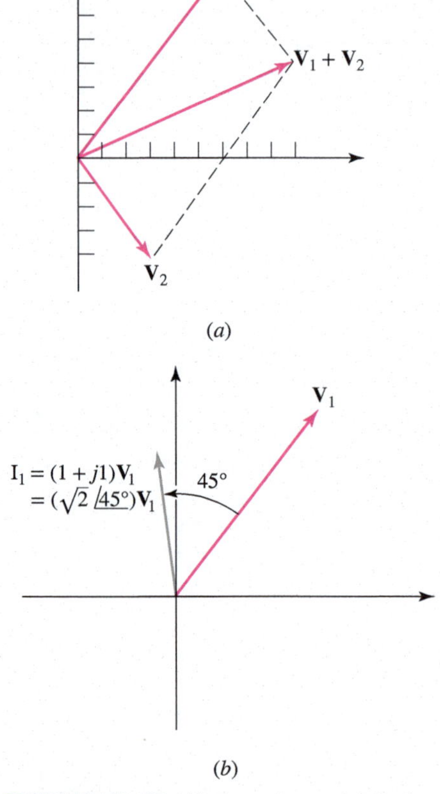

■ **FIGURE 10.38** (*a*) A phasor diagram showing the sum of $\mathbf{V}_1 = 6 + j8$ V and $\mathbf{V}_2 = 3 - j4$ V, $\mathbf{V}_1 + \mathbf{V}_2 = 9 + j4$ V $= 9.85\angle 24.0°$ V. (*b*) The phasor diagram shows $\mathbf{V}_1$ and $\mathbf{I}_1$, where $\mathbf{I}_1 = \mathbf{Y}\mathbf{V}_1$ and $\mathbf{Y} = (1 + j1)$S $= \sqrt{2}\angle 45°$ S. The current and voltage amplitude scales are different.

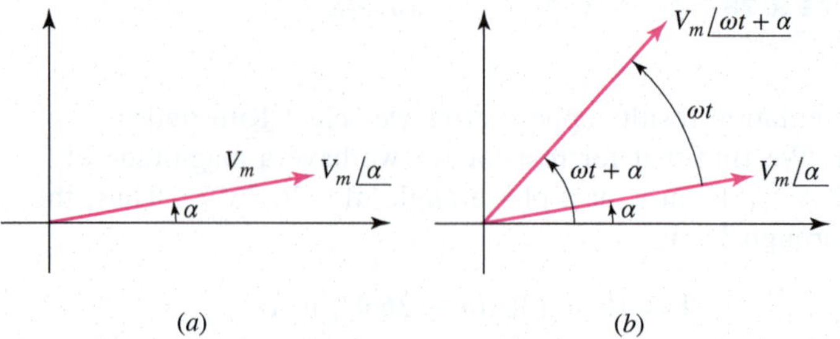

■ **FIGURE 10.39** The phasor voltage $V_m\angle\alpha$. (*b*) The complex voltage $V_m\angle\omega t + \alpha$ is shown as a phasor at a particular instant of time. This phasor leads $V_m\angle\alpha$. by ωt radians.

$V_m \underline{/\omega t + \alpha}$. The real part of this complex quantity is the projection of $V_m \underline{/\omega t + \alpha}$ on the real axis: $V_m \cos(\omega t + \alpha)$.

In summary, then, the frequency-domain phasor appears on the phasor diagram, and the transformation to the time domain is accomplished by allowing the phasor to rotate in a counterclockwise direction at an angular velocity of ω rad/s and then visualizing the projection on the real axis. It is helpful to think of the arrow representing the phasor $\mathbf{V}$ on the phasor diagram as the photographic snapshot, taken at $\omega t = 0$, of a rotating arrow whose projection on the real axis is the instantaneous voltage $v(t)$.

Let us now construct the phasor diagrams for several simple circuits. The series *RLC* circuit shown in Fig. 10.40*a* has several different voltages associated with it, but only a single current. The phasor diagram is constructed most easily by employing the single current as the reference phasor. Let us arbitrarily select $\mathbf{I} = I_m \underline{/0°}$ and place it along the real axis of the phasor diagram, Fig. 10.40*b*. The resistor, capacitor, and inductor voltages may then be calculated and placed on the diagram, where the 90° phase relationships stand out clearly. The sum of these three voltages is the source voltage, and for this circuit, which is in what we will define in a subsequent chapter as the "resonant condition" since $\mathbf{Z}_C = -\mathbf{Z}_L$, the source voltage and resistor voltage are equal. The total voltage across the resistor and inductor or resistor and capacitor is obtained from the diagram by adding the appropriate phasors as shown.

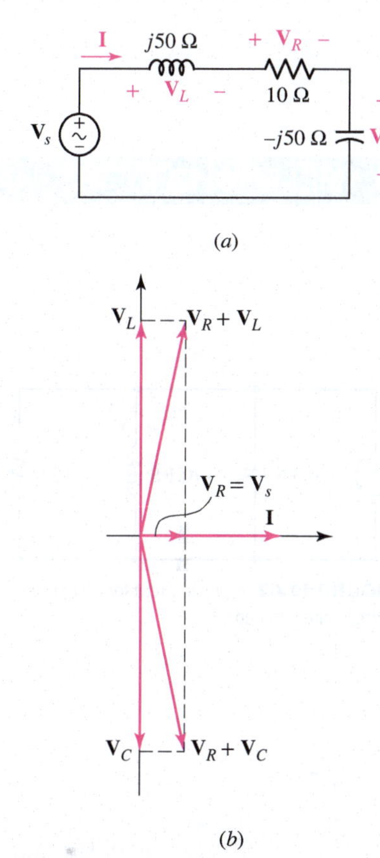

FIGURE 10.40 (*a*) A series *RLC* circuit. (*b*) The phasor diagram for this circuit; the current **I** is used as a convenient reference phasor.

Figure 10.41*a* is a simple parallel circuit in which it is logical to use the single voltage between the two nodes as a reference phasor. Suppose that $\mathbf{V} = 1\underline{/0°}$ V. The resistor current, $\mathbf{I}_R = 0.2\underline{/0°}$ A, is in phase with this voltage, and the capacitor current, $\mathbf{I}_C = j0.1$ A, leads the reference voltage by 90°. After these two currents are added to the phasor diagram, shown as Fig. 10.41*b*, they may be summed to obtain the source current. The result is $\mathbf{I}_s = 0.2 + j0.1$ A.

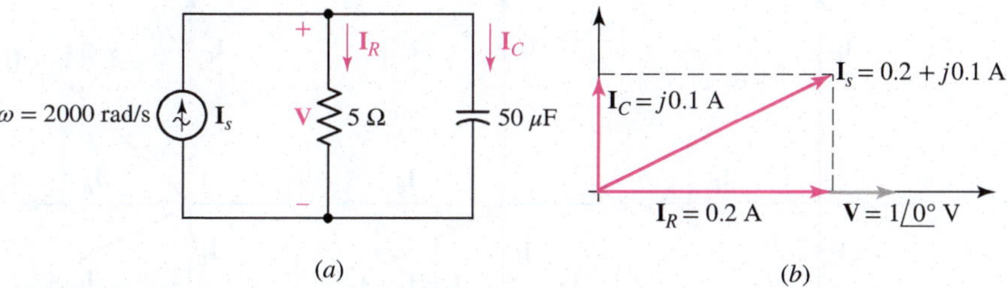

FIGURE 10.41 (*a*) A parallel *RC* circuit. (*b*) The phasor diagram for this circuit; the node voltage **V** is used as a convenient reference phasor.

If the source current is specified initially as the convenient value of $1\underline{/0°}$ A and the node voltage is not initially known, it is still convenient to begin construction of the phasor diagram by assuming a node voltage (for example, $\mathbf{V} = 1\underline{/0°}$ V. once again) and using it as the reference phasor. The diagram is then completed as before, and the source current that flows as a result of the assumed node voltage is again found to be $0.2 + j0.1$ A. The true source current is $1\underline{/0°}$ A, however, and thus the true node voltage is obtained by multiplying the assumed node voltage by $1\underline{/0°}/(0.2 + j0.1)$; the true node voltage is therefore $4 - j2$ V $= \sqrt{20}\,\underline{/-26.6°}$ V. The assumed voltage leads to a phasor diagram which differs from the true phasor diagram by a change of scale (the assumed diagram is smaller by a factor of $1/\sqrt{20}$) and an angular rotation (the assumed diagram is rotated counterclockwise through 26.6°).

EXAMPLE 10.13

Construct a phasor diagram showing $\mathbf{I}_R$, $\mathbf{I}_L$, and $\mathbf{I}_C$ for the circuit of Fig. 10.42. Combining these currents, determine the angle by which $\mathbf{I}_s$ leads $\mathbf{I}_R$, $\mathbf{I}_C$, and $\mathbf{I}_x$.

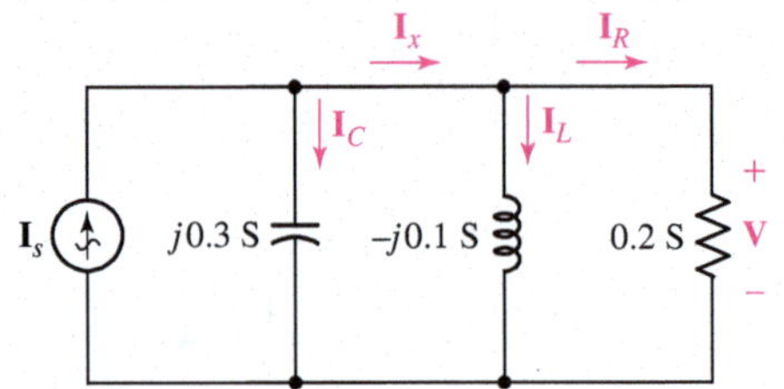

FIGURE 10.42 A simple circuit for which several currents are required.

We begin by choosing a suitable reference phasor. Upon examining the circuit and the variables to be determined, we see that once $\mathbf{V}$ is known, $\mathbf{I}_R$, $\mathbf{I}_L$, and $\mathbf{I}_C$ can be computed by simple application of Ohm's law. Thus, we select $\mathbf{V} = 1\underline{/0°}$ V for simplicity's sake, and subsequently compute

$$\mathbf{I}_R = (0.2)1\underline{/0°} \quad = 0.2\underline{/0°}\text{ A}$$
$$\mathbf{I}_L = (-j0.1)1\underline{/0°} = 0.1\underline{/-90°}\text{ A}$$
$$\mathbf{I}_C = (j0.3)1\underline{/0°} \quad = 0.3\underline{/90°}\text{ A}$$

The corresponding phasor diagram is shown in Fig. 10.43*a*. We also need to find the phasor currents $\mathbf{I}_s$ and $\mathbf{I}_x$. Figure 10.43*b* shows the determination of $\mathbf{I}_x = \mathbf{I}_L + \mathbf{I}_R = 0.2 - j0.1 = 0.224\underline{/-26.6°}$ A, and Fig. 10.43*c* shows the determination of $\mathbf{I}_s = \mathbf{I}_C + \mathbf{I}_x = 0.283\underline{/45°}$ A. From Fig. 10.43*c*, we ascertain that $\mathbf{I}_s$ leads $\mathbf{I}_R$ by 45°, $\mathbf{I}_C$ by −45°, and $\mathbf{I}_x$ by 45° + 26.6° = 71.6°. These angles are only relative, however; the exact numerical values will depend on $\mathbf{I}_s$, upon which the actual value of $\mathbf{V}$ (assumed here to be $1\underline{/0°}$ V for convenience) also depends.

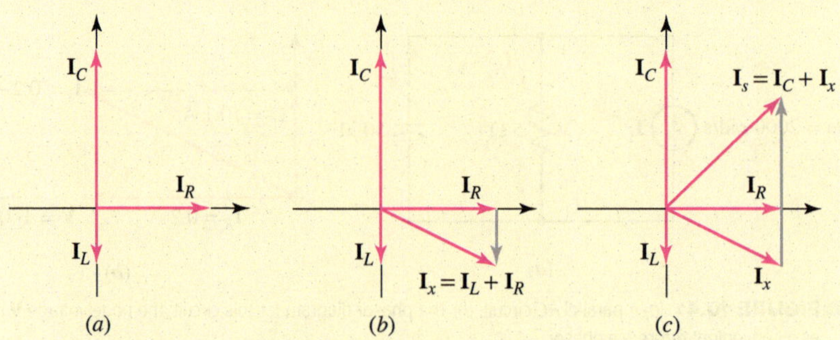

FIGURE 10.43 (*a*) Phasor diagram constructed using a reference value of $\mathbf{V} = 1\underline{/0°}$. (*b*) Graphical determination of $\mathbf{I}_x = \mathbf{I}_L + \mathbf{I}_R$. (*c*) Graphical determination of $\mathbf{I}_s = \mathbf{I}_C + \mathbf{I}_x$.

PRACTICE

10.17 Select some convenient reference value for $\mathbf{I}_C$ in the circuit of Fig. 10.44; draw a phasor diagram showing $\mathbf{V}_R$, $\mathbf{V}_2$, $\mathbf{V}_1$, and $\mathbf{V}_s$; and measure the ratio of the lengths of (*a*) $\mathbf{V}_s$ to $\mathbf{V}_1$; (*b*) $\mathbf{V}_1$ to $\mathbf{V}_2$; (*c*) $\mathbf{V}_s$ to $\mathbf{V}_R$.

Ans: 1.90; 1.00; 2.12

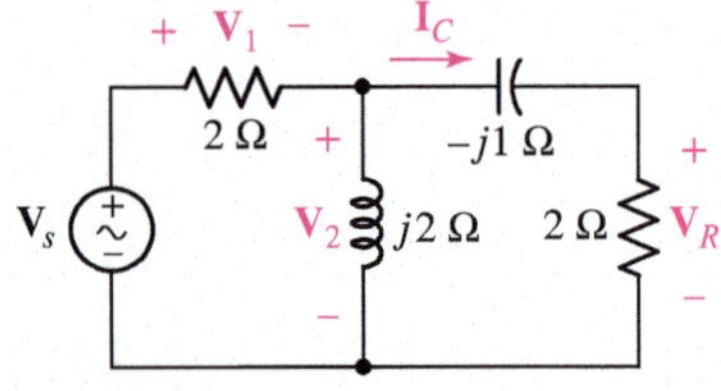

FIGURE 10.44

SUMMARY AND REVIEW

This chapter dealt with the steady-state response of circuits to sinusoidal excitation. This is a limited analysis of a circuit in some respects, as the transient behavior is completely ignored. In many situations, such an approach is more than adequate, however, and reducing the amount of information we seek about a circuit speeds up the analysis considerably. The fundamental idea behind what we did was that an *imaginary* source was added to every *real* sinusoidal source; then Euler's identity converted the combined source to a complex exponential. Since the derivative of an exponential is simply another exponential, what would otherwise be integrodifferential equations arising from mesh or nodal analysis become *algebraic equations*.

A few new terms were introduced: *lagging*, *leading*, *impedance*, *admittance*, and a particularly important one, *phasor*. Phasor relationships between current and voltage gave rise to the concept of impedance, where resistors are represented by a real number (resistance, as before), and inductors are represented by $\mathbf{Z} = j\omega L$ while capacitors are represented by $-j/\omega C$ (ω being the operating frequency of our sources). From that point forward, all the circuit analysis techniques learned in Chaps. 3 to 5 apply.

It might seem odd to have an imaginary number as part of our solution, but we found that recovering the time-domain solution to our analysis is straightforward once the voltage or current is expressed in polar form. The magnitude of our quantity of interest is the magnitude of the cosine function, the phase angle is the phase of the cosine term, and the frequency is obtained from the original circuit (it disappears from view during the analysis, but the circuits we are analyzing do not change it in any way). We concluded the chapter with an introduction to the concept of phasor diagrams. Such constructs find use in the analysis of ac power systems, as we see in subsequent chapters.

A concise list of key concepts of the chapter is presented below for the convenience of the reader, along with the corresponding example numbers.

- ❑ If two sine waves (or two cosine waves) both have positive magnitudes and the same frequency, it is possible to determine which waveform is leading and which is lagging by comparing their phase angles.
- ❑ The forced response of a linear circuit to a sinusoidal voltage or current source can always be written as a single sinusoid having the same frequency as the sinusoidal source. (Example 10.1)
- ❑ A phasor has both a magnitude and a phase angle; the frequency is understood to be that of the sinusoidal source driving the circuit. (Example 10.2)
- ❑ A phasor transform may be performed on any sinusoidal function, and vice versa: $\mathbf{V}_m \cos(\omega t + \phi) \leftrightarrow \mathbf{V}_m\underline{/\phi}$. (Example 10.3)
- ❑ When transforming a time-domain circuit into the corresponding frequency-domain circuit, resistors, capacitors, and inductors are replaced by impedances (or, occasionally, by admittances). (Examples 10.4, 10.6)
  - The impedance of a resistor is simply its resistance.
  - The impedance of a capacitor is $1/j\omega C$.
  - The impedance of an inductor is $j\omega L$.

- ❑ Impedances combine both in series and in parallel combinations in the same manner as resistors. (Example 10.6)
- ❑ All analysis techniques previously used on resistive circuits apply to circuits with capacitors and/or inductors once all elements are replaced by their frequency-domain equivalents. (Examples 10.5, 10.7, 10.8, 10.9, 10.10, 10.11)
- ❑ Phasor analysis can only be performed on single-frequency circuits. Otherwise, superposition must be invoked, and the *time-domain* partial responses added to obtain the complete response. (Example 10.12)
- ❑ The power behind phasor diagrams is evident when a convenient forcing function is used initially, and the final result scaled appropriately. (Example 10.13)

READING FURTHER

A good additional reference to phasor-based analysis techniques can be found in:

R. A. DeCarlo and P. M. Lin, *Linear Circuits,* 3rd ed. Dubuque, IA: Kendall Hunt Publishing, 2009.

Frequency-dependent transistor models are discussed from a phasor perspective in Chap. 7 of:

W. H. Hayt, Jr., and G. W. Neudeck, *Electronic Circuit Analysis and Design,* 2nd ed. New York: Wiley, 1995.

EXERCISES

10.1 Characteristics of Sinusoids

1. Evaluate the following: (*a*) $5\sin(5t - 9°)$ at $t = 0$, 0.01, and 0.1 s; (*b*) $4\cos 2t$ at $t = 0$, 1, and 1.5 s; (*c*) $3.2\cos(6t + 15°)$ at $t = 0$, 0.01, and 0.1 s.
2. (*a*) Express each of the following as a single *cosine* function: $300\sin 628t$, $4\sin(3\pi t + 30°)$, $14\sin(50t - 5°) - 10\cos 50t$. (*b*) Express each of the following as a single *sine* function: $2\cos(100t + 45°)$, $3\cos 4000t$, $5\cos(2t - 90°) + 10\sin(2t)$.
3. Determine the angle by which v_1 *leads* i_1 if $v_1 = 10\cos(10t - 45°)$ and i_1 is equal to (*a*) $5\cos 10t$; (*b*) $5\cos(10t - 80°)$; (*c*) $5\cos(10t - 40°)$; (*d*) $5\cos(10t + 40°)$; (*e*) $5\sin(10t - 19°)$.
4. Determine the angle by which v_1 *lags* i_1 if $v_1 = 3\cos(10^4 t - 5°)$ and i_1 is equal to (*a*) $5\cos 10^4 t$; (*b*) $5\cos(10^4 t - 14°)$; (*c*) $5\cos(10^4 t - 23°)$; (*d*) $5\cos(10^4 t + 23°)$; (*e*) $5\sin(10^4 t - 390°)$.
5. Determine which waveform in each of the following pairs is lagging: (*a*) $\cos 4t$, $\sin 4t$; (*b*) $\cos(4t - 80°)$, $\cos(4t)$; (*c*) $\cos(4t + 80°)$, $\cos 4t$; (*d*) $-\sin 5t$, $\cos(5t + 2°)$; (*e*) $\sin 5t + \cos 5t$, $\cos(5t - 45°)$.
6. Calculate the first three instants in time ($t > 0$) for which the following functions are zero, by first converting to a single sinusoid: (*a*) $\cos 3t - 7\sin 3t$; (*b*) $\cos(10t + 45°)$; (*c*) $\cos 5t - \sin 5t$.

7. (*a*) Determine the first two instants in time ($t > 0$) for which each of the functions in Exercise 6 are equal to unity, by first converting to a single sinusoid. (*b*) Verify your answers by plotting each waveform using an appropriate software package.

8. The concept of Fourier series is a powerful means of analyzing periodic waveforms in terms of sinusoids. For example, the triangle wave in Fig. 10.45 can be represented by the infinite sum

$$v(t) = \frac{8}{\pi^2}\left(\sin \pi t - \frac{1}{3^2}\sin 3\pi t + \frac{1}{5^2}\sin 5\pi t - \frac{1}{7^2}\sin 7\pi t + \cdots\right)$$

where in practice perhaps the first several terms provide an accurate enough approximation. (*a*) Compute the exact value of $v(t)$ at $t = 0.25$ s by first obtaining an equation for the corresponding segment of the waveform. (*b*) Compute the approximate value at $t = 0.25$ s using the first term of the Fourier series only. (*c*) Repeat part (*b*) using the first three terms. (*d*) Plot $v(t)$ using the first term only. (*e*) Plot $v(t)$ using the first two terms only. (*f*) Plot $v(t)$ using the first three terms only.

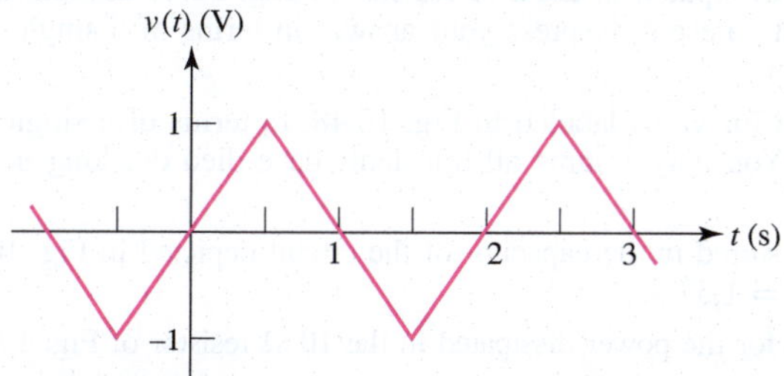

FIGURE 10.45

9. Household electrical voltages are typically quoted as either 110 V, 115 V, or 120 V. However, these values do not represent the peak ac voltage. Rather, they represent what is known as the root mean square of the voltage, defined as

$$\mathbf{V}_{\text{rms}} = \sqrt{\frac{1}{T}\int_0^T \mathbf{V}_m^2 \cos^2(\omega t)\, dt}$$

where $T =$ the period of the waveform, V_m is the peak voltage, and $\omega =$ the waveform frequency ($f = 60$ Hz in North America). (*a*) Perform the indicated integration, and show that for a sinusoidal voltage,

$$\mathbf{V}_{\text{rms}} = \frac{\mathbf{V}_m}{\sqrt{2}}$$

(*b*) Compute the peak voltages corresponding to the rms voltages of 110, 115, and 120 V.

10.2 Forced Response to Sinusoidal Functions

10. If the source v_s in Fig. 10.46 is equal to 4.53 cos 30*t* V, (*a*) obtain i_L at $t = 0$ assuming no transients are present; (*b*) obtain an expression for $v_L(t)$ in terms of a single sinusoid, valid for $t > 0$, again assuming no transients are present.

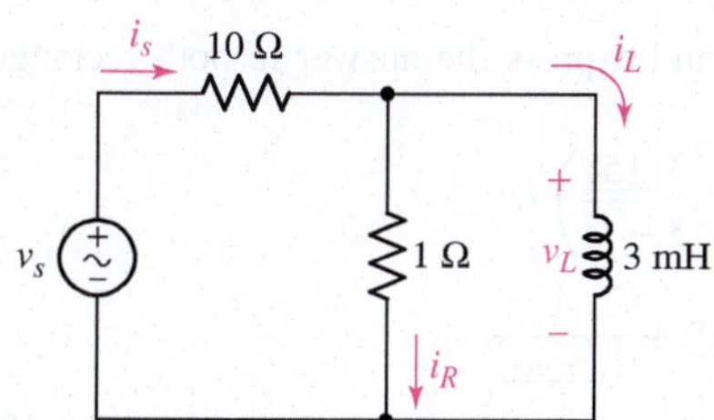

FIGURE 10.46

11. Assuming there are no longer any transients present, determine the current labeled i_L in the circuit of Fig. 10.47. Express your answer as a single sinusoid.

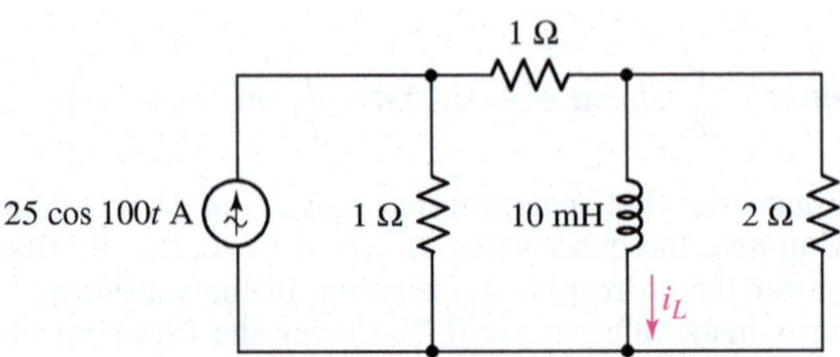

■ **FIGURE 10.47**

12. Calculate the power dissipated in the 2 Ω resistor of Fig. 10.47 assuming there are no transients present. Express your answer in terms of a single sinusoidal function.

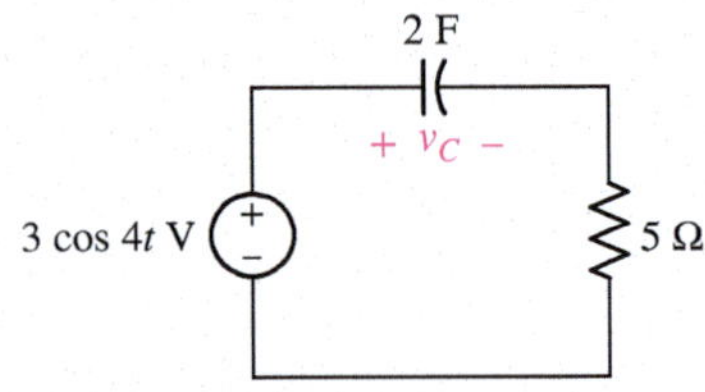

■ **FIGURE 10.48**

13. Obtain an expression for v_C as labeled in Fig. 10.48, in terms of a single sinusoidal function. You may assume all transients have died out long before $t = 0$.

14. Calculate the energy stored in the capacitor of the circuit depicted in Fig. 10.48 at $t = 785$ ms and $t = 1.57$ s.

15. Obtain an expression for the power dissipated in the 10 Ω resistor of Fig. 10.49, assuming no transients present.

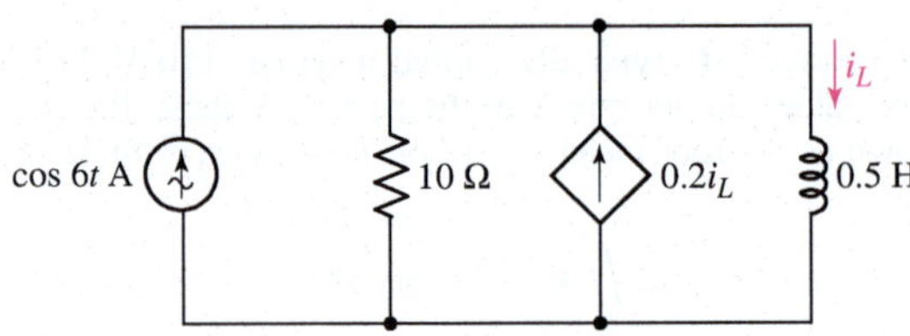

■ **FIGURE 10.49**

10.3 The Complex Forcing Function

16. Express the following complex numbers in rectangular form: (*a*) $50\underline{/-75^\circ}$; (*b*) $19e^{j30^\circ}$; (*c*) $2.5\underline{/-30^\circ} + 0.5\underline{/45^\circ}$. Convert the following to polar form: (*d*) $(2 + j2)(2 - j2)$; (*e*) $(2 + j2)(5\underline{/22^\circ})$.

17. Express the following in polar form: (*a*) $1 + e^{j45^\circ}$; (*b*) $(-j)(j^2)$; (*c*) 32. Express the following in rectangular form: (*d*) $2 - e^{j45^\circ}$; (*e*) $-j + 5\underline{/0^\circ}$.

18. Evaluate the following, and express your answer in polar form: (*a*) $4(8 - j8)$; (*b*) $4\underline{/5^\circ} - 2\underline{/15^\circ}$; (*c*) $(2 + j9) - 5\underline{/0^\circ}$; (*d*) $\dfrac{-j}{10 + 5j} - 3\underline{/40^\circ} + 2$.

19. Evaluate the following, and express your answer in rectangular form: (*a*) $3(3\underline{/30^\circ})$; (*b*) $2\underline{/25^\circ} + 5\underline{/-10^\circ}$; (*c*) $(12 + j90) - 5\underline{/30^\circ}$; (*d*) $\dfrac{10 + 5j}{8 - j} + 2\underline{/60^\circ} + 1$.

20. Perform the indicated operations, and express the answer in both rectangular and polar forms:

(*a*) $\dfrac{2 + j3}{1 + 8\underline{/90^\circ}} - 4$; (*b*) $\left(\dfrac{10\underline{/25^\circ}}{5\underline{/-10^\circ}} + \dfrac{3\underline{/15^\circ}}{3 - j5}\right) j2$;

(*c*) $\left[\dfrac{(1 - j)(1 + j) + 1\underline{/0^\circ}}{-j}\right](3\underline{/-90^\circ}) + \dfrac{j}{1\underline{/-45^\circ}}$.

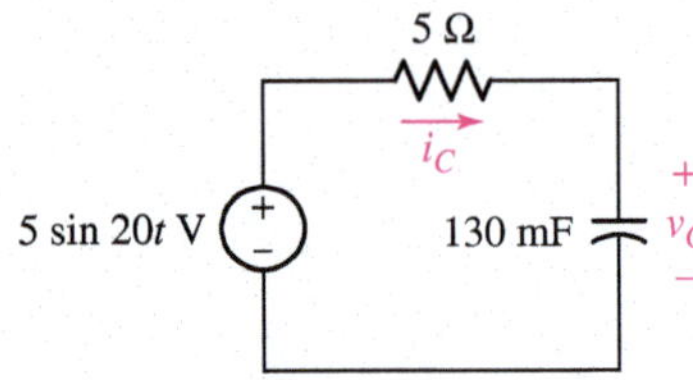

■ **FIGURE 10.50**

21. Insert an appropriate complex source into the circuit represented in Fig. 10.50, and use it to determine steady-state expressions for $i_C(t)$ and $v_C(t)$.

22. For the circuit of Fig. 10.51, if $i_s = 2 \cos 5t$ A, use a suitable complex source replacement to obtain a steady-state expression for $i_L(t)$.
23. In the circuit depicted in Fig. 10.51, if i_s is modified such that $i_L(t) = 1.8 \cos(5t + 26.6°)$ A, determine $i_s(t)$.
24. Employ a suitable complex source to determine the steady-state current i_L in the circuit of Fig. 10.52.

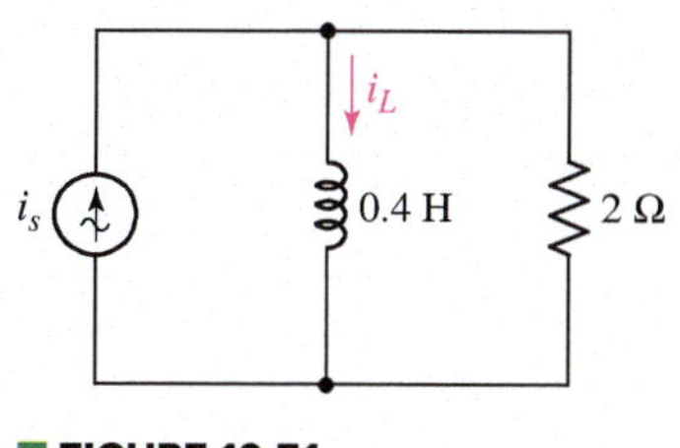

■ FIGURE 10.51

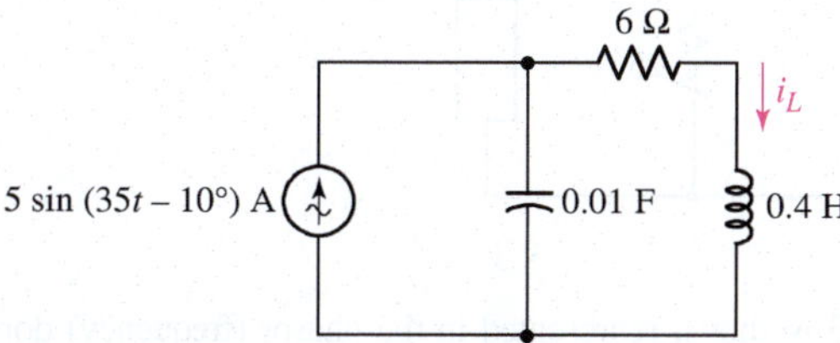

■ FIGURE 10.52

10.4 The Phasor

25. Transform each of the following into phasor form: (*a*) $28 \cos(20t)$; (*b*) $32 \sin(2t - 90°)$; (*c*) $\sin(9t + 45°)$; (*d*) $5 \cos 10t + 8 \cos(10t + 45°)$.
26. Transform each of the following into phasor form: (*a*) $11 \sin 100t$; (*b*) $11 \cos 100t$; (*c*) $11 \cos(100t - 90°)$; (*d*) $3 \cos 100t - 3 \sin 100t$.
27. Assuming an operating frequency of 1 kHz, transform the following phasor expressions into a single cosine function in the time domain: (*a*) $9\angle 65°$ V; (*b*) $\dfrac{2\angle 31°}{4\angle 25°}$ A; (*c*) $22\angle 14° - 8\angle 33°$ V.
28. The following complex voltages are written in a combination of rectangular and polar form. Rewrite each, using conventional phasor notation (i.e., a magnitude and angle): (*a*) $\dfrac{2-j}{5\angle 45°}$ V; (*b*) $\dfrac{6\angle 20°}{1000} - j$ V; (*c*) $(j)(52.5\angle -90°)$ V.
29. Assuming an operating frequency of 50 Hz, compute the instantaneous voltage at $t = 10$ ms and $t = 25$ ms for each of the phasor quantities represented in Exercise 28.
30. Assuming an operating frequency of 50 Hz, compute the instantaneous voltage at $t = 10$ ms and $t = 25$ ms for each of the quantities represented in Exercise 27.
31. Assuming the passive sign convention and an operating frequency of 5 rad/s, calculate the phasor voltage which develops across the following when driven by the phasor current $\mathbf{I} = 2\angle 0°$ mA: (*a*) a 1 kΩ resistor; (*b*) a 1 mF capacitor; (*c*) a 1 nH inductor.
32. (*a*) A series connection is formed between a 1 Ω resistor, a 1 F capacitor, and a 1 H inductor, in that order. Assuming operation at $\omega = 1$ rad/s, what are the magnitude and phase angle of the phasor current which yields a voltage of $1\angle 30°$ V across the resistor (assume the passive sign convention)? (*b*) Compute the ratio of the phasor voltage across the resistor to the phasor voltage which appears across the capacitor-inductor combination. (*c*) The frequency is doubled. Calculate the new ratio of the phasor voltage across the resistor to the phasor voltage across the capacitor-inductor combination.
33. Assuming the passive sign convention and an operating frequency of 314 rad/s, calculate the phasor voltage $\mathbf{V}$ which appears across each of the following when driven by the phasor current $\mathbf{I} = 10\angle 0°$ mA: (*a*) a 2 Ω resistor; (*b*) a 1 F capacitor; (*c*) a 1 H inductor; (*d*) a 2 Ω resistor in series with a 1 F capacitor; (*e*) a 2 Ω resistor in series with a 1 H inductor. (*f*) Calculate the instantaneous value of each voltage determined in parts (*a*) to (*e*) at $t = 0$.

34. In the circuit of Fig. 10.53, which is shown in the phasor (frequency) domain, $\mathbf{I}_{10}$ is determined to be $2\underline{/42°}$ mA; (*a*) what is the likely type of element connected to the right of the 25 Ω resistor and (*b*) what is its value, assuming the voltage source operates at a frequency of 1000 rad/s?

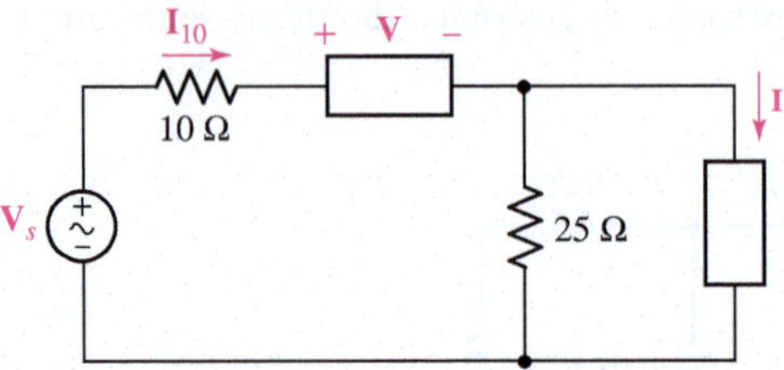

■ **FIGURE 10.53**

35. The circuit of Fig. 10.53 is shown represented in the phasor (frequency) domain. If $\mathbf{I}_{10} = 4\underline{/35°}$ A, $\mathbf{V} = 10\underline{/35°}$V, and $\mathbf{I} = 2\underline{/35°}$ A, (*a*) across what type of element does **V** appear, and what is its value? (*b*) Determine the value of $\mathbf{V}_s$.

10.5 Impedance and Admittance

36. (*a*) Obtain an expression for the equivalent impedance $\mathbf{Z}_{eq}$ of a 1 Ω resistor in series with a 2 F capacitor as a function of ω. (*b*) Plot the magnitude of $\mathbf{Z}_{eq}$ as a function of ω over the range $0.1 < \omega < 100$ rad/s (use a logarithmic scale for the frequency axis). (*c*) Plot the angle (in degrees) of $\mathbf{Z}_{eq}$ as a function of ω over the range $0.1 < \omega < 100$ rad/s (use a logarithmic scale for the frequency axis). [*Hint: semilogx*() in MATLAB is a useful plotting function.]

37. Determine the equivalent impedance of the following, assuming an operating frequency of 20 rad/s: (*a*) 1 kΩ in parallel with 1 mH; (*b*) 10 kΩ in parallel with the series combination of 1 F and 1 H.

38. (*a*) Obtain an expression for the equivalent impedance $\mathbf{Z}_{eq}$ of a 1 Ω resistor in parallel with a 10 mH inductor as a function of ω over the range $1 < \omega < 10^5$ rad/s (use a logarithmic scale for the frequency axis). (*b*) Plot the angle (in degrees) of $\mathbf{Z}_{eq}$ as a function of ω over the range $1 < \omega < 10^5$ rad/s (use a logarithmic scale for the frequency axis). [*Hint: semilogx*() in MATLAB is a useful plotting function.]

39. Determine the equivalent admittance of the following, assuming an operating frequency of 1000 rad/s: (*a*) 25 Ω in series with 20 mH; (*b*) 25 Ω in parallel with 20 mH; (*c*) 25 Ω in parallel with 20 mH in parallel with 20 mF.

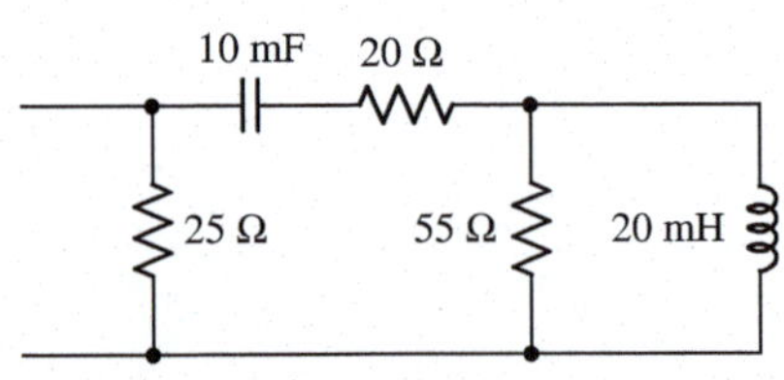

■ **FIGURE 10.54**

40. Consider the network depicted in Fig. 10.54, and determine the equivalent impedance seen looking into the open terminals if (*a*) $\omega = 1$ rad/s; (*b*) $\omega = 10$ rad/s; (*c*) $\omega = 100$ rad/s.

41. Exchange the capacitor and inductor in the network shown in Fig. 10.54, and calculate the equivalent impedance looking into the open terminals if $\omega = 25$ rad/s.

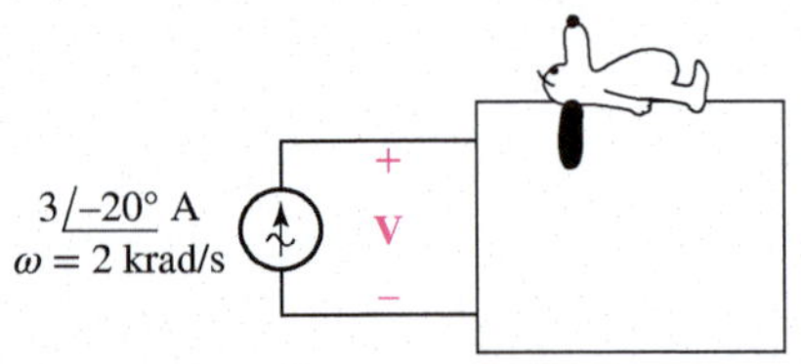

■ **FIGURE 10.55**

42. Find **V** in Fig. 10.55 if the box contains (*a*) 3 Ω in series with 2 mH; (*b*) 3 Ω in series with 125 μF; (*c*) 3 Ω, 2 mH, and 125 μF in series; (*d*) 3 Ω, 2 mH, and 125 μF in series, but $\omega = 4$ krad/s.

43. Calculate the equivalent impedance seen at the open terminals of the network shown in Fig. 10.56 if *f* is equal to (*a*) 1 Hz; (*b*) 1 kHz; (*c*) 1 MHz; (*d*) 1 GHz.

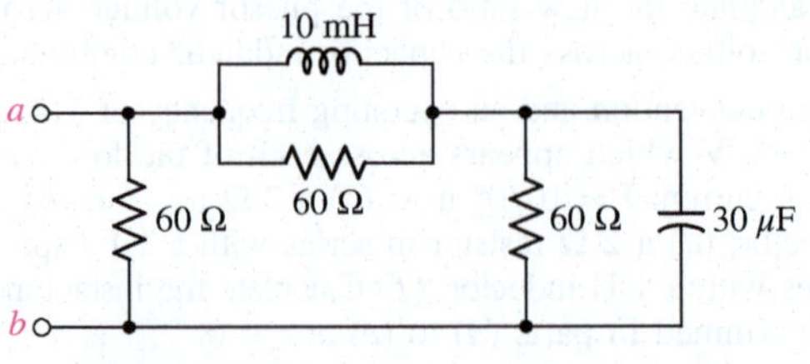

■ **FIGURE 10.56**

44. Employ phasor-based analysis to obtain an expression for $i(t)$ in the circuit of Fig. 10.57.

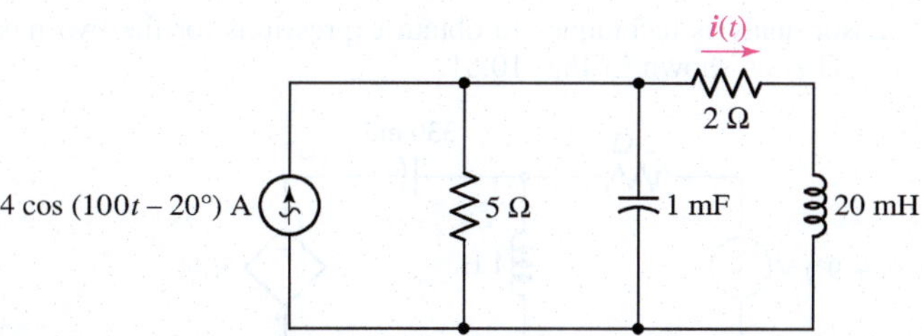

■ **FIGURE 10.57**

45. Design a suitable combination of resistors, capacitors, and/or inductors which has an equivalent impedance at $\omega = 100$ rad/s of (*a*) 1 Ω using at least one inductor; (*b*) $7\underline{/10°}$ Ω; (*c*) $3 - j4$ Ω.

46. Design a suitable combination of resistors, capacitors, and/or inductors which has an equivalent admittance at ω 10 rad/s of (*a*) 1 S using at least one capacitor; (*b*) $12\underline{/-18°}$ S; (*c*) $2 + j$ mS.

10.6 Nodal and Mesh Analysis

47. For the circuit depicted in Fig. 10.58, (*a*) redraw with appropriate phasors and impedances labeled; (*b*) employ nodal analysis to determine the two nodal voltages $v_1(t)$ and $v_2(t)$.

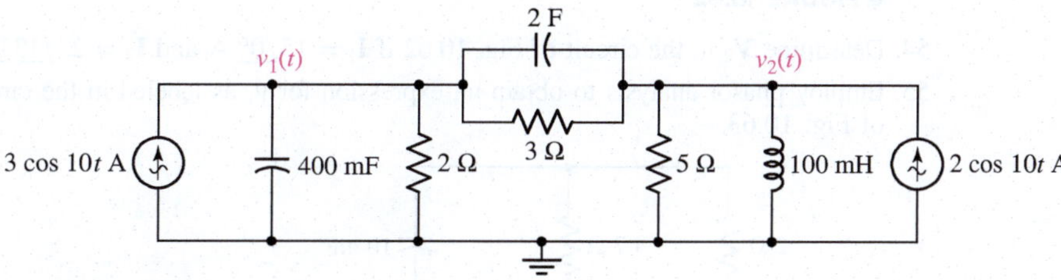

■ **FIGURE 10.58**

48. For the circuit illustrated in Fig. 10.59, (*a*) redraw, labeling appropriate phasor and impedance quantities; (*b*) determine expressions for the three time-domain mesh currents.

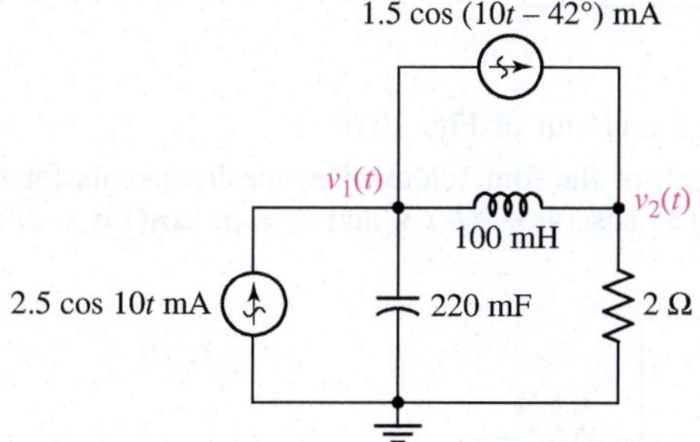

■ **FIGURE 10.59**

49. Referring to the circuit of Fig. 10.59, employ phasor-based analysis techniques to determine the two nodal voltages.

50. In the phasor-domain circuit represented by Fig. 10.60, let $\mathbf{V}_1 = 10\underline{/-80°}$ V, $\mathbf{V}_2 = 4\underline{/-0°}$ V, and $\mathbf{V}_3 = 2\underline{/-23°}$ V. Calculate $\mathbf{I}_1$ and $\mathbf{I}_2$.

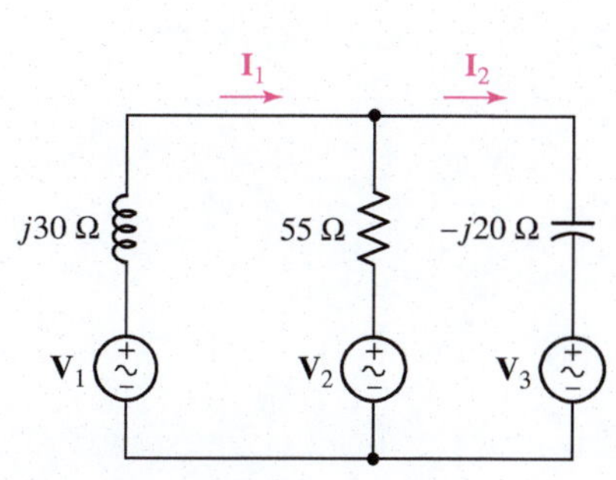

■ **FIGURE 10.60**

51. With regard to the two-mesh phasor-domain circuit depicted in Fig. 10.60, calculate the ratio of $\mathbf{I}_1$ to $\mathbf{I}_2$ if $\mathbf{V}_1 = 3\underline{/0°}$ V, $\mathbf{V}_2 = 5.5\underline{/-130°}$ V, and $\mathbf{V}_3 = 1.5\underline{/17°}$ V.
52. Employ phasor analysis techniques to obtain expressions for the two mesh currents i_1 and i_2 as shown in Fig. 10.61.

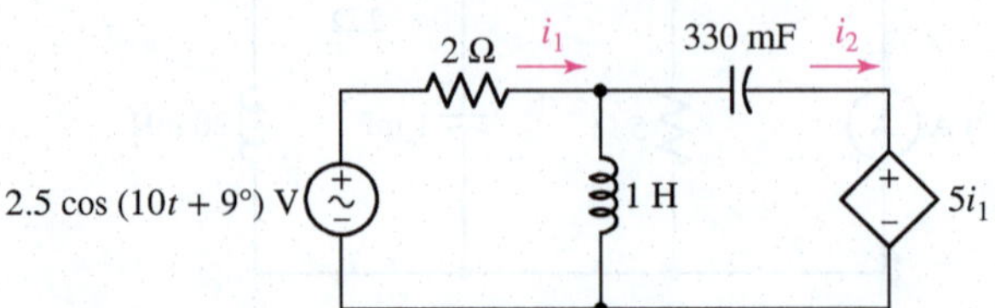

■ **FIGURE 10.61**

53. Determine $\mathbf{I}_B$ in the circuit of Fig. 10.62 if $\mathbf{I}_1 = 5\underline{/-18°}$ A and $\mathbf{I}_2 = 2\underline{/5°}$ A.

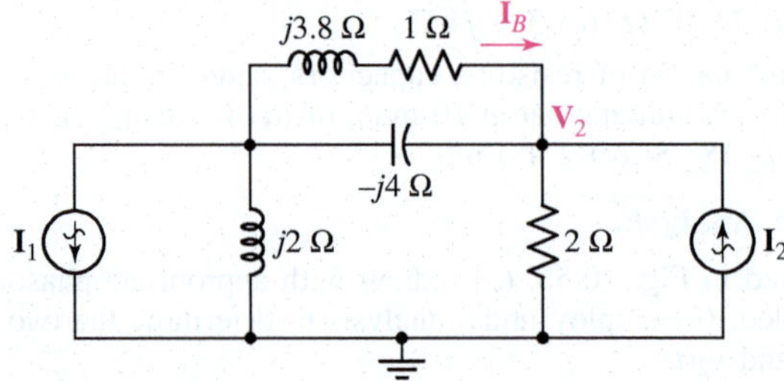

■ **FIGURE 10.62**

54. Determine $\mathbf{V}_2$ in the circuit of Fig. 10.62 if $\mathbf{I}_1 = 15\underline{/0°}$ A and $\mathbf{I}_2 = 25\underline{/131°}$ A.
55. Employ phasor analysis to obtain an expression for v_x as labeled in the circuit of Fig. 10.63.

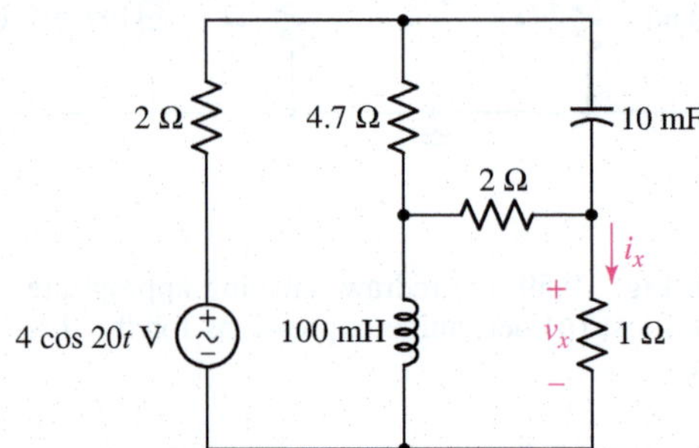

■ **FIGURE 10.63**

56. Determine the current i_x in the circuit of Fig. 10.63.
57. Obtain an expression for each of the four (clockwise) mesh currents for the circuit of Fig. 10.64 if $v_1 = 133\cos(14t + 77°)$ V and $v_2 = 55\cos(14t + 22°)$ V.

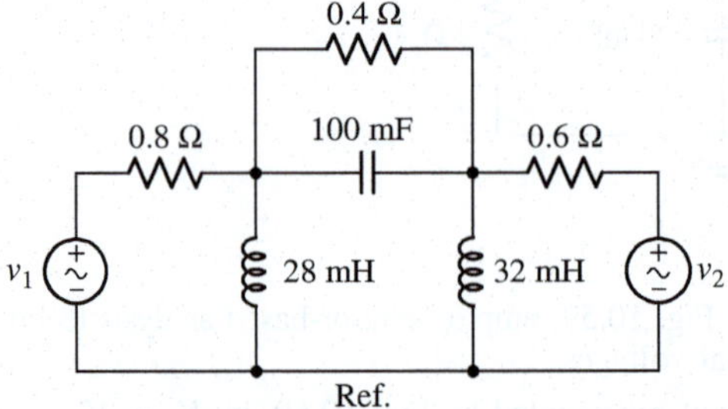

■ **FIGURE 10.64**

58. Determine the nodal voltages for the circuit of Fig. 10.64, using the bottom node as the reference node, if $v_1 = 0.009 \cos(500t + 0.5°)$ V and $v_2 = 0.004 \cos(500t + 1.5°)$ V.

59. The op amp shown in Fig. 10.65 has an infinite input impedance, zero output impedance, and a large but finite (positive, real) gain, $A = -\mathbf{V}_o/\mathbf{V}_i$. (*a*) Construct a basic differentiator by letting $\mathbf{Z}_f = \mathbf{R}_f$, find $\mathbf{V}_o/\mathbf{V}_s$, and then show that $\mathbf{V}_o/\mathbf{V}_s \to -j\omega C_1 R_f$ as $\mathbf{A} \to \infty$. (*b*) Let $\mathbf{Z}_f$ represent $\mathbf{C}_f$ and $\mathbf{R}_f$ in parallel, find $\mathbf{V}_o/\mathbf{V}_s$, and then show that $\mathbf{V}_o/\mathbf{V}_s \to -j\omega C_1 R_f/(1 + j\omega C_f R_f)$ as A $\to \infty$.

60. Obtain an expression for each of the four mesh currents labeled in the circuit of Fig. 10.66.

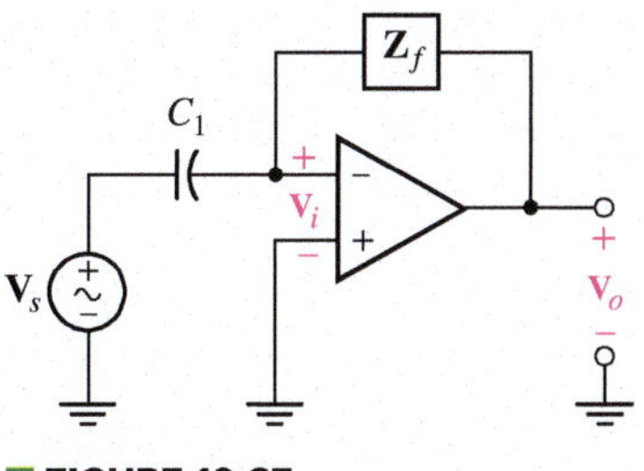

■ FIGURE 10.65

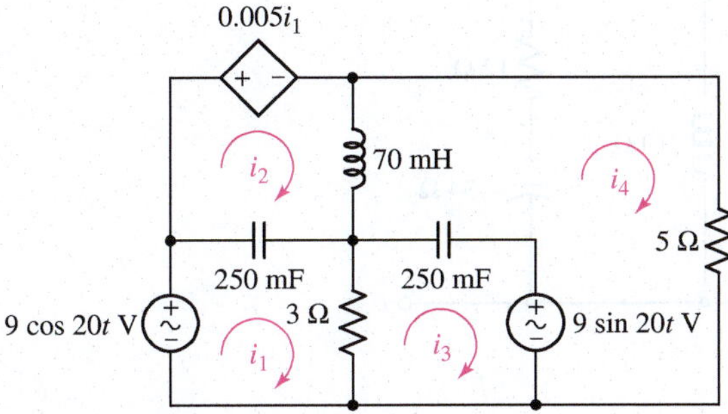

■ FIGURE 10.66

10.7 Superposition, Source Transformations, and Thévenin's Theorem

61. Determine the individual contribution each current source makes to the two nodal voltages $\mathbf{V}_1$ and $\mathbf{V}_2$ as represented in Fig. 10.67.

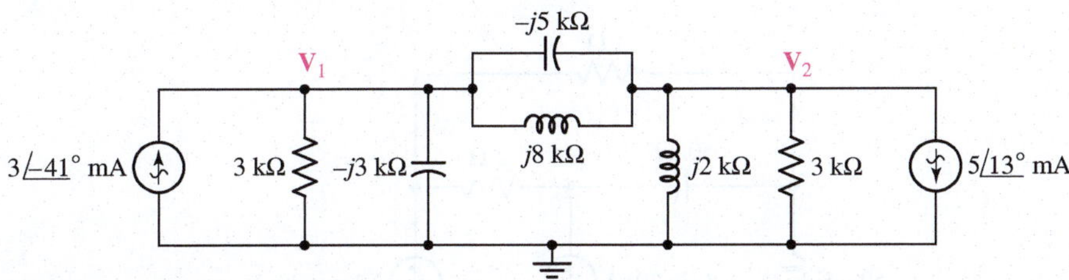

■ FIGURE 10.67

62. Determine $\mathbf{V}_1$ and $\mathbf{V}_2$ in Fig. 10.68 if $\mathbf{I}_1 = 33\angle 3°$ mA and $\mathbf{I}_2 = 51\angle -91°$ mA.

63. The phasor domain circuit of Fig. 10.68 was drawn assuming an operating frequency of 2.5 rad/s. Unfortunately, the manufacturing unit installed the wrong sources, each operating at a different frequency. If $i_1(t) = 4 \cos 40t$ mA and $i_2(t) = 4 \sin 30t$ mA, calculate $v_1(t)$ and $v_2(t)$.

64. Obtain the Thévenin equivalent seen by the $(2 - j)$ Ω impedance of Fig. 10.69, and employ it to determine the current $\mathbf{I}_1$.

■ FIGURE 10.68

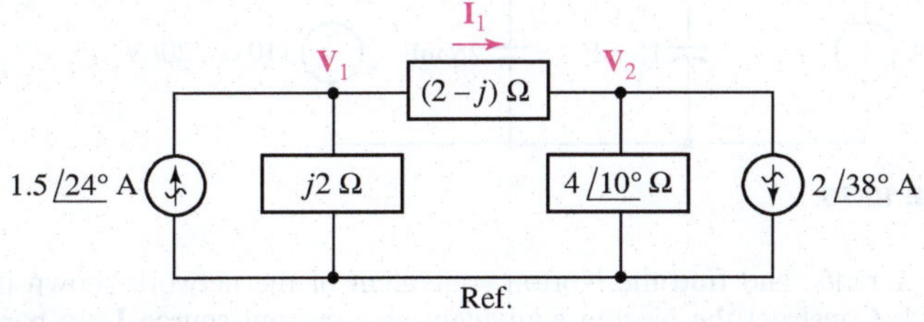

■ FIGURE 10.69

65. The $(2 - j)\ \Omega$ impedance in the circuit of Fig. 10.69 is replaced with a $(1 + j)\ \Omega$ impedance. Perform a source transformation on each source, simplify the resulting circuit as much as possible, and calculate the current flowing through the $(1 + j)\ \Omega$ impedance.

66. With regard to the circuit depicted in Fig. 10.70, (*a*) calculate the Thévenin equivalent seen looking into the terminals marked *a* and *b*; (*b*) determine the Norton equivalent seen looking into the terminals marked *a* and *b*; (*c*) compute the current flowing from *a* to *b* if a $(7 - j2)\ \Omega$ impedance is connected across them.

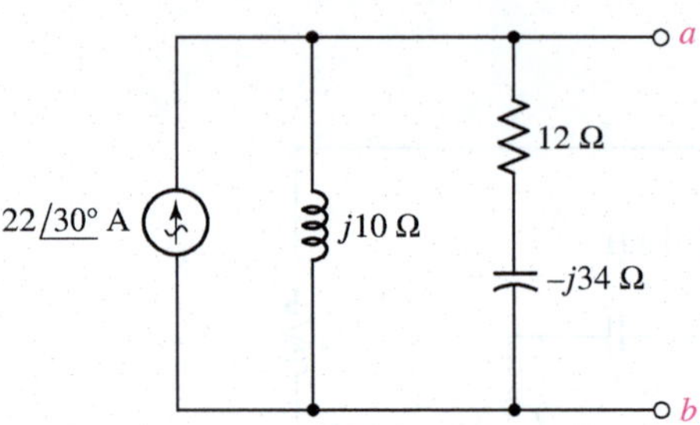

■ **FIGURE 10.70**

67. In the circuit of Fig. 10.71, $i_{s1} = 8\cos(4t - 9°)$ A, $i_{s2} = 5\cos 4t$ A, and $v_{s3} = 2\sin 4t$ V. (*a*) Redraw the circuit in the phasor domain; (*b*) reduce the circuit to a single current source with the assistance of a source transformation; (*c*) calculate $v_L(t)$. (*d*) Verify your solution with an appropriate simulation.

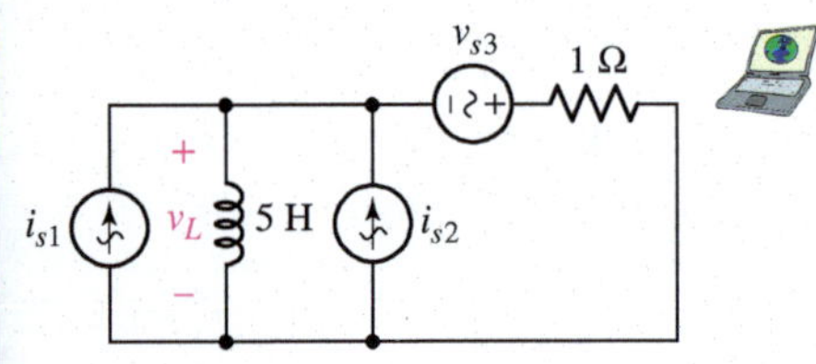

■ **FIGURE 10.71**

68. Determine the individual contribution of each source in Fig. 10.72 to the voltage $v_1(t)$.

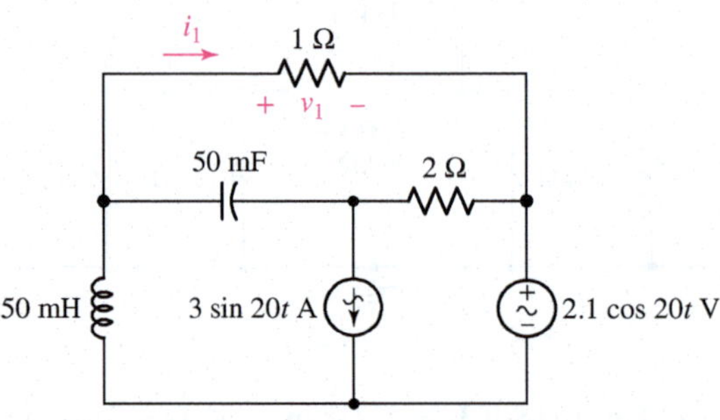

■ **FIGURE 10.72**

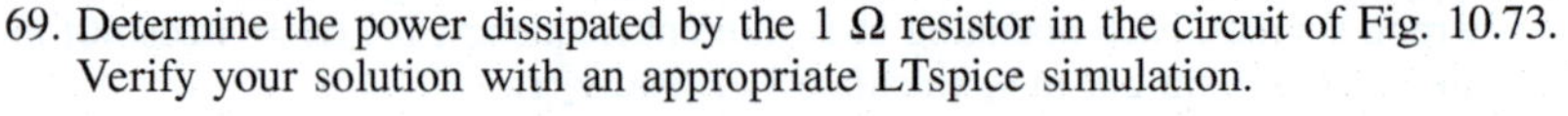

69. Determine the power dissipated by the 1 Ω resistor in the circuit of Fig. 10.73. Verify your solution with an appropriate LTspice simulation.

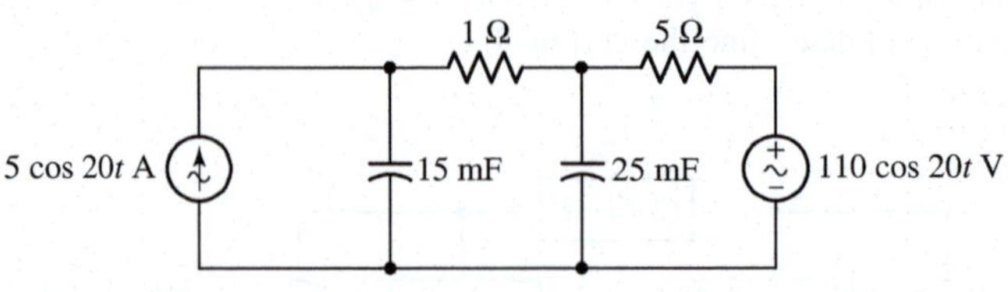

■ **FIGURE 10.73**

70. Use $\omega = 1$ rad/s, and find the Norton equivalent of the network shown in Fig. 10.74. Construct the Norton equivalent as a current source $\mathbf{I}_N$ in parallel with a resistance R_N and either an inductance L_N or a capacitance C_N.

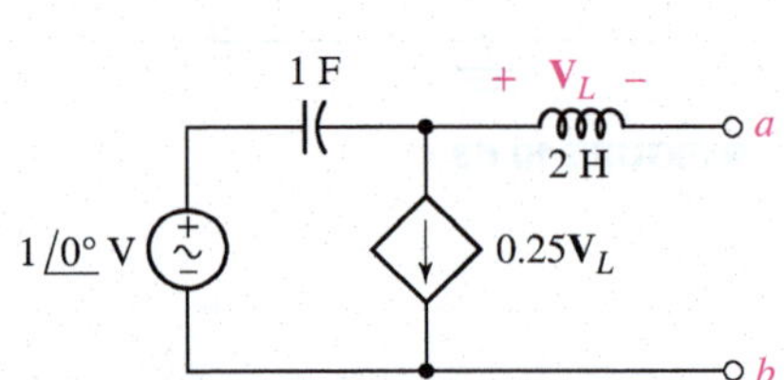

■ **FIGURE 10.74**

10.8 Phasor Diagrams

71. The source $\mathbf{I}_s$ in the circuit of Fig. 10.75 is selected such that $\mathbf{V} = 5\underline{/120°}$ V. (*a*) Construct a phasor diagram showing $\mathbf{I}_R$, $\mathbf{I}_L$, and $\mathbf{I}_C$. (*b*) Use the diagram to determine the angle by which $\mathbf{I}_s$ leads $\mathbf{I}_R$, $\mathbf{I}_C$, and $\mathbf{I}_x$.

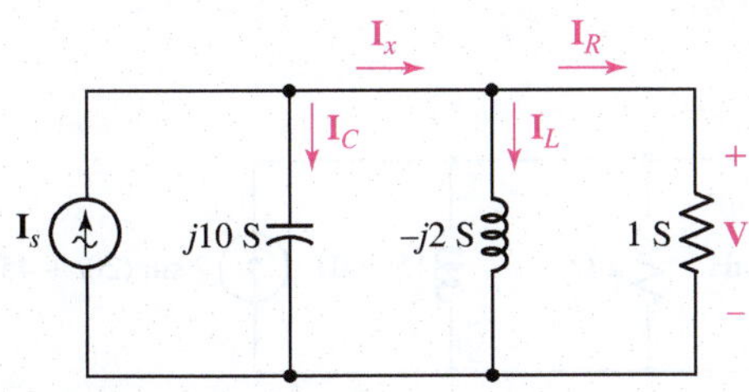

FIGURE 10.75

72. Let $\mathbf{V}_1 = 100\underline{/0°}$ V, $|\mathbf{V}_2| = 140$ V, and $|\mathbf{V}_1 + \mathbf{V}_2| = 120$ V. Use graphical methods to find two possible values for the angle of $\mathbf{V}_2$.

73. (*a*) Calculate values for $\mathbf{I}_L$, $\mathbf{I}_R$, $\mathbf{I}_C$, $\mathbf{V}_L$, $\mathbf{V}_R$, and $\mathbf{V}_C$ for the circuit shown in Fig. 10.76. (*b*) Using scales of 50 V to 1 in and 25 A to 1 in, show all seven quantities on a phasor diagram, and indicate that $\mathbf{I}_L = \mathbf{I}_R + \mathbf{I}_C$ and $\mathbf{V}_s = \mathbf{V}_L + \mathbf{V}_R$.

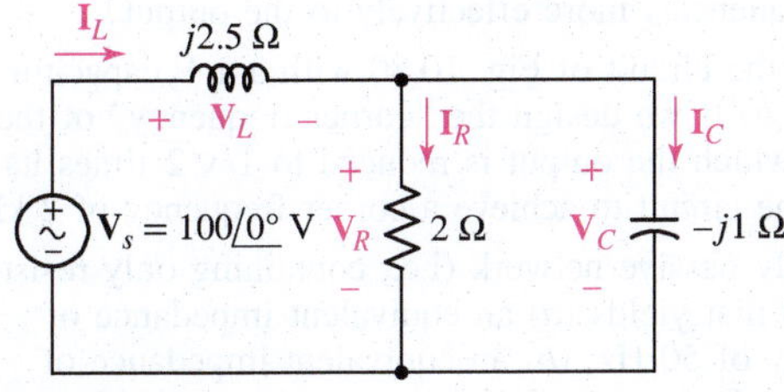

FIGURE 10.76

74. In the circuit of Fig. 10.77, (*a*) find values for $\mathbf{I}_1$, $\mathbf{I}_2$, and $\mathbf{I}_3$. (*b*) Show $\mathbf{V}_s$, $\mathbf{I}_1$, $\mathbf{I}_2$, and $\mathbf{I}_3$ on a phasor diagram (scales of 50 V/in and 2 A/in work fine). (*c*) Find $\mathbf{I}_s$ graphically and give its amplitude and phase angle.

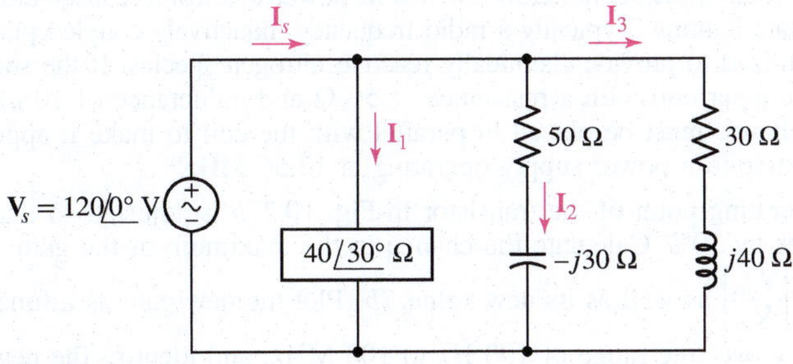

FIGURE 10.77

75. The voltage source $\mathbf{V}_s$ in Fig. 10.78 is chosen such that $\mathbf{I}_C = 1\underline{/0°}$ A. (*a*) Draw a phasor diagram showing $\mathbf{V}_1$, $\mathbf{V}_2$, $\mathbf{V}_s$, and $\mathbf{V}_R$. (*b*) Use the diagram to determine the ratio of $\mathbf{V}_2$ to $\mathbf{V}_1$.

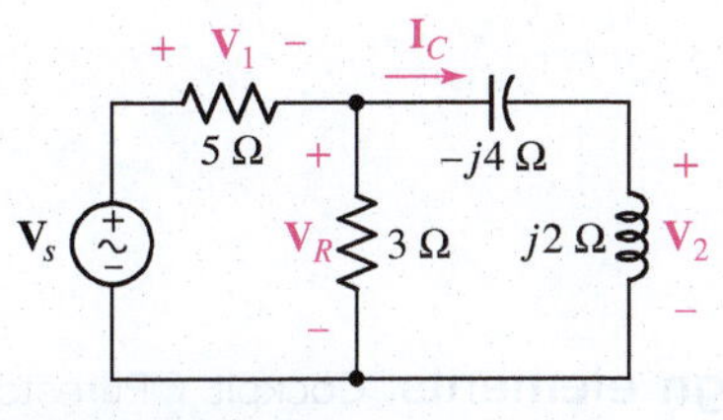

FIGURE 10.78

Chapter-Integrating Exercises

76. For the circuit shown in Fig. 10.79, (*a*) draw the phasor representation of the circuit; (*b*) determine the Thévenin equivalent seen by the capacitor, and use it to calculate $v_C(t)$. (*c*) Determine the current flowing out of the positive reference terminal of the voltage source. (*d*) Verify your solution with an appropriate LTspice simulation.

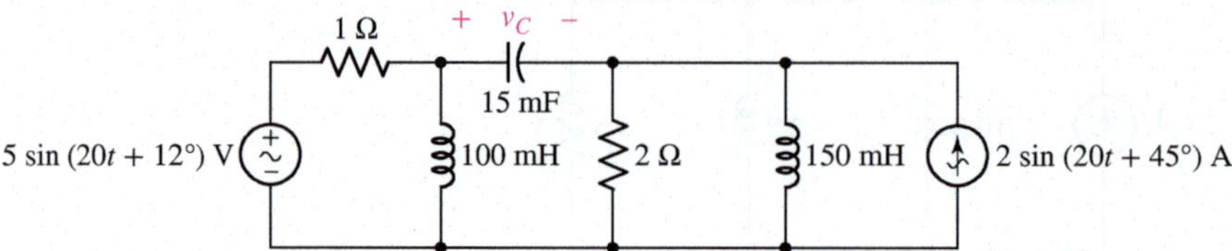

FIGURE 10.79

77. The circuit of Fig. 10.79 is unfortunately operating differently than specified; the frequency of the current source is only 19 rad/s. Calculate the actual capacitor voltage, and compare it to the expected voltage had the circuit been operating correctly.

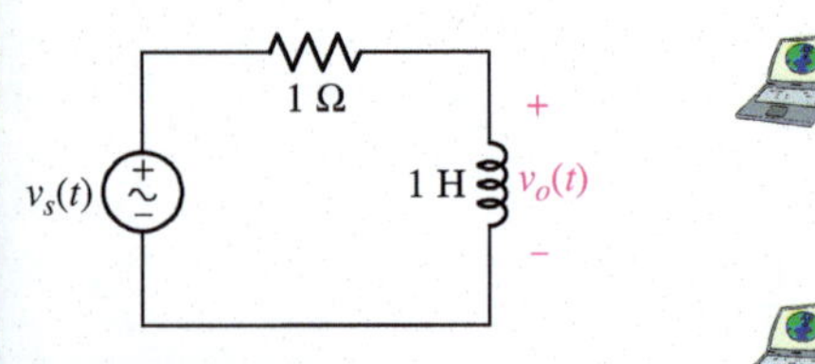

FIGURE 10.80

78. For the circuit shown in Fig. 10.80, (*a*) draw the corresponding phasor representation; (*b*) obtain an expression for $\mathbf{V}_o/\mathbf{V}_s$; (*c*) plot $|\mathbf{V}_o/\mathbf{V}_s|$, the magnitude of the phasor voltage ratio, as a function of frequency ω over the range $0.01 \leq \omega \leq 100$ rad/s (use a logarithmic x axis). (*d*) Does the circuit transfer low frequencies or high frequencies more effectively to the output?

79. (*a*) Replace the inductor in the circuit of Fig. 10.80 with a 1 F capacitor and repeat Exercise 10.78. (*b*) If we design the "corner frequency" of the circuit as the frequency at which the output is reduced to $1/\sqrt{2}$ times its maximum value, redesign the circuit to achieve a corner frequency of 2 kHz.

80. Design a two-terminal purely passive network (i.e., containing only resistors, capacitors and/or inductors) that yields (*a*) an equivalent impedance of $\mathbf{Z} = 2\angle 60^\circ\ \Omega$ at a frequency of 50 Hz; (*b*) an equivalent impedance of $\mathbf{Z} = 100.005\angle 89.427^\circ\ \Omega$ at a frequency of 100 Hz.

81. Design an RLC network which, when placed in parallel with a 100 Ω (2% tolerance) resistor, yields an equivalent parallel admittance of $\mathbf{Y} = 28.627\angle -31.900^\circ$ mS at $f = 100$ Hz when measured across the 100 Ω resistor. Use 10% tolerance (E12 standard component) values. (*b*) Verify your solution with an appropriate LTspice simulation.

82. Molecular beam epitaxy is one of several techniques available to grow gallium nitride and related semiconductors for use in power electronics, laser diodes, and solid state lighting. Typically a radio frequency inductively coupled plasma source is utilized to provide chemically reactive nitrogen species. If the source contains a copper coil with a resistance of 50 Ω and inductance of 10 μH, what capacitance must be placed in parallel with the coil to make it appear purely resistive to a power supply operating at 13.56 MHz?

83. (*a*) The operating point of the transistor in Fig. 10.27*b* is adjusted so that C_μ increases by 10%. Calculate the change in the maximum of the gain magnitude $\left|\frac{\mathbf{V}_{out}}{\mathbf{V}_{in}}\right|$ as well as its new value. (*b*) Plot the new gain as a function of frequency over the range of 100 Hz to 100 MHz, and identify the new corner frequency.

AC Circuit Power Analysis

KEY CONCEPTS

Calculating Instantaneous Power

Average Power Supplied by a Sinusoidal Source

Root-Mean-Square (RMS) Values

Reactive Power

The Relationship Between Complex, Average, and Reactive Power

Power Factor of a Load

INTRODUCTION

Often an integral part of circuit analysis is the determination of either power delivered or power absorbed (or both). In the context of ac power, we find that the rather simple approach we have taken previously does not provide a convenient picture of how a particular system is operating, so we introduce several different power-related quantities in this chapter.

We begin by considering *instantaneous* power, the product of the time-domain voltage and time-domain current associated with the element or network of interest. The instantaneous power is sometimes quite useful in its own right because its maximum value might have to be limited to avoid exceeding the safe operating range of a physical device. For example, transistor and vacuum-tube power amplifiers both produce a distorted output, and speakers give a distorted sound, when the peak power exceeds a certain limiting value. However, we are mainly interested in instantaneous power for the simple reason that it provides us with the means to calculate a more important quantity, the *average* power. In a similar way, the progress of a cross-country road trip is best described by the average velocity; our interest in the instantaneous velocity is limited to the avoidance of maximum velocities that will endanger our safety or arouse the highway patrol.

In practical problems we will deal with values of average power which range from the small fraction of a picowatt available in a telemetry signal from outer space, to the few watts of audio power supplied to the speakers in a good stereo system, to the several hundred watts required to run the morning coffee pot, or

to the 10 billion watts generated at the Grand Coulee Dam. Still, we will see that even the concept of average power has its limitations, especially when dealing with the energy exchange between reactive loads and power sources. This is easily handled by introducing the concepts of reactive power, complex power, and the power factor—all very common terms in the power industry.

11.1 INSTANTANEOUS POWER

The *instantaneous power* delivered to any device is given by the product of the instantaneous voltage across the device and the instantaneous current through it (the passive sign convention is assumed). Thus,[1]

$$p(t) = v(t)i(t) \qquad [1]$$

If the device in question is a resistor of resistance R, then the power may be expressed solely in terms of either the current or the voltage:

$$p(t) = v(t)i(t) = i^2(t)R = \frac{v^2(t)}{R} \qquad [2]$$

If the voltage and current are associated with a device that is entirely inductive, then

$$p(t) = v(t)i(t) = Li(t)\frac{di(t)}{dt} = \frac{1}{L}v(t)\int_{-\infty}^{t} v(t')\,dt' \qquad [3]$$

where we will arbitrarily assume that the voltage is zero at $t = -\infty$. In the case of a capacitor,

$$p(t) = v(t)i(t) = Cv(t)\frac{dv(t)}{dt} = \frac{1}{C}i(t)\int_{-\infty}^{t} i(t')\,dt' \qquad [4]$$

where a similar assumption about the current is made.

For example, consider the series RL circuit as shown in Fig. 11.1, excited by a step-voltage source. The familiar current response is

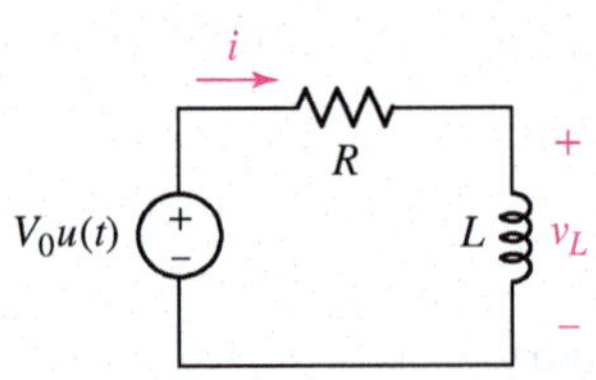

■ **FIGURE 11.1** The instantaneous power that is delivered to R is $p_R(t) = i^2(t)R = (V_0^2/R)(1 - e^{-Rt/L})^2u(t)$.

$$i(t) = \frac{V_0}{R}(1 - e^{-Rt/L})u(t)$$

and thus the total power delivered by the source or absorbed by the passive network is

$$p(t) = v(t)i(t) = \frac{V_0^2}{R}(1 - e^{-Rt/L})u(t)$$

since $v = V_0$.

The power delivered to the resistor is

$$p_R(t) = i^2(t)R = \frac{V_0^2}{R}(1 - e^{-Rt/L})^2u(t)$$

(1) Earlier, we agreed that lowercase variables in italics were understood to be functions of time, and we have carried on in this spirit up to now. However, in order to emphasize the fact that these quantities must be evaluated at a specific instant in time, we will explicitly denote the time dependence throughout this chapter.

In order to determine the power absorbed by the inductor, we first obtain the inductor voltage:

$$v_L(t) = L\frac{di(t)}{dt}$$
$$= V_0 e^{-Rt/L}u(t) + \frac{LV_0}{R}(1 - e^{-Rt/L})\frac{du(t)}{dt}$$
$$= V_0 e^{-Rt/L}u(t)$$

since $du(t)/dt$ is zero for $t > 0$ and $(1 - e^{-Rt/L})$ is zero at $t = 0$. The power absorbed by the inductor is therefore

$$p_L(t) = v_L(t)i(t) = \frac{V_0^2}{R}e^{-Rt/L}(1 - e^{-Rt/L})u(t)$$

Only a few algebraic manipulations are required to show that

$$p(t) = p_R(t) + p_L(t)$$

which serves to check the accuracy of our work; the results are sketched in Fig. 11.2.

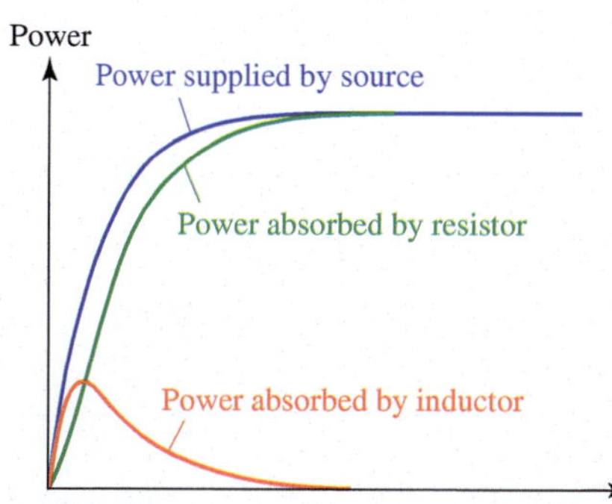

FIGURE 11.2 Sketch of $p(t)$, $p_R(t)$, and $p_L(t)$. As the transient dies out, the circuit returns to steady-state operation. Since the only source remaining in the circuit is dc, the inductor eventually acts as a short circuit, absorbing zero power.

Power Due to Sinusoidal Excitation

Let us change the voltage source in the circuit of Fig. 11.1 to the sinusoidal source $V_m \cos \omega t$. The familiar time-domain steady-state response is

$$i(t) = I_m \cos(\omega t + \phi)$$

where

$$I_m = \frac{V_m}{\sqrt{R^2 + \omega^2 L^2}} \quad \text{and} \quad \phi = -\tan^{-1}\frac{\omega L}{R}$$

The instantaneous power delivered to the entire circuit in the sinusoidal steady state is, therefore,

$$p(t) = v(t)i(t) = V_m I_m \cos(\omega t + \phi)\cos \omega t$$

which we will find convenient to rewrite in a form obtained by using the trigonometric identity for the product of two cosine functions. Thus,

$$p(t) = \frac{V_m I_m}{2}[\cos(2\omega t + \phi) + \cos \phi]$$
$$= \underbrace{\frac{V_m I_m}{2}\cos \phi}_{\text{constant}} + \underbrace{\frac{V_m I_m}{2}\cos(2\omega t + \phi)}_{\text{periodic with frequency } 2\omega}$$

The last equation possesses several characteristics that are true in general for circuits in the sinusoidal steady state. One term, the first, is not a function of time; and a second term is included which has a cyclic variation at *twice* the applied frequency. Since this term is a cosine wave, and since sine waves and cosine waves have average values which are zero (when averaged over an integral number of periods), this example suggests that the *average* power is $\frac{1}{2}V_m I_m \cos \phi$; as we will see shortly, this is indeed the case.

EXAMPLE 11.1

A voltage source, $40 + 60u(t)$V, a 5 μF capacitor, and a 200 Ω resistor form a series circuit. Find the power being absorbed by the capacitor and by the resistor at $t = 1.2$ ms.

At $t = 0^-$, no current is flowing and so 40 V appears across the capacitor. At $t = 0^+$, the voltage across the capacitor–resistor series combination jumps to 100 V. Since v_C cannot change in zero time, the resistor voltage at $t = 0^+$ is 60 V.

The current flowing through all three elements at $t = 0^+$ is therefore 60/200 = 300 mA and for $t > 0$ is given by

$$i(t) = 300e^{-t/\tau} \text{ mA}$$

where $\tau = RC = 1$ ms. Thus, the current flowing at $t = 1.2$ ms is 90.36 mA, and the power being absorbed by the resistor *at that instant* is simply

$$i^2(t)R = 1.633 \text{ W}$$

The instantaneous power absorbed by the capacitor is $i(t)v_C(t)$. Recognizing that the total voltage across both elements for $t > 0$ will always be 100 V, and that the resistor voltage is given by $60e^{-t/\tau}$,

$$v_C(t) = 100 - 60e^{-t/\tau} \text{ V}$$

and we find that $v_C(1.2 \text{ ms}) = 100 - 60e^{-1.2} = 81.93$ V. Thus, the power being absorbed by the capacitor at $t = 1.2$ ms is (90.36 mA)(81.93 V) = 7.403 W.

PRACTICE

11.1 A current source of 12 cos 2000t A, a 200 Ω resistor, and a 0.2 H inductor are in parallel. Assume steady-state conditions exist. At $t = 1$ ms, determine the power being absorbed by the (*a*) resistor; (*b*) inductor; (*c*) sinusoidal source.

Ans: 13.98 kW; −5.63 kW; −8.35 kW.

11.2 AVERAGE POWER

When we speak of an average value for the instantaneous power, the time interval over which the averaging process takes place must be clearly defined. Let us first select a general interval of time from t_1 to t_2. We may then obtain the average value by integrating $p(t)$ from t_1 to t_2 and dividing the result by the time interval $t_2 - t_1$. Thus,

$$P = \frac{1}{t_2 - t_1}\int_{t_1}^{t_2} p(t)\, dt \qquad [5]$$

The average value is denoted by the capital letter P, since it is not a function of time, and it usually appears without any specific subscripts that identify it as an average value. Although P is not a function of time, it *is* a function of t_1 and t_2, the two instants of time which define the interval of integration. This dependence of P on a specific time interval may be expressed in a simpler manner if $p(t)$ is a periodic function. We consider this important case first.

Average Power for Periodic Waveforms

Let us assume that our forcing function and the circuit responses are all periodic; a steady-state condition has been reached, although not necessarily the sinusoidal steady state. We may define a *periodic* function $f(t)$ mathematically by requiring that

$$f(t) = f(t + T) \qquad [6]$$

where T is the period. We now show that the average value of the instantaneous power as expressed by Eq. [5] may be computed over an interval of one period having an arbitrary beginning.

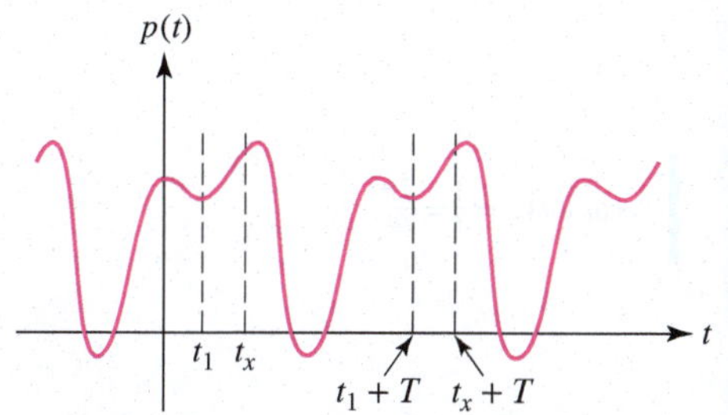

FIGURE 11.3 The average value P of a periodic function $p(t)$ is the same over any period T.

A general periodic waveform is shown in Fig. 11.3 and identified as $p(t)$. We first compute the average power by integrating from t_1 to a time t_2 which is one period later, $t_2 = t_1 + T$:

$$P_1 = \frac{1}{T}\int_{t_1}^{t_1+T} p(t)\, dt$$

and then by integrating from some other time t_x to $t_x + T$:

$$P_x = \frac{1}{T}\int_{t_x}^{t_x+T} p(t)\, dt$$

The equality of P_1 and P_x should be evident from the graphical interpretation of the integrals; the periodic nature of the curve requires the two areas to be equal. Thus, the *average power* may be computed by integrating the instantaneous power over any interval that is one period in length and then dividing by the period:

$$P = \frac{1}{T}\int_{t_x}^{t_x+T} p(t)\, dt \qquad [7]$$

It is important to note that we may also integrate over any integral number of periods, provided that we divide by the same integral number of periods. Expanding this result to the limit approaching an infinite number of periods results in the following integral in terms of the continuous variable τ:

$$P = \lim_{\tau\to\infty} \frac{1}{\tau}\int_{-\tau/2}^{\tau/2} p(t)\, dt \qquad [8]$$

We will find it convenient on several occasions to integrate periodic functions over this "infinite period."

Average Power in the Sinusoidal Steady State

Now let us obtain the general result for the sinusoidal steady state. We assume the general sinusoidal voltage

$$v(t) = V_m \cos(\omega t + \theta)$$

and current

$$i(t) = I_m \cos(\omega t + \phi)$$

associated with the device in question. The instantaneous power is

$$p(t) = V_m I_m \cos(\omega t + \theta)\cos(\omega t + \phi)$$

Again expressing the product of two cosine functions as one-half the sum of the cosine of the difference angle and the cosine of the sum angle,

$$p(t) = \underbrace{\frac{1}{2} V_m I_m \cos(\theta - \phi)}_{\text{constant}} + \underbrace{\frac{1}{2} V_m I_m \cos(2\omega t + \theta + \phi)}_{\text{periodic with frequency } 2\omega} \quad [9]$$

Recall that $T = \frac{1}{f} = \frac{2\pi}{\omega}$.

we may save ourselves some integration by an inspection of the result. The first term is a constant, independent of t. The remaining term is a cosine function; $p(t)$ is therefore periodic, and its period is $\frac{1}{2}T$. Note that the period T is associated with the given current and voltage, and not with the power; the power function has a period $\frac{1}{2}T$. However, we may integrate over an interval of T to determine the average value if we wish; it is necessary only that we also divide by T. Our familiarity with cosine and sine waves, however, shows that the average value of either over a period is zero. There is thus no need to integrate Eq. [9] formally; by inspection, the average value of the second term is zero over a period T (or $\frac{1}{2}T$), and the average value of the first term, a constant, must be that constant itself. Thus,

$$P = \frac{1}{2} V_m I_m \cos(\theta - \phi) \quad [10]$$

This important result, introduced in the previous section for a specific circuit, is therefore quite general for the sinusoidal steady state. The average power is one-half the product of the crest amplitude of the voltage, the crest amplitude of the current, and the cosine of the phase-angle difference between the current and the voltage; the sign of the difference is immaterial. The average power may also be written in phasor notation as

$$P = \frac{1}{2} Re\{\mathbf{V}\mathbf{I}^*\}$$

The notation $\mathbf{I}^*$ denotes the ***complex conjugate*** of the complex number $\mathbf{I}$. It is formed by replacing all "j"s with "$-j$"s. See Appendix 5 for more details.

Two special cases are worth isolating for consideration: the average power delivered to an ideal resistor and that to an ideal reactor (any combination of only capacitors and inductors).

EXAMPLE 11.2

Given the time-domain voltage $v = 4\cos(\pi t/6)$ V, find both the average power and an expression for the instantaneous power that result when the corresponding phasor voltage $\mathbf{V} = 4\underline{/0°}$ V is applied across an impedance $\mathbf{Z} = 2\underline{/60°}\ \Omega$.

The phasor current is $\mathbf{V}/\mathbf{Z} = 2\underline{/-60°}$ A, and so the average power is

$$P = \frac{1}{2}(4)(2)\cos 60° = 2 \text{ W}$$

We can write the time-domain voltage,

$$v(t) = 4\cos\frac{\pi t}{6} \text{ V}$$

and the time-domain current,

$$i(t) = 2\cos\left(\frac{\pi t}{6} - 60^\circ\right)\ \text{A}$$

The instantaneous power, therefore, is given by their product:

$$p(t) = 8\cos\frac{\pi t}{6}\cos\left(\frac{\pi t}{6} - 60^\circ\right)$$

$$= 2 + 4\cos\left(\frac{\pi t}{3} - 60^\circ\right)\ \text{W}$$

All three quantities are sketched on the same time axis in Fig. 11.4. Both the 2 W average value of the power and its period of 6 s, one-half the period of either the current or the voltage, are evident. The zero value of the instantaneous power at each instant when either the voltage or current is zero is also apparent.

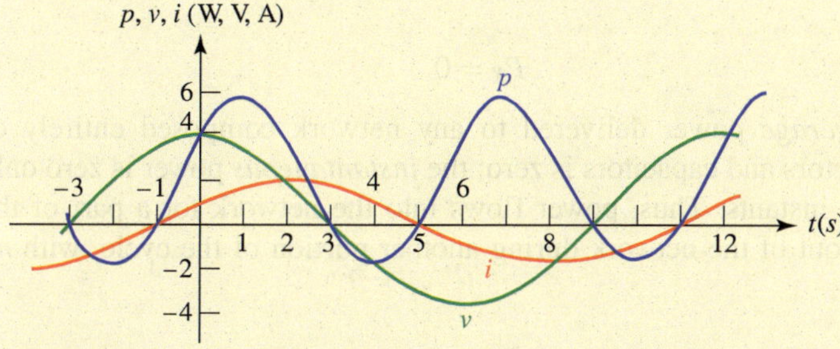

■ **FIGURE 11.4** Curves of $v(t)$, $i(t)$, and $p(t)$ are plotted as functions of time for a simple circuit in which the phasor voltage $\mathbf{V} = 4\underline{/0^\circ}$ V is applied to the impedance $\mathbf{Z} = 2\underline{/60^\circ}\ \Omega$ at $\omega = \pi/6$ rad/s.

PRACTICE

11.2 Given the phasor voltage $\mathbf{V} = 115\sqrt{2}\underline{/45^\circ}$ V across an impedance $\mathbf{Z} = 16.26\underline{/19.3^\circ}\ \Omega$, obtain an expression for the instantaneous power, and compute the average power if $\omega = 50$ rad/s.

Ans: $767.5 + 813.2\cos(100t + 70.7^\circ)$W; 767.5 W.

Average Power Absorbed by an Ideal Resistor

The phase-angle difference between the current through and the voltage across a pure resistor is zero. Thus,

$$P_R = \frac{1}{2}V_m I_m \cos 0 = \frac{1}{2}V_m I_m$$

or

$$P_R = \frac{1}{2}I_m^2 R \qquad [11]$$

or

$$P_R = \frac{V_m^2}{2R} \qquad [12]$$

The same result is obtained using the phasor form, since the value **VI*** will be a real number, and thus $P = \frac{1}{2}Re\{\mathbf{VI}^*\} = \frac{1}{2}Re\{(I_m R)I_m\underline{/0^\circ}\} = \frac{1}{2}I_m^2 R$.

Keep in mind that we are computing the ***average*** power delivered to a resistor by a sinusoidal source; take care not to confuse this quantity with the ***instantaneous*** power, which has a similar form.

The last two formulas, enabling us to determine the average power delivered to a pure resistance from a knowledge of either the sinusoidal current or voltage, are simple and important. Unfortunately, *they are often misused.* The most common error is made in trying to apply them in cases where the voltage included in Eq. [12] is *not the voltage across the resistor.* If care is taken to use the current through the resistor in Eq. [11] and the voltage across the resistor in Eq. [12], satisfactory operation is guaranteed. Also, do not forget the factor of $\frac{1}{2}$!

Average Power Absorbed by Purely Reactive Elements

The average power delivered to any device which is purely reactive (i.e., contains no resistors) must be zero. This is a direct result of the 90° phase difference which must exist between current and voltage; hence, $\cos(\theta - \phi) = \cos \pm 90° = 0$ (or in phasor notation, $\mathbf{VI}^*$ is an imaginary number)

$$P_X = 0$$

The *average* power delivered to any network composed entirely of ideal inductors and capacitors is zero; the *instantaneous* power is zero only at specific instants. Thus, power flows into the network for a part of the cycle and out of the network during another portion of the cycle, with *no* power lost.

EXAMPLE **11.3**

Find the average power being delivered to an impedance $\mathbf{Z}_L = 8 - j11\ \Omega$ by a current $\mathbf{I} = 5\underline{/20°}$ A.

We may find the solution quite rapidly by using Eq. [11]. Only the 8 Ω resistance enters the average-power calculation, since the $j11\ \Omega$ component will not absorb any *average* power. Thus,

$$P = \frac{1}{2}(5)^2 8 = 100 \text{ W}$$

We can also solve the problem directly using phasors, requiring a bit more computation:

$$\mathbf{I} = 5\underline{/20°} = 4.6985 + j1.7101 \text{ A}$$

$$\mathbf{V} = \mathbf{IZ}_L = (4.6985 + j1.7101)(8 - j11) = 56.3988 - j38.0023 \text{ V}$$

$$P = \frac{1}{2}\text{Re}\{\mathbf{VI}^*\} = \frac{1}{2}\text{Re}\{(56.3988 - j38.0023)(4.6985 - 1.7101)\}$$

$$P = 100 \text{ W}$$

PRACTICE

11.3 Calculate the average power delivered to the impedance $6\underline{/25°}\ \Omega$ by the current $\mathbf{I} = 2 + j5$ A.

Ans: 78.85 W.

EXAMPLE **11.4**

Find the average power absorbed by each of the three passive elements in Fig. 11.5, as well as the average power supplied by each source.

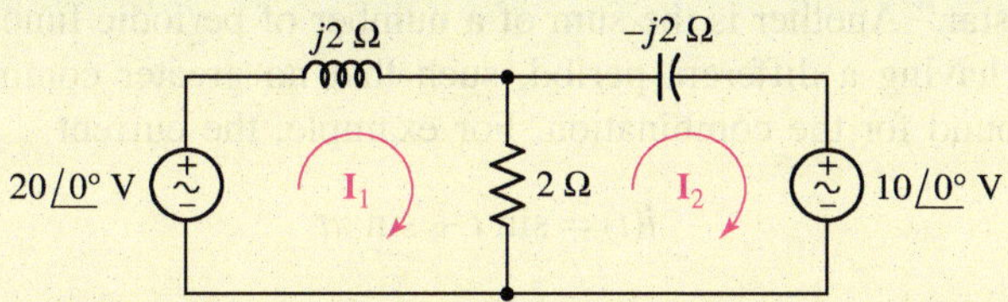

FIGURE 11.5 The average power delivered to each reactive element is zero in the sinusoidal steady state.

Without even analyzing the circuit, we already know that the average power absorbed by the two reactive elements is zero.

The values of $\mathbf{I}_1$ and $\mathbf{I}_2$ are found by any of several methods, such as mesh analysis, nodal analysis, or superposition. They are

$$\mathbf{I}_1 = 5 - j10 = 11.18\underline{/-63.43^\circ} \text{ A}$$

$$\mathbf{I}_2 = 5 - j5 = 7.071\underline{/-45^\circ} \text{ A}$$

The downward current through the 2 Ω resistor is

$$\mathbf{I}_1 - \mathbf{I}_2 = -j5 = 5\underline{/-90^\circ} \text{ A}$$

so that $I_m = 5$ A, and the average power absorbed by the resistor is found most easily by Eq. [11]:

$$P_R = \frac{1}{2}I_m^2 R = \frac{1}{2}(5)^2 2 = 25 \text{ W}$$

This result may be checked by using Eq. [10] or Eq. [12]. We next turn to the left source. The voltage $20\underline{/0^\circ}$ V and associated current $\mathbf{I}_1 = 11.18\underline{/-63.43^\circ}$ A satisfy the *active* sign convention, and thus the power *delivered* by this source is

$$P_{\text{left}} = \frac{1}{2}(20)(11.18)\cos[0^\circ - (-63.43^\circ)] = 50 \text{ W}$$

In a similar manner, we find that the right source is actually *absorbing* power according to the *passive* sign convention,

$$P_{\text{right}} = \frac{1}{2}(10)(7.071)\cos(0^\circ + 45^\circ) = 25 \text{ W}$$

Since 50 = 25 + 25, the power relations check.

PRACTICE

11.4 For the circuit of Fig. 11.6, compute the average power delivered to each of the passive elements. Verify your answer by computing the power delivered by each source.

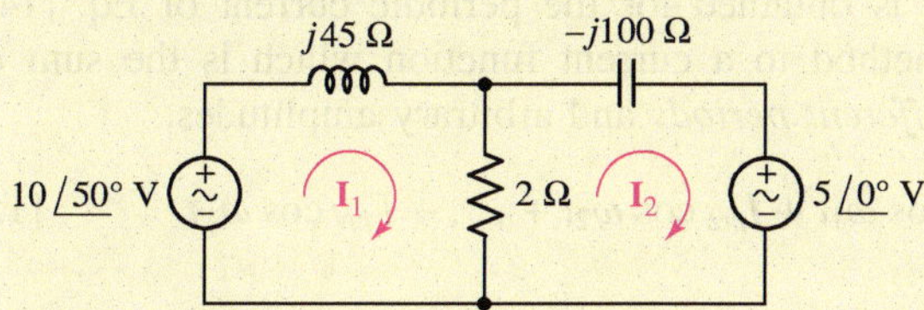

FIGURE 11.6

Ans: 0, 37.6 mW, 0, 42.0 mW, −4.4 mW

Average Power for Nonperiodic Functions

We should pay some attention to *nonperiodic* functions. One practical example of a nonperiodic power function for which an average power value is desired is the power output of a radio telescope directed toward a "radio star." Another is the sum of a number of periodic functions, each function having a different period, such that no greater common period can be found for the combination. For example, the current

$$i(t) = \sin t + \sin \pi t \tag{13}$$

is nonperiodic because the ratio of the periods of the two sine waves is an irrational number. At $t = 0$, both terms are zero and increasing. But the first term is zero and increasing only when $t = 2\pi n$, where n is an integer, and thus periodicity demands that πt or $\pi(2\pi n)$ must equal $2\pi m$, where m is also an integer. No solution (integral values for both m and n) for this equation is possible. It may be illuminating to compare the nonperiodic expression in Eq. [13] with the *periodic* function

$$i(t) = \sin t + \sin 3.14t \tag{14}$$

where 3.14 is an exact decimal expression and is *not* to be interpreted as 3.141592.... With a little effort,[2] it can be shown that the period of this current wave is 100π seconds.

The average value of the power delivered to a 1 Ω resistor by either a periodic current such as Eq. [14] or a nonperiodic current such as Eq. [13] may be found by integrating over an infinite interval. Much of the actual integration can be avoided because of our thorough knowledge of the average values of simple functions. We therefore obtain the average power delivered by the current in Eq. [13] by applying Eq. [8]:

$$P = \lim_{\tau \to \infty} \frac{1}{\tau} \int_{-\tau/2}^{\tau/2} (\sin^2 t + \sin^2 \pi t + 2 \sin t \sin \pi t)\, dt$$

We now consider P as the sum of three average values. The average value of $\sin^2 t$ over an infinite interval is found by replacing $\sin^2 t$ with $(\frac{1}{2} - \frac{1}{2}\cos 2t)$; the average is simply $\frac{1}{2}$. Similarly, the average value of $\sin^2 \pi t$ is also $\frac{1}{2}$. And the last term can be expressed as the sum of two cosine functions, each of which must certainly have an average value of zero. Thus,

$$P = \frac{1}{2} + \frac{1}{2} = 1 \text{ W}$$

An identical result is obtained for the periodic current of Eq. [14]. Applying this same method to a current function which is the sum of several sinusoids of *different periods* and arbitrary amplitudes,

$$i(t) = I_{m1} \cos \omega_1 t + I_{m2} \cos \omega_2 t + \ldots + I_{mN} \cos \omega_N t \tag{15}$$

(2) $T_1 = 2\pi$ and $T_2 = 2\pi/3.14$. Therefore, we seek integral values of m and n such that $2\pi n = 2\pi m/3.14$, or $3.14n = m$, or $(314/100)n = m$ or $157n = 50m$. Thus, the smallest integral values for n and m are $n = 50$ and $m = 157$. The period is therefore $T = 2\pi n = 100\pi$, or $T = 2\pi(157/3.14) = 100\pi$ s.

we find the average power delivered to a resistance R,

$$P = \frac{1}{2}(I_{m1}^2 + I_{m2}^2 + \cdots + I_{mN}^2)R \qquad [16]$$

The result is unchanged if an arbitrary phase angle is assigned to each component of the current. This important result is surprisingly simple when we think of the steps required for its derivation: squaring the current function, integrating, and taking the limit. The result is also just plain surprising, because it shows that, *in this special case of a current such as Eq. [15], where each term has a unique frequency, superposition is applicable to power*. Superposition is *not* applicable for a current which is the sum of two direct currents, nor is it applicable for a current which is the sum of two sinusoids of the same frequency.

EXAMPLE **11.5**

Find the average power delivered to a 4 Ω resistor by the current $i_1 = 2 \cos 10t - 3 \cos 20t$ A.

Since the two cosine terms are at *different* frequencies, the two average-power values may be calculated separately and added. Thus, this current delivers $\frac{1}{2}(2^2)4 + \frac{1}{2}(3^2)4 = 8 + 18 = 26$ W to a 4 Ω resistor.

EXAMPLE **11.6**

Find the average power delivered to a 4 Ω resistor by the current $i_2 = 2 \cos 10t - 3 \cos 10t$ A.

Here, the two components of the current are at the *same* frequency, and they must therefore be combined into a single sinusoid at that frequency. Thus, $i_2 = 2 \cos 10t - 3 \cos 10t = -\cos 10t$ delivers only $\frac{1}{2}(1^2)4 = 2$ W of average power to a 4 Ω resistor.

PRACTICE

11.5 A voltage source v_s is connected across a 4 Ω resistor. Calculate the average power absorbed by the resistor if v_s equals (*a*) $8 \sin 200t$ V; (*b*) $8 \sin 200t - 6 \cos(200t - 45°)$ V; (*c*) $8 \sin 200t - 4 \sin 100t$ V; (*d*) $8 \sin 200t - 6 \cos(200t - 45°) - 5 \sin 100t + 4$ V.

Ans: 8.00 W; 4.01 W; 10.00 W; 11.14 W.

11.3 MAXIMUM POWER TRANSFER

We previously considered the maximum power transfer theorem as it applied to resistive loads and resistive source impedances. For a Thévenin source $\mathbf{V}_{TH}$ and impedance $\mathbf{Z}_{TH} = R_{TH} + jX_{TH}$ connected to a load $\mathbf{Z}_L = R_L + jX_L$, it may be shown that the average power delivered to the load is a maximum when $R_L = R_{TH}$ and $X_L = -X_{TH}$; that is, when $\mathbf{Z}_L = \mathbf{Z}_{TH}^*$.

This result is often dignified by calling it the *maximum power transfer theorem for the sinusoidal steady state:*

> An independent voltage source in *series* with $\mathbf{Z}_{TH}$ or an independent current source in *parallel* with $\mathbf{Z}_{TH}$ delivers a maximum average power to a load $\mathbf{Z}_L$ when $\mathbf{Z}_L = \mathbf{Z}^*_{TH}$.

Average Power Delivered to Load

The details of the proof for maximum power transfer are left to the reader, but the basic approach can be understood by considering average power delivered to the load of the simple loop circuit of Fig. 11.7. The Thévenin equivalent impedance $\mathbf{Z}_{TH}$ may be written as the sum of two components, $R_{TH} + jX_{TH}$, and in a similar fashion the load impedance $\mathbf{Z}_L$ may be written as $R_L + jX_L$. The current flowing through the loop is

$$\mathbf{I}_L = \frac{\mathbf{V}_{TH}}{\mathbf{Z}_{TH} + \mathbf{Z}_L}$$

$$= \frac{\mathbf{V}_{TH}}{R_{TH} + jX_{TH} + R_L + jX_L} = \frac{\mathbf{V}_{TH}}{R_{TH} + R_L + j(X_{TH} + X_L)}$$

and

$$\mathbf{V}_L = \mathbf{V}_{TH}\frac{\mathbf{Z}_L}{\mathbf{Z}_{TH} + \mathbf{Z}_L}$$

$$= \mathbf{V}_{TH}\frac{R_L + jX_L}{R_{TH} + jX_{TH} + R_L + jX_L} = \mathbf{V}_{TH}\frac{R_L + jX_L}{R_{TH} + R_L + j(X_{TH} + X_L)}$$

The magnitude of $\mathbf{I}_L$ is

$$\frac{|\mathbf{V}_{TH}|}{\sqrt{(R_{TH} + R_L)^2 + (X_{TH} + X_L)^2}}$$

and its phase angle is

$$\underline{/\mathbf{V}_{TH}} - \tan^{-1}\left(\frac{X_{TH} + X_L}{R_{TH} + R_L}\right)$$

Similarly, the magnitude of $\mathbf{V}_L$ is

$$\frac{|\mathbf{V}_{TH}|\sqrt{R_L^2 + X_L^2}}{\sqrt{(R_{TH} + R_L)^2 + (X_{TH} + X_L)^2}}$$

and its phase angle is

$$\underline{/\mathbf{V}_{TH}} + \tan^{-1}\left(\frac{X_L}{R_L}\right) - \tan^{-1}\left(\frac{X_{TH} + X_L}{R_{TH} + R_L}\right)$$

Referring to Eq. [10], then, we find an expression for the average power P delivered to the load impedance $\mathbf{Z}_L$:

$$P = \frac{\frac{1}{2}|\mathbf{V}_{TH}|^2\sqrt{R_L^2 + X_L^2}}{(R_{TH} + R_L)^2 + (X_{TH} + X_L)^2}\cos\left(\tan^{-1}\left(\frac{X_L}{R_L}\right)\right) \quad [17]$$

FIGURE 11.7 A simple loop circuit used to illustrate the derivation of the maximum power transfer theorem as it applies to circuits operating in the sinusoidal steady state.

In order to prove that maximum average power is indeed delivered to the load when $\mathbf{Z}_L = \mathbf{Z}_{TH}^*$, we must carry out two separate steps. First, the derivative of Eq. [17] with respect to R_L must be set to zero. Second, the derivative of Eq. [17] with respect to X_L must be set to zero. The remaining details are left as an exercise for the avid reader.

EXAMPLE **11.7**

A particular circuit is composed of the series combination of a sinusoidal voltage source 3 cos(100*t* − 3°) V, a 500 Ω resistor, a 30 mH inductor, and an unknown impedance. If we are assured that the voltage source is delivering maximum average power to the unknown impedance, what is its value?

The phasor representation of the circuit is sketched in Fig. 11.8. The circuit is easily seen as an unknown impedance $\mathbf{Z}_?$ in series with a Thévenin equivalent consisting of the $3\angle{-3°}$ V source and a Thévenin impedance $500 + j3\ \Omega$.

Since the circuit of Fig. 11.8 is already in the form required to employ the maximum average power transfer theorem, we know that maximum average power will be transferred to an impedance equal to the complex conjugate of $\mathbf{Z}_{TH}$, or

$$\mathbf{Z}_? = \mathbf{Z}_{TH}^* = 500 - j3\ \Omega$$

This impedance can be constructed in several ways, the simplest being a 500 Ω resistor in series with a capacitor having impedance $-j3\ \Omega$. Since the operating frequency of the circuit is 100 rad/s, this corresponds to a capacitance of 3.33 mF.

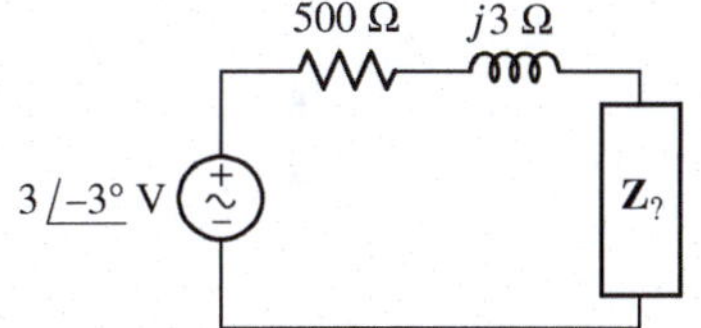

FIGURE 11.8 The phasor representation of a simple series circuit composed of a sinusoidal voltage source, a resistor, an inductor, and an unknown impedance.

PRACTICE

11.6 If the 30 mH inductor of Example 11.7 is replaced with a 10 μF capacitor, what is the value of the inductive component of the unknown impedance $\mathbf{Z}_?$ if it is known that $\mathbf{Z}_?$ is absorbing maximum power?

Ans: 10 H.

Impedance Matching

We know that maximum power transfer occurs when the load resistance is equal to the series resistance of the circuit delivering power; or similarly, the impedance of the load is the complex conjugate of the impedance of the circuit delivering power. However, the power delivery circuit typically has a different impedance than the load, and in fact this may be desirable. For example, the output resistance of a power amplifier should be less than the resistance of a speaker, due to the behavior of the speaker as a load, the mechanical response, etc. How can we maximize power delivery in such cases? For sinusoidal inputs, we can introduce

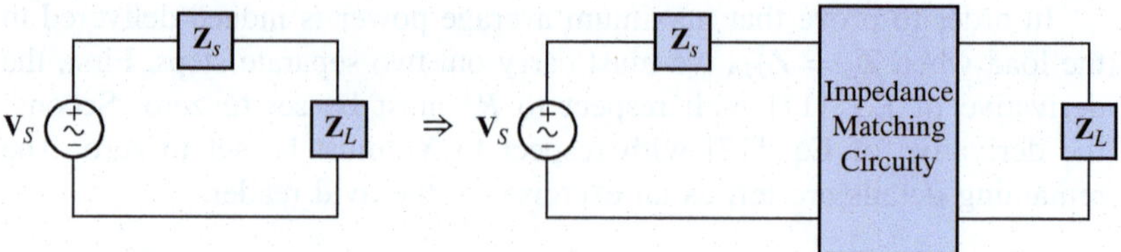

FIGURE 11.9 Illustration of impedance matching circuitry for maximum power delivery to a load.

intermediate ***impedance matching*** circuitry such that the impedance between source and load are complex conjugates, as illustrated in Fig. 11.9. Impedance matching is particularly important for applications dealing with very weak signals or where power is critical, where any power loss is a big problem. For example, impedance matching on the antenna on your radio or mobile phone will be critical in determining your range and battery life.

Consider a power amplifier at a frequency of 31.83 MHz with output resistance of 100 Ω acting as a source to power a 50 Ω antenna. Adding a 50 pF capacitor in parallel results in a load impedance of

$$\mathbf{Z}_C = \frac{1}{j\omega C} = \frac{1}{j(2\pi \times 31.83 \times 10^6)(50 \times 10^{-12})} = -j100$$

The impedance observed by the load is therefore

$$\mathbf{Z}_{eq} = \frac{R_S \mathbf{Z}_C}{R_S + \mathbf{Z}_C} = \frac{(100)(-j100)}{100 - j100} = 50 - j50$$

We have now transformed the source to have an impedance with a real part of 50 Ω, by adding a capacitor in parallel! Note that while the equivalent impedance has changed to $50 - j50$, the Thévenin equivalent source is given by

$$\mathbf{V}_{TH} = \frac{(-j100)}{100 - j100}\mathbf{V}_S = \left(\frac{1}{2} - j\frac{1}{2}\right)\mathbf{V}_S = \frac{\mathbf{V}_S}{\sqrt{2}}\ \underline{/-45^\circ}$$

To complete the impedance match, we now need to incorporate an element to cancel the imaginary part of the impedance that the load "sees." An inductor with impedance $j\omega L$ can be added in series with a value of $+j50$, requiring $L = 50/(2\pi \times 31.83 \times 10^6) = 250$ nH. We have now completed an impedance matching network that will deliver maximum power to the load. The addition of the parallel capacitor and series inductor is known as an *L matching network*. As matching requirements may vary, there are many other matching network topographies, including the *Π network* and *T network,* which are left for more advanced study of this topic. While the preceding example examines the case of a purely resistive impedance for both source and load, the impedance matching concept is also applicable for complex impedance (along with the expected feature of more complicated algebraic computation!).

EXAMPLE 11.8

Calculate and compare the average power delivered to the load for the three circuit configurations in Fig. 11.10 subject to a source with amplitude of $V_m = 5$ V.

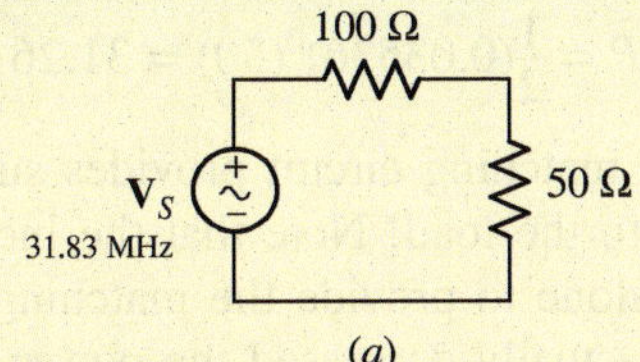

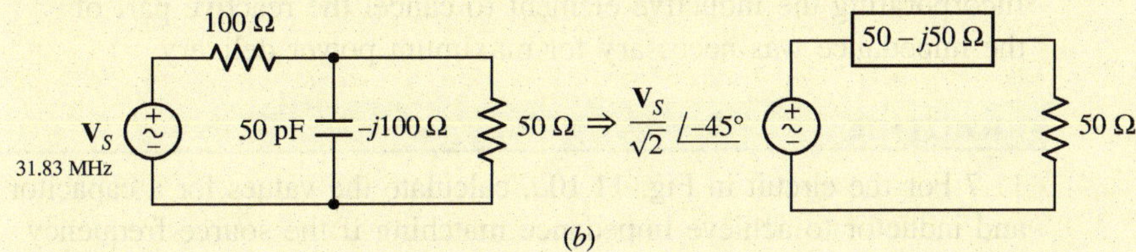

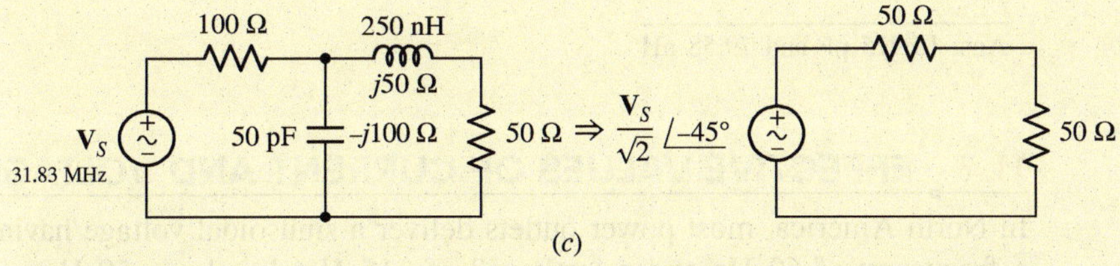

FIGURE 11.10 Impedance matching example for a circuit with sinusoidal input and source and load resistances that are not equal. (*a*) The original circuit, (*b*) addition of a capacitor in parallel, and (*c*) capacitor in parallel and inductor in series to provide impedance matching condition for the defined frequency of the sinusoidal source.

For the first case in (*a*) with no matching circuitry, the current amplitude is $I_m = 5/(100 + 50) = 1/30$ A. The power dissipated by the load is

$$P = \frac{1}{2} I_m^2 R = \frac{1}{2}\left(\frac{1}{30}\right)^2 (50) = 27.78 \text{ mW}$$

For the circuit in (*b*), the current in the load will vary due to the addition of the capacitor in parallel. From the equivalent impedance shown on the right side of (*b*),

$$\mathbf{I}_L = \frac{5\left(\frac{1}{2} - j\frac{1}{2}\right)}{100 - j50} = 0.03 - j0.01 = 0.03162\underline{/-18.435°} \text{ A}$$

resulting in an average power in the load of

$$P = \frac{1}{2}(0.03162)^2(50) = 25.00 \text{ mW}$$

(Continued on next page)

Repeating for the complete matching circuit in (c), the current and average power are

$$\mathbf{I}_L = \frac{5\left(\frac{1}{2} - j\frac{1}{2}\right)}{100} = 0.025 - j0.025 = 0.03536\underline{/-45^\circ} \text{ A}$$

$$P = \frac{1}{2}(0.03536)^2(50) = 31.26 \text{ mW}$$

We see that the matching circuit provides substantial improvement in power delivery to the load! Note that the incorporation of the parallel capacitor alone to provide the matching real part of the source impedance actually decreased the power delivered to the load. Incorporating the inductive element to cancel the reactive part of the impedance was necessary for maximum power delivery.

PRACTICE

11.7 For the circuit in Fig. 11.10a, calculate the values for a capacitor and inductor to achieve impedance matching if the source frequency is changed to 100 MHz.

Ans: 15.915 pF and 79.58 nH

11.4 EFFECTIVE VALUES OF CURRENT AND VOLTAGE

The value is also not standardized; the quoted value may be 110 V, 115, or even 120 V. Further, it varies over time depending on different loads attached to the power system, generators being added or removed to make the most effective use of resources, etc.

In North America, most power outlets deliver a sinusoidal voltage having a frequency of 60 Hz and a "voltage" of 115 V (elsewhere, 50 Hz and 240 V are typically encountered). But what is meant by "115 volts"? This is certainly not the instantaneous value of the voltage, for the voltage is not a constant. The value of 115 V is also not the amplitude which we have been symbolizing as V_m; if we displayed the voltage waveform on a calibrated oscilloscope, we would find that the amplitude of this voltage at one of our ac outlets is more like $115\sqrt{2}$, or 162.6, volts. We also cannot fit the concept of an average value to the 115 V, because the average value of the sine wave is zero. We might come a little closer by trying the magnitude of the average over a positive or negative half cycle; by using a rectifier-type voltmeter at the outlet, we should measure 103.5 V. As it turns out, however, 115 V is the ***effective value*** of this sinusoidal voltage; it is a measure of the effectiveness of a voltage source in delivering power to a resistive load.

Effective Value of a Periodic Waveform

Let us arbitrarily define effective value in terms of a current waveform, although a voltage could equally well be selected. The *effective value* of any periodic current is equal to the value of the direct current which, flowing through an R ohm resistor, delivers the same average power to the resistor as does the periodic current.

In other words, we allow the given periodic current to flow through the resistor, determine the instantaneous power $i^2 R$, and then find the average value of $i^2 R$ over a period; this is the average power. We then cause a direct current to flow through this resistor and adjust the value of the direct current until the same value of average power is obtained. The resulting magnitude of the direct current is equal to the effective value of the given periodic current. These ideas are illustrated in Fig. 11.11.

The general mathematical expression for the effective value of $i(t)$ is now easily obtained. The average power delivered to the resistor by the periodic current $i(t)$ is

$$P = \frac{1}{T}\int_0^T i^2 R\, dt = \frac{R}{T}\int_0^T i^2\, dt$$

where the period of $i(t)$ is T. The power delivered by the direct current is

$$P = I_{\text{eff}}^2 R$$

Equating the power expressions and solving for I_{eff}, we get

$$I_{\text{eff}} = \sqrt{\frac{1}{T}\int_0^T i^2\, dt} \qquad [18]$$

The result is independent of the resistance R, as it must be to provide us with a worthwhile concept. A similar expression is obtained for the effective value of a periodic voltage by replacing i and I_{eff} by v and V_{eff}, respectively.

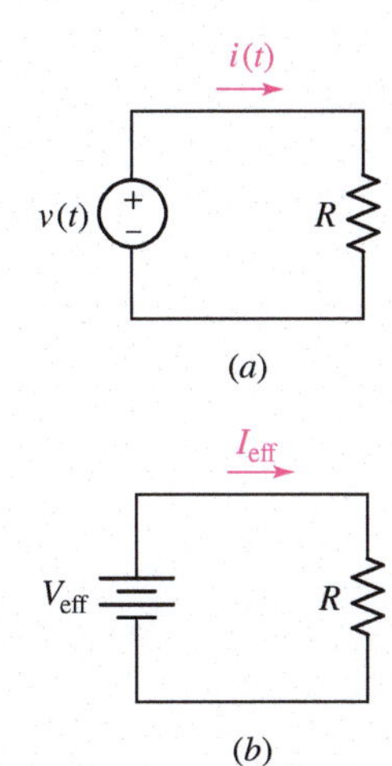

FIGURE 11.11 If the resistor receives the same average power in parts *a* and *b*, then the effective value of $i(t)$ is equal to I_{eff}, and the effective value of $v(t)$ is equal to V_{eff}.

Notice that the effective value is obtained by first squaring the time function, then taking the average value of the squared function over a period, and finally taking the square root of the average of the squared function. In short, the operation involved in finding an effective value is the (square) *root* of the *mean* of the *square;* for this reason, the effective value is often called the ***root-mean-square*** value, or simply the ***rms*** value.

Effective (RMS) Value of a Sinusoidal Waveform

The most important special case is that of the sinusoidal waveform. Let us select the sinusoidal current

$$i(t) = I_m \cos(\omega t + \phi)$$

which has a period

$$T = \frac{2\pi}{\omega}$$

and substitute in Eq. [18] to obtain the effective value

$$\begin{aligned} I_{\text{eff}} &= \sqrt{\frac{1}{T}\int_0^T I_m^2 \cos^2(\omega t + \phi) dt} \\ &= I_m \sqrt{\frac{\omega}{2\pi}\int_0^{2\pi/\omega}\left[\frac{1}{2} + \frac{1}{2}\cos(2\omega t + 2\phi)\right] dt} \\ &= I_m \sqrt{\frac{\omega}{4\pi}[t]_0^{2\pi/\omega}} \\ &= \frac{I_m}{\sqrt{2}} \end{aligned}$$

Thus the effective value of a sinusoidal current is a real quantity which is independent of the phase angle and numerically equal to $1/\sqrt{2} = 0.707$ times the amplitude of the current. A current $\sqrt{2}\cos(\omega t + \phi)$ A, therefore, has an effective value of 1 A and will deliver the **same** average power to any resistor as will a **direct** current of 1 A.

It should be noted carefully that the $\sqrt{2}$ factor that we obtained as the ratio of the amplitude of the periodic current to the effective value is applicable only when the periodic function is *sinusoidal*. For a sawtooth waveform, for example, the effective value is equal to the maximum value divided by $\sqrt{3}$. The factor by which the maximum value must be divided to obtain the effective value depends on the mathematical form of the given periodic function; it may be either rational or irrational, depending on the nature of the function.

Use of RMS Values to Compute Average Power

The use of the effective value also slightly simplifies the expression for the average power delivered by a sinusoidal current or voltage by avoiding the use of the factor $\frac{1}{2}$. For example, the average power delivered to an R ohm resistor by a sinusoidal current is

$$P = \frac{1}{2} I_m^2 R$$

Since $I_{\text{eff}} = I_m/\sqrt{2}$, the average power may be written as

$$P = I_{\text{eff}}^2 R \qquad [19]$$

The other power expressions may also be written in terms of effective values:

$$P = V_{\text{eff}} I_{\text{eff}} \cos(\theta - \phi) \qquad [20]$$

$$P = \frac{V_{\text{eff}}^2}{R} \qquad [21]$$

The fact that the effective value is defined in terms of an equivalent dc quantity provides us with average-power formulas for resistive circuits which are identical with those used in dc analysis.

Although we have succeeded in eliminating the factor $\frac{1}{2}$ from our average-power relationships, we must now take care to determine whether a sinusoidal quantity is expressed in terms of its amplitude or its effective value. In practice, the effective value is usually used in the fields of power transmission or distribution and of rotating machinery; in the areas of electronics and communications, the amplitude is more commonly used. We will assume that the amplitude is specified unless the term "rms" is explicitly used, or we are otherwise instructed.

In the sinusoidal steady state, phasor voltages and currents may be given either as effective values or as amplitudes; the two expressions differ only by a factor of $\sqrt{2}$. The voltage $50\underline{/30^\circ}$ V is expressed in terms of an amplitude; as an rms voltage, we should describe the same voltage as $35.4\underline{/30^\circ}$ V rms.

Effective Value with Multiple-Frequency Circuits

In order to determine the effective value of a periodic or nonperiodic waveform which is composed of the sum of a number of sinusoids of different frequencies, we may use the appropriate average-power relationship of Eq. [16], developed in Sec. 11.2, rewritten in terms of the effective values of the several components:

$$P = (I_{1\text{eff}}^2 + I_{2\text{eff}}^2 + \cdots + I_{N\text{eff}}^2)R \qquad [22]$$

From this we see that the effective value of a current which is composed of any number of sinusoidal currents of *different* frequencies can be expressed as

$$I_{\text{eff}} = \sqrt{I_{1\text{eff}}^2 + I_{2\text{eff}}^2 + \cdots + I_{N\text{eff}}^2} \qquad [23]$$

These results indicate that if a sinusoidal current of 5 A rms at 60 Hz flows through a 2 Ω resistor, an average power of $5^2(2) = 50$ W is absorbed by the resistor; if a second current—perhaps 3 A rms at 120 Hz, for example—is also present, the absorbed power is $3^2(2) + 50 = 68$ W. Using Eq. [23] instead, we find that the effective value of the sum of the 60 and 120 Hz currents is 5.831 A. Thus, $P = 5.831^2(2) = 68$ W as before. However, if the second current is also at 60 Hz, the effective value of the sum of the two 60 Hz currents may have any value between 2 and 8 A. Thus, the absorbed power may have *any* value between 8 W and 128 W, depending on the relative phase of the two current components.

Note that the effective value of a dc quantity K is simply K, not $\frac{K}{\sqrt{2}}$.

PRACTICE

11.8 Calculate the effective value of each of the periodic voltages: (*a*) $6 \cos 25t$; (*b*) $6 \cos 25t + 4 \sin(25t + 30°)$; (*c*) $6 \cos 25t + 5 \cos^2(25t)$; (*d*) $6 \cos 25t + 5 \sin 30t + 4$ V.

Ans: 4.24 V; 6.16 V; 5.23 V; 6.82 V.

COMPUTER-AIDED ANALYSIS

Several useful techniques are available through SPICE for calculation of power quantities. In particular, the built-in functions allow us to plot the instantaneous power and compute the average power. For example, consider the simple voltage divider circuit of Fig. 11.12,

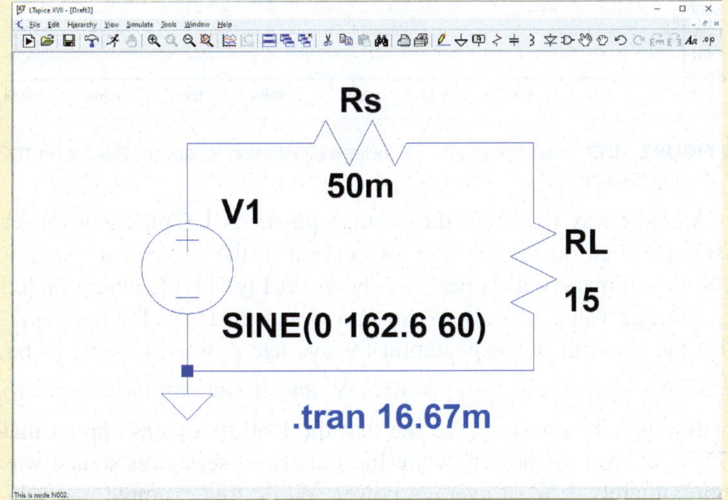

FIGURE 11.12 A simple voltage divider circuit driven by a 115 V rms source operating at 60 Hz.
Analog Devices

(Continued on next page)

which is being driven by a 60 Hz sine wave with an amplitude of $115\sqrt{2}$ V. The resistors represent a typical power supply from the wall plug, an undesired series resistance of 0.05 Ω, and a load such as a heater at 15 Ω. We begin by performing a transient simulation over one period of the voltage waveform, $\frac{1}{60}$ s(16.67 ms).

Similar to plotting current for a circuit element, the instantaneous power can be plotted by clicking on an element in the schematic after running the simulation. To plot instantaneous power in LTspice, press **alt** and click on the element (you will see a thermometer icon appear instead of a current probe when placing the cursor over the circuit element). Power waveforms can also be plotted by directly entering an expression to add to the plot (e.g., **V(N002)*I(RL)**). The waveforms for instantaneous power dissipated in resistor Rs and RL are plotted in Fig. 11.13.

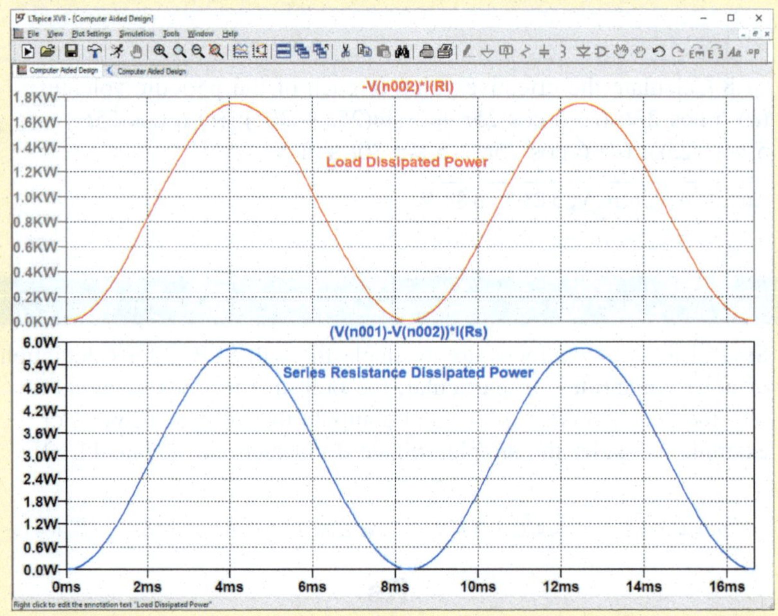

FIGURE 11.13 Instantaneous power associated with resistors Rseries and Rload of Fig. 11.12. Analog Devices

An easy way to obtain the average power in LTspice is to press **Ctrl** and click on the variable expression in the waveform window. A data summary will appear, as shown in Fig. 11.14, which includes the average value (875.41 W for Rload and 2.918 W for Rseries). This agrees with our expectation for average power in Rload to be $\frac{1}{2}\left(162.6\,\frac{15}{15+0.05}\right)\left(162.\frac{6}{15.05}\right) = 875$ W, and it can similarly be verified for Rseries. In summary, we see that the load dissipates approximately 875 W of average power, while the undesired series resistance wastes approximately 3 W of average power. While this example is a relatively simple case of a voltage divider that is easily calculated by hand, more complex circuits can benefit greatly from the use of SPICE for power analysis.

Waveform: -V(N002)*I(RL)

Interval Start:	0s
Interval End:	16.67ms
Average:	875.41W
Integral:	14.593J

FIGURE 11.14 Window indicating average power for the resistor RL.
Analog Devices

11.5 APPARENT POWER AND POWER FACTOR

Historically, the introduction of the concepts of apparent power and power factor can be traced to the electric power industry, where large amounts of electric energy must be transferred from one point to another; the efficiency with which this transfer is effected is related directly to the cost of the electric energy, which is eventually paid by the consumer. Customers who provide loads which result in a relatively poor transmission efficiency pay a greater price for each ***kilowatt hour*** (kWh) of electric energy they receive and use. In a similar way, customers who need a costlier investment in transmission and distribution equipment by the power company will also pay more for each kilowatthour unless the company is benevolent and enjoys losing money.

Let us first define ***apparent power*** and ***power factor*** and then show briefly how these terms are related to practical economic situations. We assume that the sinusoidal voltage

$$v = V_m \cos(\omega t + \theta)$$

is applied to a network and the resultant sinusoidal current is

$$i = I_m \cos(\omega t + \phi)$$

The phase angle by which the voltage leads the current is therefore $(\theta - \phi)$. The average power delivered to the network, assuming a passive sign convention at its input terminals, may be expressed either in terms of the maximum values:

$$P = \frac{1}{2} V_m I_m \cos(\theta - \phi)$$

or in terms of the effective values:

$$P = V_{\text{eff}} I_{\text{eff}} \cos(\theta - \phi)$$

Apparent power is not a concept which is limited to sinusoidal forcing functions and responses. It may be determined for any current and voltage waveshapes by simply taking the product of the effective values of the current and voltage.

If our applied voltage and current responses had been dc quantities, the average power delivered to the network would have been given simply by the product of the voltage and the current. Applying this dc technique to the sinusoidal problem, we should obtain a value for the absorbed power which is "apparently" given by the familiar product $V_{\text{eff}}I_{\text{eff}}$. However, this product of the *effective* values of the voltage and current is not the average power; we define it as the ***apparent power.*** Dimensionally, apparent power must be measured in the same units as real power, since $\cos(\theta - \phi)$ is dimensionless, but in order to avoid confusion, the term ***volt-amperes,*** or VA, is applied to the apparent power. Since $\cos(\theta - \phi)$ cannot have a magnitude greater than unity, the magnitude of the real power can never be greater than the magnitude of the apparent power.

The ratio of the real or average power to the apparent power is called the ***power factor,*** symbolized by PF. Hence,

$$\text{PF} = \frac{\text{average power}}{\text{apparent power}} = \frac{P}{V_{\text{eff}}I_{\text{eff}}}$$

In the sinusoidal case, the power factor is simply $\cos(\theta - \phi)$, where $(\theta - \phi)$ is the angle by which the voltage leads the current. This relationship is the reason why the angle $(\theta - \phi)$ is often referred to as the ***PF angle.***

For a purely resistive load, the voltage and current are in phase, $(\theta - \phi)$ is zero, and the PF is unity. In other words, the apparent power and the average power are equal. Unity PF, however, may also be achieved for loads that contain both inductance and capacitance if the element values and the operating frequency are carefully selected to provide an input impedance having a zero phase angle. A purely reactive load, that is, one containing no resistance, will cause a phase difference between the voltage and current of either plus or minus 90°, and the PF is therefore zero.

Between these two extreme cases there are the general networks for which the PF can range from zero to unity. A PF of 0.5, for example, indicates a load having an input impedance with a phase angle of either 60° or −60°; the former describes an inductive load, since the voltage leads the current by 60°, while the latter refers to a capacitive load. The ambiguity in the exact nature of the load is resolved by referring to a leading PF or a lagging PF, the terms *leading* or *lagging* referring to the *phase of the current with respect to the voltage*. Thus, an inductive load will have a lagging PF and a capacitive load a leading PF.

EXAMPLE 11.9

Calculate values for the average power delivered to each of the two loads shown in Fig. 11.15, the apparent power supplied by the source, and the power factor of the combined loads.

▶ ***Identify the goal of the problem.***

The average power refers to the power drawn by the resistive components of the load elements; the apparent power is the product of the effective voltage and the effective current of the load combination.

▶ ***Collect the known information.***

The effective voltage is 60 V rms, which appears across a combined load of $2 - j + 1 + j5 = 3 + j4\ \Omega$.

▶ ***Devise a plan.***

Simple phasor analysis will provide the current. Knowing voltage and current will enable us to calculate average power and apparent power; these two quantities can be used to obtain the power factor.

▶ ***Construct an appropriate set of equations.***

The average power P supplied to a load is given by

$$P = I_{\text{eff}}^2 R$$

where R is the real part of the load impedance. The apparent power supplied by the source is $V_{\text{eff}} I_{\text{eff}}$, where $V_{\text{eff}} = 60$ V rms.

The power factor is calculated as the ratio of these two quantities:

$$\text{PF} = \frac{\text{average power}}{\text{apparent power}} = \frac{P}{V_{\text{eff}} I_{\text{eff}}}$$

▶ ***Determine if additional information is required.***

We require I_{eff}:

$$\mathbf{I} = \frac{60\underline{/0^\circ}}{3 + j4} = 12\underline{/-53.13^\circ}\text{ A rms}$$

so $I_{\text{eff}} = 12$ A rms, and the angle of $\mathbf{I}$ is $-53.13°$.

▶ ***Attempt a solution.***

The average power delivered to the top load is given by

$$P_{\text{upper}} = I_{\text{eff}}^2 R_{\text{top}} = (12)^2(2) = 288\text{ W}$$

and the average power delivered to the right load is given by

$$P_{\text{lower}} = I_{\text{eff}}^2 R_{\text{right}} = (12)^2(1) = 144\text{ W}$$

The source itself supplies an apparent power of $V_{\text{eff}} I_{\text{eff}} = (60)(12) = 720$ VA.

Finally, the power factor of the combined loads is found by considering the voltage and current associated with the combined loads. This power factor is, of course, identical to the power factor for the source. Thus

$$\text{PF} = \frac{P}{V_{\text{eff}} I_{\text{eff}}} = \frac{432}{60(12)} = 0.6\text{ lagging}$$

since the combined load is inductive.

▶ ***Verify the solution. Is it reasonable or expected?***

The total average power delivered to the source is $288 + 144 = 432$ W. The average power supplied by the source is

$$P = V_{\text{eff}} I_{\text{eff}} \cos(\text{angle (V)} - \text{angle (I)}) = (60)(12)\cos(0 + 53.13^\circ)$$
$$= 432\text{ W}$$

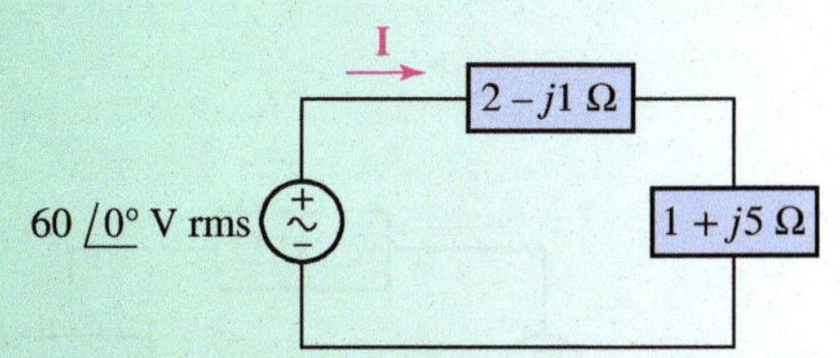

FIGURE 11.15 A circuit in which we seek the average power delivered to each element, the apparent power supplied by the source, and the power factor of the combined load.

(Continued on next page)

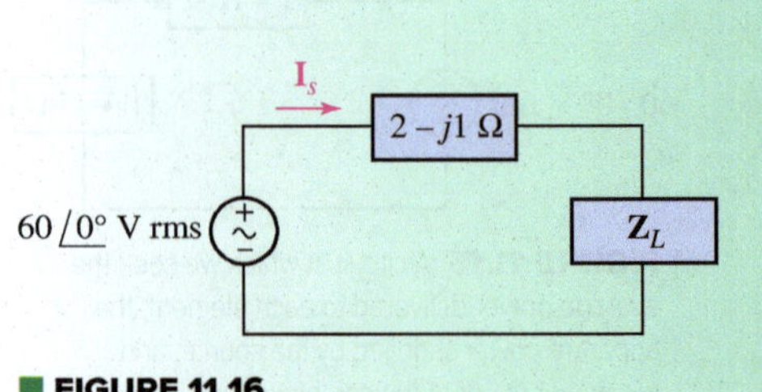

■ **FIGURE 11.16**

so we see the power balance is correct.

We might also write the combined load impedance as $5\,\underline{/53.1°}\ \Omega$, identify 53.1° as the PF angle, and thus have a PF of cos 53.1° = 0.6 *lagging*.

PRACTICE

11.9 For the circuit of Fig. 11.16, determine the power factor of the combined loads if $Z_L = 10\ \Omega$.

Ans: 0.9966 leading.

11.6 COMPLEX POWER

As we saw in Chap. 10, "complex" numbers do not actually "complicate" analysis. By allowing us to carry two pieces of information together through a series of calculations via the "real" and "imaginary" components, they often greatly simplify what might otherwise be tedious calculations. This is particularly true with power, since we have resistive as well as inductive and capacitive elements in a general load. In this section, we define ***complex power*** to allow us to calculate the various contributions to the total power in a clean, efficient fashion. The magnitude of the complex power is simply the apparent power. The real part is the average power and—as we are about to see—the imaginary part is a new quantity, termed the ***reactive power,*** which describes the rate of energy transfer into and out of reactive load components (e.g., inductors and capacitors).

We define complex power with reference to a general sinusoidal voltage $\mathbf{V}_{\text{eff}} = V_{\text{eff}}\underline{/\theta}$ across a pair of terminals and a general sinusoidal current $\mathbf{I}_{\text{eff}} = I_{\text{eff}}\underline{/\phi}$ flowing into one of the terminals in such a way as to satisfy the passive sign convention. The average power P absorbed by the two-terminal network is thus

$$P = V_{\text{eff}} I_{\text{eff}} \cos(\theta - \phi)$$

Complex nomenclature is next introduced by making use of Euler's formula in the same way we did in introducing phasors. We express P as

$$P = V_{\text{eff}} I_{\text{eff}} \text{Re}\{e^{j(\theta-\phi)}\}$$

or

$$P = \text{Re}\{V_{\text{eff}} e^{j\theta} I_{\text{eff}} e^{-j\phi}\}$$

The phasor voltage may now be recognized as the first two factors within the brackets in the preceding equation. However, the second two factors do not quite correspond to the phasor current because the angle includes a minus sign, which is not present in the expression for the phasor current. That is, the phasor current is

$$\mathbf{I}_{\text{eff}} = I_{\text{eff}} e^{j\phi}$$

and we therefore must make use of conjugate notation:

$$\mathbf{I}_{\text{eff}}^* = I_{\text{eff}} e^{-j\phi}$$

Hence

$$P = \text{Re}\{\mathbf{V}_{\text{eff}} \mathbf{I}_{\text{eff}}^*\}$$

and we may now let power become complex by defining the ***complex power* S** as

$$\mathbf{S} = \mathbf{V}_{\text{eff}} \mathbf{I}_{\text{eff}}^* \qquad [24]$$

If we first inspect the polar or exponential form of the complex power,

$$\mathbf{S} = V_{\text{eff}} I_{\text{eff}} e^{j(\theta-\phi)}$$

we see that the magnitude of **S,** $V_{\text{eff}}I_{\text{eff}}$, is the apparent power. The angle of **S,** $(\theta - \phi)$, is the PF angle (i.e., the angle by which the voltage leads the current).

In rectangular form, we have

$$\mathbf{S} = P + jQ \qquad [25]$$

where P is the average power, as before. The imaginary part of the complex power is symbolized as Q and is termed the *reactive power*. The dimensions of Q are the same as those of the real power P, the complex power **S,** and the apparent power $|\mathbf{S}|$. In order to avoid confusion with these other quantities, the unit of Q is defined as the ***volt-ampere-reactive*** (abbreviated VAR). From Eqs. [24] and [25], it is seen that

$$Q = V_{\text{eff}} I_{\text{eff}} \sin(\theta - \phi) \qquad [26]$$

The physical interpretation of reactive power is the time rate of energy flow back and forth between the source (i.e., the utility company) and the reactive components of the load (i.e., inductances and capacitances). These components alternately charge and discharge, which leads to current flow from and to the source, respectively.

The relevant quantities are summarized in Table 11.1 for convenience.

The sign of the reactive power characterizes the nature of a passive load at which V_{eff} and I_{eff} are specified. If the load is inductive, then $(\theta - \phi)$ is an angle between 0 and 90°, the sine of this angle is positive, and the reactive power is *positive*. A capacitive load results in a *negative* reactive power.

TABLE **11.1** Summary of Quantities Related to Complex Power

Quantity	Symbol	Formula	Units
Average power	P	$V_{\text{eff}} I_{\text{eff}} \cos(\theta - \phi)$	watt (W)
Reactive power	Q	$V_{\text{eff}} I_{\text{eff}} \sin(\theta - \phi)$	volt-ampere-reactive (VAR)
Complex power	**S**	$P + jQ$	
		$V_{\text{eff}} I_{\text{eff}} \underline{/\theta - \phi}$	volt-ampere (VA)
		$\mathbf{V}_{\text{eff}} \mathbf{I}_{\text{eff}}^*$	
Apparent power	$\|\mathbf{S}\|$	$V_{\text{eff}} I_{\text{eff}}$	volt-ampere (VA)

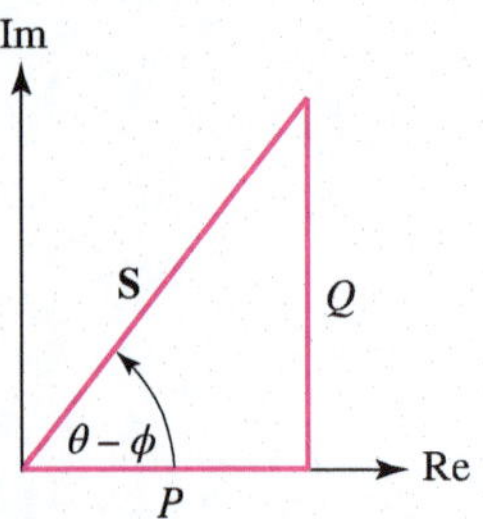

■ **FIGURE 11.17** The power triangle representation of complex power.

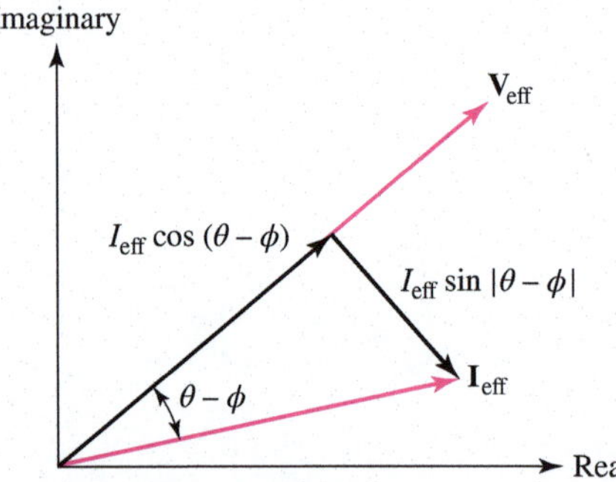

■ **FIGURE 11.18** The current phasor $\mathbf{I}_{\text{eff}}$ is resolved into two components, one in phase with the voltage phasor $\mathbf{V}_{\text{eff}}$ and the other 90° out of phase with the voltage phasor. This latter component is called a *quadrature component*.

The Power Triangle

A commonly employed graphical representation of complex power is known as the ***power triangle*** and is illustrated in Fig. 11.17. The diagram shows that only two of the three power quantities are required, as the third may be obtained by trigonometric relationships. If the power triangle lies in the first quadrant ($\theta - \phi > 0$), the power factor is *lagging* (corresponding to an inductive load); if the power triangle lies in the fourth quadrant ($\theta - \phi < 0$), the power factor is *leading* (corresponding to a capacitive load). A great deal of qualitative information concerning our load is therefore available at a glance.

Another interpretation of reactive power may be seen by constructing a phasor diagram containing $\mathbf{V}_{\text{eff}}$ and $\mathbf{I}_{\text{eff}}$ as shown in Fig. 11.18. If the phasor current is resolved into two components, one in phase with the voltage, having a magnitude $\mathbf{I}_{\text{eff}} \cos(\theta - \phi)$, and one 90° out of phase with the voltage, with magnitude equal to $\mathbf{I}_{\text{eff}} \sin|\theta - \phi|$, then it is clear that the real power is given by the product of the magnitude of the voltage phasor and the component of the phasor current which is in phase with the voltage. Moreover, the product of the magnitude of the voltage phasor and the component of the phasor current which is 90° out of phase with the voltage is the reactive power Q. It is common to speak of the component of a phasor which is 90° out of phase with some other phasor as a ***quadrature component.*** Thus Q is simply $\mathbf{V}_{\text{eff}}$ times the quadrature component of $\mathbf{I}_{\text{eff}}$. Q is also known as the ***quadrature power.***

Power Measurement

Strictly speaking, a wattmeter measures average real power P drawn by a load, and a varmeter reads the average reactive power Q drawn by a load. However, it is common to find both features in the same meter, which is often also capable of measuring apparent power and power factor (Fig. 11.19).

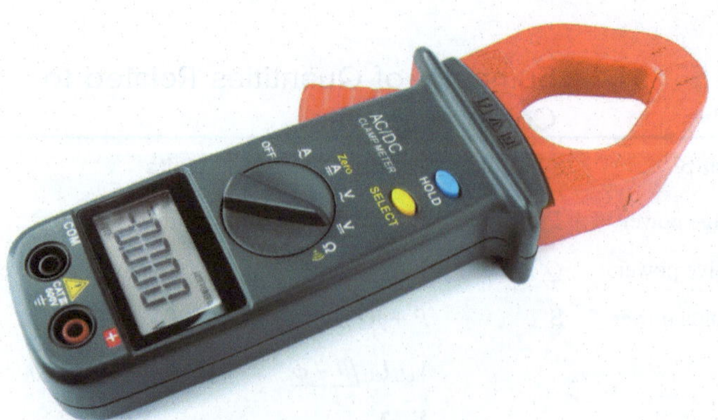

■ **FIGURE 11.19** Example of a clamp-on digital power meter capable of measuring ac currents and voltages.
KhotenkoVolodymyr/iStock/Getty Images

PRACTICAL APPLICATION

Power Factor Correction

When electric power is being supplied to large industrial consumers by a power company, the company will often include a PF clause in its rate schedules. Under this clause, an additional charge is made to the consumer whenever the PF drops below a certain specified value, usually about 0.85 lagging. Very little industrial power is consumed at leading PFs because of the nature of typical industrial loads. There are several reasons that force the power company to make this additional charge for low PFs. In the first place, larger current-carrying capacity must be built into its generators in order to provide the larger currents that go with lower-PF operation at constant power and constant voltage. Another reason is found in the increased losses in its transmission and distribution system.

In an effort to recoup losses and encourage its customers to operate at high PF, a certain utility charges a penalty of \$0.22/kVAR for each kVAR above a benchmark value computed as 0.62 times the average power demand:

$$\mathbf{S} = P + jQ = P + j0.62P = P(1 + j0.62)$$
$$= P(1.177 \underline{/31.8^\circ})$$

This benchmark targets a PF of 0.85 lagging, as cos $31.8^\circ = 0.85$ and Q is positive; this is represented graphically in Fig. 11.20. Customers with a PF smaller than the benchmark value are subject to financial penalties.

The reactive power requirement is commonly adjusted through the installation of compensation capacitors placed in parallel with the load (typically at the substation outside the customer's facility). This is analogous to the impedance matching circuitry described in Sec. 11.3. The value of the required capacitance can be shown to be

$$C = \frac{P(\tan\theta_{\text{old}} - \tan\theta_{\text{new}})}{\omega\, V_{\text{rms}}^2} \qquad [27]$$

where ω is the frequency, θ_{old} is the present PF angle, and θ_{new} is the target PF angle. For convenience, however, compensation capacitor banks are manufactured in specific increments rated in units of kVAR capacity. An example of such an installation is shown in Fig. 11.21.

Now let us consider a specific example. A particular industrial machine plant has a monthly peak demand of 5000 kW and a monthly reactive requirement of 6000 kVAR. Using the rate schedule above, what is the annual cost to this utility customer associated with PF penalties? If compensation is available through the utility company at a cost of \$2390 per 1000 kVAR increment and \$3130 per 2000 kVAR increment, what is the most cost-effective solution for the customer?

The PF of the installation is the angle of the complex power $\mathbf{S}$, which in this case is $5000 + j6000$ kVA. Thus, the angle is $\tan^{-1}(6000/5000) = 50.19^\circ$ and the PF is 0.64 lagging. The benchmark reactive power value, computed as 0.62 times the peak demand, is $0.62(5000) = 3100$ kVAR. So the plant is drawing $6000 - 3100 = 2900$ kVAR more reactive power than the utility company is willing to allow without penalty. This represents an annual assessment of $12(2900)(0.22) = \$7656$ in addition to regular electricity costs.

Reactive Power (kVAR)
0 1000 2000 3000 4000 5000 6000 7000
PF less than 0.85
Corresponds to PF 0.85 lagging
PF greater than 0.85
0 1000 2000 3000 4000 5000 6000 7000 8000 9000 10000
Average Power (kW)

■ **FIGURE 11.20** Plot showing acceptable ratio of reactive power to average power for power factor benchmark of 0.85 lagging.

■ **FIGURE 11.21** A compensation capacitor installation.
Kitja Chavanavech/jakit17/123RF

(Continued on next page)

If the customer chooses to have a single 1000 kVAR increment installed (at a cost of \$2390), the excess reactive power draw is reduced to 2900 – 1000 = 1900 kVAR, so the annual penalty is now 12(1900)(0.22) = \$5016. The total cost this year is then \$5016 + \$2390 = \$7406, for a savings of \$250. If the customer chooses to have a single 2000 kVAR increment installed (at a cost of \$3130), the excess reactive power draw is reduced to 2900 – 2000 = 900 kVAR, so the annual penalty is now 12(900)(0.22) = \$2376. The total cost this year is then \$2376 + \$3130 = \$5506, for a first-year savings of \$2150. If, however, the customer goes overboard and installs 3000 kVAR of compensation capacitors so that no penalty is assessed, it will actually cost \$14 more in the first year than if only 2000 kVAR were installed.

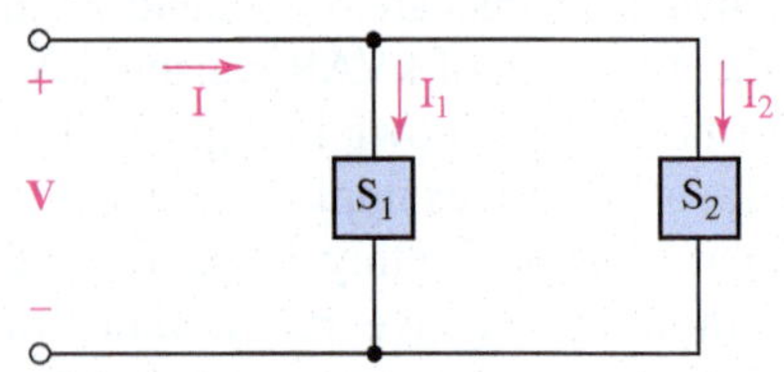

■ **FIGURE 11.22** A circuit used to show that the complex power drawn by two parallel loads is the sum of the complex powers drawn by the individual loads.

It is easy to show that the complex power delivered to several interconnected loads is the sum of the complex powers delivered to each of the individual loads, no matter how the loads are interconnected. For example, consider the two loads shown connected in parallel in Fig. 11.22. If rms values are assumed, the complex power drawn by the combined load is

$$\mathbf{S} = \mathbf{V}\mathbf{I}^* = \mathbf{V}(\mathbf{I}_1 + \mathbf{I}_2)^* = \mathbf{V}(\mathbf{I}_1^* + \mathbf{I}_2^*)$$

and thus

$$\mathbf{S} = \mathbf{V}\mathbf{I}_1^* + \mathbf{V}\mathbf{I}_2^*$$

as stated.

EXAMPLE 11.10

An industrial consumer is operating a 50 kW (67.1 hp) induction motor at a lagging PF of 0.8. The source voltage is 230 V rms. In order to obtain lower electrical rates, the customer wishes to raise the PF to 0.95 lagging. Specify a suitable solution.

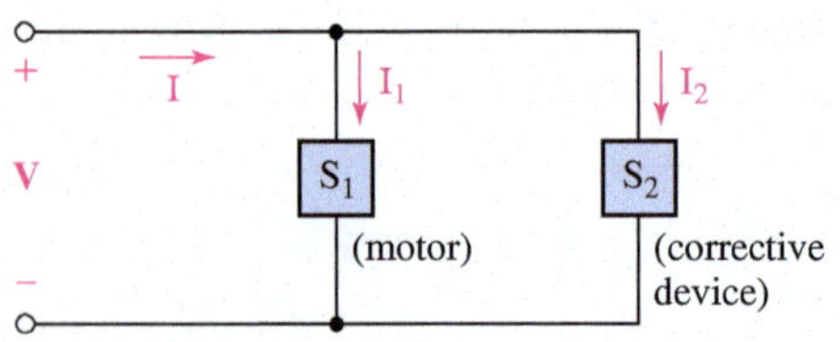

■ **FIGURE 11.23**

Although the PF might be raised by increasing the real power and maintaining the reactive power constant, this would not result in a lower bill and is not a cure that interests the consumer. A purely reactive load must be added to the system, and it is clear that it must be added in parallel, since the supply voltage to the induction motor must not change. The circuit of Fig. 11.23 is thus applicable if we interpret $\mathbf{S}_1$ as the induction motor's complex power and $\mathbf{S}_2$ as the complex power drawn by the corrective device.

The complex power supplied to the induction motor must have a real part of 50 kW and an angle of $\cos^{-1}(0.8)$, or 36.9°. Hence,

$$\mathbf{S}_1 = \frac{50\underline{/36.9^\circ}}{0.8} = 50 + j37.5 \text{ kVA}$$

In order to achieve a PF of 0.95, the total complex power must become

$$\mathbf{S} = \mathbf{S}_1 + \mathbf{S}_2 = \frac{50}{0.95}\underline{/\cos^{-1}(0.95)} = 50 + j16.43 \text{ kVA}$$

Thus, the complex power drawn by the corrective load is

$$\mathbf{S}_2 = -j21.07 \text{ kVA}$$

The necessary load impedance $\mathbf{Z}_2$ may be found in several simple steps. We select a phase angle of 0° for the voltage source, and therefore the current drawn by $\mathbf{Z}_2$ is

$$\mathbf{I}_2^* = \frac{\mathbf{S}_2}{\mathbf{V}} = \frac{-j21{,}070}{230} = -j91.6 \text{ A}$$

or

$$\mathbf{I}_2 = j91.6 \text{ A}$$

Therefore,

$$\mathbf{Z}_2 = \frac{\mathbf{V}}{\mathbf{I}_2} = \frac{230}{j91.6} = -j2.51\ \Omega$$

If the operating frequency is 60 Hz, this load can be provided by a 1056 μF capacitor connected in parallel with the motor. However, its initial cost, maintenance, and depreciation must be covered by the reduction in the electric bill.

PRACTICE

11.10 For the circuit shown in Fig. 11.24, find the complex power absorbed by the (*a*) 1 Ω resistor; (*b*) $-j10\ \Omega$ capacitor; (*c*) $5 + j10\ \Omega$ impedance; (*d*) source.

Ans: 26.6 + *j*0 VA; 0 − *j*1331VA; 532 + *j*1065 VA; −559 + *j*266 VA.

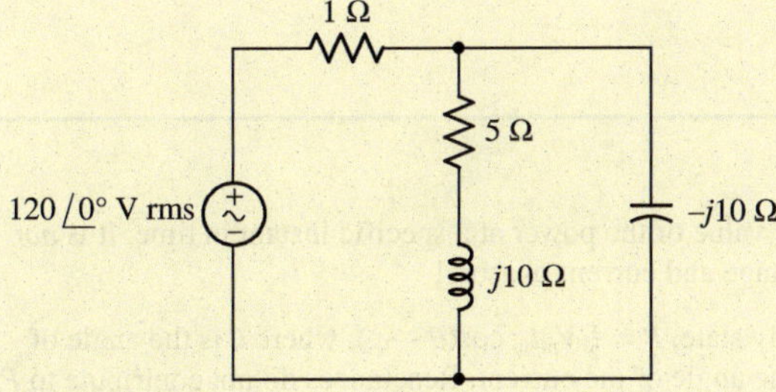

■ **FIGURE 11.24**

11.11 A 440 V rms source supplies power to a load $Z_L = 10 + j2\ \Omega$ through a transmission line having a total resistance of 1.5 Ω. Find (*a*) the average and apparent power supplied to the load; (*b*) the average and apparent power lost in the transmission line; (*c*) the average and apparent power supplied by the source; (*d*) the power factor at which the source operates.

Ans: 14.21 kW, 14.49 kVA; 2.131 kW, 2.131 kVA; 16.34 kW, 16.59 kVA; 0.985 lag.

SUMMARY AND REVIEW

In this chapter, we introduced a fair number of new power-related terms (summarized in Table 11.2), which might have come as a bit of a surprise after *watts* did so well for us up to this point. The new terminology is largely relevant to ac power systems, where voltages and currents are generally assumed to be sinusoidal (the prevalence of switched-mode power supplies in many computer systems can alter this situation, a topic of more advanced power engineering texts). After clarifying what is meant by instantaneous power, we discussed the concept of average power P. This quantity is not a function of time, but it is a strong function of the phase difference between sinusoidal voltage and current waveforms. *Maximum power transfer* to a load is found when the source and load impedance are complex conjugates, where *impedance matching* circuitry may be introduced between the source and load to provide this condition. Purely reactive elements such as ideal inductors and capacitors absorb *zero average power*. Since such elements do increase the magnitude of the current flowing between the source and load, however, two new terms find common usage: *apparent power* and *power factor*. The average power and apparent power are identical when voltage and current are in phase (i.e., associated with a purely resistive load). The power factor gives us a numerical gauge of how reactive a particular load is: a unity power factor (PF) corresponds to a purely resistive load (if inductors *are* present, they are being "canceled" by an appropriate capacitance); a zero PF indicates a purely reactive load, and the sign of the angle indicates whether the load is capacitive or inductive. Putting all of these concepts together allowed us to create a more compact representation known as *complex power*, **S**. The magnitude of **S** is the apparent power, P is the real part of **S**, and Q, the *reactive power* (zero for resistive loads), is the imaginary part of **S**.

Along the way, we paused to introduce the notion of effective values of current and voltage, often referred to as *rms values*. Care must be taken

TABLE **11.2** A Summary of AC Power Terms

Term	Symbol	Unit	Description
Instantaneous power	$p(t)$	W	$p(t) = v(t)i(t)$. It is the value of the power at a specific instant in time. It is *not* the product of the voltage and current phasors!
Average power	P	W	In the sinusoidal steady state, $P = \frac{1}{2} V_m I_m \cos(\theta - \phi)$, where θ is the angle of the voltage and ϕ is the angle of the current. Reactances do not contribute to P.
Effective or rms value	V_{rms} or I_{rms}	V or A	Defined, e.g., as $I_{\text{eff}} = \sqrt{\frac{1}{T}\int_0^T i^2\, dt}$; if $i(t)$ is sinusoidal, then $I_{\text{eff}} = I_m/\sqrt{2}$.
Apparent power	$\lvert\mathbf{S}\rvert$	VA	$\lvert\mathbf{S}\rvert = V_{\text{eff}} I_{\text{eff}}$, and is the maximum value the average power can be; $P = \lvert\mathbf{S}\rvert$ only for purely resistive loads.
Power factor	PF	None	Ratio of the average power to the apparent power. The PF is unity for a purely resistive load, and zero for a purely reactive load.
Reactive power	Q	VAR	A means of measuring the energy flow rate to and from reactive loads.
Complex power	**S**	VA	A convenient complex quantity that contains both the average power P and the reactive power Q: $\mathbf{S} = P + jQ$.

from this point forward to establish whether a particular voltage or current value is being quoted as a magnitude or its corresponding rms value because an approximate 40 percent error can be introduced. Interestingly, we also discovered an extension of the maximum power theorem encountered in Chap. 5, namely, that maximum average power is delivered to a load whose impedance $\mathbf{Z}_L$ is the *complex conjugate* of the Thévenin equivalent impedance of the network to which it is connected.

For convenience, key points of the chapter are summarized in the following list, along with corresponding example numbers.

- The instantaneous power absorbed by an element is given by the expression $p(t) = v(t)i(t)$. (Examples 11.1, 11.2)
- The average power delivered to an impedance by a sinusoidal source is $\frac{1}{2}V_m I_m \cos(\theta - \phi)$, where θ = the voltage phase angle and ϕ = the phase angle of the current. (Example 11.2)
- Only the *resistive* component of a load draws nonzero average power. The average power delivered to the *reactive* component of a load is zero. (Examples 11.3, 11.4, 11.5, 11.6)
- Maximum average power transfer occurs when the condition $\mathbf{Z}_L = \mathbf{Z}^*_{TH}$ is satisfied. (Example 11.7)
- Power delivery to a load may be improved using a matching network. (Example 11.8)
- The effective or rms value of a sinusoidal waveform is obtained by dividing its amplitude by $\sqrt{2}$. (Example 11.9)
- The power factor (PF) of a load is the ratio of its average dissipated power to the apparent power. (Example 11.9)
- A purely resistive load will have a unity power factor. A purely reactive load will have a zero power factor. (Example 11.10)
- Complex power is defined as $\mathbf{S} = P + jQ$, or $\mathbf{S} = \mathbf{V}_{\text{eff}}\mathbf{I}^*_{\text{eff}}$. It is measured in units of volt-amperes (VA). (Example 11.10)
- Reactive power Q is the imaginary component of the complex power; it is a measure of the energy flow rate into or out of the reactive components of a load. Its unit is the volt-ampere-reactive (VAR). (Example 11.10)
- Capacitors are commonly used to improve the PF of industrial loads to minimize the reactive power required from the utility company. (Example 11.10)

READING FURTHER

A good overview of ac power concepts can be found in Chap. 2 of:

B. M. Weedy, B. J. Cory, N. Jenkins, J. B. Ekanayake, Goran Strbac, *Electric Power Systems*, 5th ed. Chichester, England: Wiley, 2012.

Contemporary issues pertaining to ac power systems can be found in:

International Journal of Electrical Power & Energy Systems. Guildford, England: IPC Science and Technology Press, 1979–. ISSN: 0142-0615.

EXERCISES

11.1 Instantaneous Power

1. Determine the instantaneous power delivered to the 1 Ω resistor of Fig. 11.25 at $t = 0$, $t = 1$ s, and $t = 2$ s if v_s is equal to (*a*) 9 V; (*b*) 9 sin $2t$ V; (*c*) 9 sin$(2t + 13°)$ V; (*d*) $9e^{-t}$ V.

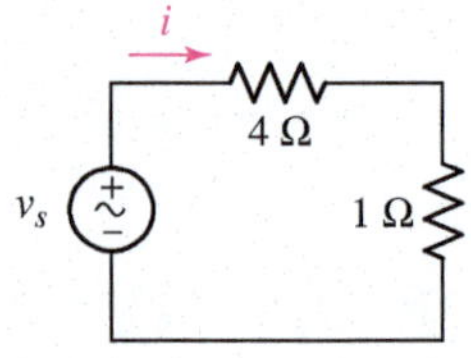

■ **FIGURE 11.25**

2. Determine the power absorbed at $t = 1.5$ ms by each of the three elements of the circuit shown in Fig. 11.26 if v_s is equal to (*a*) $30u(-t)$ V; (*b*) $10 + 20u(t)$ V.

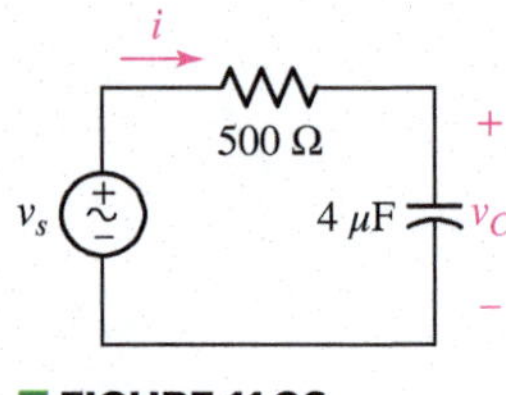

■ **FIGURE 11.26**

3. Calculate the power absorbed at $t = 0^-$, $t = 0^+$, and $t = 200$ ms by each of the elements in the circuit of Fig. 11.27 if v_s is equal to (*a*) $-10u(-t)$ V; (*b*) $20 + 5u(t)$ V.

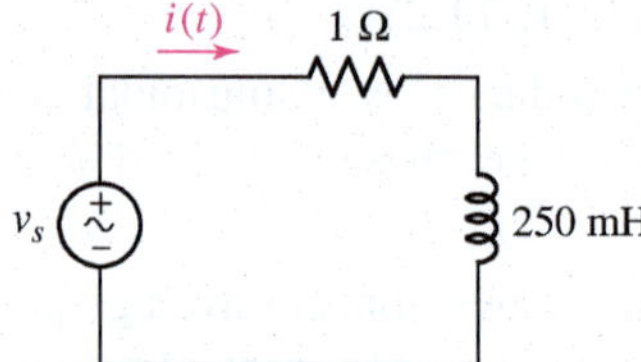

■ **FIGURE 11.27**

4. Three elements are connected in parallel: a 1 kΩ resistor, a 15 mH inductor, and a 100 cos$(2 \times 10^5 t)$ mA sinusoidal source. All transients have long since died out, so the circuit is operating in steady state. Determine the power being absorbed by each element at $t = 10$ μs.

5. Let $i_s = 4u(-t)$ A in the circuit of Fig. 11.28. (*a*) Show that, for all $t > 0$, the instantaneous power absorbed by the resistor is equal in magnitude but opposite in sign to the instantaneous power absorbed by the capacitor. (*b*) Determine the power absorbed by the resistor at $t = 60$ ms.

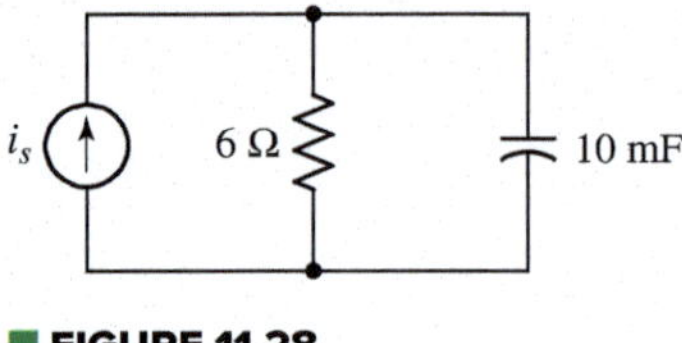

■ **FIGURE 11.28**

6. The current source in the circuit of Fig. 11.28 is given by $i_s = 8 - 7u(t)$ A. Compute the power absorbed by all three elements at $t = 0^-$, $t = 0^+$, and $t = 75$ ms.

7. Assuming no transients are present, calculate the power absorbed by each element shown in the circuit of Fig. 11.29 at $t = 0$, 10, and 20 ms.

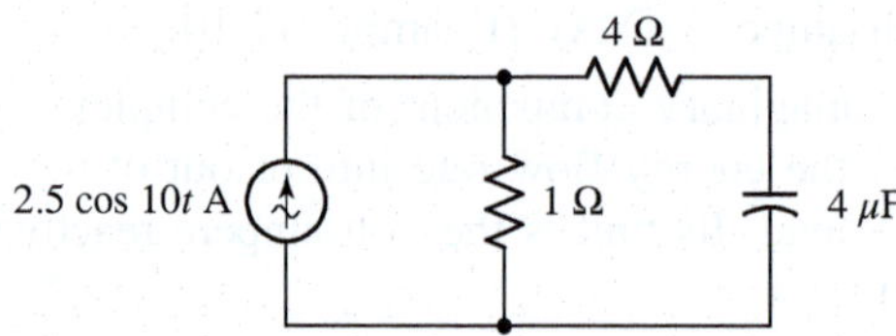

■ **FIGURE 11.29**

8. Calculate in Fig. 11.30 the power absorbed by the inductor at $t = 0$ and $t = 1$ s if $v_s = 10u(t)$ V.

9. Use SPICE to plot the instantaneous power for each circuit element of the circuit shown in Fig. 11.30 in the range of 0 to 10 s. Explain how/where power is being transferred in the circuit.

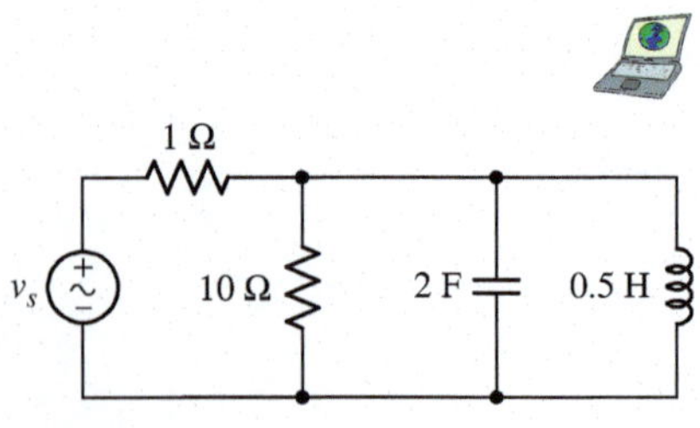

■ **FIGURE 11.30**

10. If we take a typical cloud-to-ground lightning *stroke* to represent a current of 30 kA over an interval of 150 μs, calculate (*a*) the instantaneous power delivered to a copper rod having resistance 1.2 mΩ during the stroke; (*b*) the total energy delivered to the rod.

11.2 Average Power

11. The phasor current $\mathbf{I} = 9\underline{/15°}$ mA (corresponding to a sinusoid operating at 45 rad/s) is applied to the series combination of an 18 kΩ resistor and a 1 μF capacitor. Obtain an expression for (*a*) the instantaneous power and (*b*) the average power absorbed by the combined load.

12. A phasor voltage $\mathbf{V} = 100\angle 45°$ V (the sinusoid operates at 155 rad/s) is applied to the parallel combination of a 1 Ω resistor and a 1 mH inductor. (*a*) Obtain an expression for the average power absorbed by each passive element. (*b*) Graph the instantaneous power supplied to the parallel combination, along with the instantaneous power absorbed by each element separately. (Use a single graph.)
13. Calculate the average power delivered by the current $4 - j2$ A to (*a*) $\mathbf{Z} = 9\ \Omega$; (*b*) $\mathbf{Z} = -j1000\ \Omega$; (*c*) $\mathbf{Z} = 1 - j2 + j3\ \Omega$; (*d*) $\mathbf{Z} = 6\angle 32°\ \Omega$; (*e*) $\mathbf{Z} = \frac{1.5\angle -19°}{2+j}$ kΩ.
14. With regard to the two-mesh circuit depicted in Fig. 11.31, determine the average power absorbed by each passive element and the average power supplied by each source, and verify that the total supplied average power = the total absorbed average power.

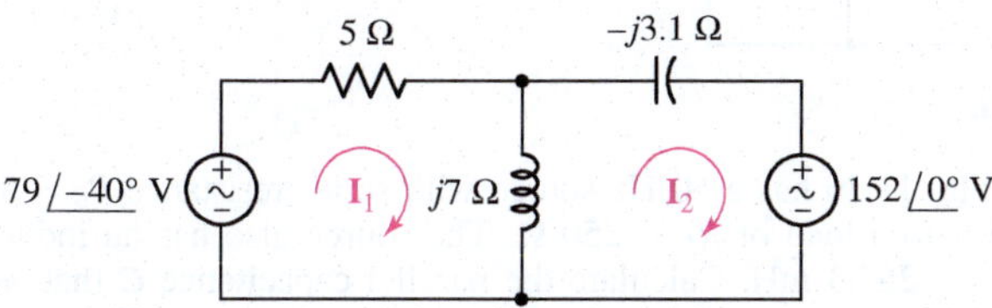

■ **FIGURE 11.31**

15. Find the average power for each element in the circuit of Fig. 11.32.

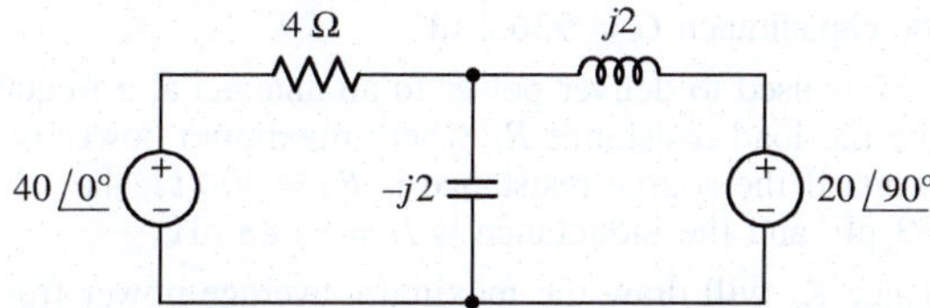

■ **FIGURE 11.32**

16. (*a*) Calculate the average power absorbed by each passive element in the circuit of Fig. 11.33, and verify that it equals the average power supplied by the source. (*b*) Check your solution with an appropriate SPICE simulation.

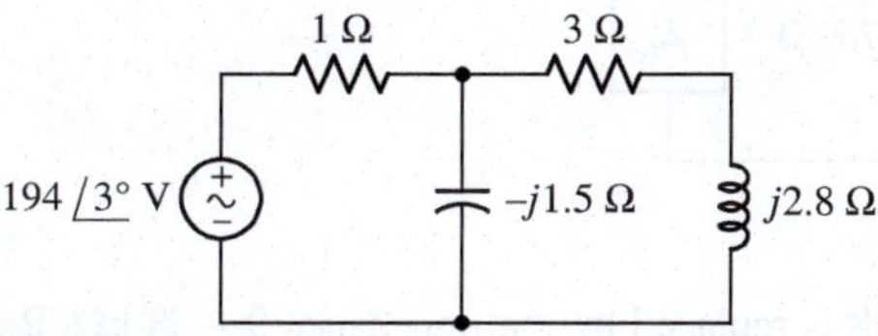

■ **FIGURE 11.33**

17. Determine the average power supplied by the dependent source in the circuit of Fig. 11.34.
18. (*a*) Calculate the average power supplied to each passive element in the circuit of Fig. 11.35. (*b*) Determine the power supplied by each source. (*c*) Replace the 8 Ω resistive load with an impedance capable of drawing maximum average power from the remainder of the circuit.

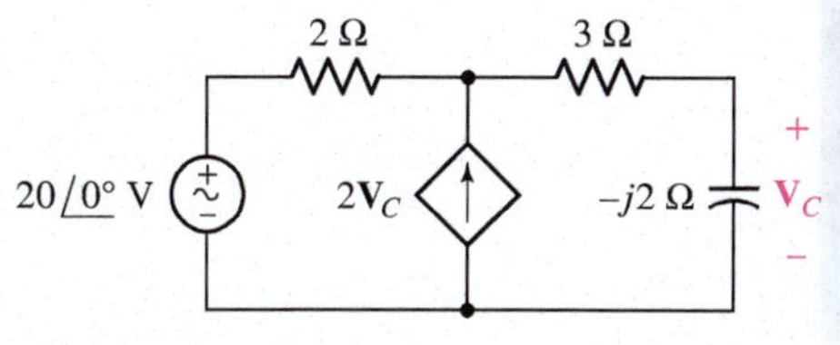

■ **FIGURE 11.34**

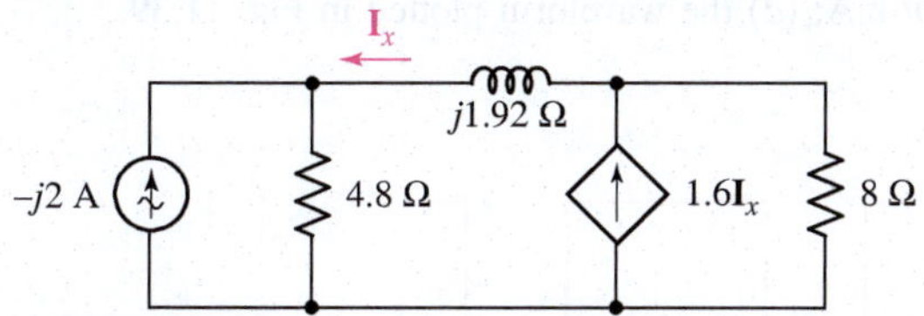

■ **FIGURE 11.35**

19. Calculate the average power delivered to a 2.2 Ω load by the voltage v_s equal to (*a*) 5 V; (*b*) 4 cos 80t − 8 sin 80t V; (*c*) 10 cos 100t + 12.5 cos(100t + 19°) V.

11.3 Maximum Power Transfer

20. The circuit in Fig. 11.36 has a series resistance of $R_S = 50\ \Omega$ and load resistance of $R_L = 82\ \Omega$. If the impedance of the inductor is $j40\ \Omega$, what would be the required impedance of the capacitor to ensure maximum power transfer to R_L?

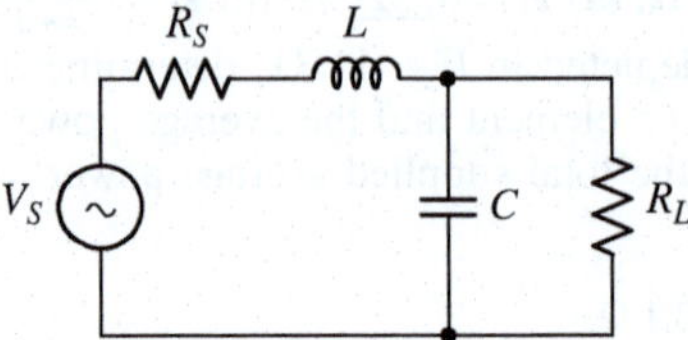

FIGURE 11.36

21. The circuit in Fig. 11.36 has a 60 Hz source with series resistance $R_S = 50\ \Omega$, delivering power to a load of $R_L = 250\ \Omega$. The source also has an inductive element with $L = 265.3$ mH. Calculate the parallel capacitance C that would provide maximum power transfer to the parallel combination of R_L and C.

22. The circuit in Fig. 11.37 is used to deliver power to a set of 8 Ω speakers (i.e., $R_L = 8\ \Omega$). Calculate the frequency (f in Hz) and value of inductance (L) where maximum power is transferred to the speakers if the source resistance is $R_S = 136\ \Omega$ and the capacitance $C = 936.2$ nF.

FIGURE 11.37

23. The circuit in Fig. 11.37 is used to deliver power to an antenna at a frequency of 300 MHz. Calculate the load resistance R_L where maximum power is transferred to the antenna if the source resistance is $R_S = 400\ \Omega$, the capacitance $C = 2.653$ pF, and the inductance is $L = 84.88$ nH.

24. (*a*) What load impedance $\mathbf{Z}_L$ will draw the maximum average power from the source shown in Fig. 11.38? (*b*) Calculate the maximum average power supplied to the load.

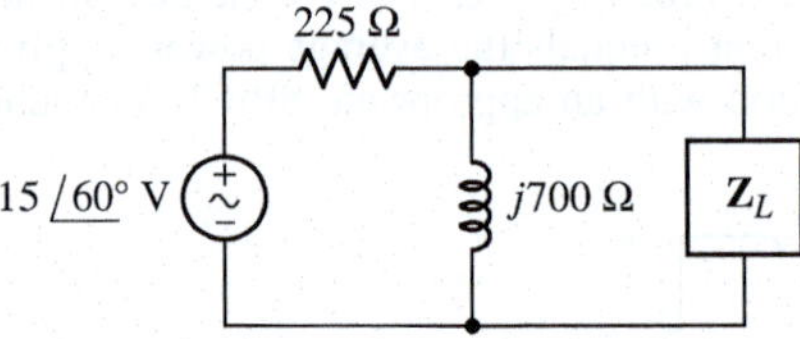

FIGURE 11.38

25. The inductance of Fig. 11.38 is replaced by the impedance 9 − j8 kΩ. Repeat Exercise 24.

11.4 Effective Values of Current and Voltage

26. Calculate the effective value of the following waveforms: (*a*) 7 sin 30t V; (*b*) 100 cos 80t mA; (*c*) $120\sqrt{2}\cos(5000t - 45°)$ V; (*d*) $\frac{100}{\sqrt{2}}\sin(2t + 72°)$ A.

27. Determine the effective value of the following waveforms: (*a*) 62.5 cos 100t mV; (*b*) 1.95 cos 2t A; (*c*) $208\sqrt{2}\cos(100\pi t + 29°)$ V; (*d*) $\frac{400}{\sqrt{2}}\sin(2000t - 14°)$ A.

28. Compute the effective value of (*a*) $i(t) = 3 \sin 4t$ A; (*b*) $v(t) = 4 \sin 20t \cos 10t$; (*c*) $i(t) = 2 - \sin 10t$ mA; (*d*) the waveform plotted in Fig. 11.39.

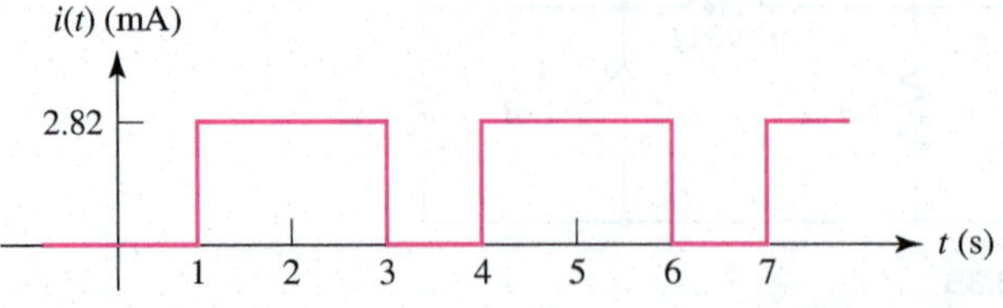

FIGURE 11.39

29. For each waveform plotted in Fig. 11.40, determine its frequency, period, and rms value.

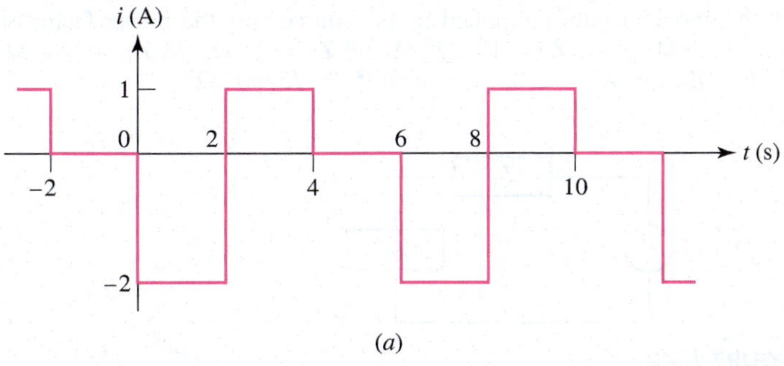

(a)

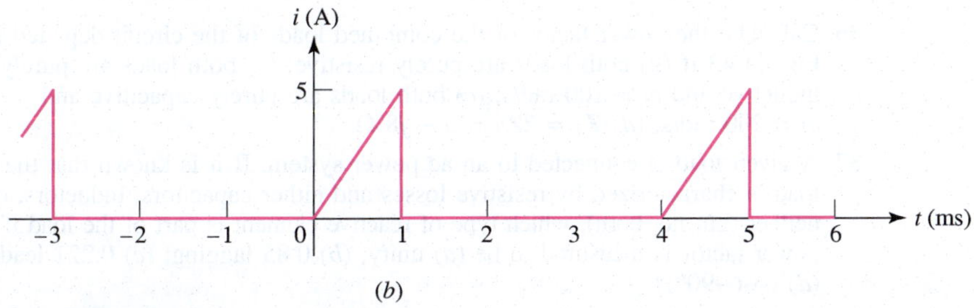

(b)

FIGURE 11.40

30. The capacitance value of the circuit shown in Fig. 11.29 is changed to 40 mF. Find the effective current and voltage for each element in the circuit following this change.

31. The series combination of a 1 kΩ resistor and a 2 H inductor must not dissipate more than 250 mW of power at any instant. Assuming a sinusoidal current with $\omega = 500$ rad/s, what is the largest rms current that can be tolerated?

32. For each of the following waveforms, determine its period, frequency, and effective value: (*a*) 5 V; (*b*) $2 \sin 80t - 7 \cos 20t + 5$ V; (*c*) $5 \cos 50t + 3 \sin 50t$ V; (*d*) $8 \cos^2 90t$ mA. (*e*) Verify your answers with an appropriate simulation.

33. With regard to the circuit of Fig. 11.41, determine whether a purely real value of *R* can result in equal rms voltages across the 14 mH inductor and the resistor *R*. If so, calculate *R* and the rms voltage across it; if not, explain why not.

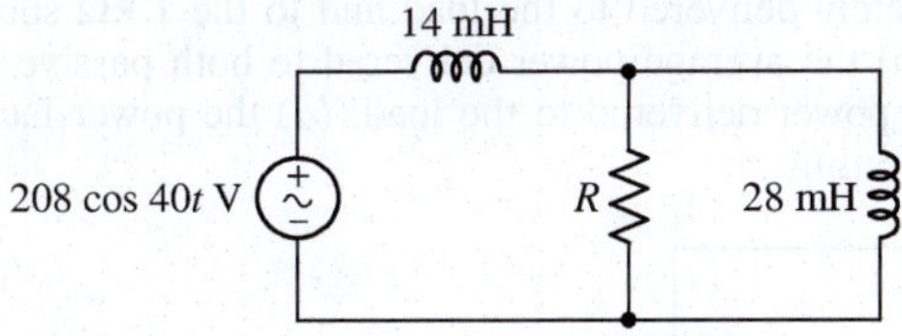

FIGURE 11.41

34. (*a*) Calculate both the average and rms values of the waveform plotted in Fig. 11.42. (*b*) Verify your solutions with appropriate SPICE simulations. (*Hint:* You may want to employ two pulse waveforms added together.)

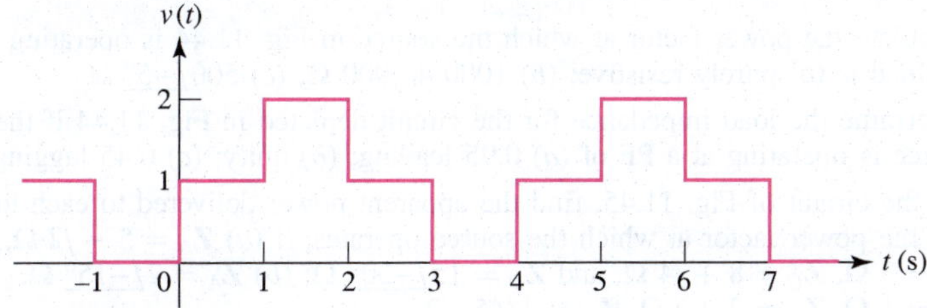

FIGURE 11.42

11.5 Apparent Power and Power Factor

35. For the circuit of Fig. 11.43, compute the average power delivered to each load, the apparent power supplied by the source, and the power factor of the combined loads if (*a*) $\mathbf{Z}_1 = 14\underline{/32^\circ}\ \Omega$ and $\mathbf{Z}_2 = 22\ \Omega$; (*b*) $\mathbf{Z}_1 = 2\underline{/0^\circ}\ \Omega$ and $\mathbf{Z}_2 = 6 - j\Omega$; (*c*) $\mathbf{Z}_1 = 100\underline{/70^\circ}\ \Omega$ and $\mathbf{Z}_2 = 75\underline{/90^\circ}\ \Omega$.

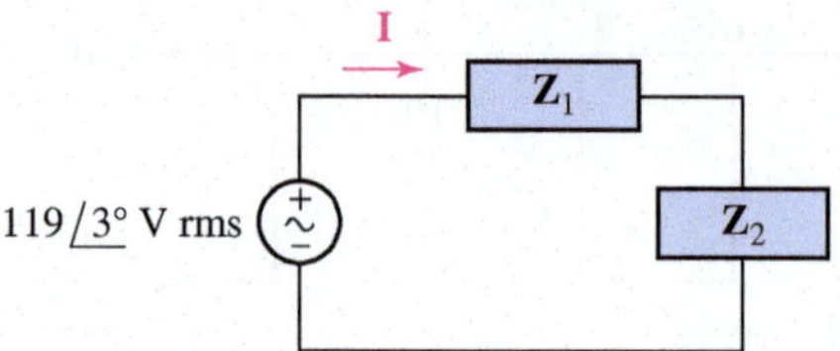

■ FIGURE 11.43

36. Calculate the power factor of the combined loads of the circuit depicted in Fig. 11.43 if (*a*) both loads are purely resistive; (*b*) both loads are purely inductive and $\omega = 100$ rad/s; (*c*) both loads are purely capacitive and $\omega = 200$ rad/s; (*d*) $\mathbf{Z}_1 = 2\mathbf{Z}_2 = 5 - j8\ \Omega$.
37. A given load is connected to an ac power system. If it is known that the load is characterized by resistive losses and either capacitors, inductors, or neither (but not both), which type of reactive element is part of the load if the power factor is measured to be (*a*) unity; (*b*) 0.85 lagging; (*c*) 0.221 leading; (*d*) $\cos(-90^\circ)$?
38. An unknown load is connected to a standard European household outlet (240 V rms, 50 Hz). Determine the phase angle difference between the voltage and current, and whether the voltage leads or lags the current, if (*a*) $\mathbf{V} = 240\underline{/243^\circ}$ V rms and $\mathbf{I} = 3\underline{/9^\circ}$ A rms; (*b*) the power factor of the load is 0.55 lagging; (*c*) the power factor of the load is 0.685 leading; (*d*) the capacitive load draws 100 W average power and 500 volt-amperes apparent power.

39. (*a*) Design a load which draws an average power of 25 W at a leading PF of 0.88 from a standard North American household outlet (120 V rms, 60 Hz). (*b*) Design a capacitor-free load which draws an average power of 150 W and an apparent power of 25 W from a household outlet in eastern Japan (110 V rms, 50 Hz).
40. Assuming an operating frequency of 40 rad/s for the circuit shown in Fig. 11.44, and a load impedance of $50\underline{/-100^\circ}\ \Omega$, calculate (*a*) the instantaneous power separately delivered to the load and to the 1 kΩ shunt resistance at $t = 20$ ms; (*b*) the average power delivered to both passive elements; (*c*) the apparent power delivered to the load; (*d*) the power factor at which the source is operating.

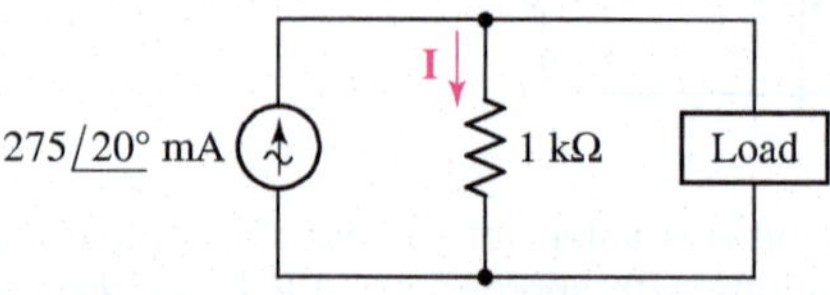

■ FIGURE 11.44

41. Calculate the power factor at which the source in Fig. 11.44 is operating if the load is (*a*) purely resistive; (*b*) $1000 + j900\ \Omega$; (*c*) $500\underline{/-5^\circ}\ \Omega$.
42. Determine the load impedance for the circuit depicted in Fig. 11.44 if the source is operating at a PF of (*a*) 0.95 leading; (*b*) unity; (*c*) 0.45 lagging.
43. For the circuit of Fig. 11.45, find the apparent power delivered to each load, and the power factor at which the source operates, if (*a*) $\mathbf{Z}_A = 5 - j2\ \Omega$, $\mathbf{Z}_B = 3\ \Omega$, $\mathbf{Z}_C = 8 + j4\ \Omega$, and $\mathbf{Z}_D = 15\underline{/-30^\circ}\ \Omega$; (*b*) $\mathbf{Z}_A = 2\underline{/-15^\circ}\ \Omega$, $\mathbf{Z}_B = 1\ \Omega$, $\mathbf{Z}_C = 2 + j\ \Omega$, $\mathbf{Z}_D = 4\underline{/45^\circ}\ \Omega$.

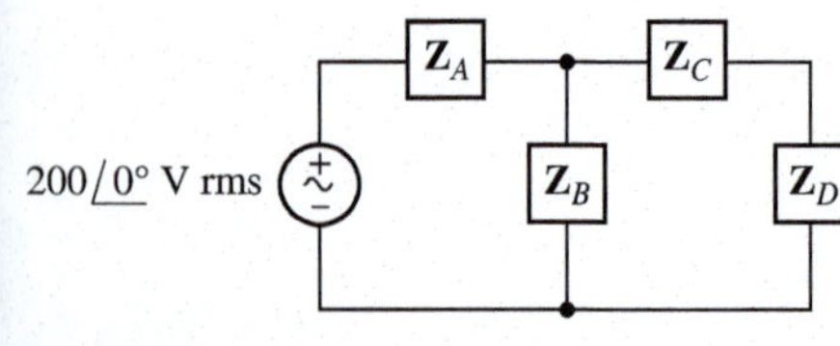

■ FIGURE 11.45

11.6 Complex Power

44. Compute the complex power **S** (in polar form) drawn by a certain load if it is known that (*a*) it draws 100 W average power at a lagging PF of 0.75; (*b*) it draws a current $\mathbf{I} = 9 + j5$ A rms when connected to the voltage $120\underline{/32°}$ V rms; (*c*) it draws 1000 W average power and 10 VAR reactive power at a leading PF; (*d*) it draws an apparent power of 450 W at a lagging PF of 0.65.

45. Calculate the apparent power, power factor, and reactive power associated with a load if it draws complex power **S** equal to (*a*) $1 + j0.5$ kVA; (*b*) 400 VA; (*c*) $150\underline{/-21°}$ VA; (*d*) $75\underline{/25°}$ VA.

46. For each power triangle depicted in Fig. 11.46, determine **S** (in polar form) and the PF.

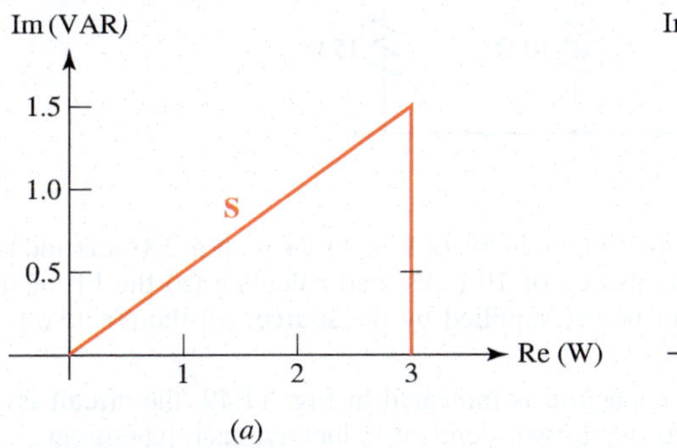

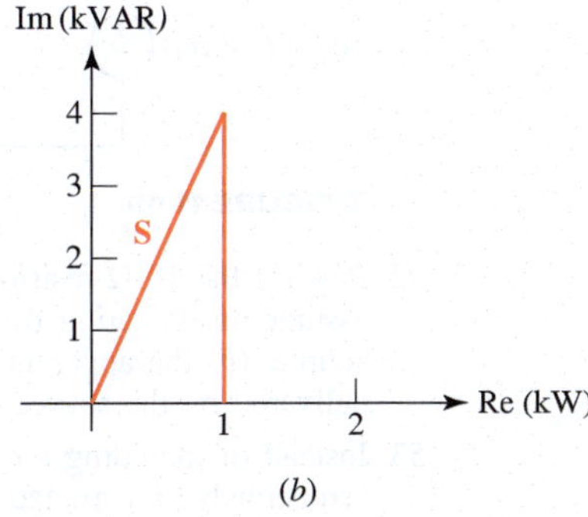

FIGURE 11.46

47. Referring to the network represented in Fig. 11.23, if the motor draws complex power $150\underline{/24°}$ VA, (*a*) determine the PF at which the source is operating; (*b*) determine the impedance of the corrective device required to change the PF of the source to 0.98 lagging. (*c*) Is it physically possible to obtain a leading PF for the source? *Explain.*

48. Determine the complex power absorbed by each passive component in the circuit of Fig. 11.47, and the power factor at which the source is operating.

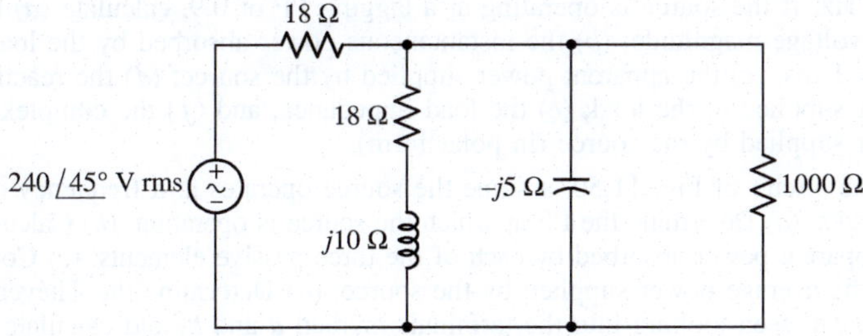

FIGURE 11.47

49. What value of capacitance must be added in parallel to the 10 Ω resistor of Fig. 11.48 to increase the PF of the source to 0.95 at 50 Hz?

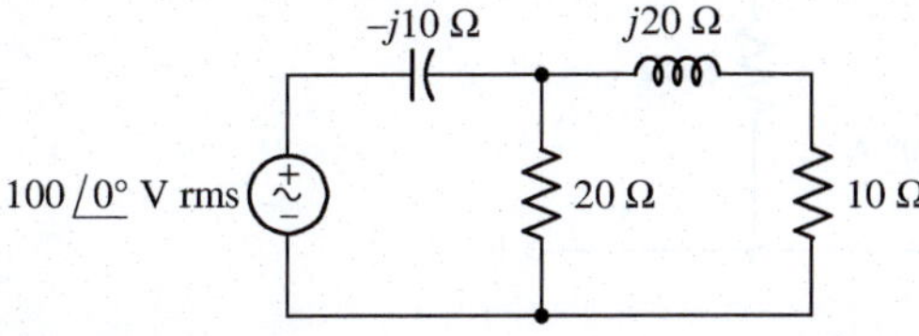

FIGURE 11.48

50. The kiln operation of a local lumberyard has a monthly average power demand of 175 kW, but associated with that is an average monthly reactive power draw of 205 kVAR. If the lumberyard's utility company charges $0.15 per kVAR for each kVAR above the benchmark value (0.7 times the peak average power demand), (*a*) estimate the annual cost to the lumberyard from PF penalties; (*b*) calculate the money saved in the first and second years, respectively, if 100 kVAR compensation capacitors are available for purchase at $75 each (installed).

51. Calculate the complex power delivered to each passive component of the circuit shown in Fig. 11.49, and determine the power factor of the source.

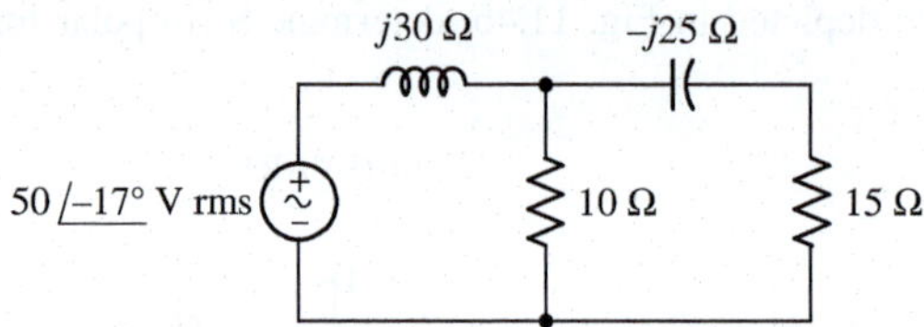

FIGURE 11.49

52. Replace the 10 Ω resistor in the circuit of Fig. 11.49 with a 200 mH inductor, assume an operating frequency of 10 rad/s, and calculate (*a*) the PF of the source; (*b*) the apparent power supplied by the source; (*c*) the reactive power delivered by the source.

53. Instead of including a capacitor as indicated in Fig. 11.49, the circuit is erroneously constructed using two identical inductors, each having an impedance of *j*30 W at the operating frequency of 50 Hz. (*a*) Compute the complex power delivered to each passive component. (*b*) Verify your solution by calculating the complex power supplied by the source. (*c*) At what power factor is the source operating?

54. Making use of the general strategy employed in Example 11.9, derive Eq. [27], which enables the corrective value of capacitance to be calculated for a general operating frequency.

Chapter-Integrating Exercises

55. A load is drawing 10 A rms when connected to a 1200 V rms supply running at 50 Hz. If the source is operating at a lagging PF of 0.9, calculate (*a*) the peak voltage magnitude; (*b*) the instantaneous power absorbed by the load at $t = 1$ ms; (*c*) the apparent power supplied by the source; (*d*) the reactive power supplied to the load; (*e*) the load impedance; and (*f*) the complex power supplied by the source (in polar form).

56. For the circuit of Fig. 11.50, assume the source operates at a frequency of 100 rad/s. (*a*) Determine the PF at which the source is operating. (*b*) Calculate the apparent power absorbed by each of the three passive elements. (*c*) Compute the average power supplied by the source. (*d*) Determine the Thévenin equivalent seen looking into the terminals marked *a* and *b*, and calculate the average power delivered to a 100 Ω resistor connected between the same terminals.

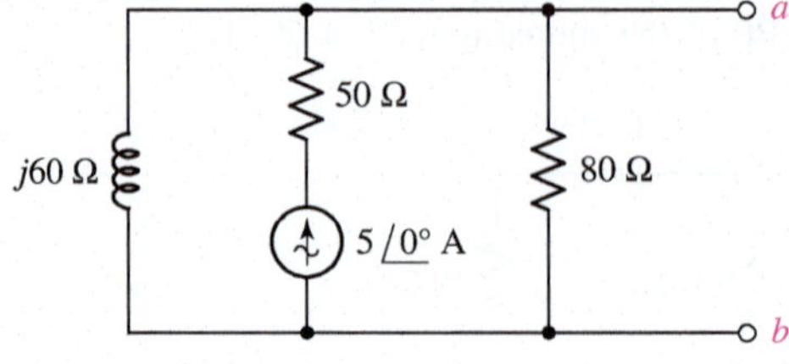

FIGURE 11.50

57. Remove the 50 Ω resistor in Fig. 11.50, assume an operating frequency of 50 Hz, and a load of 100 Ω connected between terminals *a* and *b*. (*a*) Determine the power factor at which the load is operating; (*b*) compute the average power delivered by the source; (*c*) compute the instantaneous power absorbed by the inductance at $t = 2$ ms; (*d*) determine the capacitance that must be connected between the terminals marked *a* and *b* to increase the PF of the source to 0.95.

58. A source 45 sin 32*t* V is connected in series with a 5 Ω resistor and a 20 mH inductor. Calculate (*a*) the reactive power delivered by the source; (*b*) the apparent power absorbed by each of the three elements; (*c*) the complex power **S** absorbed by each element; (*d*) the power factor at which the source is operating.

59. For the circuit of Fig. 11.40, (*a*) derive an expression for the complex power delivered by the source in terms of the unknown resistance *R* and algebraic expressions for inductor reactances X_1 and X_2; (*b*) compute the necessary capacitance that must be added in parallel to the 28 mH inductor to achieve a unity PF for the case of $R = 2$ Ω.

60. The circuit in Fig. 11.51 uses a Pi network to match the impedance between source and load for maximum power transfer. A 60 Hz source has a series impedance of 50 + *j*5 Ω. What load impedance would a Pi network with $C_S = 70.1$ μF, $C_L = 39.8$ μF, and $L = 0.225$ H correspond to for maximum power transfer?

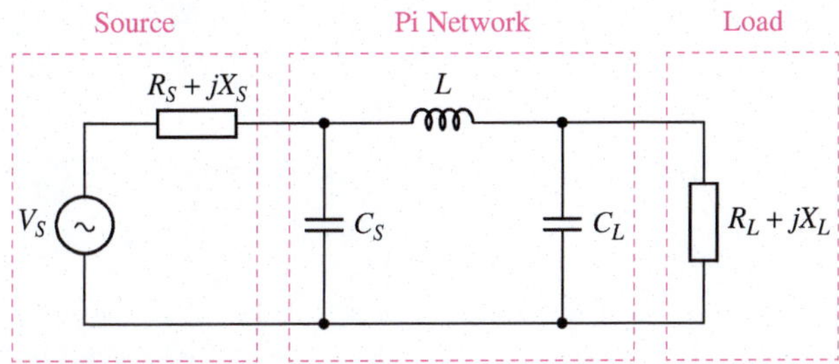

FIGURE 11.51

61. The circuit in Fig. 11.51 uses a Pi network to match the impedance between source and load for maximum power transfer. A 1 kHz source has a series impedance of 50 + *j*4 Ω. What inductance would a Pi network with $C_S =$ 506 nF, $C_L = 970$ nF require for maximum power transfer to a load with 499 + *j*23 Ω?

62. The circuit in Fig. 11.52 uses a T network to match the impedance between source and load for maximum power transfer. A 60 Hz source has a series impedance of 50 + *j*5 Ω. What load impedance would a T network with $L_S = 0.385$ H, $C = 25.4$ μF, and $L_L = 0.566$ H correspond to for maximum power transfer?

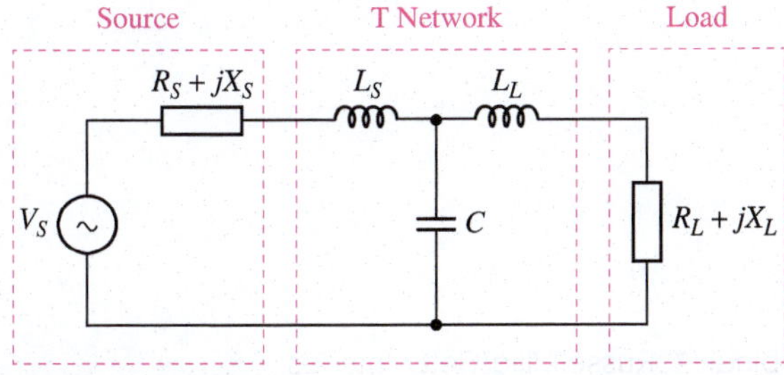

FIGURE 11.52

63. The circuit in Fig. 11.52 uses a T network to match the impedance between source and load for maximum power transfer. A 1 kHz source has a series impedance of $50 + j4\ \Omega$. What capacitance would a T network with $L_S = L_L = 20.4$ mH require for maximum power transfer to a load with $40 - j8\ \Omega$?

64. You would like to maximize power transfer to a 50 Ω antenna for VHF communications at 100 MHz. The source has an impedance of $10 + j5\ \Omega$ at this frequency. Design a T or Pi matching network for maximum power transfer (see Figs. 11.51 and 11.52). Simulate your design using SPICE, and use an appropriate supporting argument to verify maximum power transfer.

65. According to the U.S. Bureau of Reclamation (www.usbr.gov), the last three generators in the Third Power Plant of the Grand Coulee Dam (units G-22, G-23, and G-24) have been modified to increase their individual ratings from 700 MW to 805 MW at 825.6 MVA and power factor of 0.975. (*a*) If the power factor remained unchanged during the upgrade, by what percentage did the individual generator's apparent power increase? (*b*) If the power factor remains the same while the apparent power capability is reduced to 788 MVA, what is the corresponding average power delivered by such a generator? (*c*) Assuming the power factor was unchanged, estimate the apparent power capability of the three generators as originally installed. (*d*) Calculate the angle by which the voltage leads the current for one of the modernized generators.

CHAPTER 12

Polyphase Circuits

KEY CONCEPTS

Single-Phase Systems

Three-Phase Systems

Three-Phase Sources

Line Versus Phase Voltage

Line Versus Phase Current

Y-Connected Networks

Δ-Connected Networks

Balanced Loads

Per-Phase Analysis

Power Measurement in Three-Phase Systems

INTRODUCTION

The vast majority of power is supplied to consumers in the form of sinusoidal voltages and currents, typically referred to as alternating current or simply *ac*. Although there are exceptions—for example, some types of train motors—most equipment is designed to run on either 50 or 60 Hz. Most 60 Hz systems are now standardized to run on 120 V, whereas 50 Hz systems typically correspond to 240 V (both voltages being quoted in rms units). The actual voltage delivered to an appliance can vary somewhat from these values, and distribution systems employ significantly higher voltages to minimize the current and hence cable size. Originally Thomas Edison advocated a purely dc power distribution network, purportedly due to his preference for the simple algebra required to analyze such circuits. Nikola Tesla and George Westinghouse, two other pioneers in the field of electricity, proposed ac distribution systems as the achievable losses were significantly lower. Ultimately they were more persuasive, despite some rather theatrical (and questionable) demonstrations on the part of Edison.

The transient response of ac power systems is of interest when determining the peak power demand, since most equipment requires more current to start up than it does to run continuously. Often, however, it is the steady-state operation that is of primary interest, so our experience with phasor-based analysis will prove to be handy. In this chapter we introduce a new type of voltage source, the three-phase source, which can be connected in either a three- or four-wire Y configuration or a three-wire Δ configuration. Loads can also be either Y- or Δ-connected, depending on the application.

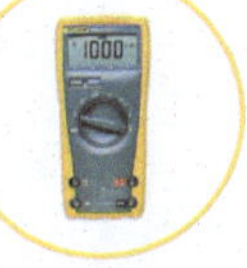

12.1 POLYPHASE SYSTEMS

In this chapter, we introduce the concept of ***polyphase*** sources, focusing on three-phase systems in particular. There are distinct advantages when using rotating machinery to generate three-phase power rather than single-phase power, and there are economical advantages in favor of the transmission of power in a three-phase system. In particular, motors used in large refrigeration systems and in machining facilities are often wired for three-phase power. For the remaining applications, once we have become familiar with the basics of polyphase systems, we will find that it is simple to obtain single-phase power by just connecting to a single "leg" of a polyphase system.

Let us look briefly at the most common polyphase system, a balanced three-phase system. The source has three terminals (not counting a ***neutral*** or ***ground*** connection), and voltmeter measurements will show that sinusoidal voltages of equal amplitude are present between any two terminals. However, these voltages are not in phase; each of the three voltages is 120° out of phase with each of the other two, the sign of the phase angle depending on the sense of the voltages. One possible set of voltages is shown in Fig. 12.1. A ***balanced load*** draws power equally from all three phases. At no instant does the instantaneous power drawn by the total load reach zero; in fact, the total instantaneous power is constant. This is an advantage in rotating machinery, for it keeps the torque on the rotor much more constant than it would be if a single-phase source were used. As a result, there is less vibration.

The use of a higher number of phases, such as 6- and 12-phase systems, is limited almost entirely to the supply of power to large ***rectifiers.*** Rectifiers convert alternating current to direct current by allowing current to flow to the load in only one direction, so that the sign of the voltage

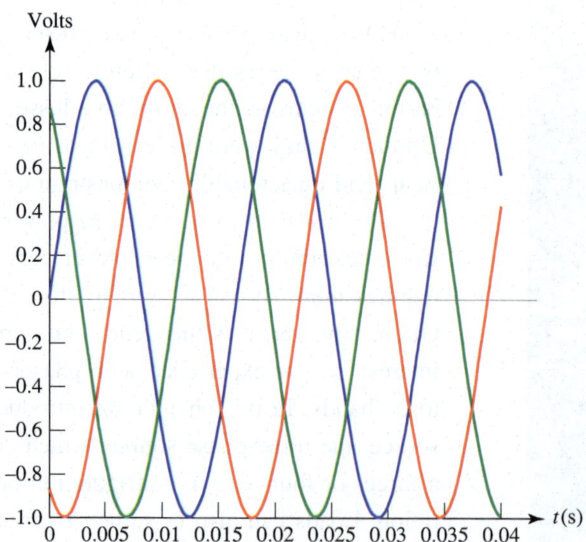

FIGURE 12.1 An example set of three voltages, each of which is 120° out of phase with the other two. As can be seen, only one of the voltages is zero at any particular instant.

across the load remains the same. The rectifier output is a direct current plus a smaller pulsating component, or ripple, which decreases as the number of phases increases.

Almost without exception, polyphase systems in practice contain sources which may be closely approximated by ideal voltage sources or by ideal voltage sources in series with small internal impedances. Three-phase current sources are extremely rare.

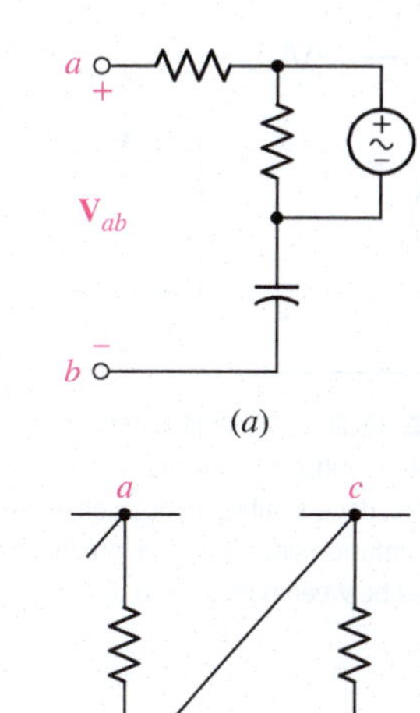

FIGURE 12.2 (a) The definition of the voltage $\mathbf{V}_{ab}$. (b) $\mathbf{V}_{ad} = \mathbf{V}_{ab} + \mathbf{V}_{bc} + \mathbf{V}_{cd} = \mathbf{V}_{ab} + \mathbf{V}_{cd}$.

Double-Subscript Notation

It is convenient to describe polyphase voltages and currents using ***double-subscript notation.*** With this notation, a voltage or current, such as $\mathbf{V}_{ab}$ or $\mathbf{I}_{aA}$, has more meaning than if it were indicated simply as $\mathbf{V}_3$ or $\mathbf{I}_x$. By definition, the voltage of point a with respect to point b is $\mathbf{V}_{ab}$. Thus, the plus sign is located at a, as indicated in Fig. 12.2a. We therefore consider the double subscripts to be *equivalent* to a plus-minus sign pair; the use of both would be redundant. With reference to Fig. 12.2b, for example, we see that $\mathbf{V}_{ad} = \mathbf{V}_{ab} + \mathbf{V}_{cd}$. The advantage of the double-subscript notation lies in the fact that Kirchhoff's voltage law requires the voltage between two points to be the same, regardless of the path chosen between the points; thus $\mathbf{V}_{ad} = \mathbf{V}_{ab} + \mathbf{V}_{bd} = \mathbf{V}_{ac} + \mathbf{V}_{cd} = \mathbf{V}_{ab} + \mathbf{V}_{bc} + \mathbf{V}_{cd}$, and so forth. The benefit of this is that KVL may be satisfied without reference to the circuit diagram; correct equations may be written even though a point, or subscript letter, is included which is not marked on the diagram. For example, we might have written $\mathbf{V}_{ad} = \mathbf{V}_{ax} + \mathbf{V}_{xd}$, where x identifies the location of any interesting point of our choice.

One possible representation of a three-phase system of voltages[1] is shown in Fig. 12.3. Let us assume that the voltages $\mathbf{V}_{an}$, $\mathbf{V}_{bn}$, and $\mathbf{V}_{cn}$ are known:

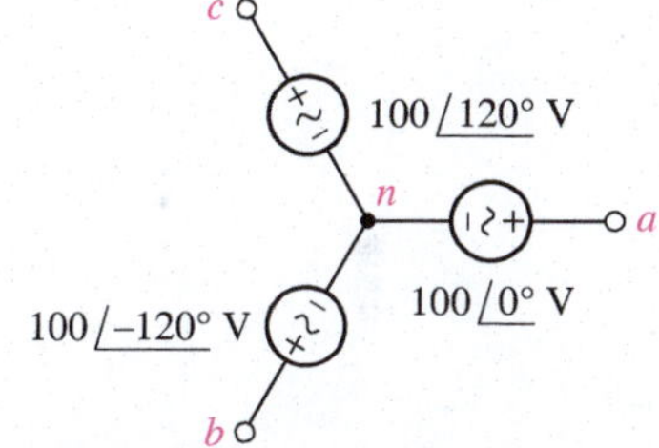

FIGURE 12.3 A network used as a numerical example of double-subscript voltage notation. The voltages are implicitly understood to be in rms units.

$$\mathbf{V}_{an} = 100\underline{/0^\circ}\ \text{V}$$
$$\mathbf{V}_{bn} = 100\underline{/-120^\circ}\ \text{V}$$
$$\mathbf{V}_{cn} = 100\underline{/-240^\circ}\ \text{V}$$

The voltage $\mathbf{V}_{ab}$ may be found, with an eye on the subscripts, as

$$\begin{aligned}\mathbf{V}_{ab} &= \mathbf{V}_{an} + \mathbf{V}_{nb} = \mathbf{V}_{an} - \mathbf{V}_{bn}\\ &= 100\underline{/0^\circ} - 100\underline{/-120^\circ}\ \text{V}\\ &= 100 - (-50 - j86.6)\ \text{V}\\ &= 173.2\underline{/30^\circ}\ \text{V}\end{aligned}$$

The three given voltages and the construction of the phasor $\mathbf{V}_{ab}$ are shown on the phasor diagram of Fig. 12.4.

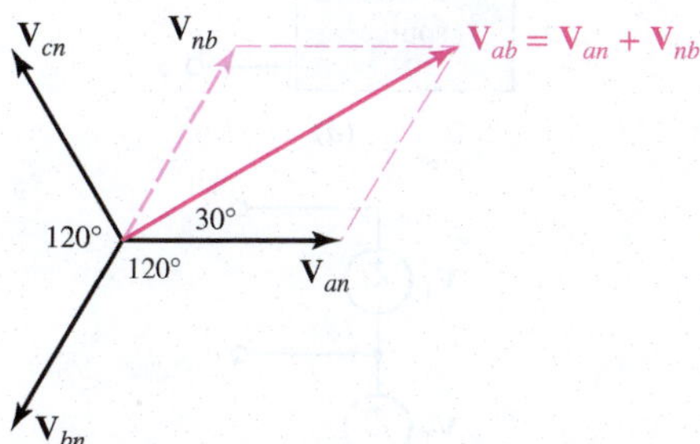

FIGURE 12.4 This phasor diagram illustrates the graphical use of the double-subscript voltage convention to obtain $\mathbf{V}_{ab}$ for the network of Fig. 12.3.

A double-subscript notation may also be applied to currents. We define the current $\mathbf{I}_{ab}$ as the current flowing from a to b by the most direct path. In every complete circuit we consider, there must of course be at least two possible paths between the points a and b, and we agree that we will not use double-subscript notation unless it is obvious that one path is much shorter, or much more direct. Usually this path is through a single element.

(1) In keeping with power industry convention, rms values for currents and voltages will be used *implicitly* throughout this chapter.

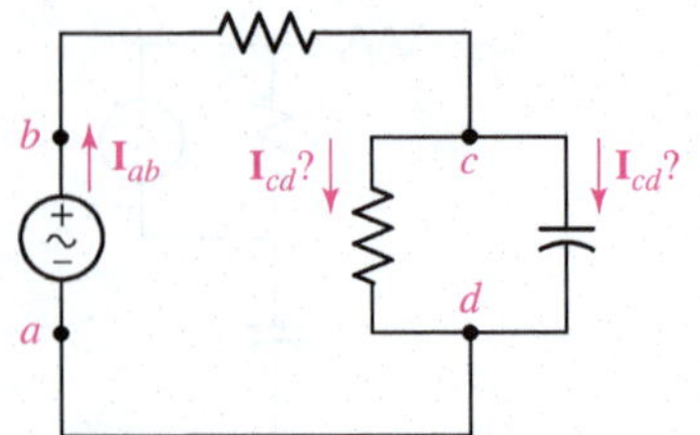

■ **FIGURE 12.5** An illustration of the use and *misuse* of the double-subscript convention for current notation. In this case, double-subscript notation to describe $\mathbf{I}_{cd}$ creates confusion since there is more than component connected between nodes *c* and *d*.

Thus, the current $\mathbf{I}_{ab}$ is correctly indicated in Fig. 12.5. In fact, we do not even need the direction arrow when talking about this current; the subscripts *tell* us the direction. However, the identification of a current as $\mathbf{I}_{cd}$ for the circuit of Fig. 12.5 would cause confusion.

PRACTICE

12.1 Let $\mathbf{V}_{ab} = 100\underline{/0°}$ V, $\mathbf{V}_{bd} = 40\underline{/80°}$ V, and $\mathbf{V}_{ca} = 70\underline{/200°}$ V. Find (*a*) $\mathbf{V}_{ad}$; (*b*) $\mathbf{V}_{bc}$; (*c*) $\mathbf{V}_{cd}$.

12.2 Refer to the circuit of Fig. 12.6 and let $\mathbf{I}_{fj} = 3$ A, $\mathbf{I}_{de} = 2$ A, and $\mathbf{I}_{hd} = -6$ A. Find (*a*) $\mathbf{I}_{cd}$; (*b*) $\mathbf{I}_{ef}$; (*c*) $\mathbf{I}_{ij}$.

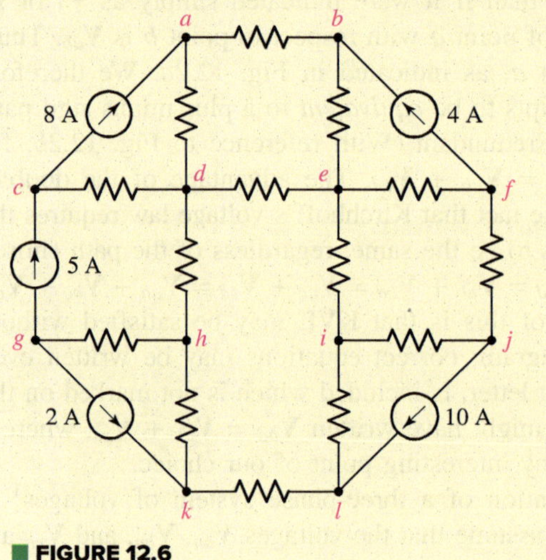

■ **FIGURE 12.6**

Ans: 12.1: $114.0\underline{/20.2°}$ V; $41.8\underline{/145.0°}$ V; $44.0\underline{/20.6°}$ V. 12.2: −3 A; 7 A; 7 A.

12.2 SINGLE-PHASE THREE-WIRE SYSTEMS

Before studying polyphase systems in detail, it can be helpful to start with a simple single-phase three-wire system. A *single-phase three-wire source* is defined as a source having three output terminals, such as *a*, *n*, and *b* in Fig. 12.7*a*, at which the phasor voltages $\mathbf{V}_{an}$ and $\mathbf{V}_{nb}$ are equal. The source may therefore be represented by the combination of two identical voltage sources; in Fig. 12.7*b*, $\mathbf{V}_{an} = \mathbf{V}_{nb} = \mathbf{V}_1$. It is apparent that $\mathbf{V}_{ab} = 2\mathbf{V}_{an} = 2\mathbf{V}_{nb}$, and we therefore have a source to which loads operating at either of two voltages may be connected. The normal North American household system is single-phase three-wire, permitting the operation of both 120 V and 240 V appliances. The higher-voltage appliances are normally those drawing larger amounts of power; operation at higher voltage results in a smaller current draw for the same power. Smaller-diameter wire may consequently be used safely in the appliance, the household distribution system, and the distribution system of the utility company, as larger-diameter wire must be used with higher currents to reduce the heat produced due to the resistance of the wire.

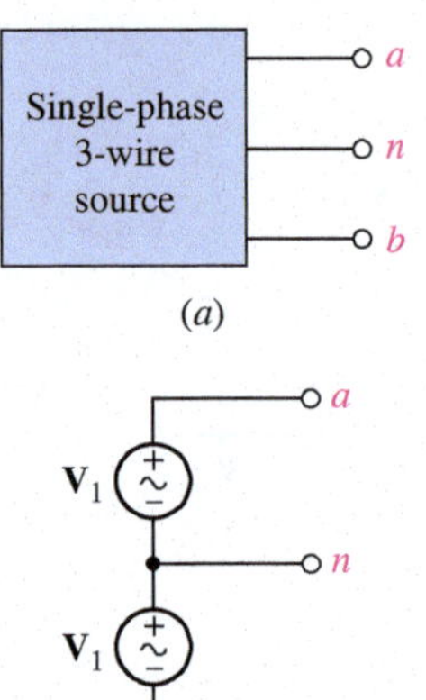

■ **FIGURE 12.7** (*a*) A single-phase three-wire source. (*b*) The representation of a single-phase three-wire source by two identical voltage sources.

The name ***single-phase*** arises because the voltages $\mathbf{V}_{an}$ and $\mathbf{V}_{nb}$, being equal, must have the same phase angle. From another viewpoint, however, the voltages between the outer wires and the central wire, which is usually referred to as the *neutral,* are exactly 180° out of phase. That is, $\mathbf{V}_{an} = -\mathbf{V}_{bn}$, and $\mathbf{V}_{an} + \mathbf{V}_{bn} = 0$. Later, we will see that balanced polyphase systems are characterized by a set of voltages of equal *amplitude* whose (phasor) sum is zero. From this viewpoint, then, the single-phase three-wire system is really a balanced two-phase system. *Two-phase,* however, is a term that is traditionally reserved for a relatively unimportant unbalanced system utilizing two voltage sources 90° out of phase.

Let us now consider a single-phase three-wire system that contains identical loads $\mathbf{Z}_p$ between each outer wire and the neutral (Fig. 12.8). We first assume that the wires connecting the source to the load are perfect conductors. Since

$$\mathbf{V}_{an} = \mathbf{V}_{nb}$$

then,

$$\mathbf{I}_{aA} = \frac{\mathbf{V}_{an}}{\mathbf{Z}_p} = \mathbf{I}_{Bb} = \frac{\mathbf{V}_{nb}}{\mathbf{Z}_p}$$

and therefore

$$\mathbf{I}_{nN} = \mathbf{I}_{Bb} + \mathbf{I}_{Aa} = \mathbf{I}_{Bb} - \mathbf{I}_{aA} = 0$$

Thus, there is no current in the neutral wire, and it could be removed without changing any current or voltage in the system. This result is achieved through the equality of the two loads and of the two sources.

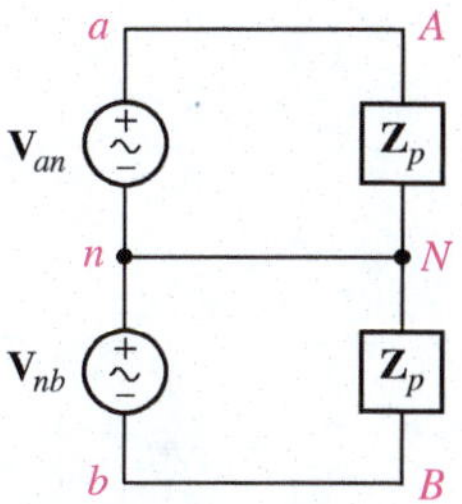

FIGURE 12.8 A simple single-phase three-wire system. The two loads are identical, and the neutral current is zero.

Effect of Finite Wire Impedance

We next consider the effect of a finite impedance in each of the wires. If lines aA and bB each have the same impedance, this impedance may be added to $\mathbf{Z}_p$, resulting in two equal loads once more, and zero neutral current. Now let us allow the neutral wire to possess some impedance $\mathbf{Z}_n$. Without carrying out any detailed analysis, superposition should show us that the symmetry of the circuit will still cause zero neutral current. Moreover, the addition of any impedance connected directly from one of the outer lines to the other outer line also yields a symmetrical circuit and zero neutral current. Thus, zero neutral current is a consequence of a balanced, or symmetrical, load; nonzero impedance in the neutral wire does not destroy the symmetry.

The most general single-phase three-wire system will contain unequal loads between each outside line and the neutral and another load directly between the two outer lines; the impedances of the two outer lines may be expected to be approximately equal, but the neutral impedance is often slightly larger. Let us consider an example of such a system, with particular interest in the current that may flow now through the neutral wire, as well as the overall efficiency with which our system is transmitting power to the unbalanced load.

EXAMPLE 12.1

Analyze the system shown in Fig. 12.9 and determine the power delivered to each of the three loads as well as the power lost in the neutral wire and each of the two lines.

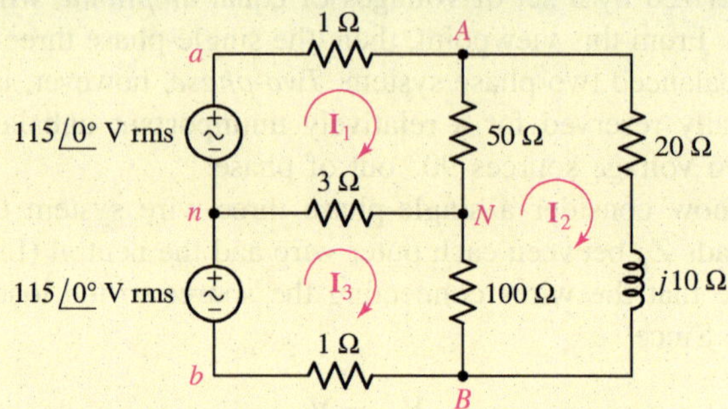

■ **FIGURE 12.9** A typical single-phase three-wire system.

▶ ***Identify the goal of the problem.***

The three loads in the circuit are the 50 Ω resistor, the 100 Ω resistor, and a $20 + j10$ Ω impedance. Each of the two lines has a resistance of 1 Ω, and the neutral wire has a resistance of 3 Ω. We need the current through each of these in order to determine power.

▶ ***Collect the known information.***

We have a single-phase three-wire system; the circuit diagram of Fig. 12.9 is completely labeled. The computed currents will be in rms units.

▶ ***Devise a plan.***

The circuit is conducive to mesh analysis, having three clearly defined meshes. The result of the analysis will be a set of mesh currents, which can then be used to compute absorbed power.

▶ ***Construct an appropriate set of equations.***

The three mesh equations are:

$$\begin{aligned} -115\underline{/0^\circ} + \mathbf{I}_1 + 50(\mathbf{I}_1 - \mathbf{I}_2) + 3(\mathbf{I}_1 - \mathbf{I}_3) &= 0 \\ (20 + j10)\mathbf{I}_2 + 100(\mathbf{I}_2 - \mathbf{I}_3) + 50(\mathbf{I}_2 - \mathbf{I}_1) &= 0 \\ -115\underline{/0^\circ} + 3(\mathbf{I}_3 - \mathbf{I}_1) + 100(\mathbf{I}_3 - \mathbf{I}_2) + \mathbf{I}_3 &= 0 \end{aligned}$$

which can be rearranged to obtain the following three equations

$$\begin{aligned} 54\,\mathbf{I}_1 \quad -50\,\mathbf{I}_2 - 3\,\mathbf{I}_3 &= 115\underline{/0^\circ} \\ -50\,\mathbf{I}_1 + (170 + j10)\mathbf{I}_2 - 100\,\mathbf{I}_3 &= 0 \\ -3\,\mathbf{I}_1 \quad -100\,\mathbf{I}_2 + 100\,\mathbf{I}_3 &= 115\underline{/0^\circ} \end{aligned}$$

▶ ***Determine if additional information is required.***

We have a set of three equations in three unknowns, so it is possible to attempt a solution at this point.

▶ ***Attempt a solution.***

Solving for the phasor currents $\mathbf{I}_1$, $\mathbf{I}_2$, and $\mathbf{I}_3$ using a scientific calculator, we find

$$\mathbf{I}_1 = 11.24\underline{/-19.83^\circ}\ \text{A}$$
$$\mathbf{I}_2 = 9.389\underline{/-24.47^\circ}\ \text{A}$$
$$\mathbf{I}_3 = 10.37\underline{/-21.80^\circ}\ \text{A}$$

The currents in the outer lines are thus

$$\mathbf{I}_{aA} = \mathbf{I}_1 = 11.24\underline{/-19.83^\circ}\ \text{A}$$

and

$$\mathbf{I}_{bB} = \mathbf{I}_3 = 10.37\underline{/158.20^\circ}\ \text{A}$$

and the smaller neutral current is

$$\mathbf{I}_{nN} = \mathbf{I}_3 - \mathbf{I}_1 = 0.9459\underline{/-177.7^\circ}\ \text{A}$$

The average power drawn by each load may thus be determined:

$$P_{50} = |\mathbf{I}_1 - \mathbf{I}_2|^2(50) = 206\ \text{W}$$
$$P_{100} = |\mathbf{I}_3 - \mathbf{I}_2|^2(100) = 117\ \text{W}$$
$$P_{20} + {}_{j10} = |\mathbf{I}_2|^2(20) = 1763\ \text{W}$$

Note that we do not need to include a factor of $\frac{1}{2}$ since we are working with rms current values.

The total load power is 2086 W. The loss in each of the wires is next found:

$$P_{aA} = |\mathbf{I}_1|^2(1) = 126\ \text{W}$$
$$P_{bB} = |\mathbf{I}_3|^2(1) = 108\ \text{W}$$
$$P_{nN} = |\mathbf{I}_{nN}|^2(3) = 3\ \text{W}$$

giving a total line loss of 237 W. The wires are evidently quite long; otherwise, the relatively high power loss in the two outer lines would cause a dangerous temperature rise.

Imagine the heat produced by two 100 W light bulbs! These outer wires must dissipate the same amount of power. In order to keep their temperature down, a large surface area is required.

▶ ***Verify the solution. Is it reasonable or expected?***

The total absorbed power is 206 + 117 + 1763 + 237, or 2323 W, which may be checked by finding the power delivered by each voltage source:

$$P_{an} = 115(11.24)\cos 19.83^\circ = 1216\ \text{W}$$
$$P_{bn} = 115(10.37)\cos 21.80^\circ = 1107\ \text{W}$$

or a total of 2323 W. The transmission efficiency for the system is

$$\eta = \frac{\text{total power delivered to load}}{\text{total power generated}} = \frac{2086}{2086 + 237} = 89.8\%$$

This value would be unbelievable for a steam engine or an internal combustion engine, but it is too low for a well-designed distribution system. Larger-diameter wires should be used if the source and the load cannot be placed closer to each other.

(Continued on next page)

A phasor diagram showing the two source voltages, the currents in the outer lines, and the current in the neutral is constructed in Fig. 12.10. The fact that $\mathbf{I}_{aA} + \mathbf{I}_{bB} + \mathbf{I}_{nN} = 0$ is indicated on the diagram.

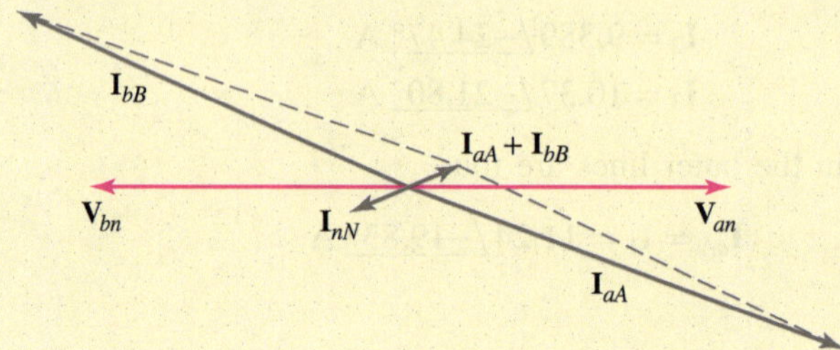

FIGURE 12.10 The source voltages and three of the currents in the circuit of Fig. 12.9 are shown on a phasor diagram. Note that $\mathbf{I}_{aA} + \mathbf{I}_{bB} + \mathbf{I}_{nN} = 0$.

PRACTICE

12.3 Modify Fig. 12.9 by adding a 1.5 Ω resistance to each of the two outer lines, and a 2.5 Ω resistance to the neutral wire. Find the average power delivered to each of the three loads.

Ans: 153.1 W; 95.8 W; 1374 W.

12.3 THREE-PHASE Y-Y CONNECTION

Three-phase sources have three terminals, called the *line* terminals, and they may or may not have a fourth terminal, the *neutral* connection. We will begin by discussing a three-phase source that does have a neutral connection. It may be represented by three ideal voltage sources connected in a Y, as shown in Fig. 12.11; terminals *a*, *b*, *c*, and *n* are available. We will consider only balanced three-phase sources, which may be defined as having

$$|\mathbf{V}_{an}| = |\mathbf{V}_{bn}| = |\mathbf{V}_{cn}|$$

and

$$\mathbf{V}_{an} + \mathbf{V}_{bn} + \mathbf{V}_{cn} = 0$$

FIGURE 12.11 A Y-connected three-phase four-wire source.

These three voltages, each existing between one line and the neutral, are called ***phase voltages.*** If we arbitrarily choose $\mathbf{V}_{an}$ as the reference, or define

$$\mathbf{V}_{an} = V_p\underline{/0°}$$

where we will consistently use V_p to represent the rms *amplitude* of any of the phase voltages, then the definition of the three-phase source indicates that either

$$\mathbf{V}_{bn} = V_p\underline{/-120°} \qquad \text{and} \qquad \mathbf{V}_{cn} = V_p\underline{/-240°}$$

or

$$\mathbf{V}_{bn} = V_p\underline{/120°} \qquad \text{and} \qquad \mathbf{V}_{cn} = V_p\underline{/240°}$$

The former is called ***positive phase sequence,*** or ***abc phase sequence,*** and is shown in Fig. 12.12*a*; the latter is termed ***negative phase sequence,*** or ***cba phase sequence,*** and is indicated by the phasor diagram of Fig. 12.12*b*.

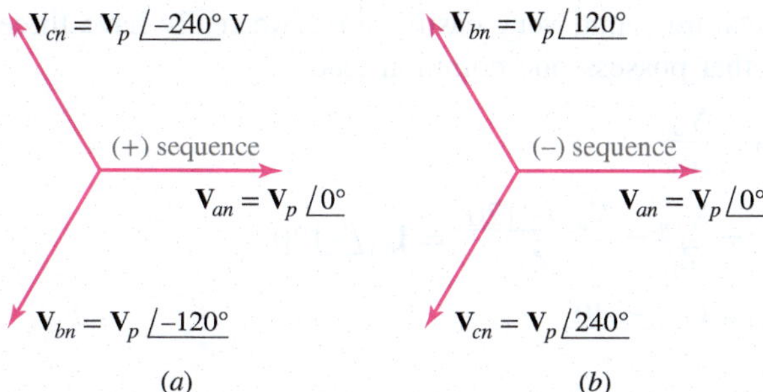

FIGURE 12.12 (*a*) Positive, or *abc*, phase sequence. (*b*) Negative, or *cba*, phase sequence.

The actual phase sequence of a physical three-phase source depends on the arbitrary choice of the three terminals to be lettered *a*, *b*, and *c*. They may always be chosen to provide positive phase sequence, and we will assume that this has been done in most of the systems we consider.

Line-to-Line Voltages

Let us next find the line-to-line voltages (often simply called the ***line voltages***) which are present when the phase voltages are those of Fig. 12.12*a*. It is easiest to do this with the help of a phasor diagram, since the angles are all multiples of 30°. The necessary construction is shown in Fig. 12.13; the results are

$$\mathbf{V}_{ab} = \sqrt{3}\, V_p \angle 30° \qquad [1]$$

$$\mathbf{V}_{bc} = \sqrt{3}\, V_p \angle -90° \qquad [2]$$

$$\mathbf{V}_{ca} = \sqrt{3}\, V_p \angle -210° \qquad [3]$$

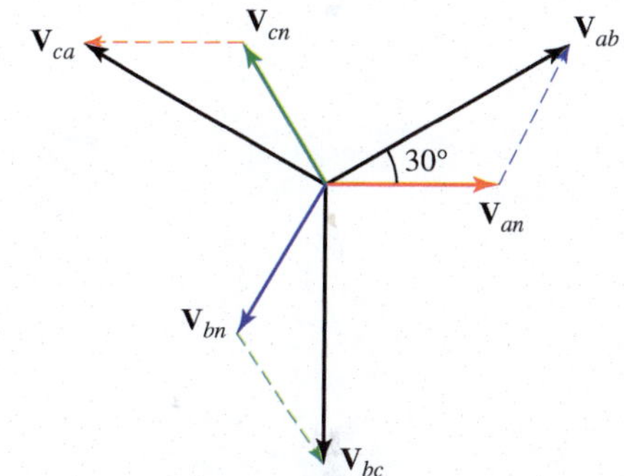

FIGURE 12.13 A phasor diagram which is used to determine the line voltages from the given phase voltages. Or, algebraically, $\mathbf{V}_{ab} = \mathbf{V}_{an} - \mathbf{V}_{bn} = V_p\angle 0° - V_p\angle -120° = V_p - V_p\cos(-120°) - jV_p\sin(-120°) = V_p(1 + \frac{1}{2} + j\sqrt{3}/2) = \sqrt{3}\,V_p\angle 30°$.

Kirchhoff's voltage law requires the sum of these three voltages to be zero; the reader is encouraged to verify this as an exercise.

If the rms amplitude of any of the line voltages is denoted by V_L, then one of the important characteristics of the Y-connected three-phase source may be expressed as

$$V_L = \sqrt{3}\, V_p$$

Note that with positive phase sequence, $\mathbf{V}_{an}$ leads $\mathbf{V}_{bn}$ and $\mathbf{V}_{bn}$ leads $\mathbf{V}_{cn}$, in each case by 120°, and also that $\mathbf{V}_{ab}$ leads $\mathbf{V}_{bc}$ and $\mathbf{V}_{bc}$ leads $\mathbf{V}_{ca}$, again by 120°. The statement is true for negative phase sequence if "lags" is substituted for "leads."

Now let us connect a balanced Y-connected three-phase load to our source, using three lines and a neutral, as drawn in Fig. 12.14. The load is represented by an impedance $\mathbf{Z}_p$ between each line and the neutral wire.

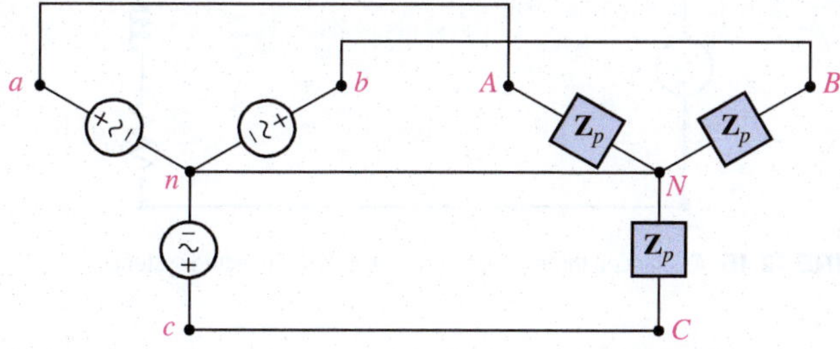

FIGURE 12.14 A balanced three-phase system, connected Y-Y and including a neutral.

The three line currents are found very easily, since we really have three single-phase circuits that possess one common lead:[2]

$$\mathbf{I}_{aA} = \frac{\mathbf{V}_{an}}{\mathbf{Z}_p}$$

$$\mathbf{I}_{bB} = \frac{\mathbf{V}_{bn}}{\mathbf{Z}_p} = \frac{\mathbf{V}_{an}\underline{/-120^\circ}}{\mathbf{Z}_p} = \mathbf{I}_{aA}\underline{/-120^\circ}$$

$$\mathbf{I}_{cC} = \mathbf{I}_{aA}\underline{/-240^\circ}$$

and therefore

$$\mathbf{I}_{Nn} = \mathbf{I}_{aA} + \mathbf{I}_{bB} + \mathbf{I}_{cC} = 0$$

Thus, the neutral carries no current if the source and load are both balanced and if the four wires have zero impedance. How will this change if an impedance $\mathbf{Z}_L$ is inserted in series with each of the three lines and an impedance $\mathbf{Z}_n$ is inserted in the neutral? The line impedances may be combined with the three load impedances; this effective load is still balanced, and a perfectly conducting neutral wire could be removed. Thus, if no change is produced in the system with a short circuit or an open circuit between n and N, any impedance may be inserted in the neutral and the neutral current will remain zero.

It follows that, if we have balanced sources, balanced loads, and balanced line impedances, a neutral wire of any impedance may be replaced by any other impedance, including a short circuit or an open circuit; the replacement will not affect the system's voltages or currents. It is often helpful to *visualize* a short circuit between the two neutral points, whether a neutral wire is actually present or not; the problem is then reduced to three single-phase problems, all identical except for the consistent difference in phase angle. We say that we thus work the problem on a "per-phase" basis.

EXAMPLE 12.2

For the circuit of Fig. 12.15, find both the phase and line currents, and the phase and line voltages throughout the circuit; then calculate the total power dissipated in the load.

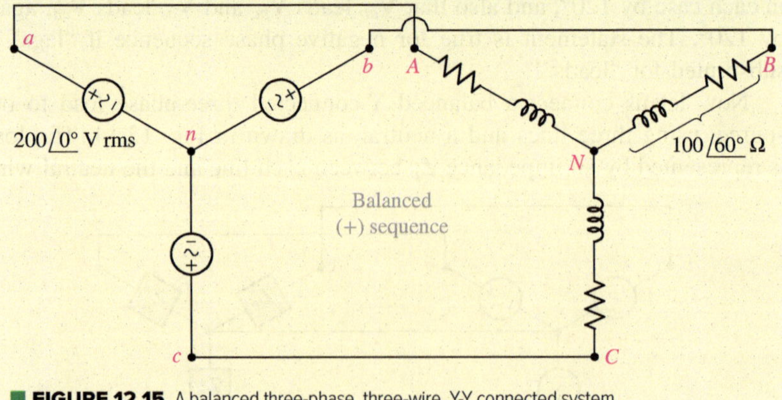

■ **FIGURE 12.15** A balanced three-phase, three-wire, Y-Y connected system.

(2) This can be seen to be true by applying superposition and looking at each phase one at a time.

Since one of the source phase voltages is given and we are told to use the positive phase sequence, the three phase voltages are:

$$\mathbf{V}_{an} = 200\underline{/0^\circ} \text{ V} \qquad \mathbf{V}_{bn} = 200\underline{/-120^\circ} \text{ V} \qquad \mathbf{V}_{cn} = 200\underline{/-240^\circ} \text{ V}$$

The line voltage is $200\sqrt{3} = 346$ V; the phase angle of each line voltage can be determined by constructing a phasor diagram, as we did in Fig. 12.13 (as a matter of fact, the phasor diagram of Fig. 12.13 is applicable), subtracting the phase voltages using a scientific calculator, or by invoking Eqs. [1] to [3]. We find that $\mathbf{V}_{ab} = 346\underline{/30^\circ}$ V, $\mathbf{V}_{bc} = 346\underline{/-90^\circ}$ V, and $\mathbf{V}_{ca} = 346\underline{/-210^\circ}$ V.

The line current for phase A is

$$\mathbf{I}_{aA} = \frac{\mathbf{V}_{an}}{\mathbf{Z}_p} = \frac{200\underline{/0^\circ}}{100\underline{/60^\circ}} = 2\underline{/-60^\circ} \text{ A}$$

Since we know this is a balanced three-phase system, we may write the remaining line currents based on $\mathbf{I}_{aA}$:

$$\mathbf{I}_{bB} = 2\underline{/(-60^\circ - 120^\circ)} = 2\underline{/-180^\circ} \text{ A}$$

$$\mathbf{I}_{cC} = 2\underline{/(-60^\circ - 240^\circ)} = 2\underline{/-300^\circ} \text{ A}$$

Finally, the average power absorbed by phase A is $\text{Re}\{\mathbf{V}_{an}\,\mathbf{I}_{aA}^*\}$, or

$$P_{AN} = 200(2)\cos(0^\circ + 60^\circ) = 200 \text{ W}$$

Thus, the total average power drawn by the three-phase load is 600 W.

The phasor diagram for this circuit is shown in Fig. 12.16. Once we knew any of the line voltage magnitudes and any of the line current magnitudes, the angles for all three voltages and all three currents could have been obtained by simply reading the diagram.

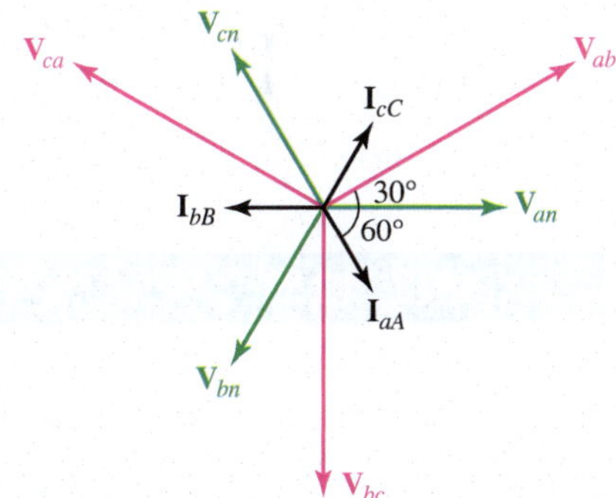

FIGURE 12.16 The phasor diagram that applies to the circuit of Fig. 12.15.

PRACTICE

12.4 A balanced three-phase three-wire system has a Y-connected load. Each phase contains three loads in parallel: $-j100\ \Omega$, $100\ \Omega$, and $50 + j50\ \Omega$. Assume positive phase sequence with $\mathbf{V}_{ab} = 400\underline{/0^\circ}$ V. Find (*a*) $\mathbf{V}_{an}$; (*b*) $\mathbf{I}_{aA}$; (*c*) the total power drawn by the load.

Ans: $231\underline{/-30^\circ}$ V; $4.62\underline{/-30^\circ}$ A; 3200 W.

Before working another example, this would be a good opportunity to quickly explore a statement made in Sec. 12.1, that even though phase voltages and currents have zero value at specific instants in time (every 1/120 s in North America), the instantaneous power delivered to the *total* load is never zero. Consider phase A of Example 12.2 once more, with the phase voltage and current written in the time domain:

The factor of $\sqrt{2}$ is required to convert from rms units.

$$v_{AN} = 200\sqrt{2}\cos(120\pi t + 0^\circ) \text{ V}$$

and

$$i_{AN} = 2\sqrt{2}\cos(120\pi t + 60^\circ) \text{ A}$$

Thus, the instantaneous power absorbed by phase A is

$$\begin{aligned} p_A(t) = v_{AN}i_{AN} &= 800\cos(120\pi t)\cos(120\pi t - 60^\circ) \\ &= 400[\cos(-60^\circ) + \cos(240\pi t - 60^\circ)] \\ &= 200 + 400\cos(240\pi t - 60^\circ)\text{ W} \end{aligned}$$

in a similar fashion,

$$p_B(t) = 200 + 400\cos(240\pi t - 300^\circ)\text{ W}$$

and

$$p_C(t) = 200 + 400\cos(240\pi t - 180^\circ)\text{ W}$$

The instantaneous power absorbed by the *total* load is therefore

$$p(t) = p_A(t) + p_B(t) + p_C(t) = 600\text{ W}$$

independent of time, and the same value as the average power computed in Example 12.2.

EXAMPLE 12.3

A balanced three-phase system with a line voltage of 300 V is supplying a balanced Y-connected load with 1200 W at a leading PF of 0.8. Find the line current and the per-phase load impedance.

The phase voltage is $300/\sqrt{3}$ V and the per-phase power is 1200/3 = 400 W. Thus the line current may be found from the power relationship

$$400 = \frac{300}{\sqrt{3}}(I_L)(0.8)$$

and the line current is therefore 2.89 A. The phase impedance magnitude is given by

$$|\mathbf{Z}_P| = \frac{V_P}{I_L} = \frac{300/\sqrt{3}}{2.89} = 60\ \Omega$$

Since the PF is 0.8 leading, the impedance phase angle is −36.9°; thus $\mathbf{Z}_p = 60\underline{/-36.9^\circ}\ \Omega$.

PRACTICE

12.5 A balanced three-phase three-wire system has a line voltage of 500 V. Two balanced Y-connected loads are present. One is a capacitive load with $7 - j2\ \Omega$ per phase, and the other is an inductive load with $4 + j2\ \Omega$ per phase. Find (*a*) the phase voltage; (*b*) the line current; (*c*) the total power drawn by the load; (*d*) the power factor at which the source is operating.

Ans: 289 V; 97.5 A; 83.0 kW; 0.983 lagging.

EXAMPLE 12.4

A balanced 600 W lighting load is added (in parallel) to the system of Example 12.3. Determine the new line current.

We first sketch a suitable per-phase circuit, as shown in Fig. 12.17. The 600 W load is assumed to be a balanced load evenly distributed among the three phases, resulting in an additional 200 W consumed by each phase.

The amplitude of the lighting current (labeled $\mathbf{I}_1$) is determined by

$$200 = \frac{300}{\sqrt{3}} |\mathbf{I}_1| \cos 0°$$

so that

$$|\mathbf{I}_1| = 1.155 \text{ A}$$

In a similar way, the amplitude of the capacitive load current (labeled $\mathbf{I}_2$) is found to be unchanged from its previous value, since the voltage across it has remained the same:

$$|\mathbf{I}_2| = 2.89 \text{ A}$$

If we assume that the phase with which we are working has a phase voltage with an angle of 0°, then since $\cos^{-1}(0.8) = 36.9°$,

$$\mathbf{I}_1 = 1.155\,\underline{/0°} \text{ A} \qquad \mathbf{I}_2 = 2.89\,\underline{/+36.9°} \text{ A}$$

and the line current is

$$\mathbf{I}_L = \mathbf{I}_1 + \mathbf{I}_2 = 3.87\,\underline{/+26.6°} \text{ A}$$

We can check our results by computing the power generated by this phase of the source

$$P_p = \frac{300}{\sqrt{3}} 3.87 \cos(+26.6°) = 600 \text{ W}$$

which agrees with the fact that the individual phase is known to be supplying 200 W to the new lighting load, as well as 400 W to the original load.

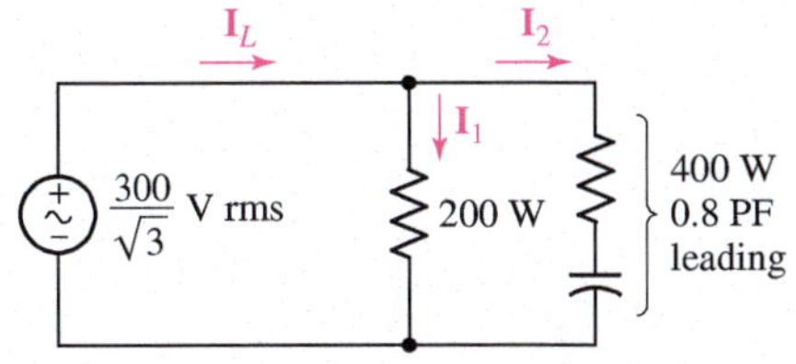

FIGURE 12.17 The per-phase circuit that is used to analyze a *balanced* three-phase example.

PRACTICE

12.6 Three balanced Y-connected loads are installed on a balanced three-phase four-wire system. Load 1 draws a total power of 6 kW at unity PF, load 2 pulls 10 kVA at PF = 0.96 lagging, and load 3 demands 7 kW at 0.85 lagging. If the phase voltage at the loads is 135 V, if each line has a resistance of 0.1 Ω, and if the neutral has a resistance of 1 Ω, find (*a*) the total power drawn by the loads; (*b*) the combined PF of the loads; (*c*) the total power lost in the four lines; (*d*) the phase voltage at the source; (*e*) the power factor at which the source is operating.

Ans: 22.6 kW; 0.954 lag; 1027 W; 140.6 V; 0.957 lagging.

If an *unbalanced* Y-connected load is present in an otherwise balanced three-phase system, the circuit may still be analyzed on a per-phase basis *if* the neutral wire is present and *if* it has zero impedance. If either of these conditions is not met, other methods must be used, such as mesh or nodal analysis. However, engineers who spend most of their time with unbalanced three-phase systems will find the use of *symmetrical components* a great time saver. We leave this topic for more specialized texts.

12.4 THE DELTA (Δ) CONNECTION

An alternative to the Y-connected load is the Δ-connected configuration, as shown in Fig. 12.18. This type of configuration is very common and does not have a neutral connection.

Let us consider a balanced Δ-connected load which consists of an impedance $\mathbf{Z}_p$ inserted between each pair of lines. With reference to Fig. 12.18, let us assume known line voltages

$$V_L = |\mathbf{V}_{ab}| = |\mathbf{V}_{bc}| = |\mathbf{V}_{ca}|$$

or known phase voltages

$$V_P = |\mathbf{V}_{an}| = |\mathbf{V}_{bn}| = |\mathbf{V}_{cn}|$$

where

$$V_L = \sqrt{3}\,V_p \qquad \text{and} \qquad \mathbf{V}_{ab} = \sqrt{3}\,V_p\underline{/30^\circ}$$

as we found previously. Because the voltage across each branch of the Δ is known, the *phase currents* are easily found:

$$\mathbf{I}_{AB} = \frac{\mathbf{V}_{ab}}{\mathbf{Z}_p} \qquad \mathbf{I}_{BC} = \frac{\mathbf{V}_{bc}}{\mathbf{Z}_p} \qquad \mathbf{I}_{CA} = \frac{\mathbf{V}_{ca}}{\mathbf{Z}_p}$$

and their differences provide us with the line currents, such as

$$\mathbf{I}_{aA} = \mathbf{I}_{AB} - \mathbf{I}_{CA}$$

Since we are working with a balanced system, the three phase currents are of equal amplitude:

$$I_p = |\mathbf{I}_{AB}| = |\mathbf{I}_{BC}| = |\mathbf{I}_{CA}|$$

The line currents are also equal in amplitude; the symmetry is apparent from the phasor diagram of Fig. 12.19. We thus have

$$I_L = |\mathbf{I}_{aA}| = |\mathbf{I}_{bB}| = |\mathbf{I}_{cC}|$$

and

$$I_L = \sqrt{3}\,I_p$$

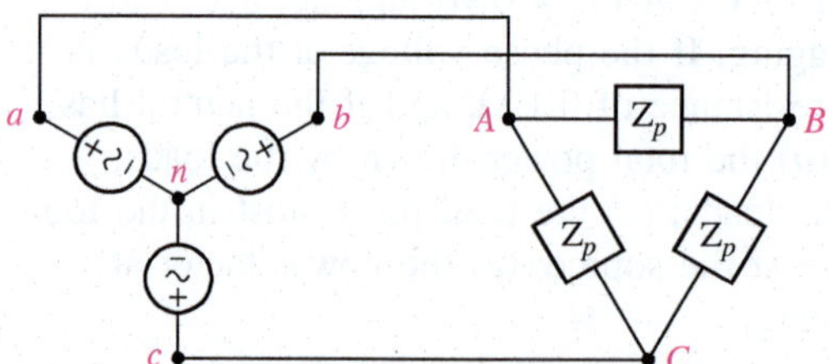

FIGURE 12.18 A balanced Δ-connected load is present on a three-wire three-phase system. The source happens to be Y-connected.

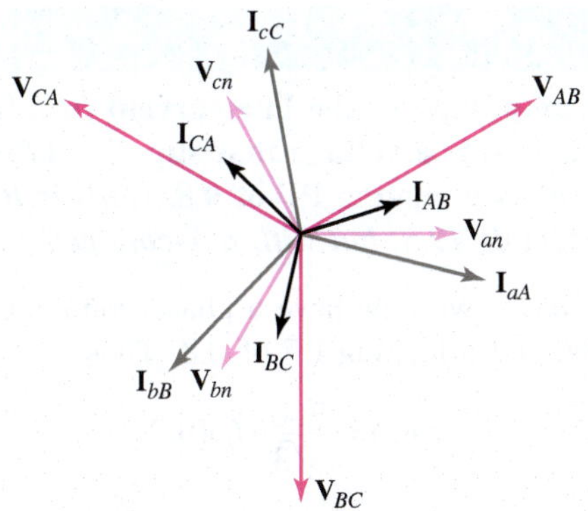

FIGURE 12.19 A phasor diagram that could apply to the circuit of Fig. 12.18 if $\mathbf{Z}_p$ were an inductive impedance.

Let us disregard the source for the moment and consider only the balanced load. If the load is Δ-connected, then the phase voltage and the line voltage are indistinguishable, but the line current is larger than the phase current by a factor of $\sqrt{3}$; with a Y-connected load, however, the phase current and the line current refer to the same current, and the line voltage is greater than the phase voltage by a factor of $\sqrt{3}$.

EXAMPLE **12.5**

Determine the amplitude of the line current in a three-phase system with a line voltage of 300 V that supplies 1200 W to a Δ-connected load at a lagging PF of 0.8; then find the phase impedance.

Let us again consider a single phase. It draws 400 W, 0.8 lagging PF, at a 300 V line voltage. Thus,

$$400 = 300(I_p)(0.8)$$

and

$$I_p = 1.667 \text{ A}$$

and the relationship between phase currents and line currents yields

$$I_L = \sqrt{3}(1.667) = 2.89 \text{ A}$$

Next, the phase angle of the load is $\cos^{-1}(0.8) = 36.9°$, and therefore the impedance in each phase must be

$$\mathbf{Z}_p = \frac{300}{1.667}\underline{/36.9°} = 180\underline{/36.9°}\ \Omega$$

Again, keep in mind that we are assuming all voltages and currents are quoted as rms values.

PRACTICE

12.7 Each phase of a balanced three-phase Δ-connected load consists of a 200 mH inductor in series with the parallel combination of a 5 μF capacitor and a 200 Ω resistance. Assume zero line resistance and a phase voltage of 200 V at ω = 400 rad/s. Find (*a*) the phase current; (*b*) the line current; (*c*) the total power absorbed by the load.

Ans: 1.158 A; 2.01 A; 693 W.

EXAMPLE 12.6

Determine the amplitude of the line current in a three-phase system with a 300 V line voltage that supplies 1200 W to a Y-connected load at a lagging PF of 0.8. (*This is the same circuit as in Example 12.5, but with a Y-connected load instead.*)

On a per-phase basis, we now have a phase voltage of $300/\sqrt{3}$ V, a power of 400 W, and a lagging PF of 0.8. Thus,

$$400 = \frac{300}{\sqrt{3}}(I_p)(0.8)$$

and

$$I_p = 2.89 \qquad \text{(and so } I_L = 2.89 \text{ A)}$$

The phase angle of the load is again 36.9°, and thus the impedance in each phase of the Y is

$$\mathbf{Z}_p = \frac{300/\sqrt{3}}{2.89}\underline{/36.9^\circ} = 60\underline{/36.9^\circ}\ \Omega$$

The $\sqrt{3}$ factor not only relates phase and line quantities but also appears in a useful expression for the total power drawn by any balanced three-phase load. If we assume a Y-connected load with a power-factor angle θ, the power taken by any phase is

$$P_p = V_p I_p \cos\theta = V_p I_L \cos\theta = \frac{V_L}{\sqrt{3}} I_L \cos\theta$$

and the total power is

$$P = 3\,P_p = \sqrt{3}\,V_L I_L \cos\theta$$

In a similar way, the power delivered to each phase of a Δ-connected load is

$$P_p = V_p I_p \cos\theta = V_L I_P \cos\theta = V_L \frac{I_L}{\sqrt{3}} \cos\theta$$

giving a total power

$$P = 3\,P_p = \sqrt{3}\,V_L I_L \cos\theta \qquad [4]$$

Thus Eq. [4] enables us to calculate the total power delivered to a balanced load from a knowledge of the magnitude of the line voltage, of the line current, and of the phase angle of the load impedance (or admittance), regardless of whether the load is Y-connected or Δ-connected. The line current in Examples 12.5 and 12.6 can now be obtained in two simple steps:

$$1200 = \sqrt{3}(300)(I_L)(0.8)$$

Therefore,

$$I_L = \frac{5}{\sqrt{3}} = 2.89 \text{ A}$$

PRACTICE

12.8 A balanced three-phase three-wire system is terminated with two Δ-connected loads in parallel. Load 1 draws 40 kVA at a lagging PF of 0.8, while load 2 absorbs 24 kW at a leading PF of 0.9. Assume no line resistance, and let $\mathbf{V}_{ab} = 440\underline{/30^\circ}$ V. Find (*a*) the total power drawn by the loads; (*b*) the phase current $\mathbf{I}_{AB1}$ for the lagging load; (*c*) $\mathbf{I}_{AB2}$; (*d*) $\mathbf{I}_{aA}$.

Ans: 56.0 kW; $30.3\underline{/-6.87^\circ}$ A; $20.2\underline{/55.8^\circ}$ A; $75.3\underline{/-12.46^\circ}$ A.

A brief comparison of phase and line voltages as well as phase and line currents is presented in Table 12.1 for both Y- and Δ-connected loads powered by a Y-connected three-phase source.

Δ-Connected Sources

The source may also be connected in a Δ configuration. This is not typical, however, for a slight unbalance in the source phases can lead to large currents circulating in the Δ loop. For example, let us call the three single-phase sources $\mathbf{V}_{ab}$, $\mathbf{V}_{bc}$, and $\mathbf{V}_{cd}$. Before closing the Δ by connecting d to a, let us determine the unbalance by measuring the sum $\mathbf{V}_{ab} + \mathbf{V}_{bc} + \mathbf{V}_{ca}$. Suppose that the amplitude of the result is only 1% of the line voltage. The circulating current is thus approximately $\frac{1}{3}$% of the line voltage divided by the internal impedance of any source. How large is this impedance apt to be? It must depend on the current that the source is expected to deliver with a negligible drop in terminal voltage. If we assume that this maximum current causes a 1% drop in the terminal voltage, then *the circulating current is one-third of the maximum current*! This reduces the useful current capacity of the source and also increases the losses in the system.

TABLE **12.1** Comparison of Y- and Δ-Connected Three-Phase Loads. V_p Is the Voltage Magnitude of Each Y-Connected *Source* Phase

Load	Phase Voltage	Line Voltage	Phase Current	Line Current	Power per Phase
Y	$\mathbf{V}_{AN} = V_p\underline{/0^\circ}$ $\mathbf{V}_{BN} = V_p\underline{/-120^\circ}$ $\mathbf{V}_{CN} = V_p\underline{/-240^\circ}$	$\mathbf{V}_{AB} = \mathbf{V}_{ab} = (\sqrt{3}\underline{/30^\circ})\mathbf{V}_{AN} = \sqrt{3}\,V_p\underline{/30^\circ}$ $\mathbf{V}_{BC} = \mathbf{V}_{bc} = (\sqrt{3}\underline{/30^\circ})\mathbf{V}_{BN} = \sqrt{3}\,V_p\underline{/-90^\circ}$ $\mathbf{V}_{CA} = \mathbf{V}_{ca} = (\sqrt{3}\underline{/30^\circ})\mathbf{V}_{CN} = \sqrt{3}\,V_p\underline{/-210^\circ}$	$\mathbf{I}_{aA} = \mathbf{I}_{AN} = \frac{\mathbf{V}_{AN}}{\mathbf{Z}_P}$ $\mathbf{I}_{bB} = \mathbf{I}_{BN} = \frac{\mathbf{V}_{BN}}{\mathbf{Z}_P}$ $\mathbf{I}_{cC} = \mathbf{I}_{CN} = \frac{\mathbf{V}_{CN}}{\mathbf{Z}_P}$	$\mathbf{I}_{aA} = \mathbf{I}_{AN} = \frac{\mathbf{V}_{AN}}{\mathbf{Z}_P}$ $\mathbf{I}_{bB} = \mathbf{I}_{BN} = \frac{\mathbf{V}_{BN}}{\mathbf{Z}_P}$ $\mathbf{I}_{cC} = \mathbf{I}_{CN} = \frac{\mathbf{V}_{CN}}{\mathbf{Z}_P}$	$V_L \frac{I_L}{\sqrt{3}} \cos\theta$
Δ	$\mathbf{V}_{AB} = \mathbf{V}_{ab} = \sqrt{3}\,V_p\underline{/30^\circ}$ $\mathbf{V}_{BC} = \mathbf{V}_{bc} = \sqrt{3}\,V_p\underline{/-90^\circ}$ $\mathbf{V}_{CA} = \mathbf{V}_{ca} = \sqrt{3}\,V_p\underline{/-120^\circ}$	$\mathbf{V}_{AB} = \mathbf{V}_{ab} = \sqrt{3}\,V_p\underline{/30^\circ}$ $\mathbf{V}_{BC} = \mathbf{V}_{bc} = \sqrt{3}\,V_p\underline{/-90^\circ}$ $\mathbf{V}_{CA} = \mathbf{V}_{ca} = \sqrt{3}\,V_p\underline{/-120^\circ}$	$\mathbf{I}_{AB} = \frac{\mathbf{V}_{AB}}{\mathbf{Z}_P}$ $\mathbf{I}_{BC} = \frac{\mathbf{V}_{BC}}{\mathbf{Z}_P}$ $\mathbf{I}_{CA} = \frac{\mathbf{V}_{CA}}{\mathbf{Z}_P}$	$\mathbf{I}_{aA} = (\sqrt{3}\underline{/-30^\circ})\frac{\mathbf{V}_{AB}}{\mathbf{Z}_P}$ $\mathbf{I}_{bB} = (\sqrt{3}\underline{/-30^\circ})\frac{\mathbf{V}_{BC}}{\mathbf{Z}_P}$ $\mathbf{I}_{cC} = (\sqrt{3}\underline{/-30^\circ})\frac{\mathbf{V}_{CA}}{\mathbf{Z}_P}$	$V_L \frac{I_L}{\sqrt{3}} \cos\theta$

PRACTICAL APPLICATION

Power-Generating Systems

Today, electrical power is generated by a rather wide variety of techniques. For example, direct conversion of solar energy into electricity using photovoltaic (solar cell) technology results in the production of dc power. Despite representing a very environmentally friendly technology, however, photovoltaic-based installations are at present more expensive than other means of producing electricity, and they require the use of inverters to convert the dc power into ac. Other technologies such as wind turbine, geothermal, hydrodynamic, nuclear, and fossil fuel–based generators are often more economical by comparison. In these systems, a shaft is rotated through the action of a ***prime mover,*** such as wind on a propeller, or water or steam on turbine blades (Fig. 12.20).

Once a prime mover has been harnessed to generate rotational movement of a shaft, there are several means of converting this mechanical energy into electrical energy. One example is a ***synchronous generator*** (Fig. 12.21). These machines are composed of two main sections: a stationary part, called the ***stator,*** and a rotating part, termed the ***rotor.*** DC current is supplied to coils of wire wound about the rotor to generate a magnetic field, which is rotated through the action of the prime mover. A set of three-phase voltages is then induced at a second set of windings around the stator. Synchronous generators get their name from the fact that the frequency of the ac voltage produced is synchronized with the mechanical rotation of the rotor.

The actual demand on a stand-alone generator can vary greatly as various loads are added or removed, such as when air conditioning units kick on or lighting is turned on or off. The voltage output of a generator should ideally be independent of the load, but this is not the case in practice. The voltage $\mathbf{E}_A$ induced in any given stator phase, often referred to as the ***internal generated voltage,*** has a magnitude given by

$$E_A = K\phi\omega$$

where K is a constant dependent on the way the machine is constructed, ϕ is the magnetic flux produced by the field windings on the rotor (and hence is independent of the load), and ω is the speed of rotation, which depends only on the prime mover and not the load attached to the generator. Thus, *changing the load does not affect the magnitude of* $\mathbf{E}_A$. The internal generated voltage can be related to the phase voltage $\mathbf{V}_\phi$ and the phase current $\mathbf{I}_A$ by

$$\mathbf{E}_A = \mathbf{V}_\phi + jX_S\mathbf{I}_A$$

■ **FIGURE 12.20** Wind-energy harvesting installation at Altamont Pass, California, which consists of over 7000 individual windmills.
Greg Balfour Evans/Alamy Stock Photo

■ **FIGURE 12.21** The 24-pole rotor of a synchronous generator as it is being lowered into position.
Courtesy of Dr. Wade Enright, Te Kura Pukaha Vira O Te Whare Wananga O Waitaha, Aotearoa

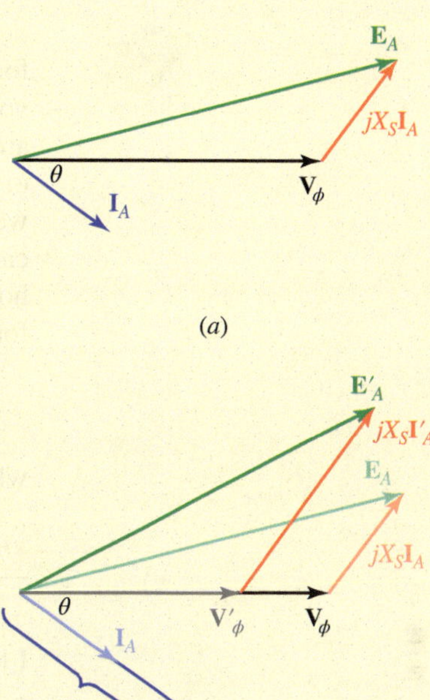

■ **FIGURE 12.22** Phasor diagrams describing the effect of loading on a stand-alone synchronous generator. (*a*) Generator connected to a load having a lagging power factor of cos θ. (*b*) An additional load is added without changing the power factor. The magnitude of the internal generated voltage $\mathbf{E}_A$ remains the same while the output current increases. Consequently, the output voltage $\mathbf{V}_\phi$ is reduced.

where X_S is the ***synchronous reactance*** of the generator. If the load is increased, then a larger current $\mathbf{I}'_A$ will be drawn from the generator. If the power factor is not changed (i.e., the angle between $\mathbf{V}_\phi$ and $\mathbf{I}_A$ remains constant), $\mathbf{V}_\phi$ will be reduced since E_A cannot change.

For example, consider the phasor diagram of Fig. 12.22*a*, which depicts the voltage–current output of a single phase of a generator connected to a load with a lagging power factor of cos θ. The internal generated voltage $\mathbf{E}_A$ is also shown. If an additional load is added without changing the power factor, as represented in Fig. 12.22*b*, the supplied current $\mathbf{I}_A$ increases to $\mathbf{I}'_A$. However, the magnitude of the internal generated voltage, formed by the sum of the phasors $jXS\mathbf{I}'_A$ and $\mathbf{V}'_\phi$, must remain unchanged. Thus, $E'_A = E_A$, and so the voltage output ($\mathbf{V}'_\phi$) of the generator will be slightly reduced, as depicted in Fig. 12.22*b*.

The ***voltage regulation*** of a generator is defined as

$$\% \text{ regulation} = \frac{V_{\text{no load}} - V_{\text{full load}}}{V_{\text{full load}}} \times 100$$

Ideally, the regulation should be as close to zero as possible, but this can only be accomplished if the dc current used to control the flux ϕ around the field winding is varied in order to compensate for changing load conditions; this can quickly become rather cumbersome. Thus, when designing a power generation facility, several smaller generators connected in parallel are usually preferable to one large generator capable of handling the peak load. Each generator can be operated at or near full load, so the voltage output is essentially constant; individual generators can be added or removed from the system depending on the demand.

We should also note that balanced three-phase sources may be transformed from Y to Δ, or vice versa, without affecting the load currents or voltages. The necessary relationships between the line and phase voltages are shown in Fig. 12.13 for the case where $\mathbf{V}_{an}$ has a reference phase angle of 0°. This transformation enables us to use whichever source connection we prefer, and all the load relationships will be correct. Of course, we cannot specify any currents or voltages within the source until we know how it is actually connected. Balanced three-phase loads may be transformed between Y- and Δ-connected configurations using the relation

$$Z_Y = \frac{Z_\Delta}{3}$$

which is probably worth remembering.

12.5 POWER MEASUREMENT IN THREE-PHASE SYSTEMS

Use of the Wattmeter

In large electrical systems, not only are voltage and current important to know, but power is quoted so often that measuring it directly proves highly valuable. This is typically performed using a device known as a ***wattmeter,*** which must have the ability to establish both the voltage and the current associated with either the source, the load, or both. Modern devices are very similar to the digital multimeter, providing a numerical display of the quantity being measured. These devices often make use of the fact that current gives rise to a magnetic field, which can be measured without breaking the circuit. However, in the field we still encounter analog versions of the multimeter, and they continue to have some advantages over digital versions, such as the ability to function without a separate power source (e.g., battery), and secondary information that comes from watching a needle move as opposed to numbers seemingly jumping around randomly on a display. Thus, in this section, we focus on power measurement using a traditional analog meter, as switching to a digital device is straightforward if one is available. Before embarking on a discussion of the specialized techniques used to measure power in three-phase systems, it is to our advantage to briefly consider how a wattmeter is used in a single-phase circuit.

Power measurement is most often accomplished at frequencies below a few hundred hertz through the use of a wattmeter that contains two separate coils. One of these coils is made of heavy wire, having a very low resistance, and is called the *current coil;* the second coil is composed of a much greater number of turns of fine wire, with relatively high resistance, and is termed the *potential coil,* or *voltage coil.* Additional resistance may also be inserted internally or externally in series with the potential coil. The torque applied to the moving system and the pointer is proportional to the instantaneous product of the currents flowing in the two coils. The mechanical inertia of the moving system, however, causes a deflection that is proportional to the *average* value of this torque.

The wattmeter is used by connecting it into a network in such a way that the current flowing in the current coil is the current flowing into the network and the voltage across the potential coil is the voltage across the

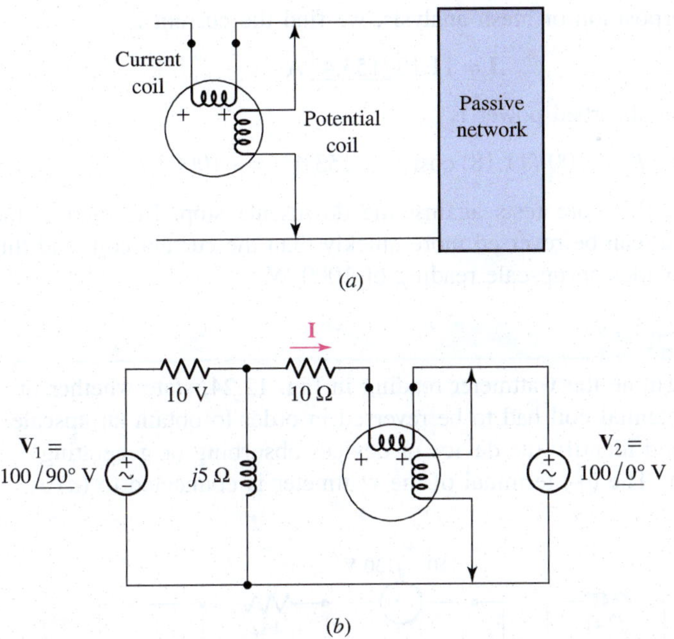

FIGURE 12.23 *(a)* A wattmeter connection that will ensure an upscale reading for the power absorbed by the passive network. *(b)* An example in which the wattmeter is installed to give an upscale indication of the power absorbed by the right source.

two terminals of the network. The current in the potential coil is thus the input voltage divided by the resistance of the potential coil.

It is apparent that the wattmeter has four available terminals, and correct connections must be made to these terminals in order to obtain an upscale reading on the meter. To be specific, let us assume that we are measuring the power absorbed by a passive network. The current coil is inserted in series with one of the two conductors connected to the load, and the potential coil is installed between the two conductors, usually on the "load side" of the current coil. The potential coil terminals are often indicated by arrows, as shown in Fig. 12.23*a*. Each coil has two terminals, and the proper relationship between the sense of the current and voltage must be observed. One end of each coil is usually marked (+), and an upscale reading is obtained if a positive current is flowing into the (+) end of the current coil while the (+) terminal of the potential coil is positive with respect to the unmarked end. The wattmeter shown in the network of Fig. 12.23*a* therefore gives an upscale deflection when the network to the right is absorbing power. A reversal of either coil, but not both, will cause the meter to try to deflect downscale; a reversal of both coils will never affect the reading.

As an example of the use of such a wattmeter in measuring average power, let us consider the circuit shown in Fig. 12.23*b*. The connection of the wattmeter is such that an upscale reading corresponds to a positive absorbed power for the network to the right of the meter, that is, the right source. The power absorbed by this source is given by

$$P = |\mathbf{V}_2||\mathbf{I}| \cos(ang\,\mathbf{V}_2 - ang\,\mathbf{I})$$

Using superposition or mesh analysis, we find the current is

$$\mathbf{I} = 11.18\underline{/153.4^\circ} \text{ A}$$

and thus the absorbed power is

$$P = (100)(11.18)\cos(0^\circ - 153.4^\circ) = -1000 \text{ W}$$

The pointer therefore rests against the downscale stop. In practice, the potential coil can be reversed more quickly than the current coil, and this reversal provides an upscale reading of 1000 W.

PRACTICE

12.9 Determine the wattmeter reading in Fig. 12.24, state whether or not the potential coil had to be reversed in order to obtain an upscale reading, and identify the device or devices absorbing or generating this power. The (+) terminal of the wattmeter is connected to (*a*) *x*; (*b*) *y*; (*c*) *z*.

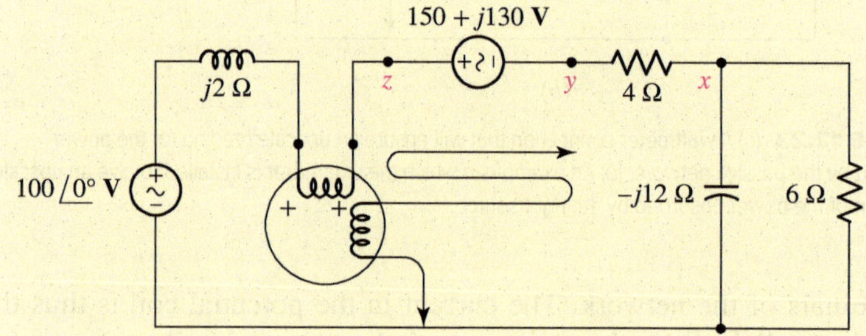

FIGURE 12.24

Ans: 1200 W, as is, $P_{6\Omega}$ (absorbed); 2200 W, as is, $P_{4\Omega} + P_{6\Omega}$ (absorbed); 500 W, reversed, absorbed by 100 V.

The Wattmeter in a Three-Phase System

At first glance, measurement of the power drawn by a three-phase load seems to be a simple problem. We need place only one wattmeter in each of the three phases and add the results. For example, the proper connections for a Y-connected load are shown in Fig. 12.25*a*. Each wattmeter has its current coil inserted in one phase of the load and its potential coil connected between the line side of that load and the neutral. In a similar way, three wattmeters may be connected as shown in Fig. 12.25*b* to measure the total power taken by a Δ-connected load. The methods are theoretically correct, but they may be useless in practice because the neutral of the Y is not always accessible and the phases of the Δ are not available. A three-phase rotating machine, for example, has only three accessible terminals, those which we have been calling *A*, *B*, and *C*.

Clearly, we have a need for a method of measuring the total power drawn by a three-phase load having only three accessible terminals; measurements may be made on the "line" side of these terminals, but not on the "load" side. Such a method is available and is capable of measuring the power taken by an *unbalanced* load from an *unbalanced* source. Let

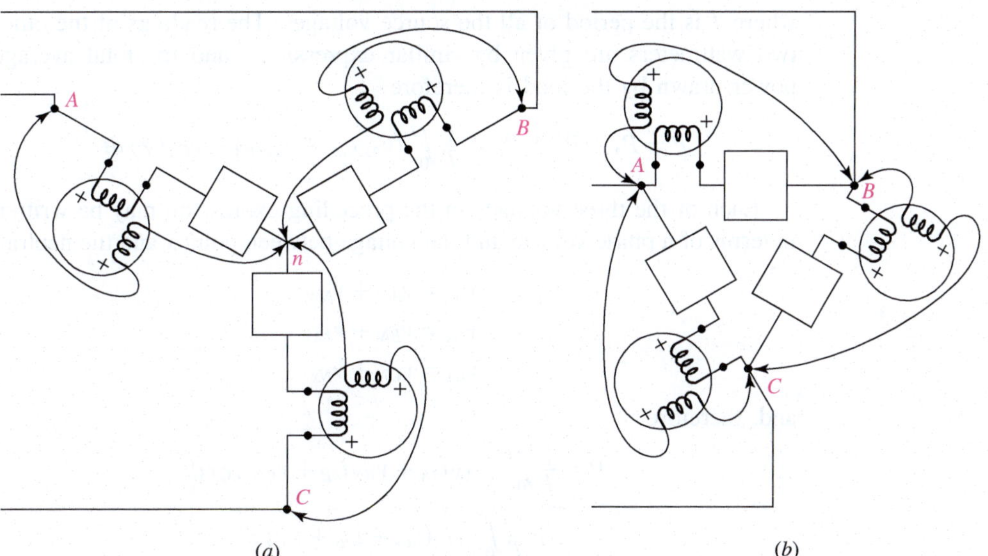

FIGURE 12.25 Three wattmeters are connected in such a way that each reads the power taken by one phase of a three-phase load, and the sum of the readings is the total power. (*a*) A Y-connected load. (*b*) A Δ-connected load. Neither the loads nor the source need be balanced.

us connect three wattmeters in such a way that each has its current coil in one line and its voltage coil between that line and some common point *x*, as shown in Fig. 12.26. Although a system with a Y-connected load is illustrated, the arguments we present are equally valid for a Δ-connected load. The point *x* may be some unspecified point in the three-phase system, or it may be merely a point in space at which the three potential coils have a common node. The average power indicated by wattmeter *A* must be

$$P_A = \frac{1}{T}\int_0^T v_{Ax} i_{aA}\, dt$$

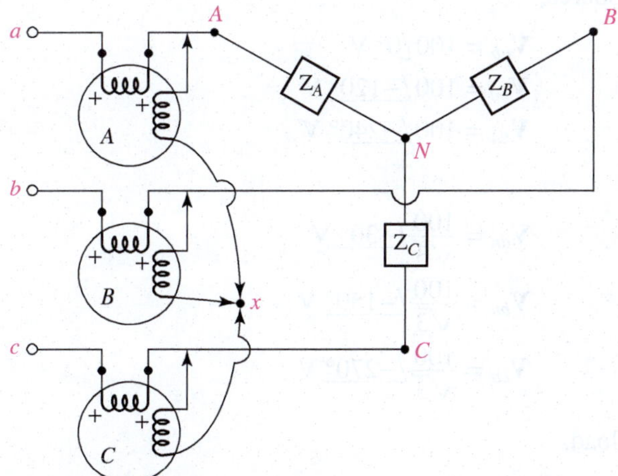

FIGURE 12.26 A method of connecting three wattmeters to measure the total power taken by a three-phase load. Only the three terminals of the load are accessible.

where T is the period of all the source voltages. The readings of the other two wattmeters are given by similar expressions, and the total average power drawn by the load is therefore

$$P = P_A + P_B + P_C = \frac{1}{T}\int_0^T (v_{Ax}i_{aA} + v_{Bx}i_{bB} + v_{Cx}i_{cC})\, dt$$

Each of the three voltages in the preceding expression may be written in terms of a phase voltage and the voltage between point x and the neutral,

$$v_{Ax} = v_{AN} + v_{Nx}$$
$$v_{Bx} = v_{BN} + v_{Nx}$$
$$v_{Cx} = v_{CN} + v_{Nx}$$

and, therefore,

$$P = \frac{1}{T}\int_0^T (v_{AN}i_{aA} + v_{BN}i_{bB} + v_{CN}i_{cC})\, dt + \frac{1}{T}\int_0^T v_{Nx}(i_{aA} + i_{bB} + i_{cC})\, dt$$

However, the entire three-phase load may be considered to be a supernode, and Kirchhoff's current law requires

$$i_{aA} + i_{bB} + i_{cC} = 0$$

Thus

$$P = \frac{1}{T}\int_0^T (v_{AN}i_{aA} + v_{BN}i_{bB} + v_{CN}i_{cC})\, dt$$

Reference to the circuit diagram shows that this sum is indeed the sum of the average powers taken by each phase of the load, and the sum of the readings of the three wattmeters therefore represents the total average power drawn by the entire load!

Let us illustrate this procedure with a numerical example before we discover that one of these three wattmeters is really superfluous. We will assume a balanced source,

$$\mathbf{V}_{ab} = 100\underline{/0^\circ}\text{ V}$$
$$\mathbf{V}_{bc} = 100\underline{/-120^\circ}\text{ V}$$
$$\mathbf{V}_{ca} = 100\underline{/-240^\circ}\text{ V}$$

or

$$\mathbf{V}_{an} = \frac{100}{\sqrt{3}}\underline{/-30^\circ}\text{ V}$$
$$\mathbf{V}_{bn} = \frac{100}{\sqrt{3}}\underline{/-150^\circ}\text{ V}$$
$$\mathbf{V}_{cn} = \frac{100}{\sqrt{3}}\underline{/-270^\circ}\text{ V}$$

and an unbalanced load,

$$\mathbf{Z}_A = -j10\ \Omega$$
$$\mathbf{Z}_B = j10\ \Omega$$
$$\mathbf{Z}_C = 10\ \Omega$$

Let us assume ideal wattmeters, connected as illustrated in Fig. 12.26, with point x located on the neutral of the source n. The three line currents may be obtained by mesh analysis,

$$\mathbf{I}_{aA} = 19.32\underline{/15^\circ}\text{ A}$$
$$\mathbf{I}_{bB} = 19.32\underline{/165^\circ}\text{ A}$$
$$\mathbf{I}_{cC} = 10\underline{/-90^\circ}\text{ A}$$

The voltage between the neutrals is

$$\mathbf{V}_{nN} = \mathbf{V}_{nb} + \mathbf{V}_{BN} = \mathbf{V}_{nb} + \mathbf{V}_{bB}(j10) = 157.7\underline{/-90^\circ}\text{ V}$$

The average power indicated by each wattmeter may be calculated,

$$P_A = V_p I_{aA}\cos(ang\ \mathbf{V}_{an} - ang\ \mathbf{I}_{aA})$$
$$= \frac{100}{\sqrt{3}}\,19.32\cos(-30^\circ - 15^\circ) = 788.7\text{ W}$$
$$P_B = \frac{100}{\sqrt{3}}\,19.32\cos(-150^\circ - 165^\circ) = 788.7\text{ W}$$
$$P_C = \frac{100}{\sqrt{3}}\,10\cos(-270^\circ + 90^\circ) = -577.4\text{ W}$$

Note that the reading of one of the wattmeters is negative. Our previous discussion on the basic use of a wattmeter indicates that an upscale reading on that meter can only be obtained after either the potential coil or the current coil is reversed.

or a total power of 1 kW. Since an rms current of 10 A flows through the *resistive* load, the total power drawn by the load is

$$P = 10^2(10) = 1\text{ kW}$$

and the two methods agree.

The Two-Wattmeter Method

We have proved that point x, the common connection of the three potential coils, may be located any place we wish without affecting the algebraic sum of the three wattmeter readings. Let us now consider the effect of placing point x, this common connection of the three wattmeters, directly on one of the lines. If, for example, one end of each potential coil is returned to B, then there is no voltage across the potential coil of wattmeter B and *this meter must read zero*. It may therefore be removed, and the algebraic sum of the remaining two wattmeter readings is still the total power drawn by the load. When the location of x is selected in this way, we describe the method of power measurement as the ***two-wattmeter*** method. The sum of the readings indicates the total power, regardless of (1) load unbalance, (2) source unbalance, (3) differences in the two wattmeters, and (4) the waveform of the periodic source. The only assumption we have made is that wattmeter corrections are sufficiently small that we can ignore them. In Fig. 12.26, for example, the current coil of each meter has passing through it the line current drawn by the load plus the current taken by the potential coil. Since the latter current is usually quite small, its effect may be estimated from a knowledge of the resistance of the potential coil and voltage across it. These two quantities enable a close estimate to be made of the power dissipated in the potential coil.

In the numerical example described previously, let us now assume that two wattmeters are used, one with current coil in line A and potential coil between lines A and B, the other with current coil in line C and potential coil between C and B. The first meter reads

$$\begin{aligned} P_1 &= V_{AB} I_{aA} \cos(ang\ V_{AB} - ang\ I_{aA}) \\ &= 100(19.32)\cos(0° - 15°) \\ &= 1866 \text{ W} \end{aligned}$$

and the second

$$\begin{aligned} P_2 &= V_{CB} I_{cC} \cos(ang\ V_{CB} - ang\ I_{cC}) \\ &= 100(10)\cos(60° + 90°) \\ &= -866 \text{ W} \end{aligned}$$

and, therefore,

$$P = P_1 + P_2 = 1866 - 866 = 1000 \text{ W}$$

as we expect from recent experience with the circuit.

In the case of a balanced load, the two-wattmeter method enables the PF angle to be determined, as well as the total power drawn by the load. Let us assume a load impedance with a phase angle θ; either a Y or Δ connection may be used, and we will assume the Δ connection shown in Fig. 12.27. The construction of a standard phasor diagram, such as that of Fig. 12.19, enables us to determine the proper phase angle between the several line voltages and line currents. We therefore determine the readings

$$\begin{aligned} P_1 &= |\mathbf{V}_{AB}||\mathbf{I}_{aA}| \cos(ang\ \mathbf{V}_{AB} - ang\ \mathbf{I}_{aA}) \\ &= V_L I_L \cos(30° + \theta) \end{aligned}$$

and

$$\begin{aligned} P_2 &= |\mathbf{V}_{CB}||\mathbf{I}_{cC}| \cos(ang\ \mathbf{V}_{CB} - ang\ \mathbf{I}_{cC}) \\ &= V_L I_L \cos(30° - \theta) \end{aligned}$$

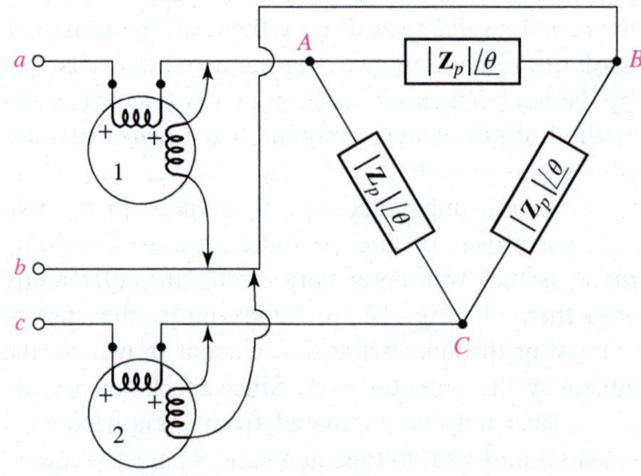

FIGURE 12.27 Two wattmeters connected to read the total power drawn by a balanced three-phase load.

The ratio of the two readings is

$$\frac{P_1}{P_2} = \frac{\cos(30° + \theta)}{\cos(30° - \theta)} \qquad [5]$$

If we expand the cosine terms, this equation is easily solved for tan θ,

$$\tan\theta = \sqrt{3}\,\frac{P_2 - P_1}{P_2 + P_1} \qquad [6]$$

Thus, equal wattmeter readings indicate a unity PF load; equal and opposite readings indicate a purely reactive load; a reading of P_2 which is (algebraically) greater than P_1 indicates an inductive impedance; and a reading of P_2 which is less than P_1 signifies a capacitive load. How can we tell which wattmeter reads P_1 and which reads P_2? It is true that P_1 is in line A, and P_2 is in line C, and our positive phase-sequence system forces V_{an} to lag V_{cn}. This is enough information to differentiate between two wattmeters, but it is confusing to apply in practice. Even if we were unable to distinguish between the two, we know the magnitude of the phase angle, but not its sign. This is often sufficient information; if the load is an induction motor, the angle must be positive and we do not need to make any tests to determine which reading is which. If no previous knowledge of the load is assumed, then there are several methods of resolving the ambiguity. Perhaps the simplest method is that which involves adding a high-impedance reactive load, say, a three-phase capacitor, across the unknown load. The load must become more capacitive. Thus, if the magnitude of tan θ (or the magnitude of θ) decreases, then the load was inductive, whereas an increase in the magnitude of tan θ signifies an original capacitive impedance.

EXAMPLE 12.7

The balanced load in Fig. 12.28 is fed by a balanced three-phase system having $\mathbf{V}_{ab} = 230\underline{/0°}$ V rms and positive phase sequence. Find the reading of each wattmeter and the total power drawn by the load.

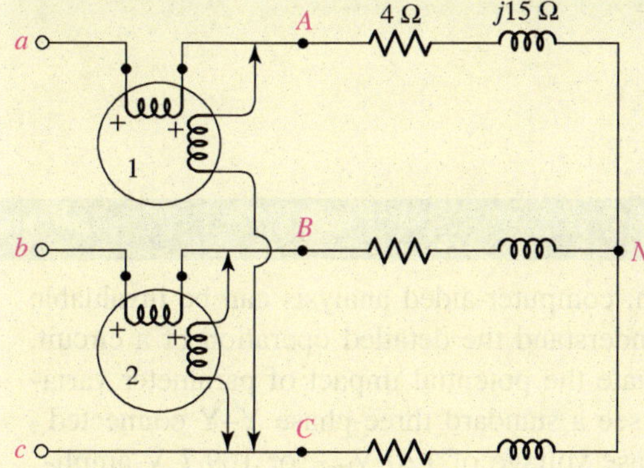

■ **FIGURE 12.28** A balanced three-phase system connected to a balanced three-phase load, the power of which is being measured using the two-wattmeter technique.

(Continued on next page)

The potential coil of wattmeter #1 is connected to measure the voltage $\mathbf{V}_{ac}$, and its current coil is measuring the phase current $\mathbf{I}_{aA}$. Since we know to use the positive phase sequence, the line voltages are

$$\mathbf{V}_{ab} = 230\underline{/0°}\text{ V}$$
$$\mathbf{V}_{bc} = 230\underline{/-120°}\text{ V}$$
$$\mathbf{V}_{ca} = 230\underline{/120°}\text{ V}$$

Note that $\mathbf{V}_{ac} = -\mathbf{V}_{ca} = 230\underline{/-60°}$ V.

The phase current $\mathbf{I}_{aA}$ is given by the phase voltage $\mathbf{V}_{an}$ divided by the phase impedance $4 + j15\ \Omega$,

$$\mathbf{I}_{aA} = \frac{\mathbf{V}_{an}}{4 + j15} = \frac{(230/\sqrt{3})\underline{/-30°}}{4 + j15}\text{ A}$$
$$= 8.554\underline{/-105.1°}\text{ A}$$

We may now compute the power measured by wattmeter #1 as

$$P_1 = |\mathbf{V}_{ac}||\mathbf{I}_{aA}| \cos(ang\ \mathbf{V}_{ac} - ang\ \mathbf{I}_{aA})$$
$$= (230)(8.554)\cos(-60° + 105.1°)\text{ W}$$
$$= 1389\text{ W}$$

In a similar fashion, we determine that

$$P_2 = |\mathbf{V}_{bc}||\mathbf{I}_{bB}| \cos(ang\ \mathbf{V}_{bc} - ang\ \mathbf{I}_{bB})$$
$$= (230)(8.554)\cos(-120° - 134.9°)\text{ W}$$
$$= -512.5\text{ W}$$

Thus, the total average power absorbed by the load is

$$P = P_1 + P_2 = 876.5\text{ W}$$

Since this measurement would result in the meter pegged at downscale, one of the coils would need to be reversed in order to take the reading.

PRACTICE

12.10 For the circuit of Fig. 12.26, let the loads be $\mathbf{Z}_A = 25\underline{/60°}\ \Omega$, $\mathbf{Z}_B = 50\underline{/-60°}\ \Omega$, $\mathbf{Z}_C = 50\underline{/60°}\ \Omega$, $\mathbf{V}_{AB} = 600\underline{/0°}$ V rms with (+) phase sequence, and locate point x at C. Find (*a*) P_A; (*b*) P_B; (*c*) P_C.

Ans: 0; 7200 W; 0.

COMPUTER-AIDED ANALYSIS

As we have already seen, computer-aided analysis can be invaluable in helping to not only understand the detailed operation of a circuit, but to also help us evaluate the potential impact of parameter variations. In Fig. 12.29, we see a standard three-phase Y–Y connected circuit operating at a phase voltage of 120 V_{rms}, or 169.7 V amplitude. The three-phase source is connected in positive, or *abc*, phase sequence, and the load could be a simple three-phase motor used to power a compressor or large pump.

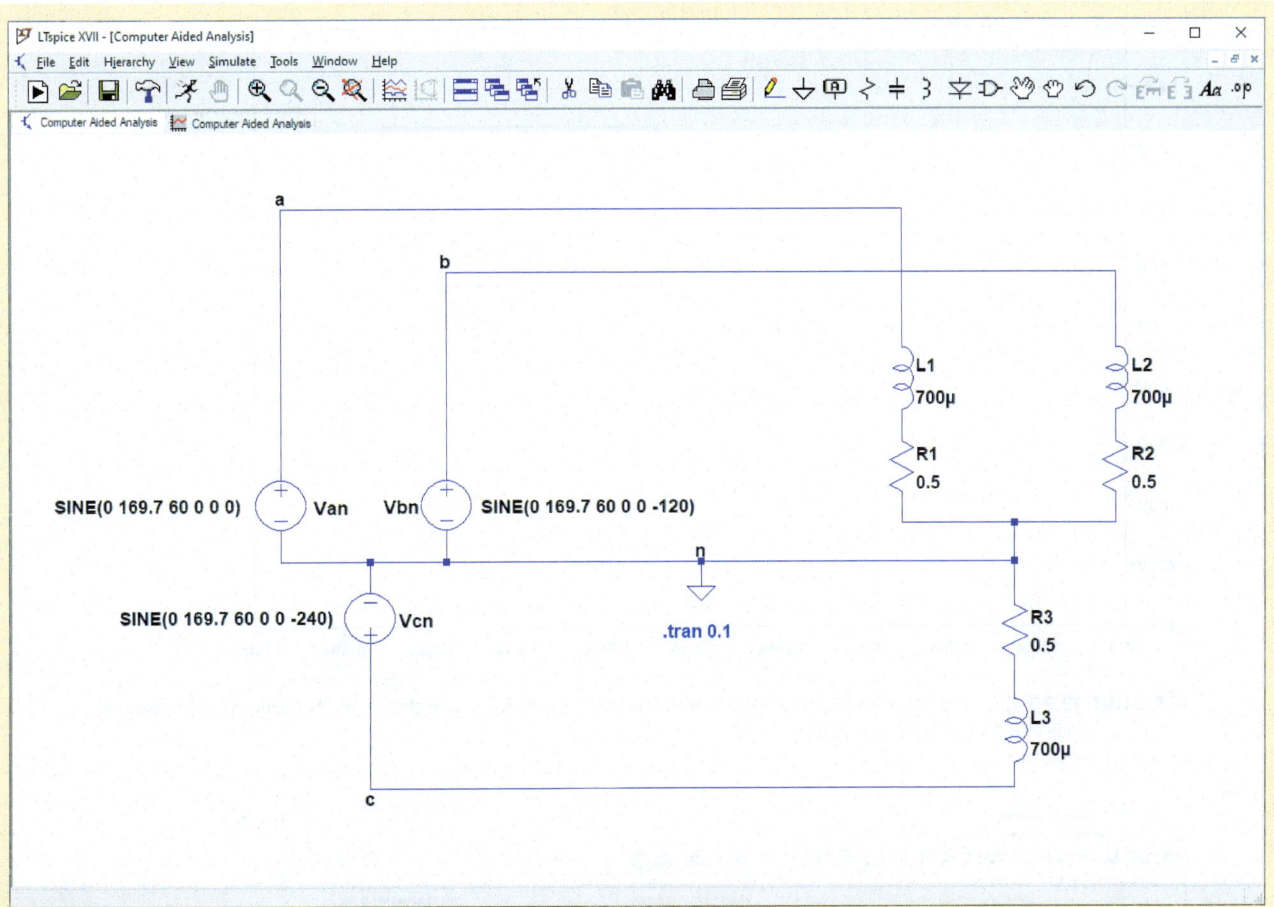

■ **FIGURE 12.29** An example Y–Y connected three-phase circuit with an inductive load which includes series resistance modeling.
Analog Devices

While the phase voltages are 120 V_{rms}, the line voltage is $\sqrt{3}$ times larger, or 208 V_{rms}, as we see in Fig. 12.30. Hence, such a source is often referred to as a 208 V three-phase system, and is often encountered when a large motor is being operated.

We now proceed to examine the current flowing through the three phases of the load, as well as the instantaneous power, which are both depicted in Fig. 12.31. As can be seen, except for an initial transient, the currents in each load phase and the instantaneous power dissipated by the series resistance of the load are identical except for a 120° phase difference as expected.

Over time, however, wear and tear occur in any mechanical system, and generally we expect some degree of difference in degradation among the three phases of the load. In the case shown in Fig. 12.32, the series resistance of the AN phase represented by R1 has increased by 10% to 550 mΩ. We see a noticeable reduction in the current which flows through that phase and a corresponding decrease in instantaneous power. The result will likely be the introduction of unbalanced vibration, which could accelerate the degradation of the load.

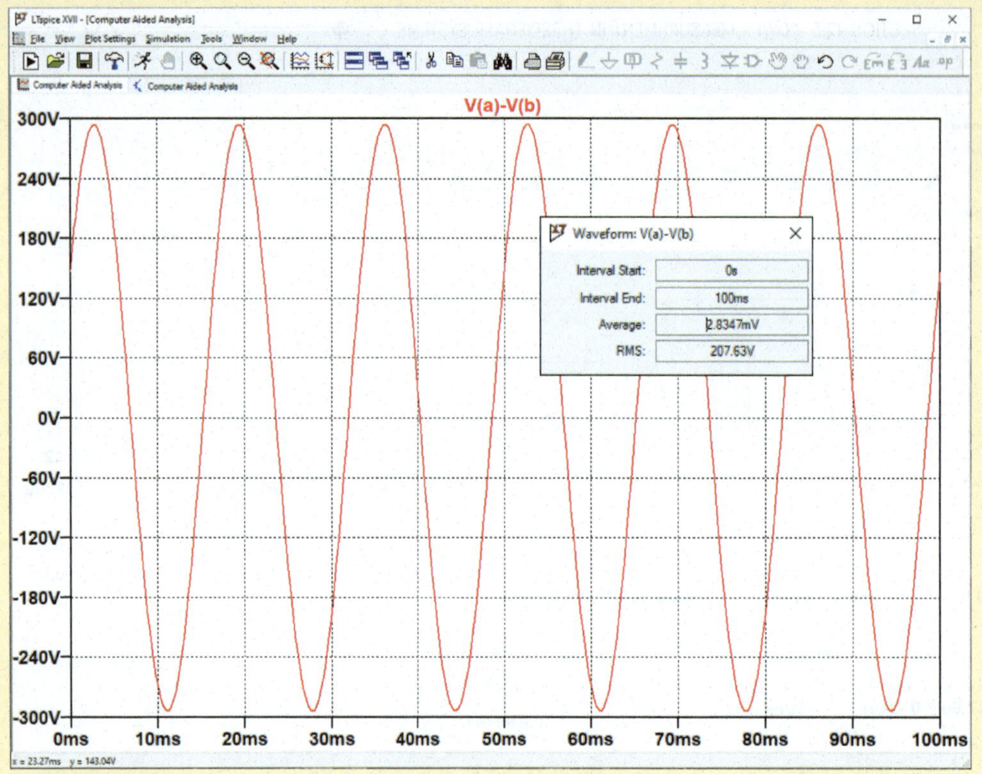

■ **FIGURE 12.30** Line voltage obtained by plotting the difference in voltage between two source phases, and then integrating the waveform. An RMS voltage of 208 V is obtained as expected.
Analog Devices

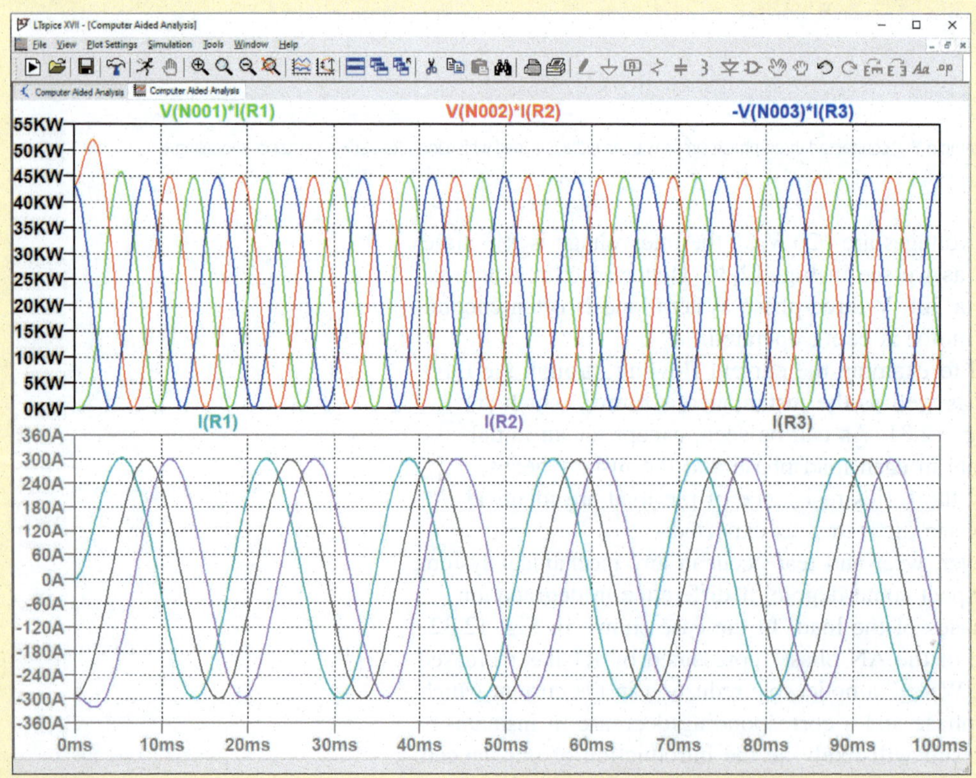

■ **FIGURE 12.31** Plot of per-phase load current (bottom) and instantaneous power (top) dissipated by the series resistance of each phase of the load.
Analog Devices

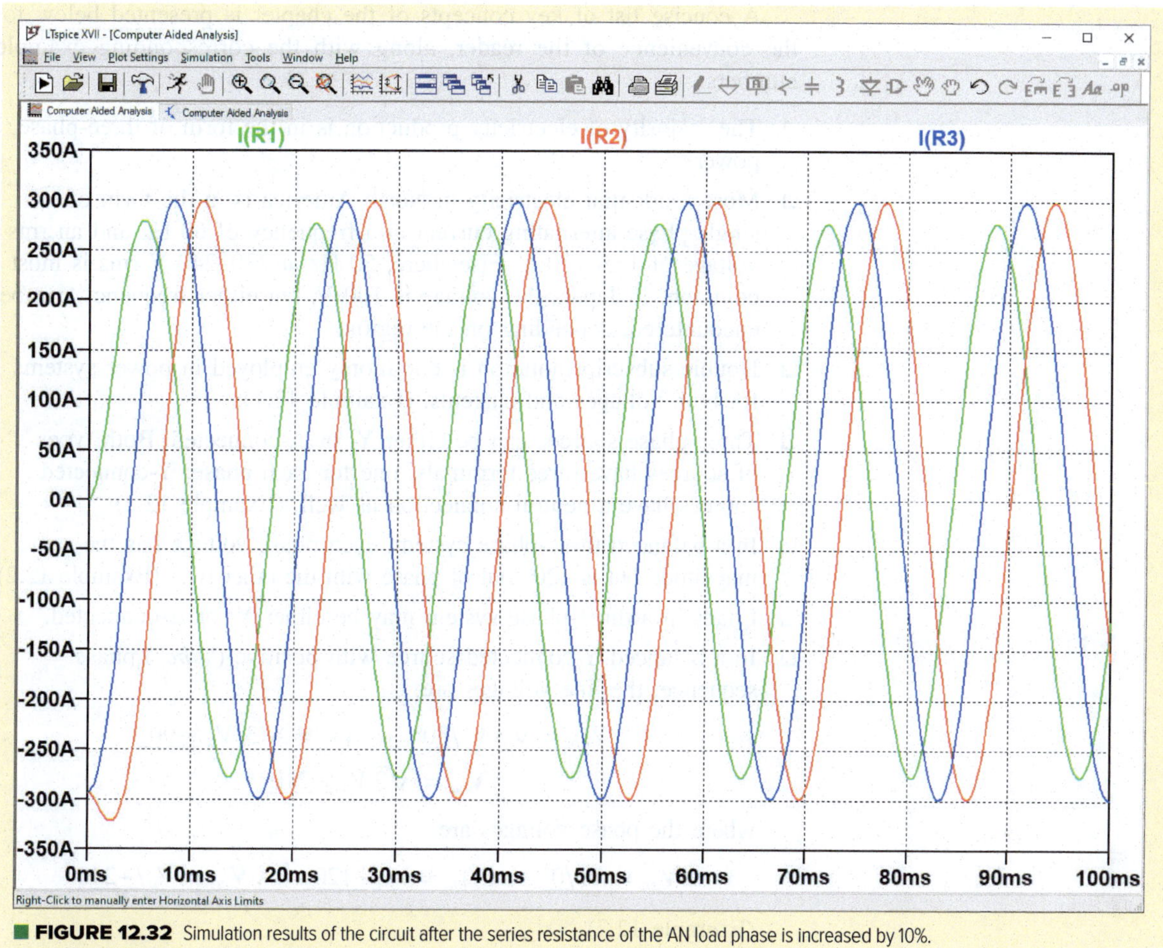

■ **FIGURE 12.32** Simulation results of the circuit after the series resistance of the AN load phase is increased by 10%.
Analog Devices

SUMMARY AND REVIEW

Polyphase circuits are not encountered directly by everyone but are part of almost every large building installation. In this chapter we studied how three voltages, each 120° out of phase with the others, can be supplied by a single generator (and hence have the same frequency) and connected to a three-component load. For the sake of convenience we introduced the double-subscript notation, which is commonly employed. A three-phase system will have at least three terminals; a neutral wire connection is not mandatory but is common at least for the source. If a Δ-connected load is employed, then there is no neutral connection to it. When a neutral wire is present, we can define *phase voltages* $\mathbf{V}_{an}$, $\mathbf{V}_{bn}$, and $\mathbf{V}_{cn}$ between each phase (a, b, or c) and neutral. Kirchhoff's voltage law requires that these three phase voltages sum to zero, regardless of whether a positive or negative phase sequence relates their angles. *Line voltages* (i.e., between phases) can be related directly to the phase voltages; for a Δ-connected load they are equal. In a similar fashion, *line currents* and *phase currents* can be directly related to one another; in a Y-connected load, they are equal. Symmetry often allows us to perform the analysis on a per-phase basis, simplifying our calculations considerably.

A concise list of key concepts of the chapter is presented below for the convenience of the reader, along with the corresponding example numbers.

- The majority of electricity production is in the form of three-phase power.
- Most residential electricity in North America is in the form of single-phase alternating current at a frequency of 60 Hz and an rms voltage of 115–120 V. Elsewhere, 50 Hz at 230–240 V rms is most common. In Japan, the voltage is 100 V, but either frequency can be encountered, depending on the region.
- Double-subscript notation is commonly employed in power systems for both voltages and currents. (Example 12.1)
- Three-phase sources can be either Y- or Δ-connected. Both types of sources have three terminals, one for each phase; Y-connected sources have a neutral connection as well. (Example 12.2)
- In a balanced three-phase system, each phase voltage has the same magnitude but is 120° out of phase with the other two. (Example 12.2)
- Loads in a three-phase system may be either Y- or Δ-connected.
- In a balanced Y-connected source with positive ("*abc*") phase sequence, the line voltages are

$$\mathbf{V}_{ab} = \sqrt{3}\,V_p\underline{/30^\circ} \qquad \mathbf{V}_{bc} = \sqrt{3}\,V_p\underline{/-90^\circ}$$
$$\mathbf{V}_{ca} = \sqrt{3}\,V_p\underline{/-210^\circ}$$

  where the phase voltages are

$$\mathbf{V}_{an} = V_p\underline{/0^\circ} \qquad \mathbf{V}_{bn} = V_p\underline{/-120^\circ} \qquad \mathbf{V}_{cn} = V_p\underline{/-240^\circ}$$

  (Example 12.2)
- In a system with a Y-connected load, the line currents are equal to the phase currents. (Examples 12.3, 12.4, 12.6)
- In a Δ-connected load, the line voltages are equal to the phase voltages. (Example 12.5)
- In a balanced system with positive phase sequence and a balanced Δ-connected load, the line currents are

$$\mathbf{I}_a = \mathbf{I}_{AB}\sqrt{3}\,\underline{/-30^\circ} \qquad \mathbf{I}_b = \mathbf{I}_{BC}\sqrt{3}\,\underline{/-150^\circ} \qquad \mathbf{I}_c = \mathbf{I}_{CA}\sqrt{3}\,\underline{/+90^\circ}$$

  where the phase currents are

$$\mathbf{I}_{AB} = \frac{\mathbf{V}_{AB}}{\mathbf{Z}_\Delta} = \frac{\mathbf{V}_{ab}}{\mathbf{Z}_\Delta} \qquad \mathbf{I}_{BC} = \frac{\mathbf{V}_{BC}}{\mathbf{Z}_\Delta} = \frac{\mathbf{V}_{bc}}{\mathbf{Z}_\Delta} \qquad \mathbf{I}_{CA} = \frac{\mathbf{V}_{CA}}{\mathbf{Z}_\Delta} = \frac{\mathbf{V}_{ca}}{\mathbf{Z}_\Delta}$$

  (Example 12.5)
- Most power calculations are performed on a per-phase basis, assuming a balanced system; otherwise, nodal/mesh analysis is always a valid approach. (Examples 12.3, 12.4, 12.5)
- The power in a three-phase system (balanced or unbalanced) can be measured with only two wattmeters. (Example 12.7)
- The instantaneous power in any balanced three-phase system is constant.

READING FURTHER

A good overview of ac power concepts can be found in Chap. 2 of:

B. M. Weedy, B. J. Cory, N. Jenkins, J. B. Ekanayake, and G. Strbac, *Electric Power Systems*, 5th ed. Chichester, England: Wiley, 2012.

Contemporary issues pertaining to ac power systems can be found in:

International Journal of Electrical Power & Energy Systems. Elsevier, 1979–. ISSN: 0142-0615.

EXERCISES

12.1 Polyphase Systems

1. An unknown three-terminal device has leads named *b*, *c*, and *e*. When installed in one particular circuit, measurements indicated that $V_{ec} = -9$ V and $V_{eb} = -0.65$ V. (*a*) Calculate V_{cb}. (*b*) Determine the power dissipated in the *b*-*e* junction if the current I_b flowing into the terminal marked *b* is equal to 1 μA.
2. A common type of transistor is known as the MESFET, which is an acronym for **me**tal-semiconductor **f**ield **e**ffect **t**ransistor. It has three terminals, named the gate (*g*), the source (*s*), and the drain (*d*). As an example, consider one particular MESFET operating in a circuit such that $V_{sg} = 0.2$ V and $V_{ds} = 3$ V. (*a*) Calculate V_{gs} and V_{dg}. (*b*) If a gate current $I_g = 100$ pA is found to be flowing into the gate terminal, compute the power lost at the gate-source junction.
3. For a certain Y-connected three-phase source, $\mathbf{V}_{an} = 400\underline{/33°}$ V, $\mathbf{V}_{bn} = 400\underline{/153°}$ V, and $\mathbf{V}_{cx} = 160\underline{/208°}$ V. Determine (*a*) $\mathbf{V}_{cn}$; (*b*) $\mathbf{V}_{an} - \mathbf{V}_{bn}$; (*c*) $\mathbf{V}_{ax}$; (*d*) $\mathbf{V}_{bx}$.
4. Describe what is meant by a "polyphase" source, state one possible advantage of such sources that might outweigh their additional complexity over single-phase sources of power, and explain the difference between "balanced" and "unbalanced" sources.
5. Several of the voltages associated with a certain circuit are given by $\mathbf{V}_{12} = 9\underline{/30°}$ V, $\mathbf{V}_{32} = 3\underline{/130°}$ V, and $\mathbf{V}_{14} = 2\underline{/10°}$ V. Determine $\mathbf{V}_{21}$, $\mathbf{V}_{13}$, $\mathbf{V}_{34}$, and $\mathbf{V}_{24}$.
6. The nodal voltages which describe a particular circuit can be expressed as $\mathbf{V}_{14} = 9 - j$ V, $\mathbf{V}_{24} = 3 + j3$ V, and $\mathbf{V}_{34} = 8$ V. Calculate $\mathbf{V}_{12}$, $\mathbf{V}_{32}$, and $\mathbf{V}_{13}$. Express your answers in phasor form.
7. In the circuit of Fig. 12.33, the resistor markings unfortunately have been omitted, but several of the currents are known. Specifically, $I_{ad} = 1$ A. (*a*) Compute I_{ab}, I_{cd}, I_{de}, I_{fe}, and I_{be}. (*b*) If $V_{ba} = 125$ V, determine the value of the resistor linking nodes *a* and *b*.
8. For the circuit shown in Fig. 12.34, (*a*) determine I_{gh}, I_{cd}, and I_{dh}. (*b*) Calculate I_{ed}, I_{ei}, and I_{jf}. (*c*) If all resistors in the circuit each have a value of 1 Ω, determine the three clockwise-flowing mesh currents.

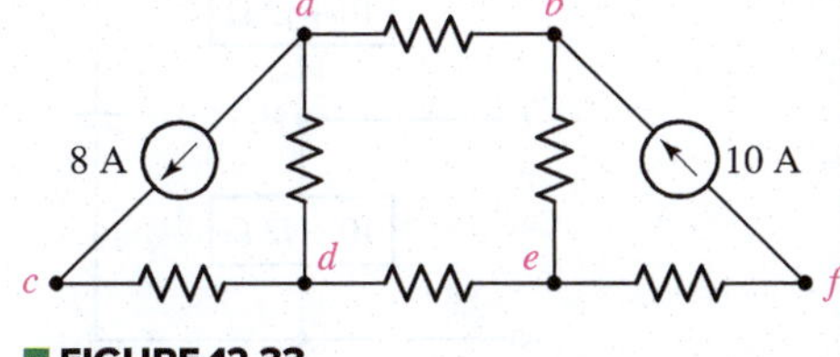

FIGURE 12.33

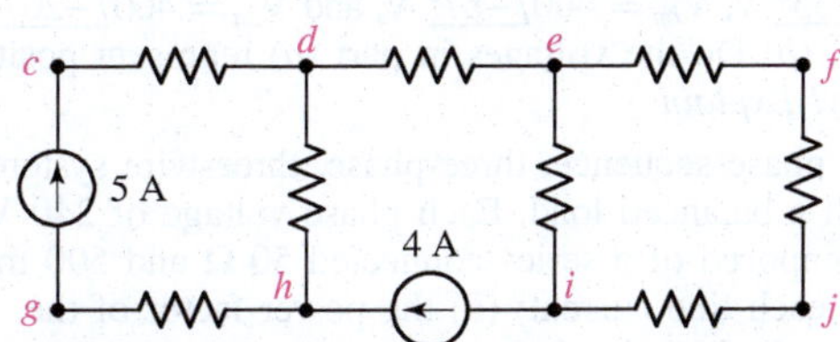

FIGURE 12.34

9. Additional resistors are added in parallel to the resistors between terminals *d* and *e*, and terminals *f* and *j*, respectively, of the circuit in Fig. 12.34. (*a*) Which voltages may still be described using double-subscript notation? (*b*) Which line currents may still be described by double-subscript notation?

12.2 Single-Phase Three-Wire Systems

10. Most consumer electronics are powered by 110 V outlets, but several types of appliances (such as clothes dryers) are powered from 220 V outlets. Lower voltages are generally safer. What, then, motivates manufacturers of some pieces of equipment to design them to run on 220 V?

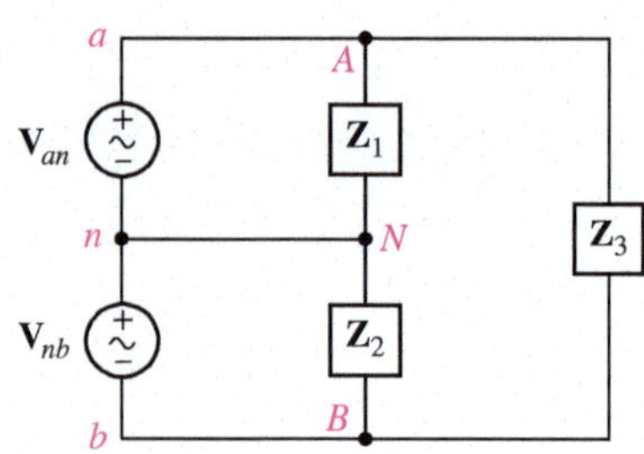

■ FIGURE 12.35

11. The single-phase three-wire system of Fig. 12.35 has three separate load impedances. If the source is balanced and $\mathbf{V}_{an} = 110 + j0$ V rms, (*a*) express $\mathbf{V}_{an}$ and $\mathbf{V}_{bn}$ in phasor notation. (*b*) Determine the phasor voltage which appears across the impedance $\mathbf{Z}_3$. (*c*) Determine the average power delivered by the two sources if $\mathbf{Z}_1 = 50 + j0\ \Omega$, $\mathbf{Z}_2 = 100 + j45\ \Omega$, and $\mathbf{Z}_3 = 100 - j90\ \Omega$. (*d*) Represent load $\mathbf{Z}_3$ by a series connection of two elements, and state their respective values if the sources operate at 60 Hz.

12. For the system represented in Fig. 12.36, the ohmic losses in the neutral wire are so small they can be neglected, and it can be adequately modeled as a short circuit. (*a*) Calculate the power lost in the two lines as a result of their nonzero resistance. (*b*) Compute the average power delivered to the load. (*c*) Determine the power factor of the total load.

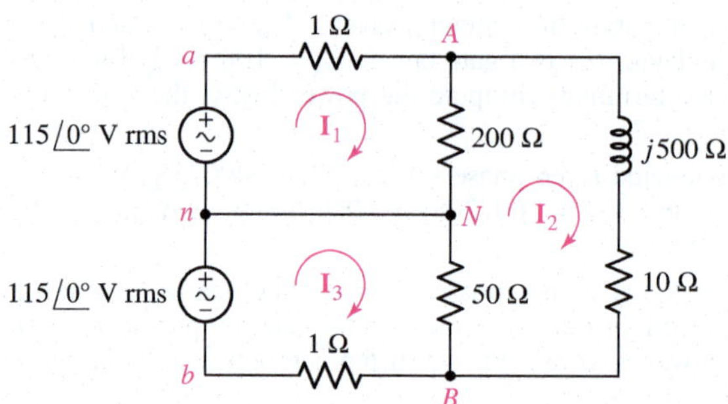

■ FIGURE 12.36

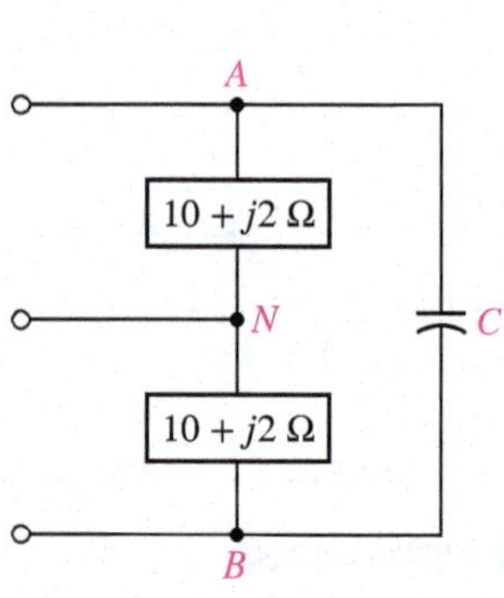

■ FIGURE 12.37

13. Referring to the balanced load represented in Fig. 12.37, if it is connected to a three-wire balanced source operating at 50 Hz such that $V_{AN} = 115$ V, (*a*) determine the power factor of the load if the capacitor is omitted; (*b*) determine the value of capacitance *C* that will achieve a unity power factor for the total load.

14. In the three-wire system of Fig. 12.36, (*a*) replace the 50 Ω resistor with a 200 Ω resistor, and calculate the current flowing through the neutral wire. (*b*) Determine a new value for the 50 Ω resistor such that the neutral wire current magnitude is 25% that of line current I_{aA}.

12.3 Three-Phase Y-Y Connection

15. (*a*) Show that if $\mathbf{V}_{an} = 400\underline{/33^\circ}$ V, $\mathbf{V}_{bn} = 400\underline{/-87^\circ}$ V, and $\mathbf{V}_{cn} = 400\underline{/-207^\circ}$ V, then $\mathbf{V}_{an} + \mathbf{V}_{bn} + \mathbf{V}_{cn} = 0$. (*b*) Do the voltages in part (*a*) represent positive or negative phase sequence? *Explain.*

16. Consider a simple positive phase sequence, three-phase, three-wire system operated at 50 Hz and with a balanced load. Each phase voltage of 240 V is connected across a load composed of a series-connected 50 Ω and 500 mH combination. Calculate (*a*) each line current; (*b*) the power factor of the load; (*c*) the total power supplied by the three-phase source.

17. Assume the system shown in Fig. 12.38 is balanced, $R_w = 0$, $\mathbf{V}_{an} = 208\underline{/0°}$ V, and a positive phase sequence applies. Calculate all phase and line currents, and all phase and line voltages, if $\mathbf{Z}_p$ is equal to (*a*) 1 kΩ; (*b*) $100 + j48$ Ω; (*c*) $100 - j48$ Ω.

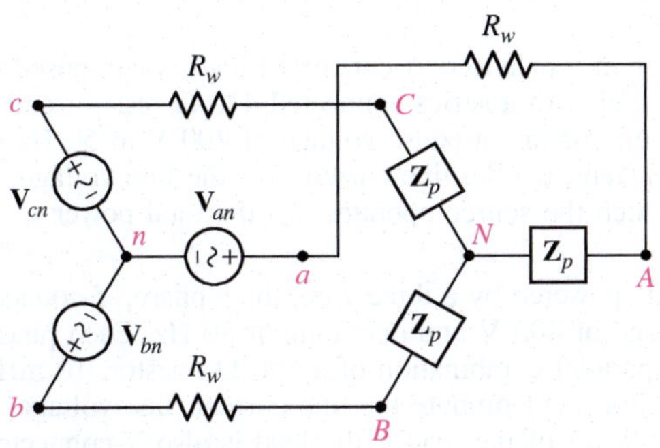

FIGURE 12.38

18. Repeat Exercise 17 with $R_w = 10$ Ω, and verify your answers with an appropriate set of simulations if the operating frequency is 60 Hz.

19. Each impedance $\mathbf{Z}_p$ in the balanced three-phase system of Fig. 12.38 is constructed using the parallel combination of a 1 mF capacitance, a 100 mH inductance, and a 10 Ω resistance. The sources have positive phase sequence and operate at 50 Hz. If $\mathbf{V}_{ab} = 208\underline{/0°}$ V, and $R_w = 0$, calculate (*a*) all phase voltages; (*b*) all line voltages; (*c*) all three line currents; (*d*) the total power drawn by the load.

20. With the assumption that the three-phase system pictured in Fig. 12.38 is balanced with a line voltage of 100 V, calculate the line current and per-phase load impedance if $R_w = 0$ and the load draws (*a*) 1 kW at a PF of 0.85 lagging; (*b*) 300 W per phase at a PF of 0.92 leading.

21. The balanced three-phase system of Fig. 12.38 is characterized by a positive phase sequence and a line voltage of 300 V. And $\mathbf{Z}_p$ is given by the parallel combination of a $5 - j3$ Ω capacitive load and a $9 + j2$ Ω inductive load. If $R_w = 0$, calculate (*a*) the power factor of the source; (*b*) the total power supplied by the source. (*c*) Repeat parts (*a*) and (*b*) if $R_w = 1$ Ω.

22. A balanced Y-connected load of $100 + j50$ Ω is connected to a balanced three-phase source. If the line current is 42 A and the source supplies 12 kW, determine (*a*) the line voltage; (*b*) the phase voltage.

23. A three-phase system is constructed from a balanced Y-connected source operating at 50 Hz and having a line voltage of 210 V, and each phase of the balanced load draws 130 W at a leading power factor of 0.75. (*a*) Calculate the line current and the total power supplied to the load. (*b*) If a purely resistive load of 1 Ω is connected in parallel with each existing load, calculate the new line current and total power supplied to the load. (*c*) Verify your answers using appropriate simulations.

24. Returning to the balanced three-phase system described in Exercise 21, determine the complex power delivered to the load for both $R_w = 0$ and $R_w = 1$ Ω.

25. Each load in the circuit of Fig. 12.38 is composed of a 1.5 H inductor in parallel with a 100 μF capacitor and a 1 kΩ resistor. The resistance is labeled $R_w = 0$ Ω. Using positive phase sequence with $\mathbf{V}_{ab} = 115\underline{/0°}$ V at $f = 60$ Hz, determine the rms line current and the total power delivered to the load. Verify your answers with an appropriate simulation.

12.4 The Delta (Δ) Connection

26. A particular balanced three-phase system is supplying a Δ-connected load with 10 kW at a leading power factor of 0.7. If the phase voltage is 208 V and the source operates at 50 Hz, (*a*) compute the line current; (*b*) determine the phase impedance; (*c*) calculate the new power factor and new total power delivered to the load if a 2.5 H inductor is connected in parallel with each phase of the load.
27. If each of the three phases in a balanced Δ-connected load is composed of a 10 mF capacitor in parallel with a series-connected 470 Ω resistor and 4 mH inductor combination, assume a phase voltage of 400 V at 50 Hz. (*a*) Calculate the phase current; (*b*) the line current; (*c*) the line voltage; (*d*) the power factor at which the source operates; (*e*) the total power delivered to the load.
28. A three-phase load is to be powered by a three-wire, three-phase, Y-connected source having phase voltage of 400 V and operating at 50 Hz. Each phase of the load consists of a parallel combination of a 500 Ω resistor, 10 mH inductor, and 1 mF capacitor. (*a*) Compute the line current, line voltage, phase current, and power factor of the load if the load is also Y-connected. (*b*) Rewire the load so that it is Δ-connected, and find the same quantities requested in part (*a*).
29. For the two situations described in Exercise 28, compute the total power delivered to each of the two loads.
30. Two Δ-connected loads are connected in parallel and powered by a balanced Y-connected system. The smaller of the two loads draws 10 kVA at a lagging PF of 0.75, and the larger draws 25 kVA at a leading PF of 0.80. The line voltage is 400 V. Calculate (*a*) the power factor at which the source is operating; (*b*) the total power drawn by the two loads; (*c*) the phase current of each load.
31. For the balanced three-phase system shown in Fig. 12.39, it is determined that 100 W is lost in each wire. If the phase voltage of the source is 400 V, and the load draws 12 kW at a lagging PF of 0.83, determine the wire resistance R_w.

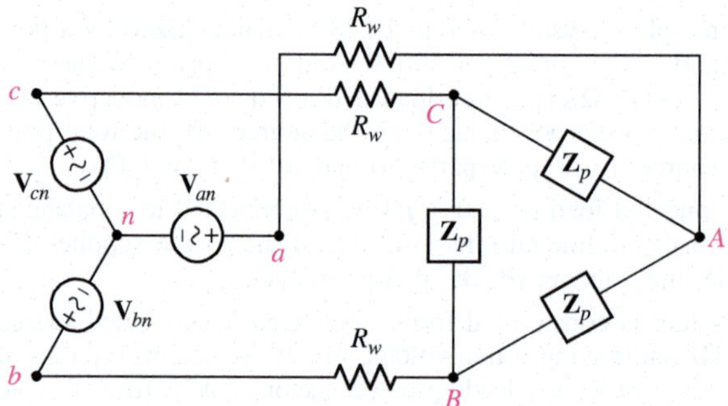

FIGURE 12.39

32. The balanced Δ-connected load in Fig. 12.39 is demanding 10 kVA at a lagging PF of 0.91. If line losses are negligible, calculate $\mathbf{I}_{bB}$ and $\mathbf{V}_{an}$ if $\mathbf{V}_{ca} = 160\underline{/30^\circ}$ V and the source voltages are described using positive phase sequence.

33. Repeat Exercise 32 if $R_w = 1\ \Omega$. Verify your solution using an appropriate simulation.
34. Compute $\mathbf{I}_{aA}$, $\mathbf{I}_{AB}$, and $\mathbf{V}_{an}$ if the Δ-connected load of Fig. 12.39 draws a total complex power of $1800 + j700$ W, $R_w = 1.2\ \Omega$, and the source generates a complex power of $1850 + j700$ W.

35. A balanced three-phase system having line voltage of 240 V rms contains a Δ-connected load of $12 + j$ kΩ per phase and also a Y-connected load of $5 + j3$ kΩ per phase. Find the line current, the power taken by the combined load, and the power factor of the load.

12.5 Power Measurement in Three-Phase Systems

36. Determine the wattmeter reading (stating whether or not the leads had to be reversed to obtain it) in the circuit of Fig. 12.40 if terminals *A* and *B*, respectively, are connected to (*a*) *x* and *y*; (*b*) *x* and *z*; (*c*) *y* and *z*.

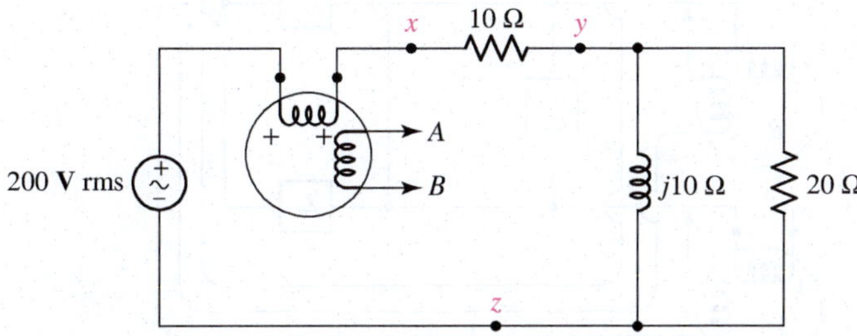

■ **FIGURE 12.40**

37. A wattmeter is connected into the circuit of Fig. 12.41 so that $\mathbf{I}_1$ enters the (+) terminal of the current coil, while $\mathbf{V}_2$ is the voltage across the potential coil. Find the wattmeter reading, and verify your solution with an appropriate simulation.

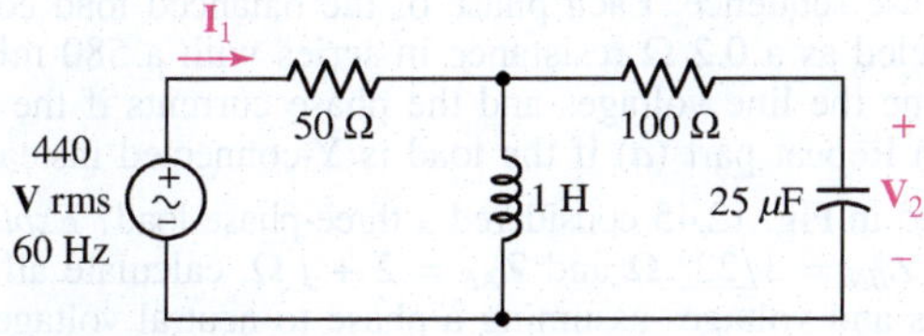

■ **FIGURE 12.41**

38. Find the reading of the wattmeter connected in the circuit of Fig. 12.42.

39. (*a*) Find both wattmeter readings in Fig. 12.43 if $\mathbf{V}_A = 100\underline{/0^\circ}$ V rms, $\mathbf{V}_B = 50\underline{/90^\circ}$ V rms, $\mathbf{Z}_A = 10 - j10\ \Omega$, $\mathbf{Z}_B = 8 + j6\ \Omega$, and $\mathbf{Z}_C = 30 + j10\ \Omega$. (*b*) Is the sum of these readings equal to the total power taken by the three loads? Verify your answer with an appropriate simulation.

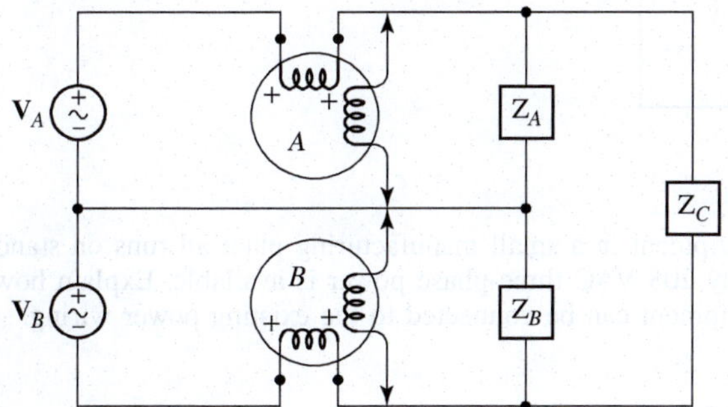

■ **FIGURE 12.43**

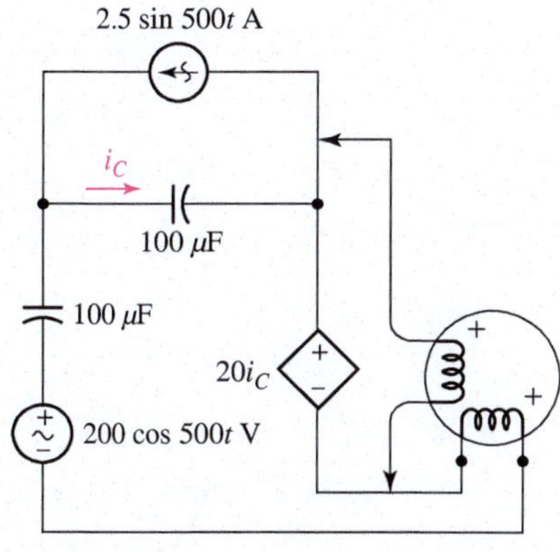

■ **FIGURE 12.42**

40. Circuit values for Fig. 12.44 are $\mathbf{V}_{ab} = 200\underline{/0^\circ}$, $\mathbf{V}_{bc} = 200\underline{/120^\circ}$, $\mathbf{V}_{ca} = 200\underline{/240^\circ}$ V rms, $\mathbf{Z}_4 = \mathbf{Z}_5 = \mathbf{Z}_6 = 25\underline{/30^\circ}\ \Omega$, $\mathbf{Z}_1 = \mathbf{Z}_2 = \mathbf{Z}_3 = 50\underline{/-60^\circ}\ \Omega$. Find the reading for each wattmeter.

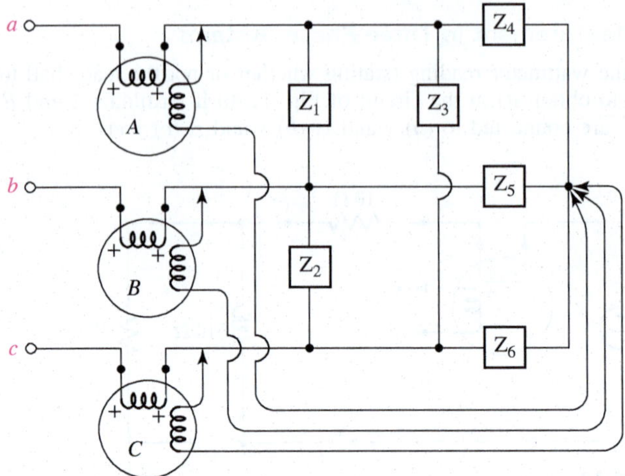

■ **FIGURE 12.44**

Chapter-Integrating Exercises

41. Explain under what circumstances a Δ-connected load might be preferred over a Y-connected load which draws the same average and complex powers.

42. A certain 208 V, 60 Hz, three-phase source is connected in a Y configuration and exhibits positive phase sequence. Each phase of the balanced load consists of a coil best modeled as a 0.2 Ω resistance in series with a 580 mH inductance. (*a*) Determine the line voltages and the phase currents if the load is Δ-connected. (*b*) Repeat part (*a*) if the load is Y-connected instead.

43. (*a*) Is the load represented in Fig. 12.45 considered a three-phase load? *Explain.* (*b*) If $\mathbf{Z}_{AN} = 1 - j7\ \Omega$, $\mathbf{Z}_{BN} = 3\underline{/22^\circ}\ \Omega$ and $\mathbf{Z}_{AB} = 2 + j\ \Omega$, calculate all phase (and line) currents and voltages assuming a phase to neutral voltage of 120 VAC (the two phases are 180° out of phase). (*c*) Under what circumstances does current flow in the neutral wire?

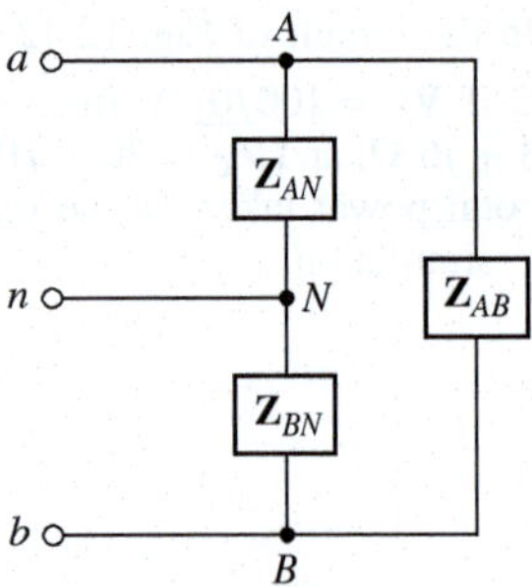

■ **FIGURE 12.45**

44. The computer equipment in a small manufacturing plant all runs on standard 120 VAC, but only 208 VAC three-phase power is available. Explain how the computer equipment can be connected to the existing power wiring.

CHAPTER 13

Magnetically Coupled Circuits

KEY CONCEPTS

Mutual Inductance

Self-Inductance

The Dot Convention

Reflected Impedance

T and Π Equivalent Networks

The Ideal Transformer

Turns Ratio of an Ideal Transformer

Impedance Matching

Voltage Level Adjustment

SPICE Analysis of Circuits with Transformers

INTRODUCTION

Whenever current flows through a conductor, whether as ac or dc, a magnetic field is generated about that conductor. In the context of circuits, we often refer to the *magnetic flux* through a loop of wire. This is the average normal component of the magnetic field density emanating from the loop multiplied by the surface area of the loop. When a time-varying magnetic field generated by one loop penetrates a second loop, a voltage is induced between the ends of the second wire. In order to distinguish this phenomenon from the "inductance" we defined earlier, more properly termed "self-inductance," we will define a new term, *mutual inductance*.

There is no such device as a "mutual inductor," but the principle forms the basis for an extremely important device—the *transformer*. A transformer consists of two coils of wire separated by a small distance, and is commonly used to convert ac voltages to higher or lower values depending on the application. Every electrical appliance that requires dc current to operate but plugs into an ac wall outlet makes use of a transformer to adjust voltage levels prior to *rectification,* a function typically performed by diodes and described in every introductory electronics text.

13.1 MUTUAL INDUCTANCE

When we defined inductance in Chap. 7, we did so by specifying the relationship between the terminal voltage and current,

$$v(t) = L\frac{di(t)}{dt}$$

where the passive sign convention is assumed. The physical basis for such a current–voltage characteristic rests upon two things:

1. The production of a ***magnetic flux*** by a current, the flux being proportional to the current in linear inductors.
2. The production of a voltage by a time-varying magnetic field; the voltage is proportional to the time rate of change of the magnetic field or the magnetic flux.

Coefficient of Mutual Inductance

Mutual inductance results from a slight extension of this same argument. A current flowing in one coil establishes a magnetic flux about that coil and also about a second coil nearby. The time-varying flux surrounding the second coil produces a voltage across the terminals of the second coil; this voltage is proportional to the time rate of change of the current flowing through the first coil. Figure 13.1*a* shows a simple model of two coils L_1 and L_2, sufficiently close together that the flux produced by a current $i_1(t)$ flowing through L_1 establishes an open-circuit voltage $v_2(t)$ across the terminals of L_2. Without considering the proper algebraic sign for the relationship at this point, we define the *coefficient of mutual inductance,* or simply ***mutual inductance,*** M_{21}, as

$$v_2(t) = M_{21}\frac{di_1(t)}{dt} \quad [1]$$

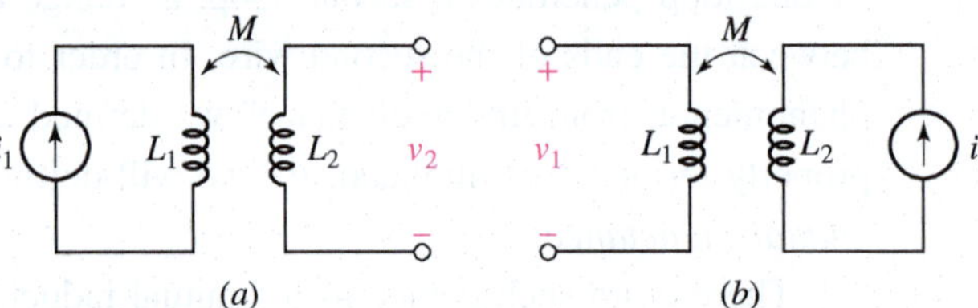

FIGURE 13.1 (*a*) A current i_1 through L_1 produces an open-circuit voltage v_2 across L_2. (*b*) A current i_2 through L_2 produces an open-circuit voltage v_1 across L_1.

The order of the subscripts on M_{21} indicates that a voltage response is produced at L_2 by a current source at L_1. If the system is reversed, as indicated in Fig. 13.1*b*, then we have

$$v_1(t) = M_{12}\frac{di_2(t)}{dt} \quad [2]$$

Two coefficients of mutual inductance are not necessary, however; we will use energy relationships a little later to prove that M_{12} and M_{21} are equal. Thus, $M_{12} = M_{21} = M$. The existence of mutual coupling between two coils is indicated by a double-headed arrow, as shown in Fig. 13.1*a* and *b*.

Mutual inductance is measured in henrys and, like resistance, inductance, and capacitance, is a positive quantity.[1] The voltage $M\,di/dt$, however, may appear as either a positive or a negative quantity depending on whether the current is increasing or decreasing at a particular instant in time.

(1) Mutual inductance is not universally assumed to be positive. It is particularly convenient to allow it to "carry its own sign" when three or more coils are involved and each coil interacts with each other coil. We will restrict our attention to the more important simple case of two coils.

Dot Convention

The inductor is a two-terminal element, and we can use the passive sign convention in order to select the correct sign for the voltage $L\ di/dt$ or $j\omega L\mathbf{I}$. If the current enters the terminal at which the positive voltage reference is located, then the positive sign is used. Mutual inductance, however, cannot be treated in exactly the same way because four terminals are involved. The choice of a correct sign is established by the use of one of several possibilities that include the ***dot convention,*** or by an examination of the particular way in which each coil is wound. We will use the dot convention and merely look briefly at the physical construction of the coils; the use of other special symbols is not necessary when only two coils are coupled.

The dot convention makes use of a large dot placed at one end of each of the two coils which are mutually coupled. We determine the sign of the mutual voltage as follows:

A current entering the *dotted* terminal of one coil produces an open-circuit voltage with a *positive* voltage reference at the *dotted* terminal of the second coil.

Thus, in Fig. 13.2*a*, i_1 enters the dotted terminal of L_1, v_2 is sensed positively at the dotted terminal of L_2, and $v_2 = M\ di_1/dt$. We have found previously that it is often not possible to select voltages or currents throughout a circuit so that the passive sign convention is everywhere satisfied; the same situation arises with mutual coupling. For example, it may be more convenient to represent v_2 by a positive voltage reference at the undotted terminal, as shown in Fig. 13.2*b*; then $v_2 = -M\ di_1/dt$. Currents that enter the dotted terminal are also not always available, as indicated by Fig. 13.2*c* and *d*. Thus, we can also say that:

A current entering the *undotted* terminal of one coil provides a voltage that is *positively* sensed at the *undotted* terminal of the second coil.

Note that the preceding discussion does not include any contribution to the voltage from self-induction, which would occur if i_2 were nonzero. We will consider this important situation in detail, but a quick example first is appropriate.

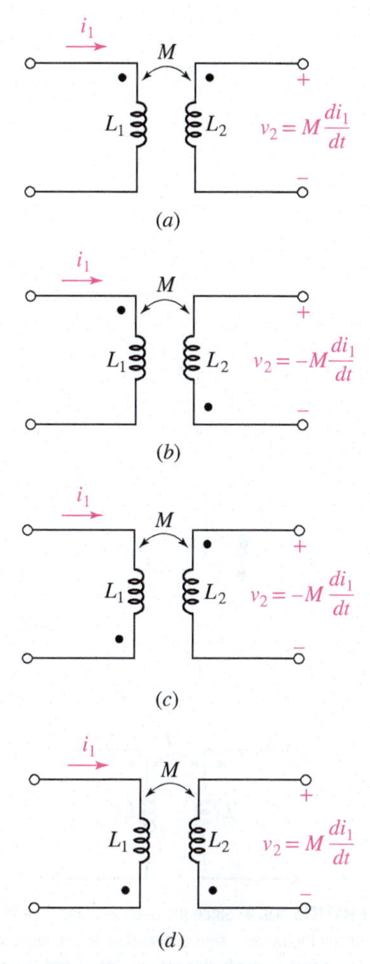

FIGURE 13.2 Current entering the dotted terminal of one coil produces a voltage that is sensed positively at the dotted terminal of the second coil. Current entering the undotted terminal of one coil produces a voltage that is sensed positively at the undotted terminal of the second coil.

EXAMPLE 13.1

For the circuit shown in Fig. 13.3, determine (*a*) v_1 if $i_2 = 5 \sin 45t$ A if $i_1 = 0$; and (*b*) v_2 if $i_1 = -8e^{-t}$ A if $i_2 = 0$.

(*a*) Since the current i_2 is entering the *undotted* terminal of the right coil, the positive reference for the voltage induced across the left coil is the undotted terminal. Thus, we have an open-circuit voltage

$$v_1 = -(2)(45)(5 \cos 45t) = -450 \cos 45t \text{ V}$$

appearing across the terminals of the left coil as a result of the time-varying magnetic flux generated by i_2 flowing into the right coil. Since no current flows through the coil on the left, there is no contribution to v_1 from self-induction.

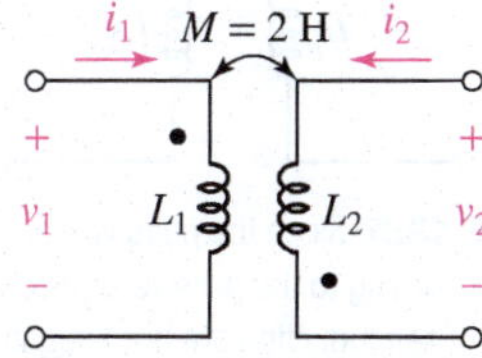

FIGURE 13.3 The dot convention provides a relationship between the terminal at which a current enters one coil and the positive voltage reference for the other coil.

(Continued on next page)

(*b*) We now have a current entering a *dotted* terminal, but v_2 has its positive reference at the *undotted* terminal. Thus,

$$v_2 = -(2)(-1)(-8e^{-t}) = -16e^{-t}\text{ V}$$

PRACTICE

13.1 Assuming $M = 10$ H, coil L_2 is open-circuited, and $i_1 = -2e^{-5t}$ A, find the voltage v_2 for (*a*) Fig. 13.2*a*; (*b*) Fig. 13.2*b*.

Ans: $100e^{-5t}$ V; $-100e^{-5t}$ V.

Combined Mutual and Self-Induction Voltage

So far, we have considered only a mutual voltage present across an *open-circuited* coil. In general, a nonzero current will be flowing in each of the two coils, and a mutual voltage will be produced in each coil because of the current flowing in the other coil. *This mutual voltage is present independently of (and in addition to) any voltage of self-induction.* In other words, the voltage across the terminals of L_1 will be composed of two terms, $L_1\, di_1/dt$ and $M\, di_2/dt$, each carrying a sign depending on the current directions, the assumed voltage sense, and the placement of the two dots. In the portion of a circuit drawn in Fig. 13.4, currents i_1 and i_2 are shown, each entering a dotted terminal. The voltage across L_1 is thus composed of two parts,

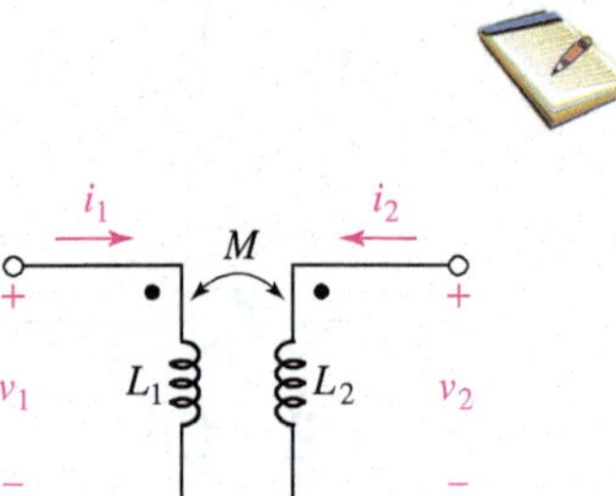

FIGURE 13.4 Since the pairs v_1, i_1 and v_2, i_2 each satisfy the passive sign convention, the voltages of self-induction are both positive; since i_1 and i_2 each enter dotted terminals, and since v_1 and v_2 are both positively sensed at the dotted terminals, the voltages of mutual induction are also both positive.

$$v_1 = L_1\frac{di_1}{dt} + M\frac{di_2}{dt}$$

as is the voltage across L_2,

$$v_2 = L_2\frac{di_2}{dt} + M\frac{di_1}{dt}$$

In Fig. 13.5 the currents and voltages are not selected with the object of obtaining all positive terms for v_1 and v_2. By inspecting only the reference symbols for i_1 and v_1, it is apparent that the passive sign convention is not satisfied and the sign of $L_1\, di_1/dt$ must therefore be negative. An identical conclusion is reached for the term $L_2\, di_2/dt$. The mutual term of v_2 is signed by inspecting the direction of i_1 and v_2; since i_1 enters the dotted terminal and v_2 is sensed positive at the dotted terminal, the sign of $M\, di/dt$ must be positive. Finally, i_2 enters the undotted terminal of L_2, and v_1 is sensed positive at the undotted terminal of L_1; hence, the mutual portion of v_1, $M\, di_2/dt$, must also be positive. Thus, we have

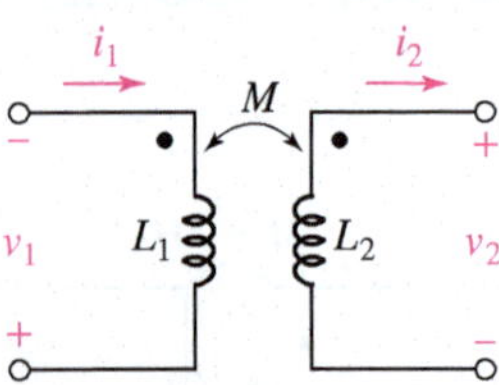

FIGURE 13.5 Since the pairs v_1, i_1 and v_2, i_2 are not sensed according to the passive sign convention, the voltages of self-induction are both negative; since i_1 enters the dotted terminal and v_2 is positively sensed at the dotted terminal, the mutual term of v_2 is positive; and since i_2 enters the undotted terminal and v_1 is positively sensed at the undotted terminal, the mutual term of v_1 is also positive.

$$v_1 = -L_1\frac{di_1}{dt} + M\frac{di_2}{dt} \quad \text{and} \quad v_2 = -L_2\frac{di_2}{dt} + M\frac{di_1}{dt}$$

The same considerations lead to identical choices of signs for excitation by a sinusoidal source operating at frequency ω

$$\mathbf{V}_1 = -j\omega L_1\mathbf{I}_1 + j\omega M\mathbf{I}_2 \quad \text{and} \quad \mathbf{V}_2 = -j\omega L_2\mathbf{I}_2 + j\omega M\mathbf{I}_1$$

Physical Basis of the Dot Convention

We can gain a more complete understanding of the dot symbolism by looking at the physical basis for the convention; the meaning of the dots is now interpreted in terms of *magnetic flux*. Two coils are shown wound on a cylindrical form in Fig. 13.6, and the direction of each winding is evident. Let us assume that the current i_1 is positive and increasing with time. The magnetic flux that i_1 produces within the form has a direction which may be found by the right-hand rule: When the right hand is wrapped around the coil with the fingers pointing in the direction of current flow, the thumb indicates the direction of the flux within the coil. Thus i_1 produces a flux which is directed downward; since i_1 is increasing with time, the flux, which is proportional to i_1, is also increasing with time. Turning now to the second coil, let us also think of i_2 as positive and increasing; the application of the right-hand rule shows that i_2 also produces a magnetic flux which is directed downward and is increasing. In other words, the assumed currents i_1 and i_2 produce *additive* fluxes.

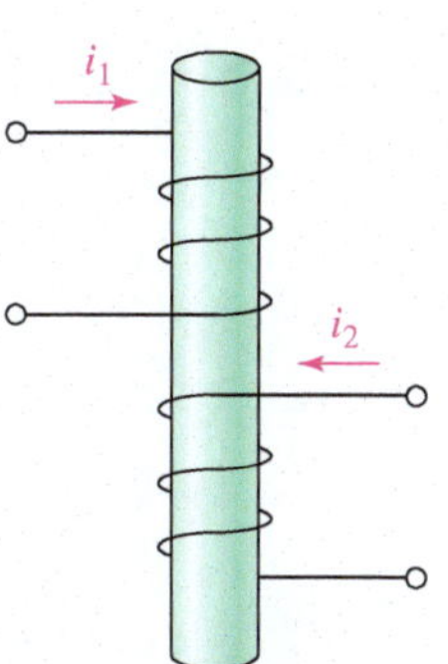

FIGURE 13.6 Example physical construction of two mutually coupled coils. From a consideration of the direction of magnetic flux produced by each coil, it can be shown that dots may be placed either on the upper terminal of each coil or on the lower terminal of each coil.

The voltage across the terminals of any coil results from the time rate of change of the flux within that coil. The voltage across the terminals of the first coil is therefore greater with i_2 flowing than it would be if i_2 were zero; i_2 induces a voltage in the first coil which has the same sense as the self-induced voltage in that coil. The sign of the self-induced voltage is known from the passive sign convention, and the sign of the mutual voltage is thus obtained.

The dot convention enables us to suppress the physical construction of the coils by placing a dot at one terminal of each coil such that currents entering dot-marked terminals produce additive fluxes. It is apparent that there are always two possible locations for the dots, because both dots may always be moved to the other ends of the coils, and additive fluxes will still result.

EXAMPLE **13.2**

For the circuit shown in Fig. 13.7*a*, find the ratio of the output voltage across the 400 Ω resistor to the source voltage, expressed using phasor notation.

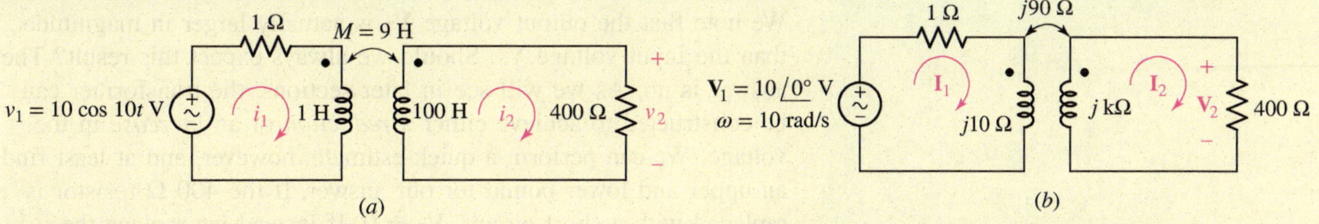

FIGURE 13.7 (*a*) A circuit containing mutual inductance in which the voltage ratio $\mathbf{V}_2/\mathbf{V}_1$ is desired. (*b*) Self- and mutual inductances are replaced by the corresponding impedances.

▶ ***Identify the goal of the problem.***

We need a numerical value for $\mathbf{V}_2$. We will then divide by $10\underline{/0°}$ V.

▶ ***Collect the known information.***

We begin by replacing the 1 H and 100 H inductances with their corresponding impedances, $j10\ \Omega$ and j kΩ, respectively (Fig. 13.7*b*). We also replace the 9 H mutual inductance with $j\omega M = j90\ \Omega$.

▶ ***Devise a plan.***

Mesh analysis is likely to be a good approach, as we have a circuit with two clearly defined meshes. Once we find $\mathbf{I}_2$, $\mathbf{V}_2$ is simply $400\mathbf{I}_2$.

(Continued on next page)

Construct an appropriate set of equations.

In the left mesh, the sign of the mutual term is determined by applying the dot convention. Since $\mathbf{I}_2$ enters the undotted terminal of L_2, the mutual voltage across L_1 must have the positive reference at the undotted terminal. Thus,

$$(1 + j10)\mathbf{I}_1 - j90\mathbf{I}_2 = 10\underline{/0^\circ}$$

Since $\mathbf{I}_1$ enters the dot-marked terminal, the mutual term in the right mesh has its (+) reference at the dotted terminal of the 100 H inductor. Therefore, we may write

$$(400 + j1000)\mathbf{I}_2 - j90\mathbf{I}_1 = 0$$

Determine if additional information is required.

We have two equations in two unknowns, $\mathbf{I}_1$ and $\mathbf{I}_2$. Once we solve for the two currents, the output voltage $\mathbf{V}_2$ may be obtained by multiplying $\mathbf{I}_2$ by 400 Ω. At this stage, we have sufficient information.

Attempt a solution.

Upon solving these two equations with a scientific calculator, we find that

$$\mathbf{I}_2 = 0.172\underline{/-16.70^\circ} \text{ A}$$

Thus,

$$\frac{\mathbf{V}_2}{\mathbf{V}_1} = \frac{400(0.172\underline{/-16.70^\circ})}{10\underline{/0^\circ}}$$
$$= 6.880\underline{/-16.70^\circ}$$

Verify the solution. Is it reasonable or expected?

We note that the output voltage $\mathbf{V}_2$ is actually larger in magnitude than the input voltage $\mathbf{V}_1$. Should we always expect this result? The answer is no. As we will see in later sections, the transformer can be constructed to achieve either a *reduction* or an *increase* in the voltage. We can perform a quick estimate, however, and at least find an upper and lower bound for our answer. If the 400 Ω resistor is replaced with a short circuit, $\mathbf{V}_2 = 0$. If instead we replace the 400 Ω resistor with an open circuit, $\mathbf{I}_2 = 0$ and hence

$$\mathbf{V}_1 = (1 + j\omega L_1)\mathbf{I}_1$$

and

$$\mathbf{V}_2 = j\omega M\mathbf{I}_1$$

Solving, we find that the maximum value we could expect for $\mathbf{V}_2/\mathbf{V}_1$ is $8.955\underline{/5.711^\circ}$. Thus, our answer at least appears reasonable.

The output voltage of the circuit in Fig. 13.7*a* is greater in magnitude than the input voltage, so a voltage gain is possible with this type of circuit. It is also interesting to consider this voltage ratio as a function of ω.

To find $\mathbf{I}_2(j\omega)$ for this particular circuit, we write the mesh equations in terms of an unspecified angular frequency ω:

$$(1 + j\omega)\mathbf{I}_1 \qquad -j\omega 9\,\mathbf{I}_2 = 10\underline{/0^\circ}$$

and

$$-j\omega 9\mathbf{I}_1 + (400 + j\omega 100)\mathbf{I}_2 = 0$$

Solving by substitution, we find that

$$\mathbf{I}_2 = \frac{j90\omega}{400 + j500\omega - 19\,\omega^2}$$

Thus, we obtain the ratio of output voltage to input voltage as a function of frequency ω

$$\frac{\mathbf{V}_2}{\mathbf{V}_1} = \frac{400\mathbf{I}_2}{10}$$

$$= \frac{j\omega 3600}{400 + j500\omega - 19\,\omega^2}$$

The magnitude of this ratio, sometimes referred to as the ***circuit transfer function,*** is plotted in Fig. 13.8 and has a peak magnitude of approximately 7 near a frequency of 4.6 rad/s. However, for very small or very large frequencies, the magnitude of the transfer function is less than unity.

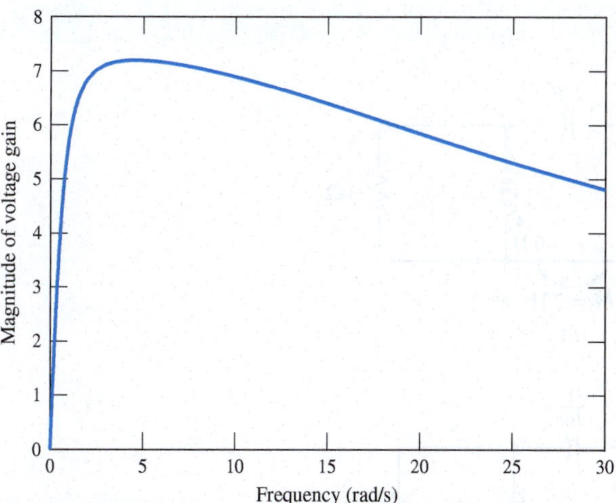

FIGURE 13.8 The voltage gain $|\mathbf{V}_2/\mathbf{V}_1|$ of the circuit shown in Fig. 13.7*a* is plotted as a function of ω using the following MATLAB script:

```
>> w = linspace(0,30,1000);
>> num = j*w*3600;
>> den = 400 + j*500*w - 19*w.*w;
>> gain = num./den;
>> plot(w,abs(gain));
>> xlabel('Frequency (rad/s)');
>> ylabel('Voltage Gain Magnitude');
```

The circuit is still passive, except for the voltage source, and the *voltage gain* must not be mistakenly interpreted as a *power gain*. At $\omega = 10$ rad/s, the voltage gain is 6.88, but the ideal voltage source, having a terminal voltage of 10 V, delivers a total power of 8.07 W, of which only 5.94 W reaches the 400 Ω resistor. The ratio of the output power to the source power, which we may define as the ***power gain,*** is thus 0.736.

PRACTICE

13.2 For the circuit of Fig. 13.9, write appropriate mesh equations for the left mesh and the right mesh if $v_s = 20e^{-1000t}$ V.

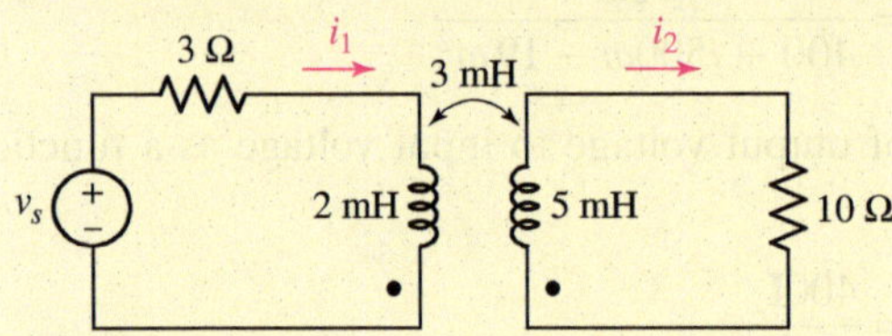

■ **FIGURE 13.9**

Ans: $20e^{-1000t} = 3i_1 + 0.002\, di_1/dt - 0.003\, di_2/dt$; $10i_2 + 0.005\, di_2/dt - 0.003\, di_1/dt = 0$.

EXAMPLE 13.3

Generate a complete set of phasor mesh equations for the circuit of Fig. 13.10*a*.

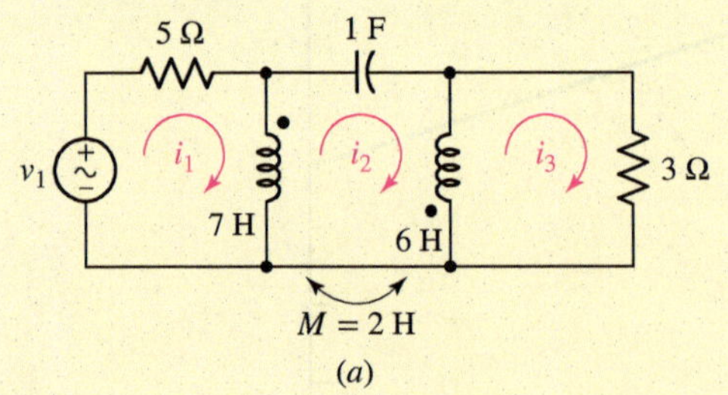

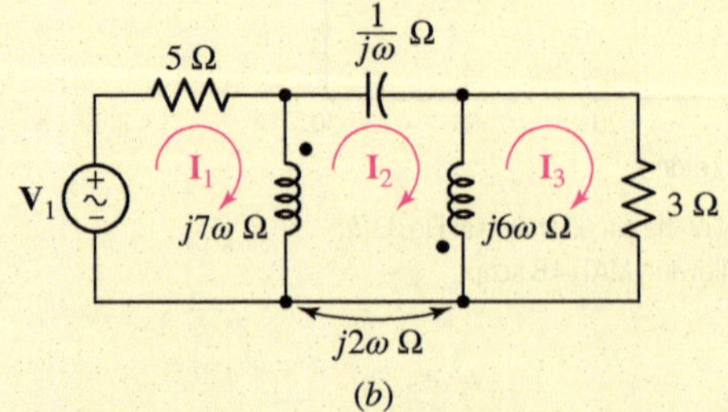

■ **FIGURE 13.10** (*a*) A three-mesh circuit with mutual coupling. (*b*) The 1 F capacitance as well as the self- and mutual inductances are replaced by their corresponding impedances.

Once again, our first step is to replace both the mutual inductance and the two self-inductances with their corresponding impedances as shown in Fig. 13.10*b*. Applying Kirchhoff's voltage law to the first mesh, a positive sign for the mutual term is ensured by selecting $(\mathbf{I}_3 - \mathbf{I}_2)$ as the current through the second coil. Thus,

$$5\mathbf{I}_1 + 7j\omega(\mathbf{I}_1 - \mathbf{I}_2) + 2j\omega(\mathbf{I}_3 - \mathbf{I}_2) = \mathbf{V}_1$$

or

$$(5 + 7j\omega)\mathbf{I}_1 - 9j\omega\mathbf{I}_2 + 2j\omega\mathbf{I}_3 = \mathbf{V}_1 \qquad [3]$$

The second mesh requires two self-inductance terms and two mutual inductance terms. Paying close attention to dots, we obtain

$$7j\omega(\mathbf{I}_2 - \mathbf{I}_1) + 2j\omega(\mathbf{I}_2 - \mathbf{I}_3) + \frac{1}{j\omega}\mathbf{I}_2 + 6j\omega(\mathbf{I}_2 - \mathbf{I}_3) + 2j\omega(\mathbf{I}_2 - \mathbf{I}_1) = 0$$

or

$$-9j\omega\mathbf{I}_1 + \left(17j\omega + \frac{1}{j\omega}\right)\mathbf{I}_2 - 8j\omega\mathbf{I}_3 = 0 \qquad [4]$$

Finally, for the third mesh,

$$6j\omega(\mathbf{I}_3 - \mathbf{I}_2) + 2j\omega(\mathbf{I}_1 - \mathbf{I}_2) + 3\mathbf{I}_3 = 0$$

or

$$2j\omega\mathbf{I}_1 - 8j\omega\mathbf{I}_2 + (3 + 6j\omega)\mathbf{I}_3 = 0 \qquad [5]$$

Equations [3] to [5] may be solved by any of the conventional methods.

PRACTICE

13.3 For the circuit of Fig. 13.11, write an appropriate mesh equation in terms of the phasor currents $\mathbf{I}_1$ and $\mathbf{I}_2$ for the (*a*) left mesh; (*b*) right mesh.

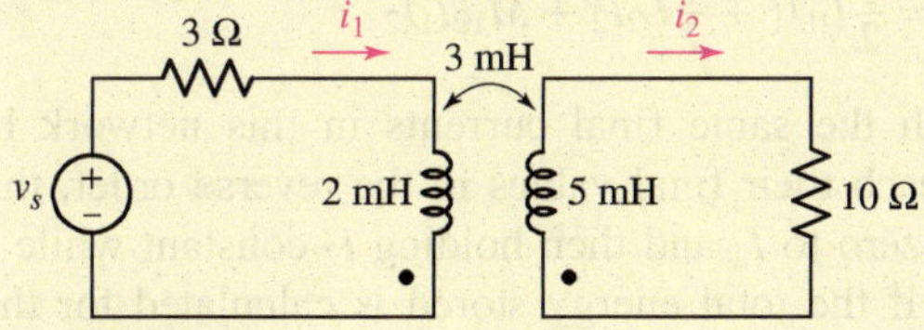

FIGURE 13.11

Ans: $\mathbf{V}_s = (3 + j10)\mathbf{I}_1 - j15\mathbf{I}_2$; $0 = -j15\mathbf{I}_1 + (10 + j25)\mathbf{I}_2$.

13.2 ENERGY CONSIDERATIONS

Let us now consider the energy stored in a pair of mutually coupled inductors. The results will be useful in several different ways. We will first justify our assumption that $M_{12} = M_{21}$, and we may then determine the maximum possible value of the mutual inductance between two given inductors.

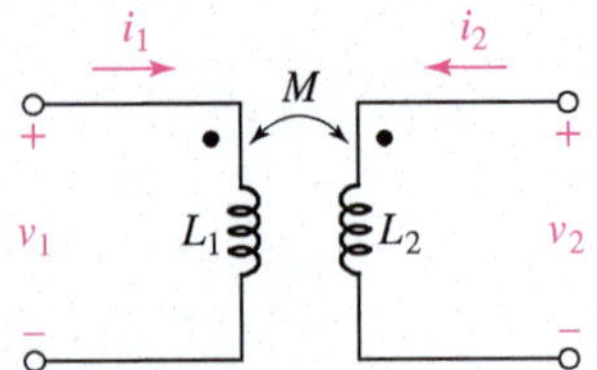

■ **FIGURE 13.12** A pair of coupled coils with a mutual inductance of $M_{12} = M_{21} = M$.

Equality of M_{12} and M_{21}

The pair of coupled coils shown in Fig. 13.12 has currents, voltages, and polarity dots indicated. In order to show that $M_{12} = M_{21}$ we begin by letting all currents and voltages be zero, thus establishing zero initial energy storage in the network. We then open-circuit the right-hand terminal pair and increase i_1 from zero to some constant (dc) value I_1 at time $t = t_1$. The power entering the network from the left at any instant is

$$v_1 i_1 = L_1 \frac{di_1}{dt} i_1$$

and the power entering from the right is

$$v_2 i_2 = 0$$

since $i_2 = 0$.

The energy stored within the network when $i_1 = I_1$ is thus

$$\int_0^{t_1} v_1 i_1 dt = \int_0^{I_1} L_1 i_1 di_1 = \frac{1}{2} L_1 I_1^2$$

We now hold i_1 constant ($i_1 = I_1$), and we let i_2 change from zero at $t = t_1$ to some constant value I_2 at $t = t_2$. The energy delivered from the right-hand source is thus

$$\int_{t_1}^{t_2} v_2 i_2 dt = \int_0^{I_2} L_2 i_2 di_2 = \frac{1}{2} L_2 I_2^2$$

However, even though the value of i_1 remains constant, the left-hand source also delivers energy to the network during this time interval:

$$\int_{t_1}^{t_2} v_1 i_1 dt = \int_{t_1}^{t_2} M_{12} \frac{di_2}{dt} i_1 dt = M_{12} I_1 \int_0^{I_2} di_2 = M_{12} I_1 I_2$$

The total energy stored in the network when both i_1 and i_2 have reached constant values is

$$W_{\text{total}} = \frac{1}{2} L_1 I_1^2 + \frac{1}{2} L_2 I_2^2 + M_{12} I_1 I_2$$

Now, we may establish the same final currents in this network by allowing the currents to reach their final values in the reverse order, that is, first increasing i_2 from zero to I_2 and then holding i_2 constant while i_1 increases from zero to I_1. If the total energy stored is calculated for this experiment, the result is found to be

$$W_{\text{total}} = \frac{1}{2} L_1 I_1^2 + \frac{1}{2} L_2 I_2^2 + M_{12} I_1 I_2$$

The only difference is the interchange of the mutual inductances M_{21} and M_{12}. The initial and final conditions in the network are the same, however, and so the two values of the stored energy must be identical. Thus,

$$M_{12} = M_{21} = M$$

and

$$W = \frac{1}{2} L_1 I_1^2 + \frac{1}{2} L_2 I_2^2 + M I_1 I_2 \qquad [6]$$

If one current enters a dot-marked terminal while the other leaves a dot-marked terminal, the sign of the mutual energy term is reversed:

$$W = \frac{1}{2}L_1I_1^2 + \frac{1}{2}L_2I_2^2 - MI_1I_2 \qquad [7]$$

Although Eqs. [6] and [7] were derived by treating the final values of the two currents as constants, these "constants" can have any value, and the energy expressions correctly represent the energy stored when the *instantaneous* values of i_1 and i_2 are i_1 and i_2, respectively. In other words, lowercase symbols might just as well be used:

$$w(t) = \frac{1}{2}L_1[i_1(t)]^2 + \frac{1}{2}L_2[i_2(t)]^2 \pm M[i_1(t)][i_2(t)] \qquad [8]$$

The only assumption upon which Eq. [8] is based is the logical establishment of a zero-energy reference level when both currents are zero.

Establishing an Upper Limit for *M*

Equation [8] may now be used to establish an upper limit for the value of M. Since $w(t)$ represents the energy stored within a *passive* network, it cannot be negative for any values of i_1, i_2, L_1, L_2, or M. Let us assume first that i_1 and i_2 are either both positive or both negative; their product is therefore positive. From Eq. [8], the only case in which the energy could possibly be negative is

$$w = \frac{1}{2}L_1i_1^2 + \frac{1}{2}L_2i_2^2 - Mi_1i_2$$

which we may write, by completing the square, as

$$w = \frac{1}{2}(\sqrt{L_1}i_1 - \sqrt{L_2}i_2)^2 + \sqrt{L_1L_2}i_1i_2 - Mi_1i_2$$

Since in reality the energy cannot be negative, the right-hand side of this equation cannot be negative. The first term, however, may be as small as zero, so we have the restriction that the sum of the last two terms cannot be negative. Hence,

$$\sqrt{L_1L_2} \geq M$$

or

$$M \leq \sqrt{L_1L_2} \qquad [9]$$

There is, therefore, an upper limit to the possible magnitude of the mutual inductance; it can be no larger than the geometric mean of the inductances of the two coils between which the mutual inductance exists. Although we have derived this inequality on the assumption that i_1 and i_2 carried the same algebraic sign, a similar development is possible if the signs are opposite; it is necessary only to select the positive sign in Eq. [8].

We might also have demonstrated the truth of inequality [9] from a physical consideration of the magnetic coupling; if we think of i_2 as being zero and the current i_1 as establishing the magnetic flux linking both L_1 and L_2, it is apparent that the flux within L_2 cannot be greater than the flux within L_1, which represents the total flux. Qualitatively, then, there is an upper limit to the magnitude of the mutual inductance possible between two given inductors.

The Coupling Coefficient

The degree to which M approaches its maximum value is described by the ***coupling coefficient,*** defined as

$$k = \frac{M}{\sqrt{L_1 L_2}} \tag{10}$$

and since $M \le \sqrt{L_1 L_2}$,

$$0 \le k \le 1$$

The larger values of the coefficient of coupling are obtained with coils which are physically closer, which are wound or oriented to provide a larger common magnetic flux, or which are provided with a common path through a material which serves to concentrate and localize the magnetic flux (a high-permeability material). Coils having a coefficient of coupling close to unity are said to be *tightly coupled.*

EXAMPLE 13.4

In Fig. 13.13, let L_1 = 0.4 H, L_2 = 2.5 H, k = 0.6, and $i_1 = 4i_2$ = 20 cos(500t − 20°) mA. Determine both $v_1(0)$ and the total energy stored in the system at t = 0.

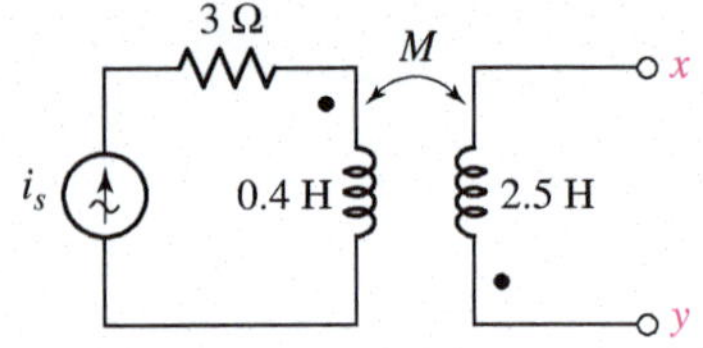

FIGURE 13.13 Two coils with a coupling coefficient of 0.6, L_1 = 0.4 H and L_2 = 2.5 H.

In order to determine the value of v_1, we need to include the contributions from both the self-inductance of coil 1 and the mutual inductance. Thus, paying attention to the dot convention,

$$v_1(t) = L_1\frac{di_1}{dt} + M\frac{di_2}{dt}$$

To evaluate this quantity, we require a value for M. This is obtained from Eq. [10],

$$M = k\sqrt{L_1 L_2} = 0.6\sqrt{(0.4)(2.5)} = 0.6 \text{ H}$$

Thus, $v_1(0) = 0.4[-10 \sin(-20°)] + 0.6[-2.5 \sin(-20°)] = 1.881$ V.

The total energy is found by summing the energy stored in each inductor, which has three separate components since the two coils are known to be magnetically coupled. Since both currents enter a "dotted" terminal,

$$w(t) = \frac{1}{2}L_1[i_1(t)]^2 + \frac{1}{2}L_2[i_2(t)]^2 + M[i_1(t)][i_2(t)]$$

Since $i_1(0) = 20 \cos(-20°) = 18.79$ mA and $i_2(0) = 0.25 i_1(0) =$ 4.698 mA, we find that the total energy stored in the two coils at t = 0 is 151.2 μJ.

PRACTICE

13.4 Let $i_s = 2 \cos 10t$ A in the circuit of Fig. 13.14, and find the total energy stored in the passive network at t = 0 if k = 0.6 and terminals x and y are (a) left open-circuited; (b) short-circuited.

Ans: 0.8 J; 0.512 J.

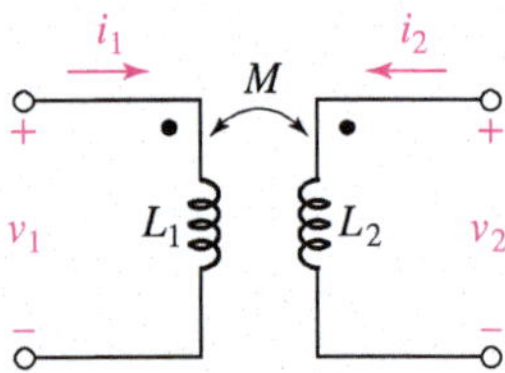

FIGURE 13.14

13.3 THE LINEAR TRANSFORMER

We are now ready to apply our knowledge of magnetic coupling to the description of two specific practical devices, each of which may be represented by a model containing mutual inductance. Both of the devices are transformers, a term which we define as a network containing two or more coils which are deliberately coupled magnetically (Fig. 13.15). In this section we consider the linear transformer, which happens to be an excellent model for devices used at radio frequencies, or higher frequencies. In Sec. 13.4 we will consider the ideal transformer, which is an idealized lossless and unity-coupled model of a linear transformer.

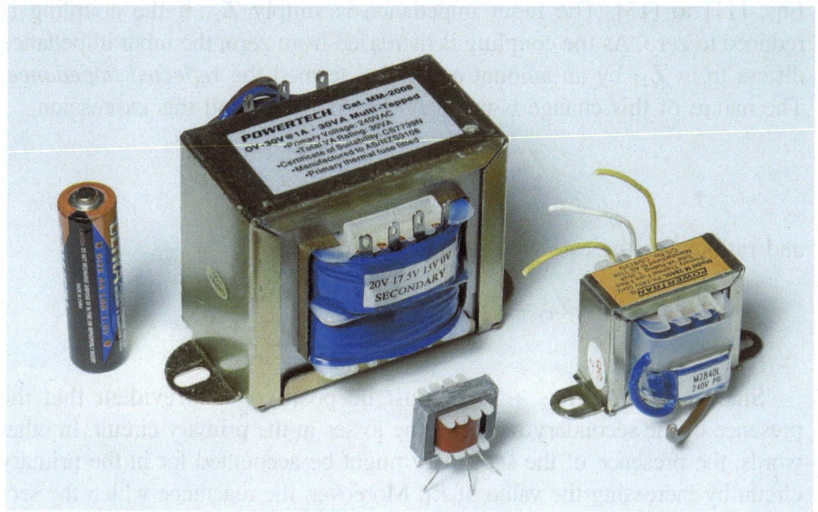

FIGURE 13.15 A selection of small transformers for use in electronic applications; the AA battery is shown for scale only.
Steve Durbin/McGraw Hill

In Fig. 13.16 a transformer is shown with two mesh currents identified. The first mesh, usually containing the source, is called the ***primary,*** while the second mesh, usually containing the load, is known as the ***secondary.*** The inductors labeled L_1 and L_2 are also referred to as the primary and secondary, respectively, of the transformer. We will assume that the transformer is *linear.* This implies that no magnetic material (which may cause a *nonlinear* flux-versus-current relationship) is employed. Without such material, however, it is difficult to achieve a coupling coefficient greater than a few tenths. The two resistors serve to account for the resistance of the wire out of which the primary and secondary coils are wound, and any other losses.

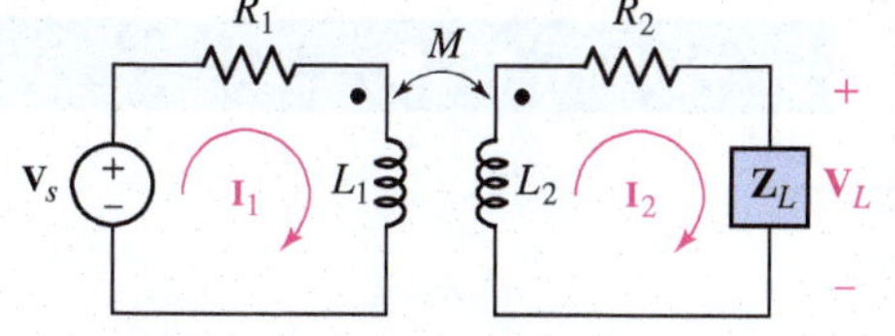

FIGURE 13.16 A linear transformer containing a source in the primary circuit and a load in the secondary circuit. Resistance is also included in both the primary and the secondary.

Reflected Impedance

Consider the input impedance offered at the terminals of the primary circuit. The two mesh equations are

$$\mathbf{V}_s = (R_1 + j\omega L_1)\mathbf{I}_1 - j\omega M\mathbf{I}_2 \qquad [11]$$

and

$$0 = -j\omega M\mathbf{I}_1 + (R_2 + j\omega L_2 + \mathbf{Z}_L)\mathbf{I}_2 \qquad [12]$$

We may simplify by defining

$$\mathbf{Z}_{11} = R_1 + j\omega L_1 \qquad \text{and} \qquad \mathbf{Z}_{22} = R_2 + j\omega L_2 + Z_L = R_{22} + jX_{22}$$

so that

$$\mathbf{V}_s = \mathbf{Z}_{11}\mathbf{I}_1 - j\omega M\mathbf{I}_2 \qquad [13]$$

$$0 = -j\omega M\mathbf{I}_1 + \mathbf{Z}_{22}\mathbf{I}_2 \qquad [14]$$

Solving the second equation for $\mathbf{I}_2$ and inserting the result in the first equation enable us to find the input impedance,

$$\mathbf{Z}_{\text{in}} = \frac{\mathbf{V}_s}{\mathbf{I}_1} = \mathbf{Z}_{11} - \frac{(j\omega)^2 M^2}{\mathbf{Z}_{22}} \qquad [15]$$

$\mathbf{Z}_{\text{in}}$ is the impedance seen looking into the primary coil of the transformer.

Before manipulating this expression any further, we can draw several exciting conclusions. The input impedance is independent of the location of the dots on either winding, which determines the sign involving M in Eqs. [11] to [15]. The input impedance is simply $\mathbf{Z}_{11}$ if the coupling is reduced to zero. As the coupling is increased from zero, the input impedance differs from $\mathbf{Z}_{11}$ by an amount $\omega^2 M^2/\mathbf{Z}_{22}$, termed the *reflected impedance*. The nature of this change is more evident if we expand this expression

$$\mathbf{Z}_{\text{in}} = \mathbf{Z}_{11} + \underbrace{\frac{\omega^2 M^2}{R_{22} + jX_{22}}}_{\text{Reflected impedance}}$$

and rationalize the reflected impedance,

$$\mathbf{Z}_{\text{in}} = \mathbf{Z}_{11} + \underbrace{\frac{\omega^2 M^2 R_{22}}{R_{22}^2 + X_{22}^2} - j\frac{\omega^2 M^2 X_{22}}{R_{22}^2 + X_{22}^2}}_{\text{Reflected impedance}}$$

Since $\omega^2 M^2 R_{22}/(R_{22}^2 + X_{22}^2)$ must be positive, it is evident that the presence of the secondary increases the losses in the primary circuit. In other words, the presence of the secondary might be accounted for in the primary circuit by increasing the value of R_1. Moreover, the reactance which the secondary reflects into the primary circuit has a sign which is opposite to that of X_{22}. This reactance X_{22} is the sum of ωL_2 and X_L; it is necessarily positive for inductive loads and either positive or negative for capacitive loads.

EXAMPLE 13.5

A linear transformer is connected in the circuit of Fig. 13.16 with $R_1 = R_2 = 2\ \Omega$, $L_1 = 4$ mH, $L_2 = 8$ mH, and $Z_L = 10\ \Omega$. For operation at $\omega = 5000$ rad/s, find M such that $\mathbf{Z}_{\text{in}}$ is purely real.

The input impedance $\mathbf{Z}_{\text{in}}$ will be all real when X_{11} is equal to the imaginary part of the reflected impedance such that the reactance will cancel.

$$X_{11} = \omega L_1 = \frac{\omega^2 M^2 X_{22}}{R_{22}^2 + X_{22}^2}$$

Solving for M,

$$M = \sqrt{\frac{L_1(R_{22}^2 + X_{22}^2)}{\omega X_{22}}}$$

Inserting values, where $R_{22} = 2 + 10 = 12\ \Omega$ and $X_{22} = \omega L_2 = 40\ \Omega$

$$M = \sqrt{\frac{(4 \times 10^{-3})(12^2 + 40^2)}{(5 \times 10^3)(40)}} = 5.906 \text{ mH}$$

PRACTICE

13.5 Element values for a certain linear transformer are $R_1 = 3\ \Omega$, $R_2 = 6\ \Omega$, $L_1 = 2$ mH, $L_2 = 10$ mH, and $M = 4$ mH. If $\omega = 5000$ rad/s, find $\mathbf{Z}_{in}$ for cases where $\mathbf{Z}_L$ is equal to (*a*) 10 Ω; (*b*) *j*20 Ω; (*c*) 10 + *j*20 Ω; (*d*) −*j*20 Ω.

Ans: 5.32 + *j*2.74 Ω; 3.49 + *j*4.33 Ω; 4.24 + *j*4.57 Ω; 5.56 − *j*2.82 Ω.

T and Π Equivalent Networks

It is often convenient to replace a transformer with an equivalent network in the form of a T or Π. If we separate the primary and secondary resistances from the transformer, only the pair of mutually coupled inductors remains, as shown in Fig. 13.17. Note that the two lower terminals of the transformer are connected together to form a three-terminal network. We do this because both of our equivalent networks are also three-terminal networks. The differential equations describing this circuit are, once again,

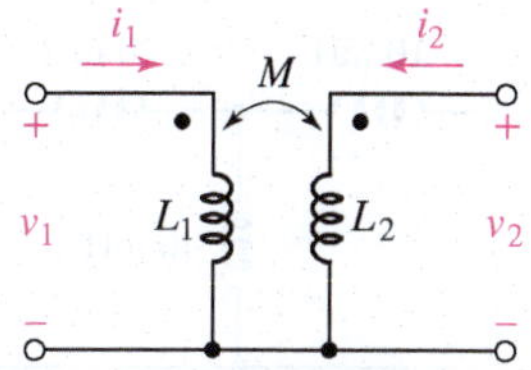

FIGURE 13.17 A given transformer which is to be replaced by an equivalent Π or T network.

$$v_1 = L_1\frac{di_1}{dt} + M\frac{di_2}{dt} \qquad [16]$$

and

$$v_2 = M\frac{di_1}{dt} + L_2\frac{di_2}{dt} \qquad [17]$$

The form of these two equations is familiar and may be easily interpreted in terms of mesh analysis with currents i_1 and i_2 where the two meshes share a common *self*-inductance M. An equivalent network consisting of $L_1 - M$, M, and $L_2 - M$ is shown in Fig. 13.18, resulting in identical pairs of equations relating v_1, i_1, v_2, and i_2 for the two networks.

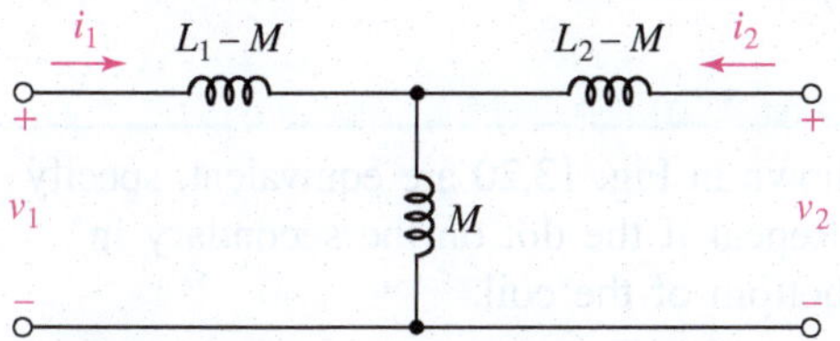

FIGURE 13.18 The T equivalent of the transformer shown in Fig. 13.17.

If either of the dots on the windings of the given transformer is placed on the opposite end of its coil, the sign of the mutual terms in Eqs. [16] and [17] will be negative. This is analogous to replacing M with $-M$, and such a replacement in the network of Fig. 13.18 leads to the correct equivalent for this case. (The three self-inductance values would then be $L_1 + M$, $-M$, and $L_2 + M$.)

The inductances in the T equivalent are all self-inductances; no mutual inductance is present. It is possible that negative values of inductance may be obtained for the equivalent circuit, but this is immaterial if our only desire is a mathematical analysis. There are times when procedures for synthesizing networks to provide a desired transfer function lead to circuits containing a T network having a negative inductance; this network may then be realized by the use of an appropriate linear transformer.

EXAMPLE 13.6

Find the T equivalent of the linear transformer shown in Fig. 13.19*a*.

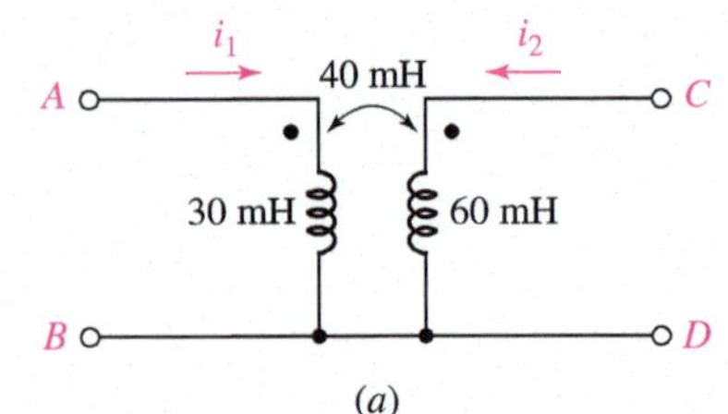

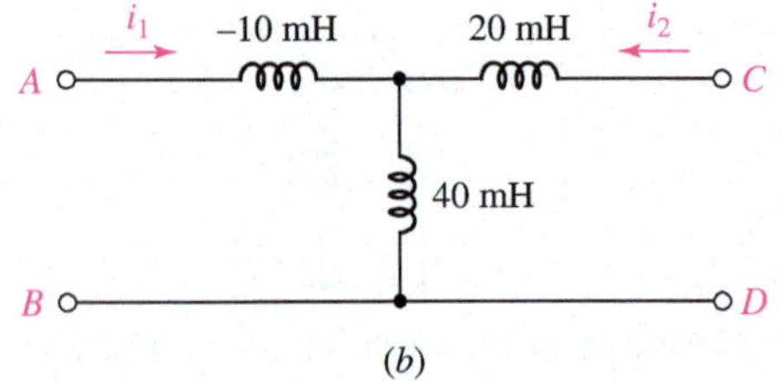

■ **FIGURE 13.19** (*a*) A linear transformer used as an example. (*b*) The T-equivalent network of the transformer.

We identify $L_1 = 30$ mH, $L_2 = 60$ mH, and $M = 40$ mH, and note that the dots are both at the upper terminals, as they are in the basic circuit of Fig. 13.17.

Hence, $L_1 - M = -10$ mH is in the upper left arm, $L_2 - M = 20$ mH is at the upper right, and the center stem contains $M = 40$ mH. The complete equivalent T is shown in Fig. 13.19*b*.

To demonstrate the equivalence, let us leave terminals *C* and *D* open-circuited and apply $v_{AB} = 10 \cos 100t$ V to the input in Fig. 13.19*a*. Thus,

$$i_1 = \frac{1}{30 \times 10^{-3}} \int 10 \cos(100t)dt = 3.33 \sin 100t \text{ A}$$

and

$$v_{CD} = M\frac{di_1}{dt} = 40 \times 10^{-3} \times 3.33 \times 100 \cos 100t$$
$$= 13.33 \cos 100t \text{ V}$$

Applying the same voltage in the T equivalent, we find that

$$i_1 = \frac{1}{(-10 + 40) \times 10^{-3}} \int 10 \cos(100t)dt = 3.33 \sin 100t \text{ A}$$

once again. Also, the voltage at *C* and *D* is equal to the voltage across the 40 mH inductor. Thus,

$$v_{CD} = 40 \times 10^{-3} \times 3.33 \times 100 \cos 100t = 13.33 \cos 100t \text{ V}$$

and the two networks yield equal results.

PRACTICE

13.6 (*a*) If the two networks shown in Fig. 13.20 are equivalent, specify values for L_x, L_y, and L_z. (*b*) Repeat if the dot on the secondary in Fig. 13.20*b* is located at the bottom of the coil.

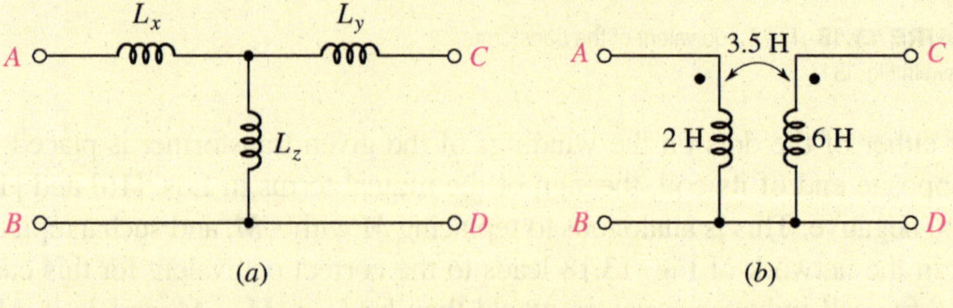

■ **FIGURE 13.20**

Ans: 1.5, 2.5, 3.5 H; 5.5, 9.5, −3.5 H

Obtaining the equivalent Π network is more complicated. The equivalent network may be represented as a pair of nodal equations, where a step-current source must be installed at each node in order to provide the proper initial conditions. Integration is required due to the relation of nodal voltages to inductance, where the factors multiplying each integral have the general form of inverses of certain equivalent inductances. The resulting equivalent Π network is shown in Fig. 13.21.

$$L_A = \frac{L_1 L_2 - M^2}{L_2 - M}$$

$$L_B = \frac{L_1 L_2 - M^2}{M}$$

$$L_C = \frac{L_1 L_2 - M^2}{L_1 - M}$$

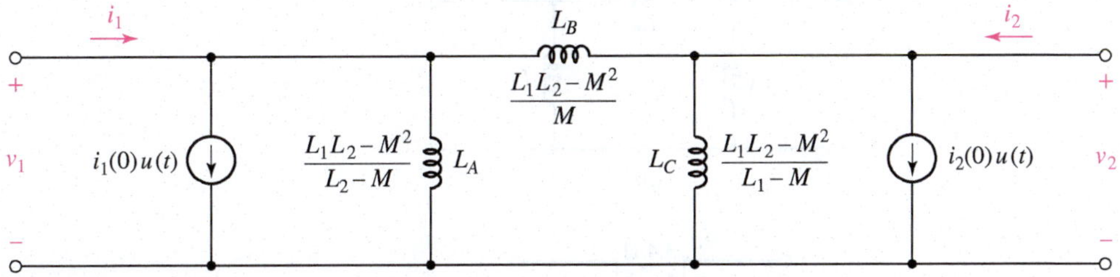

FIGURE 13.21 The Π network which is equivalent to the transformer shown in Fig. 13.17.

No magnetic coupling is present among the inductors in the equivalent Π, and the initial currents in the three *self-inductances* are zero. We may compensate for a reversal of either dot in the given transformer by merely changing the sign of M in the equivalent network. Also, just as we found in the equivalent T, negative self-inductances may appear in the equivalent Π network.

EXAMPLE 13.7

Find the equivalent Π network of the transformer in Fig. 13.19*a*, assuming zero initial currents.

Since the term $L_1 L_2 - M^2$ is common to L_A, L_B, and L_C, we begin by evaluating this quantity, obtaining

$$30 \times 10^{-3} \times 60 \times 10^{-3} - (40 \times 10^{-3})^2 = 2 \times 10^{-4}\ \text{H}^2$$

Thus,

$$L_A = \frac{L_1 L_2 - M^2}{L_2 - M} = \frac{2 \times 10^{-4}}{20 \times 10^{-3}} = 10\ \text{mH}$$

$$L_C = \frac{L_1 L_2 - M^2}{L_1 - M} = -20\ \text{mH}$$

and

$$L_B = \frac{L_1 L_2 - M^2}{M} = 5\ \text{mH}$$

The equivalent Π network is shown in Fig. 13.22.

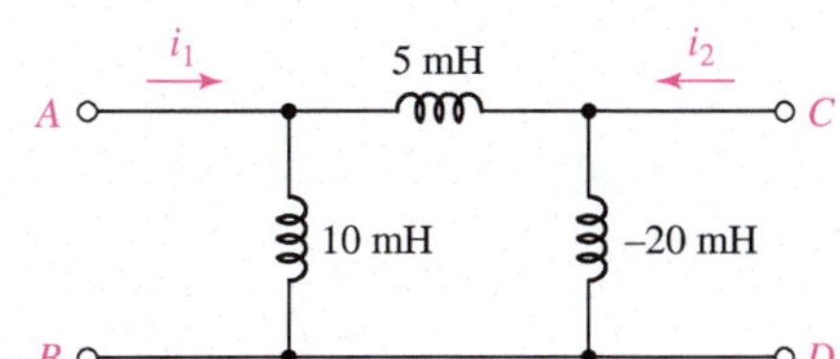

FIGURE 13.22 The Π equivalent of the linear transformer shown in Fig. 13.19*a*. It is assumed that $i_1(0) = 0$ and $i_2(0) = 0$.

(Continued on next page)

If we again check our result by letting $v_{AB} = 10 \cos 100t$ V with terminals *C-D* open-circuited, the output voltage is quickly obtained by voltage division:

$$v_{CD} = \frac{-20 \times 10^{-3}}{5 \times 10^{-3} - 20 \times 10^{-3}} 10 \cos 100t = 13.33 \cos 100t \text{ V}$$

as before. Thus, the network in Fig. 13.22 is electrically equivalent to the networks in Fig. 13.19*a* and *b*.

PRACTICE

13.7 If the networks in Fig. 13.23 are equivalent, specify values (in mH) for L_A, L_B, and L_C.

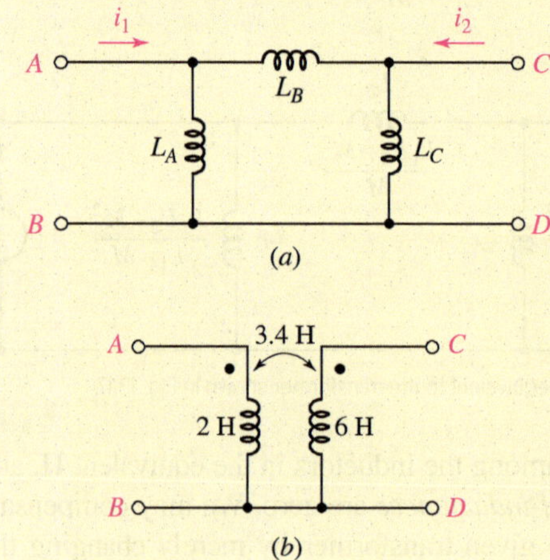

FIGURE 13.23

Ans: $L_A = 169.2$ mH, $L_B = 129.4$ mH, $L_C = -314.3$ mH.

COMPUTER-AIDED ANALYSIS

The ability to simulate circuits that contain magnetically coupled inductances is a useful skill, especially with circuit dimensions continuing to decrease. As various loops and partial loops of conductors are brought closer in new designs, various circuits and subcircuits that are intended to be isolated from one another inadvertently become coupled through stray magnetic fields and interact with one another. LTspice allows us to specify mutual inductance using a SPICE directive called a **K statement,** which links a pair of inductors in the schematic by a coupling coefficient k in the range of $0 \leq k \leq 1$.

For example, consider the circuit of Fig. 13.19*a*, which consists of two coils whose coupling is described by a mutual inductance of $M = 40$ mH, corresponding to a coupling coefficient of $k = 0.9428$. The basic circuit schematic is shown in Fig. 13.24*a*. The two coupled inductors, **L1** and **L2,** are specified along with the coupling coefficient through a SPICE directive **K1 L1 L2 0.9428,** which defines the coupling **K1** between inductors **L1** and **L2** with a coupling coefficient of $k = 0.9428$.

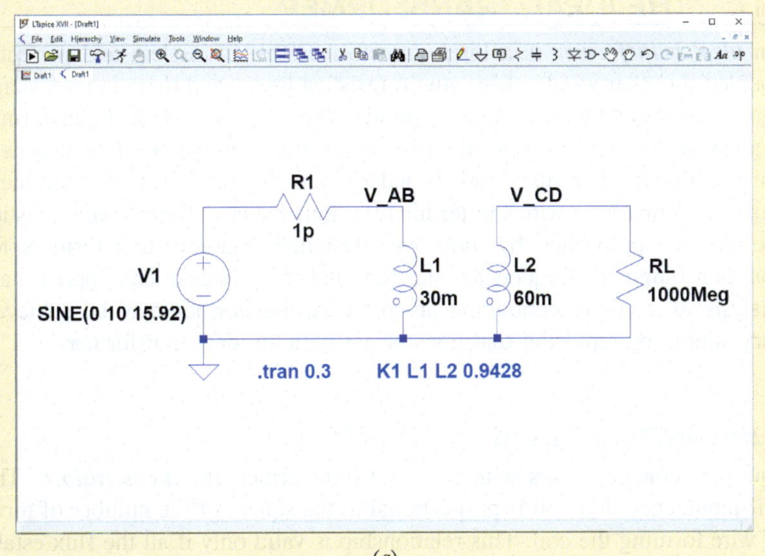

(a)

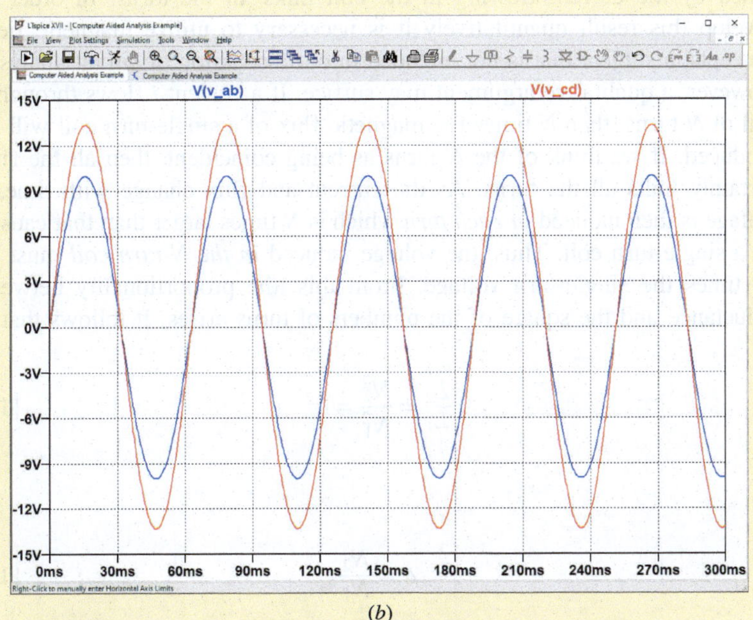

(b)

FIGURE 13.24 (*a*) The circuit of Fig. 13.19*a*, modified to meet simulation requirements. (*b*) Simulation output showing the voltage waveforms for V_{AB} and V_{CD}.
(*a*–*b*) Analog Devices

The circuit is connected to a 100 rad/s (15.92 Hz) sinusoidal voltage source. It is also necessary to add two resistors to the schematic in order for LTspice to perform the simulation without generating an error message. First, a small series resistance has been inserted between the voltage source and L1; a value of 1 pΩ was selected to minimize its effects. Second, a 1000 MΩ resistor (essentially infinite) was connected to L2. The output of the simulation is a voltage magnitude of 13.33 V that is in phase with the sinusoidal input; in agreement with the values calculated by hand in Example 13.6.

13.4 THE IDEAL TRANSFORMER

An ***ideal transformer*** is a linear transformer that is lossless and perfectly coupled. In other words, there are no resistive losses and there is no leakage flux associated with the magnetic coupling (i.e., $k = 1$). The ideal transformer is a useful approximation for the case where the coupling coefficient is near unity and both the primary and secondary inductive reactances are extremely large in comparison with the terminating impedances. These characteristics are closely approached by most well-designed iron-core transformers for common ranges of frequencies and terminal impedances. The approximate analysis of a circuit containing an iron-core transformer may be achieved very simply by replacing that transformer with an ideal transformer.

Turns Ratio of an Ideal Transformer

One new concept arises with the ideal transformer: the ***turns ratio*** a. The self-inductance of a coil is proportional to the square of the number of turns of wire forming the coil. This relationship is valid only if all the flux established by the current flowing in the coil links all the turns. In order to develop this result quantitatively it is necessary to utilize magnetic field concepts, a subject that is not included in our discussion of circuit analysis. However, a qualitative argument may suffice. If a current i flows through a coil of N turns, then N times the magnetic flux of a single-turn coil will be produced. If we think of the N turns as being coincident, then all the flux certainly links all the turns. As the current and flux change with time, a voltage is then induced *in each turn* which is N times larger than that caused by a single-turn coil. Thus, the voltage induced *in the N-turn coil* must be N^2 times the single-turn voltage. From this, the proportionality between inductance and the square of the numbers of turns arises. It follows that

$$\frac{L_2}{L_1} = \frac{N_2^2}{N_1^2} = a^2 \qquad [18]$$

or

$$a = \frac{N_2}{N_1} \qquad [19]$$

Figure 13.25 shows an ideal transformer to which a secondary load is connected. The ideal nature of the transformer is established by several conventions: the use of the vertical lines between the two coils to indicate the iron laminations present in many iron-core transformers, the unity value of the coupling coefficient, and the presence of the symbol 1:a, suggesting a turns ratio of N_1 to N_2.

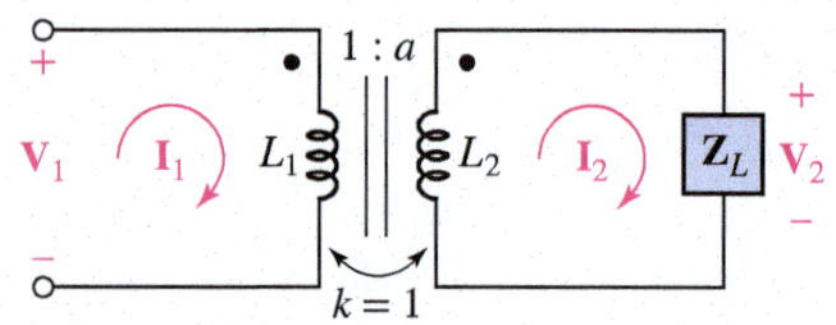

FIGURE 13.25 An ideal transformer is connected to a general load impedance.

Let us analyze this transformer in the sinusoidal steady state. The two mesh equations are

$$\mathbf{V}_1 = j\omega L_1\mathbf{I}_1 - j\omega M\mathbf{I}_2 \qquad [20]$$

$$0 = -j\omega M\mathbf{I}_1 + (\mathbf{Z}_L + j\omega L_2)\mathbf{I}_2 \qquad [21]$$

First, consider the input impedance of an ideal transformer. By solving Eq. [21] for $\mathbf{I}_2$ and substituting in Eq. [20], we obtain

$$\mathbf{V}_1 = \mathbf{I}_1 j\omega L_1 + \mathbf{I}_1 \frac{\omega^2 M^2}{\mathbf{Z}_L + j\omega L_2}$$

and

$$\mathbf{Z}_{\text{in}} = \frac{\mathbf{V}_1}{\mathbf{I}_1} = j\omega L_1 + \frac{\omega^2 M^2}{\mathbf{Z}_L + j\omega L_2}$$

Since $k = 1$, $M^2 = L_1 L_2$, so

$$\mathbf{Z}_{\text{in}} = j\omega L_1 + \frac{\omega^2 L_1 L_2}{\mathbf{Z}_L + j\omega L_2}$$

and substituting $L_2 = a^2 L_1$,

$$\mathbf{Z}_{\text{in}} = j\omega L_1 + \frac{\omega^2 a^2 L_1^2}{\mathbf{Z}_L + j\omega a^2 L_1}$$

Besides a unity coupling coefficient, another characteristic of an ideal transformer is an extremely large impedance for both the primary and secondary coils, regardless of the operating frequency. This suggests that the ideal case would be for both L_1 and L_2 to tend to infinity. Now if we let L_1 become infinite, both of the terms on the right-hand side of the preceding equation become infinite, and the result is indeterminate. Thus, it is necessary to first combine these two terms. The input impedance may be rewritten as:

$$\mathbf{Z}_{\text{in}} = \frac{j\omega L_1 \mathbf{Z}_L}{\mathbf{Z}_L + j\omega a^2 L_1} = \frac{\mathbf{Z}_L}{(\mathbf{Z}_L / j\omega L_1) + a^2}$$

Now as $L_1 \to \infty$, and for finite $\mathbf{Z}_L$, we see that $\mathbf{Z}_{\text{in}}$ becomes

$$\mathbf{Z}_{\text{in}} = \frac{\mathbf{Z}_L}{a^2}$$

The first important characteristic of the ideal transformer is therefore its ability to change the magnitude of an impedance, or to change impedance level. An ideal transformer having 100 primary turns and 10,000 secondary turns has a turns ratio of 10,000/100, or 100. Any impedance placed across the secondary then appears at the primary terminals reduced in magnitude by a factor of 100^2, or 10,000. A 20,000 Ω resistor looks like 2 Ω, a 200 mH inductor looks like 20 μH, and a 100 pF capacitor looks like 1 μF. If the primary and secondary windings are interchanged, then $a = 0.01$ and the load impedance is apparently increased in magnitude. In practice, this holds when $\mathbf{Z}_L$ is very small in comparison with $j\omega L_2$, where the ideal transformer model will become invalid if the load impedance becomes significant.

Use of Transformers for Impedance Matching

A practical example of the use of an iron-core transformer as a device for changing impedance level is in the coupling of an amplifier to a speaker system. In order to achieve maximum power transfer, we know that the resistance of the load should be equal to the internal resistance of the source; the speaker usually has an impedance magnitude (often assumed to be a resistance) of only a few ohms, while an amplifier may possess an internal resistance of several thousand ohms. Thus, an ideal transformer is required in which $N_2 < N_1$. For example, if the amplifier internal impedance is 4000 Ω and the speaker impedance is 8 Ω, then we desire that

$$\mathbf{Z}_{\text{in}} = 4000 = \frac{\mathbf{Z}_L}{a^2} = \frac{8}{a^2}$$

And solving for a,

$$a = \frac{N_2}{N_1} = \frac{1}{22.36}$$

Use of Transformers for Current Adjustment

There is a simple relationship between the primary and secondary currents $\mathbf{I}_1$ and $\mathbf{I}_2$ in an ideal transformer. From Eq. [23],

$$\frac{\mathbf{I}_2}{\mathbf{I}_1} = \frac{j\omega M}{\mathbf{Z}_L + j\omega L_2}$$

Once again we allow L_2 to become infinite, and it follows that

$$\frac{\mathbf{I}_2}{\mathbf{I}_1} = \frac{j\omega M}{j\omega\, L_2} = \sqrt{\frac{L_1}{L_2}}$$

or

$$\frac{\mathbf{I}_2}{\mathbf{I}_1} = \frac{1}{a} \qquad [22]$$

Thus, the ratio of the primary and secondary currents is the turns ratio. If we have $N_2 > N_1$, then $a > 1$, and it is apparent that the larger current flows in the winding with the smaller number of turns. In other words,

$$N_1\mathbf{I}_1 = N_2\mathbf{I}_2$$

It should also be noted that the current ratio is the *negative* of the turns ratio if either current is reversed or if either dot location is changed.

In our example in which an ideal transformer was used to change the impedance level to efficiently match a speaker to an amplifier, an rms current of 50 mA at 1000 Hz in the primary causes an rms current of 1.12 A at

1000 Hz in the secondary. The power delivered to the speaker is $(1.12)^2(8)$, or 10 W, and the power delivered to the transformer by the power amplifier is $(0.05)^2(4000)$, or 10 W. The result confirms energy conservation, since the ideal transformer contains neither an active device which can generate power nor any resistor which can absorb power.

Use of Transformers for Voltage Level Adjustment

Since the power delivered to the ideal transformer is identical with that delivered to the load, whereas the primary and secondary currents are related by the turns ratio, it should seem reasonable that the primary and secondary voltages must also be related to the turns ratio. If we define the secondary voltage, or load voltage, as

$$\mathbf{V}_2 = \mathbf{I}_2\mathbf{Z}_L$$

and the primary voltage as the voltage across L_1, then

$$\mathbf{V}_1 = \mathbf{I}_1\,\mathbf{Z}_{\text{in}} = \mathbf{I}_1\frac{\mathbf{Z}_L}{a^2}$$

The ratio of the two voltages then becomes

$$\frac{\mathbf{V}_2}{\mathbf{V}_1} = a^2\frac{\mathbf{I}_2}{\mathbf{I}_1}$$

or

$$\frac{\mathbf{V}_2}{\mathbf{V}_1} = a = \frac{N_2}{N_1} \qquad [23]$$

The ratio of the secondary to the primary voltage is equal to the turns ratio. Note that this equation is the opposite of Eq. [22], and this is a common source of error for students. This ratio may also be negative if either voltage is reversed or either dot location is changed.

Simply by choosing the turns ratio, therefore, we can now change any ac voltage to any other ac voltage. If $a > 1$, the secondary voltage will be greater than the primary voltage, and we have what is commonly referred to as a ***step-up transformer.*** If $a > 1$, the secondary voltage will be less than the primary voltage, and we have a ***step-down transformer.*** Utility companies typically generate power at a voltage in the range of 12 kV to 25 kV. Although this is a rather large voltage, transmission losses over long distances can be reduced by increasing the level to several hundred thousand volts using a step-up transformer (Fig. 13.26*a*). This voltage is then reduced to several tens of kilovolts at substations for local power distribution using step-down transformers (Fig. 13.26*b*). Additional step-down transformers are located outside buildings to reduce the voltage from the transmission voltage to the 110 V or 220 V level required to operate machinery (Fig. 13.26*c*).

Combining the voltage and current ratios, Eqs. [22] and [23],

$$\mathbf{V}_2\,\mathbf{I}_2 = \mathbf{V}_1\,\mathbf{I}_1$$

and we see that the primary and secondary complex voltamperes are equal. The magnitude of this product is usually specified as a maximum allowable value on power transformers.

(*a*)

(*b*)

(*c*)

FIGURE 13.26 (*a*) A step-up transformer used to increase the generator output voltage for transmission. (*b*) Substation transformer used to reduce the voltage from the 220 kV transmission level to several tens of kilovolts for local distribution. (*c*) Step-down transformer used to reduce the distribution voltage level to 240 V for power consumption.
Courtesy of Dr. Wade Enright, Te Kura Pukaha Vira O Te Whare Wananga O Waitaha, Aotearoa

EXAMPLE 13.8

For the circuit given in Fig. 13.27, determine the average power dissipated in the 10 kΩ resistor.

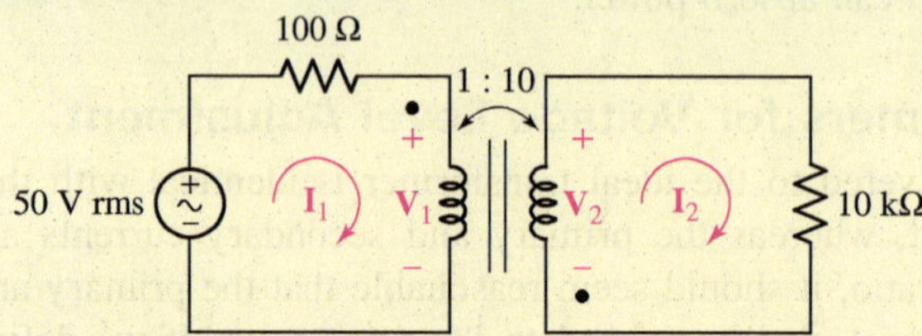

FIGURE 13.27 A simple ideal transformer circuit.

The average power dissipated by the 10 kΩ resistor is simply

$$P = 10{,}000\,|\mathbf{I}_2|^2$$

The 50 V rms source "sees" a transformer input impedance of $\mathbf{Z}_L/a^2$ or 100 Ω. Thus, we obtain

$$\mathbf{I}_1 = \frac{50}{100 + 100} = 250 \text{ mA rms}$$

From Eq. [27], $\mathbf{I}_2 = (1/a)\mathbf{I}_1 = 25$ mA rms, so we find that the 10 kΩ resistor dissipates 6.25 W.

PRACTICE

13.8 Repeat Example 13.8 using voltages to compute the dissipated power.

Ans: 6.25 W.

Voltage Relationship in the Time Domain

The characteristics of the ideal transformer that we have obtained have all been determined by phasor analysis. They are certainly true in the sinusoidal steady state, but we have no reason to believe that they are correct for the *complete* response. Actually, they are applicable in general, and the demonstration that this statement is true is much simpler than the phasor-based analysis we have just completed. Let us now determine how the time-domain quantities v_1 and v_2 are related in the ideal transformer. Returning to the circuit shown in Fig. 13.17 and the two equations, [16] and [17], describing it, we may solve the second equation for di_2/dt and substitute in the first equation:

$$v_1 = L_1\frac{di_1}{dt} + \frac{M}{L_2}v_2 - \frac{M^2}{L_2}\frac{di_1}{dt}$$

However, for unity coupling, $M^2 = L_1 L_2$, and so

$$v_1 = \frac{M}{L_2}v_2 = \sqrt{\frac{L_1}{L_2}}\,v_2 = \frac{1}{a}v_2$$

The relationship between primary and secondary voltage therefore does apply to the complete time-domain response.

PRACTICAL APPLICATION

Superconducting Transformers

For the most part, we have neglected the various types of losses that may be present in a particular transformer. When dealing with large power transformers, however, close attention must be paid to such nonidealities, despite overall efficiencies of typically 97 percent or more. Although such a high efficiency may seem nearly ideal, it can represent a great deal of wasted energy when the transformer is handling several thousand amperes. So-called i^2R (pronounced "eye-squared-R") losses represent power dissipated as heat, which can increase the temperature of the transformer coils. Wire resistance increases with temperature, so heating only leads to greater losses. High temperatures can also lead to degradation of the wire insulation, resulting in shorter transformer life. As a result, many modern power transformers employ a liquid oil bath to remove excess heat from the transformer coils. Such an approach has its drawbacks, however, including environmental impact and fire danger from leaking oil as a result of corrosion over time (Fig. 13.28).

■ **FIGURE 13.28** Fire that broke out in 2004 at the 340,000 V American Electric Power Substation near Mishawaka, Indiana.
Greg Swiercz/South Bend Tribune/AP Images

One possible means of improving the performance of such transformers is to make use of superconducting wire to replace the resistive coils of a standard transformer design. Superconductors are materials that are resistive at high temperatures but suddenly show no resistance to the flow of current below a critical temperature. Most elements are superconducting only near absolute zero, requiring expensive liquid helium–based cryogenic cooling. With the discovery in the 1980s of ceramic superconductors with critical temperatures of 90 K (−183°C) and higher, it became possible to replace helium–based equipment with significantly cheaper liquid nitrogen systems.

Figure 13.29 shows a prototype partial-core superconducting transformer being developed at the University of Canterbury. This design uses environmentally benign liquid nitrogen in place of an oil bath, and it is also significantly smaller than a comparably rated conventional transformer. The result is a measurable improvement in overall transformer efficiency, which translates into operational cost savings for the owner.

Still, all designs have disadvantages that must be weighed against their potential advantages, and superconducting transformers are no exception. The most significant obstacle at present is the relatively high cost of

■ **FIGURE 13.29** Prototype 15 kVA partial core superconducting power transformer.
Patrick Selwyn Bodger

(*Continued on next page*)

fabricating superconducting wire several kilometers in length compared to copper wire. Part of this is due to the challenge of fabricating long wires from ceramic materials, but part of it is also due to the silver tubing used to surround the superconductor to provide a low-resistance current path in the event of a cooling system failure (although less expensive than silver, copper reacts with the ceramic and is therefore not a viable alternative). The net result is that although a superconducting transformer is likely to save a utility money over a long period of time—many transformers see over 30 years of service—the initial cost is much higher than for a traditional resistive transformer. At present, many companies (including utilities) are driven by short-term cost considerations and are not always eager to make large capital investments with only long-term cost benefits.

An expression relating primary and secondary current in the time domain is most quickly obtained by dividing Eq. [16] throughout by L_1,

$$\frac{v_1}{L_1} = \frac{di_1}{dt} + \frac{M}{L_1}\frac{di_2}{dt} = \frac{di_1}{dt} + a\frac{di_2}{dt}$$

and then invoking one of the hypotheses underlying the ideal transformer: L_1 must be infinite. If we assume that v_1 is not infinite, then

$$\frac{di_1}{dt} = -a\frac{di_2}{dt}$$

Integrating,

$$i_1 = -ai_2 + A$$

where A is a constant of integration that does not vary with time. Thus, if we neglect any direct currents in the two windings and fix our attention only on the time-varying portion of the response, then

$$i_1 = -ai_2$$

The minus sign arises from the placement of the dots and the selection of the current directions in Fig. 13.17.

The same current and voltage relationships are therefore obtained in the time domain as were obtained previously in the frequency domain, provided that dc components are ignored. The time-domain results are more general, but they have been obtained by a less informative process.

Equivalent Circuits

The characteristics of the ideal transformer which we have established may be used to simplify circuits in which ideal transformers appear. Let us assume, for purposes of illustration, that everything to the left of the primary terminals has been replaced by its Thévenin equivalent, as has the network to the right of the secondary terminals. We thus consider the circuit shown in Fig. 13.30. Excitation at any frequency ω is assumed.

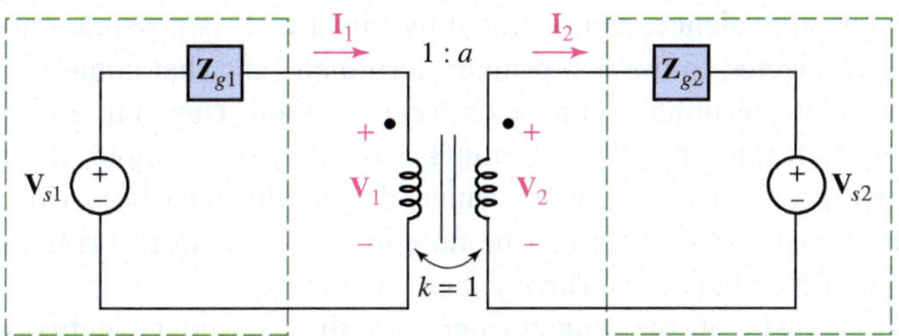

■ **FIGURE 13.30** The networks connected to the primary and secondary terminals of an ideal transformer are represented by their Thévenin equivalents.

Thévenin's or Norton's theorem may be used to achieve an equivalent circuit that does not contain a transformer. For example, let us determine the Thévenin equivalent of the network to the left of the secondary terminals. Open-circuiting the secondary, $\mathbf{I}_2 = 0$ and therefore $\mathbf{I}_1 = 0$ (remember that L_1 is infinite). No voltage appears across $\mathbf{Z}_{g1}$, and thus $\mathbf{V}_1 = \mathbf{V}_{s1}$ and $\mathbf{V}_{2oc} = a\mathbf{V}_{s1}$. The Thévenin impedance is obtained by setting V_{s1} to zero and utilizing the square of the turns ratio, being careful to use the reciprocal turns ratio, since we are looking in at the secondary terminals. Thus, $\mathbf{Z}_{TH2} = \mathbf{Z}_{g1}a^2$.

As a check on our equivalent, let us also determine the short-circuit secondary current $\mathbf{I}_{2sc}$. With the secondary short-circuited, the primary generator faces an impedance of $\mathbf{Z}_{g1}$, and, thus, $\mathbf{I}_1 = \mathbf{V}_{s1}/\mathbf{Z}_{g1}$. Therefore, $\mathbf{I}_{2sc} = \mathbf{V}_{s1}/a\mathbf{Z}_{g1}$. The ratio of the open-circuit voltage to the short-circuit current is $a^2\mathbf{Z}_{g1}$, as it should be. The Thévenin equivalent of the transformer and primary circuit is shown in the circuit of Fig. 13.31.

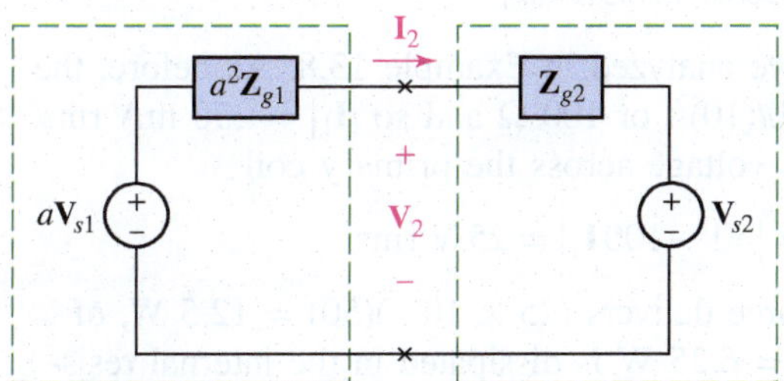

■ **FIGURE 13.31** The Thévenin equivalent of the network to the left of the secondary terminals in Fig. 13.30 is used to simplify that circuit.

Each primary voltage may therefore be multiplied by the turns ratio, each primary current divided by the turns ratio, and each primary impedance multiplied by the square of the turns ratio; and then these modified voltages, currents, and impedances replace the given voltages, currents, and impedances plus the transformer. If either dot is interchanged, the equivalent may be obtained by using the negative of the turns ratio.

Note that this equivalence, as illustrated by Fig. 13.31, is possible only if the network connected to the two primary terminals, and that connected to the two secondary terminals, can be replaced by their Thévenin equivalents. That is, each must be a two-terminal network. For example, if we cut the two primary leads at the transformer, the circuit must be divided into two separate networks; there can be no element or network bridging across the transformer between primary and secondary.

A similar analysis of the transformer and the secondary network shows that everything to the right of the primary terminals may be replaced by an identical network without the transformer, each voltage being divided by a, each current being multiplied by a, and each impedance being divided by a^2. A reversal of either winding requires the use of a turns ratio of $-a$.

EXAMPLE 13.9

For the circuit given in Fig. 13.32, determine the equivalent circuit in which the transformer and the secondary circuit are replaced, and also that in which the transformer and the primary circuit are replaced.

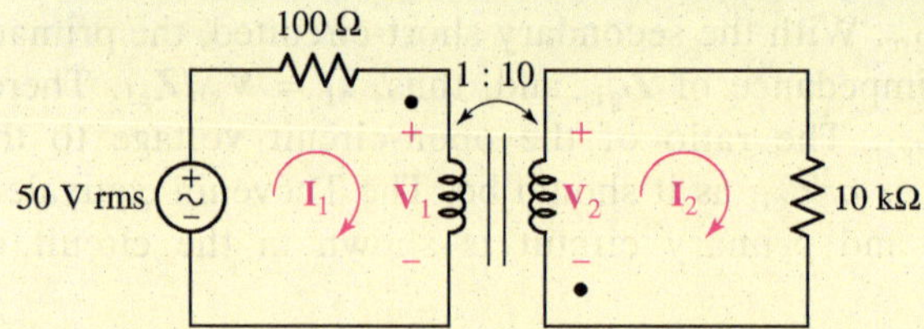

FIGURE 13.32 A simple circuit in which a resistive load is matched to the source impedance by means of an ideal transformer.

This is the same circuit we analyzed in Example 13.8. As before, the input impedance is 10,000/(10)^2, or 100 Ω and so $|\mathbf{I}_1| = 250$ mA rms. We can also compute the voltage across the primary coil

$$|\mathbf{V}_1| = |50 - 100\mathbf{I}_1| = 25 \text{ V rms}$$

and thus find that the source delivers $(25 \times 10^{-3})(50) = 12.5$ W, of which $(25 \times 10^{-3})^2(100) = 6.25$ W is dissipated in the internal resistance of the source and $12.5 - 6.25 = 6.25$ W is delivered to the load. This is the condition for maximum power transfer to the load.

If the secondary circuit and the ideal transformer are removed by the use of the Thévenin equivalent, the 50 V source and 100 Ω resistor simply see a 100 Ω impedance, and the simplified circuit of Fig. 13.33a is obtained. The primary current and voltage are now immediately evident.

If, instead, the network to the left of the secondary terminals is replaced by its Thévenin equivalent, we find (keeping in mind the location of the dots) $\mathbf{V}_{TH} = -10(50) = -500$ V rms, and $\mathbf{Z}_{TH} = (-10)^2(100) = 10$ kΩ; the resulting circuit is shown in Fig. 13.33b.

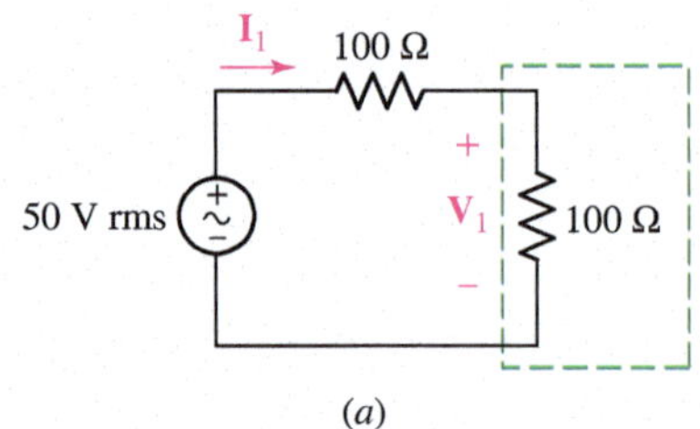

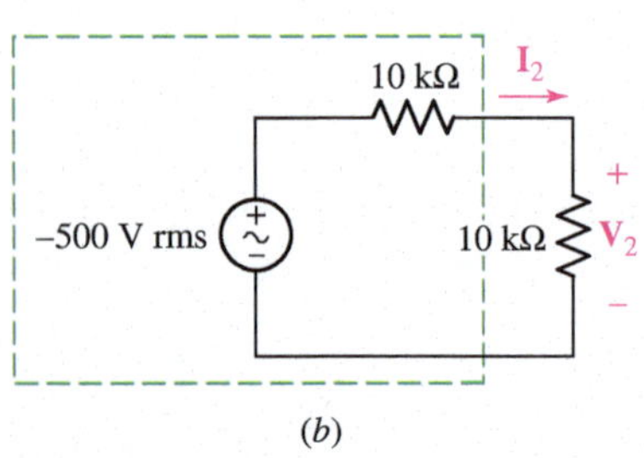

FIGURE 13.33 The circuit of Fig. 13.32 is simplified by replacing (a) the transformer and secondary circuit with the Thévenin equivalent or (b) the transformer and primary circuit with the Thévenin equivalent.

PRACTICE

13.9 Let $N_1 = 1000$ turns and $N_2 = 5000$ turns in the ideal transformer shown in Fig. 13.34. If $\mathbf{Z}_L = 500 - j400\ \Omega$, find the average power delivered to $\mathbf{Z}_L$ for (*a*) $\mathbf{I}_2 = 1.4\underline{/20^\circ}$ A rms; (*b*) $\mathbf{V}_2 = 900\underline{/40^\circ}$ V rms; (*c*) $\mathbf{V}_1 = 80\underline{/100^\circ}$ V rms; (*d*) $\mathbf{I}_1 = 6\underline{/45^\circ}$ A rms; (*e*) $\mathbf{V}_s = 200\underline{/0^\circ}$ V rms.

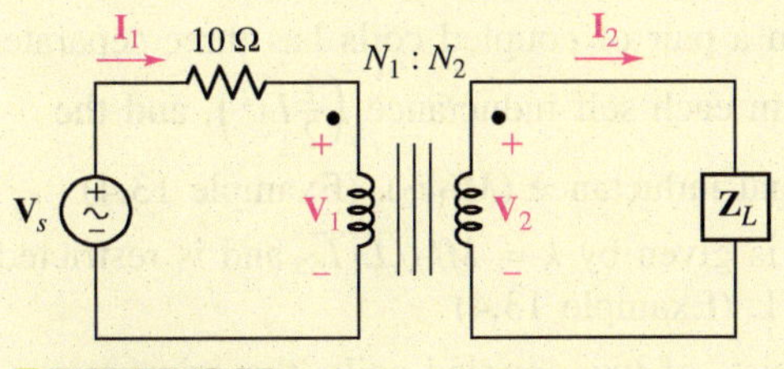

FIGURE 13.34

Ans: 980 W; 988 W; 195.1 W; 720 W; 692 W.

SUMMARY AND REVIEW

Transformers play a critical role in the power industry, allowing voltages to be stepped up for transmission and stepped down to the level required for individual pieces of equipment. In this chapter, we studied transformers in the broader context of magnetically coupled circuits, where the magnetic flux associated with current can link two or more elements in a circuit (or even neighboring circuits). This is most easily understood by extending the concept of inductance studied in Chap. 7 to introduce the idea of mutual inductance (also having units of henrys). We saw that the coefficient M of mutual inductance is limited to less than the geometric mean of the two inductances being coupled (i.e., $M \leq \sqrt{L_1 L_2}$), and we used the dot convention to determine the polarity of the voltage induced across one inductance as a result of current flowing through the other. When the two inductances are not particularly close, M might be rather small. However, in the case of a well-designed transformer, it might approach its maximum value. To describe such situations, we introduced the concept of the coupling coefficient k. When dealing with a linear transformer, analysis may be assisted by representing the element with an equivalent T (or, less commonly, Π) network, but a great deal of circuit analysis is performed assuming an ideal transformer. In such cases we no longer concern ourselves with M or k, but rather the turns ratio a. We saw that the voltages across the primary and secondary coils, as well as their individual currents, are related by this parameter. This approximation is very useful for both analysis and design. We concluded the chapter with a brief discussion of how Thévenin's theorem can be applied to circuits with ideal transformers.

We could continue, as the study of inductively coupled circuits is an interesting and important topic, but at this point it might be appropriate to list some of the key concepts we have already discussed, along with corresponding example numbers.

- ❑ Mutual inductance describes the voltage induced at the ends of a coil due to the magnetic field generated by a second coil. (Example 13.1)

- The dot convention allows a sign to be assigned to the mutual inductance term. (Example 13.1)
- According to the dot convention, a current entering the *dotted* terminal of one coil produces an open-circuit voltage with a positive voltage reference at the *dotted* terminal of the second coil. (Examples 13.1, 13.2, 13.3)
- The total energy stored in a pair of coupled coils has three separate terms: the energy stored in each self-inductance $\left(\frac{1}{2}Li^2\right)$, and the energy stored in the mutual inductance $(Mi_1 i_2)$. (Example 13.4)
- The coupling coefficient is given by $k = M/\sqrt{L_1 L_2}$ and is restricted to values between 0 and 1. (Example 13.4)
- A linear transformer consists of two coupled coils: the primary winding and the secondary winding. (Examples 13.5, 13.6, 13.7)
- An ideal transformer is a useful approximation for practical iron-core transformers. The coupling coefficient is taken to be unity, and the inductance values are assumed to be infinite. (Examples 13.8, 13.9)
- The turns ratio $a = N_2/N_1$ of an ideal transformer relates the primary and secondary coil voltages: $\mathbf{V}_2 = a\mathbf{V}_1$. (Example 13.9)
- The turns ratio a also relates the currents in the primary and secondary coils: $\mathbf{I}_1 = a\mathbf{I}_2$. (Examples 13.8, 13.9)

READING FURTHER

Almost everything you ever wanted to know about transformers can be found in:

M. Heathcote, *J&P Transformer Book,* 13th ed. Oxford: Reed Educational and Professional Publishing Ltd., 2007.

Another comprehensive transformer title is:

W. T. McLyman, *Transformer and Inductor Design Handbook,* 4th ed. New York: Marcel Dekker, 2011.

A good transformer book with a strong economic focus is:

B. K. Kennedy, *Energy Efficient Transformers.* New York: McGraw Hill, 1998.

EXERCISES

13.1 Mutual Inductance

1. Consider the two inductances depicted in Fig. 13.35. Set $L_1 = 10$ mH, $L_2 = 5$ mH, and $M = 1$ mH. Determine the steady-state expression for (*a*) v_1 if $i_1 = 0$ and $i_2 = 5 \cos 8t$ A; (*b*) v_2 if $i_1 = 3 \sin 100t$ A and $i_2 = 0$; (*c*) v_2 if $i_1 = 5 \cos(8t - 40°)$ A and $i_2 = 4 \sin 8t$ A.

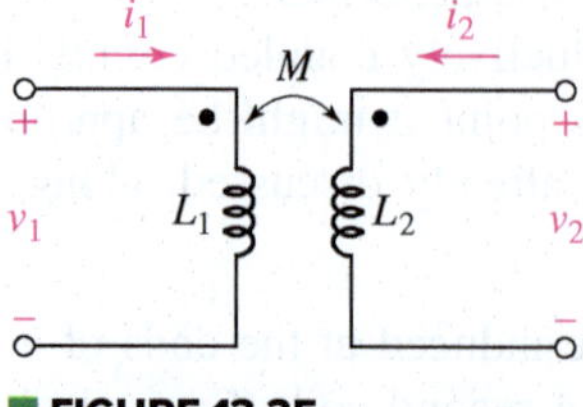

FIGURE 13.35

2. With respect to Fig. 13.36, assume $L_1 = 500$ mH, $L_2 = 250$ mH, and $M = 20$ mH. Determine the steady-state expression for (*a*) v_1 if $i_1 = 0$ and $i_2 = 3 \cos 80t$ A; (*b*) v_2 if $i_1 = 4 \cos(30t - 15°)$ A and $i_2 = 0$. (*c*) Repeat parts (*a*) and (*b*) if M is increased to 200 mH.
3. The circuit in Fig. 13.36 has a sinusoidal input at $\omega = 2{,}000$ rad/s with $\mathbf{I}_1 = 2\underline{/30°}$ A and 100 Ω resistor attached across the terminals labeled v_2. For the case where $L_1 = 400$ mH, $L_2 = 100$ mH, and M = 50 mH, determine $\mathbf{V}_1$, $\mathbf{I}_2$, and $\mathbf{V}_2$ in phasor form.
4. In Fig. 13.37, set $L_1 = 1$ μH, $L_2 = 2$ μH, and $M = 150$ nH. Obtain a steady-state expression for (*a*) v_1 if $i_2 = -\cos 70t$ mA and $i_1 = 0$; (*b*) v_2 if $i_1 = 55 \cos(5t - 30°)$ A; (*c*) v_2 if $i_1 = 6 \sin 5t$ A and $i_2 = 3 \sin 5t$ A.

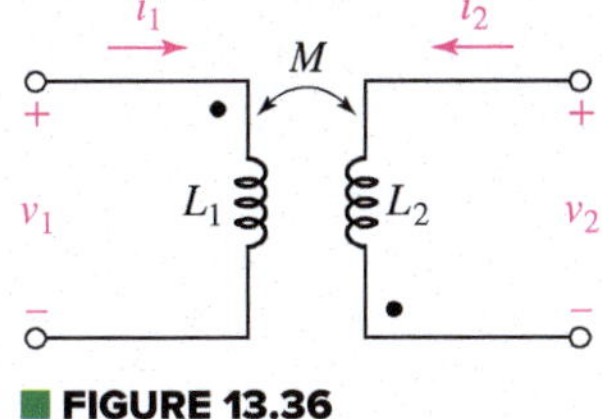

■ **FIGURE 13.36**

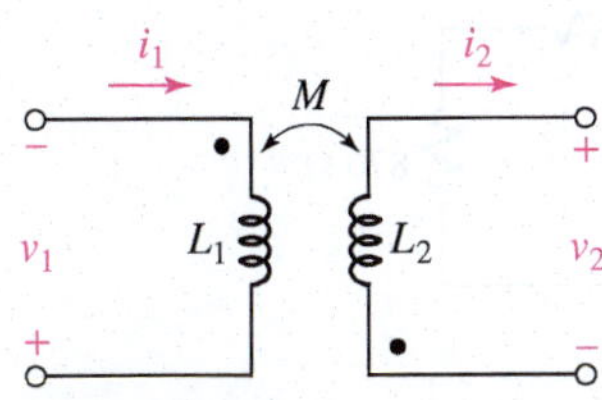

■ **FIGURE 13.37**

5. For the configuration of Fig. 13.38, $L_1 = 0.5L_2 = 1$ mH and $M = 0.85\sqrt{L_1 L_2}$. Calculate $v_2(t)$ if (*a*) $i_2 = 0$ and $i_1 = 5e^{-t}$ mA; (*b*) $i_2 = 0$ and $i_1 = 5 \cos 10t$ mA; (*c*) $i_2 = 5 \cos 70t$ mA and $i_1 = 0.5i_2$.
6. The circuit in Fig. 13.38 has a sinusoidal input at $\omega = 1{,}000$ rad/s with $\mathbf{I}_1 = 3\underline{/45°}$ A and a 100 Ω resistor attached across the terminals labeled v_2. For the case where $L_1 = 50$ mH, $L_2 = 250$ mH, and $M = 0.75\sqrt{L_1 L_2}$, determine $\mathbf{V}_1$, $\mathbf{I}_2$, and $\mathbf{V}_2$ in phasor form.
7. The physical construction of three pairs of coupled coils is shown in Fig. 13.39. Show the two different possible locations for the two dots on each pair of coils.

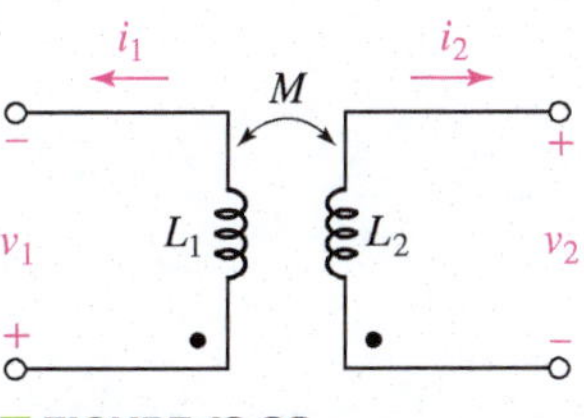

■ **FIGURE 13.38**

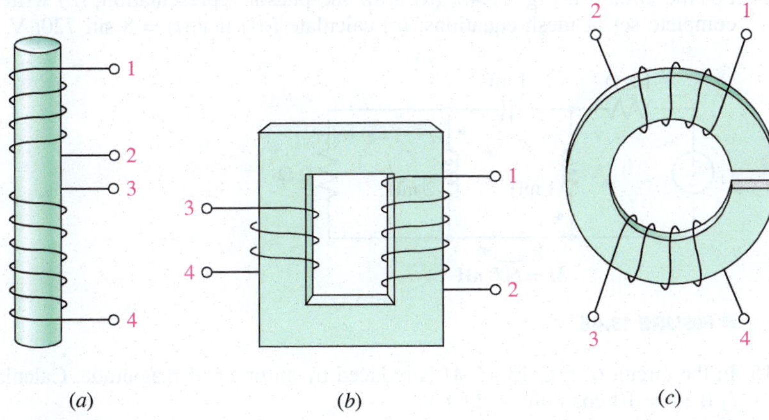

■ **FIGURE 13.39**

8. In the circuit of Fig. 13.40, $i_1 = 5 \cos(100t - 80°)$ mA, $L_1 = 1$ H, and $L_2 = 2$ H. If $v_2 = 250 \sin(100t - 80°)$ mV, calculate M.

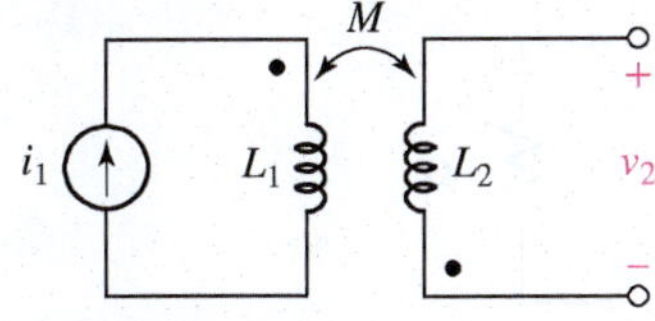

■ **FIGURE 13.40**

9. In the circuit represented in Fig. 13.40, determine i_1 if $v_2(t) = 4 \cos 5t$ V, $L_1 = 1$ mH, $L_2 = 4$ mH, and $M = 1.5$ mH.
10. Calculate v_1 and v_2 if $i_1 = 5 \sin 40t$ mA and $i_2 = 5 \cos 40t$ mA, $L_1 = 1$ mH, $L_2 = 3$ mH, and $M = 0.5$ mH, for the coupled inductances shown in (*a*) Fig. 13.37; (*b*) Fig. 13.38.
11. Calculate v_1 and v_2 if $i_1 = 3 \cos(2000t + 13°)$ mA and $i_2 = 5 \sin 400t$ mA, $L_1 = 1$ mH, $L_2 = 3$ mH, and $M = 200$ nH, for the coupled inductances shown in (*a*) Fig. 13.35; (*b*) Fig. 13.36.
12. For the circuit of Fig. 13.41, calculate $\mathbf{I}_1$, $\mathbf{I}_2$, $\mathbf{V}_2/\mathbf{V}_1$, and $\mathbf{I}_2/\mathbf{I}_1$.

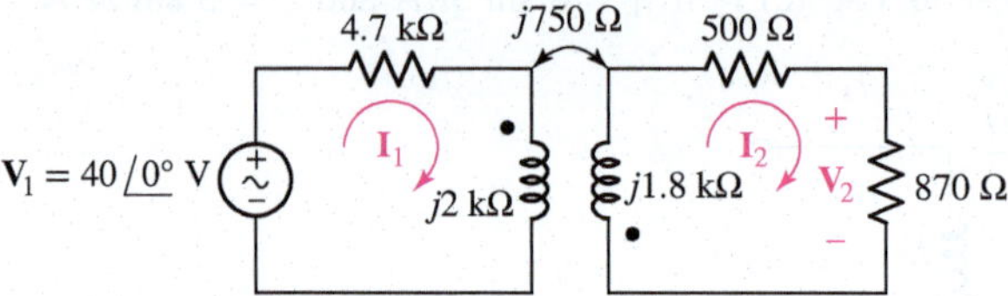

■ FIGURE 13.41

13. For the circuit of Fig. 13.42, plot the magnitude of $\mathbf{V}_2/\mathbf{V}_1$ as a function of frequency ω, over the range $0 \leq \omega \leq 2$ rad/s.

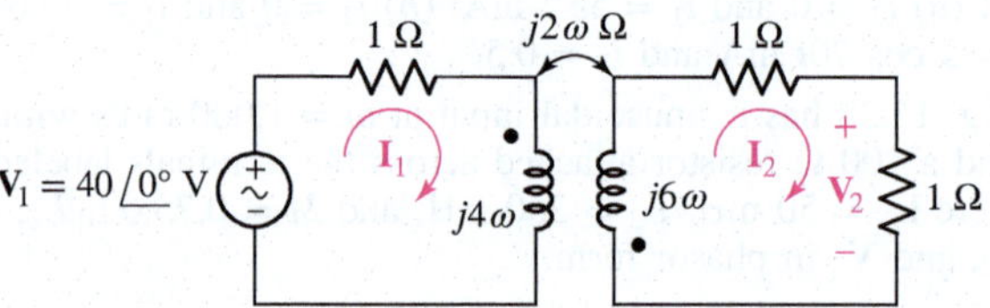

■ FIGURE 13.42

14. For the circuit of Fig. 13.43, (*a*) draw the phasor representation; (*b*) write a complete set of mesh equations; (*c*) calculate $i_2(t)$ if $v_1(t) = 8 \sin 720t$ V.

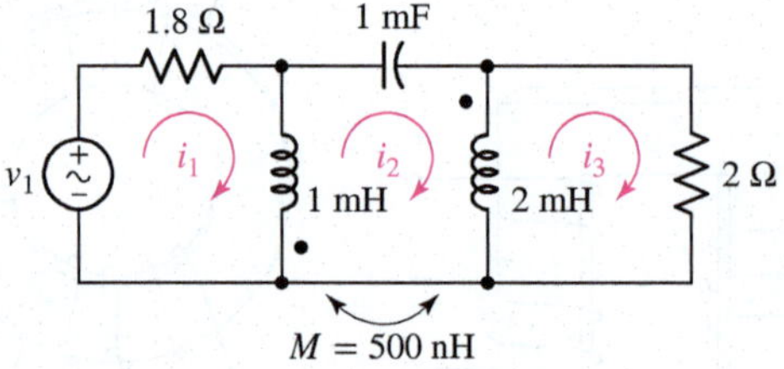

■ FIGURE 13.43

15. In the circuit of Fig. 13.43, M is reduced by an order of magnitude. Calculate i_3 if $v_1 = 10 \cos(800t - 20°)$ V.
16. In the circuit shown in Fig. 13.44, find the average power absorbed by (*a*) the source; (*b*) each of the two resistors; (*c*) each of the two inductances; (*d*) the mutual inductance.

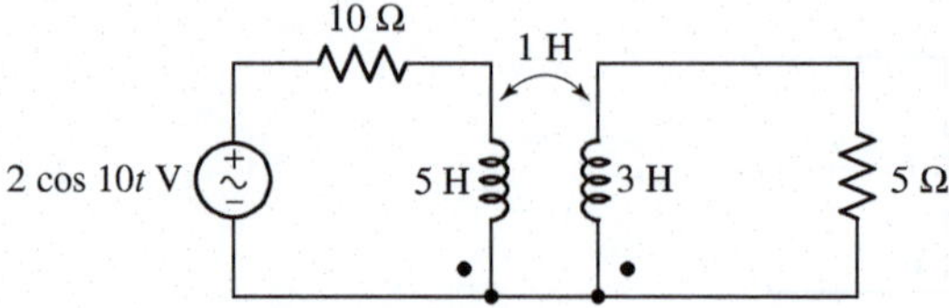

■ FIGURE 13.44

17. The circuit of Fig. 13.45 is designed to drive a simple 8 Ω speaker. What value of M results in 1 W of average power being delivered to the speaker?

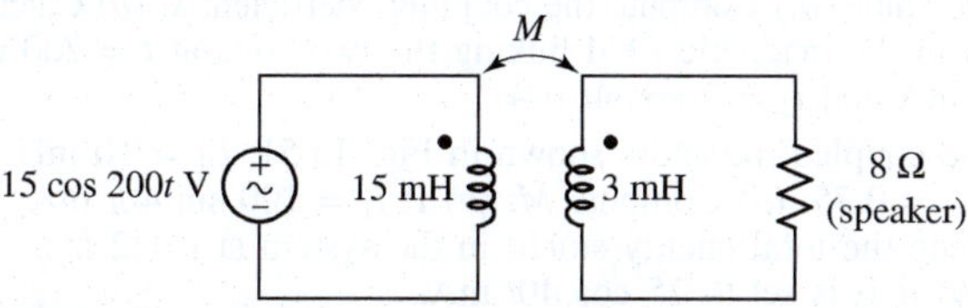

■ **FIGURE 13.45**

18. Determine an expression for $i_c(t)$ valid for $t > 0$ in the circuit of Fig. 13.46, if $v_s(t) = 10t^2u(t)/(t^2 + 0.01)$ V.

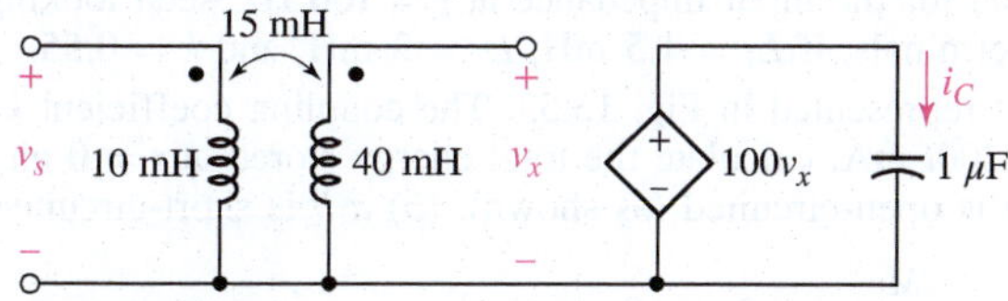

■ **FIGURE 13.46**

19. For the coupled inductor network of Fig. 13.47*a*, set $L_1 = 20$ mH, $L_2 = 30$ mH, $M = 10$ mH, and obtain equations for v_A and v_B if (*a*) $i_1 = 0$ and $i_2 = 5 \sin 10t$; (*b*) $i_1 = 5 \cos 20t$ and $i_2 = 2 \cos(20t - 100°)$ mA. (*c*) Express $\mathbf{V}_1$ and $\mathbf{V}_2$ as functions of $\mathbf{I}_A$ and $\mathbf{I}_B$ for the network shown in Fig. 13.47*b*.

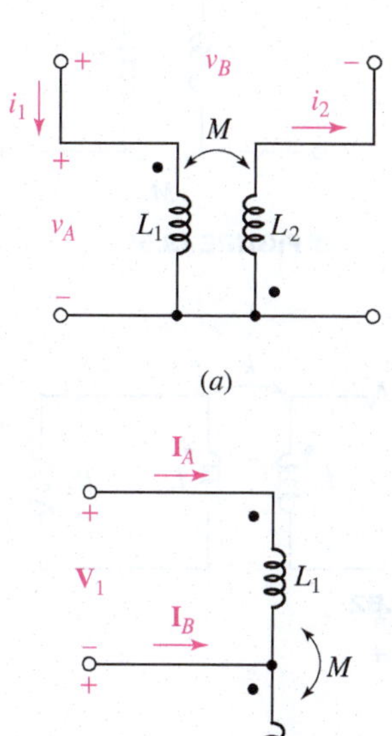

■ **FIGURE 13.47**

20. Note that there is no mutual coupling between the 5 H and 6 H inductors in the circuit of Fig. 13.48. (*a*) Write a set of equations in terms of $\mathbf{I}_1(j\omega)$, $\mathbf{I}_2(j\omega)$, and $\mathbf{I}_3(j\omega)$. (*b*) Find $\mathbf{I}_3(j\omega)$, if $\omega = 2$ rad/s.

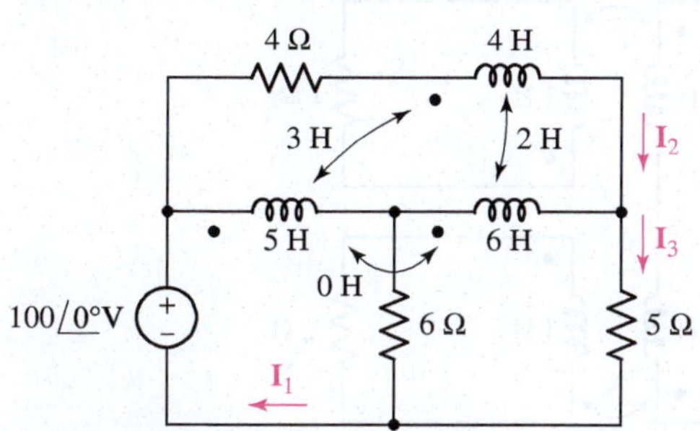

■ **FIGURE 13.48**

21. Find $\mathbf{V}_1(j\omega)$ and $\mathbf{V}_2(j\omega)$ in terms of $\mathbf{I}_1(j\omega)$ and $\mathbf{I}_2(j\omega)$ for each circuit of Fig. 13.49.

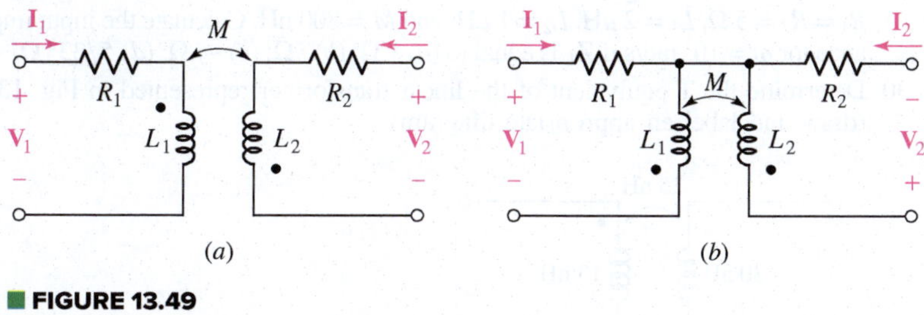

■ **FIGURE 13.49**

22. (*a*) Find $\mathbf{Z}_{in}(j\omega)$ for the network of Fig 13.50. (*b*) Plot $\mathbf{Z}_{in}$ over the frequency range of $0 \leq \omega \leq 1000$ rad/s. (*c*) Find $\mathbf{Z}_{in}(j\omega)$ for $\omega = 50$ rad/s.

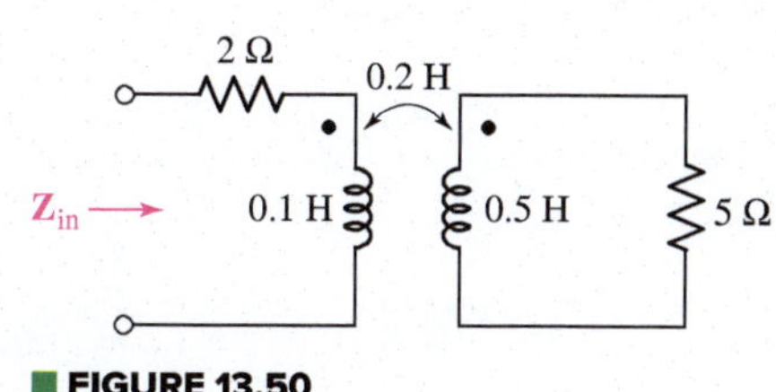

■ **FIGURE 13.50**

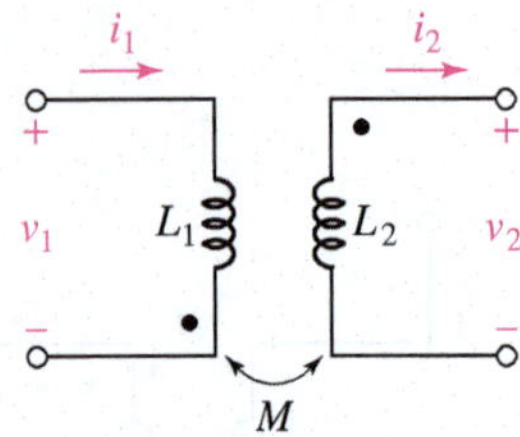

■ FIGURE 13.51

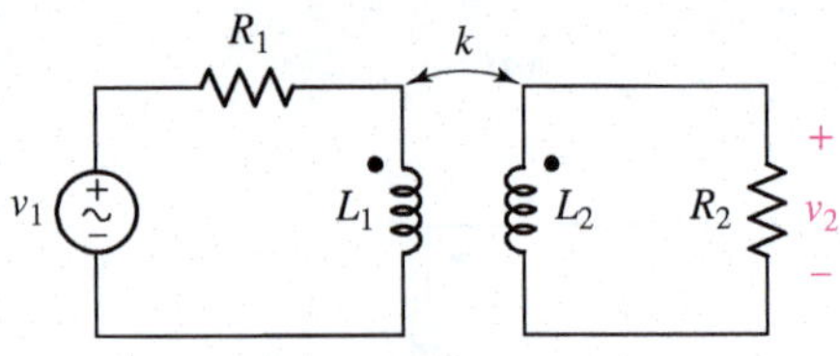

■ FIGURE 13.52

13.2 Energy Considerations

23. For the coupled coils of Fig. 13.51, $L_1 = L_2 = 10$ H, and M is equal to its maximum possible value. (*a*) Compute the coupling coefficient k. (*b*) Calculate the energy stored in the magnetic field linking the two coils at $t = 200$ ms if $i_1 = 10 \cos 4t$ mA and $i_2 = 2 \cos 4t$ mA.
24. With regard to the coupled inductors shown in Fig. 13.51, $L_1 = 10$ mH, $L_2 = 5$ mH, and $k = 0.75$. (*a*) Compute M. (*b*) If $i_1 = 100 \sin 40t$ mA, and $i_2 = 0$, compute the total energy stored in the system at $t = 2$ ms. (*c*) Repeat part (*b*) if i_2 is set to $75 \cos 40t$ mA.
25. For the circuit of Fig. 13.52, $L_1 = 4$ mH, $L_2 = 12$ mH, $R_1 = 1\ \Omega$, $R_2 = 10\ \Omega$, and $v_1 = 2 \cos 8t$ V. (*a*) Obtain an equation for the phasor $\mathbf{V}_2$ as a function of k and circuit parameters. (*b*) Plot the magnitude and phase angle of $\mathbf{V}_2$ as a function of k.
26. Connect a load $\mathbf{Z}_L = 5\underline{/33^\circ}\ \Omega$ to the right-hand terminals of Fig. 13.51. Derive an expression for the input impedance at $f = 100$ Hz, seen looking into the left-hand terminals, if $L_1 = 1.5$ mH, $L_2 = 3$ mH, and $k = 0.55$.
27. Consider the circuit represented in Fig. 13.53. The coupling coefficient $k = 0.75$. If $i_s = 5 \cos 200t$ mA, calculate the total energy stored at $t = 0$ and $t = 5$ ms if (*a*) *a-b* is open-circuited (as shown); (*b*) *a-b* is short-circuited.

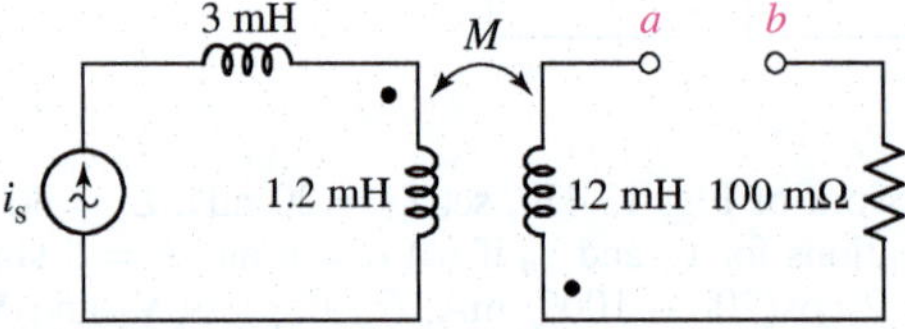

■ FIGURE 13.53

28. Compute v_1, v_2, and the average power delivered to each resistor in the circuit of Fig. 13.54.

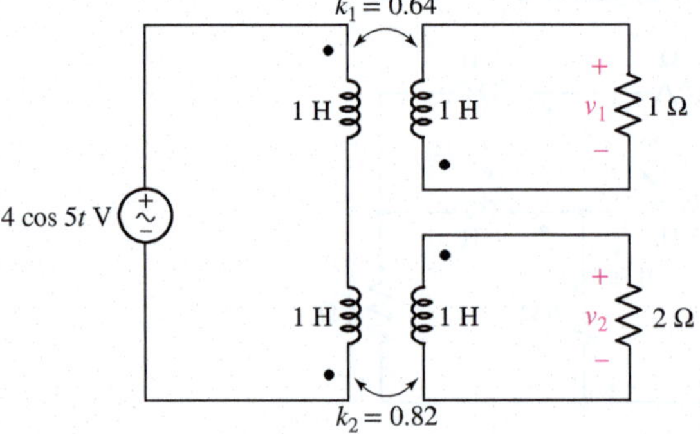

■ FIGURE 13.54

13.3 The Linear Transformer

29. Assume the following values for the circuit depicted schematically in Fig. 13.16: $R_1 = R_2 = 5\ \Omega$, $L_1 = 2\ \mu$H, $L_2 = 1\ \mu$H, and $M = 800$ nH. Calculate the input impedance for $\omega = 10^7$ rad/s if $\mathbf{Z}_L$ is equal to (*a*) 1 Ω; (*b*) j Ω; (*c*) $-j$ Ω; (*d*) $5\underline{/33^\circ}$ Ω.
30. Determine the T equivalent of the linear transformer represented in Fig. 13.55 (draw and label an appropriate diagram).

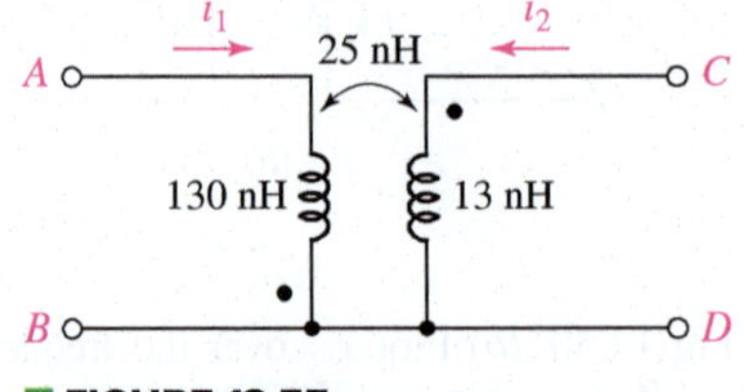

■ FIGURE 13.55

31. (*a*) Draw and label an appropriate diagram of a T equivalent network for the linear transformer shown in Fig. 13.56. (*b*) Calculate the open circuit voltage phasor $\mathbf{V_{CD}}$ for the case where $\mathbf{V_{AB}} = 5\angle 45^\circ$ V(frequency of 60 Hz).

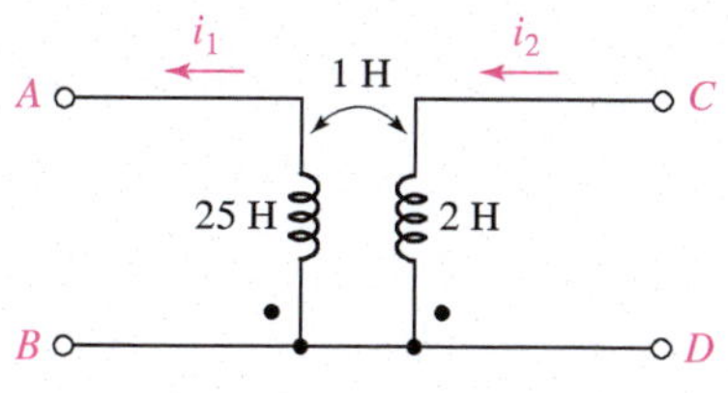

■ **FIGURE 13.56**

32. Represent the T network shown in Fig. 13.57 as an equivalent linear transformer if (*a*) $L_x = 1$ H, $L_y = 2$ H, and $L_z = 4$ H; (*b*) $L_x = 10$ mH, $L_y = 50$ mH, and $L_z = 22$ mH.

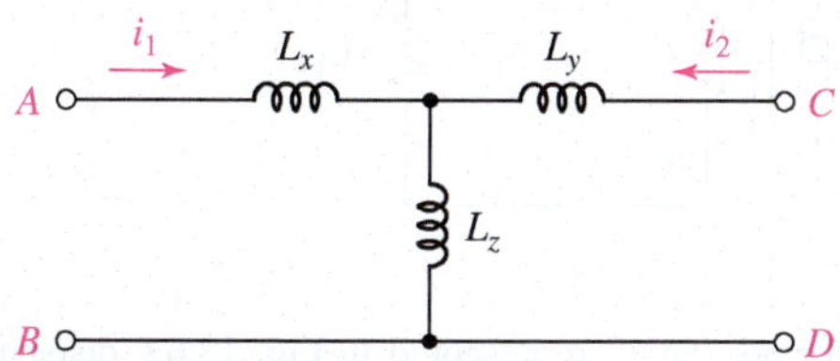

■ **FIGURE 13.57**

33. Assuming zero initial currents, obtain an equivalent Π network of the transformer depicted in Fig. 13.55.

34. (*a*) Draw and label a suitable equivalent Π network of the linear transformer shown in Fig. 13.56, assuming zero initial currents. (*b*) Verify their equivalence with an appropriate simulation.

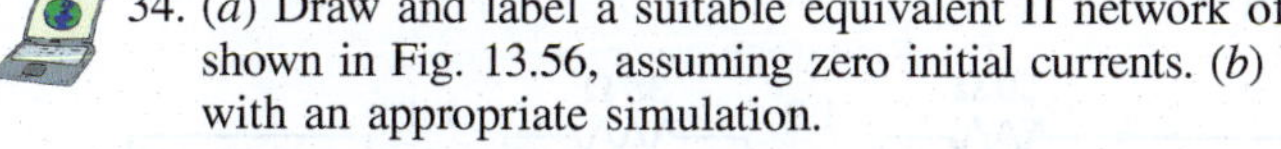

35. Represent the Π network of Fig. 13.58 as an equivalent linear transformer with zero initial currents if (*a*) $L_A = 1$ H, $L_B = 2$ H, and $L_C = 4$ H; (*b*) $L_A = 10$ mH, $L_B = 50$ mH, and $L_C = 22$ mH.

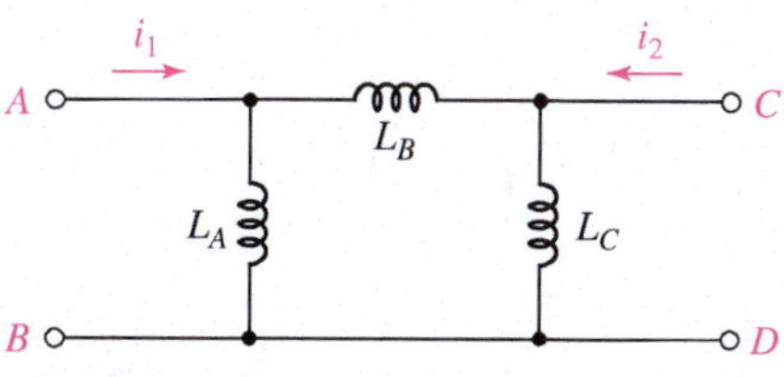

■ **FIGURE 13.58**

36. For the circuit of Fig. 13.59, determine an expression for (*a*) $\mathbf{I}_L/\mathbf{V}_s$; (*b*) $\mathbf{V}_1/\mathbf{V}_s$.

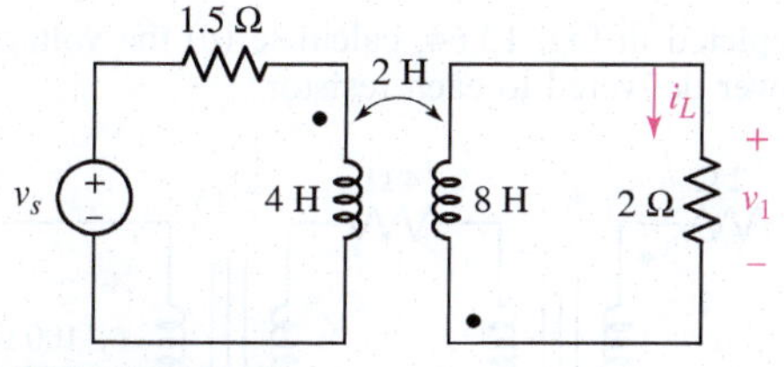

■ **FIGURE 13.59**

37. (*a*) For the circuit of Fig. 13.60, if $v_s = 8 \cos 1000t$ V, calculate v_o. (*b*) Verify your solution with an appropriate LTspice simulation.

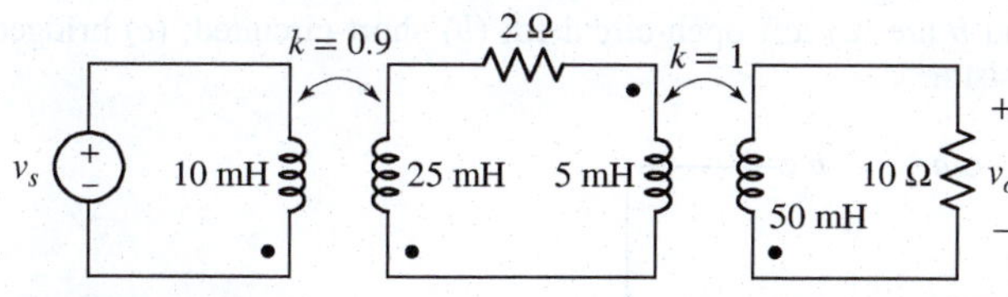

■ **FIGURE 13.60**

38. For the circuit of Fig. 13.60, redraw using equivalent T networks. Calculate v_o for the case where $v_s = 12 \sin 500t$ V.

13.4 The Ideal Transformer

39. Calculate $\mathbf{I}_2$ and $\mathbf{V}_2$ for the ideal transformer circuit of Fig. 13.61 if (*a*) $\mathbf{V}_1 = 4\angle 32^\circ$ V and $\mathbf{Z}_L = 1 - j\ \Omega$; (*b*) $\mathbf{V}_1 = 4\angle 32^\circ$ V and $\mathbf{Z}_L = 0$; (*c*) $\mathbf{V}_1 = 2\angle 118^\circ$ V and $\mathbf{Z}_L = 1.5\angle 10^\circ\ \Omega$.

40. With respect to the ideal transformer circuit depicted in Fig. 13.61, calculate $\mathbf{I}_2$ and $\mathbf{V}_2$ if (*a*) $\mathbf{I}_1 = 244\angle 0^\circ$ mA and $\mathbf{Z}_L = 5 - j2\ \Omega$; (*b*) $\mathbf{I}_1 = 100\angle 10^\circ$ mA and $\mathbf{Z}_L = j2\ \Omega$.

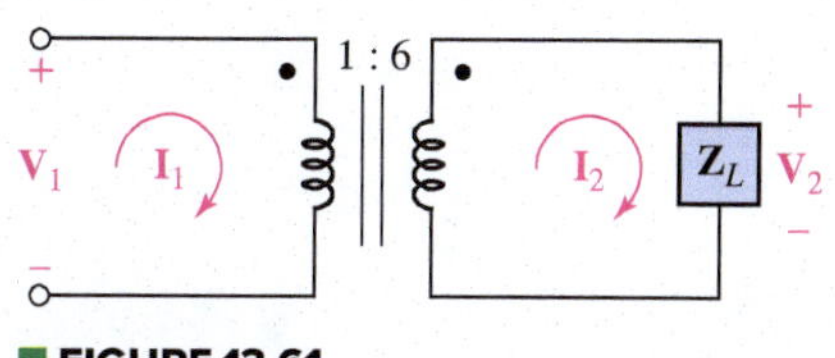

■ **FIGURE 13.61**

41. Calculate the average power delivered to the 400 mΩ and 21 Ω resistors, respectively, in the circuit of Fig. 13.62.

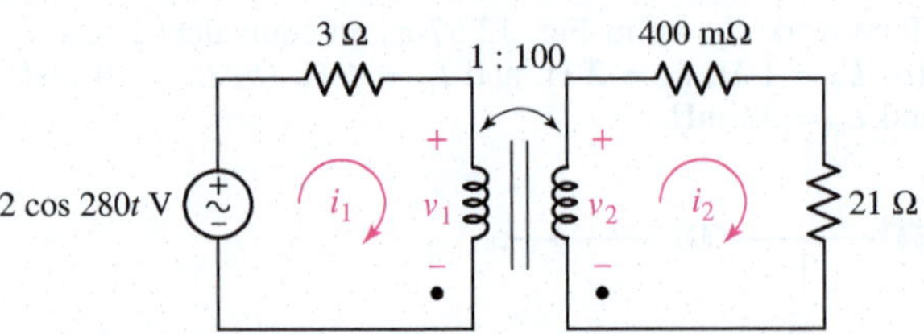

■ **FIGURE 13.62**

42. With regard to the ideal transformer circuit represented in Fig. 13.62, determine an equivalent circuit in which (*a*) the transformer and primary circuit are replaced, so that $\mathbf{V}_2$ and $\mathbf{I}_2$ are unchanged; (*b*) the transformer and secondary circuit are replaced, so that $\mathbf{V}_1$ and $\mathbf{I}_1$ are unchanged.
43. Calculate the average power delivered to each resistor shown in Fig. 13.63.

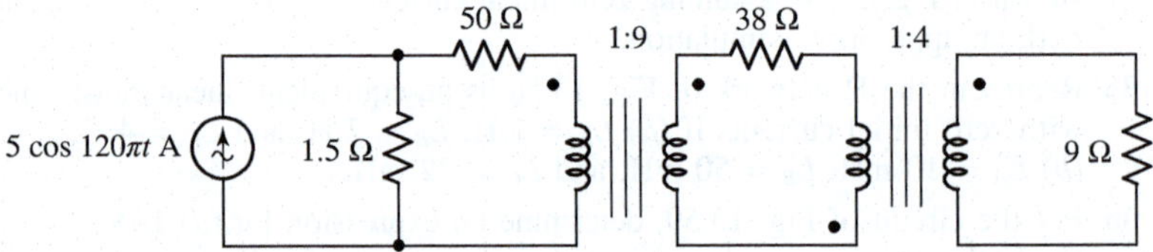

■ **FIGURE 13.63**

44. With respect to the circuit depicted in Fig. 13.64, calculate (*a*) the voltages v_1 and v_2; (*b*) the average power delivered to each resistor.

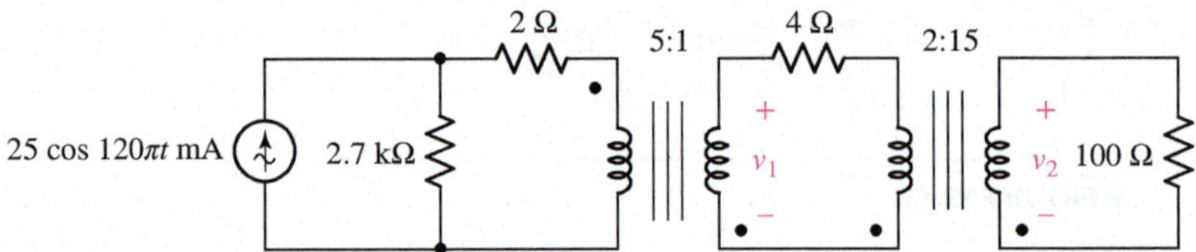

■ **FIGURE 13.64**

45. For the circuit of Fig. 13.65, $v_s = 117 \sin 500t$ V. Calculate v_2 if the terminals marked *a* and *b* are (*a*) left open-circuited; (*b*) short-circuited; (*c*) bridged by a 2 Ω resistor.

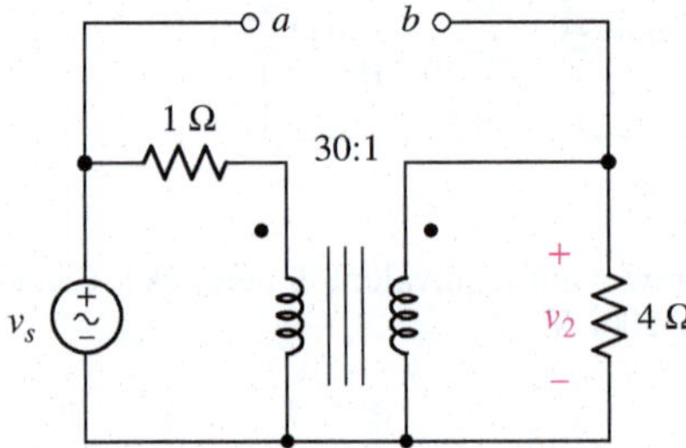

■ **FIGURE 13.65**

46. The turns ratio of the ideal transformer in Fig. 13.65 is changed from 30:1 to 1:3. Take $v_s = 720 \cos 120\pi\, t$ V, and calculate v_2 if terminals *a* and *b* are (*a*) short-circuited; (*b*) bridged by a 10 Ω resistor; (*c*) bridged by a 1 MΩ resistor.

47. For the circuit of Fig. 13.66, $R_1 = 1\ \Omega$, $R_2 = 4\ \Omega$, and $R_L = 1\ \Omega$. Select a and b to achieve a peak voltage of 200 mV magnitude across R_L.

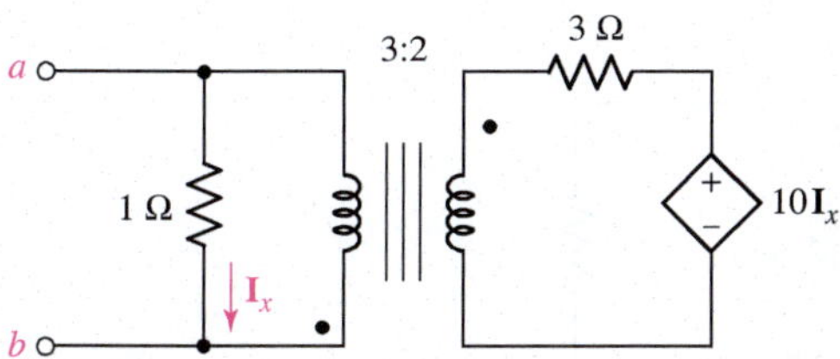

■ **FIGURE 13.66**

48. Calculate v_x for the circuit of Fig. 13.66 if $a = 0.01$, $b = 1$, $R_1 = 300\ \Omega$, $R_2 = 14\ \Omega$, and $R_L = 1\ \text{k}\Omega$.

49. (*a*) Referring to the ideal transformer circuit in Fig. 13.68, determine the load current i_L if $b = 0.25$, $a = 1$, $R_1 = 2.2\ \Omega$, $R_2 = 3.1\ \Omega$, and $R_L = 200\ \Omega$. (*b*) Verify your solution with an appropriate LTspice simulation.

50. Determine the equivalent impedance of the network in Fig. 13.67 as seen looking into terminals a and b. Find the current through the circuit if it is connected to the wall plug with 120 Vrms/60 Hz.

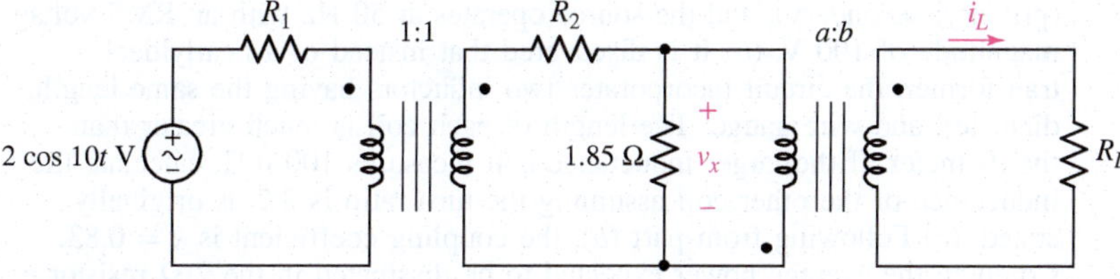

■ **FIGURE 13.67**

Chapter-Integrating Exercises

51. A transformer whose nameplate reads 2300/230 V, 25 kVA operates with primary and secondary voltages of 2300 V and 230 V rms, respectively, and can supply 25 kVA from its secondary winding. If this transformer is supplied with 2300 V rms and is connected to secondary loads requiring 8 kW at unity PF and 15 kVA at 0.8 PF lagging, (*a*) what is the primary current? (*b*) How many kilowatts can the transformer still supply to a load operating at 0.95 PF lagging?

52. A friend brings a vintage stereo system back from a recent trip to Warnemünde, unaware that it was designed to operate on twice the supply voltage (240 VAC) available at American household outlets. Design a circuit to allow your friend to listen to the stereo in the United States, assuming the operating frequency (50 Hz in Germany, 60 Hz in the United States) difference can be neglected.

53. As the lead singer in the local rock band, you just secured your first "gig" to perform at the local spring festival. As you are setting up your sound system, you notice that the output of the power amplifier (50 V peak) has a step up transformer marked 70.7 Vrms and 50 Ω, and that this output is supplied to three speakers in parallel: two speakers marked 8 ohm and one speaker marked 4 Ω, with a step down transformer for each speaker. (*a*) Explain why the transformer configuration is used. (*b*) Sketch a schematic diagram of the audio circuit, indicating desired turns ratio for all step up/step down transformers.

54. Obtain an expression for $\mathbf{V}_2/\mathbf{V}_s$ in the circuit of Fig. 13.68 if (*a*) $L_1 = 100$ mH, $L_2 = 500$ mH, and M is its maximum possible value; (*b*) $L_1 = 5L_2 = 1.4$ H

and $k = 87\%$ of its maximum possible value; (*c*) the two coils can be treated as an ideal transformer, the left-hand coil having 500 turns and the right-hand coil having 10,000 turns.

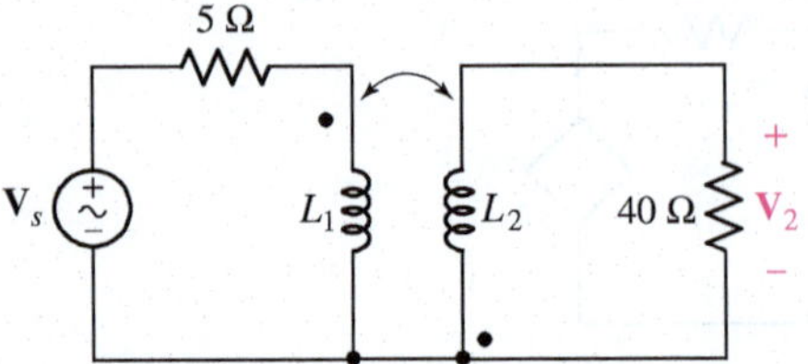

■ **FIGURE 13.68**

55. You notice your neighbor has installed a large coil of wire in close proximity to the power line coming into your house (underground cables are not available in your neighborhood). (*a*) What is the likely intention of your neighbor? (*b*) Is the plan likely to succeed? *Explain.* (*c*) When confronted, your neighbor simply shrugs and claims there's no way it can cost you anything, anyway, since the wire is in no way touching anything on your property. True or not? *Explain.*

56. Consider the circuit shown in Fig. 13.69. (*a*) Calculate the average power delivered to the 2 Ω resistor if the transformer is ideal with a 3:2 turns ratio (primary: secondary), and the source operates at 50 Hz with an RMS voltage magnitude of 100 V. (*b*) It is discovered that instead of a nearlyideal transformer, the circuit incorporates two inductors having the same length, diameter, and wire gauge. The length of each coil is much greater than the diameter. If the larger inductance coil measures 100 mH, calculate the inductance of the other coil assuming the turn ratio is 3:2 as originally stated. (*c*) Following from part (*b*), the coupling coefficient is $k = 0.83$. Compute the average power expected to be dissipated in the 2 Ω resistor. (*d*) Verify your solution with an appropriate simulation.

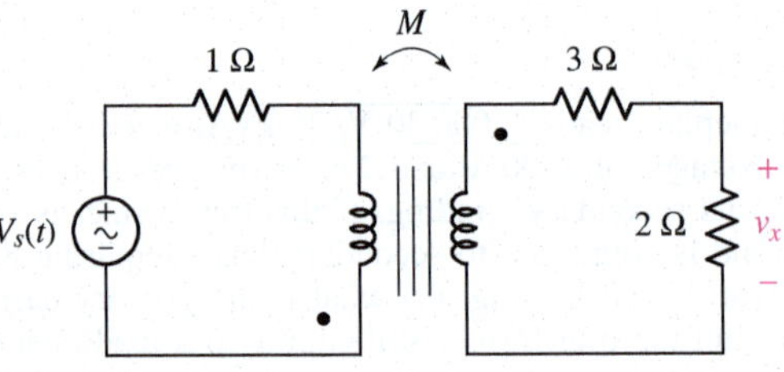

■ **FIGURE 13.69**

CHAPTER 14

Circuit Analysis in the s-Domain

KEY CONCEPTS

- Complex Frequency
- The Laplace Transform and Inverse Transforms
- Use of Transform Tables
- Method of Residuals
- Initial Value and Final Value Theorems
- Impedance in the **s**-Domain
- Modeling Initial Conditions with Ideal Sources
- Circuit Analysis in the **s**-Domain
- Identifying Poles and Zeros in Transfer Functions
- Impulse Response of a Circuit
- Using Convolution to Determine System Response

INTRODUCTION

When faced with time-varying sources, or a circuit with switches, we have several choices with respect to the analysis approach. Chapters 7 through 9 detail direct differential equation–based analysis, which is particularly useful when examining turn-on or turn-off transients of simple circuits. In contrast, Chaps. 10 to 13 describe analysis situations where sinusoidal excitation is assumed, and transients are of no interest. Unfortunately, not all sources are sinusoidal, and there are times when both transient and steady-state responses are required. In such instances, the Laplace transform proves to be a highly valuable tool.

Many textbooks simply launch straight into the Laplace transform integral, but such an approach conveys no intuitive understanding. For this reason, we have chosen to first introduce what may strike the reader at first as a somewhat unexpected concept—the notion of a "complex" frequency. Simply a mathematical convenience, complex frequency allows us to manipulate both periodic and nonperiodic time-varying quantities in parallel, greatly simplifying the analysis. After getting a feel for the basic idea, we develop it as a specific circuit analysis method.

14.1 COMPLEX FREQUENCY

We introduce the notion of ***complex frequency*** by considering a (purely real) exponentially damped sinusoidal function, such as the voltage

$$v(t) = V_m e^{\sigma t} \cos(\omega t + \theta) \qquad [1]$$

where σ (sigma) is a real quantity and is usually negative. Although we often refer to this function as being "damped," it is conceivable that we might occasionally encounter a situation where $\sigma > 0$ and hence the sinusoidal amplitude is increasing. (In Chap. 9, our study of the natural response of the *RLC* circuit also indicates that σ is the negative of the exponential damping coefficient.)

We may construct a constant voltage from Eq. [1] by letting $\sigma = \omega = 0$:

$$v(t) = V_m \cos(\theta) = V_0 \qquad [2]$$

If we set only σ equal to zero, then we obtain a general sinusoidal voltage

$$v(t) = V_m \cos(\omega t + \theta) \qquad [3]$$

and if $\omega = 0$, we have the exponential voltage

$$v(t) = V_m \cos\theta e^{\sigma t} = V_0 e^{\sigma t} \qquad [4]$$

Thus, the damped sinusoid of Eq. [1] includes as special cases the dc Eq. [2], sinusoidal Eq. [3], and exponential Eq. [4] functions.

Some additional insight into the significance of σ can be obtained by comparing the exponential function of Eq. [4] with the complex representation of a sinusoidal function with a zero-degree phase angle,

$$v(t) = V_0 e^{j\omega t} \qquad [5]$$

The neper itself was named after the Scottish philosopher and mathematician John Napier (1550–1617) and his napierian logarithm system; the spelling of his name is historically uncertain (see, for example, H. A. Wheeler, *IRE Transactions on Circuit Theory* 2, 1955, p. 219).

It is apparent that the two functions shown in Eqs. [4] and [5] have much in common. The only difference is that the exponent in Eq. [4] is real and the one in Eq. [5] is imaginary. The similarity between the two functions is emphasized by describing σ as a "frequency." This choice of terminology will be discussed in detail in the following sections, but for now we need merely note that σ is specifically termed the *real part* of the complex frequency. It should not be called the "real frequency," however, for this is a term that is more suitable for f (or, loosely, for ω). We can also refer to σ as the ***neper frequency,*** the name arising from the dimensionless unit of the exponent of e. Thus, given e^{7t} the dimensions of $7t$ are ***nepers*** (Np), and 7 is the neper frequency in nepers per second.

The General Form

The forced response of a network to a general forcing function of the form of Eq. [1] can be found very simply by using a method almost identical with that used in phasor-based analysis. Once we are able to find the forced response to this damped sinusoid, we will also have found the forced response to a dc voltage, an exponential voltage, and a sinusoidal voltage. First we consider σ and ω as the real and imaginary parts of a complex frequency.

We suggest that any function that may be written in the form

$$f(t) = \mathbf{K}e^{\mathbf{s}t} \qquad [6]$$

where $\mathbf{K}$ and $\mathbf{s}$ are complex constants (independent of time) is characterized by the complex frequency $\mathbf{s}$. The complex frequency $\mathbf{s}$ is therefore simply the factor that multiplies t in this complex exponential representation. Until we can determine the complex frequency of a given function by inspection, it is necessary to write the function in the form of Eq. [6].

The DC Case

We apply this definition first to the more familiar forcing functions. For example, a constant voltage

$$v(t) = V_0$$

may be written in the form

$$v(t) = V_0 e^{(0)t}$$

Therefore, we conclude that the complex frequency of a dc voltage or current is zero (i.e., $\mathbf{s} = 0$).

The Exponential Case

The next simple case is the exponential function

$$v(t) = V_0 e^{\sigma t}$$

which is already in the required form. The complex frequency of this voltage is therefore σ (i.e., $\mathbf{s} = \sigma + j0$).

The Sinusoidal Case

Now let us consider a sinusoidal voltage, one that may provide a slight surprise. Given

$$v(t) = V_m \cos(\omega t + \theta)$$

we want to find an equivalent expression in terms of the complex exponential. Drawing on our past experience, we use the formula we derived from Euler's identity

$$\cos(\omega t + \theta) = \frac{1}{2}[e^{j(\omega t+\theta)} + e^{-j(\omega t+\theta)}]$$

and obtain

$$v(t) = \frac{1}{2}V_m[e^{j(\omega t+\theta)} + e^{-j(\omega t+\theta)}]$$

$$= \left(\frac{1}{2}V_m e^{j\theta}\right)e^{j\omega t} + \left(\frac{1}{2}V_m e^{-j\theta}\right)e^{-j\omega t}$$

or

$$v(t) = \mathbf{K}_1 e^{\mathbf{s}_1 t} + \mathbf{K}_2 e^{\mathbf{s}_2 t}$$

We have the *sum* of two complex exponentials, and *two* complex frequencies are therefore present, one for each term. The complex frequency of the first term is $\mathbf{s} = \mathbf{s}_1 = j\omega$, and that of the second term is $\mathbf{s} = \mathbf{s}_2 = -j\omega$. These two values of $\mathbf{s}$ are ***conjugates*** (i.e., $\mathbf{s}_2 = \mathbf{s}_1^*$) and the two values of $\mathbf{K}$ are also conjugates: $\mathbf{K}_1 = \frac{1}{2}V_m e^{j\theta}$ and $\mathbf{K}_2 = \mathbf{K}_1^* = \frac{1}{2}V_m e^{-j\theta}$. The entire first term and the entire second term are therefore conjugates, which we might have expected inasmuch as their sum must be a real quantity, $v(t)$.

The complex conjugate of any number can be obtained by simply replacing all occurrences of "j" with "$-j$." The concept arises from our arbitrary choice of $j = +\sqrt{-1}$. However, the negative root is just as valid, which leads us to the definition of a complex conjugate.

The Exponentially Damped Sinusoidal Case

Finally, let us determine the complex frequency or frequencies associated with the exponentially damped sinusoidal function, Eq. [1]. We again use Euler's formula to obtain a complex exponential representation:

$$v(t) = V_m e^{\sigma t}\cos(\omega t + \theta)$$

$$= \frac{1}{2}V_m e^{\sigma t}\left[e^{j(\omega t+\theta)} + e^{-j(\omega t+\theta)}\right]$$

and thus

$$v(t) = \frac{1}{2}V_m e^{j\theta}e^{j(\sigma+j\omega)t} + \frac{1}{2}V_m e^{-j\theta}e^{j(\sigma-j\omega)t}$$

We therefore find that a conjugate complex pair of frequencies, $\mathbf{s}_1 = \sigma + j\omega$ and $\mathbf{s}_2 = \mathbf{s}_1^* = \sigma - j\omega$, is also required to describe the exponentially damped sinusoid. In general, neither σ nor ω is zero, and the exponentially varying sinusoidal waveform is the general case; the constant, sinusoidal, and exponential waveforms are special cases.

The Relationship of s to Reality

A positive real value of **s,** e.g., $\mathbf{s} = 5 + j0$, identifies an exponentially increasing function $\mathbf{K}e^{+5t}$, where **K** must be real if the function is to be a physical one. A negative real value for **s**, such as $\mathbf{s} = -5 + j0$, refers to an exponentially decreasing function $\mathbf{K}e^{-5t}$.

A purely imaginary value of **s**, such as $j10$, can never be associated with a purely real quantity. The functional form is $\mathbf{K}e^{j10t}$, which can also be written as $\mathbf{K}(\cos 10t + j\sin 10t)$; it obviously has both a real and an imaginary part, each of which is sinusoidal. To construct a real function, we must consider conjugate values of **s**, such as $\mathbf{s}_{1,2} = \pm j10$, with which we associate conjugate values of **K.** Loosely speaking, however, we may identify either of the complex frequencies $\mathbf{s}_1 = +j10$ or $\mathbf{s}_2 = -j10$ with a sinusoidal voltage at the radian frequency of 10 rad/s; the presence of the conjugate complex frequency is understood. The amplitude and phase angle of the sinusoidal voltage will depend on the choice of **K** for each of the two frequencies. Thus, selecting $\mathbf{s}_1 = j10$ and $\mathbf{K}_1 = 6 - j8$, where

Note that $|6 - j8| = 10$, so that $V_m = 2|\mathbf{K}| = 20$. Also, $ang(6 - j8) = -53.13°$.

$$v(t) = \mathbf{K}_1 e^{\mathbf{s}_1 t} + \mathbf{K}_2 e^{\mathbf{s}_2 t} \qquad \mathbf{s}_2 = \mathbf{s}_1^* \qquad \text{and} \qquad \mathbf{K}_2 = \mathbf{K}_1^*$$

we obtain the real sinusoid $20\cos(10t - 53.1°)$.

In a similar manner, a general value for **s**, such as $3 - j5$, can be associated with a real quantity only if it is accompanied by its conjugate, $3 + j5$. Speaking loosely again, we may think of either of these two conjugate frequencies as describing an exponentially increasing sinusoidal function, $e^{3t}\cos 5t$; the specific amplitude and phase angle will again depend on the values of the conjugate complex **K** terms.

By now we should have achieved some appreciation of the physical nature of the complex frequency **s**; in general, it describes an exponentially varying sinusoid. The real part of **s** is associated with the exponential variation; if it is negative, the function decays as t increases; if it is positive, the function increases; and if it is zero, the sinusoidal amplitude is constant. The larger the *magnitude* of the real part of **s**, the greater is

Large magnitudes for the real part of **s**, the imaginary part of **s**, or the magnitude of **s** indicate a rapidly varying function.

the rate of exponential increase or decrease. The imaginary part of **s** describes the sinusoidal variation; it is specifically the radian frequency. A large magnitude for the imaginary part of **s** indicates a more rapidly changing function of time.

It is customary to use the letter σ to designate the real part of **s**, and ω (*not* $j\omega$) to designate the imaginary part:

$$\mathbf{s} = \sigma + j\omega \qquad [7]$$

The radian frequency is sometimes referred to as the "real frequency," but this terminology can be very confusing when we find that we must then say that "the real frequency is the imaginary part of the complex frequency"! When we need to be specific, we will call **s** the complex frequency, σ the neper frequency, ω the radian frequency, and $f = \omega/2\pi$ the cyclic frequency; when no confusion seems likely, it is permissible to use "frequency" to refer to any of these four quantities. The *neper frequency* is measured in nepers per second, *radian frequency* is measured in radians per second, and *complex frequency* **s** is measured in units which are variously termed complex nepers per second or complex radians per second.

PRACTICE

14.1 Identify all the complex frequencies present in these real functions: (*a*) $(2e^{-100t} + e^{-200t}) \sin 2000t$; (*b*) $(2 - e^{-10t}) \cos(4t + \phi)$; (*c*) $e^{-10t} \cos 10t \sin 40t$.

14.2 Use real constants A, B, C, ϕ, and so forth, to construct the general form of the real function of time for a current having components at these frequencies: (*a*) 0, 10, -10 s^{-1}; (*b*) -5, $j8$, $-5 - j8$ s^{-1}; (*c*) -20, 20, $-20 + j20$, $20 - j20$ s^{-1}.

Ans: 14.1: $-100 + j2000$, $-100 - j2000$, $-200 + j2000$, $-200 - j2000$ s^{-1}; $j4$, $-j4$, $-10 + j4$, $-10 - j4$ s^{-1}; $-10 + j30$, $-10 - j30$, $-10 + j50$, $-10 - j50$ s^{-1}.
14.2: $A + Be^{10t} + Ce^{-10t}$; $Ae^{-5t} + B\cos(8t + \phi_1) + Ce^{-5t} \times \cos(8t + \phi_2)$; $Ae^{-20t} + Be^{20t} + Ce^{-20t}\cos(20t + \phi_1) + De^{20t}\cos(20t + \phi_2)$

14.2 DEFINITION OF THE LAPLACE TRANSFORM

Shortly after being exposed to the use of the sinusoidal forcing function in circuits with energy storage elements, the tedium and complexity of solving the integrodifferential equations caused us to begin casting about for an easier way to work problems. The phasor transform was the result, and we might remember that we were led to it through consideration of a complex forcing function of the form $V_0 e^{j\theta} e^{j\omega t}$. As soon as we concluded that we did not need the factor containing t, we were left with the phasor $V_0 e^{j\theta}$; we had arrived at the *frequency domain.*

Now a little flexing of our cerebral cortex has caused us to apply a forcing function of the form $V_0 e^{j\theta} e^{(\sigma + j\omega)t}$, leading to the invention of the complex frequency **s**, and thereby relegating all our previous functional forms to special cases: dc ($\mathbf{s} = 0$), exponential ($\mathbf{s} = \sigma$), sinusoidal ($\mathbf{s} = j\omega$), and exponential sinusoid ($\mathbf{s} = \sigma + j\omega$). By analogy to our previous experience with phasors, we saw that in these cases we may omit the factor containing t and once again obtain a solution by working in the frequency domain.

The Laplace Transform

We know that sinusoidal forcing functions lead to sinusoidal responses, and also that exponential forcing functions lead to exponential responses. However, as practicing engineers we will encounter many waveforms that are neither sinusoidal nor exponential, such as square waves, sawtooth waveforms, and pulses beginning at arbitrary instants of time. When such forcing functions are applied to a linear circuit, we will see that the response is neither similar to the form of the excitation waveform nor exponential. As a result, we cannot eliminate the terms containing t to form a frequency-domain response. This is rather unfortunate, as working in the frequency domain has proved to be rather useful.

There is a solution, however, which makes use of a technique that allows us to expand any function into a *sum* of exponential waveforms, each with its own complex frequency, and thus building on what we have already learned. Since we are considering linear circuits, we know that the total response of our circuit can be obtained by simply adding the individual response to each exponential waveform (i.e., by applying superposition). And, in dealing with each exponential waveform, we may once again neglect any terms containing t and work instead in the *frequency* domain. It unfortunately takes an infinite number of exponential terms to accurately represent a general time function, so taking a brute-force approach and applying superposition to the exponential series might be somewhat ill-advised (to put it mildly). Instead, we will sum these terms by performing an integration, leading to a frequency-domain function.

We formalize this approach using what is known as a ***Laplace transform,*** defined for a general function $f(t)$ as

$$\mathbf{F}(\mathbf{s}) = \int_{-\infty}^{\infty} e^{-\mathbf{s}t} f(t)dt \qquad [8]$$

The mathematical derivation of this integral operation requires an understanding of Fourier series and the Fourier transform, which are discussed in Chap. 18. The fundamental concept behind the Laplace transform, however, can be understood based on our discussion of complex frequency, as well as our prior experience with both phasors and converting back and forth between the time domain and the frequency domain. In fact, that is precisely what the Laplace transform does: it converts the general time-domain function $f(t)$ into a corresponding frequency-domain representation, $\mathbf{F}(\mathbf{s})$.

Equation [8] defines the two-sided, or bilateral, Laplace transform of $f(t)$. The term *two-sided* or *bilateral* is used to emphasize the fact that both positive and negative values of t are included in the range of integration. The inverse operation, often referred to as the inverse Laplace transform, is also defined as an integral expression[1]

$$f(t) = \frac{1}{2\pi j}\int_{\sigma_0 - j\infty}^{\sigma_0 + j\infty} e^{\mathbf{s}t}\mathbf{F}(\mathbf{s})d\mathbf{s} \qquad [9]$$

where the real constant σ_0 is included in the limits to ensure convergence of this improper integral; the two equations [8] and [9] constitute the two-sided Laplace transform pair. The good news is that Eq. [9] need never

(1) If we ignore the distracting factor of $1/2\pi j$ and view the integral as a summation over all frequencies such that $f(t) \propto \Sigma[\mathbf{F}(\mathbf{s})\, d\mathbf{s}]e^{\mathbf{s}t}$, this reinforces the notion that $f(t)$ is indeed a sum of complex frequency terms having a magnitude proportional to $\mathbf{F}(\mathbf{s})$.

be invoked in the study of circuit analysis: there is a quick and easy alternative to look forward to learning.

The One-Sided Laplace Transform

In many of our circuit analysis problems, the forcing and response functions do not exist forever in time, but rather they are initiated at some specific instant that we usually select as $t = 0$. Thus, for time functions that do not exist for $t < 0$, or for those time functions whose behavior for $t < 0$ is of no interest, the time-domain description can be thought of as $v(t)u(t)$. The defining integral for the Laplace transform is taken with the lower limit at $t = 0^-$ in order to include the effect of any discontinuity at $t = 0$, such as an impulse or a higher-order singularity. The corresponding Laplace transform is then

$$\mathbf{F(s)} = \int_{-\infty}^{\infty} e^{-st} f(t)u(t)dt = \int_{0^-}^{\infty} e^{-st} f(t)dt$$

This defines the *one-sided* Laplace transform of $f(t)$, or simply the *Laplace transform* of $f(t)$, one-sided being understood. (Eq. 8 then is referred to as the *two-sided* Laplace transform.)

The inverse transform expression remains unchanged, but when evaluated, it is understood to be valid only for $t > 0$. Here then is the definition of the Laplace transform pair that we will use from now on:

$$\mathbf{F(s)} = \int_{0^-}^{\infty} e^{-st} f(t)dt \quad [10]$$

$$f(t) = \frac{1}{2\pi j}\int_{\sigma_0 - j\infty}^{\sigma_0 + j\infty} e^{st}\mathbf{F(s)}ds \qquad f(t) \Leftrightarrow \mathbf{F(s)} \quad [11]$$

The script $\mathscr{L}$ may also be used to indicate the direct or inverse Laplace transform operation:

$$\mathbf{F(s)} = \mathscr{L}\{f(t)\} \quad \text{and} \quad f(t) = \mathscr{L}^{-1}\{\mathbf{F(s)}\}$$

EXAMPLE **14.1**

Compute the Laplace transform of the function $f(t) = 2u(t - 3)$.

In order to find the one-sided Laplace transform of $f(t) = 2u(t - 3)$, we must evaluate the integral

$$\begin{aligned}\mathbf{F(s)} &= \int_{0^-}^{\infty} e^{-st} f(t)dt \\ &= \int_{0^-}^{\infty} e^{-st} 2u(t-3)dt \\ &= 2\int_{3}^{\infty} e^{-st} dt\end{aligned}$$

Simplifying, we find

$$\mathbf{F(s)} = -\frac{2}{\mathbf{s}}e^{-st}\bigg|_3^{\infty} = -\frac{2}{\mathbf{s}}(0 - e^{-3s}) = \frac{2}{\mathbf{s}}e^{-3s}$$

(Continued on next page)

PRACTICE

14.3 Let $f(t) = -6e^{-2t}\,[u(t+3) - u(t-2)]$. Find the (*a*) two-sided $\mathbf{F(s)}$; (*b*) one-sided $\mathbf{F(s)}$.

Ans: $\frac{6}{2+s}[e^{-4-2s} - e^{6+3s}]$; $\frac{6}{2+s}[e^{-4-2s} - 1]$

14.3 LAPLACE TRANSFORMS OF SIMPLE TIME FUNCTIONS

In this section we will begin to build up a catalog of Laplace transforms for those time functions most often encountered in circuit analysis; we will assume for now that the function of interest is a voltage, although such a choice is completely arbitrary. We will create this catalog, at least initially, by using the definition,

$$\mathbf{V(s)} = \int_{0^-}^{\infty} e^{-st}\, v(t)dt = \mathscr{L}\{v(t)\}$$

which, along with the expression for the inverse transform,

$$v(t) = \frac{1}{2\pi j}\int_{\sigma_0 - j\infty}^{\sigma_0 + j\infty} e^{st}\, \mathbf{V(s)}ds = \mathscr{L}^{-1}\{\mathbf{V(s)}\}$$

establishes a one-to-one correspondence between $v(t)$ and $\mathbf{V(s)}$. That is, for every $v(t)$ for which $\mathbf{V(s)}$ exists, there is a unique $\mathbf{V(s)}$. At this point, we may be looking with some trepidation at the rather ominous form given for the inverse transform. Fear not! As we will see shortly, *an introductory study of Laplace transform theory does not require actual evaluation of this integral*. By going from the time domain to the frequency domain and taking advantage of the uniqueness just mentioned, we will be able to generate a catalog of transform pairs that will already contain the corresponding time function for nearly every transform that we wish to invert.

Before we continue, however, we should pause to consider whether there is any chance that the transform may not even exist for some $v(t)$ that concerns us. A set of conditions sufficient to ensure the absolute convergence of the Laplace integral for $\text{Re}\{\mathbf{s}\} > \sigma_0$ is

1. The function $v(t)$ is integrable in every finite interval $t_1 < t < t_2$, where $0 \le t_1 < t_2 < \infty$.

2. $\lim_{t\to\infty} e^{-\sigma_0 t}|v(t)|$ exists for some value of σ_0.

Time functions that do not satisfy these conditions are seldom encountered by the circuit analyst.[2]

(2) Examples of such functions are e^{t^2} and e^{e^t}, but not t^n or n^t. For a somewhat more detailed discussion of the Laplace transform and its applications, refer to Clare D. McGillem and George R. Cooper, *Continuous and Discrete Signal and System Analysis,* 3d ed. Oxford University Press, North Carolina: 1991, Chap. 5.

The Unit-Step Function $u(t)$

We first examine the Laplace transform of the unit-step function $u(t)$. From the defining equation, we may write

$$\mathscr{L}\{u(t)\} = \int_{0^-}^{\infty} e^{-\mathbf{s}t} u(t) dt = \int_{0}^{\infty} e^{-\mathbf{s}t}\, dt$$

$$= -\frac{1}{\mathbf{s}} e^{-\mathbf{s}t}\Big|_0^{\infty} = \frac{1}{\mathbf{s}}$$

for Re$\{\mathbf{s}\} > 0$, to satisfy condition 2. Thus,

$$u(t) \Leftrightarrow \frac{1}{\mathbf{s}} \qquad [12]$$

and our first Laplace transform pair has been established with ease.

The double arrow notation is commonly used to indicate Laplace transform pairs.

The Unit-Impulse Function $\delta(t - t_0)$

A singularity function whose transform is of considerable interest is the unit-impulse function $\delta(t - t_0)$. This function, plotted in Fig. 14.1, seems rather strange at first but is enormously useful in practice. The unit-impulse function is defined to have an area of unity, so that

$$\delta(t - t_0) = 0 \qquad t \neq t_0$$

$$\int_{t_0-\varepsilon}^{t_0+\varepsilon} \delta(t - t_0) dt = 1$$

where ε is a small constant. Thus, this "function" (a naming that makes many purist mathematicians cringe) has a nonzero value only at the point t_0. For $t_0 > 0^-$, we therefore find the Laplace transform to be

$$\mathscr{L}\{\delta(t - t_0)\} = \int_{0^-}^{\infty} e^{-\mathbf{s}t} \delta(t - t_0) dt = e^{-\mathbf{s}t_0}$$

$$= \delta(t - t_0) \Leftrightarrow e^{-\mathbf{s}t_0} \qquad [13]$$

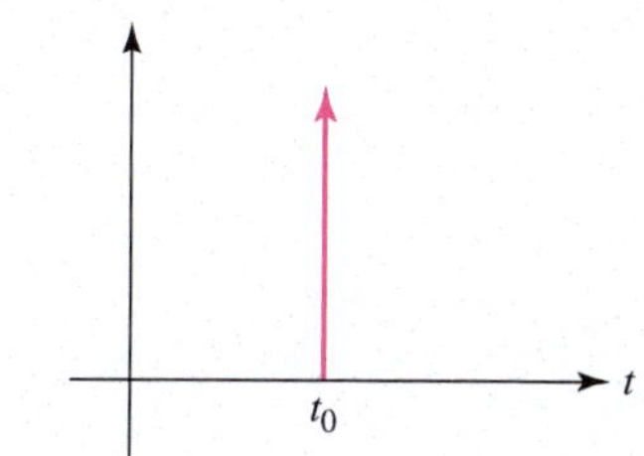

FIGURE 14.1 The unit-impulse function $\delta(t - t_0)$. This function is often used to approximate a signal pulse whose duration is very short compared to circuit time constants.

In particular, note that we obtain

$$\delta(t) \Leftrightarrow 1 \qquad [14]$$

for $t_0 = 0$.

An interesting feature of the unit-impulse function is known as the ***sifting property.*** Consider the integral of the impulse function multiplied by an arbitrary function $f(t)$:

$$\int_{-\infty}^{\infty} f(t)\delta(t - t_0)dt$$

Since the function $\delta(t - t_0)$ is zero everywhere except at $t = t_0$, the value of this integral is simply $f(t_0)$. The property turns out to be *very* useful in simplifying integral expressions containing the unit-impulse function.

The Exponential Function $e^{-\alpha t}$

Recalling our past interest in the exponential function, we examine its transform,

$$\mathscr{L}\{e^{-\alpha t}u(t)\} = \int_{0^-}^{\infty} e^{-\alpha t} e^{-\mathbf{s}t}\, dt$$

$$= -\frac{1}{\mathbf{s} + \alpha} e^{-(\mathbf{s}+\alpha)t}\Big|_0^{\infty} = \frac{1}{\mathbf{s} + \alpha}$$

and therefore,

$$e^{-\alpha t}\,u(t) \Leftrightarrow \frac{1}{\mathbf{s}+\alpha} \qquad [15]$$

It is understood that $\text{Re}\{\mathbf{s}\} > -\alpha$.

The Ramp Function *tu(t)*

As a final example, for the moment, let us consider the ramp function $tu(t)$. We obtain

$$\mathscr{L}\{tu(t)\} = \int_{0^-}^{\infty} te^{-\mathbf{s}t}\,dt = \frac{1}{\mathbf{s}^2}$$

$$tu(t) \Leftrightarrow \frac{1}{\mathbf{s}^2} \qquad [16]$$

either by a straightforward integration by parts or from a table of integrals.

So what of the function $te^{-\alpha t}u(t)$? We leave it to the reader to show that

$$te^{-\alpha t}u(t) \Leftrightarrow \frac{1}{(\mathbf{s}+\alpha)^2} \qquad [17]$$

There are, of course, quite a few additional time functions worth considering, but it may be best if we pause for the moment to consider the reverse of the process—the inverse Laplace transform—before returning to add to our list.

PRACTICE

14.4 Determine $\mathbf{V}(\mathbf{s})$ if $v(t)$ equals (*a*) $4\delta(t) - 3u(t)$; (*b*) $4\delta(t-2) - 3tu(t)$; (*c*) $[u(t)][u(t-2)]$.

14.5 Determine $v(t)$ if $\mathbf{V}(\mathbf{s})$ equals (*a*) 10; (*b*) $10/\mathbf{s}$; (*c*) $10/\mathbf{s}^2$; (*d*) $10/[\mathbf{s}(\mathbf{s}+10)]$; (*e*) $10\mathbf{s}/(\mathbf{s}+10)$.

Ans: 14.4: $(4\mathbf{s} - 3)/\mathbf{s}$; $4e^{-2\mathbf{s}} - (3/\mathbf{s}^2)$; $e^{-2\mathbf{s}}/\mathbf{s}$. 14.5: $10\delta(t)$; $10u(t)$; $10tu(t)$; $u(t) - e^{-10t}u(t)$; $10\delta(t) - 100e^{-10t}u(t)$.

14.4 INVERSE TRANSFORM TECHNIQUES

The Linearity Theorem

Although we mentioned that Eq. [9] can be applied to convert an **s**-domain expression into a time-domain expression, we also alluded to the fact that this is far more work than required—if we're willing to exploit the uniqueness of any Laplace transform pair. In order to fully capitalize on this fact, we must first introduce one of several helpful and well-known Laplace transform theorems—the ***linearity theorem.*** This theorem states that the Laplace transform of the sum of two or more time functions is equal to the sum of the transforms of the individual time functions. For two time functions we have

$$\begin{aligned}\mathscr{L}\{f_1(t)+f_2(t)\} &= \int_{0^-}^{\infty} e^{-\mathbf{s}t}[f_1(t)+f_2(t)]dt \\ &= \int_{0^-}^{\infty} e^{-\mathbf{s}t} f_1(t)dt + \int_{0^-}^{\infty} e^{-\mathbf{s}t} f_2(t)dt \\ &= \mathbf{F}_1(\mathbf{s}) + \mathbf{F}_2(\mathbf{s})\end{aligned}$$

This is known as the "additive property" of the Laplace transform.

As an example of the use of this theorem, suppose that we have a Laplace transform $\mathbf{V}(\mathbf{s})$ and want to know the corresponding time function $v(t)$. It will often be possible to decompose $\mathbf{V}(\mathbf{s})$ into the sum of two or more functions, e.g., $\mathbf{V}_1(\mathbf{s})$ and $\mathbf{V}_2(\mathbf{s})$, whose inverse transforms, $v_1(t)$ and $v_2(t)$, are already tabulated. It then becomes a simple matter to apply the linearity theorem and write

$$\begin{aligned}v(t) = \mathscr{L}^{-1}\{\mathbf{V}(\mathbf{s})\} &= \mathscr{L}^{-1}\{\mathbf{V}_1(\mathbf{s})+\mathbf{V}_2(\mathbf{s})\} \\ &= \mathscr{L}^{-1}\{\mathbf{V}_1(\mathbf{s})\} + \mathscr{L}^{-1}\{\mathbf{V}_2(\mathbf{s})\} = v_1(t)+v_2(t)\end{aligned}$$

Another important consequence of the linearity theorem is evident by studying the definition of the Laplace transform. Since we are working with an integral, *the Laplace transform of a constant times a function is equal to the constant times the Laplace transform of the function.* In other words,

$$\mathscr{L}\{kv(t)\} = k\mathscr{L}\{v(t)\}$$

or

$$kv(t) \Leftrightarrow k\mathbf{V}(\mathbf{s})$$

This is known as the "homogeneity property" of the Laplace transform.

where k is a constant of proportionality. This result is extremely handy in many situations that arise from circuit analysis, as we will see.

EXAMPLE 14.2

Given the function G(s) = (7/s) − 31/(s + 17), obtain *g*(*t*).

This **s**-domain function is composed of the sum of two terms, 7/**s** and −31/(**s** + 17). Through the linearity theorem we know that $g(t)$ will be composed of two terms as well, each the inverse Laplace transform of one of the two **s**-domain terms:

$$g(t) = \mathscr{L}^{-1}\left\{\frac{7}{\mathbf{s}}\right\} - \mathscr{L}^{-1}\left\{\frac{31}{\mathbf{s}+17}\right\}$$

Let's begin with the first term. The homogeneity property of the Laplace transform allows us to write that

$$\mathscr{L}^{-1}\left\{\frac{7}{\mathbf{s}}\right\} = 7\mathscr{L}^{-1}\left\{\frac{1}{\mathbf{s}}\right\} = 7u(t)$$

Thus, we have made use of the known transform pair $u(t) \Leftrightarrow 1/\mathbf{s}$ and the homogeneity property to find this first component of $g(t)$. In a similar fashion, we find that $\mathscr{L}^{-1}\left\{\frac{31}{\mathbf{s}+17}\right\} = 31e^{-17t}u(t)$. Putting these two terms together,

$$g(t) = [7 - 31e^{-17t}]u(t)$$

(Continued on next page)

PRACTICE

14.6 Given $\mathbf{H(s)} = \frac{2}{\mathbf{s}} - \frac{4}{\mathbf{s}^2} + \frac{3.5}{(\mathbf{s}+10)(\mathbf{s}+10)}$, obtain $h(t)$.

Ans: $h(t) = [2 - 4t + 3.5te^{-10t}]u(t)$.

Inverse Transform Techniques for Rational Functions

In analyzing circuits with multiple energy storage elements, we will often encounter **s**-domain expressions that are ratios of **s**-polynomials. We thus expect to routinely encounter expressions of the form

$$\mathbf{V(s)} = \frac{\mathbf{N(s)}}{\mathbf{D(s)}}$$

where **N(s)** and **D(s)** are polynomials in **s**. The values of **s** which lead to **N(s)** = 0 are referred to as ***zeros*** of **V(s)**, and those values of **s** which lead to the situation where **D(s)** = 0 are referred to as ***poles*** of **V(s)**.

In practice, it is seldom necessary to ever invoke Eq. [9] for functions encountered in circuit analysis, provided that we are clever in using the various techniques presented in this chapter.

Rather than rolling up our sleeves and invoking Eq. [9] each time we need to find an inverse transform, it is often possible to decompose these expressions using the method of residues into simpler terms whose inverse transforms are already known. The criterion for this is that **V(s)** must be a ***rational function*** for which the degree of the numerator **N(s)** must be less than that of the denominator **D(s)**. If it is not, we must first perform a simple division step, as shown in the following example. The result will include an impulse function (assuming the degree of the numerator is the same as that of the denominator) and a rational function. The inverse transform of the first is simple; the straightforward method of residues applies to the rational function if its inverse transform is not already known.

EXAMPLE 14.3

Calculate the inverse transform of $\mathbf{F(s)} = 2\frac{\mathbf{s}+2}{\mathbf{s}}$.

Since the degree of the numerator is equal to the degree of the denominator, **F(s)** is *not* a rational function. Thus, we begin by performing long division:

$$\mathbf{F(s)} = \mathbf{s}\overline{)2\mathbf{s}+4}^{\;2} \qquad \begin{array}{r} 2\mathbf{s} \\ \hline 4 \end{array}$$

so that **F(s)** = 2 + (4/**s**). By the linearity theorem,

$$\mathscr{L}^{-1}\{\mathbf{F(s)}\} = \mathscr{L}^{-1}\{2\} + \mathscr{L}^{-1}\left\{\frac{4}{\mathbf{s}}\right\} = 2\delta(t) + 4u(t)$$

(It should be noted that this particular function can be simplified without the process of long division; such a route was chosen to provide an example of the basic process.)

PRACTICE

14.7 Given the function $\mathbf{Q(s)} = \frac{3\mathbf{s}^2 - 4}{\mathbf{s}^2}$, find $q(t)$.

Ans: $q(t) = 3\delta(t) - 4tu(t)$.

In employing the method of residues, essentially performing a partial fraction expansion of $\mathbf{V(s)}$, we focus our attention on the roots of the denominator. Thus, it is first necessary to factor the $\mathbf{s}$-polynomial that comprises $\mathbf{D(s)}$ into a product of binomial terms. The roots of $\mathbf{D(s)}$ may be any combination of distinct or repeated roots, and they may be real or complex. It is worth noting, however, that complex roots always occur as conjugate pairs, provided that the coefficients of $\mathbf{D(s)}$ are real.

Distinct Poles and the Method of Residues

As a specific example, let us determine the inverse Laplace transform of

$$\mathbf{V(s)} = \frac{1}{(\mathbf{s}+\alpha)(\mathbf{s}+\beta)}$$

The denominator has been factored into two distinct roots, $-\alpha$ and $-\beta$. Although it is possible to substitute this expression in the defining equation for the inverse transform, it is much easier to use the linearity theorem. Using partial-fraction expansion, we can split the given transform into the sum of two simpler transforms,

$$\mathbf{V(s)} = \frac{A}{\mathbf{s}+\alpha} + \frac{B}{\mathbf{s}+\beta}$$

where A and B may be found by any of several methods. Perhaps the quickest solution is obtained by recognizing that

$$A = \lim_{\mathbf{s}\to-\alpha}\left[(\mathbf{s}+\alpha)\mathbf{V(s)} - \frac{(\mathbf{s}+\alpha)}{(\mathbf{s}+\beta)}B\right]$$

$$= \lim_{\mathbf{s}\to-\alpha}\left[\frac{1}{\mathbf{s}+\beta} - 0\right] = \frac{1}{\beta-\alpha}$$

In this equation, we use the single-fraction (i.e., nonexpanded) version of $\mathbf{V(s)}$.

Recognizing that the second term is always zero, in practice we would simply write

$$A = (\mathbf{s}+\alpha)\mathbf{V(s)}|_{\mathbf{s}=-\alpha}$$

Similarly,

$$B = (\mathbf{s}+\beta)\mathbf{V(s)}|_{\mathbf{s}=-\beta} = \frac{1}{\alpha-\beta}$$

and therefore,

$$\mathbf{V(s)} = \frac{1/(\beta-\alpha)}{\mathbf{s}+\alpha} + \frac{1/(\alpha-\beta)}{\mathbf{s}+\beta}$$

We have already evaluated inverse transforms of this form, and so

$$v(t) = \frac{1}{\beta-\alpha}e^{-\alpha t}u(t) + \frac{1}{\alpha-\beta}e^{-\beta t}u(t)$$

$$= \frac{1}{\beta-\alpha}(e^{-\alpha t} - e^{-\beta t})u(t)$$

If we wished, we could now include this as a new entry in our catalog of Laplace pairs,

$$\frac{1}{\beta-\alpha}(e^{-\alpha t} - e^{-\beta t})u(t) \Leftrightarrow \frac{1}{(\mathbf{s}+\alpha)(\mathbf{s}+\beta)}$$

This approach is easily extended to functions whose denominators are higher-order **s**-polynomials, although the operations can become somewhat tedious. It should also be noted that we did not specify that the constants A and B must be real. However, in situations where α and β are complex, we will find that α and β are complex conjugates (this is not required mathematically, but it is required for physical circuits). In such instances, we will also find that $A = B^*$; in other words, the coefficients will be complex conjugates as well.

EXAMPLE **14.4**

Determine the inverse transform of $\mathbf{P(s)} = \dfrac{7\mathbf{s} + 5}{\mathbf{s}^2 + \mathbf{s}}$

We see that **P(s)** is a rational function (the degree of the numerator is *one,* whereas the degree of the denominator is *two*), so we begin by factoring the denominator and write:

$$\mathbf{P(s)} = \frac{7\mathbf{s}+5}{\mathbf{s}(\mathbf{s}+1)} = \frac{a}{\mathbf{s}} + \frac{b}{\mathbf{s}+1}$$

where our next step is to determine values for a and b. Applying the method of residues,

$$a = \left.\frac{7\mathbf{s}+5}{\mathbf{s}+1}\right|_{\mathbf{s}=0} = 5 \quad \text{and} \quad b = \left.\frac{7\mathbf{s}+5}{\mathbf{s}}\right|_{\mathbf{s}=-1} = 2$$

We may now write **P(s)** as

$$\mathbf{P(s)} = \frac{5}{\mathbf{s}} + \frac{2}{\mathbf{s}+1}$$

the inverse transform of which is simply $p(t) = [5 + 2e^{-t}]u(t)$.

PRACTICE

14.8 Given the function $\mathbf{Q(s)} = \dfrac{11\mathbf{s} + 30}{\mathbf{s}^2 + 3\mathbf{s}}$, find $q(t)$.

Ans: $q(t) = [10 + e^{-3t}]u(t)$.

Repeated Poles

A closely related situation is that of repeated poles. Consider the function

$$\mathbf{V(s)} = \frac{\mathbf{N(s)}}{(\mathbf{s} - p)^n}$$

which we want to expand into

$$\mathbf{V(s)} = \frac{a_n}{(\mathbf{s}-p)^n} + \frac{a_{n-1}}{(\mathbf{s}-p)^{n-1}} + \cdots + \frac{a_1}{(\mathbf{s}-p)}$$

To determine each constant, we first multiply the nonexpanded version of **V(s)** by $(\mathbf{s} - p)^n$. The constant a_n is found by evaluating the resulting expression at $\mathbf{s} = p$. The remaining constants are found by differentiating the expression $(\mathbf{s} - p)^n\mathbf{V(s)}$ the appropriate number of times prior to evaluating at $\mathbf{s} = p$, and dividing by a factorial term. The differentiation procedure removes the constants previously found, and evaluating at $\mathbf{s} = p$ removes the remaining constants.

For example, a_{n-2} is found by evaluating

$$\frac{1}{2!}\frac{d^2}{d\mathbf{s}^2}[(\mathbf{s} - p)^n \mathbf{V}(\mathbf{s})]_{\mathbf{s}=p}$$

and the term a_{n-k} is found by evaluating

$$\frac{1}{k!}\frac{d^k}{d\mathbf{s}^k}[(\mathbf{s} - p)^n \mathbf{V}(\mathbf{s})]_{\mathbf{s}=p}$$

To illustrate the basic procedure, let's find the inverse Laplace transform of a function having a combination of both situations: one pole at $\mathbf{s} = 0$ and two poles at $\mathbf{s} = -6$.

EXAMPLE 14.5

Compute the inverse transform of the function $\mathbf{V}(\mathbf{s}) = \dfrac{2}{\mathbf{s}^3 + 12\mathbf{s}^2 + 36\mathbf{s}}$

We note that the denominator can be easily factored, leading to

$$\mathbf{V}(\mathbf{s}) = \frac{2}{\mathbf{s}(\mathbf{s}+6)(\mathbf{s}+6)} = \frac{2}{\mathbf{s}(\mathbf{s}+6)^2}$$

As promised, we see that there are indeed three poles, one at $\mathbf{s} = 0$, and two at $\mathbf{s} = -6$. Next, we expand the function into

$$\mathbf{V}(\mathbf{s}) = \frac{a_1}{(\mathbf{s}+6)^2} + \frac{a_2}{(\mathbf{s}+6)} + \frac{a_3}{\mathbf{s}}$$

and apply our new procedure to obtain the unknown constants a_1 and a_2; we will find a_3 using the previous procedure. Thus,

$$a_1 = \left[(\mathbf{s}+6)^2 \frac{2}{\mathbf{s}(\mathbf{s}+6)^2}\right]_{\mathbf{s}=-6} = \frac{2}{\mathbf{s}}\bigg|_{\mathbf{s}=-6} = -\frac{1}{3}$$

and

$$a_2 = \frac{d}{d\mathbf{s}}\left[(\mathbf{s}+6)^2 \frac{2}{\mathbf{s}(\mathbf{s}+6)^2}\right]_{\mathbf{s}=-6} = \frac{d}{d\mathbf{s}}\left(\frac{2}{\mathbf{s}}\right)\bigg|_{\mathbf{s}=-6} = -\frac{2}{\mathbf{s}^2}\bigg|_{\mathbf{s}=-6} = -\frac{1}{18}$$

The remaining constant a_3 is found using the procedure for distinct poles

$$a_3 = \mathbf{s}\frac{2}{\mathbf{s}(\mathbf{s}+6)^2}\bigg|_{\mathbf{s}=0} = \frac{2}{6^2} = \frac{1}{18}$$

Thus, we may now write $\mathbf{V}(\mathbf{s})$ as

$$\mathbf{V}(\mathbf{s}) = \frac{-\frac{1}{3}}{(\mathbf{s}+6)^2} + \frac{-\frac{1}{18}}{(\mathbf{s}+6)} + \frac{\frac{1}{18}}{\mathbf{s}}$$

Using the linearity theorem, the inverse transform of $\mathbf{V}(\mathbf{s})$ can now be found by simply determining the inverse transform of each term. We see that the first term on the right is of the form

$$\frac{\mathbf{K}}{(\mathbf{s}+\alpha)^2}$$

(Continued on next page)

and making use of Eq. [17], we find that its inverse transform is $-\frac{1}{3}te^{-6t}u(t)$. In a similar fashion, we find that the inverse transform of the second term is $-\frac{1}{18}e^{-6t}u(t)$, and that of the third term is $\frac{1}{18}u(t)$. Thus,

$$v(t) = -\frac{1}{3}te^{-6t}\,u(t) - \frac{1}{18}e^{-6t}\,u(t) + \frac{1}{18}u(t)$$

or, more compactly,

$$v(t) = \frac{1}{18}[1 - (1 + 6t)e^{-6t}]u(t)$$

PRACTICE

14.9 Determine $g(t)$ if $\mathbf{G(s)} = \dfrac{3}{\mathbf{s}^3 + 5\mathbf{s}^2 + 8\mathbf{s} + 4}$.

Ans: $g(t) = 3[e^{-t} - te^{-2t} - e^{-2t}]u(t)$.

COMPUTER-AIDED ANALYSIS

MATLAB can be used to assist in the solution of equations arising from the analysis of circuits with time-varying excitation in several different ways. The most straightforward technique makes use of ordinary differential equation (ODE) solver routines *ode*23() and *ode*45(). These two routines are based on numerical methods of solving differential equations, with *ode*45() having greater accuracy. The solution is determined only at discrete points, however, and therefore is not known for all values of time. For many applications this is adequate, provided a sufficient density of points is used.

The Laplace transform technique provides us with the means of obtaining an exact expression for the solution of differential equations, and as such has many advantages over the use of numerical ODE solution techniques. Another significant advantage to the Laplace transform technique will become apparent in subsequent chapters when we study the significance of the form of **s**-domain expressions, particularly once we factor the denominator polynomials.

As we have already seen, lookup tables can be handy when working with Laplace transforms, although the method of residues can become somewhat tedious for functions with higher-order polynomials in their denominators. In these instances MATLAB can also be of assistance, as it contains several useful functions for the manipulation of polynomial expressions.

In MATLAB, the polynomial

$$p(x) = a_n x^n + a_{n-1}x^{n-1} + \cdots + a_1 x + a_0$$

is stored as the vector $[a_n a_{n-1} \ldots a_1 a_0]$. Thus, to define the polynomials $\mathbf{N(s)} = 2$ and $\mathbf{D(s)} = \mathbf{s}^3 + 12\mathbf{s}^2 + 36\mathbf{s}$ we write

```
>> N = [2];
>> D = [1 12 36 0];
```

The roots of either polynomial can be obtained by invoking the function *roots*(**p**), where **p** is a vector containing the coefficients of the polynomial. For example,

```
>> q = [1 8 16];
>> roots (q)
```

yields

```
ans =
   -4
   -4
```

MATLAB also enables us to determine the residues of the rational function **N(s)/D(s)** using the function *residue*(). For example,

```
>> [r p y] = residue (N, D);
```

returns three vectors **r**, **p**, and **y**, such that

$$\frac{\mathbf{N(s)}}{\mathbf{D(s)}} = \frac{r_1}{x - p_1} + \frac{r_2}{x - p_2} + \cdots + \frac{r_n}{x - p_n} + \mathbf{y(s)}$$

in the case of no multiple poles; in the case of n multiple poles,

$$\frac{\mathbf{N(s)}}{\mathbf{D(s)}} = \frac{r_1}{(x - p)} + \frac{r_2}{(x - p)^2} + \cdots + \frac{r_n}{(x - p)^n} + \mathbf{y(s)}$$

Note that as long as the order of the numerator polynomial is less than the order of the denominator polynomial, the vector **y(s)** will always be empty.

Executing the command without the semicolon results in the output

```
r =
  -0.0556
  -0.3333
   0.0556
p =
  -6
  -6
   0
y =
   [ ]
```

which agrees with the answer found in Example 14.5.

14.5 BASIC THEOREMS FOR THE LAPLACE TRANSFORM

We can now consider two theorems that might be considered collectively the *raison d'être* for Laplace transforms in circuit analysis—the time differentiation and integration theorems. These will help us transform the derivatives and integrals appearing in the time-domain circuit equations.

Time Differentiation Theorem

Let us look at time differentiation first by considering a time function $v(t)$ whose Laplace transform $\mathbf{V}(\mathbf{s})$ is known to exist. We want the transform of the first derivative of $v(t)$,

$$\mathscr{L}\left\{\frac{dv}{dt}\right\} = \int_{0^-}^{\infty} e^{-\mathbf{s}t}\frac{dv}{dt}dt$$

This can be integrated by parts:

$$U = e^{-\mathbf{s}t} \qquad dV = \frac{dv}{dt}dt$$

with the result

$$\mathscr{L}\left\{\frac{dv}{dt}\right\} = v(t)e^{-\mathbf{s}t}\Big|_{0^-}^{\infty} + \mathbf{s}\int_{0^-}^{\infty} e^{-\mathbf{s}t}v(t)dt$$

The first term on the right must approach zero as t increases without limit; otherwise $\mathbf{V}(\mathbf{s})$ would not exist. Hence,

$$\mathscr{L}\left\{\frac{dv}{dt}\right\} = 0 - v(0^-) + \mathbf{s}\mathbf{V}(\mathbf{s})$$

and

$$\frac{dv}{dt} \Leftrightarrow \mathbf{s}\mathbf{V}(\mathbf{s}) - v(0^-)$$

Similar relationships may be developed for higher-order derivatives:

$$\frac{d^2v}{dt^2} \Leftrightarrow \mathbf{s}^2\mathbf{V}(\mathbf{s}) - \mathbf{s}v(0^-) - v'(0^-) \qquad [18]$$

$$\frac{d^3v}{dt^3} \Leftrightarrow \mathbf{s}^3\mathbf{V}(\mathbf{s}) - \mathbf{s}^2v(0^-) - \mathbf{s}v'(0^-) - v''(0^-) \qquad [19]$$

where $v'(0^-)$ is the value of the first derivative of $v(t)$ evaluated at $t = 0^-$, $v''(0^-)$ is the initial value of the second derivative of $v(t)$, and so forth. When all initial conditions are zero, we see that differentiating once with respect to t in the time domain corresponds to multiplication by $\mathbf{s}$ in the frequency domain; differentiating twice in the time domain corresponds to multiplication by $\mathbf{s}^2$ in the frequency domain, and so on. Thus, *differentiation in the time domain is equivalent to multiplication in the frequency domain*. This is a substantial simplification! We also notice that, when the initial conditions are not zero, their presence is still accounted for.

EXAMPLE 14.6

Given the series RL circuit shown in Fig. 14.2, calculate the current through the 4 Ω resistor, given the initial condition shown.

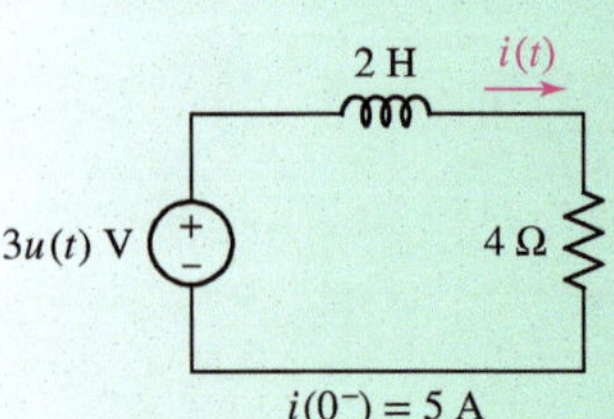

FIGURE 14.2 A circuit that is analyzed by transforming the differential equation $2\,di/dt + 4i = 3u(t)$ into $2[\mathbf{sI}(\mathbf{s}) - i(0^-)] + 4\mathbf{I}(\mathbf{s}) = 3/\mathbf{s}$.

▶ ***Identify the goal of the problem.***

We need to find an expression for the current labeled $i(t)$.

▶ ***Collect the known information.***

The network is driven by a step voltage, and we are given an initial value of the current (at $t = 0^-$) of 5 A.

▶ ***Devise a plan.***

Applying KVL to the circuit will result in a differential equation with $i(t)$ as the unknown. Taking the Laplace transform of both sides of this equation will convert it to the **s**-domain. Solving the resulting algebraic equation for **I**(**s**), the inverse Laplace transform will yield $i(t)$.

▶ *Construct an appropriate set of equations.*

Using KVL to write the single-loop equation in the time domain,

$$2\frac{di}{dt} + 4i = 3u(t)$$

Now, we take the Laplace transform of each term, so that

$$2[\mathbf{sI(s)} - i(0^-)] + 4\mathbf{I(s)} = \frac{3}{\mathbf{s}}$$

▶ *Determine if additional information is required.*

We have an equation that may be solved for the frequency-domain representation **I**(**s**) of our goal, $i(t)$.

▶ *Attempt a solution.*

We next solve for **I**(**s**), substituting $i(0^-) = 5$:

$$(2\mathbf{s} + 4)\mathbf{I(s)} = \frac{3}{\mathbf{s}} + 10$$

and

$$\mathbf{I(s)} = \frac{1.5}{\mathbf{s(s+2)}} + \frac{5}{\mathbf{s}+2}$$

Applying the method of residues to the first term,

$$\left.\frac{1.5}{\mathbf{s}+2}\right|_{\mathbf{s}=0} = 0.75 \quad \text{and} \quad \left.\frac{1.5}{\mathbf{s}}\right|_{\mathbf{s}=-2} = -0.75$$

so that

$$\mathbf{I(s)} = \frac{0.75}{\mathbf{s}} + \frac{4.25}{\mathbf{s}+2}$$

We then use our known transform pairs to invert:

$$\begin{aligned} i(t) &= 0.75u(t) + 4.25e^{-2t}u(t) \\ &= (0.75 + 4.25e^{-2t})u(t) \text{ A} \end{aligned}$$

▶ *Verify the solution. Is it reasonable or expected?*

Based on our previous experience with this type of circuit, we expected a dc forced response plus an exponentially decaying natural response. At $t = 0$, we obtain $i(0) = 5$ A, as required, and as $t \to \infty$, $i(t) \to \frac{3}{4}$A as we would expect.

Our solution for $i(t)$ is therefore complete. Both the forced response $0.75u(t)$ and the natural response $4.25e^{-2t}u(t)$ are present, and the initial condition was automatically incorporated into the solution. The method illustrates a very painless way of obtaining the complete solution to many differential equations.

PRACTICE

14.10 Use Laplace transform methods to find $i(t)$ in the circuit of Fig. 14.3.

Ans: $(0.25 + 4.75e^{-20t})u(t)$ A.

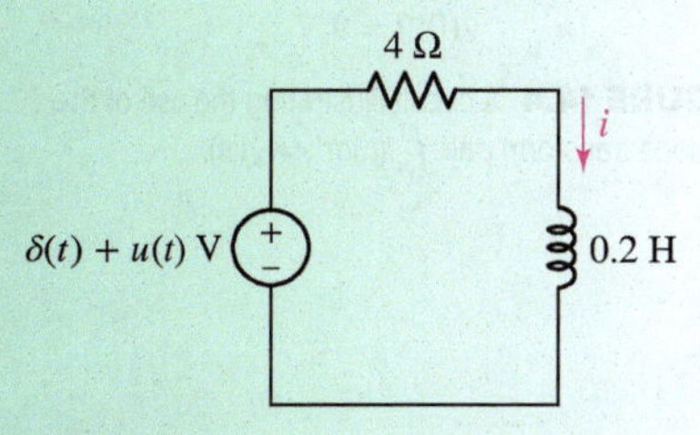

FIGURE 14.3

Time-Integration Theorem

The same kind of simplification can be accomplished when we meet the operation of integration with respect to time in our circuit equations. Let us determine the Laplace transform of the time function described by $\int_{0^-}^{t} v(x)dx$,

$$\mathscr{L}\left\{\int_{0^-}^{t} v(x)dx\right\} = \int_{0^-}^{\infty} e^{-\mathbf{s}t}\left[\int_{0^-}^{t} v(x)dx\right]dt$$

Integrating by parts, we let

$$u = \int_{0^-}^{t} v(x)dx \qquad dv = e^{-\mathbf{s}t}dt$$

$$du = v(t)dt \qquad v = -\frac{1}{\mathbf{s}}e^{-\mathbf{s}t}$$

Then

$$\mathscr{L}\left\{\int_{0^-}^{t} v(x)dx\right\} = \left\{\left[\int_{0^-}^{t} v(x)dx\right]\left[-\frac{1}{\mathbf{s}}e^{-\mathbf{s}t}\right]\right\}_{t=0^-}^{t=\infty} - \int_{0^-}^{\infty} -\frac{1}{\mathbf{s}}e^{-\mathbf{s}t}v(t)dt$$

$$= \left[-\frac{1}{\mathbf{s}}e^{-\mathbf{s}t}\int_{0^-}^{t} v(x)dx\right]_{0^-}^{\infty} + \frac{1}{\mathbf{s}}\mathbf{V}(\mathbf{s})$$

But, since $e^{-\mathbf{s}t} \to 0$ as $t \to \infty$, the first term on the right vanishes at the upper limit, and when $t \to 0^-$, the integral in this term likewise vanishes. This leaves only the $\mathbf{V}(\mathbf{s})/\mathbf{s}$ term, so that

$$\int_{0^-}^{t} v(x)dx \Leftrightarrow \frac{\mathbf{V}(\mathbf{s})}{\mathbf{s}} \qquad [20]$$

and thus *integration in the time domain corresponds to division by* **s** *in the frequency domain.* Once more, a relatively complicated calculus operation in the time domain simplifies to an algebraic operation in the frequency domain.

EXAMPLE 14.7

Determine $i(t)$ for $t > 0$ in the series RC circuit shown in Fig. 14.4.

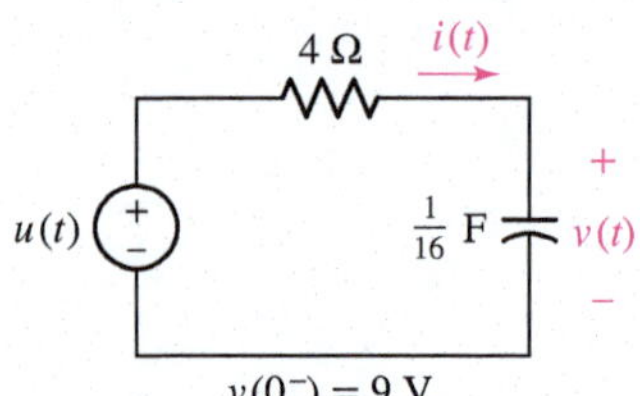

FIGURE 14.4 A circuit illustrating the use of the Laplace transform pair $\int_{0^-}^{t} i(t')dt' \Leftrightarrow \frac{1}{\mathbf{s}}\mathbf{I}(\mathbf{s})$.

We first write the single-loop equation,

$$u(t) - 4i(t) + 16\int_{-\infty}^{t} i(t')dt'$$

In order to apply the time-integration theorem, we must arrange for the lower limit of integration to be 0^-. Thus, we set

$$16\int_{-\infty}^{t} i(t')dt' = 16\int_{-\infty}^{0^-} i(t')dt' + 16\int_{0^-}^{t} i(t')dt'$$

$$= v(0^-) + 16\int_{0^-}^{t} i(t')dt'$$

Therefore,

$$u(t) = 4i(t) = v(0^-) + 16\int_{0^-}^{t} i(t')dt'$$

We next take the Laplace transform of both sides of this equation. Since we are using the one-sided transform, $\mathscr{L}\{v(0^-)\}$ is simply $\mathscr{L}\{v(0^-)u(t)\}$, and thus

$$\frac{1}{\mathbf{s}} = 4\mathbf{I(s)} + \frac{9}{\mathbf{s}} + \frac{16}{\mathbf{s}}\mathbf{I(s)}$$

Solving for $\mathbf{I(s)}$,

$$\mathbf{I(s)} = -\frac{2}{\mathbf{s}+4}$$

the desired result is immediately obtained,

$$i(t) = -2e^{-4t}u(t) \text{ A}$$

EXAMPLE **14.8**

Find $v(t)$ for the same circuit, repeated as Fig. 14.5 for convenience.

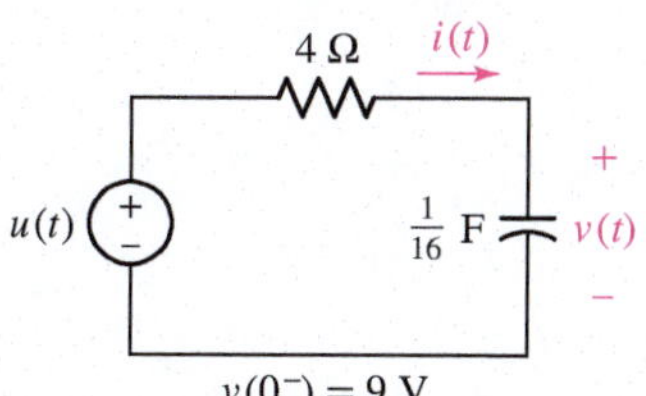

FIGURE 14.5 The circuit of Fig. 14.4 repeated, in which the voltage $v(t)$ is required.

This time we write a single nodal equation,

$$\frac{v(t) - u(t)}{4} + \frac{1}{16}\frac{dv}{dt} = 0$$

Taking the Laplace transform, we obtain

$$\frac{\mathbf{V(s)}}{4} - \frac{1}{4\mathbf{s}} + \frac{1}{16}\mathbf{s}\mathbf{V(s)} - \frac{v(0^-)}{16} = 0$$

or

$$\mathbf{V(s)}\left(1 + \frac{\mathbf{s}}{4}\right) = \frac{1}{\mathbf{s}} + \frac{9}{4}$$

Thus,

$$\mathbf{V(s)} = \frac{4}{\mathbf{s}(\mathbf{s}+4)} + \frac{9}{\mathbf{s}+4}$$
$$= \frac{1}{\mathbf{s}} - \frac{1}{\mathbf{s}+4} + \frac{9}{\mathbf{s}+4}$$
$$= \frac{1}{\mathbf{s}} + \frac{8}{\mathbf{s}+4}$$

and taking the inverse transform,

$$v(t) = (1 + 8e^{-4t})u(t) \text{ V}$$

To check this result, we note that $\left(\frac{1}{16}\right)dv/dt$ should yield the previous expression for $i(t)$. For $t > 0$,

$$\frac{1}{16}\frac{dv}{dt} = \frac{1}{16}(-32)e^{-4t} = -2e^{-4t}$$

which is in agreement with what was found in Example 14.7.

PRACTICE

14.11 Find $v(t)$ at $t = 800$ ms for the circuit of Fig. 14.6.

Ans: 802 mV.

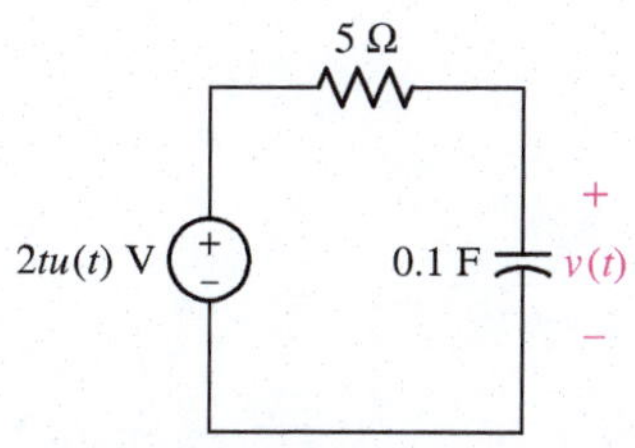

FIGURE 14.6

Laplace Transforms of Sinusoids

To illustrate the use of both the linearity theorem and the time-differentiation theorem, not to mention the addition of a most important pair to our forthcoming Laplace transform table, let us establish the Laplace transform of $\sin \omega t\ u(t)$. We could use the defining integral expression with integration by parts, but this is needlessly difficult. Instead, we use the relationship

$$\sin \omega t = \frac{1}{2j}(e^{j\omega t} - e^{-j\omega t})$$

The transform of the sum of these two terms is just the sum of the transforms, and each term is an exponential function for which we already have the transform. We may immediately write

$$\mathscr{L}\{\sin \omega t\ u(t)\} = \frac{1}{2j}\left(\frac{1}{\mathbf{s} - j\omega} - \frac{1}{\mathbf{s} + j\omega}\right) = \frac{\omega}{\mathbf{s}^2 + \omega^2}$$

$$\sin \omega t\ u(t) \Leftrightarrow \frac{\omega}{\mathbf{s}^2 + \omega^2} \qquad [21]$$

We next use the time-differentiation theorem to determine the transform of $\cos \omega t u(t)$, which is proportional to the derivative of $\cos \omega t$. That is,

$$\mathscr{L}\{\cos \omega t\ u(t)\} = \mathscr{L}\left\{\frac{1}{\omega}\frac{d}{dt}[\sin \omega t\ u(t)]\right\} = \frac{1}{\omega}\mathbf{s}\frac{\omega}{\mathbf{s}^2 + \omega^2}$$

$$\cos \omega t\ u(t) \Leftrightarrow \frac{\mathbf{s}}{\mathbf{s}^2 + \omega^2} \qquad [22]$$

Note that we have made use of the fact that $\sin \omega t|_{t=0} = 0$.

The Time-Shift Theorem

As we have seen in some of our earlier transient problems, not all forcing functions begin at $t = 0$. What happens to the transform of a time function if that function is simply shifted in time by some known amount? In particular, if the transform of $f(t)u(t)$ is the known function $\mathbf{F}(\mathbf{s})$, then what is the transform of $f(t - a)u(t - a)$, the original time function delayed by a seconds (and not existing for $t < a$)? Working directly from the definition of the Laplace transform, we get

$$\mathscr{L}\{f(t - a)u(t - a)\} = \int_{0^-}^{\infty} e^{-\mathbf{s}t} f(t - a)u(t - a)dt$$

$$= \int_{a^-}^{\infty} e^{-\mathbf{s}t} f(t - a)dt$$

for $t \geq a^-$. Choosing a new variable of integration, $\tau = t - a$, we obtain

$$\mathscr{L}\{f(t - a)u(t - a)\} = \int_{0^-}^{\infty} e^{-\mathbf{s}(\tau + a)} f(\tau)d\tau = e^{-a\mathbf{s}}\mathbf{F}(\mathbf{s})$$

Therefore,

$$f(t - a)u(t - a) \Leftrightarrow e^{-a\mathbf{s}}\mathbf{F}(\mathbf{s}) \qquad (a \geq 0) \qquad [23]$$

This result is known as the *time-shift theorem,* and it simply states that if a time function is delayed by a time a in the time domain, the result in the frequency domain is a multiplication by e^{-as}.

EXAMPLE **14.9**

Determine the transform of the rectangular pulse $v(t) = u(t-2) - u(t-5)$.

This pulse, shown plotted in Fig. 14.7, has unit value for the time interval $2 < t < 5$, and zero value elsewhere. We know that the transform of $u(t)$ is just $1/\mathbf{s}$, and since $u(t-2)$ is simply $u(t)$ delayed by 2 s, the transform of this delayed function is $e^{-2\mathbf{s}}/\mathbf{s}$. Similarly, the transform of $u(t-5)$ is $e^{-5\mathbf{s}}/\mathbf{s}$. It follows, then, that the desired transform is

$$\mathbf{V}(\mathbf{s}) = \frac{e^{-2\mathbf{s}}}{\mathbf{s}} - \frac{e^{-5\mathbf{s}}}{\mathbf{s}} = \frac{e^{-2\mathbf{s}} - e^{-5\mathbf{s}}}{\mathbf{s}}$$

It was not necessary to revert to the definition of the Laplace transform in order to determine $\mathbf{V}(\mathbf{s})$.

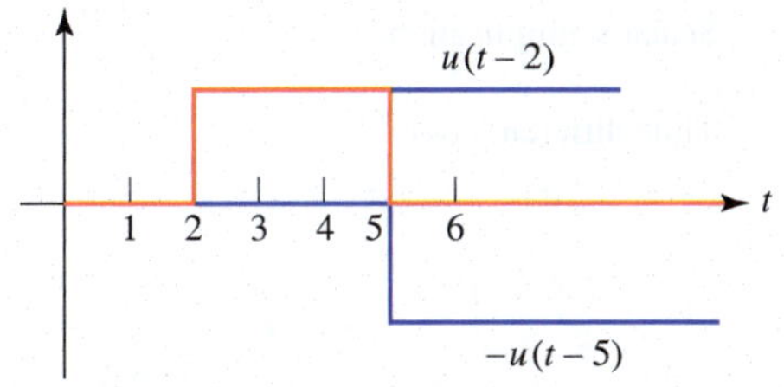

■ **FIGURE 14.7** Plot of $u(t-2) - u(t-5)$.

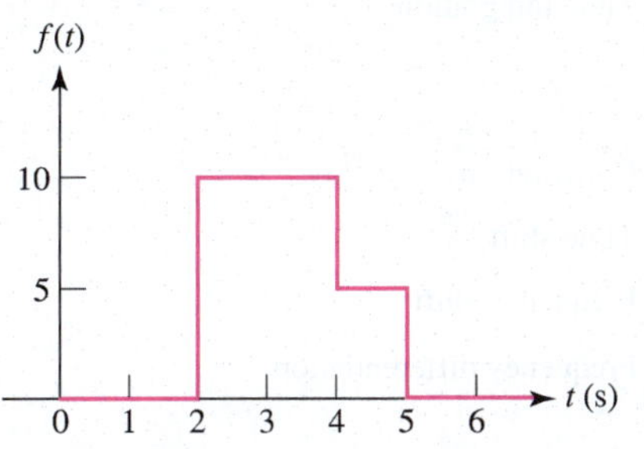

■ **FIGURE 14.8**

PRACTICE

14.12 Obtain the Laplace transform of the time function shown in Fig. 14.8.

Ans: $(5/\mathbf{s})(2e^{-2\mathbf{s}} - e^{-4\mathbf{s}} - e^{-5\mathbf{s}})$.

At this point we have obtained a number of entries for the catalog of Laplace transform pairs that we agreed to construct earlier. Included are the transforms of the *impulse function,* the *step function,* the *exponential function,* the *ramp function,* the *sine* and *cosine functions,* and the sum of two exponentials. In addition, we have noted the consequences in the **s** domain of the time-domain operations of addition, multiplication by a constant, differentiation, and integration. These results are collected in Tables 14.1 and 14.2; several others which are derived in Appendix 7 are also included.

TABLE **14.1** Laplace Transform Pairs

$f(t) = \mathscr{L}^{-1}\{\mathbf{F}(\mathbf{s})\}$	$\mathbf{F}(\mathbf{s}) = \mathscr{L}\{f(t)\}$	$f(t) = \mathscr{L}^{-1}\{\mathbf{F}(\mathbf{s})\}$	$\mathbf{F}(\mathbf{s}) = \mathscr{L}\{f(t)\}$
$\delta(t)$	1	$\frac{1}{\beta-\alpha}(e^{-\alpha t} - e^{-\beta t})u(t)$	$\frac{1}{(\mathbf{s}+\alpha)(\mathbf{s}+\beta)}$
$u(t)$	$\frac{1}{\mathbf{s}}$	$\sin \omega t\, u(t)$	$\frac{\omega}{\mathbf{s}^2+\omega^2}$
$tu(t)$	$\frac{1}{\mathbf{s}^2}$	$\cos \omega t\, u(t)$	$\frac{\mathbf{s}}{\mathbf{s}^2+\omega^2}$
$\frac{t^{n-1}}{(n-1)!}u(t), n = 1, 2, \ldots$	$\frac{1}{\mathbf{s}^n}$	$\sin(\omega t + \theta)\, u(t)$	$\frac{\mathbf{s}\sin\theta + \omega\cos\theta}{\mathbf{s}^2+\omega^2}$
$e^{-\alpha t}u(t)$	$\frac{1}{\mathbf{s}+\alpha}$	$\cos(\omega t + \theta)\, u(t)$	$\frac{\mathbf{s}\cos\theta - \omega\sin\theta}{\mathbf{s}^2+\omega^2}$
$te^{-\alpha t}u(t)$	$\frac{1}{(\mathbf{s}+\alpha)^2}$	$e^{-\alpha t}\sin \omega t\, u(t)$	$\frac{\omega}{(\mathbf{s}+\alpha)^2+\omega^2}$
$\frac{t^{n-1}}{(n-1)!}e^{-\alpha t}u(t), n = 1, 2, \ldots$	$\frac{1}{(\mathbf{s}+\alpha)^n}$	$e^{-\alpha t}\cos \omega t\, u(t)$	$\frac{\mathbf{s}+\alpha}{(\mathbf{s}+\alpha)^2+\omega^2}$

TABLE 14.2 Laplace Transform Operations

Operation	$f(t)$	$\mathbf{F}(\mathbf{s})$
Addition	$f_1(t) \pm f_2(t)$	$\mathbf{F}_1(\mathbf{s}) \pm \mathbf{F}_2(\mathbf{s})$
Scalar multiplication	$kf(t)$	$k\mathbf{F}(\mathbf{s})$
Time differentiation	$\frac{df}{dt}$	$\mathbf{s}\mathbf{F}(\mathbf{s}) - f(0^-)$
	$\frac{d^2f}{dt^2}$	$\mathbf{s}^2\mathbf{F}(\mathbf{s}) - \mathbf{s}f(0^-) - f'(0^-)$
	$\frac{d^3f}{dt^3}$	$\mathbf{s}^3\mathbf{F}(\mathbf{s}) - \mathbf{s}^2f(0^-) - \mathbf{s}f'(0^-) - f''(0^-)$
Time integration	$\int_{0^-}^{t} f(t)dt$	$\frac{1}{\mathbf{s}}\mathbf{F}(\mathbf{s})$
	$\int_{-\infty}^{t} f(t)dt$	$\frac{1}{\mathbf{s}}F(\mathbf{s}) + \frac{1}{\mathbf{s}}\int_{-\infty}^{0^-} f(t)dt$
Convolution	$f_1(t) * f_2(t)$	$\mathbf{F}_1(\mathbf{s})\mathbf{F}_2(\mathbf{s})$
Time shift	$f(t-a)u(t-a), a \geq 0$	$e^{-a\mathbf{s}}\mathbf{F}(\mathbf{s})$
Frequency shift	$f(t)e^{-at}$	$\mathbf{F}(\mathbf{s}+a)$
Frequency differentiation	$tf(t)$	$-\frac{d\mathbf{F}(\mathbf{s})}{d\mathbf{s}}$
Frequency integration	$\frac{f(t)}{t}$	$\int_{\mathbf{s}}^{\infty} \mathbf{F}(\mathbf{s})d\mathbf{s}$
Scaling	$f(at), a \geq 0$	$\frac{1}{a}\mathbf{F}\left(\frac{\mathbf{s}}{a}\right)$
Initial value	$f(0^+)$	$\lim_{\mathbf{s}\to\infty} \mathbf{s}\mathbf{F}(\mathbf{s})$
Final value	$f(\infty)$	$\lim_{\mathbf{s}\to 0} \mathbf{s}\mathbf{F}(\mathbf{s})$, all poles of $\mathbf{s}\mathbf{F}(\mathbf{s})$ in LHP
Time periodicity	$f(t) = f(t+nT), n = 1, 2, \ldots$	$\frac{1}{1-e^{-T\mathbf{s}}}\mathbf{F}_1(\mathbf{s})$, where $\mathbf{F}_1(\mathbf{s}) = \int_{0^-}^{T} f(t)e^{-\mathbf{s}t}\,dt$

14.6 THE INITIAL-VALUE AND FINAL-VALUE THEOREMS

The last two fundamental theorems that we will discuss are known as the initial-value and final-value theorems. They will enable us to evaluate $f(0^+)$ and $f(\infty)$ by examining the limiting values of $\mathbf{s}\mathbf{F}(\mathbf{s})$. Such an ability can be invaluable; if only the initial and final values are needed for a particular function of interest, there is no need to take the time to perform an inverse transform operation.

The Initial-Value Theorem

To derive the initial-value theorem, we consider the Laplace transform of the derivative once again,

$$\mathscr{L}\left\{\frac{df}{dt}\right\} = \mathbf{s}\mathbf{F}(\mathbf{s}) - f(0^-) = \int_{0^-}^{\infty} e^{-\mathbf{s}t}\frac{df}{dt}\,dt$$

We now let **s** approach infinity. By breaking the integral into two parts,

$$\lim_{\mathbf{s}\to\infty}[\mathbf{sF(s)} - f(0^-)] = \lim_{\mathbf{s}\to\infty}\left(\int_{0^-}^{0^+} e^0 \frac{df}{dt}dt + \int_{0^+}^{\infty} e^{-\mathbf{s}t}\frac{df}{dt}dt\right)$$

we see that the second integral must approach zero in the limit, since the integrand itself approaches zero. Also, $f(0^-)$ is not a function of **s**, and it may be removed from the left limit:

$$-f(0^-) + \lim_{\mathbf{s}\to\infty}[\mathbf{sF(s)}] = \lim_{\mathbf{s}\to\infty}\int_{0^-}^{0^+} df = \lim_{\mathbf{s}\to\infty}[f(0^+) - f(0^-)]$$
$$= f(0^+) - f(0^-)$$

and finally,

$$f(0^+) = \lim_{\mathbf{s}\to\infty}[\mathbf{sF(s)}]$$

or

$$\lim_{t\to 0^+} f(t) = \lim_{\mathbf{s}\to\infty}[\mathbf{sF(s)}] \qquad [24]$$

This is the mathematical statement of the ***initial-value theorem.*** It states that the initial value of the time function $f(t)$ can be obtained from its Laplace transform **F(s)** by first multiplying the transform by **s** and then letting **s** approach infinity. Note that the initial value of $f(t)$ we obtain is the limit from the right.

The initial-value theorem, along with the final-value theorem that we will consider in a moment, is useful in checking the results of a transformation or an inverse transformation. For example, when we first calculated the transform of $\cos(\omega_0 t)u(t)$, we obtained $\mathbf{s}/(\mathbf{s}^2 + \omega_0^2)$. After noting that $f(0^+) = 1$, we can make a partial check on the validity of this result by applying the initial-value theorem:

$$\lim_{\mathbf{s}\to\infty}\left(\mathbf{s}\frac{\mathbf{s}}{\mathbf{s}^2 + \omega_0^2}\right) = 1$$

and the check is accomplished.

The Final-Value Theorem

The ***final-value theorem*** is not quite as useful as the initial-value theorem, for it can be used only with a certain class of transforms. In order to determine whether a transform fits into this class, the denominator of **F(s)** must be evaluated to find all values of **s** for which it is zero, i.e., the ***poles*** of **F(s)**. Only those transforms **F(s)** whose poles lie entirely within the left half of the **s** plane (i.e., $\sigma < 0$), except for a simple pole at $\mathbf{s} = 0$, are suitable for use with the final-value theorem. We again consider the Laplace transform of df/dt,

$$\int_{0^-}^{\infty} e^{-\mathbf{s}t}\frac{df}{dt}dt = \mathbf{sF(s)} - f(0^-)$$

this time in the limit as **s** approaches zero,

$$\lim_{\mathbf{s}\to\infty}\int_{0^-}^{\infty} e^{-\mathbf{s}t}\frac{df}{dt}dt = \lim_{\mathbf{s}\to\infty}[\mathbf{sF(s)} - f(0^-)] = \int_{0^-}^{\infty}\frac{df}{dt}dt$$

We assume that both $f(t)$ and its first derivative are transformable. Now, the last term of this equation is readily expressed as a limit,

$$\int_{0^-}^{\infty} \frac{df}{dt}dt = \lim_{t\to\infty} \int_{0^-}^{t} \frac{df}{dt}dt$$
$$= \lim_{t\to\infty} [f(t) - f(0^-)]$$

By recognizing that $f(0^-)$ is a constant, a comparison of the last two equations shows us that

$$\lim_{t\to\infty} f(t) = \lim_{\mathbf{s}\to 0} [\mathbf{sF(s)}]$$

which is the ***final-value theorem.*** In applying this theorem, it is necessary to know that $f(\infty)$, the limit of $f(t)$ as t becomes infinite, exists or—what amounts to the same thing—that the poles of $\mathbf{F(s)}$ all lie *within* the left half of the $\mathbf{s}$ plane except for (possibly) a simple pole at the origin. The product $\mathbf{sF(s)}$ thus has all of its poles lying within the left half plane.

EXAMPLE 14.10

Use the final-value theorem to determine $f(\infty)$ for the function $(1 - e^{-at})u(t)$, where $a > 0$.

Without even using the final-value theorem, we see immediately that $f(\infty) = 1$. The transform of $f(t)$ is

$$\mathbf{F(s)} = \frac{1}{\mathbf{s}} - \frac{1}{\mathbf{s}+a}$$
$$= \frac{a}{\mathbf{s}(\mathbf{s}+a)}$$

The poles of $\mathbf{F(s)}$ are $\mathbf{s} = 0$ and $\mathbf{s} = -a$. Thus, the nonzero pole of $\mathbf{F(s)}$ lies in the left-hand $\mathbf{s}$-plane, as we were assured that $a > 0$; we find that we may indeed apply the final-value theorem to this function. Multiplying by $\mathbf{s}$ and letting $\mathbf{s}$ approach zero, we obtain

$$\lim_{\mathbf{s}\to 0}[\mathbf{sF(s)}] = \lim_{\mathbf{s}\to 0} \frac{a}{\mathbf{s}+a} = 1$$

which agrees with $f(\infty)$.

If $f(t)$ is a sinusoid, however, so that $\mathbf{F(s)}$ has poles on the $j\omega$ axis, then a blind use of the final-value theorem might lead us to conclude that the final value is zero. We know, however, that the final value of either $\sin\omega_0 t$ or $\cos\omega_0 t$ is indeterminate. So, beware of $j\omega$-axis poles!

PRACTICE

14.13 Without finding $f(t)$ first, determine $f(0^+)$ and $f(\infty)$ for each of the following transforms: (*a*) $4e^{-2\mathbf{s}}(\mathbf{s} + 50)/\mathbf{s}$; (*b*) $(\mathbf{s}^2 + 6)/(\mathbf{s}^2 + 7)$; (*c*) $(5\mathbf{s}^2 + 10)/[2\mathbf{s}(\mathbf{s}^2 + 3\mathbf{s} + 5)]$.

Ans: 0, 200; ∞, indeterminate (poles lie on the $j\omega$ axis); 2.5, 1.

14.7 Z(s) AND Y(s)

Having been introduced to the concept of complex frequency and to the Laplace transform technique, we now are ready to see the details of how circuit analysis in the **s**-domain actually works. As the reader might suspect, particularly if Chap. 10 has already been studied, in fact several shortcuts are routinely applied. The first of these is to create a new way of viewing capacitors and inductors, so that **s**-domain nodal and mesh equations can be written directly. As part of this method, we will learn how to take care to account for initial conditions. Another "shortcut" is the concept of a circuit transfer function. This general function can be exploited to predict the response of a circuit to various inputs, its stability, and even its frequency-selective response.

The key concept that makes phasors so useful in the analysis of sinusoidal steady-state circuits is the transformation of resistors, capacitors, and inductors into *impedances*. Circuit analysis then proceeds using the basic techniques of nodal or mesh analysis, superposition, source transformation, as well as Thévenin or Norton equivalents. This concept can be extended to the **s**-domain, since the sinusoidal steady state is included in **s**-domain analysis as a special case (where $\sigma = 0$).

Resistors in the Frequency Domain

Let's begin with the simplest situation: a resistor connected to a voltage source $v(t)$. Ohm's law specifies that

$$v(t) = Ri(t)$$

Taking the Laplace transform of both sides,

$$\mathbf{V}(\mathbf{s}) = R\mathbf{I}(\mathbf{s})$$

Thus, the ratio of the frequency-domain representation of the voltage to the frequency-domain representation of the current is simply the resistance, R. Since we are working in the frequency domain, we refer to this quantity as an *impedance* for the sake of clarity, but we still assign it the unit ohms (Ω):

$$\mathbf{Z}(\mathbf{s}) \equiv \frac{\mathbf{V}(\mathbf{s})}{\mathbf{I}(\mathbf{s})} = R \qquad [25]$$

Just as we found in working with phasors in the sinusoidal steady state, the impedance of a resistor does not depend on frequency. The *admittance* $\mathbf{Y}(\mathbf{s})$ of a resistor, defined as the ratio of $\mathbf{I}(\mathbf{s})$ to $\mathbf{V}(\mathbf{s})$, is simply $1/R$; the unit of admittance is the siemen (S).

Inductors in the Frequency Domain

Next, we consider an inductor connected to some time-varying voltage source $v(t)$, as shown in Fig. 14.9*a*. Since

$$v(t) = L\frac{di}{dt}$$

taking the Laplace transform of both sides of this equation yields

$$\mathbf{V}(\mathbf{s}) = L[\mathbf{s}\mathbf{I}(\mathbf{s}) - i(0^-)] \qquad [26]$$

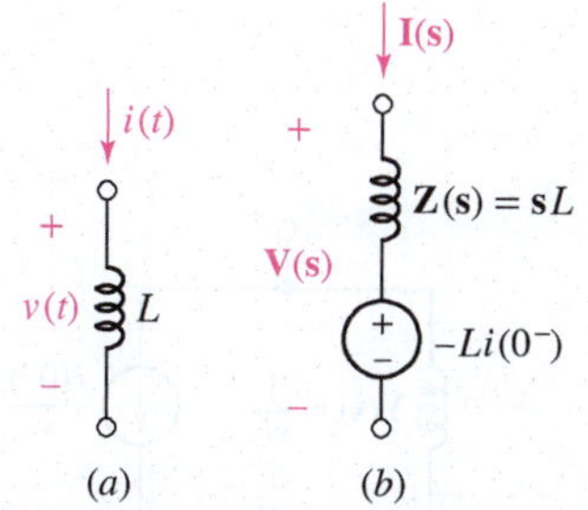

FIGURE 14.9 (*a*) Inductor in the time domain. (*b*) The complete model for an inductor in the frequency domain, consisting of an impedance **s***L* and a voltage source $-Li(0^-)$ that incorporates the effects of nonzero initial conditions on the element.

We now have two terms: $\mathbf{s}L\mathbf{I}(\mathbf{s})$ and $Li(0^-)$. In situations where we have zero initial energy stored in the inductor (i.e., $i(0^-) = 0$), then

$$\mathbf{V}(\mathbf{s}) = \mathbf{s}L\mathbf{I}(\mathbf{s})$$

so that

$$\mathbf{Z}(\mathbf{s}) \equiv \frac{\mathbf{V}(\mathbf{s})}{\mathbf{I}(\mathbf{s})} = \mathbf{s}L \qquad [27]$$

Equation [27] may be further simplified if we are only interested in the sinusoidal steady-state response. It is permissible to neglect the initial conditions in such instances as they only affect the nature of the transient response. Thus, we substitute $\mathbf{s} = j\omega$ and find

$$\mathbf{Z}(j\omega) = j\omega L$$

as was obtained previously in Chap. 10.

Modeling Inductors in the s-Domain

Although we refer to the quantity in Eq. [27] as the impedance of an inductor, we must remember that it was obtained by assuming zero initial current. In the more general situation where energy is stored in the element at $t = 0^-$, this quantity is not sufficient to represent the inductor in the frequency domain. Fortunately, it is possible to include the initial condition by modeling an inductor as an impedance in combination with either a voltage or current source. To do this, we begin by rearranging Eq. [26] as

$$\mathbf{V}(\mathbf{s}) = \mathbf{s}L\mathbf{I}(\mathbf{s}) - Li(0^-) \qquad [28]$$

The second term on the right will be a constant: the inductance L in henrys multiplied by the initial current $i(0^-)$ in amperes. The result is a constant voltage term that is subtracted from the frequency-dependent term $\mathbf{s}L\mathbf{I}(\mathbf{s})$. A small leap of intuition at this point leads us to the realization that we can model a single inductor L as a two-component frequency-domain element, as shown in Fig. 14.9*b*.

The frequency-domain inductor model shown in Fig. 14.9*b* consists of an impedance $\mathbf{s}L$ and a voltage source $Li(0^-)$. The voltage across the impedance $\mathbf{s}L$ is given by Ohm's law as $\mathbf{s}L\mathbf{I}(\mathbf{s})$. Since the two-element combination in Fig. 14.9*b* is linear, every circuit analysis technique previously explored can be brought to bear in the s-domain as well. For example, it is possible to perform a source transformation on the model in order to obtain an impedance $\mathbf{s}L$ in parallel with a current source $[-Li(0^-)]/\mathbf{s}L = -i(0^-)/\mathbf{s}$. This can be verified by taking Eq. [28] and solving for $\mathbf{I}(\mathbf{s})$:

$$\mathbf{I}(\mathbf{s}) = \frac{\mathbf{V}(\mathbf{s}) + Li(0^-)}{\mathbf{s}L}$$
$$= \frac{\mathbf{V}(\mathbf{s})}{\mathbf{s}L} + \frac{i(0^-)}{\mathbf{s}} \qquad [29]$$

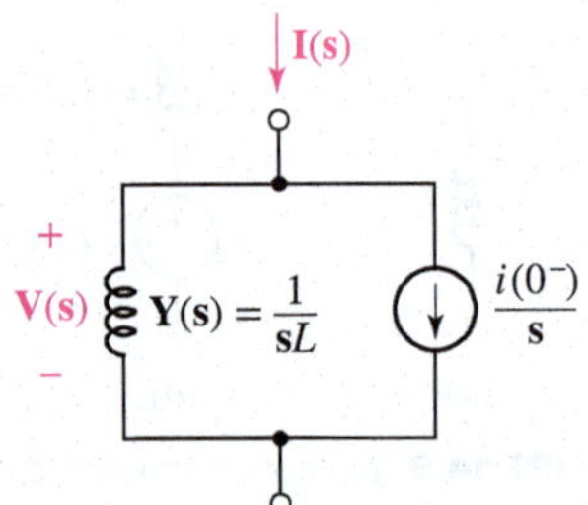

FIGURE 14.10 An alternative frequency-domain model for the inductor, consisting of an admittance $1/\mathbf{s}L$ and a current source $i(0^-)/\mathbf{s}$.

We are once again left with two terms. The first term on the right is simply an admittance $1/\mathbf{s}L$ times the voltage $\mathbf{V}(\mathbf{s})$. The second term on the right is a current, although it has units of ampere·seconds. Thus, we can model this equation with two separate components: an admittance $1/\mathbf{s}L$ in parallel with a current source $i(0^-)/\mathbf{s}$; the resulting model is shown in Fig. 14.10. The choice of whether to use the model of Fig. 14.9*b* or that shown in Fig. 14.10

is usually made depending on which one will result in simpler equations. Note that although Fig. 14.10 shows the inductor symbol labeled with an admittance $\mathbf{Y(s)} = 1/\mathbf{s}L$, it can also be viewed as an impedance $\mathbf{Z(s)} = \mathbf{s}L$; again, the choice of which to use is generally based on personal preference.

A brief comment on units is in order. When we take the Laplace transform of a current $i(t)$, we are integrating over time. Thus, the units of $\mathbf{I(s)}$ are technically ampere·seconds; in a similar fashion, the units of $\mathbf{V(s)}$ are volt·seconds. However, it is the convention to drop the seconds and assign $\mathbf{I(s)}$ the units of amperes, and to measure $\mathbf{V(s)}$ in volts. This convention does not present any problems until we scrutinize an equation such as Eq. [29] and see a term like $i(0^-)/\mathbf{s}$ seemingly in conflict with the units of $\mathbf{I(s)}$ on the left-hand side. Although we will continue to measure these phasor quantities in "amperes" and "volts," when checking the units of an equation to verify algebra, we must remember the seconds!

EXAMPLE 14.11

Calculate the voltage $v(t)$ shown in Fig. 14.11a, given an initial current $i(0^-) = 1$ A.

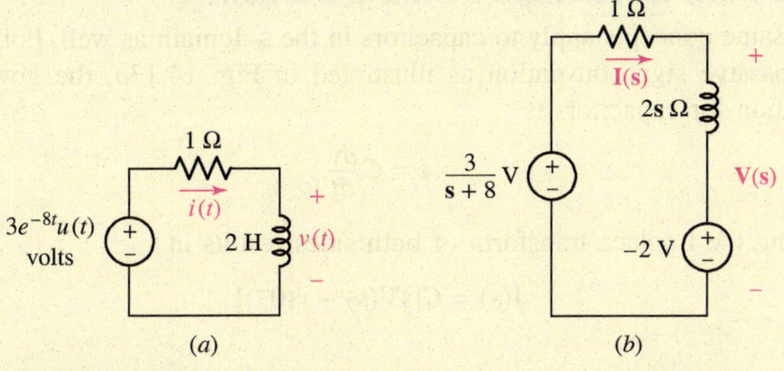

■ **FIGURE 14.11** (a) A simple resistor-inductor circuit for which the voltage $v(t)$ is desired. (b) The equivalent frequency-domain circuit, including the initial current in the inductor through the use of a series voltage source $-Li(0^-)$.

We begin by first converting the circuit in Fig. 14.11a to its frequency-domain equivalent, shown in Fig. 14.11b; the inductor has been replaced with a two-component model: an impedance $\mathbf{s}L = 2\mathbf{s}\ \Omega$ and an independent voltage source $-Li(0^-) = -2$ V.

We seek the quantity labeled $\mathbf{V(s)}$, as its inverse transform will result in $v(t)$. Note that $\mathbf{V(s)}$ appears across the *entire* inductor model and not just the impedance component.

Taking a straightforward route, we write

$$\mathbf{I(s)} = \frac{\dfrac{3}{\mathbf{s}+8}+2}{1+2\mathbf{s}} = \frac{\mathbf{s}+9.5}{(\mathbf{s}+8)(\mathbf{s}+0.5)}$$

and

$$\mathbf{V(s)} = 2\mathbf{s}\,\mathbf{I(s)} - 2$$

so that

$$\mathbf{V(s)} = \frac{2\mathbf{s}(\mathbf{s}+9.5)}{(\mathbf{s}+8)(\mathbf{s}+0.5)} - 2$$

(Continued on next page)

Before attempting to take the inverse Laplace transform of this expression, it is well worth a little time and effort to simplify it first. Thus,

$$\mathbf{V(s)} = \frac{2\mathbf{s} - 8}{(\mathbf{s}+8)(\mathbf{s}+0.5)}$$

Employing the technique of partial-fraction expansion (on paper or with the assistance of MATLAB), we find that

$$\mathbf{V(s)} = \frac{3.2}{\mathbf{s}+8} - \frac{1.2}{\mathbf{s}+0.5}$$

Referring to Table 14.1, then, the inverse transform is found to be

$$v(t) = [3.2e^{-8t} - 1.2e^{-0.5t}]u(t) \text{ V}$$

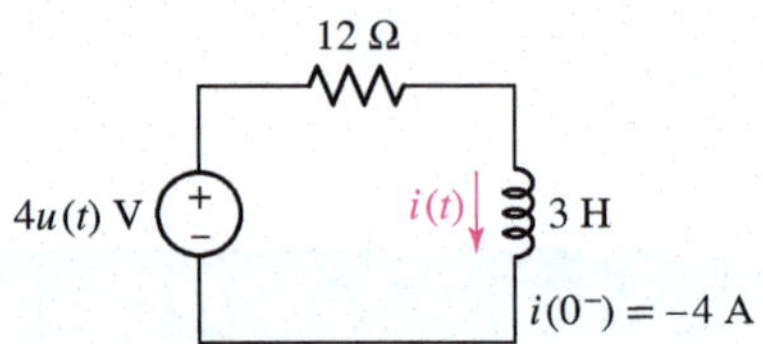

■ **FIGURE 14.12**

PRACTICE

14.14 Determine the current $i(t)$ in the circuit of Fig. 14.12.

Ans: $\frac{1}{3}[1 - 13e^{-4t}]u(t)$ A.

Modeling Capacitors in the s-Domain

The same concepts apply to capacitors in the **s**-domain as well. Following the passive sign convention as illustrated in Fig. 14.13*a*, the governing equation for capacitors is

$$i = C\frac{dv}{dt}$$

Taking the Laplace transform of both sides results in

$$\mathbf{I(s)} = C[\mathbf{sV(s)} - v(0^-)]$$

or

$$\mathbf{I(s)} = \mathbf{s}C\mathbf{V(s)} - Cv(0^-) \qquad [30]$$

which can be modeled as an admittance $\mathbf{s}C$ in parallel with a current source $Cv(0^-)$ as shown in Fig. 14.13*b*. Performing a source transformation on this circuit (taking care to follow the passive sign convention) results in an equivalent model for the capacitor consisting of an impedance $1/\mathbf{s}C$ in series with a voltage source $v(0^-)/\mathbf{s}$, as shown in Fig. 14.13*c*.

In working with these **s**-domain equivalents, we should be careful not to be confused with the independent sources being used to include initial

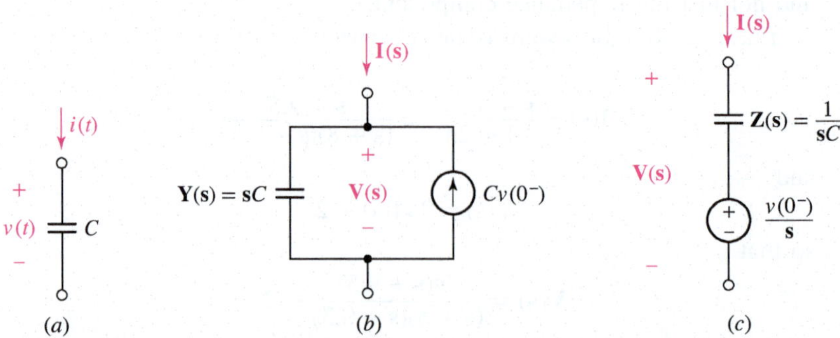

■ **FIGURE 14.13** (*a*) Capacitor in the time domain, with $v(t)$ and $i(t)$ labeled. (*b*) Frequency-domain model of a capacitor with initial voltage $v(0^-)$. (*c*) An equivalent model obtained by performing a source transformation.

conditions. The initial condition for an inductor is given as $i(0^-)$; this term may appear as part of either a voltage source or a current source, depending on which model is chosen. The initial condition for a capacitor is given as $v(0^-)$; this term may thus appear as part of either a voltage source or a current source. A very common mistake for students working with **s**-domain analysis for the first time is to always use $v(0^-)$ for the voltage source component of the model, even when dealing with an inductor.

EXAMPLE 14.12

Determine $v_C(t)$ in the circuit of Fig. 14.14*a*, given an initial voltage $v_C(0^-) = -2$ V.

▶ ***Identify the goal of the problem.***

An expression for the capacitor voltage, $v_C(t)$.

▶ ***Collect the known information.***

The problem specifies an initial capacitor voltage of −2 V.

▶ ***Devise a plan.***

Our first step is to draw the equivalent **s**-domain circuit; in doing so, we must choose between the two possible capacitor models. With no clear benefit in choosing one over the other, we select the current-source-based model, as in Fig. 14.14*b*.

▶ ***Construct an appropriate set of equations.***

We proceed with the analysis by writing a single nodal equation:

$$-1 = \frac{\mathbf{V}_C}{2/\mathbf{s}} + \frac{\mathbf{V}_C - 9/\mathbf{s}}{3}$$

▶ ***Determine if additional information is required.***

We have one equation in one unknown, the frequency-domain representation of the desired capacitor voltage.

▶ ***Attempt a solution.***

Solving for $\mathbf{V}_C$, we find that

$$\mathbf{V}_C = \frac{18/\mathbf{s} - 6}{3\mathbf{s} + 2} = -2\frac{\mathbf{s} - 3}{\mathbf{s}(\mathbf{s} + 2/3)}$$

Partial fraction expansion yields

$$\mathbf{V}_C = \frac{9}{\mathbf{s}} - \frac{11}{\mathbf{s} + 2/3}$$

We obtain $v_C(t)$ by taking the inverse Laplace transform of this expression, resulting in

$$v_C(t) = 9u(t) - 11e^{-2t/3}u(t) \text{ V}$$

or, more compactly,

$$v_C(t) = [9 - 11e^{-2t/3}]u(t) \text{ V}$$

▶ ***Verify the solution. Is it reasonable or expected?***

A quick check for $t = 0$ yields $v_C(t) = -2$ V, as it should based on our knowledge of the initial condition. Also, as $t \to \infty$, $v_C(t) \to 9$ V, as we would expect from Fig. 14.14*a* once the transient has died out.

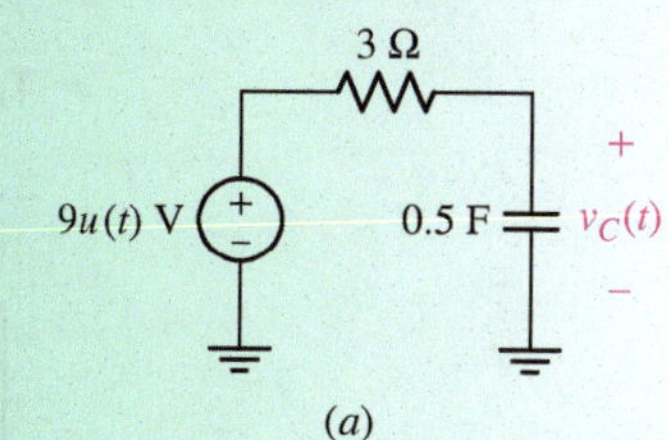

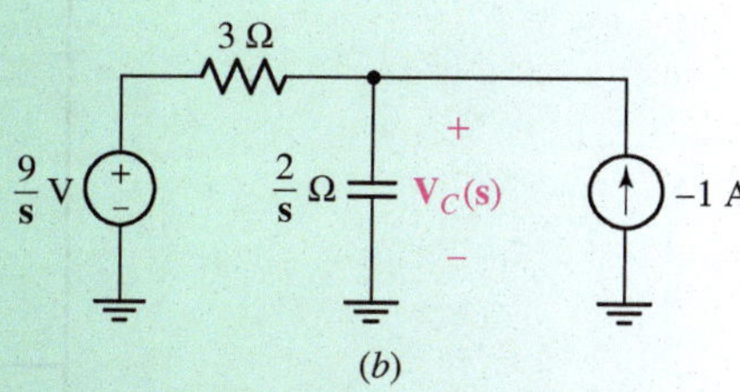

■ **FIGURE 14.14** (*a*) A circuit for which the current $v_C(t)$ is required. (*b*) Frequency-domain equivalent circuit, employing the current-source-based model to account for the initial condition of the capacitor.

(Continued on next page)

PRACTICE

14.15 Repeat Example 14.12 using the voltage-source–based capacitor model.

Ans: $[9 - 11e^{-2t/3}]u(t)$ V.

The results of this section are summarized in Table 14.3. Note that in each case, we have assumed the passive sign convention to relate voltage polarity and current direction.

TABLE **14.3** Summary of Element Representations in the Time and Frequency Domains

Time Domain	Frequency Domain	
Resistor $v(t) = Ri(t)$ 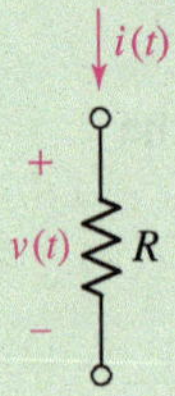	$\mathbf{V(s)} = R\mathbf{I(s)}$; 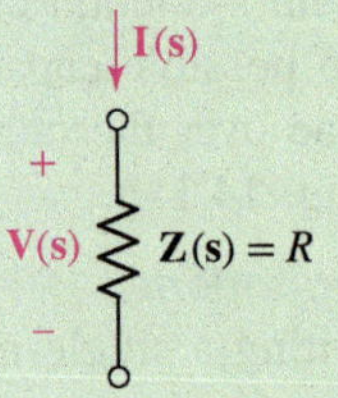 $\mathbf{Z(s)} = R$	$\mathbf{I(s)} = \frac{1}{R}\mathbf{V(s)}$; 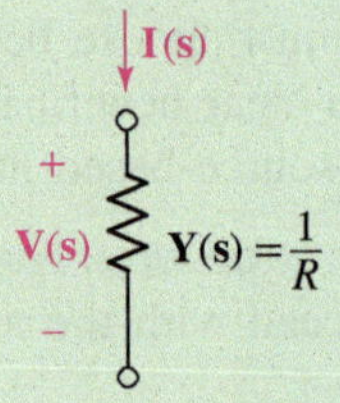 $\mathbf{Y(s)} = \frac{1}{R}$
Inductor $v(t) = L\frac{di}{dt}$ 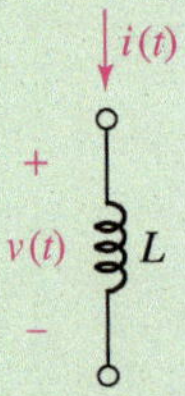	$\mathbf{V(s)} = \mathbf{s}L\mathbf{I(s)} - Li(0^-)$; 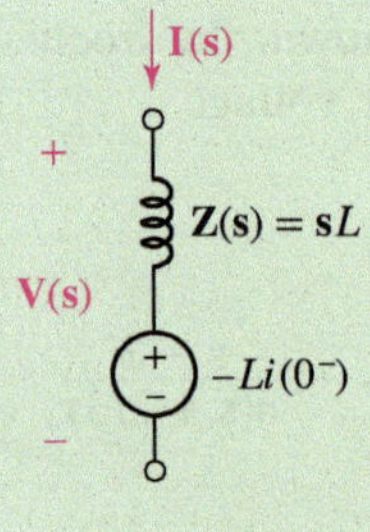 $\mathbf{Z(s)} = \mathbf{s}L$; source $-Li(0^-)$	$\mathbf{I(s)} = \frac{\mathbf{V(s)}}{\mathbf{s}L} + \frac{i(0^-)}{\mathbf{s}}$; $\mathbf{Y(s)} = \frac{1}{\mathbf{s}L}$; source $\frac{i(0^-)}{\mathbf{s}}$ 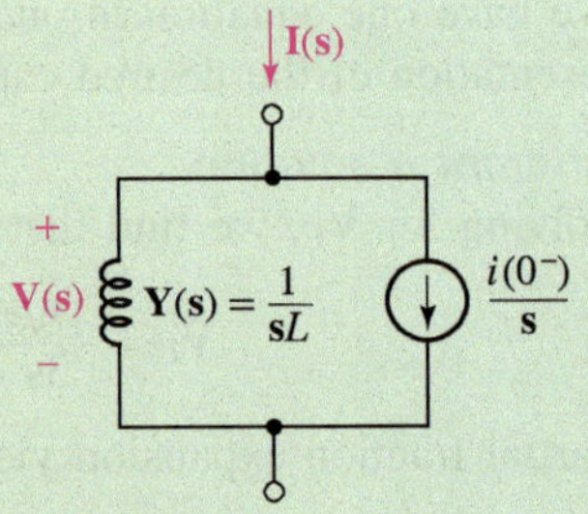
Capacitor $i(t) = C\frac{dv}{dt}$ 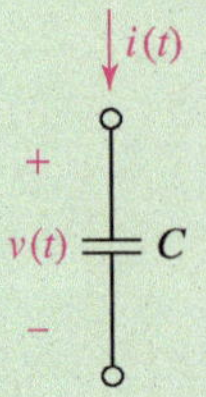	$\mathbf{V(s)} = \frac{\mathbf{I(s)}}{\mathbf{s}C} + \frac{v(0^-)}{\mathbf{s}}$; 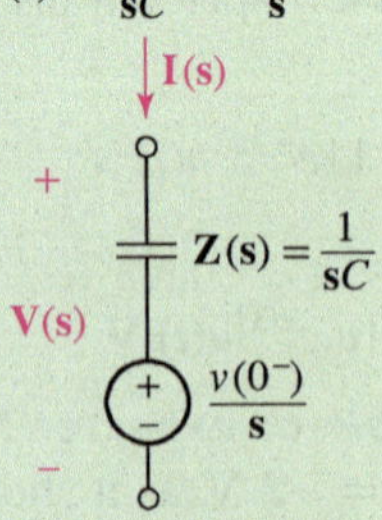 $\mathbf{Z(s)} = \frac{1}{\mathbf{s}C}$; source $\frac{v(0^-)}{\mathbf{s}}$	$\mathbf{I(s)} = \mathbf{s}C\mathbf{V(s)} - Cv(0^-)$; 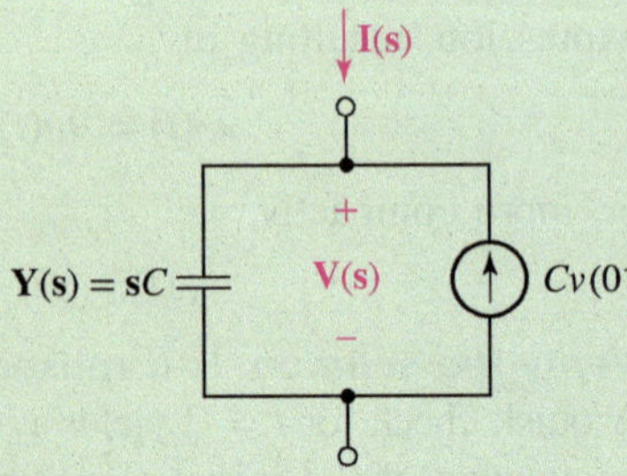 $\mathbf{Y(s)} = \mathbf{s}C$; source $Cv(0^-)$

14.8 NODAL AND MESH ANALYSIS IN THE s-DOMAIN

In Chap. 10, we learned how to transform time-domain circuits driven by sinusoidal sources into their frequency-domain equivalents. The benefits of this transformation were immediately evident, as we were no longer required to solve integrodifferential equations. Nodal and mesh analysis of such circuits (restricted to determining only the steady-state response) resulted in algebraic expressions in terms of $j\omega$, where ω is the frequency of the sources.

We have now seen that the concept of impedance can be extended to the more general case of complex frequency ($\mathbf{s} = \sigma + j\omega$). Once we transform circuits from the time domain into the frequency domain, performing nodal or mesh analysis will once again result in purely algebraic expressions, this time in terms of the complex frequency **s**. Solution of the resulting equations requires the use of variable substitution, or software capable of symbolic algebra manipulation (e.g., MATLAB). In this section, we present two examples of reasonable complexity so that we may examine these issues in greater detail. First, however, we consider how MATLAB can be used to assist us in such endeavors.

COMPUTER-AIDED ANALYSIS

In Sec. 14.4, we saw that MATLAB can be used to determine the residues of rational functions in the **s**-domain, making the inverse Laplace transform process significantly easier. However, the software package is actually much more powerful, having many built-in routines for the manipulation of algebraic expressions. In fact, as we will see in this example, MATLAB is even capable of performing inverse Laplace transforms directly from the rational functions we obtain through circuit analysis.

Let's begin by seeing how MATLAB can be used to work with algebraic expressions. These expressions are stored as character strings, with the apostrophe symbol (') used in the defining expression. For example, we previously represented the polynomial $\mathbf{p(s)} = \mathbf{s}^3 - 12\mathbf{s} + 6$ as a vector:

```
≫ p = [1 0 -12 6];
```

However, we can also represent it symbolically:

```
SYMS p s
≫ p = s^3 -12*s + 6;
```

These two representations are not equal in MATLAB; they are two distinct concepts. When we wish to manipulate an algebraic expression *symbolically,* the second representation is necessary. This ability is especially useful in working with simultaneous equations.

(Continued on next page)

Consider the set of equations

$$(2\mathbf{s} + 10)\mathbf{I}_1 - 10\mathbf{I}_2 = \frac{4}{\mathbf{s} + 2}$$

$$-10\mathbf{I}_1 + (4\mathbf{s} + 10)\mathbf{I}_2 = -\frac{2}{\mathbf{s} + 1}$$

Using MATLAB's symbolic notation, we define two string variables:

```
>> syms eqn1 eqn2 I1 I2 s
>> eqn1 = (3*s + 10)*I1 – 10*I2 = 4/(s + 2);
>> eqn2 = –10*I1 + (4*s + 10)*I2 = –2/(s + 1);
```

Note that the entire equation has been included in each string; our goal is to solve the two equations for the variables I1 and I2. MATLAB provides a special routine, *solve*(), that can manipulate the equations for us. It is invoked by listing the separate equations (defined as strings), followed by a list of the unknowns (also defined as strings):

```
>> solution = solve(eqn1, eqn2, I1, I2);
```

The answer is stored in the variable *solution,* although in a somewhat unexpected form. MATLAB returns the answer in what is termed a structure, a construct that will be familiar to C programmers. At this stage, however, all we need to know is how to extract our answer. If we type

```
>> solution.I1
```

we obtain the response

```
(4*s + 9)/((2*s + 15)*(s + 1)*(s + 2))
```

indicating that an s-polynomial expression has been assigned to the variable I1. You can also express the result in a nicer-looking format using the function *pretty*().

```
>> pretty(solution.I1)
(4 s + 9)
-----------------------------
(2s + 15) (s + 1) (s + 2)
```

A similar operation is used for the variable I2.

We can now proceed directly to determining the inverse Laplace transform using the function *ilaplace*():

```
>> i1 = ilaplace(solution.I1)
i1 = (5*exp(-t))/13 - exp(-2*t)/11 - (42*exp(-(15*t)/2))/143
```

In this manner, we can quickly obtain the solution to simultaneous equations resulting from nodal or mesh analysis, and we can also obtain the inverse Laplace transforms. The command *ezplot*(i1) allows us to see what the solution looks like, if we're so inclined. It should be noted that complicated expressions sometimes may confuse MATLAB; in such situations, *ilaplace*() may not return a useful answer.

It is worth mentioning a few related functions, as they can also be used to quickly check answers obtained by hand. The function *numden*() converts a rational function into two separate variables: one containing the numerator and the other containing the denominator. For example,

```
≫ [N,D] = numden(solution.I1)
```

returns two algebraic expressions stored in N and D, respectively:

```
N =
4*s + 9
D = (2*s + 15)(s + 1)*(s + 2)
```

In order to apply our previous experience with the function *residue*(), we need to convert each symbolic (string) expression into a vector containing the coefficients of the polynomial. This is achieved using the command *sym2poly*():

```
≫ n = sym2poly (N)
n =
4 9
and
≫ d = sym2poly (D)
d =
2 21 49 30
```

after which we can determine the residues:

```
≫ [r p y] =residue (n, d)
r =
  -0.2937
  -0.0909
   0.3846
p =
  -7.5000
  -2.0000
  -1.0000
y =
[ ]
```

which is in agreement with what we obtained using *ilaplace*().

With these new MATLAB skills, (or a deep-seated desire to try an alternative approach such as Cramer's rule or direct substitution), we are ready to proceed to analyze a few circuits.

EXAMPLE 14.13

Determine the two mesh currents i_1 and i_2 in the circuit of Fig. 14.15*a*. There is no energy initially stored in the circuit.

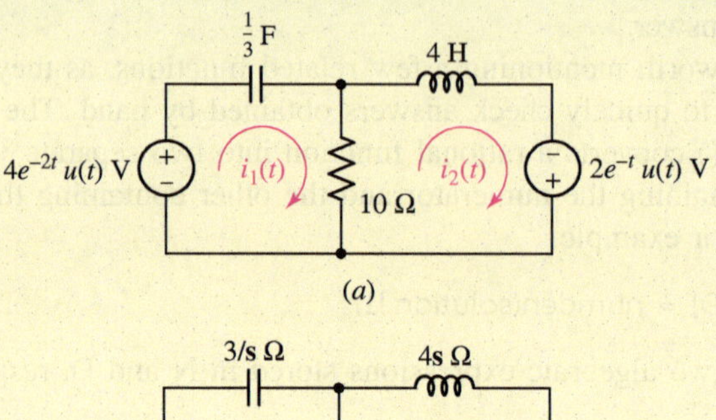

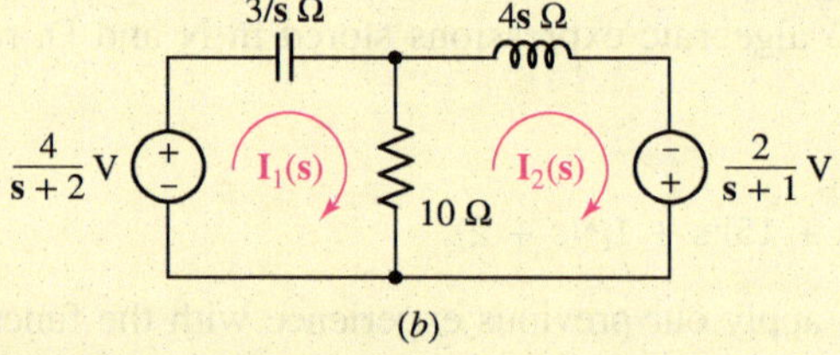

■ **FIGURE 14.15** (*a*) A two-mesh circuit for which the individual mesh currents are desired. (*b*) The frequency-domain equivalent circuit.

As always, our first step is to draw the appropriate frequency-domain equivalent circuit. Since we have zero energy stored in the circuit at $t = 0^-$, we replace the $\frac{1}{3}$F capacitor with a 3/s Ω impedance, and the 4 H inductor with a 4s Ω impedance, as shown in Fig. 14.15*b*.

Next, we write two mesh equations just as we have before:

$$-\frac{4}{\mathbf{s}+2} + \frac{3}{\mathbf{s}}\mathbf{I}_1 + 10\mathbf{I}_1 - 10\mathbf{I}_2 = 0$$

or

$$\left(\frac{3}{\mathbf{s}} + 10\right)\mathbf{I}_1 - 10\mathbf{I}_2 = \frac{4}{\mathbf{s}+2} \qquad \text{(mesh 1)}$$

and

$$\frac{-2}{\mathbf{s}+1} + 10\mathbf{I}_2 - 10\mathbf{I}_1 + 4\mathbf{s}\mathbf{I}_2 = 0$$

or

$$-10\mathbf{I}_1 + (4\mathbf{s} + 10)\mathbf{I}_2 = \frac{2}{\mathbf{s}+1} \qquad \text{(mesh 2)}$$

Solving for $\mathbf{I}_1$ and $\mathbf{I}_2$, we find that

$$\mathbf{I}_1 = \frac{2\mathbf{s}(4\mathbf{s}^2 + 19\mathbf{s} + 20)}{20\mathbf{s}^4 + 66\mathbf{s}^3 + 73\mathbf{s}^2 + 57\mathbf{s} + 30} \text{ A}$$

and

$$\mathbf{I}_2 = \frac{30\mathbf{s}^2 + 43\mathbf{s} + 6}{(\mathbf{s}+2)(20\mathbf{s}^3 + 26\mathbf{s}^2 + 21\mathbf{s} + 15)} \text{ A}$$

All that remains is for us to take the inverse Laplace transform of each function, after which we find that

$$i_1(t) = -96.39e^{-2t} - 344.8e^{-t} + 841.2e^{-0.15t}\cos 0.8529t$$
$$= +197.7e^{-0.15t}\sin 0.8529t \text{ mA}$$

and

$$i_2(t) = -481.9e^{-2t} - 241.4e^{-t} + 723.3e^{-0.15t}\cos 0.8529t$$
$$= +472.8e^{-0.15t}\sin 0.8529t \text{ mA}$$

We were (indirectly) told that no current flows through the inductor at $t = 0^-$. Therefore, $i_2(0^-) = 0$, and consequently $i_2(0^+)$ must be zero as well. Does this result hold true for our answer?

PRACTICE

14.16 Find the mesh currents i_1 and i_2 in the circuit of Fig. 14.16. You may assume no energy is stored in the circuit at $t = 0^-$.

Ans: $i_1 = e^{-2t/3}\cos\left(\frac{4}{3}\sqrt{2}t\right) + (\sqrt{2}/8)e^{-2t/3}\sin\left(\frac{4}{3}\sqrt{2}t\right)$ A; $i_2 = -\frac{2}{3} + \frac{2}{3}e^{-2t/3}\cos\left(\frac{4}{3}\sqrt{2}t\right) + (13\sqrt{2}/24)e^{-2t/3}\sin\left(\frac{4}{3}\sqrt{2}t\right)$ A.

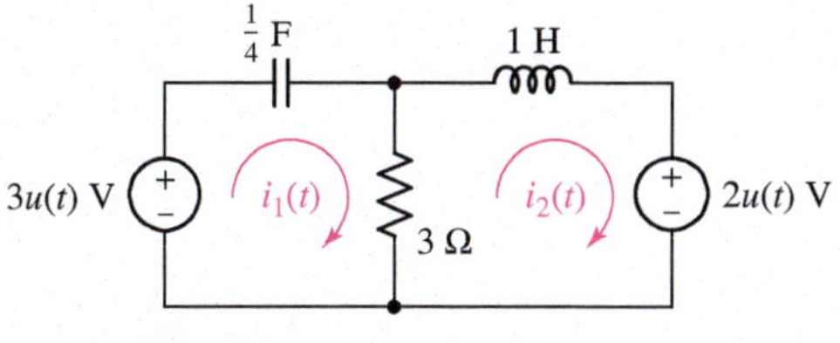

■ **FIGURE 14.16**

EXAMPLE 14.14

Calculate the voltage v_x in the circuit of Fig. 14.17 using nodal analysis techniques.

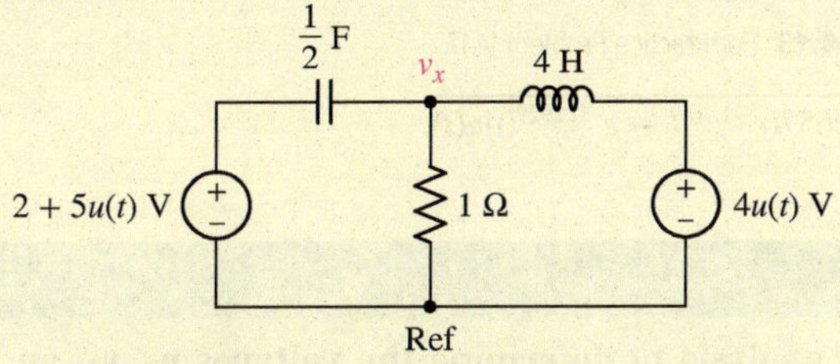

■ **FIGURE 14.17** A simple four-node circuit containing two energy storage elements.

The first step is to draw the corresponding **s**-domain circuit. We see that the $\frac{1}{2}$**F** capacitor has an initial voltage of 2 V across it at $t = 0^-$, requiring that we employ one of the two models of Fig. 14.13. Since we are to use nodal analysis, perhaps the model of Fig. 14.13*b* is the better route. The resulting circuit is shown in Fig. 14.18.

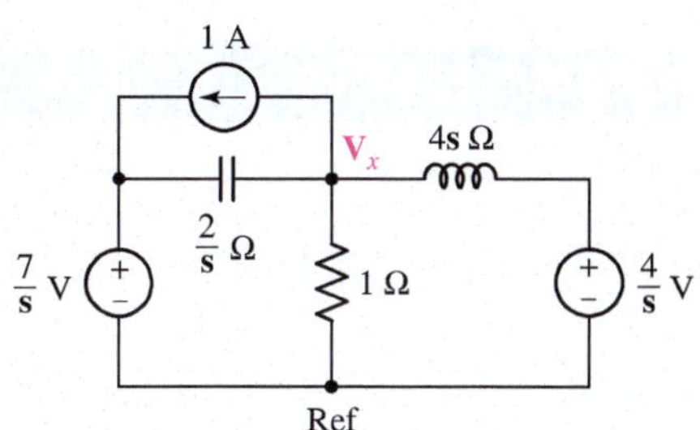

■ **FIGURE 14.18** The **s**-domain equivalent circuit of Fig. 14.17.

With two of the three nodal voltages specified, we have only one nodal equation to write:

$$-1 = \frac{\mathbf{V}_x - 7/\mathbf{s}}{2/\mathbf{s}} + \mathbf{V}_x + \frac{\mathbf{V}_x - 4/\mathbf{s}}{4\mathbf{s}}$$

so that

$$\mathbf{V}_x = \frac{10\mathbf{s}^2 + 4}{\mathbf{s}(2\mathbf{s}^2 + 4\mathbf{s} + 1)} = \frac{5\mathbf{s}^2 + 2}{\mathbf{s}\left(\mathbf{s} + 1 + \frac{\sqrt{2}}{2}\right)\left(\mathbf{s} + 1 - \frac{\sqrt{2}}{2}\right)}$$

The nodal voltage v_x is found by taking the inverse Laplace transform, and we find that

$$v_x = [4 + 6.864e^{+1.707t} - 5.864e^{-0.2929t}]\,u(t)$$

(Continued on next page)

or

$$v_x = \left[4 - e^{-t}\left(9\sqrt{2}\,\sinh\frac{\sqrt{2}}{2}t - \cosh\frac{\sqrt{2}}{2}t\right)\right]u(t)$$

Is our answer correct? One way to check is to evaluate the capacitor voltage at $t = 0$, since we know it to be 2 V. Thus,

$$\mathbf{V}_C = \frac{7}{\mathbf{s}} - \mathbf{V}_x = \frac{4\mathbf{s}^2 + 28\mathbf{s} + 3}{\mathbf{s}(2\mathbf{s}^2 + 4\mathbf{s} + 1)}$$

Multiplying $\mathbf{V}_C$ by $\mathbf{s}$ and taking the limit as $\mathbf{s} \to \infty$, we find that

$$v_c(0^+) = \lim_{\mathbf{s}\to\infty}\left[\frac{4\mathbf{s}^2 + 28\mathbf{s} + 3}{2\mathbf{s}^2 + 4\mathbf{s} + 1}\right] = 2\ \text{V}$$

as expected.

PRACTICE

14.17 Employ nodal analysis to calculate $v_x(t)$ for the circuit of Fig. 14.19.

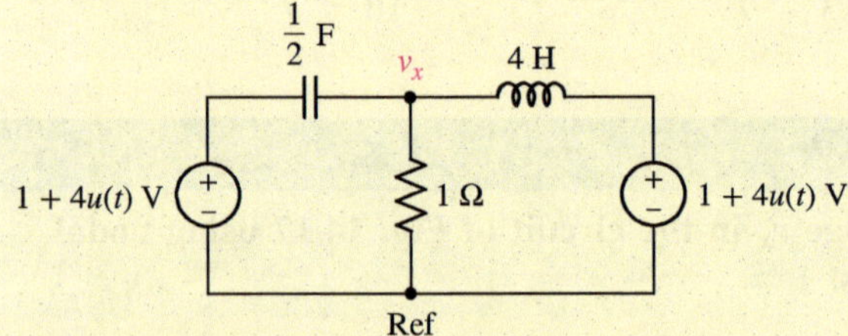

FIGURE 14.19 For Practice Problem 14.17.

Ans: $[5 + 5.657(e^{-1.707t} - e^{-0.2929t})]u(t)$.

EXAMPLE 14.15

Use nodal analysis to determine the voltages v_1, v_2, and v_3 in the circuit of Fig. 14.20*a*. No energy is stored in the circuit at $t = 0^-$.

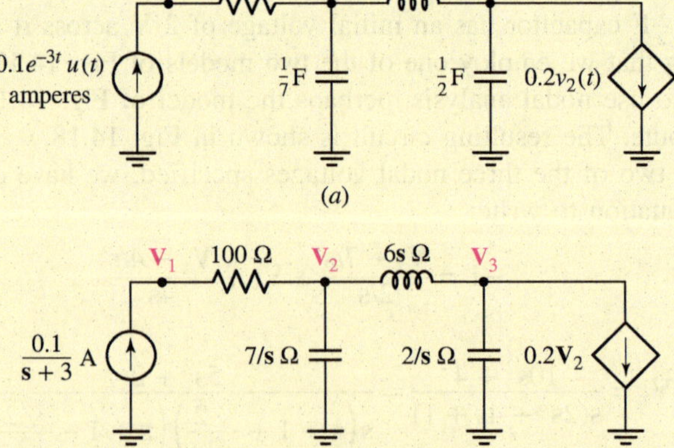

FIGURE 14.20 (*a*) A four-node circuit containing two capacitors and one inductor, none of which are storing energy at $t = 0^-$. (*b*) The frequency-domain equivalent circuit.

This circuit consists of three separate energy storage elements, none of which is storing any energy at $t = 0^-$. Thus, each may be replaced by its corresponding impedance as shown in Fig. 14.20*b*. We also note the presence of a dependent current source controlled by the nodal voltage $v_2(t)$.

Beginning at node 1, we write the following equation:

$$\frac{0.1}{\mathbf{s}+3} = \frac{\mathbf{V}_1 - \mathbf{V}_2}{100}$$

or

$$\frac{10}{\mathbf{s}+3} = \mathbf{V}_1 - \mathbf{V}_2 \qquad \text{(node 1)}$$

and at node 2,

$$0 = \frac{\mathbf{V}_2 - \mathbf{V}_1}{100} + \frac{\mathbf{V}_2}{7/\mathbf{s}} + \frac{\mathbf{V}_2 - \mathbf{V}_3}{6\mathbf{s}}$$

or

$$-42\mathbf{s}\mathbf{V}_1 + (600\mathbf{s}^2 + 42\mathbf{s} + 700)\mathbf{V}_2 - 700\mathbf{V}_3 = 0 \qquad \text{(node 2)}$$

and finally, at node 3,

$$-0.2\mathbf{V}_2 = \frac{\mathbf{V}_3 - \mathbf{V}_2}{6\mathbf{s}} + \frac{\mathbf{V}_3}{2/\mathbf{s}}$$

or

$$(1.2\mathbf{s} - 1)\mathbf{V}_2 + (3\mathbf{s}^2 + 1)\mathbf{V}_3 = 0$$

Solving this set of equations for the nodal voltages, we obtain

$$\mathbf{V}_1 = 3\frac{100\mathbf{s}^3 + 7\mathbf{s}^2 + 150\mathbf{s} + 49}{(\mathbf{s}+3)(30\mathbf{s}^3 + 45\mathbf{s} + 14)}$$

$$\mathbf{V}_2 = 7\frac{3\mathbf{s}^2 + 1}{(\mathbf{s}+3)(30\mathbf{s}^3 + 45\mathbf{s} + 14)}$$

$$\mathbf{V}_3 = -1.4\frac{6\mathbf{s} - 5}{(\mathbf{s}+3)(30\mathbf{s}^3 + 45\mathbf{s} + 14)}$$

The only remaining step is to take the inverse Laplace transform of each voltage, so that, for $t > 0$,

$$v_1(t) = 9.789e^{-3t} + 0.06173e^{-0.2941t} + 0.1488e^{0.1471t}\cos(1.25\,1t) + 0.05172e^{0.1471t}\sin(1.251t)\text{ V}$$

$$v_2(t) = -0.2105e^{-3t} + 0.06173e^{-0.2941t} + 0.1488e^{0.1471t}\cos(1.25\,1t) + 0.05172e^{0.1471t}\sin(1.251t)\text{ V}$$

$$v_3(t) = -0.03459e^{-3t} + 0.06631e^{-0.2941t} - 0.03172e^{0.1471t}\cos(1.25\,1t) - 0.06362e^{0.1471t}\sin(1.251t)\text{ V}$$

Note that the response grows exponentially as a result of the action of the dependent current source. In essence, the circuit is "running away," indicating that at some point a component will melt, explode, or fail in some related fashion. Although analyzing such circuits can evidently entail a great deal of work, the advantages to **s**-domain techniques are clear once we contemplate performing the analysis in the time domain!

(Continued on next page)

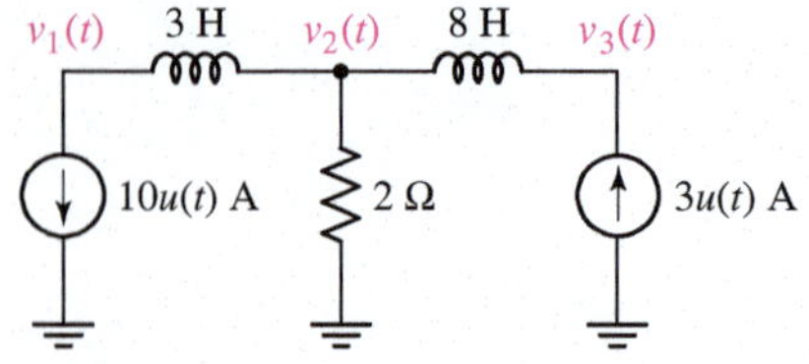

■ **FIGURE 14.21**

PRACTICE

14.18 Use nodal analysis to determine the voltages v_1, v_2, and v_3 in the circuit of Fig. 14.21. Assume there is zero energy stored in the inductors at $t = 0^-$.

Ans: $v_1(t) = -30\delta(t) - 14u(t)$ V; $v_2(t) = -14u(t)$ V; $v_3(t) = 24\delta(t) - 14u(t)$ V.

14.9 ADDITIONAL CIRCUIT ANALYSIS TECHNIQUES

Depending on the specific goal in analyzing a particular circuit, we often find that we can simplify our task by carefully choosing our analysis technique. For example, it is seldom desirable to apply superposition to a circuit containing 215 independent sources, as such an approach requires analysis of 215 separate circuits! By treating passive elements such as capacitors and inductors as impedances, however, we are free to apply any of the circuit analysis techniques studied in Chaps. 3, 4, and 5 to circuits that have been transformed to their **s**-domain equivalents.

Thus, superposition, source transformations, Thévenin's theorem, and Norton's theorem all apply in the **s**-domain.

EXAMPLE 14.16

Simplify the circuit of Fig. 14.22*a* using source transformations, and determine an expression for the voltage $v(t)$.

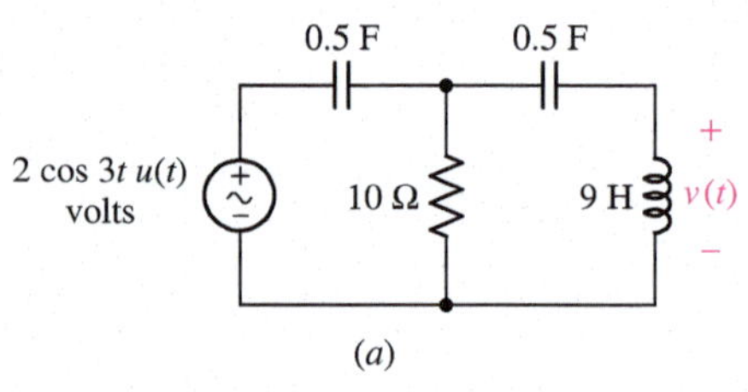

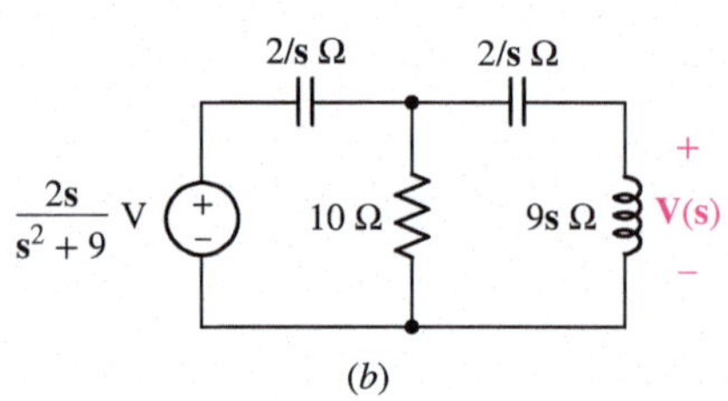

■ **FIGURE 14.22** (*a*) Circuit to be simplified using source transformations. (*b*) Frequency-domain representation.

With no initial currents or voltages specified, and a $u(t)$ multiplying the voltage source, we conclude that there is no energy initially stored in the circuit. Thus, we draw the frequency-domain circuit as shown in Fig. 14.22*b*.

Our strategy will be to perform several source transformations in succession in order to combine the two 2/**s** Ω impedances and the 10 Ω resistor; we must leave the 9**s** Ω impedance alone as the desired quantity **V**(**s**) appears across its terminals. We may now transform the voltage source and the leftmost 2/**s** Ω impedance into a current source

$$\mathbf{I}(\mathbf{s}) = \left(\frac{2\mathbf{s}}{\mathbf{s}^2+9}\right)\left(\frac{\mathbf{s}}{2}\right) = \frac{\mathbf{s}^2}{\mathbf{s}^2+9}\ \text{A}$$

in parallel with a 2/**s** Ω impedance.

As depicted in Fig. 14.23*a*, after this transformation, we have $\mathbf{Z}_1 \equiv (2/\mathbf{s})\|10 = 20/(10\mathbf{s}+2)$ Ω facing the current source. Performing another source transformation, we obtain a voltage source $\mathbf{V}_2(\mathbf{s})$ such that

$$\mathbf{V}_2(\mathbf{s}) = \left(\frac{\mathbf{s}^2}{\mathbf{s}^2+9}\right)\left(\frac{20}{10\mathbf{s}+2}\right)$$

This voltage source is in series with $\mathbf{Z}_1$ and also with the remaining 2/**s** impedance; combining $\mathbf{Z}_1$ and 2/**s** into a new impedance $\mathbf{Z}_2$ yields

$$\mathbf{Z}_2 = \frac{20}{10\mathbf{s}+2} + \frac{2}{\mathbf{s}} = \frac{40\mathbf{s}+4}{\mathbf{s}(10\mathbf{s}+2)}\ \Omega$$

The resulting circuit is shown in Fig. 14.23*b*. At this stage, we are now ready to obtain an expression for the voltage **V(s)** using simple voltage division:

$$\mathbf{V(s)} = \left(\frac{\mathbf{s}^2}{\mathbf{s}^2+9}\right)\left(\frac{20}{10\mathbf{s}+2}\right)\frac{9\mathbf{s}}{9\mathbf{s}+\left[\dfrac{40\mathbf{s}+4}{\mathbf{s}(10\mathbf{s}+2)}\right]}$$

$$= \frac{180\mathbf{s}^4}{(\mathbf{s}^2+9)(90\mathbf{s}^3+18\mathbf{s}^2+40\mathbf{s}+4)}$$

Both terms in the denominator have complex roots. Employing MATLAB to expand the denominator and then determine the residues,

```
>> syms s;
>> d1=s^2 + 9;
>> d2=90*s^3 + 18*s^2 + 40*s + 4;
>> d=d1*d2;
>> denominator=expand(d);
>> den=sym2poly(denominator)
den =
90 18 850 166 360 36
>> num=[180 0 0 0 0];
>> [r p y]=residue(num,den)
>> [Th,R] = cart2pol(real(r),imag(r))
```

we find

$$\mathbf{V(s)} = \frac{1.047+j0.0716}{\mathbf{s}-j3} + \frac{1.047-j0.0716}{\mathbf{s}+j3} - \frac{0.0471+j0.0191}{\mathbf{s}+0.04885-j0.6573}$$

$$= \frac{0.0471-j0.0191}{\mathbf{s}+0.04885+j0.6573} + \frac{5.590\times10^{-5}}{\mathbf{s}+0.1023}$$

Taking the inverse transform of each term, writing $1.047 + j0.0716$ as $1.049e^{j3.912^\circ}$ and $0.0471 + j0.0191$ as $0.05083e^{j157.9^\circ}$ results in

$$\begin{aligned} v(t) = {} & 1.049e^{j3.912^\circ}\, e^{j3t}\, u(t) + 1.049e^{-j3.912^\circ}\, e^{-j3t}\, u(t) \\ & + 0.05083e^{-j157.9^\circ}\, e^{-0.04885t}\, e^{-j0.6573t}\, u(t) \\ & + 0.05083e^{+j157.9^\circ}\, e^{-0.04885t}\, e^{+j0.6573t}\, u(t) \\ & + 5.590\times10^{-5}\, e^{-0.1023t}\, u(t) \end{aligned}$$

Converting the complex exponentials to sinusoids then allows us to write a slightly simplified expression for our voltage

$$\begin{aligned} v(t) = {} & [5.590\times10^{-5}\, e^{-0.1023t} + 2.098\cos(3t+3.912^\circ)] \\ & [+0.1017e^{-0.04885t}\cos(0.6573t+157.9^\circ)]u(t)\ \text{V} \end{aligned}$$

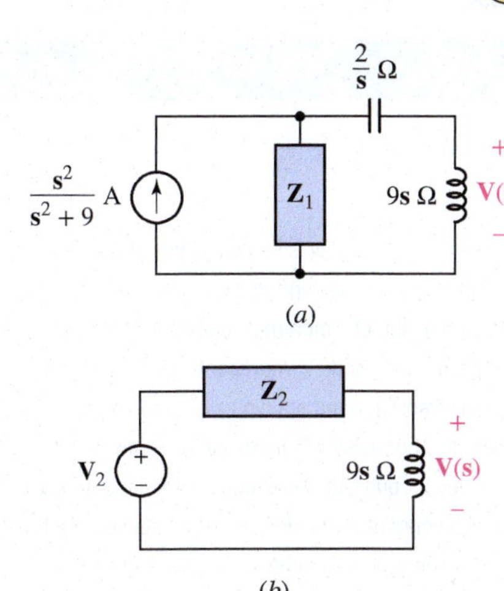

FIGURE 14.23 (*a*) Circuit after first source transformation. (*b*) Final circuit to be analyzed for **V(s)**.

Note that each term having a complex pole has a companion term that is its complex conjugate. For any physical system, complex poles will always occur in conjugate pairs.

PRACTICE

14.19 Using the method of source transformation, reduce the circuit of Fig. 14.24 to a single **s**-domain current source in parallel with a single impedance.

Ans: $\mathbf{I}_s = \frac{35}{\mathbf{s}^2(18\mathbf{s}+63)}$ A, $\mathbf{Z}_s = \frac{72\mathbf{s}^2+252\mathbf{s}}{18\mathbf{s}^3+63\mathbf{s}^2+12\mathbf{s}+28}$ Ω.

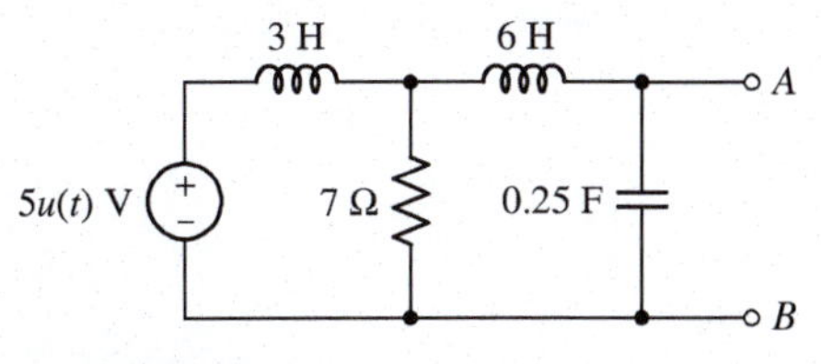

FIGURE 14.24

EXAMPLE 14.17

Find the frequency-domain Thévenin equivalent of the highlighted network shown in Fig. 14.25*a*.

This particular circuit is known as the "hybrid π" model for a special type of single-transistor circuit known as the common base amplifier. The two capacitors, C_π and C_μ, represent capacitances internal to the transistor and are typically on the order of a few pF. The resistor R_L in the circuit represents the Thévenin equivalent resistance of the output device, which could be a speaker or even a semiconductor laser. The voltage source v_s and the resistor R_s together represent the Thévenin equivalent of the input device, which may be a microphone, a light-sensitive resistor, or possibly a radio antenna.

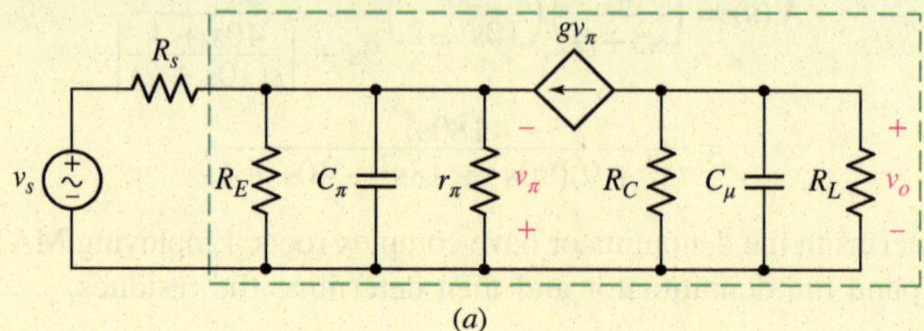

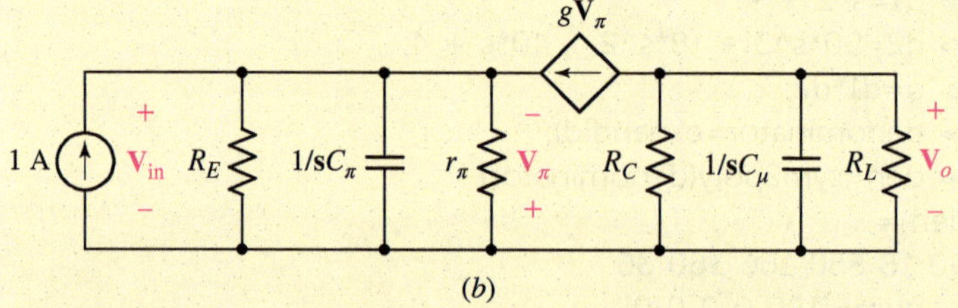

FIGURE 14.25 (*a*) An equivalent circuit for the "common base" transistor amplifier. (*b*) The frequency-domain equivalent circuit with a 1 A test source substituted for the input source represented by v_s and R_s.

We are being asked to determine the Thévenin equivalent of the circuit connected to the input device; this quantity is often referred to as the ***input impedance*** of the amplifier circuit. After converting the circuit to its frequency-domain equivalent, we replace the input device (v_s and R_s) with a 1 A "test" source, as shown in Fig. 14.25*b*. The input impedance $\mathbf{Z}_{in}$ is then

$$\mathbf{Z}_{in} = \frac{\mathbf{V}_{in}}{1}$$

or simply $\mathbf{V}_{in}$. We must find an expression for this quantity in terms of the 1 A source, resistors and capacitors, and/or the dependent source parameter g.

Writing a single nodal equation at the input, then, we find that

$$1 + g\mathbf{V}_\pi = \frac{\mathbf{V}_{in}}{\mathbf{Z}_{eq}}$$

where

$$\mathbf{Z}_{eq} \equiv R_E \left\| \frac{1}{\mathbf{s}C_\pi} \right\| r_\pi = \frac{R_E r_\pi}{r_\pi + R_E + \mathbf{s}R_E r_\pi C_\pi}$$

Since $\mathbf{V}_\pi = -\mathbf{V}_{in}$, we find that

$$\mathbf{Z}_{in} = \mathbf{V}_{in} = \frac{R_E r_\pi}{r_\pi + R_E + \mathbf{s}R_E r_\pi C_\pi + gR_E r_\pi}$$

PRACTICE

14.20 Working in the **s**-domain, find the Norton equivalent connected to the 1 Ω resistor in the circuit of Fig. 14.26.

Ans: $\mathbf{I}_{sc} = 3(\mathbf{s} + 1)/4\mathbf{s}$ A; $\mathbf{Z}_{th} = 4/(\mathbf{s} + 1)$ Ω.

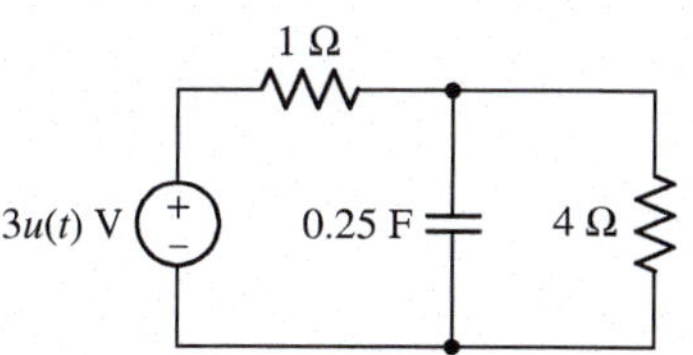

FIGURE 14.26

14.10 POLES, ZEROS, AND TRANSFER FUNCTIONS

In this section, we define terminology that will be very useful in examining system response (applicable to circuits and many other fields of study such as control systems), namely, *poles, zeros,* and *transfer functions*.

Consider the simple circuit in Fig. 14.27*a*. The **s**-domain equivalent is given in Fig. 14.27*b*, and nodal analysis yields

$$0 = \frac{\mathbf{V}_{\text{out}}}{1/\mathbf{s}C} + \frac{\mathbf{V}_{\text{out}} - \mathbf{V}_{\text{in}}}{R}$$

Rearranging and solving for $\mathbf{V}_{\text{out}}$, we find

$$\mathbf{V}_{\text{out}} = \frac{\mathbf{V}_{\text{in}}}{1 + \mathbf{s}RC}$$

or

$$\mathbf{H}(\mathbf{s}) \equiv \frac{\mathbf{V}_{\text{out}}}{\mathbf{V}_{\text{in}}} = \frac{1}{1 + \mathbf{s}RC} \qquad [31]$$

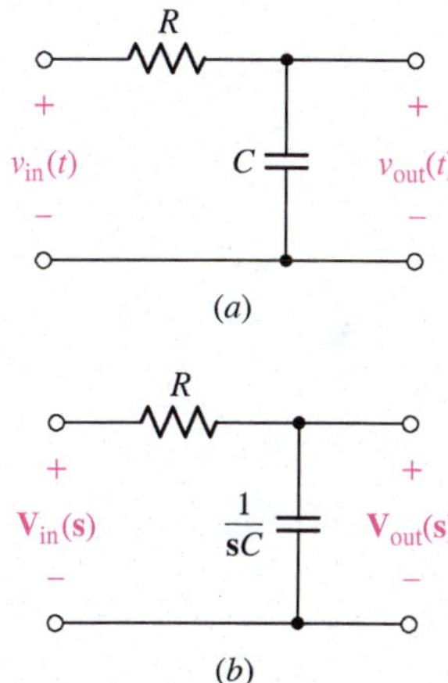

FIGURE 14.27 (*a*) A simple resistor-capacitor circuit, with an input voltage and output voltage specified. (*b*) The **s**-domain equivalent circuit.

where **H(s)** is the ***transfer function*** of the circuit, defined as the ratio of the output to the input. We could just as easily specify a particular current as either the input or output quantity, leading to a different transfer function for the same circuit. Circuit schematics are typically read from left to right, so designers often place the input of a circuit on the left of the schematic and the output terminals on the right, at least to the extent where it is possible.

The concept of a transfer function is very important, both in terms of circuit analysis and in other areas of engineering. There are two reasons for this. First, once we know the transfer function of a particular circuit, we can easily find the output that results from *any* input. All we need to do is multiply **H(s)** by the input quantity and take the inverse transform of the resulting expression. Second, the form of the transfer function contains a great deal of information about the behavior we might expect from a particular circuit (or system).

In order to evaluate the stability of a system, it is necessary to determine the poles and zeros of the transfer function **H(s)**. Writing Eq. [31] as

$$\mathbf{H}(\mathbf{s}) = \frac{1/RC}{\mathbf{s} + 1/RC} \qquad [32]$$

we see that the magnitude of this function approaches zero as $\mathbf{s} \to \infty$. Thus, we say that **H(s)** has a ***zero*** at $\mathbf{s} = \infty$. The function approaches infinity

When computing magnitude, it is customary to consider $+\infty$ and $-\infty$ as being the same frequency. The phase angle of the response at very large positive and negative values of ω need not be the same, however.

at $\mathbf{s} = -1/RC$; we therefore say that $\mathbf{H}(\mathbf{s})$ has a ***pole*** at $\mathbf{s} = -1/RC$. These frequencies are termed ***critical frequencies***.

Pole-Zero Constellations

There is a useful method for visualizing poles and zeroes of a transfer function with an eye toward evaluating system stability. Let us conceptualize the **s** plane as a floor, and then imagine a large elastic sheet laid on it. We now fix our attention on all the poles and zeros of the response. At each zero, the response is zero, the height of the sheet must be zero, and we therefore tack the sheet to the floor. At the value of **s** corresponding to each pole, we may prop up the sheet with a thin vertical rod. Zeros and poles at infinity must be treated by using a large-radius clamping ring or a high circular fence, respectively. If we have used an infinitely large, weightless, perfectly elastic sheet, tacked down with vanishingly small tacks, and propped up with infinitely long, zero-diameter rods, then the elastic sheet assumes a height that is exactly proportional to the magnitude of the response.

These comments may be illustrated by considering the configuration of the poles and zeros, sometimes called a ***pole-zero constellation,*** that locates all the critical frequencies of a frequency-domain quantity, for example, an impedance $\mathbf{Z}(\mathbf{s})$. A pole-zero constellation for an example impedance is shown in Fig. 14.28; in such a diagram, poles are denoted by crosses and zeros by circles. If we visualize an elastic-sheet model, tacked down at $\mathbf{s} = -2 + j0$ and propped up at $\mathbf{s} = -1 + j5$ and at $\mathbf{s} = -1 - j5$, we should see a terrain whose distinguishing features are two mountains and one conical crater or depression. The portion of the model for the upper LHP is shown in Fig. 14.28*b*.

Let us now build up the expression for $\mathbf{Z}(\mathbf{s})$ that leads to this pole-zero configuration. The zero requires a factor of $(\mathbf{s} + 2)$ in the numerator, and the two poles require the factors $(\mathbf{s} + 1 - j5)$ and $(\mathbf{s} + 1 + j5)$ in the denominator. Except for a multiplying constant k, we now know the form of $\mathbf{Z}(\mathbf{s})$:

$$\mathbf{Z}(\mathbf{s}) = k\frac{\mathbf{s}+2}{(\mathbf{s}+1-j5)(\mathbf{s}+1+j5)}$$

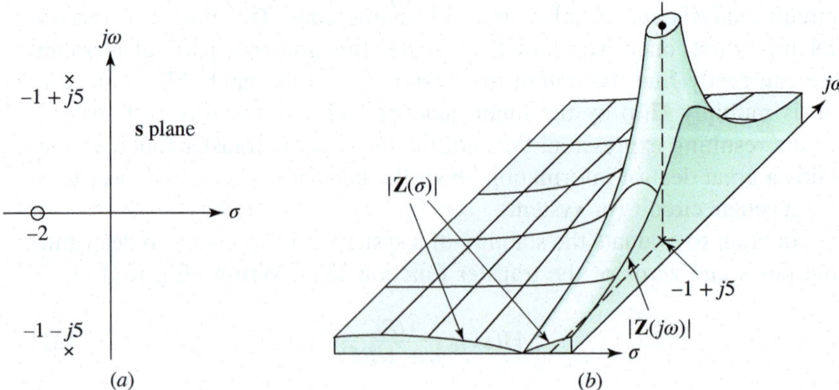

FIGURE 14.28 (*a*) The pole-zero constellation of some impedance $\mathbf{Z}(\mathbf{s})$. (*b*) A portion of the elastic-sheet model of the magnitude of $\mathbf{Z}(\mathbf{s})$.

or

$$\mathbf{Z}(\mathbf{s}) = k\frac{\mathbf{s}+2}{\mathbf{s}^2+2\mathbf{s}+26} \qquad [33]$$

In order to determine k, we require a value for $\mathbf{Z}(\mathbf{s})$ at some $\mathbf{s}$ other than a critical frequency. For this function, let us suppose we are told $\mathbf{Z}(0) = 1$. By direct substitution in Eq. [33], we find that k is 13, and therefore

$$\mathbf{Z}(\mathbf{s}) = 13\frac{\mathbf{s}+2}{\mathbf{s}^2+2\mathbf{s}+26} \qquad [34]$$

The plots $|\mathbf{Z}(\sigma)|$ versus σ and $|\mathbf{Z}(j\omega)|$ versus ω may be obtained exactly from Eq. [34], but the general form of the function is apparent from the pole-zero configuration and the elastic-sheet analogy. Portions of these two curves appear at the sides of the model shown in Fig. 14.28*b*.

PRACTICE

14.21 The parallel combination of 0.25 mH and 5 Ω is in series with the parallel combination of 40 μF and 5 Ω. (*a*) Find $\mathbf{Z}_{in}(\mathbf{s})$, the input impedance of the series combination. (*b*) Specify all the zeros of $\mathbf{Z}_{in}(\mathbf{s})$. (*c*) Specify all the poles of $\mathbf{Z}_{in}(\mathbf{s})$. (*d*) Draw the pole-zero configuration.

Ans: $5(\mathbf{s}^2 + 10{,}000\mathbf{s} + 10^8)/(\mathbf{s}^2 + 25{,}000\mathbf{s} + 10^8)$ Ω; $-5 \pm j8.66$ krad/s; $-5, -20$ krad/s.

14.11 CONVOLUTION

The **s**-domain techniques we have developed up to this point are very useful in determining the current and voltage response of a particular circuit. However, in practice we are often faced with circuits to which arbitrary sources can be connected, and we need an efficient way to determine the new output each time. This is easily accomplished if we can characterize the basic circuit by a transfer function called the ***system function.***

The analysis can proceed in either the time domain or the frequency domain, although it is generally more useful to work in the frequency domain. In such situations, we have a simple four-step process:

1. Determine the circuit system function (if not already known);
2. Obtain the Laplace transform of the forcing function to be applied;
3. Multiply this transform and the system function; and finally
4. Perform an inverse transform operation on the product to find the output.

By these means some relatively complicated integral expressions will be reduced to simple functions of **s**, and the mathematical operations of integration and differentiation will be replaced by the simpler operations of algebraic multiplication and division.

The Impulse Response

Consider a linear electrical network N, without initial stored energy, to which a forcing function $x(t)$ is applied. At some point in this circuit, a response function $y(t)$ is present. We show this in block diagram form in Fig. 14.29*a* along with sketches of generic time functions. The forcing function is shown to exist only in the interval $a < t < b$. Thus, $y(t)$ exists only for $t > a$.

The question that we now wish to answer is this: *"If we know the form of $x(t)$, how is $y(t)$ described?"* To answer this question, we need to know something about N, such as its response when the forcing function is a unit impulse $\delta(t)$. That is, we are assuming that we know $h(t)$, the response function resulting when a unit impulse is supplied as the forcing function at $t = 0$, as shown in Fig. 14.29*b*. The function $h(t)$ is commonly called the unit-impulse response function, or the ***impulse response.***

Based on our knowledge of Laplace transforms, we can view this from a slightly different perspective. Transforming $x(t)$ into $\mathbf{X(s)}$ and $y(t)$ into $\mathbf{Y(s)}$, we define the system transfer function $\mathbf{H(s)}$ as

$$\mathbf{H(s)} \equiv \frac{\mathbf{Y(s)}}{\mathbf{X(s)}}$$

If $x(t) = \delta(t)$, then according to Table 14.1, $\mathbf{X(s)} = 1$. Thus, $\mathbf{H(s)} = \mathbf{Y(s)}$ and so *in this instance* $h(t) = y(t)$.

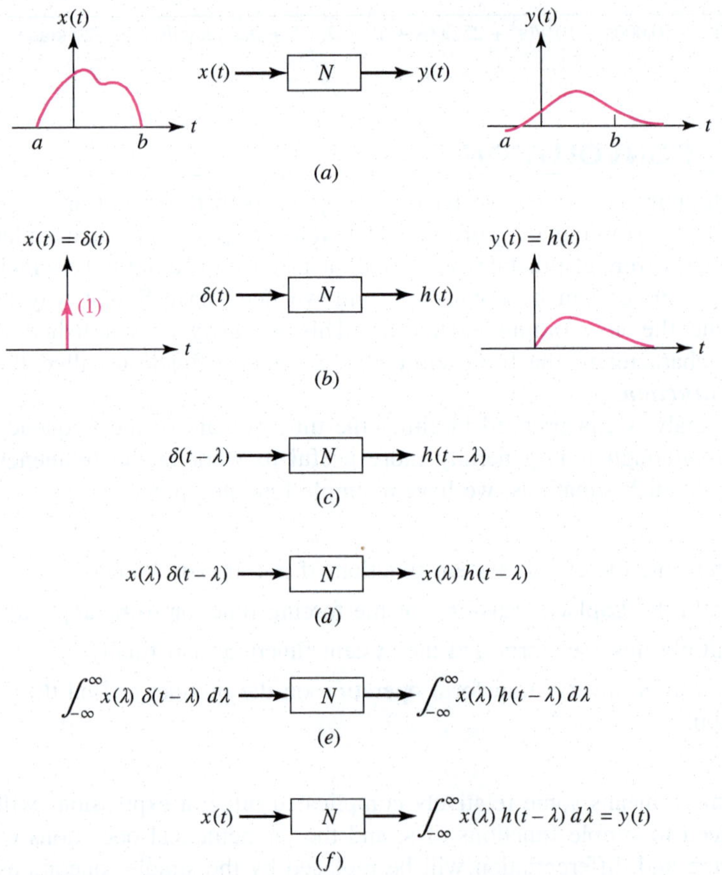

FIGURE 14.29 A conceptual development of the convolution integral.

Instead of applying the unit impulse at time $t = 0$, let us now suppose that it is applied at time $t = \lambda$ (lambda). We see that the only change in the output is a time delay. Thus, the output becomes $h(t - \lambda)$ when the input is $\delta(t - \lambda)$, as shown in Fig. 14.29*c*. Next, suppose that the input impulse has some strength other than unity. Specifically, let the strength of the impulse be numerically equal to the value of $x(t)$ when $t = \lambda$. This value $x(\lambda)$ is a constant; we know that the multiplication of a single forcing function in a *linear* circuit by a constant simply causes the response to change proportionately. Thus, if the input is changed to $x(\lambda)\delta(t - \lambda)$, then the response becomes $x(\lambda)h(t - \lambda)$, as shown in Fig. 14.29*d*.

Now let us sum this latest input over all possible values of λ and use the result as a forcing function for N. Linearity decrees that the output must be equal to the sum of the responses resulting from the use of all possible values of λ. Loosely speaking, the integral of the input produces the integral of the output, as shown in Fig. 14.29*e*. But what is the input now? Given the sifting property[3] of the unit impulse, we see that the input is simply $x(t)$, the original input. Thus, Fig. 14.29*e* may be represented as in Fig. 14.29*f*.

The Convolution Integral

If the input to our system N is the forcing function $x(t)$, we know the output must be the function $x(t)$ as depicted in Fig. 14.29*a*. Thus, from Fig. 14.29*f* we conclude that

$$y(t) = \int_{-\infty}^{\infty} x(\lambda)h(t - \lambda)d\lambda \qquad [35]$$

where $h(t)$ is the impulse response of N. This important relationship is known far and wide as the ***convolution integral.*** In words, this last equation states that *the output is equal to the input convolved with the impulse response.* It is often abbreviated by means of

$$y(t) = x(t) * h(t)$$

where the asterisk is read "convolved with."

Be careful not to confuse this new notation with multiplication!

Equation [35] sometimes appears in a slightly different but equivalent form. If we let $z = t - \lambda$, then $d\lambda = -dz$, and the expression for $y(t)$ becomes

$$y(t) = \int_{\infty}^{-\infty} -x(t - z)h(z)dz = \int_{-\infty}^{\infty} x(t - z)h(z)dz$$

and since the symbol that we use for the variable of integration is unimportant, we can modify Eq. [35] to write

$$\begin{aligned} y(t) = x(t) * h(t) &= \int_{-\infty}^{\infty} x(z)h(t - z)dz \\ &= \int_{-\infty}^{\infty} x(t - z)h(z)dz \end{aligned} \qquad [36]$$

Convolution and Realizable Systems

The result that we have in Eq. [36] is very general; it applies to any linear system. However, we are usually interested in ***physically realizable systems,*** those that *do* exist or *could* exist, and such systems have a property that modifies the convolution integral slightly. That is, *the response of the*

(3) The sifting property of the impulse function, described in Section 14.3, states that $\int_{-\infty}^{\infty} f(t)\delta(t - t_0)dt = f(t_0)$.

system cannot begin before the forcing function is applied. In particular, $h(t)$ is the response of the system resulting from the application of a unit impulse at $t = 0$. Therefore, $h(t)$ cannot exist for $t < 0$. It follows that, in the second integral of Eq. [36], the integrand is zero when $z < 0$ in the first integral, the integrand is zero when $(t - z)$ is negative, or when $z > t$. Therefore, for *realizable* systems the limits of integration change in the convolution integrals:

$$y(t) = x(t) * h(t) = \int_{-\infty}^{t} x(z)h(t - z)dz = \int_{0}^{\infty} x(t - z)h(z)dz \qquad [37]$$

Equations [36] and [37] are both valid, but the latter is more specific when we are speaking of *realizable* linear systems, and well worth memorizing.

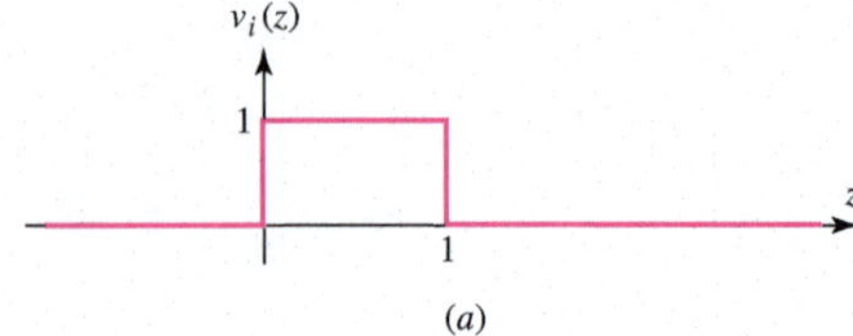

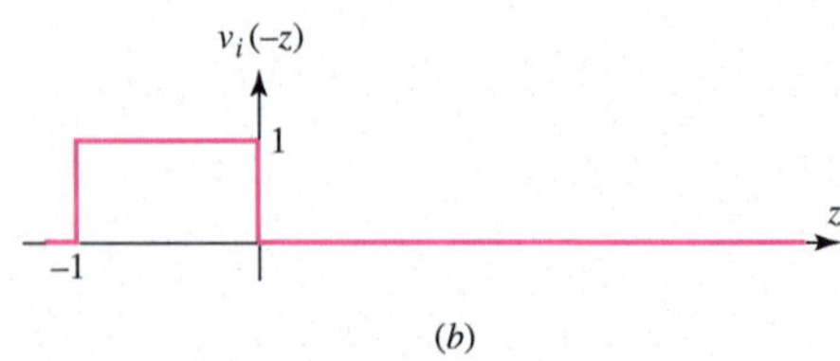

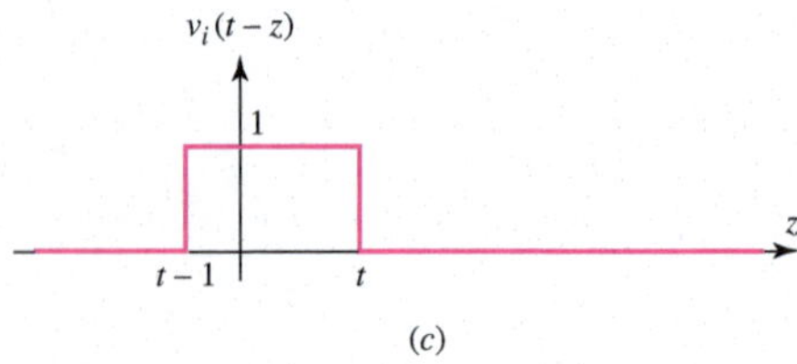

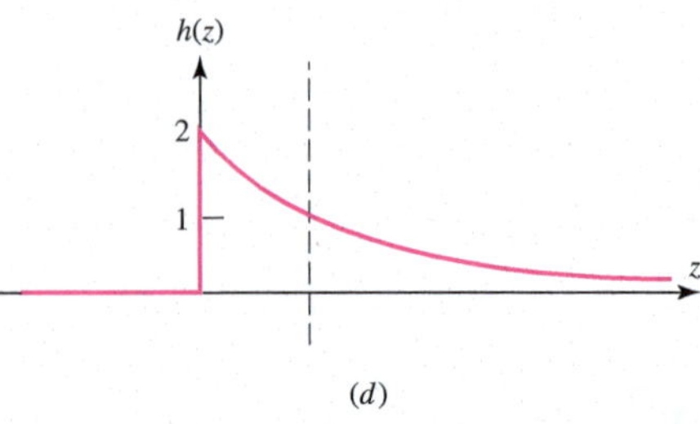

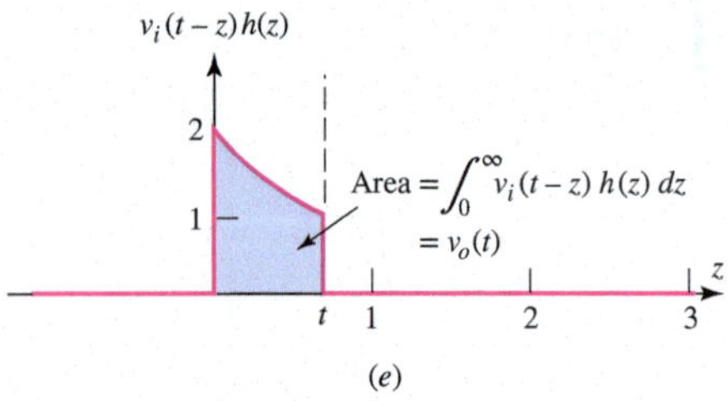

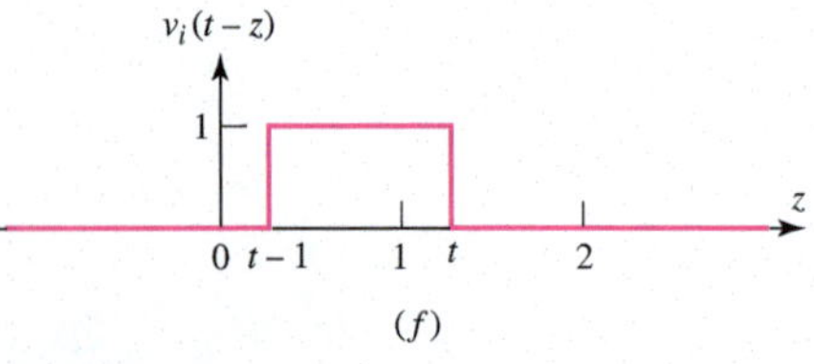

FIGURE 14.30 Graphical concepts in evaluating a convolution integral.

Graphical Method of Convolution

Before discussing the significance of the impulse response of a circuit any further, let us consider a numerical example that will give us some insight into just how the convolution integral can be evaluated. Although the expression itself is simple enough, the evaluation is sometimes troublesome, especially with regard to the values used as the limits of integration.

Suppose that the input is a rectangular voltage pulse that starts at $t = 0$, has a duration of 1 second, and is 1 V in amplitude:

$$x(t) = v_i(t) = u(t) - u(t - 1)$$

Suppose also that this voltage pulse is applied to a circuit whose impulse response is known to be an exponential function of the form:

$$h(t) = 2e^{-t}u(t)$$

We wish to evaluate the output voltage $v_o(t)$, and we can write the answer immediately in integral form,

$$y(t) = v_o(t) = v_i(t) * h(t) = \int_0^\infty v_i(t - z)h(z)dz$$
$$= \int_0^\infty [u(t - z) - u(t - z - 1)][2e^{-z}u(z)]dz$$

Obtaining this expression for $v_o(t)$ is simple enough, but the presence of the many unit-step functions tends to make its evaluation confusing and possibly even a little obnoxious. Careful attention must be paid to the determination of those portions of the range of integration in which the integrand is zero.

Let us use some graphical assistance to help us understand what the convolution integral says. We begin by drawing several z axes lined up one above the other, as shown in Fig. 14.30. We know what $v_i(t)$ looks like, and so we know what $v_i(z)$ looks like also; this is plotted as Fig. 14.30*a*. The function $v_i(-z)$ is simply $v_i(z)$ run backward with respect to z, or rotated about the ordinate axis; it is shown in Fig. 14.30*b*. Next we wish to represent $v_i(t - z)$, which is $v_i(-z)$ after it is shifted to the right by an amount $z = t$ as shown in Fig. 14.30*c*. On the next z axis, in Fig. 14.30*d*, our impulse response $h(z) = 2e^{-z}u(z)$ is plotted.

The next step is to multiply the two functions $v_i(t - z)$ and $h(z)$; the result for an arbitrary value of $t < 1$ is shown in Fig. 14.30*e*. We seek a value for the output $v_o(t)$, which is given by the *area* under the product curve (shown shaded in the figure).

First consider $t < 0$. There is no overlap between $v_i(t - z)$ and $h(z)$, so $v_o = 0$. As we increase t, we slide the pulse shown in Fig. 14.30*c* to the right, leading to an overlap with $h(z)$ once $t > 0$. The area under the corresponding curve of Fig. 14.30*e* continues to increase as we increase the value of t until we reach $t = 1$. As t increases above this value, a gap opens up between $z = 0$ and the leading edge of the pulse, as shown in Fig. 14.30*f*. As a result, the overlap with $h(z)$ decreases.

In other words, for values of t that lie between zero and unity, we must integrate from $z = 0$ to $z = t$; for values of t that exceed unity, the range of integration is $t - 1 < z < t$. Thus, we may write

$$v_0(t) = \begin{cases} 0 & t < 0 \\ \int_0^t 2e^{-z}\,dz = 2(1 - e^{-t}) & 0 \le t \le 1 \\ \int_{t-1}^t 2e^{-z}\,dz = 2(e - 1)e^{-t} & t > 1 \end{cases}$$

This function is shown plotted versus the time variable t in Fig. 14.31, and our solution is completed.

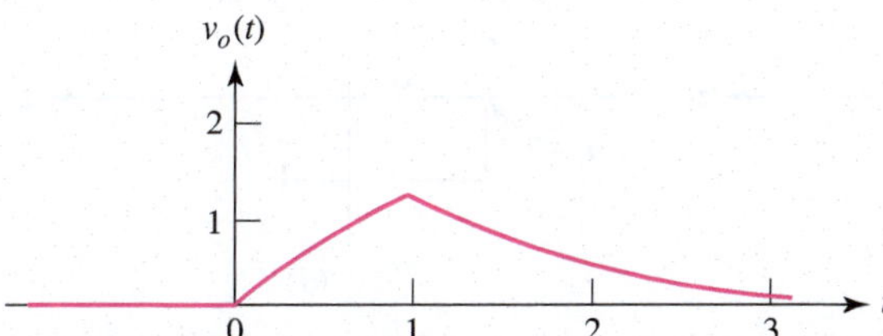

FIGURE 14.31 The output function v_o obtained by graphical convolution.

EXAMPLE **14.18**

Apply a unit-step function, $x(t) = u(t)$, as the input to a system whose impulse response is $h(t) = u(t) - 2u(t - 1) + u(t - 2)$, and determine the corresponding output $y(t) = x(t) * h(t)$.

Our first step is to plot both $x(t)$ and $h(t)$, as shown in Fig. 14.32.

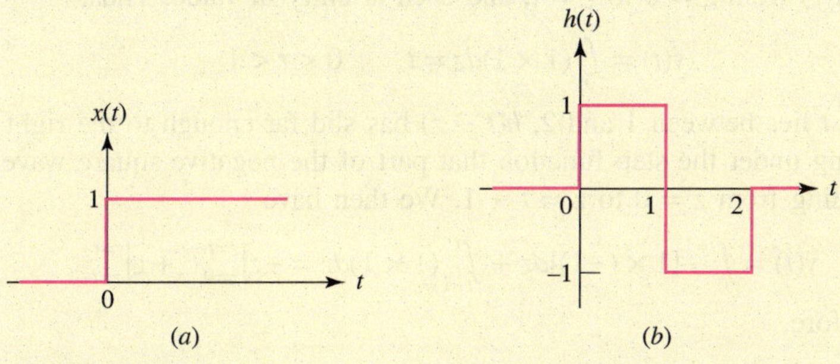

FIGURE 14.32 Sketches of (*a*) the input signal $x(t) = u(t)$ and (*b*) the unit-impulse response $h(t) = u(t) - 2u(t - 1) + u(t - 2)$, for a linear system.

(Continued on next page)

We arbitrarily choose to evaluate the first integral of Eq. [37],

$$y(t) = \int_{-\infty}^{t} x(z)h(t - z)dz$$

and prepare a sequence of sketches to help select the correct limits of integration. Figure 14.33 shows these functions in order: the input $x(z)$ as a function of z; the impulse response $h(z)$; the curve of $h(-z)$, which is just $h(z)$ rotated about the vertical axis; and $h(t - z)$, obtained by sliding $h(-z)$ to the right t units. For this sketch, we have selected t in the range $0 < t < 1$.

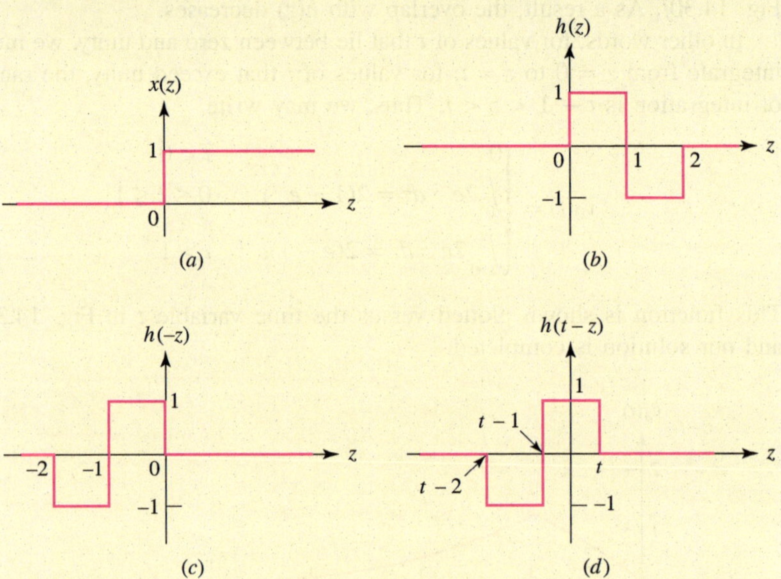

FIGURE 14.33 (*a*) The input signal and (*b*) the impulse response are plotted as functions of *z*. (*c*) *h*(−*z*) is obtained by flipping *h*(*z*) about the vertical axis, and (*d*) *h*(*t* − *z*) results when *h*(−*z*) is slid *t* units to the right.

It is now possible to visualize the product of the first graph, $x(z)$, and the last, $h(t - z)$, for the various ranges of t. When t is less than zero, there is no overlap, and

$$y(t) = 0 \qquad t < 0$$

For the case sketched in Fig. 14.33*d*, $h(t - z)$ has a nonzero overlap with $x(z)$ from $z = 0$ to $z = t$, and each is unity in value. Thus,

$$y(t) = \int_{0}^{t} (1 \times 1)dz = t \qquad 0 < t < 1$$

When t lies between 1 and 2, $h(t - z)$ has slid far enough to the right to bring under the step function that part of the negative square wave extending from $z = 0$ to $z = t - 1$. We then have

$$y(t) = \int_{0}^{t-1} [1 \times (-1)]dz + \int_{t-1}^{t} (1 \times 1)dz = -z\Big|_{z=0}^{z=t-1} + z\Big|_{z=t-1}^{z=t}$$

Therefore,

$$y(t) = -(t - 1) + t - (t - 1) = 2 - t \qquad 1 < t < 2$$

Finally, when t is greater than 2, $h(t - z)$ has slid far enough to the right that it lies entirely to the right of $z = 0$. The intersection with the unit step is complete, and

$$y(t) = \int_{t-2}^{t-1} [1 \times (-1)]dz + \int_{t-1}^{t} (1 \times 1)dz = -z\Big|_{z=t-2}^{z=t-1} + z\Big|_{z=t-1}^{z=t}$$

or

$$y(t) = -(t - 1) + (t - 2) + t - (t - 1) = 0 \qquad t > 2$$

These four segments of $y(t)$ are collected as a continuous curve in Fig. 14.34.

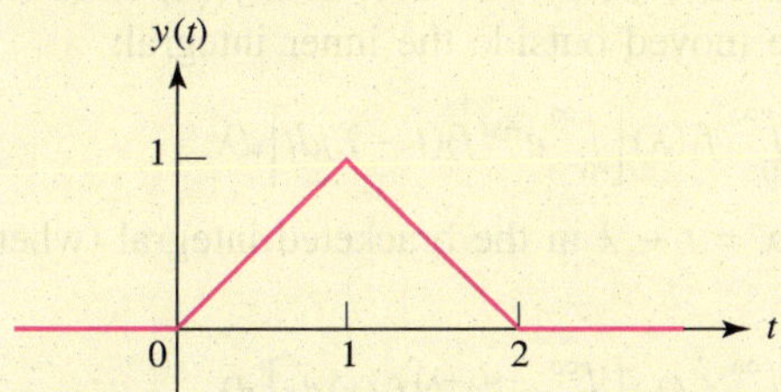

FIGURE 14.34 The result of convolving $x(t)$ and $h(t)$ as shown in Fig. 14.32.

PRACTICE

14.22 Repeat Example 14.18 using the *second* integral of Eq. [37].

14.23 The impulse response of a network is given by $h(t) = 5u(t - 1)$. If an input signal $x(t) = 2[u(t) - u(t - 3)]$ is applied, determine the output $y(t)$ at t equal to (*a*) −0.5; (*b*) 0.5; (*c*) 2.5; (*d*) 3.5.

Ans: 14.23: 0; 0; 15; 25.

Convolution and the Laplace Transform

Convolution has applications in a wide variety of disciplines beyond linear circuit analysis, including image processing, communications, and semiconductor transport theory. It is often helpful therefore to have a graphical intuition of the basic process, even if the integral expressions of Eqs. [36] and [37] are not always the best solution route. One powerful alternative approach makes use of properties of the Laplace transform—hence our introduction to convolution in this chapter.

Let $\mathbf{F}_1(\mathbf{s})$ and $\mathbf{F}_2(\mathbf{s})$ be the Laplace transforms of $f_1(t)$ and $f_2(t)$, respectively, and consider the Laplace transform of $f_1(t) * f_2(t)$,

$$\mathscr{L}\{f_1(t) * f_2(t)\} = L\left\{\int_{-\infty}^{\infty} f_1(\lambda) f_2(t - \lambda)d\lambda\right\}$$

One of these time functions will typically be the forcing function that is applied at the input terminals of a linear circuit, and the other will be the unit-impulse response of the circuit.

Since we are now dealing with time functions that do not exist prior to $t = 0^-$ (the definition of the Laplace transform forces us to assume this), the lower limit of integration can be changed to 0^-. Then, using the definition of the Laplace transform, we get

$$\mathscr{L}\{f_1(t) * f_2(t)\} = \int_{0^-}^{\infty} e^{-st}\left[\int_{0^-}^{\infty} f_1(\lambda)f_2(t-\lambda)d\lambda\right]dt$$

Since e^{-st} does not depend upon λ, we can move this factor inside the inner integral. If we do this and also reverse the order of integration, the result is

$$\mathscr{L}\{f_1(t) * f_2(t)\} = \int_{0^-}^{\infty} \left[\int_{0^-}^{\infty} e^{-st}f_1(\lambda)f_2(t-\lambda)dt\right]d\lambda$$

Continuing with the same type of trickery, we note that $f_1(\lambda)$ does not depend upon t, and so it can be moved outside the inner integral:

$$\mathscr{L}\{f_1(t) * f_2(t)\} = \int_{0^-}^{\infty} f_1(\lambda)\left[\int_{0^-}^{\infty} e^{-st}f_2(t-\lambda)dt\right]d\lambda$$

We then make the substitution $x = t - \lambda$ in the bracketed integral (where we may treat λ as a constant):

$$\begin{aligned}\mathscr{L}\{f_1(t) * f_2(t)\} &= \int_{0^-}^{\infty} f_1(\lambda)\left[\int_{-\lambda}^{\infty} e^{-\mathbf{s}(x+\lambda)}f_2(x)dx\right]d\lambda \\ &= \int_{0^-}^{\infty} f_1(\lambda)e^{-\mathbf{s}\lambda}\left[\int_{-\lambda}^{\infty} e^{-\mathbf{s}x}f_2(x)dx\right]d\lambda \\ &= \int_{0^-}^{\infty} f_1(\lambda)e^{-\mathbf{s}\lambda}[\mathbf{F}_2(\mathbf{s})]d\lambda \\ &= \mathbf{F}_2(\mathbf{s})\int_{0^-}^{\infty} f_1(\lambda)e^{-\mathbf{s}\lambda}d\lambda\end{aligned}$$

Since the remaining integral is simply $\mathbf{F}_1(\mathbf{s})$, we find that

$$\mathscr{L}\{f_1(t) * f_2(t)\} = \mathbf{F}_1(\mathbf{s}) \cdot \mathbf{F}_2(\mathbf{s}) \qquad [38]$$

Stated slightly differently, we may conclude that the inverse transform of the product of two transforms is the convolution of the individual inverse transforms, a result that is sometimes useful in obtaining inverse transforms.

EXAMPLE 14.19

Find $v(t)$ by applying convolution techniques, given that $\mathbf{V}(\mathbf{s}) = 1/[(\mathbf{s} + \alpha)(\mathbf{s} + \beta)]$.

We obtained the inverse transform of this particular function in Sec. 14.4 using a partial-fraction expansion. We now identify $\mathbf{V}(\mathbf{s})$ as the product of two transforms,

$$\mathbf{V}_1(\mathbf{s}) = \frac{1}{\mathbf{s} + \alpha}$$

and

$$\mathbf{V}_2(\mathbf{s}) = \frac{1}{\mathbf{s} + \beta}$$

where

$$v_1(t) = e^{-\alpha t}u(t)$$

and

$$v_2(t) = e^{-\beta t}u(t)$$

The desired $v(t)$ can be expressed as

$$v(t) = \mathscr{L}^{-1}\{\mathbf{V}_1(\mathbf{s})\mathbf{V}_2(\mathbf{s})\} = v_1(t) * v_2(t) = \int_{0^-}^{\infty} v_1(\lambda)v_2(t-\lambda)d\lambda$$

$$= \int_{0^-}^{\infty} e^{-\alpha\lambda}u(\lambda)e^{-\beta(t-\lambda)}u(t-\lambda)d\lambda = \int_{0^-}^{t} e^{-\alpha\lambda}e^{-\beta t}e^{\beta\lambda}d\lambda$$

$$= e^{-\beta t}\int_{0^-}^{t} e^{(\beta-\alpha)\lambda}d\lambda = e^{-\beta t}\frac{e^{(\beta-\alpha)t}-1}{\beta-\alpha}u(t)$$

or, more compactly,

$$v(t) = \frac{1}{\beta-\alpha}(e^{-\alpha t} - e^{-\beta t})u(t)$$

which is the same result that we obtained before using partial-fraction expansion. Note that it is necessary to insert the unit step $u(t)$ in the result because all (one-sided) Laplace transforms are valid only for nonnegative time.

Was the result easier to obtain by this method? Not unless one is in love with convolution integrals! The partial-fraction-expansion method is usually simpler, assuming that the expansion itself is not too cumbersome. However, the operation of convolution is easier to perform in the **s**-domain, since it only requires multiplication.

PRACTICE

14.24 Repeat Example 14.18, performing the convolution in the s-domain.

Further Comments on Transfer Functions

As we have noted several times before, the output $v_o(t)$ at some point in a linear circuit can be obtained by convolving the input $v_i(t)$ with the unit-impulse response $h(t)$. However, we must remember that the impulse response results from the application of a unit impulse at $t = 0$ *with all initial conditions zero.* Under these conditions, the Laplace transform of $v_o(t)$ is

$$\mathscr{L}\{v_o(t)\} = \mathbf{V}_0(\mathbf{s}) = \mathscr{L}\{v_i(t) * h(t)\} = \mathbf{V}_i(\mathbf{s})[\mathscr{L}\{h(t)\}]$$

Thus, the ratio $\mathbf{V}_o(\mathbf{s})/\mathbf{V}_i(\mathbf{s})$ is equal to the transform of the impulse response, which we shall denote by $\mathbf{H}(\mathbf{s})$,

$$\mathscr{L}\{h(t)\} = \mathbf{H}(\mathbf{s}) = \frac{\mathbf{V}_o(\mathbf{s})}{\mathbf{V}_i(\mathbf{s})} \qquad [39]$$

From Eq. [39] we see that the impulse response and the transfer function make up a Laplace transform pair,

$$h(t) \Leftrightarrow \mathbf{H}(\mathbf{s})$$

EXAMPLE 14.20

Determine the impulse response of the circuit in Fig. 14.35*a*, and use this to compute the forced response $v_o(t)$ if the input $v_{in}(t) = 6e^{-t}\,u(t)$ V.

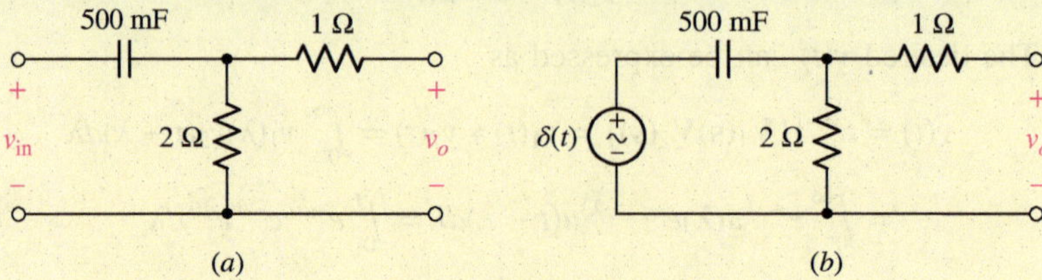

■ **FIGURE 14.35** (*a*) A simple circuit to which an exponential input is applied at $t = 0$. (*b*) Circuit used to determine $h(t)$.

We first connect an impulse voltage pulse $\delta(t)$ V to the circuit as shown in Fig. 14.35*b*. Although we may work in either the time domain with $h(t)$ or the s-domain with $\mathbf{H}(\mathbf{s})$, we choose the latter, so we next consider the s-domain representation of Fig. 14.35*b* as depicted in Fig. 14.36.

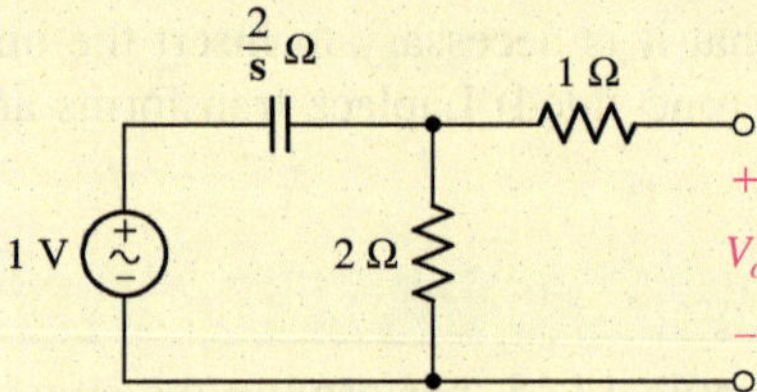

■ **FIGURE 14.36** Circuit used to find $\mathbf{H}(\mathbf{s})$.

The impulse response $\mathbf{H}(\mathbf{s})$ is given by

$$\mathbf{H}(\mathbf{s}) = \frac{\mathbf{V}_o}{1}$$

so our immediate goal is to find $\mathbf{V}_o$—a task easily performed by simple voltage division:

$$\mathbf{V}_o\big|_{v_{in}} = \partial(t) = \frac{2}{2/\mathbf{s} + 2} = \frac{\mathbf{s}}{\mathbf{s}+1} = \mathbf{H}(\mathbf{s})$$

We may now find $v_o(t)$ when $v_{in} = 6e^{-t}u(t)$ using convolution, as

$$v_{in} = \mathscr{L}^{-1}\{\mathbf{V}_{in}(\mathbf{s}) \cdot \mathbf{H}(\mathbf{s})\}$$

Since $\mathbf{V}_{in}(\mathbf{s}) = 6/(\mathbf{s} + 1)$,

$$\mathbf{V}_o = \frac{6\mathbf{s}}{(\mathbf{s}+1)^2} = \frac{6}{\mathbf{s}+1} - \frac{6}{(\mathbf{s}+1)^2}$$

Taking the inverse Laplace transform, we find that

$$v_o(t) = 6e^{-t}(1 - t)u(t) \text{ V}$$

PRACTICE

14.25 Referring to the circuit of Fig. 14.35*a*, use convolution to obtain $v_o(t)$ if $v_{in} = tu(t)$ V.

Ans: $v_o(t) = (1 - e^{-t})u(t)$ V.

14.12 A TECHNIQUE FOR SYNTHESIZING THE VOLTAGE RATIO $\mathbf{H(s)} = \mathbf{V}_{out}/\mathbf{V}_{in}$

Much of the discussion in this chapter has been related to the poles and zeros of a transfer function. Now let us see how we might determine a network that can provide a desired transfer function. We consider only a small part of the general problem, working with a transfer function of the form $\mathbf{H(s)} = \mathbf{V}_{out}(\mathbf{s})/\mathbf{V}_{in}(\mathbf{s})$, as indicated in Fig. 14.37. For simplicity, we restrict $\mathbf{H(s)}$ to critical frequencies on the negative σ axis (including the origin). Thus, we will consider transfer functions such as

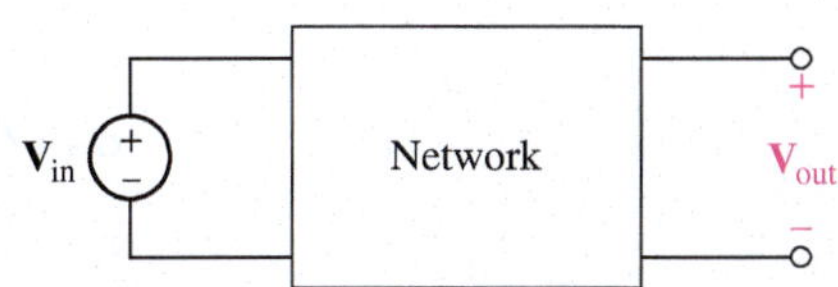

FIGURE 14.37 Given $\mathbf{H(s)} = \mathbf{V}_{out}/\mathbf{V}_{in}$, we seek a network having a specified $\mathbf{H(s)}$.

$$\mathbf{H}_1(\mathbf{s}) = \frac{10(\mathbf{s}+2)}{\mathbf{s}+5}$$

or

$$\mathbf{H}_2(\mathbf{s}) = \frac{-5\mathbf{s}}{(\mathbf{s}+8)^2}$$

or

$$\mathbf{H}_3(\mathbf{s}) = 0.1\mathbf{s}(\mathbf{s}+2)$$

Let us begin by finding the voltage gain of the network of Fig. 14.38, which contains an ideal op amp. The voltage between the two input terminals of the op amp is essentially zero, and the input impedance of the op amp is essentially infinite. We therefore may set the sum of the currents entering the inverting input terminal equal to zero:

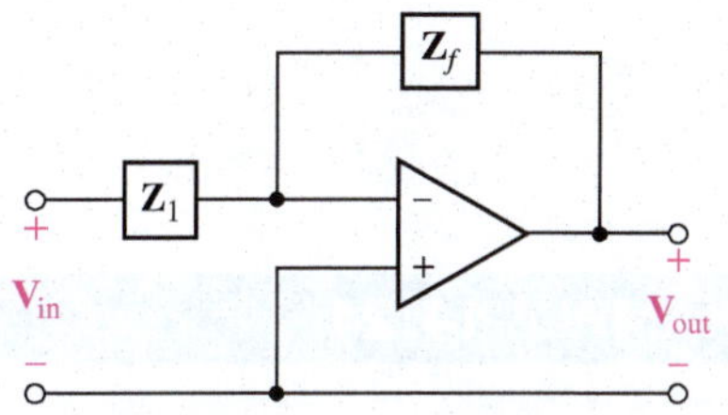

FIGURE 14.38 For an ideal op amp, $\mathbf{H(s)} = \mathbf{V}_{out}/\mathbf{V}_{in} = -\mathbf{Z}_f/\mathbf{Z}_1$.

$$\frac{\mathbf{V}_{in}}{\mathbf{Z}_1} + \frac{\mathbf{V}_{out}}{\mathbf{Z}_f} = 0$$

or

$$\frac{\mathbf{V}_{out}}{\mathbf{V}_{in}} = -\frac{\mathbf{Z}_f}{\mathbf{Z}_1}$$

If $\mathbf{Z}_f$ and $\mathbf{Z}_1$ are both resistances, the circuit acts as an inverting amplifier, or possibly an ***attenuator*** (if the ratio is less than unity). Our present interest, however, lies with those cases in which one of these impedances is a resistance while the other is an *RC* network.

In Fig. 14.39*a*, we let $\mathbf{Z}_1 = R_1$, while $\mathbf{Z}_f$ represents the impedance of the parallel combination of R_f and C_f. Therefore,

$$\mathbf{Z}_f = \frac{R_f/\mathbf{s}C_f}{R_f + (1+\mathbf{s}C_f)} = \frac{R_f}{1+\mathbf{s}C_fR_f} = \frac{1/C_f}{\mathbf{s}+(1/R_fC_f)}$$

and

$$\mathbf{H(s)} = \frac{\mathbf{V}_{out}}{\mathbf{V}_{in}} = -\frac{\mathbf{Z}_f}{\mathbf{Z}_1} = -\frac{1/R_1C_f}{\mathbf{s}+(1/R_fC_f)}$$

We have a transfer function with a single (finite) critical frequency, a pole at $\mathbf{s} = -1/R_fC_f$.

Moving on to Fig. 14.39*b*, we now let $\mathbf{Z}_f$ be resistive while $\mathbf{Z}_1$ is an *RC* parallel combination:

$$\mathbf{Z}_1 = \frac{1/C_1}{\mathbf{s}+(1/R_1C_1)}$$

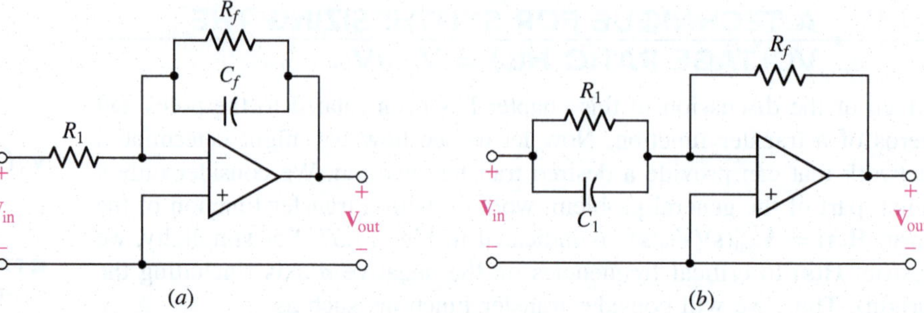

FIGURE 14.39 (a) The transfer function $\mathbf{H(s)} = \mathbf{V}_{out}/\mathbf{V}_{in}$ has a pole at $\mathbf{s} = -1/R_f C_f$. (b) Here, there is a zero at $\mathbf{s} = -1/R_1 C_1$.

and

$$\mathbf{H(s)} = \frac{\mathbf{V}_{out}}{\mathbf{V}_{in}} = -\frac{\mathbf{Z}_f}{\mathbf{Z}_1} = -R_f C_1\left(\mathbf{s} + \frac{1}{R_1 C_1}\right)$$

The only finite critical frequency is a zero at $\mathbf{s} = -1/R_1C_1$.

For our ideal op amps, the output or Thévenin impedance is zero and therefore $\mathbf{V}_{out}$ and $\mathbf{V}_{out}/\mathbf{V}_{in}$ are not functions of any load $\mathbf{Z}_L$ that may be placed across the output terminals. This includes the input to another op amp as well, and therefore we may connect circuits having poles and zeros at specified locations in ***cascade,*** where the output of one op amp is connected directly to the input of the next, and thus generates any desired transfer function.

EXAMPLE 14.21

Synthesize a circuit that will yield the transfer function $\mathbf{H(s)} = \mathbf{V}_{out}/\mathbf{V}_{in} = 10(\mathbf{s} + 2)/(\mathbf{s} + 5)$.

The pole at $\mathbf{s} = -5$ may be obtained by a network of the form of Fig. 14.39*a*. Calling this network *A*, we have $1/R_{fA}C_{fA} = 5$. We arbitrarily select $R_{fA} = 100\ \text{k}\Omega$; therefore, $C_{fA} = 2\ \mu\text{F}$. For this portion of the complete circuit,

$$\mathbf{H}_A(\mathbf{s}) = -\frac{1/R_{1A}C_{fA}}{\mathbf{s} + (1/R_{fA}C_{fA})} = -\frac{5 \times 10^5/R_{1A}}{\mathbf{s} + 5}$$

Next, we consider the zero at $\mathbf{s} = -2$. From Fig. 14.39*b*, $1/R_{1B}C_{1B} = 2$, and, with $R_{1B} = 100\ \text{k}\Omega$, we have $C_{1B} = 5\ \mu\text{F}$. Thus

$$\mathbf{H}_B(\mathbf{s}) = -R_{fB}C_{1B}\left(\mathbf{s} + \frac{1}{R_{1B}C_{1B}}\right)$$
$$= -5 \times 10^{-6}\, R_{fB}(\mathbf{s} + 2)$$

and

$$\mathbf{H(s)} = \mathbf{H}_A(\mathbf{s})\mathbf{H}_B(\mathbf{s}) = 2.5\frac{R_{fB}}{R_{1A}}\frac{\mathbf{s} + 2}{\mathbf{s} + 5}$$

We complete the design by letting $R_{fB} = 100\ \text{k}\Omega$ and $R_{1A} = 25\ \text{k}\Omega$. The result is shown in Fig. 14.40. The capacitors in this circuit are fairly large, but this is a direct consequence of the low frequencies selected for the pole and zero of $\mathbf{H(s)}$.

If $\mathbf{H(s)}$ were changed to $10(\mathbf{s} + 2000)/(\mathbf{s} + 5000)$, we could use 2 and 5 nF values.

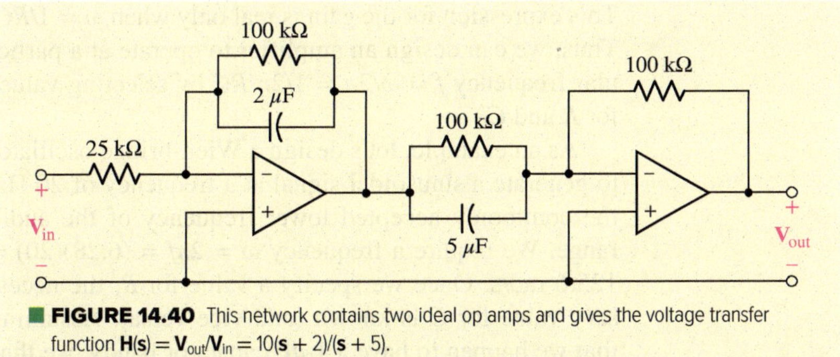

■ **FIGURE 14.40** This network contains two ideal op amps and gives the voltage transfer function $\mathbf{H(s)} = \mathbf{V}_{out}/\mathbf{V}_{in} = 10(\mathbf{s} + 2)/(\mathbf{s} + 5)$.

PRACTICE

14.26 Specify suitable element values for $\mathbf{Z}_1$ and $\mathbf{Z}_f$ in each of three cascaded stages to realize the transfer function $\mathbf{H(s)} = -20\mathbf{s}^2/(\mathbf{s} + 1000)$.

Ans: 1 μF || ∞, 1 MΩ; 1 μF || ∞, 1 MΩ; 100 kΩ || 10 nF, 5 MΩ.

PRACTICAL APPLICATION

Design of Oscillator Circuits

At several points throughout this book, we have investigated the behavior of various circuits responding to sinusoidal excitation. The creation of sinusoidal waveforms, however, is an interesting topic in itself. Generation of large sinusoidal voltages and currents is straightforward using magnets and rotating coils of wire, for example, but such an approach is not easily scaled down for the creation of small signals. Instead, low-current applications typically make use of what is known as an ***oscillator,*** which exploits the concept of ***positive feedback*** using an appropriate amplifier circuit. Oscillator circuits are an integral component of many consumer products, such as the touch screen in Fig. 14.41. These touch screens are often based on capacitive sensing, where touching the screen will change the ac coupling to an oscillator circuit, thereby changing the oscillation frequency.

One straightforward but useful oscillator circuit is known as the ***Wien-bridge oscillator,*** shown in Fig. 14.42.

The circuit resembles a noninverting op amp circuit, with a resistor R_1 connected between the inverting input pin and ground, and a resistor R_f connected between the output and the inverting input pin. The resistor R_f supplies what is referred to as a ***negative feedback path,*** since it connects the output of the amplifier to the inverting input. Any increase $\Delta\mathbf{V}_o$ in the output then leads to a reduction of the input, which in turn leads to a smaller output; this process increases the stability of the output voltage $\mathbf{V}_o$. The ***gain*** of the op amp, defined as the ratio of $\mathbf{V}_o$ to $\mathbf{V}_i$, is determined by the relative sizes of R_1 and R_f.

■ **FIGURE 14.41** Many consumer electronic products, such as GPS receivers, rely on oscillator circuits to provide a reference frequency. Aleksandra Suzi/Shutterstock

(Continued on next page)

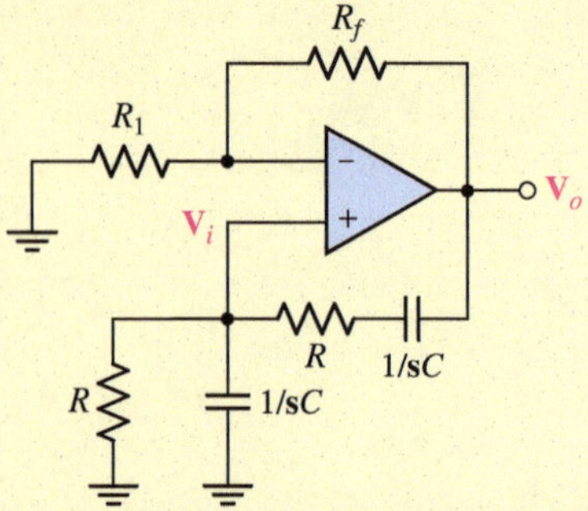

■ **FIGURE 14.42** A Wien-bridge oscillator circuit.

The *positive* feedback loop consists of two separate resistor-capacitor combinations, defined as $\mathbf{Z}_s = R + 1/\mathbf{s}C$ and $\mathbf{Z}_p = R\|(1/\mathbf{s}C)$. The values we choose for R and C allow us to design an oscillator having a specific frequency (*the internal capacitances of the op amp itself will limit the maximum frequency that can be obtained*). In order to determine the relationship between R, C, and the oscillation frequency, we seek an expression for the amplifier gain, $\mathbf{V}_o/\mathbf{V}_i$.

Recalling the two ideal op amp rules as discussed in Chap. 6 and examining the circuit in Fig. 14.42 closely, we recognize that $\mathbf{Z}_p$ and $\mathbf{Z}_s$ form a voltage divider such that

$$\mathbf{V}_i = \mathbf{V}_o \frac{\mathbf{Z}_p}{\mathbf{Z}_p + \mathbf{Z}_s} \qquad [40]$$

Simplifying the expressions for $\mathbf{Z}_p = R\|(1/\mathbf{s}C) = R/(1 + \mathbf{s}RC)$, and $\mathbf{Z}_s = R + 1/\mathbf{s}C = (1 + \mathbf{s}RC)/\mathbf{s}C$, we find that

$$\frac{\mathbf{V}_i}{\mathbf{V}_o} = \frac{\dfrac{R}{1+\mathbf{s}RC}}{\dfrac{1+\mathbf{s}RC}{\mathbf{s}C} + \dfrac{R}{1+\mathbf{s}RC}} = \frac{\mathbf{s}RC}{1 + 3\mathbf{s}RC + \mathbf{s}^2R^2C^2} \qquad [41]$$

Since we are interested in the sinusoidal steady-state operation of the amplifier, we replace $\mathbf{s}$ with $j\omega$, so that

$$\frac{\mathbf{V}_i}{\mathbf{V}_o} = \frac{j\omega RC}{1 + 3j\omega RC + (j\omega)^2R^2C^2} = \frac{j\omega RC}{1 - \omega^2R^2C^2 + 3j\omega RC} \qquad [42]$$

This expression for the gain is real only when $\omega = 1/RC$. Thus, we can design an amplifier to operate at a particular frequency $f = \omega/2\pi = 1/2\pi\, RC$ by selecting values for R and C.

As an example, let's design a Wien-bridge oscillator to generate a sinusoidal signal at a frequency of 20 Hz, the commonly accepted lower frequency of the audio range. We require a frequency $\omega = 2\pi f = (6.28)(20) = 125.6$ rad/s. Once we specify a value for R, the necessary value for C is known (and vice versa). Assuming that we happen to have a 1 μF capacitor handy, we thus compute a required resistance of $R = 7962\ \Omega$. Since this is not a standard resistor value, we will likely have to use several resistors in series and/or parallel combinations to obtain the necessary value. Referring to Fig. 14.42 in preparation for simulating the circuit using SPICE, however, we notice that no values for R_f or R_1 have been specified.

Although Eq. [40] correctly specifies the relationship between $\mathbf{V}_o$ and $\mathbf{V}_i$, we may also write another equation relating these quantities:

$$0 = \frac{\mathbf{V}_i}{R_1} + \frac{\mathbf{V}_i - \mathbf{V}_o}{R_f}$$

which can be rearranged to obtain

$$\frac{\mathbf{V}_o}{\mathbf{V}_i} = 1 + \frac{R_f}{R_1} \qquad [43]$$

Setting $\omega = 1/RC$ in Eq. [42] results in

$$\frac{\mathbf{V}_i}{\mathbf{V}_o} = \frac{1}{3}$$

Therefore, we need to select values of R_1 and R_f such that $R_f/R_1 = 2$. Unfortunately, if we proceed to perform a transient SPICE analysis on the circuit selecting $R_f = 2$ kΩ and $R_1 = 1$ kΩ, for example, we will likely be disappointed in the outcome. In order to ensure that the circuit is indeed unstable (*a necessary condition for oscillations to begin*), it is necessary to have R_f/R_1 slightly greater than 2. The simulated output of our final design ($R = 7962\ \Omega$, $C = 1\ \mu$F, $R_f = 2.1$ kΩ, and $R_1 = 1$ kΩ) is shown in Fig. 14.43. Note that the magnitude of the oscillations is increasing in the plot; in practice, nonlinear circuit elements are required to stabilize the voltage magnitude of the oscillator circuit. We should also note that the output is easier to obtain if the transient analysis is set so that dc sources start at 0 V (.tran 2 startup).

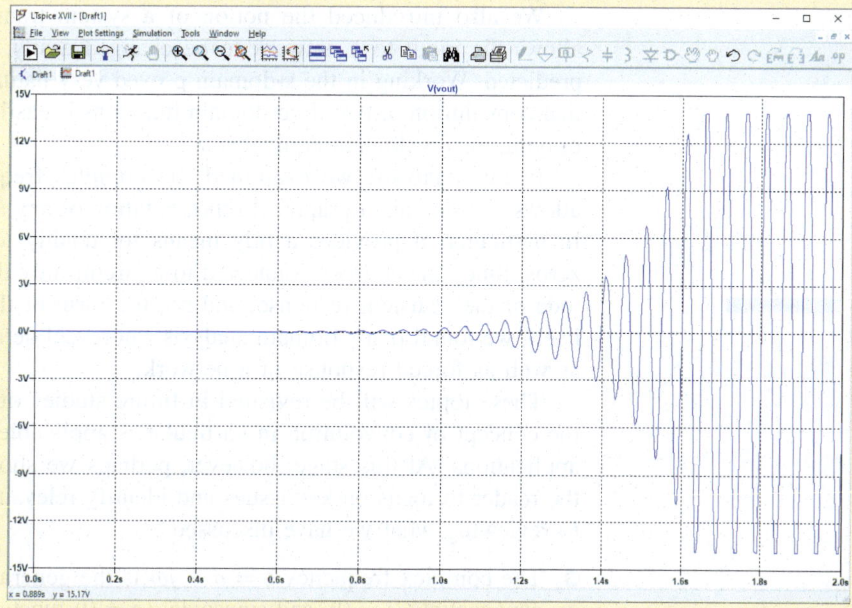

■ **FIGURE 14.43** Simulated output of the Wien-bridge oscillator designed for operation at 20 Hz. An LT1001 precision op amp was used to perform this transient simulation. Note the clipping evident in the waveform, since there is no voltage amplitude stabilization in our simple circuit.
Analog Devices

SUMMARY AND REVIEW

The primary topic of this chapter was circuit analysis using the Laplace transform, a mathematical tool for converting well-behaved time-domain functions into frequency-domain expressions. Before introducing the transform, we first considered the notion of a complex frequency, which we referred to as **s**. This convenient term has both a real (σ) and imaginary (ω) component, so it can be written as $\mathbf{s} = \sigma + j\omega$.

One of the most surprising things is that day-to-day circuit analysis does not technically require direct implementation of either the Laplace transform integral or its corresponding inverse integral! Instead, look-up tables are routinely employed, and the **s**-polynomials which result from analyzing circuits in the **s**-domain are factored into smaller, easily recognizable terms. This works because each Laplace transform pair is unique. There are several theorems associated with Laplace transforms which do see daily usage, however. These include the linearity theorem, the time-differentiation theorem, and the time-integration theorem. The time-shift as well as initial-value and final-value theorems are also commonly employed.

Ultimately, we exploit these techniques in the **s**-domain by replacing capacitors and inductors with appropriate impedances, which can be written in such a way as to include initial conditions. The concept of impedance (or admittance) allows us to directly construct **s**-domain equations which describe nodal voltages, mesh currents, etc., without having to rely on taking the Laplace transform of each term of an integrodifferential equation.

We also introduced the notion of a system transfer function, which allows the input to a network to be changed easily, and the new output predicted. Working in the **s**-domain proved very natural here, and we saw that convolution of two time-domain functions is easily performed by multiplying their **s**-domain equivalents.

In our analysis, we examined the complex-frequency plane, which allows us to create a graphical representation of any **s**-domain expression. In particular, it provides a tidy means for readily identifying poles and zeros. Since the sources connected to a circuit only determine the magnitude of the transient response, and not the form of the transient response itself, we found that **s**-domain analysis can reveal details about the natural as well as forced response of a network.

These topics will be revisited in future studies of signal analysis, and the concept of convolution in particular is applicable to a broad range of applications. At this stage, however, perhaps we should pause and allow the reader to focus on key issues and identify relevant examples as a start to reviewing what we have discussed.

- ❑ The complex frequency $\mathbf{s} = \sigma + j\omega$ is the general case; dc($\mathbf{s} = 0$), exponential ($\omega = 0$), and sinusoidal ($\sigma = 0$) functions are special cases.
- ❑ In circuit analysis problems, we convert time-domain functions into the frequency domain using the one-sided Laplace transform:

  $\mathbf{F}(\mathbf{s}) = \int_{0^-}^{\infty} e^{-\mathbf{s}t} f(t)dt$. (Example 14.1)
- ❑ The inverse Laplace transform converts frequency-domain expressions into the time domain. However, it is seldom needed due to the existence of tables listing Laplace transform pairs. (Example 14.2, Table 14.1)
- ❑ The unit-impulse function is a common approximation to pulses with very narrow widths compared to circuit time constants. It is nonzero only at a single point, and has unity area.
- ❑ $\mathscr{L}\{f_1(t) + f_2(t)\} = \mathscr{L}\{f_1(t)\} + \mathscr{L}\{f_2(t)\}$ *(additive property)*
- ❑ $\mathscr{L}\{kf(t)\} = k\mathscr{L}\{f(t)\}$, k = constant *(homogeneity property)*
- ❑ Inverse transforms are typically found using a combination of partial-fraction expansion techniques and various operations (Table 14.2) to simplify **s**-domain quantities into expressions that can be found in transform tables such as Table 14.1. (Examples 14.3, 14.4, 14.5, 14.9)
- ❑ The differentiation and integration theorems allow us to convert integrodifferential equations in the time domain into simple algebraic equations in the frequency domain. (Examples 14.6, 14.7, 14.8)
- ❑ The initial-value and final-value theorems are useful when only the specific values $f(t = 0^+)$ or $f(t \to \infty)$ are desired. (Example 14.10)
- ❑ Resistors may be represented in the frequency domain by an impedance having the same magnitude. (Example 14.11)
- ❑ Inductors may be represented in the frequency domain by an impedance $\mathbf{s}L$. If the initial current is nonzero, then the impedance must be placed in series with a voltage source $-Li(0^-)$ or in parallel with a current source $i(0^-)/\mathbf{s}$. (Example 14.11)

- Capacitors may be represented in the frequency domain by an impedance $1/\mathbf{s}C$. If the initial voltage is nonzero, then the impedance must be placed in series with a voltage source $v(0^-)/\mathbf{s}$ or in parallel with a current source $Cv(0^-)$. (Example 14.12)
- Circuits may be analyzed using **s**-domain techniques to determine its transient response. (Examples 14.11, 14.12, 14.13, 14.14, 14.15, 14.16)
- Nodal and mesh analysis in the **s**-domain lead to simultaneous equations in terms of **s**-polynomials. MATLAB is a particularly useful tool for solving such systems of equations. (Examples 14.13, 14.14, 14.15)
- Superposition, source transformation, and the Thévenin and Norton theorems all apply in the **s**-domain. (Examples 14.16, 14.17)
- A circuit transfer function $\mathbf{H}(\mathbf{s})$ is defined as the ratio of the **s**-domain output to the **s**-domain input. Either quantity may be a voltage or a current. (Examples 14.18, 14.20)
- The zeros of $\mathbf{H}(\mathbf{s})$ are those values that result in zero magnitude. The poles of $\mathbf{H}(\mathbf{s})$ are those values that result in infinite magnitude.
- Convolution provides us with both an analytic and a graphical means of determining the output of a circuit from its impulse response $h(t)$. (Examples 14.18, 14.19, 14.20)
- There are several graphical approaches to representing **s**-domain expressions in terms of poles and zeros. Such plots can be used to synthesize a circuit to obtain a desired response. (Example 14.21)
- Single op amp stages can be used to synthesize transfer functions having either a zero or a pole. More complex functions can be synthesized by cascading multiple stages. (Example 14.21)

READING FURTHER

An easily readable development of the Laplace transform and some of its key properties can be found in Chap. 4 of:

A. Pinkus and S. Zafrany, *Fourier Series and Integral Transforms.* Cambridge, United Kingdom: Cambridge University Press, 1997.

A much more detailed treatment of integral transforms and their application to science and engineering problems can be found in:

B. Davies, *Integral Transforms and Their Applications,* 3rd ed. New York: Springer-Verlag, 2002.

Stability and the Routh test are discussed in Chap. 5 of:

K. Ogata, *Modern Control Engineering,* 4th ed. Englewood Cliffs, N.J.: Prentice-Hall, 2002.

More details regarding **s**-domain analysis of systems, use of Laplace transforms, and properties of transfer functions can be found in:

K. Ogata, *Modern Control Engineering,* 4th ed. Englewood Cliffs, N.J.: Prentice-Hall, 2002.

A good discussion of various types of oscillator circuits can be found in:

R. Mancini, *Op Amps for Everyone,* 2nd ed. Amsterdam: Newnes, 2003.

and

G. Clayton and S. Winder, *Operational Amplifiers,* 5th ed. Amsterdam: Newnes, 2003.

EXERCISES

14.1 Complex Frequency

1. Determine the conjugate of each of the following: (*a*) $8 - j$; (*b*) $8e^{-9}t$; (*c*) 22.5; (*d*) $4e^{j9}$; (*e*) $j2e^{-j11}$.
2. Compute the complex conjugate of each of the following expressions: (*a*) -1; (*b*) $\frac{-j}{5\angle 20^\circ}$; (*c*) $5e^{-j5} + 2e^{j3}$; (*d*) $(2 + j)(8\angle 30^\circ)e^{j2t}$.
3. Several real voltages are written down on a piece of paper, but coffee spills across half of each one. Complete the voltage expression if the legible part is (*a*) $5e^{-j50t}$; (*b*) $(2 + j)e^{j9t}$; (*c*) $(1 - j)e^{j78t}$; (*d*) $-je^{-5t}$. Assume the units of each voltage are volts (V).
4. State the complex frequency or frequencies associated with each function: (*a*) $f(t) = \sin 100t$; (*b*) $f(t) = 10$; (*c*) $g(t) = 5e^{-7t}\cos 80t$; (*d*) $f(t) = 5e^{8t}$; (*e*) $g(t) = (4e^{-2t} - e^{-t})\cos(4t - 95^\circ)$.
5. For each of the following functions, determine the complex frequency **s** as well as **s***: (*a*) $7e^{-9t}\sin(100t + 9^\circ)$; (*b*) $\cos 9t$; (*c*) $2\sin 45t$; (*d*) $e^{7t}\cos 7t$.
6. Use real constants A, B, θ, ϕ, etc. to construct the general form of a real-time function characterized by the following frequency components: (*a*) $10 - j3$ $\mathbf{s}^{-1}$; (*b*) 0.25 $\mathbf{s}^{-1}$; (*c*) 0, 1, $-j$, $1 + j$ (all $\mathbf{s}^{-1}$).
7. The following voltage sources $Ae^{Bt}\cos(Ct + \theta)$ are connected (one at a time) to a 280 Ω resistor. Calculate the resulting current at $t = 0$, 0.1, and 0.5 s, assuming the passive sign convention: (*a*) $A = 1$ V, $B = 0.2$ Hz, $C = 0$, $\theta = 45^\circ$; (*b*) $A = 285$ mV, $B = -1$ Hz, $C = 2$ rad/s, $\theta = -45^\circ$.
8. Your neighbor's cell phone interferes with your laptop speaker system whenever the phone is connecting to the local network. Connecting an oscilloscope to the output jack of your computer, you observe a voltage waveform that can be described by a complex frequency $\mathbf{s} = -1 + j200\pi$ $\mathbf{s}^{-1}$. (*a*) What can you deduce about your neighbor's movements? (*b*) The imaginary part of the complex frequency starts to decrease suddenly. Alter your deduction as appropriate.
9. Compute the real part of each of the following complex functions: (*a*) $\mathbf{v}(t) = 9e^{-j4t}$ V; (*b*) $\mathbf{v}(t) = 12 - j9$ V; (*c*) $5\cos 100t - j43\sin 100t$ V; (*d*) $(2 + j)e^{j3t}$ V.
10. Your new assistant has measured the signal coming from a piece of test equipment, writing $v(t) = \mathbf{V}_x e^{(-2+j60)t}$, where $\mathbf{V}_x = 8 - j100$ V. (*a*) There is a missing term. What is it, and how can you tell it's missing? (*b*) What is the complex frequency of the signal? (*c*) What is the significance of the fact that $\text{Im}\{\mathbf{V}_x\} > \text{Re}\{\mathbf{V}_x\}$? (*d*) What is the significance of the fact that $|\text{Re}\{\mathbf{s}\}| < |\text{Im}\{\mathbf{s}\}|$?

14.2 Definition of the Laplace Transform

11. Calculate, with the assistance of Eq. [10] (and showing intermediate steps), the Laplace transform of the following: (*a*) $2.1u(t)$; (*b*) $2u(t - 1)$; (*c*) $5u(t - 2) - 2u(t)$; (*d*) $3u(t - b)$, where $b > 0$.
12. Employ the one-sided Laplace transform integral (with intermediate steps explicitly included) to compute the **s**-domain expressions which correspond to the following: (*a*) $5u(t - 6)$; (*b*) $2e^{-t}u(t)$; (*c*) $2e - tu(t - 1)$; (*d*) $e^{-2t}\sin 5t\, u(t)$.
13. With the assistance of Eq. [10] and showing appropriate intermediate steps, compute the one-sided Laplace transform of the following: (*a*) $(t - 1)u(t - 1)$; (*b*) $t^2u(t)$; (*c*) $\sin 2t\, u(t)$; (*d*) $\cos 100t\, u(t)$.
14. The Laplace transform of $tf(t)$, assuming $\mathscr{L}\{f(t)\} = \mathbf{F}(\mathbf{s})$, is given by $-\frac{d}{ds}\mathbf{F}(\mathbf{s})$. Test this by comparing the predicted result to what is found by directly employing Eq. [10] for (*a*) $tu(t)$; (*b*) $t^2u(t)$; (*c*) $t^3u(t)$; (*d*) $te^{-t}u(t)$.

14.3 Laplace Transforms of Simple Time Functions

15. For the following functions, specify the range of σ_0 for which the one-sided Laplace transform exists: (*a*) $t + 4$; (*b*) $(t + 1)(t - 2)$; (*c*) $e^{-t/2}u(t)$; (*d*) $\sin 10t\, u(t + 1)$.

16. Show, with the assistance of Eq. [10], that $\mathscr{L}\{f(t) + g(t) + h(t)\} = \mathscr{L}\{f(t)\} + \mathscr{L}\{g(t)\} + \mathscr{L}\{h(t)\}$.

17. Determine $\mathbf{F(s)}$ if $f(t)$ is equal to (*a*) $3u(t-2)$; (*b*) $3e^{-2t}u(t) + 5u(t)$; (*c*) $\delta(t) + u(t) - tu(t)$; (*d*) $5\delta(t)$.

18. Obtain an expression for $\mathbf{G(s)}$ if $g(t)$ is given by (*a*) $[5u(t)]^2 - u(t)$; (*b*) $2u(t) - 2u(t-2)$; (*c*) $tu(2t)$.

19. Without recourse to Eq. [11], obtain an expression for $f(t)$ if $\mathbf{F(s)}$ is given by (*a*) $\frac{1}{\mathbf{s}}$; (*b*) $1.55 - \frac{2}{\mathbf{s}}$; (*c*) $\frac{1}{\mathbf{s}+1.5}$; (*d*) $\frac{5}{\mathbf{s}^2} + \frac{5}{\mathbf{s}} + 5$.

20. Obtain an expression for $g(t)$ without employing the inverse Laplace transform integral, if $\mathbf{G(s)}$ is known to be (*a*) $\frac{1.5}{(\mathbf{s}+9)^2}$; (*b*) $\frac{2}{\mathbf{s}} - 0$; (*c*) π; (*d*) $\frac{a}{(\mathbf{s}+1)^2} - a,\ a > 0$.

21. Evaluate the following: (*a*) $\delta(t)$ at $t = 1$; (*b*) $5\delta(t+1) + u(t+1)$ at $t = 0$; (*c*) $\int_{-1}^{2} \delta(t)dt$; (*d*) $3 - \int_{-1}^{2} 2\delta(t)dt$.

22. Evaluate the following: (*a*) $[\delta(2t)]^2$ at $t = 1$; (*b*) $2\delta(t-1) + u(-t+1)$ at $t = 0$; (*c*) $\frac{1}{3}\int_{-0.001}^{0.003} \delta(t)dt$; (*d*) $\frac{1}{2}\left[\int_{-\infty}^{\infty} \delta(t-1)dt\right]^2$.

23. Evaluate the following expressions at $t = 0$:

(*a*) $\int_{-\infty}^{+\infty} 2\delta(t-1)dt$; (*b*) $\frac{\int_{-\infty}^{+\infty} \delta(t+1)dt}{u(t+1)}$; (*c*) $\frac{\sqrt{3\int_{-\infty}^{+\infty} \delta(t-2)dt}}{[u(1-t)]^3} - \sqrt{u(t+2)}$;

(*d*) $\left[\frac{\int_{-\infty}^{+\infty} \delta(t-1)dt}{\int_{-\infty}^{+\infty} \delta(t+1)dt}\right]^2$.

24. Evaluate the following:

(*a*) $\int_{-\infty}^{+\infty} e^{-100}\delta\left(t-\frac{1}{5}\right)dt$; (*b*) $\int_{-\infty}^{+\infty} 4t\delta(t-2)dt$; (*c*) $\int_{-\infty}^{+\infty} 4t^2\delta(t-1.5)dt$;

(*d*) $\frac{\int_{-\infty}^{+\infty} (4-t)\delta(t-1)dt}{\int_{-\infty}^{+\infty} (4-t)\delta(t+1)}$.

14.4 Inverse Transform Techniques

25. Determine the inverse transform of $\mathbf{F(s)}$ equal to (*a*) $5 + \frac{5}{\mathbf{s}^2} - \frac{5}{(\mathbf{s}+1)}$;

(*b*) $\frac{1}{\mathbf{s}} + \frac{5}{(0.1\mathbf{s}+4)} - 3$; (*c*) $-\frac{1}{2\mathbf{s}} + \frac{1}{(0.5\mathbf{s})^2} + \frac{4}{(\mathbf{s}+5)(\mathbf{s}+5)} + 2$;

(*d*) $\frac{4}{(\mathbf{s}+5)(\mathbf{s}+5)} + \frac{2}{\mathbf{s}+1} + \frac{1}{\mathbf{s}+3}$.

26. Obtain an expression for $g(t)$ if $\mathbf{G(s)}$ is given by (*a*) $\frac{3(\mathbf{s}+1)}{(\mathbf{s}+1)^2} + \frac{2\mathbf{s}}{\mathbf{s}^2} - \frac{1}{(\mathbf{s}+2)^2}$;

(*b*) $-\frac{10}{(\mathbf{s}+3)^3}$; (*c*) $19 - \frac{8}{(\mathbf{s}+3)^2} + \frac{18}{\mathbf{s}^2+6\mathbf{s}+9}$.

27. Reconstruct the time-domain function if its transform is (*a*) $\frac{\mathbf{s}}{\mathbf{s}(\mathbf{s}+2)}$; (*b*) 1;

(*c*) $3\frac{\mathbf{s}+2}{(\mathbf{s}^2+2\mathbf{s}+4)}$; (*d*) $4\frac{\mathbf{s}}{2\mathbf{s}+3}$.

28. Determine the inverse transform of $\mathbf{V(s)}$ equal to (*a*) $\frac{\mathbf{s}+2}{2\mathbf{s}}$; (*b*) $\frac{\mathbf{s}+8}{\mathbf{s}} + \frac{2}{\mathbf{s}^2}$;

(*c*) $\frac{\mathbf{s}+1}{\mathbf{s}(\mathbf{s}+2)} + \frac{2\mathbf{s}^2-1}{\mathbf{s}^2}$; (*d*) $\frac{\mathbf{s}^2+4\mathbf{s}+4}{\mathbf{s}}$.

29. Obtain the time-domain expression which corresponds to each of the following $\mathbf{s}$-domain functions: (*a*) $2\frac{3\mathbf{s}+\frac{1}{2}}{\mathbf{s}^2+3\mathbf{s}}$; (*b*) $7 - \frac{\mathbf{s}+\frac{1}{2}}{\mathbf{s}^2+3\mathbf{s}+2}$; (*c*) $\frac{2}{\mathbf{s}^2}$;

(*d*) $\frac{2}{(\mathbf{s}+1)(\mathbf{s}+1)}$; (*e*) $\frac{14}{(\mathbf{s}+1)(\mathbf{s}+4)(\mathbf{s}+5)}$.

30. Find the inverse Laplace transform of the following: (*a*) $\frac{1}{\mathbf{s}^2+9\mathbf{s}+20}$; (*b*) $\frac{4}{\mathbf{s}^3+18\mathbf{s}^2+17\mathbf{s}}$; (*c*) $\frac{3}{\mathbf{s}(\mathbf{s}+1)(\mathbf{s}+4)(\mathbf{s}+5)(\mathbf{s}+2)}$. (*d*) Verify your answers with MATLAB.

31. Determine the inverse Laplace transform of each of the following **s**-domain expressions: (*a*) $\frac{1}{(\mathbf{s}+2)^2(\mathbf{s}+1)}$; (*b*) $\frac{\mathbf{s}}{(\mathbf{s}^2+4\mathbf{s}+4)(\mathbf{s}+2)}$; (*c*) $\frac{1}{\mathbf{s}^2+8\mathbf{s}+7}$. (*d*) Verify your answers with MATLAB.

32. Given the following expressions in the **s**-domain, determine the corresponding time-domain functions: (*a*) $\frac{1}{3\mathbf{s}}-\frac{1}{2\mathbf{s}+1}+\frac{3}{\mathbf{s}^3+8\mathbf{s}^2+16\mathbf{s}}-1$; (*b*) $\frac{1}{(3\mathbf{s}+5)^2}$; (*c*) $\frac{2\mathbf{s}}{(\mathbf{s}+a)^2}$.

33. Compute $\mathcal{L}^{-1}\{\mathbf{G(s)}\}$ if $\mathbf{G(s)}$ is given by (*a*) $\frac{3\mathbf{s}}{(\mathbf{s}/2+2)^2(\mathbf{s}+2)}$; (*b*) $3-3\frac{\mathbf{s}}{(2\mathbf{s}^2+24\mathbf{s}+70)(\mathbf{s}+5)}$; (*c*) $2-\frac{1}{\mathbf{s}+100}+\frac{\mathbf{s}}{\mathbf{s}^2+100}$; (*d*) $\mathcal{L}\{tu(2t)\}$.

34. Obtain the time-domain expression which corresponds to the following **s**-domain functions: (*a*) $\frac{1}{(\mathbf{s}+2)^2}$; (*b*) $\frac{4}{(\mathbf{s}+1)^2}$; (*c*) $\frac{1}{\mathbf{s}(\mathbf{s}+4)(\mathbf{s}+6)}$. (*d*) Verify your solutions with MATLAB.

14.5 Basic Theorems for the Laplace Transform

35. Take the Laplace transform of the following equations: (*a*) $5\,di/dt-7\,d^2i/dt^2+9i=4$; (*b*) $m\frac{d^2p}{dt^2}+\mu_f\frac{dp}{dt}+kp(t)=0$, the equation that describes the "force-free" response of a simple shock absorber system; (*c*) $\frac{d\Delta n_p}{dt}=-\frac{\Delta n_p}{\tau}+G_L$, with τ = constant, which describes the recombination rate of excess electrons (Δn_p) in *p*-type silicon under optical illumination (G_L is a constant proportional to the light intensity).

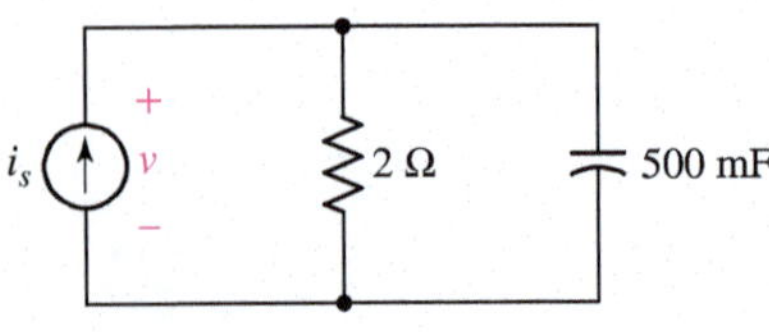

■ **FIGURE 14.44**

36. With regard to the circuit depicted in Fig. 14.44, the initial voltage across the capacitor is $v(0^-)=1.5$ V and the current source is $\mathbf{i}_s=700u(t)$ mA. (*a*) Write the differential equation which arises from KCL, in terms of the nodal voltage $v(t)$. (*b*) Take the Laplace transform of the differential equation. (*c*) Determine the frequency-domain representation of the nodal voltage. (*d*) Solve for the time-domain voltage $v(t)$.

37. For the circuit of Fig. 14.44, if $\mathbf{I}_s=\frac{2}{\mathbf{s}+1}$ mA, (*a*) write the time-domain nodal equation in terms of $v(t)$; (*b*) solve for $\mathbf{V(s)}$; (*c*) determine the time-domain voltage $v(t)$.

38. The voltage source in the circuit of Fig. 14.3 is replaced with one whose **s**-domain equivalent is $\frac{2}{\mathbf{s}}-\frac{1}{\mathbf{s}+1}$ V. The initial condition is unchanged. (*a*) Write the **s**-domain KVL equation in terms of $\mathbf{I(s)}$. (*b*) Solve for $i(t)$.

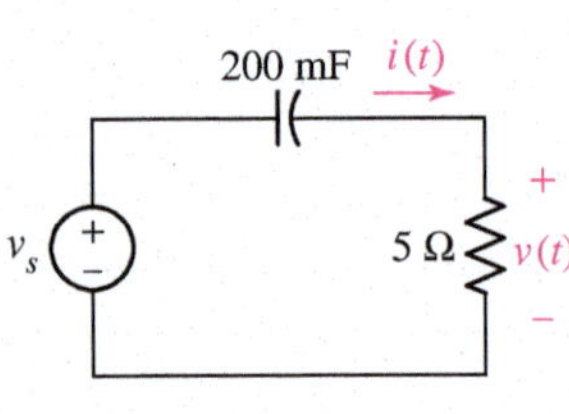

■ **FIGURE 14.45**

39. For the circuit of Fig. 14.45, $v_s(t)=2u(t)$ V and the capacitor initially stores zero energy. (*a*) Write the time-domain loop equation in terms of the current $i(t)$. (*b*) Obtain the **s**-domain representation of this integral equation. (*c*) Solve for $i(t)$.

40. The **s**-domain representation of the voltage source in Fig. 14.45 is $\mathbf{V}_s(\mathbf{s})=\frac{2}{\mathbf{s}+1}$ V. The initial voltage across the capacitor, defined using the passive sign convention in terms of the current i, is 4.5 V. (*a*) Write the time-domain integral equation that arises from application of KVL. (*b*) By first solving for $\mathbf{I(s)}$, determine the time-domain current $i(t)$.

41. If the current source of Fig. 14.46 is given by $450u(t)$ mA, and $i_x(0) =$ 150 mA, work initially in the s-domain to obtain an expression for $v(t)$ valid for $t > 0$.

42. Obtain, through purely legitimate means, an s-domain expression which corresponds to the time-domain waveform plotted in Fig. 14.47.

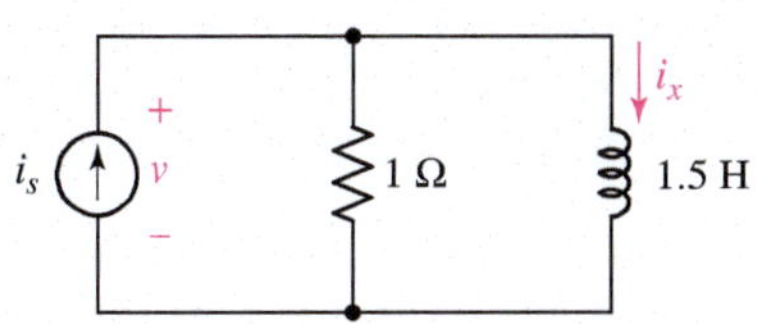

■ FIGURE 14.46

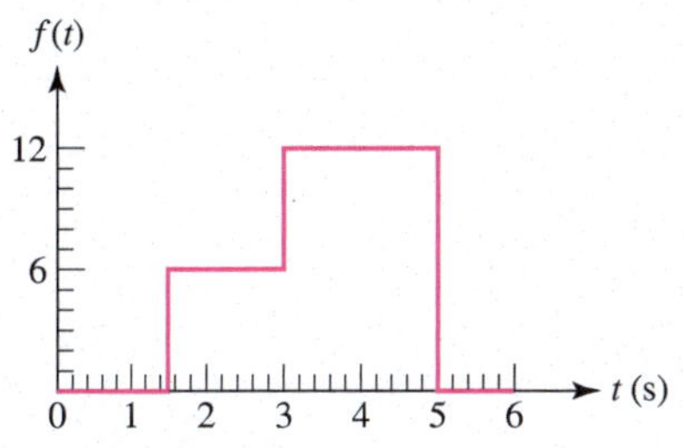

■ FIGURE 14.47

14.6 The Initial-Value and Final-Value Theorems

43. Employ the initial-value theorem to determine the initial value of each of the following time-domain functions: (*a*) $2u(t)$; (*b*) $2e^{-t}u(t)$; (*c*) $u(t-6)$; (*d*) $\cos 5t\, u(t)$.

44. Employ the initial-value theorem to determine the initial value of each of the following time-domain functions: (*a*) $u(t-3)$; (*b*) $2e^{-(t-2)}u(t-2)$; (*c*) $\frac{u(t-2)+[u(t)]^2}{2}$; (*d*) $\sin 5t\, e^{-2t}u(t)$.

45. Make use of the final-value theorem (if appropriate) to ascertain $f(\infty)$ for (*a*) $\frac{1}{\mathbf{s}+2} - \frac{2}{\mathbf{s}}$; (*b*) $\frac{2\mathbf{s}}{(\mathbf{s}+2)(\mathbf{s}+1)}$; (*c*) $\frac{1}{(\mathbf{s}+2)(\mathbf{s}+4)} + \frac{2}{\mathbf{s}}$; (*d*) $\frac{1}{(\mathbf{s}^2+\mathbf{s}-6)(\mathbf{s}+9)}$.

46. Without recourse to $f(t)$, determine $f(0^+)$ and $f(\infty)$ (or show they do not exist) for each of the following s-domain expressions: (*a*) $\frac{1}{\mathbf{s}+18}$; (*b*) $10\left(\frac{1}{\mathbf{s}^2}+\frac{3}{\mathbf{s}}\right)$; (*c*) $\frac{\mathbf{s}^2-4}{\mathbf{s}^3+8\mathbf{s}^2+4\mathbf{s}}$; (*d*) $\frac{\mathbf{s}^2+2}{\mathbf{s}^3+3\mathbf{s}^2+5\mathbf{s}}$.

47. Apply the initial- or final-value theorems as appropriate to determine $f(0^+)$ and $f(\infty)$ for the following functions: (*a*) $\frac{\mathbf{s}+2}{\mathbf{s}^2+8\mathbf{s}+4}$; (*b*) $\frac{1}{\mathbf{s}^2(\mathbf{s}+4)^2(\mathbf{s}+6)^3}$; (*c*) $\frac{4\mathbf{s}^2+1}{(\mathbf{s}+1)^2(\mathbf{s}+2)^2}$.

48. Determine which of the following functions are appropriate for the final-value theorem: (*a*) $\frac{1}{(\mathbf{s}-1)}$; (*b*) $\frac{10}{\mathbf{s}^2-4\mathbf{s}+4}$; (*c*) $\frac{13}{\mathbf{s}^3-5\mathbf{s}^2+8\mathbf{s}-6}$; (*d*) $\frac{3}{2\mathbf{s}^3-10\mathbf{s}^2+16\mathbf{s}-12}$.

14.7 Z(s) and Y(s)

49. Draw an s-domain equivalent of the circuit depicted in Fig. 14.48, if the only quantity of interest is $v(t)$. (*Hint:* Omit the source, but don't ignore it.)

50. For the circuit represented in Fig. 14.49, draw an s-domain equivalent and analyze it to obtain a value for $i(t)$ if $i(0)$ is equal to (*a*) 0; (*b*) −2 A.

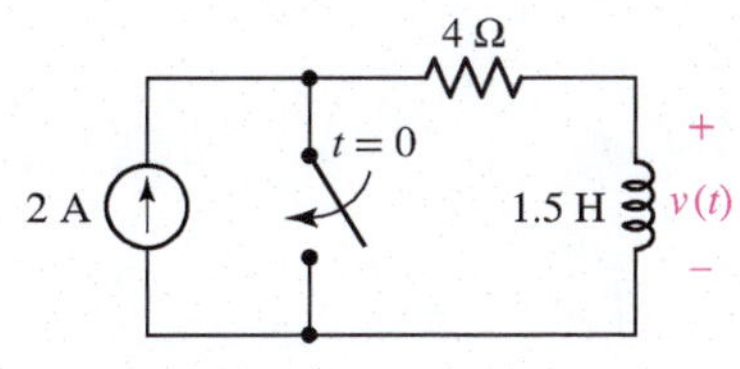

■ FIGURE 14.48

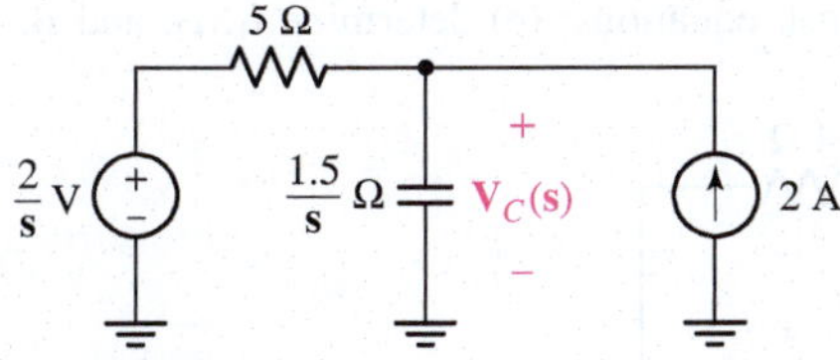

■ FIGURE 14.49

51. For the circuit of Fig. 14.49, draw an s-domain equivalent and analyze it to obtain a value for $v(t)$ if $i(0)$ is equal to (*a*) 0; (*b*) 3 A.

52. With respect to the s-domain circuit drawn in Fig. 14.50, (*a*) calculate $\mathbf{V}_C(\mathbf{s})$; (*b*) determine $v_C(t)$, $t > 0$; (*c*) draw the time-domain representation of the circuit.

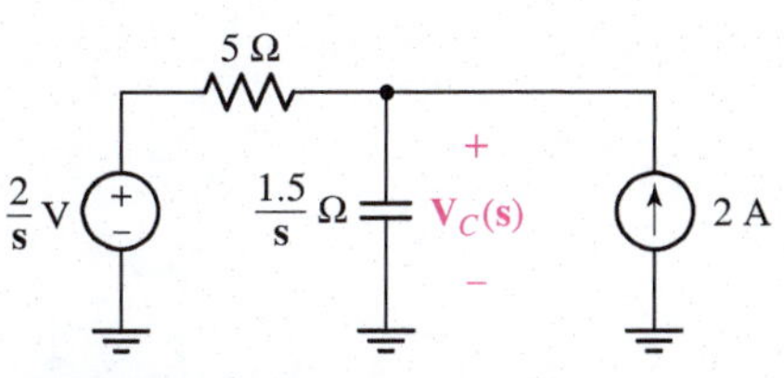

■ FIGURE 14.50

53. Determine $v(t)$ for $t > 0$ for the circuit shown in Fig. 14.51.

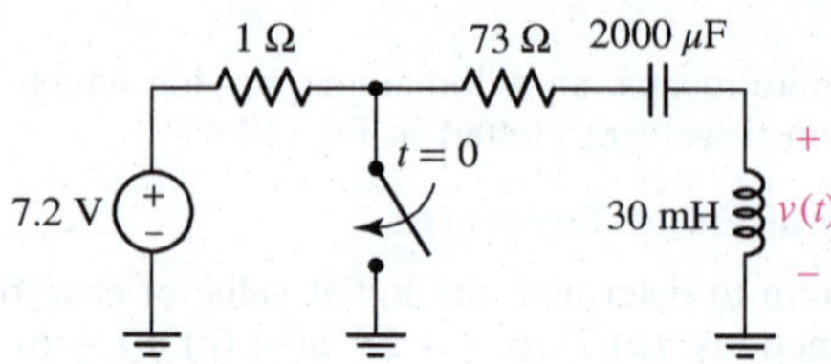

■ FIGURE 14.51

54. Determine the input impedance $\mathbf{Z}_{in}(\mathbf{s})$ seen looking into the terminals of the network depicted in Fig. 14.52. Express your answer as a ratio of two **s**-polynomials.

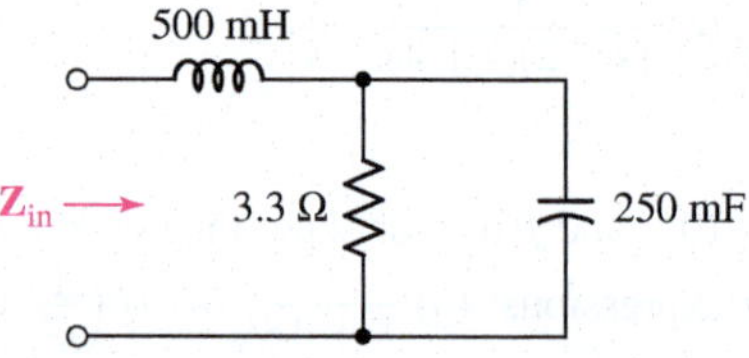

■ FIGURE 14.52

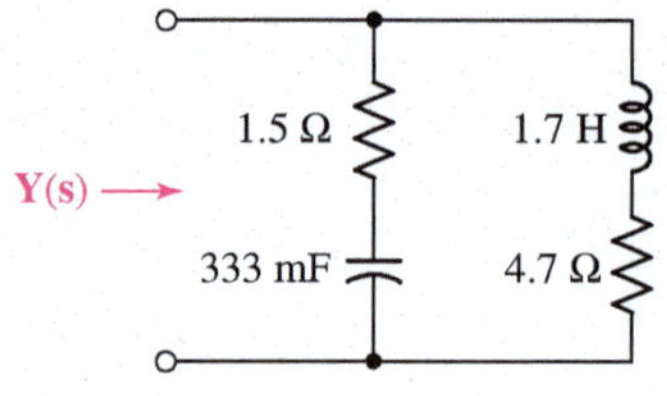

■ FIGURE 14.53

55. With respect to the network of Fig. 14.53, obtain an expression for the input admittance $\mathbf{Y}(\mathbf{s})$ as labeled. Express your answer as a ratio of two **s**-polynomials.

56. For the circuit of Fig. 14.54, (*a*) draw both **s**-domain equivalent circuits; (*b*) choose one and solve for $\mathbf{V}(\mathbf{s})$; (*c*) determine $v(t)$.

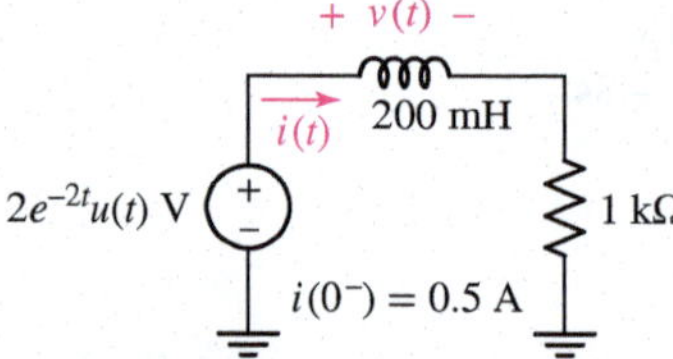

■ FIGURE 14.54

14.8 Nodal and Mesh Analysis in the s-Domain

57. For the circuit given in Fig. 14.55, (*a*) draw the **s**-domain equivalent; (*b*) write the three **s**-domain mesh equations; (*c*) determine i_1, i_2, and i_3.

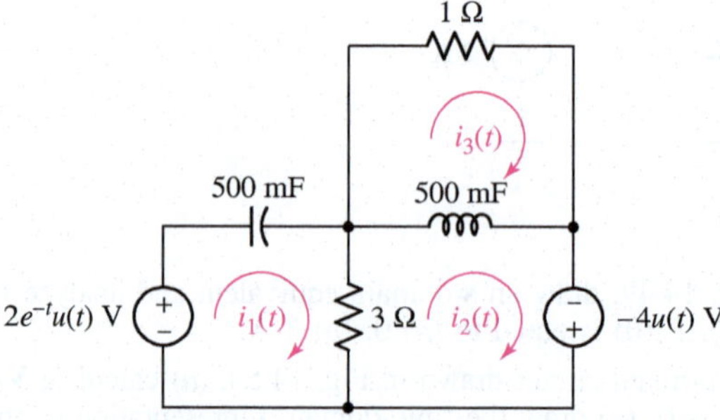

■ FIGURE 14.55

58. For the circuit shown in Fig. 14.56, (*a*) write an **s**-domain nodal equation for $\mathbf{V}_x(\mathbf{s})$; (*b*) solve for $v_x(t)$.

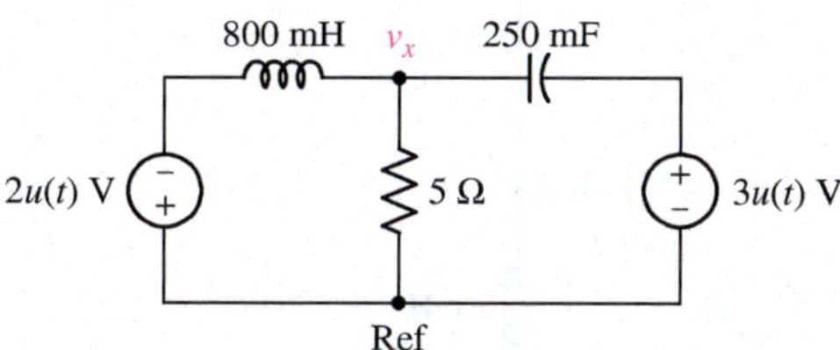

■ **FIGURE 14.56**

59. Determine v_1 and v_2 for the circuit of Fig. 14.57 using nodal analysis in the **s**-domain.

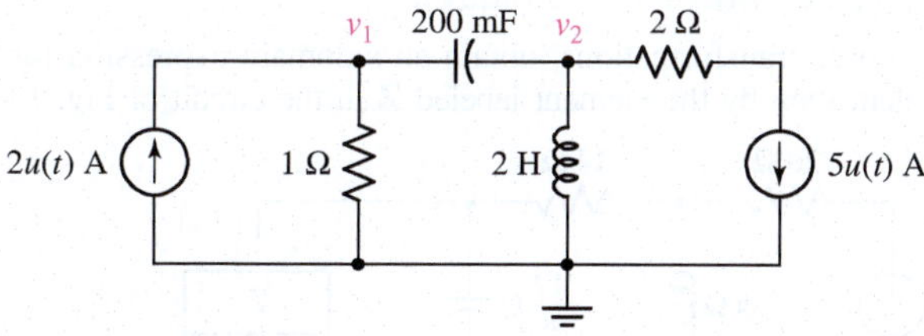

■ **FIGURE 14.57**

60. The $2u(t)$ A source in Fig. 14.57 is replaced with a $4e^{-t}u(t)$ A source. Employ **s**-domain analysis to determine the power dissipated by the 1 Ω resistor.

61. For the circuit shown in Fig. 14.58, let $i_{s1} = 3u(t)$ A and $i_{s2} = 5 \sin 2t$ A. Working initially in the **s**-domain, obtain an expression for $v_x(t)$.

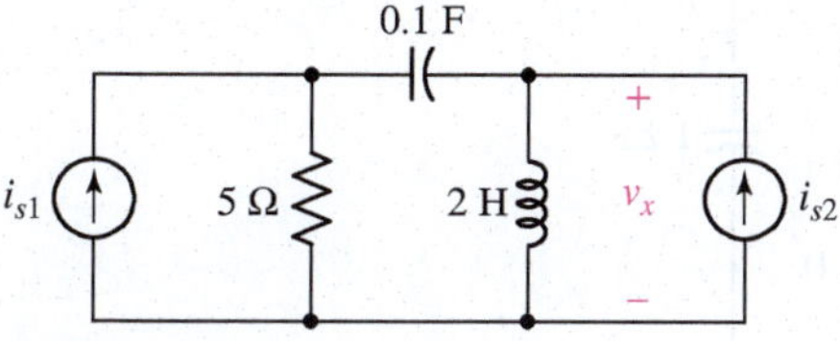

■ **FIGURE 14.58**

62. Determine the mesh current $i_1(t)$ in Fig. 14.59 if the current through the 1 mH inductor ($i_2 - i_4$) is 1 A at $t = 0^-$.

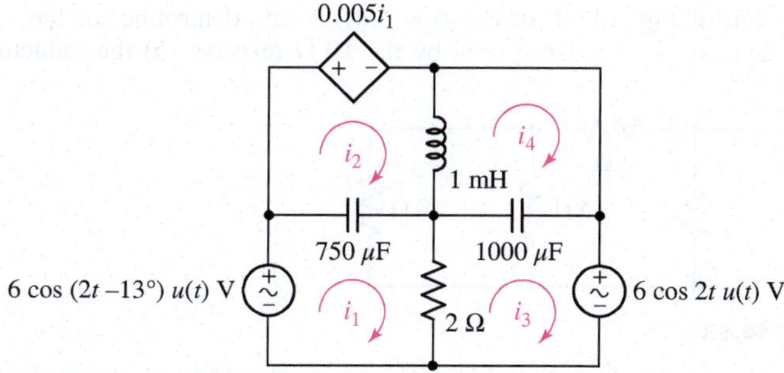

■ **FIGURE 14.59**

63. Assuming no energy initially stored in the circuit of Fig. 14.60, determine the value of v_2 at t equal to (*a*) 1 ms; (*b*) 100 ms; (*c*) 10 s.

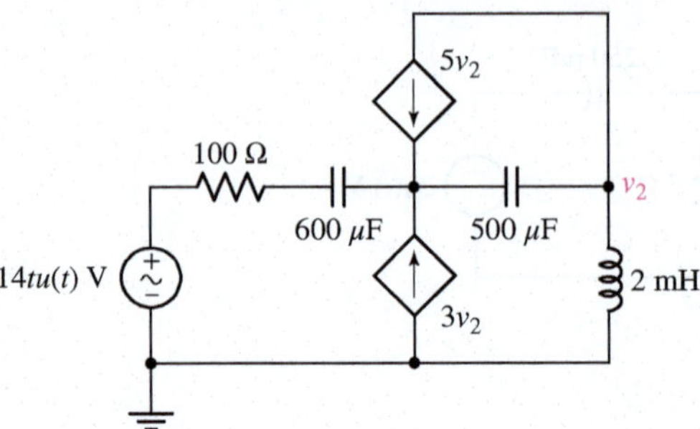

■ **FIGURE 14.60**

14.9 Additional Circuit Analysis Techniques

64. Using repeated source transformations, obtain an **s**-domain expression for the Thévenin equivalent seen by the element labeled **Z** in the circuit of Fig. 14.61.

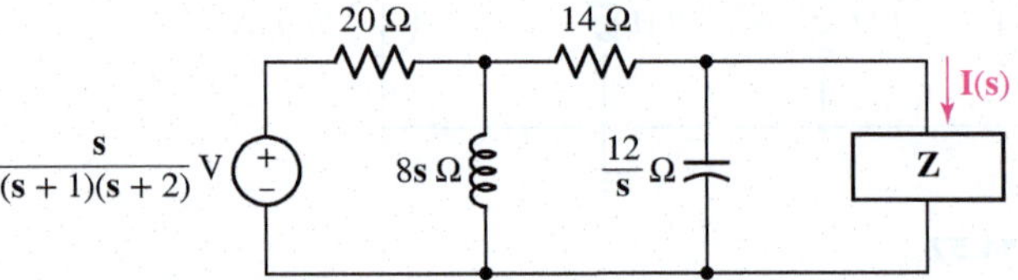

■ **FIGURE 14.61**

65. For the circuit shown in Fig. 14.62, determine the **s**-domain Thévenin equivalent seen by the (*a*) 2 Ω resistor; (*b*) 4 Ω resistor; (*c*) 1.2 F capacitor; (*d*) current source.

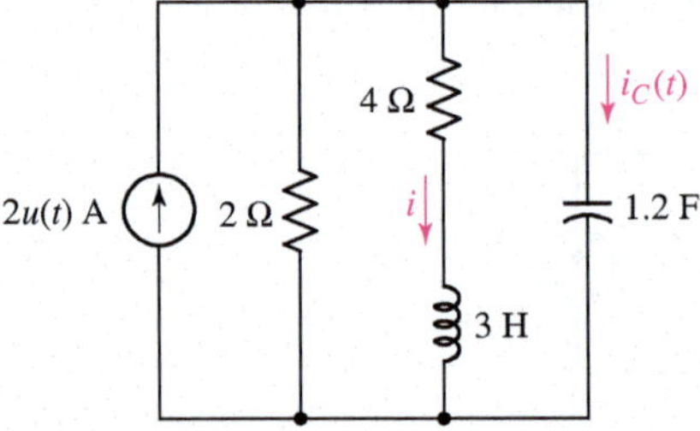

■ **FIGURE 14.62**

66. Use **s**-domain techniques and MATLAB to determine the capacitor current $i_C(t)$ for the circuit of Fig. 14.62. Compare results using the residue and ilaplace MATLAB functions.

67. For the circuit of Fig. 14.63, take $i_s(t) = 5u(t)$ A and determine (*a*) the Thévenin equivalent impedance seen by the 10 Ω resistor; (*b*) the inductor current $i_L(t)$.

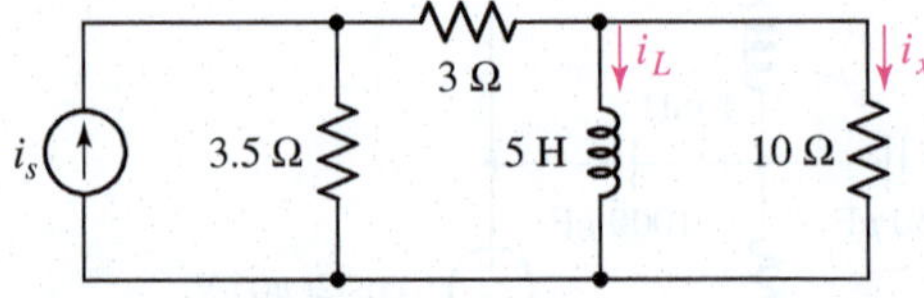

■ **FIGURE 14.63**

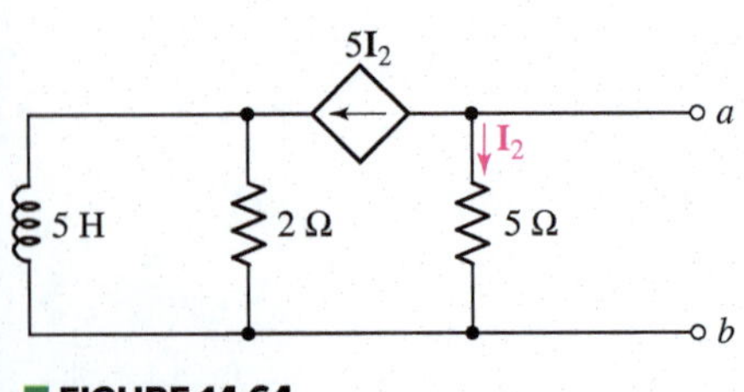

■ **FIGURE 14.64**

68. For the **s**-domain circuit of Fig. 14.64, determine the Thévenin equivalent seen looking into the terminals marked a and b.

14.10 Poles, Zeros, and Transfer Functions

69. Determine the poles and zeros of the following **s**-domain functions:
$(a)\ \frac{\mathbf{s}}{\mathbf{s}+12.5}$; $(b)\ \frac{\mathbf{s}(\mathbf{s}+1)}{(\mathbf{s}+5)(\mathbf{s}+3)}$; $(c)\ \frac{\mathbf{s}+4}{\mathbf{s}^2+8\mathbf{s}+7}$; $(d)\ \frac{\mathbf{s}^2-\mathbf{s}-2}{3\mathbf{s}^3+24\mathbf{s}^2+21\mathbf{s}}$.

70. Use appropriate means to ascertain the poles and zeros of (a) $\mathbf{s}+4$; $(b)\ \frac{2\mathbf{s}}{\mathbf{s}^2-8\mathbf{s}+16}$; $(c)\ \frac{4}{\mathbf{s}^2+8\mathbf{s}+7}$; $(d)\ \frac{\mathbf{s}-5}{\mathbf{s}^3-7\mathbf{s}+6}$.

71. Consider the following expressions and determine the critical frequencies of each:
$(a)\ 5+\mathbf{s}^{-1}$; $(b)\ \frac{\mathbf{s}(\mathbf{s}+1)(\mathbf{s}+4)}{(\mathbf{s}+5)(\mathbf{s}+3)^2}$; $(c)\ \frac{1}{\mathbf{s}^2+4}$; $(d)\ \frac{0.5\mathbf{s}^2-18}{\mathbf{s}^2+1}$.

72. For the network represented schematically in Fig. 14.65, (a) write the transfer function $\mathbf{H}(\mathbf{s}) \equiv \mathbf{V}_{\text{out}}(\mathbf{s})/\mathbf{V}_{\text{in}}(\mathbf{s})$; (b) determine the poles and zeros of $\mathbf{H}(\mathbf{s})$.

73. For each of the two networks represented schematically in Fig. 14.66, (a) write the transfer function $\mathbf{H}(\mathbf{s}) \equiv \mathbf{V}_{\text{out}}(\mathbf{s})/\mathbf{V}_{\text{in}}(\mathbf{s})$; (b) determine the poles and zeros of $\mathbf{H}(\mathbf{s})$.

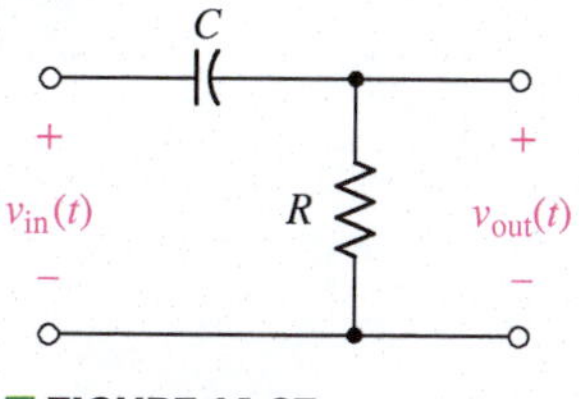

FIGURE 14.65

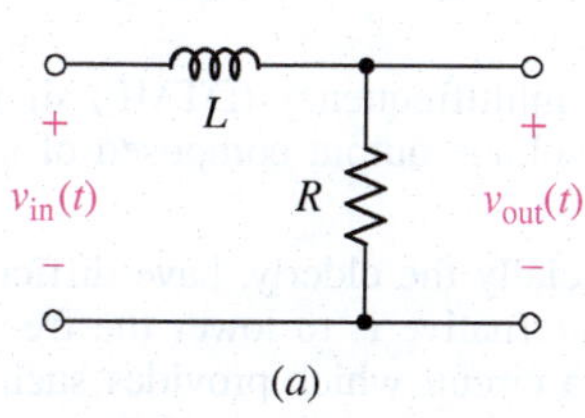

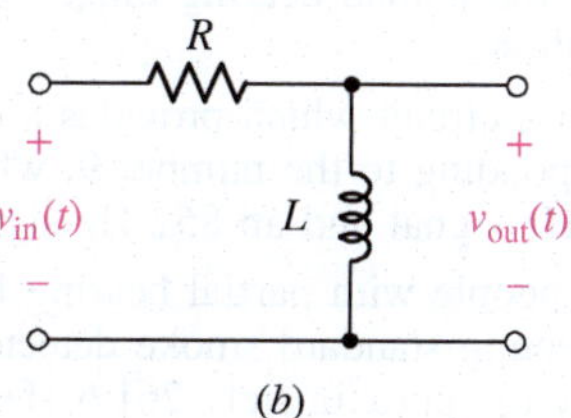

FIGURE 14.66

74. Specify the poles and zeros of $\mathbf{Y}(\mathbf{s})$ as defined by Fig. 14.53.

75. If a network is found to have the transfer function $\mathbf{H}(\mathbf{s}) = \frac{\mathbf{s}}{\mathbf{s}^2+8\mathbf{s}+7}$, determine the **s**-domain output voltage for $v_{\text{in}}(t)$ equal to (a) $3u(t)$ V; (b) $25e^{-2t}u(t)$ V; (c) $4u(t+1)$ V; (d) $2\sin 5t\ u(t)$ V.

14.11 Convolution

76. Referring to Fig. 14.67, employ Eq. [36] to obtain $x(t) * y(t)$.

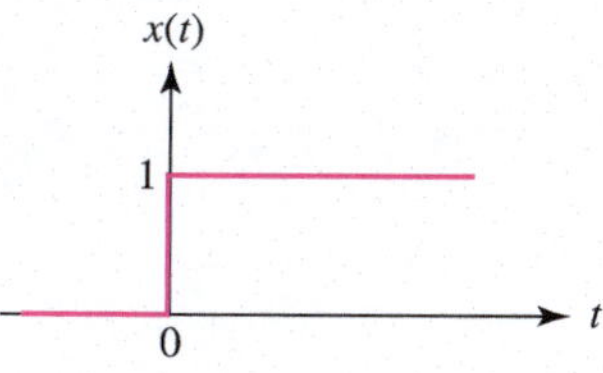

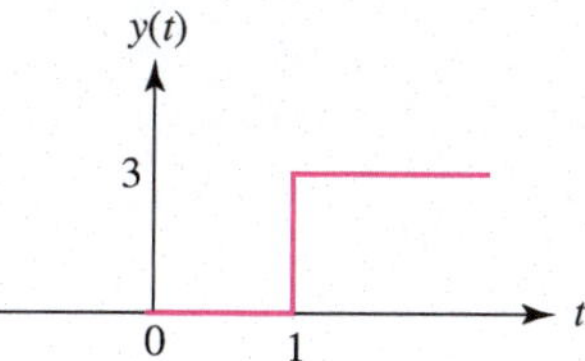

FIGURE 14.67

77. With respect to the functions $x(t)$ and $y(t)$ as plotted in Fig. 14.67, use Eq. [36] to obtain (a) $x(t) * x(t)$; (b) $y(t) * \delta(t)$.

78. Employ graphical convolution techniques to determine $f * g$ if $f(t) = 5u(t)$ and $g(t) = 2u(t) - 2u(t-2) + 2u(t-4) - 2u(t-6)$.

79. Let $h(t) = 2e^{-3t}u(t)$ and $x(t) = u(t) - \delta(t)$. Find $y(t) = h(t) * x(t)$ by (a) using convolution in the time domain; (b) finding $\mathbf{H}(\mathbf{s})$ and $\mathbf{X}(\mathbf{s})$ and then obtaining $\mathscr{L}^{-1}\{\mathbf{H}(\mathbf{s})\mathbf{X}(\mathbf{s})\}$.

80. (a) Determine the impulse response $h(t)$ of the network shown in Fig. 14.68. (b) Use convolution to determine $v_o(t)$ if $v_{\text{in}}(t) = 8u(t)$ V.

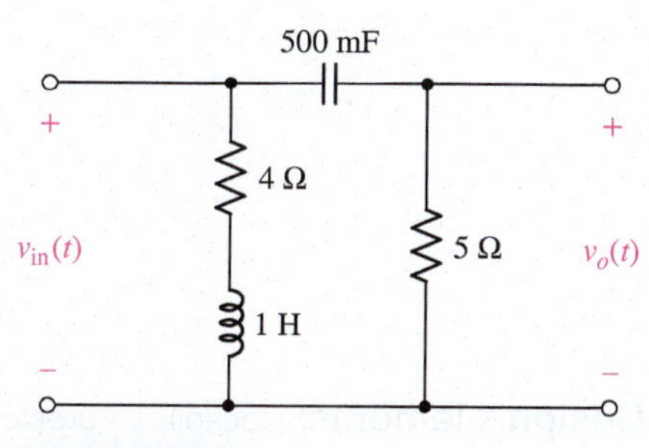

FIGURE 14.68

14.12 A Technique for Synthesizing the Voltage Ratio $\mathbf{H}(\mathbf{s}) = \mathbf{V}_{out}/\mathbf{V}_{in}$

81. Design a circuit which produces the transfer function $\mathbf{H}(\mathbf{s}) = \mathbf{V}_{out}/\mathbf{V}_{in}$ equal to (*a*) $5(\mathbf{s}+1)$; (*b*) $\dfrac{5}{(\mathbf{s}+1)}$; (*c*) $5\dfrac{\mathbf{s}+1}{\mathbf{s}+2}$.

82. Design a circuit which produces the transfer function $\mathbf{H}(\mathbf{s}) = \mathbf{V}_{out}/\mathbf{V}_{in}$ equal to (*a*) $2(\mathbf{s}+1)^2$; (*b*) $\dfrac{3}{(\mathbf{s}+500)(\mathbf{s}+100)}$.

83. Design a circuit which produces the transfer function

$$\mathbf{H}(\mathbf{s}) = \frac{\mathbf{V}_{out}}{\mathbf{V}_{in}} = 3\frac{\mathbf{s}+50}{(\mathbf{s}+75)^2}.$$

84. Find $\mathbf{H}(\mathbf{s}) = \mathbf{V}_{out}/\mathbf{V}_{in}$ as a ratio of polynomials in $\mathbf{s}$ for the op-amp circuit of Fig. 14.38, given the impedance values (in Ω): (*a*) $\mathbf{Z}_1(\mathbf{s}) = 10^3 + (10^8/\mathbf{s})$, $\mathbf{Z}_f(\mathbf{s}) = 5000$; (*b*) $\mathbf{Z}_1(\mathbf{s}) = 5000$, $\mathbf{Z}_f(\mathbf{s}) = 10^3 + (10^8/\mathbf{s})$; (*c*) $\mathbf{Z}_1(\mathbf{s}) = 10^3 + (10^8/\mathbf{s})$, $\mathbf{Z}_f(\mathbf{s}) = 10^4 + (10^8/\mathbf{s})$.

Chapter-Integrating Exercises

85. Design a circuit that provides a frequency of 16 Hz, which is near the lower end of the human hearing range. Verify your design with an appropriate simulation.

86. Design a circuit which provides a dual-tone multifrequency (DTMF) signal corresponding to the number 9, which is a voltage output composed of a 1477 Hz signal and an 852 Hz signal.

87. Many people with partial hearing loss, especially the elderly, have difficulty in detecting standard smoke detectors. An alternative is to lower the frequency to approximately 261.6 Hz. Design a circuit which provides such a signal.

88. (*a*) A standard acoustic guitar is composed of six strings which are traditionally tuned to achieve the notes E2, A2, D3, G3, B3, and E4. The note A2 corresponds to 110 Hz. Design a circuit using 10% tolerance component values (E12 series preferred numbers) that generates a reference signal that can be used to guide the tuning of the guitar's A string. (*b*) Evaluate your design's performance using an appropriate simulation. Comment on how close your output frequency is to the target value of 110 Hz.

CHAPTER 15

Frequency Response

KEY CONCEPTS

Resonant Frequency of Circuits with Inductors and Capacitors

Transfer Function

Bode Diagram Techniques

Quality Factor

Bandwidth

Frequency and Magnitude Scaling

Low- and High-Pass Filters

Bandpass Filter Design

Active Filters

Butterworth Filter Design

INTRODUCTION

We have already been introduced to the concept of frequency response, meaning that the behavior of our circuit can change dramatically depending on the frequency (or frequencies) of operation—a radical departure from our first experiences with simple dc circuits. In this chapter we take the topic to a more refined level, as even simple circuits designed for specific frequency response can be enormously useful in a wide variety of applications. In fact, we make use of frequency-selective circuits throughout the day, probably without even realizing it. For example, switching to our favorite radio station is in fact tuning our radio to selectively amplify a narrow band of signal frequencies; heating microwave popcorn is possible while watching television or talking on a cell phone because the frequencies of each device can be isolated from one another. In addition, studying frequency response and filters can be enjoyable because it gives us the opportunity to push past the analysis of existing circuits to design complex circuits from scratch to meet sometimes stringent specifications. We'll start this journey with a short discussion of *resonance, loss, quality factor,* and *bandwidth*—important concepts for filters as well as any circuit (or system, for that matter) containing energy storage elements.

15.1 TRANSFER FUNCTION

We have seen that the characteristics of a network can vary drastically at different frequencies. For example, a capacitor looks like an open circuit (high impedance) at low frequency and like a short

circuit (low impedance) at high frequency. An important way of describing the frequency response of a network is through the ***transfer function*** **H(s)** describing the output of the network divided by the input as a function of complex frequency **s**, as discussed in Chap. 14. Since we are interested in studying the dependence of the network on frequency, we will focus our attention on the case where $\mathbf{s} = j\omega$ and therefore the transfer function $\mathbf{H(s)} = \mathbf{H}(j\omega)$. The network can consist of a voltage or current as the input, where the corresponding transfer function can be defined as a voltage gain, current gain, impedance, or admittance. Thus, the transfer function will take one of the following forms:

$$\mathbf{H}(j\omega) = \frac{\mathbf{V}_{\text{out}}(j\omega)}{\mathbf{V}_{\text{in}}(j\omega)} = \text{Voltage gain}$$

$$\mathbf{H}(j\omega) = \frac{\mathbf{I}_{\text{out}}(j\omega)}{\mathbf{I}_{\text{in}}(j\omega)} = \text{Current gain}$$

$$\mathbf{H}(j\omega) = \frac{\mathbf{V}_{\text{out}}(j\omega)}{\mathbf{I}_{\text{in}}(j\omega)} = \text{Transfer impedance}$$

$$\mathbf{H}(j\omega) = \frac{\mathbf{I}_{\text{out}}(j\omega)}{\mathbf{V}_{\text{in}}(j\omega)} = \text{Transfer admittance}$$

The transfer function is extremely useful because it provides a way of describing the output of a network for any input. For our case of frequency response, the transfer function can be represented in phasor form $\mathbf{H}(j\omega) = H(j\omega)\underline{/\phi(j\omega)}$, where $H(j\omega)$ and $\phi(j\omega)$ are the frequency-dependent magnitude and phase, respectively. The character of the network will defined by the magnitude and phase frequency response.

EXAMPLE 15.1

Determine the transfer function of the *RC* circuit in Fig. 15.1, defined as $\mathbf{H} = \mathbf{V}_{\text{out}}/\mathbf{V}_{\text{in}}$. Construct a plot of the magnitude and phase as a function of frequency.

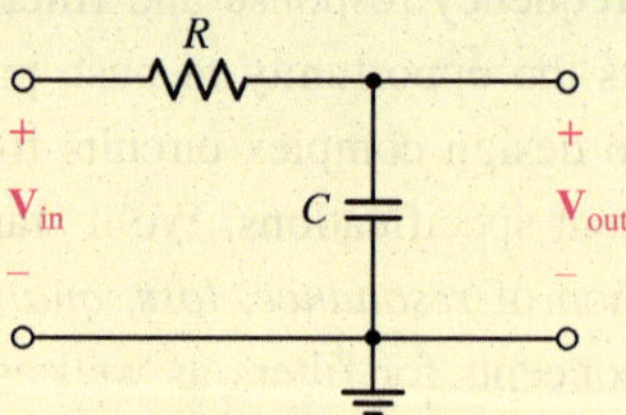

■ **FIGURE 15.1** Example RC network, the transfer function of which is of interest.

As we have observed in the circuit of Fig. 14.27, the circuit is a voltage divider with transfer function given by

$$\mathbf{H}(j\omega) = \frac{\mathbf{V}_{\text{out}}(j\omega)}{\mathbf{V}_{\text{in}}(j\omega)} = \frac{\dfrac{1}{j\omega C}}{R + \dfrac{1}{j\omega C}} = \frac{1}{1 + j\omega CR}$$

The circuit has a pole at $\omega = j/CR$, which is at the natural frequency $\omega_0 = 1/CR$. It is useful to rewrite the transfer function in terms of ω_0

$$\mathbf{H}(j\omega) = \frac{1}{1 + j\omega/\omega_0}$$

The magnitude and phase are given by

$$H = \frac{1}{\sqrt{1 + (\omega/\omega_0)^2}}$$

and

$$\phi = -\tan^{-1}\left(\frac{\omega}{\omega_0}\right)$$

The resulting plot is shown in Fig. 15.2, produced in MATLAB assisted by the following code to describe magnitude and phase:

```
omega=linspace(0,10,100); % define frequency vector omega
for i=1:100; % step through all points in frequency
  H(i)=1/sqrt(1+omega(i)^2);
  phi(i)=-atan(omega(i))*180/pi;
end
```

We see that the transfer function magnitude is unity and phase is zero at low frequency, since the capacitor has a very high impedance, behaving as an open circuit. As frequency increases, the capacitor impedance decreases and eventually reaches the short-circuit condition, where the transfer function magnitude and phase become zero and −90°, respectively. The natural frequency ω_0 is a point of interest, where the magnitude and phase are $1/\sqrt{2}$ and −45°, respectively.

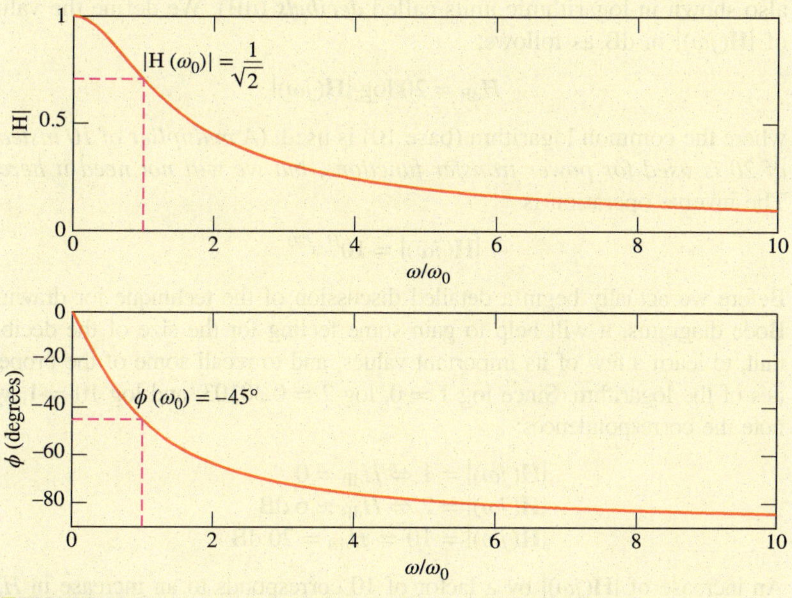

FIGURE 15.2 Frequency response of the magnitude and phase of RC network.

(Continued on next page)

PRACTICE

15.1 Write an expression for the transfer function of the RC network in Fig 15.1 after switching the positions of the resistor and capacitor such that $\mathbf{V}_{out}$ is now the voltage drop across the resistor. Evaluate at $\omega = 0$, $\omega = \omega_0 = 1/CR$, and $\omega = \infty$.

Ans: $j\omega CR/(j\omega CR + 1)$. At $\omega = 0$, $H = 0$ and $\phi = 90°$. At $\omega = \omega_0$, $H = 1/\sqrt{(2)}$ and $\phi = 45°$. At $\omega = \infty$, $H = 1$ and $\phi = 0$.

15.2 BODE DIAGRAMS

In this section we will discover a quick method of obtaining an *approximate* picture of the amplitude and phase variation of a given transfer function as functions of ω. Accurate curves may, of course, be plotted after calculating values with a programmable calculator or a computer; curves may also be produced directly on the computer. Our object here, however, is to obtain a better picture of the response than we could visualize from a pole-zero plot, but yet not mount an all-out computational offensive. Furthermore, we will be using logarithmic scales for frequency and magnitude, providing an improved snapshot of frequency response over a wide range of values.

The Decibel (dB) Scale

The approximate response curve we construct is called an asymptotic plot, or a ***Bode plot,*** or a ***Bode diagram,*** after its developer, Hendrik W. Bode, who was an electrical engineer and mathematician with the Bell Telephone Laboratories. Both the magnitude and phase curves are shown using a logarithmic frequency scale for the abscissa, and the magnitude itself is also shown in logarithmic units called ***decibels*** (dB). We define the value of $|\mathbf{H}(j\omega)|$ in dB as follows:

The decibel is named in honor of Alexander Graham Bell.

$$H_{dB} = 20 \log |\mathbf{H}(j\omega)|$$

where the common logarithm (base 10) is used. (*A multiplier of 10 instead of 20 is used for power transfer functions, but we will not need it here.*) The inverse operation is

$$|\mathbf{H}(j\omega)| = 10^{H_{dB}/20}$$

Before we actually begin a detailed discussion of the technique for drawing Bode diagrams, it will help to gain some feeling for the size of the decibel unit, to learn a few of its important values, and to recall some of the properties of the logarithm. Since $\log 1 = 0$, $\log 2 = 0.30103$, and $\log 10 = 1$, we note the correspondences:

$$|\mathbf{H}(j\omega)| = 1 \Leftrightarrow H_{dB} = 0$$
$$|\mathbf{H}(j\omega)| = 2 \Leftrightarrow H_{dB} \approx 6 \text{ dB}$$
$$|\mathbf{H}(j\omega)| = 10 \Leftrightarrow H_{dB} = 20 \text{ dB}$$

An increase of $|\mathbf{H}(j\omega)|$ by a factor of 10 corresponds to an increase in H_{dB} by 20 dB. Moreover, $\log 10^n = n$, and thus $10^n \Leftrightarrow 20n$ dB, so that 1000 corresponds to 60 dB, while 0.01 is represented as −40 dB. Using only the values already given, we may also note that 20 log 5 = 20 log $\frac{10}{2}$ = 20 log 10 − 20 log 2 = 20 − 6 = 14 dB, and thus 5 ⇔ 14 dB. Also, log $\sqrt{x} = \frac{1}{2}\log x$, and therefore $\sqrt{2}$ ⇔ 3 dB and $1/\sqrt{2}$ ⇔ −3 dB.[1]

We will write our transfer functions in terms of **s**, substituting $\mathbf{s} = j\omega$ when we are ready to find the magnitude or phase angle. If desired, the magnitude may be written in terms of dB at that point.

PRACTICE

15.2 Calculate H_{db} at ω = 146 rad/s if **H**(**s**) equals (*a*) 20/(**s** + 100); (*b*) 20(**s** + 100); (*c*) 20**s**. Calculate $|\mathbf{H}(j\omega)|$ if H_{dB} equals (*d*) 29.2 dB; (*e*) −15.6 dB; (*f*) −0.318 dB.

Ans: −18.94 dB; 71.0 dB; 69.3 dB; 28.8; 0.1660; 0.964.

Determination of Asymptotes

Our next step is to factor **H**(**s**) to display its poles and zeros. We first consider a zero at $\mathbf{s} = -a$, written in a standardized form as

$$\mathbf{H}(\mathbf{s}) = 1 + \frac{\mathbf{s}}{a} \qquad [1]$$

The Bode diagram for this function consists of the two asymptotic curves approached by H_{dB} for very large and very small values of ω. Thus, we begin by finding

$$|\mathbf{H}(j\omega)| = \left|1 + \frac{j\omega}{a}\right| = \sqrt{1 + \frac{\omega^2}{a^2}}$$

and thus

$$H_{dB} = 20 \log \left|1 + \frac{j\omega}{a}\right| = 20 \log \sqrt{1 + \frac{\omega^2}{a^2}}$$

When $\omega \ll a$,

$$H_{dB} \approx 20 \log 1 = 0 \qquad (\omega \ll a)$$

This simple asymptote is shown in Fig. 15.3. It is drawn as a solid line for $\omega < a$ and as a dashed line for $\omega > a$.

When $\omega \gg a$,

$$H_{dB} \approx 20 \log \frac{\omega}{a} \qquad (\omega \gg a)$$

At $\omega = a$, $H_{dB} = 0$; at $\omega = 10a$, H_{dB} = 20 dB; and at $\omega = 100a$, H_{dB} = 40 dB. Thus, the value of H_{dB} increases 20 dB for every tenfold increase in frequency. The asymptote therefore has a slope of 20 dB/decade. Since H_{dB} increases by 6 dB when ω doubles, an alternate value for the slope is 6 dB/octave. The high-frequency asymptote is also shown in Fig. 15.3, a solid line for $\omega > a$, and a broken line for $\omega < a$. Note that the two asymptotes intersect at $\omega = a$, the frequency of the zero. This frequency is also described as the ***corner, break, 3 dB,*** or ***half-power frequency.***

A ***decade*** refers to a range of frequencies defined by a factor of 10, such as 3 Hz to 30 Hz, or 12.5 MHz to 125 MHz. An ***octave*** refers to a range of frequencies defined by a factor of 2, such as 7 GHz to 14 GHz.

(1) Note that we are being slightly dishonest here by using 20 log $\sqrt{2}$ = 6 dB rather than 6.02 dB. It is customary, however, to represent 2 as 3 dB; since the dB scale is inherently logarithmic, the small inaccuracy is seldom significant.

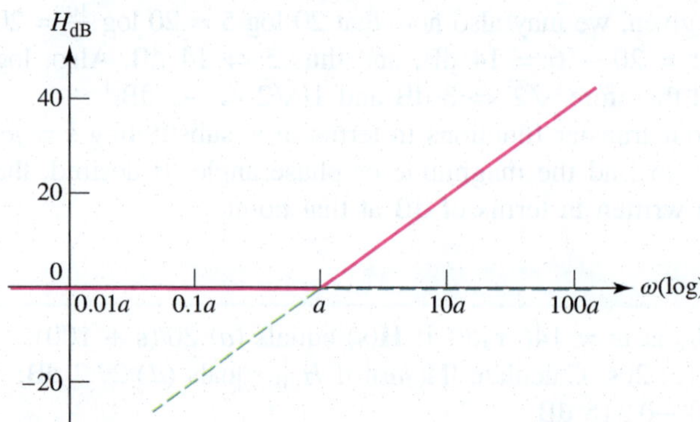

FIGURE 15.3 The Bode amplitude plot for **H**(**s**) = 1 + **s**/*a* consists of the low- and high-frequency asymptotes, shown as dashed lines. They intersect on the abscissa at the corner frequency. The Bode plot represents the response in terms of two asymptotes, both straight lines and both easily drawn.

Smoothing Bode Plots

Let us see how much error is embodied in our asymptotic response curve. At the corner frequency ($\omega = a$),

$$H_{dB} = 20 \log \sqrt{1 + \frac{a^2}{a^2}} = 3 \text{ dB}$$

as compared with an asymptotic value of 0 dB. At $\omega = 0.5a$, we have

$$H_{dB} = 20 \log \sqrt{1.25} \approx 1 \text{ dB}$$

Note that we continue to abide by the convention of taking $\sqrt{2}$ as corresponding to 3 dB.

Thus, the exact response is represented by a smooth curve that lies 3 dB above the asymptotic response at $\omega = a$, and 1 dB above it at $\omega = 0.5a$ (and also at $\omega = 2a$). This information can always be used to smooth out the corner if a more exact result is desired.

Multiple Terms

Most transfer functions will consist of more than a simple zero (or simple pole). This, however, is easily handled by the Bode method, since we are in fact working with logarithms. For example, consider a function

$$\mathbf{H}(\mathbf{s}) = K\left(1 + \frac{\mathbf{s}}{s_1}\right)\left(1 + \frac{\mathbf{s}}{s_2}\right)$$

where $K =$ constant, and $-s_1$ and $-s_2$ represent the two zeros of our function $\mathbf{H}(\mathbf{s})$. For this function H_{dB} may be written as

$$H_{dB} = 20 \log \left| K\left(1 + \frac{j\omega}{s_1}\right)\left(1 + \frac{j\omega}{s_2}\right) \right|$$

$$= 20 \log \left[K\sqrt{1 + \left(\frac{\omega}{s_1}\right)^2} \sqrt{1 + \left(\frac{\omega}{s_2}\right)^2} \right]$$

or

$$H_{dB} = 20 \log K + 20 \log \sqrt{1 + \left(\frac{\omega}{s_1}\right)^2} + 20 \log \sqrt{1 + \left(\frac{\omega}{s_2}\right)^2}$$

which is simply the sum of a constant (frequency-independent) term 20 log K and two simple zero terms of the form previously considered. In other words, *we may construct a sketch of H_{dB} by simply graphically adding the plots of the separate terms.* We explore this in the following example.

EXAMPLE 15.2

Obtain the Bode plot of the input impedance of the network shown in Fig. 15.4.

We begin by writing the input impedance,

$$\mathbf{Z}_{in}(\mathbf{s}) = \mathbf{H}(\mathbf{s}) = 20 + 0.2\mathbf{s}$$

Putting this in standard form, we obtain

$$\mathbf{H}(\mathbf{s}) = 20\left(1 + \frac{\mathbf{s}}{100}\right)$$

The two factors constituting $\mathbf{H}(\mathbf{s})$ are a zero at $\mathbf{s} = -100$, leading to a break frequency of $\omega = 100$ rad/s, and a constant equivalent to 20 log 20 = 26 dB. Each of these is sketched lightly in Fig. 15.5*a*. Since we are working with the logarithm of $|\mathbf{H}(j\omega)|$, we next add together the Bode plots corresponding to the individual factors. The resultant magnitude plot appears as Fig. 15.5*b*. No attempt has been made to smooth out the corner with a +3 dB correction at $\omega = 100$ rad/s; this is left to the reader as a quick exercise.

■ **FIGURE 15.4** If $\mathbf{H}(\mathbf{s})$ is selected as $\mathbf{Z}_{in}(\mathbf{s})$ for this network, then the Bode plot for $\mathbf{H}_{dB}$ is constructed as shown in Fig. 15.5.

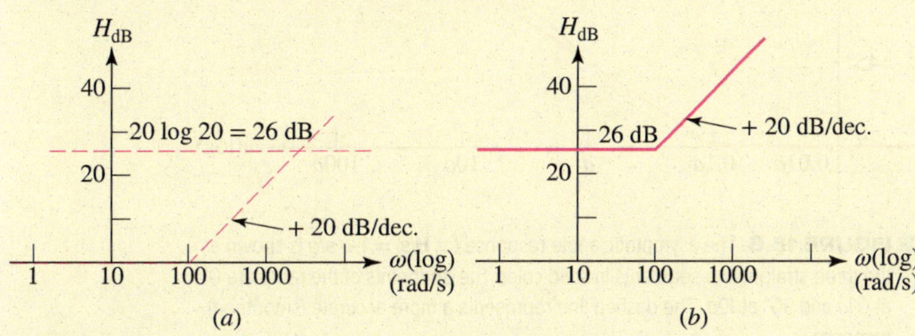

■ **FIGURE 15.5** (*a*) The Bode plots for the factors of $\mathbf{H}(\mathbf{s}) = 20(1 + \mathbf{s}/100)$ are sketched individually. (*b*) The composite Bode plot is shown as the sum of the plots of part (*a*).

PRACTICE

15.3 Construct a Bode magnitude plot for $\mathbf{H}(\mathbf{s}) = 50 + \mathbf{s}$.

Ans: 34 dB, $\omega < 50$ rad/s; slope = +20 dB/decade $\omega > 50$ rad/s.

Phase Response

Returning to the transfer function of Eq. [1], we would now like to determine the ***phase response*** for the simple zero,

$$\text{ang}\,\mathbf{H}(j\omega) = \text{ang}\left(1 + \frac{j\omega}{a}\right) = \tan^{-1}\frac{\omega}{a}$$

This expression is also represented by its asymptotes, although three straight-line segments are required. For $\omega \ll a$, ang $\mathbf{H}(j\omega) \approx 0°$, and we use this as our asymptote when $\omega < 0.1a$:

$$\text{ang } \mathbf{H}(j\omega) = 0° \qquad (\omega < 0.1a)$$

At the high end, $\omega \gg a$, we have ang $\mathbf{H}(j\omega) \approx 90°$, and we use this above $\omega = 10a$:

$$\text{ang } \mathbf{H}(j\omega) = 90° \qquad (\omega > 10a)$$

Since the angle is 45° at $\omega = a$, we now construct the straight-line asymptote extending from 0° at $\omega = 0.1a$ through 45° at $\omega = a$, to 90° at $\omega = 10a$. This straight line has a slope of 45°/decade. It is shown as a solid curve in Fig. 15.6, while the exact angle response is shown as a broken line. The maximum differences between the asymptotic and true responses are $\pm 5.71°$ at $\omega = 0.1a$ and $10a$. Errors of $\mp 5.29°$ occur at $\omega = 0.394a$ and $2.54a$; the error is zero at $\omega = 0.159a$, a, and $6.31a$. The phase angle plot is typically left as a straight-line approximation, although smooth curves can also be drawn in a manner similar to that depicted in Fig. 15.6 by the dashed line.

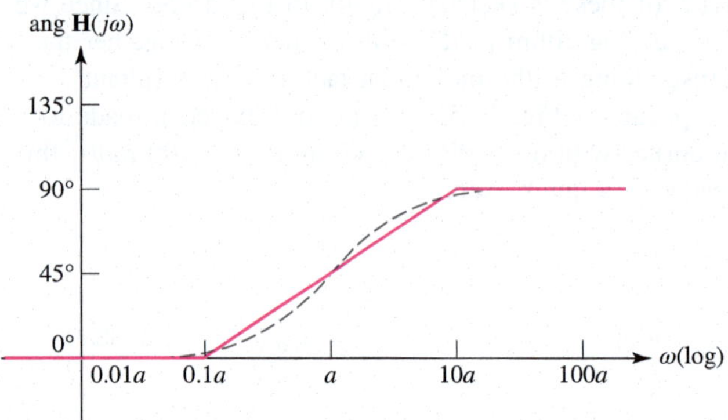

FIGURE 15.6 The asymptotic angle response for $\mathbf{H(s)} = 1 + \mathbf{s}/a$ is shown as the three straight-line segments in solid color. The endpoints of the ramp are 0° at 0.1a and 90° at 10a. The dashed line represents a more accurate (smoothed) response.

It is worth pausing briefly here to consider what the phase plot is telling us. In the case of a simple zero at $\mathbf{s} = a$, we see that for frequencies much less than the corner frequency, the phase of the response function is 0°. For high frequencies, however ($\omega \gg a$), the phase is 90°. In the vicinity of the corner frequency, the phase of the transfer function varies somewhat rapidly. The actual phase angle imparted to the response can therefore be selected through the design of the circuit (which determines a).

PRACTICE

15.4 Draw the Bode phase plot for the transfer function of Example 15.2.

Ans: 0°, $\omega \leq 10$; 90°, $\omega \geq 1000$; 45°, $\omega = 100$; 45°/dec slope, $10 < \omega < 1000$. (ω in rad/s).

Additional Considerations in Creating Bode Plots

We next consider a simple pole, similar to the *RC* network shown in Example 15.1:

$$\mathbf{H(s)} = \frac{1}{1 + \mathbf{s}/a} \qquad [2]$$

Since this is the reciprocal of a zero, the logarithmic operation leads to a Bode plot which is the *negative* of that obtained previously. The amplitude is 0 dB up to $\omega = a$, and then the slope is -20 dB/decade for $\omega > a$. The angle plot is 0° for $\omega < 0.1a$, $-90°$ for $\omega > 10a$, and $-45°$ at $\omega = a$, and it has a slope of $-45°$/decade when $0.1a < \omega < 10a$. The reader is encouraged to generate the Bode plot for this function by working directly with Eq. [2].

Another term that can appear in $\mathbf{H(s)}$ is a factor of $\mathbf{s}$ in the numerator or denominator. If $\mathbf{H(s)} = \mathbf{s}$, then

$$H_{\text{dB}} = 20 \log |\omega|$$

Thus, we have an infinite straight line passing through 0 dB at $\omega = 1$ and having a slope everywhere of 20 dB/decade. This is shown in Fig. 15.7*a*. If the $\mathbf{s}$ factor occurs in the denominator, a straight line is obtained having a slope of -20 dB/decade and passing through 0 dB at $\omega = 1$, as shown in Fig. 15.7*b*.

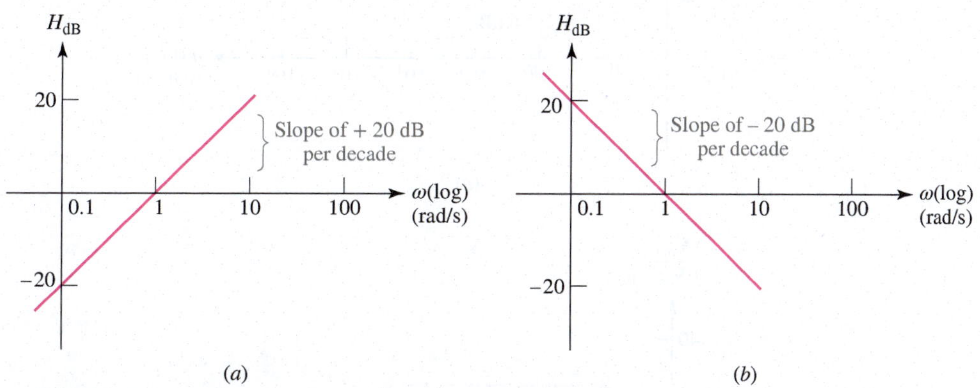

FIGURE 15.7 The asymptotic diagrams are shown for (*a*) $\mathbf{H(s)} = \mathbf{s}$ and $\mathbf{H(s)} = 1/\mathbf{s}$. Both are infinitely long straight lines passing through 0 dB at $\omega = 1$ and having slopes of ± 20 dB/decade.

Another simple term found in $\mathbf{H(s)}$ is the multiplying constant K. This yields a Bode plot which is a horizontal straight line lying $20 \log |K|$ dB above the abscissa. It will actually be below the abscissa if $|K| < 1$.

EXAMPLE **15.3**

Obtain the Bode plot for the gain of the circuit shown in Fig. 15.8.

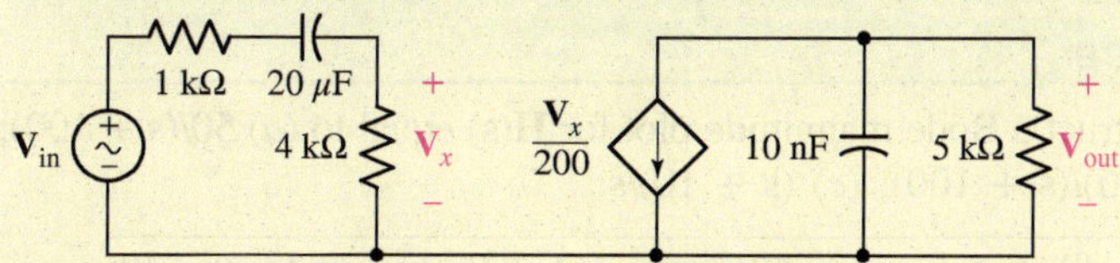

FIGURE 15.8 If $\mathbf{H(s)} = \mathbf{V}_{out}/\mathbf{V}_{in}$, this amplifier is found to have the Bode amplitude plot shown in Fig. 15.9*b*, and the phase plot shown in Fig. 15.10.

(*Continued on next page*)

We work from left to right through the circuit and write the expression for the voltage gain,

$$\mathbf{H(s)} = \frac{\mathbf{V}_{out}}{\mathbf{V}_{in}} = \frac{4000}{5000 + 10^6/20\mathbf{s}}\left(-\frac{1}{200}\right)\frac{5000(10^8/\mathbf{s})}{5000 + 10^8/\mathbf{s}}$$

which simplifies (mercifully) to

$$\mathbf{H(s)} = \frac{-2\mathbf{s}}{(1 + \mathbf{s}/10)(1 + \mathbf{s}/20{,}000)} \qquad [3]$$

We see a constant, $20 \log |-2| = 6$ dB, break points at $\omega = 10$ rad/s and $\omega = 20{,}000$ rad/s, and a linear factor **s**. Each of these is sketched in Fig. 15.9*a*, and the four sketches are added to give the Bode magnitude plot in Fig. 15.9*b*.

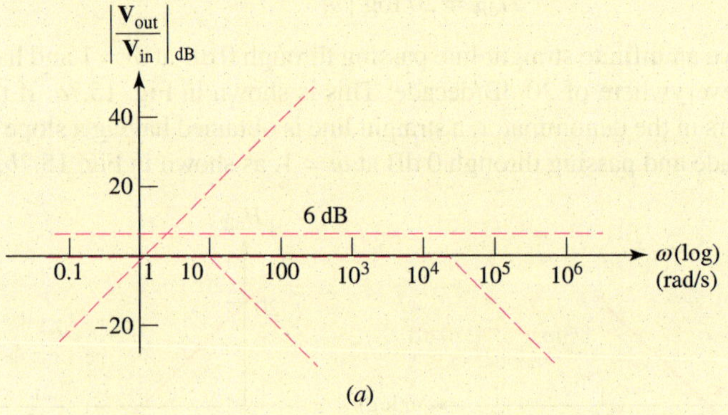

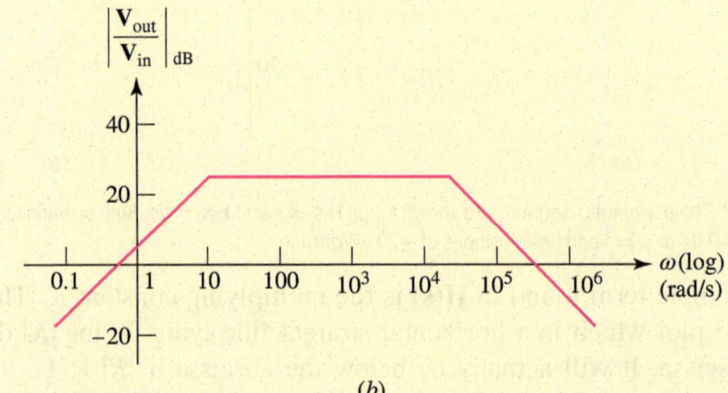

FIGURE 15.9 (*a*) Individual Bode magnitude sketches are made for the factors (−2), (**s**), $(1 + \mathbf{s}/10)^{-1}$, and $(1 + \mathbf{s}/20{,}000)^{-1}$. (*b*) The four separate plots of part (*a*) are added to give the Bode magnitude plots for the amplifier of Fig. 15.8.

PRACTICE

15.5 Construct a Bode magnitude plot for **H(s)** equal to (*a*) 50/(**s** + 100); (*b*) (**s** + 10)/(**s** + 100); (*c*) (**s** + 10)/**s**.

Ans: (*a*) −6 dB, $\omega < 100$; −20 dB/decade, $\omega > 100$; (*b*) −20 dB, $\omega < 10$; +20 dB/decade, $10 < \omega < 100$; 0 dB, $\omega > 100$; (*c*) 0 dB, $\omega > 10$; −20 dB/decade, $\omega < 10$. All ω in rad/s.

Before we construct the phase plot for the amplifier of Fig. 15.8, let us take a few moments to investigate several of the details of the magnitude plot.

First, it is wise not to rely too heavily on graphical addition of the individual magnitude plots. Instead, the exact value of the combined magnitude plot may be found easily at selected points by considering the asymptotic value of each factor of **H**(**s**) at the point in question. For example, in the flat region of Fig. 15.9*a* between $\omega = 10$ and $\omega = 20{,}000$, we are below the corner at $\omega = 20{,}000$, and so we represent $(1 + \mathbf{s}/20{,}000)$ by 1; but we are above $\omega = 10$, so $(1 + \mathbf{s}/10)$ is represented as $\omega/10$. Hence,

$$H_{\text{dB}} = 20 \log \left| \frac{-2\omega}{(\omega/10)(1)} \right|$$

$$= 20 \log 20 = 26 \text{ dB} \qquad (10 < \omega < 20{,}000)$$

We might also wish to know the frequency at which the asymptotic response crosses the abscissa at the high end. The two factors are expressed here as $\omega/10$ and $\omega/20{,}000$; thus

$$H_{\text{dB}} = 20 \log \left| \frac{-2\omega}{(\omega/10)(\omega/20{,}000)} \right| = 20 \log \left| \frac{400{,}000}{\omega} \right|$$

Since $H_{\text{dB}} = 0$ at the abscissa crossing, $400{,}000/\omega = 1$, and therefore $\omega = 400{,}000$ rad/s.

Many times we do not need an accurate Bode plot drawn on printed semilog paper. Instead we construct a rough logarithmic frequency axis on simple lined paper. After selecting the interval for a decade—say, a distance L extending from $\omega = \omega_1$ to $\omega = 10\omega_1$ to (where ω_1 is usually an integral power of 10)—we let x locate the distance that ω lies to the right of ω_1, so that $x/L = \log(\omega/\omega_1)$. Of particular help is the knowledge that $x = 0.3L$ when $\omega = 2\omega_1$, $x = 0.6L$ at $\omega = 4\omega_1$, and $x = 0.7L$ at $\omega = 5\omega_1$.

EXAMPLE 15.4

Draw the phase plot for the transfer function given by Eq. [3], H(s) = −2s/[(1 + s/10)(1 + s/20,000)].

We begin by inspecting $\mathbf{H}(j\omega)$:

$$\mathbf{H}(j\omega) = \frac{-j2\omega}{(1 + j\omega/10)(1 + j\omega/20{,}000)} \qquad [4]$$

The angle of the numerator is a constant, −90°.

The remaining factors are represented as the sum of the angles contributed by break points at $\omega = 10$ and $\omega = 20{,}000$. These three terms appear as broken-line asymptotic curves in Fig. 15.10, and their sum is shown as the solid curve. An equivalent representation is obtained if the curve is shifted upward by 360°.

Exact values can also be obtained for the asymptotic phase response. For example, at $\omega = 10^4$ rad/s, the angle in Fig. 15.10 is obtained from the numerator and denominator terms in Eq. [4]. The numerator angle

(Continued on next page)

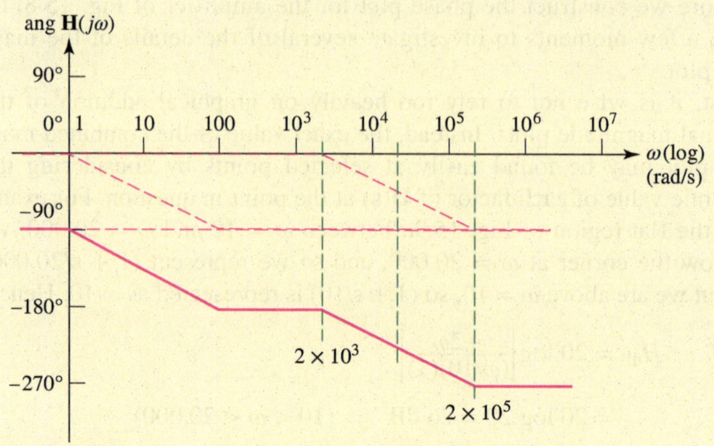

■ **FIGURE 15.10** The solid curve displays the asymptotic phase response of the amplifier shown in Fig. 15.8.

is −90°. The angle for the pole at $\omega = 10$ is −90°, since ω is greater than 10 times the corner frequency. Between 0.1 and 10 times the corner frequency, we recall that the slope is −45° per decade for a simple pole. For the break point at 20,000 rad/s, we therefore calculate the angle, $-45° \log(\omega/0.1a) = -45° \log[10{,}000/(0.1 \times 20{,}000)] = -31.5°$.

The algebraic sum of these three contributions is $-90° - 90° - 31.5° = -211.5°$, a value that appears to be moderately near the asymptotic phase curve of Fig. 15.10.

PRACTICE

15.6 Draw the Bode phase plot for $\mathbf{H}(\mathbf{s})$ equal to (*a*) $50/(\mathbf{s} + 100)$; (*b*) $(\mathbf{s} + 10)/(\mathbf{s} + 100)$; (*c*) $(\mathbf{s} + 10)/\mathbf{s}$.

Ans: (*a*) 0°, $\omega < 10$; −45°/decade, $10 < \omega < 1000$; −90°, $\omega > 1000$; (*b*) 0°, $\omega < 1$; +45°/decade, $1 < \omega < 10$; 45°, $10 < \omega < 100$; −45°/decade, $100 < \omega < 1000$; 0°, $\omega > 1000$; (*c*) −90°, $\omega < 1$; +45°/decade, $1 < \omega < 100$; 0°, $\omega > 100$.

Higher-Order Terms

The zeros and poles that we have been considering are all first-order terms, such as $\mathbf{s}^{\pm 1}$, $(1 + 0.2\mathbf{s})^{\pm 1}$, and so forth. We may extend our analysis to higher-order poles and zeros very easily, however. A term $\mathbf{s}^{\pm n}$ yields a magnitude response that passes through $\omega = 1$ with a slope of $+20n$ dB/decade; the phase response is a constant angle of $\pm 90n°$. Also, a multiple zero, $(1 + \mathbf{s}/a)^n$, must represent the sum of n of the magnitude-response curves, or n of the phase-response curves of the simple zero. We therefore obtain an asymptotic magnitude plot that is 0 dB for $\omega < a$ and has a slope of $20n$ dB/ decade when $\omega < a$; the error is $-3n$ dB at $\omega = a$, and $-n$ dB at $\omega = 0.5a$ and $2a$. The phase plot is 0° for $\omega < 0.1a$, $90n°$ for $\omega > 10a$, $45n°$ at $\omega = a$, and a straight line with a slope of $45n°$/decade for $0.1a < \omega < 10a$, and it has errors as large as $\pm 5.71n°$ at two frequencies.

The asymptotic magnitude and phase curves associated with a factor such as $(1 + \mathbf{s}/20)^{-3}$ may be drawn quickly, but the relatively large errors associated with the higher powers should be kept in mind.

Complex Conjugate Pairs

The last type of factor we should consider represents a conjugate complex pair of poles or zeros. We adopt the following as the standard form for a pair of zeros:

$$\mathbf{H(s)} = 1 + 2\zeta\left(\frac{\mathbf{s}}{\omega_0}\right) + \left(\frac{\mathbf{s}}{\omega_0}\right)^2$$

The paired complex zeros or poles are a defining characteristic of ***resonance*** in circuits, which will be covered in the following sections of this chapter. The quantity ζ is the ***damping factor,*** and we will see shortly that ω_0 is the corner frequency of the asymptotic response. The damping factor is analogous to the damping coefficient we have studied in the *RLC* circuits of Chap. 9. In the field of system theory or automatic control theory, it is traditional to describe damping in a form that utilizes the dimensionless parameter ζ (zeta) and the characteristic equation

$$\mathbf{s}^2 + 2\zeta\omega_0\mathbf{s} + \omega_0^2$$

If $\zeta = 1$, we see that $\mathbf{H(s)} = 1 + 2(\mathbf{s}/\omega_0) + (\mathbf{s}/\omega_0)^2 = (1 + \mathbf{s}/\omega_0)^2$, a second-order zero such as we have just considered. If $\zeta > 1$, then $\mathbf{H(s)}$ may be factored to show two simple zeros. Thus, if $\zeta = 1.25$, then $\mathbf{H(s)} = 1 + 2.5(\mathbf{s}/\omega_0) + (\mathbf{s}/\omega_0)^2 = (1 + \mathbf{s}/2\omega_0)(1 + \mathbf{s}/0.5\omega_0)$, and we again have a familiar situation.

A new case arises when $0 \le \zeta \le 1$. There is no need to find values for the conjugate complex pair of roots. Instead, we determine the low- and high-frequency asymptotic values for both the magnitude and phase response, and we then apply a correction that depends on the value of ζ.

For the magnitude response, we have

$$H_{\text{dB}} = 20 \log |\mathbf{H}(j\omega)| = 20 \log \left|1 + j2\zeta\left(\frac{\omega}{\omega_0}\right) - \left(\frac{\omega}{\omega_0}\right)^2\right| \qquad [5]$$

When $\omega \ll \omega_0$, $\mathbf{H}_{\text{dB}} = 20 \log |1| = 0$ dB. This is the low-frequency asymptote. Next, if $\omega \gg \omega_0$, only the squared term is important, and $\mathbf{H}_{\text{dB}} = 20 \log |-(\omega/\omega_0)^2| = 40 \log(\omega/\omega_0)$. We have a slope of +40 dB/decade. This is the high-frequency asymptote, and the two asymptotes intersect at 0 dB, $\omega = \omega_0$. The solid curve in Fig. 15.11 shows this asymptotic representation of the magnitude response. However, a correction must be applied in the neighborhood of the corner frequency. We let $\omega = \omega_0$ in Eq. [5] and have

$$H_{\text{dB}} = 20 \log \left|j2\zeta\left(\frac{\omega}{\omega_0}\right)\right| = 20 \log(2\zeta) \qquad [6]$$

If $\zeta = 1$, a limiting case, the correction is +6 dB; for $\zeta = 0.5$, no correction is required; and if $\zeta = 0.1$, the correction is -14 dB. Knowing this one correction value is often sufficient to draw a satisfactory asymptotic magnitude response. Figure 15.11 shows more accurate curves for $\zeta = 1$, 0.5, 0.25, and 0.1, as calculated from Eq. [5]. For example, if $\zeta = 0.25$, then the exact value of H_{dB} at $\omega = 0.5\omega_0$ is

$$H_{\text{dB}} = 20 \log |1 + j0.25 - 0.25| = 20 \log \sqrt{0.75^2 + 0.25^2} = -2.0 \text{ dB}$$

The negative peaks do not show a minimum value exactly at $\omega = \omega_0$, as we can see by the curve for $\zeta = 0.5$. The valley is always found at a slightly lower frequency.

If then $\zeta = 0$, then $\mathbf{H}(j\omega_0) = 0$ and $H_{\text{dB}} = -\infty$. Bode plots are not usually drawn for this situation.

Our last task is to draw the asymptotic phase response for $\mathbf{H}(j\omega) = 1 + j2\zeta(\omega/\omega_0) - (\omega/\omega_0)^2$. Below $\omega = 0.1\omega_0$, we let ang $\mathbf{H}(j\omega) = 0°$; above $\omega = 10\omega_0$,

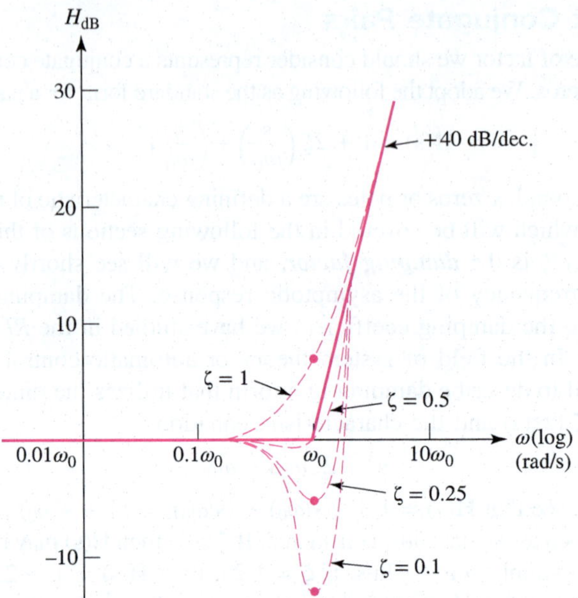

FIGURE 15.11 Bode amplitude plots are shown for $\mathbf{H}(\mathbf{s}) = 1 + 2\zeta(\mathbf{s}/\omega_0) + (\mathbf{s}/\omega_0)^2$ for several values of the damping factor ζ.

we have ang $\mathbf{H}(j\omega)$ = ang $[-(\omega/\omega_0)^2] = 180°$. At the corner frequency, ang $\mathbf{H}(j\omega_0) = \text{ang}(j2\zeta) = 90°$. In the interval $0.1\omega_0 < \omega < 10\omega_0$, we begin with the straight line shown as a solid curve in Fig. 15.12. It extends from $(0.1\omega_0, 0°)$, through $(\omega_0, 90°)$, and terminates at $(10\omega_0, 180°)$; it has a slope of 90°/decade.

We must now provide some correction to this basic curve for various values of ζ. From Eq. [5], we may obtain

$$\text{ang } \mathbf{H}(j\omega) = \tan^{-1}\frac{2\zeta(\omega/\omega_0)}{1-(\omega/\omega_0)^2}$$

One accurate value above and one below $\omega = \omega_0$ may be sufficient to give an approximate shape to the curve. If we take $\omega = 0.5\omega_0$, we find ang $\mathbf{H} = (j0.5\omega_0) = \tan^{-1}(4\zeta/3)$, while the angle is $180° - \tan^{-1}(4\zeta/3)$ at $\omega = 2\omega_0$. Phase curves are shown as broken lines in Fig. 15.12 for $\zeta = 1, 0.5, 0.25$, and 0.1; heavy dots identify accurate values at $\omega = 0.5\omega_0$ and $\omega = 2\omega_0$.

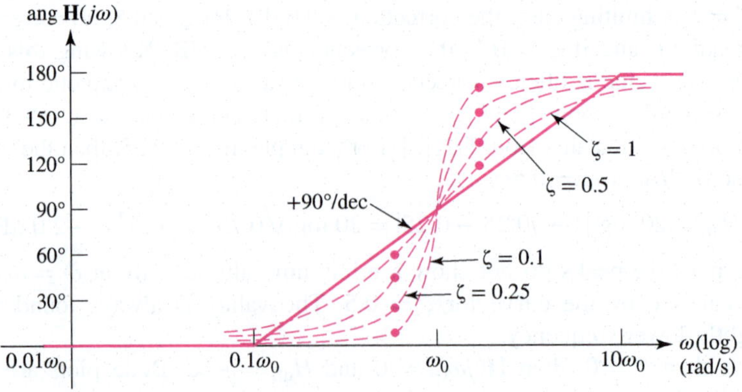

FIGURE 15.12 The straight-line approximation to the phase characteristic for $\mathbf{H}(j\omega) = 1 + j2\zeta(\omega/\omega_0) - (\omega/\omega_0)^2$ is shown as a solid curve, and the true phase response is shown for $\zeta = 1, 0.5, 0.25$, and 0.1 as broken lines.

If the quadratic factor appears in the denominator, both the magnitude and phase curves are the *negatives* of those just discussed. We conclude with an example that contains both linear and quadratic factors.

EXAMPLE 15.5

Construct the Bode plot for the transfer function H(s) = 100,000s/[(s + 1)(10,000 + 20s + s²)].

Let's consider the quadratic factor first and arrange it in a form such that we can see the value of ζ. We begin by dividing the second-order factor by its constant term, 10,000:

$$\mathbf{H}(\mathbf{s}) = \frac{10\mathbf{s}}{(1+\mathbf{s})(1+0.002\mathbf{s}+0.0001\,\mathbf{s}^2)}$$

An inspection of the $\mathbf{s}^2$ term next shows that $\omega_0 = \sqrt{1/0.0001} = 100$. Then the linear term of the quadratic is written to display the factor 2, the factor $(\mathbf{s}/\omega_0)$, and finally the factor ζ:

$$\mathbf{H}(\mathbf{s}) = \frac{10\mathbf{s}}{(1+\mathbf{s})[1+(2)(0.1)(\mathbf{s}/100)+(\mathbf{s}/100)^2]}$$

We see that $\zeta = 0.1$.

The asymptotes of the magnitude-response curve are sketched in lightly in Fig. 15.13: 20 dB for the factor of 10, an infinite straight line through $\omega = 1$ with a +20 dB/decade slope for the **s** factor, a corner at $\omega = 1$ for the simple pole, and a corner at $\omega = 100$ with a slope of −40 dB/decade for the second-order term in the denominator. Adding these four curves and supplying a correction of +14 dB for the quadratic factor lead to the heavy curve of Fig. 15.13.

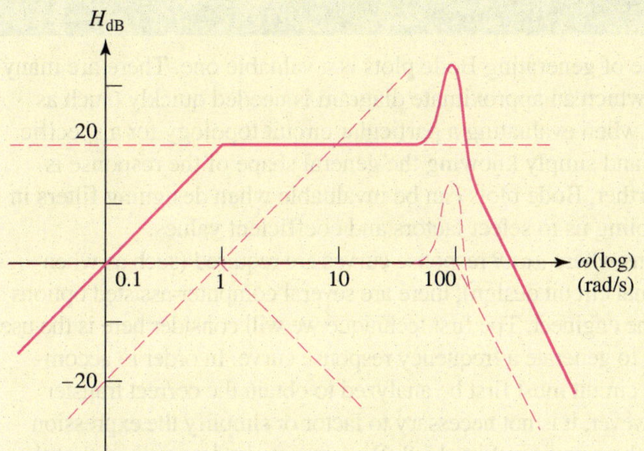

■ **FIGURE 15.13** The Bode magnitude plot of the transfer function $\mathbf{H}(\mathbf{s}) = \dfrac{100{,}000\mathbf{s}}{(\mathbf{s}+1)(10{,}000+20\mathbf{s}+\mathbf{s}^2)}$.

The phase response contains three components: +90° for the factor **s**; 0° for $\omega < 0.1$, −90° for $\omega > 10$, and −45°/decade for the simple pole; and 0° for $\omega < 10$, −180° for $\omega > 1000$, and −90° per decade for the quadratic factor. The addition of these three asymptotes

(Continued on next page)

plus some improvement for $\zeta = 0.1$ is shown as the solid curve in Fig. 15.14.

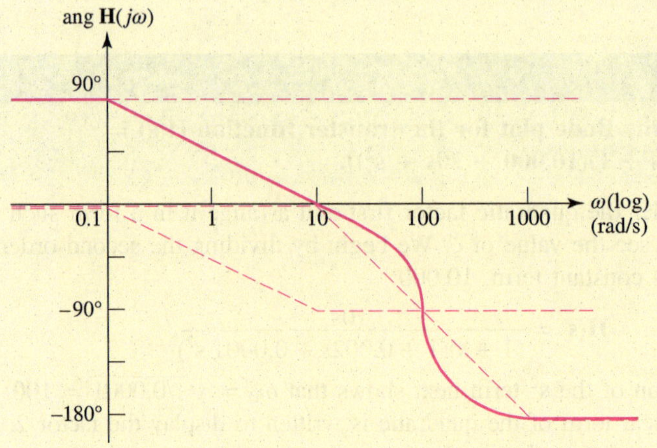

■ **FIGURE 15.14** The Bode magnitude plot of the transfer function $H(s) = \frac{100{,}000s}{(s+1)(10{,}000+20s+s^2)}$

PRACTICE

15.7 If $\mathbf{H}(\mathbf{s}) = 1000\mathbf{s}^2/(\mathbf{s}^2 + 5\mathbf{s} + 100)$, sketch the Bode amplitude plot and calculate a value for (*a*) ω when $H_{dB} = 0$; (*b*) H_{dB} at $\omega = 1$; (*c*) H_{dB} as $\omega \to \infty$.

Ans: 0.316 rad/s; 20 dB; 60 dB.

COMPUTER-AIDED ANALYSIS

The technique of generating Bode plots is a valuable one. There are many situations in which an approximate diagram is needed quickly (such as on exams, or when evaluating a particular circuit topology for a specific application), and simply knowing the general shape of the response is adequate. Further, Bode plots can be invaluable when designing filters in terms of enabling us to select factors and coefficient values.

In situations where *exact* response curves are required (such as when verifying a final circuit design), there are several computer-assisted options available to the engineer. The first technique we will consider here is the use of MATLAB to generate a frequency response curve. In order to accomplish this, the circuit must first be analyzed to obtain the correct transfer function. However, it is not necessary to factor or simplify the expression.

Consider the circuit in Fig. 15.8. We previously determined that the transfer function for this circuit can be expressed as

$$\mathbf{H}(\mathbf{s}) = \frac{-2\mathbf{s}}{(1+\mathbf{s}/10)(1+\mathbf{s}/20{,}000)}$$

We seek a detailed graph of this function over the frequency range of 100 mrad/s to 1 Mrad/s. Since the final graph will be plotted on a logarithmic scale, there is no need to uniformly space our discrete frequencies. Instead, we use the MATLAB function *logspace*() to generate a frequency vector, where the first two arguments represent the power of 10 for starting and ending frequencies, respectively (−1 and 6 in the present example), and the third argument is the total number of points desired. Thus, our MATLAB script is

```
>> w = logspace(-1,6,100);
>> denom = (1 + j*w/10) .* (1 + j*w/20000);
>> H = -2*j*w ./denom;
>> Hdb = 20*log10(abs(H));
>> semilogx(w,Hdb);
>> xlabel('Frequency (rad/s)')
>> ylabel('|H(jw)|(dB)')
```

which yields the graph depicted in Fig. 15.15.

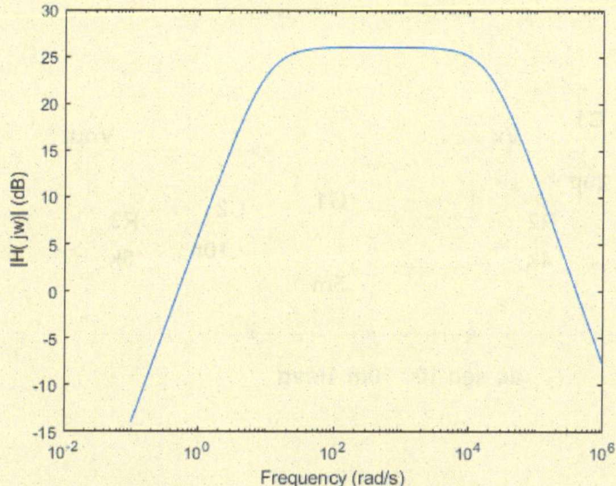

FIGURE 15.15 Plot of H_{dB} generated using MATLAB.

A few comments about the MATLAB code are warranted. First, note that we have substituted **s** $= j\omega$ in our expression for **H**(**s**). Also, MATLAB treats the variable *w* as a vector, or one-dimensional matrix. As such, this variable can cause difficulties in the denominator of an expression as MATLAB will try to apply matrix algebra rules to any expression. Thus, the denominator of **H**($j\omega$) is computed in a separate line, and the operator ".*" is required instead of "*" to multiply the two terms. This operator is equivalent to the following MATLAB code:

```
>> for k = 1:100
denom = (1 + j*w(k)/10) * (1 + j*w(k)/20000);
end
```

In a similar fashion, the new operator "./" is used in the subsequent line of code. The results are desired in dB, so the function *log*10() is invoked; *log*() represents the natural logarithm in MATLAB. Finally, the new plot command *semilogx*() is used to generate a graph with the *x* axis having a logarithmic scale. The reader is encouraged at this point to return to previous examples and use these techniques to generate exact curves for comparison to the corresponding Bode plots.

(Continued on next page)

LTspice is also commonly used to generate frequency response curves, especially to evaluate a final design. Figure 15.16*a* depicts the circuit of Fig. 15.8, where the voltage across the resistor R3 represents the desired output voltage. The source component V1 has been defined with an AC amplitude of 1 V (edit the voltage source to define **Small Signal Parameters (.AC)** to have an AC amplitude of 1). The resulting Vout will then correspond to the transfer function due to our convenient choice of an input of 1 V. An ac sweep simulation is required to determine the frequency response of our circuit; Fig. 15.16*b* was generated using 100 points on a decade scale ranging from 10 mHz to 1 MHz using the SPICE directive **.ac dec 100 10m 1meg.** (Note the simulation has been performed in Hz, not rad/s.)

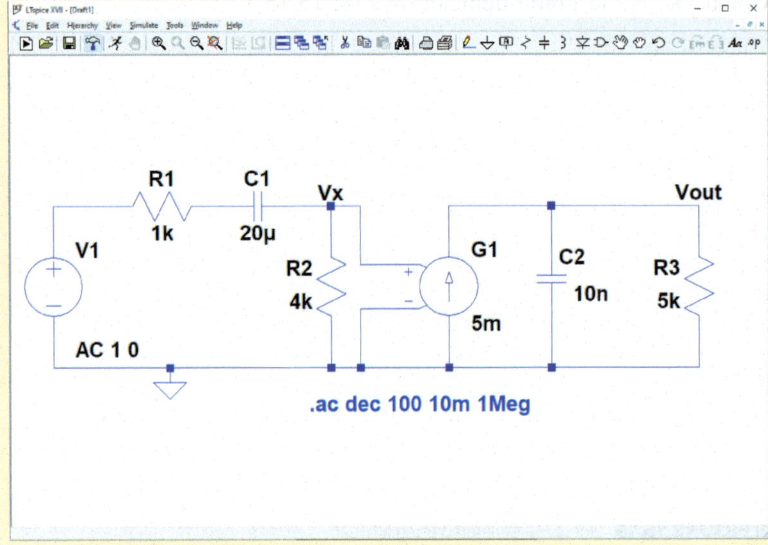

(*a*)

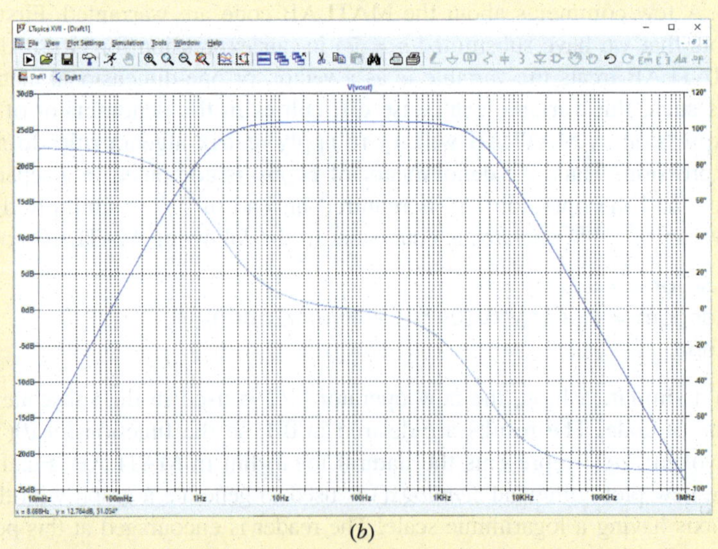

(*b*)

■ **FIGURE 15.16** (*a*) The circuit of Fig. 15.8. (*b*) Frequency response of the circuit in terms of transfer function magnitude (solid line, axis in dB) and phase angle (dashed line, axis in degrees). (*a–b*) Analog Devices

15.3 PARALLEL RESONANCE

Suppose that a certain forcing function is found to contain sinusoidal *components* having frequencies within the range of 10 to 100 Hz. Now let us imagine that this forcing function is applied to a network that has the property that all sinusoidal voltages with frequencies from zero to 200 Hz applied at the input terminals appear doubled in magnitude at the output terminals, with no change in phase angle. The output function is therefore an undistorted replica of the input function, but with twice the amplitude. If, however, the network has a frequency response such that the magnitudes of input sinusoids between 10 and 50 Hz are multiplied by a different factor than are those between 50 and 100 Hz, then the output would in general be distorted; it would no longer be a magnified version of the input. This distorted output might be desirable in some cases and undesirable in others, such as in tuning circuits for radio transmitters/receivers. That is, the network frequency response might be chosen *deliberately* to reject some frequency components of the forcing function, or to emphasize others.

Such behavior is characteristic of tuned circuits or resonant circuits, as we will see in this chapter. In discussing resonance we will be able to apply all the methods we have discussed in presenting frequency response.

Resonance

In this section we will begin the study of a very important phenomenon that may occur in circuits that contain both inductors and capacitors. The phenomenon is called ***resonance,*** and it may be loosely described as the condition existing in any physical system when a fixed-amplitude sinusoidal forcing function produces a response of maximum amplitude. However, we often speak of resonance as occurring even when the forcing function is not sinusoidal. The resonant system may be electrical, mechanical, hydraulic, acoustic, or some other kind, but we will restrict our attention, for the most part, to electrical systems.

Resonance is a familiar phenomenon. Jumping up and down on the bumper of an automobile, for example, can put the vehicle into rather large oscillatory motion if the jumping is done at the proper frequency (about one jump per second), and if the shock absorbers are somewhat decrepit. However, if the jumping frequency is increased or decreased, the vibrational response of the automobile will be considerably less than it was before. A further illustration is furnished in the case of an opera singer who can shatter crystal goblets by means of a well-formed note at the proper frequency. In each of these examples, we are thinking of frequency as being adjusted until resonance occurs; it is also possible to adjust the size, shape, and material of the mechanical object being vibrated, but this may not be so easily accomplished physically. The condition of resonance may or may not be desirable, depending upon the purpose which the physical system is to serve. In the automotive example, a large amplitude of vibration may help to separate locked bumpers, but it would be somewhat disagreeable at 65 mph (105 km/h).

Let us now define resonance more carefully. In a two-terminal electrical network containing at least one inductor and one capacitor, we define

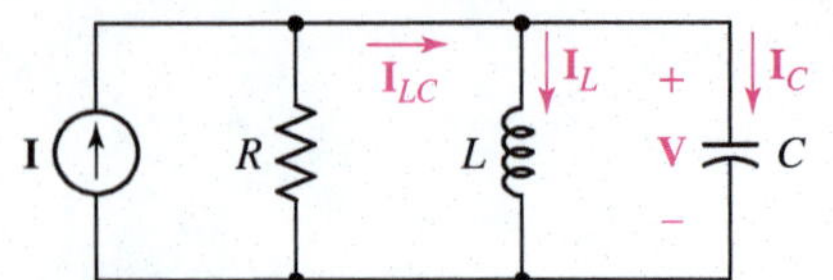

■ **FIGURE 15.17** The parallel combination of a resistor, an inductor, and a capacitor, often referred to as a *parallel resonant circuit*.

resonance as the condition which exists when the input impedance of the network is purely resistive. Thus,

a network is in resonance (or resonant) when the voltage and current at the network input terminals are in phase.

We will also find that a maximum-amplitude response is produced in the network when it is in the resonant condition.

We first apply the definition of resonance to a parallel *RLC* network driven by a sinusoidal current source as shown in Fig. 15.17. In many practical situations, this circuit is a very good approximation to the circuit we might build in the laboratory by connecting a physical inductor in parallel with a physical capacitor, where the parallel combination is driven by an energy source having a very high output impedance. The steady-state admittance offered to the ideal current source is

$$\mathbf{Y} = \frac{1}{R} + j\left(\omega C - \frac{1}{\omega L}\right) \qquad [7]$$

Resonance occurs when the voltage and current at the input terminals are in phase. This corresponds to a purely real admittance, so that the necessary condition is given by

$$\omega C - \frac{1}{\omega L} = 0$$

The resonant condition may be achieved by adjusting L, C, or ω; we will devote our attention to the case for which ω is the variable. Hence, the resonant frequency ω_0 is

$$\omega_0 = \frac{1}{\sqrt{LC}} \qquad \text{rad/s} \qquad [8]$$

or

$$f_0 = \frac{1}{2\pi\sqrt{LC}} \qquad \text{Hz} \qquad [9]$$

This resonant frequency ω_0 is identical to the resonant frequency defined in Eq. [10], Chap. 9.

The pole-zero configuration of the admittance function can also be used to considerable advantage here. Given $\mathbf{Y}(\mathbf{s})$,

$$\mathbf{Y}(\mathbf{s}) = \frac{1}{R} + \frac{1}{\mathbf{s}L} + \mathbf{s}C$$

or

$$\mathbf{Y}(\mathbf{s}) = C\frac{\mathbf{s}^2 + \mathbf{s}/RC + 1/LC}{\mathbf{s}} \qquad [10]$$

we may display the zeros of $\mathbf{Y}(\mathbf{s})$ by factoring the numerator:

$$\mathbf{Y}(\mathbf{s}) = C\frac{(\mathbf{s} + \alpha - j\omega_d)(\mathbf{s} + \alpha + j\omega_d)}{\mathbf{s}}$$

where α and ω_d represent the same quantities that they did when we discussed the natural response of the parallel *RLC* circuit in Sec. 9.4. That is, α is the *exponential damping coefficient,*

$$\alpha = \frac{1}{2RC}$$

and ω_d is the *natural resonant frequency* (not the resonant frequency ω_0),

$$\omega_d = \sqrt{\omega_0^2 - \alpha^2}$$

The pole-zero constellation shown in Fig. 15.18*a* follows directly from the factored form.

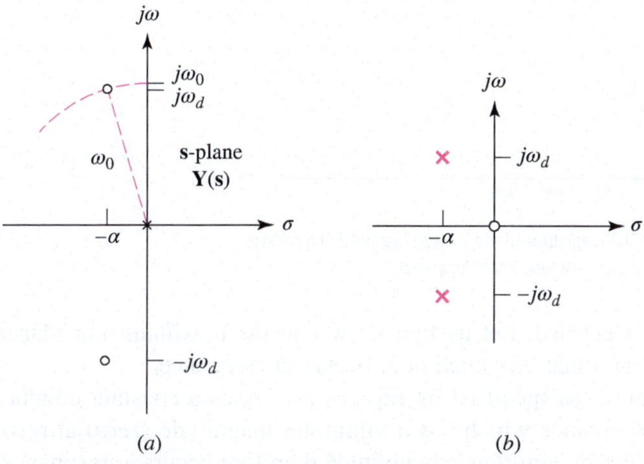

FIGURE 15.18 (*a*) The pole-zero constellation of the input admittance of a parallel resonant circuit is shown on the **s**-plane; $\omega_0^2 = \alpha^2 + \omega_d^2$. (*b*) The pole-zero constellation of the input impedance.

In view of the relationship among α, ω_d, and ω_0, it is apparent that the distance from the origin of the **s** plane to one of the admittance zeros is numerically equal to ω_0. Given the pole-zero configuration, the resonant frequency may therefore be obtained by purely graphical methods. We merely swing an arc, using the origin of the **s** plane as a center, through one of the zeros. The intersection of this arc and the positive $j\omega$ axis locates the point $\mathbf{s} = j\omega_0$. It is evident that ω_0 is slightly greater than the natural resonant frequency ω_d, but their ratio approaches unity as the ratio of ω_d to α increases.

Resonance and the Voltage Response

Next let us examine the magnitude of the response, the voltage **V(s)** indicated in Fig. 15.17, as the frequency ω of the forcing function is varied. If we assume a constant-amplitude sinusoidal current source, the voltage response is proportional to the input impedance. This response can be obtained from the pole-zero plot of the impedance

$$\mathbf{Z(s)} = \frac{\mathbf{s}/C}{(\mathbf{s} + \alpha - j\omega_d)(\mathbf{s} + \alpha + j\omega_d)}$$

shown in Fig. 15.18*b*. The response of course starts at zero, reaches a maximum value in the vicinity of the natural resonant frequency, and then drops again to zero as ω becomes infinite. The frequency response is sketched in Fig. 15.19. The maximum value of the response is indicated as R times the amplitude of the source current, implying that the maximum magnitude of the circuit impedance is R; moreover, the response maximum is shown to occur exactly at the resonant frequency ω_0. Two additional frequencies, ω_1 and ω_2, which we will later use as a measure of the width of the response

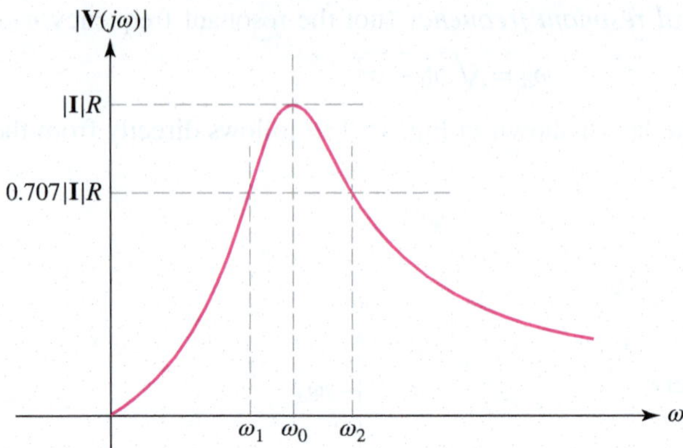

FIGURE 15.19 The magnitude of the voltage response of a parallel resonant circuit is shown as a function of frequency.

curve, are also identified. Let us first show that the maximum impedance magnitude is R and that this maximum occurs at resonance.

The admittance, as specified by Eq. [7], possesses a constant conductance and a susceptance which has a minimum magnitude (zero) at resonance. The minimum admittance magnitude therefore occurs at resonance, and it is $1/R$. Hence, the maximum impedance magnitude is R, *and it occurs at resonance.*

At the resonant frequency, therefore, the voltage across the parallel resonant circuit of Fig. 15.17 is simply $\mathbf{I}R$, and the *entire* source current $\mathbf{I}$ flows through the resistor. However, current is also present in L and C. For the inductor, $\mathbf{I}_{L,0} = \mathbf{V}_{L,0}/j\omega_0 L = \mathbf{I}R/j\omega_0 L$, and the capacitor current at resonance is $\mathbf{I}_{C,0} = (j\omega_0 C)\mathbf{V}_{C,0} = j\omega_0 CR\mathbf{I}$. Since $1/\omega_0 C = \omega_0 L$ at resonance, we find that

$$\mathbf{I}_{C,0} = -\mathbf{I}_{L,0} = j\omega_0 CR\mathbf{I} \qquad [11]$$

and

$$\mathbf{I}_{C,0} + \mathbf{I}_{L,0} = \mathbf{I}_{LC} = 0$$

Thus, the *net* current flowing into the LC combination is zero. The maximum value of the response magnitude and the frequency at which it occurs are not always found so easily. In less standard resonant circuits, we may find it necessary to express the magnitude of the response in analytical form, usually as the square root of the sum of the real part squared and the imaginary part squared; then we should differentiate this expression with respect to frequency, equate the derivative to zero, solve for the frequency of maximum response, and finally substitute this frequency in the magnitude expression to obtain the maximum-amplitude response. The procedure may be carried out for this simple case merely as a corroborative exercise but, as we have seen, it is not necessary.

Quality Factor

It should be emphasized that, although the *height* of the response curve of Fig. 15.19 depends only upon the value of R for constant-amplitude excitation, the width of the curve or the steepness of the sides depends upon the other two element values also. We will shortly relate the "width of the response curve" to a more carefully defined quantity, the *bandwidth,* but it is helpful to express this relationship in terms of a very important parameter, the *quality factor Q*.

We should be very careful not to confuse the quality factor with charge or reactive power, all of which unfortunately are represented by the letter Q.

We will find that the sharpness of the response curve of any resonant circuit is determined by the maximum amount of energy that can be stored in the circuit, compared with the energy that is lost during one complete period of the response.

We define Q as

$$Q = \text{quality factor} \equiv 2\pi \frac{\text{maximum energy stored}}{\text{total energy lost per period}} \qquad [12]$$

The proportionality constant 2π is included in the definition in order to simplify the more useful expressions for Q which we will now obtain. Since energy can be stored only in the inductor and the capacitor, and can be lost only in the resistor, we may express Q in terms of the instantaneous energy associated with each of the reactive elements and the average power P_R dissipated in the resistor:

$$Q = 2\pi \frac{[w_L(t) + w_C(t)]_{\text{max}}}{P_R T}$$

where T is the period of the sinusoidal frequency at which Q is evaluated.

Now let us apply this definition to the parallel RLC circuit of Fig. 15.17 and determine the value of Q at the resonant frequency; this value of Q is denoted by Q_0. We select the current forcing function

$$i(t) = I_m \cos \omega_0 t$$

and obtain the corresponding voltage response at resonance,

$$v(t) = Ri(t) = RI_m \cos \omega_0 t$$

The energy stored in the capacitor is then

$$w_C(t) = \frac{1}{2} C v^2 = \frac{I_m^2 R^2 C}{2} \cos^2 \omega_0 t$$

and the instantaneous energy stored in the inductor is given by

$$w_L(t) = \frac{1}{2} L i_L^2 = \frac{1}{2} L \left(\frac{1}{L} \int v\, dt\right)^2 = \frac{1}{2L} \left[\frac{RI_m}{\omega_0} \sin \omega_0 t\right]^2$$

so that

$$w_L(t) = \frac{I_m^2 R^2 C}{2} \sin^2 \omega_0 t$$

The total *instantaneous* stored energy is therefore constant:

$$w(t) = w_L(t) + w_C(t) = \frac{I_m^2 R^2 C}{2}$$

and this constant value must also be the maximum value. In order to find the energy lost in the resistor in one period, we take the average power absorbed by the resistor (see Sec. 11.2),

$$P_R = \frac{1}{2} I_m^2 R$$

and multiply by one period, obtaining

$$P_R T = \frac{1}{2 f_0} I_m^2 R$$

We thus find the quality factor at resonance:

$$Q_0 = 2\pi \frac{I_m^2 R^2 C/2}{I_m^2 R/2 f_0}$$

or

$$Q_0 = 2\pi f_0 RC = \omega_0 RC \qquad [13]$$

This equation holds only for the parallel *RLC* circuit of Fig. 15.17. Equivalent expressions for Q_0 which are often quite useful may be obtained by substitution:

$$Q_0 = R\sqrt{\frac{C}{L}} = \frac{R}{|X_{C,0}|} = \frac{R}{|X_{L,0}|} \qquad [14]$$

So we see that for this specific circuit, decreasing the resistance decreases Q_0; the lower the resistance, the greater the amount of energy lost in the element. Intriguingly, increasing the capacitance *increases* Q_0, but increasing the inductance leads to a *reduction* in Q_0. These statements, of course, apply to operation of the circuit at the resonant frequency.

Other Interpretations of Q

Another useful interpretation of Q is obtained when we inspect the inductor and capacitor currents at resonance, as given by Eq. [11],

$$\mathbf{I}_{C,0} = -\mathbf{I}_{L,0} = j\omega_0 CR\mathbf{I} = jQ_0\mathbf{I} \qquad [15]$$

Note that each is Q_0 times the source current in amplitude and that each is 180° out of phase with the other. Thus, if we apply 2 mA at the resonant frequency to a parallel resonant circuit with a Q_0 of 50, we find 2 mA in the resistor, and 100 mA in both the inductor and the capacitor. A parallel resonant circuit can therefore act as a current amplifier, but not, of course, as a power amplifier, since it is a passive network.

Resonance, by definition, is fundamentally associated with the forced response, since it is defined in terms of a (purely resistive) input impedance, a sinusoidal steady-state concept. The two most important parameters of a resonant circuit are perhaps the resonant frequency ω_0 and the quality factor Q_0. Both the exponential damping coefficient and the natural resonant frequency may be expressed in terms of ω_0 and Q_0:

$$\alpha = \frac{1}{2RC} = \frac{1}{2(Q_0/\omega_0 C)C}$$

or

$$\alpha = \frac{\omega_0}{2Q_0} \qquad [16]$$

and

$$\omega_d = \sqrt{\omega_0^2 - \alpha^2}$$

or

$$\omega_d = \omega_0\sqrt{1 - \left(\frac{1}{2Q_0}\right)^2} \qquad [17]$$

Damping Factor

For future reference it may be helpful to note one additional relationship involving ω_0 and Q_0. The quadratic factor appearing in the numerator of Eq. [10],

$$\mathbf{s}^2 + \frac{1}{RC}\mathbf{s} + \frac{1}{LC}$$

may be written in terms of α and ω_0:

$$\mathbf{s}^2 + 2\alpha\mathbf{s} + \omega_0^2$$

or in the form with the damping factor ζ (zeta)

$$\mathbf{s}^2 + 2\zeta\omega_0\mathbf{s} + \omega_0^2$$

Comparison of these expressions allows us to relate ζ to other parameters:

$$\zeta = \frac{\alpha}{\omega_0} = \frac{1}{2Q_0} \qquad [18]$$

EXAMPLE 15.6

Consider a parallel RLC circuit such that $L = 2$ mH, $Q_0 = 5$, and $C = 10$ nF. Determine the value of R and the magnitude of the steady-state admittance at $0.1\omega_0$, ω_0, and $1.1\omega_0$.

We derived several expressions for Q_0, a parameter directly related to energy loss, and hence the resistance in our circuit. Rearranging the expression in Eq. [14], we calculate

$$R = Q_0\sqrt{\frac{L}{C}} = 2.236 \text{ k}\Omega$$

Next, we compute ω_0, a term we may recall from Chap. 9,

$$\omega_0 = \frac{1}{\sqrt{LC}} = 223.6 \text{ krad/s}$$

or, alternatively, we may exploit Eq. [13] and obtain the same answer,

$$\omega_0 = \frac{Q_0}{RC} = 223.6 \text{ krad/s}$$

The admittance of any parallel RLC network is simply

$$\mathbf{Y} = \frac{1}{R} + j\omega C + \frac{1}{j\omega L}$$

and hence

$$|\mathbf{Y}| = \frac{1}{R} + j\omega C + \frac{1}{j\omega L}$$

evaluated at the three designated frequencies is equal to

$$|\mathbf{Y}(0.9\,\omega_0)| = 6.504 \times 10^{-4} \text{ S} \qquad |\mathbf{Y}(\omega_0)| = 4.472 \times 10^{-4} \text{ S}$$

$$|\mathbf{Y}(1.1\,\omega_0)| = 6.182 \times 10^{-4} \text{ S}$$

(Continued on next page)

We thus obtain a minimum impedance at the resonant frequency, or a *maximum voltage response* to a particular input current. If we quickly compute the reactance at these three frequencies, we find

$$X(0.9\,\omega_0) = -4.72 \times 10^{-4}\ \text{S} \qquad X(1.1\,\omega_0) = 4.72 \times 10^{-4}\ \text{S}$$
$$X(\omega_0) = -1.36 \times 10^{-7}\ \text{S}$$

We leave it to the reader to show that our value for $X(\omega_0)$ is nonzero only as a result of rounding error.

PRACTICE

15.8 A parallel resonant circuit is composed of the elements $R = 8$ kΩ, $L = 50$ mH, and $C = 80$ nF. Compute (*a*) ω_0; (*b*) Q_0; (*c*) ωd; (*d*) α; (*e*) ζ.

15.9 Determine the values of R, L, and C in a parallel resonant circuit for which $\omega_0 = 1000$ rad/s, $\omega_d = 998$ rad/s, and $\mathbf{Y}_{\text{in}} = 1$ mS at resonance.

Ans: 15.8: 15.811 krad/s; 10.12; 115.792 krad/s; 781 Np/s; 0.0494. 15.9: 1000 Ω: 126.4 mH; 7.91 μF.

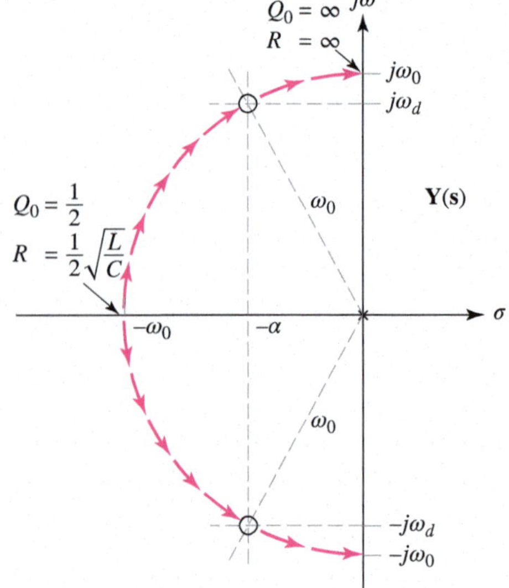

■ **FIGURE 15.20** The two zeros of the admittance $\mathbf{Y}(\mathbf{s})$, located at $\mathbf{s} = -\alpha \pm j\omega d$, provide a semicircular locus as R increases from $\frac{1}{2}\sqrt{L/C}$ to ∞.

Now let us interpret Q_0 in terms of the pole-zero locations of the admittance $\mathbf{Y}(\mathbf{s})$ of the parallel *RLC* circuit. We will keep ω_0 constant; this may be done, for example, by changing R while holding L and C constant. As Q_0 is increased, the relationships relating α, Q_0, and ω_0 indicate that the two zeros must move closer to the $j\omega$ axis. These relationships also show that the zeros must simultaneously move away from the σ axis. The exact nature of the movement becomes clearer when we remember that the point at which $\mathbf{s} = j\omega_0$ could be located on the $j\omega$ axis by swinging an arc, centered at the origin, through one of the zeros and over to the positive $j\omega$ axis; since ω_0 is to be held constant, the radius must be constant, and the zeros must therefore move along this arc toward the positive $j\omega$ axis as Q_0 increases.

The two zeros are indicated in Fig. 15.20, and the arrows show the path they take as R increases. When R is infinite, Q_0 is also infinite, and the two zeros are found at $\mathbf{s} = \pm j\omega_0$ on the $j\omega$ axis. As R decreases, the zeros move toward the σ axis along the circular locus, joining to form a double zero on the σ axis at $\mathbf{s} = -\omega_0$ when $R = \frac{1}{2}\sqrt{L/C}$ or $Q_0 = \frac{1}{2}$. This condition may be recalled as that for critical damping, so that $\omega d = 0$ and $\alpha = \omega_0$. Lower values of R and lower values of Q_0 cause the zeros to separate and move in opposite directions on the negative σ axis, but these low values of Q_0 are not really typical of resonant circuits and we need not track them any further.

Later, we will use the criterion $Q_0 \geq 5$ to describe a high-Q circuit. When $Q_0 = 5$, the zeros are located at $\mathbf{s} = -0.1\,\omega_0 \pm j0.995\omega_0$, and thus ω_0 and ω_d differ by only one-half of 1 percent.

15.4 BANDWIDTH AND HIGH-Q CIRCUITS

We continue our discussion of parallel resonance by defining half-power frequencies and bandwidth, and then we will make good use of these new concepts in obtaining approximate response data for high-Q circuits. The "width" of a resonance response curve, such as the one shown in Fig. 15.19, may now be defined more carefully and related to Q_0. Let us first define the two half-power frequencies ω_1 and ω_2 as those frequencies at which the magnitude of the input admittance of a parallel resonant circuit is greater than the magnitude at resonance by a factor of $\sqrt{2}$. Since the response curve of Fig. 15.19 displays the voltage produced across the parallel circuit by a sinusoidal current source as a function of frequency, the half-power frequencies also locate those points at which the voltage response is $1/\sqrt{2}$, or 0.707, times its maximum value. A similar relationship holds for the impedance magnitude. We will designate ω_1 as the ***lower half-power frequency*** and ω_2 as the ***upper half-power frequency.***

These names arise from the fact that a voltage which is $1/\sqrt{2}$ times the resonant voltage is equivalent to a squared voltage which is one-half the squared voltage at resonance. Thus, at the half-power frequencies, the resistor absorbs one-half the power that it does at resonance.

Bandwidth

The (half-power) ***bandwidth*** of a resonant circuit is defined as the difference of these two half-power frequencies.

$$\mathcal{B} \equiv \omega_2 - \omega_1 \qquad [19]$$

We tend to think of bandwidth as the "width" of the response curve, even though the curve actually extends from $\omega = 0$ to $\omega = \infty$. More exactly, the half-power bandwidth is measured by that portion of the response curve which is equal to or greater than 70.7 percent of the maximum value, as illustrated in Fig. 15.21.

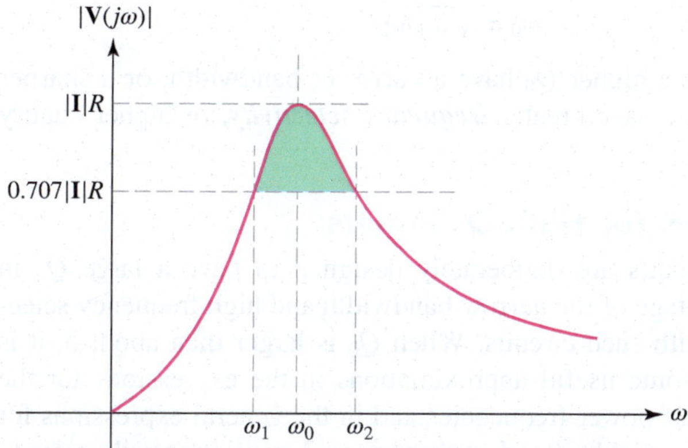

FIGURE 15.21 The bandwidth of the circuit response is highlighted in green; it corresponds to the portion of the response curve greater than or equal to 70.7 percent of the maximum value.

We can express the bandwidth in terms of Q_0 and the resonant frequency. In order to do so, we first express the admittance of the parallel RLC circuit,

$$\mathbf{Y} = \frac{1}{R} + j\left(\omega C - \frac{1}{\omega L}\right)$$

in terms of Q_0:

$$\mathbf{Y} = \frac{1}{R} + j\frac{1}{R}\left(\frac{\omega\omega_0 CR}{\omega_0} - \frac{\omega_0 R}{\omega\omega_0 L}\right)$$

or

$$\mathbf{Y} = \frac{1}{R}\left[1 + jQ_0\left(\frac{\omega}{\omega_0} - \frac{\omega_0}{\omega}\right)\right] \qquad [20]$$

We note again that the magnitude of the admittance at resonance is 1/R, and we then realize that an admittance magnitude of $\sqrt{2}/R$ can occur only when a frequency is selected such that the imaginary part of the bracketed quantity has a magnitude of unity. Thus

Keep in mind that $\omega_2 > \omega_0$, while $\omega_1 < \omega_0$.

$$Q_0\left(\frac{\omega_1}{\omega_0} - \frac{\omega_0}{\omega_1}\right) = -1 \quad \text{and} \quad Q_0\left(\frac{\omega_2}{\omega_0} - \frac{\omega_0}{\omega_2}\right) = 1$$

Solving, we have

$$\omega_1 = \omega_0\left[\sqrt{1 + \left(\frac{1}{2Q_0}\right)^2} - \frac{1}{2Q_0}\right] \qquad [21]$$

$$\omega_2 = \omega_0\left[\sqrt{1 + \left(\frac{1}{2Q_0}\right)^2} + \frac{1}{2Q_0}\right] \qquad [22]$$

Although these expressions are somewhat unwieldy, their difference provides a very simple formula for the bandwidth:

$$\mathcal{B} = \omega_2 - \omega_1 = \frac{\omega_0}{Q_0}$$

Equations [15] and [16] may be multiplied by each other to show that ω_0 is exactly equal to the geometric mean of the half-power frequencies:

$$\omega_0^2 = \omega_1 \omega_2$$

or

$$\omega_0 = \sqrt{\omega_1 \omega_2}$$

Circuits possessing a higher Q_0 have a narrower bandwidth, or a sharper response curve; they have greater ***frequency selectivity,*** or higher quality (factor).

Approximations for High-Q Circuits

Many resonant circuits are deliberately designed to have a large Q_0 in order to take advantage of the narrow bandwidth and high frequency selectivity associated with such circuits. When Q_0 is larger than about 5, it is possible to make some useful approximations in the expressions for the upper and lower half-power frequencies and in the general expressions for the response in the neighborhood of resonance. Let us arbitrarily refer to a "high-Q circuit" as one for which Q_0 is equal to or greater than 5. The pole-zero configuration of $\mathbf{Y}(\mathbf{s})$ for a parallel RLC circuit having a Q_0 of about 5 is shown in Fig. 15.22. Since

$$\alpha = \frac{\omega_0}{2Q_0}$$

then

$$\alpha = \frac{1}{2}\mathcal{B}$$

and the locations of the two zeros $\mathbf{s}_1$ and $\mathbf{s}_2$ may be approximated:

$$\mathbf{s}_{1,2} = -\alpha \pm j\omega_d$$
$$\approx -\frac{1}{2}\mathcal{B} \pm j\omega_0$$

Moreover, the locations of the two half-power frequencies (on the positive $j\omega$ axis) may also be determined in a concise approximate form:

$$\omega_{1,2} = \omega_0\left[\sqrt{1 + \left(\frac{1}{2Q_0}\right)^2} \mp \frac{1}{2Q_0}\right] \approx \omega_0\left(1 \mp \frac{1}{2Q_0}\right)$$

or

$$\omega_{1,2} \approx \omega_0 \mp \frac{1}{2}\mathcal{B} \qquad [23]$$

In a high-Q circuit, therefore, each half-power frequency is located approximately one-half bandwidth from the resonant frequency; this is indicated in Fig. 15.22.

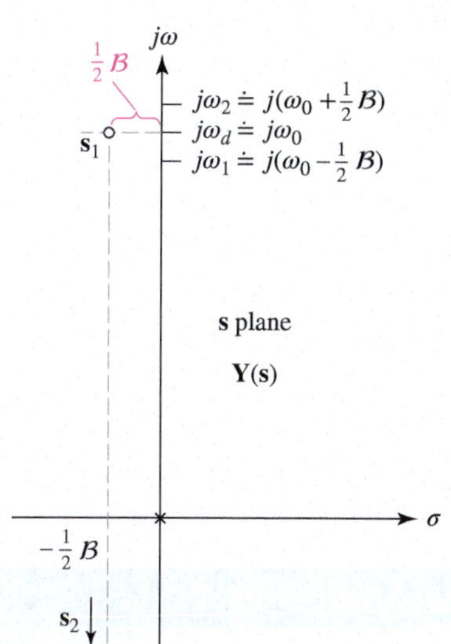

FIGURE 15.22 The pole-zero constellation of **Y(s)** for a parallel *RLC* circuit. The two zeros are exactly $\frac{1}{2}\mathcal{B}$ **Np/s** (or rad/s) to the left of the $j\omega$ axis and approximately $j\omega_0$ rad/s (or Np/s) from the σ axis. The upper and lower half-power frequencies are separated exactly $\mathcal{B}$ rad/s, and each is approximately $\frac{1}{2}\mathcal{B}$ **rad/s** away from the resonant frequency and the natural resonant frequency.

The approximate relationships for ω_1 and ω_2 in Eq. [17] may be added to show that ω_0 is approximately equal to the arithmetic mean of ω_1 and ω_2 in high-Q circuits:

$$\omega_0 \approx \frac{1}{2}(\omega_1 + \omega_2)$$

Now let us visualize a test point slightly above $j\omega_0$ on the $j\omega$ axis. In order to determine the admittance offered by the parallel RLC network at this frequency, we construct the three vectors from the critical frequencies to the test point. If the test point is close to $j\omega_0$, then the vector from the pole is approximately $j\omega_0$ and that from the lower zero is nearly $j2\omega_0$. The admittance is therefore given approximately by

$$\mathbf{Y(s)} \approx C\frac{(j2\omega_0)(\mathbf{s} - \mathbf{s}_1)}{j\omega_0} \approx 2C(\mathbf{s} - \mathbf{s}_1) \qquad [24]$$

where C is the capacitance, as shown in Eq. [10]. In order to determine a useful approximation for the vector $(\mathbf{s} - \mathbf{s}_1)$, let us consider an enlarged view of that portion of the $\mathbf{s}$ plane in the neighborhood of the zero $\mathbf{s}_1$ (Fig. 15.23).

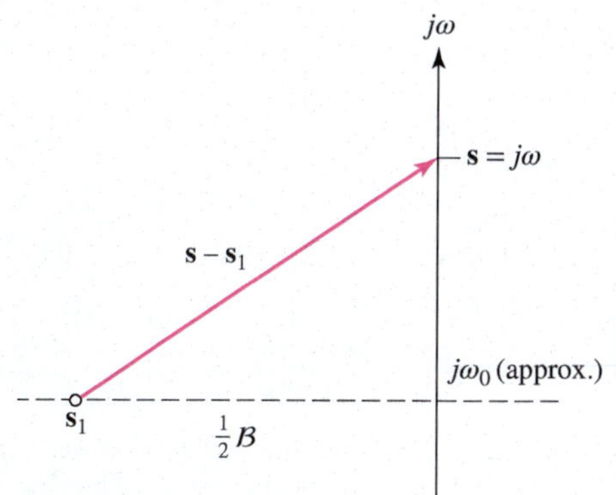

FIGURE 15.23 An enlarged portion of the pole-zero constellation for **Y(s)** of a high-Q_0 parallel *RLC* circuit.

In terms of its cartesian components, we see that

$$\mathbf{s} - \mathbf{s}_1 \approx \frac{1}{2}\mathcal{B} + j(\omega - \omega_0)$$

where this expression would be exact if ω_0 were replaced by ω_d. We now substitute this equation in the approximation for **Y(s)**, Eq. [24], and factor out $\frac{1}{2}\mathcal{B}$:

$$\mathbf{Y(s)} \approx 2C\left(\frac{1}{2}\mathcal{B}\right)\left(1 + j\frac{\omega - \omega_0}{\frac{1}{2}\mathcal{B}}\right)$$

or

$$\mathbf{Y(s)} \approx \frac{1}{R}\left(1 + j\frac{\omega - \omega_0}{\frac{1}{2}\mathcal{B}}\right)$$

The fraction $(\omega - \omega_0)/\left(\frac{1}{2}\mathcal{B}\right)$ may be interpreted as the "number of half-bandwidths off resonance" and abbreviated by N. Thus,

$$\mathbf{Y(s)} \approx \frac{1}{R}(1 + jN) \qquad [25]$$

where

$$N = \frac{\omega - \omega_0}{\frac{1}{2}\mathcal{B}} \qquad [26]$$

At the upper half-power frequency, $\omega_2 \approx \omega_0 + \frac{1}{2}\mathcal{B}$, $N = +1$, and we are one half-bandwidth above resonance. For the lower half-power frequency, $\omega_1 \approx \omega_0 - \frac{1}{2}\mathcal{B}$, so that $N = -1$, locating us one half-bandwidth below resonance.

Equation [25] is much easier to use than the exact relationships we have had up to now. It shows that the magnitude of the admittance is

$$|\mathbf{Y}(j\omega)| \approx \frac{1}{R}\sqrt{1+N^2}$$

while the angle of $\mathbf{Y}(j\omega)$ is given by the inverse tangent of N:

$$\text{ang } \mathbf{Y}(j\omega) \approx \tan^{-1} N$$

EXAMPLE 15.7

Estimate the location of the two half-power frequencies of the voltage response of a parallel *RLC* network for which $R = 40$ kΩ, $L = 1$ H, and $C = \frac{1}{64}\,\mu$F, and determine the approximate value of the admittance for an operating frequency of 8200 rad/s.

▶ ***Identify the goal of the problem.***

We seek the lower and upper half-power frequencies of the voltage response as well as $\mathbf{Y}(\omega_0)$. Since we are asked to "estimate" and "approximate," the implication is that this is a high-Q circuit, an assumption we should verify.

▶ ***Collect the known information.***

Given R, L, and C, we can compute ω_0 and Q_0. If $Q_0 \geq 5$, we may invoke approximate expressions for half-power frequencies and admittance near resonance, but we could compute these quantities exactly if required.

▶ ***Devise a plan.***

To use approximate expressions, we must first determine Q_0, the quality factor at resonance, as well as the bandwidth.

The resonant frequency ω_0 is given as $1/\sqrt{LC} = 8$ krad/s. Thus, $Q_0 = \omega_0 RC = 5$, and the bandwidth is $\omega_0/Q_0 = 1.6$ krad/s. The value of Q_0 for this circuit is sufficient to employ "high-Q" approximations.

▶ ***Construct an appropriate set of equations.***

The bandwidth is simply

$$\mathcal{B} = \frac{\omega_0}{Q_0} = 1600 \text{ rad/s}$$

and so

$$\omega_1 \approx \omega_0 - \frac{\mathcal{B}}{2} = 7200 \text{ rad/s} \qquad \omega_1 \approx \omega_0 + \frac{\mathcal{B}}{2} = 8800 \text{ rad/s}$$

Equation [25] states that

$$\mathbf{Y(s)} \approx \frac{1}{R}(1 + jN)$$

so

$$|\mathbf{Y}(j\omega)| \approx \frac{1}{R}\sqrt{1+N^2} \qquad \text{and} \qquad \text{ang } \mathbf{Y}(j\omega) \approx \tan^{-1} N$$

▶ ***Determine if additional information is required.***

We still require N, which tells us how many half-bandwidths ω is from the resonant frequency ω_0:

$$N = (8.2 - 8)/0.8 = 0.25$$

▶ ***Attempt a solution.***

Now we are ready to employ our approximate relationships for the magnitude and angle of the network admittance,

$$\text{ang } \mathbf{Y} \approx \tan^{-1} 0.25 = 14.04^\circ$$

and

$$|\mathbf{Y}| \approx 25\sqrt{1 + (0.25)^2} = 25.77\ \mu\text{S}$$

▶ ***Verify the solution. Is it reasonable or expected?***

An exact calculation of the admittance using Eq. [7] shows that

$$\mathbf{Y}(j8200) = 25.75\underline{/13.87^\circ}\ \mu\text{S}$$

The approximate method therefore leads to values of admittance magnitude and angle that are reasonably accurate (better than 2 percent) for this frequency. We leave it to the reader to judge the accuracy of our prediction for ω_1 and ω_2.

PRACTICE

15.10 A marginally high-Q parallel resonant circuit has $f_0 = 440$ Hz with $Q_0 = 6$. Use Eqs. [21] and [22] to obtain accurate values for (*a*) f_1; (*b*) f_2. Now use Eq. [23] to calculate approximate values for (*c*) f_1; (*d*) f_2.

Ans: 404.9 Hz; 478.2 Hz; 403.3 Hz; 476.7 Hz.

We conclude our coverage of the *parallel* resonant circuit by reviewing some key conclusions we have reached:

- The resonant frequency ω_0 is the frequency at which the imaginary part of the input admittance becomes zero, or the angle of the admittance becomes zero. For this circuit, $\omega_0 = 1/\sqrt{LC}$.
- The circuit's figure of merit Q_0 is defined as 2π times the ratio of the maximum energy stored in the circuit to the energy lost each period in the circuit. For this circuit, $Q_0 = \omega_0 RC$.
- We defined two half-power frequencies, ω_1 and ω_2, as the frequencies at which the admittance magnitude is $\sqrt{2}$ times the minimum admittance magnitude. (These are also the frequencies at which the voltage response is 70.7 percent of the maximum response.)
- The exact expressions for ω_1 and ω_2 are

$$\omega_{1,2} = \omega_0 \left[\sqrt{1 + \left(\frac{1}{2Q_0}\right)^2} \mp \frac{1}{2Q_0}\right]$$

- The approximate (high-Q_0) expressions for ω_1 and ω_2 are

$$\omega_{1,2} \approx \omega_0 \mp \frac{1}{2}\mathcal{B}$$

- The half-power bandwidth $\mathcal{B}$ is given by

$$\mathcal{B} = \omega_2 - \omega_1 = \frac{\omega_0}{Q_0}$$

- The input admittance may also be expressed in approximate form for high-Q circuits:

$$\mathbf{Y} \approx \frac{1}{R}(1 + jN) = \frac{1}{R}\sqrt{1 + N^2}\,\underline{/\tan^{-1} N}$$

where N is defined as the number of half-bandwidths off resonance, or

$$N = \frac{\omega - \omega_0}{\frac{1}{2}\mathcal{B}}$$

This approximation is valid for $0.9\omega_0 \le \omega \le 1.1\omega_0$.

15.5 SERIES RESONANCE

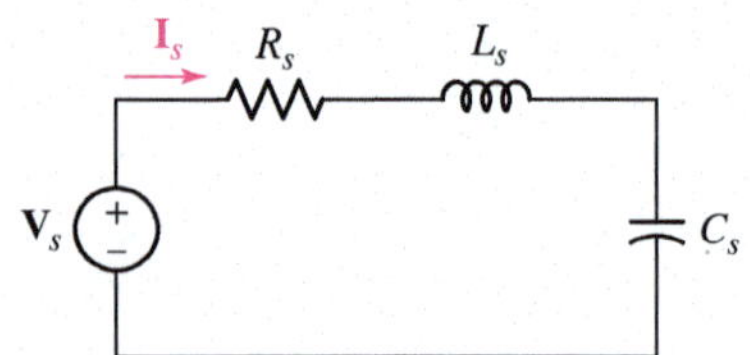

FIGURE 15.24 A series resonant circuit.

Although we probably find less use for the series *RLC* circuit than we do for the parallel *RLC* circuit, it is still worthy of our attention. We will consider the circuit shown in Fig. 15.24. It should be noted that the various circuit elements are given the subscript *s* (for series) for the time being in order to avoid confusing them with the parallel elements when the circuits are compared.

Our discussion of parallel resonance occupied two sections of considerable length. We could now give the series *RLC* circuit the same kind of treatment, but it is much cleverer to avoid such needless repetition and use the concept of duality. For simplicity, let us concentrate on the conclusions presented in the last paragraph of the preceding section on parallel resonance. The important results are contained there, and the use of dual language enables us to transcribe this paragraph to present the important results for the series *RLC* circuit.

Again, this paragraph is the same as the last paragraph of Sec. 15.4, with the parallel *RLC* language converted to series *RLC* language using duality (hence the quotation marks).

"We conclude our coverage of the *series* resonant circuit by reviewing some key conclusions we have reached:

- The resonant frequency ω_0 is the frequency at which the imaginary part of the input impedance becomes zero, or the angle of the impedance becomes zero. For this circuit, $\omega_0 = 1/\sqrt{C_s L_s}$.
- The circuit's figure of merit Q_0 is defined as 2π times the ratio of the maximum energy stored in the circuit to the energy lost each period in the circuit. For this circuit, $Q_0 = \omega_0 L_s/R_s$.
- We defined two half-power frequencies, ω_1 and ω_2, as the frequencies at which the impedance magnitude is $\sqrt{2}$ times the minimum

impedance magnitude. (These are also the frequencies at which the current response is 70.7 percent of the maximum response.)

- The exact expressions for ω_1 and ω_2 are

$$\omega_{1,2} = \omega_0\left[\sqrt{1 + \left(\frac{1}{2Q_0}\right)^2} \mp \frac{1}{2Q_0}\right]$$

- The approximate (high-Q_0) expressions for ω_1 and ω_2 are

$$\omega_{1,2} \approx \omega_0 \mp \frac{1}{2}\mathcal{B}$$

- The half-power bandwidth $\mathcal{B}$ is given by

$$\mathcal{B} = \omega_2 - \omega_1 = \frac{\omega_0}{Q_0}$$

- The input admittance may also be expressed in approximate form for high-Q circuits:

$$\mathbf{Y} \approx \frac{1}{R}(1 + jN) = \frac{1}{R}\sqrt{1 + N^2}\,\underline{/\tan^{-1} N}$$

where N is defined as the number of half-bandwidths off resonance, or

$$N = \frac{\omega - \omega_0}{\frac{1}{2}\mathcal{B}}$$

This approximation is valid for $0.9\omega_0 \leq \omega \leq 1.1\omega_0$."

From this point on, we will no longer identify series resonant circuits by use of the subscript s, unless clarity requires it.

EXAMPLE 15.8

The voltage $v_s = 100 \cos \omega t$ mV is applied to a series resonant circuit composed of a 10 Ω resistance, a 200 nF capacitance, and a 2 mH inductance. Use both exact and approximate methods to calculate the current amplitude if $\omega = 48$ krad/s.

The resonant frequency of the circuit is given by

$$\omega_0 = \frac{1}{\sqrt{LC}} = \frac{1}{\sqrt{(2 \times 10^{-3})(200 \times 10^{-9})}} = 50 \text{ krad/s}$$

Since we are operating at $\omega = 48$ krad/s, which is within 10 percent of the resonant frequency, it is reasonable to apply our approximate relationships to estimate the equivalent impedance of the network provided that we find that we are working with a high-Q circuit:

$$\mathbf{Z}_{\text{eq}} \approx R\sqrt{1 + N^2}\,\underline{/\tan^{-1} N}$$

(Continued on next page)

where N is computed once we determine Q_0. This is a series circuit, so

$$Q_0 = \frac{\omega_0 L}{R} = \frac{(50 \times 10^3)(2 \times 10^{-3})}{10} = 10$$

which qualifies as a high-Q circuit. Thus,

$$\mathcal{B} = \frac{\omega_0}{Q_0} = \frac{50 \times 10^3}{10} = 5 \text{ krad/s}$$

The number of half-bandwidths off resonance (N) is therefore

$$N = \frac{\omega - \omega_0}{\mathcal{B}/2} = \frac{48 - 50}{2.5} = -0.8$$

Thus,

$$\mathbf{Z}_{\text{eq}} \approx R\sqrt{1 + N^2}\,\underline{/\tan^{-1} N} = 12.81\,\underline{/-38.66^\circ}\ \Omega$$

The approximate current magnitude is then

$$\frac{|\mathbf{V}_s|}{|\mathbf{Z}_{\text{eq}}|} = \frac{100}{12.81} = 7.806 \text{ mA}$$

Using the exact expressions, we find that $\mathbf{I} = 7.746\,\underline{/39.24^\circ}$ mA and thus

$$|\mathbf{I}| = 7.746 \text{ mA}$$

PRACTICE

15.11 A series resonant circuit has a bandwidth of 100 Hz and contains a 20 mH inductance and a 2 μF capacitance. Determine (*a*) f_0; (*b*) Q_0; (*c*) $\mathbf{Z}_{\text{in}}$ at resonance; (*d*) f_2.

Ans: 796 Hz; 7.96; 12.57 + $j0$ Ω; 846 Hz (approx.).

The series resonant circuit is characterized by a minimum impedance at resonance, whereas the parallel resonant circuit produces a maximum resonant impedance. The latter circuit provides inductor currents and capacitor currents at resonance which have amplitudes Q_0 times as great as the source current; the series resonant circuit provides inductor voltages and capacitor voltages which are greater than the source voltage by the factor Q_{0s}. The series circuit thus provides voltage amplification at resonance.

A comparison of our results for series and parallel resonance, as well as the exact and approximate expressions we have developed, appears in Table 15.1.

TABLE **15.1** A Short Summary of Resonance

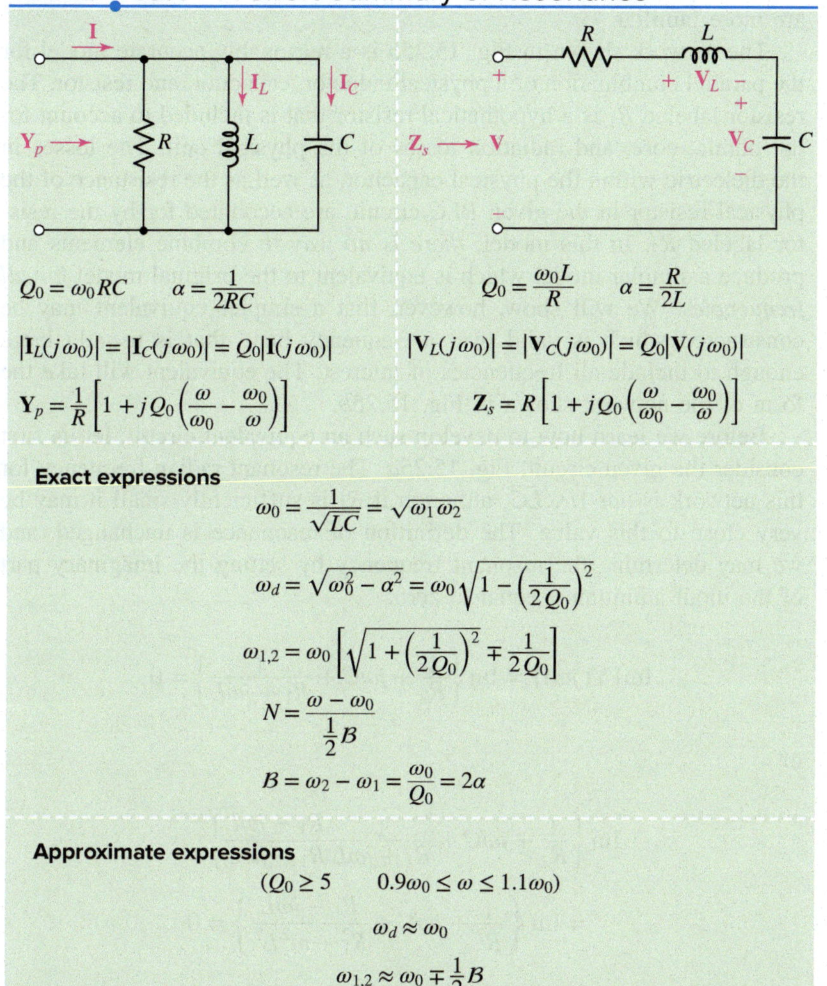

Parallel	Series
$Q_0 = \omega_0 RC \qquad \alpha = \dfrac{1}{2RC}$	$Q_0 = \dfrac{\omega_0 L}{R} \qquad \alpha = \dfrac{R}{2L}$
$\lvert \mathbf{I}_L(j\omega_0)\rvert = \lvert \mathbf{I}_C(j\omega_0)\rvert = Q_0\lvert \mathbf{I}(j\omega_0)\rvert$	$\lvert \mathbf{V}_L(j\omega_0)\rvert = \lvert \mathbf{V}_C(j\omega_0)\rvert = Q_0\lvert \mathbf{V}(j\omega_0)\rvert$
$\mathbf{Y}_p = \dfrac{1}{R}\left[1 + jQ_0\left(\dfrac{\omega}{\omega_0} - \dfrac{\omega_0}{\omega}\right)\right]$	$\mathbf{Z}_s = R\left[1 + jQ_0\left(\dfrac{\omega}{\omega_0} - \dfrac{\omega_0}{\omega}\right)\right]$

Exact expressions

$$\omega_0 = \frac{1}{\sqrt{LC}} = \sqrt{\omega_1 \omega_2}$$

$$\omega_d = \sqrt{\omega_0^2 - \alpha^2} = \omega_0\sqrt{1 - \left(\frac{1}{2Q_0}\right)^2}$$

$$\omega_{1,2} = \omega_0\left[\sqrt{1 + \left(\frac{1}{2Q_0}\right)^2} \mp \frac{1}{2Q_0}\right]$$

$$N = \frac{\omega - \omega_0}{\frac{1}{2}B}$$

$$B = \omega_2 - \omega_1 = \frac{\omega_0}{Q_0} = 2\alpha$$

Approximate expressions

$$(Q_0 \geq 5 \qquad 0.9\omega_0 \leq \omega \leq 1.1\omega_0)$$

$$\omega_d \approx \omega_0$$

$$\omega_{1,2} \approx \omega_0 \mp \frac{1}{2}B$$

$$\omega_0 \approx \frac{1}{2}(\omega_1 + \omega_2)$$

$$\mathbf{Y}_p \approx \frac{\sqrt{1 + N^2}}{R}\underline{/\tan^{-1} N}$$

$$\mathbf{Z}_s \approx R\sqrt{1 + N^2}\,\underline{/\tan^{-1} N}$$

15.6 OTHER RESONANT FORMS

The parallel and series *RLC* circuits of the previous two sections represent *idealized* resonant circuits. The degree of accuracy with which the idealized model fits an *actual* circuit depends on the operating frequency range, the Q of the circuit, the materials present in the physical elements, the element sizes, and many other factors. We are not studying the techniques for determining the best model of a given physical circuit, for this requires some knowledge of electromagnetic field theory and the properties of materials; we are, however, concerned with the problem of reducing a

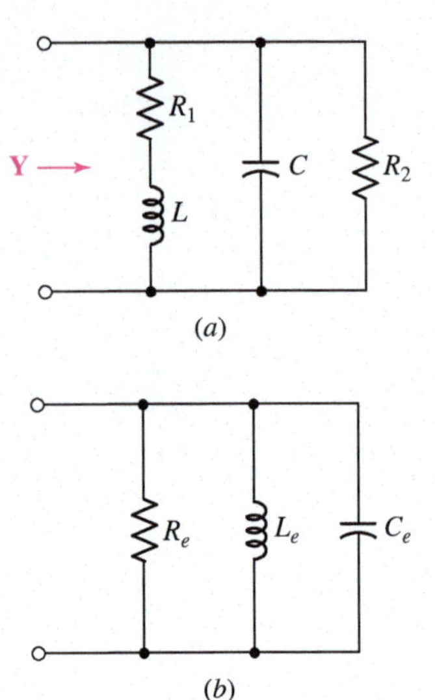

FIGURE 15.25 (*a*) A useful model of a physical network which consists of a physical inductor, capacitor, and resistor in parallel. (*b*) A network which can be equivalent to part (*a*) over a narrow frequency band.

more complicated model to one of the two simpler models with which we are more familiar.

The network shown in Fig. 15.25*a* is a reasonably accurate model for the parallel combination of a physical inductor, capacitor, and resistor. The resistor labeled R_1 is a hypothetical resistor that is included to account for the ohmic, core, and radiation losses of the physical coil. The losses in the dielectric within the physical capacitor, as well as the resistance of the physical resistor in the given *RLC* circuit, are accounted for by the resistor labeled R_2. In this model, *there is no way* to combine elements and produce a simpler model which is equivalent to the original model *for all frequencies*. We will show, however, that a simpler equivalent may be constructed which is valid over a frequency band that is usually large enough to include all frequencies of interest. The equivalent will take the form of the network shown in Fig. 15.25*b*.

Before we learn how to develop such an equivalent circuit, let us first consider the given circuit, Fig. 15.25*a*. The resonant radian frequency for this network is *not* $1/\sqrt{LC}$, although if R_1 is sufficiently small it may be very close to this value. The definition of resonance is unchanged, and we may determine the resonant frequency by setting the imaginary part of the input admittance equal to zero:

$$\text{Im}\{\mathbf{Y}(j\omega)\} = \text{Im}\left\{\frac{1}{R_2} + j\omega C + \frac{1}{R_1 + j\omega L}\right\} = 0$$

or

$$\text{Im}\left\{\frac{1}{R_2} + j\omega C + \frac{1}{R_1 + j\omega L}\frac{R_1 - j\omega L}{R_1 - j\omega L}\right\}$$

$$= \text{Im}\left\{\frac{1}{R^2} + j\omega C + \frac{R_1 - j\omega L}{R_1^2 + \omega^2 L^2}\right\} = 0$$

Thus, we have the resonance condition that

$$C = \frac{L}{R_1^2 + \omega^2 L^2}$$

and so

$$\omega_0 = \sqrt{\frac{1}{LC} - \left(\frac{R_1}{L}\right)^2} \qquad [27]$$

We note that ω_0 is less than $1/\sqrt{LC}$, but sufficiently small values of the ratio R_1/L may result in a negligible difference between ω_0 and $1/\sqrt{LC}$.

The maximum magnitude of the input impedance also deserves consideration. It is not R_2, and it does not occur at ω_0 (or at $\omega = 1/\sqrt{LC}$). The proof of these statements will not be shown, because the expressions soon become algebraically cumbersome; the theory, however, is straightforward. Let us be content with a numerical example.

EXAMPLE 15.9

Using the values $R_1 = 2\ \Omega$, $L = 1$ H, $C = 125$ mF, and $R_2 = 3\ \Omega$ for Fig. 15.25*a*, determine the resonant frequency and the impedance at resonance.

Substituting the appropriate values in Eq. [27], we find

$$\omega_0 = \sqrt{8 - 2^2} = 2 \text{ rad/s}$$

and this enables us to calculate the input admittance,

$$\mathbf{Y} = \frac{1}{3} + j2\left(\frac{1}{8}\right) + \frac{1}{2 + j(2)(1)} = \frac{1}{3} + \frac{1}{4} = 0.583 \text{ S}$$

and then the input impedance at resonance:

$$\mathbf{Z}(j2) = \frac{1}{0.583} = 1.714\ \Omega$$

At the frequency which would be the resonant frequency if R_1 were zero,

$$\frac{1}{\sqrt{LC}} = 2.83 \text{ rad/s}$$

the input impedance would be

$$\mathbf{Z}(j2.83) = 1.947\underline{/-13.26^\circ}\ \Omega$$

As can be seen in Fig. 15.26, however, the frequency at which the *maximum* impedance magnitude occurs, indicated by ω_m, can be determined to be $\omega_m = 3.26$ rad/s, and the *maximum* impedance magnitude is

$$\mathbf{Z}(j3.26) = 1.980\underline{/-21.4^\circ}\ \Omega$$

The impedance magnitude at resonance and the maximum magnitude differ by about 16 percent. Although it is true that such an error may be neglected occasionally in practice, it is too large to neglect on an exam. (The later work in this section will show that the Q of the inductor-resistor combination at 2 rad/s is unity; this low value accounts for the 16 percent discrepancy.)

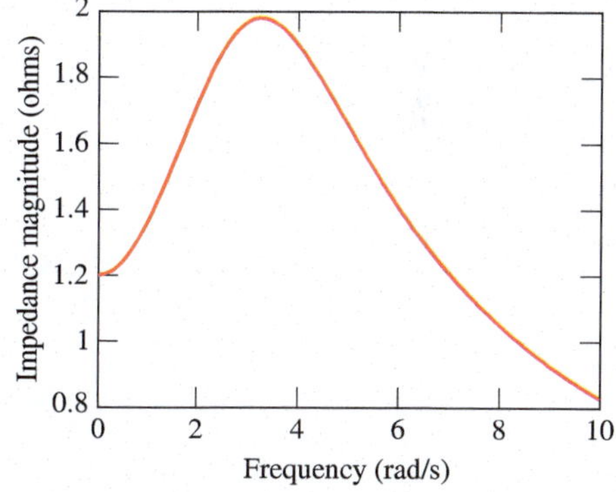

FIGURE 15.26 Plot of |**Z**| vs. ω, generated using the following MATLAB script:

```
>> omega = linspace(0,10,100);
>> for i = 1:100
Y(i) = 1/3 + j*omega(i)/8 + 1/(2 + j*omega(i));
Z(i) = 1/Y(i);
end
>> plot(omega,abs(Z));
>> xlabel('Frequency (rad/s)');
>> ylabel('Impedance magnitude (ohms)');
```

PRACTICE

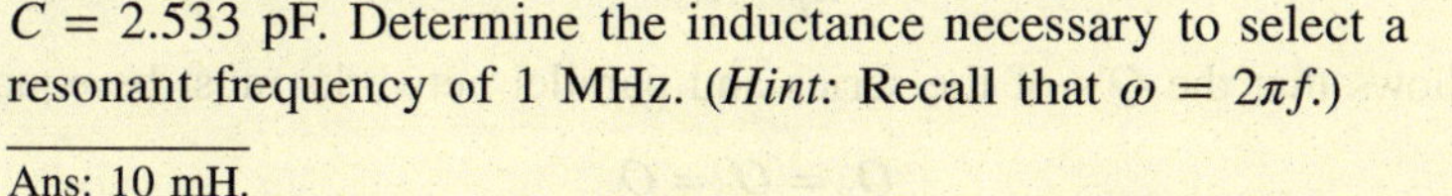

15.12 Referring to the circuit of Fig. 15.25*a*, let $R_1 = 1$ kΩ and $C = 2.533$ pF. Determine the inductance necessary to select a resonant frequency of 1 MHz. (*Hint:* Recall that $\omega = 2\pi f$.)

Ans: 10 mH.

Equivalent Series and Parallel Combinations

In order to transform the given circuit of Fig. 15.25*a* into an equivalent of the form shown in Fig. 15.25*b*, we must discuss the Q of a simple series or parallel combination of a resistor and a reactor (inductor or capacitor). We first consider the series circuit shown in Fig. 15.27*a*. The Q of this network

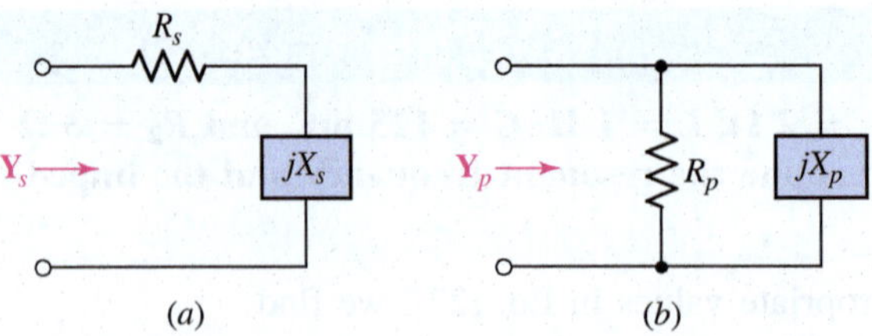

FIGURE 15.27 (*a*) A series network which consists of a resistance R_S and an inductive or capacitive reactance X_S may be transformed into (*b*) a parallel network such that $\mathbf{Y}_S = \mathbf{Y}_p$ at one specific frequency. The reverse transformation is equally possible.

is again defined as 2π times the ratio of the maximum stored energy to the energy lost each period, but the Q may be evaluated at any frequency we choose. In other words, Q is a function of ω. It is true that we will choose to evaluate it at a frequency which is, or apparently is, the resonant frequency of some network of which the series arm is a part. This frequency, however, is not known until a more complete circuit is available. The reader is encouraged to show that the Q of this series arm is $|X_s|/R_s$, whereas the Q of the parallel network of Fig. 15.27*b* is $R_p/|X_p|$.

Let us now carry out the details necessary to find values for R_p and X_p so that the parallel network of Fig. 15.27*b* is equivalent to the series network of Fig. 15.27*a* at some single specific frequency. We equate $\mathbf{Y}_s$ and $\mathbf{Y}_p$,

$$\mathbf{Y}_s = \frac{1}{R_s + jX_s} = \frac{R_s - jX_s}{R_s^2 + X_s^2}$$

$$= Y_p = \frac{1}{R_p} - j\frac{1}{X_p}$$

and obtain

$$R_p = \frac{R_s^2 + X_s^2}{R_s}$$

$$X_p = \frac{R_s^2 + X_s^2}{X_s}$$

Dividing these two expressions, we find

$$\frac{R_p}{X_p} = \frac{X_s}{R_s}$$

It follows that the Q's of the series and parallel networks must be equal:

$$Q_p = Q_s = Q$$

The transformation equations may therefore be simplified:

$$R_p = R_s(1 + Q^2) \qquad [28]$$

$$X_p = X_s\left(1 + \frac{1}{Q^2}\right) \qquad [29]$$

Also R_s and X_s may be found if R_p and X_p are the given values; the transformation in either direction may be performed.

If $Q \geq 5$, little error is introduced by using the approximate relationships

$$R_p \approx Q^2 R_s \qquad [30]$$

$$X_p \approx X_s \qquad (C_p \approx C_s \quad \text{or} \quad L_p \approx L_s) \qquad [31]$$

EXAMPLE 15.10

Find the parallel equivalent of the series combination of a 100 mH inductor and a 5 Ω resistor at a frequency of 1000 rad/s. Details of the network to which this series combination is connected are unavailable.

At ω = 1000 rad/s, $X_s = 1000(100 \times 10^{-3}) = 100\ \Omega$. The Q of this series combination is

$$Q = \frac{X_s}{R_s} = \frac{100}{5} = 20$$

Since the Q is sufficiently high (20 is much greater than 5), we use Eqs. [30] and [31] to obtain

$$R_p \approx Q^2 R_s = 2000\ \Omega \qquad \text{and} \qquad L_p \approx L_s = 100\ \text{mH}$$

Our assertion here is that a 100 mH inductor in series with a 5 Ω resistor provides *essentially the same* input impedance as does a 100 mH inductor in parallel with a 2000 Ω resistor at the frequency 1000 rad/s.

To check the accuracy of the equivalence, let us evaluate the input impedance for each network at 1000 rad/s. We find

$$\mathbf{Z}_s(j1000) = 5 + j100 = 100.1\,\underline{/87.1^\circ}\ \Omega$$

$$\mathbf{Z}_p(j1000) = \frac{2000(j100)}{2000 + j100} = 99.9\,\underline{/87.1^\circ}\ \Omega$$

and conclude that the accuracy of our approximation at the transformation frequency is pretty impressive. The accuracy at 900 rad/s is also reasonably good, because

$$\mathbf{Z}_s(j900) = 90.1\,\underline{/86.8^\circ}\ \Omega$$

$$\mathbf{Z}_p(j900) = 89.9\,\underline{/87.4^\circ}\ \Omega$$

PRACTICE

15.13 At ω = 1000 rad/s, find a parallel network that is equivalent to the series combination in Fig. 15.28*a*.

15.14 Find a series equivalent for the parallel network shown in Fig. 15.28*b*, assuming ω = 1000 rad/s.

Ans: 15.13: 8 H, 640 kΩ; 15.14: 5 H, 250 Ω.

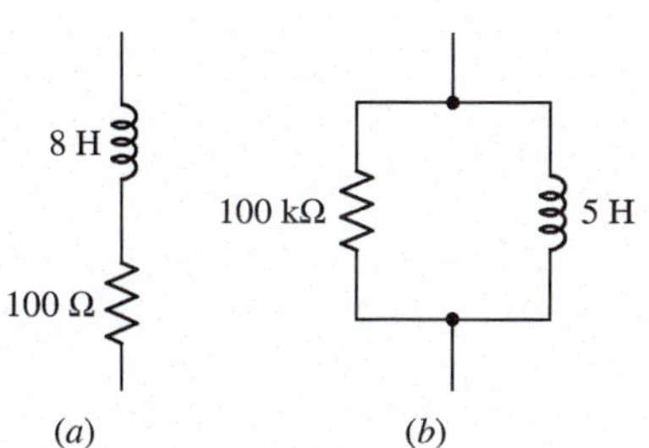

FIGURE 15.28 (*a*) A series network for which an equivalent parallel network (at ω = 1000 rad/s) is needed. (*b*) A parallel network for which an equivalent series network (at ω = 1000 rad/s) is needed.

As a further example of the replacement of a more complicated resonant circuit by an equivalent series or parallel *RLC* circuit, let us consider a problem in electronic instrumentation. The simple series *RLC* network in Fig. 15.29*a* is excited by a sinusoidal voltage source at the network's

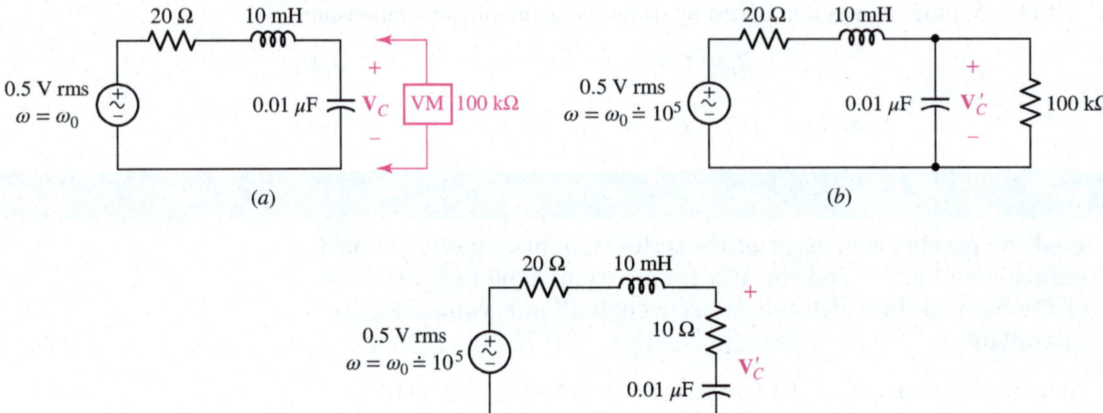

FIGURE 15.29 *(a)* A given series resonant circuit in which the capacitor voltage is to be measured by a nonideal electronic voltmeter. *(b)* The effect of the voltmeter is included in the circuit; it reads $\mathbf{V}'_C$. *(c)* A series resonant circuit is obtained when the parallel *RC* network in part *(b)* is replaced by the series *RC* network which is equivalent at 10^5 rad/s.

An "ideal" meter is an instrument that measures a particular quantity of interest without disturbing the circuit being tested.

resonant frequency. The effective (rms) value of the source voltage is 0.5 V, and we wish to measure the effective value of the voltage across the capacitor with an electronic voltmeter (VM) having an internal resistance of 100,000 Ω. That is, an equivalent representation of the voltmeter is an ideal voltmeter in parallel with a 100 kΩ resistor.

Before the voltmeter is connected, we compute that the resonant frequency is 10^5 rad/s, $Q_0 = 50$, the current is 25 mA, and the rms capacitor voltage is 25 V. (As indicated at the end of Sec. 15.5, this voltage is Q_0 times the applied voltage.) Thus, if the voltmeter were ideal, it would read 25 V when connected across the capacitor.

However, when the actual voltmeter is connected, the circuit shown in Fig. 15.29*b* results. In order to obtain a series *RLC* circuit, it is now necessary to replace the parallel *RC* network with a series *RC* network. Let us assume that the *Q* of this *RC* network is sufficiently high that the equivalent series capacitor will be the same as the given parallel capacitor. We do this in order to approximate the resonant frequency of the final series *RLC* circuit. Thus, if the series *RLC* circuit also contains a 0.01 μF capacitor, the resonant frequency remains 10^5 rad/s. We need to know this estimated resonant frequency in order to calculate the *Q* of the parallel *RC* network; it is

$$Q = \frac{R_p}{|X_p|} = \omega R_p C_p = 10^5(10^5)(10^{-8}) = 100$$

Since this value is greater than 5, our vicious circle of assumptions is justified, and the equivalent series *RC* network consists of the capacitor $C_s = 0.01\ \mu\text{F}$ and the resistor

$$R_s \approx \frac{R_p}{Q^2} = 10\ \Omega$$

Hence, the equivalent circuit of Fig. 15.29*c* is obtained. The resonant *Q* of this circuit is now only 33.3, and thus the voltage across the capacitor in the circuit of Fig. 15.29*c* is $16\frac{2}{3}$V. But we need to find $|\mathbf{V}'_C|$, the voltage across the series *RC* combination; we obtain

$$|\mathbf{V}'_C| = \frac{0.5}{30}|10 - j1000| = 16.67 \text{ V}$$

The capacitor voltage and $|\mathbf{V}'_C|$ are essentially equal, since the voltage across the 10 Ω resistor is quite small.

The final conclusion must be that an apparently good voltmeter may still produce a severe effect on the response of a high-*Q* resonant circuit. A similar effect may occur when a nonideal ammeter is inserted in the circuit.

We wrap up this section with a technical fable.

Once upon a time there was a student named Sean, who had a professor identified simply as Dr. Abel.

In the laboratory one afternoon, Dr. Abel gave Sean three practical circuit devices: a resistor, an inductor, and a capacitor, having nominal element values of 20 Ω, 20 mH, and 1 μF. The student was asked to connect a variable-frequency voltage source to the series combination of these three elements, to measure the resultant voltage across the resistor as a function of frequency, and then to calculate numerical values for the resonant frequency, the *Q* at resonance, and the half-power bandwidth. The student was also asked to predict the results of the experiment before making the measurements.

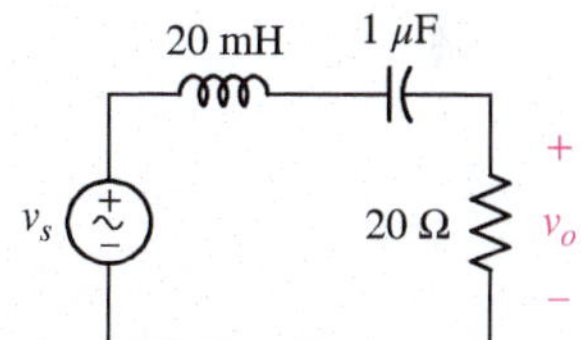

FIGURE 15.30 A first model for a 20 mH inductor, a 1 μF capacitor, and a 20 Ω resistor in series with a voltage generator.

Sean first drew an equivalent circuit for this problem that was like the circuit of Fig. 15.30 and then calculated

$$f_0 = \frac{1}{2\pi\sqrt{LC}} = \frac{1}{2\pi\sqrt{20 \times 10^{-3} \times 10^{-6}}} = 1125 \text{ Hz}$$

$$Q_0 = \frac{\omega_0 L}{R} = 7.07$$

$$\mathcal{B} = \frac{f_0}{Q_0} = 159 \text{ Hz}$$

Next, Sean made the measurements that Dr. Abel requested, compared them with the predicted values, and then felt a strong urge to transfer to the business school. The results were

$$f_0 = 1000 \text{ Hz} \quad Q_0 = 0.625 \quad \mathcal{B} = 1600 \text{ Hz}$$

Sean knew that discrepancies of this magnitude could not be characterized as being "within engineering accuracy" or "due to meter errors." Sadly, the results were handed to the professor.

Remembering many past errors in judgment, some of which were even (possibly) self-made, Dr. Abel smiled kindly and called Sean's attention to the *Q*-meter (or impedance bridge) which is present in most well-equipped laboratories, and suggested that it might be used to find out what these practical circuit elements really looked like at some convenient frequency near resonance, 1000 Hz, for example.

Upon doing so, Sean discovered that the resistor had a measured value of 18 Ω and the inductor was 21.4 mH with a *Q* of 1.2, while the capacitor had a capacitance of 1.41 μF and a dissipation factor (the reciprocal of *Q*) equal to 0.123.

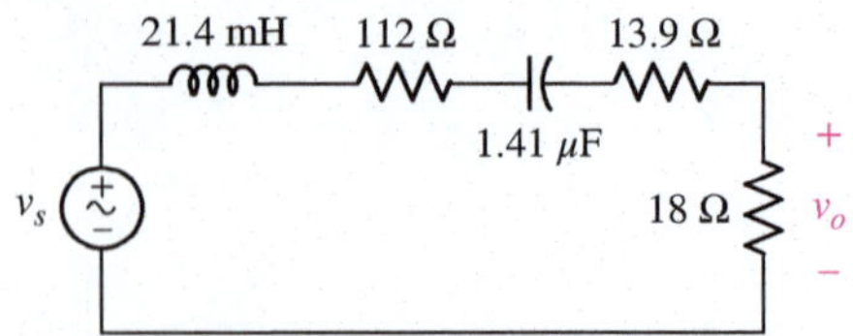

FIGURE 15.31 An improved model in which more accurate values are used and the losses in the inductor and capacitor are acknowledged.

So, with the hope that springs eternal within the heart of every engineering undergraduate, Sean reasoned that a better model for the practical inductor would be 21.4 mH in series with $\omega L/Q$ = 112 Ω, while a more appropriate model for the capacitor would be 1.41 μF in series with $1/\omega C\ Q$ = 13.9 Ω. Using these data, Sean prepared the modified circuit model shown as Fig. 15.31 and calculated a new set of predicted values:

$$f_0 = \frac{1}{2\pi\sqrt{21.4 \times 10^{-3} \times 1.41 \times 10^{-6}}} = 916 \text{ Hz}$$

$$Q_0 = \frac{2\pi \times 916 \times 21.4 \times 10^{-3}}{143.9} = 0.856$$

$$\mathcal{B} = 916/0.856 = 1070 \text{ Hz}$$

Since these results were much closer to the measured values, Sean was much happier. Dr. Abel, however, being a stickler for detail, pondered the differences in the predicted and measured values for both Q_0 and the bandwidth. *"Have you,"* Dr. Abel asked, *"given any consideration to the output impedance of the voltage source?" "Not yet,"* said Sean, trotting back to the laboratory bench.

It turned out that the output impedance in question was 50 Ω, and so Sean added this value to the circuit diagram, as shown in Fig. 15.32. Using the new equivalent resistance value of 193.9 Ω, improved values for Q_0 and $\mathcal{B}$ were then obtained:

$$Q_0 = 0.635 \quad \mathcal{B} = 1442 \text{ Hz}$$

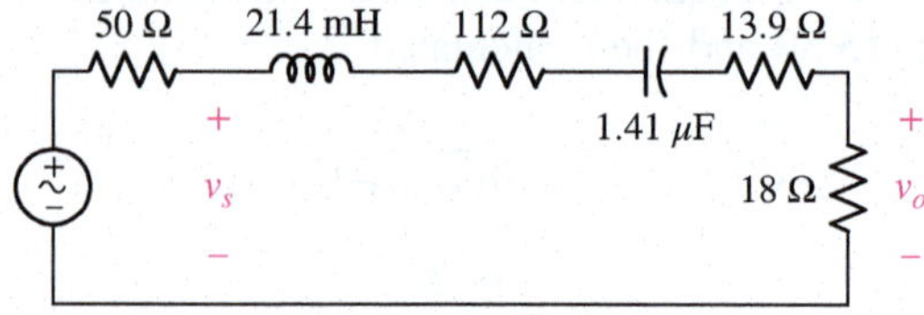

FIGURE 15.32 The final model also contains the output resistance of the voltage source.

Since all the theoretical and experimental values now agreed within 10 percent, Sean was once again an enthusiastic, confident engineering student, motivated to start homework early and read the textbook prior to class.[2] Dr. Abel simply nodded her head agreeably as she moralized:

When using real devices,
Watch the models that you choose;
Think well before you calculate,
And mind your Z's and Q's!

PRACTICE

15.15 The series combination of 10 Ω and 10 nF is in parallel with the series combination of 20 Ω and 10 mH. (*a*) Find the approximate resonant frequency of the parallel network. (*b*) Find the Q of the RC branch. (*c*) Find the Q of the RL branch. (*d*) Find the three-element equivalent of the original network.

Ans: 10^5 rad/s; 100; 50; 10 nF ∥ 10 mH ∥ 33.3 kΩ.

(2) Okay, this last part is a bit much. Sorry about that. Got a bit carried away.

15.7 SCALING

Some of the examples and problems that we have been solving have involved circuits containing passive element values ranging around a few ohms, a few henrys, and a few farads. The applied frequencies were a few radians per second. These particular numerical values were used not because they are those commonly met in practice, but because arithmetic manipulations are so much easier than they would be if it were necessary to carry along various powers of 10 throughout the calculations. The scaling procedures that will be discussed in this section enable us to analyze networks composed of practical-sized elements by scaling the element values to permit more convenient numerical calculations. We will consider both ***magnitude scaling*** and ***frequency scaling.***

Let us select the parallel resonant circuit shown in Fig. 15.33*a* as our example. The impractical element values lead to the unlikely response curve drawn as Fig. 15.33*b*; the maximum impedance is 2.5 Ω, the resonant frequency is 1 rad/s, Q_0 is 5, and the bandwidth is 0.2 rad/s. These numerical values are much more characteristic of the electrical analog of some mechanical system than they are of any basically electrical device. We have convenient numbers with which to calculate, but an impractical circuit to construct.

Our goal is to scale this network in such a way as to provide an impedance maximum of 5000 Ω at a resonant frequency of 5×10^6 rad/s, or 796 kHz. In other words, we may use the same response curve shown in Fig. 15.33*b* if every number on the *ordinate* scale is increased by a factor of 2000 and every number on the *abscissa* scale is increased by a factor of 5×10^6. We will treat this as two problems: (1) scaling in magnitude by a factor of 2000 and (2) scaling in frequency by a factor of 5×10^6.

Recall that "ordinate" refers to the vertical axis and "abscissa" refers to the horizontal axis.

Magnitude scaling is defined as the process by which the impedance of a two-terminal network is increased by a factor of K_m, the frequency remaining constant. The factor K_m is real and positive; it may be greater or smaller than unity. We will understand that the shorter statement *"the network is scaled in magnitude by a factor of 2"* indicates that the impedance of the new network is to be twice that of the old network at any frequency. Let us now determine how we must scale each type of passive element. To increase the input impedance of a network by a factor of K_m, it is sufficient to increase the impedance of each element in the network by this same factor. Thus, a resistance R must be replaced by a resistance K_mR. Each inductance must also exhibit an impedance which is K_m times as great at any frequency. In order to increase an impedance sL by a factor of K_m when **s** remains constant, the inductance L must be replaced by an inductance K_mL. In a similar manner, each capacitance C must be replaced by a capacitance C/K_m. In summary, these changes will produce a network which is scaled in magnitude by a factor of K_m:

$$\left.\begin{aligned} R &\rightarrow K_m R \\ L &\rightarrow K_m L \\ C &\rightarrow \frac{C}{K_m} \end{aligned}\right\} \quad \text{magnitude scaling}$$

When each element in the network of Fig. 15.33*a* is scaled in magnitude by a factor of 2000, the network shown in Fig. 15.34*a* results. The response curve shown in Fig. 15.34*b* indicates that no change in the previously drawn response curve need be made other than a change in the scale of the ordinate.

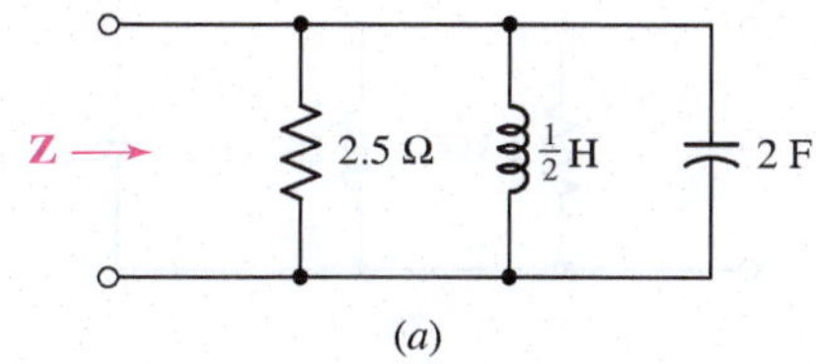

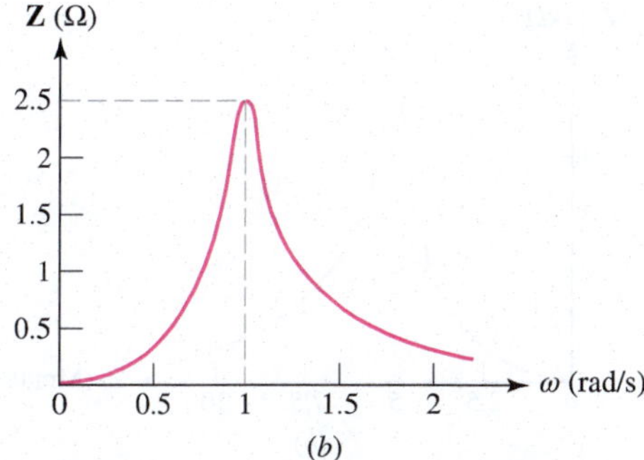

FIGURE 15.33 (*a*) A parallel resonant circuit used as an example to illustrate magnitude and frequency scaling. (*b*) The magnitude of the input impedance is shown as a function of frequency.

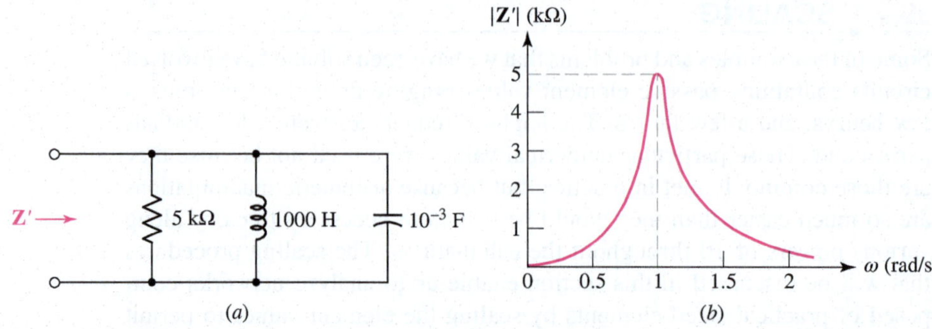

FIGURE 15.34 (a) The network of Fig. 15.33a after being scaled in magnitude by a factor $K_m = 2000$. (b) The corresponding response curve.

Let us now take this new network and scale it in frequency. We define frequency scaling as the process by which the frequency at which any impedance occurs is increased by a factor of K_f. Again, we will make use of the shorter expression *"the network is scaled in frequency by a factor of 2"* to indicate that the same impedance is now obtained at a frequency twice as great. Frequency scaling is accomplished by scaling each passive element in frequency. It is apparent that no resistor is affected. The impedance of any inductor is sL, and if this same impedance is to be obtained at a frequency K_f times as great, then the inductance L must be replaced by an inductance of L/K_f. Similarly, a capacitance C is to be replaced by a capacitance C/K_f. Thus, if a network is to be scaled in frequency by a factor of K_f, then the changes necessary in each passive element are

$$\left.\begin{aligned} R &\rightarrow R \\ L &\rightarrow \frac{L}{K_f} \\ C &\rightarrow \frac{C}{K_f} \end{aligned}\right\} \quad \text{frequency scaling}$$

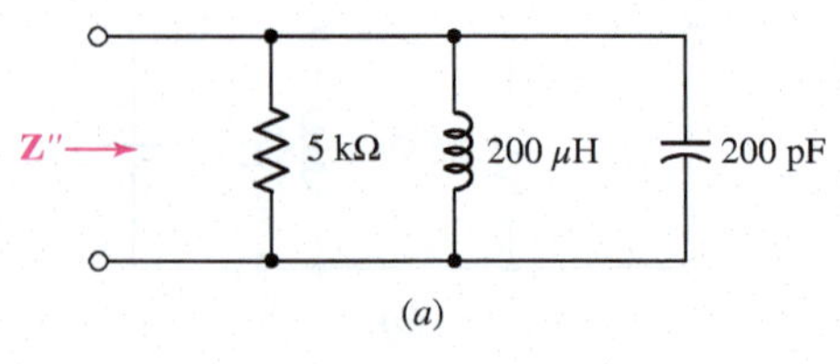

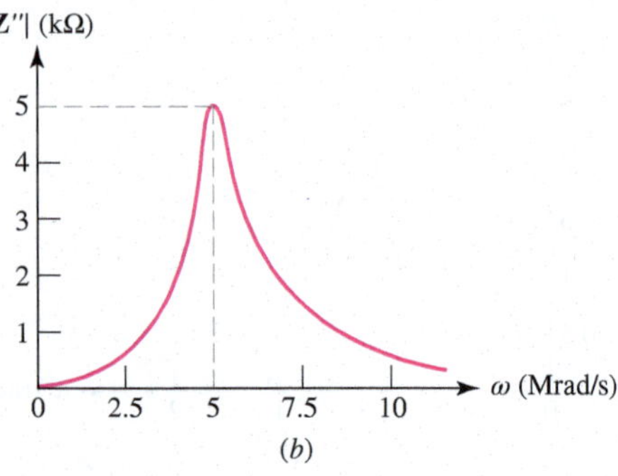

FIGURE 15.35 (a) The network of Fig. 15.34a after being scaled in frequency by a factor $K_f = 5 \times 10^6$. (b) The corresponding response curve.

When each element of the magnitude-scaled network of Fig. 15.34a is scaled in frequency by a factor of 5×10^6, the network of Fig. 15.35a is obtained. The corresponding response curve is shown in Fig. 15.35b.

The circuit elements in this last network have values which are easily achieved in physical circuits; the network can actually be built and tested. It follows that, if the original network of Fig. 15.33a were actually an analog of some mechanical resonant system, we could have scaled this analog in both magnitude and frequency in order to achieve a network which we might construct in the laboratory; tests that are expensive or inconvenient to run on the mechanical system could then be made on the scaled electrical system, and the results should then be "unscaled" and converted into mechanical units to complete the analysis.

An impedance that is given as a function of **s** may also be scaled in magnitude or frequency, and this may be done without any knowledge of the specific elements of which the two-terminal network is composed. In order to scale **Z(s)** in magnitude, the definition of magnitude scaling shows that it is necessary only to multiply **Z(s)** by K_m in order to obtain the magnitude-scaled impedance. Hence, the impedance **Z′(s)** of the magnitude-scaled network is

$$\mathbf{Z}'(\mathbf{s}) = K_m\,\mathbf{Z}(\mathbf{s})$$

If $\mathbf{Z}'(\mathbf{s})$ is now to be scaled in frequency by a factor of 5×10^6, then $\mathbf{Z}''(\mathbf{s})$ and $\mathbf{Z}'(\mathbf{s})$ are to provide identical values of impedance if $\mathbf{Z}''(\mathbf{s})$ is evaluated at a frequency K_f times that at which $\mathbf{Z}'(\mathbf{s})$ is evaluated, or

$$\mathbf{Z}''(\mathbf{s}) = \mathbf{Z}'\left(\frac{\mathbf{s}}{K_f}\right)$$

Although scaling is a process normally applied to passive elements, dependent sources may also be scaled in magnitude and frequency. We assume that the output of any source is given as $k_x v_x$ or $k_y i_y$, where k_x has the dimensions of an admittance for a dependent current source and is dimensionless for a dependent voltage source, while k_y has the dimensions of ohms for a dependent voltage source and is dimensionless for a dependent current source. If the network containing the dependent source is scaled in magnitude by K_m, then it is necessary only to treat k_x or k_y as if it were the type of element consistent with its dimensions. That is, if k_x (or k_y) is dimensionless, it is left unchanged; if it is an admittance, it is divided by K_m; and if it is an impedance, it is multiplied by K_m. *Frequency scaling does not affect the dependent sources.*

EXAMPLE 15.11

Scale the network shown in Fig. 15.36 by $K_m = 20$ and $K_f = 50$, and then find $\mathbf{Z}_{in}(\mathbf{s})$ for the scaled network.

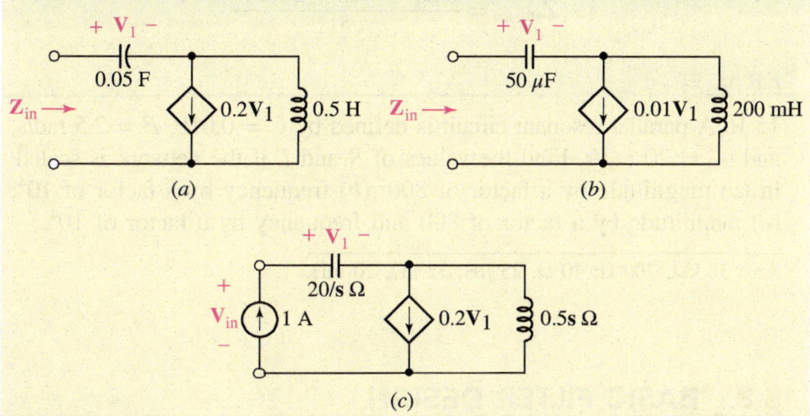

■ **FIGURE 15.36** (*a*) A network to be magnitude-scaled by a factor of 20, and frequency-scaled by a factor of 50. (*b*) The scaled network. (*c*) A 1 A test source is applied to the input terminals in order to obtain the impedance of the unscaled network in part (*a*).

Magnitude scaling of the capacitor is accomplished by dividing 0.05 F by the scaling factor $K_m = 20$, and frequency scaling is accomplished by dividing by $K_f = 50$. Carrying out both operations simultaneously,

$$C_{\text{scaled}} = \frac{0.05}{(20)(50)} = 50\ \mu\text{F}$$

The inductor is also scaled:

$$L_{\text{scaled}} = \frac{(20)(0.5)}{50} = 200\ \text{mH}$$

In scaling the dependent source, only magnitude scaling need be considered, as frequency scaling does not affect dependent sources. Since

(Continued on next page)

this is a *voltage*-controlled *current* source, the multiplying constant 0.2 has units of A/V, or S. Since the factor has units of admittance, we divide by K_m, so the new term is 0.01 $\mathbf{V}_1$. The resulting (scaled) network is shown in Fig. 15.36*b*.

To find the impedance of the new network, we need to apply a 1 A test source at the input terminals. We may work with either circuit; however, let's proceed by first finding the impedance of the *unscaled* network in Fig. 15.36*a*, and then scaling the result.

Referring to Fig. 15.36*c*,

$$\mathbf{V}_{\text{in}} = \mathbf{V}_1 + 0.5\mathbf{s}(1 - 0.2\,\mathbf{V}_1)$$

Also,

$$\mathbf{V}_1 = \frac{20}{\mathbf{s}}(1)$$

Performing the indicated substitution followed by a little algebraic manipulation yields

$$\mathbf{Z}_{\text{in}} = \frac{\mathbf{V}_{\text{in}}}{1} = \frac{\mathbf{s}^2 - 4\mathbf{s} + 40}{2\mathbf{s}}$$

To scale this quantity to correspond to the circuit of Fig. 15.36*b*, we multiply by $K_m = 20$ and replace $\mathbf{s}$ with $\mathbf{s}/\mathbf{K}_f = \mathbf{s}/50$. Thus,

$$\mathbf{Z}_{\text{in}_{\text{scaled}}} = \frac{0.2\mathbf{s}^2 - 40\mathbf{s} + 20{,}000}{\mathbf{s}}$$

PRACTICE

15.16 A parallel resonant circuit is defined by $C = 0.01\text{F}$, $\mathcal{B} = 2.5$ rad/s, and $\omega_0 = 20$ rad/s. Find the values of R and L if the network is scaled in (*a*) magnitude by a factor of 800; (*b*) frequency by a factor of 10^4; (*c*) magnitude by a factor of 800 and frequency by a factor of 10^4.

Ans: 32 kΩ, 200 H; 40 Ω, 25 μH; 32 kΩ, 20 mH.

15.8 BASIC FILTER DESIGN

The design of filters is a very practical (and interesting) subject, worthy of a separate textbook in its own right. In this section, we introduce some of the basic concepts of filtering and explore both passive and active filter circuits. These circuits may be very simple, consisting of a single capacitor or inductor whose addition to a given network leads to improved performance. They may also be fairly sophisticated, consisting of many resistors, capacitors, inductors, and op amps in order to obtain the precise response curve required for a given application. Filters are used in modern electronics to obtain dc voltages in power supplies, eliminate noise in communication channels, separate radio and television channels from the multiplexed signal provided by antennas, and boost the bass signal in a car stereo, to name just a few applications.

The underlying concept of a filter is that it selects the frequencies that may pass through a network. There are several varieties, depending on the needs of a particular application. A ***low-pass filter,*** the response of which

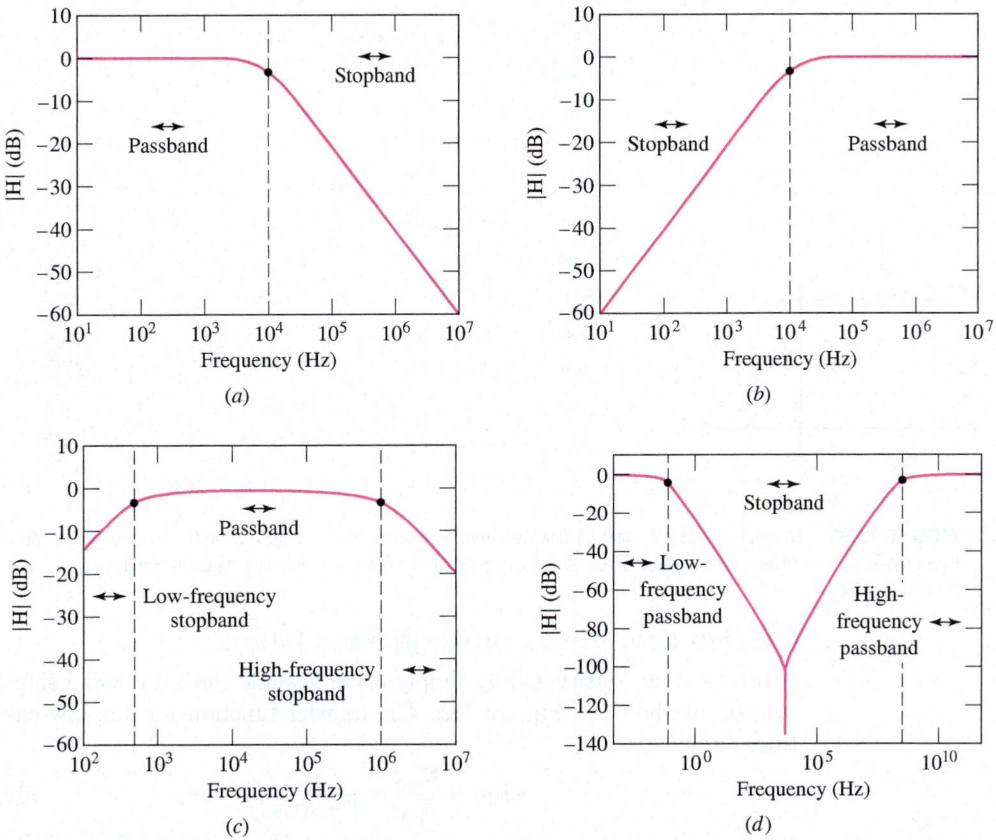

FIGURE 15.37 Frequency response curves for *(a)* a low-pass filter; *(b)* a high-pass filter; *(c)* a bandpass filter; *(d)* a bandstop filter. In each diagram, a solid dot corresponds to −3 dB.

is illustrated in Fig. 15.37*a*, passes frequencies below a cutoff frequency, while significantly damping frequencies above that cutoff. A ***high-pass filter,*** on the other hand, does just the opposite, as shown in Fig. 15.37*b*. The chief figure of merit of a filter is the sharpness of the cutoff, or the steepness of the curve in the vicinity of the corner frequency. In general, steeper response curves require more complex circuits.

Combining a low-pass and a high-pass filter can lead to what is known as a ***bandpass filter,*** as illustrated by the response curve shown in Fig. 15.37*c*. In this type of filter, the region between the two corner frequencies is referred to as the ***passband;*** the region outside the passband is referred to as the ***stopband.*** These terms may also be applied to the low- and high-pass filters, as indicated in Fig. 15.37*a* and *b*. We can also create a ***bandstop filter,*** which allows both high and low frequencies to pass but attenuates any signal with a frequency between the two corner frequencies (Fig. 15.37*d*).

The ***notch filter*** is a specialized bandstop filter, designed with a narrow response characteristic that blocks a single frequency component of a signal. ***Multiband filters*** are also possible; these are filter circuits which have multiple passbands and stopbands. The design of such filters is straightforward, but it is beyond the range of this book.

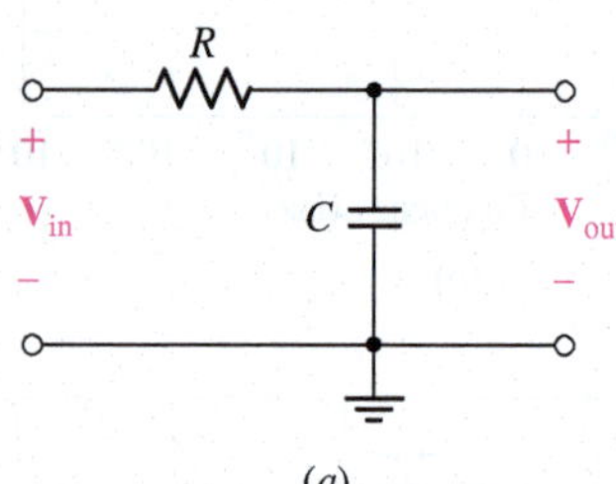

(a)

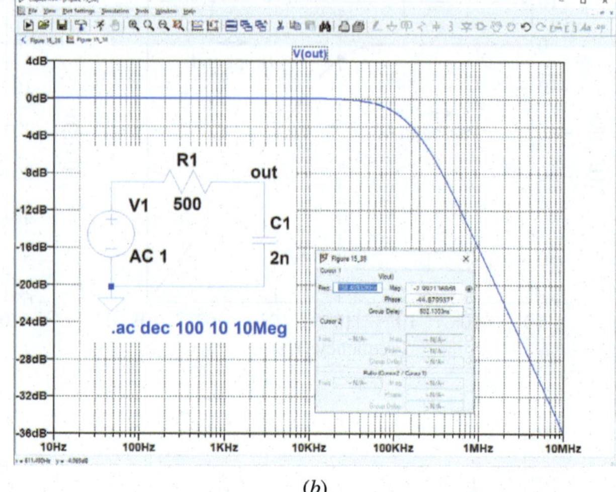

(b)

FIGURE 15.38 (a) A simple low-pass RC filter. (b) Simulated frequency response for $R = 500\ \Omega$ and $C = 2$ nF, showing a corner frequency at 159 kHz. Inset is the simulated circuit as well as the cursor window used to identify the corner frequency.
(b) Analog Devices

Passive Low-Pass and High-Pass Filters

A filter can be constructed by simply using a single capacitor and a single resistor, as shown in Fig. 15.38*a*. The transfer function for this low-pass filter circuit is

$$\mathbf{H(s)} \equiv \frac{\mathbf{V}_{out}}{\mathbf{V}_{in}} = \frac{1}{1 + RC\mathbf{s}} \qquad [32]$$

$\mathbf{H(s)}$ has a single corner frequency, which occurs at $\omega = 1/RC$, and a zero at $\mathbf{s} = \infty$, leading to its "low-pass" filtering behavior. Low frequencies ($\mathbf{s} \rightarrow 0$) result in $|\mathbf{H(s)}|$ near its maximum value (unity, or 0 dB), and high frequencies ($\mathbf{s} \rightarrow 0$) result in $|\mathbf{H(s)}| \rightarrow 0$. This behavior can be understood qualitatively by considering the impedance of the capacitor: as the frequency increases, the capacitor begins to act like a short-circuit to ac signals, leading to a reduction in the output voltage. An example response curve for such a filter with $R = 500\ \Omega$ and $C = 2$ nF is shown in Fig. 15.38*b*; the corner frequency of 159 kHz (1 Mrad/s) can be found by moving the cursor to −3 dB. The sharpness of the response curve in the vicinity of the cutoff frequency can be improved by moving to a circuit containing additional reactive (i.e., capacitive and/or inductive) elements.

A high-pass filter can be constructed by simply swapping the locations of the resistor and capacitor in Fig. 15.38*a*, as we see in the next example.

EXAMPLE 15.12

Design a high-pass filter with a corner frequency of 3 kHz.

We begin by selecting a circuit topology. Since no requirements as to the sharpness of the response are given, we choose the simple circuit of Fig. 15.39.

The transfer function of this circuit is easily found to be

$$\mathbf{H(s)} \equiv \frac{\mathbf{V}_{out}}{\mathbf{V}_{in}} = \frac{RC\mathbf{s}}{1 + RC\mathbf{s}}$$

which has a zero at $\mathbf{s} = 0$ and a pole at $\mathbf{s} = -1/RC$, leading to "high-pass" filter behavior (i.e., $|\mathbf{H}| \to 0$ as $\omega \to \infty$).

The corner frequency of the filter circuit is $\omega_c = 1/RC$, and we seek a value of $\omega_c = 2\pi f_c = 2\pi(3000) = 18.85$ krad/s. Again, we must select a value for either R or C. In practice, our decision would most likely be based on the values of resistors and capacitors at hand, but since no such information has been provided here, we are free to make arbitrary choices.

We therefore choose the standard resistor value 4.7 kΩ for R, leading to a requirement of $C = 11.29$ nF.

The only remaining step is to verify our design with an LTspice simulation; the predicted frequency response curve is shown in Fig. 15.40.

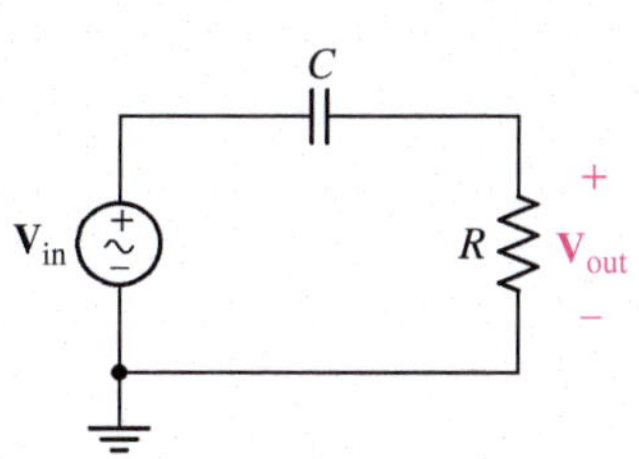

■ **FIGURE 15.39** A simple high-pass filter circuit, for which values for R and C must be selected to obtain a cutoff frequency of 3 kHz.

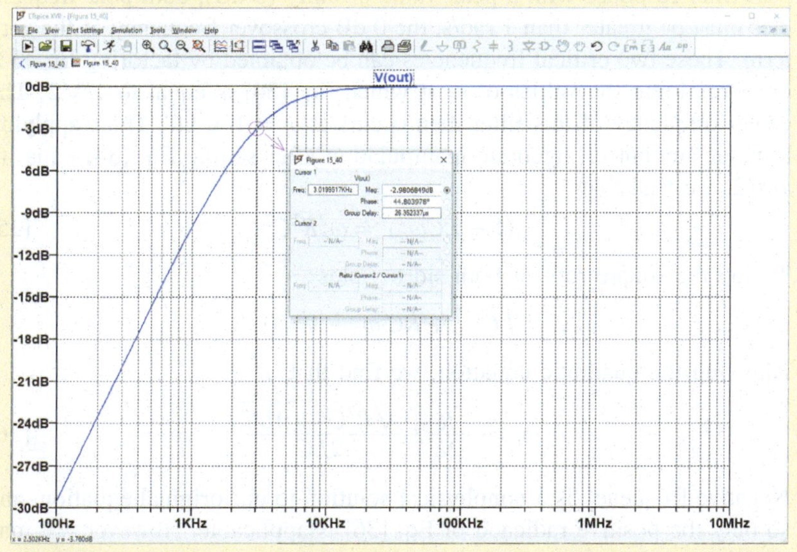

■ **FIGURE 15.40** Simulated frequency response of the final filter design, showing a cutoff (3 dB) frequency of 3 kHz as expected.
Analog Devices

PRACTICE

15.17 Design a high-pass filter with a cutoff frequency of 13.56 MHz, a common RF power supply frequency.

Ans: One possibility is the circuit of Fig. 15.39 where $RC = 1.174 \times 10^{-8}$ s. Example values are $C = 2$ pF and $R = 5.87$ kΩ.

Bandpass Filters

We have already seen several circuits earlier in this chapter which could be classified as "bandpass" filters (e.g., Figs. 15.17 and 15.24). Consider the simple circuit of Fig. 15.41, in which the output is taken across the resistor. The transfer function of this circuit is easily found to be

$$\mathbf{A}_V = \frac{\mathbf{s}RC}{LC\mathbf{s}^2 + RC\mathbf{s} + 1} \qquad [33]$$

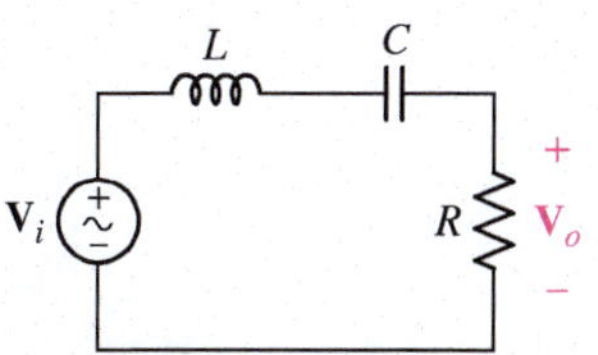

■ **FIGURE 15.41** A simple bandpass filter, constructed using a series RLC circuit.

The magnitude of this function is (after a few algebraic maneuvers)

$$|\mathbf{A}_V| = \frac{\omega RC}{\sqrt{(1-\omega^2 LC)^2 + \omega^2 R^2 C^2}} \qquad [34]$$

which, in the limit of $\omega \to 0$, becomes

$$|\mathbf{A}_V| \approx \omega RC \to 0$$

and in the limit of $\omega \to \infty$ becomes

$$|\mathbf{A}_V| \approx \frac{R}{\omega L} \to 0$$

We know from our experience with Bode plots that Eq. [33] represents three critical frequencies: one zero and two poles. In order to obtain a bandpass filter response with a peak value of unity (0 dB), both pole frequencies must be greater than 1 rad/s, the 0 dB crossover frequency of the zero term. These two critical frequencies can be obtained by factoring Eq. [33] or determining the values of ω at which Eq. [34] is equal to $1/\sqrt{2}$. The center frequency of this filter then occurs at $\omega = 1/\sqrt{LC}$. Thus, applying a minor amount of algebraic manipulation after setting Eq. [34] equal to $1/\sqrt{2}$, we find that

$$(1 - LC\omega_c^2)^2 = \omega_c^2 R^2 C^2 \qquad [35]$$

Taking the square root of both sides yields

$$LC\omega_c^2 + RC\,\omega_c - 1 = 0$$

Applying the quadratic equation, we find that

$$\omega_c = -\frac{R}{2L} \pm \frac{\sqrt{R^2C^2 + 4LC}}{2LC} \qquad [36]$$

Negative frequency is a nonphysical solution to our original equation, and so only the positive radicand of Eq. [36] is applicable. However, we may have been a little too hasty in taking the positive square root of both sides of Eq. [35]. Considering the negative square root as well, which is equally valid, we also obtain

$$\omega_c = \frac{R}{2L} \pm \frac{\sqrt{R^2C^2 + 4LC}}{2LC} \qquad [37]$$

from which it can be shown that only the positive radicand is physical. Thus, we obtain ω_L from Eq. [36] and ω_H from Eq. [37]; since $\omega_H - \omega_L = \mathcal{B}$, simple algebra shows that $\mathcal{B} = R/L$.

EXAMPLE 15.13

Design a bandpass filter characterized by a bandwidth of 1 MHz and a high-frequency cutoff of 1.1 MHz.

We choose the circuit topology of Fig. 15.41, and we begin by determining the corner frequencies required. The bandwidth is given by $f_H - f_L$, so

$$f_L = 1.1 \times 10^6 - 1 \times 10^6 = 100 \text{ kHz}$$

and

$$\omega_L = 2\pi f_L = 628.3 \text{ krad/s}$$

The high-frequency cutoff (ω_H) is simply 6.912 Mrad/s.

In order to proceed to design a circuit with these characteristics, it is necessary to obtain an expression for each frequency in terms of the variables R, L, and C.

Setting Eq. [37] equal to $2\pi(1.1 \times 10^6)$ allows us to solve for $1/LC$, as we already know that $\mathcal{B} = 2\pi(f_H - f_L) = 6.283 \times 10^6$.

$$\frac{1}{2}\mathcal{B} + \left[\frac{1}{4}\mathcal{B}^2 + \frac{1}{LC}\right]^{1/2} = 2\pi(1.1 \times 10^6)$$

Solving, we find that $1/LC = 4.343 \times 10^{12}$. Arbitrarily selecting $L = 50$ mH, we obtain $R = 314$ kΩ and $C = 4.6$ pF. It should be noted that there is no unique solution for this "design" problem—R, L, or C can be selected as a starting point.

LTspice verification of our design is shown in Fig. 15.42.

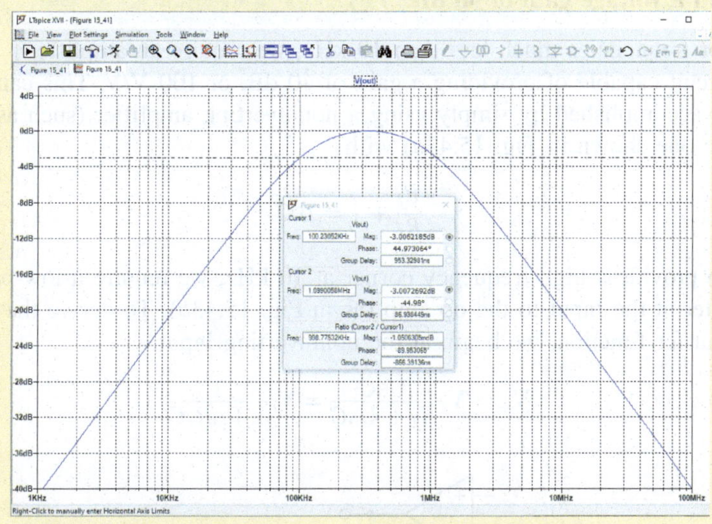

FIGURE 15.42 Simulated response of the bandpass filter design showing a bandwidth of 1 MHz and a high-frequency cutoff of 1.1 MHz as desired.
Analog Devices

PRACTICE

15.18 Design a bandpass filter with a low-frequency cutoff of 100 rad/s and a high-frequency cutoff of 10 krad/s.

Ans: One possible answer of many: $R = 990\ \Omega$, $L = 100$ mH, and $C = 10\ \mu$F.

The type of circuit we have been considering is known as a ***passive filter,*** as it is constructed of only passive components (i.e., no transistors, op amps, or other "active" elements). Although passive filters are relatively common, they are not well suited to all applications. The gain (defined as the output voltage divided by the input voltage) of a passive filter can be difficult to set, and amplification is often desirable in filter circuits.

Active Filters

The use of an active element such as the op amp in filter design can overcome many of the shortcomings of passive filters. As we saw in Chap. 6, op amp circuits can easily be designed to provide gain. Op amp circuits can also exhibit inductorlike behavior through the strategic location of capacitors.

The internal circuitry of an op amp contains very small capacitances (typically on the order of 100 pF), and these limit the maximum frequency at which the op amp will function properly. Thus, any op amp circuit will behave as a low-pass filter, with a cutoff frequency for modern devices of perhaps 20 MHz or more (depending on the circuit gain).

EXAMPLE 15.14

Design an active low-pass filter with a cutoff frequency of 10 kHz and a voltage gain of 40 dB.

For frequencies much less than 10 kHz, we require an amplifier circuit capable of providing a gain of 40 dB, or 100 V/V. This can be accomplished by simply using a noninverting amplifier (such as the one shown in Fig. 15.43*a*) with

$$\frac{R_f}{R_1} + 1 = 100$$

To provide a high-frequency corner at 10 kHz, we require a low-pass filter at the input to the op amp (as in Fig. 15.43*b*). To derive the transfer function, we begin at the noninverting input,

$$\mathbf{V}_+ = \mathbf{V}_i \frac{1/\mathbf{s}C}{R_2 + 1/\mathbf{s}C} = \mathbf{V}_i \frac{1}{1 + \mathbf{s}R_2C}$$

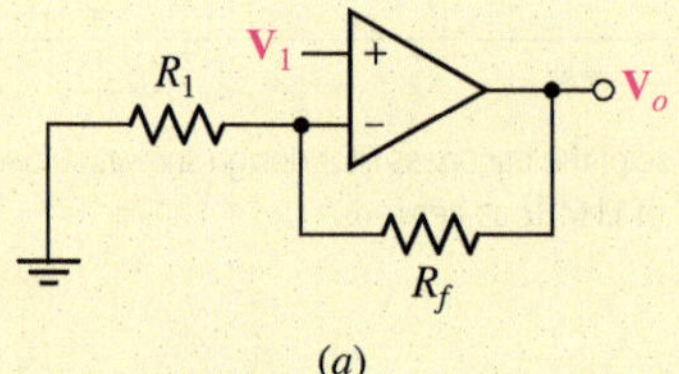

(*a*)

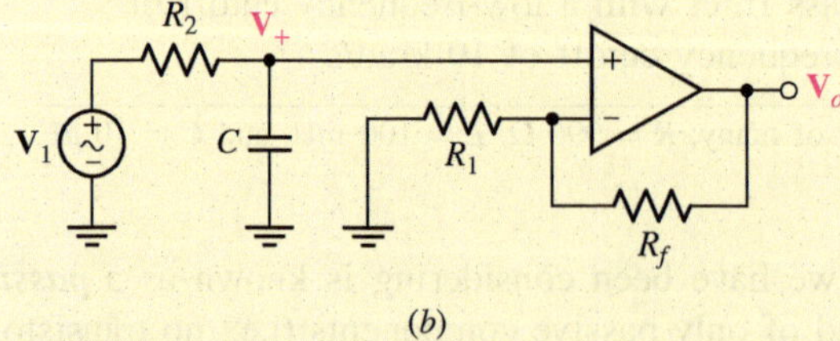

(*b*)

FIGURE 15.43 (*a*) A simple noninverting op amp circuit. (*b*) A low-pass filter consisting of a resistor R_2 and a capacitor C has been added to the input.

At the inverting input we have

$$\frac{\mathbf{V}_o - \mathbf{V}_+}{R_f} = \frac{\mathbf{V}_+}{R_1}$$

Combining these two equations and solving for $\mathbf{V}_o$, we find that

$$\mathbf{V}_o = \mathbf{V}_i\left(\frac{1}{1 + \mathbf{s}R_2C}\right)\left(1 + \frac{R_f}{R_1}\right)$$

The maximum value of the gain $\mathbf{A}V = \mathbf{V}_o/\mathbf{V}_i$ is $1 + R_f/R_1$, so we set this quantity equal to 100. Since neither resistor appears in the expression for the corner frequency $(R_2C)^{-1}$, either may be selected first. We thus choose $R_1 = 1\ \text{k}\Omega$, so $R_f = 99\ \text{k}\Omega$.

Arbitrarily selecting $C = 1\ \mu\text{F}$, we find that

$$R_2 = \frac{1}{2\pi(10 \times 10^3)C} = 15.9\ \Omega$$

The simulated frequency response of this circuit is shown in Fig. 15.44 is far enough away from 10 kHz that it does not affect our design.

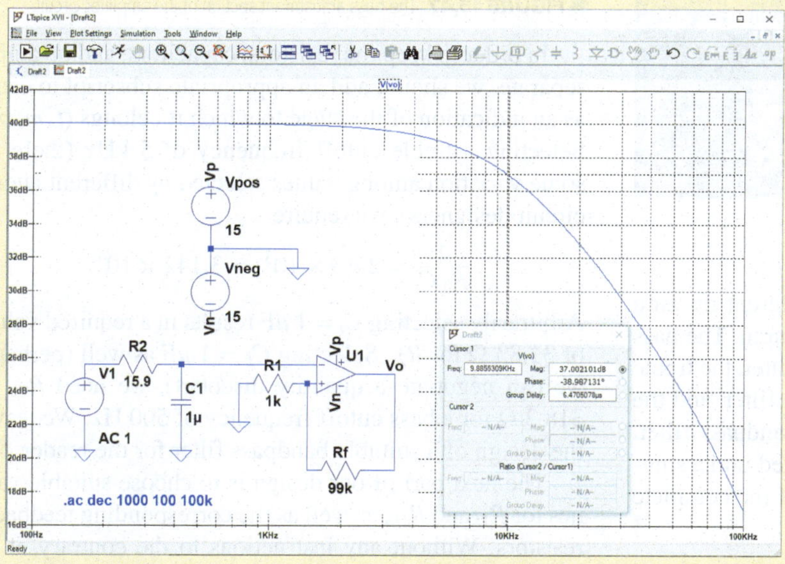

■ **FIGURE 15.44** The frequency response of the op amp-based filter (inset), which has a cutoff (3 dB) frequency of 9.9 kHz, is close to the design goal of 10 kHz.
Analog Devices

PRACTICE

15.19 Design a low-pass filter circuit with a gain of 30 dB and a cutoff frequency of 1 kHz.

Ans: One possible answer of many: $R_1 = 100\ \text{k}\Omega$, $R_f = 3.062\ \text{M}\Omega$, $R_2 = 79.58\ \Omega$, and $C = 2\ \mu\text{F}$.

PRACTICAL APPLICATION

Bass, Treble, and Midrange Adjustment

We often wish to be able to independently adjust the bass, treble, and midrange settings on a sound system, even for inexpensive equipment. The audio frequency range (at least for the human ear) is commonly accepted to be 20 Hz to 20 kHz, with bass corresponding to lower frequencies (< 500 Hz or so) and treble corresponding to higher frequencies (> 5 kHz or thereabouts).

Designing a simple graphic equalizer is a relatively straightforward endeavor, although a system such as that shown in Fig. 15.45 requires a bit more effort. In the bass, midrange, treble type of equalizer common on many portable radios, the main signal (provided by the radio receiver circuit, or perhaps a CD player) consists of a wide spectrum of frequencies having a bandwidth of approximately 20 kHz.

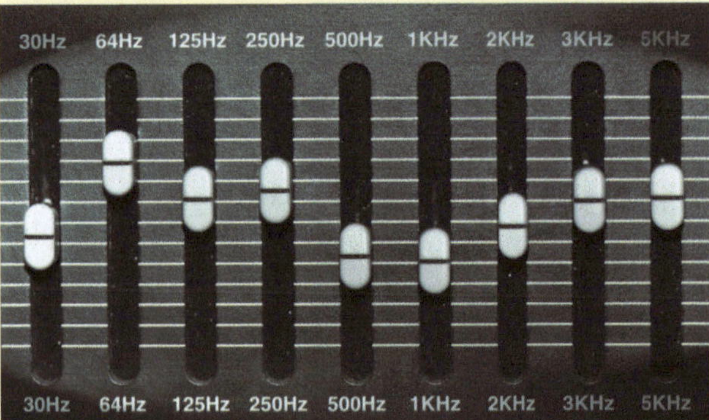

FIGURE 15.45 An example of a graphic equalizer. winnond/Shutterstock

This signal must be sent to three different op amp circuits, each with a different filter at the input. The bass adjustment circuit will require a low-pass filter, the treble adjustment circuit will require a high-pass filter, and the midrange adjustment circuit requires a bandpass filter. The output of each op amp circuit is then fed into a summing amplifier circuit; a block diagram of the complete circuit is shown in Fig. 15.46.

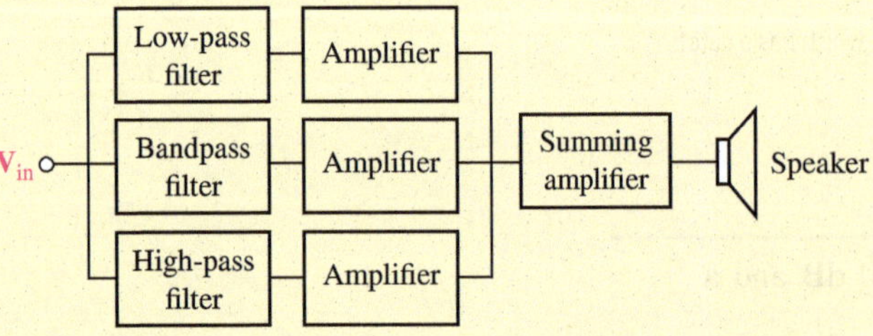

FIGURE 15.46 Block diagram of a simple graphic equalizer circuit.

Our basic building block is shown in Fig. 15.47. This circuit consists of a noninverting op amp circuit characterized by a voltage gain of $1 + R_f/R_1$, and a simple low-pass filter composed of a resistor R_2 and a capacitor C. The feedback resistor R_f is a variable resistor (sometimes referred to as a *potentiometer*), and it allows the gain to be varied through the rotation of a knob; the layperson would call this resistor the volume control. The low-pass filter network restricts the frequencies that will enter the op amp and hence be amplified; the corner frequency is simply $(R_2C)^{-1}$. If the circuit designer needs to allow the user to also select the break frequency for the filter, R_2 may be replaced by a potentiometer, or, alternatively, C could be replaced by a variable capacitor. The remaining stages are constructed in essentially the same way, but with a different filter network at the input. For example, the high-pass filter for treble will swap positions for R_2 and C.

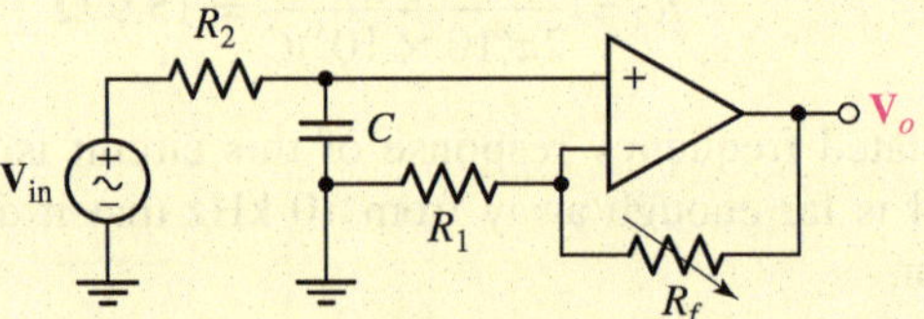

FIGURE 15.47 The bass adjustment section of the amplifier circuit.

In order to keep the resistors, capacitors, and op amps separate, we should add an appropriate subscript to each as an indication of the stage to which it belongs (t, m, b). Selecting a treble cutoff frequency of 5 kHz (there is some variation among values selected by different audio circuit designers), we require

$$\frac{1}{R_{2t}C_t} = 2\pi(5 \times 10^3) = 3.142 \times 10^4$$

Arbitrarily selecting $C_t = 1\ \mu$F results in a required value of 31.83 Ω for R_{2t}. Selecting $C_b = 1\ \mu$F as well (perhaps we can negotiate a quantity discount), we need $R_{2b} =$ 318.3 Ω for a bass cutoff frequency of 500 Hz. We leave the design of a suitable bandpass filter for the reader.

The next part of our design is to choose suitable values for R_{1t} and R_{1b}, as well as the corresponding feedback resistors. Without any instructions to the contrary, it is probably simplest to make both stages identical. Therefore, we arbitrarily select both R_{1t} and R_{1b} as 1 kΩ, and R_{ft} and R_{fb} as 10 kΩ potentiometers (meaning that the range will be from 0 to 10 kΩ). This allows the volume of one signal to be up to 11 times louder than the other.

Now that the design of the filter stage is complete, we are ready to consider the design of the summing stage. We use an inverting op amp configuration, with the output of each of the filter op amp stages fed directly into its own 1 kΩ resistor. The other terminal of each 1 kΩ resistor is then connected to the inverting input of the summing amplifier stage. The appropriate potentiometer for the summing amplifier stage must be selected in order to prevent saturation, so knowledge of both the input voltage range and the output speaker wattage is required. To illustrate the design, a simulation of the circuit with treble (low pass) and bass (high pass) components (without the mid-range bandpass filter) is shown in Fig. 15.48, along with the resulting summing output with gain of 10.

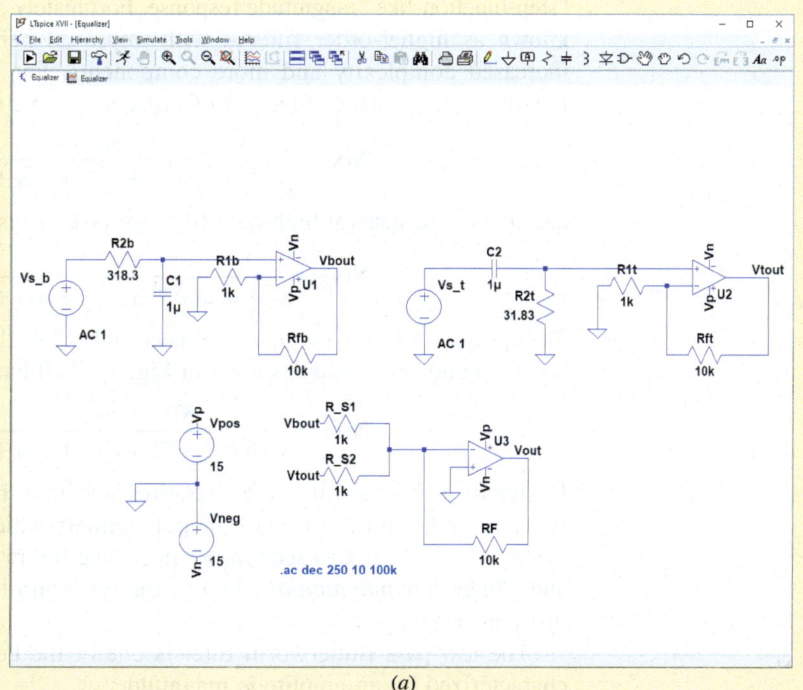

(*a*)

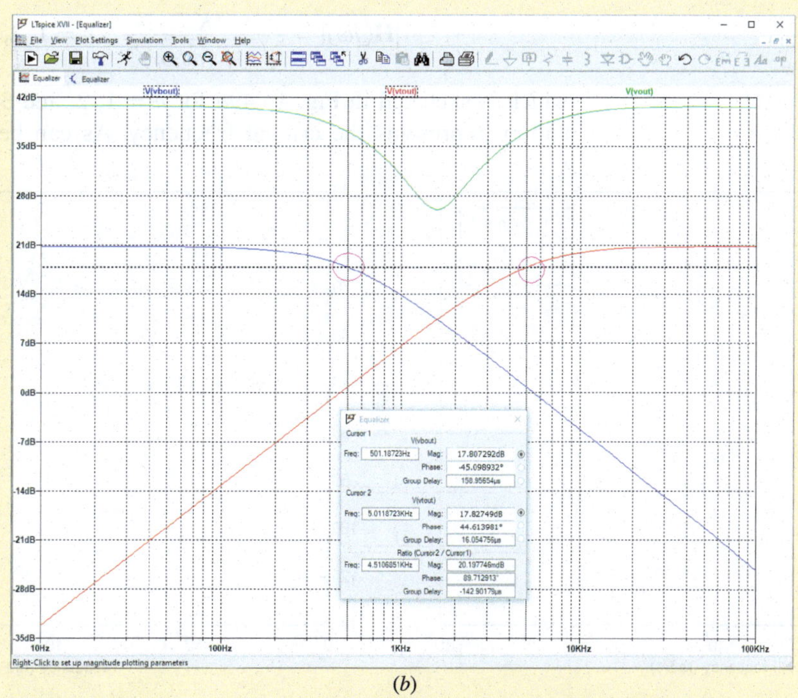

(*b*)

FIGURE 15.48 (*a*) Simulated low- and high-pass filter stages along with summing stage. (*b*) Simulated frequency response of the low-pass, high-pass, and summing output of the audio equalizer circuit. The use of the UniversalOpamp2 model, as well as the 250 points per decade, introduce a slight difference in the measured corner frequencies, but within reasonable limits. (*a–b*) Analog Devices

15.9 ADVANCED FILTER DESIGN

Although the basic filters we have encountered so far function adequately for a number of applications, their characteristics are far from an ideal "step-function-like" magnitude response. Fortunately, we have alternatives—known as higher-order filters—with improved behavior, at the cost of increased complexity and more components. For example, the general low-pass filter transfer function of order n may be written as

$$\mathbf{N(s)} = \frac{Ka_0}{\mathbf{s}^n + a_{n-1}\mathbf{s}^{n-1} + \cdots + a_1\mathbf{s} + a_0}$$

and that of the general high-pass filter (of order n) is only subtly different:

$$\mathbf{N(s)} = \frac{K\mathbf{s}^n}{\mathbf{s}^n + a_{n-1}\mathbf{s}^{n-1} + \cdots + a_1\mathbf{s} + a_0}$$

To represent a bandpass filter, we need only alter the numerator to $K\mathbf{s}^{n/2}$, and the band-reject filter (shown in Fig. 15.37d) has the transfer function

$$\mathbf{N(s)} = \frac{K(\mathbf{s}^2 + \omega_0^2)^{n/2}}{\mathbf{s}^n + a_{n-1}\mathbf{s}^{n-1} + \cdots + a_1\mathbf{s} + a_0}$$

Design of a specific filter, then, requires selecting the appropriate transfer function and choosing a class of polynomials which specify the coefficients a_1, a_2, etc. In this section, we introduce filters based on ***Butterworth*** and ***Chebyshev polynomials,*** two of the types most commonly employed in filter design.

The low-pass Butterworth filter is one of the best-known filters. It is characterized by an amplitude magnitude

$$|\mathbf{H}(j\omega)| = \frac{K}{\sqrt{1 + (\omega/\omega_c)^{2n}}} \qquad n = 1, 2, 3, \ldots$$

which is sketched in Fig. 15.49a for $n = 1$, 2, and 3; K is a real constant, and ω_c represents the critical frequency. As can be seen, the magnitude

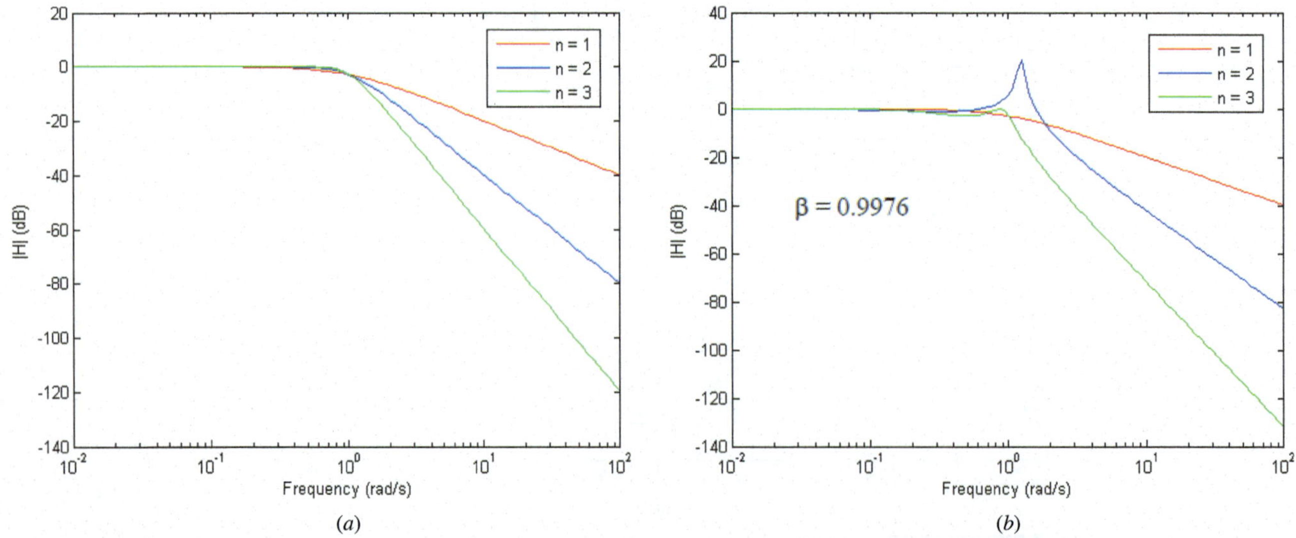

FIGURE 15.49 Plot of $|\mathbf{H}(j\omega)|$ for first-, second-, and third-order low-pass (a) Butterworth filters and (b) Chebyshev filters. All filters were normalized to a corner frequency of 1 rad/s.

TABLE **15.2** Coefficients for Low-Pass Butterworth and Chebyshev ($\beta = 0.9976$, or 3 dB) Filter Functions, Normalized to $\omega_C = 1$

Butterworth					
n	a_0	a_1	a_2	a_3	a_4
1	1.0000				
2	1.0000	1.4142			
3	1.0000	2.0000	2.0000		
4	1.0000	2.6131	3.4142	2.6131	
5	1.0000	3.2361	5.2361	5.2361	3.2361
Chebyshev ($\beta = 0.9976$)					
n	a_0	a_1	a_2	a_3	a_4
1	1.0024				
2	0.7080	0.6449			
3	0.2506	0.9284	0.5972		
4	0.1770	0.4048	1.1691	0.5816	
5	0.0626	0.4080	0.5489	1.4150	0.5744

approaches a step-function-like shape as the order n increases. In contrast, the low-pass Chebyshev filter is characterized by rather prominent ripples in the passband, the number of which depends upon the order of the filter as illustrated in Fig. 15.49*b*. Its magnitude response is described by

$$|\mathbf{H}(j\omega)| = \frac{K}{\sqrt{1 + \beta^2 C_n^2(\omega/\omega_c)}} \qquad n = 1, 2, 3, \ldots$$

where β is a real constant known as the ***ripple factor*** and $C_n(\omega/\omega_c)$ denotes the Chebyshev polynomial of the first kind of degree n. For convenience, selected coefficients of both polynomial types are listed in Table 15.2.

The Sallen-Key Amplifier

As seen in Sec. 14.12, we may create an op-amp-based filter circuit having a double pole simply by cascading two circuits such as the one shown in Fig. 14.39*a*, in which case we obtain a transfer function

$$\mathbf{H(s)} = \frac{(1/R_1C_f)^2}{\mathbf{s}^2 + 2/R_fC_f + (1/R_fC_f)^2} \qquad [38]$$

If we wish to improve upon this basic approach, a circuit worth considering is known as the Sallen-Key amplifier, shown in Fig. 15.50, configured as a low-pass filter. Analysis of this circuit by nodal analysis is straightforward. We first define the gain G of the noninverting amplifier as

$$G \equiv \frac{R_A + R_B}{R_B} \qquad [39]$$

Then voltage division yields

$$\mathbf{V}_y = \mathbf{V}_x \frac{1}{1 + R_2C_2\mathbf{s}} \qquad [40]$$

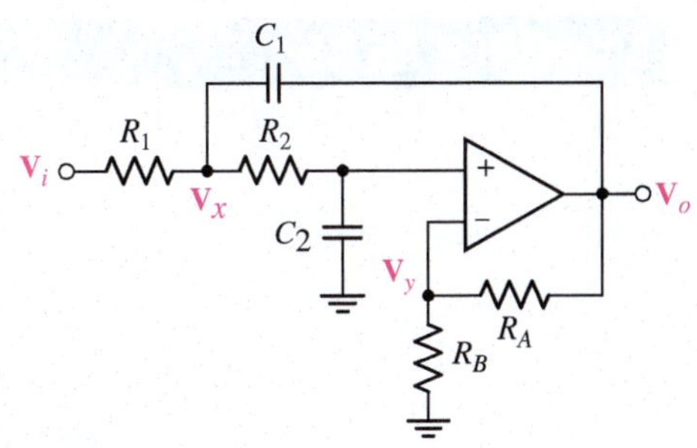

FIGURE 15.50 General low-pass Sallen-Key filter circuit.

and we may write a single nodal equation

$$0 = \frac{\mathbf{V}_x - \mathbf{V}_i}{R_1} + \frac{\mathbf{V}_x - \mathbf{V}_y}{R_2} + \frac{\mathbf{V}_x - \mathbf{V}_o}{1/\mathbf{s}C_1} \qquad [41]$$

Substituting Eqs. [39] and [40] into Eq. [41] and performing a few algebraic maneuvers, we arrive at an expression for the transfer function of the amplifier,

$$\frac{\mathbf{V}_o}{\mathbf{V}_i} = \frac{G/R_1R_2C_1C_2}{\mathbf{s}^2 + \left[\frac{1}{R_1C_1} + \frac{1}{R_2C_1} + \frac{1-G}{R_2C_2}\right]\mathbf{s} + \frac{1}{R_1R_2C_1C_2}} \qquad [42]$$

Noting that the coefficients in Table 15.2 represent filters with a cutoff frequency of 1 rad/s, so that when finished we will need to make use of the simple scaling techniques described in Sec. 15.7, we are now ready to explore the design of a second-order Butterworth low-pass filter.

Design of high-pass filters based on the Sallen-Key model is similarly straightforward; the only modification required is to replace capacitors C_1 and C_2 with resistors, and resistors R_1 and R_2 with capacitors. The remainder of the circuit remains unchanged. Nodal analysis of the resulting circuit with $C_1 = C_2 = C$ and $R_1 = R_2 = R$ yields

$$a_0 = \frac{1}{R^2C^2} \qquad [43]$$

and

$$a_1 = \frac{3-G}{RC} \qquad [44]$$

as we found for the low-pass filter.

Higher-order filters can be realized by cascading appropriate op amp stages. For example, Butterworth filters of odd order (e.g., 3, 5, …) require an additional pole at $\mathbf{s} = -1$. Thus, a third-order Butterworth filter is constructed using a Sallen-Key stage which provides a transfer function denominator $\mathbf{D(s)}$ of

$$(\mathbf{s}+1)\overline{)\mathbf{s}^3 + 2\mathbf{s}^2 + 2\mathbf{s} + 1}\quad \text{quotient: } \mathbf{s}^2 + \mathbf{s} + 1$$

or

$$\mathbf{D(s)} = \mathbf{s}^2 + \mathbf{s} + 1 \qquad [45]$$

with an additional op amp stage such as the one in Fig. 14.39*a* to provide the term $(\mathbf{s} + 1)$.

EXAMPLE 15.15

Design a second-order low-pass Butterworth filter having a gain of 4 and a corner frequency at 1400 rad/s.

We begin by selecting the Sallen-Key prototype shown in Fig. 15.50, and we opt for the simplification which arises when we set $R_1 = R_2 = R$ and $C_1 = C_2 = C$. With a second-order Butterworth filter we expect from Table 15.2 to have a denominator polynomial

$$\mathbf{s}^2 + 1.4142\mathbf{s} + 1$$

and comparing to Eq. [44]

$$RC = 1$$

and

$$\frac{2}{RC} + \frac{1-G}{RC} = 1.414$$

hence

$$G = \frac{R_A + R_B}{R_B} = 1.586$$

We first set values for the two resistors in our gain network (which do not need to undergo scaling) by arbitrarily choosing $R_B = 1\ \text{k}\Omega$, so that $R_A = 586\ \Omega$.

Next, we note that if $C = 1$ F, then $R = 1\ \Omega$, neither of which is a particularly conventional value. Instead we select $C' = 1\ \mu\text{F}$; this requires scaling the resistor by 10^6. We also need to frequency-scale to 1400 rad/s. Thus,

$$10^{-6}\ \text{F} = \frac{1\ \text{F}}{k_m k_f} = \frac{1\ \text{F}}{1400\ k_m}$$

and $k_m = 714\ \Omega$. Consequently, $R' = k_m R = 714\ \Omega$.

Unfortunately, our design is not quite done. We were constrained to an amplifier gain of 1.586, or 4 dB, but the specifications called for a gain of 4, or 12 dB. The only option available to us is to feed the output of our circuit into a noninverting amplifier such as the one in Fig. 6.7*a*. Choosing 1 kΩ and 1.52 kΩ for R_1 (output stage) and R_f completes the design.

PRACTICE

15.20 Design a second-order Butterworth low-pass filter having a gain of 10 dB and a cutoff frequency of 1000 Hz.

Ans: A two-stage circuit, with the output of the circuit of Fig. 15.50 fed into the input of a noninverting amplifier, with component values $C_1 = C_2 = 1\ \mu\text{F}$, $R_1 = R_2 = 159\ \Omega$, $R_A = 586\ \Omega$, $R_B = 1\ \text{k}\Omega$ (stage 1) and $R_1 = 1\ \text{k}\Omega$, $R_f = 994\ \Omega$ (stage 2).

EXAMPLE 15.16

Design a third-order low-pass Butterworth filter having a voltage gain magnitude of 4 and a corner frequency at 2000 rad/s.

We begin by again selecting the Sallen-Key prototype shown in Fig. 15.50, and we opt for the simplification which arises when we set $R_1 = R_2 = R$ and $C_1 = C_2 = C$. We will also add an input stage of the form shown in Fig. 14.39*a* to add the necessary pole. The basic design is shown in Fig. 15.51.

(Continued on next page)

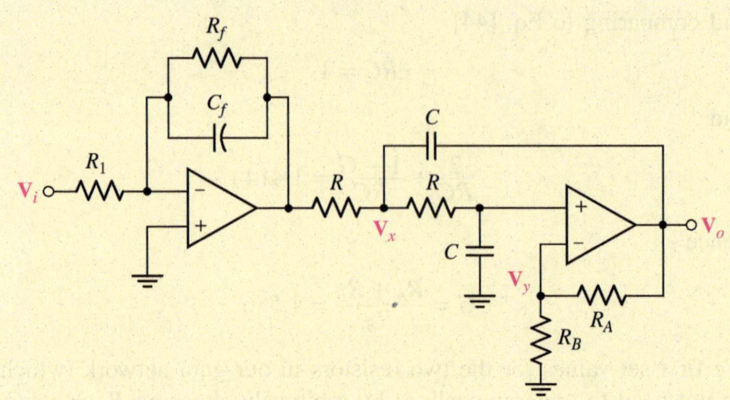

FIGURE 15.51 Basic structure of the proposed third-order low-pass Butterworth filter, with component values still to be chosen.

Comparing Eqs. [41], [42], and [43], we determine that our design must ensure that

$$1 = \frac{1}{R^2C^2}$$

and

$$1 = \frac{3 - G}{RC}$$

Consequently, $RC = 1$ and $G = 4$. If we choose $R_A = 3\ \text{k}\Omega$, it follows that $R_B = 1\ \text{k}\Omega$. We may scale these values later if we choose when adjusting for operation at 2000 rad/s, but this is unnecessary as the dc gain is set by the ratio of the two resistors.

Initially we design for $R = 1\ \Omega$ and $C = 1$ F as this automatically satisfies the $RC = 1$ requirement. Neither value being easy to locate, however, we select a more reasonable capacitor value of 0.1 μF, which combined with our frequency scaling factor $k_f = 2000$, results in a resistor scaling factor $k_m = 5000$. Thus, $R = 5\ \text{k}\Omega$ in our final design.

All that remains is to select values for R_1, R_f, and C_f in our front-end stage. Recall that the transfer function of this stage is

$$-\frac{1/R_1C_f}{\mathbf{s} + (1/R_fC_f)}$$

Setting $R_f = 1\ \Omega$ and $C_f = 1$ F initially allows the pole to be properly located prior to scaling operations, which dictate that we build the circuit with $R_f = 5\ \text{k}\Omega$ and $C_f = 0.1$ F. Our only remaining choice, then, is to ensure that R_1 allows us to meet our gain requirement of 4. Since we have already achieved this with our Sallen-Key stage, R_1 must be equal to R_f, or 5 kΩ.

Design of Chebyshev filters proceeds along the same lines as that of Butterworth filters, except we have more choices now with the ripple factor. Also, for filters not having a 3 dB ripple factor, the critical frequency is where the ripple channel in the passband terminates, which is slightly different than what we have specified previously. Filters with order $n > 2$ are constructed by cascading stages, either multiple Sallen-Key stages for even orders, or a simple stage such as Fig. 14.39*a* in conjunction with the appropriate number of Sallen-Key stages for odd orders. For filters with a specific gain requirement, an op amp stage containing only resistors is typically required at the output.

SUMMARY AND REVIEW

We began this chapter with a short discussion of *resonance*. Of course the reader was likely to already have an intuitive understanding of the basic concept—timing when to kick our legs on a swing as a child; watching videos of crystal glasses shattering under the power of a trained soprano's voice; instinctively slowing down when driving over a corrugated surface. In the context of linear circuit analysis, we found (perhaps surprisingly) that a frequency can be chosen even for networks with capacitors and inductors such that the voltage and current are in phase (hence the network appears purely resistive at that particular frequency). How quickly our circuit response changes as we move "off resonance" was related to a new term—the *quality factor* (Q) of our circuit. After defining what is meant by *critical frequencies* for our circuit response, we introduced the concept of *bandwidth,* and we discovered that our expressions may be simplified rather dramatically for high-Q ($Q > 5$) circuits. We briefly extended this discussion to consider the differences between series and parallel circuits near resonance, along with more practical networks which cannot be classed as either.

The remainder of this chapter dealt with the analysis and design of filter circuits. Prior to launching into that discussion, the topic of "scaled" circuit components dealt with both *frequency* and *magnitude scaling* as a convenient design tool. We also introduced the handy method of *Bode plots,* which allows us to quickly sketch a reasonable approximation to the response of a filter circuit as a function of frequency. We next considered both *passive* and *active filters,* starting with simple designs using a single capacitor to achieve either low-pass or high-pass behavior. Shortly thereafter, *bandpass filter* design was studied. Although they are straightforward to work with, the response of such simple circuits is not particularly abrupt. As an alternative, filter designs based on either Butterworth or Chebyshev polynomials were examined, with higher-order filters yielding sharper magnitude response at the expense of increased complexity.

- ❑ Transfer functions describe the input/output relations of a circuit, and can be written in terms of various combinations of input current or voltage and output current or voltage. (Example 15.1)
- ❑ A Bode plot is a useful representation of the transfer function, where the magnitude (in dB) and phase are plotted on a logarithmic frequency scale. Linear approximations based on location of poles and zeros may be used to approximate Bode plots to quickly determine the behavior of a transfer function. (Examples 15.2–15.5)

- Resonance is the condition in which a fixed-amplitude sinusoidal forcing function produces a response of maximum amplitude. Resonant behavior can be defined by the quality factor, half-power frequency, and bandwidth. Examples 15.6, 15.7, 15.8, 15.9, 15.10
- In a high-Q circuit, each half-power frequency is located approximately one-half bandwidth from the resonant frequency. (Example 15.7)
- A series resonant circuit is characterized by a *low* impedance at resonance, whereas a parallel resonant circuit is characterized by a *high* impedance at resonance. (Examples 15.6 and 15.8)
- Impractical values for components often make design easier. The transfer function of a network may be scaled in magnitude or frequency using appropriate replacement values for components. (Example 15.11)
- The four basic types of filters are low-pass, high-pass, bandpass, and bandstop. (Examples 15.12 and 15.13)
- Passive filters use only resistors, capacitors, and inductors; active filters are based on op amps or other active elements. (Example 15.14)
- Butterworth and Chebyshev filters can be designed based on the simple Sallen-Key amplifier. Filter gain typically must be adjusted by adding a purely resistor-based amplifier circuit at the output. (Examples 15.15 and 15.16)

READING FURTHER

A good discussion of a large variety of filters can be found in:

J. T. Taylor and Q. Huang, eds., *CRC Handbook of Electrical Filters*. Boca Raton, Fla.: CRC Press, 1997.

A comprehensive compilation of various active filter circuits and design procedures is given in:

D. Lancaster, *Lancaster's Active Filter Cookbook,* 2nd ed. Burlington, Mass.: Newnes, 1996.

Additional filter references which the reader might find useful include:

D. E. Johnson and J. L. Hilburn, *Rapid Practical Design of Active Filters*. New York: John Wiley & Sons, Inc., 1975.

J. V. Wait, L. P. Huelsman, and G. A. Korn, *Introduction to Operational Amplifier Theory and Applications,* 2nd ed. New York: McGraw Hill, 1992.

EXERCISES

15.1 Transfer Function

1. For the *RL* circuit in Fig. 15.52, (*a*) determine the transfer function defined as $\mathbf{H}(j\omega) = v_{out}/v_{in}$; (*b*) for the case of $R = 200\ \Omega$ and $L = 5$ mH, construct a plot of the magnitude and phase as a function of frequency; and (*c*) evaluate the magnitude and phase at a frequency of 10 kHz.

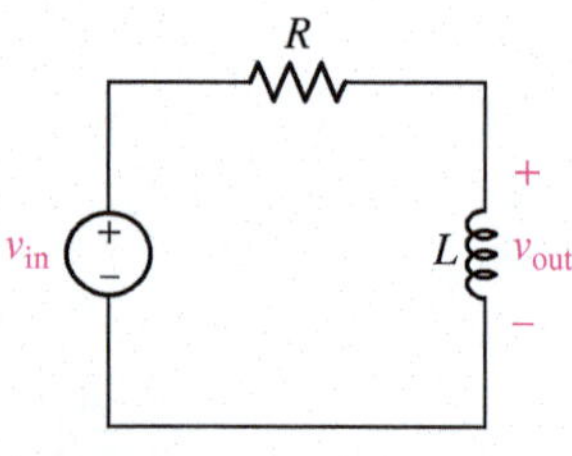

FIGURE 15.52

2. For the RL circuit in Fig. 15.52, switch the positions of the resistor and inductor such that v_{out} is the voltage drop across the resistor. (*a*) Write an expression or the transfer function, defined as $\mathbf{H}(j\omega) = v_{out}/v_{in}$; (*b*) for the case of $R = 200\ \Omega$ and $L = 5$ mH, construct a plot of the magnitude and phase as a function of frequency; and (*c*) evaluate the magnitude and phase at a frequency of 10 kHz.
3. Examine the series RLC circuit in Fig. 15.53, with $R = 100\ \Omega$, $L = 5$ mH, and $C = 2\ \mu$F. Calculate the magnitude of the transfer function $\mathbf{H}(j\omega) = v_{out}/v_{in}$ at frequencies of 0, 2 kHz, and ∞ for the three cases where (*a*) $v_{out} = v_R$, (*b*) $v_{out} = v_L$, and $v_{out} = v_C$.

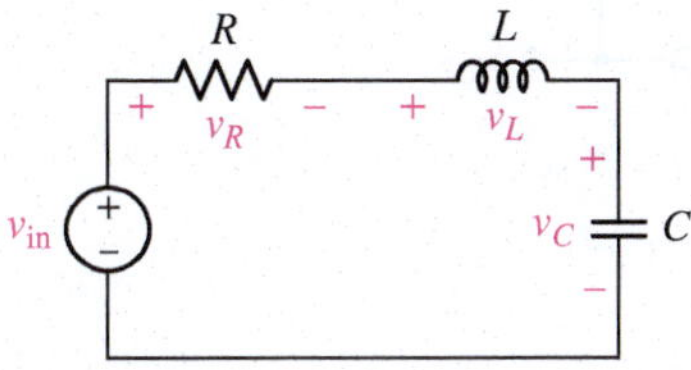

FIGURE 15.53

4. For the circuit in Fig. 15.54, (*a*) derive an algebraic expression for the transfer function $\mathbf{H}(j\omega) = v_{out}/i_{in}$ terms of circuit components R_1, R_2, C_1, and C_2; and (*b*) evaluate the magnitude of $\mathbf{H}$ at frequencies of 100 Hz, 10 kHz, and 1 MHz for case where $R_1 = 20$ kΩ, $R_2 = 5$ kΩ, $C_1 = 10$ nF, and $C_2 = 40$ nF.

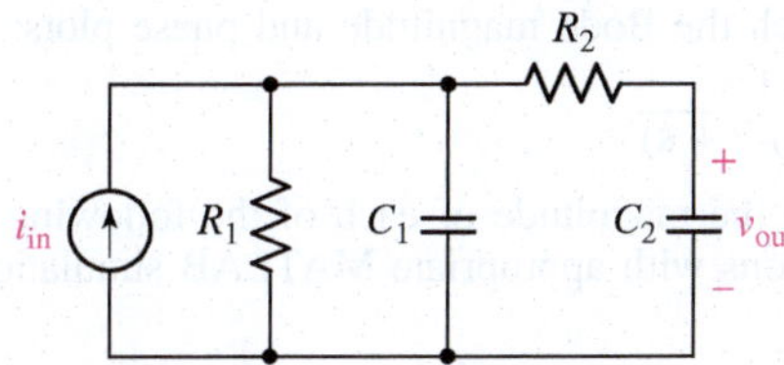

FIGURE 15.54

5. For the circuit in Fig. 15.55, (*a*) derive an algebraic expression for the transfer function $\mathbf{H}(j\omega) = i_{out}/v_{in}$ in terms of circuit components R_1, R_2, L_1, and L_2; (*b*) evaluate the magnitude of $\mathbf{H}(j\omega)$ at frequencies of 10 kHz, 1 MHz, and 100 MHz for the case where $R_1 = 3$ kΩ, $R_2 = 12$ kΩ, $L_1 = 5$ mH, and $L_2 = 8$ mH; (*c*) qualitatively, explain the behavior of the transfer function magnitude frequency response.

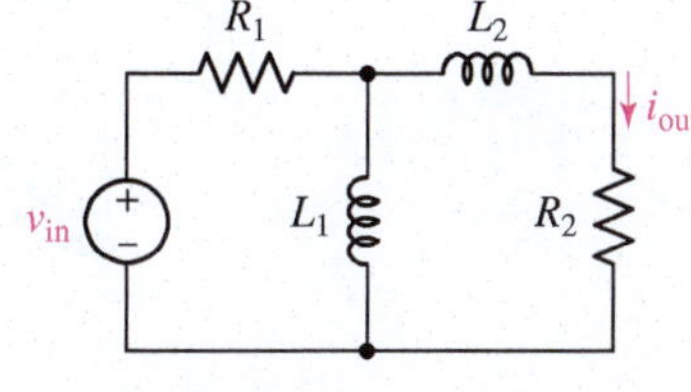

FIGURE 15.55

6. For the circuit in Fig. 15.56, (*a*) determine the transfer function $\mathbf{H}(j\omega) = \mathbf{V}_{out}/\mathbf{V}_{in}$ in terms of circuit parameters R_1, R_2, and C; (*b*) determine the magnitude and phase of the transfer function at $\omega = 0$, 3×10^4 rad/s, and as $\omega \to \infty$ for the case where circuit values are $R_1 = 500\ \Omega$, $R_2 = 40$ kΩ, and $C = 10$ nF.

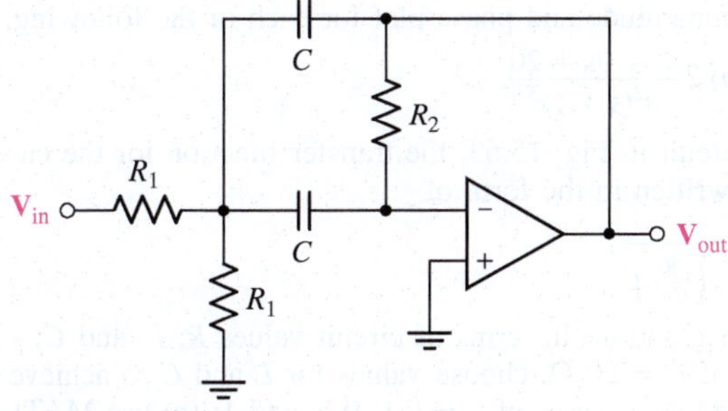

FIGURE 15.56

7. For the circuit in Fig. 15.57, (*a*) determine the transfer function $\mathbf{H(s)} = \mathbf{V}_{out}/\mathbf{V}_{in}$ in terms of circuit parameters R_1, R_2, R_3, L_1, and L_2; (*b*) determine the magnitude and phase of the transfer function at $\omega = 0$, 3×10^3 rad/s, and as $\omega \to \infty$ for the case where circuit values are $R_1 = 2$ kΩ, $R_2 = 2$ kΩ, $R_3 = 20$ kΩ, $L_1 = 2$ H, and $L_2 = 2$ H.

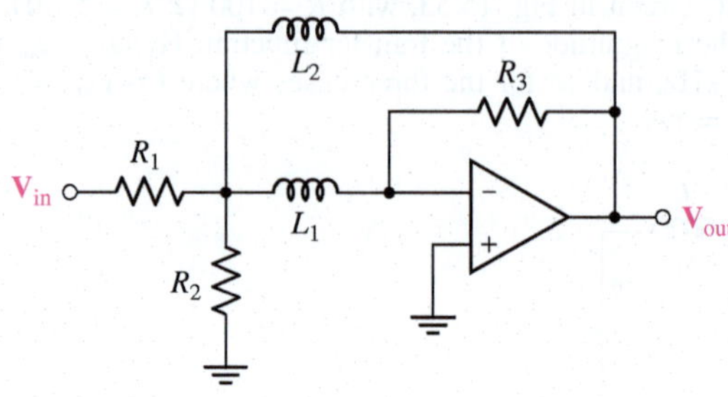

■ **FIGURE 15.57**

15.2 Bode Diagrams

8. Sketch the Bode magnitude and phase plots for the following functions: (*a*) $3 + 4\mathbf{s}$; (*b*) $\dfrac{1}{3 + 4\mathbf{s}}$.

9. For the following functions, sketch the Bode magnitude and phase plots: (*a*) $25\left(1 + \frac{\mathbf{s}}{3}\right)(5 + \mathbf{s})$; (*b*) $\dfrac{0.1}{(1 + 5\mathbf{s})(2 + \mathbf{s})}$.

10. Use the Bode approach to sketch the magnitude of each of the following responses, then verify your solutions with appropriate MATLAB simulations: (*a*) $3\dfrac{\mathbf{s}}{\mathbf{s}^2 + 7\mathbf{s} + 10}$; (*b*) $\dfrac{4}{\mathbf{s}^3 + 7\mathbf{s}^2 + 12\mathbf{s}}$.

11. If a particular network is described by transfer function $\mathbf{H(s)}$, use MATLAB to plot the magnitude and phase Bode plot for $\mathbf{H(s)}$ equal to (*a*) $\dfrac{\mathbf{s} + 300}{\mathbf{s}(5\mathbf{s} + 8)}$; (*b*) $\dfrac{\mathbf{s}(\mathbf{s}^2 + 7\mathbf{s} + 7)}{\mathbf{s}(2\mathbf{s} + 4)^2}$.

12. Use MATLAB to plot the magnitude and phase Bode plot for each of the following transfer functions: (*a*) $\dfrac{\mathbf{s} + 1}{\mathbf{s}(\mathbf{s} + 2)^2}$; (*b*) $5\dfrac{\mathbf{s}^2 + \mathbf{s}}{\mathbf{s} + 2}$.

13. Determine the Bode magnitude plot for the following transfer functions, and compare to what is predicted using MATLAB: (*a*) $\mathbf{s}^2 + 0.2\mathbf{s} + 1$; (*b*) $\left(\frac{\mathbf{s}}{4}\right)^2 + 0.1\left(\frac{\mathbf{s}}{4}\right) + 1$.

14. Determine the Bode magnitude and phase plot for each of the following: (*a*) $\dfrac{3 + 0.1\mathbf{s} + \mathbf{s}^2/3}{\mathbf{s}^2 + 1}$; (*b*) $2\dfrac{\mathbf{s}^2 + 9\mathbf{s} + 20}{\mathbf{s}^2(\mathbf{s} + 1)^3}$.

15. For the series *RLC* circuit in Fig. 15.53, the transfer function for the case of $\mathbf{H(s)} = v_c/v_{in}$ can be written in the form of

$$\mathbf{H(s)} = \frac{1}{1 + 2\zeta\left(\frac{\mathbf{s}}{\omega_0}\right) + \left(\frac{\mathbf{s}}{\omega_0}\right)^2}.$$

(*a*) Find the values of ζ and ω_0 in terms of circuit values *R*, *L*, and *C*; (*b*) for a fixed value of $R = 50$ Ω, choose values for *L* and *C* to achieve $\omega_0 = 2 \times 10^3$ rad/s and three cases of $\zeta = 0.1$, 0.5, and 1; (*c*) use MATLAB to construct magnitude Bode plots for the three cases in part (*b*).

16. Repeat Exercise 15 for the series RLC circuit in the case where $\mathbf{H(s)} = v_L/v_{\text{in}}$. (*a*) Determine the new form of the transfer function in terms of parameters ζ and ω_0, relating each to circuit values R, L, and C; (*b*) for a fixed value of $R = 100\ \Omega$, choose values for L and C to achieve $\omega_0 = 5 \times 10^3$ rad/s and three cases of $\zeta = 0.1$, 0.5, and 1; (*c*) use MATLAB to construct magnitude Bode plots for the three cases in part (*b*).

17. For the circuit of Fig. 15.56, construct a magnitude and phase Bode plot for the transfer function $\mathbf{H(s)} = \mathbf{V}_{\text{out}}/\mathbf{V}_{\text{in}}$ and circuit values of $R_1 = 500\ \Omega$, $R_2 = 40\ \text{k}\Omega$, and $C_1 = 10$ nF.

18. Construct a magnitude and phase Bode plot for the transfer function $\mathbf{H(s)} = \mathbf{V}_{\text{out}}/\mathbf{V}_{\text{in}}$ for the circuit shown in Fig. 15.57 and circuit values of $R_1 = R_2 = 2\ \text{k}\Omega$, $R_3 = 20\ \text{k}\Omega$, and $L_1 = L_2 = 2$ H.

19. For the circuit in Fig. 15.54, use LTspice to construct a Bode plot of the frequency response for the case where $R_1 = 20\ \text{k}\Omega$, $R_2 = 5\ \text{k}\Omega$, $C_1 = 10$ nF, and $C_2 = 40$ nF. Use your plot to estimate locations of poles and zeros.

20. For the circuit in Fig. 15.55, use LTspice to construct a Bode plot of the frequency response for the case where $R_1 = 3\ \text{k}\Omega$, $R_2 = 12\ \text{k}\Omega$, $L_1 = 5$ mH, and $L_2 = 8$ mH. Use your plot to estimate locations of poles and zeros.

15.3 Parallel Resonance

21. Compute Q_0 and ζ for a simple parallel RLC network if $R = 1\ \text{k}\Omega$, $C = 10$ mF, and $L = 1$ H; (*b*) $R = 1\ \Omega$, $C = 10$ mF, and $L = 1$ H; (*c*) $R = 1\ \text{k}\Omega$, $C = 1$ F, and $L = 1$ H; (*d*) $R = 1\ \Omega$, $C = 1$ F, and $L = 1$ H.

22. A certain parallel RLC circuit is built using component values $L = 50$ mH and $C = 33$ mF. If $Q_0 = 10$, determine the value of R, and sketch the magnitude of the steady-state impedance over the range of $2 < \omega < 40$ rad/s.

23. A parallel RLC network is constructed using $R = 5\ \Omega$, $L = 100$ mH, and $C = 1$ mF. (*a*) Compute Q_0. (*b*) Determine at which frequencies the impedance magnitude drops to 90% of its maximum value.

24. For the network of Fig. 15.58, derive an expression for the steady-state input impedance and determine the frequency at which it has maximum amplitude.

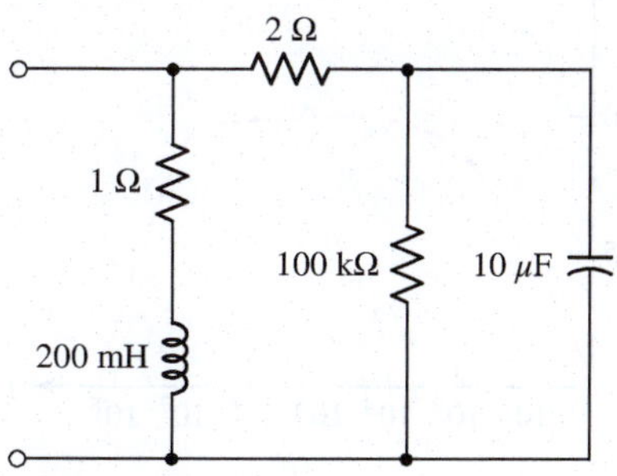

FIGURE 15.58

25. Plot the input admittance of the network depicted in Fig. 15.58 using a logarithmic frequency and magnitude scale over the range $0.1\omega_0 < \omega_0 < 10\omega_0$, and determine the resonant frequency and the bandwidth of the network.

26. Delete the 2 Ω resistor in the network of Fig. 15.58 and determine (*a*) the magnitude of the input impedance at resonance and (*b*) the resonant frequency.

27. Delete the 1 Ω resistor in the network of Fig. 15.58 and determine (*a*) the magnitude of the input impedance at resonance and (*b*) the resonant frequency.
28. A varactor is a semiconductor device whose reactance may be varied by applying a bias voltage. The quality factor can be expressed[3] as

$$Q \approx \frac{\omega C_J R_P}{1 + \omega^2 C_J^2 R_P R_S}$$

where C_J is the junction capacitance (which depends on the voltage applied to the device), R_S is the series resistance of the device, and R_P is an equivalent parallel resistance term. (*a*) If $C_J = 3.77$ pF at 1.5 V, $R_P = 1.5$ MΩ, and $R_S = 2.8$ Ω plot the quality factor as a function of frequency ω. (*b*) Differentiate the expression for Q to obtain both ω_0 and Q_{max}.

15.4 Bandwidth and High-Q Circuits

29. The circuit of Fig. 15.17 is built using component values $L = 1$ mH and $C = 100$ μF. If $Q_0 = 15$, determine the bandwidth and estimate the magnitude and angle of the input impedance for operation at (*a*) 3162 rad/s; (*b*) 3000 rad/s; (*c*) 3200 rad/s; (*d*) 2000 rad/s.
30. A parallel *RLC* network is constructed with a 5 mH inductor, and the remaining component values are chosen such that $Q_0 = 6.5$ and $\omega_0 = 1000$ rad/s. Determine the approximate value of the input impedance magnitude for operation at (*a*) 500 rad/s; (*b*) 750 rad/s; (*c*) 900 rad/s; (*d*) 1100 rad/s. (*e*) Plot your estimates along with the exact result using a linear frequency (rad/s) axis.
31. A parallel *RLC* network is constructed with a 200 μH inductor, and the remaining component values are chosen such that $Q_0 = 8$ and $\omega_0 = 5000$ rad/s. Use approximate expressions to estimate the input impedance angle for operation at (*a*) 2000 rad/s; (*b*) 3000 rad/s; (*c*) 4000 rad/s; (*d*) 4500 rad/s. (*e*) Plot your estimates along with the exact result using a linear frequency (rad/s) axis.
32. Find the bandwidth of each of the response curves shown in Fig. 15.59.

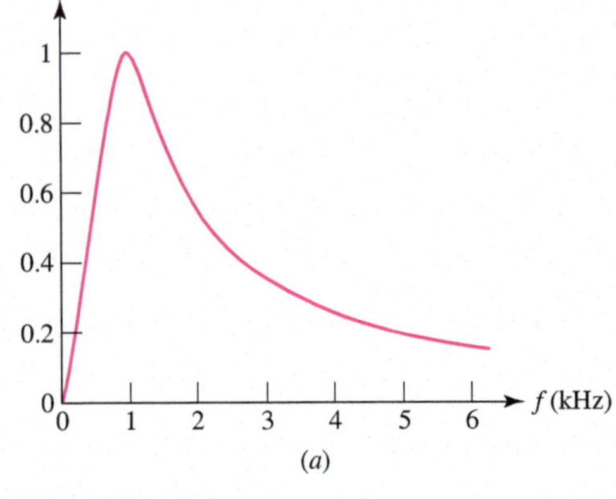

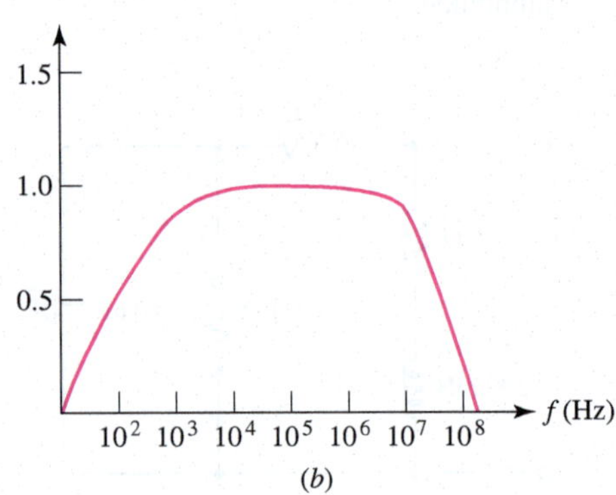

FIGURE 15.59

33. A parallel *RLC* circuit is constructed such that it has the impedance magnitude characteristic plotted in Fig. 15.60. (*a*) Determine the resistor value. (*b*) Determine the capacitor value if a 1 H inductor was used. (*c*) Obtain values for the bandwidth, Q_0, and both the low half-power frequency and the high half-power frequency.

(3) S. M. Sze, *Physics of Semiconductor Devices,* 2d ed. New York: Wiley, 1981, p. 116.

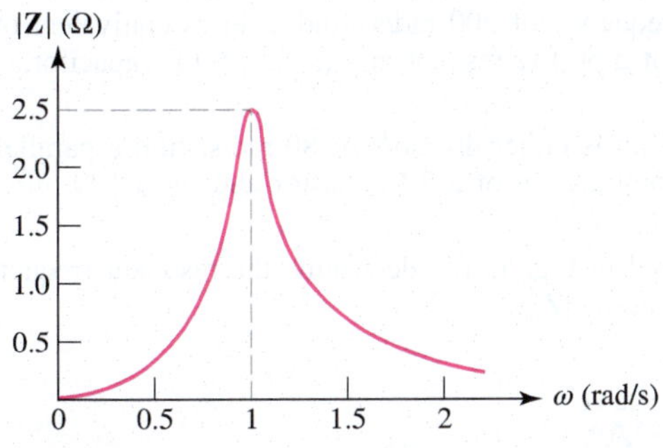

■ **FIGURE 15.60**

15.5 Series Resonance

34. A series RLC circuit is constructed using component values $R = 100\ \Omega$ and $L = 1.5$ mH along with a sinusoidal voltage source v_s. If $Q_0 = 7$, determine (*a*) the magnitude of the impedance at 500 Mrad/s; (*b*) the current which flows in response to a voltage $v_s = 2.5\cos(425 \times 10^6 t)$ V.

35. With regard to the series RLC circuit described in Exercise 34, adjust the resistor value such that Q_0 is reduced to 5, and (*a*) estimate the angle of the impedance at 90 krad/s, 100 krad/s, and 110 krad/s. (*b*) Determine the percent error in the estimated values, compared to the exact expression.

36. An RLC circuit is constructed using $R = 5\ \Omega$, $L = 20$ mH, and $C = 1$ mF. Calculate Q_0, the bandwidth, and the magnitude of the impedance at $0.95\omega_0$ if the circuit is (*a*) parallel-connected; (*b*) series-connected. (*c*) Verify your solutions using appropriate LTspice simulations.

37. Inspect the circuit of Fig. 15.61, noting the amplitude of the source voltage. Now decide whether you would be willing to put your bare hands across the capacitor if the circuit were actually built in the lab. Plot $|\mathbf{V}_C|$ versus ω to justify your answer.

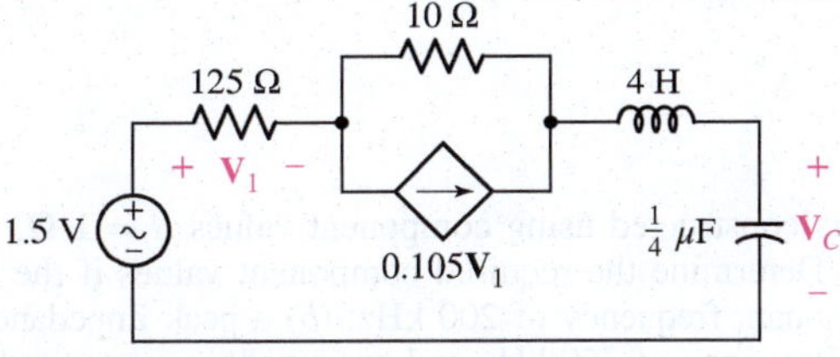

■ **FIGURE 15.61**

38. After deriving $\mathbf{Z}_{in}(\mathbf{s})$ in Fig. 15.62, find (*a*) ω_0; (*b*) Q_0.

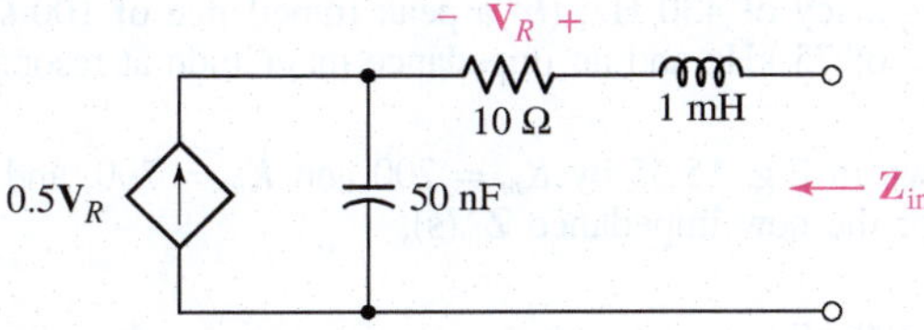

■ **FIGURE 15.62**

15.6 Other Resonant Forms

39. For the network of Fig. 15.25*a*, $R_1 = 100\ \Omega$, $R_2 = 150\ \Omega$, $L = 30$ mH, and C is chosen so that $\omega_0 = 750$ rad/s. Calculate the impedance magnitude at (*a*) the frequency corresponding to resonance when $R_1 = 0$; (*b*) 700 rad/s; (*c*) 800 rad/s.

40. Assuming an operating frequency of 200 rad/s, find a series equivalent of the parallel combination of a 500 Ω resistor and (*a*) a 1.5 μF capacitor; (*b*) a 200 mH inductor.
41. If the frequency of operation is either 40 rad/s or 80 rad/s, find a parallel equivalent of the series combination of a 2 Ω resistor and (*a*) a 100 mF capacitor; (*b*) a 3 mH inductor.
42. For the network represented in Fig. 15.63, determine the resonant frequency and the corresponding value of $|\mathbf{Z}_{in}|$.

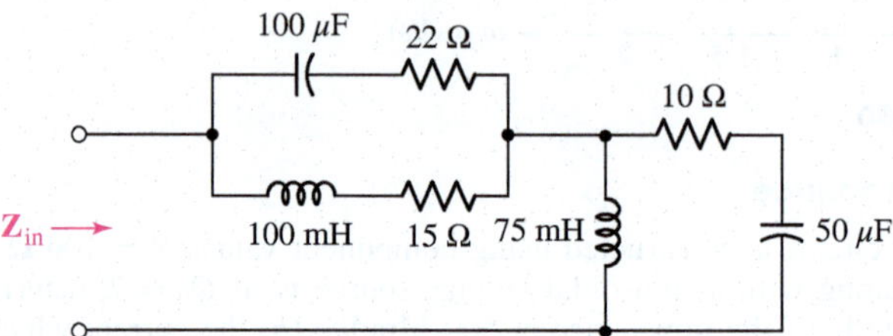

■ **FIGURE 15.63**

43. For the circuit shown in Fig. 15.64, the voltage source has magnitude 1 V and phase angle 0°. Determine the resonant frequency $\omega°$ and the value of $\mathbf{V}_x$ at $0.95\omega_0$.

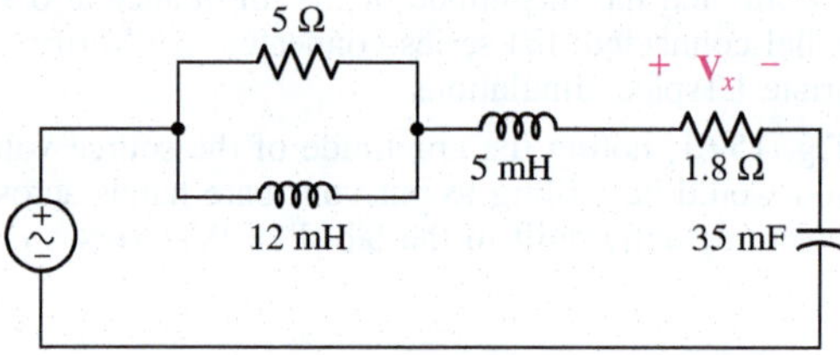

■ **FIGURE 15.64**

15.7 Scaling

44. A parallel *RLC* circuit is constructed using component values $R = 1\ \Omega$, $C = 3$ F, and $L = \frac{1}{3}$ **H**. Determine the required component values if the network is to have (*a*) a resonant frequency of 200 kHz; (*b*) a peak impedance of 500 kΩ; (*c*) a resonant frequency of 750 kHz and an impedance magnitude at resonance of 25 Ω.
45. A series *RLC* circuit is constructed using component values $R = 1\ \Omega$, $C = 5$ F, and $L = \frac{1}{5}$ **H**. Determine the required component values if the network is to have (*a*) a resonant frequency of 430 Hz; (*b*) a peak impedance of 100 Ω; (*c*) a resonant frequency of 75 kHz and an impedance magnitude at resonance of 15 kΩ.
46. Scale the network shown in Fig. 15.65 by $K_m = 200$ and $K_m = 700$, and obtain an expression for the new impedance $\mathbf{Z}_{in}(\mathbf{s})$.

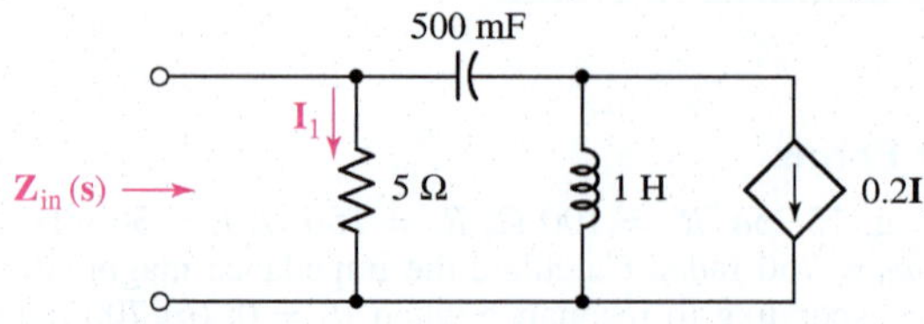

■ **FIGURE 15.65**

47. The filter shown in Fig. 15.66*a* has the response curve shown in Fig. 15.66*b*. (*a*) Scale the filter so that it operates between a 50 Ω source and a 50 Ω load and has a cutoff frequency of 20 kHz. (*b*) Draw the new response curve.

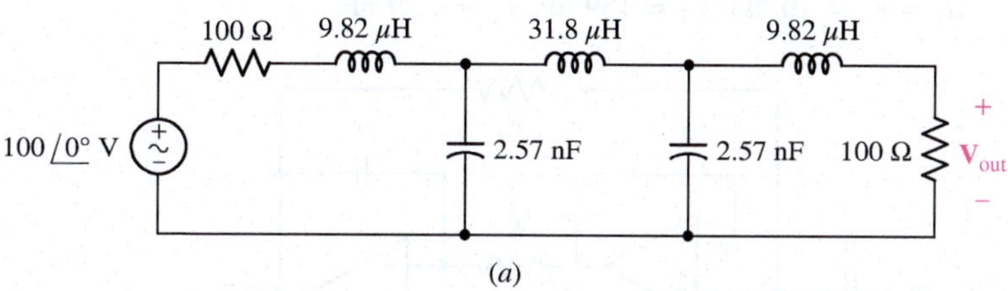

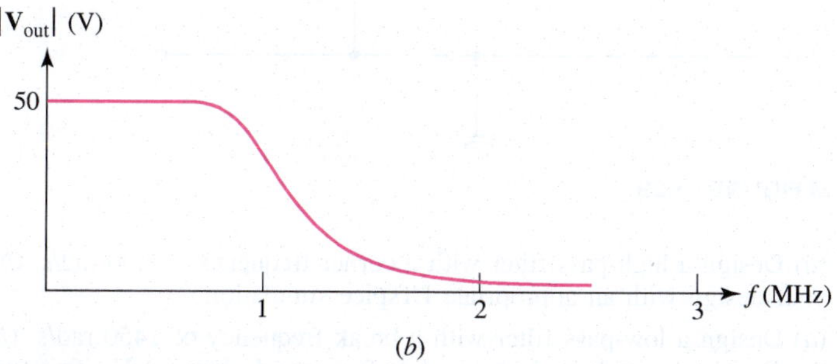

■ **FIGURE 15.66**

48. (*a*) Draw the new configuration for Fig. 15.67 after the network is scaled by $K_m = 250$ and $K_f = 400$. (*b*) Determine the Thévenin equivalent of the scaled network at $\omega = 1$ krad/s.

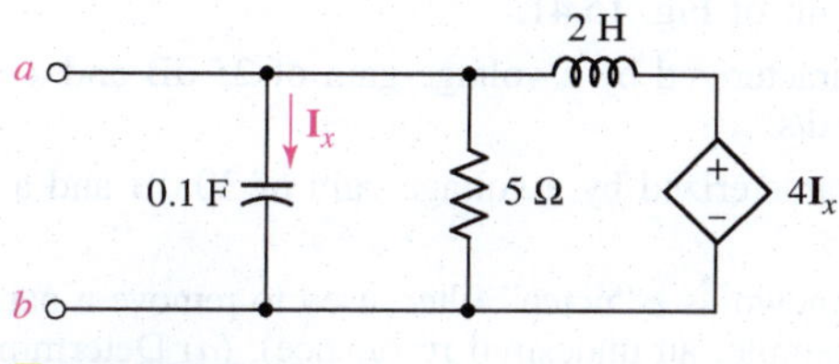

■ **FIGURE 15.67**

15.8 Basic Filter Design

49. Examine the filter for the circuit in Fig. 15.68. (*a*) Without going through a full mathematical analysis of the circuit, determine what kind of filter this is. (*b*) Determine an expression for the transfer function $\mathbf{H(s)} = v_{out}/v_{in}$. (*c*) Use MATLAB to construct a Bode plot (with frequency in Hz) for $R_1 = R_2 = 50$ Ω, $C_1 = 50$ nF, $C_2 = 225$ nF, $L_1 = 563$ μH, and $L_2 = 125$ μH.

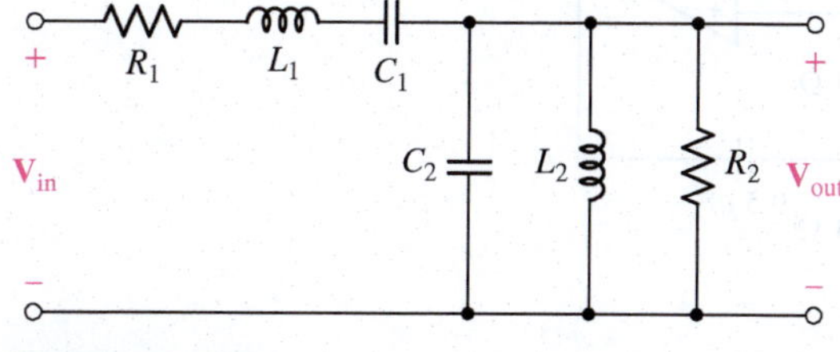

■ **FIGURE 15.68**

50. Examine the filter for the circuit in Fig. 15.69. (*a*) Without going through a full mathematical analysis of the circuit, determine what kind of filter this is. (*b*) Determine an expression for the transfer function $\mathbf{H(s)} = v_{out}/v_{in}$. (*c*) Use MATLAB to construct a Bode plot (with frequency in Hz) for $R_1 = R_2 = 10\ \text{k}\Omega$, $C_1 = 159$ nF, $C_2 = 1.59$ nF.

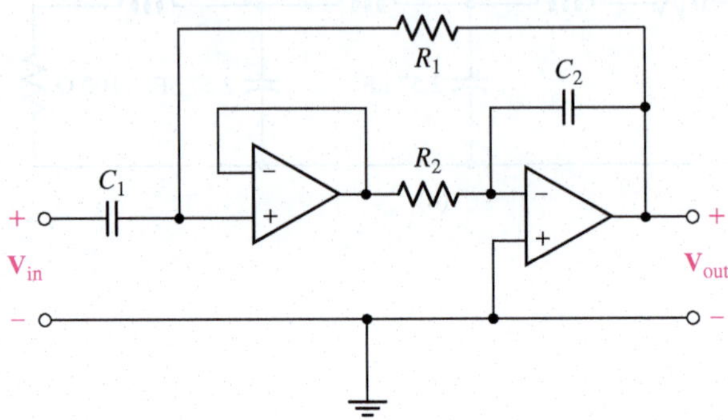

FIGURE 15.69

51. (*a*) Design a high-pass filter with a corner frequency of 100 rad/s. (*b*) Verify your design with an appropriate LTspice simulation.
52. (*a*) Design a low-pass filter with a break frequency of 1450 rad/s. (*b*) Sketch the Bode magnitude and phase plots for your design. (*c*) Verify your filter performance with an appropriate simulation.
53. (*a*) Design a bandpass filter characterized by a bandwidth of 1000 rad/s and a low-frequency corner of 250 Hz. (*b*) Verify your design with an appropriate LTspice simulation.

54. Design a notch filter which removes 60 Hz "noise" from power line influences on a particular signal by taking the output across the inductor-capacitor series connection in the circuit of Fig. 15.41.

55. Design a low-pass filter characterized by a voltage gain of 25 dB and a corner frequency of 5000 rad/s.

56. Design a high-pass filter characterized by a voltage gain of 30 dB and a corner frequency of 50 rad/s.
57. The circuit in Fig. 15.70 is known as a "notch" filter, used to remove a narrow range of frequencies (for example, an undesired resonance). (*a*) Determine the transfer function for this circuit; (*b*) plot the magnitude Bode plot for this filter; (*c*) determine the center frequency for the notch filter and the reduction in magnitude at the center frequency (in dB).

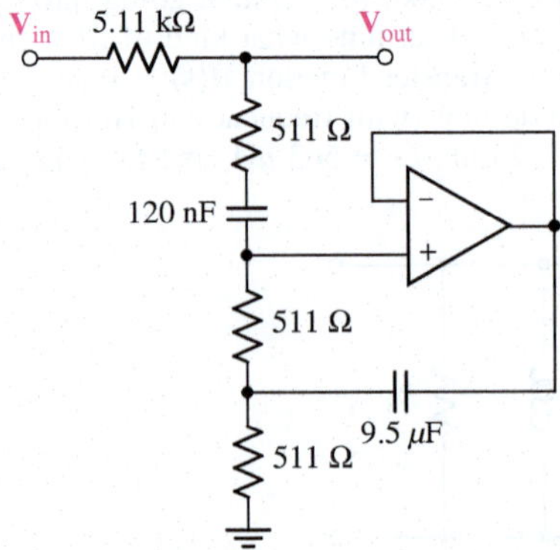

FIGURE 15.70

58. (*a*) Design a two-stage op amp filter circuit with a bandwidth of 1000 rad/s, a low-frequency cutoff of 100 rad/s, and a voltage gain of 20 dB. (*b*) Verify your design with an appropriate LTspice simulation.

59. Design a circuit which removes the entire audio frequency range (approximately 20 Hz to 20 kHz, for human hearing) but amplifies the signal voltage of all other frequencies by a factor of 15.

15.9 Advanced Filter Design

60. The circuit in Fig. 15.71 is a low-pass Butterworth filter, whose primary virtue is to have an extremely flat passband. (*a*) Determine the transfer function $\mathbf{H(s)} = \mathbf{V}_{out}(s)/\mathbf{V}_{in}(s)$, (*b*) plot the magnitude Bode plot, using frequency in Hz rather than rad/s, and (*c*) determine the cutoff frequency and attenuation for the filter (in dB/decade).

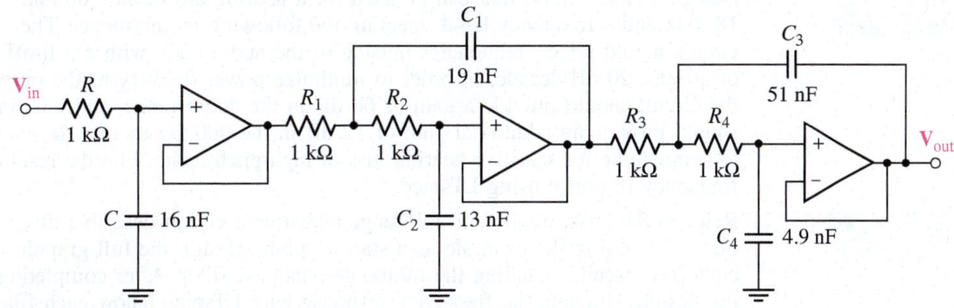

■ **FIGURE 15.71**

61. Design a second-order low-pass filter having a voltage gain of 5 dB and a cutoff frequency of 1700 kHz based on (*a*) Butterworth polynomials; (*b*) Chebyshev polynomials for a 3 dB ripple factor.

62. If a high-pass filter is required having gain of 6 dB and a cutoff frequency of 350 Hz, design a suitable second-order Butterworth-based solution.

63. (*a*) Design a second-order high-pass Butterworth filter with a cutoff frequency of 2000 Hz and a voltage gain of 4.5 dB. (*b*) Verify your design with an appropriate simulation.

64. (*a*) Design a third-order low-pass Butterworth filter having a gain of 13 dB and a corner frequency at 1800 Hz. (*b*) Compare your filter response to that of a Chebyshev filter with the same specifications.

65. Design a fourth-order high-pass Butterworth filter having a minimum gain of 15 dB and a corner frequency of 1100 rad/s.

66. Choose parameters for the circuit described by Eq. [36] such that it has a cutoff frequency at 450 rad/s, and compare its performance to a comparable second-order Butterworth filter.

67. (*a*) Design a Sallen-Key low-pass filter with a corner frequency of 10 kHz and $Q = 0.5$. (*b*) Simulate the frequency response of the circuit using LTspice.

68. (*a*) Design a Sallen-Key high-pass filter with a corner frequency of 2 kHz and $Q = 0.5$. (*b*) Simulate the frequency response of the circuit using LTspice.

Chapter-Integrating Exercises

69. A piezoelectric sensor has an equivalent circuit representation as shown in Fig. 15.72. Determine (*a*) the transfer function $\mathbf{H(s)} = \mathbf{V}_{out}/\mathbf{V}_{in}$; (*b*) plot the frequency response (Bode plot) for the case where $C_e = 100$ nF, $C_0 = 200$ nF, $L_m = 20\ \mu$H, and $R_i = 50$ kΩ, and (*c*) comment on the approximate frequency range where the sensor would be useful.

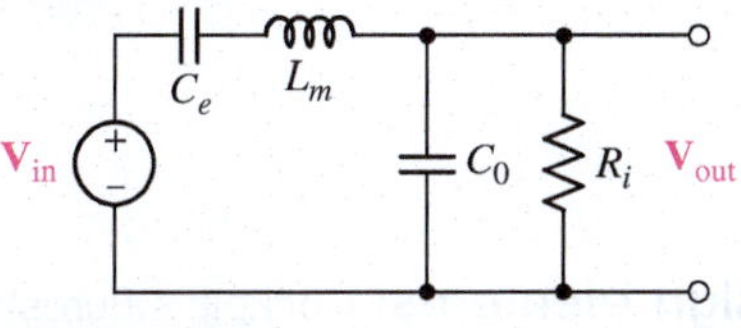

■ **FIGURE 15.72**

70. Design a parallel resonant circuit for an AM radio so that a variable inductor can adjust the resonant frequency over the AM broadcast band, 535 to 1605 kHz, with $Q_0 = 45$ at one end of the band and $Q_0 \leq 45$ throughout the band. Let $R = 20$ kΩ, and specify values for C, L_{min}, and L_{max}.

71. The network of Fig. 15.72 was implemented as a low-pass filter designed with a corner frequency of 1250 rad/s. Its performance is inadequate in two respects: (1) a voltage gain of at least 2 dB is required, and (2) the magnitude of the output voltage does not decrease quickly enough in the stopband. Design a better alternative if only one op amp is available and only two 1 μF capacitors can be located.

72. Determine the effect of component tolerance on the circuit designed in Example 15.15 if each component is specified to be only within 10% of its stated value.

73. Design a filter circuit that can be used for a hearing aid in the 100 and 18 kHz audio frequency band, meeting the following requirements: The circuit should reject frequencies outside of the audio band with a roll-off of at least 20 dB/decade, in order to minimize power delivery to the output, the circuit should provide a gain of 60 dB in the audio range, and the circuit should provide an additional gain of 12 dB in the 900 Hz to 18 kHz range to compensate for selective hearing loss of high-pitch audio. Plot the resulting frequency response using LTspice.

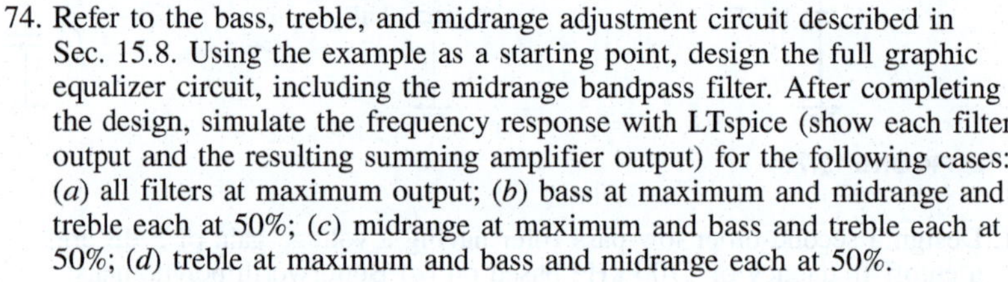

74. Refer to the bass, treble, and midrange adjustment circuit described in Sec. 15.8. Using the example as a starting point, design the full graphic equalizer circuit, including the midrange bandpass filter. After completing the design, simulate the frequency response with LTspice (show each filter output and the resulting summing amplifier output) for the following cases: (*a*) all filters at maximum output; (*b*) bass at maximum and midrange and treble each at 50%; (*c*) midrange at maximum and bass and treble each at 50%; (*d*) treble at maximum and bass and midrange each at 50%.

75. (*a*) Modify the circuit of Fig. 15.48*a* making use of simulated potentiometers instead of fixed resistors so that the peak output voltage of either the bass or treble stage may be 10 times that of the other stage. The output voltage magnitudes should be independent of one another. (*b*) Plot the output with a summing amplifier stage gain of 10, showing the range of obtainable outputs using a few well-chosen example values. Ensure your plot covers the frequency range of 10 Hz to 100 kHz.

CHAPTER

16 Two-Port Networks

KEY CONCEPTS

One-Port and Two-Port Networks

Admittance (**y**) Parameters

Impedance (**z**) Parameters

Hybrid (**h**) Parameters

Transmission (**t**) Parameters

Transformation Between **y, z, h,** and **t** Parameters

Circuit Analysis Using Network Parameters

INTRODUCTION

A general network having two pairs of terminals, one often labeled the "input terminals" and the other the "output terminals," is an important building block in many types of systems, including electronic, communication, automatic control, transmission, and distribution systems, as well as other systems in which an electrical signal or electric energy enters the input terminals, is acted upon by the network, and leaves via the output terminals. The output terminal pair may very well connect with the input terminal pair of another network. When we studied the concept of Thévenin and Norton equivalent networks in Chap. 5, we were introduced to the idea that it is not always necessary to know the detailed workings of part of a circuit. This chapter extends such concepts to linear networks, resulting in parameters that allow us to predict how any network will interact with other networks. The terms in this chapter are also often used in the characterization of electronic components.

16.1 ONE-PORT NETWORKS

A pair of terminals at which a signal may enter or leave a network is called a ***port;*** a network having only one such pair of terminals is called a ***one-port network,*** or simply a *one-port*. No connections may be made to any other nodes internal to the one-port, and therefore i_a must equal i_b in the one-port shown in Fig. 16.1*a*. When more than one pair of terminals is present, the network is known as a ***multiport network.*** The two-port network to which this chapter is principally devoted is shown in Fig. 16.1*b*. The currents in the two leads making up each port must be equal, and so it follows that $i_a = i_b$ and $i_c = i_d$ in the two-port of Fig. 16.1*b*. Sources and loads must be connected directly across the two terminals of

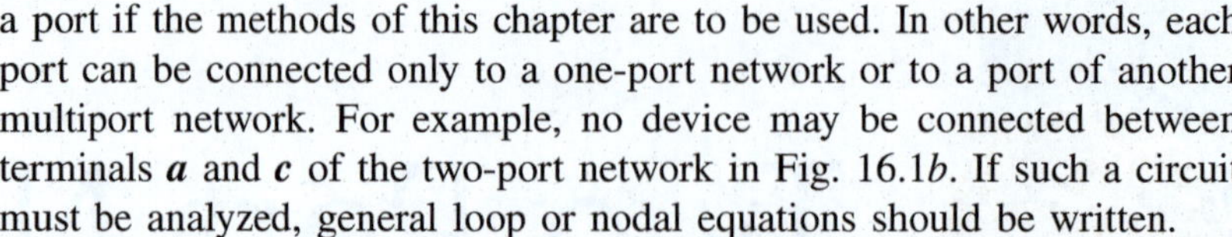

a port if the methods of this chapter are to be used. In other words, each port can be connected only to a one-port network or to a port of another multiport network. For example, no device may be connected between terminals ***a*** and ***c*** of the two-port network in Fig. 16.1*b*. If such a circuit must be analyzed, general loop or nodal equations should be written.

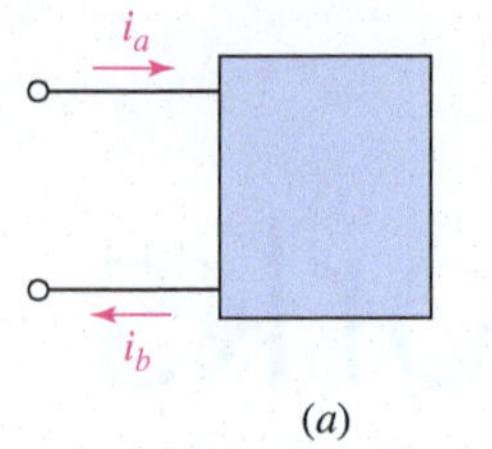

(*a*)

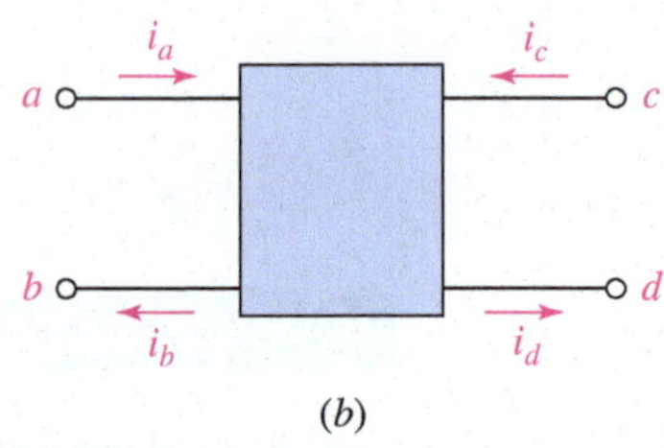

(*b*)

FIGURE 16.1 (*a*) A one-port network. (*b*) A two-port network. Connections between terminals *a* and *b*, or *c* and *d*, are allowed; if connections between *a* and *c* are required, for example, then alternative analysis techniques are needed.

Some of the introductory study of one- and two-port networks is accomplished best by using a generalized network notation and the abbreviated nomenclature for determinants introduced in Appendix 2. Thus, if we write a set of loop equations for a passive network,

$$\begin{array}{c} \mathbf{Z}_{11}\mathbf{I}_1 + \mathbf{Z}_{12}\mathbf{I}_2 + \mathbf{Z}_{13}\mathbf{I}_3 + \cdots + \mathbf{Z}_{1N}\mathbf{I}_N = \mathbf{V}_1 \\ \mathbf{Z}_{21}\mathbf{I}_1 + \mathbf{Z}_{22}\mathbf{I}_2 + \mathbf{Z}_{23}\mathbf{I}_3 + \cdots + \mathbf{Z}_{2N}\mathbf{I}_N = \mathbf{V}_2 \\ \mathbf{Z}_{31}\mathbf{I}_1 + \mathbf{Z}_{32}\mathbf{I}_2 + \mathbf{Z}_{33}\mathbf{I}_3 + \cdots + \mathbf{Z}_{3N}\mathbf{I}_N = \mathbf{V}_3 \\ \cdots\cdots\cdots\cdots\cdots\cdots\cdots\cdots\cdots\cdots \\ \mathbf{Z}_{N1}\mathbf{I}_1 + \mathbf{Z}_{N2}\mathbf{I}_2 + \mathbf{Z}_{N3}\mathbf{I}_3 + \cdots + \mathbf{Z}_{NN}\mathbf{I}_N = \mathbf{V}_N \end{array} \qquad [1]$$

then the coefficient of each current will be an impedance $\mathbf{Z}_{ij}(\mathbf{s})$, and the circuit determinant, or determinant of the coefficients, is

$$\Delta_\mathbf{Z} = \begin{vmatrix} \mathbf{Z}_{11} & \mathbf{Z}_{12} & \mathbf{Z}_{13} & \cdots & \mathbf{Z}_{1N} \\ \mathbf{Z}_{21} & \mathbf{Z}_{22} & \mathbf{Z}_{23} & \cdots & \mathbf{Z}_{2N} \\ \mathbf{Z}_{31} & \mathbf{Z}_{32} & \mathbf{Z}_{33} & \cdots & \mathbf{Z}_{3N} \\ \cdots & \cdots & \cdots & \cdots & \cdots \\ \mathbf{Z}_{N1} & \mathbf{Z}_{N2} & \mathbf{Z}_{N3} & \cdots & \mathbf{Z}_{NN} \end{vmatrix} \qquad [2]$$

Here N loops have been assumed, the currents appear in subscript order in each equation, and the order of the equations is the same as that of the currents. We also assume that KVL is applied so that the sign of each $\mathbf{Z}_{ii}$ term ($\mathbf{Z}_{11}$, $\mathbf{Z}_{22}$, . . . , $\mathbf{Z}_{NN}$) is positive; the sign of any $\mathbf{Z}_{ij}(i \neq j)$ or mutual term may be either positive or negative, depending on the reference directions assigned to $\mathbf{I}_i$ and $\mathbf{I}_j$.

If there are dependent sources within the network, then it is possible that not all the coefficients in the loop equations represent resistances or impedances. Even so, we will continue to refer to the circuit determinant as Δ_Z.

The use of minor notation (Appendix 2) allows for the input or driving-point impedance at the terminals of a *one-port* network to be expressed very concisely. The result is also applicable to a *two-port* network if one of the two ports is terminated in a passive impedance, including an open or a short circuit.

Cramer's rule is reviewed in Appendix 2.

Let us suppose that the one-port network shown in Fig. 16.2 is composed entirely of passive elements and dependent sources; linearity is also assumed. An ideal voltage source $\mathbf{V}_1$ is connected to the port, and the source current is identified as the current in loop 1. Employing Cramer's rule, then,

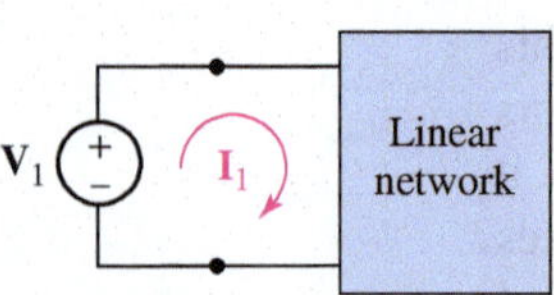

FIGURE 16.2 An ideal voltage source $\mathbf{V}_1$ is connected to the single port of a linear one-port network containing no independent sources; $\mathbf{Z}_{in} = \Delta_\mathbf{Z}/\Delta_{11}$.

$$\mathbf{I}_1 = \frac{\begin{vmatrix} \mathbf{V}_1 & \mathbf{Z}_{12} & \mathbf{Z}_{13} & \cdots & \mathbf{Z}_{1N} \\ 0 & \mathbf{Z}_{22} & \mathbf{Z}_{23} & \cdots & \mathbf{Z}_{2N} \\ 0 & \mathbf{Z}_{32} & \mathbf{Z}_{33} & \cdots & \mathbf{Z}_{3N} \\ \cdots & \cdots & \cdots & \cdots & \cdots \\ 0 & \mathbf{Z}_{N2} & \mathbf{Z}_{N3} & \cdots & \mathbf{Z}_{NN} \end{vmatrix}}{\begin{vmatrix} \mathbf{Z}_{11} & \mathbf{Z}_{12} & \mathbf{Z}_{13} & \cdots & \mathbf{Z}_{1N} \\ \mathbf{Z}_{21} & \mathbf{Z}_{22} & \mathbf{Z}_{23} & \cdots & \mathbf{Z}_{2N} \\ \mathbf{Z}_{31} & \mathbf{Z}_{32} & \mathbf{Z}_{33} & \cdots & \mathbf{Z}_{3N} \\ \cdots & \cdots & \cdots & \cdots & \cdots \\ \mathbf{Z}_{N1} & \mathbf{Z}_{N2} & \mathbf{Z}_{N3} & \cdots & \mathbf{Z}_{NN} \end{vmatrix}}$$

or, more concisely,

$$\mathbf{I}_1 = \frac{\mathbf{V}_1 \Delta_{11}}{\Delta_\mathbf{Z}}$$

Thus,

$$\mathbf{Z}_{\text{in}} = \frac{\mathbf{V}_1}{\mathbf{I}_1} = \frac{\Delta \mathbf{Z}}{\Delta_{11}} \qquad [3]$$

EXAMPLE 16.1

Calculate the input impedance for the one-port resistive network shown in Fig. 16.3.

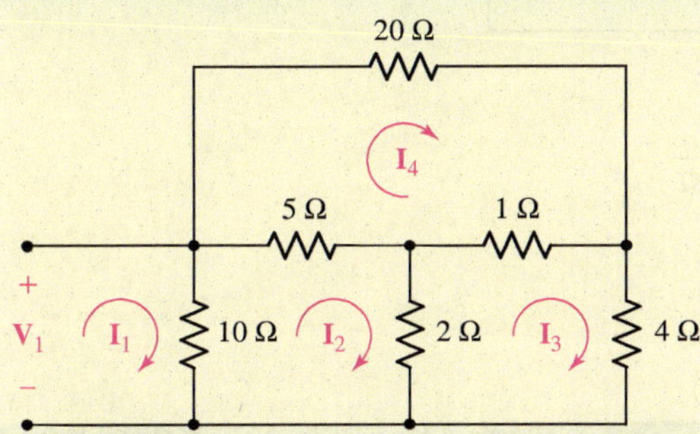

■ **FIGURE 16.3** An example one-port network containing only resistive elements.

We first assign the four mesh currents as shown and write the corresponding mesh equations by inspection:

$$\begin{aligned}
\mathbf{V}_1 &= 10\mathbf{I}_1 - 10\mathbf{I}_2 \\
0 &= -10\mathbf{I}_1 + 17\mathbf{I}_2 - 2\mathbf{I}_3 - 5\mathbf{I}_4 \\
0 &= \quad - 2\mathbf{I}_2 + 7\mathbf{I}_3 - \mathbf{I}_4 \\
0 &= \quad - 5\mathbf{I}_2 - \mathbf{I}_3 + 26\mathbf{I}_4
\end{aligned}$$

The circuit determinant is then given by

$$\Delta_\mathbf{Z} = \begin{vmatrix} 10 & -10 & 0 & 0 \\ -10 & 17 & -2 & -5 \\ 0 & -2 & 7 & -1 \\ 0 & -5 & -1 & 26 \end{vmatrix}$$

and has the value 9680 Ω^4. Eliminating the first row and first column, we have

$$\Delta_{11} = \begin{vmatrix} 17 & -2 & -5 \\ -2 & 7 & -1 \\ -5 & -1 & 26 \end{vmatrix} = 2778\ \Omega^3$$

Thus, Eq. [3] provides the value of the input impedance,

$$\mathbf{Z}_{\text{in}} = \frac{9680}{2778} = 3.485\ \Omega$$

PRACTICE

16.1 Find the input impedance of the network shown in Fig. 16.4 if it is formed into a one-port network by breaking it at terminals (*a*) *a* and *a*′; (*b*) *b* and *b*′; (*c*) *c* and *c*′.

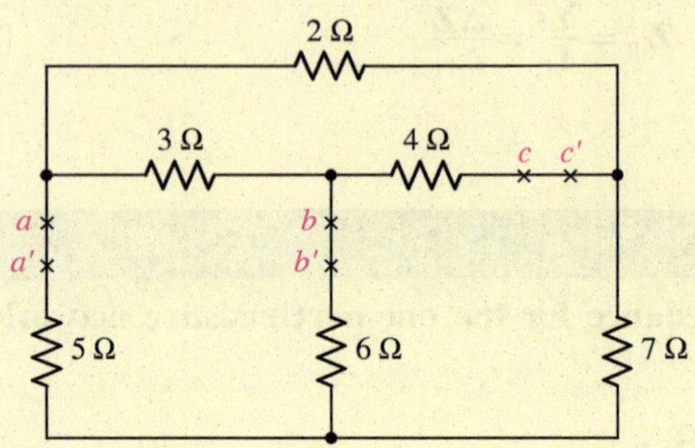

■ **FIGURE 16.4**

Ans: 9.47 Ω; 10.63 Ω; 7.58 Ω.

EXAMPLE 16.2

Determine the input impedance of the network shown in Fig. 16.5.

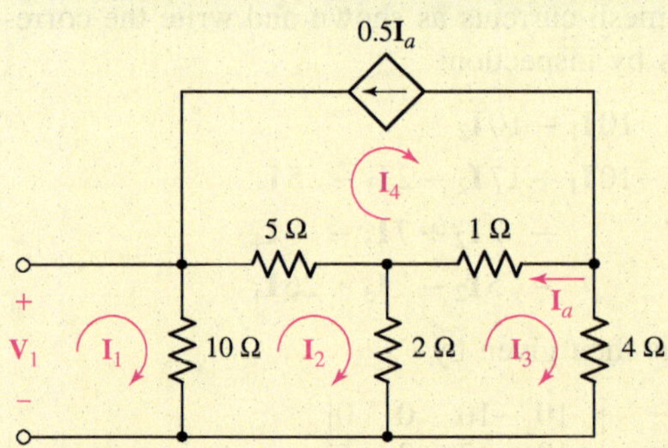

■ **FIGURE 16.5** A one-port network containing a dependent source.

The four mesh equations are written in terms of the four assigned mesh currents:

$$\begin{aligned} 10\mathbf{I}_1 - 10\mathbf{I}_2 \qquad\qquad\qquad &= \mathbf{V}_1 \\ -10\mathbf{I}_1 + 17\mathbf{I}_2 - 2\mathbf{I}_3 - 5\mathbf{I}_4 &= 0 \\ -\ 2\mathbf{I}_2 + 7\mathbf{I}_3 - \ \mathbf{I}_4 &= 0 \end{aligned}$$

and

$$\mathbf{I}_4 = -0.5\,\mathbf{I}_a = -0.5(\mathbf{I}_4 - \mathbf{I}_3)$$

or

$$-0.5\mathbf{I}_3 + 1.5\mathbf{I}_4 = 0$$

Thus we can write

$$\Delta_\mathbf{Z} = \begin{vmatrix} 10 & -10 & 0 & 0 \\ -10 & 17 & -2 & -5 \\ 0 & -2 & 7 & -1 \\ 0 & 0 & -0.5 & 1.5 \end{vmatrix} = 590\ \Omega^4$$

while

$$\Delta_{11} = \begin{vmatrix} 17 & -2 & -5 \\ -2 & 7 & -1 \\ 0 & -0.5 & 1.5 \end{vmatrix} = 159\ \Omega^3$$

giving

$$\mathbf{Z}_{\text{in}} = \frac{590}{159} = 3.711\ \Omega$$

We may also select a similar procedure using nodal equations, yielding the input admittance:

$$\mathbf{Y}_{\text{in}} = \frac{1}{\mathbf{Z}_{\text{in}}} = \frac{\Delta_\mathbf{Y}}{\Delta_{11}} \qquad [4]$$

where Δ_{11} now refers to the minor of $\Delta_\mathbf{Y}$.

PRACTICE

16.2 Write a set of nodal equations for the circuit of Fig. 16.6, calculate $\Delta_\mathbf{Y}$, and then find the input admittance seen between (*a*) node 1 and the reference node; (*b*) node 2 and the reference.

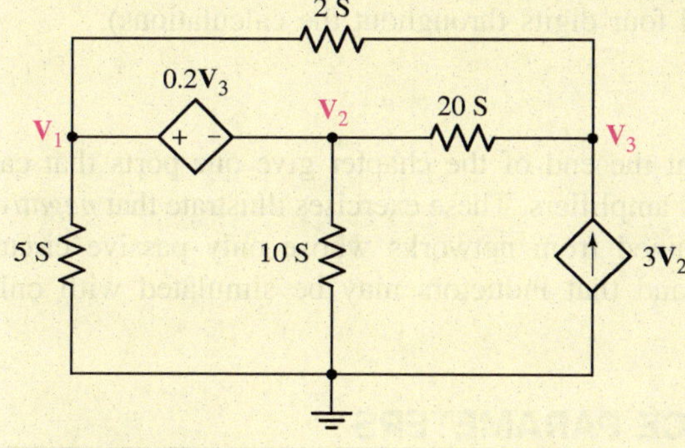

FIGURE 16.6

Ans: 10.68 S; 13.16 S.

EXAMPLE 16.3

Use Eq. [4] to again determine the input impedance of the network shown in Fig. 16.3, repeated here as Fig. 16.7.

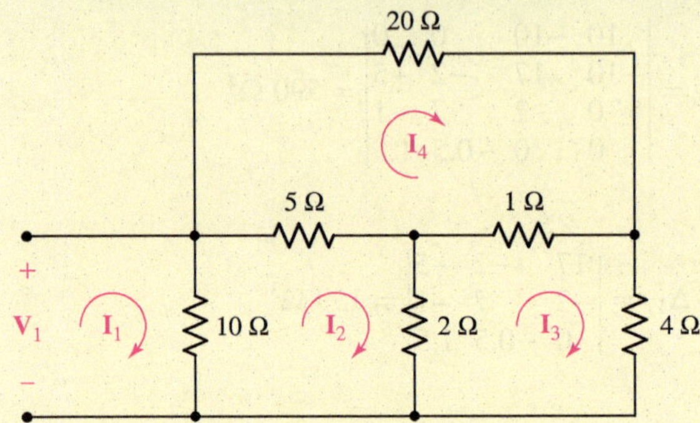

■ **FIGURE 16.7** The circuit from Example 16.1, repeated for convenience.

We first order the node voltages $\mathbf{V}_1$, $\mathbf{V}_2$, and $\mathbf{V}_3$ from left to right, select the reference at the bottom node, and then write the system admittance matrix by inspection:

$$\Delta_{\mathbf{Y}} = \begin{vmatrix} 0.35 & -0.2 & -0.05 \\ -0.2 & 1.7 & -1 \\ -0.05 & -1 & 1.3 \end{vmatrix} = 0.3473 \text{ S}^3$$

$$\Delta_{11} = \begin{vmatrix} 1.7 & -1 \\ -1 & 1.3 \end{vmatrix} = 1.21 \text{ S}^2$$

so that

$$\mathbf{Y}_{\text{in}} = \frac{0.3473}{1.21} = 0.2870 \text{ S}$$

which corresponds to

$$\mathbf{Z}_{\text{in}} = \frac{1}{0.287} = 3.484 \ \Omega$$

which agrees with our previous answer to within expected rounding error (we only retained four digits throughout the calculations).

Exercises 9 and 10 at the end of the chapter give one-ports that can be built using operational amplifiers. These exercises illustrate that *negative* resistances may be obtained from networks whose only passive circuit elements are resistors, and that inductors may be simulated with only resistors and capacitors.

16.2 ADMITTANCE PARAMETERS

Let us now turn our attention to two-port networks. We will assume in all that follows that the network is composed of linear elements and contains no independent sources; *dependent* sources are permissible, however. Further conditions will also be placed on the network in some special cases.

We will consider the two-port as it is shown in Fig. 16.8; the voltage and current at the input terminals are $\mathbf{V}_1$ and $\mathbf{I}_1$, and $\mathbf{V}_2$ and $\mathbf{I}_2$ are specified at the output port. The directions of $\mathbf{I}_1$ and $\mathbf{I}_2$ are both customarily selected as *into* the network at the upper conductors (and out at the lower conductors). Since the network is linear and contains no independent sources within it, $\mathbf{I}_1$ may be considered to be the superposition of two components, one caused by $\mathbf{V}_1$ and the other by $\mathbf{V}_2$. When the same argument is applied to $\mathbf{I}_2$, we may begin with the set of equations

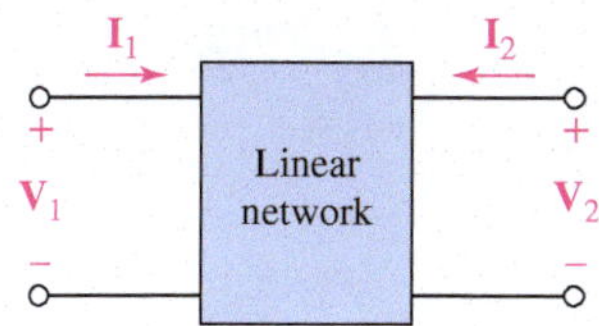

FIGURE 16.8 A general two-port with terminal voltages and currents specified. The two-port is composed of linear elements, possibly including dependent sources, but not containing any independent sources.

$$\mathbf{I}_1 = \mathbf{y}_{11}\mathbf{V}_1 + \mathbf{y}_{12}\mathbf{V}_2 \qquad [5]$$

$$\mathbf{I}_2 = \mathbf{y}_{21}\mathbf{V}_1 + \mathbf{y}_{22}\mathbf{V}_2 \qquad [6]$$

where the $\mathbf{y}$'s are no more than proportionality constants, or unknown coefficients, for the present. However, it should be clear that their dimensions must be A/V, or S. They are therefore called the $\mathbf{y}$ (or admittance) parameters, and are defined by Eqs. [5] and [6].

The $\mathbf{y}$ parameters, as well as other sets of parameters we will define later in the chapter, are represented concisely as matrices. Here, we define the (2 × 1) column matrix $\mathbf{I}$,

$$\mathbf{I} = \begin{bmatrix}\mathbf{I}_1\\ \mathbf{I}_2\end{bmatrix} \qquad [7]$$

the (2 × 2) square matrix of the $\mathbf{y}$ parameters,

$$\mathbf{y} = \begin{bmatrix}\mathbf{y}_{11} & \mathbf{y}_{12}\\ \mathbf{y}_{21} & \mathbf{y}_{22}\end{bmatrix} \qquad [8]$$

and the (2 × 1) column matrix $\mathbf{V}$,

$$\mathbf{V} = \begin{bmatrix}\mathbf{V}_1\\ \mathbf{V}_2\end{bmatrix} \qquad [9]$$

The notation adopted in this text to represent a matrix is standard, but it also can be easily confused with our previous notation for phasors or general complex quantities. The nature of any such symbol should be clear from the context in which it is used.

Thus, we may write the matrix equation $\mathbf{I} = \mathbf{yV}$, or

$$\begin{bmatrix}\mathbf{I}_1\\ \mathbf{I}_2\end{bmatrix} = \begin{bmatrix}\mathbf{y}_{11} & \mathbf{y}_{12}\\ \mathbf{y}_{21} & \mathbf{y}_{22}\end{bmatrix}\begin{bmatrix}\mathbf{V}_1\\ \mathbf{V}_2\end{bmatrix}$$

and matrix multiplication of the right-hand side gives us the equality

$$\begin{bmatrix}\mathbf{I}_1\\ \mathbf{I}_2\end{bmatrix} = \begin{bmatrix}\mathbf{y}_{11}\mathbf{V}_1 + \mathbf{y}_{12}\mathbf{V}_2\\ \mathbf{y}_{21}\mathbf{V}_1 + \mathbf{y}_{22}\mathbf{V}_2\end{bmatrix}$$

These (2 × 1) matrices must be equal, element by element, and thus we are led to the defining equations, [5] and [6].

The most useful and informative way to attach a physical meaning to the $\mathbf{y}$ parameters is through a direct inspection of Eqs. [5] and [6]. Consider Eq. [5], for example; if we let $\mathbf{V}_2$ be zero, then we see that $\mathbf{y}_{11}$ must be given by the ratio of $\mathbf{I}_1$ to $\mathbf{V}_1$. We therefore describe $\mathbf{y}_{11}$ as the admittance measured at the input terminals with the output terminals *short-circuited* ($\mathbf{V}_2 = 0$). Since there can be no question which terminals are short-circuited, $\mathbf{y}_{11}$ is best described as the *short-circuit input admittance*. Alternatively, we might describe $\mathbf{y}_{11}$ as the reciprocal of the input impedance measured with the output terminals short-circuited, but a description as an admittance is obviously more direct. It is not the *name* of the parameter that is important. Rather, it is the conditions which must be applied to Eq. [5] or [6], and hence to the network, that are most meaningful; when the conditions are determined, the parameter can be found directly from an analysis of the circuit (or by experiment on the physical circuit). Each of the **y** parameters may be described as a current–voltage ratio with either $\mathbf{V}_1 = 0$ (the input terminals short-circuited) or $\mathbf{V}_2 = 0$ (the output terminals short-circuited):

$$\mathbf{y}_{11} = \left.\frac{\mathbf{I}_1}{\mathbf{V}_1}\right|_{\mathbf{V}_2=0} \quad [10]$$

$$\mathbf{y}_{12} = \left.\frac{\mathbf{I}_1}{\mathbf{V}_2}\right|_{\mathbf{V}_1=0} \quad [11]$$

$$\mathbf{y}_{21} = \left.\frac{\mathbf{I}_2}{\mathbf{V}_1}\right|_{\mathbf{V}_2=0} \quad [12]$$

$$\mathbf{y}_{22} = \left.\frac{\mathbf{I}_2}{\mathbf{V}_2}\right|_{\mathbf{V}_1=0} \quad [13]$$

Because each parameter is an admittance which is obtained by short-circuiting either the output or the input port, the **y** parameters are known as the **short-circuit admittance parameters.** The specific name of $\mathbf{y}_{11}$ is the **short-circuit input admittance,** $\mathbf{y}_{22}$ is the **short-circuit output admittance,** and $\mathbf{y}_{12}$ and $\mathbf{y}_{21}$ are the **short-circuit transfer admittances.**

EXAMPLE 16.4

Find the four short-circuit admittance parameters for the resistive two-port shown in Fig. 16.9.

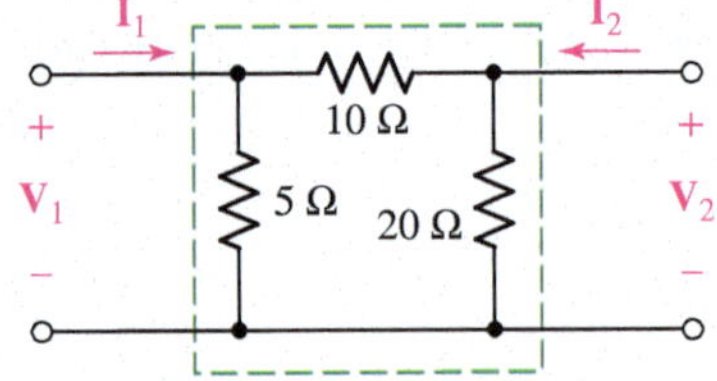

■ **FIGURE 16.9** A resistive two-port.

The values of the parameters may be easily established by applying Eqs. [10] to [13], which we obtained directly from the defining equations, [5] and [6]. To determine $\mathbf{y}_{11}$, we short-circuit the output and find the ratio of $\mathbf{I}_1$ to $\mathbf{V}_1$. This may be done by letting $\mathbf{V}_1 = 1$ V, for then $\mathbf{y}_{11} = \mathbf{I}_1$. By inspection of Fig. 16.9, it is apparent that 1 V applied at the input with the output short-circuited will cause an input current of $\left(\frac{1}{5} + \frac{1}{10}\right)$, or 0.3 A. Hence,

$$\mathbf{y}_{11} = 0.3 \text{ S}$$

In order to find $\mathbf{y}_{12}$, we short-circuit the input terminals and apply 1 V at the output terminals. The input current flows through the short circuit and is $\mathbf{I}_1 = -\frac{1}{10}$ A. Thus

$$\mathbf{y}_{12} = -0.1 \text{ S}$$

By similar methods,

$$\mathbf{y}_{21} = -0.1 \text{ S} \qquad \mathbf{y}_{22} = 0.15 \text{ S}$$

The describing equations for this two-port in terms of the admittance parameters are, therefore,

$$\mathbf{I}_1 = 0.3\,\mathbf{V}_1 - 0.1\,\mathbf{V}_2 \quad [14]$$

$$\mathbf{I}_2 = -0.1\,\mathbf{V}_1 + 0.15\,\mathbf{V}_2 \quad [15]$$

and

$$\mathbf{y} = \begin{bmatrix} 0.3 & -0.1 \\ -0.1 & 0.15 \end{bmatrix} \quad \text{(all S)}$$

It is not necessary to find these parameters one at a time by using Eqs. [10] to [13], however. We may find them all at once—as shown in the next example.

EXAMPLE 16.5

Assign node voltages $\mathbf{V}_1$ and $\mathbf{V}_2$ in the two-port of Fig. 16.9 and write the expressions for $\mathbf{I}_1$ and $\mathbf{I}_2$ in terms of them.

We have

$$\mathbf{I}_1 = \frac{\mathbf{V}_1}{5} + \frac{\mathbf{V}_1 - \mathbf{V}_2}{10} = 0.3\ \mathbf{V}_1 - 0.1\ \mathbf{V}_2$$

and

$$\mathbf{I}_2 = \frac{\mathbf{V}_2 - \mathbf{V}_1}{10} + \frac{\mathbf{V}_2}{20} = -0.1\ \mathbf{V}_1 + 0.15\ \mathbf{V}_2$$

These equations are identical with Eqs. [14] and [15], and the four **y** parameters may be read from them *directly*.

PRACTICE

16.3 By applying the appropriate 1 V sources and short circuits to the circuit shown in Fig. 16.10, find (*a*) $\mathbf{y}_{11}$; (*b*) $\mathbf{y}_{21}$; (*c*) $\mathbf{y}_{22}$; (*d*) $\mathbf{y}_{12}$.

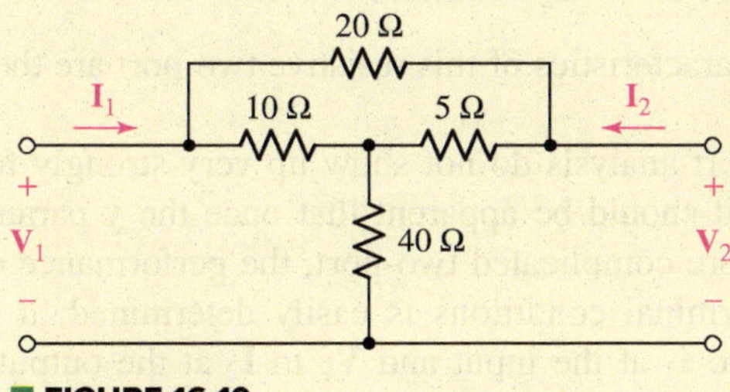

■ **FIGURE 16.10**

Ans: 0.1192 S; −0.1115 S; 0.1269 S; −0.1115 S.

In general, it is easier to use Eq. [10], [11], [12], or [13] when only one parameter is desired. If we need all of them, however, it is usually easier to assign $\mathbf{V}_1$ and $\mathbf{V}_2$ to the input and output nodes, to assign other node-to-reference voltages at any interior nodes, and then to carry through with the general solution.

In order to see what use might be made of such a system of equations, let us now terminate each port with some specific one-port network. Consider the simple two-port network of Example 16.4, shown in Fig. 16.11

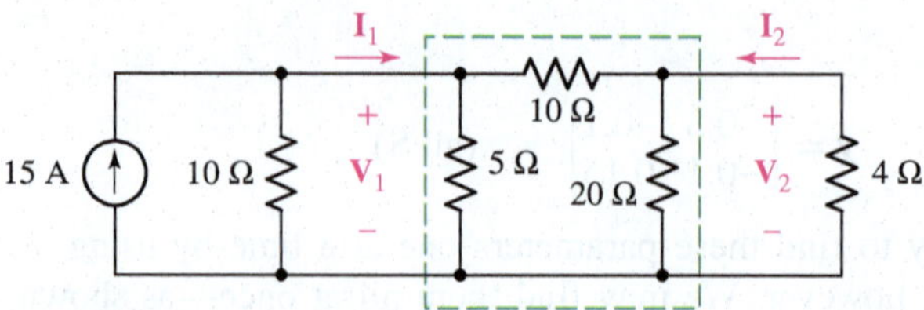

FIGURE 16.11 The resistive two-port network of Fig. 16.9, terminated with specific one-port networks.

with a practical current source connected to the input port and a resistive load connected to the output port. A relationship must now exist between $\mathbf{V}_1$ and $\mathbf{I}_1$ that is independent of the two-port network. This relationship may be determined solely from this external circuit. If we apply KCL (or write a single nodal equation) at the input,

$$\mathbf{I}_1 = 15 - 0.1\,\mathbf{V}_1$$

For the output, Ohm's law yields

$$\mathbf{I}_2 = -0.25\,\mathbf{V}_2$$

Substituting these expressions for $\mathbf{I}_1$ and $\mathbf{I}_2$ in Eqs. [14] and [15], we have

$$15 = 0.4\,\mathbf{V}_1 - 0.1\,\mathbf{V}_2$$
$$0 = -0.1\,\mathbf{V}_1 + 0.4\,\mathbf{V}_2$$

from which are obtained

$$\mathbf{V}_1 = 40 \text{ V} \qquad \mathbf{V}_2 = 10 \text{ V}$$

The input and output currents are also easily found:

$$\mathbf{I}_1 = 11 \text{ A} \qquad \mathbf{I}_2 = -2.5 \text{ A}$$

and the complete terminal characteristics of this resistive two-port are then known.

The advantages of two-port analysis do not show up very strongly for such a simple example, but it should be apparent that once the **y** parameters are determined for a more complicated two-port, the performance of the two-port for different terminal conditions is easily determined; it is necessary only to relate $\mathbf{V}_1$ to $\mathbf{I}_1$ at the input and $\mathbf{V}_2$ to $\mathbf{I}_2$ at the output.

In the example just concluded, $\mathbf{y}_{12}$ and $\mathbf{y}_{21}$ were both found to be −0.1 S. It is not difficult to show that this equality is also obtained if three general impedances $\mathbf{Z}_A$, $\mathbf{Z}_B$, and $\mathbf{Z}_C$ are contained in this Π network. It is somewhat more difficult to determine the specific conditions which are necessary in order that $\mathbf{y}_{12} = \mathbf{y}_{21}$, but the use of determinant notation is of some help. Let us see if the relationships of Eqs. [10] to [13] can be expressed in terms of the impedance determinant and its minors.

Since our concern is with the two-port and not with the specific networks with which it is terminated, we will let $\mathbf{V}_1$ and $\mathbf{V}_2$ be represented by two ideal voltage sources. Equation [10] is applied by letting $\mathbf{V}_2 = 0$ (thus short-circuiting the output) and finding the input admittance. The network now, however, is simply a one-port, and the input impedance of a one-port was found in Sec. 16.1. We select loop **1** to include the input terminals, and let $\mathbf{I}_1$ be that loop's current; we identify $(-\mathbf{I}_2)$ as the loop current in loop **2** and assign the remaining loop currents in any convenient manner. Thus,

$$\mathbf{Z}_{\text{in}}|_{\mathbf{V}_2=0} = \frac{\Delta_\mathbf{Z}}{\Delta_{11}}$$

and, therefore,

$$\mathbf{y}_{11} = \frac{\Delta_{11}}{\Delta_\mathbf{Z}}$$

Similarly,

$$\mathbf{y}_{22} = \frac{\Delta_{22}}{\Delta_\mathbf{Z}}$$

In order to find $\mathbf{y}_{12}$, we let $\mathbf{V}_1 = 0$ and find $\mathbf{I}_1$ as a function of $\mathbf{V}_2$. We find that $\mathbf{I}_1$ is given by the ratio

$$\mathbf{I}_1 = \frac{\begin{vmatrix} 0 & \mathbf{Z}_{12} & \cdots & \mathbf{Z}_{1N} \\ -\mathbf{V}_2 & \mathbf{Z}_{22} & \cdots & \mathbf{Z}_{2N} \\ 0 & \mathbf{Z}_{32} & \cdots & \mathbf{Z}_{3N} \\ \cdots & \cdots & \cdots & \cdots \\ 0 & \mathbf{Z}_{N2} & \cdots & \mathbf{Z}_{NN} \end{vmatrix}}{\begin{vmatrix} \mathbf{Z}_{11} & \mathbf{Z}_{12} & \cdots & \mathbf{Z}_{1N} \\ \mathbf{Z}_{21} & \mathbf{Z}_{22} & \cdots & \mathbf{Z}_{2N} \\ \mathbf{Z}_{31} & \mathbf{Z}_{32} & \cdots & \mathbf{Z}_{3N} \\ \cdots & \cdots & \cdots & \cdots \\ \mathbf{Z}_{N1} & \mathbf{Z}_{N2} & \cdots & \mathbf{Z}_{NN} \end{vmatrix}}$$

Thus,

$$\mathbf{I}_1 = -\frac{(-\mathbf{V}_2)\Delta_{21}}{\Delta_\mathbf{Z}}$$

and

$$\mathbf{y}_{12} = \frac{\Delta_{21}}{\Delta_\mathbf{Z}}$$

In a similar manner, we may show that

$$\mathbf{y}_{21} = \frac{\Delta_{12}}{\Delta_\mathbf{Z}}$$

The equality of $\mathbf{y}_{12}$ and $\mathbf{y}_{21}$ is thus contingent on the equality of the two minors of $\Delta_\mathbf{Z}$–Δ_{12} and Δ_{21}. These two minors are

$$\Delta_{21} = \begin{vmatrix} \mathbf{Z}_{12} & \mathbf{Z}_{13} & \mathbf{Z}_{14} & \cdots & \mathbf{Z}_{1N} \\ \mathbf{Z}_{32} & \mathbf{Z}_{33} & \mathbf{Z}_{34} & \cdots & \mathbf{Z}_{3N} \\ \mathbf{Z}_{42} & \mathbf{Z}_{43} & \mathbf{Z}_{44} & \cdots & \mathbf{Z}_{4N} \\ \cdots & \cdots & \cdots & \cdots & \cdots \\ \mathbf{Z}_{N2} & \mathbf{Z}_{N3} & \mathbf{Z}_{N4} & \cdots & \mathbf{Z}_{NN} \end{vmatrix}$$

and

$$\Delta_{12} = \begin{vmatrix} \mathbf{Z}_{21} & \mathbf{Z}_{23} & \mathbf{Z}_{24} & \cdots & \mathbf{Z}_{2N} \\ \mathbf{Z}_{31} & \mathbf{Z}_{33} & \mathbf{Z}_{34} & \cdots & \mathbf{Z}_{3N} \\ \mathbf{Z}_{41} & \mathbf{Z}_{43} & \mathbf{Z}_{44} & \cdots & \mathbf{Z}_{4N} \\ \cdots & \cdots & \cdots & \cdots & \cdots \\ \mathbf{Z}_{N1} & \mathbf{Z}_{N3} & \mathbf{Z}_{N4} & \cdots & \mathbf{Z}_{NN} \end{vmatrix}$$

Their equality is shown by first interchanging the rows and columns of one minor (for example, Δ_{21}), an operation which any college algebra book proves is valid, and then letting every mutual impedance $\mathbf{Z}_{ij}$ be replaced by $\mathbf{Z}_{ji}$. Thus, we set

$$\mathbf{Z}_{12} = \mathbf{Z}_{21} \qquad \mathbf{Z}_{23} = \mathbf{Z}_{32} \qquad \text{etc.}$$

This equality of $\mathbf{Z}_{ij}$ and $\mathbf{Z}_{ji}$ is evident for the three familiar passive elements—the resistor, capacitor, and inductor—and it is also true for mutual inductance. However, it is *not* true for *every* type of device which we may wish to include inside a two-port network. Specifically, it is not true in general for a dependent source, and it is not true for the gyrator, a useful model for Hall-effect devices and for waveguide sections containing ferrites. Over a narrow range of radian frequencies, the gyrator provides an additional phase shift of 180° for a signal passing from the output to the input over that for a signal in the forward direction, and thus $\mathbf{y}_{12} = -\mathbf{y}_{21}$. A common type of passive element leading to the inequality of $\mathbf{Z}_{ij}$ and $\mathbf{Z}_{ji}$, however, is a nonlinear element.

Any device for which $\mathbf{Z}_{ij} = \mathbf{Z}_{ji}$ is called a *bilateral element,* and a circuit which contains only bilateral elements is called a *bilateral circuit*. We have therefore shown that an important property of a bilateral two-port is

$$\mathbf{y}_{12} = \mathbf{y}_{21}$$

and this property is glorified by stating it as the *reciprocity theorem:*

Another way of stating the theorem is to say that the interchange of an ideal voltage source and an ideal ammeter in any passive, linear, bilateral circuit will not change the ammeter reading.

In any passive linear bilateral network, if the single voltage source $\mathbf{V}_x$ in branch x produces the current response $\mathbf{I}_y$ in branch y, then the removal of the voltage source from branch x and its insertion in branch y will produce the current response $\mathbf{I}_y$ in branch x.

If we had been working with the admittance determinant of the circuit and had proved that the minors Δ_{21} and Δ_{12} of the admittance determinant $\Delta_\mathbf{Y}$ were equal, then we should have obtained the reciprocity theorem in its dual form:

In other words, the interchange of an ideal current source and an ideal voltmeter in any passive linear bilateral circuit will not change the voltmeter reading.

In any passive linear bilateral network, if the single current source $\mathbf{I}_x$ between nodes x and x' produces the voltage response $\mathbf{V}_y$ between nodes y and y', then the removal of the current source from nodes x and x' and its insertion between nodes y and y' will produce the voltage response $\mathbf{V}_y$ between nodes x and x'.

In other words, the interchange of an ideal current source and an ideal voltmeter in any passive linear bilateral circuit will not change the voltmeter reading.

PRACTICE

16.4 In the circuit of Fig. 16.10, let $\mathbf{I}_1$ and $\mathbf{I}_2$ represent ideal current sources. Assign the node voltage $\mathbf{V}_1$ at the input, $\mathbf{V}_2$ at the output, and $\mathbf{V}_x$ from the central node to the reference node. Write three nodal equations, eliminate $\mathbf{V}_x$ to obtain two equations, and then rearrange these equations into the form of Eqs. [5] and [6] so that all four $\mathbf{y}$ parameters may be read directly from the equations.

16.5 Find **y** for the two-port shown in Fig. 16.12.

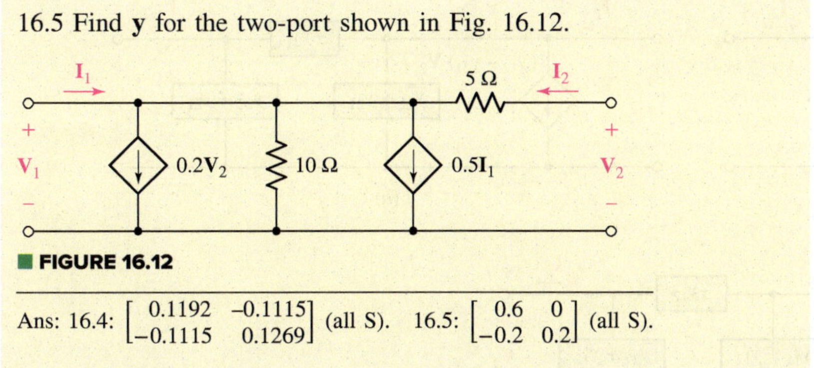

FIGURE 16.12

Ans: 16.4: $\begin{bmatrix} 0.1192 & -0.1115 \\ -0.1115 & 0.1269 \end{bmatrix}$ (all S). 16.5: $\begin{bmatrix} 0.6 & 0 \\ -0.2 & 0.2 \end{bmatrix}$ (all S).

16.3 SOME EQUIVALENT NETWORKS

When analyzing electronic circuits, it is often necessary to replace the active device (and perhaps some of its associated passive circuitry) with an equivalent two-port containing only three or four impedances. The validity of the equivalent may be restricted to small signal amplitudes and a single frequency, or perhaps a limited range of frequencies. The equivalent is also a linear approximation of a nonlinear circuit. However, if we are faced with a network containing a number of resistors, capacitors, and inductors, plus a transistor labeled 2N3823, then we cannot analyze the circuit by any of the techniques we have studied previously; the transistor must first be replaced by a linear model, just as we replaced the op amp by a linear model in Chap. 6. The **y** parameters provide one such model in the form of a two-port network that is often used at high frequencies. Another common linear model for a transistor appears in Sec. 16.5.

The two equations that determine the short-circuit admittance parameters,

$$\mathbf{I}_1 = \mathbf{y}_{11}\,\mathbf{V}_1 + \mathbf{y}_{12}\,\mathbf{V}_2 \qquad [16]$$

$$\mathbf{I}_2 = \mathbf{y}_{21}\,\mathbf{V}_1 + \mathbf{y}_{22}\,\mathbf{V}_2 \qquad [17]$$

have the form of a pair of nodal equations written for a circuit containing two nonreference nodes. The determination of an equivalent circuit that leads to Eqs. [16] and [17] is made more difficult by the inequality, in general, of $\mathbf{y}_{12}$ and $\mathbf{y}_{21}$; it helps to resort to a little trickery in order to obtain a pair of equations that possess equal mutual coefficients. Let us both add and subtract $\mathbf{y}_{12}\mathbf{V}_1$ (the term we would like to see present on the right side of Eq. [17]):

$$\mathbf{I}_2 = \mathbf{y}_{12}\,\mathbf{V}_1 + \mathbf{y}_{22}\,\mathbf{V}_2 + (\mathbf{y}_{21} - \mathbf{y}_{12})\mathbf{V}_1 \qquad [18]$$

or

$$\mathbf{I}_2 - (\mathbf{y}_{21} - \mathbf{y}_{12})\mathbf{V}_1 = \mathbf{y}_{12}\,\mathbf{V}_1 + \mathbf{y}_{22}\,\mathbf{V}_2 \qquad [19]$$

The right-hand sides of Eqs. [16] and [19] now show the proper symmetry for a bilateral circuit; the left-hand side of Eq. [19] may be interpreted as the algebraic sum of two current sources, one an independent source $\mathbf{I}_2$ entering node **2**, and the other a dependent source $(\mathbf{y}_{21} - \mathbf{y}_{12})\mathbf{V}_1$ leaving node **2**.

Let us now "read" the equivalent network from Eqs. [16] and [19]. We first provide a reference node, and then a node labeled $\mathbf{V}_1$ and one

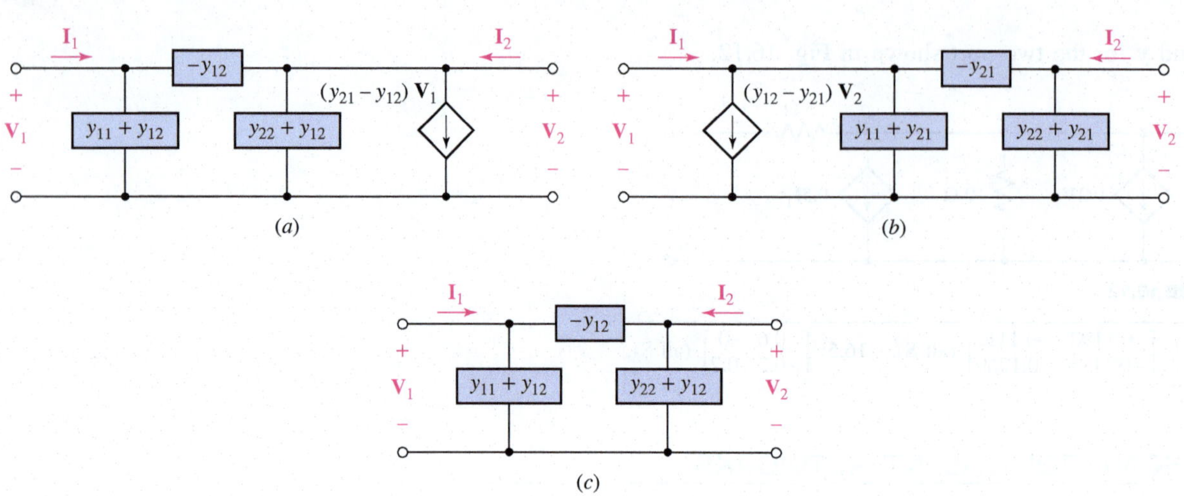

■ **FIGURE 16.13** (*a*, *b*) Two-ports which are equivalent to any general linear two-port. The dependent source in part (*a*) depends on $\mathbf{V}_1$, and that in part (*b*) depends on $\mathbf{V}_2$. (*c*) An equivalent for a bilateral network.

labeled $\mathbf{V}_2$. From Eq. [16], we establish the current $\mathbf{I}_1$ flowing into node **1**, we supply a mutual admittance $(-\mathbf{y}_{12})$ between nodes **1** and **2**, and we supply an admittance of $(\mathbf{y}_{11} + \mathbf{y}_{12})$ between node **1** and the reference node. With $\mathbf{V}_2 = 0$, the ratio of $\mathbf{I}_1$ to $\mathbf{V}_1$ is then $\mathbf{y}_{11}$, as it should be. Now consider Eq. [19]; we cause the current $\mathbf{I}_2$ to flow into the second node, we cause the current $(\mathbf{y}_{21} - \mathbf{y}_{12})\mathbf{V}_1$ to leave the node, we note that the proper admittance $(-\mathbf{y}_{12})$ exists between the nodes, and we complete the circuit by installing the admittance $(\mathbf{y}_{22} + \mathbf{y}_{12})$ from node **2** to the reference node. The completed circuit is shown in Fig. 16.13*a*.

Another form of equivalent network is obtained by subtracting and adding $\mathbf{y}_{21}\,\mathbf{V}_2$ in Eq. [16]; this equivalent circuit is shown in Fig. 16.13*b*. If the two-port is bilateral, then $\mathbf{y}_{12} = \mathbf{y}_{21}$, and either of the equivalents reduces to a simple passive Π network. The dependent source disappears. This equivalent of the bilateral two-port is shown in Fig. 16.13*c*.

There are several uses to which these equivalent circuits may be put. In the first place, we have succeeded in showing that an equivalent of any complicated linear two-port *exists*. It does not matter how many nodes or loops are contained within the network; the equivalent is no more complex than the circuits of Fig. 16.13. One of these may be much simpler to use than the given circuit if we are interested only in the terminal characteristics of the given network.

The three-terminal network shown in Fig. 16.14*a* is often referred to as a Δ of impedances, while that in Fig. 16.14*b* is called a Y. One network may be replaced by the other if certain specific relationships between the impedances are satisfied, and these interrelationships may be established by the use of the **y** parameters. We find that

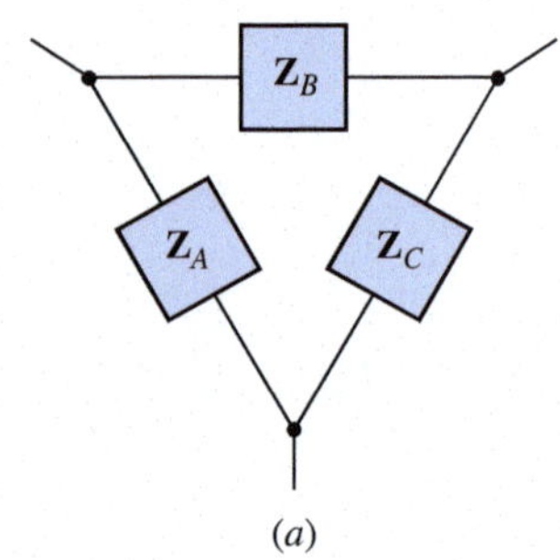

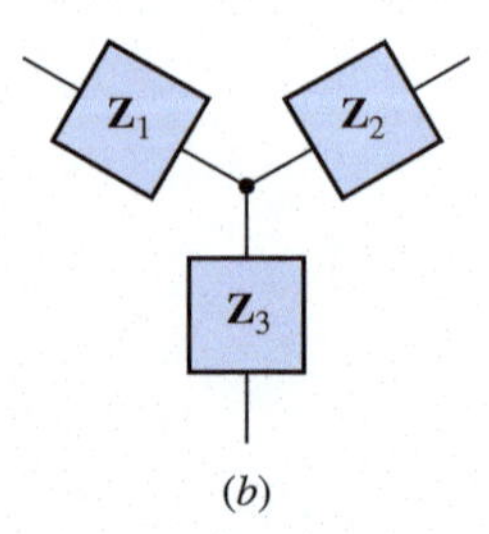

■ **FIGURE 16.14** The three-terminal Δ network (*a*) and the three-terminal Y network (*b*) are equivalent if the six impedances satisfy the conditions of the Y-Δ (or Π-T) transformation, Eqs. [20] to [25].

$$\mathbf{y}_{11} = \frac{1}{\mathbf{Z}_A} + \frac{1}{\mathbf{Z}_B} = \frac{1}{\mathbf{Z}_1 + \mathbf{Z}_2\mathbf{Z}_3/(\mathbf{Z}_2 + \mathbf{Z}_3)}$$

$$\mathbf{y}_{12} = \mathbf{y}_{21} = -\frac{1}{\mathbf{Z}_B} = \frac{-\mathbf{Z}_3}{\mathbf{Z}_1\mathbf{Z}_2 + \mathbf{Z}_2\mathbf{Z}_3 + \mathbf{Z}_3\mathbf{Z}_1}$$

$$\mathbf{y}_{22} = \frac{1}{\mathbf{Z}_C} + \frac{1}{\mathbf{Z}_B} = \frac{1}{\mathbf{Z}_2 + \mathbf{Z}_1\mathbf{Z}_3/(\mathbf{Z}_1 + \mathbf{Z}_3)}$$

These equations may be solved for $\mathbf{Z}_A$, $\mathbf{Z}_B$, and $\mathbf{Z}_C$ in terms of $\mathbf{Z}_1$, $\mathbf{Z}_2$, and $\mathbf{Z}_3$:

$$\mathbf{Z}_A = \frac{\mathbf{Z}_1\mathbf{Z}_2 + \mathbf{Z}_2\mathbf{Z}_3 + \mathbf{Z}_3\mathbf{Z}_1}{\mathbf{Z}_2} \qquad [20]$$

$$\mathbf{Z}_B = \frac{\mathbf{Z}_1\mathbf{Z}_2 + \mathbf{Z}_2\mathbf{Z}_3 + \mathbf{Z}_3\mathbf{Z}_1}{\mathbf{Z}_3} \qquad [21]$$

$$\mathbf{Z}_C = \frac{\mathbf{Z}_1\mathbf{Z}_2 + \mathbf{Z}_2\mathbf{Z}_3 + \mathbf{Z}_3\mathbf{Z}_1}{\mathbf{Z}_1} \qquad [22]$$

or, for the inverse relationships:

$$\mathbf{Z}_1 = \frac{\mathbf{Z}_A\mathbf{Z}_B}{\mathbf{Z}_A + \mathbf{Z}_B + \mathbf{Z}_C} \qquad [23]$$

$$\mathbf{Z}_2 = \frac{\mathbf{Z}_B\mathbf{Z}_C}{\mathbf{Z}_A + \mathbf{Z}_B + \mathbf{Z}_C} \qquad [24]$$

$$\mathbf{Z}_3 = \frac{\mathbf{Z}_C\mathbf{Z}_A}{\mathbf{Z}_A + \mathbf{Z}_B + \mathbf{Z}_C} \qquad [25]$$

These equations enable us to transform easily between the equivalent Y and Δ networks, a process known as the Y–Δ transformation (or Π–T transformation if the networks are drawn in the forms of those letters). In going from Y to Δ, Eqs. [20] to [22], first find the value of the common numerator as the sum of the products of the impedances in the Y taken two at a time. Each impedance in the Δ is then found by dividing the numerator by the impedance of that element in the Y which has no common node with the desired Δ element. Conversely, given the Δ, first take the sum of the three impedances around the Δ; then divide the product of the two Δ impedances having a common node with the desired Y element by that sum.

The reader may recall these useful relationships from Chap. 5, where their derivation was described.

These transformations are often useful in simplifying passive networks, particularly resistive ones, thus avoiding the need for any mesh or nodal analysis.

EXAMPLE **16.6**

Find the input resistance of the circuit shown in Fig. 16.15*a*.

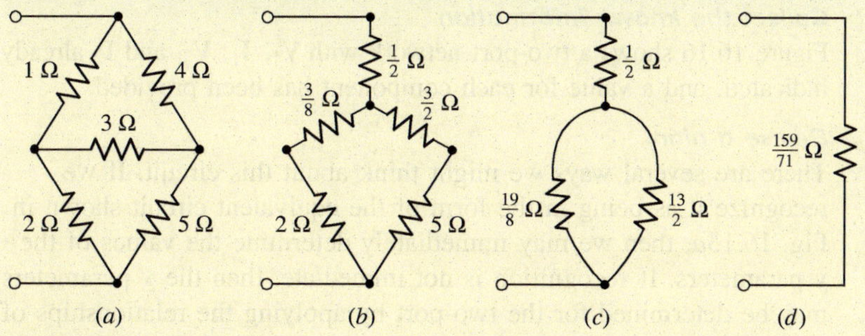

■ **FIGURE 16.15** (*a*) A resistive network whose input resistance is desired. This example is repeated from Chap. 5. (*b*) The upper Δ is replaced by an equivalent Y. (*c*, *d*) Series and parallel combinations give the equivalent input resistance $\frac{159}{71}$ Ω.

(Continued on next page)

We first make a Δ–Y transformation on the upper Δ appearing in Fig. 16.15*a*. The sum of the three resistances forming this Δ is $1 + 4 + 3 = 8\ \Omega$. The product of the two resistors connected to the top node is $1 \times 4 = 4\ \Omega^2$. Thus, the upper resistor of the **Y** is $\frac{4}{8}$, or $\frac{1}{2}\ \Omega$. Repeating this procedure for the other two resistors, we obtain the network shown in Fig. 16.15*b*.

We next make the series and parallel combinations indicated, obtaining in succession Fig. 16.15*c* and *d*. Thus, the input resistance of the circuit in Fig. 16.15*a* is found to be $\frac{159}{71}$ or 2.24 Ω.

Now let us tackle a slightly more complicated example, shown as Fig. 16.16. We note that the circuit contains a dependent source, and thus the Y–Δ transformation is not applicable.

EXAMPLE 16.7

The circuit shown in Fig. 16.16 is an approximate linear equivalent of a transistor amplifier in which the emitter terminal is the bottom node, the base terminal is the upper input node, and the collector terminal is the upper output node. A 2000 Ω resistor is connected between collector and base for some special application and makes the analysis of the circuit more difficult. Determine the y parameters for this circuit.

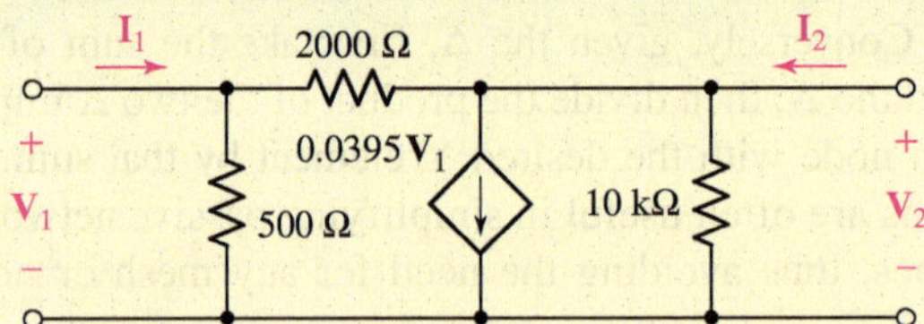

FIGURE 16.16 The linear equivalent circuit of a transistor in common emitter configuration with resistive feedback between collector and base.

▶ ***Identify the goal of the problem.***

Cutting through the problem-specific jargon, we realize that we have been presented with a two-port network and require the **y** parameters.

▶ ***Collect the known information.***

Figure 16.16 shows a two-port network with $\mathbf{V}_1$, $\mathbf{I}_1$, $\mathbf{V}_2$, and $\mathbf{I}_2$ already indicated, and a value for each component has been provided.

▶ ***Devise a plan.***

There are several ways we might think about this circuit. If we recognize it as being in the form of the equivalent circuit shown in Fig. 16.13*a*, then we may immediately determine the values of the **y** parameters. If recognition is not immediate, then the **y** parameters may be determined for the two-port by applying the relationships of Eqs. [10] to [13]. We also might avoid any use of two-port analysis methods and write equations directly for the circuit as it stands. The first option seems best in this case.

▶ ***Construct an appropriate set of equations.***

By inspection, we find that $-\mathbf{y}_{21}$ corresponds to the admittance of our 2 kΩ resistor, that $\mathbf{y}_{11} + \mathbf{y}_{12}$ corresponds to the admittance of the 500 Ω resistor, the gain of the dependent current source corresponds to $\mathbf{y}_{21} - \mathbf{y}_{12}$, and finally that $\mathbf{y}_{22} + \mathbf{y}_{12}$ corresponds to the admittance of the 10 kΩ resistor. Hence we may write

$$\mathbf{y}_{12} = -\frac{1}{2000}$$

$$\mathbf{y}_{11} = \frac{1}{500} - \mathbf{y}_{12}$$

$$\mathbf{y}_{21} = 0.0395 + \mathbf{y}_{12}$$

$$\mathbf{y}_{22} = \frac{1}{10{,}000} - \mathbf{y}_{12}$$

▶ ***Determine if additional information is required.***

With the equations written as they are, we see that once $\mathbf{y}_{12}$ is computed, the remaining $\mathbf{y}$ parameters may also be obtained.

▶ ***Attempt a solution.***

Plugging the numbers into a calculator, we find that

$$\mathbf{y}_{12} = -\frac{1}{2000} = -0.5 \text{ mS}$$

$$\mathbf{y}_{11} = \frac{1}{500} - \left(-\frac{1}{2000}\right) = 2.5 \text{ mS}$$

$$\mathbf{y}_{22} = \frac{1}{10{,}000} - \left(-\frac{1}{2000}\right) = 0.6 \text{ mS}$$

and

$$\mathbf{y}_{21} = 0.0395 + \left(-\frac{1}{2000}\right) = 39 \text{ mS}$$

The following equations must then apply:

$$\mathbf{I}_1 = 2.5\,\mathbf{V}_1 - 0.5\,\mathbf{V}_2 \qquad [26]$$

$$\mathbf{I}_2 = 39\,\mathbf{V}_1 + 0.6\,\mathbf{V}_2 \qquad [27]$$

where we are now using units of mA, V, and mS or kΩ.

▶ ***Verify the solution. Is it reasonable or expected?***

Writing two nodal equations directly from the circuit, we find

$$\mathbf{I}_1 = \frac{\mathbf{V}_1 - \mathbf{V}_2}{2} + \frac{\mathbf{V}_1}{0.5} \quad \text{or} \quad \mathbf{I}_1 = 2.5\,\mathbf{V}_1 - 0.5\,\mathbf{V}_2$$

and

$$-39.5\,\mathbf{V}_1 + \mathbf{I}_2 = \frac{\mathbf{V}_2 - \mathbf{V}_1}{2} + \frac{\mathbf{V}_2}{10} \quad \text{or} \quad \mathbf{I}_2 = 39\,\mathbf{V}_1 + 0.6\,\mathbf{V}_2$$

which agree with Eqs. [26] and [27] obtained directly from the $\mathbf{y}$ parameters.

Now let us make use of Eqs. [26] and [27] by analyzing the performance of the two-port in Fig. 16.16 under several different operating conditions. We first provide a current source of $1\underline{/0^\circ}$ mA at the input and connect a 0.5 kΩ(2 mS) load to the output. The terminating networks are therefore both one-ports and give us the following specific information relating $\mathbf{I}_1$ to $\mathbf{V}_1$ and $\mathbf{I}_2$ to $\mathbf{V}_2$:

$$\mathbf{I}_1 = 1 \text{ (for any } \mathbf{V}_1) \qquad \mathbf{I}_2 = -2\,\mathbf{V}_2$$

We now have four equations in the four variables, $\mathbf{V}_1$, $\mathbf{V}_2$, $\mathbf{I}_1$, and $\mathbf{I}_2$. Substituting the two one-port relationships in Eqs. [26] and [27], we obtain two equations relating $\mathbf{V}_1$ and $\mathbf{V}_2$:

$$1 = 2.5\,\mathbf{V}_1 - 0.5\,\mathbf{V}_2 \qquad 0 = 39\,\mathbf{V}_1 + 2.6\,\mathbf{V}_2$$

Solving, we find that

$$\mathbf{V}_1 = 0.1 \text{ V} \qquad \mathbf{V}_2 = -1.5 \text{ V}$$

$$\mathbf{I}_1 = 1 \text{ mA} \qquad \mathbf{I}_2 = 3 \text{ mA}$$

These four values apply to the two-port operating with a prescribed input ($\mathbf{I}_1 = 1$ mA) and a specified load ($R_L = 0.5$ kΩ).

The performance of an amplifier is often described by giving a few specific values. Let us calculate four of these values for this two-port with its terminations. We will define and evaluate the voltage gain, the current gain, the power gain, and the input impedance.

The *voltage gain* $\mathbf{G}_V$ is

$$\mathbf{G}_V = \frac{\mathbf{V}_2}{\mathbf{V}_1}$$

From the numerical results, it is easy to see that $\mathbf{G}_V = -15$.

The *current gain* $\mathbf{G}_I$ is defined as

$$\mathbf{G}_I = \frac{\mathbf{I}_2}{\mathbf{I}_1}$$

and we have

$$\mathbf{G}_I = 3$$

Let us define and calculate the *power gain* $\mathbf{G}_P$ for an assumed sinusoidal excitation. We have

$$\mathbf{G}_P = \frac{P_{\text{out}}}{P_{\text{in}}} = \frac{\text{Re}\left\{-\frac{1}{2}\mathbf{V}_2\mathbf{I}_2^*\right\}}{\text{Re}\left\{\frac{1}{2}\mathbf{V}_1\mathbf{I}_1^*\right\}} = 45$$

The device might be termed either a voltage, a current, or a power amplifier, since all the gains are greater than unity. If the 2 kΩ resistor were removed, the power gain would rise to 354.

The input and output impedances of the amplifier are often desired in order that maximum power transfer may be achieved to or from an adjacent two-port. We define the *input impedance* $\mathbf{Z}_{\text{in}}$ as the ratio of input voltage to current:

$$\mathbf{Z}_{\text{in}} = \frac{\mathbf{V}_1}{\mathbf{I}_1} = 0.1 \text{ k}\Omega$$

This is the impedance offered to the current source when the 500 Ω load is connected to the output. (With the output short-circuited, the input impedance is necessarily $1/\mathbf{y}_{11}$, or 400 Ω.)

It should be noted that the input impedance *cannot* be determined by replacing every source with its internal impedance and then combining resistances or conductances. In the given circuit, such a procedure would yield a value of 416 Ω. The error, of course, comes from treating the *dependent* source as an *independent* source. If we think of the input impedance as being numerically equal to the input voltage produced by an input current of 1 A, the application of the 1 A source produces some input voltage $\mathbf{V}_1$, and the strength of the dependent source ($0.0395\,\mathbf{V}_1$) cannot be zero. We should recall that when we obtain the Thévenin equivalent impedance of a circuit containing a dependent source along with one or more independent sources, we must replace the independent sources with short circuits or open circuits, but a dependent source must not be deactivated. Of course, if the voltage or current on which the dependent source depends is zero, then the dependent source will itself be inactive; occasionally a circuit may be simplified by recognizing such an occurrence.

Besides $\mathbf{G}_V$, $\mathbf{G}_I$, $\mathbf{G}_P$, and $\mathbf{Z}_{\text{in}}$, there is one other performance parameter that is quite useful. This is the *output impedance* $\mathbf{Z}_{\text{out}}$, and it is determined for a different circuit configuration.

The output impedance is just another term for the Thévenin impedance appearing in the Thévenin equivalent circuit of that portion of the network faced by the load. In our circuit, which we have assumed is driven by a $1\underline{/0^\circ}$ mA current source, we therefore replace this independent source with an open circuit, leave the dependent source alone, and seek the *input* impedance seen looking to the left from the output terminals (with the load removed). Thus, we define

$$\mathbf{Z}_{\text{out}} = \mathbf{V}_2\big|_{\mathbf{I}_2=1\text{ A with all other independent sources deactivated } and\ R_L \text{ removed}}$$

We therefore remove the load resistor, apply $1\underline{/0^\circ}$ mA (since we are working in V, mA, and kΩ) at the output terminals, and determine $\mathbf{V}_2$. We place these requirements on Eqs. [26] and [27] and obtain

$$0 = 2.5\,\mathbf{V}_1 - 0.5\,\mathbf{V}_2 \qquad 1 = 39\,\mathbf{V}_1 + 0.6\,\mathbf{V}_2$$

Solving,

$$\mathbf{V}_2 = 0.1190\text{ V}$$

and thus

$$\mathbf{Z}_{\text{out}} = 0.1190\text{ k}\Omega$$

An alternative procedure might be to find the open-circuit output voltage and the short-circuit output current. That is, the Thévenin impedance is the output impedance:

$$\mathbf{Z}_{\text{out}} = \mathbf{Z}_{\text{th}} = -\frac{\mathbf{V}_{2\text{oc}}}{\mathbf{I}_{2\text{sc}}}$$

Carrying out this procedure, we first rekindle the independent source so that $\mathbf{I}_1 = 1$ mA, and then open-circuit the load so that $\mathbf{I}_2 = 0$. We have

$$1 = 2.5\,\mathbf{V}_1 - 0.5\,\mathbf{V}_2 \qquad 0 = 39\,\mathbf{V}_1 + 0.6\,\mathbf{V}_2$$

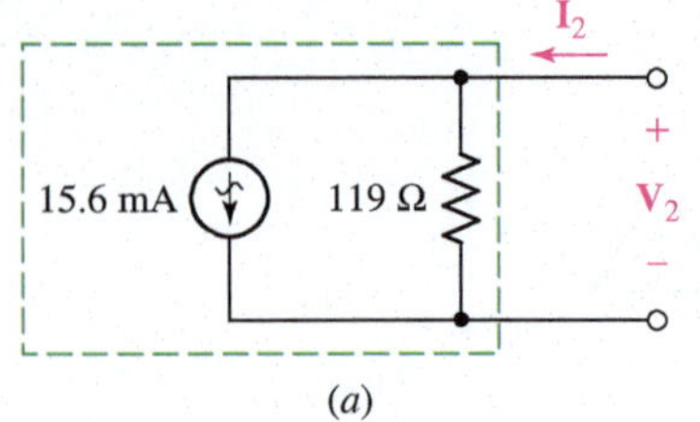

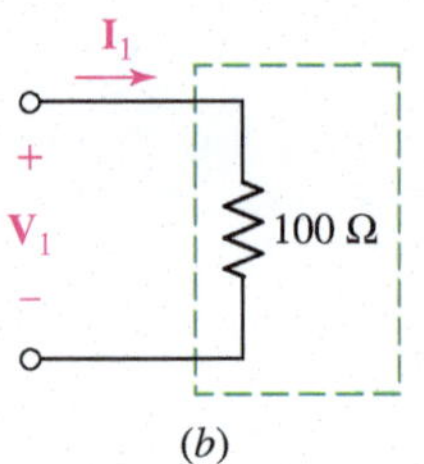

FIGURE 16.17 (a) The Norton equivalent of the network in Fig. 16.16 to the left of the output terminal, $\mathbf{I}_1 = 1\underline{/0°}$ mA. (b) The Thévenin equivalent of that portion of the network to the right of the input terminals, if $\mathbf{I}_2 = -2\mathbf{V}_2$ mA.

and thus

$$\mathbf{V}_{2oc} = -1.857 \text{ V}$$

Next, we apply short-circuit conditions by setting $\mathbf{V}_2 = 0$ and again let $\mathbf{I}_1 = 1$ mA. We find that

$$\mathbf{I}_1 = 1 = 2.5\,\mathbf{V}_1 - 0 \qquad \mathbf{I}_2 = 39\,\mathbf{V}_1 + 0$$

and thus

$$\mathbf{I}_{2sc} = 15.6 \text{ mA}$$

The assumed directions of $\mathbf{V}_2$ and $\mathbf{I}_2$ therefore result in a Thévenin or output impedance

$$\mathbf{Z}_{out} = -\frac{\mathbf{V}_{2oc}}{\mathbf{I}_{2sc}} = -\frac{-1.857}{15.6} = 0.1190 \text{ k}\Omega$$

as before.

We now have enough information to enable us to draw the Thévenin or Norton equivalent of the two-port of Fig. 16.16 when it is driven by a $1\underline{/0°}$ mA current source and terminated in a 500 Ω load. Thus, the Norton equivalent presented to the load must contain a current source equal to the short-circuit current $\mathbf{I}_{2sc}$ in parallel with the output impedance; this equivalent is shown in Fig. 16.17*a*. Also, the Thévenin equivalent offered to the $1\underline{/0°}$ mA input source must consist solely of the input impedance, as drawn in Fig. 16.17*b*.

Before leaving the **y** parameters, we should recognize their usefulness in describing the parallel connection of two-ports, as indicated in Fig. 16.18. When we first defined a port in Sec. 16.1, we noted that the currents entering and leaving the two terminals of a port had to be equal, and there could be no external connections made that bridged between ports. Apparently the parallel connection shown in Fig. 16.18 violates this condition. However, if each two-port has a reference node that is common to its input and output port, and if the two-ports are connected in parallel so that they have a common reference node, then all ports remain ports after the connection. Thus, for the *A* network,

$$\mathbf{I}_A = \mathbf{y}_A \mathbf{V}_A$$

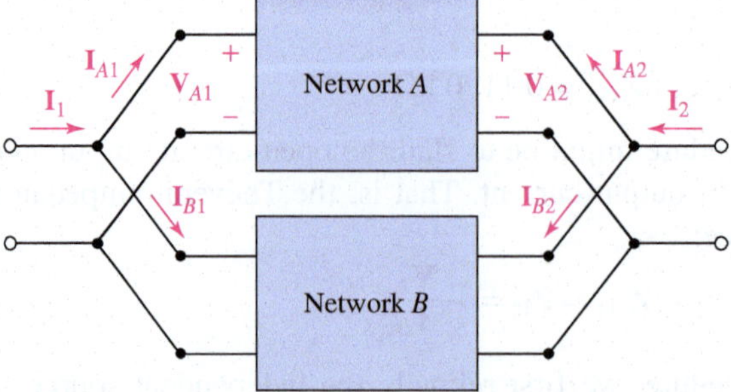

FIGURE 16.18 The parallel connection of two two-port networks. If both inputs and outputs have the same reference node, then the admittance matrix $\mathbf{y} = \mathbf{y}_A + \mathbf{y}_B$.

where

$$\mathbf{I}_A = \begin{bmatrix} \mathbf{I}_{A1} \\ \mathbf{I}_{A2} \end{bmatrix} \quad \text{and} \quad \mathbf{V}_A = \begin{bmatrix} \mathbf{V}_{A1} \\ \mathbf{V}_{A2} \end{bmatrix}$$

and for the B network

$$\mathbf{I}_B = \mathbf{y}_B \mathbf{V}_B$$

But

$$\mathbf{V}_A = \mathbf{V}_B = \mathbf{V} \quad \text{and} \quad \mathbf{I} = \mathbf{I}_A + \mathbf{I}_B$$

Thus,

$$\mathbf{I} = (\mathbf{y}_A + \mathbf{y}_B)\mathbf{V}$$

and we see that each $\mathbf{y}$ parameter of the parallel network is given as the sum of the corresponding parameters of the individual networks,

$$\mathbf{y} = \mathbf{y}_A + \mathbf{y}_B \qquad [28]$$

This may be extended to any number of two-ports connected in parallel.

PRACTICE

16.6 Find $\mathbf{y}$ and $\mathbf{Z}_{out}$ for the terminated two-port shown in Fig. 16.19.

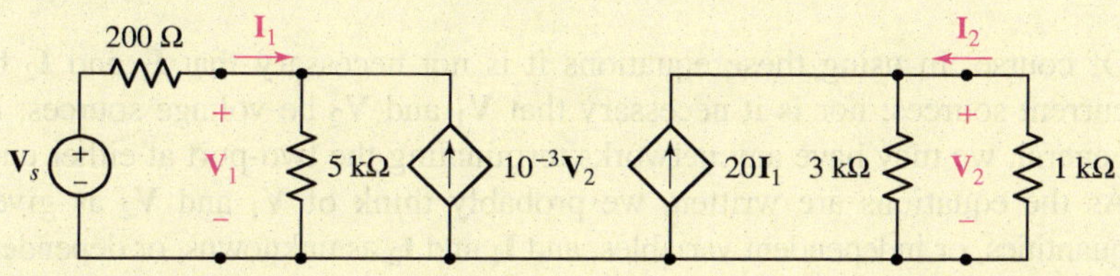

■ **FIGURE 16.19**

16.7 Use Δ−Y and Y−Δ transformations to determine R_{in} for the network shown in (*a*) Fig. 16.20*a*; (*b*) Fig. 16.20*b*.

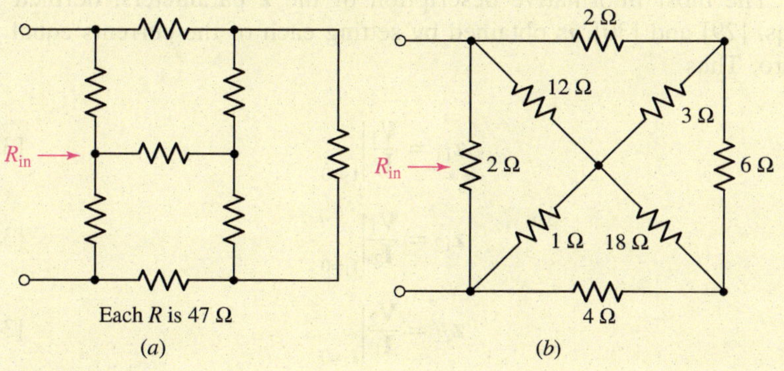

■ **FIGURE 16.20**

Ans: 16.6: $\begin{bmatrix} 2 \times 10^{-4} & -10^{-3} \\ -4 \times 10^{-3} & 20.3 \times 10^{-3} \end{bmatrix}$ (S); 51.1 Ω. 16.7: 53.71 Ω, 1.311 Ω.

16.4 IMPEDANCE PARAMETERS

The concept of two-port parameters has been introduced in terms of the short-circuit admittance parameters. There are other sets of parameters, however, and each set is associated with a particular class of networks for which its use provides the simplest analysis. We will consider three other types of parameters: the open-circuit impedance parameters, which are the subject of this section; and the hybrid and the transmission parameters, which are discussed in following sections.

We begin again with a general linear two-port that does not contain any independent sources; the currents and voltages are assigned as before (Fig. 16.8). Now let us consider the voltage $\mathbf{V}_1$ as the response produced by two current sources $\mathbf{I}_1$ and $\mathbf{I}_2$. We thus write for $\mathbf{V}_1$

$$\mathbf{V}_1 = \mathbf{z}_{11}\mathbf{I}_1 + \mathbf{z}_{12}\mathbf{I}_2 \qquad [29]$$

and for $\mathbf{V}_2$

$$\mathbf{V}_2 = \mathbf{z}_{21}\mathbf{I}_1 + \mathbf{z}_{22}\mathbf{I}_2 \qquad [30]$$

or

$$\mathbf{V} = \begin{bmatrix}\mathbf{V}_1\\ \mathbf{V}_2\end{bmatrix} = \mathbf{zI} = \begin{bmatrix}\mathbf{z}_{11} & \mathbf{z}_{12}\\ \mathbf{z}_{21} & \mathbf{z}_{22}\end{bmatrix}\begin{bmatrix}\mathbf{I}_1\\ \mathbf{I}_2\end{bmatrix} \qquad [31]$$

Of course, in using these equations it is not necessary that $\mathbf{I}_1$ and $\mathbf{I}_2$ be current sources; nor is it necessary that $\mathbf{V}_1$ and $\mathbf{V}_2$ be voltage sources. In general, we may have any networks terminating the two-port at either end. As the equations are written, we probably think of $\mathbf{V}_1$ and $\mathbf{V}_2$ as given quantities, or independent variables, and $\mathbf{I}_1$ and $\mathbf{I}_2$ as unknowns, or dependent variables.

The six ways in which two equations may be written to relate these four quantities define the different systems of parameters. We study the four most important of these six systems of parameters.

The most informative description of the **z** parameters, defined in Eqs. [29] and [30], is obtained by setting each of the currents equal to zero. Thus

$$\mathbf{z}_{11} = \left.\frac{\mathbf{V}_1}{\mathbf{I}_1}\right|_{\mathbf{I}_2=0} \qquad [32]$$

$$\mathbf{z}_{12} = \left.\frac{\mathbf{V}_1}{\mathbf{I}_2}\right|_{\mathbf{I}_1=0} \qquad [33]$$

$$\mathbf{z}_{21} = \left.\frac{\mathbf{V}_2}{\mathbf{I}_1}\right|_{\mathbf{I}_2=0} \qquad [34]$$

$$\mathbf{z}_{22} = \left.\frac{\mathbf{V}_2}{\mathbf{I}_2}\right|_{\mathbf{I}_1=0} \qquad [35]$$

Since zero current results from an open-circuit termination, the **z** parameters are known as the *open-circuit impedance parameters*. They are easily related to the short-circuit admittance parameters by solving Eqs. [29] and [30] for $\mathbf{I}_1$ and $\mathbf{I}_2$:

$$\mathbf{I}_1 = \frac{\begin{vmatrix} \mathbf{V}_1 & \mathbf{z}_{12} \\ \mathbf{V}_2 & \mathbf{z}_{22} \end{vmatrix}}{\begin{vmatrix} \mathbf{z}_{11} & \mathbf{z}_{12} \\ \mathbf{z}_{21} & \mathbf{z}_{22} \end{vmatrix}}$$

or

$$\mathbf{I}_1 = \left(\frac{\mathbf{z}_{22}}{\mathbf{z}_{11}\mathbf{z}_{22} - \mathbf{z}_{12}\mathbf{z}_{21}}\right)\mathbf{V}_1 - \left(\frac{\mathbf{z}_{12}}{\mathbf{z}_{11}\mathbf{z}_{22} - \mathbf{z}_{12}\mathbf{z}_{21}}\right)\mathbf{V}_2$$

Using determinant notation, and being careful that the subscript is a lowercase **z**, we assume that $\Delta\mathbf{Z} \neq 0$ and obtain

$$\mathbf{y}_{11} = \frac{\Delta_{11}}{\Delta_\mathbf{z}} = \frac{\mathbf{z}_{22}}{\Delta_\mathbf{z}} \qquad \mathbf{y}_{12} = -\frac{\Delta_{21}}{\Delta_\mathbf{z}} = -\frac{\mathbf{z}_{12}}{\Delta_\mathbf{z}}$$

and from solving for $\mathbf{I}_2$,

$$\mathbf{y}_{21} = -\frac{\Delta_{12}}{\Delta_\mathbf{z}} = -\frac{\mathbf{z}_{21}}{\Delta_\mathbf{z}} \qquad \mathbf{y}_{22} = \frac{\Delta_{22}}{\Delta_\mathbf{z}} = \frac{\mathbf{z}_{11}}{\Delta_\mathbf{z}}$$

In a similar manner, the **z** parameters may be expressed in terms of the admittance parameters. Transformations of this nature are possible between any of the various parameter systems, and quite a collection of occasionally useful formulas may be obtained. Transformations between the **y** and **z** parameters (as well as the **h** and **t** parameters which we will consider in the following sections) are given in Table 16.1 as a helpful reference.

TABLE **16.1** Transformations Between **y, z, h,** and **t** Parameters

	y		z		h		t	
y	$\mathbf{y}_{11}$	$\mathbf{y}_{12}$	$\frac{\mathbf{z}_{22}}{\Delta_\mathbf{z}}$	$\frac{-\mathbf{z}_{12}}{\Delta_\mathbf{z}}$	$\frac{1}{\mathbf{h}_{11}}$	$\frac{-\mathbf{h}_{12}}{\mathbf{h}_{11}}$	$\frac{\mathbf{t}_{22}}{\mathbf{t}_{12}}$	$\frac{-\Delta_t}{\mathbf{t}_{12}}$
	$\mathbf{y}_{21}$	$\mathbf{y}_{22}$	$\frac{-\mathbf{z}_{21}}{\Delta_\mathbf{z}}$	$\frac{\mathbf{z}_{11}}{\Delta_\mathbf{z}}$	$\frac{\mathbf{h}_{21}}{\mathbf{h}_{11}}$	$\frac{\Delta_\mathbf{h}}{\mathbf{h}_{11}}$	$\frac{-1}{\mathbf{t}_{12}}$	$\frac{\mathbf{t}_{11}}{\mathbf{t}_{12}}$
z	$\frac{\mathbf{y}_{22}}{\Delta_\mathbf{y}}$	$\frac{-\mathbf{y}_{12}}{\Delta_\mathbf{y}}$	$\mathbf{z}_{11}$	$\mathbf{z}_{12}$	$\frac{\Delta_\mathbf{h}}{\mathbf{h}_{22}}$	$\frac{\mathbf{h}_{12}}{\mathbf{h}_{22}}$	$\frac{\mathbf{t}_{11}}{\mathbf{t}_{21}}$	$\frac{\Delta_t}{\mathbf{t}_{21}}$
	$\frac{-\mathbf{y}_{21}}{\Delta_\mathbf{y}}$	$\frac{\mathbf{y}_{11}}{\Delta_\mathbf{y}}$	$\mathbf{z}_{21}$	$\mathbf{z}_{22}$	$\frac{-\mathbf{h}_{21}}{\mathbf{h}_{22}}$	$\frac{1}{\mathbf{h}_{22}}$	$\frac{1}{\mathbf{t}_{21}}$	$\frac{\mathbf{t}_{22}}{\mathbf{t}_{21}}$
h	$\frac{1}{\mathbf{y}_{11}}$	$\frac{-\mathbf{y}_{12}}{\mathbf{y}_{11}}$	$\frac{\Delta_\mathbf{z}}{\mathbf{z}_{22}}$	$\frac{\mathbf{z}_{12}}{\mathbf{z}_{22}}$	$\mathbf{h}_{11}$	$\mathbf{h}_{12}$	$\frac{\mathbf{t}_{12}}{\mathbf{t}_{22}}$	$\frac{\Delta_\mathbf{t}}{\mathbf{t}_{22}}$
	$\frac{\mathbf{y}_{21}}{\mathbf{y}_{11}}$	$\frac{\Delta_\mathbf{y}}{\mathbf{y}_{11}}$	$\frac{-\mathbf{z}_{21}}{\mathbf{z}_{22}}$	$\frac{1}{\mathbf{z}_{22}}$	$\mathbf{h}_{21}$	$\mathbf{h}_{22}$	$\frac{-1}{\mathbf{t}_{22}}$	$\frac{\mathbf{t}_{21}}{\mathbf{t}_{22}}$
t	$\frac{-\mathbf{y}_{22}}{\mathbf{y}_{21}}$	$\frac{-1}{\mathbf{z}_{21}}$	$\frac{\mathbf{z}_{11}}{\mathbf{z}_{21}}$	$\frac{\Delta_\mathbf{z}}{\mathbf{z}_{21}}$	$\frac{-\Delta_\mathbf{h}}{\mathbf{h}_{21}}$	$\frac{-\mathbf{h}_{11}}{\mathbf{h}_{21}}$	$\mathbf{t}_{11}$	$\mathbf{t}_{12}$
	$\frac{-\Delta_\mathbf{y}}{\mathbf{y}_{21}}$	$\frac{-\mathbf{y}_{11}}{\mathbf{y}_{21}}$	$\frac{1}{\mathbf{z}_{21}}$	$\frac{\mathbf{z}_{22}}{\mathbf{z}_{21}}$	$\frac{-\mathbf{h}_{22}}{\mathbf{h}_{21}}$	$\frac{-1}{\mathbf{h}_{21}}$	$\mathbf{t}_{21}$	$\mathbf{t}_{22}$

For all parameter sets: $\Delta_\mathbf{p} = \mathbf{p}_{11}\mathbf{p}_{22} - \mathbf{p}_{12}\mathbf{p}_{21}$.

If the two-port is a bilateral network, reciprocity is present; it is easy to show that this results in the equality of $\mathbf{z}_{12}$ and $\mathbf{z}_{21}$.

Equivalent circuits may again be obtained from an inspection of Eqs. [29] and [30]; their construction is facilitated by adding and subtracting either $\mathbf{z}_{12}\mathbf{I}_1$ in Eq. [30] or $\mathbf{z}_{21}\mathbf{I}_2$ in Eq. [29]. Each of these equivalent circuits contains a dependent voltage source.

Let us leave the derivation of such an equivalent to some leisure moment, and consider next an example of a rather general nature. Can we construct a general Thévenin equivalent of the two-port, as viewed from the output terminals? It is necessary first to assume a specific input circuit configuration, and we will select an independent voltage source $\mathbf{V}_s$ (positive sign at top) in series with a generator impedance $\mathbf{Z}_g$. Thus

$$\mathbf{V}_s = \mathbf{V}_1 + \mathbf{I}_1\mathbf{Z}g$$

Combining this result with Eqs. [29] and [30], we may eliminate $\mathbf{V}_1$ and $\mathbf{I}_1$ and obtain

$$\mathbf{V}_2 = \frac{\mathbf{z}_{21}}{\mathbf{z}_{11} + \mathbf{Z}_g}\mathbf{V}_s + \left(\mathbf{z}_{22} - \frac{\mathbf{z}_{12}\mathbf{z}_{21}}{\mathbf{z}_{11} + \mathbf{Z}_g}\right)\mathbf{I}_2$$

The Thévenin equivalent circuit may be drawn directly from this equation; it is shown in Fig. 16.21. The output impedance, expressed in terms of the $\mathbf{z}$ parameters, is

$$\mathbf{Z}_{\text{out}} = \mathbf{z}_{22} - \frac{\mathbf{z}_{12}\mathbf{z}_{21}}{\mathbf{z}_{11} + \mathbf{Z}_g}$$

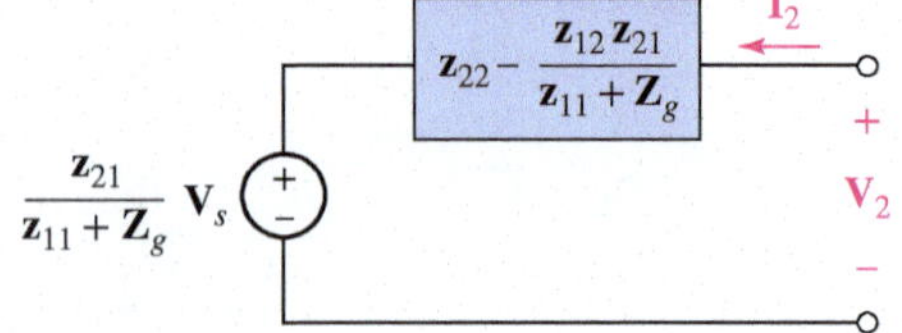

FIGURE 16.21 The Thévenin equivalent of a general two-port, as viewed from the output terminals, expressed in terms of the open-circuit impedance parameters.

If the generator impedance is zero, the simpler expression

$$\mathbf{Z}_{\text{out}} = \frac{\mathbf{z}_{11}\mathbf{z}_{22} - \mathbf{z}_{12}\mathbf{z}_{21}}{\mathbf{z}_{11}} = \frac{\Delta_{\mathbf{z}}}{\Delta_{22}} = \frac{1}{\mathbf{y}_{22}} \qquad (\mathbf{Z}_g = 0)$$

is obtained. For this special case, the output *admittance* is identical to $\mathbf{y}_{22}$, as indicated by the basic relationship of Eq. [13].

EXAMPLE 16.8

Given the set of impedance parameters

$$\mathbf{z} = \begin{bmatrix} 10^3 & 10 \\ -10^6 & 10^4 \end{bmatrix} \qquad \text{(all } \Omega)$$

which is representative of a bipolar junction transistor operating in the common-emitter configuration, determine the voltage, current, and power gains, as well as the input and output impedances. The two-port is driven by an ideal sinusoidal voltage source $\mathbf{V}_s$ in series with a 500 Ω resistor, and terminated in a 10 kΩ load resistor.

The two describing equations for the two-port are

$$\mathbf{V}_1 = 10^3\mathbf{I}_1 + 10\,\mathbf{I}_2 \qquad [36]$$

$$\mathbf{V}_2 = -10^6\mathbf{I}_1 + 10^4\mathbf{I}_2 \qquad [37]$$

and the characterizing equations of the input and output networks are

$$\mathbf{V}_s = 500\mathbf{I}_1 + \mathbf{V}_1 \quad [38]$$

$$\mathbf{V}_2 = -10^4\mathbf{I}_2 \quad [39]$$

From these last four equations, we may easily obtain expressions for $\mathbf{V}_1$, $\mathbf{I}_1$, $\mathbf{V}_2$, and $\mathbf{I}_2$ in terms of $\mathbf{V}_s$:

$$\mathbf{V}_1 = 0.75\mathbf{V}_s \qquad \mathbf{I}_1 = \frac{\mathbf{V}_s}{2000}$$

$$\mathbf{V}_2 = -250\mathbf{V}_s \qquad \mathbf{I}_2 = \frac{\mathbf{V}_s}{40}$$

From this information, it is simple to determine the voltage gain,

$$\mathbf{G_V} = \frac{\mathbf{V}_2}{\mathbf{V}_1} = -333$$

the current gain,

$$\mathbf{G_I} = \frac{\mathbf{I}_2}{\mathbf{I}_1} = 50$$

the power gain,

$$G_P = \frac{\text{Re}\left\{-\frac{1}{2}\mathbf{V}_2\mathbf{I}_2^*\right\}}{\text{Re}\left\{\frac{1}{2}\mathbf{V}_1\mathbf{I}_1^*\right\}} = 16{,}670$$

and the input impedance,

$$\mathbf{Z}_{\text{in}} = \frac{\mathbf{V}_1}{\mathbf{I}_1} = 1500\ \Omega$$

The output impedance may be obtained by referring to Fig. 16.21:

$$\mathbf{Z}_{\text{out}} = \mathbf{z}_{22} - \frac{\mathbf{z}_{12}\mathbf{z}_{21}}{\mathbf{z}_{11} + \mathbf{Z}_g} = 16.67\ \text{k}\Omega$$

In accordance with the predictions of the maximum power transfer theorem, the power gain reaches a maximum value when $\mathbf{Z}_L = \mathbf{Z}_{\text{out}}^* =$ 16.67 kΩ; that maximum value is 17,045.

The **y** parameters are useful when two-ports are interconnected in parallel, and, in a dual manner, the **z** parameters simplify the problem of a series connection of networks, shown in Fig. 16.22. Note that the series connection is *not* the same as the cascade connection that we will discuss later in connection with the transmission parameters. If each two-port has a common reference node for its input and output, and if the references are connected together as indicated in Fig. 16.22, then $\mathbf{I}_1$ flows through the input ports of the two networks in series. A similar statement holds for $\mathbf{I}_2$. Thus, ports remain ports after the interconnection. It follows that $\mathbf{I} = \mathbf{I}_A = \mathbf{I}_B$ and

$$\mathbf{V} = \mathbf{V}_A + \mathbf{V}_B = \mathbf{z}_A\mathbf{I}_A + \mathbf{z}_B\mathbf{I}_B$$
$$= (\mathbf{z}_A + \mathbf{z}_B)\mathbf{I} = \mathbf{z}\mathbf{I}$$

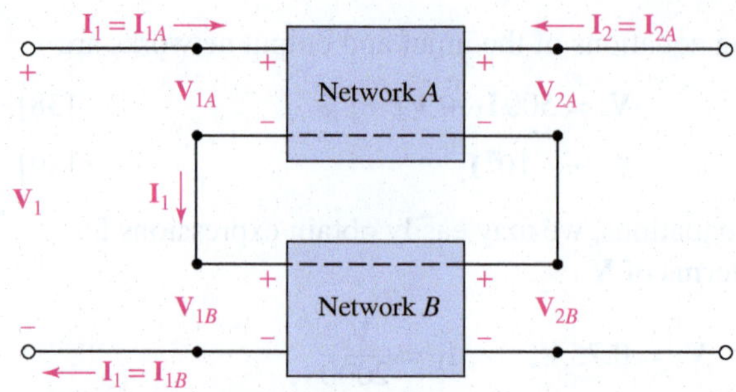

■ **FIGURE 16.22** The series connection of two two-port networks is made by connecting the four common reference nodes together; then the matrix $\mathbf{Z} = \mathbf{Z}_A + \mathbf{Z}_B$.

where

$$\mathbf{z} = \mathbf{z}_A + \mathbf{z}_B$$

so that $\mathbf{Z}_{11} = \mathbf{Z}_{11A} + \mathbf{Z}_{11B}$, and so forth.

PRACTICE

16.8 Find **z** for the two-port shown in (*a*) Fig. 16.23*a*; (*b*) Fig. 16.23*b*.

16.9 Find **z** for the two-port shown in Fig. 16.23*c*.

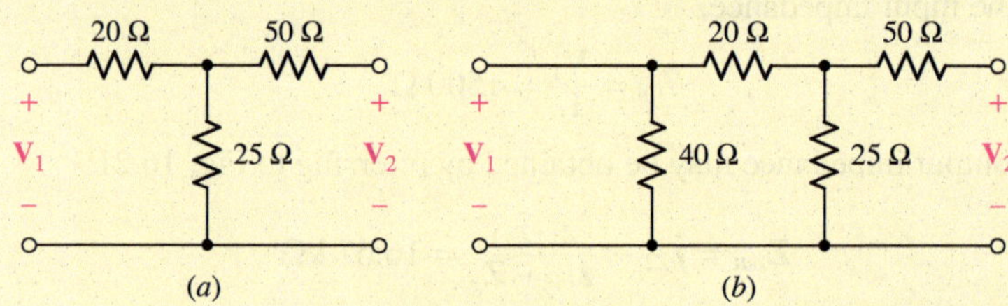

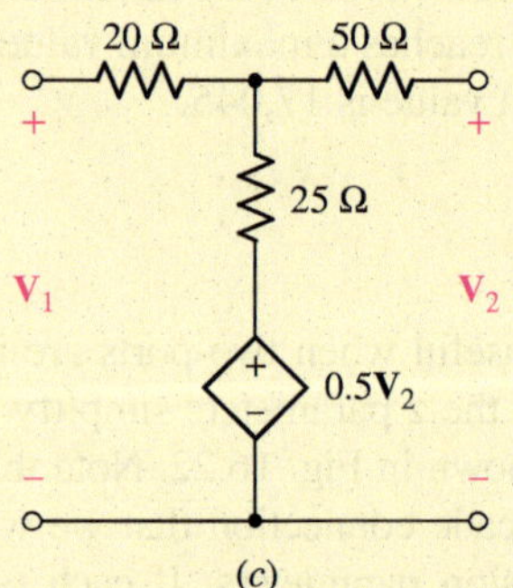

■ **FIGURE 16.23**

Ans: 16.8: $\begin{bmatrix} 45 & 25 \\ 25 & 75 \end{bmatrix}$ (Ω), $\begin{bmatrix} 21.2 & 11.76 \\ 11.76 & 67.6 \end{bmatrix}$ (Ω). 16.9: $\begin{bmatrix} 70 & 100 \\ 50 & 150 \end{bmatrix}$ (Ω).

16.5 HYBRID PARAMETERS

The difficulty in measuring quantities such as the open-circuit impedance parameters arises when a parameter such as $\mathbf{z}_{21}$ must be measured. A known sinusoidal current is easily supplied at the input terminals, but because of the exceedingly high output impedance of the transistor circuit, it is difficult to open-circuit the output terminals and yet supply the necessary dc biasing voltages and measure the sinusoidal output voltage. A short-circuit current measurement at the output terminals is much simpler to implement.

The hybrid parameters are defined by writing the pair of equations relating $\mathbf{V}_1$, $\mathbf{I}_1$, $\mathbf{V}_2$, and $\mathbf{I}_2$ as if $\mathbf{V}_1$ and $\mathbf{I}_2$ were the independent variables:

$$\mathbf{V}_1 = \mathbf{h}_{11}\mathbf{I}_1 + \mathbf{h}_{12}\mathbf{V}_2 \qquad [40]$$

$$\mathbf{I}_2 = \mathbf{h}_{21}\mathbf{I}_1 + \mathbf{h}_{22}\mathbf{V}_2 \qquad [41]$$

or

$$\begin{bmatrix} \mathbf{V}_1 \\ \mathbf{I}_2 \end{bmatrix} = \mathbf{h} \begin{bmatrix} \mathbf{I}_1 \\ \mathbf{V}_2 \end{bmatrix} \qquad [42]$$

The nature of the parameters is made clear by first setting $\mathbf{V}_2 = 0$. Thus,

$$\mathbf{h}_{11} = \left.\frac{\mathbf{V}_1}{\mathbf{I}_1}\right|_{\mathbf{V}_2=0} = \text{short-circuit input impedance}$$

$$\mathbf{h}_{21} = \left.\frac{\mathbf{I}_2}{\mathbf{I}_1}\right|_{\mathbf{V}_2=0} = \text{short-circuit forward current gain}$$

Letting $\mathbf{I}_1 = 0$, we obtain

$$\mathbf{h}_{12} = \left.\frac{\mathbf{V}_1}{\mathbf{V}_2}\right|_{\mathbf{I}_1=0} = \text{open-circuit reverse voltage gain}$$

$$\mathbf{h}_{22} = \left.\frac{\mathbf{I}_2}{\mathbf{V}_2}\right|_{\mathbf{I}_1=0} = \text{open-circuit output admittance}$$

Since the parameters represent an impedance, an admittance, a voltage gain, and a current gain, they are called the "hybrid" parameters.

The subscript designations for these parameters are often simplified when they are applied to transistors. Thus, $\mathbf{h}_{11}$, $\mathbf{h}_{12}$, $\mathbf{h}_{21}$, and $\mathbf{h}_{22}$ become $\mathbf{h}_i$, $\mathbf{h}_r$, $\mathbf{h}_f$, and $\mathbf{h}_o$, respectively, where the subscripts denote input, reverse, forward, and output.

EXAMPLE 16.9

Find h for the bilateral resistive circuit drawn in Fig. 16.24.

With the output short-circuited ($\mathbf{V}_2 = 0$), the application of a 1 A source at the input ($\mathbf{I}_1 = 1$ A) produces an input voltage of 3.4 V($\mathbf{V}_1 = 3.4$ V); hence, $\mathbf{h}_{11} = 3.4\ \Omega$. Under these same conditions, the output current is easily obtained by current division: $\mathbf{I}_2 = -0.4$ A; thus, $\mathbf{h}_{21} = -0.4$.

The remaining two parameters are obtained with the input open-circuited ($\mathbf{I}_1 = 0$). We apply 1 V to the output terminals ($\mathbf{V}_2 = 1$ V).

(Continued on next page)

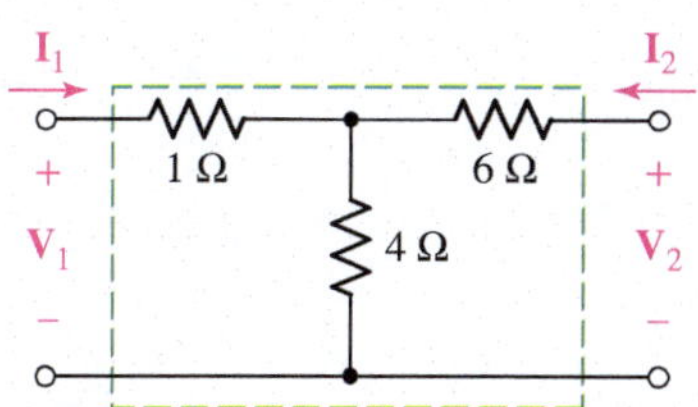

FIGURE 16.24 A bilateral network for which the **h** parameters are found: $\mathbf{h}_{12} = -\mathbf{h}_{21}$.

The response at the input terminals is 0.4 V($\mathbf{V}_1$ = 0.4 V), and thus $\mathbf{h}_{12} = 0.4$. The current delivered by this source at the output terminals is 0.1 A($\mathbf{I}_2$ = 0.1 A), and therefore $\mathbf{h}_{22} = 0.1$ S.

We therefore have $\mathbf{h} = \begin{bmatrix} 3.4\ \Omega & 0.4 \\ -0.4 & 0.1\ \text{S} \end{bmatrix}$. It is a consequence of the reciprocity theorem that $\mathbf{h}_{12} = -\mathbf{h}_{21}$ for a bilateral network.

PRACTICE

16.10 Find **h** for the two-port shown in (*a*) Fig. 16.25*a*; (*b*) Fig. 16.25*b*.

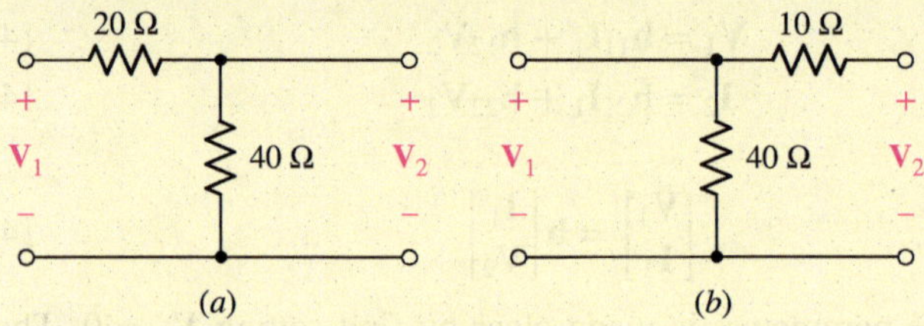

FIGURE 16.25

16.11 If $\mathbf{h} = \begin{bmatrix} 5\ \Omega & 2 \\ -0.5 & 0.1\ \text{S} \end{bmatrix}$, find (*a*) **y**; (*b*) **z**.

Ans: 16.10: $\begin{bmatrix} 20\ \Omega & 1 \\ -1 & 25\ \text{mS} \end{bmatrix}$, $\begin{bmatrix} 8\ \Omega & 0.8 \\ -0.8 & 20\ \text{mS} \end{bmatrix}$. 16.11: $\begin{bmatrix} 0.2 & -0.4 \\ -0.1 & 0.3 \end{bmatrix}$ (S), $\begin{bmatrix} 15 & 20 \\ 5 & 10 \end{bmatrix}$ (Ω).

The circuit shown in Fig. 16.26 is a direct translation of the two defining equations, [40] and [41]. The first represents KVL about the input loop, while the second is obtained from KCL at the upper output node. This circuit is also a popular transistor equivalent circuit. Let us assume some reasonable values for the common-emitter configuration: $\mathbf{h}_{11} = 1200\ \Omega$, $\mathbf{h}_{12} = 2 \times 10^{-4}$, $\mathbf{h}_{21} = 50$, $\mathbf{h}_{22} = 50 \times 10^{-6}$ S, a voltage generator of $1\underline{/0^\circ}$ mV in series with 800 Ω, and a 5 kΩ load. For the input,

$$10^{-3} = (1200 + 800)\mathbf{I}_1 + 2 \times 10^{-4}\,\mathbf{V}_2$$

and at the output,

$$\mathbf{I}_2 = -2 \times 10^{-4}\,\mathbf{V}_2 = 50\mathbf{I}_1 + 50 \times 10^{-6}\,\mathbf{V}_2$$

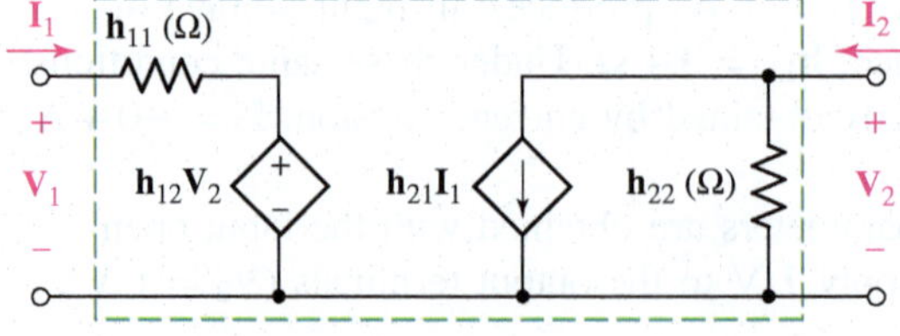

FIGURE 16.26 The four **h** parameters are referred to a two-port. The pertinent equations are $\mathbf{V}_1 = \mathbf{h}_{11}\mathbf{I}_1 + \mathbf{h}_{12}\mathbf{V}_2$ and $\mathbf{I}_2 = \mathbf{h}_{21}\mathbf{I}_1 + \mathbf{h}_{22}\mathbf{V}_2$.

PRACTICAL APPLICATION

Characterizing Transistors

Parameter values for bipolar junction transistors are commonly quoted in terms of **h** parameters. Invented in the late 1940s by researchers at Bell Laboratories (Fig. 16.27), the transistor is a nonlinear three-terminal passive semiconductor device that forms the basis for almost all amplifiers and digital logic circuits.

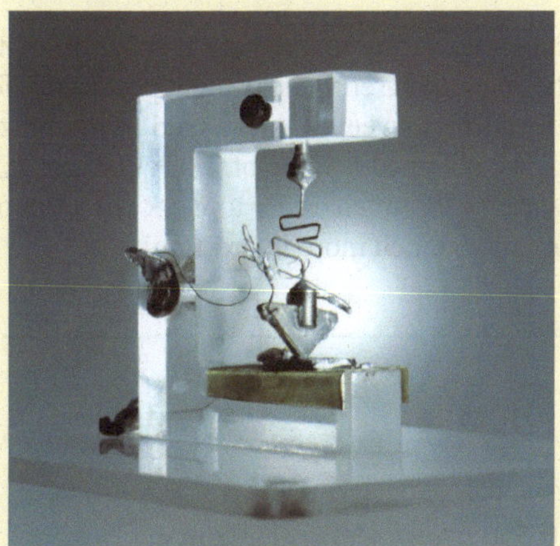

■ **FIGURE 16.27** Photograph of the first demonstrated bipolar junction transistor ("bjt").
©Lucent Technologies Inc./Bell Labs

The three terminals of a transistor are labeled the *base* (*b*), *collector* (*c*), and *emitter* (*e*) as shown in Fig. 16.28, and are named after their roles in the transport of charge within the device. The **h** parameters of a bipolar junction transistor are typically measured with the emitter terminal grounded, also known as the *common-emitter* configuration; the base is then designated as the input and the collector as the output. As mentioned previously, however, the transistor is a nonlinear device, and so definition of **h** parameters which are valid for all voltages and currents is not possible. Therefore, it is common practice to quote **h** parameters at a specific value of collector current I_C and collector-emitter voltage V_{CE}. Another consequence

■ **FIGURE 16.28** Schematic of a bjt showing currents and voltages defined using the IEEE convention.

of the nonlinearity of the device is that ac **h** parameters and dc **h** parameters are often quite different in value.

There are many types of instruments which may be employed to obtain the **h** parameters for a particular transistor. One example is a semiconductor parameter analyzer, shown in Fig. 16.29. This instrument sweeps the desired current (plotted on the vertical axis) against a specified voltage (plotted on the horizontal axis). A "family" of curves is produced by varying a third parameter, often the base current, in discrete steps.

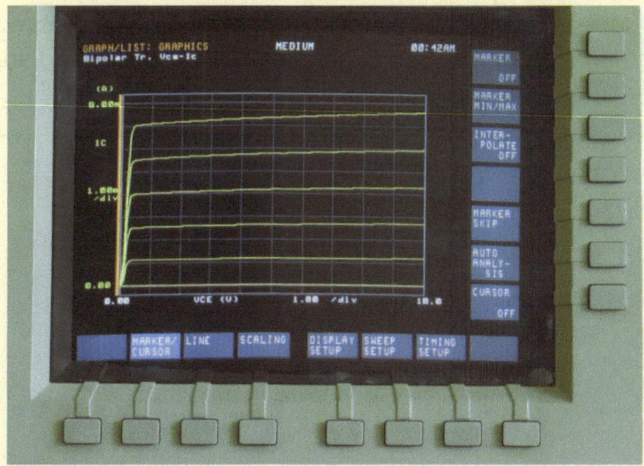

■ **FIGURE 16.29** Display snapshot of an HP 4155A Semiconductor Parameter Analyzer used to measure the **h** parameters of a 2N3904 bipolar junction transistor (bjt).
Steve Durbin/McGraw Hill

As an example, the manufacturer of the 2N3904 NPN silicon transistor quotes **h** parameters as indicated in Table 16.2; note that the specific parameters are given alternative designations (h_{ie}, h_{re}, etc.) by transistor engineers. The measurements were made at $I_C = 1.0$ mA, $V_{CE} = 10$ V dc, and $f = 1.0$ kHz.

Just for fun, one of the authors and a friend decided to measure these parameters for themselves. Grabbing an inexpensive device off the shelf and using the instrument in Fig. 16.29, they found

$$h_{oe} = 3.3\ \mu\text{S} \qquad h_{fe} = 109$$
$$h_{ie} = 3.02\ \text{k}\Omega \qquad h_{re} = 4 \times 10^{-3}$$

the first three of which were all well within the manufacturer's published tolerances, although much closer to the minimum values than to the maximum values. The value for h_{re}, however, was an order of magnitude larger than the maximum value specified by the manufacturer's

(Continued on next page)

TABLE 16.2 Summary of 2N3904 AC Parameters

Parameter	Name	Specification	Units
$h_{ie}(h_{11})$	Input impedance	1.0–10	kΩ
$h_{re}(h_{12})$	Voltage feedback ratio	$0.5\text{–}8.0 \times 10^{-4}$	–
$h_{fe}(h_{21})$	Small-signal current gain	100–400	–
$h_{oe}(h_{22})$	Output admittance	1.0–40	μS

datasheet! This was rather disconcerting, as we thought we were doing pretty well up to that point.

Upon further reflection, we realized that the experimental setup allowed the device to heat up during the measurement, as we were sweeping below and above $I_C = 1$ mA. Transistors, unfortunately, can change their properties rather dramatically as a function of temperature; the manufacturer values were specifically for 25°C. Once the current sweep was changed to minimize device heating, we obtained a value of 2.0×10^{-4} for h_{re}. Linear circuits are by far much easier to work with, but nonlinear circuits can be much more interesting!

Solving,

$$\mathbf{I}_1 = 0.510\ \mu\text{A} \qquad \mathbf{V}_1 = 0.592\ \text{mV}$$
$$\mathbf{I}_2 = 20.4\ \mu\text{A} \qquad \mathbf{V}_2 = -102\ \text{mV}$$

Through the transistor we have a current gain of 40, a voltage gain of −172, and a power gain of 6880. The input impedance to the transistor is 1160 Ω, and a few more calculations show that the output impedance is 22.2 kΩ.

Hybrid parameters may be added directly when two-ports are connected in series at the input and in parallel at the output. This is called a series-parallel interconnection, and it is not used very often.

16.6 TRANSMISSION PARAMETERS

The last two-port parameters that we will consider are called the **t** ***parameters,*** the ***ABCD parameters,*** or simply the ***transmission parameters.*** They are defined by

$$\mathbf{V}_1 = \mathbf{t}_{11}\mathbf{V}_2 - \mathbf{t}_{12}\mathbf{I}_2 \qquad [43]$$

and

$$\mathbf{I}_1 = \mathbf{t}_{21}\mathbf{V}_2 - \mathbf{t}_{22}\mathbf{I}_2 \qquad [44]$$

or

$$\begin{bmatrix} \mathbf{V}_1 \\ \mathbf{I}_1 \end{bmatrix} = \mathbf{t}\begin{bmatrix} V_2 \\ -\mathbf{I}_2 \end{bmatrix} \qquad [45]$$

where $\mathbf{V}_1$, $\mathbf{V}_2$, $\mathbf{I}_1$, and $\mathbf{I}_2$ are defined as usual (Fig. 16.8). The minus signs that appear in Eqs. [43] and [44] should be associated with the output current, as $(-\mathbf{I}_2)$. Thus, both $\mathbf{I}_1$ and $-\mathbf{I}_{12}$ are directed to the right, the direction of energy or signal transmission.

Other widely used nomenclature for this set of parameters is

$$\begin{bmatrix} \mathbf{t}_{11} & \mathbf{t}_{12} \\ \mathbf{t}_{21} & \mathbf{t}_{22} \end{bmatrix} = \begin{bmatrix} \mathbf{A} & \mathbf{B} \\ \mathbf{C} & \mathbf{D} \end{bmatrix} \qquad [46]$$

Note that there are no minus signs in the **t** or **ABCD** matrices.

Looking again at Eqs. [43] to [45], we see that the quantities on the left, often thought of as the given or independent variables, are the input voltage and current, $\mathbf{V}_1$ and $\mathbf{I}_1$; the dependent variables, $\mathbf{V}_2$ and $\mathbf{I}_2$, are the output quantities. Thus, the transmission parameters provide a direct relationship between input and output. Their major use arises in transmission-line analysis and in cascaded networks.

Let us find the **t** parameters for the bilateral resistive two-port of Fig. 16.30*a*. To illustrate one possible procedure for finding a single parameter, consider

$$\mathbf{t}_{12} = \left.\frac{\mathbf{V}_1}{-\mathbf{I}_2}\right|_{\mathbf{V}_2=0}$$

We therefore short-circuit the output ($\mathbf{V}_2 = 0$) and set $\mathbf{V}_1 = 1$ V, as shown in Fig. 16.30*b*. Note that we cannot set the denominator equal to unity by placing a 1 A current source at the output; we already have a short circuit there. The equivalent resistance offered to the 1 V source is $R_{eq} = 2 + (4 \parallel 10)\ \Omega$, and we then use current division to get

$$-\mathbf{I}_2 = \frac{1}{2 + (4 \parallel 10)} \times \frac{10}{10+4} = \frac{5}{34}\ \text{A}$$

Hence,

$$\mathbf{t}_{12} = \frac{1}{-\mathbf{I}_2} = \frac{34}{5} = 6.8\ \Omega$$

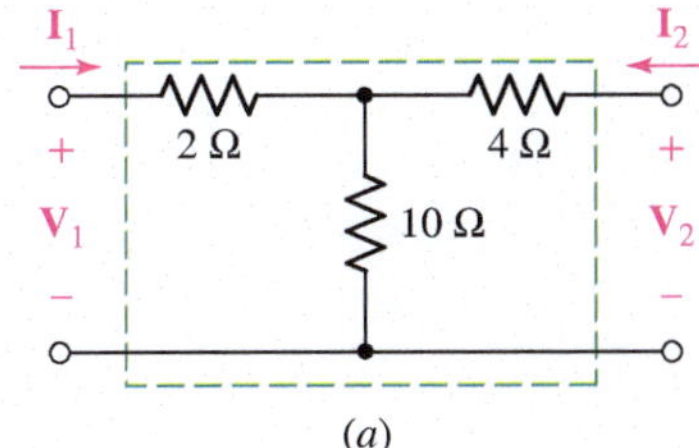

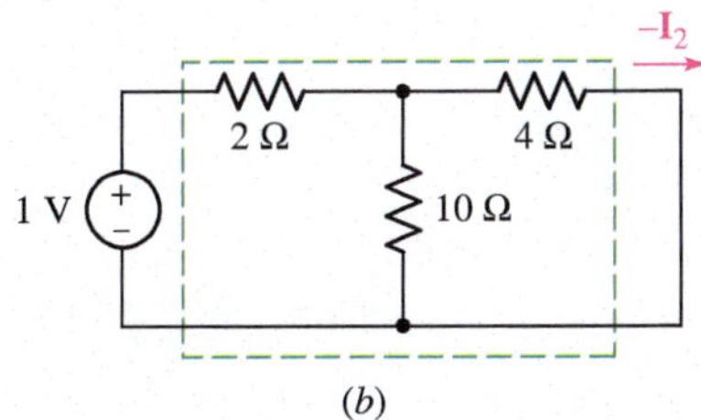

■ **FIGURE 16.30** (*a*) A two-port resistive network for which the **t** parameters are to be found. (*b*) To find $\mathbf{t}_{12}$, set $\mathbf{V}_1 = 1$ V with $\mathbf{V}_2 = 0$; then $\mathbf{t}_{12} = 1/(-\mathbf{I}_2) = 6.8\ \Omega$.

If it is necessary to find all four parameters, we write any convenient pair of equations using all four terminal quantities, $\mathbf{V}_1$, $\mathbf{V}_2$, $\mathbf{I}_1$, and $\mathbf{I}_2$. From Fig. 16.30*a*, we have two mesh equations:

$$\mathbf{V}_1 = 12\mathbf{I}_1 + 10\mathbf{I}_2 \qquad [47]$$

$$\mathbf{V}_2 = 10\mathbf{I}_1 + 14\mathbf{I}_2 \qquad [48]$$

Solving Eq. [48] for $\mathbf{I}_1$, we get

$$\mathbf{I}_1 = 0.1\,\mathbf{V}_2 - 1.4\,\mathbf{I}_2$$

so that $\mathbf{t}_{21} = 0.1$ S and $\mathbf{t}_{22} = 1.4$. Substituting the expression for $\mathbf{I}_1$ in Eq. [47], we find

$$\mathbf{V}_1 = 12(0.1\,\mathbf{V}_2 - 1.4\,\mathbf{I}_2) + 10\,\mathbf{I}_2 = 1.2\,\mathbf{V}_2 - 6.8\,\mathbf{I}_2$$

and $\mathbf{t}_{11} = 1.2$ and $\mathbf{t}_{12} = 6.8\ \Omega$, once again.

For reciprocal networks, the determinant of the **t** matrix is equal to unity:

$$\Delta_t = \mathbf{t}_{11}\mathbf{t}_{22} - \mathbf{t}_{12}\mathbf{t}_{21} = 1$$

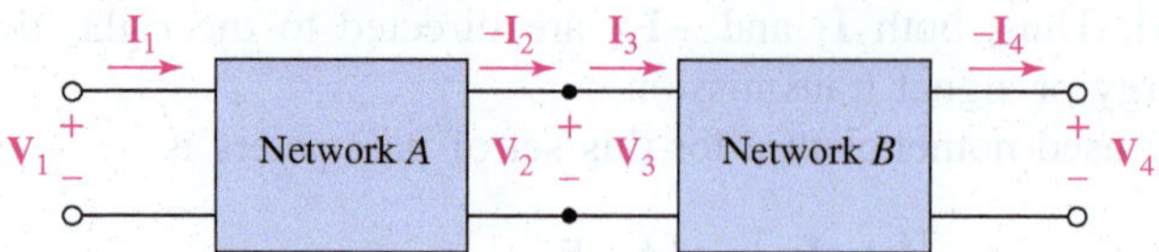

■ **FIGURE 16.31** When two-port networks *A* and *B* are cascaded, the **t** parameter matrix for the combined network is given by the matrix product $\mathbf{t} = \mathbf{t}_A\mathbf{t}_B$.

In the resistive example of Fig. 16.30, $\Delta_t = 1.2 \times 1.4 - 6.8 \times 0.1 = 1$. Good!

We conclude our two-port discussion by connecting two two-ports in cascade, as illustrated for two networks in Fig. 16.31. Terminal voltages and currents are indicated for each two-port, and the corresponding **t** parameter relationships are, for network *A*,

$$\begin{bmatrix} \mathbf{V}_1 \\ \mathbf{I}_1 \end{bmatrix} = \mathbf{t}_A \begin{bmatrix} \mathbf{V}_2 \\ -\mathbf{I}_2 \end{bmatrix} = \mathbf{t}_A \begin{bmatrix} \mathbf{V}_3 \\ \mathbf{I}_3 \end{bmatrix}$$

and for network *B*,

$$\begin{bmatrix} \mathbf{V}_3 \\ \mathbf{I}_3 \end{bmatrix} = \mathbf{t}_B \begin{bmatrix} \mathbf{V}_4 \\ -\mathbf{I}_4 \end{bmatrix}$$

Combining these results, we have

$$\begin{bmatrix} \mathbf{V}_1 \\ \mathbf{I}_1 \end{bmatrix} = \mathbf{t}_A\mathbf{t}_B \begin{bmatrix} \mathbf{V}_4 \\ -\mathbf{I}_4 \end{bmatrix}$$

Therefore, the **t** parameters for the cascaded networks are found by the matrix product,

$$\mathbf{t} = \mathbf{t}_A\mathbf{t}_B$$

This product is *not* obtained by multiplying corresponding elements in the two matrices. If necessary, review the correct procedure for matrix multiplication in Appendix 2.

EXAMPLE 16.10

Find the t parameters for the cascaded networks shown in Fig. 16.32.

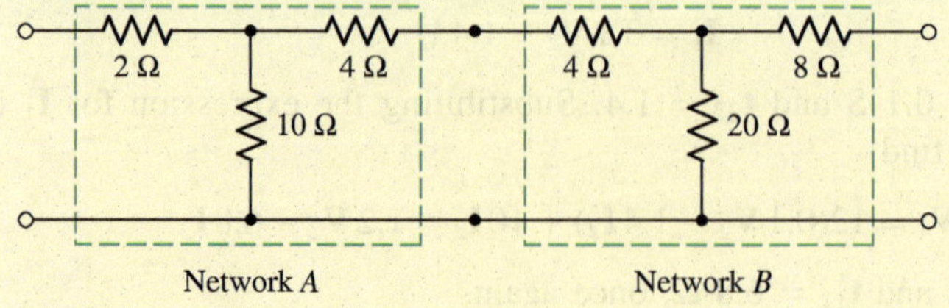

■ **FIGURE 16.32** A cascaded connection.

Network A is the two-port of Fig. 16.32, and, therefore

$$\mathbf{t}_A = \begin{bmatrix} 1.2 & 6.8\ \Omega \\ 0.1\ \text{S} & 1.4 \end{bmatrix}$$

while network B has resistance values twice as large, so that

$$\mathbf{t}_B = \begin{bmatrix} 1.2 & 13.6\ \Omega \\ 0.05\ \text{S} & 1.4 \end{bmatrix}$$

For the combined network,

$$\mathbf{t} = \mathbf{t}_A \mathbf{t}_B = \begin{bmatrix} 1.2 & 6.8 \\ 0.1 & 1.4 \end{bmatrix} \begin{bmatrix} 1.2 & 13.6 \\ 0.05 & 1.4 \end{bmatrix}$$
$$= \begin{bmatrix} 1.2 \times 1.2 + 6.8 \times 0.05 & 1.2 \times 13.6 + 6.8 \times 1.4 \\ 0.1 \times 1.2 \times 1.4 \times 0.05 & 0.1 \times 13.6 + 1.4 \times 1.4 \end{bmatrix}$$

and

$$\mathbf{t} = \begin{bmatrix} 1.78 & 25.84\ \Omega \\ 0.19\ \text{S} & 3.32 \end{bmatrix}$$

PRACTICE

16.12 Given $\mathbf{t} = \begin{bmatrix} 3.2 & 8\ \Omega \\ 0.2\ \text{S} & 4 \end{bmatrix}$, find (*a*) $\mathbf{z}$; (*b*) $\mathbf{t}$ for two identical networks in cascade; (*c*) $\mathbf{z}$ for two identical networks in cascade.

Ans: $\begin{bmatrix} 16 & 56 \\ 5 & 20 \end{bmatrix}$ (Ω); $\begin{bmatrix} 11.84 & 57.6\Omega \\ 1.44\text{S} & 14.6 \end{bmatrix}$; $\begin{bmatrix} 8.22 & 87.1 \\ 0.694 & 12.22 \end{bmatrix}$ (Ω).

COMPUTER-AIDED ANALYSIS

The characterization of two-port networks using **t** parameters creates the opportunity for vastly simplified analysis of cascaded two-port network circuits. As seen in this section, where, for example,

$$\mathbf{t}_A = \begin{bmatrix} 1.2 & 6.8\ \Omega \\ 0.1\ \text{S} & 1.4 \end{bmatrix}$$

and

$$\mathbf{t}_B = \begin{bmatrix} 1.2 & 13.6\ \Omega \\ 0.05\ \text{S} & 1.4 \end{bmatrix}$$

we found that the **t** parameters characterizing the cascaded network can be found by simply multiplying $\mathbf{t}_A$ and $\mathbf{t}_B$:

$$\mathbf{t} = \mathbf{t}_A \cdot \mathbf{t}_B$$

(*Continued on next page*)

Such matrix operations are easily carried out using scientific calculators or software packages such as MATLAB. The MATLAB script, for example, would be

```
>> tA = [1.2 6.8; 0.1 1.4];
>> tB = [1.2 13.6; 0.05 1.4];
>> t = tA*tB

t =

   1.7800    25.8400
   0.1900     3.3200
```

as we found in Example 16.10.

In terms of entering matrices in MATLAB, each has a case-sensitive variable name (tA, tB, and t in this example). Matrix elements are entered a row at a time, beginning with the top row; rows are separated by a semicolon. Again, the reader should always be careful to remember that the order of operations is critical when performing matrix algebra. For example, tB*tA results in a totally different matrix than the one we sought:

$$\mathbf{t}_B \cdot \mathbf{t}_A = \begin{bmatrix} 2.8 & 27.2 \\ 0.2 & 2.3 \end{bmatrix}$$

For simple matrices such as seen in this example, a scientific calculator is just as handy (if not more so). However, larger cascaded networks are more easily handled on a computer, where it is more convenient to see all arrays on the screen simultaneously.

SUMMARY AND REVIEW

In this chapter we encountered a somewhat abstract way to represent networks. This new approach is especially useful if the network is *passive* and if it will either be connected somehow to other networks at some point or perhaps component values will often be changed. We introduced the concept through the idea of a *one-port network,* where all we really did was determine the Thévenin equivalent resistance (or impedance, more generally speaking). Our first exposure to the idea of a two-port network (perhaps one port is an input, the other an output?) was through admittance parameters, also called **y** parameters. The result is a matrix which, when multiplied by the vector containing the terminal voltages, yields a vector with the currents into each port. A little manipulation yielded what we called Δ−Y equivalents in Chap. 5. The direct counterpart to **y** parameters are **z** parameters, where each matrix element is the ratio of a voltage to a current. Occasionally **y** and **z** parameters are not particularly convenient, so we also introduced "hybrid" or **h** parameters, as well as "transmission" or **t** parameters, also referred to as *ABCD* parameters.

Table 16.1 summarizes the conversion process between **y**, **z**, **h**, and **t** parameters; having one set of parameters which completely describes a network is sufficient regardless of what type of matrix we prefer for a particular analysis.

As a convenience to the reader, we will now proceed directly to a list of key concepts in the chapter, along with corresponding examples.

- In order to employ the analysis methods described in this chapter, it is critical to remember that each port can only be connected to either a one-port network or a port of another multiport network.
- The input impedance of a one-port (*passive*) linear network can be obtained using either nodal or mesh analysis; in some instances the set of coefficients can be written directly by inspection. (Examples 16.1, 16.2, 16.3)
- The defining equations for analyzing a two-port network in terms of its admittance (**y**) parameters are:

$$\mathbf{I}_1 = \mathbf{y}_{11}\mathbf{V}_1 + \mathbf{y}_{12}\mathbf{V}_2 \quad \text{and} \quad \mathbf{I}_2 = \mathbf{y}_{21}\mathbf{V}_1 + \mathbf{y}_{22}\mathbf{V}_2$$

  where

$$\mathbf{y}_{11} = \left.\frac{\mathbf{I}_1}{\mathbf{V}_1}\right|_{\mathbf{V}_2=0} \qquad \mathbf{y}_{12} = \left.\frac{\mathbf{I}_1}{\mathbf{V}_2}\right|_{\mathbf{V}_1=0}$$

$$\mathbf{y}_{21} = \left.\frac{\mathbf{I}_2}{\mathbf{V}_1}\right|_{\mathbf{V}_2=0} \quad \text{and} \quad \mathbf{y}_{22} = \left.\frac{\mathbf{I}_2}{\mathbf{V}_2}\right|_{\mathbf{V}_1=0}$$

  (Examples 16.4, 16.5, 16.7)
- The defining equations for analyzing a two-port network in terms of its impedance (**z**) parameters are:

$$\mathbf{V}_1 = \mathbf{z}_{11}\mathbf{I}_1 + \mathbf{z}_{12}\mathbf{I}_2 \quad \text{and} \quad \mathbf{V}_2 = \mathbf{z}_{21}\mathbf{I}_1 + \mathbf{z}_{22}\mathbf{I}_2$$

  (Example 16.8)
- The defining equations for analyzing a two-port network in terms of its hybrid (**h**) parameters are:

$$\mathbf{V}_1 = \mathbf{h}_{11}\mathbf{I}_1 + \mathbf{h}_{12}\mathbf{V}_2 \quad \text{and} \quad \mathbf{I}_2 = \mathbf{h}_{21}\mathbf{I}_1 + \mathbf{h}_{22}\mathbf{V}_2$$

  (Example 16.9)
- The defining equations for analyzing a two-port network in terms of its transmission (**t**) parameters (also called the **ABCD** parameters) are:

$$\mathbf{V}_1 = \mathbf{t}_{11}\mathbf{V}_2 - \mathbf{t}_{12}\mathbf{I}_2 \quad \text{and} \quad \mathbf{I}_1 = \mathbf{t}_{21}\mathbf{V}_2 - \mathbf{t}_{22}\mathbf{I}_2$$

  (Example 16.10)
- It is straightforward to convert between **h**, **z**, **t**, and **y** parameters, depending on circuit analysis needs; the transformations are summarized in Table 16.1. (Example 16.6)

READING FURTHER

Further details of matrix methods for circuit analysis can be found in:

R. A. DeCarlo and P. M. Lin, *Linear Circuit Analysis,* 2nd ed. New York: Oxford University Press, 2001.

Analysis of transistor circuits using network parameters is described in:

W. H. Hayt, Jr., and G. W. Neudeck, *Electronic Circuit Analysis and Design,* 2nd ed. New York: Wiley, 1995.

EXERCISES

16.1 One-Port Networks

1. Consider the following system of equations, which represents a resistive two-port network:

$$\begin{aligned} 2\mathbf{I}_1 \qquad\quad - \mathbf{I}_3 &= 15 \\ -3\mathbf{I}_1 + 2\mathbf{I}_2 + 7\mathbf{I}_3 &= -2 \\ 4\mathbf{I}_1 - 7\mathbf{I}_2 + 2\mathbf{I}_3 &= 0 \end{aligned}$$

2. For the following system of equations, (*a*) write the set of equations in matrix form. (*b*) Use Δ_Y to calculate $\mathbf{V}_2$ only.

$$\begin{aligned} 100\mathbf{V}_1 - 45\mathbf{V}_2 + 30\mathbf{V}_3 &= 0.2 \\ 75\mathbf{V}_1 \qquad\qquad + 80\mathbf{V}_3 &= -0.1 \\ 48\mathbf{V}_1 + 200\mathbf{V}_2 + 42\mathbf{V}_3 &= 0.5 \end{aligned}$$

3. With regard to the passive network depicted in Fig. 16.33, (*a*) obtain the three mesh equations; (*b*) compute Δ_Z; and (*c*) calculate the input impedance.

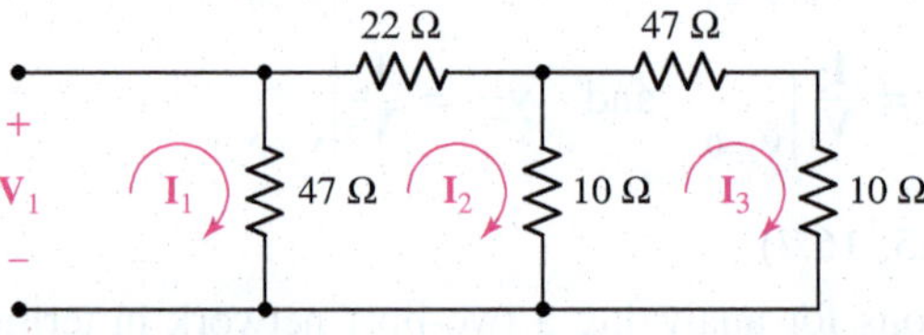

■ **FIGURE 16.33**

4. Determine the input impedance of the network shown in Fig. 16.34 after first calculating Δ_Z.

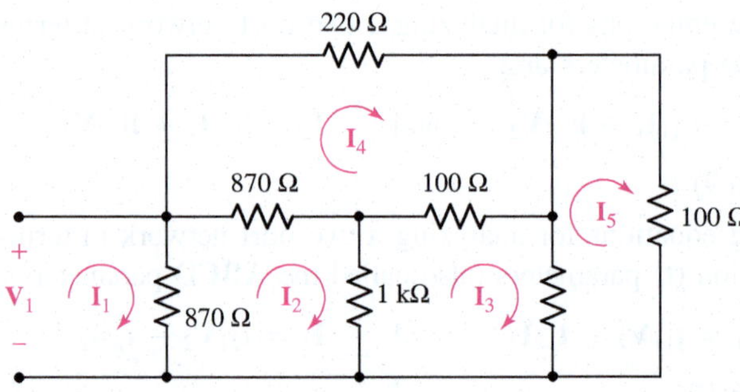

■ **FIGURE 16.34**

5. For the one-port network represented schematically in Fig. 16.35, choose the bottom node as the reference; name the junction between the 3, 10, and 20 S conductances $\mathbf{V}_2$ and the remaining node $\mathbf{V}_3$. (*a*) Write the three nodal equations. (*b*) Compute Δ_Y. (*c*) Calculate the input admittance.

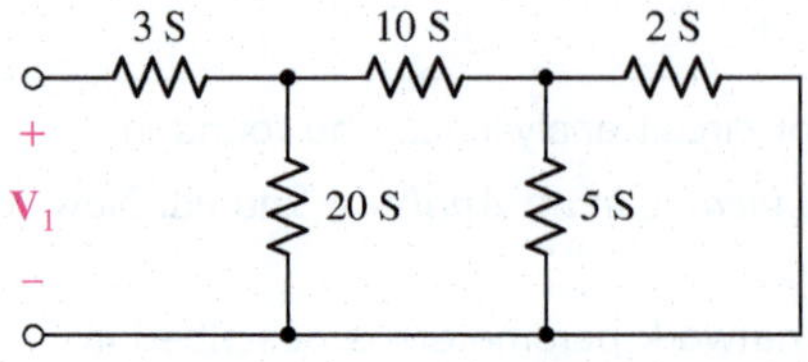

■ **FIGURE 16.35**

6. Calculate $\Delta_\mathbf{Z}$ and $\mathbf{Z}_{in}$ for the network of Fig. 16.36 if ω is equal to (*a*) 1 rad/s; (*b*) 320 krad/s.
7. Set $\omega = 100\pi$ rad/s in the one-port of Fig. 16.36. (*a*) Calculate $\Delta_\mathbf{Y}$ and the input admittance at ω, $\mathbf{Y}_{in}(\omega)$. (*b*) A sinusoidal current source having magnitude 100 A, frequency 100π rad/s, and 0° phase is connected to the network. Calculate the voltage across the current source (express answer as a phasor).
8. With reference to the one-port of Fig. 16.37, which contains a dependent current source controlled by a resistor voltage, (*a*) calculate $\Delta_\mathbf{Z}$; (*b*) compute $\mathbf{Z}_{in}$.

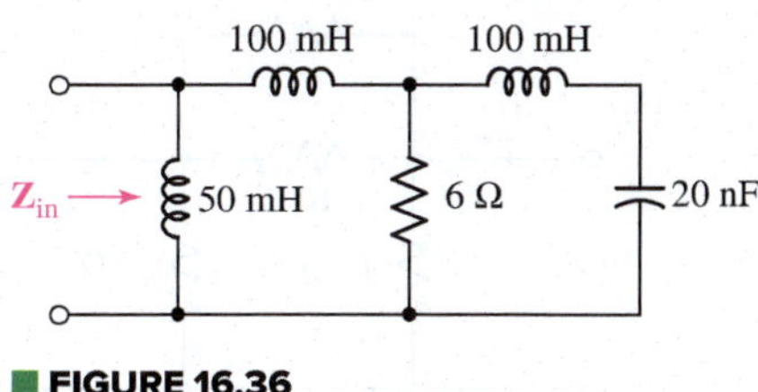

FIGURE 16.36

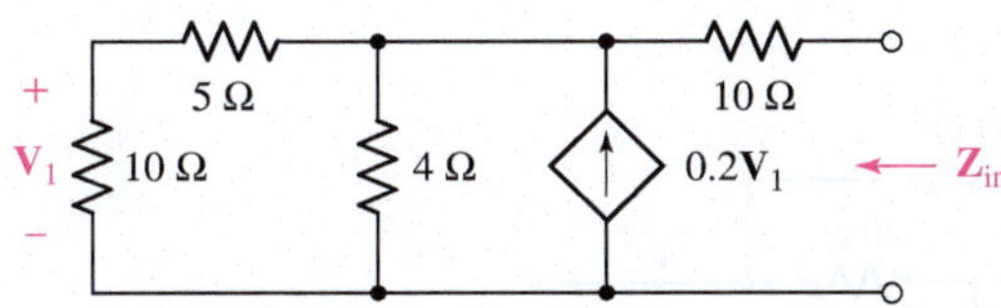

FIGURE 16.37

9. For the ideal op amp circuit represented in Fig. 16.38, the input resistance is defined by looking between the positive input terminal of the op amp and ground. (*a*) Write the appropriate nodal equations for the one-port. (*b*) Obtain an expression for R_{in}. Is your answer somewhat unexpected? Explain.
10. (*a*) If both the op amps shown in the circuit of Fig. 16.39 are assumed to be ideal ($R_i = \infty$, $R_o = 0$, and $A = \infty$), find $\mathbf{Z}_{in}$. (*b*) $R_1 = 4$ kΩ, $R_2 = 10$ kΩ, $R_3 = 10$ kΩ, $R_4 = 1$ kΩ, and $C = 200$ pF; show that $\mathbf{Z}_{in} = j\omega L_{in}$, where $L_{in} = 0.8$ mH.

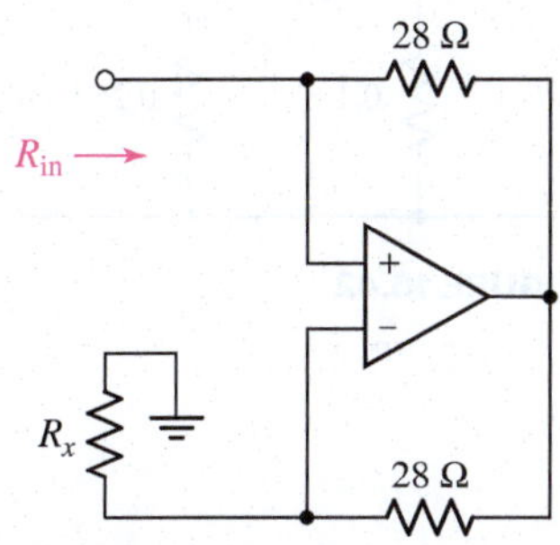

FIGURE 16.38

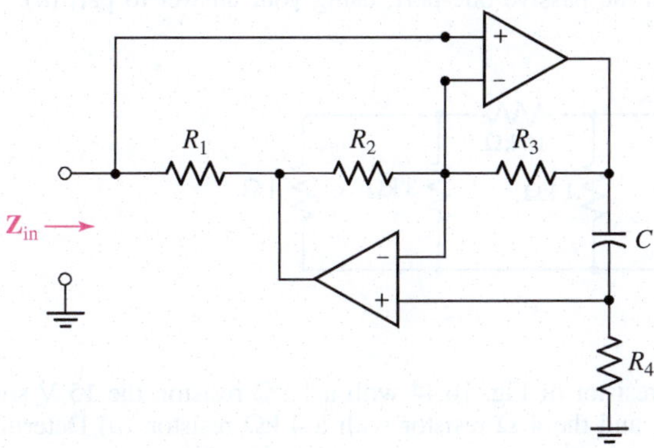

FIGURE 16.39

16.2 Admittance Parameters

11. Obtain a complete set of **y** parameters which describes the two-port shown in Fig. 16.40.

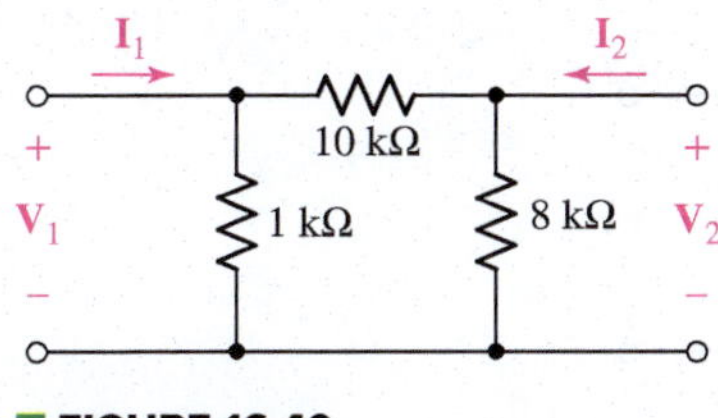

FIGURE 16.40

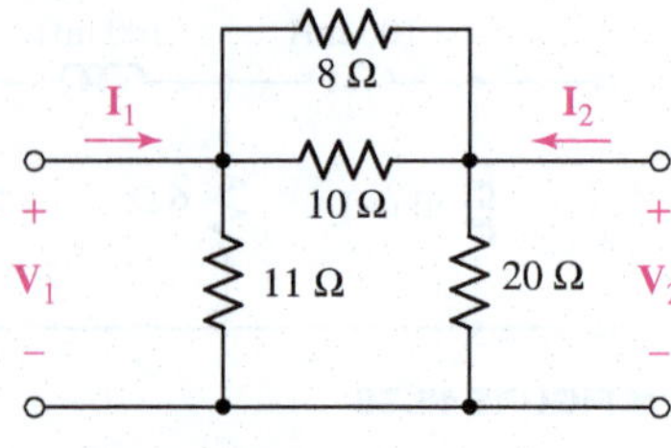

■ **FIGURE 16.41**

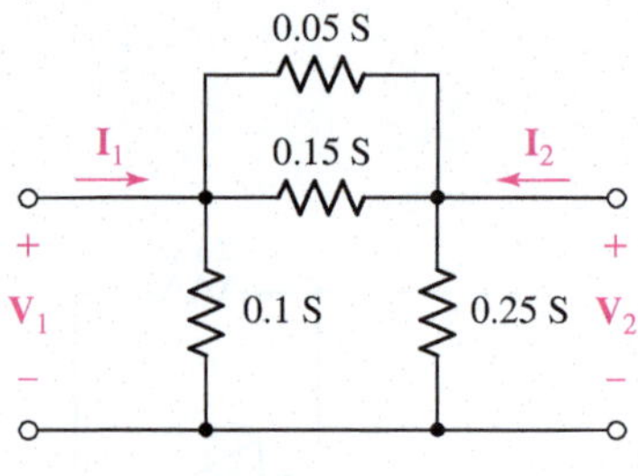

■ **FIGURE 16.42**

12. (*a*) Determine the short-circuit admittance parameters which completely describe the two-port network of Fig. 16.41. (*b*) If $\mathbf{V}_1 = 3$ **V** and $\mathbf{V}_2 = -2$ **V**, use your answer in part (*a*) to compute $\mathbf{I}_1$ and $\mathbf{I}_2$.
13. (*a*) Determine the **y** parameters for the two-port of Fig. 16.42. (*b*) Define the bottom node of Fig. 16.42 as the reference node, and apply nodal analysis to obtain expressions for $\mathbf{I}_1$ and $\mathbf{I}_2$ in terms of $\mathbf{V}_1$ and $\mathbf{V}_2$. Use these expressions to write down the admittance matrix. (*c*) If $\mathbf{V}_1 = 2\mathbf{V}_2 = 10$ V, calculate the power dissipated in the 100 mS conductance.
14. Obtain a complete set of **y** parameters to describe the two-port network depicted in Fig. 16.43.

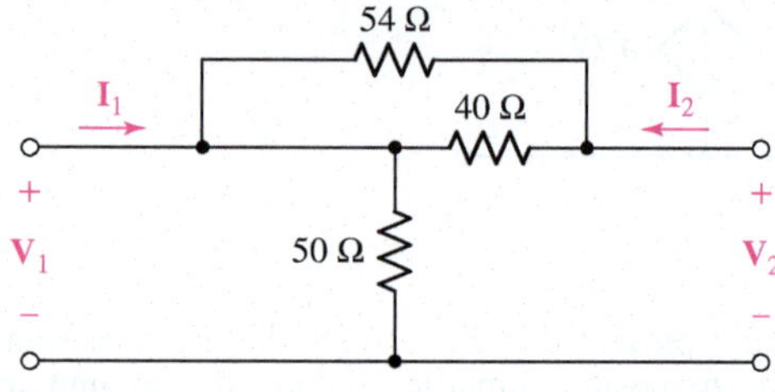

■ **FIGURE 16.43**

15. The circuit of Fig. 16.44 is simply the two-port of Fig. 16.40 terminated by a passive one-port and a separate one-port consisting of a voltage source in series with a resistor. (*a*) Determine the complete set of admittance parameters which describes the two-port network. (*Hint:* Draw the two-port by itself, properly labeled with a voltage and current at each port.) (*b*) Calculate the power dissipated in the passive one-port, using your answer to part (*a*).

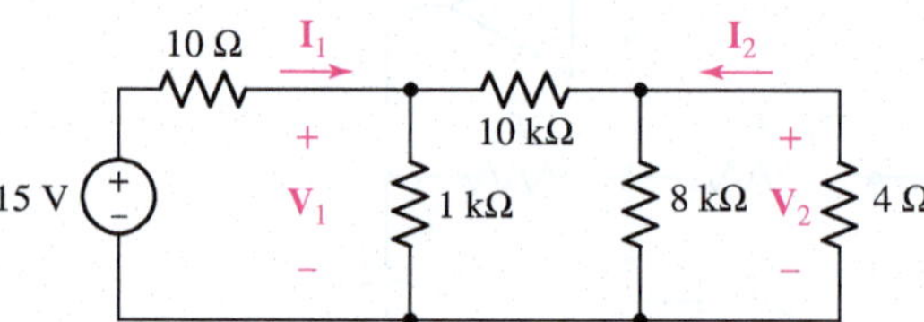

■ **FIGURE 16.44**

16. Replace the 10 Ω resistor of Fig. 16.44 with a 1 kΩ resistor, the 15 V source with a 9 V source, and the 4 Ω resistor with a 4 kΩ resistor. (*a*) Determine the complete set of admittance parameters which describes the two-port network consisting of the 1 kΩ, 10 kΩ, and 8 kΩ resistors. (*Hint:* Draw the two-port by itself, properly labeled with a voltage and current at each port.) (*b*) Calculate the power dissipated in the passive one-port, using your answer to part (*a*).
17. Calculate the admittance parameters which describe the two-port depicted in Fig. 16.45.

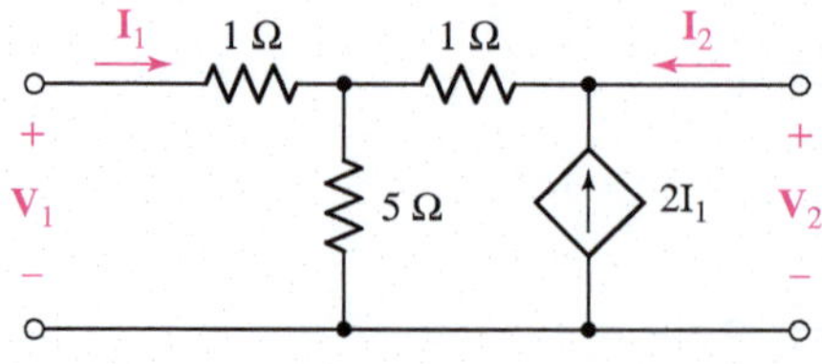

■ **FIGURE 16.45**

18. Obtain the **y** parameter for the network shown in Fig. 16.46 and use it to determine $\mathbf{I}_1$ and $\mathbf{I}_2$ if (*a*) $\mathbf{V}_1 = 0$, $\mathbf{V}_2 = 1$ **V**; (*b*) $\mathbf{V}_1 = -8$ **V**, $\mathbf{V}_2 = 3$ **V**; (*c*) $\mathbf{V}_1 = \mathbf{V}_2 = 5$ **V**.

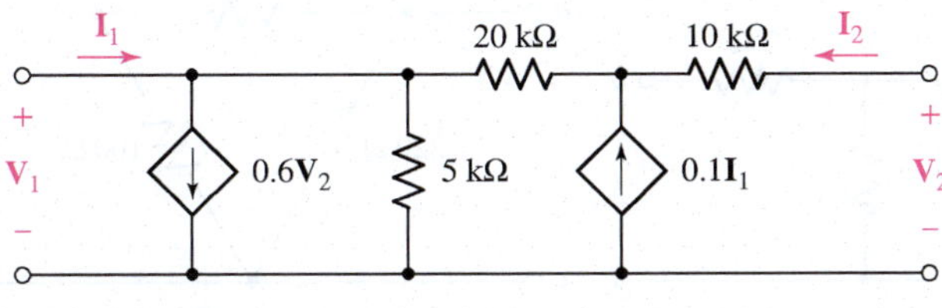

FIGURE 16.46

19. Employ an appropriate method to obtain **y** for the network of Fig. 16.47.

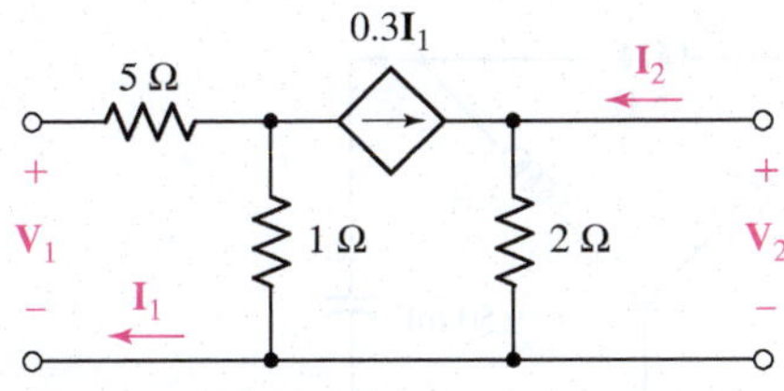

FIGURE 16.47

20. The metal-oxide-semiconductor field effect transistor (MOSFET), a three-terminal nonlinear element used in many electronics applications, is often specified in terms of its **y** parameters. The ac parameters are strongly dependent on the measurement conditions and are commonly named y_{is}, y_{rs}, y_{fs}, and y_{os}, as in

$$I_g = y_{is}V_{gs} + y_{rs}V_{ds} \qquad [49]$$

$$I_d = y_{fs}V_{gs} + y_{os}V_{ds} \qquad [50]$$

where I_g is the transistor gate current, I_d is the transistor drain current, and the third terminal (the source) is common to the input and output during the measurement. Thus, V_{gs} is the voltage between the gate and the source, and V_{ds} is the voltage between the drain and the source. The typical high-frequency model used to approximate the behavior of a MOSFET is shown in Fig. 16.48.

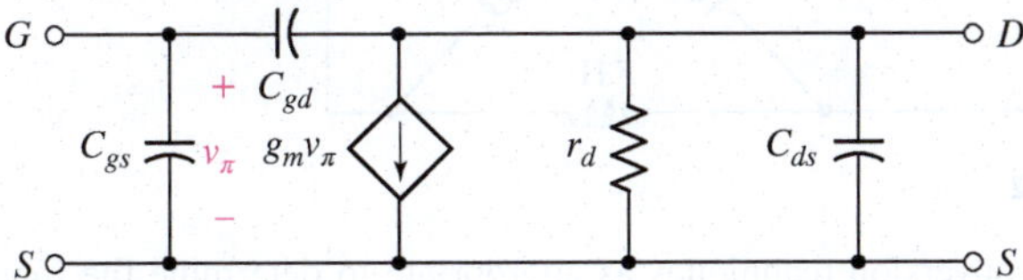

FIGURE 16.48

(*a*) For the configuration stated above, which transistor terminal is used as the input, and which terminal is used as the output? (*b*) Derive expressions for the parameters y_{is}, y_{rs}, y_{fs}, and y_{os} defined in Eqs. [49] and [50], in terms of the model parameters C_{gs}, C_{gd}, g_m, r_d, and C_{ds} of Fig. 16.48. (*c*) Compute y_{is}, y_{rs}, y_{fs}, and y_{os} if $g_m = 4.7$ mS, $C_{gs} = 3.4$ pF, $C_{gd} = 1.4$ pF, $C_{ds} = 0.4$ pF, and $r_d = 10$ kΩ.

16.3 Some Equivalent Networks

21. For the two-port displayed in Fig. 16.49, (*a*) determine the input resistance; (*b*) compute the power dissipated by the network if connected in parallel with a 1 A current source.

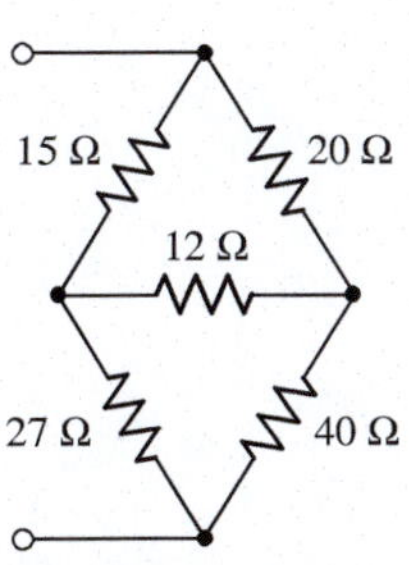

FIGURE 16.49

22. With reference to the two networks in Fig. 16.50, convert the Δ-connected network to a Y-connected network, and vice versa.

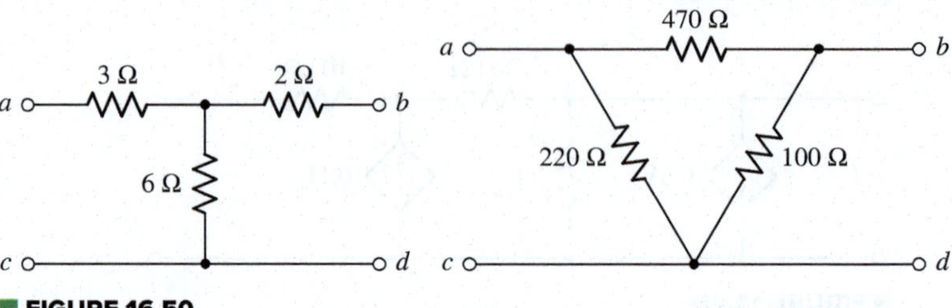

■ **FIGURE 16.50**

23. Determine the input impedance $\mathbf{Z}_{in}$ of the one-port shown in Fig. 16.51 if ω is equal to (*a*) 50 rad/s; (*b*) 1000 rad/s.

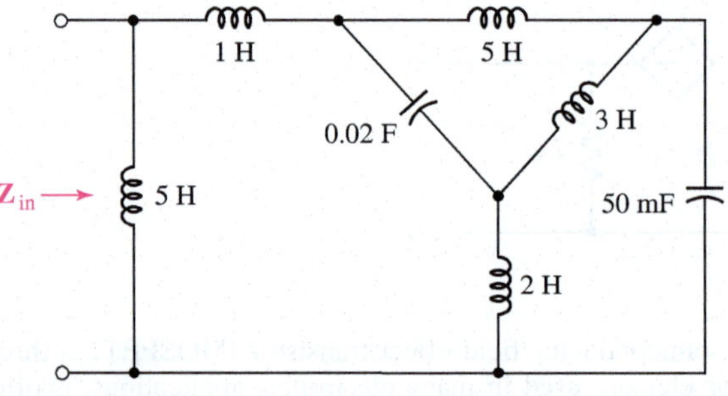

■ **FIGURE 16.51**

24. Determine the input impedance $\mathbf{Z}_{in}$ of the one-port shown in Fig. 16.52 if ω is equal to (*a*) 50 rad/s; (*b*) 1000 rad/s.

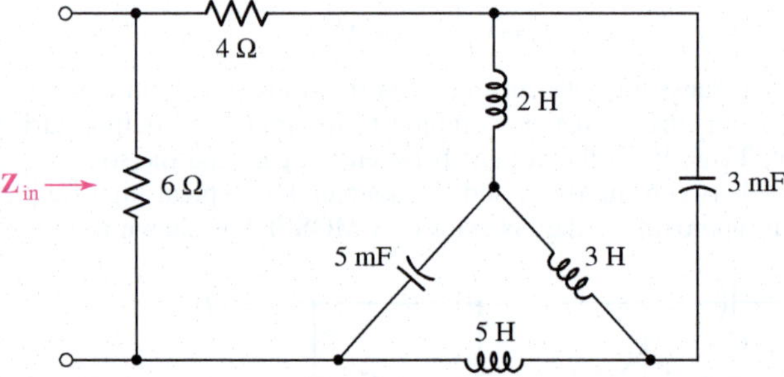

■ **FIGURE 16.52**

25. Employ Δ–Y conversion techniques as appropriate to determine the input resistance R_{in} of the one-port shown in Fig. 16.53.

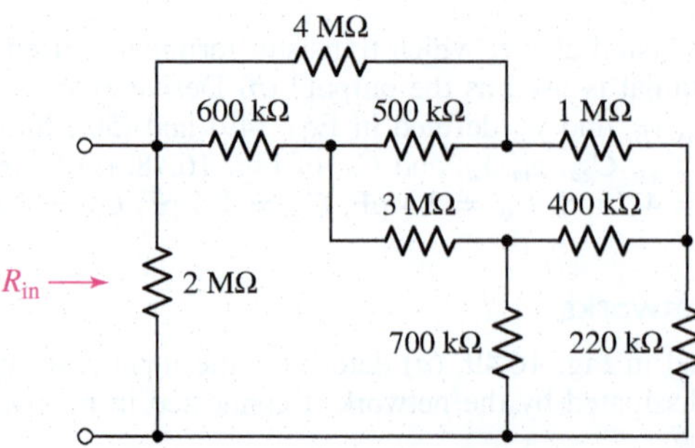

■ **FIGURE 16.53**

26. Employ appropriate techniques to find a value for the input resistance of the one-port network represented by the schematic of Fig. 16.54.

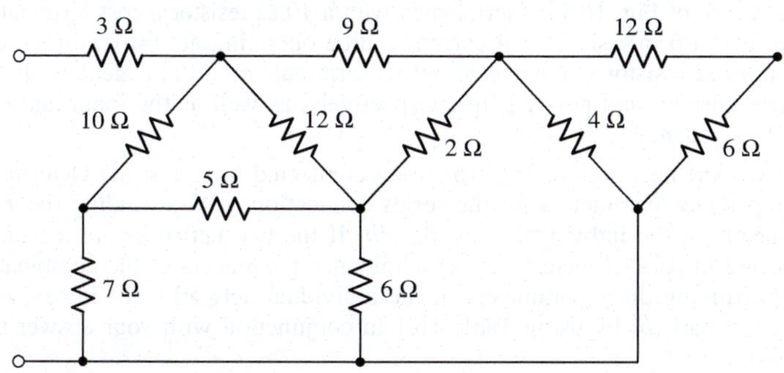

FIGURE 16.54

27. (*a*) Determine the parameter values required to model the network of Fig. 16.43 with the alternative network shown in Fig. 16.13*a*. (*b*) Verify that the two networks are in fact equivalent by computing the power dissipated in a 2 Ω resistor connected to the right of each network and connecting a 1 A current source to the left-hand terminals.

28. The network of Fig. 16.13*b* is equivalent to the network of Fig. 16.43, assuming the appropriate parameter values are chosen. (*a*) Compute the necessary parameter values. (*b*) Verify the equivalence of the two networks by terminating each with a 1 Ω resistor (across their $\mathbf{V}_2$ terminals), connecting a 10 mA source to the other terminals, and showing that $\mathbf{I}_1$, $\mathbf{V}_1$, $\mathbf{I}_2$, and $\mathbf{V}_2$ are equal for both networks.

29. Compute the three parameter values necessary to construct an equivalent network for Fig. 16.43 modeled after the network of Fig. 16.13*c*. Verify their equivalence with an appropriate computer simulation. (*Hint:* Connect some type of source(s) and load(s).)

30. It is possible to construct an alternative two-port to the one shown in Fig. 16.47 by selecting appropriate parameter values as labeled on the diagram in Fig. 16.13. (*a*) Construct such an equivalent network. (*b*) Verify their equivalence with an appropriate computer simulation. (*Hint:* Connect some type of source(s) and load(s).)

31. Let $\mathbf{y} = \begin{bmatrix} 0.1 & -0.05 \\ -0.5 & 0.2 \end{bmatrix}$ (S) for the two-port of Fig. 16.55. Find (*a*) $\mathbf{G}_V$; (*b*) $\mathbf{G}_I$; (*c*) G_P; (*d*) $\mathbf{Z}_{\text{in}}$; (*e*) $\mathbf{Z}_{\text{out}}$. (*f*) If the reverse voltage gain $G_{v,\text{rev}}$ is defined as $\mathbf{V}_1/\mathbf{V}_2$ with $\mathbf{V}_s = 0$ and R_L removed, calculate $G_{V,\text{rev}}$. (*g*) If the insertion power gain G_{ins} is defined as the ratio of $P_{5\Omega}$ with the two-port in place to $P_{5\Omega}$ with the two-port replaced by jumpers connecting each input terminal to the corresponding output terminal, calculate G_{ins}.

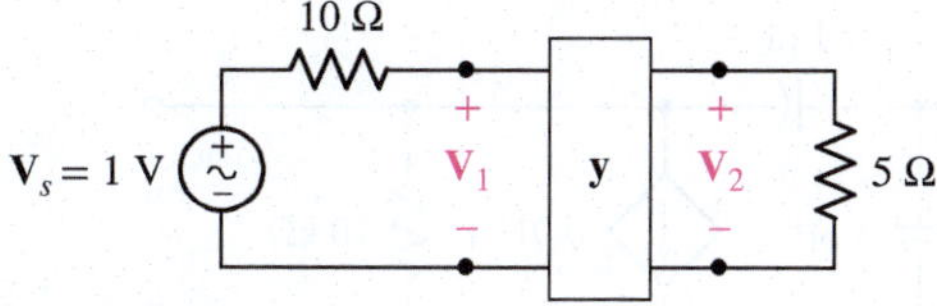

FIGURE 16.55

16.4 Impedance Parameters

32. Convert the following **z** parameters to **y** parameters, or vice versa, as appropriate:

$$\mathbf{z} = \begin{bmatrix} 2 & 3 \\ 5 & 2 \end{bmatrix} \Omega \quad \mathbf{z} = \begin{bmatrix} 100 & 37 \\ 25 & 90 \end{bmatrix} \Omega$$

$$\mathbf{y} = \begin{bmatrix} 1 & 5 \\ 6 & 3 \end{bmatrix} \text{S} \quad \mathbf{y} = \begin{bmatrix} 1 & 2 \\ -1 & 3 \end{bmatrix} \text{S}$$

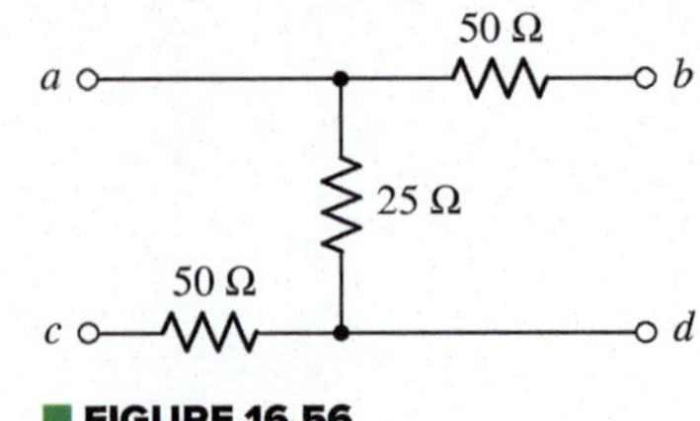

■ **FIGURE 16.56**

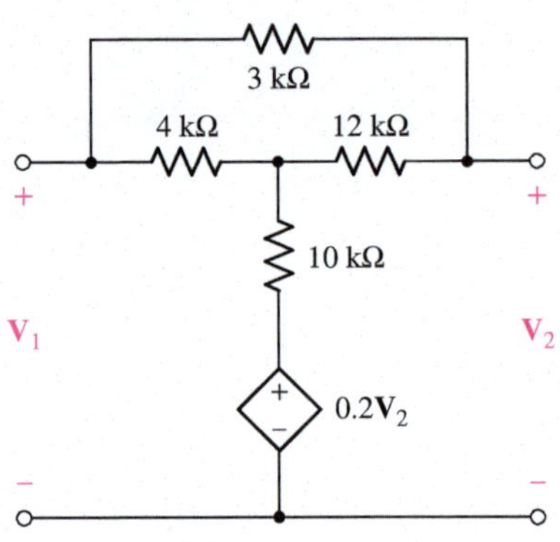

■ **FIGURE 16.57**

33. Determine the complete set of **z** parameters for the network represented in Fig. 16.56.
34. The network of Fig. 16.56 is terminated with a 10 Ω resistor across terminals *b* and *d*, and a 6 mA sinusoidal current source operating at 100 Hz in parallel with a 50 Ω resistor is connected across terminals *a* and *c*. Calculate the voltage, current, and power gains, respectively, as well as the input and output impedance.
35. The two-port networks of Fig. 16.50 are connected in series. (*a*) Determine the impedance parameters for the series connection by first finding the **z** parameters of the individual networks. (*b*) If the two networks are instead connected in parallel, determine the admittance parameters of the combination by first finding the **y** parameters of the individual networks. (*c*) Verify your answer to part (*b*) by using Table 16.1 in conjunction with your answer to part (*a*).
36. (*a*) Use an appropriate method to obtain the impedance parameters which describe the network illustrated in Fig. 16.57. (*b*) If a 1 V source in series with a 1 kΩ resistor is connected to the left-hand port such that the negative reference terminal of the source is connected to the common terminal of the network, and a 5 kΩ load is connected across the right-hand terminals, compute the current, voltage, and power gain.
37. Determine the impedance parameters for the two-port exhibited in Fig. 16.58.

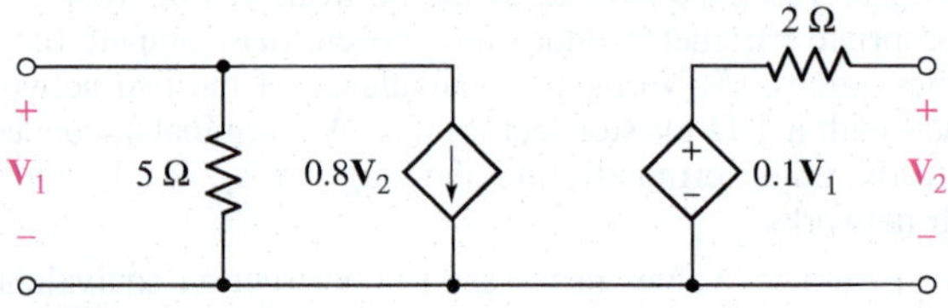

■ **FIGURE 16.58**

38. Obtain both the impedance and admittance parameters for the two-port network of Fig. 16.59.

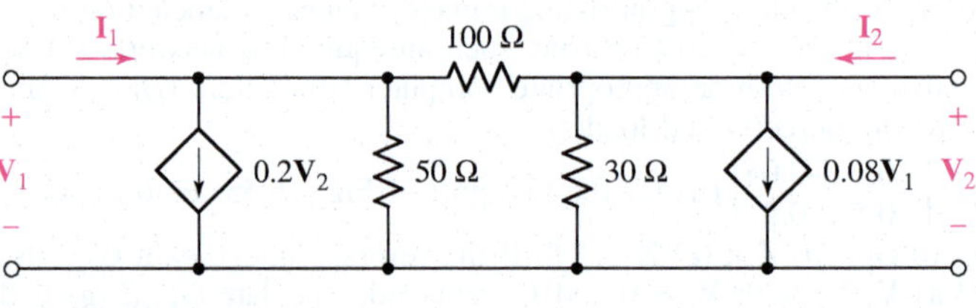

■ **FIGURE 16.59**

39. Find the four **z** parameters at $\omega = 10^8$ rad/s for the transistor high-frequency equivalent circuit shown in Fig. 16.60.

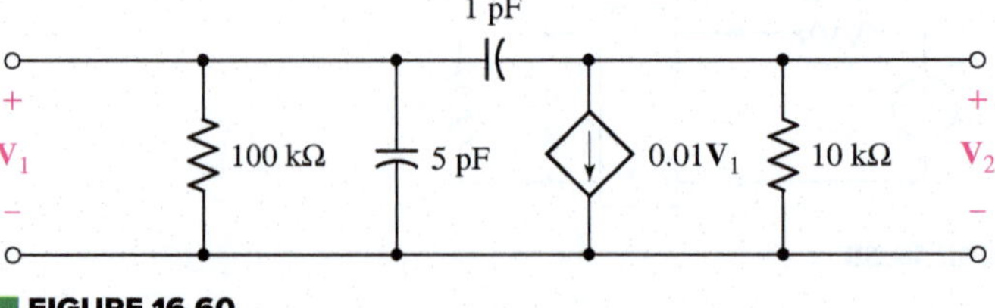

■ **FIGURE 16.60**

16.5 Hybrid Parameters

40. Determine the **h** parameters which describe the purely resistive network shown in Fig. 16.56 by connecting appropriate 1 V, 1 A, and short circuits to terminals as required.

41. (*a*) Obtain the **h** parameters of the two-ports of Fig. 16.61. (*b*) Repeat for the left-hand network, if both resistors are replaced with 1 Ω resistors.

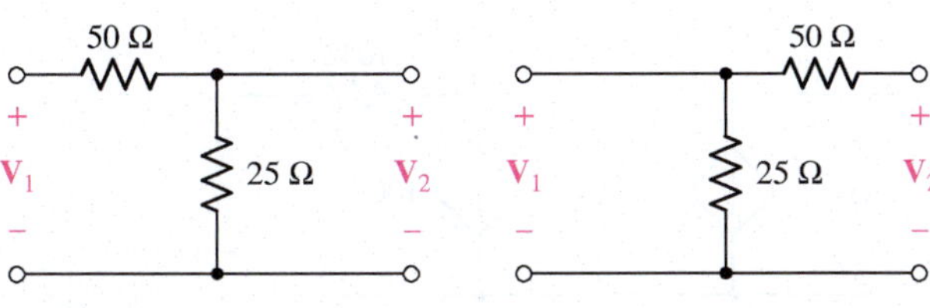

FIGURE 16.61

42. If **h** for some particular two-port is given by $\mathbf{h} = \begin{bmatrix} 2\text{ k}\Omega & -3 \\ 5 & 0.01\text{ S} \end{bmatrix}$, calculate (*a*) **z**; (*b*) **y**.

43. The hybrid parameters $\mathbf{h} = \begin{bmatrix} 75\ \Omega & -2 \\ 5 & 0.1\text{ S} \end{bmatrix}$ describe a particular network. Determine the new **h** parameters if a 17 Ω resistor is connected in parallel with the input terminals.

44. A bipolar junction transistor is connected in common-emitter configuration and found to have **h** parameters $h_{11} = 5\text{ k}\Omega$, $h_{12} = 0.55 \times 10^{-4}$, $h_{21} = 300$, and $h_{22} = 39\ \mu\text{S}$. (*a*) Write **h** in matrix form. (*b*) Determine the small-signal current gain. (*c*) Determine the output resistance in kΩ. (*d*) If a sinusoidal voltage source having frequency 100 rad/s and amplitude 5 mV in series with a 100 Ω resistor is connected to the input terminals, calculate the peak voltage which appears across the output terminals.

45. The two-port which plays a central role in the circuit of Fig. 16.62 can be characterized by hybrid parameters $\mathbf{h} = \begin{bmatrix} 1\Omega & -1 \\ 2 & 0.5\text{ S} \end{bmatrix}$. Determine $\mathbf{I}_1$, $\mathbf{I}_2$, $\mathbf{V}_1$, and $\mathbf{V}_2$.

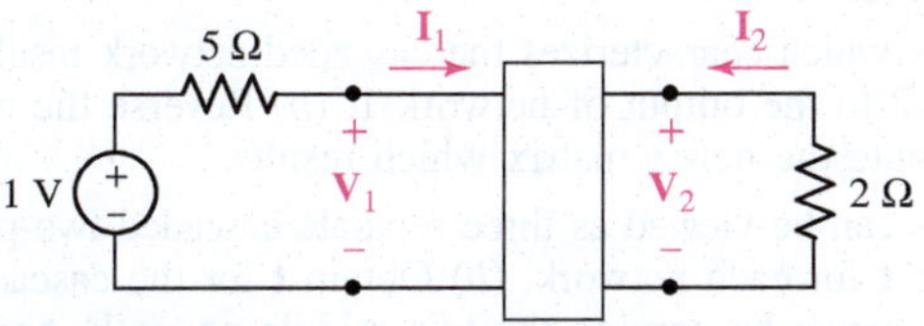

FIGURE 16.62

46. The two networks of Fig. 16.61 are connected in series by connecting the terminals as illustrated in Fig. 16.22 (assume the left-hand network of Fig. 16.61 is network *A*). Determine the new set of **h** parameters which describe the series connection.

47. The two networks of Fig. 16.61 are connected in parallel by tying the corresponding input terminals together, and then tying the corresponding output terminals together. Determine the new set of **h** parameters which describe the parallel connection.

48. Find **y**, **z**, and **h** for both of the two-ports shown in Fig. 16.63. If any parameter is infinite, skip that parameter set.

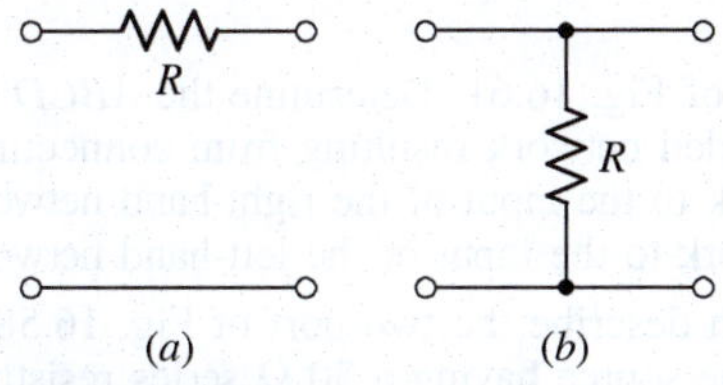

FIGURE 16.63

49. (*a*) Find $\mathbf{h}$ for the two-port of Fig. 16.64. (*b*) Find $\mathbf{Z}_{out}$ if the input contains $\mathbf{V}_S$ in series with $R_S = 200\ \Omega$.

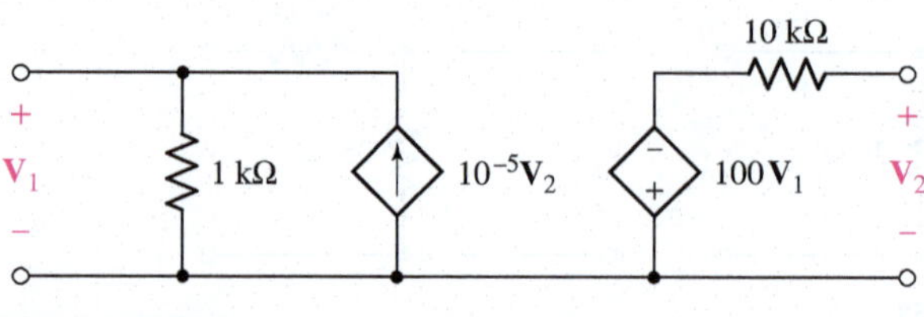

FIGURE 16.64

16.6 Transmission Parameters

50. (*a*) With the assistance of appropriate mesh equations, determine the *ABCD* matrix which represents the two-port shown in Fig. 16.9. (*b*) Convert your answer to $\mathbf{h}$.

51. (*a*) Employ suitably written mesh equations to obtain the $\mathbf{t}$ parameters which characterize the network of Fig. 16.57. (*b*) If currents $\mathbf{I}_1$ and $\mathbf{I}_2$ are defined as flowing into the (+) reference terminals of $\mathbf{V}_1$ and $\mathbf{V}_2$, respectively, compute the voltages if $\mathbf{I}_1 = 2\mathbf{I}_2 = 3$ mA.

52. Consider the following matrices:

$$\mathbf{a} = \begin{bmatrix} 5 & 2 \\ 4 & 1 \end{bmatrix} \quad \mathbf{b} = \begin{bmatrix} 1.5 & 1 \\ 1 & 0.5 \end{bmatrix} \quad \mathbf{c} = \begin{bmatrix} -4 \\ 2 \end{bmatrix}$$

Calculate (*a*) $\mathbf{a} \cdot \mathbf{b}$; (*b*) $\mathbf{b} \cdot \mathbf{a}$; (*c*) $\mathbf{a} \cdot \mathbf{c}$; (*d*) $\mathbf{b} \cdot \mathbf{c}$; (*e*) $\mathbf{b} \cdot \mathbf{a} \cdot \mathbf{c}$; (*f*) $\mathbf{a} \cdot \mathbf{a}$.

53. Two networks are represented by the following impedance matrices:

$$\mathbf{z}_1 = \begin{bmatrix} 4 & 5 \\ 8 & 1 \end{bmatrix}\ \Omega \text{ and } \mathbf{z}_2 = \begin{bmatrix} 1.1 & 2.2 \\ 0.89 & 1.8 \end{bmatrix}\ \Omega, \text{ respectively.}$$

(*a*) Determine the $\mathbf{t}$ matrix which characterizes the cascaded network resulting from connecting network 2 to the output of network 1. (*b*) Reverse the order of the networks and compute the new $\mathbf{t}$ matrix which results.

54. The two-port of Fig. 16.65 can be viewed as three separate cascaded two-ports *A*, *B*, and *C*. (*a*) Compute $\mathbf{t}$ for each network. (*b*) Obtain $\mathbf{t}$ for the cascaded network. (*c*) Verify your answer by naming the two middle nodes V_x and V_y, respectively, writing nodal equations, obtaining the admittance parameters from your nodal equations, and converting to $\mathbf{t}$ parameters using Table 16.1.

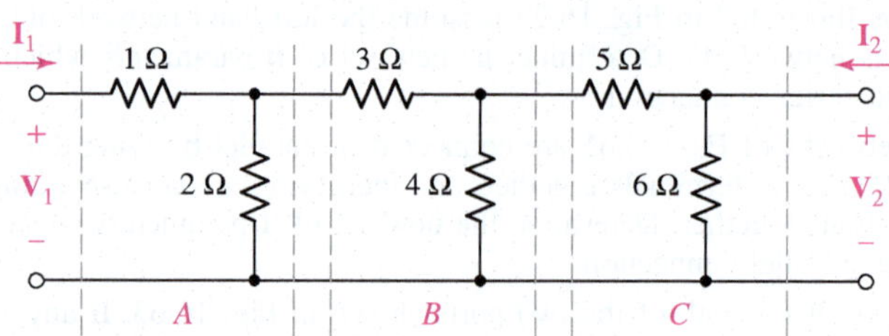

FIGURE 16.65

55. Consider the two separate two-ports of Fig. 16.61. Determine the *ABCD* matrix which characterizes the cascaded network resulting from connecting (*a*) the output of the left-hand network to the input of the right-hand network; (*b*) the output of the right-hand network to the input of the left-hand network.

56. (*a*) Determine the $\mathbf{t}$ parameters which describe the two-port of Fig. 16.58. (*b*) Compute $\mathbf{Z}_{out}$ if a practical voltage source having a 30 Ω series resistance is connected to the input terminals of the network.

57. Three identical networks to the one depicted in Fig. 16.56 are cascaded together. Determine the **t** parameters which fully represent the result.

58. (*a*) Find $\mathbf{t}_a$, $\mathbf{t}_b$, and $\mathbf{t}_c$ for the networks shown in Fig. 16.66*a*, *b*, and *c*. (*b*) By using the rules for interconnecting two-ports in cascade, find **t** for the network of Fig. 16.66*d*.

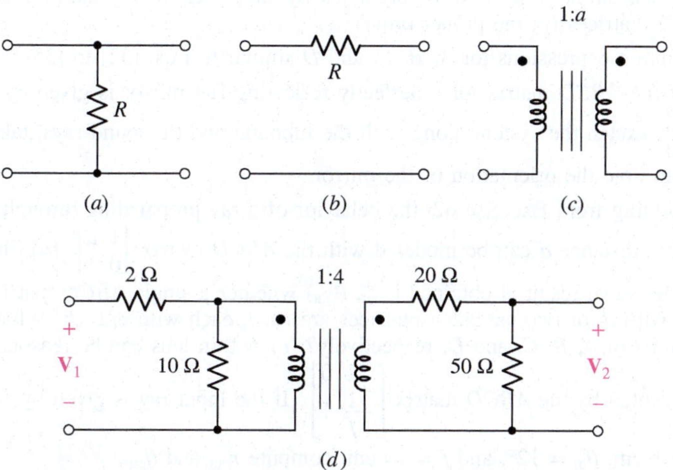

FIGURE 16.66

Chapter-Integrating Exercises

59. (*a*) Obtain **y**, **z**, **h**, and **t** parameters for the network shown in Fig. 16.67 using either the defining equations or mesh/nodal equations. (*b*) Verify your answers using the relationships in Table 16.1.

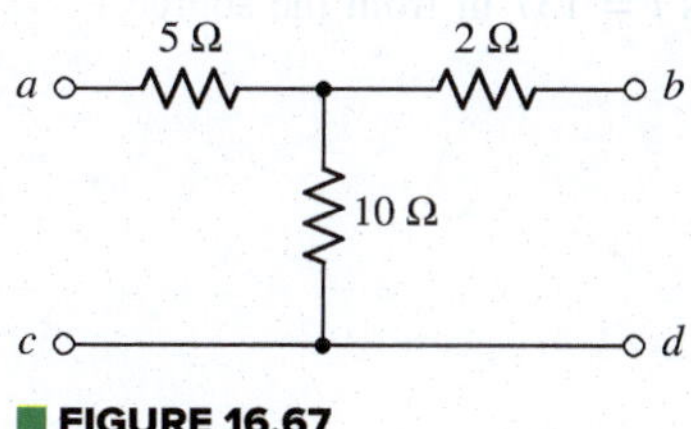

FIGURE 16.67

60. Four networks, each identical to the one depicted in Fig. 16.67, are connected in parallel such that all terminals labeled *a* are tied together, all terminals designated *b* are tied together, and all terminals labeled *c* and *d* are connected. Obtain the **y**, **z**, **h**, and **t** parameters which describe the parallel-connected network.

61. A cascaded 12-element network is formed using four two-port networks identical to the one shown in Fig. 16.67. Determine the **y**, **z**, **h**, and **t** parameters which describe the result.

62. The concept of *ABCD* matrices extends to systems beyond electrical circuits. For example, they are commonly employed for ray-tracing calculations in optical systems. In that case, we envision parallel input and output planes in *xy*, skewered by an optical axis *z*. An inbound ray crosses the input plane a distance $x = r_{\text{in}}$ from the optical axis, making an angle θ_{in}. The corresponding

parameters r_{out}, θ_{out} for the outbound ray crossing the output plane are then given by the *ABCD* matrix such that

$$\begin{bmatrix} r_{out} \\ \theta_{out} \end{bmatrix} = \begin{bmatrix} A & B \\ C & D \end{bmatrix} \begin{bmatrix} r_{in} \\ \theta_{in} \end{bmatrix}$$

Each type of optical element (e.g., mirror, lens, or even propagation through free space) has its own *ABCD* matrix. If the ray passes through several elements, the net effect can be predicted by simply cascading the individual *ABCD* matrices (in the proper order).

(*a*) Obtain expressions for *A, B, C,* and *D* similar to Eqs. [32] to [35].

(*b*) If the *ABCD* matrix for a perfectly reflecting flat mirror is given by $\begin{bmatrix} 1 & 0 \\ 0 & 1 \end{bmatrix}$, sketch the system along with the inbound and outbound rays, taking care to note the orientation of the mirror.

63. Continuing from Exercise 62, the behavior of a ray propagating through free space a distance *d* can be modeled with the *ABCD* matrix $\begin{bmatrix} 1 & d \\ 0 & 1 \end{bmatrix}$. (*a*) Show that the same result is obtained (r_{out}, θ_{out}) whether a single *ABCD* matrix is used with *d*, or two cascaded matrices are used, each with *d*/2. (*b*) What are the units of *A, B, C,* and *D,* respectively? (*c*) A thin lens can be reasonably represented by the *ABCD* matrix $\begin{bmatrix} 1 & 0 \\ -\frac{1}{f} & 1 \end{bmatrix}$. If the input ray is given by $r_{in} = 1$ cm, $\theta_{in} = 12°$, and $f = 10$ cm, compute r_{out} and θ_{out}.

64. Another application for *ABCD* matrices is the modeling of power transmission lines. We denote the input side as $\mathbf{V}_1$, $\mathbf{I}_1$, and the output side (what is delivered by the transmission line) as $\mathbf{V}_2$, $\mathbf{I}_2$; here $\mathbf{I}_1$ is defined as flowing into the transmission line, and $\mathbf{I}_2$ is defined as flowing out of the transmission line. Then $A = \mathbf{V}_1/\mathbf{V}_2$ = voltage ratio; $B = \mathbf{V}_1/\mathbf{I}_2$ = short circuit resistance; $C = \mathbf{I}_1/\mathbf{V}_2$ = open circuit conductance; and $D = \mathbf{I}_1/\mathbf{I}_2$ = current ratio. (*a*) State the units of each parameter. (*b*) If the *ABCD* matrix is given by $\begin{bmatrix} \cos\beta l & jZ_0 \sin\beta l \\ jY_0 \sin\beta l & \cos\beta l \end{bmatrix}$ and the phase constant $\beta = 0.8\pi$ radians per meter, compute the voltage ratio if the load is $l = 137$ m from the source.

CHAPTER 17

Fourier Circuit Analysis

KEY CONCEPTS

Representing Periodic Functions as a Sum of Sines and Cosines

Harmonic Frequencies

Even and Odd Symmetry

Half-Wave Symmetry

Complex Form of the Fourier Series

Discrete Line Spectra

Fourier Transform

Using Fourier Series and Fourier Transform Techniques in Circuit Analysis

System Response and Convolution in the Frequency Domain

INTRODUCTION

In this chapter we continue our introduction to circuit analysis by studying periodic functions in both the time and frequency domains. Specifically, we will consider forcing functions which are *periodic* and have functional natures which satisfy certain mathematical restrictions that are characteristic of any function which we can generate in the laboratory. Such functions may be represented as the sum of an infinite number of sine and cosine functions which are harmonically related. Therefore, since the forced response to each sinusoidal component can be determined easily by sinusoidal steady-state analysis, the response of the linear network to the general periodic forcing function may be obtained by superposing the partial responses.

The topic of Fourier series is of vital importance in a number of fields, particularly communications. The use of Fourier-based techniques to assist in circuit analysis, however, had been slowly falling out of fashion for a number of years. Now as we face an increasingly larger fraction of global power usage coming from equipment employing pulse-modulated power supplies (e.g., computers), the subject of harmonics in power systems and power electronics is rapidly becoming a serious problem in even large-scale generation plants. It is only with Fourier-based analysis that the underlying problems and possible solutions can be understood.

17.1 TRIGONOMETRIC FORM OF THE FOURIER SERIES

We know that the complete response of a linear circuit to an arbitrary forcing function is composed of the sum of a *forced response* and a *natural response*. The natural response has been considered both in the time domain (Chaps. 7, 8, and 9) and in the frequency domain (Chaps. 14 and 15). The forced response has also been considered from several perspectives, including the phasor-based techniques of Chap. 10. As we have discovered, in some cases we need *both* components of the total response of a particular circuit, while in others we need only the natural or the forced response. In this section, we refocus our attention on forcing functions that are *sinusoidal* in nature, and we discover how to write a general periodic function as a *sum* of such functions—leading us into a discussion of a new set of circuit analysis procedures.

Harmonics

Some feeling for the validity of representing a general *periodic* function by an infinite sum of sine and cosine functions may be gained by considering a simple example. Let us first assume a cosine function of radian frequency ω_0,

$$v_1(t) = 2\cos\omega_0 t$$

where

$$\omega_0 = 2\pi f_0$$

and the period T is

$$T = \frac{1}{f_0} = \frac{2\pi}{\omega_0}$$

Although T does not usually carry a zero subscript, it is the period of the fundamental frequency. The ***harmonics*** of this sinusoid have frequencies $n\omega_0$, where ω_0 is the fundamental frequency and $n = 1, 2, 3, \ldots$. The frequency of the first harmonic is the ***fundamental frequency.***

Next let us combine this with a third-harmonic voltage given by

$$v_3(t) = \cos 3\omega_0 t$$

The fundamental $v_1(t)$, the third harmonic $v_3(t)$, and the sum of these two waves $v(t)$ are shown as functions of time in Fig. 17.1*a*. Note that the sum is also periodic, with period $T = 2\pi/\omega_0$.

The form of the resultant periodic function changes as the phase and amplitude of the third-harmonic component change. Thus, Fig. 17.1*b* shows the effect of combining $v_1(t)$ and a third harmonic of slightly larger amplitude,

$$v_3(t) = 2\cos 3\omega_0 t$$

By shifting the phase of the third harmonic by 90 degrees to give

$$v_3(t) = \sin 3\omega_0 t$$

the sum, shown in Fig. 17.1*c*, takes on a still different character. In all cases, the period of the resultant waveform is the same as the period of the fundamental waveform. The nature of the waveform depends on the amplitude and phase of every possible harmonic component, and we will find that we can generate waveforms which have extremely nonsinusoidal characteristics by an appropriate combination of sinusoidal functions.

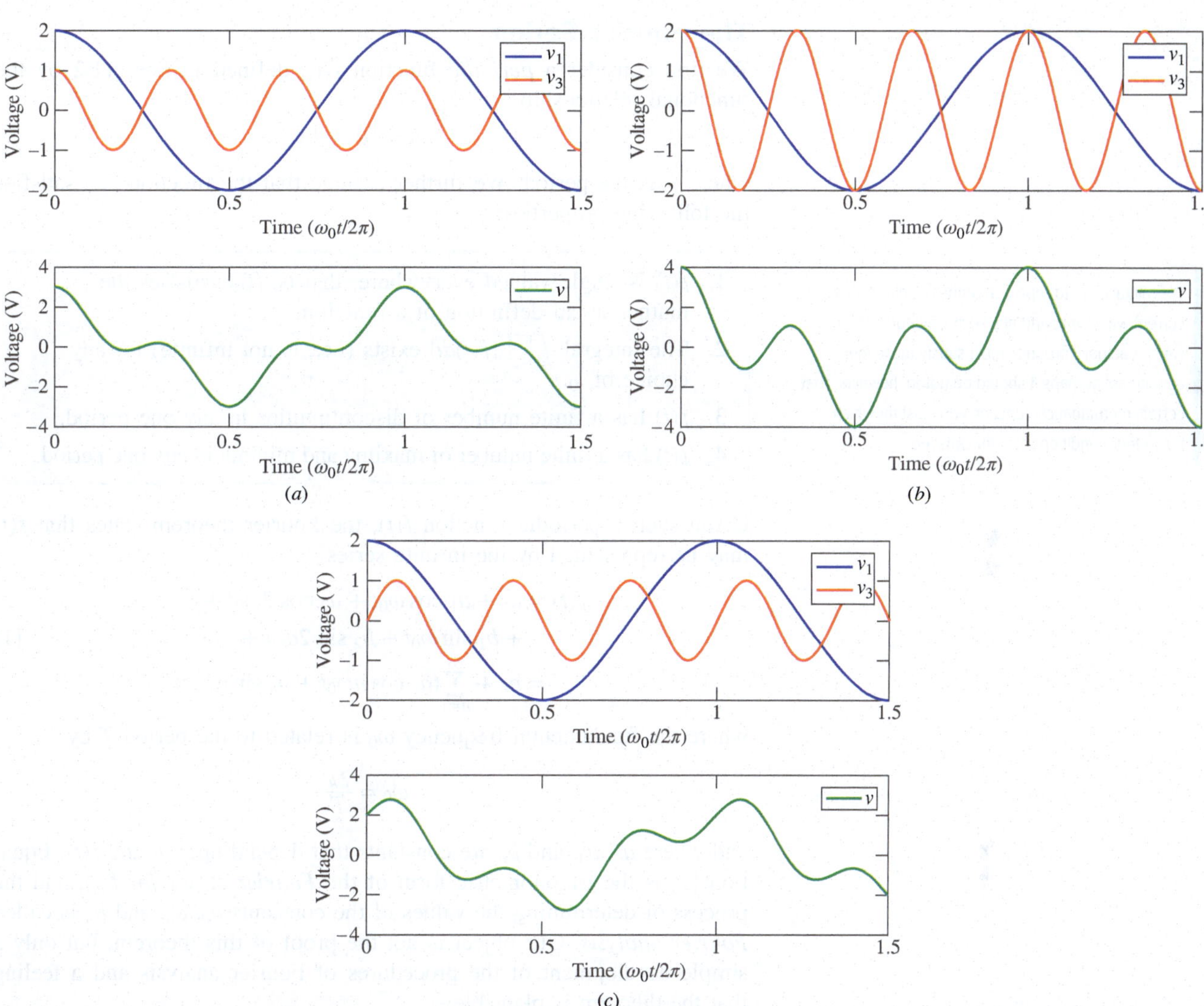

■ **FIGURE 17.1** Several example waveforms combining a fundamental and a third harmonic. The fundamental is $v_1(t) = 2\cos\omega_0 t$, and the third harmonic is (a) $v_3(t) = \cos 3\omega_0 t$; (b) $v_3(t) = 2\cos 3\omega_0 t$; (c) $v_3(t) = \sin 3\omega_0 t$.

After we have become familiar with the use of the sum of an infinite number of sine and cosine functions to represent a periodic waveform, we will consider the frequency-domain representation of a general nonperiodic waveform in a manner similar to the Laplace transform.

PRACTICE

17.1 Let a third-harmonic voltage be added to the fundamental to yield $v = 2\cos\omega_0 t + V_{m3}\sin 3\omega_0 t$, the waveform shown in Fig. 17.1*c* for $V_{m3} = 1$. (*a*) Find the value of V_{m3} so that $v(t)$ will have zero slope at $\omega_0 t = 2\pi/3$. (*b*) Evaluate $v(t)$ at $\omega_0 t = 2\pi/3$.

Ans: 0.577; −1.000.

The Fourier Series

We first consider a *periodic* function $f(t)$, defined in Sec. 11.2 by the functional relationship

$$f(t) = f(t + T)$$

where T is the period. We further assume that the function $f(t)$ satisfies the following properties:

We will take $f(t)$ to represent either a voltage or a current waveform, and any such waveform which we can actually produce must satisfy these four conditions; perhaps it should be noted, however, that certain mathematical functions do exist for which these four conditions are not satisfied.

1. $f(t)$ is single-valued everywhere; that is, $f(t)$ satisfies the mathematical definition of a function.
2. The integral $\int_{t_0}^{t_0+T} |f(t)dt|$ exists (i.e., is not infinite) for any choice of t_0.
3. $f(t)$ has a finite number of discontinuities in any one period.
4. $f(t)$ has a finite number of maxima and minima in any one period.

Given such a periodic function $f(t)$, the Fourier theorem states that $f(t)$ may be represented by the infinite series

$$\begin{aligned} f(t) &= a_0 + a_1 \cos \omega_0 t + a_2 \cos 2\omega_0 t + \cdots \\ &\quad + b_1 \sin \omega_0 t + b_2 \sin 2\omega_0 t + \cdots \\ &= a_0 + \sum_{n=1}^{\infty} (a_n \cos n\omega_0 t + b_n \sin n\omega_0 t) \end{aligned} \qquad [1]$$

where the fundamental frequency ω_0 is related to the period T by

$$\omega_0 = \frac{2\pi}{T}$$

and where a_0, a_n, and b_n are constants that depend upon n and $f(t)$. Equation [1] is the trigonometric form of the ***Fourier series** for* $f(t)$, and the process of determining the values of the constants a_0, a_n, and b_n is called *Fourier analysis*. Our object is not the proof of this theorem, but only a simple development of the procedures of Fourier analysis and a feeling that the theorem is plausible.

Some Useful Trigonometric Integrals

Before we discuss the evaluation of the constants appearing in the Fourier series, let us collect a set of useful trigonometric integrals. We let both n and k represent any element of the set of integers 1, 2, 3, ... In the following integrals, 0 and T are used as the integration limits, but it is understood that any interval of one period is equally correct.

$$\int_0^T \sin n\omega_0 t \, dt = 0 \qquad [2]$$

$$\int_0^T \cos n\omega_0 t \, dt = 0 \qquad [3]$$

$$\int_0^T \sin k\omega_0 t \cos n\omega_0 t \, dt = 0 \qquad [4]$$

$$\int_0^T \sin k\omega_0 t \sin n\omega_0 t \, dt = 0 \qquad (k \neq n) \qquad [5]$$

$$\int_0^T \cos k\omega_0 t \cos n\omega_0 t \, dt = 0 \qquad (k \neq n) \qquad [6]$$

Those cases which are excepted in Eqs. [5] and [6] are also easily evaluated; we obtain

$$\int_0^T \sin^2 n\omega_0 t\, dt = \frac{T}{2} \qquad [7]$$

$$\int_0^T \cos^2 n\omega_0 t\, dt = \frac{T}{2} \qquad [8]$$

Evaluation of the Fourier Coefficients

The evaluation of the unknown constants in the Fourier series may now be accomplished readily. We first attack a_0. If we integrate each side of Eq. [1] over a full period, we obtain

$$\int_0^T f(t)dt = \int_0^T a_0 dt + \int_0^T \sum_{n=1}^{\infty} (a_n \cos n\omega_0 t + b_n \sin n\omega_0 t)dt$$

But every term in the summation is of the form of Eq. [2] or [3], and thus

$$\int_0^T f(t)dt = a_0 T$$

or

$$a_0 = \frac{1}{T}\int_0^T f(t)dt \qquad [9]$$

This constant a_0 is simply the average value of $f(t)$ over a period, and we therefore describe it as the dc component of $f(t)$.

To evaluate one of the cosine coefficients, for example, a_k, the coefficient of $\cos k\omega_0 t$, we first multiply each side of Eq. [1] by $\cos k\omega_0 t$ and then integrate both sides of the equation over a full period:

$$\begin{aligned}\int_0^T f(t)\cos k\omega_0 t\, dt &= \int_0^T a_0 \cos k\omega_0 t\, dt \\ &+ \int_0^T \sum_{n=1}^{\infty} a_n \cos k\omega_0 t \cos n\omega_0 t\, dt \\ &+ \int_0^T \sum_{n=1}^{\infty} b_n \cos k\omega_0 t \sin n\omega_0 t\, dt\end{aligned}$$

From Eqs. [3], [4], and [6] we note that every term on the right-hand side of this equation is zero except for the single a_n term where $k = n$. We evaluate that term using Eq. [8], and in so doing we find a_k, or a_n:

$$a_n = \frac{2}{T}\int_0^T f(t)\cos n\omega_0 t\, dt \qquad [10]$$

This result is *twice* the average value of the product $f(t)\cos n\omega_0 t$ over a period.

In a similar way, we obtain b_k by multiplying by $\sin k\omega_0 t$ integrating over a period, noting that all but one of the terms on the right-hand side are zero, and performing that single integration by Eq. [7]. The result is

$$b_n = \frac{2}{T}\int_0^T f(t)\sin n\omega_0 t\, dt \qquad [11]$$

which is *twice* the average value of $f(t)\sin n\omega_0 t$ over a period.

Equations [9] to [11] now enable us to determine values for a_0 and all the a_n and b_n in the Fourier series, Eq. [1], as summarized below:

$$f(t) = a_0 + \sum_{n=1}^{\infty}(a_n \cos n\omega_0 t + b_n \sin n\omega_0 t) \qquad [1]$$

$$\omega_0 = \frac{2\pi}{T} = 2\pi f_0$$

$$a_0 = \frac{1}{T}\int_0^T f(t)\,dt \qquad [9]$$

$$a_n = \frac{2}{T}\int_0^T f(t)\cos n\omega_0 t\,dt \qquad [10]$$

$$b_n = \frac{2}{T}\int_0^T f(t)\sin n\omega_0 t\,dt \qquad [11]$$

EXAMPLE 17.1

The sawtooth waveform shown in Fig. 17.2 represents the periodic voltage response of an integrator circuit, such as that obtained in digital imaging. Find the Fourier series representation of this waveform, and plot for the cases of $n = 3$ and $n = 30$.

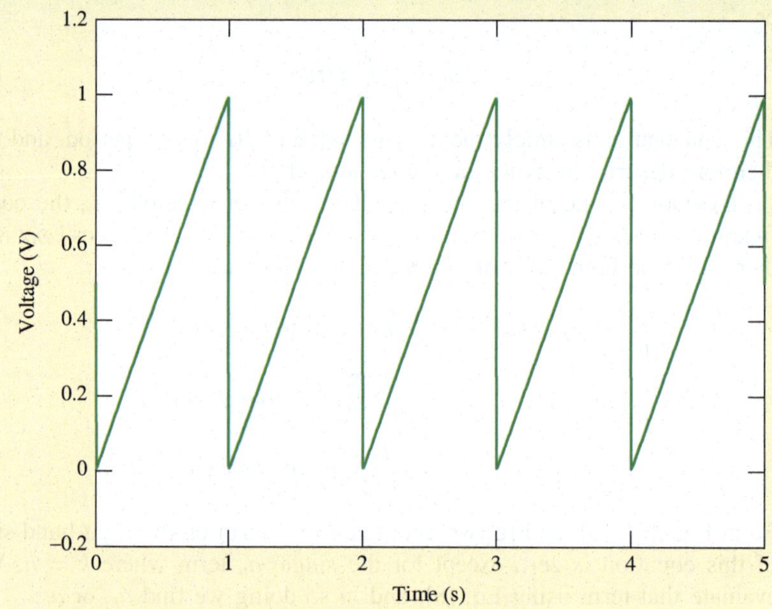

■ **FIGURE 17.2** A periodic sawtooth function of an integrator circuit.

▶ ***Identify the goal of the problem.***

We are presented with a periodic function and are asked to find the Fourier series representation.

▶ ***Collect the known information.***

In order to represent this voltage as a Fourier series, we must first determine the period and then express the graphical voltage as an analytical function of time. From the graph, the period is seen to be

$$T = 1 \text{ s}$$

and

$$\omega_0 = 2\pi \text{ rad/s}$$

▶ ***Devise a plan.***

The most straightforward approach is to apply Eqs. [9] to [11] to calculate the set of coefficients a_0, a_n, and b_n. To do this, we need a functional expression for $v(t)$, which is a linear function that repeats each period

$$v(t) = \frac{t}{T} \qquad 0 \le t \le 1$$

▶ ***Construct an appropriate set of equations.***

The zero-frequency component is easily obtained:

$$a_0 = \frac{1}{T}\int_0^T \frac{t}{T}dt = \frac{1}{T^2}\left(\frac{t^2}{2}\right)\Big|_0^T = \frac{1}{2}$$

Notice that integration over an entire period must be broken up into subintervals of the period, in each of which the functional form of $v(t)$ is known.

The amplitude of a *general* cosine term is

$$a_n = \frac{2}{T}\int_0^T \frac{t}{T}\cos(n\omega_0 t)dt$$

$$a_n = \frac{2}{T^2}\left[\frac{1}{(n\omega_0)^2}\cos(n\omega_0 t) + \frac{t}{n\omega_0}\sin(n\omega_0 t)\right]\Bigg|_0^T$$

$$a_n = 0!$$

and the amplitude of a general sine term is

$$b_n = \frac{2}{T}\int_0^T \frac{t}{T}\sin(n\omega_0 t)dt$$

$$b_n = \frac{2}{T^2}\left[-\frac{t}{n\omega_0}\cos(n\omega_0 t) + \frac{1}{(n\omega_0)^2}\sin(n\omega_0 t)\right]\Bigg|_0^T$$

$$b_n = -\frac{1}{n\pi}$$

▶ ***Determine if additional information is required.***

We now have the required coefficients for the Fourier series representation.

▶ ***Attempt a solution.***

Combining terms, we have

$$v(t) = \frac{1}{2} - \sum_{n=1}^{\infty} \frac{1}{n\pi}\sin(n\omega_0 t)$$

▶ ***Verify the solution. Is it reasonable or expected?***

Our solution can be checked by plotting the function for an increasing number of terms, as shown in Fig. 17.3. As can be seen, as more terms are included, the more the plot resembles that of Fig. 17.3. For $n = 3$, there are only a few sinusoids to work with, resulting in a relatively inaccurate representation of the sawtooth function.

(Continued on next page)

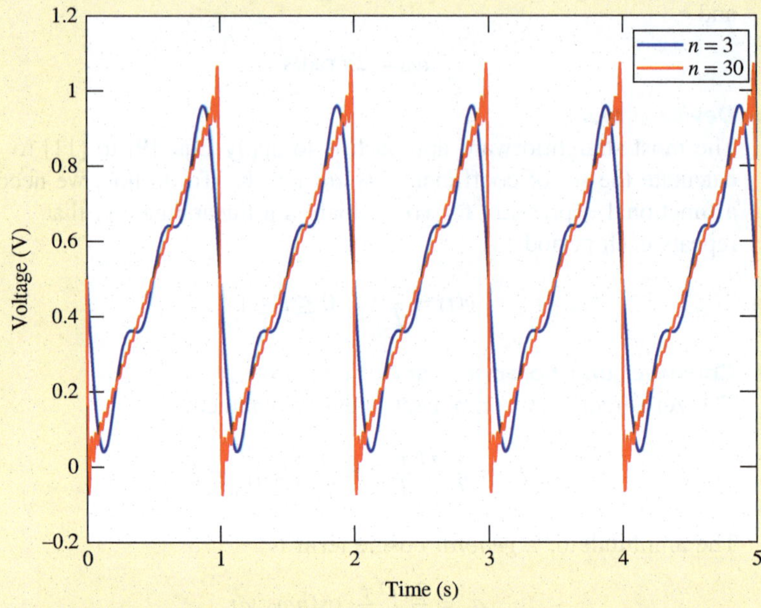

FIGURE 17.3 Solution truncated after $n = 3$ and $n = 30$ terms, showing convergence to the sawtooth function $v(t)$.

For $n = 30$, the superposition of a large number of sinusoids of varying frequency provides a close representation of the sawtooth signal.

```
t=linspace(0,5,1000); % vector for time over 1000 points
T=1; % Period
w0=2*pi/T; % natural frequency
a0=0.5; % constant
for i=1:1000;
  sum=0; % begin sum
  for k=1:3; % loop for n=3
    sum=sum-1/k/pi*sin(k*w0*t(i));
  end
  f3(i)=a0+sum; % function for n=3
  sum=0;
  for k=1:30; % loop for n=30
    sum=sum-1/k/pi*sin(k*w0*t(i));
  end
  f30(i)=a0+sum; % function for n=30
end
figure(1)
plot(t,f3,t,f30,'LineWidth',1.0)
xlabel('Time (s)')
ylabel('Voltage (V)')
legend('n=3','n=30')
```

PRACTICE

17.2 A periodic waveform $f(t)$ is described as follows: $f(t) = -4$, $0 < t < 0.3$; $f(t) = 6$, $0.3 < t < 0.4$; $f(t) = 0$, $0.4 < t < 0.5$; $T = 0.5$. Evaluate (*a*) a_0; (*b*) a_3; (*c*) b_1.

17.3 Write the Fourier series for the three voltage waveforms shown in Fig. 17.4.

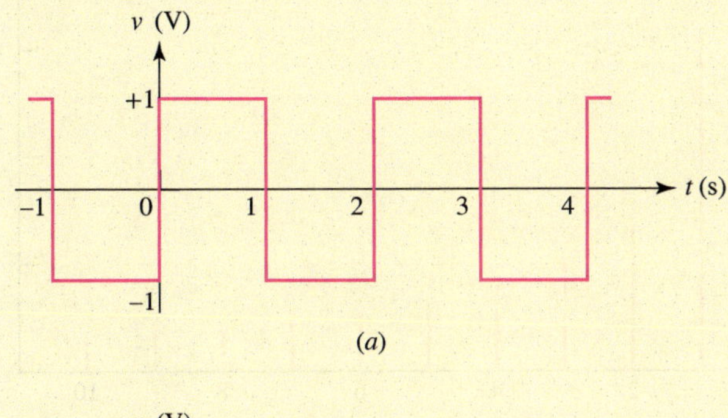

(*a*)

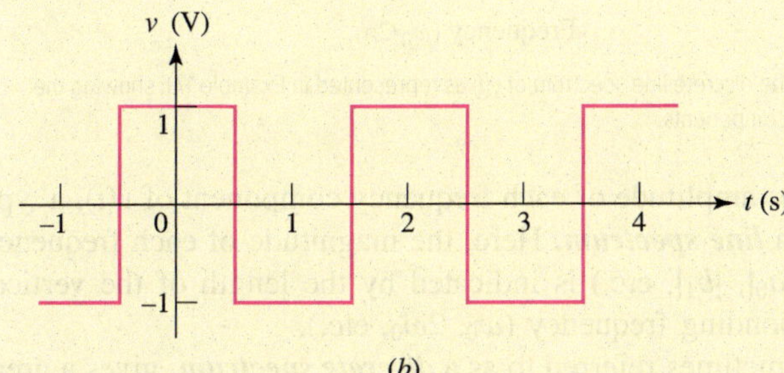

(*b*)

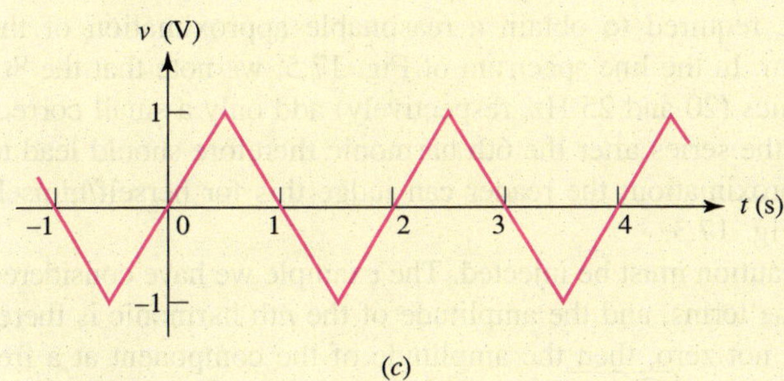

(*c*)

FIGURE 17.4

Ans: 17.2: −1.200; 1.383; −4.44. 17.3: $(4/\pi)(\sin \pi t + \frac{1}{3}\sin 3\pi t + \frac{1}{5}\sin 5\pi t + \cdots)$ V; $(4/\pi)(\cos \pi t - \frac{1}{3}\cos 3\pi t + \frac{1}{5}\cos 5\pi t - \cdots)$ V; $(8/\pi^2)(\sin \pi t - \frac{1}{9}\sin 3\pi t + \frac{1}{25}\sin 5\pi t - \cdots)$.

Line and Phase Spectra

We depicted the function $v(t)$ of Example 17.1 , graphically and analytically—both representations being in the time domain. The Fourier series representation of $v(t)$ is also a time-domain expression, but it may be transformed into a *frequency-domain* representation as well. For example,

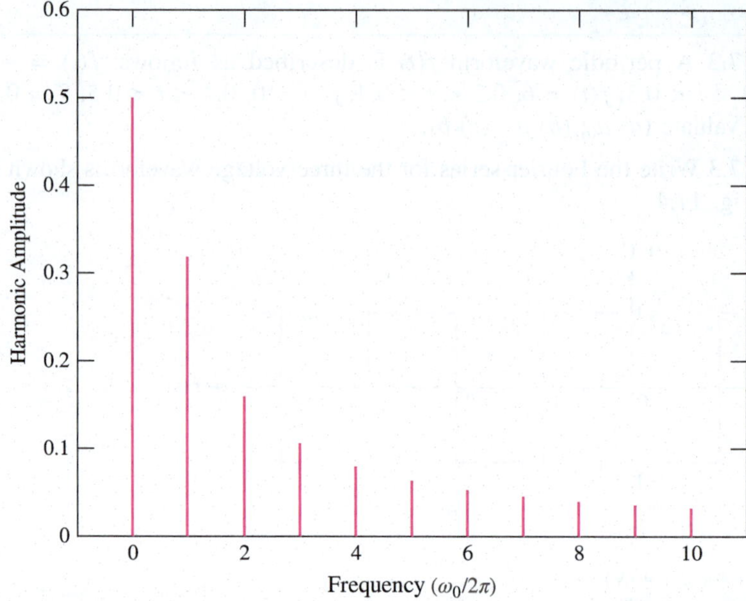

FIGURE 17.5 The discrete line spectrum of $v(t)$ as represented in Example 17.1, showing the first ten frequency components.

Fig. 17.5 shows the amplitude of each frequency component of $v(t)$, a type of plot known as a ***line spectrum.*** Here, the magnitude of each frequency component (i.e., $|a_0|$, $|b_1|$, etc.) is indicated by the length of the vertical line at the corresponding frequency (ω_0, $2\omega_0$, etc.).

Such a plot, sometimes referred to as a ***discrete spectrum,*** gives a great deal of information at a glance. In particular, we can see how many terms of the series are required to obtain a reasonable approximation of the original waveform. In the line spectrum of Fig. 17.5, we note that the 8th and 10th harmonics (20 and 25 Hz, respectively) add only a small correction. Truncating the series after the 6th harmonic therefore should lead to a reasonable approximation; the reader can judge this for herself/himself by considering Fig. 17.3.

One note of caution must be injected. The example we have considered contains no cosine terms, and the amplitude of the nth harmonic is therefore $|b_n|$. If a_n is not zero, then the amplitude of the component at a frequency $n\omega_0$ must be $\sqrt{a_n^2 + b_n^2}$. This is the general quantity which we must show in a line spectrum. When we discuss the complex form of the Fourier series, we will see that this amplitude is obtained more directly.

In addition to the amplitude spectrum, we may construct a discrete ***phase spectrum.*** At any frequency $n\omega_0$, we combine the cosine and sine terms to determine the phase angle ϕ_n:

$$a_n \cos n\omega_0 t + b_n \sin n\omega_0 t = \sqrt{a_n^2 + b_n^2}\cos\left(n\omega_0 t + \tan^{-1}\frac{-b_n}{a_n}\right)$$

$$= \sqrt{a_n^2 + b_n^2}\cos(n\omega_0 t + \phi_n)$$

or

$$\phi_n = \tan^{-1}\frac{-b_n}{a_n}$$

The Fourier series obtained for this example includes no cosine terms. It is possible to anticipate the absence of certain terms in a Fourier series, before any integrations are performed, by an inspection of the symmetry of the given time function. We will investigate the use of symmetry in Sec. 17.2.

17.2 THE USE OF SYMMETRY

Even and Odd Symmetry

The two types of symmetry which are most readily recognized are *even-function symmetry* and *odd-function symmetry*, or simply *even symmetry* and *odd symmetry*. We say that $f(t)$ possesses the property of even symmetry if

$$f(t) = f(-t) \qquad [12]$$

Such functions as t^2, $\cos 3t$, $\ln(\cos t)$, $\sin^2 7t$, and a constant C all possess even symmetry; the replacement of t with $(-t)$ does not change the value of any of these functions. This type of symmetry may also be recognized graphically, for if $f(t) = f(-t)$, then mirror symmetry exists about the $f(t)$ axis. The function shown in Fig. 17.6*a* possesses even symmetry; if the figure were to be folded along the $f(t)$ axis, then the portions of the graph for positive and negative time would fit exactly, one on top of the other.

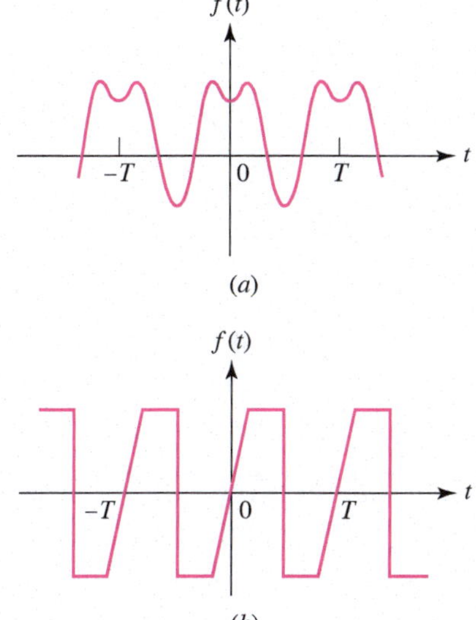

FIGURE 17.6 (*a*) A waveform showing even symmetry. (*b*) A waveform showing odd symmetry.

We define odd symmetry by stating that if odd symmetry is a property of $f(t)$, then

$$f(t) = -f(-t) \qquad [13]$$

In other words, if t is replaced by $(-t)$, then the negative of the given function is obtained; for example, t, $\sin t$, $t \cos 70t$, $t\sqrt{1 + t^2}$, and the function sketched in Fig. 17.6*b* are all odd functions and possess odd symmetry. The graphical characteristics of odd symmetry are apparent if the portion of $f(t)$ for $t > 0$ is rotated about the positive t axis and the resultant figure is then rotated about the $f(t)$ axis; the two curves will fit exactly, one on top of the other. That is, we now have symmetry about the origin, rather than about the $f(t)$ axis as we did for even functions.

Having definitions for even and odd symmetry, we should note that the product of two functions with even symmetry, or of two functions with odd symmetry, yields a function with even symmetry. Furthermore, the product of an even and an odd function gives a function with odd symmetry.

Symmetry and Fourier Series Terms

Now let us investigate the effect that even symmetry produces in a Fourier series. If we think of the expression which equates an even function $f(t)$ and the sum of an infinite number of sine and cosine functions, then it is apparent that the sum must also be an even function. A sine wave, however, is an odd function, and *no sum of sine waves can produce any even function other than zero* (which is both even and odd). It is thus plausible that the Fourier series of any even function is composed of only a constant and cosine functions. Let us now show carefully that $b_n = 0$. We have

$$b_n = \frac{2}{T}\int_{-T/2}^{T/2} f(t)\sin n\omega_0 t\, dt$$

$$= \frac{2}{T}\left[\int_{-T/2}^{0} f(t)\sin n\omega_0 t\, dt + \int_{0}^{T/2} f(t)\sin n\omega_0 t\, dt\right]$$

We replace the variable t in the first integral with $-\tau$, or $\tau = -t$, and make use of the fact that $f(t) = f(-t) = f(\tau)$:

$$b_n = \frac{2}{T}\left[\int_{T/2}^{0} f(-\tau)\sin(-n\omega_0\tau)(-d\tau) + \int_0^{T/2} f(t)\sin n\omega_0 t\, dt\right]$$

$$= \frac{2}{T}\left[-\int_0^{T/2} f(\tau)\sin n\omega_0\tau\, d\tau + \int_0^{T/2} f(t)\sin n\omega_0 t\, dt\right]$$

but the symbol we use to identify the variable of integration cannot affect the value of the integral. Thus,

$$\int_0^{T/2} f(\tau)\sin n\omega_0\tau\, d\tau = \int_0^{T/2} f(t)\sin n\omega_0 t\, dt$$

and

$$b_n = 0 \qquad \text{(even sym.)} \qquad [14]$$

No sine terms are present. Therefore, if $f(t)$ shows even symmetry, then $b_n = 0$; conversely, if $b_n = 0$, then $f(t)$ must have even symmetry.

A similar examination of the expression for a_n leads to an integral over the *half period* extending from $t = 0$ to $t = \frac{1}{2}T$:

$$a_n = \frac{4}{T}\int_0^{T/2} f(t)\cos n\omega_0 t\, dt \qquad \text{(even sym.)} \qquad [15]$$

The fact that a_n may be obtained for an even function by taking "twice the integral over half the range" should seem logical.

A function having odd symmetry can contain no constant term or cosine terms in its Fourier expansion. Let us prove the second part of this statement. We have

$$a_n = \frac{2}{T}\int_{-T/2}^{T/2} f(t)\cos n\omega_0 t\, dt$$

$$= \frac{2}{T}\left[\int_{-T/2}^{0} f(t)\cos n\omega_0 t\, dt + \int_0^{T/2} f(t)\cos n\omega_0 t\, dt\right]$$

and we now let $t = -\tau$ in the first integral:

$$a_n = \frac{2}{T}\left[\int_{T/2}^{0} f(-\tau)\cos(-n\omega_0\tau)(-d\tau) + \int_0^{T/2} f(t)\cos n\omega_0 t\, dt\right]$$

$$= \frac{2}{T}\left[\int_0^{T/2} f(-\tau)\cos n\omega_0\tau\, d\tau + \int_0^{T/2} f(t)\cos n\omega_0 t\, dt\right]$$

But $f(-\tau) = -f(\tau)$, and therefore

$$a_n = 0 \qquad \text{(odd sym.)} \qquad [16]$$

A similar, but simpler, proof shows that

$$a_0 = 0 \qquad \text{(odd sym.)}$$

With odd symmetry, therefore, $a_n = 0$ and $a_0 = 0$; conversely, if $a_n = 0$ and $a_0 = 0$, odd symmetry is present.

The values of b_n may again be obtained by integrating over half the range:

$$b_n = \frac{4}{T}\int_0^{T/2} f(t)\sin n\omega_0 t\, dt \qquad \text{(odd sym.)} \qquad [17]$$

Half-Wave Symmetry

The Fourier series for both of these square waves have one other interesting characteristic: neither contains any even *harmonics*.[1] That is, the only frequency components present in the series have frequencies which are odd multiples of the fundamental frequency; a_n and b_n are zero for even values of n. This result is caused by another type of symmetry, called half-wave symmetry. We will say that $f(t)$ possesses *half-wave symmetry* if

$$f(t) = -f\left(t - \frac{1}{2}T\right)$$

or the equivalent expression,

$$f(t) = -f\left(t - \frac{1}{2}T\right)$$

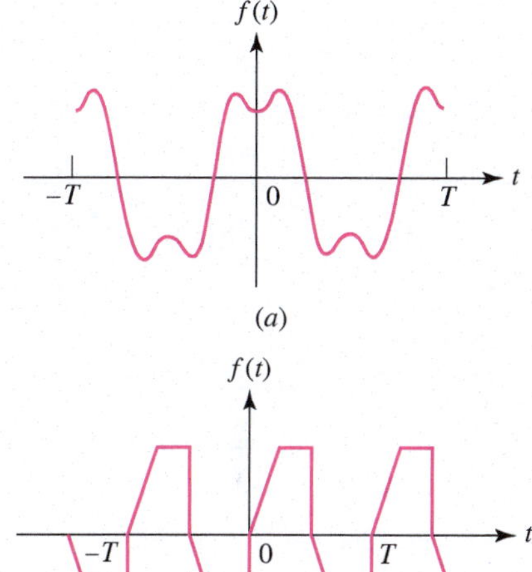

FIGURE 17.7 (*a*) A waveform somewhat similar to the one shown in Fig. 17.6*a* but possessing half-wave symmetry. (*b*) A waveform somewhat similar to the one shown in Fig. 17.6*b* but possessing half-wave symmetry.

Except for a change of sign, each half cycle is like the adjacent half cycles. Half-wave symmetry, unlike even and odd symmetry, is not a function of the choice of the point $t = 0$. Thus, we can state that the square wave (Fig. 17.4*a* or *b*) shows half-wave symmetry. Neither waveform shown in Fig. 17.6 has half-wave symmetry, but the two somewhat similar functions plotted in Fig. 17.7 do possess half-wave symmetry.

It may be shown that the Fourier series of any function which has half-wave symmetry contains only odd harmonics. Let us consider the coefficients a_n. We have again

$$a_n = \frac{2}{T}\int_{-T/2}^{T/2} f(t)\cos n\omega_0 t\, dt$$
$$= \frac{2}{T}\left[\int_{-T/2}^{0} f(t)\cos n\omega_0 t\, dt + \int_0^{T/2} f(t)\cos n\omega_0 t\, dt\right]$$

which we may represent as

$$a_n = \frac{2}{T}(I_1 + I_2)$$

Now we substitute the new variable $\tau = t + \frac{1}{2}T$ in the integral I_1:

$$I_1 = \int_0^{T/2} f\left(\tau - \frac{1}{2}T\right)\cos n\omega_0\left(\tau - \frac{1}{2}T\right) d\tau$$
$$= \int_0^{T/2} -f(\tau)\left(\cos n\omega_0\tau \cos\frac{n\omega_0 T}{2} + \sin n\omega_0\tau \sin\frac{n\omega_0 T}{2}\right) d\tau$$

(1) Constant vigilance is required to avoid confusion between an even function and an even harmonic, or between an odd function and an odd harmonic. For example, b_{10} is the coefficient of an even harmonic, and it is zero if $f(t)$ is an even function.

But $\omega_0 T$ is 2π, and thus

$$\sin\frac{n\omega_0 T}{2} = \sin n\pi = 0$$

Hence

$$I_1 = -\cos n\pi \int_0^{T/2} f(\tau)\cos n\omega_0\tau \, d\tau$$

After noting the form of I_2, we therefore may write

$$a_n = \frac{2}{T}(1 - \cos n\pi)\int_0^{T/2} f(t)\cos n\omega_0 t \, dt$$

The factor $(1 - \cos n\pi)$ indicates that a_n is zero if n is even. Thus,

$$a_n = \begin{cases} \frac{4}{T}\int_0^{T/2} f(t)\cos n\omega_0 t \, dt & n \text{ odd} \\ 0 & n \text{ even} \end{cases} \qquad \left(\tfrac{1}{2}\text{-wave sym.}\right) \qquad [18]$$

A similar investigation shows that b_n is also zero for all even n, and therefore

$$b_n = \begin{cases} \frac{4}{T}\int_0^{T/2} f(t)\sin n\omega_0 t \, dt & n \text{ odd} \\ 0 & n \text{ even} \end{cases} \qquad \left(\tfrac{1}{2}\text{-wave sym.}\right) \qquad [19]$$

It should be noted that half-wave symmetry may be present in a waveform which also shows odd symmetry or even symmetry. The waveform sketched in Fig. 17.7*a*, for example, possesses both even symmetry and half-wave symmetry. When a waveform possesses half-wave symmetry and either even or odd symmetry, then it is possible to reconstruct the waveform if the function is known over any quarter-period interval. The value of a_n or b_n may also be found by integrating over any quarter period. Thus,

$$\left.\begin{aligned} a_n &= \frac{8}{T}\int_0^{T/4} f(t)\cos n\omega_0 t \, dt && n \text{ odd} \\ a_n &= 0 && n \text{ even} \\ b_n &= 0 && \text{all } n \end{aligned}\right\} \qquad \left(\tfrac{1}{2}\text{-wave and even sym.}\right) \qquad [20]$$

It is *always* worthwhile to spend a few moments investigating the symmetry of a function for which a Fourier series is to be determined.

$$\left.\begin{aligned} a_n &= 0 && \text{all } n \\ b_n &= \frac{8}{T}\int_0^{T/4} f(t)\sin n\omega_0 t \, dt && n \text{ odd} \\ b_n &= 0 && n \text{ even} \end{aligned}\right\} \qquad \left(\tfrac{1}{2}\text{-wave and odd sym.}\right) \qquad [21]$$

Table 17.1 provides a short summary of the simplifications arising from the various types of symmetry discussed.

TABLE **17.1** Summary of Symmetry-Based Simplifications in Fourier Series

Symmetry Type	Characteristic	Simplification
Even	$f(t) = -f(t)$	$b_n = 0$
Odd	$f(t) = -f(-t)$	$a_n = 0$
Half-Wave	$f(t) = -f\left(t - \frac{T}{2}\right)$ or $f(t) = -f\left(t + \frac{T}{2}\right)$	$a_n = \begin{cases} \frac{4}{T}\int_0^{T/2} f(t)\cos n\omega_0 t\, dt & n \text{ odd} \\ 0 & n \text{ even} \end{cases}$ $b_n = \begin{cases} \frac{4}{T}\int_0^{T/2} f(t)\sin n\omega_0 t\, dt & n \text{ odd} \\ 0 & n \text{ even} \end{cases}$
Half-Wave and Even	$f(t) = -f\left(t - \frac{T}{2}\right)$ and $f(t) = -f(t)$ or $f(t) = -f\left(t + \frac{T}{2}\right)$ and $f(t) = -f(t)$	$a_n = \begin{cases} \frac{8}{T}\int_0^{T/4} f(t_1)\cos n\omega_0 t\, dt & n \text{ odd} \\ 0 & n \text{ even} \end{cases}$ $b_n = 0$ all n
Half-Wave and Odd	$f(t) = -f\left(t - \frac{T}{2}\right)$ and $f(t) = -f(-t)$ or $f(t) = -f\left(t + \frac{T}{2}\right)$ and $f(t) = -f(-t)$	$a_n = 0$ all n $b_n = \begin{cases} \frac{8}{T}\int_0^{T/4} f(t)\sin n\omega_0 t\, dt & n \text{ odd} \\ 0 & n \text{ even} \end{cases}$

PRACTICE

17.4 Sketch each of the functions described; state whether or not even symmetry, odd symmetry, and half-wave symmetry are present; and give the period: (*a*) $v = 0$, $-2 < t < 0$ and $2 < t < 4$; $v = 5$, $0 < t < 2$; $v = -5$, $4 < t < 6$; repeats; (*b*) $v = 10$, $1 < t < 3$; $v = 0$, $3 < t < 7$; $v = -10$, $7 < t < 9$; repeats; (*c*) $v = 8t$, $-1 < t < 1$; $v = 0$, $1 < t < 3$; repeats.

17.5 Determine the Fourier series for the waveforms of Practice Problem 17.4*a* and *b*.

Ans: 17.4: No, no, yes, 8; no, no, no, 8; no, yes, no, 4.

17.5: $\sum_{n=1(\text{odd})}^{\infty} \frac{10}{n\pi}\left(\sin\frac{n\pi}{2}\cos\frac{n\pi t}{4} + \sin\frac{n\pi t}{4}\right)$;

$\sum_{n=1}^{\infty} \frac{10}{n\pi}\left[\left(\sin\frac{3n\pi}{4} - 3\sin\frac{n\pi}{4}\right)\cos\frac{n\pi t}{4} + \left(\cos\frac{n\pi}{4} - \cos\frac{3n\pi}{4}\right)\sin\frac{n\pi t}{4}\right]$

17.3 COMPLETE RESPONSE TO PERIODIC FORCING FUNCTIONS

Through the use of the Fourier series, we may now express an arbitrary periodic forcing function as the sum of an infinite number of sinusoidal forcing functions. The forced response to each of these functions may be determined by conventional steady-state analysis, and the form of the natural response may be determined from the poles of an appropriate network transfer function. The initial conditions existing throughout the network, including the initial value of the forced response, enable the amplitude of the natural response to be selected; the complete response is then obtained as the sum of the forced and natural responses.

EXAMPLE 17.2

For the circuit of Fig. 17.8*a*, determine the periodic response *i*(*t*) corresponding to the forcing function shown in Fig. 17.8*b* if *i*(0) = 0.

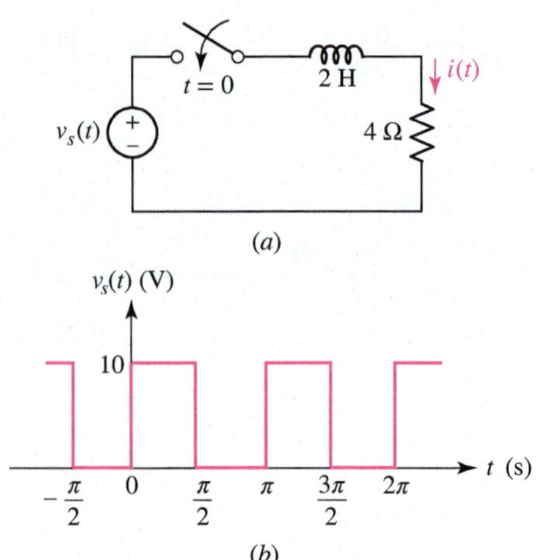

FIGURE 17.8 (*a*) A simple series *RL* circuit subjected to a periodic forcing function $v_s(t)$. (*b*) The form of the forcing function.

The forcing function has a fundamental frequency $\omega_0 = 2$ rad/s, and its Fourier series may be written down by comparison with the Fourier series developed for the waveform of Fig. 17.4*b* in the solution of Practice Problem 17.3,

$$v_s(t) = 5 + \frac{20}{\pi} \sum_{n=1(\text{odd})}^{\infty} \frac{\sin 2nt}{n}$$

We will find the forced response for the *n*th harmonic by working in the frequency domain. Thus,

$$v_{sn}(t) = \frac{20}{n\pi} \sin 2nt$$

and

$$\mathbf{V}_{sn} = \frac{20}{n\pi}\angle{-90^\circ} = -j\,\frac{20}{n\pi}$$

Recall that $V_m \sin \omega t$ is equal to $V_m \cos(\omega t - 90^\circ)$, corresponding to $V_m\angle{-90^\circ} = -jV_m$.

The impedance offered by the *RL* circuit at this frequency is

$$\mathbf{Z}_n = 4 + j(2n)(2) = 4 + j4n$$

and thus the component of the forced response at this frequency is

$$\mathbf{I}_{fn} = \frac{\mathbf{V}_{sn}}{\mathbf{Z}_n} = \frac{-j5}{n\pi(1 + jn)}$$

Transforming to the time domain, we have

$$\begin{aligned} i_{fn} &= \frac{5}{n\pi} \frac{1}{\sqrt{1+n^2}} \cos\left(2nt - 90^\circ - \tan^{-1} n\right) \\ &= \frac{5}{\pi(1+n^2)} \left(\frac{\sin 2nt}{n} - \cos 2nt \right) \end{aligned}$$

Since the response to the dc component is simply 5 V/4Ω = 1.25 A, the forced response may be expressed as the summation

$$i_f(t) = 1.25 + \frac{5}{\pi} \sum_{n=1(\text{odd})}^{\infty} \left[\frac{\sin 2nt}{n(1+n^2)} - \frac{\cos 2nt}{1+n^2} \right]$$

The familiar natural response of this simple circuit is the single exponential term [characterizing the single pole of the transfer function, $\mathbf{I}_f/\mathbf{V}_s = 1/(4 + 2\mathbf{s})$]

$$i_n(t) = Ae^{-2t}$$

The *complete* response is therefore the sum

$$i(t) = i_f(t) + i_n(t)$$

Letting $t = 0$, we find A using $i(0) = 0$:

$$A = -1.25 + \frac{5}{\pi} \sum_{n=1(\text{odd})}^{\infty} \frac{1}{1+n^2}$$

Although correct, it is more convenient to use the numerical value of the summation. The sum of the first 5 terms of $\Sigma 1/(1 + n^2)$ is 0.671, the sum of the first 10 terms is 0.695, the sum of the first 20 terms is 0.708, and the exact sum is 0.720 to three significant figures. Thus

$$A = -1.25 + \frac{5}{\pi}(0.720) = -0.104$$

and

$$i(t) = -0.104e^{-2t} + 1.25 + \frac{5}{\pi} \sum_{n=1(\text{odd})}^{\infty} \left[\frac{\sin 2nt}{n(1+n^2)} - \frac{\cos 2nt}{1+n^2} \right] \text{ amperes}$$

In obtaining this solution, we have had to use many of the most general concepts introduced in this and the preceding 16 chapters. Some we did not have to use because of the simple nature of this particular circuit, but their places in the general analysis were indicated. In this sense, we may look upon the solution of this problem as a significant achievement in our introductory study of circuit analysis. In spite of this glorious feeling of accomplishment, however, it must be pointed out that the complete response, as obtained in Example 17.2 in analytical form, is not of much value as it stands; it furnishes no clear picture of the nature of the response. What we really need is a sketch of $i(t)$ as a function of time. This may be obtained by a laborious calculation at a sufficient number of instants of time; a desktop computer or a programmable calculator can be of great assistance here. The sketch may be approximated by the graphical addition of the natural response, the dc term, and the first few harmonics; this is an unrewarding task.

When all is said and done, the most informative solution of this problem is probably obtained by making a repeated transient analysis. That is, the form of the response can certainly be calculated in the interval from $t = 0$ to $t = \pi/2$ s; it is an exponential rising toward 2.5 A. After determining the

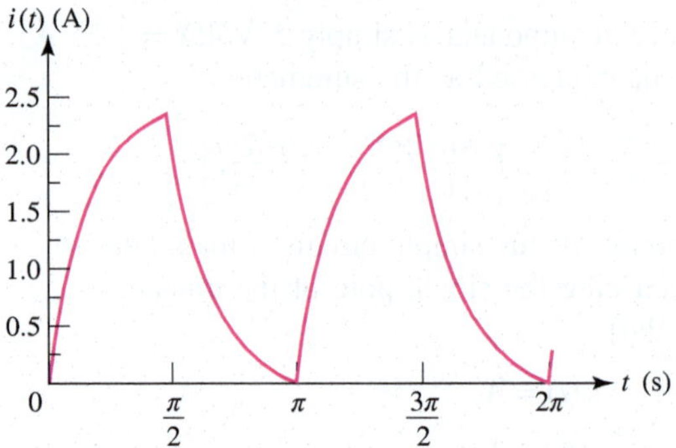

FIGURE 17.9 The initial portion of the complete response of the circuit of Fig. 17.8*a* to the forcing function of Fig. 17.8.

value at the end of this first interval, we have an initial condition for the next ($\pi/2$)-second interval. The process is repeated until the response assumes a generally periodic nature. The method is eminently suitable to this example, for there is negligible change in the current waveform in the successive periods $\pi/2 < t < 3\pi/2$ and $3\pi/2 < t < 5\pi/2$. The complete current response is sketched in Fig. 17.9.

PRACTICE

17.6 Use the methods of Chap. 8 to determine the value of the current sketched in Fig. 17.9 at t equal to (*a*) $\pi/2$; (*b*) π; $3\pi/2$.

Ans: 2.392 A; 0.1034 A; 2.396 A.

17.4 COMPLEX FORM OF THE FOURIER SERIES

In obtaining a frequency spectrum, we have seen that the amplitude of each frequency component depends on both a_n and b_n; that is, the sine term and the cosine term both contribute to the amplitude. The exact expression for this amplitude is $\sqrt{a_n^2 + b_n^2}$. It is also possible to obtain the amplitude directly by using a form of Fourier series in which each term is a cosine function with a phase angle; the amplitude and phase angle are functions of $f(t)$ and n. An even more convenient and concise form of the Fourier series is obtained if the sines and cosines are expressed as exponential functions with complex multiplying constants.

Let us first take the trigonometric form of the Fourier series:

$$f(t) = a_0 + \sum_{n=1}^{\infty}(a_n \cos n\omega_0 t + b_n \sin n\omega_0 t)$$

The reader may recall the identities

$$\sin\alpha = \frac{e^{j\alpha} - e^{-j\alpha}}{j2}$$

and

$$\cos\alpha = \frac{e^{j\alpha} + e^{-j\alpha}}{2}$$

and then substitute the exponential forms for the sine and cosine. After rearranging,

$$f(t) = a_0 + \sum_{n=1}^{\infty}\left(e^{jn\omega_0 t}\frac{a_n - jb_n}{2} + e^{-jn\omega_0 t}\frac{a_n + jb_n}{2}\right)$$

We now define a complex constant $\mathbf{c}_n$:

$$\mathbf{c}_n = \frac{1}{2}(a_n - jb_n) \qquad (n = 1, 2, 3, \ldots) \tag{22}$$

The values of a_n, b_n, and $\mathbf{c}_n$ all depend on n and $f(t)$. Suppose we now replace n with $(-n)$; how do the values of the constants change? The coefficients a_n and b_n are defined by Eqs. [10] and [11], and it is evident that

$$a_{-n} = a_n$$

but

$$b_{-n} = -b_n$$

From Eq. [22], then,

$$\mathbf{c}_{-n} = \frac{1}{2}(a_n + jb_n) \qquad (n = 1, 2, 3, \ldots) \tag{23}$$

Thus,

$$\mathbf{c}_n = \mathbf{c}^*_{-n}$$

We also let

$$\mathbf{c}_0 = a_0$$

We may therefore express $f(t)$ as

$$f(t) = \mathbf{c}_0 + \sum_{n=1}^{\infty} \mathbf{c}_n\, e^{jn\omega_0 t} + \sum_{n=1}^{\infty} \mathbf{c}_{-n} e^{-jn\omega_0 t}$$

or

$$f(t) = \sum_{n=0}^{\infty} \mathbf{c}_n e^{jn\omega_0 t} + \sum_{n=1}^{\infty} \mathbf{c}_{-n} e^{-jn\omega_0 t}$$

Finally, instead of summing the second series over the positive integers from 1 to ∞, let us sum over the negative integers from -1 to $-\infty$:

$$f(t) = \sum_{n=0}^{\infty} \mathbf{c}_n e^{jn\omega_0 t} + \sum_{n=-1}^{-\infty} \mathbf{c}_n e^{jn\omega_0 t}$$

or

$$f(t) = \sum_{n=-\infty}^{\infty} \mathbf{c}_n e^{jn\omega_0 t} \tag{24}$$

By agreement, a summation from $-\infty$ to ∞ is understood to include a term for $n = 0$.

Equation [24] is the *complex form* of the Fourier series for $f(t)$; its conciseness is one of the most important reasons for its use. In order to obtain the expression by which a particular complex coefficient $\mathbf{c}_n$ may be evaluated, we substitute Eqs. [10] and [11] in Eq. [22]:

$$\mathbf{c}_n = \frac{1}{T}\int_{-T/2}^{T/2} f(t)\cos n\omega_0 t\, dt - j\frac{1}{T}\int_{-T/2}^{T/2} f(t)\sin n\omega_0 t\, dt$$

and then we use the exponential equivalents of the sine and cosine and simplify:

$$\mathbf{c}_n = \frac{1}{T}\int_{-T/2}^{T/2} f(t)e^{-jn\omega_0 t}\,dt \qquad [25]$$

Thus, a single concise equation serves to replace the two equations required for the trigonometric form of the Fourier series. Instead of evaluating two integrals to find the Fourier coefficients, only one integration is required; moreover, it is almost always a simpler integration. It should be noted that the integral of Eq. [25] contains the multiplying factor $1/T$, whereas the integrals for a_n and b_n both contain the factor $2/T$.

Collecting the two basic relationships for the exponential form of the Fourier series, we have

$$f(t) = \sum_{n=-\infty}^{\infty} \mathbf{c}_n e^{jn\omega_0 t} \qquad [24]$$

$$\mathbf{c}_n = \frac{1}{T}\int_{-T/2}^{T/2} f(t)e^{-jn\omega_0 t}\,dt \qquad [25]$$

where $\omega_0 = 2\pi/T$ as usual.

The amplitude of the component of the exponential Fourier series at $\omega = n\omega_0$, where $n = 0, \pm1, \pm2, \ldots$, is $|\mathbf{c}_n|$. We may plot a discrete frequency spectrum giving $|\mathbf{c}_n|$ versus $n\omega_0$ or nf_0, using an abscissa that shows both positive and negative values; and when we do this, the graph is symmetrical about the origin, since Eqs. [22] and [23] show that $|\mathbf{c}_n| = |\mathbf{c}_{-n}|$.

We note also from Eqs. [24] and [25] that the amplitude of the sinusoidal component at $\omega = n\omega_0$, where $n = 1, 2, 3, \ldots$, is $\sqrt{a_n^2 + b_n^2} = 2|\mathbf{c}_n| = 2|\mathbf{c}_{-n}| = |\mathbf{c}_n| + |\mathbf{c}_{-n}|$. For the dc component, $a_0 = \mathbf{c}_0$.

The exponential Fourier coefficients, given by Eq. [25], are also affected by the presence of certain symmetries in $f(t)$. Thus, appropriate expressions for $\mathbf{c}_n$ are

$$\mathbf{c}_n = \frac{2}{T}\int_0^{T/2} f(t)\cos n\omega_0 t\,dt \qquad \text{(even sym.)} \qquad [26]$$

$$\mathbf{c}_n = \frac{-j2}{T}\int_0^{T/2} f(t)\sin n\omega_0 t\,dt \qquad \text{(odd sym.)} \qquad [27]$$

$$\mathbf{c}_n = \begin{cases} \dfrac{2}{T}\displaystyle\int_0^{T/2} f(t)e^{-jn\omega_0 t}\,dt & (n \text{ odd, } \tfrac{1}{2}\text{-wave sym.}) \quad [28a] \\ 0 & (n \text{ even, } \tfrac{1}{2}\text{-wave sym.}) \quad [28b] \end{cases}$$

$$\mathbf{c}_n = \begin{cases} \dfrac{4}{T}\displaystyle\int_0^{T/4} f(t)\cos n\omega_0 t\,dt & (n \text{ odd, } \tfrac{1}{2}\text{-wave and even sym.}) \quad [29a] \\ 0 & (n \text{ even, } \tfrac{1}{2}\text{-wave and even sym.}) \quad [29b] \end{cases}$$

$$\mathbf{c}_n = \begin{cases} \dfrac{-j4}{T}\displaystyle\int_0^{T/4} f(t)\sin n\omega_0 t\,dt & (n \text{ odd, } \tfrac{1}{2}\text{-wave and odd sym.}) \quad [30a] \\ 0 & (n \text{ even, } \tfrac{1}{2}\text{-wave and odd sym.}) \quad [30b] \end{cases}$$

EXAMPLE 17.3

Determine $\mathbf{c}_n$ for the square wave of Fig. 17.10.

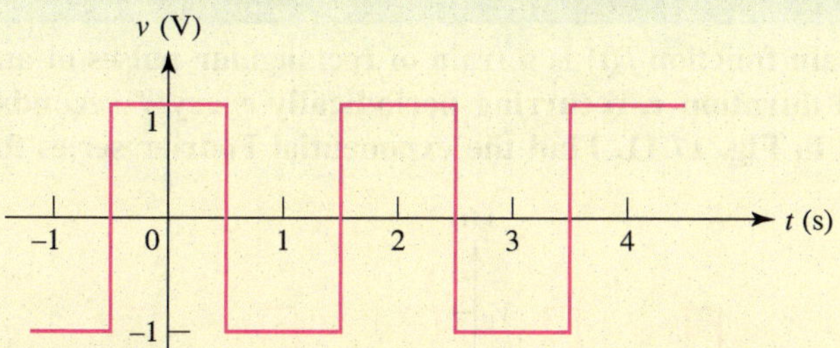

FIGURE 17.10 A square wave function possessing both even and half-wave symmetry.

This square wave possesses both even and half-wave symmetry. If we ignore the symmetry and use our general equation [25], with $T = 2$ and $\omega_0 = 2\pi/2 = \pi$, we have

$$\mathbf{c}_n = \frac{1}{T}\int_{-T/2}^{T/2} f(t)e^{-jn\omega_0 t}\,dt$$

$$= \frac{1}{2}\left[\int_{-1}^{-0.5} -e^{-jn\pi t}\,dt + \int_{-0.5}^{0.5} e^{-jn\pi t}\,dt - \int_{0.5}^{1} e^{-jn\pi t}\,dt\right]$$

$$= \frac{1}{2}\left[\frac{-1}{-jn\pi}(e^{-jn\pi t})\Big|_{-1}^{-0.5} + \frac{1}{-jn\pi}(e^{-jn\pi t})\Big|_{-0.5}^{0.5} + \frac{-1}{-jn\pi}(e^{-jn\pi t})\Big|_{0.5}^{1}\right]$$

$$= \frac{1}{j2n\pi}(e^{jn\pi/2} - e^{jn\pi} - e^{-jn\pi/2} + e^{jn\pi/2} + e^{-jn\pi} - e^{-jn\pi/2})$$

$$= 2\frac{e^{jn\pi/2} - e^{-jn\pi/2}}{j2n\pi} - \frac{e^{jn\pi} - e^{-jn\pi}}{j2n\pi}$$

$$= \frac{1}{n\pi}\left[2\sin\frac{n\pi}{2} - \sin n\pi\right]$$

We thus find that $\mathbf{c}_0 = 0$, $\mathbf{c}_1 = 2/\pi$, $\mathbf{c}_2 = 0$, $\mathbf{c}_3 = -2/3\pi$, $\mathbf{c}_4 = 0$, $\mathbf{c}_5 = 2/5\pi$, and so forth. These values agree with the trigonometric Fourier series given as the answer we obtained in Practice Problem 17.3 for the same waveform shown in Fig. 17.4*b* if we remember that $a_n = 2\mathbf{c}_n$ when $b_n = 0$.

Utilizing the symmetry of the waveform (even and half-wave), there is less work when we apply Eqs. [29*a*] and [29*b*], leading to

$$\mathbf{c}_n = \frac{4}{T}\int_0^{T/4} f(t)\cos n\omega_0 t\,dt$$

$$= \frac{4}{2}\int_0^{0.5} \cos n\pi t\,dt = \frac{2}{n\pi}(\sin n\pi t)\Big|_0^{0.5}$$

$$= \begin{cases} \dfrac{2}{n\pi}\sin\dfrac{n\pi}{2} & (n \text{ odd}) \\ 0 & (n \text{ even}) \end{cases}$$

These results are the same as those we just obtained when we did not take the symmetry of the waveform into account.

Now let us consider a more difficult, more interesting example.

EXAMPLE **17.4**

A certain function $f(t)$ is a train of rectangular pulses of amplitude v_o and duration τ, recurring periodically every T seconds, as shown in Fig. 17.11. Find the exponential Fourier series for $f(t)$.

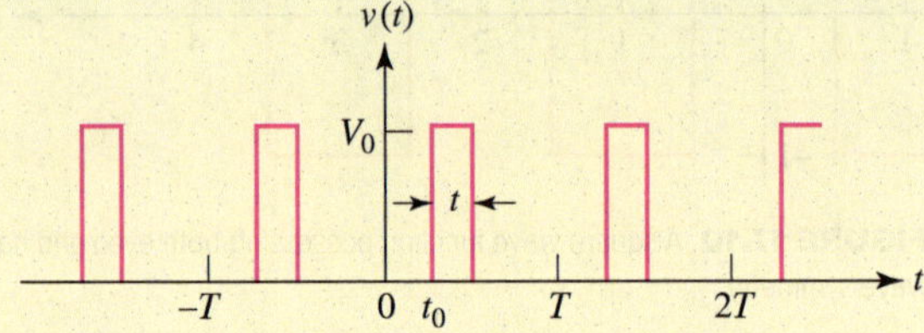

FIGURE 17.11 A periodic sequence of rectangular pulses.

The fundamental frequency is $f_0 = 1/\mathrm{T}$. No symmetry is present, and the value of a general complex coefficient is found from Eq. [25]:

$$\mathbf{c}_n = \frac{1}{T}\int_{-T/2}^{T/2} f(t)e^{-jn\omega_0 t}\,dt = \frac{V_0}{T}\int_{t_0}^{t_0+\tau} e^{-jn\omega_0 t}\,dt$$

$$= \frac{V_0}{-jn\omega_0 T}\left(e^{-jn\omega_0(t_0+\tau)} - e^{-jn\omega_0 t_0}\right)$$

$$= \frac{2V_0}{n\omega_0 T}e^{-jn\omega_0(t_0+\tau/2)}\sin\left(\frac{1}{2}n\omega_0\tau\right)$$

$$= \frac{V_0\tau}{T}\,\frac{\sin\left(\frac{1}{2}n\omega_0\tau\right)}{\frac{1}{2}n\omega_0\tau}\,e^{-jn\omega_0(t_0+\tau/2)}$$

The magnitude of $\mathbf{c}_n$ is therefore

$$|\mathbf{c}_n| = \frac{V_0\tau}{T}\left|\frac{\sin\left(\frac{1}{2}n\omega_0\tau\right)}{\frac{1}{2}n\omega_0\tau}\right| \qquad [31]$$

and the angle of $\mathbf{c}_n$ is

$$\text{ang } \mathbf{c}_n = -n\omega_0\left(t_0 + \frac{\tau}{2}\right) \quad \text{(possibly plus 180°)} \qquad [32]$$

Equations [31] and [32] represent our solution to this exponential Fourier series problem.

The Sampling Function

The trigonometric factor in Eq. [31] occurs frequently in modern communication theory, and it is called the ***sampling function.*** The "sampling" refers to the time function of Fig. 17.11 from which the sampling function is derived. The product of this sequence of pulses and any other function

$f(t)$ represents *samples* of $f(t)$ every T seconds if τ is small and $v_o = 1$. We define

$$\text{Sa}(x) = \frac{\sin x}{x}$$

Because of the way in which it helps to determine the amplitude of the various frequency components in $f(t)$, it is worth our while to discover the important characteristics of this function. First, we note that Sa(x) is zero whenever x is an integral multiple of π; that is,

$$\text{Sa}(n\pi) = 0 \qquad n = 1, 2, 3, \ldots$$

When x is zero, the function is indeterminate, but it is easy to show that its value is unity:

$$\text{Sa}(0) = 1$$

The magnitude of Sa(x) therefore decreases from unity at $x = 0$ to zero at $x = \pi$. As x increases from π to 2π, $|\text{Sa}(x)|$ increases from zero to a maximum less than unity, and then decreases to zero once again. As x continues to increase, the successive maxima continually become smaller because the numerator of Sa(x) cannot exceed unity and the denominator is continually increasing. Also, Sa(x) shows even symmetry.

Now let us construct the line spectrum. We first consider $|\mathbf{c}_n|$, writing Eq. [31] in terms of the fundamental cyclic frequency f_0:

$$|\mathbf{c}_n| = \frac{V_0\tau}{T}\left|\frac{\sin(n\pi f_0\tau)}{n\pi f_0\tau}\right| \qquad [33]$$

The amplitude of any $\mathbf{c}_n$ is obtained from Eq. [33] by using the known values τ and $T = 1/f_0$ and selecting the desired value of n, $n = 0, \pm 1, \pm 2, \ldots$. Instead of evaluating Eq. [33] at these discrete frequencies, let us sketch the *envelope* of $|\mathbf{c}_n|$ by considering the frequency nf_0 to be a continuous variable. That is, f, which is nf_0, can actually take on only the discrete values of the harmonic frequencies 0, $\pm f_0$, $\pm 2f_0$, $\pm 3f_0$, and so forth, but we may think of n for the moment as a continuous variable. When f is zero, $|\mathbf{c}_n|$ is evidently $v_o\tau/\text{T}$, and when f has increased to $1/\tau$, $|\mathbf{c}_n|$ is zero. The resultant envelope is first sketched as in Fig. 17.12*a*. The line spectrum is then obtained by simply erecting a vertical line at each harmonic frequency, as shown in the sketch. The amplitudes shown are those of the $\mathbf{c}_n$. The particular case sketched applies to the case where $\tau/T = 1/(1.5\pi) = 0.212$. In this example, it happens that there is no harmonic exactly at that frequency at which the envelope amplitude is zero; another choice of τ or T could produce such an occurrence, however.

In Fig. 17.12*b*, the amplitude of the sinusoidal component is plotted as a function of frequency. Note again that $a_0 = \mathbf{c}_0$ and $\sqrt{a_n^2 + b_n^2} = |\mathbf{c}_n| + |\mathbf{c}_{-n}|$.

There are several observations and conclusions which we may make about the line spectrum of a periodic sequence of rectangular pulses, as given in Fig. 17.12*b*. With respect to the envelope of the discrete spectrum, it is evident that the "width" of the envelope depends upon τ, and not upon T. As a matter of fact, the shape of the envelope is not a function

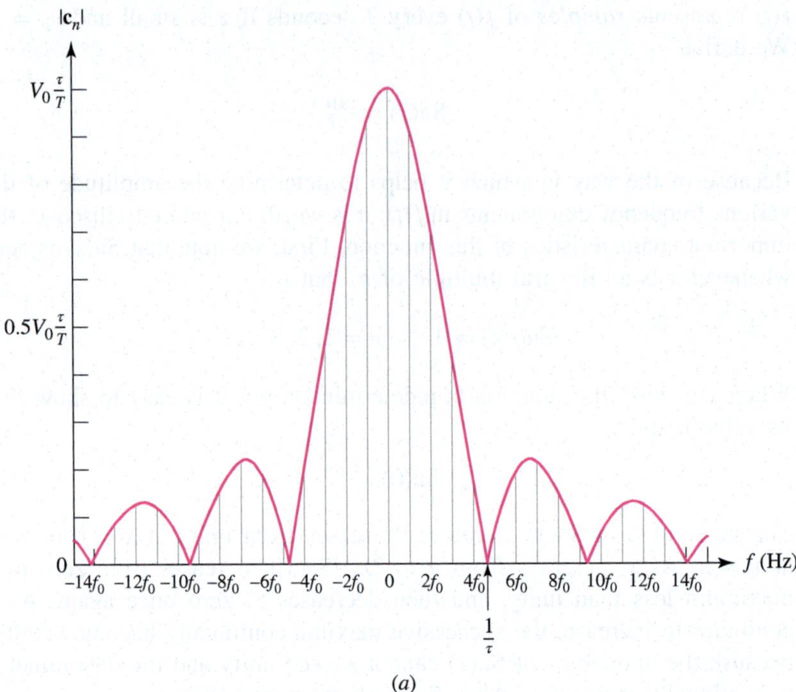

(*a*)

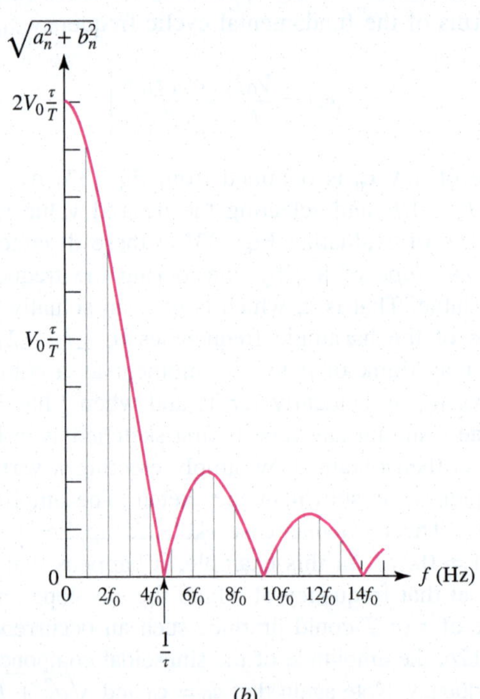

(*b*)

FIGURE 17.12 (*a*) The discrete line spectrum of $|\mathbf{c}_n|$ versus $f = nf_0$, $n = 0, \pm1, \pm2, \ldots$, corresponding to the pulse train shown in Fig. 17.11. (*b*) $\sqrt{a^2 + b^2}$ versus $f = nf_0$, $n = 0, 1, 2, \ldots$, for the same pulse train.

of T. It follows that the bandwidth of a filter which is designed to pass the periodic pulses is a function of the pulse width τ, but not of the pulse period T; an inspection of Fig. 17.12*b* indicates that the required bandwidth is about $1/\tau$ Hz. If the pulse period T is increased (or the pulse repetition frequency f_0 is decreased), the bandwidth $1/\tau$ does not change, but the number of spectral lines between zero frequency and $1/\tau$ Hz increases, albeit discontinuously; the amplitude of each line is inversely proportional to T. Finally, a shift in the time origin does not change the line spectrum; that is, $|\mathbf{c}_n|$ is not a function of t_0. The relative phases of the frequency components do change with the choice of t_0.

PRACTICE

17.7 Determine the general coefficient $\mathbf{c}_n$ in the complex Fourier series for the waveform shown in (*a*) Fig. 17.4*a*; (*b*) Fig. 17.4*c*.

Ans: $-j2/(n\pi)$ for n odd, 0 for n even; $-j[4/(n^2\pi^2)]\sin n\pi/2$ for all n.

17.5 DEFINITION OF THE FOURIER TRANSFORM

Now that we are familiar with the basic concepts of the Fourier series representation of periodic functions, let us proceed to define the Fourier transform by first recalling the spectrum of the periodic train of rectangular pulses we obtained in Sec. 17.4. That was a *discrete* line spectrum, which is the type that we must always obtain for periodic functions of time. The spectrum was discrete in the sense that it was not a smooth or continuous function of frequency; instead, it had nonzero values only at specific frequencies.

There are many important forcing functions, however, that are not periodic functions of time, such as a single rectangular pulse, a step function, a ramp function, or the somewhat strange type of function called the *impulse function* defined in Chap. 14. Frequency spectra may be obtained for such nonperiodic functions, but they will be *continuous* spectra in which some energy, in general, may be found in any nonzero frequency interval, no matter how small.

We will develop this concept by beginning with a periodic function and then letting the period become infinite. Our experience with periodic rectangular pulses should indicate that the envelope will decrease in amplitude without otherwise changing shape, and that more and more frequency components will be found in any given frequency interval. In the limit, we should expect an envelope of vanishingly small amplitude, filled with an infinite number of frequency components separated by vanishingly small frequency intervals. The number of frequency components between 0 and 100 Hz, for example, becomes infinite, but the amplitude of each one approaches zero. At first thought, a spectrum of zero amplitude is a puzzling concept. We know that the line spectrum of a periodic forcing function shows the amplitude of each frequency component. But what does the zero-amplitude continuous spectrum of a nonperiodic forcing function signify? That question will be answered in the following section; now we proceed to carry out the limiting procedure just suggested.

We begin with the exponential form of the Fourier series:

$$f(t) = \sum_{n=-\infty}^{\infty} \mathbf{c}_n e^{jn\omega_0 t} \qquad [34]$$

where

$$\mathbf{c}_n = \frac{1}{T} \int_{-T/2}^{T/2} f(t) e^{-jn\omega_0 t} \, dt \qquad [35]$$

and

$$\omega_0 = \frac{2\pi}{T} \qquad [36]$$

We now let

$$T \to \infty$$

and thus, from Eq. [36], ω_0 must become vanishingly small. We represent this limit by a differential:

$$\omega_0 \to d\omega$$

Thus

$$\frac{1}{T} = \frac{\omega_0}{2\pi} \to \frac{d\omega}{2\pi} \qquad [37]$$

Finally, the frequency of any "harmonic" $n\omega_0$ must now correspond to the general frequency variable which describes the continuous spectrum. In other words, n must tend to infinity as ω_0 approaches zero, so that the product is finite:

$$n\omega_0 \to \omega \qquad [38]$$

When these four limiting operations are applied to Eq. [35], we find that $\mathbf{c}_n$ must approach zero, as we had previously presumed. If we multiply each side of Eq. [35] by the period T and then undertake the limiting process, a nontrivial result is obtained:

$$\mathbf{c}_n T \to \int_{-\infty}^{\infty} f(t) e^{-j\omega t} \, dt$$

The right-hand side of this expression is a function of ω (and *not* of t), and we represent it by $\mathbf{F}(j\omega)$:

$$\mathbf{F}(j\omega) = \int_{-\infty}^{\infty} f(t) e^{-j\omega t} \, dt \qquad [39]$$

Now let us apply the limiting process to Eq. [34]. We begin by multiplying and dividing the summation by T,

$$f(t) = \sum_{n=-\infty}^{\infty} \mathbf{c}_n T e^{jn\omega_0 t} \frac{1}{T}$$

next replacing $\mathbf{c}_n T$ with the new quantity $\mathbf{F}(j\omega)$, and then making use of expressions [42] and [43]. In the limit, the summation becomes an integral, and

$$f(t) = \frac{1}{2\pi} \int_{-\infty}^{\infty} \mathbf{F}(j\omega) e^{j\omega t} \, d\omega \qquad [40]$$

Equations [39] and [40] are collectively called the *Fourier transform pair*. The function $\mathbf{F}(j\omega)$ is the *Fourier transform* of $f(t)$, and $f(t)$ is the *inverse* Fourier transform of $\mathbf{F}(j\omega)$.

This transform-pair relationship is very important! We should memorize it, draw arrows pointing to it, and mentally keep it on the conscious level. We emphasize the importance of these relations by repeating them in boxed form:

$$\mathbf{F}(j\omega) = \int_{-\infty}^{\infty} e^{-j\omega t} f(t)dt \qquad [41a]$$

$$f(t) = \frac{1}{2\pi}\int_{-\infty}^{\infty} e^{j\omega t}\, \mathbf{F}(j\omega)d\omega \qquad [41b]$$

The exponential terms in these two equations carry opposite signs for the exponents. To keep them straight, it may help to note that the positive sign is associated with the expression for $f(t)$, as it is with the complex Fourier series, Eq. [34].

It is appropriate to raise one question at this time. For the Fourier transform relationships of Eq. [41], can we obtain the Fourier transform of *any* arbitrarily chosen $f(t)$? It turns out that the answer is affirmative for almost any voltage or current that we can actually produce. A sufficient condition for the existence of $\mathbf{F}(j\omega)$ is that

$$\int_{-\infty}^{\infty} |f(t)|dt < \infty$$

This condition is not *necessary,* however, because some functions that do not meet it still have a Fourier transform; the step function is one such example. Furthermore, we will see later that $f(t)$ does not even need to be nonperiodic in order to have a Fourier transform; the Fourier series representation for a periodic time function is just a special case of the more general Fourier transform representation.

As we indicated earlier, the Fourier transform-pair relationship is unique. For a given $f(t)$ there is one specific $\mathbf{F}(j\omega)$; and for a given $\mathbf{F}(j\omega)$ there is one specific $f(t)$.

The reader may have already noticed a few similarities between the Fourier transform and the Laplace transform. Key differences between the two include the fact that initial energy storage is not easily incorporated in circuit analysis using Fourier transforms, while it is very easily incorporated in the case of Laplace transforms. Also, there are several time functions (e.g., the *increasing* exponential) for which a Fourier transform does not exist. However, if it is spectral information as opposed to transient response in which we are primarily concerned, the Fourier transform is the ticket.

EXAMPLE 17.5

Use the Fourier transform to obtain the continuous spectrum of the single rectangular pulse in Fig. 17.13*a*.

The pulse is a truncated version of the sequence considered previously in Fig. 17.11 and is described by

$$f(t) = \begin{cases} V_o & t_0 < t < t_0 + \tau \\ 0 & t < t_0 \text{ and } t > t_0 + \tau \end{cases}$$

The Fourier transform of $f(t)$ is found from Eq. [41*a*]:

$$\mathbf{F}(j\omega) = \int_{t_0}^{t_0+\tau} V_0 e^{-j\omega t}\, dt$$

(Continued on next page)

and this may be easily integrated and simplified:

$$\mathbf{F}(j\omega) = V_0\tau \frac{\sin\frac{1}{2}\omega\tau}{\frac{1}{2}\omega\tau} e^{-j\omega(t_0+\tau/2)}$$

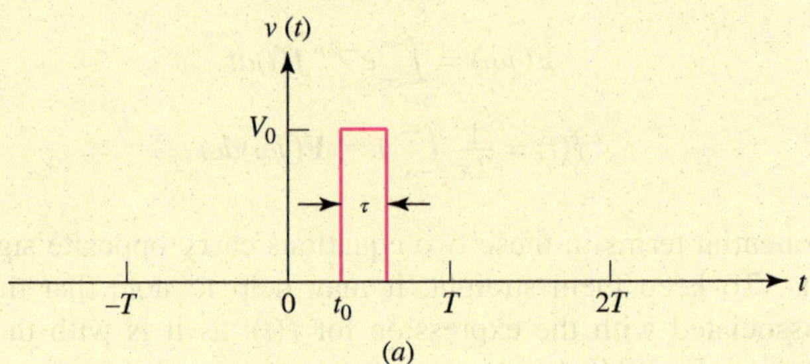

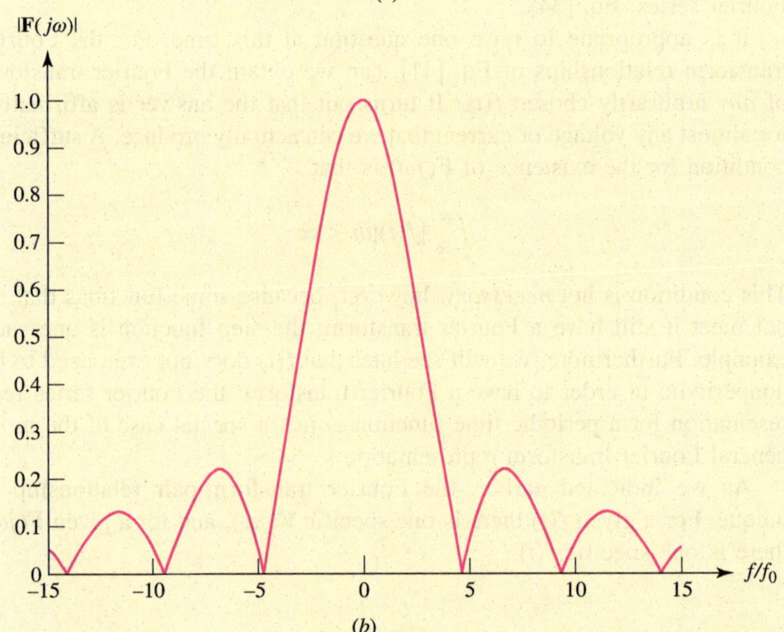

FIGURE 17.13 (*a*) A single rectangular pulse identical to those of the sequence in Fig. 17.11. (*b*) A plot of $|\mathbf{F}(j\omega)|$ corresponding to the pulse, with $v_o = 1$, $\tau = 1$, and $t_0 = 0$. The frequency axis has been normalized to the value of $f_0 = 1/1.5\pi$ corresponding to Fig. 17.12*a* to allow comparison; note that f_0 has no meaning or relevance in the context of $\mathbf{F}(j\omega)$.

The magnitude of $\mathbf{F}(j\omega)$ yields the continuous frequency spectrum, and it is of the form of the sampling function. The value of $\mathbf{F}(0)$ is $v_o\tau$. The shape of the spectrum is identical with the envelope in Fig. 17.12*b*. A plot of $|\mathbf{F}(j\omega)|$ as a function of ω does *not* indicate the magnitude of the voltage present at any given frequency. What is it, then? Examination of Eq. [40] shows that, if $f(t)$ is a voltage waveform, then $\mathbf{F}(j\omega)$ is dimensionally "volts per unit frequency."

PRACTICE

17.8 If $f(t) = -10$ V, $-0.2 < t < -0.1$ s, $f(t) = 10$ V, $0.1 < t < 0.2$ s, and $f(t) = 0$ for all other t, evaluate $\mathbf{F}(j\omega)$ for ω equal to (*a*) 0; (*b*) 10π rad/s; (*c*) -10π rad/s; (*d*) 15π rad/s; (*e*) -20π rad/s.

17.9 If $\mathbf{F}(j\omega) = -10$ V/(rad/s) for $-4 < \omega < -2$ rad/s, + 10 V/(rad/s) for $2 < \omega < 4$ rad/s, and 0 for all other ω, find the numerical value of $f(t)$ at t equal to (*a*) 10^{-4} s; (*b*) 10^{-2} s; (*c*) $\pi/4$ s; (*d*) $\pi/2$ s; (*e*) π s.

Ans: 17.8: 0; $j1.273$ V/(rad/s); $-j1.273$ V/(rad/s); $-j0.424$ V/(rad/s); 0.
17.9: $j1.9099 \times 10^{-3}$ V; $j0.1910$ V; $j4.05$ V; $-j4.05$ V; 0.

17.6 SOME PROPERTIES OF THE FOURIER TRANSFORM

Our object in this section is to establish several of the mathematical properties of the Fourier transform and, even more important, to understand its physical significance. We begin by using Euler's identity to replace $e^{-j\omega t}$ in Eq. [41*a*]:

$$\mathbf{F}(j\omega) = \int_{-\infty}^{\infty} f(t)\cos\omega t\,dt - j\int_{-\infty}^{\infty} f(t)\sin\omega t\,dt \qquad [42]$$

Since $f(t)$, $\cos\omega t$, and $\sin\omega t$ are all real functions of time, both the integrals in Eq. [42] are real functions of ω. Thus, by letting

$$\mathbf{F}(j\omega) = A(\omega) + jB(\omega) = |\mathbf{F}(j\omega)|\, e^{j\phi(\omega)} \qquad [43]$$

we have

$$A(\omega) = -\int_{-\infty}^{\infty} f(t)\cos\omega t\,dt \qquad [44]$$

$$B(\omega) = -\int_{-\infty}^{\infty} f(t)\sin\omega t\,dt \qquad [45]$$

$$|\mathbf{F}(j\omega)| = \sqrt{A^2(\omega) + B^2(\omega)} \qquad [46]$$

and

$$\phi(\omega) = \tan^{-1}\frac{B(\omega)}{A(\omega)} \qquad [47]$$

Replacing ω with $-\omega$ shows that $A(\omega)$ and $|\mathbf{F}(j\omega)|$ are both even functions of ω, while $B(\omega)$ and $\phi(\omega)$ are both odd functions of ω.

Now, if $f(t)$ is an even function of t, then the integrand of Eq. [45] is an odd function of t, and the symmetrical limits force $B(\omega)$ to be zero; thus, if $f(t)$ is even, its Fourier transform $\mathbf{F}(j\omega)$ is a real, even function of ω, and the phase function $\phi(\omega)$ is zero or π for all ω. However, if $f(t)$ is an odd function of t, then $A(\omega) = 0$ and $\mathbf{F}(j\omega)$ is both odd and a pure imaginary function of ω; $\phi(\omega)$ is $\pm\pi/2$. In general, however, $\mathbf{F}(j\omega)$ is a complex function of ω.

Finally, we note that the replacement of ω by $-\omega$ in Eq. [42] forms the *conjugate* of $\mathbf{F}(j\omega)$. Thus,

$$\mathbf{F}(-j\omega) = A(\omega) - jB(\omega) = \mathbf{F}^*(j\omega)$$

and we have

$$\mathbf{F}(j\omega)\mathbf{F}(-j\omega) = \mathbf{F}(j\omega)\mathbf{F}^*(j\omega) = A^2(\omega) + B^2(\omega) = |\mathbf{F}(j\omega)|^2$$

Physical Significance of the Fourier Transform

With these basic mathematical properties of the Fourier transform in mind, we are now ready to consider its physical significance. Let us suppose that $f(t)$ is either the voltage across or the current through a 1 Ω resistor, so that $f^2(t)$ is the instantaneous power delivered to the 1 Ω resistor by $f(t)$. Integrating this power over all time, we obtain the total energy delivered by $f(t)$ to the 1 Ω resistor,

$$W_{1\Omega} = \int_{-\infty}^{\infty} f^2(t)dt \qquad [48]$$

Now let us resort to a little trickery. Thinking of the integrand in Eq. [48] as $f(t)$ times itself, we replace one of those functions with Eq. [41*b*]:

$$W_{1\Omega} = \int_{-\infty}^{\infty} f(t)\left[\frac{1}{2\pi}\int_{-\infty}^{\infty} e^{j\omega t}\mathbf{F}(j\omega)d\omega\right]dt$$

Since $f(t)$ is not a function of the variable of integration ω, we may move it inside the bracketed integral and then interchange the order of integration:

$$W_{1\Omega} = \frac{1}{2\pi}\int_{-\infty}^{\infty}\left[\int_{-\infty}^{\infty}\mathbf{F}(j\omega)e^{j\omega t}f(t)dt\right]d\omega$$

Next we shift $\mathbf{F}(j\omega)$ outside the inner integral, causing that integral to become $\mathbf{F}(-j\omega)$:

$$W_{1\Omega} = \frac{1}{2\pi}\int_{-\infty}^{\infty}\mathbf{F}(j\omega)\mathbf{F}(-j\omega)d\omega = \frac{1}{2\pi}\int_{-\infty}^{\infty}|\mathbf{F}(j\omega)|^2 d\omega$$

Collecting these results,

$$\int_{-\infty}^{\infty} f^2(t)dt = \frac{1}{2\pi}\int_{-\infty}^{\infty}|\mathbf{F}(j\omega)|^2 d\omega \qquad [49]$$

Marc Antoine Parseval-Deschenes was a somewhat obscure French mathematician, geographer, and occasional poet who published these results in 1805, seventeen years before Fourier published his theorem.

Equation [49] is a very useful expression known as Parseval's theorem. This theorem, along with Eq. [48], tells us that the energy associated with $f(t)$ can be obtained either from an integration over all time in the time domain or by $1/(2\pi)$ times an integration over all (radian) frequency in the frequency domain.

Parseval's theorem also leads us to a greater understanding and interpretation of the meaning of the Fourier transform. Consider a voltage $v(t)$ with Fourier transform $\mathbf{F}_v(j\omega)$ and 1 Ω energy $W_{1\Omega}$:

$$W_{1\Omega} = \frac{1}{2\pi}\int_{-\infty}^{\infty}|\mathbf{F}_v(j\omega)|^2 d\omega = \frac{1}{\pi}\int_{0}^{\infty}|\mathbf{F}_v(j\omega)|^2 d\omega$$

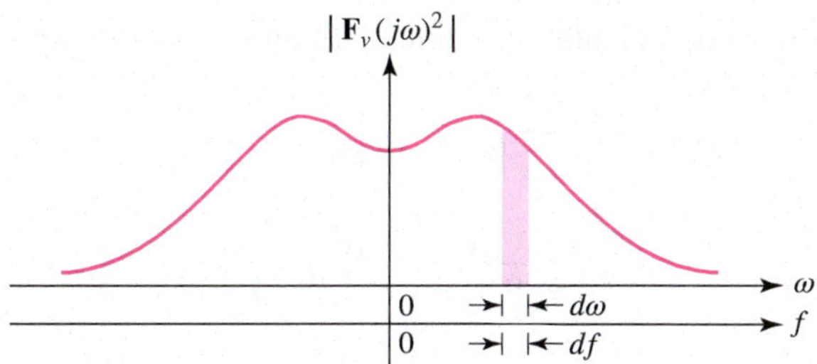

FIGURE 17.14 The area of the slice $|\mathbf{F}_v(j\omega)|^2$ is the 1 Ω energy associated with $v(t)$ lying in the bandwidth df.

where the rightmost equality follows from the fact that $|\mathbf{F}_v(j\omega)|^2$ is an even function of ω. Then, since $\omega = 2\pi f$, we can write

$$W_{1\Omega} = \int_{-\infty}^{\infty} |\mathbf{F}_v(j\omega)|^2 df = 2\int_{0}^{\infty} |\mathbf{F}_v(j\omega)|^2 df \qquad [50]$$

Figure 17.14 illustrates a typical plot of $|\mathbf{F}_v(j\omega)|^2$ as a function of both ω and f. If we divide the frequency scale up into vanishingly small increments df, Eq. [50] shows us that the area of a differential slice under the $|\mathbf{F}_v(j\omega)|^2$ curve, having a width df, is $|\mathbf{F}_v(j\omega)|^2 df$. This area is shown shaded. The sum of all such areas, as f ranges from minus to plus infinity, is the total 1 Ω energy contained in $v(t)$. Thus, $|\mathbf{F}_v(j\omega)|^2$ is the (1 Ω) ***energy density*** or energy per unit bandwidth (J/Hz) of $v(t)$, and this energy density is always a real, even, nonnegative function of ω. By integrating $|\mathbf{F}_v(j\omega)|^2$ over an appropriate frequency interval, we can calculate that portion of the total energy lying within the chosen interval. Note that the energy density is not a function of the phase of $\mathbf{F}_v(j\omega)$, and thus there are an infinite number of time functions and Fourier transforms that possess identical energy-density functions.

EXAMPLE **17.6**

The one-sided [i.e., $v(t) = 0$ for $t < 0$] exponential pulse $v(t) = 4e^{-3t}u(t)$ V is applied to the input of an ideal bandpass filter. If the filter passband is defined by $1 < |f| < 2$ Hz, calculate the total output energy.

We call the filter output voltage $v_o(t)$. The energy in $v_o(t)$ will therefore be equal to the energy of that part of $v(t)$ having frequency components in the intervals $1 < f < 2$ and $-2 < f < -1$. We determine the Fourier transform of $v(t)$,

$$\mathbf{F}_v(j\omega) = 4\int_{-\infty}^{\infty} e^{-j\omega t} e^{-3t}\, u(t)dt$$

$$= 4\int_{0}^{\infty} e^{-(3+j\omega)t}\, dt = \frac{4}{3 + j\omega}$$

(Continued on next page)

and then we may calculate the total 1 Ω energy in the input signal by either

$$W_{1\Omega} = \frac{1}{2\pi}\int_{-\infty}^{\infty} |\mathbf{F}_v(j\omega)|^2\, d\omega$$

$$= \frac{8}{\pi}\int_{-\infty}^{\infty} \frac{d\omega}{9+\omega^2} = \frac{16}{\pi}\int_{0}^{\infty} \frac{d\omega}{9+\omega^2} = \frac{8}{3}\text{ J}$$

or

$$W_{1\Omega} = \int_{-\infty}^{\infty} v^2(t)dt = 16\int_{0}^{\infty} e^{-6t}\, dt = \frac{8}{3}\text{ J}$$

The total energy in $v_o(t)$, however, is smaller:

$$W_{o1} = \frac{1}{2\pi}\int_{-4\pi}^{-2\pi} \frac{16\, d\omega}{9+\omega^2} + \frac{1}{2\pi}\int_{2\pi}^{4\pi} \frac{16\, d\omega}{9+\omega^2}$$

$$= \frac{16}{\pi}\int_{2\pi}^{4\pi} \frac{d\omega}{9+\omega^2} = \frac{16}{3\pi}\left(\tan^{-1}\frac{4\pi}{3} - \tan^{-1}\frac{2\pi}{3}\right) = 358\text{ mJ}$$

In general, we see that an ideal bandpass filter enables us to remove energy from prescribed frequency ranges while still retaining the energy contained in other frequency ranges. The Fourier transform helps us to describe the filtering action quantitatively without actually evaluating $v_o(t)$, although we will see later that the Fourier transform can also be used to obtain the expression for $v_o(t)$ if we wish to do so.

PRACTICE

17.10 If $i(t) = 10e^{20t}\,[u(t+0.1) - u(t-0.1)]$ A, find (*a*) $\mathbf{F}_i(j0)$; (*b*) $\mathbf{F}_i(j10)$; (*c*) $A_i(10)$; (*d*) $B_i(10)$; (*e*) $\phi_i(10)$.

17.11 Find the 1 Ω energy associated with the current $i(t) = 20e^{-10t}u(t)$ A in the interval (*a*) $-0.1 < t < 0.1$ s; (*b*) $-10 < \omega < 10$ rad/s; (*c*) $10 < \omega < \infty$ rad/s.

Ans: 17.10: 3.63 A/(rad/s); $3.33\underline{/-31.7^\circ}$ A/(rad/s); 2.83 A/(rad/s); −1.749 A/(rad/s); −31.7°. 17.11: 17.29 J; 10 J; 5 J.

17.7 FOURIER TRANSFORM PAIRS FOR SOME SIMPLE TIME FUNCTIONS

The Unit-Impulse Function

We now seek the Fourier transform of the unit impulse $\delta(t - t_0)$, a function we introduced in Sec. 14.3. That is, we are interested in the spectral properties or frequency-domain description of this singularity function. If we use the notation $\mathcal{F}\{\}$ to symbolize "Fourier transform of {}," then

$$\mathcal{F}\{\delta(t-t_0)\} = \int_{-\infty}^{\infty} e^{-j\omega t}\delta(t-t_0)dt$$

From our earlier discussion of this type of integral, we have

$$\mathcal{F}\{\delta(t-t_0)\} = e^{-j\omega t_0} = \cos\omega t_0 - j\sin\omega t_0$$

This complex function of ω leads to the 1 Ω energy-density function,

$$|\mathcal{F}\{\delta(t - t_0)\}|^2 = \cos^2 \omega t_0 + \sin^2 \omega t_0 = 1$$

This remarkable result says that the (1 Ω) energy per unit bandwidth is unity *at all frequencies,* and that the total energy in the unit impulse is infinitely large. No wonder, then, that we must conclude that the unit impulse is "impractical" in the sense that it cannot be generated in the laboratory. Moreover, even if one were available to us, it must appear distorted after being subjected to the finite bandwidth of any practical laboratory instrument.

Since there is a unique one-to-one correspondence between a time function and its Fourier transform, we can say that the inverse Fourier transform of $e^{-j\omega t_0}$ is $\delta(t - t_0)$. Utilizing the symbol $\mathcal{F}^{-1}\{\}$ for the inverse transform, we have

$$\mathcal{F}^{-1}\{e^{-j\omega t_0}\} = \delta(t - t_0)$$

Thus, we now know that

$$\frac{1}{2\pi}\int_{-\infty}^{\infty} e^{j\omega t}e^{-j\omega t_0}d\omega = \delta(t - t_0)$$

even though we would fail in an attempt at the direct evaluation of this improper integral. Symbolically, we may write

$$\delta(t - t_0) \Leftrightarrow e^{-j\omega t_0} \qquad [51]$$

where $\Leftrightarrow$ indicates that the two functions constitute a Fourier transform pair.

Continuing with our consideration of the unit-impulse function, let us consider a Fourier transform in that form,

$$\mathbf{F}(j\omega) = \delta(\omega - \omega_0)$$

which is a unit impulse *in the frequency domain* located at $\omega = \omega_0$. Then $f(t)$ must be

$$f(t) = \mathcal{F}^{-1}\{\mathbf{F}(j\omega)\} = \frac{1}{2\pi}\int_{-\infty}^{\infty} e^{j\omega t}\delta(\omega - \omega_0)d\omega = \frac{1}{2\pi}e^{j\omega_0 t}$$

where we have used the sifting property of the unit impulse. Thus we may now write

$$\frac{1}{2\pi}e^{j\omega_0 t} \Leftrightarrow \delta(\omega - \omega_0)$$

or

$$e^{j\omega_0 t} \Leftrightarrow 2\pi\delta(\omega - \omega_0) \qquad [52]$$

Also, by a simple sign change we obtain

$$e^{-j\omega_0 t} \Leftrightarrow 2\pi\delta(\omega + \omega_0) \qquad [53]$$

Clearly, the time function is complex in both expressions [52] and [53], and does not exist in the real world of the laboratory.

However, we know that

$$\cos \omega_0 t = \frac{1}{2}e^{j\omega_0 t} + \frac{1}{2}e^{-j\omega_0 t}$$

and it is easily seen from the definition of the Fourier transform that

$$\mathscr{F}\{f_1(t)\} + \mathscr{F}\{f_2(t)\} = \mathscr{F}\{f_1(t) + f_2(t)\} \qquad [54]$$

Therefore,

$$\mathscr{F}\{\cos \omega_0 t\} = \mathscr{F}\left\{\frac{1}{2}e^{j\omega_0 t}\right\} + \mathscr{F}\left\{\frac{1}{2}e^{-j\omega_0 t}\right\}$$
$$= \pi\delta(\omega - \omega_0) + \pi\delta(\omega + \omega_0)$$

which indicates that the frequency-domain description of $\cos \omega_0 t$ shows a *pair* of impulses, located at $\omega = \pm\omega_0$. This should not be a great surprise, for in our first discussion of complex frequency in Chap. 14, we noted that a sinusoidal function of time was always represented by a pair of imaginary frequencies located at $\mathbf{s} = \pm j\omega_0$. We have, therefore,

$$\cos \omega_0 t \Leftrightarrow \pi[\delta(\omega + \omega_0) + \delta(\omega - \omega_0)] \qquad [55]$$

The Constant Forcing Function

To find the Fourier transform of a constant function of time, $f(t) = K$, our first inclination might be to substitute this constant in the defining equation for the Fourier transform and evaluate the resulting integral. If we did, we would find ourselves with an indeterminate expression on our hands. Fortunately, however, we have already solved this problem, for from expression [53],

$$e^{-j\omega_0 t} \Leftrightarrow 2\pi\delta(\omega + \omega_0)$$

We see that if we simply let $\omega_0 = 0$, then the resulting transform pair is

$$1 \Leftrightarrow 2\pi\delta(\omega) \qquad [56]$$

from which it follows that

$$K \Leftrightarrow 2\pi K\delta(\omega) \qquad [57]$$

and our problem is solved. The frequency spectrum of a constant function of time consists only of a component at $\omega = 0$, which we knew all along.

The Signum Function

As another example, let us obtain the Fourier transform of a singularity function known as the ***signum function,*** $\text{sgn}(t)$, defined by

$$\text{sgn}(t) = \begin{cases} -1 & t < 0 \\ 1 & t > 0 \end{cases} \qquad [58]$$

or

$$\text{sgn}(t) = u(t) - u(-t)$$

Again, if we should try to substitute this time function in the defining equation for the Fourier transform, we would face an indeterminate expression upon substitution of the limits of integration. This same problem will arise every time we try to obtain the Fourier transform of a time function that does not approach zero as $|t|$ approaches infinity. Fortunately, we can avoid this situation by using the *Laplace transform*, as it contains a built-in convergence factor that cures many of the inconvenient ills associated with the evaluation of certain Fourier transforms.

Along those lines, the signum function under consideration can be written as

$$\text{sgn}(t) = \lim_{a\to 0}[e^{-at}u(t) - e^{at}u(-t)]$$

Notice that the expression within the brackets *does* approach zero as $|t|$ gets very large. Using the definition of the Fourier transform, we obtain

$$\mathscr{F}\{\text{sgn}(t)\} = \lim_{a\to 0}\left[\int_0^{\infty} e^{-j\omega t}e^{-at}\,dt - \int_{-\infty}^{0} e^{-j\omega t}e^{at}\,dt\right]$$

$$= \lim_{a\to 0}\frac{-j2\omega}{\omega^2 + a^2} = \frac{2}{j\omega}$$

The real component is zero, since sgn(t) is an odd function of t. Thus,

$$\text{sgn}(t) \Leftrightarrow \frac{2}{j\omega} \qquad [59]$$

The Unit-Step Function

As a final example in this section, let us look at the familiar unit-step function, $u(t)$. Making use of our work on the signum function in the preceding paragraphs, we represent the unit step by

$$u(t) = \frac{1}{2} + \frac{1}{2}\text{sgn}(t)$$

and obtain the Fourier transform pair

$$u(t) \Leftrightarrow \left[\pi\delta(\omega) + \frac{1}{j\omega}\right] \qquad [60]$$

Table 17.2 presents the conclusions drawn from the examples discussed in this section, along with a few others that have not been detailed here.

EXAMPLE 17.7

Use Table 17.2 to find the Fourier transform of the time function $3e^{-t}\cos 4t\, u(t)$.

From the next to the last entry in the table, we have

$$e^{-\alpha t}\cos\omega_d t\, u(t) \Leftrightarrow \frac{\alpha + j\omega}{(\alpha + j\omega)^2 + \omega_d^2}$$

We therefore identify α as 1 and ω_d as 4, and have

$$\mathbf{F}(j\omega) = 3\frac{1 + j\omega}{(1 + j\omega)^2 + 16}$$

PRACTICE

17.12 Evaluate the Fourier transform at $\omega = 12$ for the time function (*a*) $4u(t) - 10\delta(t)$; (*b*) $5e^{-8t}\,u(t)$; (*c*) $4\cos 8tu(t)$; (*d*) $-4\,\text{sgn}(t)$.

17.13 Find $f(t)$ at $t = 2$ if $\mathbf{F}(j\omega)$ is equal to (*a*) $5e^{-j3\omega} - j(4/\omega)$; (*b*) $8[\delta(\omega - 3) + \delta(\omega + 3)]$; (*c*) $(8/\omega)\sin 5\omega$.

Ans: 17.12: $10.01\underline{/-178.1^\circ}$; $0.347\underline{/-56.3^\circ}$; $-j0.6$; $j0.667$. 17.13: 2.00; 2.45; 4.00.

TABLE 17.2 A Summary of Some Fourier Transform Pairs

f(t)	f(t)	$\mathscr{F}\{f(t)\} = \mathbf{F}(j\omega)$	$\lvert\mathbf{F}(j\omega)\rvert$
(1) t_0 t	$\delta(t - t_0)$	$e^{-j\omega t_0}$	1 ω
Complex t	$e^{j\omega_0 t}$	$2\pi\delta(\omega - \omega_0)$	(2π) ω_0 ω
1 −1 t	$\cos\omega_0 t$	$\pi[\delta(\omega + \omega_0) + \delta(\omega - \omega_0)]$	(π) (π) $-\omega_0$ ω_0 ω
1 t	1	$2\pi\delta(\omega)$	(2π) ω
1 −1 t	$\mathrm{sgn}(t)$	$\frac{2}{j\omega}$	ω
1 t	$u(t)$	$\pi\delta(\omega) + \frac{1}{j\omega}$	(π) ω
1 t	$e^{-\alpha t}u(t)$	$\frac{1}{\alpha + j\omega}$	$\frac{1}{\alpha}$ ω
t	$[e^{-\alpha t}\cos\omega_d t]u(t)$	$\frac{\alpha + j\omega}{(\alpha + j\omega)^2 + \omega_d^2}$	$-\omega_d$ ω_d ω
1 $-\frac{T}{2}$ $\frac{T}{2}$ t	$u\left(t + \frac{1}{2}T\right) - u\left(t - \frac{1}{2}T\right)$	$T\frac{\sin\frac{\omega T}{2}}{\frac{\omega T}{2}}$	$-\frac{2\pi}{T}$ $\frac{2\pi}{T}$ ω

17.8 THE FOURIER TRANSFORM OF A GENERAL PERIODIC TIME FUNCTION

In Sec. 17.5 we remarked that we would be able to show that periodic time functions, as well as nonperiodic functions, possess Fourier transforms. Let us now establish this fact on a rigorous basis. Consider a periodic time function $f(t)$ with period T and Fourier series expansion, as outlined by Eqs. [34], [35], and [36], repeated here for convenience:

$$f(t) = \sum_{n=-\infty}^{\infty} \mathbf{c}_n e^{jn\omega_0 t} \qquad [34]$$

$$\mathbf{c}_n = \frac{1}{T}\int_{-T/2}^{T/2} f(t)e^{-jn\omega_0 t}\,dt \qquad [35]$$

and

$$\omega_0 = \frac{2\pi}{T} \qquad [36]$$

Bearing in mind that the Fourier transform of a sum is just the sum of the transforms of the terms in the sum, and that $\mathbf{c}_n$ is not a function of time, we can write

$$\mathscr{F}\{f(t)\} = \mathscr{F}\left\{\sum_{n=-\infty}^{\infty} \mathbf{c}_n e^{jn\omega_0 t}\right\} = \sum_{n=-\infty}^{\infty} \mathbf{c}_n \mathscr{F}\{e^{jn\omega_0 t}\}$$

After obtaining the transform of $e^{jn\omega_0 t}$ from expression [52], we have

$$f(t) \Leftrightarrow 2\pi \sum_{n=-\infty}^{\infty} \mathbf{c}_n \delta(\omega - n\omega_0) \qquad [61]$$

This shows that $f(t)$ has a discrete spectrum consisting of impulses located at points on the ω axis given by $\omega = n\omega_0$, $n = \ldots, -2, -1, 0, 1, \ldots$. The strength of each impulse is 2π times the value of the corresponding Fourier coefficient appearing in the complex form of the Fourier series expansion for $f(t)$.

As a check on our work, let us see whether the inverse Fourier transform of the right side of expression [61] is once again $f(t)$. This inverse transform can be written as

$$\mathscr{F}^{-1}\{\mathbf{F}(j\omega)\} = \frac{1}{2\pi}\int_{-\infty}^{\infty} e^{j\omega t}\left[2\pi \sum_{n=-\infty}^{\infty} \mathbf{c}_n \delta(\omega - n\omega_0)\right] d\omega \stackrel{?}{=} f(t)$$

Since the exponential term does not contain the index of summation n, we can interchange the order of the integration and summation operations:

$$\mathscr{F}^{-1}\{\mathbf{F}(j\omega)\} = \sum_{n=-\infty}^{\infty} \int_{-\infty}^{\infty} \mathbf{c}_n e^{j\omega t}\delta(\omega - n\omega_0)\,d\omega \stackrel{?}{=} f(t)$$

Because it is not a function of the variable of integration, $\mathbf{c}_n$ can be treated as a constant. Then, using the sifting property of the impulse, we obtain

$$\mathscr{F}^{-1}\{\mathbf{F}(j\omega)\} = \sum_{n=-\infty}^{\infty} \mathbf{c}_n e^{jn\omega_0 t} \stackrel{?}{=} f(t)$$

which is exactly the same as Eq. [34], the complex Fourier series expansion for $f(t)$. The question marks in the preceding equations can now be removed, and the existence of the Fourier transform for a periodic time function is established. This should come as no great surprise, however. In the last section we evaluated the Fourier transform of a cosine function, which is certainly periodic, although we made no direct reference to its periodicity. However, we did use a backhanded approach in getting the transform. But now we have a mathematical tool by which the transform can be obtained more directly. To demonstrate this procedure, consider $f(t) = \cos \omega_0 t$ once more. First we evaluate the Fourier coefficients $\mathbf{c}_n$:

$$\mathbf{c}_n = \frac{1}{T}\int_{-T/2}^{T/2} \cos \omega_0 t \, e^{-jn\omega_0 t}\, dt = \begin{cases} \frac{1}{2} & n = \pm 1 \\ 0 & \text{otherwise} \end{cases}$$

Then

$$\mathscr{F}\{f(t)\} = 2\pi \sum_{n=-\infty}^{\infty} \mathbf{c}_n \delta(\omega - n\omega_0)$$

This expression has values that are nonzero only when $n = \pm 1$, and it follows, therefore, that the entire summation reduces to

$$\mathscr{F}\{\cos \omega_0 t\} = \pi[\delta(\omega - \omega_0) + \delta(\omega + \omega_0)]$$

which is precisely the expression that we obtained before. What a relief!

PRACTICE

17.14 Find (*a*) $\mathscr{F}\{5 \sin^2 3t\}$; (*b*) $\mathscr{F}\{A \sin \omega_0 t\}$; (*c*) $\mathscr{F}\{6 \cos(8t + 0.1\pi)\}$.

Ans: $2.5\pi[2\delta(\omega) - \delta(\omega + 6) - \delta(\omega - 6)]$; $j\pi A[\delta(\omega + \omega_0) - \delta(\omega - \omega_0)]$; $[18.85\underline{/18^\circ}]\delta(\omega - 8) + [18.85\underline{/-18^\circ}]\delta(\omega + 8)$.

17.9 THE SYSTEM FUNCTION AND RESPONSE IN THE FREQUENCY DOMAIN

In Sec. 14.11, the problem of determining the output of a physical system in terms of the input and the impulse response was solved by using the convolution integral and initially working in the time domain. The input, the output, and the impulse response are all time functions. Subsequently, we found that it was often more convenient to perform such operations in the frequency domain, as the Laplace transform of the convolution of two functions is simply the product of each function in the frequency domain. Along the same lines, we find the same is true when working with Fourier transforms.

To do this we examine the Fourier transform of the system output. Assuming arbitrarily that the input and output are voltages, we apply the basic definition of the Fourier transform and express the output by the convolution integral:

$$\mathscr{F}\{v_0(t)\} = \mathbf{F}_0(j\omega) = \int_{-\infty}^{\infty} e^{-j\omega t}\left[\int_{-\infty}^{\infty} v_i(t-z)h(z)dz\right]dt$$

where we again assume no initial energy storage. At first glance this expression may seem rather formidable, but it can be reduced to a result that is surprisingly simple. We may move the exponential term inside the inner integral because it does not contain the variable of integration z. Next we reverse the order of integration, obtaining

$$\mathbf{F}_0(j\omega) = \int_{-\infty}^{\infty}\left[\int_{-\infty}^{\infty} e^{-j\omega t} v_i(t-z)h(z)dt\right]dz$$

Since it is not a function of t, we can extract $h(z)$ from the inner integral and simplify the integration with respect to t by a change of variable, $t - z = x$:

$$\mathbf{F}_0(j\omega) = \int_{-\infty}^{\infty} h(z)\left[\int_{-\infty}^{\infty} e^{-j\omega(x+z)} v_i(x)dx\right]dz$$

$$= \int_{-\infty}^{\infty} e^{-j\omega z}h(z)\left[\int_{-\infty}^{\infty} e^{-j\omega x} v_i(x)dx\right]dz$$

But now the sun is starting to break through, for the inner integral is merely the Fourier transform of $v_i(t)$. Furthermore, it contains no z terms and can be treated as a constant in any integration involving z. Thus, we can move this transform, $\mathbf{F}_i(j\omega)$, completely outside all the integral signs:

$$\mathbf{F}_0(j\omega) = \mathbf{F}_i(j\omega)\int_{-\infty}^{\infty} e^{-j\omega z}\, h(z)dz$$

Finally, the remaining integral exhibits our old friend once more, another Fourier transform! This one is the Fourier transform of the impulse response, which we will designate by the notation $\mathbf{H}(j\omega)$. Therefore, all our work has boiled down to the simple result:

$$\mathbf{F}_0(j\omega) = \mathbf{F}_i(j\omega)\mathbf{H}(j\omega) = \mathbf{F}_i(j\omega)\mathscr{F}\{h(t)\}$$

This is another important result: it defines the *system function* $\mathbf{H}(j\omega)$ as the ratio of the Fourier transform of the response function to the Fourier transform of the forcing function. Moreover, the system function and the impulse response constitute a Fourier transform pair:

$$h(t) \Leftrightarrow \mathbf{H}(j\omega) \qquad [62]$$

The development in the preceding paragraph also serves to prove the general statement that the Fourier transform of the convolution of two time functions is the product of their Fourier transforms,

$$\mathscr{F}\{f(t) * g(t)\} = \mathbf{F}_f(j\omega)\mathbf{F}_g(j\omega) \qquad [63]$$

To recapitulate, if we know the Fourier transforms of the forcing function and the impulse response, then the Fourier transform of the response function can be obtained as their product. The result is a description of the response function in the frequency domain; the time-domain description of the response function is obtained by simply taking the inverse Fourier transform. Thus we see that the process of convolution in the time domain is equivalent to the relatively simple operation of multiplication in the frequency domain.

The foregoing comments might make us wonder once again why we would ever choose to work in the time domain at all, but we must always remember that we seldom get something for nothing. A poet once said, "*Our sincerest laughter/with some pain is fraught.*"[2] The pain herein is the occasional difficulty in obtaining the inverse Fourier transform of a response function, for reasons of mathematical complexity. On the other hand, a simple desktop computer can convolve two time functions with magnificent celerity. For that matter, it can also obtain an FFT (fast Fourier transform) quite rapidly. Consequently there is no clear-cut advantage between working in the time domain and in the frequency domain. A decision must be made each time a new problem arises; it should be based on the information available and on the computational facilities at hand.

Consider a forcing function of the form

$$v_i(t) = u(t) - u(t-1)$$

and a unit-impulse response defined by

$$h(t) = 2e^{-t}u(t)$$

We first obtain the corresponding Fourier transforms. The forcing function is the difference between two unit-step functions. These two functions are identical, except that one is initiated 1 s after the other. We will evaluate the response due to $u(t)$; the response due to $u(t-1)$ is the same, but delayed in time by 1 s. The difference between these two partial responses will be the total response due to $v_i(t)$.

The Fourier transform of $u(t)$ was obtained in Sec. 17.7:

$$\mathscr{F}\{u(t)\} = \pi\delta(\omega) + \frac{1}{j\omega}$$

The system function is obtained by taking the Fourier transform of $h(t)$, listed in Table 17.2,

$$\mathscr{F}\{h(t)\} = \mathbf{H}(j\omega) = \mathscr{F}\{2e^{-t}u(t)\} = \frac{2}{1+j\omega}$$

The inverse transform of the product of these two functions yields that component of $v_o(t)$ caused by $u(t)$,

$$v_{o1}(t) = \mathscr{F}^{-1}\left\{\frac{2\pi\delta(\omega)}{1+j\omega} + \frac{2}{j\omega(1+j\omega)}\right\}$$

Using the sifting property of the unit impulse, the inverse transform of the first term is just a constant equal to unity. Thus,

$$v_{o1}(t) = 1 + \mathscr{F}^{-1}\left\{\frac{2}{j\omega(1+j\omega)}\right\}$$

The second term contains a product of terms in the denominator, each of the form $(\alpha + j\omega)$, and its inverse transform is found most easily by making use of the partial-fraction expansion that we developed in Sec. 14.4. Let us select a technique for obtaining a partial-fraction expansion that

(2) P. B. Shelley, "To a Skylark," 1821.

has one big advantage—it always works, although faster methods are usually available for most situations. We assign an unknown quantity in the numerator of each fraction, here two in number,

$$\frac{2}{j\omega(1+j\omega)} = \frac{A}{j\omega} + \frac{B}{1+j\omega}$$

and then substitute a corresponding number of simple values for $j\omega$. Here we let $j\omega = 1$:

$$1 = A + \frac{B}{2}$$

and then let $j\omega = -2$:

$$1 = -\frac{A}{2} - B$$

This leads to $A = 2$ and $B = -2$. Thus,

$$\mathscr{F}^{-1}\left\{\frac{2}{j\omega(1+j\omega)}\right\} = \mathscr{F}^{-1}\left\{\frac{2}{j\omega} - \frac{2}{1+j\omega}\right\} = \text{sgn}(t) - 2e^{-t}\,u(t)$$

so that

$$\begin{aligned} v_{o1}(t) &= 1 + \text{sgn}(t) - 2e^{-t}\,u(t) \\ &= 2u(t) - 2e^{-t}u(t) \\ &= 2(1 - e^{-t})u(t) \end{aligned}$$

It follows that $v_{o2}(t)$, the component of $v_o(t)$, produced by $u(t-1)$, is

$$v_{o2}(t) = 2(1 - e^{-(t-1)})u(t-1)$$

Therefore,

$$\begin{aligned} v_o(t) &= v_{o1}(t) - v_{o2}(t) \\ &= 2(1 - e^{-t})u(t) - 2(1 - e^{-t+1})u(t-1) \end{aligned}$$

The discontinuities at $t = 0$ and $t = 1$ dictate a separation into three time intervals:

$$v_o(t) = \begin{cases} 0 & t < 0 \\ 2(1 - e^{-t}) & 0 < t < 1 \\ 2(e - 1)e^{-t} & t > 1 \end{cases}$$

PRACTICE

17.15 The impulse response of a certain linear network is $h(t) = 6e^{-20t}\,u(t)$. The input signal is $3e^{-6t}\,u(t)$ V. Find (*a*) $\mathbf{H}(j\omega)$; (*b*) $\mathbf{V}_i(j\omega)$; (*c*) $V_o(j\omega)$; (*d*) $v_o(0.1)$; (*e*) $v_o(0.3)$; (*f*) $v_{o,\text{max}}$.

Ans: $6/(20 + j\omega)$; $3/(6 + j\omega)$; $18/[(20 + j\omega)(6 + j\omega)]$; 0.532 V; 0.209 V; 0.5372.

COMPUTER-AIDED ANALYSIS

The material presented in this chapter forms the foundation for many advanced fields of study, including signal processing, communications, and controls. We can only introduce some of the more fundamental concepts within the context of an introductory circuits text, but even at this point some of the power of Fourier-based analysis can be brought to bear. As a first example, consider the op amp circuit of Fig. 17.15 constructed in LTspice using UniversalOpamp2.

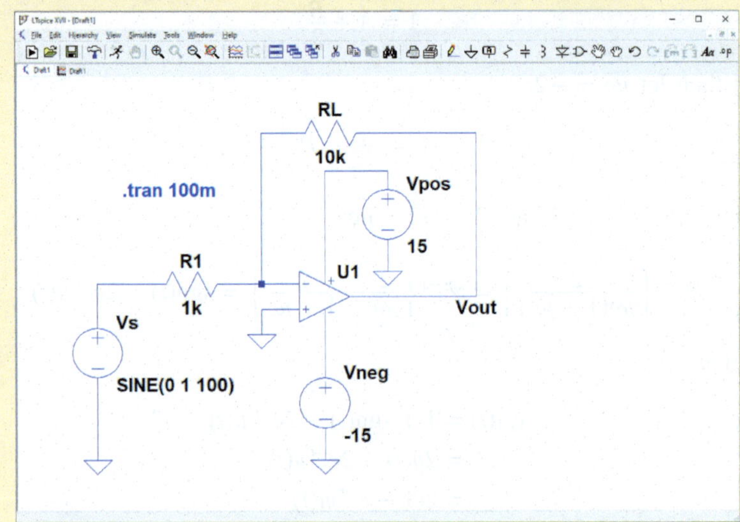

■ **FIGURE 17.15** An inverting amplifier circuit with a voltage gain of −10, driven by a sinusoidal input operating at 100 Hz.
Analog Devices

The circuit has a voltage gain of −10, and so we would expect a sinusoidal output of 10 V amplitude. This is indeed what we obtain from a transient analysis of the circuit, as shown in Fig. 17.16.

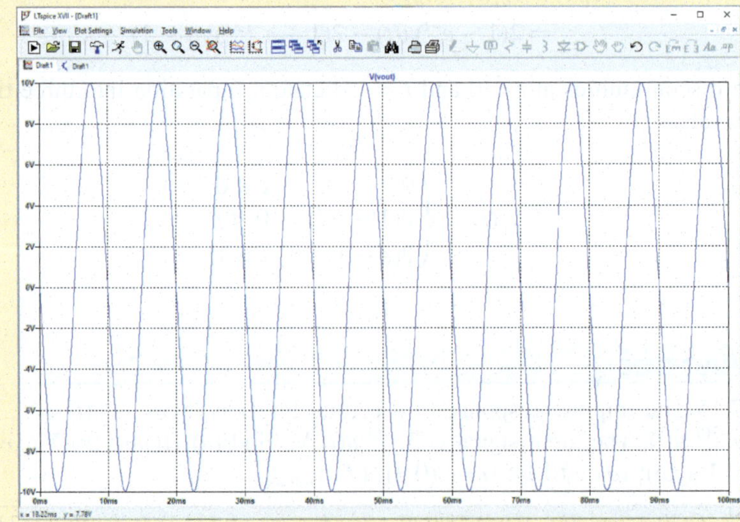

■ **FIGURE 17.16** Simulated output voltage of the amplifier circuit shown in Fig. 17.15.
Analog Devices

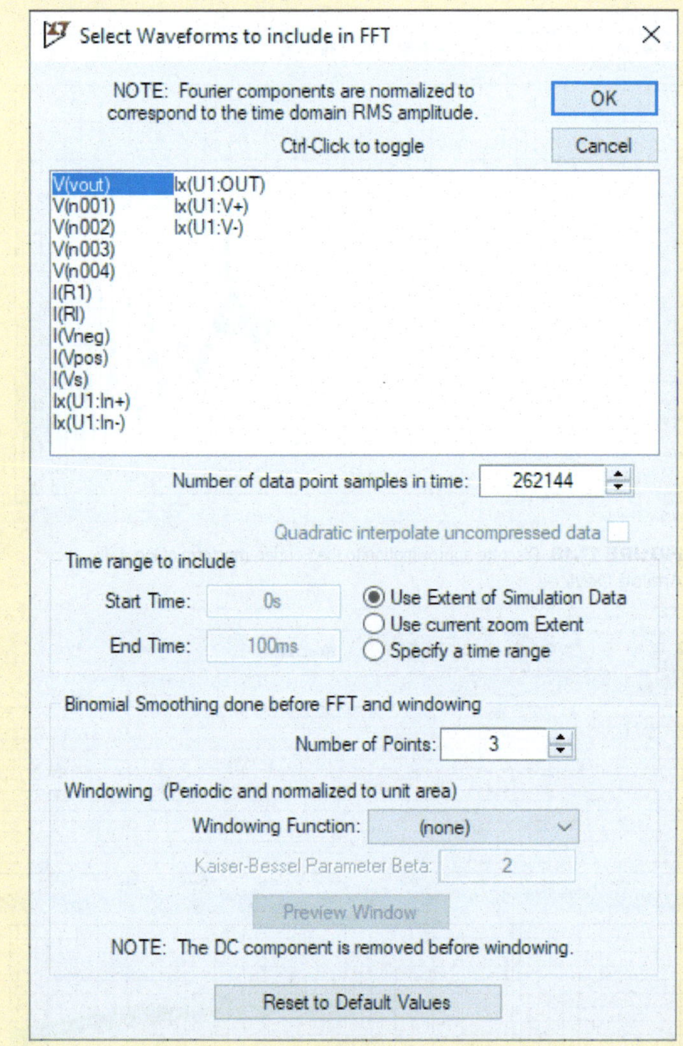

FIGURE 17.17 Pop-up menu for FFT analysis performed on waveform plot.
Analog Devices

LTspice allows us to determine the frequency spectrum of the output voltage through what is known as a fast Fourier transform (FFT), a discrete-time approximation to the exact Fourier transform of the signal. In the plot window, select **FFT** from under **View.** A pop-up window appears (Fig. 17.17), which provides multiple options to adjust the plot. The FFT is shown in Fig. 17.18 after adjusting the *x*-axis limit to 100 kHz. Note that several plotting options are available, including dB scale and log frequency. It is often useful to examine both the dB scale and log scale for frequency, as we have for Bode plots. The result of our amplifier shows a dominant single feature at a frequency of 100 Hz. Other components are observed, which are a combination of noise from numerical analysis and response of the op amp according to the SPICE model.

(Continued on next page)

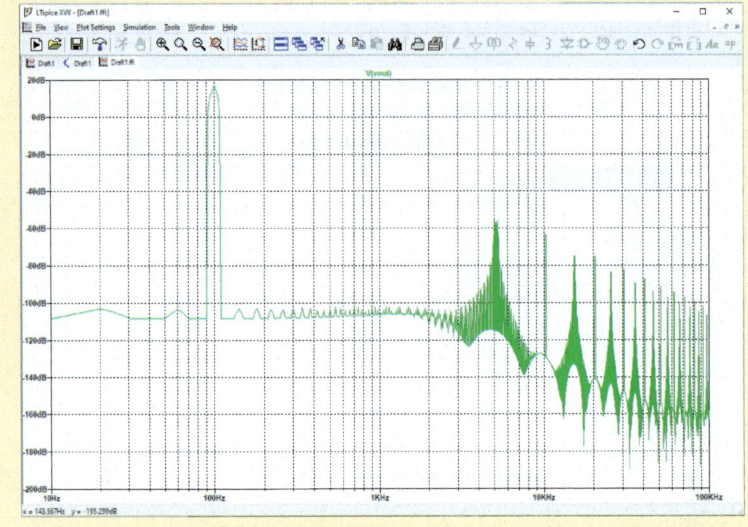

■ **FIGURE 17.18** Discrete approximation to the Fourier transform of Fig. 17.16.
Analog Devices

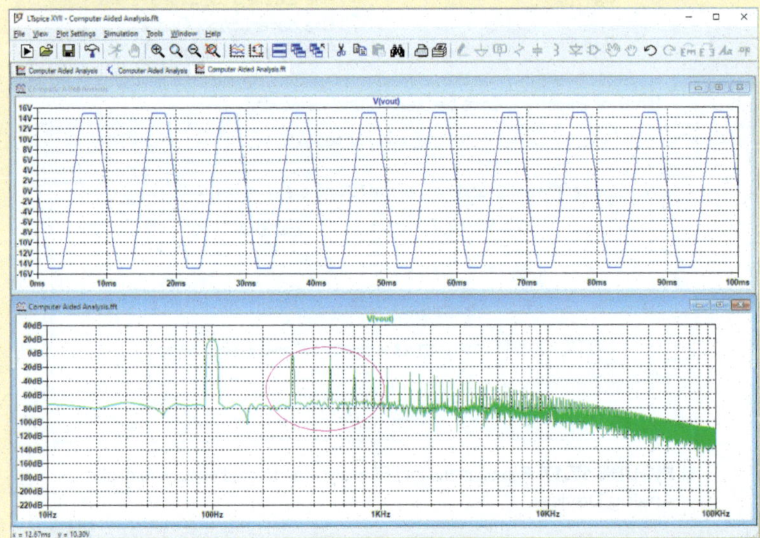

■ **FIGURE 17.19** *Top graph:* Transient analysis simulation results for the amplifier circuit when the input voltage magnitude is increased to 1.8 V. Saturation effects manifest themselves in the plot as clipped waveform extrema. *Bottom graph:* corresponding FFT. The first several harmonics introduced by clipping are circled.
Analog Devices

As the input voltage magnitude is increased, the output of the amplifier approaches the saturation condition determined by the positive and negative dc supply voltages (±15 V in this example). This behavior is evident in the simulation result of Fig. 17.19, which corresponds to an input voltage magnitude of 1.8 V. A key feature of interest is that the output voltage waveform is no longer a pure sinusoid. As a result, we expect nonzero values at harmonic frequencies to appear in the frequency spectrum of the function, as is the case in the FFT plot in Fig. 17.19. The effect of reaching saturation in the amplifier circuit is a distortion of the signal; if connected to a speaker, we do not hear a "clean" 100 Hz waveform. Instead, we now hear a superposition of waveforms which include not only the 100 Hz fundamental frequency, but significant harmonic components at 300 and 500 Hz as well. Further distortion of the waveform would increase the amount of energy in harmonic frequencies, so contributions from higher-frequency harmonics would become more significant. This is evident in the simulation results of Fig. 17.20 and 17.21, which show the output voltage in the time and frequency domains, respectively.

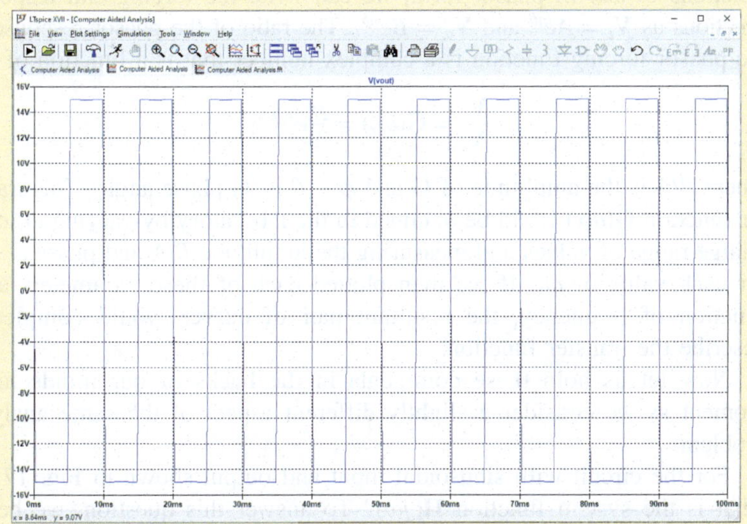

FIGURE 17.20 Severe effects of amplifier saturation are observed in the simulated response to a 15 V sinusoidal input.
Analog Devices

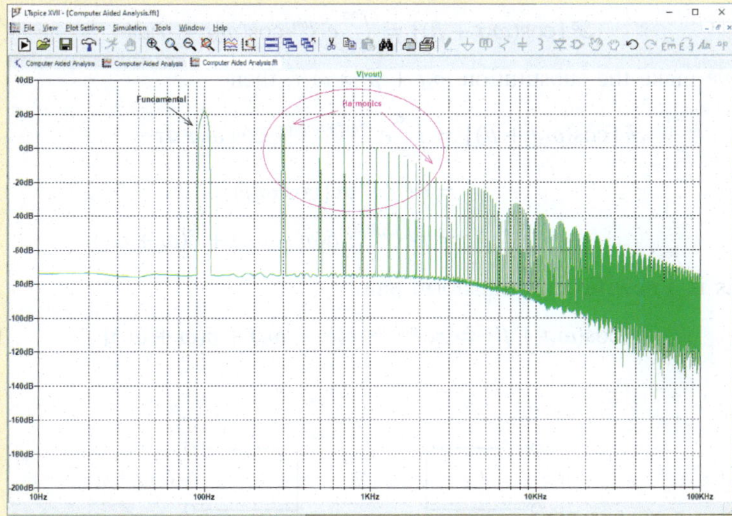

FIGURE 17.21 An FFT of the waveform shows a significant increase in the fraction of energy present in harmonics as opposed to the fundamental frequency of 100 Hz.
Analog Devices

17.10 THE PHYSICAL SIGNIFICANCE OF THE SYSTEM FUNCTION

In this section we will try to connect several aspects of the Fourier transform with work we completed in earlier chapters.

Given a general linear two-port network N without any initial energy storage, we assume sinusoidal forcing and response functions, arbitrarily taken to be voltages, as shown in Fig. 17.22. We let the input voltage be simply $A\cos(\omega_x t + \theta)$, and the output can be described in general terms as $v_o(t) = B\cos(\omega_x t + \phi)$, where the amplitude B and phase angle ϕ are functions of ωx. In phasor form, we can write the forcing and response functions as $\mathbf{V}_i = Ae^{j\theta}$ and $\mathbf{V}_o = Be^{j\phi}$. The ratio of the phasor response to the phasor forcing function is a complex number that is a function of ω_x:

$$\frac{\mathbf{V}_o}{\mathbf{V}_i} = \mathbf{G}(\omega_x) = \frac{B}{A}e^{j(\phi-\theta)}$$

where B/A is the amplitude of $\mathbf{G}$ and $\phi - \theta$ is its phase angle. This transfer function $\mathbf{G}(\omega x)$ could be obtained in the laboratory by varying ω_x over a large range of values and measuring the amplitude B/A and phase $\phi - \theta$ for each value of ω_x. If we then plotted each of these parameters as a function of frequency, the resultant pair of curves would completely describe the transfer function.

Now let us hold these comments in the backs of our minds for a moment as we consider a slightly different aspect of the same analysis problem.

For the circuit with sinusoidal input and output shown in Fig. 17.22, what is the system function $\mathbf{H}(j\omega)$? To answer this question, we begin with the definition of $\mathbf{H}(j\omega)$ as the ratio of the Fourier transforms of the output and the input. Both of these time functions involve the functional form $\cos(\omega_x t + \beta)$, whose Fourier transform we have not evaluated as yet, although we can handle $\cos\omega_x t$. The transform we need is

$$\mathscr{F}\{\cos(\omega_x t + \beta)\} = \int_{-\infty}^{\infty} e^{-j\omega t}\cos(\omega_x t + \beta)dt$$

If we make the substitution $\omega_x t + \beta = \omega_x \tau$, then

$$\mathscr{F}\{\cos(\omega_x t + \beta)\} = \int_{-\infty}^{\infty} e^{-j\omega\tau + j\omega\beta/\omega_x}\cos\omega_x\tau\, d\tau$$

$$= e^{j\omega\beta/\omega_x}\mathscr{F}\{\cos\omega_x t\}$$

$$= \pi e^{j\omega\beta/\omega_x}[\delta(\omega - \omega_x) + \delta(\omega + \omega_x)]$$

This is a new Fourier transform pair,

$$\cos(\omega_x t + \beta) \Leftrightarrow \pi e^{j\omega\beta/\omega_x}[\delta(\omega - \omega_x) + \delta(\omega + \omega_x)] \qquad [64]$$

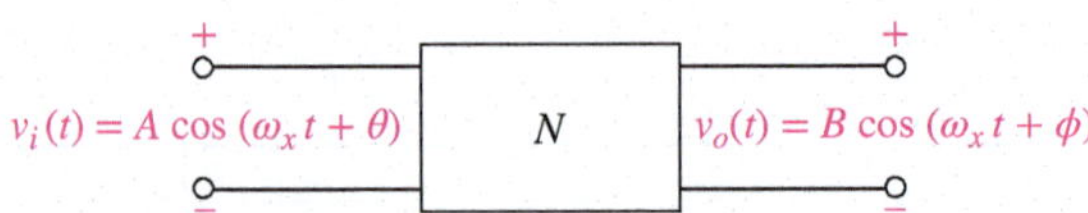

FIGURE 17.22 Sinusoidal analysis can be used to determine the transfer function $\mathbf{H}(j\omega_x) = (B/A)e^{j(\phi-\theta)}$, where B and ϕ are functions of ω_x.

which we can now use to evaluate the desired system function,

$$\begin{aligned}\mathbf{H}(j\omega) &= \frac{\mathcal{F}\{B\cos(\omega_x t+\phi)\}}{\mathcal{F}\{A\cos(\omega_x t+\theta)\}}\\ &= \frac{\pi B e^{j\omega\phi/\omega_x}[\delta(\omega-\omega_x)+\delta(\omega+\omega_x)]}{\pi A e^{j\omega\theta/\omega_x}[\delta(\omega-\omega_x)+\delta(\omega+\omega_x)]}\\ &= \frac{B}{A}e^{j\omega(\phi-\theta)/\omega_x}\end{aligned}$$

Now we recall the expression for $\mathbf{G}(\omega_x)$,

$$\mathbf{G}(\omega_x) = \frac{B}{A}e^{j(\phi-\theta)}$$

where B and ϕ were evaluated at $\omega = \omega x$, and we see that evaluating at $\mathbf{H}(j\omega)$ at $\omega = \omega_x$ gives

$$\mathbf{H}(\omega_x) = \mathbf{G}(\omega_x) = \frac{B}{A}e^{j(\phi-\theta)}$$

Since there is nothing special about the x subscript, we conclude that the system function and the transfer function are identical:

$$\mathbf{H}(j\omega) = \mathbf{G}(\omega) \qquad [65]$$

The fact that one argument is ω while the other is indicated by $j\omega$ is immaterial and arbitrary; the j merely makes possible a more direct comparison between the Fourier and Laplace transforms.

Equation [65] represents a direct connection between Fourier transform techniques and sinusoidal steady-state analysis. Our previous work on steady-state sinusoidal analysis using phasors was but a special case of the more general techniques of Fourier transform analysis. It was "special" in the sense that the inputs and outputs were sinusoids, whereas the use of Fourier transforms and system functions enables us to handle nonsinusoidal forcing functions and responses.

Thus, to find the system function $\mathbf{H}(j\omega)$ for a network, all we need to do is to determine the corresponding sinusoidal transfer function as a function of ω (or $j\omega$).

EXAMPLE **17.8**

Find the voltage across the inductor of the circuit shown in Fig. 17.23*a* when the input voltage is a simple exponentially decaying pulse, as indicated.

We need the system function; but it is not necessary to apply an impulse, find the impulse response, and then determine its inverse transform. Instead we use Eq. [65] to obtain the system function $\mathbf{H}(j\omega)$ by assuming that the input and output voltages are both sinusoids described by their corresponding phasors, as shown in Fig. 17.23*b*.

Using voltage division, we have

$$\mathbf{H}(j\omega) = \frac{\mathbf{V}_o}{\mathbf{V}_i} = \frac{j2\omega}{4+j2\omega}$$

(Continued on next page)

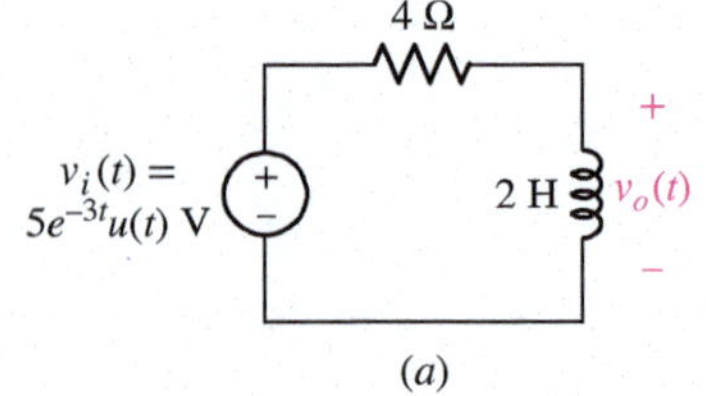

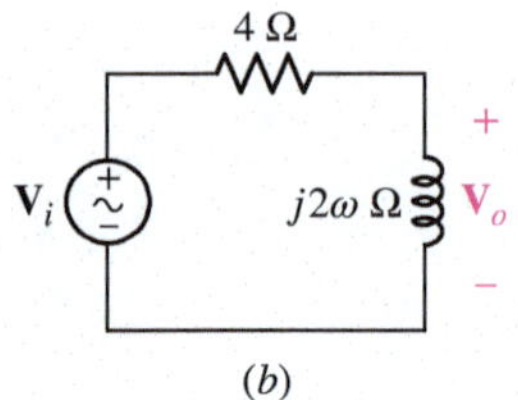

FIGURE 17.23 (a) The response $v_o(t)$ caused by $v_i(t)$ is desired. (b) The system function $\mathbf{H}(j\omega)$ may be determined by sinusoidal steady-state analysis: $\mathbf{H}(j\omega) = \mathbf{V}_o/\mathbf{V}_i$.

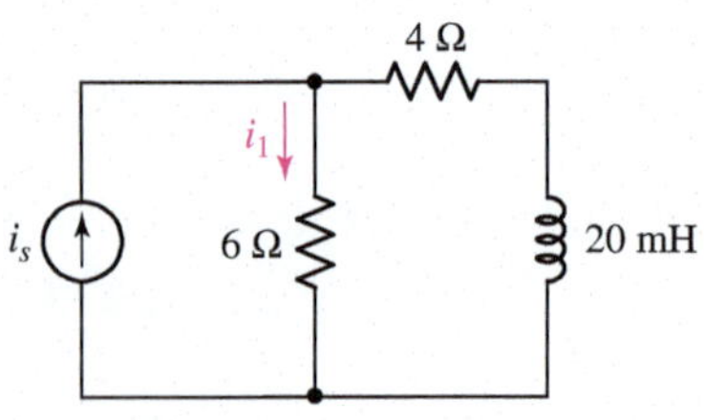

FIGURE 17.24

The transform of the forcing function is

$$\mathscr{F}\{v_i(t)\} = \frac{5}{3 + j\omega}$$

and thus the transform of $v_o(t)$ is given as

$$\begin{aligned}\mathscr{F}\{v_o(t)\} &= \mathbf{H}(j\omega)\mathscr{F}\{v_i(t)\} \\ &= \frac{j2\omega}{4 + j2\omega}\frac{5}{3 + j\omega} \\ &= \frac{15}{3 + j\omega} - \frac{10}{2 + j\omega}\end{aligned}$$

where the partial fractions appearing in the last step help to determine the inverse Fourier transform

$$\begin{aligned}v_o(t) &= \mathscr{F}^{-1}\left\{\frac{15}{3 + j\omega} - \frac{10}{2 + j\omega}\right\} \\ &= 15e^{-3t}\,u(t) - 10e^{-2t}\,u(t) \\ &= 5(3e^{-3t} - 2e^{-2t})u(t) \text{ V}\end{aligned}$$

Our problem is completed without fuss, convolution, or differential equations.

PRACTICE

17.16 Use Fourier transform techniques on the circuit of Fig. 17.24 to find $i_1(t)$ at $t = 1.5$ ms if i_s equals (a) $\delta(t)$ A; (b) $u(t)$ A; (c) $\cos 500t$ A.

Ans: −141.7 A; 0.683 A; 0.308 A.

PRACTICAL APPLICATION

Image Processing

Although a great deal of progress has been made toward developing a complete understanding of the function of muscle, there remain many open questions. A great deal of research in this field has been carried out using vertebrate skeletal muscle, in particular the *sartorius* or leg muscle of the frog (Fig. 17.25).

Of the many analytical techniques scientists use, one of the most common is electron microscopy. Figure 17.26 shows an electron micrograph of frog sartorius muscle tissue, sectioned in such a fashion as to highlight the regular arrangement of *myosin,* a filamentary type of contractile protein. Of interest to structural biologists are the periodicity and disorder of these proteins over a large area of muscle tissue. In order to develop a model for

FIGURE 17.25 Close-up of a frog against an orange background.
Simon Murrell/Alamy Stock Photo

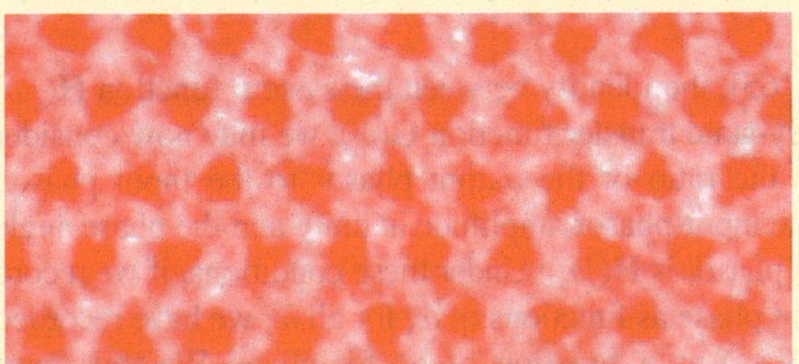

■ **FIGURE 17.26** Electron micrograph of a region of frog sartorius muscle tissue. False color has been employed for clarity.
Prof. John M. Squire

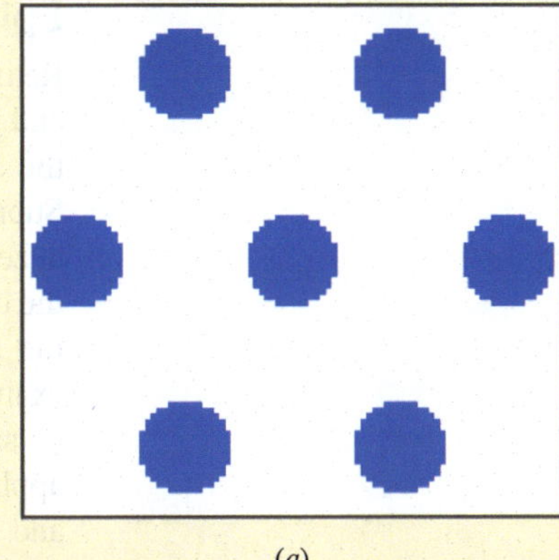

(*a*)

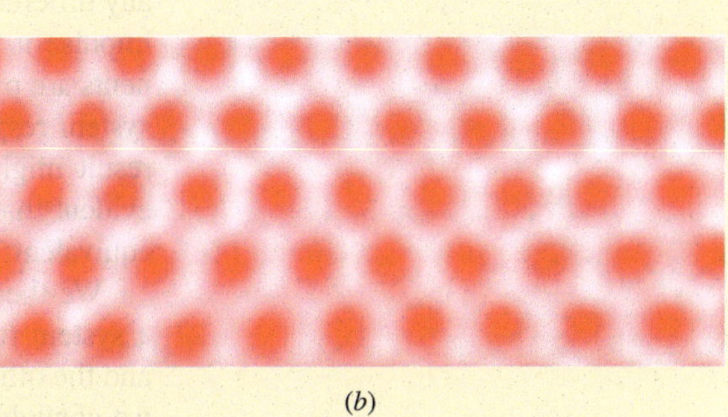

(*b*)

■ **FIGURE 17.27** (*a*) Spatial filter having hexagonal symmetry. (*b*) Image after convolution and inverse discrete Fourier transform are performed, showing a reduction in background noise.
(*b*) Prof. John M. Squire

these characteristics, a numerical approach is preferable, where the analysis of such images can be automated. As can be seen in the figure, however, the image produced by the electron microscope can be contaminated by a high level of background noise, making automated identification of the myosin filaments prone to error.

Introduced with the intent of ultimately assisting us in the analysis of time-varying linear circuits, the Fourier-based techniques of this chapter are in fact very powerful general methods which find application in many other situations. Among these, the field of *image processing* makes frequent use of Fourier techniques, especially through the fast Fourier transform and related numerical methods. The image of Fig. 17.26 can be described by a spatial function $f(x, y)$ where $f(x, y) = 0$ corresponds to white, $f(x, y) = 1$ corresponds to red, and (x, y) denotes a pixel location in the image. Defining a filter function $h(x, y)$ that has the appearance of Fig. 17.27*a*, the convolution operation

$$g(x, y) = f(x, y) * h(x, y)$$

results in the image of Fig. 17.27*b* in which the myosin filaments (viewed on end) are more clearly identifiable.

In practice, this image processing is performed in the frequency domain, where the FFT of both f and h are calculated, and the resulting matrices multiplied together.

An inverse FFT operation then produces the filtered image of Fig. 17.27*b*. Why does this convolution equate to a filtering operation? The myosin filament arrangement possesses hexagonal symmetry, as does the filter function $h(x, y)$—in a sense, both the myosin filament arrangement and the filter function possess the same spatial frequencies. The convolution of f with h results in a reinforcement of the hexagonal pattern within the original image and the removal of noise pixels (which do not possess hexagonal symmetry). This can be understood qualitatively if we model a horizontal row of Fig. 17.26 as a sinusoidal function $f(x) = \cos \omega_0 t$, which has the Fourier transform shown in Fig. 17.28*a*—a matched pair of impulse functions separated by $2\omega_0$. If we convolve this function with a filter function $h(x) = \cos \omega_1 t$, the Fourier transform of which is depicted in Fig. 17.28*b*, we get zero if $\omega_1 \neq \omega_0$; the frequencies (periodicities) of the two functions do not match. If, instead, we choose a filter function with the same frequency as $f(x)$, the convolution has a nonzero value at $\omega = \pm\omega_0$.

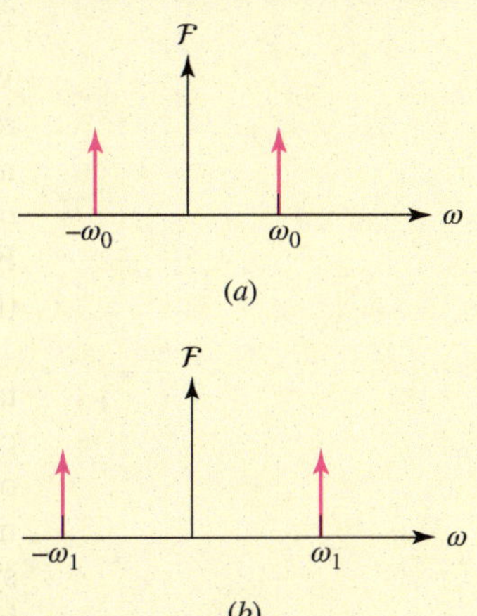

■ **FIGURE 17.28** (*a*) Fourier transform of $f(x) = \cos \omega_0 t$. (*b*) Fourier transform of $h(x) = \cos \omega_1 t$.

Epilogue

Returning again to Eq. [65], the identity between the system function $\mathbf{H}(j\omega)$ and the sinusoidal steady-state transfer function $\mathbf{G}(\omega)$, we may now consider the system function as the ratio of the output phasor to the input phasor. Suppose that we hold the input-phasor amplitude at unity and the phase angle at zero. Then the output phasor is $\mathbf{H}(j\omega)$. Under these conditions, if we record the output amplitude and phase as functions of ω, for all ω, we have recorded the system function $\mathbf{H}(j\omega)$ as a function of ω, for all ω. We thus have examined the system response under the condition that an infinite number of sinusoids, all with unity amplitude and zero phase, were successively applied at the input. Now suppose that our input is a single unit impulse, and look at the impulse response $h(t)$. Is the information we examine really any different from what we just obtained? The Fourier transform of the unit impulse is a constant equal to unity, indicating that all frequency components are present, all with the same magnitude, and all with zero phase. Our system response is the sum of the responses to all these components. The result might be viewed at the output on a cathode-ray oscilloscope. It is evident that the system function and the impulse-response function contain equivalent information regarding the response of the system.

We therefore have two different methods of describing the response of a system to a general forcing function; one is a time-domain description, and the other a frequency-domain description. Working in the time domain, we convolve the forcing function with the impulse response of the system to obtain the response function. As we saw when we first considered convolution, this procedure may be interpreted by thinking of the input as a continuum of impulses of different strengths and times of application; the output which results is a continuum of impulse responses.

In the frequency domain, however, we determine the response by multiplying the Fourier transform of the forcing function by the system function. In this case we interpret the transform of the forcing function as a frequency spectrum, or a continuum of sinusoids. Multiplying this by the system function, we obtain the response function, also as a continuum of sinusoids.

SUMMARY AND REVIEW

Whether we choose to think of the output as a continuum of impulse responses or as a continuum of sinusoidal responses, the linearity of the network and the superposition principle enable us to determine the total output as a time function by summing over all frequencies (the inverse Fourier transform), or as a frequency function by summing over all time (the Fourier transform).

Unfortunately, both of these techniques have some difficulties or limitations associated with their use. In using convolution, the integral itself can often be rather difficult to evaluate when complicated forcing functions or impulse-response functions are present. Furthermore, from the experimental point of view, we cannot really measure the impulse response of a system because we cannot actually generate an impulse. Even if we approximated the impulse by a narrow high-amplitude pulse, we would probably drive our system into saturation and out of its linear operating range.

With regard to the frequency domain, we encounter one absolute limitation in that we may easily hypothesize forcing functions that we would like to apply theoretically that do not possess Fourier transforms. Moreover, if we wish to find the time-domain description of the response function, we must evaluate an inverse Fourier transform, and some of these inversions can be extremely difficult.

Finally, neither of these techniques offers a very convenient method of handling initial conditions. For this, the Laplace transform is clearly superior.

The greatest benefits derived from the use of the Fourier transform arise through the abundance of useful information it provides about the spectral properties of a signal, particularly the energy or power per unit bandwidth. Some of this information is also easily obtained through the Laplace transform; we must leave a detailed discussion of the relative merits of each to more advanced signals and systems courses.

So, why has this all been withheld until now? The best answer is probably that these powerful techniques can overcomplicate the solution of simple problems and tend to obscure the physical interpretation of the performance of the simpler networks. For example, if we are interested only in the forced response, then there is little point in using the Laplace transform and obtaining both the forced and natural response after laboring through a difficult inverse transform operation.

Well, we could go on, but all good things must come to an end. Best of luck to you in your future studies.

- The harmonic frequencies of a sinusoid having the fundamental frequency ω_0 are $n\omega_0$, where n is an integer. (Examples 17.1, 17.2)
- The Fourier theorem states that provided a function $f(t)$ satisfies certain key properties, it may be represented by the infinite series $a_0 + \sum_{n=1}^{\infty}(a_n \cos n\omega_0 t + b_n \sin n\omega_0 t)$, where $a_0 = (1/T)\int_0^T f(t)dt$, $a_n = (2/T)\int_0^T f(t)\cos n\omega_0 t\, dt$, and $b_n = (2/T)\int_0^T f(t)\sin n\omega_0 t\, dt$. (Example 17.1)
- A function $f(t)$ possesses *even* symmetry if $f(t) = f(-t)$.
- A function $f(t)$ possesses *odd* symmetry if $f(t) = -f(-t)$.
- A function $f(t)$ possesses *half-wave* symmetry if $f(t) = -f\left(t - \frac{1}{2}T\right)$.
- The Fourier series of an even function is composed of only a constant and cosine functions.
- The Fourier series of an odd function is composed of only sine functions.
- The Fourier series of any function possessing half-wave symmetry contains only odd harmonics.
- The Fourier series of a function may also be expressed in complex or exponential form, where $f(t) = \sum_{n=-\infty}^{\infty} c_n e^{jn\omega_0 t}$ and $\mathbf{c}_n = (1/T)\int_{-T/2}^{T/2} f(t)e^{-jn\omega_0 t}\, dt$. (Examples 17.3, 17.4)
- The Fourier transform allows us to represent time-varying functions in the frequency domain, in a manner similar to that of the Laplace transform. The defining equations are $\mathbf{F}(j\omega) = \int_{-\infty}^{\infty} e^{-j\omega t} f(t)dt$ and $f(t) = (1/2\pi)\int_{-\infty}^{\infty} e^{j\omega t}\mathbf{F}(j\omega)d\omega$. (Examples 17.5, 17.6, 17.7)
- Fourier transform analysis can be implemented to analyze circuits containing resistors, inductors, and/or capacitors in a manner similar to what is done using Laplace transforms. (Example 17.8)

READING FURTHER

A very readable treatment of Fourier analysis can be found in:

A. Pinkus and S. Zafrany, *Fourier Series and Integral Transforms*. Cambridge: Cambridge University Press, 1997.

Finally, for those interested in learning more about muscle research, including electron microscopy of tissue, an excellent treatment can be found in:

J. Squire, *The Structural Basis of Muscular Contraction*. New York: Plenum Press, 1981.

EXERCISES

17.1 Trigonometric Form of the Fourier Series

1. Determine the fundamental frequency, fundamental radian frequency, and period of the following: (*a*) $5 \sin 9t$; (*b*) $200 \cos 70t$; (*c*) $4 \sin(4t - 10°)$; (*d*) $4 \sin(4t + 10°)$.
2. Plot multiple periods of the first, third, and fifth harmonics on the same graph of each of the following periodic waveforms (three separate graphs in total are desired): (*a*) $3 \sin t$; (*b*) $40 \cos 100t$; (*c*) $2 \cos(10t - 90°)$.
3. Calculate a_0 for the following: (*a*) $4 \sin 4t$; (*b*) $4 \cos 4t$; (*c*) $4 + \cos 4t$; (*d*) $4 \cos(4t + 40°)$.
4. Compute a_0, a_1, and b_1 for the following functions: (*a*) $2 \cos 3t$; (*b*) $3 - \cos 3t$; (*c*) $4 \sin(4t - 35°)$.
5. Calculate the Fourier coefficients a_0, a_1, a_2, a_3, b_1, b_2, and b_3 for the periodic function $f(t) = 2u(t) - 2u(t + 1) + 2u(t + 2) - 2u(t + 3) + \cdots$.
6. (*a*) Compute the Fourier coefficients a_0, a_1, a_2, a_3, a_4, b_1, b_2, b_3, and b_4 for the periodic function $g(t)$ partially sketched in Fig. 17.29. (*b*) Plot $g(t)$ along with the Fourier series representation truncated after $n = 4$.

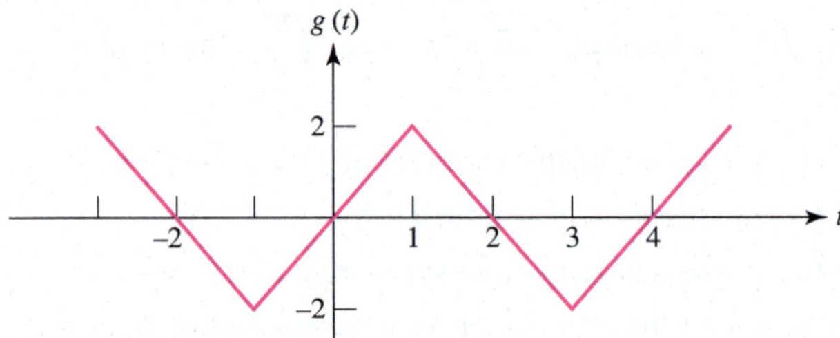

■ FIGURE 17.29

7. For the periodic waveform $f(t)$ represented in Fig. 17.30, calculate a_1, a_2, a_3 and b_1, b_2, b_3.

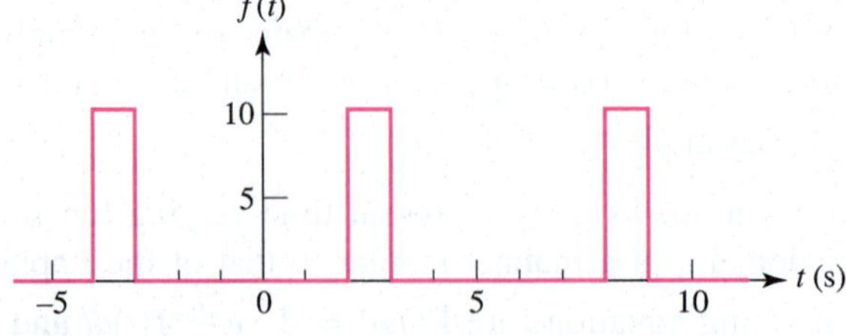

■ FIGURE 17.30

8. With respect to the periodic waveform sketched in Fig. 17.30, let $g_n(t)$ represent the Fourier series representation of $f(t)$ truncated at n. [For example, if $n = 1$, $g_1(t)$ has three terms, defined through a_0, a_1 and b_1.] (*a*) Sketch $g_2(t)$, $g_3(t)$, and $g_5(t)$, along with $f(t)$. (*b*) Calculate $f(2.5)$, $g_2(2.5)$, $g_3(2.5)$, and $g_5(2.5)$.

9. With respect to the periodic waveform $g(t)$ sketched in Fig. 17.29, define $y_n(t)$ which represents the Fourier series representation truncated at n. (For example, $y_2(t)$ has five terms, defined through a_0, a_1, a_2, b_1, and b_2.) (*a*) Plot $y_3(t)$ and $y_5(t)$ along with $g(t)$. (*b*) Compute $y_1(0.5)$, $y_2(0.5)$, $y_3(0.5)$, and $g(0.5)$.
10. Determine expressions for a_n and b_n for $g(t - 1)$ if the periodic waveform $g(t)$ is defined as sketched in Fig. 17.29.
11. A "half-sinusoidal" waveform is shown in Fig. 17.31, which is the output of a half-wave rectifier used to help convert a sinusoidal input to dc. Find the Fourier series representation and plot the signal and Fourier series representation for $n = 10$ terms.

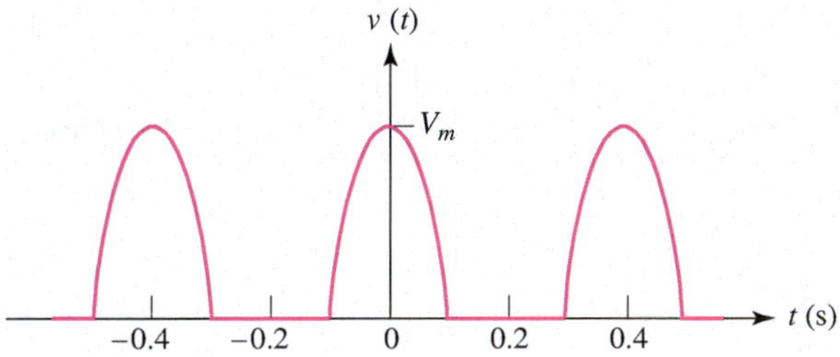

FIGURE 17.31

12. Plot the line spectrum (limited to the six largest terms) for the waveform shown in Fig. 17.4*a*.
13. Plot the line spectrum (limited to the five largest terms) for the waveform of Fig. 17.4*b*.
14. Plot the line spectrum (limited to the five largest terms) for the waveform represented by the graph of Fig. 17.4*c*.
15. Plot the line spectrum for the waveform shown in Fig. 17.31.

17.2 The Use of Symmetry

16. State whether the following exhibit odd symmetry, even symmetry, and/or half-wave symmetry: (*a*) $4 \sin 100t$; (*b*) $4 \cos 100t$; (*c*) $4 \cos(4t + 70°)$; (*d*) $4 \cos 100t + 4$; (*e*) each waveform in Fig. 17.4.
17. Determine whether the following exhibit odd symmetry, even symmetry, and/or half-wave symmetry: (*a*) the waveform in Fig. 17.29; (*b*) $g(t - 1)$, if $g(t)$ is represented in Fig. 17.29; (*c*) $g(t + 1)$, if $g(t)$ is represented in Fig. 17.29; (*d*) the waveform of Fig. 17.30.
18. A periodic function has the form $f(t) = t^2$ over the period $-\pi < t < \pi$. Determine if the function has even or odd symmetry, and evaluate the Fourier coefficients for $n = 1$, 2, and 3.
19. The nonperiodic waveform $g(t)$ is defined in Fig. 17.32. Use it to create a new function $y(t)$ such that $y(t)$ is identical to $g(t)$ over the range of $0 < t < 4$ and also is characterized by a period $T = 8$ and has (*a*) odd symmetry; (*b*) even symmetry; (*c*) both even and half-wave symmetry; (*d*) both odd and half-wave symmetry.
20. Calculate a_0, a_1, a_2, a_3 and b_1, b_2, b_3 for the periodic waveform $v(t)$ represented in Fig. 17.33.

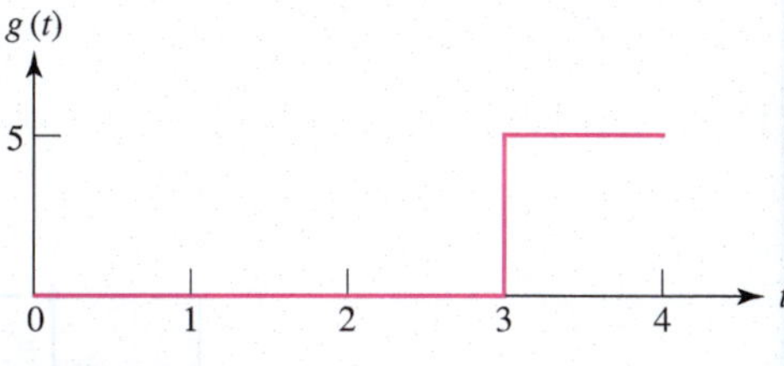

FIGURE 17.32

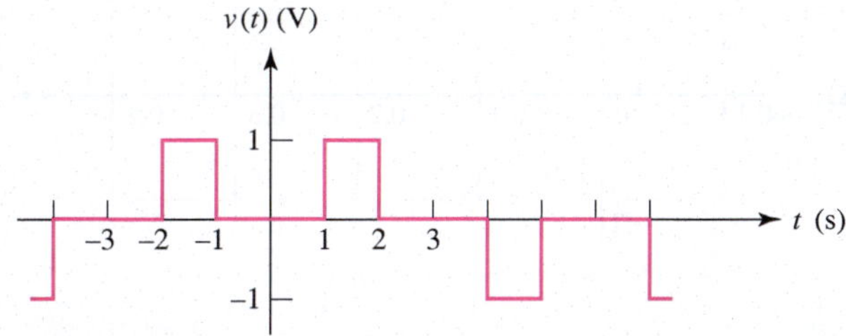

FIGURE 17.33

21. Design a triangular waveform having a peak magnitude of 3, a period of 2 s, and characterized by (*a*) half-wave and even symmetry; (*b*) half-wave and odd symmetry.

22. Make use of symmetry as much as possible to obtain numerical values for a_0, a_n, and b_n, $1 \leq n \leq 10$, for the waveform shown in Fig. 17.34.

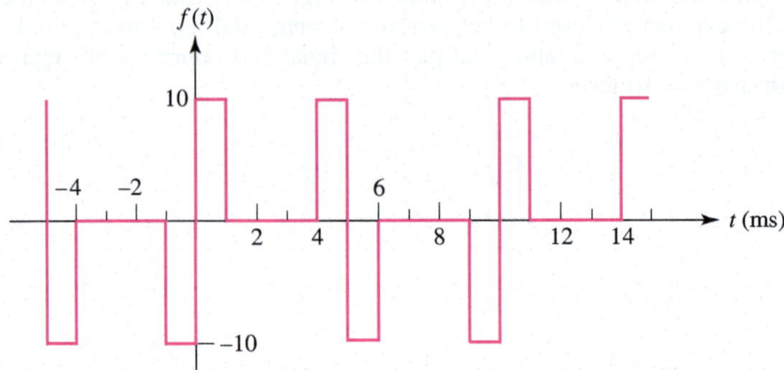

FIGURE 17.34

17.3 Complete Response to Periodic Forcing Functions

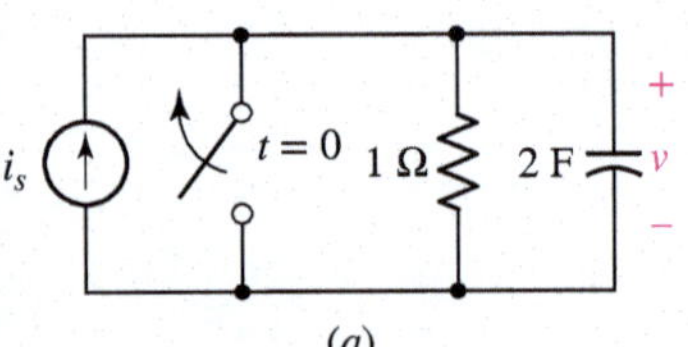

(*a*)

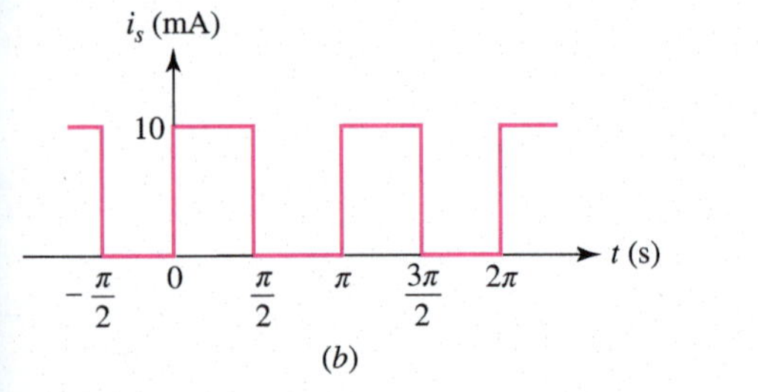

(*b*)

FIGURE 17.35

23. For the circuit of Fig. 17.35*a*, calculate $v(t)$ in a Fourier series representation if $i_s(t)$ is given by Fig. 17.35*b* and $v(0) = 0$.

24. If the waveform shown in Fig. 17.36 is applied to the circuit of Fig. 17.8*a*, calculate $i(t)$ in a Fourier series representation.

FIGURE 17.36

25. The circuit of Fig. 17.37*a* is subjected to the waveform depicted in Fig. 17.37*b*. Determine the steady-state voltage $v(t)$ in a Fourier series representation.

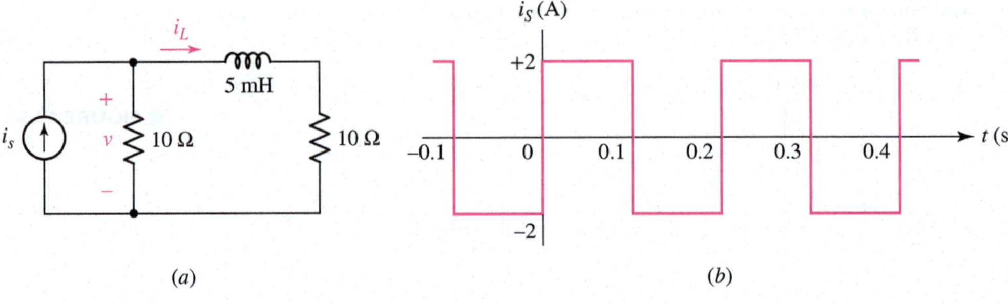

FIGURE 17.37

26. Apply the waveform of Fig. 17.38 to the circuit of Fig. 17.37*a*, and calculate the steady-state current $i_L(t)$ in a Fourier series representation.

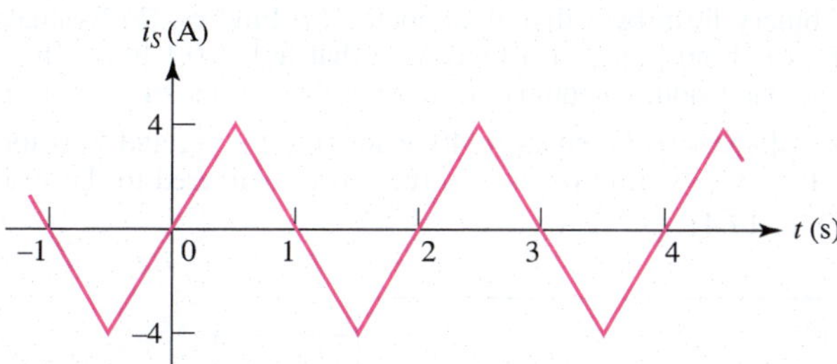

FIGURE 17.38

17.4 Complex Form of the Fourier Series

27. Let the function $v(t)$ be defined as indicated in Fig. 17.10. Determine $\mathbf{c}_n$ for (*a*) $v(t + 0.5)$; (*b*) $v(t - 0.5)$.
28. Calculate $\mathbf{c}_0$, $\mathbf{c}_{\pm1}$, and $\mathbf{c}_{\pm2}$ for the waveform of Fig. 17.38.
29. Determine the first five terms of the exponential Fourier series representation of the waveform graphed in Fig. 17.35*b*.
30. For the periodic waveform shown in Fig. 17.39, determine (*a*) the period T; (*b*) $\mathbf{c}_0$, $\mathbf{c}_{\pm1}$, $\mathbf{c}_{\pm2}$, and $\mathbf{c}_{\pm3}$.

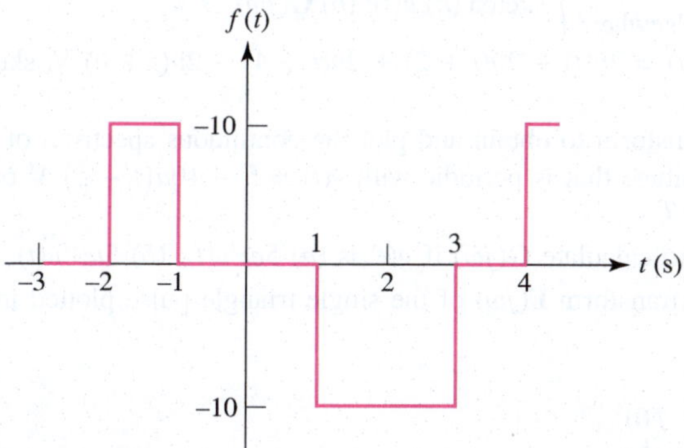

FIGURE 17.39

31. For the periodic waveform represented in Fig. 17.40, calculate (*a*) the period T; (*b*) $\mathbf{c}_1$ and $\mathbf{c}_2$.

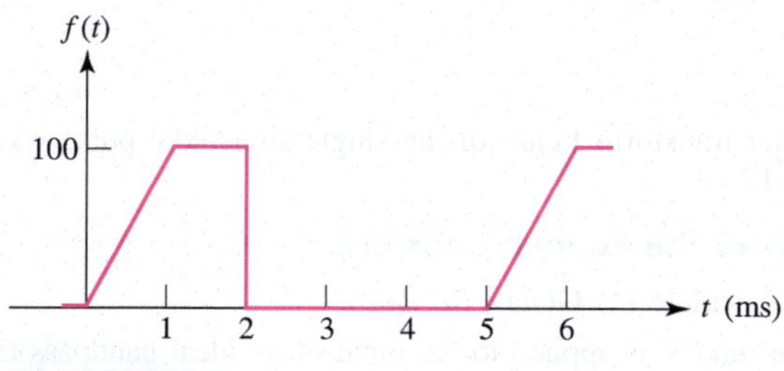

FIGURE 17.40

32. Determine the Fourier coefficients c_i for the sawtooth signal in Fig 17.2 in Example 17.1. Plot the resulting $v(t)$ for $n = 50$.

33. A pulse sequence has a period of 5 μs, an amplitude of unity for $-0.6 < t < -0.4$ μs and for $0.4 < t < 0.6$ μs, and zero amplitude elsewhere in the period interval. This series of pulses might represent the decimal number 3 being transmitted in binary form by a digital computer. (*a*) Find $\mathbf{c}_n$. (*b*) Evaluate $\mathbf{c}_4$. (*c*) Evaluate $\mathbf{c}_0$. (*d*) Find $|\mathbf{c}_n|_{\max}$. (*e*) Find N so that $|\mathbf{c}_n| \leq 0.1\ |\mathbf{c}_n|_{\max}$ for all $n > N$. (*f*) What bandwidth is required to transmit this portion of the spectrum?

34. Let a periodic voltage $v_s(t)$ be equal to 40 V for $0 < t < \frac{1}{96}$, and to 0 for $\frac{1}{96} < t < \frac{1}{16}$ s. If $T = \frac{1}{16}$ s, find (*a*) $\mathbf{c}_3$; (*b*) the power delivered to the load in the circuit of Fig. 17.41.

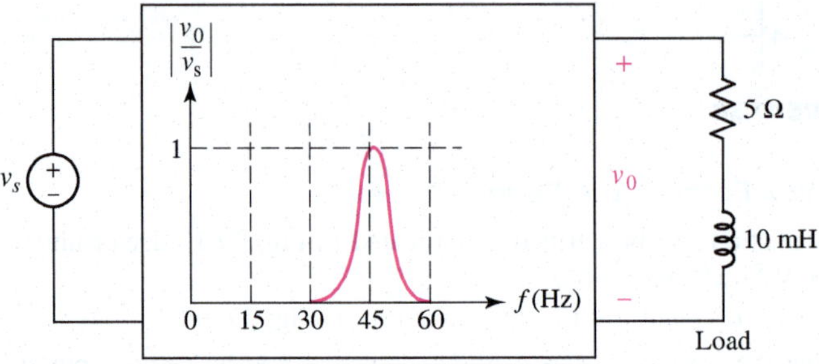

■ **FIGURE 17.41**

17.5 Definition of the Fourier Transform

35. Given $g(t) = \begin{Bmatrix} 5 & -1 < t < 1 \\ 0 & \text{elsewhere} \end{Bmatrix}$ sketch (*a*) $g(t)$; (*b*) $\mathbf{G}(j\omega)$.

36. For the function $v(t) = 2u(t) - 2u(t+2) + 2u(t+4) - 2u(t+6)$ V, sketch (*a*) $v(t)$; (*b*) $\mathbf{V}(j\omega)$.

37. Use the Fourier transform to obtain and plot the continuous spectrum of the square wave voltage that is periodic with $v(t) = 5 - 10u\left(t - \frac{T}{2}\right)$ V over the period $0 < t < T$.

38. Employ Eq. [41*a*] to calculate $\mathbf{G}(j\omega)$ if $g(t)$ is (*a*) $5e^{-t}u(t)$; (*b*) $5te^{-t}u(t)$.

39. Obtain the Fourier transform $\mathbf{F}(j\omega)$ of the single triangle pulse plotted in Fig. 17.42.

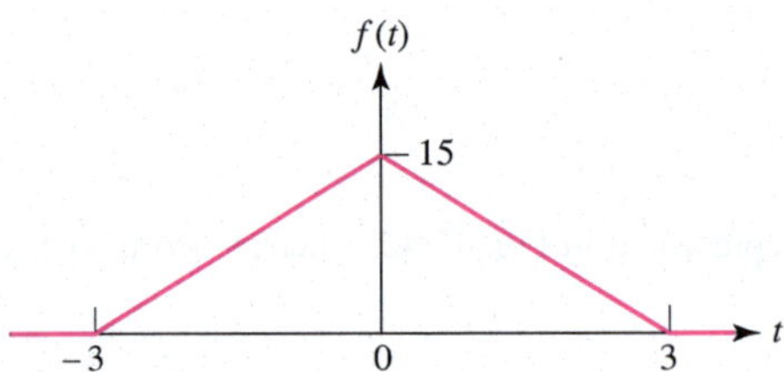

■ **FIGURE 17.42**

40. Determine the Fourier transform $\mathbf{F}(j\omega)$ of the single sinusoidal pulse waveform shown in Fig. 17.43.

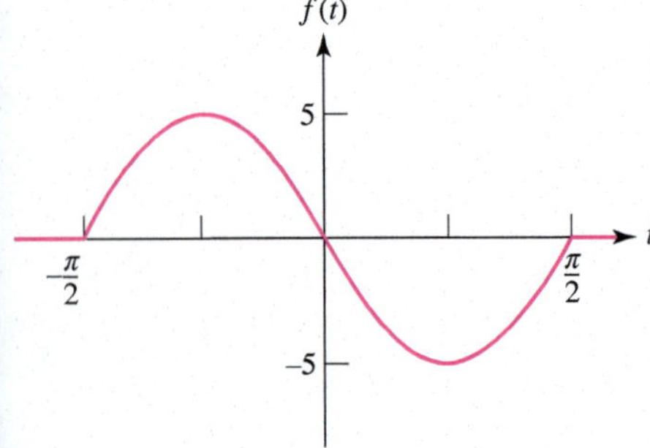

■ **FIGURE 17.43**

17.6 Some Properties of the Fourier Transform

41. For $g(t) = 3e^{-t}u(t)$, calculate (*a*) $\mathbf{G}(j\omega)$; (*b*) $\phi(\omega)$.

42. The voltage pulse $2e^{-t}u(t)$ V is applied to the input of an ideal bandpass filter. The passband of the filter is defined by $100 < |f| < 500$ Hz. Calculate the total output energy.

43. Given that $v(t) = 4e^{-|t|}$ V, calculate the frequency range in which 85% of the 1 Ω energy lies.

44. Use the definition of the Fourier transform to prove the following results, where $\mathscr{F}\{f(t)\} = \mathbf{F}(j\omega)$: (*a*) $\mathscr{F}\{f(t - t_0)\} = e^{-j\omega t0}\mathscr{F}\{f(t)\}$; (*b*) $\mathscr{F}\{df(t)/dt\} = j\omega\mathscr{F}\{f(t)\}$; (*c*) $\mathscr{F}\{f(kt)\} = (1/|k|)\mathbf{F}(j\omega/k)$; (*d*) $\mathscr{F}\{f(-t)\} = \mathbf{F}(-j\omega)$; (*e*) $\mathscr{F}\{tf(t)\} = j\,d[\mathbf{F}(j\omega)]/d\omega$.

17.7 Fourier Transform Pairs for Some Simple Time Functions

45. Determine the Fourier transform of the following: (*a*) $5u(t) - 2\,\text{sgn}(t)$; (*b*) $2\cos 3t - 2$; (*c*) $4e^{-j3t} + 4e^{j3t} + 5u(t)$.
46. Find the Fourier transform of each of the following: (*a*) $85u(t + 2) - 50u(t - 2)$; (*b*) $5\delta(t) - 2\cos 4t$.
47. Find $\mathbf{F}(j\omega)$ if $f(t)$ is given by (*a*) $2\cos 10t$; (*b*) $e^{-4t}u(t)$; (*c*) $5\,\text{sgn}(t)$.
48. Determine $f(t)$ if $\mathbf{F}(j\omega)$ is given by (*a*) $4\delta(\omega)$; $2/(5000 + j\omega)$; (*c*) $e^{-j120\omega}$.
49. Obtain an expression for $f(t)$ if $\mathbf{F}(j\omega)$ is given by (*a*) $-j\frac{231}{\omega}$; (*b*) $\frac{1 + j2}{1 + j4}$; (*c*) $5\delta(\omega) + \frac{1}{2 + j10}$.

17.8 The Fourier Transform of a General Periodic Time Function

50. Calculate the Fourier transform of the following functions: (*a*) $2\cos^2 5t$; (*b*) $7\sin 4t\cos 3t$; (*c*) $3\sin(4t - 40°)$.
51. Determine the Fourier transform of the periodic function $g(t)$, which is defined over the range $0 < t < 10$ s by $(t) = 2u(t) - 3u(t - 4) + 2u(t - 8)$.
52. If $\mathbf{F}(j\omega) = 20\sum_{n=1}^{\infty}[1/(|n|! + 1)]\delta(\omega - 20n)$, find the value of $f(0.05)$.
53. Given the periodic waveform shown in Fig. 17.44, determine its Fourier transform.

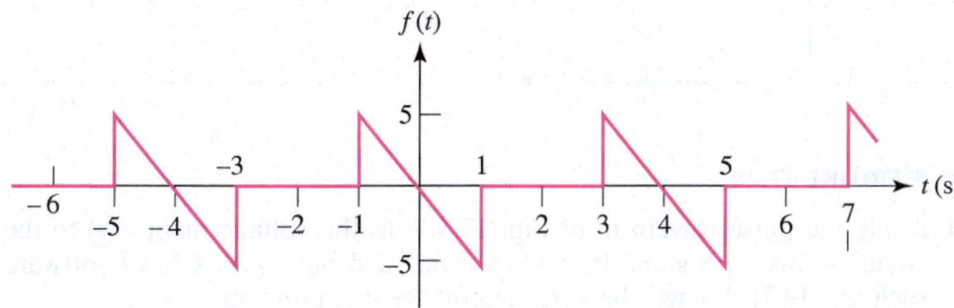

FIGURE 17.44

17.9 The System Function and Response in the Frequency Domain

54. If a system is described by transfer function $h(t) = 2u(t) + 2u(t - 1)$, use convolution to calculate the output (time domain) if the input is (*a*) $2u(t)$; (*b*) $2te^{-2t}u(t)$.
55. Given the input function $x(t) = 5e^{-5t}u(t)$, employ convolution to obtain a time-domain output if the system transfer function $h(t)$ is given by (*a*) $3u(t + 1)$; (*b*) $10te^{-t}u(t)$.

56. (*a*) Design a noninverting amplifier having a gain of 10. If the circuit is constructed using an op amp powered by ±15 V supplies, determine the FFT of the output through appropriate simulations if the input voltage operates at 1 kHz and has magnitude (*b*) 10 mV; (*c*) 1 V; (*d*) 2 V.

57. (*a*) Design an inverting amplifier having a gain of 5. If the circuit is constructed using an op amp powered by ±10 V supplies, perform appropriate simulations to determine the FFT of the output voltage if the input voltage has a frequency of 10 kHz and magnitude (*b*) 500 mV; (*c*) 1.8 V; (*d*) 3 V.

17.10 The Physical Significance of the System Function

58. With respect to the circuit of Fig. 17.45, calculate $v_o(t)$ using Fourier techniques if $v_i(t) = 2te^{-t}u(t)$ V.
59. After the inductor of Fig. 17.45 is surreptitiously replaced with a 2 F capacitor, calculate $v_o(t)$ using Fourier techniques if $v_i(t)$ is equal to (*a*) $5u(t)$ V; (*b*) $3e^{-4t}u(t)$ V.

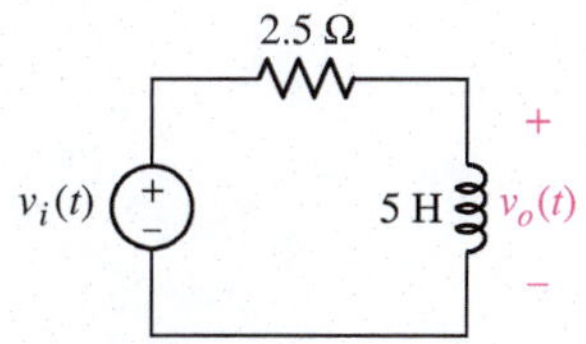

FIGURE 17.45

60. Employ Fourier-based techniques to calculate $v_C(t)$ as labeled in Fig. 17.46 if $v_i(t)$ is equal to (*a*) $2u(t)$ V; (*b*) $2\delta(t)$ V.

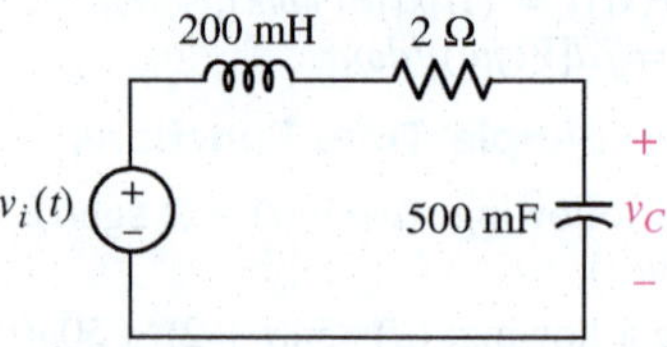

■ **FIGURE 17.46**

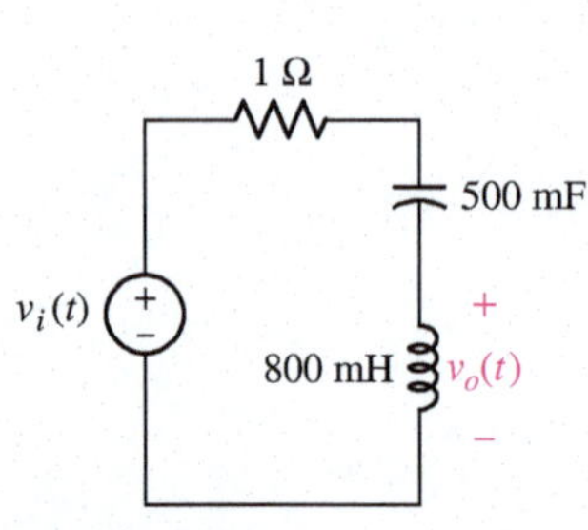

■ **FIGURE 17.47**

61. Employ Fourier-based techniques to calculate $v_o(t)$ as labeled in Fig. 17.47 if $v_i(t)$ is equal to (*a*) $5u(t - 1)$V; (*b*) $3\delta(t)$V.

62. Employ Fourier-based techniques to calculate $v_o(t)$ as labeled in Fig. 17.47 if $v_i(t)$ is equal to (*a*) $5u(t - 1)$V; (*b*) $2 + 8e^{-t}u(t)$V.

Chapter-Integrating Exercises

63. Apply the pulse waveform of Fig. 17.48*a* as the voltage input $v_i(t)$ to the circuit shown in Fig. 17.45, and calculate $v_C(t)$.

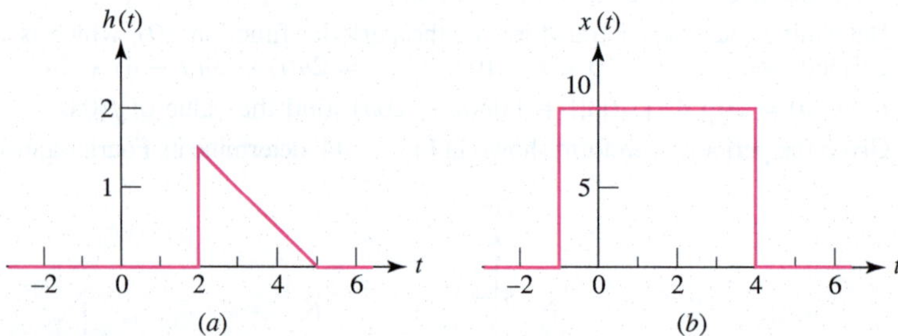

■ **FIGURE 17.48**

64. Apply the pulse waveform of Fig. 17.48*b* as the voltage input $v_i(t)$ to the circuit shown in Fig. 17.46, and calculate and plot $v_C(t)$. Use of software such as MATLAB will be very helpful for this problem!

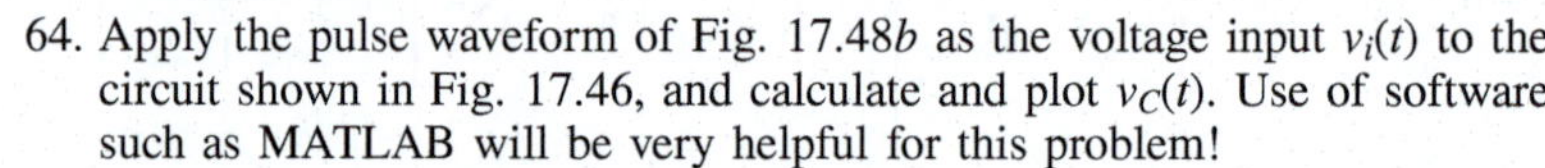

65. Apply the pulse waveform of Fig. 17.48*b* as the voltage input $v_i(t)$ to the circuit shown in Fig. 17.47, and calculate and plot $v_o(t)$. Use of software such as MATLAB will be very helpful for this problem!

66. Design an audio amplifier with gain of 10, using power supplies of $V_{CC} = \pm 12$ V, and choose an appropriate op amp for audio applications, such as the OP27. Use a SPICE simulation to determine the maximum in harmonic distortion, defined as $P_{HD} = P_{harmonic} - P_{fundamental}$ (in dB), for the cases of input with amplitude of 1.2 V at frequencies (*a*) 250 Hz (bass), (*b*) 1 kHz (mid-range), and (*c*) 4 kHz (treble).

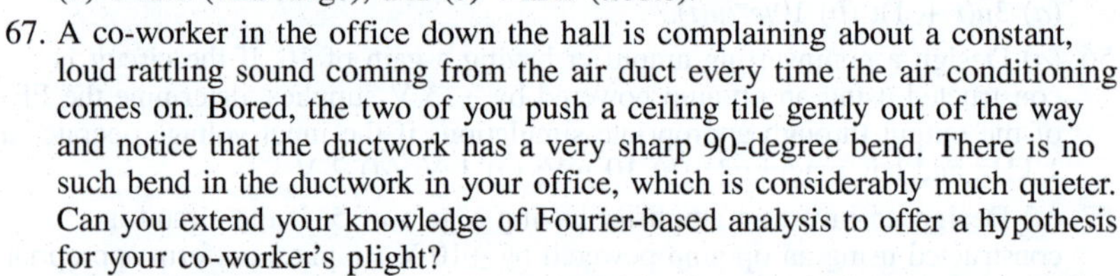

67. A co-worker in the office down the hall is complaining about a constant, loud rattling sound coming from the air duct every time the air conditioning comes on. Bored, the two of you push a ceiling tile gently out of the way and notice that the ductwork has a very sharp 90-degree bend. There is no such bend in the ductwork in your office, which is considerably much quieter. Can you extend your knowledge of Fourier-based analysis to offer a hypothesis for your co-worker's plight?

68. Atomic force microscopy is an extremely high-resolution technique for imaging surfaces at submicron length scales. Either two- or three-dimensional images can be generated, with "height" over the imaged area being the typical output. Just for fun, you apply an FFT to an example periodic image and see only a single row of dots (the *x*-*y*-axes of the FFT plot, in this case, of course, represent spatial frequency). A friend shows you an FFT of an image they acquired, which shows again a row of dots, but this time spaced further apart. What do the two images have in common, and what is their primary difference?

State-Variable Analysis

KEY CONCEPTS

Introducing the Concept of State Variables and Normal-Form Equations

Learning How to Write a Complete Set of Normal-Form Equations for a Given Circuit

Matrix-Based Solution of the Circuit Equations

Development of a General Technique Applicable to Higher-Order Problems

INTRODUCTION

Up to this point, we have seen several different methods by which circuits might be analyzed. The resistive circuit came first, and for it we wrote a set of algebraic equations, often cast in the form of nodal or mesh equations. However, in Appendix 1 we learn that we can choose other, more convenient voltage or current variables after drawing an appropriate tree for the network. The tree sprouts up again in this chapter, in the selection of circuit variables.

We next added inductors and capacitors to our networks, and this produced equations containing derivatives and integrals with respect to time. Except for simple first- and second-order systems that either were source-free or contained only dc sources, we did not attempt solving these equations. The results we obtained were found by time-domain methods. Subsequently, we explored the use of phasors to determine the sinusoidal steady-state response of such circuits, and a little later on we were introduced to the concept of complex frequency and the Laplace transform method.

In this last chapter, we return to the time domain and introduce the use of state variables. Once again we will not obtain many explicit solutions for circuits of even moderate complexity, but we will write sets of equations compatible with numerical analysis techniques.

18.1 STATE VARIABLES AND NORMAL-FORM EQUATIONS

State-variable analysis, or state-space analysis, as it is sometimes called, is a procedure that can be applied both to linear and, with some modifications, to nonlinear circuits, as well as to circuits containing

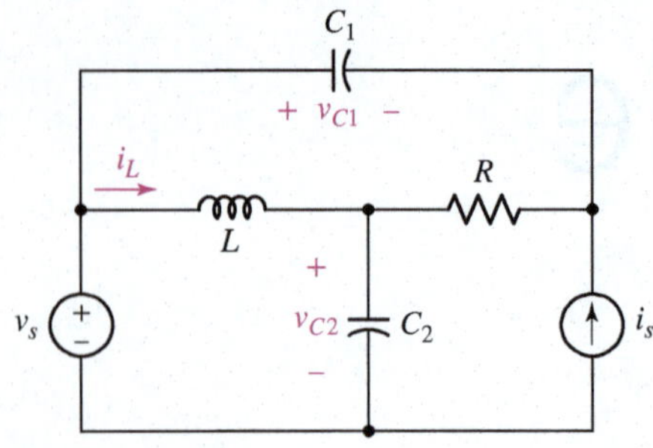

FIGURE 18.1 A four-node *RLC* circuit.

time-varying parameters, such as the capacitance $C = 50 \cos 20t$ pF. Our attention, however, will be restricted to time-invariant linear circuits.

We introduce some of the ideas underlying state variables by looking at a general *RLC* circuit drawn in Fig. 18.1. When we write equations for this circuit, we could use nodal analysis, the two dependent variables being the node voltages at the central and right nodes. We could also opt for mesh analysis and use two currents as the variables, or, alternatively, we could draw a tree first and then select a set of tree-branch voltages or link currents as the dependent variables. It is possible that each approach could lead to a different number of variables, although two seems to be the most likely number for this circuit.

The set of variables we will select in state-variable analysis is a hybrid set that may include both currents and voltages. They are the *inductor currents* and the *capacitor voltages*. Each of these quantities may be used directly to express the energy stored in the inductor or capacitor at any instant of time. That is, they collectively describe the *energy state* of the system, and for that reason, they are called the *state variables*.

Let us try to write a set of equations for the circuit of Fig. 18.1 in terms of the state variables i_L, v_{C1}, and v_{C2}, as defined on the circuit diagram. The method we use will be outlined more formally in the following section, but for the present let us try to use KVL once for each inductor, and KCL once for each capacitor.

Beginning with the inductor, we set the sum of voltages around the lower left mesh equal to zero:

Here we are using a "prime" symbol to denote a derivative with respect to time.

$$Li'_L + v_{C2} - v_s = 0 \qquad [1]$$

We presume that the source voltage v_s and the source current i_s are known, and we therefore have one equation in terms of our chosen state variables.

Next, we consider the capacitor C_1. Since the left terminal of C_1 is also one terminal of a voltage source, it will become part of a supernode. Therefore, we select the right terminal of C_1 as the node to which we apply KCL. The current through the capacitor branch is $C_1v'_{C1}$, the upward source current is i_s, and the current in R is obtained by noting that the voltage across R, positive reference on the left, is $(v_{C2} - v_s + v_{C1})$ and, therefore, the current to the right in R is $(v_{C2} - v_s + v_{C1})$. Thus,

$$C_1v'_{C1} + \frac{1}{R}(v_{C2} - v_s + v_{C1}) + i_s = 0 \qquad [2]$$

Again we have been able to write an equation without introducing any new variables, although we might not have been able to express the current through R directly in terms of the state variables if the circuit had been any more complicated.

Finally, we apply KCL to the upper terminal of C_2:

$$C_2v'_{C2} - i_L + \frac{1}{R}(v_{C2} - v_s + v_{C1}) = 0 \qquad [3]$$

Equations [1] to [3] are written solely in terms of the three state variables, the known element values, and the two known forcing functions. They are not, however, written in the standardized form which state-variable analysis demands. The state equations are said to be in *normal form* when the derivative of each state variable is expressed as a linear combination of all the state variables and forcing functions. The ordering of the equations defining the derivatives and the order in which the state variables appear in every equation must be the same. Let us arbitrarily select the order i_L, v_{C1}, v_{C2}, and rewrite Eq. [1] as

$$i_L' = -\frac{1}{L}v_{C2} + \frac{1}{L}v_s \qquad [4]$$

Then Eq. [2] is rewritten as

$$v_{C1}' = -\frac{1}{RC_1}v_{C1} - \frac{1}{RC_1}v_{C2} + \frac{1}{RC_1}v_s - \frac{1}{C_1}i_s \qquad [5]$$

while Eq. [3] becomes

$$v_{C2}' = \frac{1}{C_2}i_L - \frac{1}{RC_2}v_{C1} - \frac{1}{RC_2}v_{C2} + \frac{1}{RC_2}v_s \qquad [6]$$

Note that these equations define, in order, i_L', v_{C1}', and v_{C2}', with a corresponding order of variables on the right sides, i_L, v_{C1}, and v_{C2}. The forcing functions come last and may be written in any convenient order.

As another example of the determination of a set of normal-form equations, we look at the circuit shown in Fig. 18.2*a*. Since the circuit has one capacitor and one inductor, we expect two state variables, the capacitor voltage and the inductor current. To facilitate writing the normal-form equations, let us construct a tree for this circuit which follows all the rules for tree construction discussed in Appendix 1 and in addition requires all capacitors to be located in the tree and all inductors in the cotree. This is usually possible, and it leads to a *normal tree*. In those few exceptional cases where it is not possible to draw a normal tree, we use a slightly different method which will be considered at the end of Sec. 18.2. Here, we can place C in the tree and L and i_s in the cotree, as shown in Fig. 18.2*b*. This is the only normal tree possible for this circuit. The source quantities and the state variables are indicated on the tree and cotree.

Next, we determine the current in every link and the voltage across each tree branch in terms of the state variables. For a simple circuit such as this, it is possible to do so by beginning with any resistor for which either the current or the voltage is obvious. The results are shown on the tree in Fig. 18.2*c*.

We may now write the normal-form equations by invoking KCL at the upper terminal of the capacitor:

$$0.4v_C' + 0.2v_C - i_s + i_L = 0$$

or, in normal form,

$$v_C' = -0.5v_C - 2.5i_L + 2.5i_s \qquad [7]$$

Around the outer loop, we have

$$2i_L' - v_C + 7i_L = 0$$

or

$$i_L' = 0.5v_C - 3.5i_L \qquad [8]$$

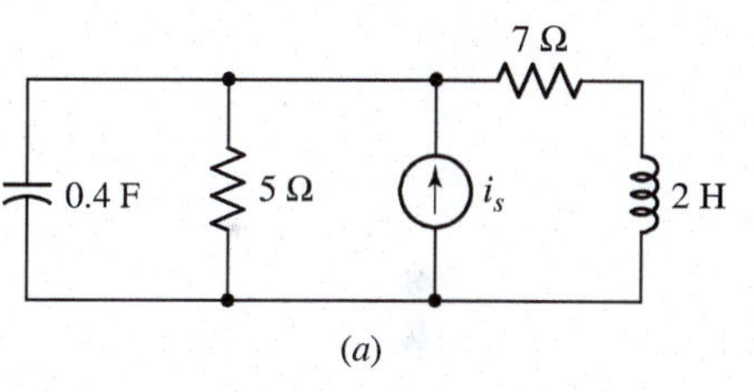

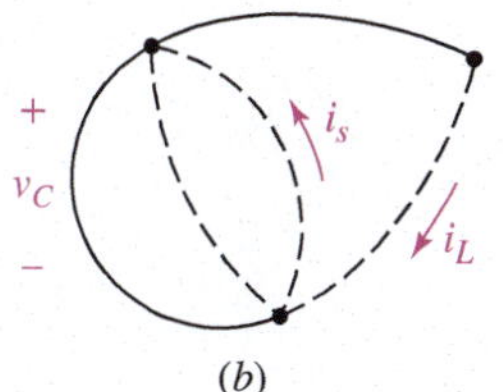

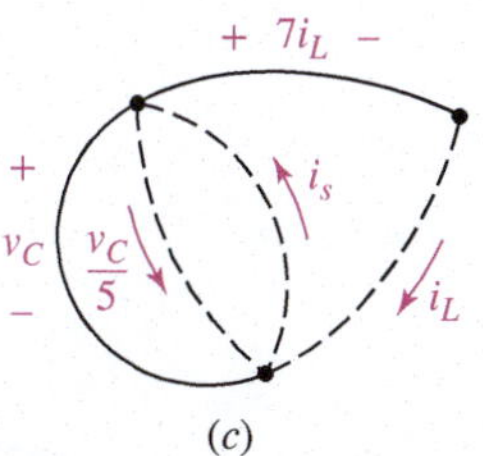

FIGURE 18.2 (*a*) An *RLC* circuit requiring two state variables. (*b*) A normal tree showing the state variables v_C and i_L. (*c*) The current in every link and the voltage across every tree branch is expressed in terms of the state variables.

Equations [7] and [8] are the desired normal-form equations. Their solution will yield all the information necessary for a complete analysis of the given circuit. Of course, explicit expressions for the state variables can be obtained only if a specific function is given for $i_s(t)$. For example, it will be shown later that if

$$i_s(t) = 12 + 3.2e^{-2t}u(t) \quad \text{A} \qquad [9]$$

then

$$v_C(t) = 35 + (10e^{-t} - 12e^{-2t} + 2e^{-3t})u(t) \quad \text{V} \qquad [10]$$

and

$$i_L(t) = 5 + (2e^{-t} - 4e^{-2t} + 2e^{-3t})u(t) \quad \text{A} \qquad [11]$$

The solutions, however, are far from obvious, and we will develop the technique for obtaining them from the normal-form equations in Sec. 18.6.

PRACTICE

18.1 Write a set of normal-form equations for the circuit shown in Fig. 18.3. Order the state variables as i_{L1}, i_{L2}, and v_C.

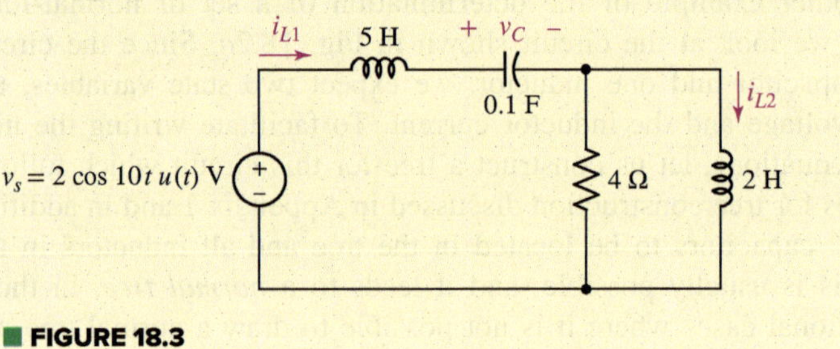

FIGURE 18.3

Ans: $i'_{L1} = -0.8i_{L1} + 0.8i_{L2} - 0.2v_C + 0.2v_s$; $i'_{L2} = 2i_{L1} - 2i_{L2}$; $v'_C = 10i_{L1}$.

18.2 WRITING A SET OF NORMAL-FORM EQUATIONS

In the two examples considered in Sec. 18.1, the methods whereby we obtained a set of normal-form equations may have seemed to be more of an art than a science. In order to bring a little order into our chaos, let's try to follow the procedure used when we were studying nodal analysis, mesh analysis, and the use of trees in general loop and general nodal analysis. We seek a set of guidelines that will systematize the procedure. Then we will apply these rules to three new examples, each a little more involved than the preceding one.

Here are the six steps that we have been following:

1. **Establish a normal tree.** Place capacitors and voltage sources in the tree, and inductors and current sources in the cotree; place control voltages in the tree and control currents in the cotree if possible. More than one normal tree may be possible. Certain types of networks do not permit any normal tree to be drawn; these exceptions are considered at the end of this section.

2. **Assign voltage and current variables.** Assign a voltage (with polarity reference) to every capacitor and a current (with arrow) to every inductor; these voltages and currents are the state variables. Indicate the voltage across every tree branch and the current through every link in terms of the source voltages, the source currents, and the state variables, if possible; otherwise, assign a new voltage or current variable to that resistive tree branch or link.
3. **Write the *C* equations.** Use KCL to write one equation for each capacitor. Set Cv'_C equal to the sum of link currents obtained by considering the node (or supernode) at either end of the capacitor. The supernode is identified as the set of all tree branches connected to that terminal of the capacitor. Do not introduce any new variables.
4. **Write the *L* equations.** Use KVL to write one equation for each inductor. Set Li'_L equal to the sum of tree-branch voltages obtained by considering the single closed path consisting of the link in which L lies and a convenient set of tree branches. Do not introduce any new variables.
5. **Write the *R* equations (if necessary).** If any new voltage variables were assigned to resistors in step 2, use KCL to set v_R/R equal to a sum of link currents. If any new current variables were assigned to resistors in step 2, use KVL to set $i_R R$ equal to a sum of tree-branch voltages. Solve these resistor equations simultaneously to obtain explicit expressions for each v_R and i_R in terms of the state variables and source quantities.
6. **Write the normal-form equations.** Substitute the expressions for each v_R and i_R into the equations obtained in steps 3 and 4, thus eliminating all resistor variables. Put the resultant equations in normal form.

EXAMPLE 18.1

Obtain the normal-form equations for the circuit of Fig. 18.4*a*, a four-node circuit containing two capacitors, one inductor, and two independent sources.

Following step 1, we draw a normal tree. Note that only one such tree is possible here, as shown in Fig. 18.4*b*, since the two capacitors and the voltage source must be in the tree and the inductor and the current source must be in the cotree.

We next define the voltage across the $\frac{1}{6}$ F capacitor as v_{C1}, the voltage across the $\frac{1}{7}$ F capacitor as v_{C2}, and the current through the inductor as i_L. The source voltage is indicated across its tree branch, the source current is marked on its link, and only the resistor link remains without an assigned variable. The current directed to the left through that link is the voltage $v_{C1} - v_s$ divided by 3 Ω, and we thus find it unnecessary to introduce any additional variables. These tree-branch voltages and link currents are shown in Fig. 18.4*c*.

(Continued on next page)

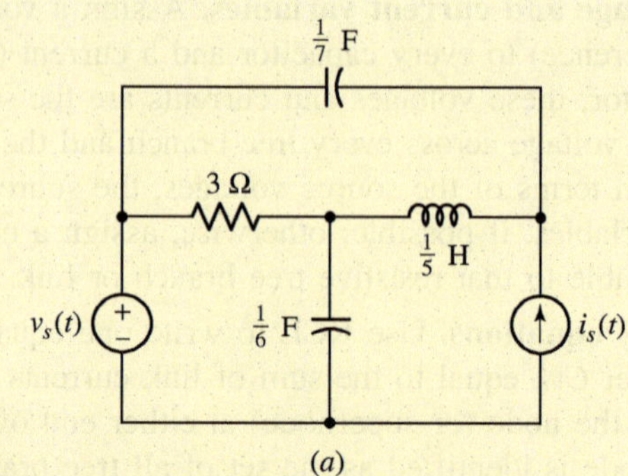

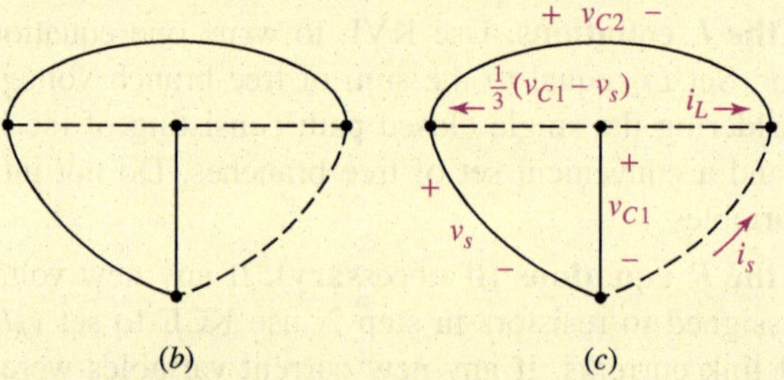

■ **FIGURE 18.4** (*a*) A given circuit for which normal-form equations are to be written. (*b*) Only one normal tree is possible. (*c*) Tree-branch voltages and link currents are assigned.

Two equations must be written for step 3. For the $\frac{1}{6}$ F capacitor, we apply KCL to the central node:

$$\frac{v'_{C1}}{6} + i_L + \frac{1}{3}(v_{C1} - v_s) = 0$$

while the right-hand node is most convenient for the $\frac{1}{7}$ F capacitor:

$$\frac{v'_{C2}}{7} + i_L + i_s = 0$$

Moving on to step 4, KVL is applied to the inductor link and the entire tree in this case:

$$\frac{i'_L}{5} - v_{C2} + v_s - v_{C1} = 0$$

Since there were no new variables assigned to the resistor, we skip step 5 and simply rearrange the three preceding equations to obtain the desired normal-form equations,

$$v'_{C1} = -2v_{C1} - 6i_L + 2v_s \qquad [12]$$

$$v'_{C2} = -7i_L - 7i_s \qquad [13]$$

$$i'_L = 5v_{C1} + 5v_{C2} - 5v_s \qquad [14]$$

The state variables have arbitrarily been ordered as v_{C1}, v_{C2}, and i_L. If the order i_L, v_{C1}, v_{C2} had been selected instead, the three normal-form equations would be

$$i'_L = 5v_{C1} + 5v_{C2} - 5v_s$$

$$v'_{C1} = -6i_L - 2v_{C1} + 2v_s$$

$$v'_{C2} = -7i_L - 7i_s$$

Note that the ordering of the terms on the right-hand sides of the equations has been changed to agree with the ordering of the equations.

EXAMPLE 18.2

Write a set of normal-form equations for the circuit of Fig. 18.5*a*.

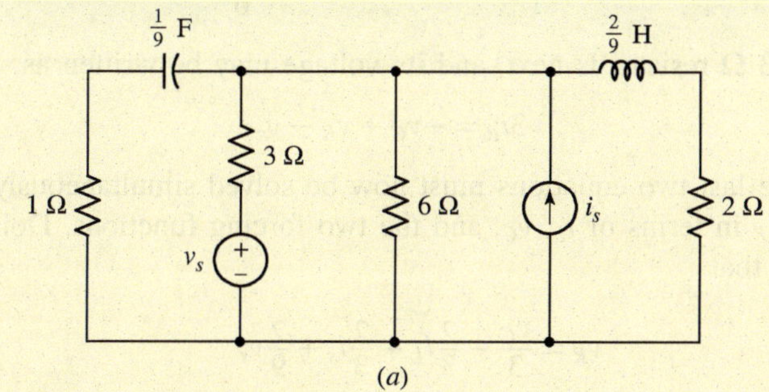

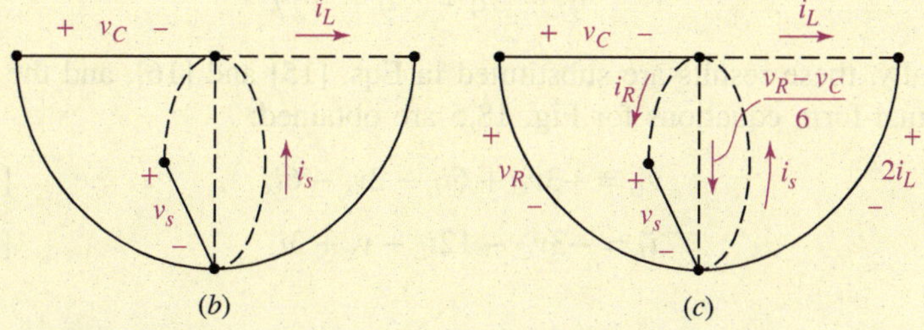

■ **FIGURE 18.5** (*a*) The given circuit. (*b*) One of many possible normal trees. (*c*) The assigned voltages and currents.

The circuit contains several resistors, and this time it will be necessary to introduce resistor variables. Many different normal trees can be constructed for this network, and it is often worthwhile to sketch several possible trees to see whether resistor voltage and current variables can be avoided by a judicious selection. The tree we shall use is shown in Fig. 18.5*b*. The state variables v_C and i_L and the forcing functions v_s and i_s are indicated on the graph.

Although we might study this circuit and the tree for a few minutes and arrive at a method to avoid the introduction of any new variables, let us plead temporary stupidity and assign the tree-branch voltage v_R and link current i_R for the 1 Ω and 3 Ω branches, respectively.

(Continued on next page)

The voltage across the 2 Ω resistor is easily expressed as $2i_L$, while the downward current through the 6 Ω resistor becomes $(v_R - v_C)/6$. All the link currents and tree-branch voltages are marked on Fig. 18.5*c*.

The capacitor equation may now be written

$$\frac{v'_C}{9} = i_R + i_L - i_s + \frac{v_R - v_C}{6} \qquad [15]$$

and that for the inductor is

$$\frac{2}{9}i'_L = -v_C + v_R - 2i_L \qquad [16]$$

We begin step 5 with the 1 Ω resistor. Since it is in the tree, we must equate its current to a sum of link currents. Both terminals of the capacitor collapse into a supernode, and we have

$$\frac{v_R}{1} = i_s - i_R - i_L - \frac{v_R - v_C}{6}$$

The 3 Ω resistor is next, and its voltage may be written as

$$3i_R = -v_C + v_R - v_s$$

These last two equations must now be solved simultaneously for v_R and i_R in terms of i_L, v_C, and the two forcing functions. Doing so, we find that

$$v_R = \frac{v_C}{3} - \frac{2}{3}i_L + \frac{2}{3}i_s + \frac{2}{9}v_s$$

$$i_R = -\frac{2}{9}v_C - \frac{2}{9}i_L + \frac{2}{9}i_s - \frac{7}{27}v_s$$

Finally, these results are substituted in Eqs. [15] and [16], and the normal-form equations for Fig. 18.5 are obtained:

$$v'_C = -3v_C + 6i_L - 2v_s - 6i_s \qquad [17]$$

$$i'_L = -3v_C - 12i_L + v_s + 3i_s \qquad [18]$$

Up to now we have discussed only those circuits in which the voltage and current sources were independent sources. At this time we will look at a circuit containing a dependent source.

EXAMPLE 18.3

Find the set of normal-form equations for the circuit of Fig. 18.6.

The only possible normal tree is that shown in Fig. 18.6*b*, and it might be noted that it was not possible to place the controlling current i_x in a link. The tree-branch voltages and link currents are shown on the linear graph, and they are unchanged from the earlier example except for the additional source voltage, $18i_x$.

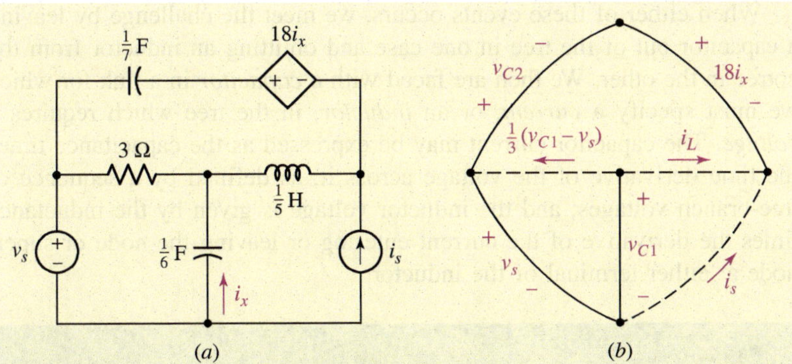

FIGURE 18.6 (a) A circuit containing a dependent voltage source. (b) The normal tree for this circuit with tree-branch voltages and link currents assigned.

For the $\frac{1}{6}$ F capacitor, we again find that

$$\frac{v'_{C1}}{6} + i_L + \frac{1}{3}(v_{C1} - v_s) = 0 \qquad [19]$$

Letting the dependent voltage source shrink into a supernode, we also find the relationship unchanged for the $\frac{1}{7}$ F capacitor,

$$\frac{v'_{C2}}{7} + i_L + i_s = 0 \qquad [20]$$

Our previous result for the inductor changes, however, because there is an added branch in the tree:

$$\frac{i'_L}{5} - 18i_x - v_{C2} + v_s - v_{C1} = 0$$

Finally, we must write a control equation that expresses i_x in terms of our tree-branch voltages and link currents. It is

$$i_x = i_L + \frac{v_{C1} - v_s}{3}$$

and thus the inductor equation becomes

$$\frac{i'_L}{5} - 18i_L - 6v_{C1} + 6v_s - v_{C2} + v_s - v_{C1} = 0 \qquad [21]$$

When Eqs. [19] to [21] are written in normal form, we have

$$v'_{C1} = -2v_{C1} - 6i_L + 2v_s \qquad [22]$$

$$v'_{C2} = -7i_L - 7i_s \qquad [23]$$

$$i'_L = 35v_{C1} + 5v_{C2} + 90i_L - 35v_s \qquad [24]$$

Our equation-writing process must be modified slightly if we cannot construct a normal tree for the circuit, and there are two types of networks for which this occurs. One contains a loop in which every element is a capacitor or voltage source, thus making it impossible to place them all in the tree. The other occurs if there is a node or supernode that is connected to the remainder of the circuit only by inductors and current sources.

Remember, a tree, by definition, contains no closed paths (loops), and every node must have at least one tree branch connected to it.

When either of these events occurs, we meet the challenge by leaving a capacitor out of the tree in one case and omitting an inductor from the cotree in the other. We then are faced with a *capacitor* in a link for which we must specify a *current*, or an *inductor*, in the tree which requires a *voltage*. The capacitor current may be expressed as the capacitance times the time derivative of the voltage across it, as defined by a sequence of tree-branch voltages; and the inductor voltage is given by the inductance times the derivative of the current entering or leaving the node or supernode at either terminal of the inductor.

EXAMPLE 18.4

Write a set of normal-form equations for the network shown in Fig. 18.7*a*.

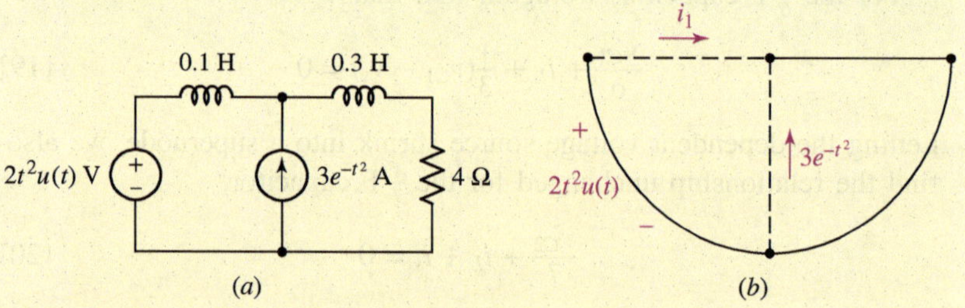

FIGURE 18.7 (*a*) A circuit for which a normal tree cannot be drawn. (*b*) A tree is constructed in which one inductor must be a tree branch.

The only elements connected to the upper central node are the two inductors and the current source, and we therefore must place one inductor in the tree, as illustrated by Fig. 18.7*b*. The two forcing functions and the single state variable, the current i_1, are shown on the graph. Note that the current in the upper right-hand inductor is known in terms of i_1 and $3e^{-t^2}$ A. Thus we could not specify the energy states of the two inductors independently of each other, and this system requires only the single state variable i_1. Of course, if we had placed the 0.1 H inductor in the tree (instead of the 0.3 H inductor), the current in the right-hand inductor would have ended up as the single state variable.

We must still assign voltages to the remaining two tree branches in Fig. 18.7*b*. Since the current directed to the right in the 0.3 H inductor is $i_1 + 3e^{-t^2}$, the voltages appearing across the inductor and the 4 Ω resistor are $0.3d(i_1 + 3e^{-t^2})/dt = 0.3i_1' - 1.8te^{-t^2}$ and $4i_1 + 12e^{-t^2}$, respectively.

The single normal-form equation is obtained in step 4 of our procedure:

$$0.1i_1' + 0.3i_1' - 1.8te^{-t^2} + 4i_1 + 12e^{-t^2} - 2t^2u(t) = 0$$

or

$$i_1' = -10i_1 + e^{-t^2}(4.5t - 30) + 5t^2u(t) \qquad [25]$$

Note that one of the terms on the right-hand side of the equation is proportional to the derivative of one of the source functions.

In this example, the two inductors and the current source were all connected to a common node. With circuits containing more branches and more nodes, the connection may be to a supernode. As an indication of such a network, Fig. 18.8 shows a slight rearrangement of two elements appearing in Fig. 18.7*a*. Note that, once again, the energy states of the two inductors cannot be specified independently; with one specified, and the source current given, the other is then specified also. The method of obtaining the normal-form equation is the same.

The exception created by a loop of capacitors and voltage sources is treated by a similar (dual) procedure, and the wary student should have only the most minor troubles.

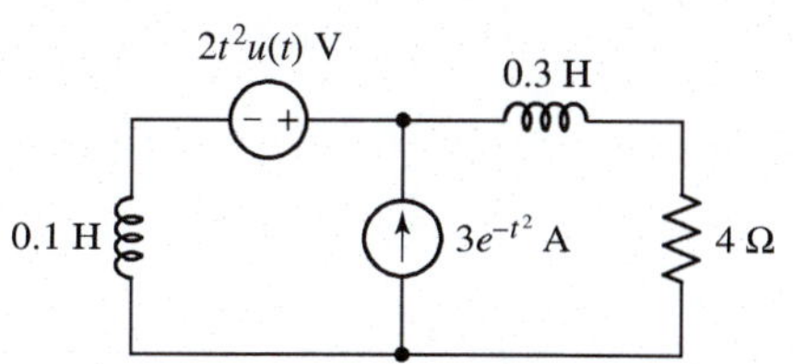

FIGURE 18.8 The circuit of Fig. 18.7*a* is redrawn in such a way as to cause the supernode containing the voltage source to be connected to the remainder of the network only through the two inductors and the current source.

PRACTICE

18.2 Write normal-form equations for the circuit of Fig. 18.9*a*.

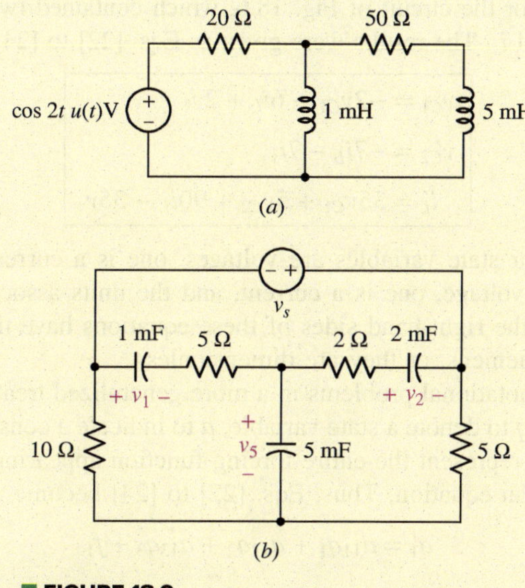

FIGURE 18.9

18.3 Write the normal-form equations for the circuit of Fig. 18.9*b* using the state-variable order v_1, v_2, v_5.

18.4 Find the normal-form equation for the circuit shown in Fig. 18.10 using the state variable: (*a*) v_{C1}; (*b*) v_{C2}.

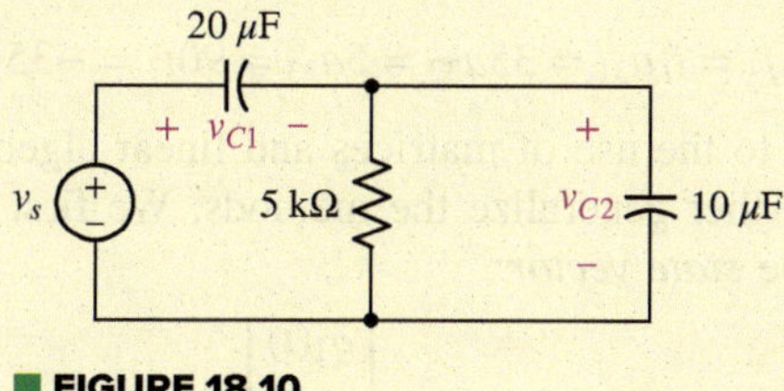

FIGURE 18.10

Ans: 18.2: $i_1' = -20{,}000i_1 - 20{,}000i_5 + 1000\cos 2tu(t)$; $i_5' = -4000i_1 - 14{,}000i_5 + 200\cos 2tu(t)$ [using $i_1 \downarrow$ and $i_5 \downarrow$]. 18.3: $v_1' = -160v_1 - 100v_2 - 60v_5 - 140v_s$; $i_2' = -50v_1 - 125v_2 + 75v_5 - 75v_s$; $i_5' = -12v_1 + 30v_2 - 42v_5 + 2v_s$. 18.4: $v_{C1}' = -6.67v_{C1} + 6.67v_s + 0.333v_s'$; $v_{C2}' = -6.67v_{C2} + 0.667v_s'$.

18.3 THE USE OF MATRIX NOTATION

In the examples that we studied in the previous two sections, the state variables selected were the capacitor voltages and the inductor currents, except for the final case, where it was not possible to draw a normal tree and only one inductor current could be selected as a state variable. As the number of inductors and capacitors in a network increases, it is apparent that the number of state variables will increase. More complicated circuits thus require a greater number of state equations, each of which contains a greater array of state variables on the right-hand side. Not only does the solution of such a set of equations require computer assistance,[1] but also the sheer effort of writing down all the equations can easily lead to writer's cramp.

In this section we will establish a useful symbolic notation that minimizes the equation-writing effort.

Let us introduce this method by recalling the normal-form equations that we obtained for the circuit of Fig. 18.6, which contained two independent sources, v_s and i_s. The results were given as Eqs. [22] to [24] of Sec. 18.2:

$$v'_{C1} = -2v_{C1} - 6i_L + 2v_s \qquad [22]$$

$$v'_{C2} = -7i_L - 7i_s \qquad [23]$$

$$i'_L = 35v_{C1} + 5v_{C2} + 90i_L - 35v_s \qquad [24]$$

Two of our state variables are voltages, one is a current, one source function is a voltage, one is a current, and the units associated with the constants on the right-hand sides of these equations have the dimensions of ohms, or siemens, or they are dimensionless.

To avoid notational problems in a more generalized treatment, we will use the letter q to denote a state variable, a to indicate a constant multiplier of q, and f to represent the entire forcing function appearing on the right-hand side of an equation. Thus, Eqs. [22] to [24] become

$$q'_1 = a_{11}q_1 + a_{12}q_2 + a_{13}q_3 + f_1 \qquad [26]$$

$$q'_2 = a_{21}q_1 + a_{22}q_2 + a_{23}q_3 + f_2 \qquad [27]$$

$$q'_3 = a_{31}q_1 + a_{32}q_2 + a_{33}q_3 + f_3 \qquad [28]$$

where

$$q_1 = v_{C1} \quad a_{11} = -2 \quad a_{12} = 0 \quad a_{13} = -6 \quad f_1 = 2v_s$$

$$q_2 = v_{C2} \quad a_{21} = 0 \quad a_{22} = 0 \quad a_{23} = -7 \quad f_2 = -7i_s$$

$$q_3 = i_L \quad a_{31} = 35 \quad a_{32} = 5 \quad a_{33} = 90 \quad f_3 = -35v_s$$

For a review of matrices and matrix notation refer to Appendix 2.

We now turn to the use of matrices and linear algebra to simplify our equations and further generalize the methods. We first define a matrix **q** which we call the *state vector:*

$$\mathbf{q}(t) = \begin{bmatrix} q_1(t) \\ q_2(t) \\ \vdots \\ q_n(t) \end{bmatrix} \qquad [29]$$

(1) Although SPICE can be used to analyze any of the circuits we meet in this chapter, other computer programs are used specifically to solve normal-form equations.

The derivative of a matrix is obtained by taking the derivative of each element of the matrix. Thus,

$$\mathbf{q}'(t) = \begin{bmatrix} q_1'(t) \\ q_2'(t) \\ \vdots \\ q_n'(t) \end{bmatrix}$$

We shall represent all matrices and vectors in this chapter by lowercase boldface letters, such as **q** or $\mathbf{q}(t)$, with the single exception of the identity matrix **I**, to be defined in Sec. 18.5. The elements of any matrix are scalars, and they are symbolized by lowercase italic letters, such as q_1 or $q_1(t)$.

We also define the set of forcing functions, $f_1, f_2, \ldots, f_n$, as a matrix **f** and call it the *forcing-function matrix:*

$$\mathbf{f}(t) = \begin{bmatrix} f_1(t) \\ f_2(t) \\ \vdots \\ f_n(t) \end{bmatrix} \qquad [30]$$

Now let us turn our attention to the coefficients a_{ij}, which represent elements in the $(n \times n)$ square matrix **a**,

$$\mathbf{a} = \begin{bmatrix} a_{11} & a_{12} & \cdots & a_{1n} \\ a_{21} & a_{22} & \cdots & a_{2n} \\ \cdots & \cdots & \cdots & \cdots \\ a_{n1} & a_{n2} & \cdots & a_{nn} \end{bmatrix} \qquad [31]$$

The matrix **a** is termed the *system matrix.*

Using the matrices we have defined in the preceding paragraphs, we can combine these results to obtain a concise, compact representation of the state equations,

$$\mathbf{q}' = \mathbf{aq} + \mathbf{f} \qquad [32]$$

The matrices $\mathbf{q}'$ and **f**, and the matrix product **aq**, are all $(n \times 1)$ column matrices.

The advantages of this representation are obvious, because a system of 100 equations in 100 state variables has exactly the same form as one equation in one state variable.

For the example of Eqs. [22] to [24], the four matrices in Eq. [32] may be written out explicitly as

$$\begin{bmatrix} v_{C1}' \\ v_{C2}' \\ i_L' \end{bmatrix} = \begin{bmatrix} -2 & 0 & -6 \\ 0 & 0 & -7 \\ 35 & 5 & 90 \end{bmatrix} \begin{bmatrix} v_{C1} \\ v_{C2} \\ i_L \end{bmatrix} + \begin{bmatrix} 2v_s \\ -7i_s \\ -35v_s \end{bmatrix} \qquad [33]$$

Everyone except full-blooded matrix experts should take a few minutes off to expand Eq. [33] and then check the results with Eqs. [22] to [24]; three identical equations should result.

What now is our status with respect to state-variable analysis? Given a circuit, we should be able to construct a normal tree, specify a set of state variables, order them as the state vector, write a set of normal-form equations, and finally specify the system matrix and the forcing-function vector from the equations.

The next problem facing us is to obtain the explicit functions of time that the state variables represent.

PRACTICE

18.5 (*a*) Using the state vector $\mathbf{q} = \begin{bmatrix} i \\ v \end{bmatrix}$, determine the system matrix and the forcing-function vector for the circuit of Fig. 18.11. (*b*) Repeat for the state vector $\mathbf{q} = \begin{bmatrix} v \\ i \end{bmatrix}$.

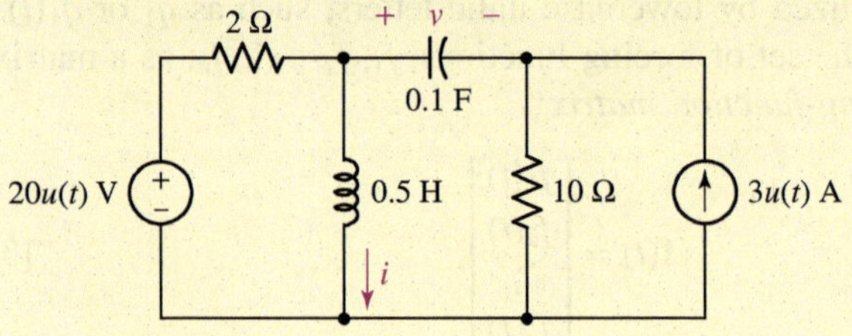

■ **FIGURE 18.11**

Ans: $\begin{bmatrix} -3.33 & 0.3333 \\ -1.667 & -0.833 \end{bmatrix}, \begin{bmatrix} 43.3u(t) \\ -8.33u(t) \end{bmatrix}; \begin{bmatrix} -0.833 & -1.667 \\ 0.333 & -3.33 \end{bmatrix}, \begin{bmatrix} -8.33u(t) \\ 43.3u(t) \end{bmatrix}$

18.4 SOLUTION OF THE FIRST-ORDER EQUATION

The matrix equation representing the set of normal-form equations for a general nth-order system was obtained as Eq. [32] in the previous section, and it is repeated here for convenience as Eq. [34]:

$$\mathbf{q}' = \mathbf{aq} + \mathbf{f} \tag{34}$$

The $(n \times n)$ system matrix **a** is composed of constant elements for our time-invariant circuits, and $\mathbf{q}'$, **q**, and **f** are all $(n \times 1)$ column matrices. We need to solve this matrix equation for **q**, whose elements are q_1, q_2, ..., q_n. Each must be found as a function of time. Remember that these are the state variables, the collection of which enables us to specify every voltage and current in the given circuit.

If each matrix in Eq. [34] has only one row and one column, then we may write the matrix equation as

$$\begin{aligned}[q_1'(t)] &= [a_{11}][q_1(t)] + [f_1(t)] \\ &= [a_{11}q_1(t)] + [f_1(t)] \\ &= [a_{11}q_1(t) + f_1(t)]\end{aligned}$$

and therefore, we have the first-order equation

$$q_1'(t) = a_{11}q_1(t) + f_1(t) \tag{35}$$

or

$$q_1'(t) - a_{11}q_1(t) = f_1(t) \tag{36}$$

We proceed by multiplying each side of the equation by the integrating factor $e^{-ta_{11}}$:

$$e^{-ta_{11}}q_1'(t) - e^{-ta_{11}}q_1(t) = e^{-ta_{11}}f_1(t)$$

The left-hand side of this equation is again an exact derivative, and so we have

$$\frac{d}{dt}[e^{-ta_{11}}q_1(t)] = e^{-ta_{11}}f_1(t) \qquad [37]$$

The order in which the various factors in Eq. [37] have been written may seem a little strange because a term which is the product of a constant and a time function is usually written with the constant appearing first and the time function following it. In scalar equations, multiplication is commutative, and so the order in which the factors appear is of no consequence. But in the matrix equations which we will be considering next, the corresponding factors will be matrices, and matrix multiplication is *not* commutative.

That is, we know that

$$\begin{bmatrix} 2 & 4 \\ 6 & 8 \end{bmatrix}\begin{bmatrix} a & b \\ c & d \end{bmatrix} = \begin{bmatrix} (2a+4c)(2b+4d) \\ (6a+8c)(6b+8d) \end{bmatrix}$$

while

$$\begin{bmatrix} a & b \\ c & d \end{bmatrix}\begin{bmatrix} 2 & 4 \\ 6 & 8 \end{bmatrix} = \begin{bmatrix} (2a+6b)(4a+8b) \\ (2c+6d)(4c+8d) \end{bmatrix}$$

Different results are obtained, and we therefore need to be careful later about the order in which we write matrix factors.

Continuing with Eq. [37], let us integrate each side with respect to time from $-\infty$ to a general time t:

$$e^{-ta_{11}}q_1(t) = \int_{-\infty}^{t} e^{-za_{11}}f_1(z)dz \qquad [38]$$

where z is simply a dummy variable of integration, and where we have assumed that $e^{-ta_{11}}q_1(t)$ approaches zero as t approaches $-\infty$. We now multiply (premultiply, if this were a matrix equation) each side of Eq. [38] by the exponential factor $e^{ta_{11}}$, obtaining

$$q_1(t) = e^{ta_{11}}\int_{-\infty}^{t} e^{-za_{11}}f_1(z)dz \qquad [39]$$

which is the desired expression for the single unknown state variable.

In many circuits, however, particularly those in which switches are present and the circuit is reconfigured at some instant of time (often $t = 0$), we do not know the forcing function or the normal-form equation prior to that instant. We therefore incorporate all of the past history in an integral from $-\infty$ to that instant, here assumed to be $t = 0$, by letting $t = 0$ in Eq. [39]:

$$q_1(0) = \int_{-\infty}^{0} e^{-za_{11}}f_1(z)dz$$

We then use this initial value in the general solution for $q_1(t)$:

$$q_1(t) = e^{ta_{11}}q_1(0) + e^{ta_{11}}\int_{0}^{t} e^{-za_{11}}f_1(z)dz \qquad [40]$$

This last expression shows that the state-variable time function may be interpreted as the sum of two terms. The first is the response that would arise if the forcing function were zero [$f_1(t) = 0$], and in the language of state-variable analysis it is called the *zero-input response*. It has the *form* of the natural response, although it may not have the same amplitude as the term we have been calling the natural response. The zero-input response also is the solution to the homogeneous normal-form equation, obtained by letting $f_1(t) = 0$ in Eq. [36].

The second part of the solution would represent the complete response if $q_1(0)$ were zero, and it is termed the *zero-state response*. We will see in the following example that what we have termed the forced response appears as a part of the zero-state response.

EXAMPLE 18.5

Find $v_C(t)$ for the circuit of Fig. 18.12.

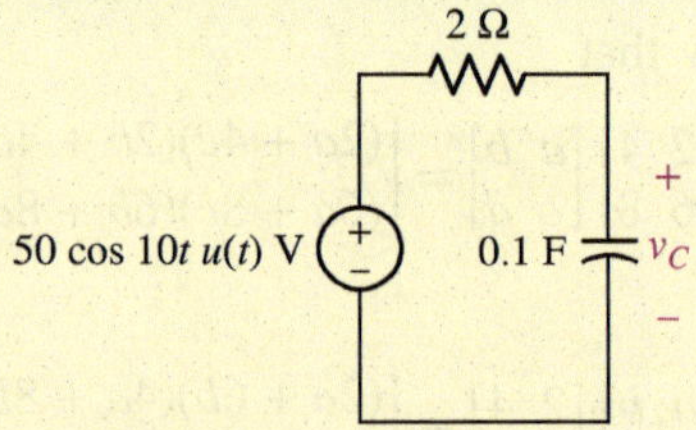

FIGURE 18.12 A first-order circuit for which $v_C(t)$ is to be found through the methods of state-variable analysis.

The normal-form equation is easily found to be

$$v_C' = -5v_C + 250 \cos 10t\, u(t)$$

Thus, $a_{11} = -5$, $f_1(t) = 250 \cos 10t\, ut(t)$, and we may substitute directly in Eq. [39] to obtain the solution:

$$v_C(t) = e^{-5t} \int_{-\infty}^{t} e^{5z}\, 250 \cos 10z\, u(z)dz$$

The unit-step function inside the integral may be replaced by $u(t)$ outside the integral if the lower limit is changed to zero:

$$v_C(t) = e^{-5t}u(t) \int_0^t e^{5z}\, 250 \cos 10z\, dz$$

Integrating, we have

$$v_C(t) = e^{-5t}u(t)\left[\frac{250e^{5z}}{5^2 + 10^2}(5 \cos 10z + 10 \sin 10z)\right]_0^t$$

or

$$v_C(t) = [-10e^{-5t} + 10(\cos 10t + 2 \sin 10t)]u(t) \qquad [41]$$

The same result may be obtained through the use of Eq. [40]. Since $v_C(0) = 0$, we have

$$v_C(t) = e^{-5t}(0) + e^{-5t} \int_0^t e^{5z}\, 250 \cos 10z\, u(z)dz$$

$$= e^{-5t}u(t) \int_0^t e^{5z}\, 250 \cos 10z\, dz$$

and this leads to the same solution as before, of course. However, we also see that the entire solution for $v_C(t)$ is the zero-state response, and there is no zero-input response. It is interesting to note that, if we had solved this problem by the methods of Chap. 17, we would have set up a natural response,

$$v_{C,n}(t) = Ae^{-5t}$$

and computed the forced response by frequency-domain methods,

$$\mathbf{V}_{C,f} = \frac{50}{2 - j1}(-j1) = 10 - j20$$

so that

$$v_{C,f}(t) = 10 \cos 10t + 20 \sin 10t$$

Thus,

$$v_C(t) = Ae^{-5t} + 10(\cos 10t + 2 \sin 10t) \quad (t > 0)$$

and the application of the initial condition, $v_C(0) = 0$, leads to $A = -10$ and an expression identical to Eq. [41] once again. Looking at the partial responses obtained in the two methods, we therefore find that

$$\begin{aligned} v_{C,\text{zero-input}} &= 0 \\ v_{C,\text{zero-state}} &= [-10e^{-5t} + 10(\cos 10t + 2 \sin 10t)]u(t) \\ v_{C,n} &= -10e^{-5t}u(t) \\ v_{C,f} &= 10(\cos 10t + 2 \sin 10t)u(t) \end{aligned}$$

EXAMPLE **18.6**

The switch in the circuit of Fig. 18.13 is thrown at $t = 0$, and the form of the circuit changes at that instant. Find $i_L(t)$ for $t > 0$.

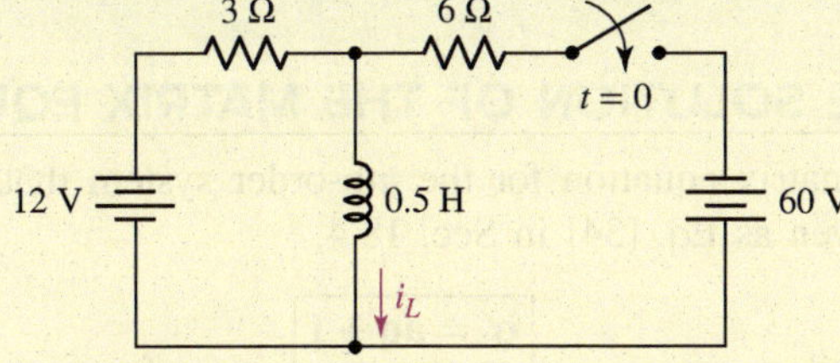

■ **FIGURE 18.13** A first-order example in which the form of the circuit changes at $t = 0$.

We represent everything before $t = 0$ by the statement that $i_L(0) = 4$ A, and we obtain the normal-form equation for the circuit in the configuration it has $t = 0$. It is

$$i_L'(t) = -4i_L(t) - 24$$

This time, we must use Eq. [40], since we do not have a single normal-form equation that is valid for all time. The result is

$$i_L(t) = e^{-4t}(4) + e^{-4t}\int_0^t e^{+4z}(-24)dz$$

(Continued on next page)

or

$$i_L(t) = 4e^{-4t} - 6(1 - e^{-4t}) \quad (t > 0)$$

The several components of the response are now identified:

$$i_{L,\text{zero-input}} = 4e^{-4t} \qquad (t > 0)$$
$$i_{L,\text{zero-state}} = -6(1 - e^{-4t})(t > 0)$$

while our earlier analytical methods would have led to

$$i_{L,n} = 10e^{-4t}(t > 0)$$
$$i_{L,f} = -6 \quad (t > 0)$$

These first-order networks certainly do not require the use of state variables for their analyses. However, the method by which we solved the single normal-form equation does offer a few clues to how the *n*th-order solution might be obtained. We follow this exciting trail in the next section.

PRACTICE

18.6 (*a*) Write the normal-form equation for the circuit of Fig. 18.14. Find $v_C(t)$ for $t > 0$ by (*b*) Eq. [39]; (*c*) Eq. [40].

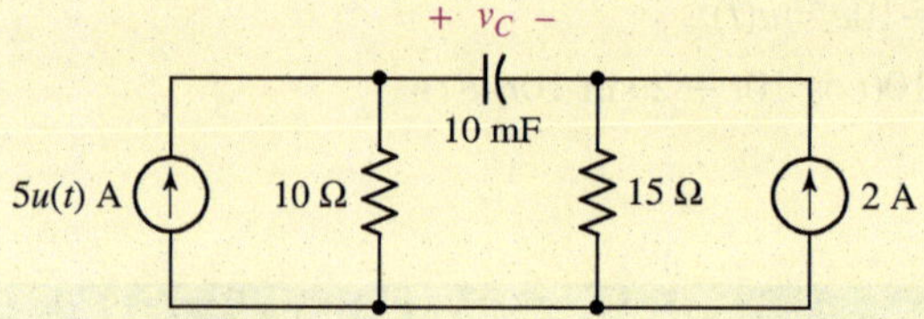

■ **FIGURE 18.14**

Ans: $v_C' = -4v_C - 120 + 200u(t)$; $20 - 50e^{-4t}$ V; $20 - 50e^{-4t}$ V.

18.5 THE SOLUTION OF THE MATRIX EQUATION

The general matrix equation for the *n*th-order system that we now wish to solve is given as Eq. [34] in Sec. 18.4,

$$\boxed{\mathbf{q}' = \mathbf{aq} + \mathbf{f}} \qquad [34]$$

where **a** is an $(n \times n)$ square matrix of constants and the other three matrices are all $(n \times 1)$ column matrices whose elements are, in general, time functions. In the most general case, all the matrices would be composed of time functions.

In this section, we obtain the matrix solution for this equation. In Sec. 18.6 we will interpret our results and indicate how we might obtain a useful solution for **q**.

We begin by subtracting the matrix product **aq** from each side of Eq. [34]:

$$\mathbf{q}' - \mathbf{aq} = \mathbf{f} \qquad [42]$$

Recalling our case of the integrating factor $e^{-ta_{11}}$ in the first-order case, let us premultiply each side of Eq. [42] by $e^{-ta_{11}}$:

$$e^{-t\mathbf{a}}\mathbf{q}' - e^{-t\mathbf{a}}\mathbf{a}\mathbf{q} = e^{-t\mathbf{a}}\mathbf{f} \qquad [43]$$

Although the presence of a matrix as an exponent may seem to be somewhat strange, the function e^{-ta} may be defined in terms of its infinite power series expansion in t,

$$e^{-t\mathbf{a}} = \mathbf{I} - t\mathbf{a} + \frac{t^2}{2!}(\mathbf{a})^2 - \frac{t^3}{3!}(\mathbf{a})^3 + \cdots \qquad [44]$$

We identify $\mathbf{I}$ as the $(n \times n)$ identity matrix,

$$\mathbf{I} = \begin{bmatrix} 1 & 0 & \cdots & 0 \\ 0 & 1 & \cdots & 0 \\ \cdots & \cdots & \cdots & \cdots \\ 0 & 0 & \cdots & 1 \end{bmatrix}$$

such that

$$\mathbf{Ia} = \mathbf{aI} = \mathbf{a}$$

The products $(\mathbf{a})^2$, $(\mathbf{a})^3$, and so forth in Eq. [44] may be obtained by repeated multiplication of the matrix $\mathbf{a}$ by itself, and therefore each term in the expansion is again an $(n \times n)$ matrix. Thus, it is apparent that $e^{-t\mathbf{a}}$ is also an $(n \times n)$ square matrix, but its elements are all functions of time in general.

Again following the first-order procedure, we would now like to show that the left-hand side of Eq. [43] is equal to the time derivative of $e^{-t\mathbf{a}}\mathbf{q}$. Since this is a product of two functions of time, we have

$$\frac{d}{dt}(e^{-t\mathbf{a}}\mathbf{q}) = e^{-t\mathbf{a}}\frac{d}{dt}(\mathbf{q}) + \left[\frac{d}{dt}(e^{-t\mathbf{a}})\right]\mathbf{q}$$

The derivative of e^{-ta} is obtained by again considering the infinite series of Eq. [44], and we find that it is given by $-\mathbf{a}e^{-t\mathbf{a}}$. The series expansion can also be used to show that $-\mathbf{a}e^{-t\mathbf{a}} = -e^{-t\mathbf{a}}\mathbf{a}$. Thus

$$\frac{d}{dt}(e^{-t\mathbf{a}}\mathbf{q}) = e^{-t\mathbf{a}}\mathbf{q}' - e^{-t\mathbf{a}}\mathbf{aq}$$

and Eq. [43] helps to simplify this expression into the form

$$\frac{d}{dt}(e^{-t\mathbf{a}}\mathbf{q}) = e^{-t\mathbf{a}}\mathbf{f}$$

Multiplying by dt and integrating from $-\infty$ to t, we have

$$e^{-t\mathbf{a}}\mathbf{q} = \int_{-\infty}^{t} e^{-z\mathbf{a}}\mathbf{f}(z)dz \qquad [45]$$

To solve for $\mathbf{q}$, we must premultiply the left-hand side of Eq. [45] by the matrix inverse of $e^{-t\mathbf{a}}$. That is, any square matrix $\mathbf{b}$ has an inverse $\mathbf{b}^{-1}$ such that $\mathbf{b}^{-1}\mathbf{b} = \mathbf{bb}^{-1} = \mathbf{I}$. In this case, another power series expansion shows that the inverse of $e^{-t\mathbf{a}}$ is $e^{t\mathbf{a}}$, or

$$e^{t\mathbf{a}}e^{-t\mathbf{a}}\mathbf{q} = \mathbf{Iq} = \mathbf{q}$$

and we may therefore write our solution as

$$\mathbf{q} = e^{t\mathbf{a}} \int_{-\infty}^{t} e^{-z\mathbf{a}}\mathbf{f}(z)dz \qquad [46]$$

In terms of the initial value of the state vector,

$$\mathbf{q} = e^{t\mathbf{a}}\mathbf{q}(0) + e^{t\mathbf{a}} \int_0^t e^{-z\mathbf{a}}\mathbf{f}(z)dz \qquad [47]$$

The function $e^{t\mathbf{a}}$ is a very important quantity in state-space analysis. It is called the *state-transition matrix,* for it describes how the state of the system changes from its zero state to its state at time t. Eqs. [46] and [47] are the nth-order matrix equations that correspond to the first-order results we numbered [39] and [40]. Although these expressions represent the "solutions" for $\mathbf{q}$, the fact that we express $e^{t\mathbf{a}}$ and $e^{t\mathbf{a}}$ only as infinite series is a serious deterrent to our making any effective use of these results. We would have infinite power series in t for each $q_i(t)$, and while a computer might find this procedure compatible with its memory and computational speed, we would probably have other, more pressing chores to handle than carrying out the summation by ourselves.

PRACTICE

18.7 Let $\mathbf{a} = \begin{bmatrix} 0 & 2 \\ -1 & 1 \end{bmatrix}$ and $t = 0.1$ s. Use the power series expansion to find the (*a*) matrix $e^{-t\mathbf{a}}$; (*b*) matrix $e^{t\mathbf{a}}$; (*c*) value of the determinant of $e^{-t\mathbf{a}}$; (*d*) value of the determinant of $e^{t\mathbf{a}}$; (*e*) product of the last two results.

Ans: $\begin{bmatrix} 0.9903 & -0.1897 \\ 0.0948 & 0.8955 \end{bmatrix}$; $\begin{bmatrix} 0.9897 & 0.2097 \\ -0.1048 & 1.0945 \end{bmatrix}$; 0.9048; 1.1052; 1.0000.

18.6 A FURTHER LOOK AT THE STATE-TRANSITION MATRIX

In this section, we seek a more satisfactory representation for $e^{t\mathbf{a}}$ and $e^{-t\mathbf{a}}$. If the system matrix $\mathbf{a}$ is an $(n \times n)$ square matrix, then each of these exponentials is an $(n \times n)$ square matrix of time functions, and one of the consequences of a theorem developed in linear algebra, known as the Cayley–Hamilton theorem, shows that such a matrix may be expressed as an $(n - 1)^{\text{st}}$-degree polynomial in the matrix $\mathbf{a}$. That is,

$$e^{t\mathbf{a}} = u_0\mathbf{I} + u_1\mathbf{a} + u_2(\mathbf{a})^2 + \cdots + u_{n-1}(\mathbf{a})^{n-1} \qquad [48]$$

where each of the u_i is a scalar function of time that is still to be determined; the $\mathbf{a}^i$ are constant $(n \times n)$ matrices. The theorem also states that Eq. [48] remains an equality if $\mathbf{I}$ is replaced by unity and $\mathbf{a}$ is replaced by any one of the scalar roots s_i of the nth-degree scalar equation,

$$\det(\mathbf{a} - s\mathbf{I}) = 0 \qquad [49]$$

The expression $(\mathbf{a} - s\mathbf{I})$ indicates the determinant of the matrix $(\mathbf{a} - s\mathbf{I})$. This determinant is an nth-degree polynomial in s. We assume that the n roots are all different. Eq. [49] is called the *characteristic equation of* the matrix $\mathbf{a}$, and the values of s which are the roots of the equation are known as the *eigenvalues* of $\mathbf{a}$.

These values of s are identical with the natural frequencies that we dealt with in Chap. 15 as poles of an appropriate transfer function. That is, if our state variable is $v_{C1}(t)$, then the poles of $\mathbf{H(s)} = \mathbf{V}_{C1}(\mathbf{s})/\mathbf{I}_s$ or of $\mathbf{V}_{C1}(\mathbf{s})/\mathbf{V}_s$ are also the eigenvalues of the characteristic equation.

Thus, this is the procedure we shall follow to obtain a simpler form for $e^{t\mathbf{a}}$:

1. Given **a**, form the matrix $(\mathbf{a} - s\mathbf{I})$.
2. Set the determinant of this square matrix equal to zero.
3. Solve the resultant nth-degree polynomial for its n roots, s_1, s_2, . . . , s_n.
4. Write the n scalar equations of the form

$$e^{ts_i} = u_0 + u_1 s_i + \cdots + u_{n-1} s_i^{n-1} \qquad [50]$$

5. Solve for the n time functions u_0, u_1, . . . , u_{n-1}.
6. Substitute these time functions in Eq. [48] to obtain the $(n \times n)$ matrix $e^{t\mathbf{a}}$.

To illustrate this procedure, let us use the system matrix that corresponds to Eqs. [7] and [8] of Sec. 18.1 and to the circuit shown in Fig. 18.2*a*.

$$\mathbf{a} = \begin{bmatrix} -0.5 & -2.5 \\ 0.5 & -3.5 \end{bmatrix}$$

Therefore,

$$(\mathbf{a} - s\mathbf{I}) = \begin{bmatrix} -0.5 & -2.5 \\ 0.5 & -3.5 \end{bmatrix} - s\begin{bmatrix} 1 & 0 \\ 0 & 1 \end{bmatrix}$$

$$= \begin{bmatrix} (-0.5 - s) & -2.5 \\ 0.5 & (-3.5 - s) \end{bmatrix}$$

and the expansion of the corresponding (2×2) determinant gives

$$\det\begin{bmatrix} (-0.5 - s) & -2.5 \\ 0.5 & (-3.5 - s) \end{bmatrix} = \begin{vmatrix} (-0.5 - s) & -2.5 \\ 0.5 & (-3.5 - s) \end{vmatrix}$$

$$= (-0.5 - s)(-3.5 - s) + 1.25$$

so that

$$\det(\mathbf{a} - s\mathbf{I}) = s^2 + 4s + 3$$

The roots of this polynomial are $s_1 = -1$ and $s_2 = -3$ and we substitute each of these values in Eq. [50], obtaining the two equations

$$e^{-t} = u_0 - u_1 \quad \text{and} \quad e^{-3t} = u_0 - 3u_1$$

Subtracting, we find that

$$u_1 = 0.5e^{-t} - 0.5e^{-3t}$$

and, therefore,

$$u_0 = 1.5e^{-t} - 0.5e^{-3t}$$

Note that each u_i has the general form of the natural response.

When these two functions are installed in Eq. [48], so that

$$e^{t\mathbf{a}} = (1.5e^{-t} - 0.5e^{-3t})\mathbf{I} + (0.5e^{-t} - 0.5e^{-3t})\begin{bmatrix} -0.5 & -2.5 \\ 0.5 & -3.5 \end{bmatrix}$$

we may carry out the indicated operations to find

$$e^{t\mathbf{a}} = \begin{bmatrix} (1.25e^{-t} - 0.25e^{-3t})(-1.25e^{-t} + 1.25e^{-3t}) \\ (0.25e^{-t} - 0.25e^{-3t})(-0.25e^{-t} + 1.25e^{-3t}) \end{bmatrix} \qquad [51]$$

Having $e^{t\mathbf{a}}$, we form $e^{-t\mathbf{a}}$ by replacing each t in Eq. [51] with $-t$. To complete this example, we may identify $\mathbf{q}$ and $\mathbf{f}$ from Eqs. [7], [8], and [9]:

$$\mathbf{q} = \begin{bmatrix} v_C \\ i_L \end{bmatrix}$$

$$\mathbf{f} = \begin{bmatrix} 30 + 8e^{-2t}u(t) \\ 0 \end{bmatrix} \qquad [52]$$

Since part of the forcing-function vector is present for $t < 0$, we may as well use Eq. [46] to solve for the state vector:

$$\mathbf{q} = e^{t\mathbf{a}} \int_{-\infty}^{t} e^{-z\mathbf{a}}\mathbf{f}(z)dz \qquad [46]$$

The matrix product $e^{-z\mathbf{a}}\mathbf{f}(z)$ is next formed by replacing t with $-z$ in Eq. [51] and then postmultiplying by Eq. [52] with z replacing t. With only moderate labor, we find that

$$e^{-z\mathbf{a}}\mathbf{f} = \begin{bmatrix} 37.5e^{z} - 7.5e^{3z} + (10e^{-z} - 2e^{z})u(z) \\ 7.5e^{z} - 7.5e^{3z} + (2e^{-z} - 2e^{z})u(z) \end{bmatrix}$$

Integrating the first two terms of each element from $-\infty$ to t and the last two terms from 0 to t, we have

$$\int_{-\infty}^{t} e^{-z\mathbf{a}}\mathbf{f}\,dz = \begin{bmatrix} 35.5e^{t} - 2.5e^{3t} - 10e^{-t} + 12 \\ 5.5e^{t} - 2.5e^{3t} - 2e^{-t} + 4 \end{bmatrix} \quad (t > 0)$$

Finally, we must premultiply this matrix by Eq. [51], and this matrix multiplication involves a large number of scalar multiplications and algebraic additions of like time functions. The result is the desired state vector

$$\mathbf{q} = \begin{bmatrix} 35 + 10e^{-t} - 12e^{-2t} + 2e^{-3t} \\ 5 + 2e^{-t} - 4e^{-2t} + 2e^{-3t} \end{bmatrix} \quad (t > 0)$$

These are the expressions that were given offhandedly in Eqs. [10] and [11] at the end of Sec. 18.1.

We now have a general technique that we could apply to higher-order problems. Such a procedure is theoretically possible, but the labor involved soon becomes monumental. Instead, we utilize computer programs that work directly from the normal-form equations and the initial values of the state variables. Knowing $\mathbf{q}(0)$, we compute $\mathbf{f}(0)$ and solve the equations for $\mathbf{q}'(0)$. Then $\mathbf{q}(\Delta t)$ may be approximated from the initial value and its derivative: $\mathbf{q}(\Delta t) = \mathbf{q}(0) + q'(0)\Delta t$. With this value for $\mathbf{q}(\Delta t)$, the normal-form equations are used to determine a value for $\mathbf{q}'(\Delta t)$, and the process moves along by time increments of Δt. Smaller values for Δt lead to greater accuracy for $\mathbf{q}'(t)$ at any time $t > 0$, but with attendant penalties in computational time and computer storage requirements.

We accomplished several things in this chapter. First, and perhaps most important, we have learned some of the terms and ideas of this branch of system analysis, and this should make future study of this field more meaningful and pleasurable.

Another accomplishment is the general solution for the first-order case in matrix form.

We have also obtained the matrix solution for the general case. This introduction to the use of matrices in circuit and system analysis is a tool which becomes increasingly necessary in more advanced work in these areas.

Finally, we have indicated how a numerical solution might be obtained using numerical methods, hopefully by a digital computer.

PRACTICE

18.8 Given $\mathbf{q} = \begin{bmatrix} v_C \\ i_L \end{bmatrix}$, $\mathbf{a} = \begin{bmatrix} 0 & -27 \\ \frac{1}{3} & -10 \end{bmatrix}$, $\mathbf{f} = \begin{bmatrix} 243u(t) \\ 40u(t) \end{bmatrix}$, and $\mathbf{q}(0) = \begin{bmatrix} 150 \\ 5 \end{bmatrix}$, use a Δt of 0.001 s and calculate: *(a)* $v_C(0.001)$; *(b)* $i_L(0.001)$; *(c)* $v'_C(0.001)$; *(d)* $i'_L(0.001)$; *(e)* $v_C(0.002)$; *(f)* $i_L(0.002)$.

Ans: 150.108 V; 5.04 A; 106.92 V/s; 39.636 A/s; 150.2149 V; 5.0796 A.

SUMMARY AND REVIEW

- The term *state variable* in circuit analysis refers to either an inductor current or a capacitor voltage, as they can be used to describe the *energy state* of a system (i.e., a circuit).
- The six steps for constructing a set of equations for state-variable analysis are:
  1. Establish a normal tree.
  2. Assign voltage and current variables.
  3. Write the C equations.
  4. Write the L equations.
  5. Write the R equations (if necessary).
  6. Write the normal-form equations.
- The *state vector* is a matrix which contains the state variables q_i.
- The *forcing-function matrix* is a matrix which contains the set of forcing functions f_i.
- The *system matrix* **a** contains the coefficients a_{ij} of the state equations.

- If we write the first-order matrix equation representation of the set of normal-form state variable equations as $\mathbf{q}' = \mathbf{aq} + \mathbf{f}$, the single unknown state variable $q_1(t)$ is given as the sum of the zero-input response $e^{ta_{11}}q_1(0)$ and the zero-state response $e^{ta_{11}}\int_0^t e^{-za_{11}}f_1(z)dz$.
- The general solution to the matrix equation, expressed in terms of the state-transition matrix e^{ta}, is $\mathbf{q} = e^{t\mathbf{a}}\int_{-\infty}^{t} e^{-z\mathbf{a}}\mathbf{f1}(z)dz$.

EXERCISES

18.1 State Variables and Normal-Form Equations

1. Using the order i_1, i_2, i_3, write the following equations as a set of normal-form equations: $-2i_1' - 6i_3' = 5 + 2\cos 10t - 3i_1 + 2i_2$, $4i_2 = 0.05i_1' - 0.15i_2' + 0.25i_3'$, $i_2 = -2i_1 - 5i_3 + 0.4\int_0^t (i_1 - i_3)dt + 8$.
2. Given the two linear differential equations $x' + y' = x + y + 1$ and $x' - 2y' = 2x - y - 1$: (*a*) write the two normal-form equations, using the order x, y; (*b*) obtain a single differential equation involving only $x(t)$ and its derivatives. (*c*) If $x(0) = 2$ and $y(0) = -5$ find $x'(0)$, $x''(0)$, and $x'''(0)$.
3. Write the following equations in normal form, using the order x, y, z: $x' - 2y - 3z' = f_1(t)$, $2x' + 5z = 3$, $z' - 2y' - x = 0$.
4. If $x' = -2x - 3y + 4$ and $y' = 5x - 6y + 7$, let $x(0) = 2$ and $y(0) = \frac{1}{3}$, and find: (*a*) $x''(0)$; (*b*) $y''(0)$; (*c*) $y'''(0)$.

18.2 Writing a Set of Normal-Form Equations

5. If $v_s = 100\cos 120\pi t$ V, write a set of normal-form equations for the circuit shown in Fig. 18.15. Use i_L and v_C as the state variables.

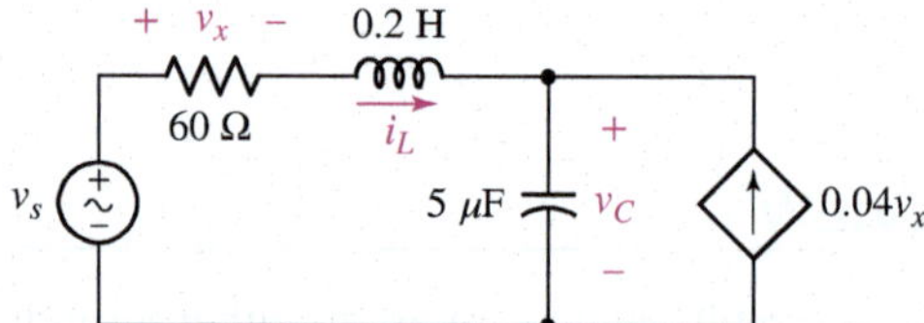

FIGURE 18.15

6. (*a*) Draw a normal tree for the circuit of Fig. 18.16 and assign the necessary state variables (for uniformity, put + references at the top or left side of an element, and direct arrows down or to the right). (*b*) Specify every link current and tree-branch voltage in terms of the sources, element values, and state variables. (*c*) Write the normal-form equations using the order i_L, v_x, v_{2F}.

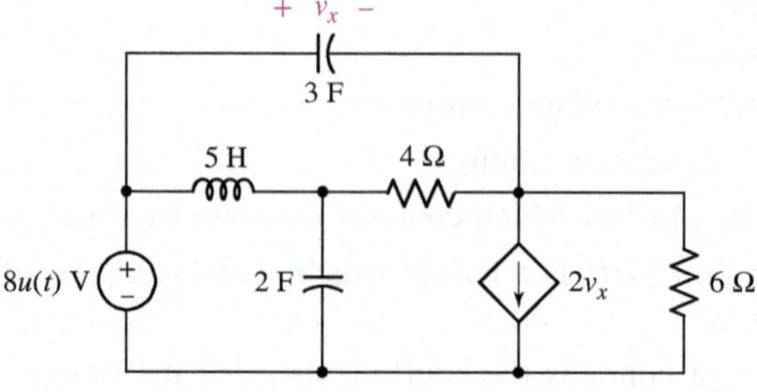

FIGURE 18.16

7. Write a set of normal-form equations for each circuit shown in Fig. 18.17. Use the state-variable order given below the circuit.

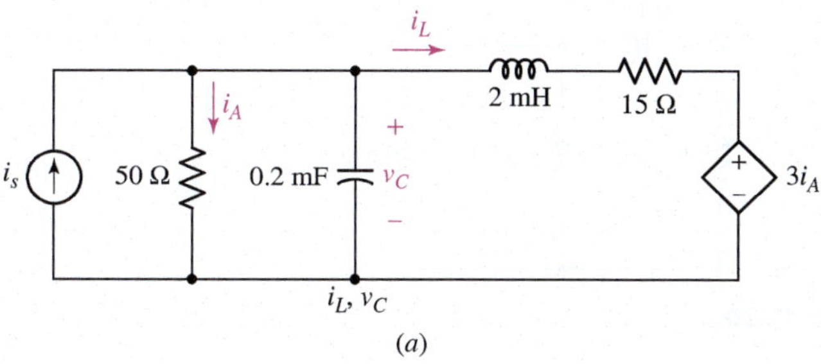

(a)

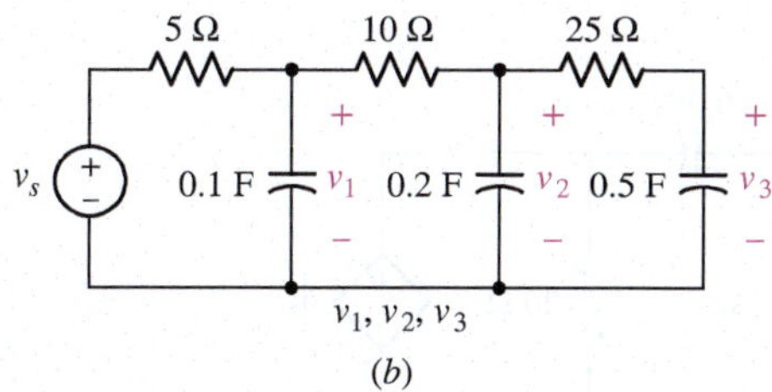

(b)

■ **FIGURE 18.17**

8. A 12.5 Ω resistor is placed in series with the 0.2 mF capacitor in the circuit of Fig. 18.17*a*. Using one current and one voltage, in that order, write a set of normal-form equations for this circuit.

9. (*a*) Write normal-form equations in the state variables v_1, v_2, i_1, and i_2 for Fig. 18.18. (*b*) Repeat if the current source is replaced with a 3 Ω.

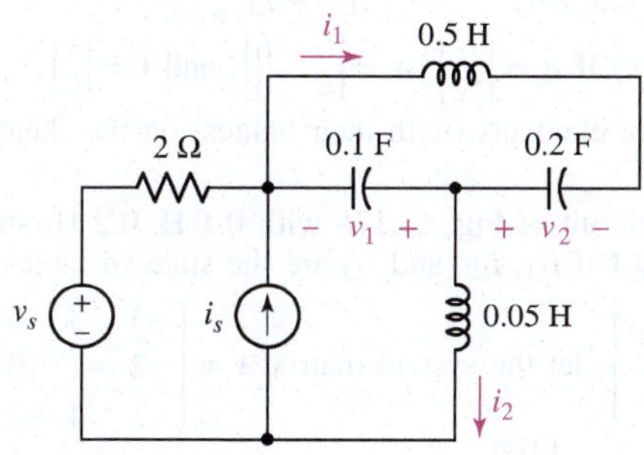

■ **FIGURE 18.18**

10. Write a set of normal-form equations in the order i_L, v_C for each circuit shown in Fig. 18.19.

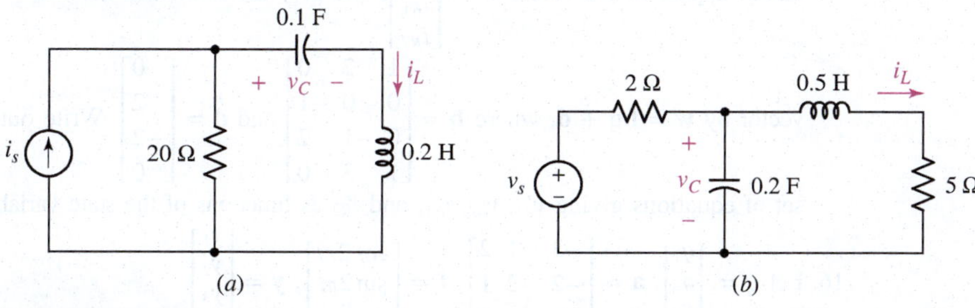

■ **FIGURE 18.19**

11. Write a set of normal-form equations for the circuit shown in Fig. 18.20. Use the variable order v, i.

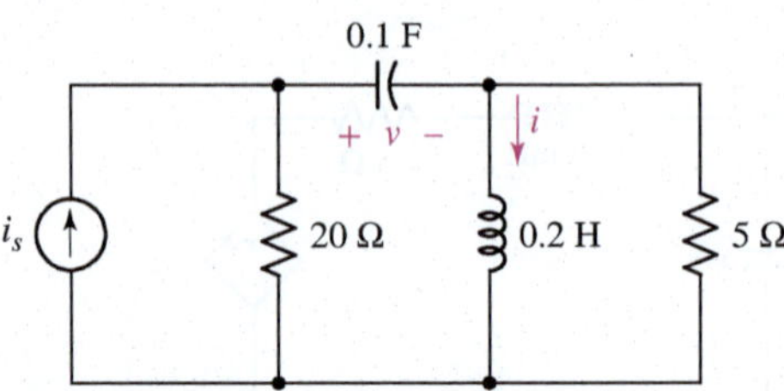

■ **FIGURE 18.20**

12. If $v_s = 5 \sin 2t\, u(t)$ V, write a set of normal-form equations for the circuit of Fig. 18.21.

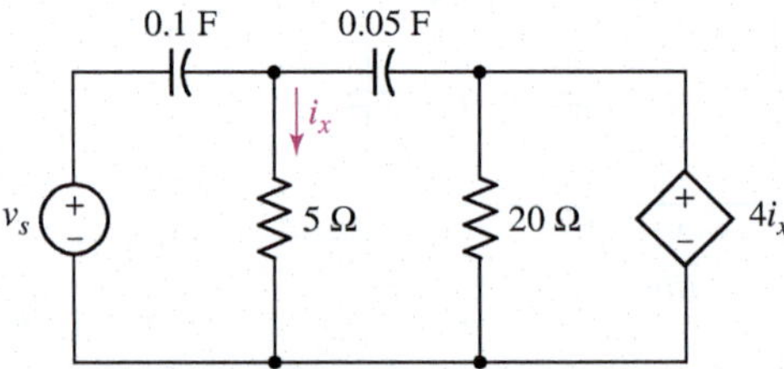

■ **FIGURE 18.21**

18.3 The Use of Matrix Notation

■ **FIGURE 18.22**

13. (*a*) Given $\mathbf{q} = \begin{bmatrix} i_{L1} \\ i_{L2} \\ v_C \end{bmatrix}$, $\mathbf{a} = \begin{bmatrix} -1 & -2 & -3 \\ 4 & -5 & 6 \\ 7 & -8 & -9 \end{bmatrix}$, and $\mathbf{f} = \begin{bmatrix} 2t \\ 3t^2 \\ 1+t \end{bmatrix}$, write out the three normal-form equations. (*b*) If $\mathbf{q} = \begin{bmatrix} i_L \\ v_C \end{bmatrix}$, $\mathbf{a} = \begin{bmatrix} 0 & -6 \\ 4 & 0 \end{bmatrix}$, and $\mathbf{f} = \begin{bmatrix} 1 \\ 0 \end{bmatrix}$, show the state variables and the elements (with their values) on the skeleton circuit diagram of Fig. 18.22.

14. Replace the capacitors in the circuit of Fig. 18.17*b* with 0.1 H, 0.2 H, and 0.5 H inductors and find **a** and **f** if i_{L1}, i_{L2}, and i_{L3} are the state variables.

15. Given the state vector $\mathbf{q} = \begin{bmatrix} v_{C1} \\ v_{C2} \\ i_L \end{bmatrix}$, let the system matrix $\mathbf{a} = \begin{bmatrix} -3 & 3 & 0 \\ 2 & -1 & 0 \\ 1 & 4 & -2 \end{bmatrix}$, and the forcing function vector $\mathbf{f} = \begin{bmatrix} 10 \\ 0 \\ 0 \end{bmatrix}$. Other voltages and currents in the circuit appear in the vector $\mathbf{w} = \begin{bmatrix} v_{o1} \\ v_{o2} \\ i_{R1} \\ i_{R2} \end{bmatrix}$, and they are related to the state vector by $\mathbf{w} = \mathbf{bq} + \mathbf{d}$, where $\mathbf{b} = \begin{bmatrix} 1 & 2 & 0 \\ 0 & 0 & -1 \\ 0 & -1 & 2 \\ 1 & 3 & 0 \end{bmatrix}$ and $\mathbf{d} = \begin{bmatrix} 0 \\ 2 \\ -2 \\ 0 \end{bmatrix}$. Write out the set of equations giving v'_{o1}, v'_{o2}, i'_{R1}, and i'_{R2} as functions of the state variables.

16. Let $\mathbf{q} = \begin{bmatrix} q_1 \\ q_2 \\ q_3 \end{bmatrix}$, $\mathbf{a} = \begin{bmatrix} -3 & 1 & 2 \\ -2 & -2 & 1 \\ -1 & 3 & 0 \end{bmatrix}$, $\mathbf{f} = \begin{bmatrix} \cos 2\pi t \\ \sin 2\pi t \\ 0 \end{bmatrix}$, $\mathbf{y} = \begin{bmatrix} y_1 \\ y_2 \\ y_3 \\ y_3 \end{bmatrix}$,

$\mathbf{b} = \begin{bmatrix} 1 & 2 & 0 & 3 \\ 0 & -1 & 1 & 1 \\ 2 & -1 & -1 & 3 \end{bmatrix}$, and $\mathbf{d} = \begin{bmatrix} 2 \\ 1 \\ 3 \end{bmatrix}$. Then if $\mathbf{q}' = \mathbf{aq} + \mathbf{f}$ and $\mathbf{q} = \mathbf{by} + \mathbf{d}$, determine $\mathbf{q}(0)$ and $\mathbf{q}'(0)$ if $\mathbf{y}(0) = \begin{bmatrix} 10 \\ -10 \\ -5 \\ 5 \end{bmatrix}$.

18.4 Solution of the First-Order Equation

17. Find **a** and **f** for the circuit shown in Fig. 18.23 if $\mathbf{q} = \begin{bmatrix} v_C \\ i_L \end{bmatrix}$.

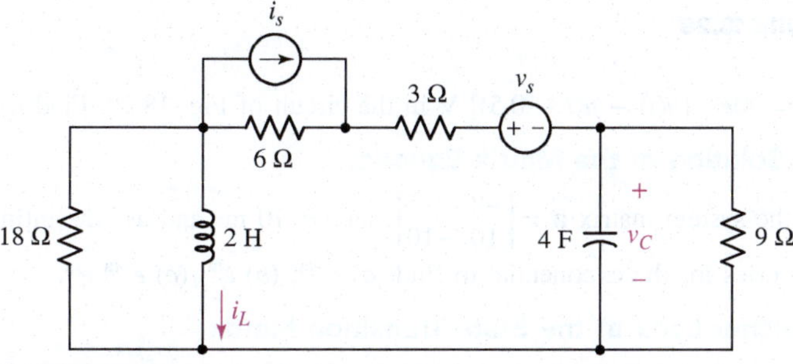

■ **FIGURE 18.23**

18. For the circuit shown in Fig. 18.24: (*a*) write the normal-form equation for i, $t > 0$; (*b*) solve this equation for i; (*c*) identify the zero-state and zero-input responses; (*d*) find i by the methods of Chap. 8 and specify the natural and forced responses.

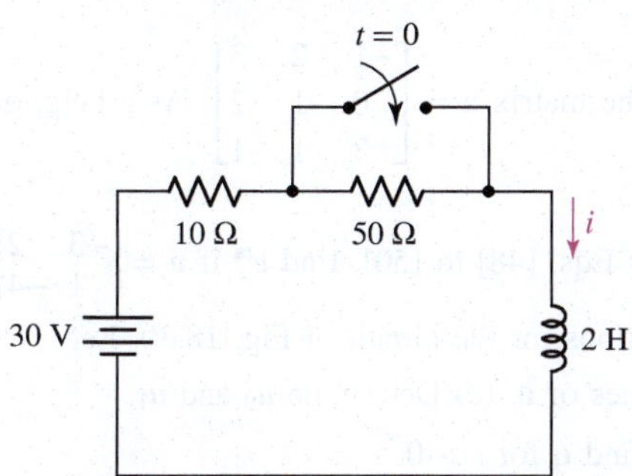

■ **FIGURE 18.24**

19. Let $v_s = 2tu(t)$ V in the circuit shown in Fig. 18.25. Find $i_2(t)$.

20. (*a*) Use state-variable methods to find $i_L(t)$ for all t in the circuit shown in Fig. 18.26. (*b*) Identify the zero-input, zero-state, natural, and forced responses.

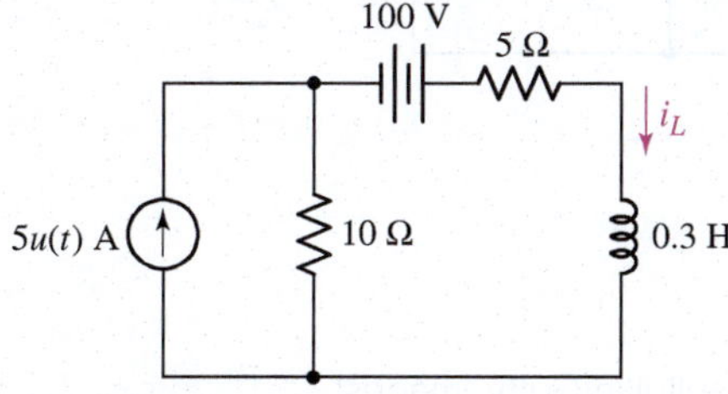

■ **FIGURE 18.26**

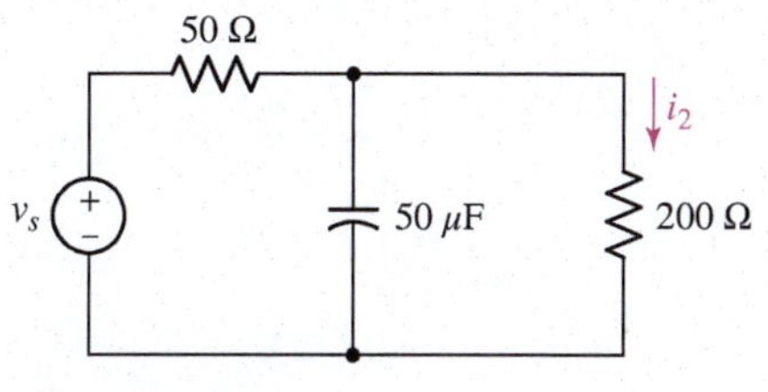

■ **FIGURE 18.25**

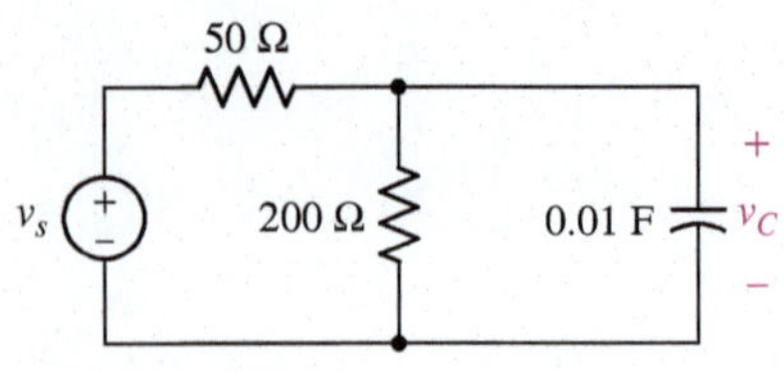

■ **FIGURE 18.27**

21. Let $v_s = 100[u(t) - u(t - 0.5)] \cos \pi t$ V in Fig. 18.27. Find v_C for $t > 0$.
22. (*a*) Write the normal-form equation for the circuit shown in Fig. 18.28. (*b*) Find $v(t)$ for $t < 0$ and $t > 0$. (*c*) Identify the forced response, the natural response, the zero-state response, and the zero-input response.

■ **FIGURE 18.28**

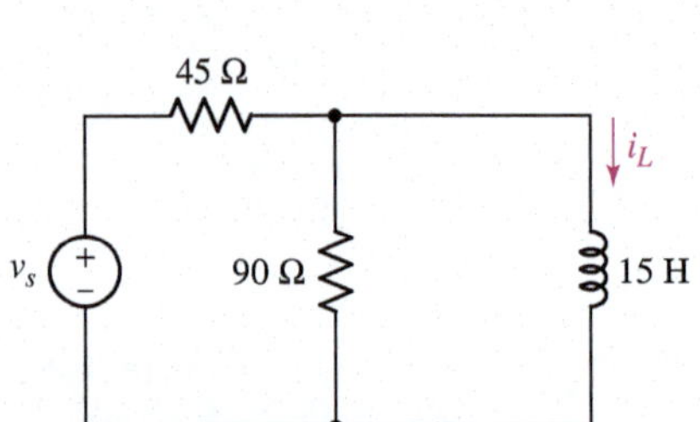

■ **FIGURE 18.29**

23. Let $v_s = 90e^{-t}[u(t) - u(t - 0.5)]$ V in the circuit of Fig. 18.29. Find $i_L(t)$.

18.5 The Solution of the Matrix Equation

24. Given the system matrix $\mathbf{a} = \begin{bmatrix} -8 & 5 \\ 10 & -10 \end{bmatrix}$, let $t = 10$ ms and use the infinite power series for the exponential to find (*a*) $e^{-t\mathbf{a}}$; (*b*) $e^{t\mathbf{a}}$; (*c*) $e^{-t\mathbf{a}}\, e^{t\mathbf{a}}$.

18.6 A Further Look at the State-Transition Matrix

25. (*a*) If $\mathbf{q} = \begin{bmatrix} x \\ y \\ z \end{bmatrix}$, $\mathbf{f} = \begin{bmatrix} u(t) \\ \cos t \\ -u(t) \end{bmatrix}$, and $\mathbf{a} = \begin{bmatrix} 1 & 2 & -1 \\ 0 & -1 & 3 \\ -2 & -3 & -1 \end{bmatrix}$, write the set of normal-form equations. (*b*) If $\mathbf{q}(0) = \begin{bmatrix} 2 \\ -3 \\ 1 \end{bmatrix}$, estimate $q(0.1)$ using $\Delta t = 0.1$. (*c*) Repeat for $\Delta t = 0.05$.
26. Determine the eigenvalues of the matrix $\mathbf{a} = \begin{bmatrix} -1 & 2 & 3 \\ 0 & -1 & 2 \\ 3 & 1 & -1 \end{bmatrix}$. As a help, one root is near $s = -3.5$.
27. Using the method described by Eqs. [48] to [50], find $e^{t\mathbf{a}}$ if $\mathbf{a} = \begin{bmatrix} -3 & 2 \\ 1 & -4 \end{bmatrix}$.
28. (*a*) Find the normal-form equations for the circuit of Fig. 18.30. Let $\mathbf{q} = \begin{bmatrix} i \\ v \end{bmatrix}$. (*b*) Find the eigenvalues of $\mathbf{a}$. (*c*) Determine u_0 and u_1. (*d*) Specify $e^{t\mathbf{a}}$. (*e*) Use $e^{t\mathbf{a}}$ to find $\mathbf{q}$ for $t > 0$.

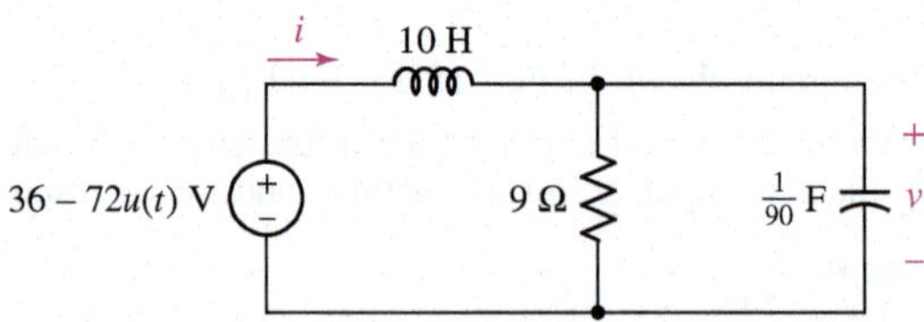

■ **FIGURE 18.30**

AN INTRODUCTION TO NETWORK TOPOLOGY

After working a good number of circuits problems, it slowly becomes evident that many of the circuits we encounter have a lot in common, at least in terms of the arrangement of components. From this realization, it is possible to create a more abstract view of circuits which we call *network topology,* a subject we introduce in this appendix.

A1.1 TREES AND GENERAL NODAL ANALYSIS

We now plan to generalize the method of nodal analysis that we have come to know and love. Since nodal analysis is applicable to any network, we cannot promise that we will be able to solve a wider class of circuit problems. We can, however, look forward to being able to select a general nodal analysis method for any particular problem that may result in fewer equations and less work.

We must first extend our list of definitions relating to network topology. We begin by defining *topology* itself as a branch of geometry which is concerned with those properties of a geometrical figure which are unchanged when the figure is twisted, bent, folded, stretched, squeezed, or tied in knots, with the provision that no parts of the figure are to be cut apart or to be joined together. A sphere and a tetrahedron are topologically identical, as are a square and a circle. In terms of electric circuits, then, we are not now concerned with the particular types of elements appearing in the circuit, but only with the way in which branches and nodes are arranged. As a matter of fact, we usually suppress the nature of the elements and simplify the drawing of the circuit by showing the elements as lines. The resultant drawing is called a linear graph, or simply a graph. A circuit and its graph are shown in Fig. A1.1. Note that all nodes are identified by heavy dots in the graph.

Since the topological properties of the circuit or its graph are unchanged when it is distorted, the three graphs shown in Fig. A1.2 are all topologically identical with the circuit and graph of Fig. A1.1.

Topological terms that we already know and have been using correctly are

Node: A point at which two or more elements have a common connection.

Path: A set of elements that may be traversed in order without passing through the same node twice.

Branch: A single path, containing one simple element, which connects one node to any other node.

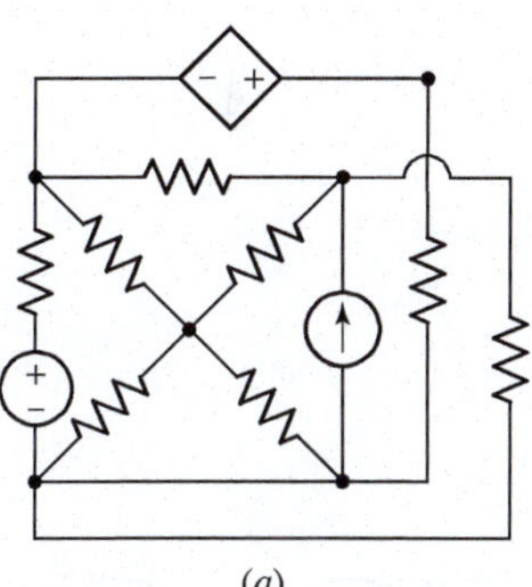

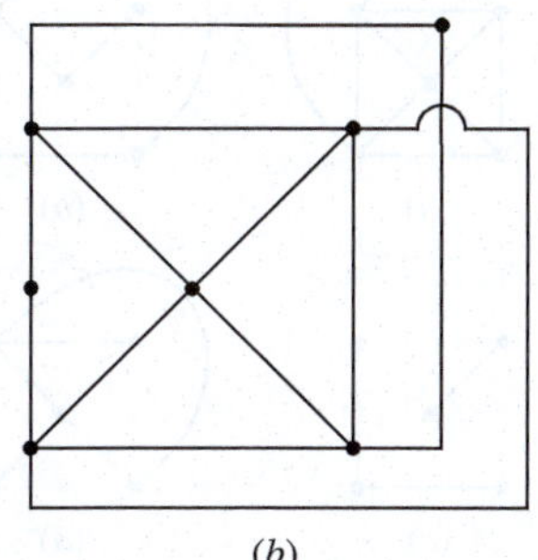

FIGURE A1.1 (*a*) A given circuit. (*b*) The linear graph of this circuit.

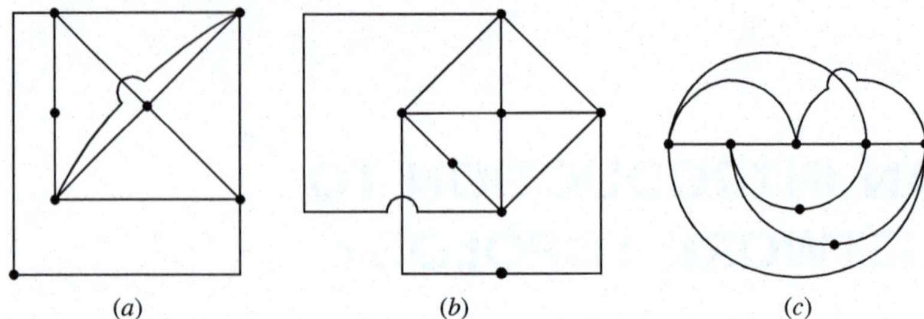

FIGURE A1.2 (*a, b, c*) Alternative linear graphs of the circuit of Fig. A1.1.

Loop: A closed path.
Mesh: A loop which does not contain any other loops within it.
Planar circuit: A circuit which may be drawn on a plane surface in such a way that no branch passes over or under any other branch.
Nonplanar circuit: Any circuit which is not planar.

The graphs of Fig. A1.2 each contain 12 branches and 7 nodes.

Three new properties of a linear graph must now be defined—a *tree,* a *cotree,* and a *link*. We define a *tree* as any set of branches which does not contain any loops and yet connects every node to every other node, not necessarily directly. There are usually a number of different trees which may be drawn for a network, and the number increases rapidly as the complexity of the network increases. The simple graph shown in Fig. A1.3*a* has eight possible trees, four of which are shown by heavy lines in Fig. A1.3*b*, *c*, *d*, and *e*.

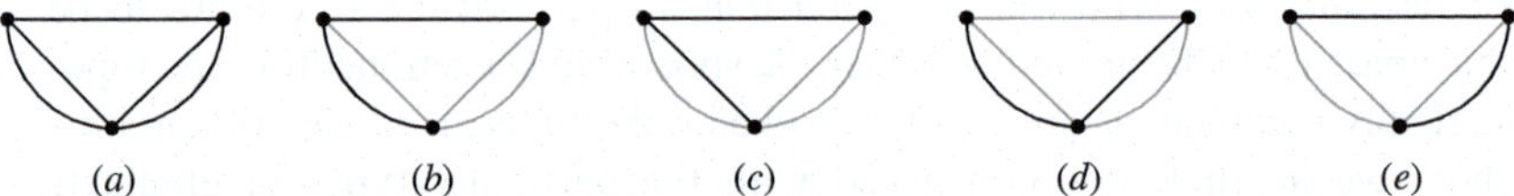

FIGURE A1.3 (*a*) The linear graph of a three-node network. (*b, c, d, e*) Four of the eight different trees which may be drawn for this graph are shown by the black lines.

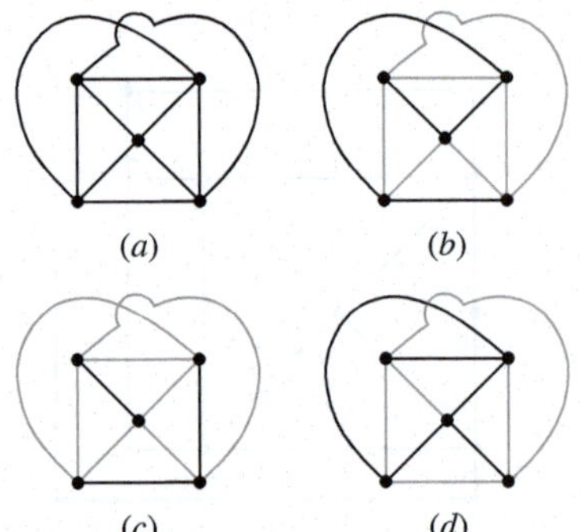

FIGURE A1.4 (*a*) A linear graph. (*b*) A possible tree for this graph. (*c, d*) These sets of branches do not satisfy the definition of a tree.

In Fig. A1.4*a* a more complex graph is shown. Figure A1.4*b* shows one possible tree, and Fig. A1.4*c* and *d* show sets of branches which are not trees because neither set satisfies the definition.

After a tree has been specified, those branches that are not part of the tree form the *cotree,* or complement of the tree. The lightly drawn branches in Fig. A1.3*b* to *e* show the cotrees that correspond to the heavier trees.

Once we understand the construction of a tree and its cotree, the concept of the *link* is very simple, for a link is any branch belonging to the cotree. It is evident that any particular branch may or may not be a link, depending on the particular tree which is selected.

The number of links in a graph may easily be related to the number of branches and nodes. If the graph has N nodes, then exactly $(N - 1)$ branches are required to construct a tree because the first branch chosen connects two nodes and each additional branch includes one more node.

Thus, given B branches, the number of links L must be

$$L = B - (N - 1)$$

or

$$L = B - N + 1 \qquad [1]$$

There are L branches in the cotree and $(N - 1)$ branches in the tree.

In any of the graphs shown in Fig. A1.3, we note that $3 = 5 - 3 + 1$, and in the graph of Fig. A1.4*b*, $6 = 10 - 5 + 1$. A network may be in several disconnected parts, and Eq. [1] may be made more general by replacing $+1$ with $+S$, where S is the number of separate parts. However, it is also possible to connect two separate parts by a single conductor, thus causing two nodes to form one node; no current can flow through this single conductor. This process may be used to join any number of separate parts, and thus we will not suffer any loss of generality if we restrict our attention to circuits for which $S = 1$.

We are now ready to discuss a method by which we may write a set of nodal equations that are independent and sufficient. The method will enable us to obtain many different sets of equations for the same network, and all the sets will be valid. However, the method does not provide us with every possible set of equations. Let us first describe the procedure, illustrate it by three examples, and then point out the reason that the equations are independent and sufficient.

Given a network, we should:

1. Draw a graph and then identify a tree.
2. Place all voltage sources in the tree.
3. Place all current sources in the cotree.
4. Place all control-voltage branches for voltage-controlled dependent sources in the tree, if possible.
5. Place all control-current branches for current-controlled dependent sources in the cotree, if possible.

The last four steps effectively associate voltages with the tree and currents with the cotree.

We now assign a voltage variable (with its plus-minus pair) across each of the $(N - 1)$ branches in the tree. A branch containing a voltage source (dependent or independent) should be assigned that source voltage, and a branch containing a controlling voltage should be assigned that controlling voltage. The number of new variables that we have introduced is therefore equal to the number of branches in the tree $(N - 1)$, reduced by the number of voltage sources in the tree, and reduced also by the number of control voltages we were able to locate in the tree. In Example A1.3, we will find that the number of new variables required may be zero.

Having a set of variables, we now need to write a set of equations that are sufficient to determine these variables. The equations are obtained through the application of KCL. Voltage sources are handled in the same way that they were in our earlier attack on nodal analysis; each voltage source and the two nodes at its terminals constitute a supernode or a part of a supernode. Kirchhoff's current law is then applied at all but one of the remaining nodes and supernodes. We set the sum of the currents leaving the node in all of the branches connected to it equal to zero. Each current is expressed in terms of the voltage variables we just assigned. One node may be ignored, just as was the case earlier for the reference node. Finally, in case there are current-controlled dependent sources, we must write an equation for each control current that relates it to the voltage variables; this also is no different from the procedure used before with nodal analysis.

Let us try out this process on the circuit shown in Fig. A1.5*a*. It contains four nodes and five branches, and its graph is shown in Fig. A1.5*b*.

EXAMPLE A1.1

Find the value of v_x in the circuit of Fig. A1.5*a*.

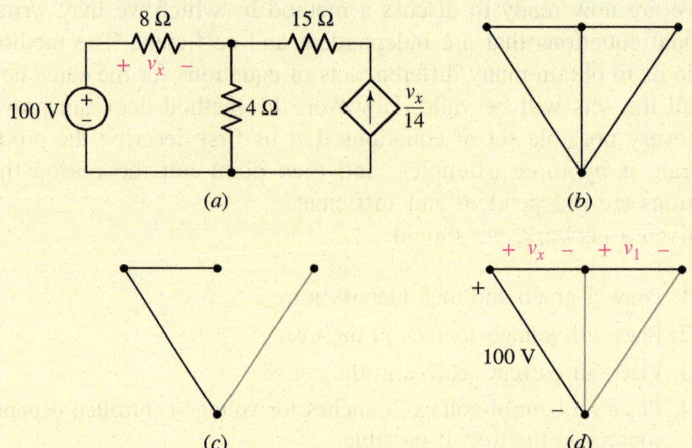

FIGURE A1.5 (*a*) A circuit used as an example for general nodal analysis. (*b*) The graph of the given circuit. (*c*) The voltage source and the control voltage are placed in the tree, while the current source goes in the cotree. (*d*) The tree is completed and a voltage is assigned across each tree branch.

In accordance with steps 2 and 3 of the tree-drawing procedure, we place the voltage source in the tree and the current source in the cotree. Following step 4, we see that the v_x branch may also be placed in the tree, since it does not form any loop which would violate the definition of a tree. We have now arrived at the two tree branches and the single link shown in Fig. A1.5*c*, and we see that we do not yet have a tree, since the right node is not connected to the others by a path through tree branches. The only possible way to complete the tree is shown in Fig. A1.5*d*. The 100 V source voltage, the control voltage v_x, and a new voltage variable v_1 are next assigned to the three tree branches as shown.

We therefore have two unknowns, v_x and v_1, and we need to obtain two equations in terms of them. There are four nodes, but the presence of the voltage source causes two of them to form a single supernode. Kirchhoff's current law may be applied at any two of the three remaining nodes or supernodes. Let's attack the right node first. The current leaving to the left is $-v_1/15$, while that leaving downward is $-v_x/14$. Thus, our first equation is

$$-\frac{v_1}{15} + \frac{-v_x}{14} = 0$$

The central node at the top looks easier than the supernode, and so we set the sum of the current to the left ($-v_x/8$), the current to the right ($-v_1/15$), and the downward current through the 4 Ω resistor equal to zero. This latter current is given by the voltage across the resistor divided by 4 Ω, but there is no voltage labeled on that link. However, when a tree is constructed according to the definition, there is a path through it from any node to any other node. Then, since every branch in the tree is assigned a voltage, we may express the voltage across any link in terms of the tree-branch voltages. This downward current is therefore $(-v_x + 100)/4$, and we have the second equation,

$$-\frac{v_x}{8} + \frac{v_1}{15} + \frac{-v_x + 100}{4} = 0$$

The simultaneous solution of these two nodal equations gives

$$v_1 = -60 \text{ V} \qquad v_x = 56 \text{ V}$$

EXAMPLE A1.2

Find the values of v_x and v_y in the circuit of Fig. A1.6*a*.

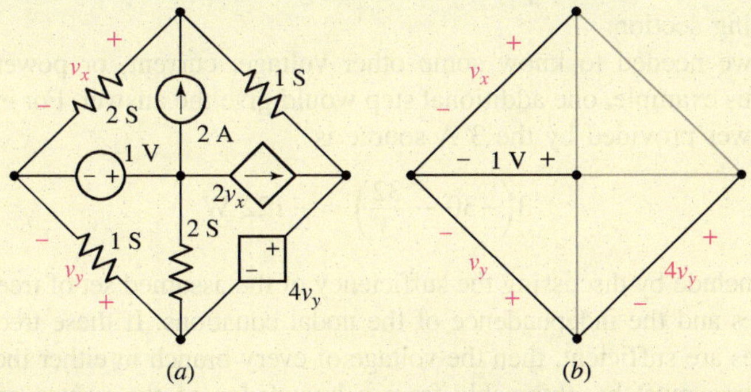

■ **FIGURE A1.6** (*a*) A circuit with five nodes. (*b*) A tree is chosen such that both voltage sources and both control voltages are tree branches.

We draw a tree so that both voltage sources and both control voltages appear as tree-branch voltages and, hence, as assigned variables. As it happens, these four branches constitute a tree, Fig. A1.6*b*, and tree-branch voltages v_x, 1, v_y, and $4v_y$ are chosen, as shown.

(Continued on next page)

Both voltage sources define supernodes, and we apply KCL twice, once to the top node,

$$2v_x + 1(v_x - v_y - 4v_y) = 2$$

and once to the supernode consisting of the right node, the bottom node, and the dependent voltage source,

$$1v_y + 2(v_y - 1) + 1(4v_y + v_y - v_x) = 2v_x$$

Instead of the four equations we would expect using previously studied techniques, we have only two, and we find easily that $v_x = \frac{26}{9}$ V and $v_y = \frac{4}{3}$ V.

EXAMPLE A1.3

Find the value of v_x in the circuit of Fig. A1.7*a*.

The two voltage sources and the control voltage establish the three-branch tree shown in Fig. A1.7*b*. Since the two upper nodes and the lower right node all join to form one supernode, we need write only one KCL equation. Selecting the lower left node, we have

$$-1 - \frac{v_x}{4} + 3 + \frac{-v_x + 30 + 6v_x}{5} = 0$$

and it follows that $v_x = -\frac{32}{3}$ V. In spite of the apparent complexity of this circuit, the use of general nodal analysis has led to an easy solution. Employing mesh currents or node-to-reference voltages would require more equations and more effort.

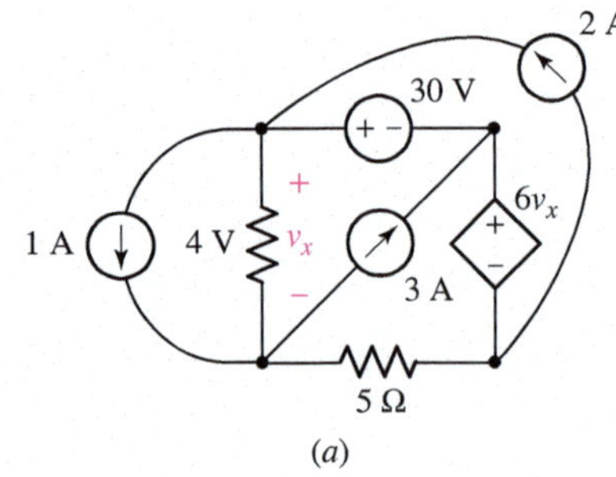

(*a*)

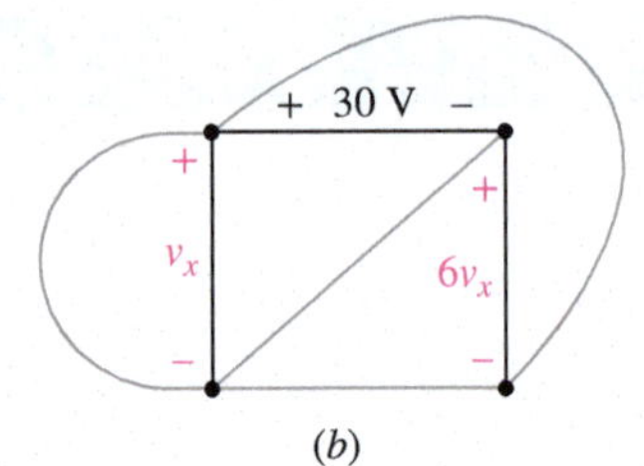

(*b*)

FIGURE A1.7 (*a*) A circuit for which only one general nodal equation need be written. (*b*) The tree and the tree-branch voltages used.

We will discuss the problem of finding the best analysis scheme in the following section.

If we needed to know some other voltage, current, or power in the previous example, one additional step would give the answer. For example, the power provided by the 3 A source is

$$3\left(-30 - \frac{32}{3}\right) = -122 \text{ W}$$

We conclude by discussing the sufficiency of the assumed set of tree-branch voltages and the independence of the nodal equations. If these tree-branch voltages are sufficient, then the voltage of every branch in either the tree or the cotree must be obtainable from a knowledge of the values of all the tree-branch voltages. This is certainly true for those branches in the tree. For the links we know that each link extends between two nodes, and, by definition, the tree must also connect those two nodes. Hence, every link voltage may also be established in terms of the tree-branch voltages.

Once the voltage across every branch in the circuit is known, then all the currents may be found by using either the given value of the current if the branch consists of a current source, by using Ohm's law if it is a resistive branch, or by using KCL and these current values if the branch happens to be a voltage source. Thus, all the voltages and currents are determined, and sufficiency is demonstrated.

To demonstrate independency, let us satisfy ourselves by assuming the situation where the only sources in the network are independent current sources. As we have noticed earlier, independent voltage sources in the circuit result in fewer equations, while dependent sources usually necessitate a greater number of equations. With independent current sources only, there will then be precisely $(N - 1)$ nodal equations written in terms of $(N - 1)$ tree-branch voltages. To show that these $(N - 1)$ equations are independent, visualize the application of KCL to the $(N - 1)$ different nodes. Each time we write the KCL equation, there is a new tree branch involved—the one which connects that node to the remainder of the tree. Since that circuit element has not appeared in any previous equation, we must obtain an independent equation. This is true for each of the $(N - 1)$ nodes in turn, and hence we have $(N - 1)$ independent equations.

PRACTICE

A1.1 (*a*) How many trees may be constructed for the circuit of Fig. A1.8 that follow all five of the tree-drawing suggestions listed earlier? (*b*) Draw a suitable tree, write two equations in two unknowns, and find i_3. (*c*) What power is supplied by the dependent source?

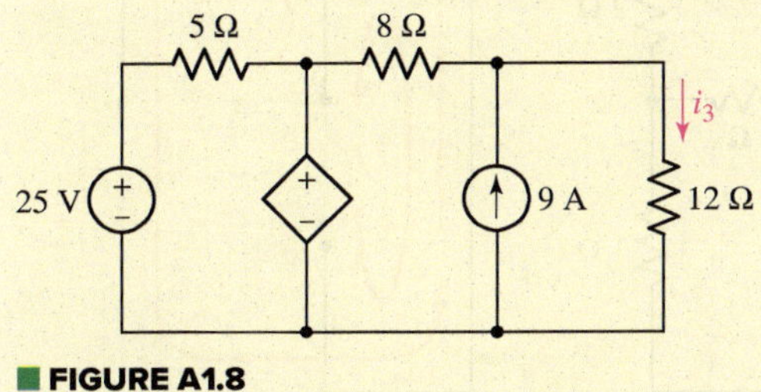

FIGURE A1.8

Ans: 1; 7.2 A; 547 W.

A1.2 LINKS AND LOOP ANALYSIS

Now we will consider the use of a tree to obtain a suitable set of loop equations. In some respects this is the *dual* of the method of writing nodal equations. Again it should be pointed out that, although we are able to guarantee that any set of equations we write will be both sufficient and independent, we should not expect that the method will lead directly to every possible set of equations.

We again begin by constructing a tree, and we use the same set of rules as we did for general nodal analysis. The objective for either nodal or loop analysis is to place voltages in the tree and currents in the cotree; this is a mandatory rule for sources and a desirable rule for controlling quantities.

Now, however, instead of assigning a voltage to each branch in the tree, we assign a current (including reference arrow, of course) to each element in the cotree or to each link. If there were 10 links, we would assign exactly 10 link currents. Any link that contains a current source is assigned that source current as the link current. Note that each link current may also be thought of as a loop current, for the link must extend between two specific nodes, and there must also be a path between those same two nodes through the tree. Thus, with each link there is associated a single specific loop that includes that one link and a unique path through the tree. It is evident that the assigned current may be thought of either as a loop current or as a link current. The link connotation is most helpful at the time the currents are being defined, for one must be established for each link; the loop interpretation is more convenient at equation-writing time, because we will apply KVL around each loop.

Let us try out this process of defining link currents by considering the circuit shown in Fig. A1.9*a*. The tree selected is one of several that might be constructed for which the voltage source is in a tree branch and the current source is in a link. Let us first consider the link containing the current source. The loop associated with this link is the left-hand mesh, and so we show our link current flowing about the perimeter of this mesh (Fig. A1.9*b*). An obvious choice for the symbol for this link current is "7 A." Remember that no other current can flow through this particular link, and thus its value must be exactly the strength of the current source.

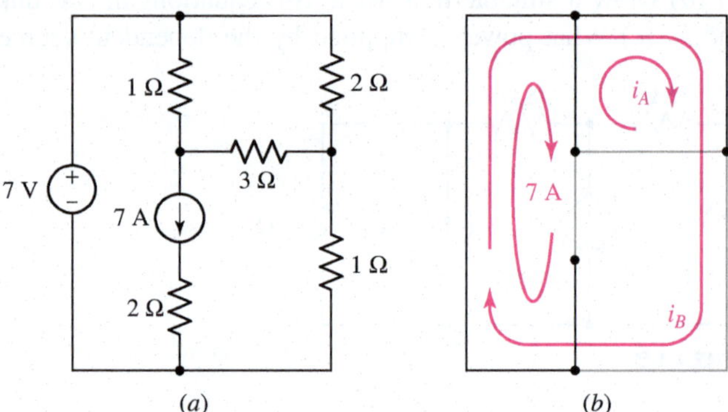

FIGURE A1.9 (*a*) A simple circuit. (*b*) A tree is chosen such that the current source is in a link and the voltage source is in a tree branch.

We next turn our attention to the 3 Ω resistor link. The loop associated with it is the upper right-hand mesh, and this loop (or mesh) current is defined as i_A and also shown in Fig. A1.9*b*. The last link is the lower 1 Ω resistor, and the only path between its terminals through the tree is around the perimeter of the circuit. That link current is called i_B, and the arrow indicating its path and reference direction appears in Fig. A1.9*b*. It is not a mesh current.

Note that each link has only one current present in it, but a tree branch may have any number from 1 to the total number of link currents assigned. The use of long, almost closed, arrows to indicate the loops helps to indicate which loop currents flow through which tree branch and what their reference directions are.

A KVL equation must now be written around each of these loops. The variables used are the assigned link currents. Since the voltage across a current source cannot be expressed in terms of the source current, and since we have already used the value of the source current as the link current, we discard any loop containing a current source.

EXAMPLE A1.4

For the example of Fig. A1.9, find the values of i_A and i_B.

We first traverse the i_A loop, proceeding clockwise from its lower left corner. The current going our way in the 1 Ω resistor is $(i_A - 7)$, in the 2 Ω element it is $(i_A + i_B)$, and in the link it is simply i_B. Thus

$$1(i_A - 7) + 2(i_A + i_B) + 3i_A = 0$$

For the i_B link, clockwise travel from the lower left corner leads to

$$-7 + 2(i_A + i_B) + 1i_B = 0$$

Traversal of the loop defined by the 7 A link is not required. Solving, we have $i_A = 0.5$ A, $i_B = 2$ A, once again. The solution has been achieved with one less equation than before!

EXAMPLE A1.5

Evaluate i_1 in the circuit shown in Fig. A1.10*a*.

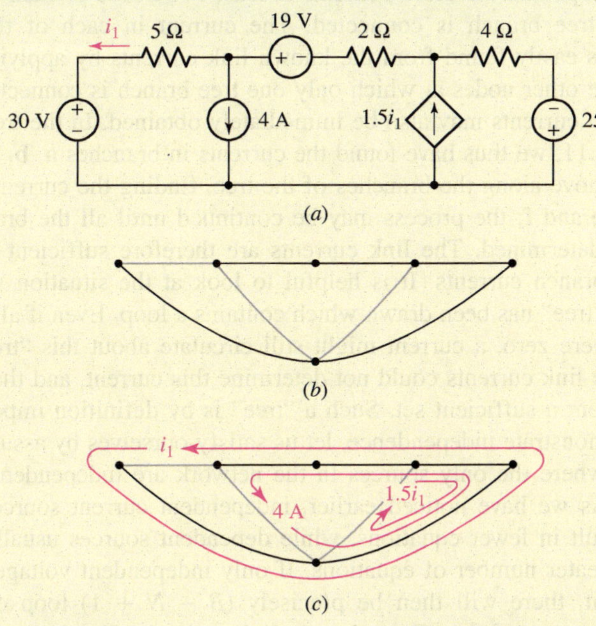

■ **FIGURE A1.10** (*a*) A circuit for which i_1 may be found with one equation using general loop analysis. (*b*) The only tree that satisfies the rules outlined in Sec. A1.1. (*c*) The three link currents are shown with their loops.

This circuit contains six nodes, and its tree therefore must have five branches. Since there are eight elements in the network, there are three links in the cotree. If we place the three voltage sources in the tree and the two current sources and the current control in the cotree, we are led to the tree shown in Fig. A1.10*b*. The source current of 4 A defines a

(*Continued on next page*)

loop as shown in Fig. A1.10*c*. The dependent source establishes the loop current $1.5i_1$ around the right mesh, and the control current i_1 gives us the remaining loop current about the perimeter of the circuit. Note that all three currents flow through the 4 Ω resistor.

We have only one unknown quantity, i_1, and after discarding the loops defined by the two current sources, we apply KVL around the outside of the circuit:

$$-30 + 5(-i_1) + 19 + 2(-i_1 - 4) + 4(-i_1 - 4 + 1.5i_1) - 25 = 0$$

Besides the three voltage sources, there are three resistors in this loop. The 5 Ω resistor has one loop current in it, since it is also a link; the 2 Ω resistor contains two loop currents; and the 4 Ω resistor has three. A carefully drawn set of loop currents is a necessity if errors in skipping currents, utilizing extra ones, or erring in choosing the correct direction are to be avoided. The foregoing equation is guaranteed, however, and it leads to $i_1 = -12$ A.

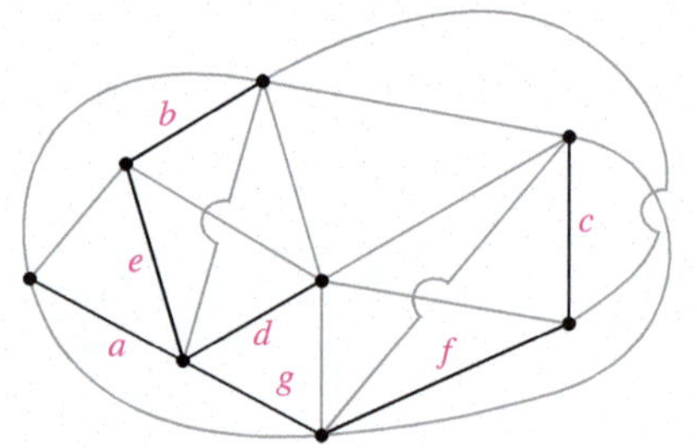

■ **FIGURE A1.11** A tree that is used as an example to illustrate the sufficiency of the link currents.

How may we demonstrate sufficiency? Let us visualize a tree. It contains no loops and therefore contains at least two nodes to each of which only one tree branch is connected. The current in each of these two branches is easily found from the known link currents by applying KCL. If there are other nodes at which only one tree branch is connected, these tree-branch currents may also be immediately obtained. In the tree shown in Fig. A1.11, we thus have found the currents in branches **a**, **b**, **c**, and **d**. Now we move along the branches of the tree, finding the currents in tree branches **e** and **f**; the process may be continued until all the branch currents are determined. The link currents are therefore sufficient to determine all branch currents. It is helpful to look at the situation where an incorrect "tree" has been drawn which contains a loop. Even if all the link currents were zero, a current might still circulate about this "tree loop." Hence, the link currents could not determine this current, and they would not represent a sufficient set. Such a "tree" is by definition impossible.

To demonstrate independence, let us satisfy ourselves by assuming the situation where the only sources in the network are independent voltage sources. As we have noticed earlier, independent current sources in the circuit result in fewer equations, while dependent sources usually necessitate a greater number of equations. If only independent voltage sources are present, there will then be precisely $(B - N + 1)$ loop equations written in terms of the $(B - N + 1)$ link currents. To show that these $(B - N + 1)$ loop equations are independent, it is necessary only to point out that each represents the application of KVL around a loop which contains one link not appearing in any other equation. We might visualize a different resistance $R_1, R_2, \ldots, R_{B-N+1}$ in each of these links, and it is then apparent that one equation can never be obtained from the others, since each contains one coefficient not appearing in any other equation.

Hence, the link currents are sufficient to enable a complete solution to be obtained, and the set of loop equations which we use to find the link currents is a set of independent equations.

Having looked at both general nodal analysis and loop analysis, we should now consider the advantages and disadvantages of each method so that an intelligent choice of a plan of attack can be made on a given analysis problem.

The nodal method in general requires $(N - 1)$ equations, but this number is reduced by 1 for each independent or dependent voltage source in a tree branch, and increased by 1 for each dependent source that is voltage-controlled by a link voltage, or current-controlled.

The loop method basically involves $(B - N + 1)$ equations. However, each independent or dependent current source in a link reduces this number by 1, while each dependent source that is current-controlled by a tree-branch current, or is voltage-controlled, increases the number by 1.

As a grand finale for this discussion, let us inspect the T-equivalent circuit model for a transistor shown in Fig. A1.12, to which is connected a sinusoidal source, 4 sin 1000t mV, and a 10 kΩ load.

EXAMPLE A1.6

Find the input (emitter) current i_e and the load voltage v_L in the circuit of Fig. A1.12, assuming typical values for the emitter resistance $r_e = 50\ \Omega$; the base resistance $r_b = 500\ \Omega$; the collector resistance $r_c = 20\ \Omega$; and the common-base forward-current-transfer ratio $\alpha = 0.99$.

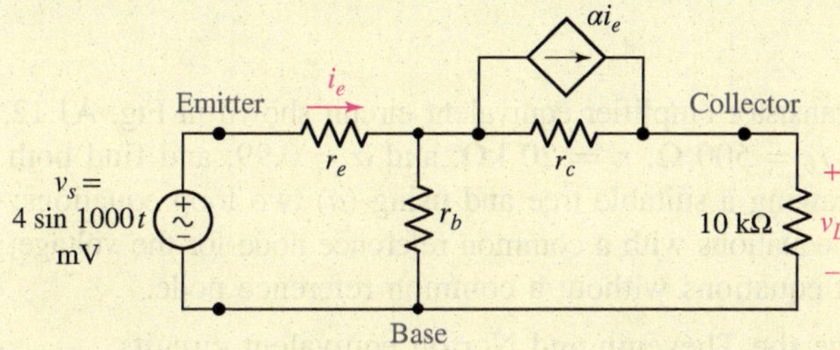

FIGURE A1.12 A sinusoidal voltage source and a 10 kΩ load are connected to the T-equivalent circuit of a transistor. The common connection between the input and output is at the base terminal of the transistor, and the arrangement is called the *common-base* configuration.

Although the details are requested in the practice problems that follow, we should see readily that the analysis of this circuit might be accomplished by drawing trees requiring three general nodal equations $(N - 1 - 1 + 1)$ or two loop equations $(B - N + 1 - 1)$. We might also note that three equations are required in terms of node-to-reference voltages, as are three mesh equations.

No matter which method we choose, these results are obtained for this specific circuit:

$$i_e = 18.42 \sin 1000t \quad \mu\text{A}$$
$$v_L = 122.6 \sin 1000t \quad \text{mV}$$

(Continued on next page)

and we therefore find that this transistor circuit provides a voltage gain (v_L/v_s) of 30.6, a current gain ($v_L/10{,}000i_e$) of 0.666, and a power gain equal to the product 30.6(0.666) = 20.4. Higher gains could be secured by operating this transistor in a common-emitter configuration.

PRACTICE

A1.2 Draw a suitable tree and use general loop analysis to find i_{10} in the circuit of (*a*) Fig. A1.13*a* by writing just one equation with i_{10} as the variable; (*b*) Fig. A1.13*b* by writing just two equations with i_{10} and i_3 as the variables.

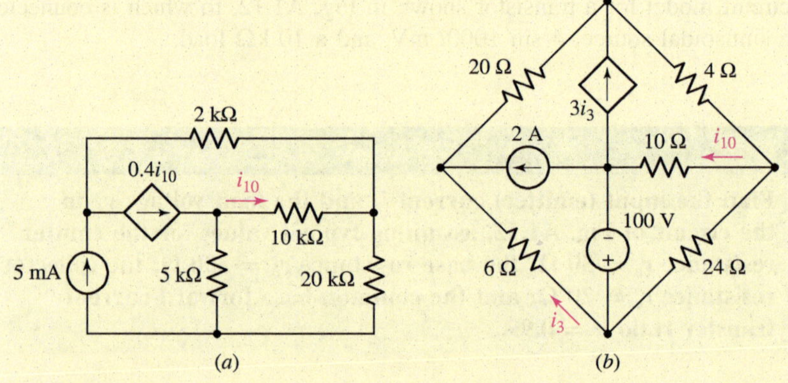

FIGURE A1.13

A1.3 For the transistor amplifier equivalent circuit shown in Fig. A1.12, let $r_e = 50\ \Omega$, $r_b = 500\ \Omega$, $r_c = 20\ \text{k}\Omega$, and $\alpha = 0.99$, and find both i_e and v_L by drawing a suitable tree and using (*a*) two loop equations; (*b*) three nodal equations with a common reference node for the voltage; (*c*) three nodal equations without a common reference node.

A1.4 Determine the Thévenin and Norton equivalent circuits presented to the 10 kΩ load in Fig. A1.12 by finding (*a*) the open-circuit value of v_L; (*b*) the (downward) short-circuit current; (*c*) the Thévenin equivalent resistance. All circuit values are given in Practice Problem A1.3.

Ans: A1.2: −4.00 mA; 4.69 A. A1.3: 18.42 sin 1000t μA; 122.6 sin 1000t mV. A1.4: 147.6 sin 1000t mV; 72.2 sin 1000t μA; 2.05 kΩ.

SOLUTION OF SIMULTANEOUS EQUATIONS

Consider the simple system of equations

$$7v_1 - 3v_2 - 4v_3 = -11 \qquad [1]$$

$$-3v_1 + 6v_2 - 2v_3 = 3 \qquad [2]$$

$$-4v_1 - 2v_2 + 11v_3 = 25 \qquad [3]$$

This set of equations *could* be solved by a systematic elimination of the variables. Such a procedure is lengthy, however, and may never yield answers if done unsystematically for a greater number of simultaneous equations. Fortunately, there are many more options available to us, some of which we will explore in this appendix.

The Scientific Calculator

Perhaps the most straightforward approach when faced with a system of equations such as Eqs. [1] to [3], in which we have numerical coefficients and are only interested in the specific values of our unknowns (as opposed to algebraic relationships), is to employ any of the various scientific calculators currently on the market. For example, on a Texas Instruments TI-84, we can employ the Polynomial Root Finder and Simultaneous Equation Solver (you may need to install the application using TI Connect™). Pressing the **APPS** key and scrolling down, find the application named **PLYSmlt2**. Running and continuing past the welcome screen shows the Main Menu of Fig. A2.1*a*. Selecting the second menu item results in the screen shown in Fig. A2.1*b*, where we have chosen three equations in

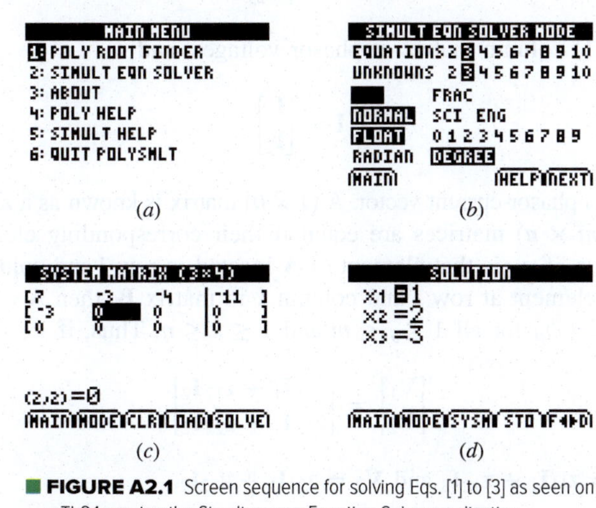

■ **FIGURE A2.1** Screen sequence for solving Eqs. [1] to [3] as seen on a TI-84 running the Simultaneous Equation Solver application.

three unknowns. After pressing **NEXT**, we are presented with a screen similar to that shown in Fig. A2.1*c*. After we have finished entering all coefficients, pressing the **SOLVE** button yields the Solution screen depicted in Fig. A2.1*d*. Since we did not name the variables, a slight mental conversion is required to realize $X_1 = v_1$, $X_2 = v_2$, etc.

It should be noted that each calculator capable of solving simultaneous equations has its own procedure for entering the required information—therefore, it is a good idea not to throw away anything marked "Owner's Manual" or "Instructions," no matter how tempting such an action might be.

Matrices

Another powerful approach to the solution of a system of equations is based on the concept of matrices. Consider Eqs. [1], [2], and [3]. The array of the constant coefficients of the equations,

$$\mathbf{G} = \begin{bmatrix} 7 & -3 & -4 \\ -3 & 6 & -2 \\ -4 & -2 & 11 \end{bmatrix}$$

is called a ***matrix;*** the symbol **G** has been selected since each element of the matrix is a conductance value. A matrix itself has no "value"; it is merely an ordered array of elements. We use a letter in boldface type to represent a matrix, and we enclose the array itself by square brackets.

A matrix having m rows and n columns is called an $(m \times n)$ (pronounced "m by n") matrix. Thus,

$$\mathbf{A} = \begin{bmatrix} 2 & 0 & 5 \\ -1 & 6 & 3 \end{bmatrix}$$

is a (2×3) matrix, and the **G** matrix of our example is a (3×3) matrix. An $(n \times n)$ matrix is a ***square matrix*** of order n.

An $(m \times 1)$ matrix is called a *column matrix,* or a ***vector.*** Thus,

$$\mathbf{V} = \begin{bmatrix} \mathbf{V}_1 \\ \mathbf{V}_2 \end{bmatrix}$$

is a (2×1) column matrix of phasor voltages, and

$$\mathbf{I} = \begin{bmatrix} \mathbf{I}_1 \\ \mathbf{I}_2 \end{bmatrix}$$

is a (2×1) phasor-current vector. A $(1 \times n)$ matrix is known as a *row vector.*

Two $(m \times n)$ matrices are equal if their corresponding elements are equal. Thus, if a_{jk} is that element of **A** located in row j and column k and b_{jk} is the element at row j and column k in matrix **B**, then $\mathbf{A} = \mathbf{B}$ *if and only if* $a_{jk} = b_{jk}$ for all $1 \le j \le m$ and $1 \le k \le n$. Thus, if

$$\begin{bmatrix} \mathbf{V}_1 \\ \mathbf{V}_2 \end{bmatrix} = \begin{bmatrix} \mathbf{z}_{11}\mathbf{I}_1 + \mathbf{z}_{12}\mathbf{I}_2 \\ \mathbf{z}_{12}\mathbf{I}_1 + \mathbf{z}_{22}\mathbf{I}_2 \end{bmatrix}$$

then $\mathbf{V}_1 = \mathbf{z}_{11}\mathbf{I}_1 + \mathbf{z}_{12}\mathbf{I}_2$ and $\mathbf{V}_2 = \mathbf{z}_{21}\mathbf{I}_1 + \mathbf{z}_{22}\mathbf{I}_2$.

Two $(m \times n)$ matrices may be added by adding corresponding elements. Thus,

$$\begin{bmatrix} 2 & 0 & 5 \\ -1 & 6 & 3 \end{bmatrix} + \begin{bmatrix} 1 & 2 & 3 \\ -3 & -2 & -1 \end{bmatrix} = \begin{bmatrix} 3 & 2 & 8 \\ -4 & 4 & 2 \end{bmatrix}$$

Next let us consider the matrix product **AB**, where **A** is an $(m \times n)$ matrix and **B** is a $(p \times q)$ matrix. If $n = p$, the matrices are said to be *conformal,* and their product exists. That is, matrix multiplication is defined only for the case where the number of columns of the first matrix in the product is equal to the number of rows in the second matrix.

The formal definition of matrix multiplication states that the product of the $(m \times n)$ matrix **A** and the $(n \times q)$ matrix **B** is an $(m \times q)$ matrix having elements c_{jk}, $1 \le j \le m$ and $1 \le k \le q$, where

$$c_{jk} = a_{j1}b_{1k} + a_{j2}b_{2k} + \cdots + a_{jn}b_{nk}$$

That is, to find the element in the second row and third column of the product, we multiply each of the elements in the second row of **A** by the corresponding element in the third column of **B** and then add the n results. For example, given the (2×3) matrix **A** and the (3×2) matrix **B**,

$$\begin{bmatrix} a_{11} & a_{12} & a_{13} \\ a_{21} & a_{22} & a_{23} \end{bmatrix} \begin{bmatrix} b_{11} & b_{12} \\ b_{21} & b_{22} \\ b_{31} & b_{32} \end{bmatrix} =$$

$$\begin{bmatrix} (a_{11}b_{11} + a_{12}b_{21} + a_{13}b_{31}) & (a_{11}b_{12} + a_{12}b_{22} + a_{13}b_{32}) \\ (a_{21}b_{11} + a_{22}b_{21} + a_{23}b_{31}) & (a_{21}b_{12} + a_{22}b_{22} + a_{23}b_{32}) \end{bmatrix}$$

The result is a (2×2) matrix.

As a numerical example of matrix multiplication, we take

$$\begin{bmatrix} 3 & 2 & 1 \\ -2 & -2 & 4 \end{bmatrix} \begin{bmatrix} 2 & 3 \\ -2 & -1 \\ 4 & -3 \end{bmatrix} = \begin{bmatrix} 6 & 4 \\ 16 & -16 \end{bmatrix}$$

where $6 = (3)(2) + (2)(-2) + (1)(4)$, $4 = (3)(3) + (2)(-1) + (1)(-3)$, and so forth.

Matrix multiplication is not commutative. For example, given the (3×2) matrix **C** and the (2×1) matrix **D**, it is evident that the product **CD** may be calculated, but the product **DC** is not even defined.

As a final example, let

$$\mathbf{t}_A = \begin{bmatrix} 2 & 3 \\ -1 & 4 \end{bmatrix}$$

and

$$\mathbf{t}_B = \begin{bmatrix} 3 & 1 \\ 5 & 0 \end{bmatrix}$$

so that both $\mathbf{t}_A\mathbf{t}_B$ and $\mathbf{t}_B\mathbf{t}_A$ are defined. However,

$$\mathbf{t}_A\mathbf{t}_B = \begin{bmatrix} 21 & 2 \\ 17 & -1 \end{bmatrix}$$

while

$$\mathbf{t}_B\mathbf{t}_A = \begin{bmatrix} 5 & 13 \\ 10 & 15 \end{bmatrix}$$

PRACTICE

A2.1 Given $\mathbf{A} = \begin{bmatrix} 1 & -3 \\ 3 & 5 \end{bmatrix}$, $\mathbf{B} = \begin{bmatrix} 4 & -1 \\ -2 & 3 \end{bmatrix}$, $\mathbf{C} = \begin{bmatrix} 50 \\ 30 \end{bmatrix}$, and $\mathbf{V} = \begin{bmatrix} \mathbf{V}_1 \\ \mathbf{V}_2 \end{bmatrix}$, find (*a*) $\mathbf{A} + \mathbf{B}$; (*b*) $\mathbf{AB}$; (*c*) $\mathbf{BA}$; (*d*) $\mathbf{AV} + \mathbf{BC}$; (*e*) $\mathbf{A}^2 = \mathbf{AA}$.

Ans: $\begin{bmatrix} 5 & -4 \\ 1 & 8 \end{bmatrix}$; $\begin{bmatrix} 10 & -10 \\ 2 & 12 \end{bmatrix}$; $\begin{bmatrix} 1 & -17 \\ 7 & 21 \end{bmatrix}$; $\begin{bmatrix} V_1 - 3V_2 + 170 \\ 3V_1 + 5V_2 - 10 \end{bmatrix}$; $\begin{bmatrix} -8 & -18 \\ 18 & 16 \end{bmatrix}$

Matrix Inversion

If we write our system of equations using matrix notation,

$$\begin{bmatrix} 7 & -3 & -4 \\ -3 & 6 & -2 \\ -4 & -2 & 11 \end{bmatrix} \begin{bmatrix} v_1 \\ v_2 \\ v_3 \end{bmatrix} = \begin{bmatrix} -11 \\ 3 \\ 25 \end{bmatrix} \qquad [4]$$

we may solve for the voltage vector by multiplying both sides of Eq. [4] by the inverse of our matrix **G**:

$$\mathbf{G}^{-1} \begin{bmatrix} 7 & -3 & -4 \\ -3 & 6 & -2 \\ -4 & -2 & 11 \end{bmatrix} \begin{bmatrix} v_1 \\ v_2 \\ v_3 \end{bmatrix} = \mathbf{G}^{-1} \begin{bmatrix} -11 \\ 3 \\ 25 \end{bmatrix} \qquad [5]$$

This procedure makes use of the identity $\mathbf{G}^{-1}\mathbf{G} = \mathbf{I}$, where **I** is the identity matrix, a square matrix of the same size as **G**, with zeros everywhere except on the diagonal. Each element on the diagonal of an identity matrix is unity. Thus, Eq. [5] becomes

$$\begin{bmatrix} 1 & 0 & 0 \\ 0 & 1 & 0 \\ 0 & 0 & 1 \end{bmatrix} \begin{bmatrix} v_1 \\ v_2 \\ v_3 \end{bmatrix} = \mathbf{G}^{-1} \begin{bmatrix} -11 \\ 3 \\ 25 \end{bmatrix}$$

which may be simplified to

$$\begin{bmatrix} v_1 \\ v_2 \\ v_3 \end{bmatrix} = \mathbf{G}^{-1} \begin{bmatrix} -11 \\ 3 \\ 25 \end{bmatrix}$$

since the identity matrix times any vector is simply equal to that vector (the proof is left to the reader as a 30-second exercise). The solution of our system of equations has therefore been transformed into the problem of obtaining the inverse matrix of **G**. Many scientific calculators provide the means of performing matrix algebra.

Returning to the TI-84, we press **2ND** and **MATRIX** to obtain the screen shown in Fig. A2.2*a*. Scrolling horizontally to **EDIT**, we press the **ENTER** key, and select a 3 × 3 matrix, resulting in a screen similar to that shown in Fig. A2.2*b*. Once we have finished entering the matrix, we press **2ND** and **QUIT**. Returning to the **MATRIX** editor, we create a 3 × 1 vector named **B**, as shown in Fig. A2.2*c*. We are now (finally) ready to solve for the solution vector. Pressing **2ND** and **MATRIX**, under **NAMES** we select **[A]** and press **ENTER**, followed by the x^{-1} key. Next, we select **[B]** the same way (we could have pressed the multiplication key in between but it is not necessary). The result of our calculation is shown in Fig. A2.2*d*, and agrees with our earlier exercise.

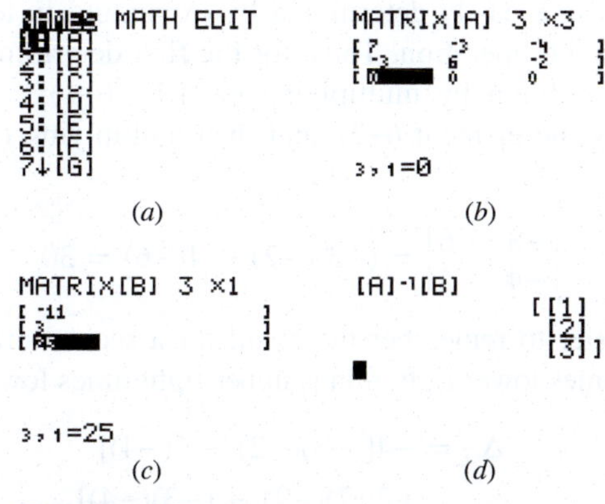

(*a*) (*b*) (*c*) (*d*)

FIGURE A2.2 Sequence of screens for matrix solution. (*a*) Matrix editor screen; (*b*) entering terms; (*c*) creating right-hand side vector; (*d*) solving matrix equation.

Determinants

Although a matrix *itself* has no "value," the ***determinant*** of a square matrix *does* have a value. To be precise, we should say that the determinant of a matrix is a value, but common usage enables us to speak of both the array itself and its value as the determinant. We shall symbolize a determinant by Δ, and employ a suitable subscript to denote the matrix to which the determinant refers. Thus,

$$\Delta_G = \begin{vmatrix} 7 & -3 & -4 \\ -3 & 6 & -2 \\ -4 & -2 & 11 \end{vmatrix}$$

Note that simple vertical lines are used to enclose the determinant.

The value of any determinant is obtained by expanding it in terms of its minors. To do this, we select any row j or any column k, multiply each element in that row or column by its minor and by $(-1)^{j+k}$, and then add the products. The minor of the element appearing in both row j and column k is the determinant obtained when row j and column k are removed; it is indicated by Δ_{jk}.

As an example, let us expand the determinant Δ_G along column 3. We first multiply the (−4) at the top of this column by $(-1)^{1+3} = 1$ and then by its minor:

$$(-4)(-1)^{1+3}\begin{vmatrix} -3 & 6 \\ -4 & -2 \end{vmatrix}$$

and then repeat for the other two elements in column 3, adding the results:

$$-4\begin{vmatrix} -3 & 6 \\ -4 & -2 \end{vmatrix} + 2\begin{vmatrix} 7 & -3 \\ -4 & -2 \end{vmatrix} + 11\begin{vmatrix} 7 & -3 \\ -3 & 6 \end{vmatrix}$$

The minors contain only two rows and two columns. They are of order 2, and their values are easily determined by expanding in terms of minors again, here a trivial operation. Thus, for the first determinant, we expand along the first column by multiplying (−3) by $(-1)^{1+1}$ and its minor, which is merely the element (−2), and then multiplying (−4) by $(-1)^{2+1}$ and by 6. Thus,

$$\begin{vmatrix} -3 & 6 \\ -4 & -2 \end{vmatrix} = (-3)(-2) - 4(-6) = 30$$

It is usually easier to remember the result for a second-order determinant as "upper left times lower right minus upper right times lower left." Finally,

$$\begin{aligned} \Delta_G &= -4[(-3)(-2) - 6(-4)] \\ &\quad +2[(7)(-2) - (-3)(-4)] \\ &\quad +11[(7)(6) - (-3)(-3)] \\ &= -4(30) + 2(-26) + 11(33) \\ &= 191 \end{aligned}$$

For practice, let us expand this same determinant along the first row:

$$\begin{aligned} \Delta_G &= 7\begin{vmatrix} 6 & -2 \\ -2 & 11 \end{vmatrix} - (-3)\begin{vmatrix} -3 & -2 \\ -4 & 11 \end{vmatrix} + (-4)\begin{vmatrix} -3 & 6 \\ -4 & -2 \end{vmatrix} \\ &= 7(62) + 3(-41) - 4(30) \\ &= 191 \end{aligned}$$

The expansion by minors is valid for a determinant of any order.

Repeating these rules for evaluating a determinant in more general terms, we would say, given a matrix $\mathbf{a}$,

$$\mathbf{a} = \begin{bmatrix} a_{11} & a_{12} & \cdots & a_{1N} \\ a_{21} & a_{22} & \cdots & a_{2N} \\ \cdots & \cdots & \cdots & \cdots \\ a_{N1} & a_{N2} & \cdots & a_{NN} \end{bmatrix}$$

that Δ_a may be obtained by expansion in terms of minors along any row j:

$$\begin{aligned} \Delta_a &= a_{j1}(-1)^{j+1}\Delta_{j1} + a_{j2}(-1)^{j+2}\Delta_{j2} + \cdots + a_{jN}(-1)^{j+N}\Delta_{jN} \\ &= \sum_{n=1}^{N} a_{jn}(-1)^{j+n}\Delta_{jn} \end{aligned}$$

or along any column k:

$$\Delta_a = a_{1k}(-1)^{1+k}\Delta_{1k} + a_{2k}(-1)^{2+k}\Delta_{2k} + \cdots + a_{Nk}(-1)^{N+k}\Delta_{Nk}$$
$$= \sum_{n=1}^{N} a_{nk}(-1)^{n+k}\Delta_{nk}$$

The cofactor C_{jk} of the element appearing in both row j and column k is simply $(-1)^{j+k}$ times the minor Δ_{jk}. Thus, $C_{11} = \Delta_{11}$, but $C_{12} = -\Delta_{12}$. We may now write

$$\Delta_a = \sum_{n=1}^{N} a_{jn}C_{jn} = \sum_{n=1}^{N} a_{nk}C_{nk}$$

As an example, let us consider this fourth-order determinant:

$$\Delta = \begin{vmatrix} 2 & -1 & -2 & 0 \\ -1 & 4 & 2 & -3 \\ -2 & -1 & 5 & -1 \\ 0 & -3 & 3 & 2 \end{vmatrix}$$

We find

$$\Delta_{11} = \begin{vmatrix} 4 & 2 & -3 \\ -1 & 5 & -1 \\ -3 & 3 & 2 \end{vmatrix} = 4(10 + 3) + 1(4 + 9) - 3(-2 + 15) = 26$$

$$\Delta_{12} = \begin{vmatrix} -1 & 2 & -3 \\ -2 & 5 & -1 \\ 0 & 3 & 2 \end{vmatrix} = -1(10 + 3) + 2(4 + 9) + 0 = 13$$

and $C_{11} = 26$, whereas $C_{12} = -13$. Finding the value of Δ for practice, we have

$$\Delta = 2C_{11} + (-1)C_{12} + (-2)C_{13} + 0$$
$$= 2(26) + (-1)(-13) + (-2)(3) + 0 = 59$$

Cramer's Rule

We next consider Cramer's rule, which enables us to find the values of the unknown variables. It is also useful in solving systems of equations where numerical coefficients have not yet been specified, thus confounding our calculators. Let us again consider Eqs. [1], [2], and [3]; we define the determinant Δ_1 as that determinant which is obtained when the first column of Δ_G is replaced by the three constants on the right-hand sides of the three equations. Thus,

$$\Delta_1 = \begin{vmatrix} -11 & -3 & -4 \\ 3 & 6 & -2 \\ 25 & -2 & 11 \end{vmatrix}$$

We expand along the first column:

$$\Delta_1 = -11\begin{vmatrix} 6 & -2 \\ -2 & 11 \end{vmatrix} - 3\begin{vmatrix} -3 & -4 \\ -2 & 11 \end{vmatrix} + 25\begin{vmatrix} -3 & -4 \\ 6 & -2 \end{vmatrix}$$
$$= -682 + 123 + 750 = 191$$

Cramer's rule then states that

$$v_1 = \frac{\Delta_1}{\Delta_G} = \frac{191}{191} = 1 \text{ V}$$

and

$$v_2 = \frac{\Delta_2}{\Delta_G} = \begin{vmatrix} 7 & -11 & -4 \\ -3 & 3 & -2 \\ -4 & 25 & 11 \end{vmatrix} = \frac{581 - 63 - 136}{191} = 2 \text{ V}$$

and finally,

$$v_3 = \frac{\Delta_3}{\Delta_G} = \begin{vmatrix} 7 & -3 & -11 \\ -3 & 6 & 3 \\ -4 & -2 & 25 \end{vmatrix} = \frac{1092 - 291 - 228}{191} = 3 \text{ V}$$

Cramer's rule is applicable to a system of N simultaneous linear equations in N unknowns; for the ith variable v_i:

$$v_i = \frac{\Delta_i}{\Delta_G}$$

PRACTICE

A2.2 Evaluate

(*a*) $\begin{vmatrix} 2 & -3 \\ -2 & 5 \end{vmatrix}$; (*b*) $\begin{vmatrix} 1 & -1 & 0 \\ 4 & 2 & -3 \\ 3 & -2 & 5 \end{vmatrix}$; (*c*) $\begin{vmatrix} 2 & -3 & 1 & 5 \\ -3 & 1 & -1 & 0 \\ 0 & 4 & 2 & -3 \\ 6 & 3 & -2 & 5 \end{vmatrix}$.

(*d*) Find i_2 if $5i_1 - 2i_2 - i_3 = 100$, $-2i_1 + 6i_2 - 3i_3 - i_4 = 0$, $-i_1 - 3i_2 + 4i_3 - i_4 = 0$, and $-i_2 - i_3 = 0$.

Ans: 4; 33; –411; 1.266.

A PROOF OF THÉVENIN'S THEOREM

Here we prove Thévenin's theorem in the same form in which it is stated in Sec. 5.4 of Chap. 5:

> Given any linear circuit, rearrange it in the form of two networks A and B connected by two wires. Define a voltage v_{oc} as the open-circuit voltage which appears across the terminals of A when B is disconnected. Then all currents and voltages in B will remain unchanged if all *independent* voltage and current sources in A are "zeroed out," and an independent voltage source v_{oc} is connected, with proper polarity, in series with the dead (inactive) A network.

We will effect our proof by showing that the original A network and the Thévenin equivalent of the A network both cause the same current to flow into the terminals of the B network. If the currents are the same, then the voltages must be the same; in other words, if we apply a certain current, which we might think of as a current source, to the B network, then the current source and the B network constitute a circuit that has a specific input voltage as a response. Thus, the current determines the voltage. Alternatively we could, if we wished, show that the terminal voltage at B is unchanged, because the voltage also determines the current uniquely. If the input voltage and current to the B network are unchanged, then it follows that the currents and voltages *throughout* the B network are also unchanged.

Let us first prove the theorem for the case where the B network is inactive (no independent sources). After this step has been accomplished, we may then use the superposition principle to extend the theorem to include B networks that contain independent sources. Each network may contain dependent sources, provided that their control variables are in the same network.

The current i, flowing in the upper conductor from the A network to the B network in Fig. A3.1a, is therefore caused entirely by the independent

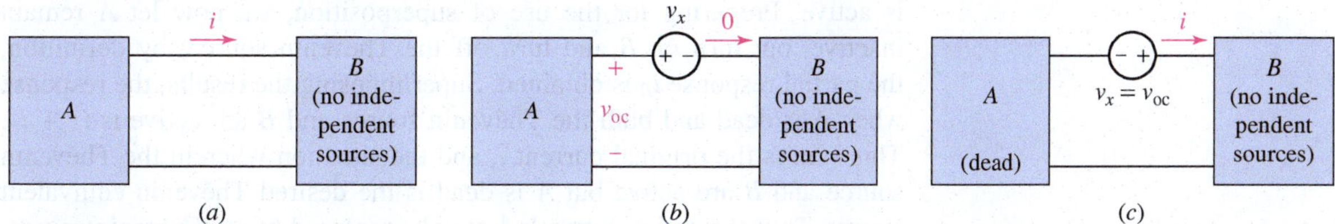

FIGURE A3.1 (*a*) A general linear network A and a network B that contains no independent sources. Controls for dependent sources must appear in the same part of the network. (*b*) The Thévenin source is inserted in the circuit and adjusted until $i = 0$. No voltage appears across network B and thus $v_x = v_{oc}$. The Thévenin source thus produces a current $-i$ while network A provides i. (*c*) The Thévenin source is reversed and network A is zeroed out. The current is therefore i.

sources present in the A network. Suppose now that we add an additional voltage source v_x, which we shall call the Thévenin source, in the conductor in which i is measured, as shown in Fig. A3.1*b*, and then adjust the magnitude and time variation of v_x until the current is reduced to zero. By our definition of v_{oc}, then, the voltage across the terminals of A must be v_{oc}, since $i = 0$. Network B contains no independent sources, and no current is entering its terminals; therefore, there is no voltage across the terminals of the B network, and by Kirchhoff's voltage law the voltage of the Thévenin source is v_{oc} volts, $v_x = v_{oc}$. Moreover, since the Thévenin source and the A network jointly deliver no current to B, and since the A network by itself delivers a current i, superposition requires that the Thévenin source acting by itself must deliver a current of $-i$ to B. The source acting alone in a reversed direction, as shown in Fig. A3.1*c*, therefore produces a current i in the upper lead. This situation, however, is the same as the conclusion reached by Thévenin's theorem: the Thévenin source v_{oc} acting in series with the inactive A network is equivalent to the given network.

Now let us consider the case where the B network may be an active network. We now think of the current i, flowing from the A network to the B network in the upper conductor, as being composed of two parts, i_A and i_B, where i_A is the current produced by A acting alone and the current i_B is due to B acting alone. Our ability to divide the current into these two components is a direct consequence of the applicability of the superposition principle to these two *linear* networks; the complete response and the two partial responses are indicated by the diagrams of Fig. A3.2.

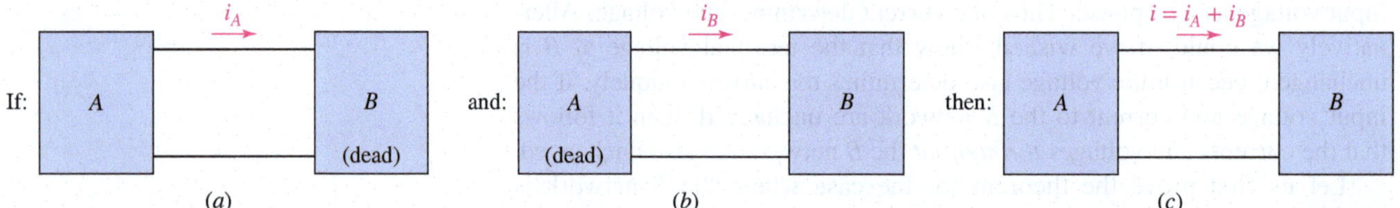

FIGURE A3.2 Superposition enables the current *i* to be considered as the sum of two partial responses.

The partial response i_A has already been considered; if network B is inactive, we know that network A may be replaced by the Thévenin source and the inactive A network. In other words, of the three sources which we must keep in mind—those in A, those in B, and the Thévenin source—the partial response i_A occurs when A and B are dead and the Thévenin source is active. Preparing for the use of superposition, we now let A remain inactive, but turn on B and turn off the Thévenin source; by definition, the partial response i_B is obtained. Superimposing the results, the response when A is dead and both the Thévenin source and B are active is $i_A + i_B$. This sum is the original current i, and the situation wherein the Thévenin source and B are active but A is dead is the desired Thévenin equivalent circuit. Thus the active network A may be replaced by its Thévenin source, the open-circuit voltage, in series with the inactive A network, regardless of the status of the B network; it may be either active or inactive.

AN LTspice® TUTORIAL

SPICE is an acronym for Simulation Program with Integrated Circuit Emphasis. A powerful program, it is an industry standard and used throughout the world for a variety of circuit analysis applications. SPICE was originally developed in the early 1970s by Donald O. Peterson and coworkers at the University of California at Berkeley. Interestingly, Peterson advocated free and unhindered distribution of knowledge created in university labs, choosing to make an impact as opposed to profiting financially. There are now several variations of SPICE available on a variety of computing platforms, as well as competing software products.

The goal of this appendix is to simply introduce the basics of computer-aided circuit analysis using LTspice (www.analog.com); additional details are presented in the main text as well as in the references listed under Reading Further. The student (and instructor) should not feel limited to any one package—the authors have selected this one simply as an example, in part because it is both freeware and supported on both Windows and Mac OS X. Advanced topics covered in the references include how to determine the sensitivity of an output variable to changes in a specific component value; how to obtain plots of the output versus a source value; determining ac output as a function of source frequency; methods for performing noise and distortion analyses; nonlinear component models; and how to model temperature effects on specific types of circuits.

Getting Started

The first step is to download and install the software, available at http://www.linear.com/designtools/software/#LTspice. The website also includes a "Getting Started Guide" and shortcuts for Mac OS X.

A computer-aided circuit analysis consists of three separate steps: (1) drawing the schematic, (2) simulating the circuit, and (3) extracting the desired information from the simulation output. Begin by launching LTspice and selecting **New Schematic** under **File**. A blank schematic window will open, where you will place components, connect components with wires, and define simulation parameters. To add, move, or delete components and wires, we have two options: right-click on the schematic or use the menu options, as shown in Fig. A4.1.

Mac OS X users: The user interface is somewhat different from the Windows version, so you will need to become familiar with shortcuts and a way to right-click. Right-clicking can usually be accomplished using **Ctrl**+click with a mouse or a two-finger click with a track pad.

We begin by analyzing the basic series circuit with a voltage source, resistor, and a lamp (a light bulb that has a circuit equivalent of a resistor). The objective is to find the current through the lamp and the voltage across the lamp it. Begin by creating a new schematic. Add components and wire them together using the following procedure.

- Add new components using the appropriate choice under **Edit**. For this example, we will choose voltage source **(voltage)** by typing *voltage*

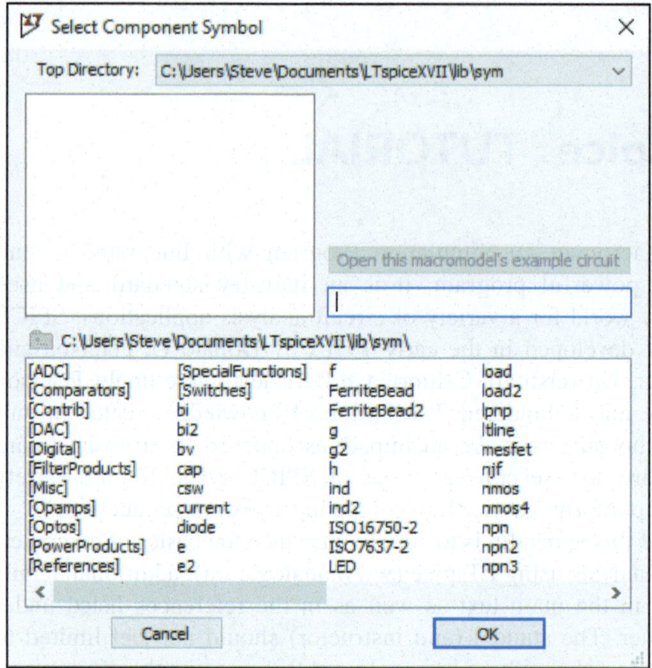

■ **FIGURE A4.1** **Select Component** menu, obtained through **Component** under **Edit**.
Analog Devices

in the pop-up menu from **Component**, then selecting **Resistor** in a separate step (again, under **Edit**). Use edit tools to move or delete components, and the shortcut **Ctrl+R** to rotate objects. Note that the top (vertical) node of the resistor as it first appears is the default + reference terminal for the device, and this is the node about which it rotates.

- Add a reference ground node, either by selecting the ground symbol from the menu bar, or by selecting **Place GND** under **Edit**.
- Connect the components with wires either by selecting the pencil symbol from the menu bar, or **Draw Wire** under **Edit**. *Be careful not to wire across a component by accident!* Note that crossing wires does not automatically create a node at that junction.
- Define component values by right-clicking on each component and changing the value.

Prior to simulating our circuit, we save it by selecting **Save** from the **File** menu. Your circuit should look very similar to Fig. A4.2, with the exception of the .op, which we are now ready to obtain.

To run a simulation, we need to first define the type of simulation. In our case, we are simply interested in the dc voltage and current. To define the simulation to find the dc values, select **SPICE Directive'S'** under **Edit**. In the dialog box that appears, type the command **.op**, which refers

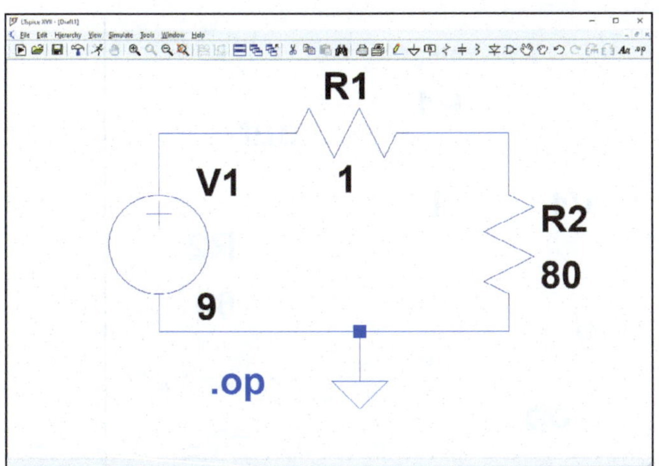

FIGURE A4.2 Schematic of series circuit with a lamp (represented by an 80 Ω resistor) drawn in LTspice. Note that units are not shown; prefixes such as k for kilo, m for milli, and Meg for mega are acceptable alternatives to decimal or scientific notation.
Analog Devices

to the DC operating point of the circuit, and place the text anywhere on the schematic. There are six different types of analyses that can be performed, each with a different analysis command:

- **.tran** Transient analysis
- **.ac** Small signal AC
- **.dc** DC sweep
- **.noise** Noise
- **.tf** DC transfer function
- **.op** DC operating point

We can access these options also by selecting **SPICE Analysis** under **Edit**, which helps us identify the parameter values we need to set. Now run the simulation by clicking **Run** (the running person in the upper left corner). In Windows you can also select **Run** under **Simulate**.

Following a successful simulation, you will see a new window. Resulting values from the simulation will be shown in a new window:

> On the circuit schematic, we select the symbol with the letter A in a box, which lies between the ground and resistor symbols on the menu bar. Place the label on a node of interest, and the DC operating point will appear. Note that this only works for node voltage, not current (see Fig. A4.3).

The resulting output using these three methods is shown in Fig. A4.3, where we find $I_{out} = 0.1111$ A and $V_{out} = 8.8889$ V.

Fortunately, our simulation yields the expected result, where most of the voltage from the 9 V source is dropped across the lamp, and only a small fraction of the voltage is dropped across the 1 Ω series resistance.

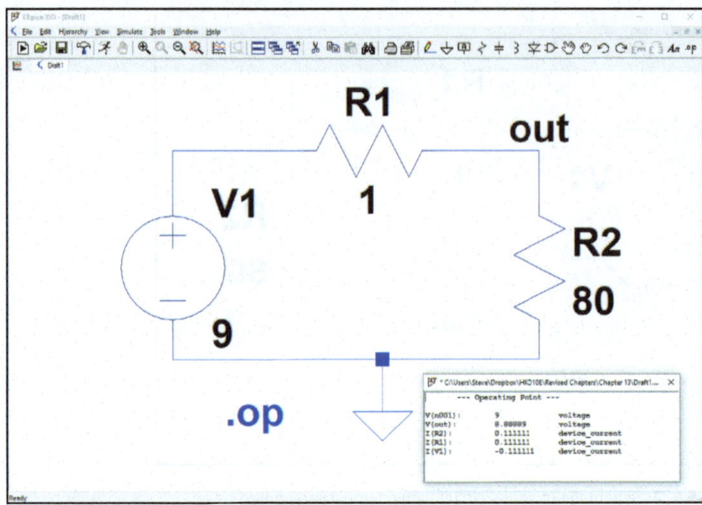

FIGURE A4.3 LTspice dc operating point results shown after naming an output node "out."
Analog Devices

This example of simulating a lamp circuit is intended to simply get you started in using LTspice. More sophisticated analysis and functions are described throughout the text in the highlighted Computer-Aided Analysis sections as new circuits concepts are introduced.

READING FURTHER

Some good resources devoted to LTspice simulation are:

The LTwiki online at http://ltwiki.org/?title=Main_Page

G. Brocard, *The LTSpice IV Simulator: Manual, Methods and Applications,* Künzelsau: Swiridoff Verlag, 2013.

A. K. Singh and R. Singh, *Electronics Circuit SPICE Simulations with LTspice: A Schematic Based Approach,* CreateSpace Independent Publishing, 2015.

An interesting history of circuit simulators, as well as Donald Peterson's contributions to the field, can be found in:

T. Perry, "Donald O. Peterson [electronic engineering biography]," *IEEE Spectrum* **35**, 1998, pp. 22–27.

COMPLEX NUMBERS

This appendix includes sections covering the definition of a complex number, the basic arithmetic operations for complex numbers, Euler's identity, and the exponential and polar forms of the complex number.

A5.1 THE COMPLEX NUMBER

Our early training in mathematics dealt exclusively with real numbers, such as 4, $-\frac{2}{7}$, and π. Soon, however, we began to encounter algebraic equations, such as $x^2 = -3$, which could not be satisfied by any real number. Such an equation can be solved only through the introduction of the *imaginary unit,* or the *imaginary operator,* which we shall designate by the symbol j. By definition, $j^2 = -1$, and thus $j = \sqrt{-1}$, $j^3 = -j$, $j^4 = 1$, and so forth. The product of a real number and the imaginary operator is called an *imaginary number,* and the sum of a real number and an imaginary number is called a *complex number*. Thus, a number having the form $a + jb$, where a and b are real numbers, is a complex number.

Mathematicians designate the imaginary operator by the symbol *i*, but it is customary to use *j* in electrical engineering in order to avoid confusion with the symbol for current.

The choice of the words *imaginary* and *complex* is unfortunate. They are used here and in the mathematical literature as technical terms to designate a class of numbers. To interpret imaginary as "not pertaining to the physical world" or complex as "complicated" is neither justified nor intended.

We shall designate a complex number by means of a special single symbol; thus, $\mathbf{A} = a + jb$. The complex nature of the number is indicated by the use of boldface type; in handwritten material, a bar over the letter is customary. The complex number **A** just shown is described as having a *real component* or *real part a* and an *imaginary component* or *imaginary part b*. This is also expressed as

$$\text{Re}\{\mathbf{A}\} = a$$
$$\text{Im}\{\mathbf{A}\} = b$$

The imaginary component of **A** is *not jb*. By definition, the imaginary component is a real number.

It should be noted that all real numbers may be regarded as complex numbers having imaginary parts equal to zero. The real numbers are therefore included in the system of complex numbers, and we may now consider them as a special case. When we define the fundamental arithmetic operations for complex numbers, we should therefore expect them to reduce to the corresponding definitions for real numbers if the imaginary part of every complex number is set equal to zero.

Since any complex number is completely characterized by a pair of real numbers, such as a and b in the previous example, we can obtain some visual assistance by representing a complex number graphically on a rectangular, or Cartesian, coordinate system. By providing ourselves with a real axis and an imaginary axis, as shown in Fig. A5.1, we form a *complex plane,* or *Argand diagram,* on which any complex number can be represented as a single point. The complex numbers $\mathbf{M} = 3 + j1$ and $\mathbf{N} = 2 - j2$ are indicated. It is important to understand that this complex

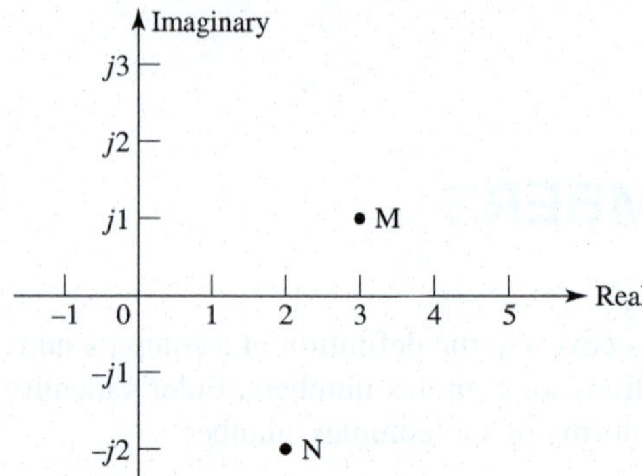

FIGURE A5.1 The complex numbers $\mathbf{M} = 3 + j1$ and $\mathbf{N} = 2 - j2$ are shown on the complex plane. Often, just "j" is written instead of "$j1$," although either is correct.

plane is only a visual aid; it is not at all essential to the mathematical statements which follow.

We define two complex numbers as being equal if, and only if, their real parts are equal and their imaginary parts are equal. Graphically, then, to each point in the complex plane there corresponds only one complex number, and conversely, to each complex number there corresponds only one point in the complex plane. Thus, suppose we are given the two complex numbers

$$\mathbf{A} = a + jb \quad \text{and} \quad \mathbf{B} = c + jd$$

Then, if

$$\mathbf{A} = \mathbf{B}$$

it is necessary that

$$a = c \quad \text{and} \quad b = d$$

A complex number expressed as the sum of a real number and an imaginary number, such as $\mathbf{A} = a + jb$, is said to be in *rectangular* or *Cartesian* form. Other forms for a complex number will appear shortly.

Let us now define the fundamental operations of addition, subtraction, multiplication, and division for complex numbers. The sum of two complex numbers is defined as the complex number whose real part is the sum of the real parts of the two complex numbers and whose imaginary part is the sum of the imaginary parts of the two complex numbers. Thus,

$$(a + jb) + (c + jd) = (a + c) + j(b + d)$$

For example,

$$(3 + j4) + (4 - j2) = 7 + j2$$

The difference of two complex numbers is taken in a similar manner; for example,

$$(3 + j4) - (4 - j2) = -1 + j6$$

Addition and subtraction of complex numbers may also be accomplished graphically on the complex plane. Each complex number is represented as a vector, or directed line segment, and the sum is obtained by completing the parallelogram, illustrated by Fig. A5.2*a*, or by connecting the vectors in a head-to-tail manner, as shown in Fig. A5.2*b*. A graphical sketch is often useful as a check for a more exact numerical solution.

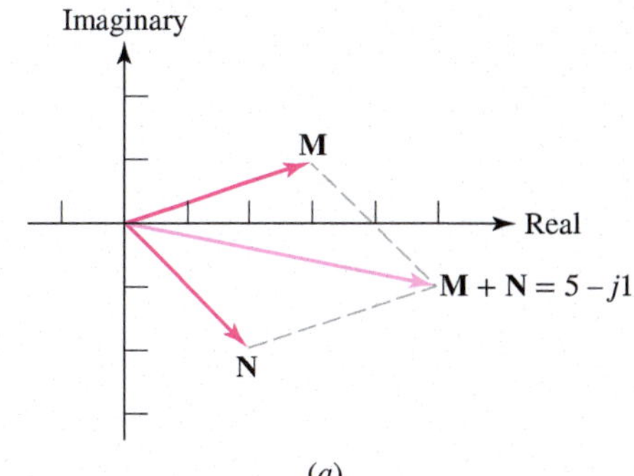

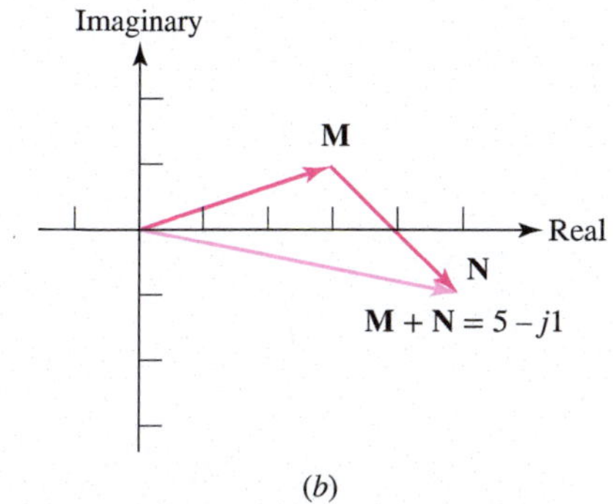

FIGURE A5.2 (*a*) The sum of the complex numbers $\mathbf{M} = 3 + j1$ and $\mathbf{N} = 2 - j2$ is obtained by constructing a parallelogram. (*b*) The sum of the same two complex numbers is found by a head-to-tail combination.

The product of two complex numbers is defined by

$$(a + jb)(c + jd) = (ac - bd) + j(bc + ad)$$

This result may be easily obtained by a direct multiplication of the two binomial terms, using the rules of the algebra of real numbers, and then simplifying the result by letting $j^2 = -1$. For example,

$$\begin{aligned}(3 + j4)(4 - j2) &= 12 - j6 + j16 - 8j^2\\ &= 12 + j10 + 8\\ &= 20 + j10\end{aligned}$$

It is easier to multiply the complex numbers by this method, particularly if we immediately replace j^2 with -1, than it is to substitute in the general formula that defines the multiplication.

Before defining the operation of division for complex numbers, we should define the conjugate of a complex number. The *conjugate* of the complex number $\mathbf{A} = a + jb$ is $a - jb$ and is represented as $\mathbf{A}^*$. The conjugate of any complex number is therefore easily obtained by merely changing the sign of the imaginary part of the complex number. Thus, if

$$\mathbf{A} = 5 + j3$$

then

$$\mathbf{A}^* = 5 - j3$$

It is evident that the conjugate of any complicated complex expression may be found by replacing every complex term in the expression by its conjugate, which may be obtained by replacing every j in the expression by $-j$.

The definitions of addition, subtraction, and multiplication show that the following statements are true: the sum of a complex number and its conjugate is a real number; the difference of a complex number and its conjugate is an imaginary number; and the product of a complex number and its conjugate is a real number. It is also evident that if $\mathbf{A}^*$ is the conjugate of $\mathbf{A}$, then $\mathbf{A}$ is the conjugate of $\mathbf{A}^*$; in other words, $\mathbf{A} = (\mathbf{A}^*)^*$. A complex number and its conjugate are said to form a *conjugate complex pair* of numbers.

In any physical problem, a complex number will somehow be accompanied by its conjugate.

We now define the quotient of two complex numbers:

$$\frac{\mathbf{A}}{\mathbf{B}} = \frac{(\mathbf{A})(\mathbf{B}^*)}{(\mathbf{B})(\mathbf{B}^*)}$$

and thus

$$\frac{a + jb}{c + jd} = \frac{(ac + bd) + j(bc - ad)}{c^2 + d^2}$$

We multiply numerator and denominator by the conjugate of the denominator in order to obtain a denominator which is real; this process is called *rationalizing the denominator.* As a numerical example,

$$\frac{3+j4}{4-j2} = \frac{(3+j4)(4+j2)}{(4-j2)(4+j2)}$$

$$= \frac{4+j22}{16+4} = 0.2 + j1.1$$

The addition or subtraction of two complex numbers which are each expressed in rectangular form is a relatively simple operation; multiplication or division of two complex numbers in rectangular form, however, is a rather unwieldy process. These latter two operations will be found to be much simpler when the complex numbers are given in either exponential or polar form. These forms will be introduced in Secs. A5.3 and A5.4.

PRACTICE

A5.1 Let $\mathbf{A} = -4 + j5$, $\mathbf{B} = 3 - j2$, and $\mathbf{C} = -6 - j5$, and find (*a*) $\mathbf{C} - \mathbf{B}$; (*b*) $2\mathbf{A} - 3\mathbf{B} + 5\mathbf{C}$; (*c*) $j^5\mathbf{C}^2(\mathbf{A} + \mathbf{B})$; (*d*) $\mathbf{B}\,\text{Re}\{\mathbf{A}\} + \mathbf{A}\,\text{Re}\{\mathbf{B}\}$.

A5.2 Using the same values for **A**, **B**, and **C** as in the previous problem, find (*a*) $[(\mathbf{A} - \mathbf{A}^*)(\mathbf{B} + \mathbf{B}^*)^*]^*$; (*b*) $(1/\mathbf{C}) - (1/\mathbf{B})^*$; (*c*) $(\mathbf{B} + \mathbf{C})/(2\mathbf{BC})$.

Ans: A5.1: $-9 - j3$; $-47 - j9$; $27 - j191$; $-24 + j23$.
A5.2: $-j60$; $-0.329 + j0.236$; $0.0662 + j0.1179$.

A5.2 EULER'S IDENTITY

In Chap. 9 we encounter functions of time which contain complex numbers, and we are concerned with the differentiation and integration of these functions with respect to the real variable t. We differentiate and integrate such functions with respect to t by exactly the same procedures we use for real functions of time. That is, the complex constants are treated just as though they were real constants when performing the operations of differentiation or integration. If $\mathbf{f}(t)$ is a complex function of time, such as

$$\mathbf{f}(t) = a\cos ct + jb\sin ct$$

then

$$\frac{d\mathbf{f}(t)}{dt} = -ac\sin ct + jbc\cos ct$$

and

$$\int \mathbf{f}(t)dt = \frac{a}{c}\sin ct - j\frac{b}{c}\cos ct + \text{C}$$

where the constant of integration **C** is a complex number in general.

It is sometimes necessary to differentiate or integrate a function of a complex variable with respect to that complex variable. In general, the successful accomplishment of either of these operations requires that the function which is to be differentiated or integrated satisfy certain conditions. All our functions do meet these conditions, and integration or differentiation with respect to a complex variable is achieved by using methods identical to those used for real variables.

At this time we must make use of a very important fundamental relationship known as Euler's identity (pronounced "oilers"). We shall prove this identity, for it is extremely useful in representing a complex number in a form other than rectangular form.

The proof is based on the power series expansions of $\cos\theta$, $\sin\theta$, and e^z, given somewhere (hopefully) in your favorite college calculus text:

$$\cos\theta = 1 - \frac{\theta^2}{2!} + \frac{\theta^4}{4!} - \frac{\theta^6}{6!} + \cdots$$

$$\sin\theta = \theta - \frac{\theta^3}{3!} + \frac{\theta^5}{5!} - \frac{\theta^7}{7!} + \cdots$$

or

$$\cos\theta + j\sin\theta = 1 + j\theta - \frac{\theta^2}{2!} - j\frac{\theta^3}{3!} + \frac{\theta^4}{4!} + j\frac{\theta^5}{5!} - \cdots$$

and

$$e^z = 1 + z + \frac{z^2}{2!} + \frac{z^3}{3!} + \frac{z^4}{4!} + \frac{z^5}{5!} + \cdots$$

so that

$$e^{j\theta} = 1 + j\theta - \frac{\theta^2}{2!} - j\frac{\theta^3}{3!} + \frac{\theta^4}{4!} + \cdots$$

We conclude that

$$e^{j\theta} = \cos\theta + j\sin\theta \qquad [1]$$

or, if we let $z = -j\theta$, we find that

$$e^{-j\theta} = \cos\theta - j\sin\theta \qquad [2]$$

By adding and subtracting Eqs. [1] and [2], we obtain the two expressions which we use without proof in our study of the underdamped natural response of the parallel and series *RLC* circuits,

$$\cos\theta = \frac{1}{2}(e^{j\theta} + e^{-j\theta}) \qquad [3]$$

$$\sin\theta = -j\frac{1}{2}(e^{j\theta} + e^{-j\theta}) \qquad [4]$$

PRACTICE

A5.3 Use Eqs. [1] through [4] to evaluate (*a*) e^{-j1}; (*b*) e^{1-j1}; (*c*) $\cos(-j1)$; (*d*) $\sin(-j1)$.

A5.4 Evaluate at $t = 0.5$: (*a*) $(d/dt)(3\cos 2t - j2\sin 2t)$; (*b*) $\int_0^t (3\cos 2t - j2\sin 2t)dt$. Evaluate at $\mathbf{s} = 1 + j2$: (*c*) $\int_{\mathbf{s}}^{\infty} \mathbf{s}^{-3} d\mathbf{s}$; (*d*) $(d/d\mathbf{s})[3/(\mathbf{s} + 2)]$.

Ans: A5.3: 0.540 − *j*0.841; 1.469 − *j*2.29; 1.543; −*j*1.175. A5.4: −5.05 − *j*2.16; 1.262 − *j*0.460; −0.06 − *j*0.08; −0.0888 + *j*0.213.

A5.3 THE EXPONENTIAL FORM

Let us now take Euler's identity

$$e^{j\theta} = \cos\theta + j\sin\theta$$

and multiply each side by the real positive number C:

$$Ce^{j\theta} = C\cos\theta + jC\sin\theta \qquad [5]$$

The right-hand side of Eq. [5] consists of the sum of a real number and an imaginary number and thus represents a complex number in rectangular form; let us call this complex number $\mathbf{A}$, where $\mathbf{A} = a + jb$. By equating the real parts

$$a = C\cos\theta \qquad [6]$$

and the imaginary parts

$$b = C\sin\theta \qquad [7]$$

then squaring and adding Eqs. [6] and [7],

$$a^2 + b^2 = C^2$$

or

$$C = +\sqrt{a^2 + b^2} \qquad [8]$$

and dividing Eq. [7] by Eq. [6]:

$$\frac{b}{a} = \tan\theta$$

or

$$\theta = \tan^{-1}\frac{b}{a} \qquad [9]$$

we obtain the relationships of Eqs. [8] and [9], which enable us to determine C and θ from a knowledge of a and b. For example, if $\mathbf{A} = 4 + j2$, then we identify a as 4 and b as 2 and find C and θ:

$$C = \sqrt{4^2 + 2^2} = 4.47$$

$$\theta = \tan^{-1}\frac{2}{4} = 26.6^\circ$$

We could use this new information to write $\mathbf{A}$ in the form

$$\mathbf{A} = 4.47\cos 26.6^\circ + j4.47\sin 26.6^\circ$$

but it is the form of the left-hand side of Eq. [5] which will prove to be the more useful:

$$A = Ce^{j\theta} = 4.47e^{j26.6^\circ}$$

A complex number expressed in this manner is said to be in *exponential form*. The real positive multiplying factor C is known as the *amplitude* or *magnitude,* and the real quantity θ appearing in the exponent is called the *argument* or *angle*. A mathematician would always express θ in radians and would write

$$\mathbf{A} = 4.47e^{j0.464}$$

but engineers customarily work in terms of degrees. The use of the degree symbol (°) in the exponent should make confusion impossible.

To recapitulate, if we have a complex number which is given in rectangular form,

$$\mathbf{A} = a + jb$$

and wish to express it in exponential form,

$$\mathbf{A} = Ce^{j\theta}$$

we may find C and θ by Eqs. [8] and [9]. If we are given the complex number in exponential form, then we may find a and b by Eqs. [6] and [7].

When **A** is expressed in terms of numerical values, the transformation between exponential (or polar) and rectangular forms is available as a built-in operation on most handheld scientific calculators.

One question will be found to arise in the determination of the angle θ by using the arctangent relationship of Eq. [9]. This function is multivalued, and an appropriate angle must be selected from various possibilities. One method by which the choice may be made is to select an angle for which the sine and cosine have the proper signs to produce the required values of a and b from Eqs. [6] and [7]. For example, let us convert

$$\mathbf{V} = 4 - j3$$

to exponential form. The amplitude is

$$C = \sqrt{4^2 + (-3)^2} = 5$$

and the angle is

$$\theta = \tan^{-1}\left(\frac{-3}{4}\right) \qquad [10]$$

A value of θ has to be selected which leads to a positive value for $\cos\theta$, since $4 = 5\cos\theta$, and a negative value for $\sin\theta$, since $-3 = 5\sin\theta$. We therefore obtain $\theta = -36.9°$, $323.1°$, $-396.9°$, and so forth. Any of these angles is correct, and we usually select that one which is the simplest, here, $-36.9°$. We should note that the alternative solution of Eq. [10], $\theta = 143.1°$, is not correct, because $\cos\theta$ is negative and $\sin\theta$ is positive.

A simpler method of selecting the correct angle is available if we represent the complex number graphically in the complex plane. Let us first select a complex number, given in rectangular form, $\mathbf{A} = a + jb$, which lies in the first quadrant of the complex plane, as illustrated in Fig. A5.3. If we draw a line from the origin to the point which represents the complex number, we shall have constructed a right triangle whose hypotenuse is evidently the amplitude of the exponential representation of the complex number. In other words, $C = \sqrt{a^2 + b^2}$. Moreover, the counterclockwise angle which the line makes with the positive real axis is seen to be the angle θ of the exponential representation, because $a = C\cos\theta$ and $b = C\sin\theta$. Now if we are given the rectangular form of a complex number which lies in another quadrant, such as $\mathbf{V} = 4 - j3$, which is depicted in Fig. A5.4, the correct angle is graphically evident, either $-36.9°$ or $323.1°$ for this example. The sketch may often be visualized and need not be drawn.

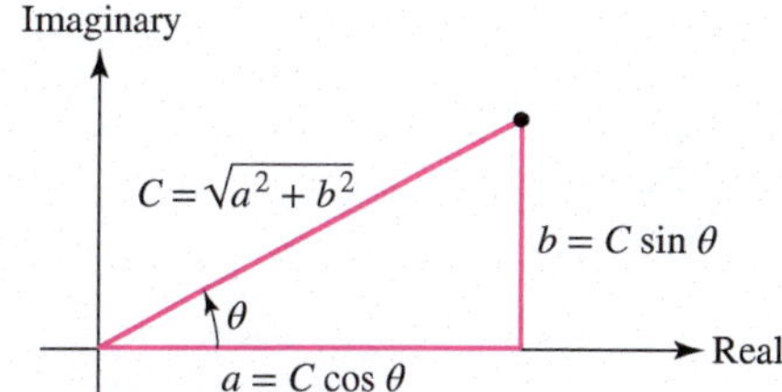

FIGURE A5.3 A complex number may be represented by a point in the complex plane by choosing the correct real and imaginary parts from the rectangular form, or by selecting the magnitude and angle from the exponential form.

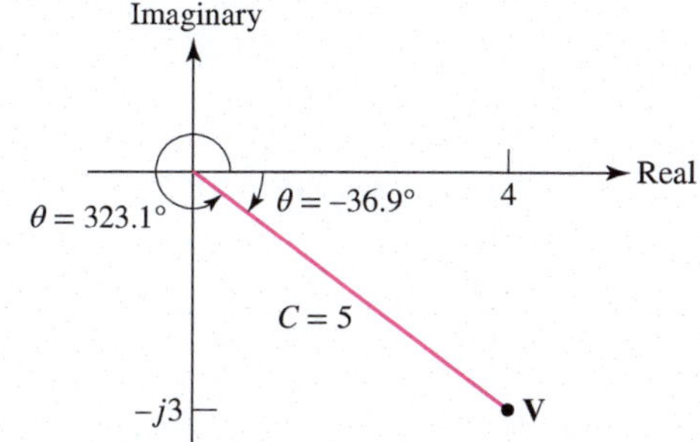

FIGURE A5.4 The complex number $\mathbf{V} = 4 - j3 = 5e^{-j36.9°}$ is represented in the complex plane.

If the rectangular form of the complex number has a negative real part, it is often easier to work with the negative of the complex number, thus avoiding angles greater than 90° in magnitude. For example, given

$$\mathbf{I} = -5 + j2$$

we write

$$\mathbf{I} = -(5 - j2)$$

and then transform $(5 - j2)$ to exponential form:

$$\mathbf{I} = -Ce^{j\theta}$$

where

$$C = \sqrt{29} = 5.39 \quad \text{and} \quad \theta = \tan^{-1}\left(\frac{-2}{5}\right) = -21.8°$$

We therefore have

$$\mathbf{I} = -5.39e^{-j21.8°}$$

The negative sign may be removed from the complex number by increasing or decreasing the angle by 180°, as shown by reference to a sketch in the complex plane. Thus, the result may be expressed in exponential form as

$$\mathbf{I} = 5.39e^{j158.2°} \quad \text{or} \quad \mathbf{I} = 5.39e^{-j201.8°}$$

Note that use of an electronic calculator in the inverse tangent mode always yields angles having magnitudes less than 90°. Thus, both $\tan^{-1}[(-3)/4]$ and $\tan^{-1}[3/(-4)]$ come out as −36.9°. Calculators capable of performing rectangular-to-polar conversion, however, give the correct angle in all cases.

One last remark about the exponential representation of a complex number should be made. Two complex numbers, both written in exponential form, are equal if, and only if, their amplitudes are equal and their angles are equivalent. Equivalent angles are those which differ by multiples of 360°.

For example, if $\mathbf{A} = Ce^{j\theta}$ and $\mathbf{B} = De^{j\phi}$, then if $\mathbf{A} = \mathbf{B}$, it is necessary that $C = D$ and $\theta = \phi \pm (360°)n$, where $n = 0, 1, 2, 3, \ldots$.

PRACTICE

A5.5 Express each of the following complex numbers in exponential form, using an angle lying in the range $-180° < \theta \leq 180°$: (*a*) $-18.5 - j26.1$; (*b*) $17.9 - j12.2$; (*c*) $-21.6 + j31.2$.

A5.6 Express each of these complex numbers in rectangular form: (*a*) $61.2e^{-j111.1°}$; (*b*) $-36.2e^{j108°}$; (*c*) $5e^{-j2.5}$.

Ans: A5.5: $32.0e^{-j125.3°}$; $21.7e^{-j34.3°}$; $37.9e^{j124.7°}$. A5.6: $-22.0 - j57.1$; $11.19 - j34.4$; $-4.01 - j2.99$.

A5.4 THE POLAR FORM

The third (and last) form in which we may represent a complex number is essentially the same as the exponential form, except for a slight difference in symbolism. We use an angle sign ($\angle$) to replace the combination e^j. Thus, the exponential representation of a complex number **A**,

$$\mathbf{A} = Ce^{j\theta}$$

may be written somewhat more concisely as

$$\mathbf{A} = C\underline{/\theta}$$

The complex number is now said to be expressed in *polar* form, a name which suggests the representation of a point in a (complex) plane through the use of polar coordinates.

It is apparent that transformation from rectangular to polar form or from polar form to rectangular form is basically the same as transformation between rectangular and exponential form. The same relationships exist between C, θ, a, and b.

The complex number

$$\mathbf{A} = -2 + j5$$

is thus written in exponential form as

$$\mathbf{A} = 5.39e^{j111.8^\circ}$$

and in polar form as

$$\mathbf{A} = 5.39\underline{/111.8^\circ}$$

In order to appreciate the utility of the exponential and polar forms, let us consider the multiplication and division of two complex numbers represented in exponential or polar form. If we are given

$$\mathbf{A} = 5\underline{/53.1^\circ} \quad \text{and} \quad \mathbf{B} = 15\underline{/-36.9^\circ}$$

then the expression of these two complex numbers in exponential form

$$\mathbf{A} = 5e^{j53.1^\circ} \quad \text{and} \quad \mathbf{B} = 15e^{-j36.9^\circ}$$

enables us to write the product as a complex number in exponential form whose amplitude is the product of the amplitudes and whose angle is the algebraic sum of the angles, in accordance with the normal rules for multiplying two exponential quantities:

$$(\mathbf{A})(\mathbf{B}) = (5)(15)e^{j(53.1^\circ - 36.9^\circ)}$$

or

$$\mathbf{AB} = 75e^{j16.2^\circ} = 75\underline{/16.2^\circ}$$

From the definition of the polar form, it is evident that

$$\frac{\mathbf{A}}{\mathbf{B}} = 0.333\underline{/90^\circ}$$

Addition and subtraction of complex numbers are accomplished most easily by operating on complex numbers in rectangular form, and the addition or subtraction of two complex numbers given in exponential or polar form should begin with the conversion of the two complex numbers to rectangular form. The reverse situation applies to multiplication and division; two numbers given in rectangular form should be transformed to polar form, unless the numbers happen to be small integers. For example, if we wish to multiply $(1 - j3)$ by $(2 + j1)$, it is easier to multiply them directly as they stand and obtain $(5 - j5)$. If the numbers can be multiplied mentally, then time is wasted in transforming them to polar form.

We should now try to become familiar with the three different forms in which complex numbers may be expressed and with the rapid conversion from one form to another. The relationships among the three forms seem almost endless, and the following lengthy equation summarizes the various interrelationships

$$\begin{aligned}\mathbf{A} &= a + jb \\ &= \mathrm{Re}\{\mathbf{A}\} + j\,\mathrm{Im}\{\mathbf{A}\} \\ &= Ce^{j\theta} \\ &= \sqrt{a^2+b^2}\,e^{j\tan^{-1}(b/a)} \\ &= \sqrt{a^2+b^2}\,\underline{/\tan^{-1}(b/a)}\end{aligned}$$

Most of the conversions from one form to another can be done quickly with the help of a calculator, and many calculators are equipped to solve linear equations with complex numbers.

We will find that complex numbers are a convenient mathematical tool which facilitates the analysis of real physical situations.

PRACTICE

A5.7 Express the result of each of these complex-number manipulations in polar form, using six significant figures just for the pure joy of calculating (*a*) $[2 - (1\underline{/-41^\circ})]/(0.3\underline{/41^\circ})$; (*b*) $50/(2.87\underline{/83.6^\circ} + 5.16\underline{/63.2^\circ})$; (*c*) $4\underline{/18^\circ} - 6\underline{/-75^\circ} + 5\underline{/28^\circ}$.

A5.8 Find $\mathbf{Z}$ in rectangular form if (*a*) $\mathbf{Z} + j2 = 3/\mathbf{Z}$; (*b*) $\mathbf{Z} = 2\ln(2 - j3)$; (*c*) $\sin \mathbf{Z} = 3$.

Ans: A5.7: $4.69179\underline{/-13.2183^\circ}$; $6.318\,33\underline{/-70.4626^\circ}$; $11.5066\underline{/54.5969^\circ}$. A5.8: $\pm 1.414 - j1$; $2.56 - j1.966$; $1.571 \pm j1.763$.

A BRIEF MATLAB® TUTORIAL

The intention of this tutorial is to provide a brief introduction to some basic concepts required to use a powerful software package known as MATLAB. The use of MATLAB is a completely optional part of the material in this textbook, but as it is becoming an increasingly common tool in all areas of electrical engineering, we felt that it was worthwhile to provide students with the opportunity to begin exploring some of its features, particularly in plotting 2D and 3D functions, performing matrix operations, solving simultaneous equations, and manipulating algebraic expressions. While some institutions provide the full version of MATLAB for their students, at the time of this writing, a student version is available at reduced cost from The MathWorks, Inc. (http://www.mathworks.com/academia/student_version/).

Getting Started

MATLAB is launched by clicking on the program icon; the typical opening window is shown in Fig. A6.1. Programs may be run from files or by

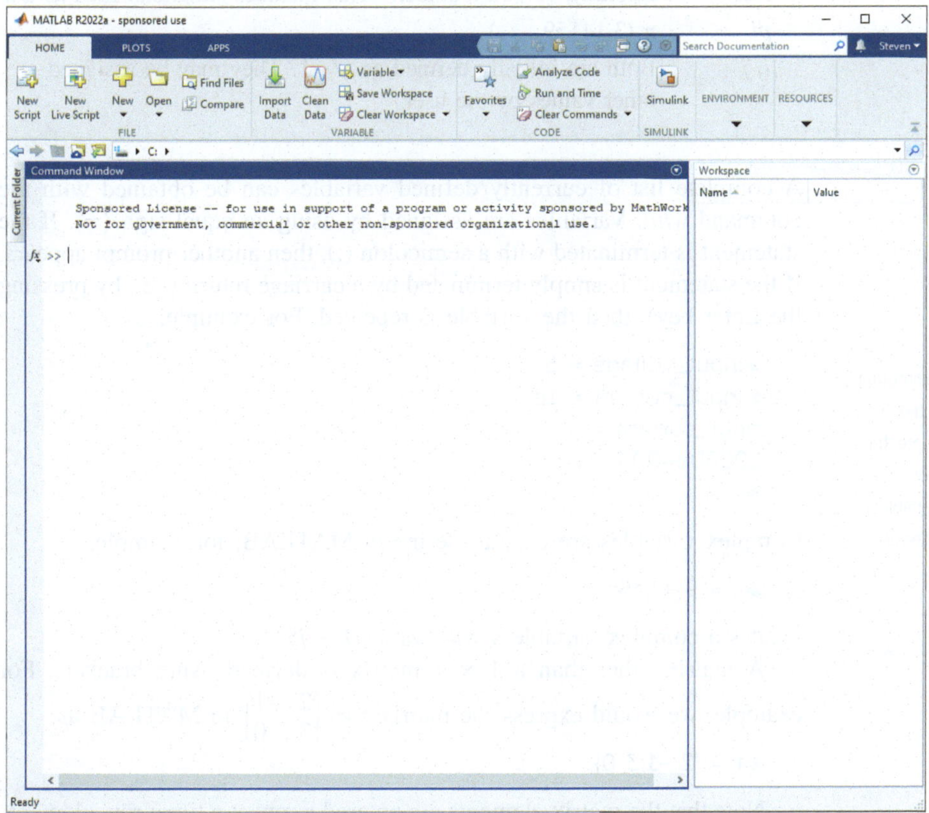

FIGURE A6.1 MATLAB command window upon start-up.
MATLAB

directly entering commands in the window. MATLAB also has extensive online help resources, useful for both beginners and advanced users. Typical MATLAB programs very much resemble programs written in C, although familiarity with this language is by no means required.

Variables and Mathematical Operations

MATLAB makes a great deal more sense once the user realizes that all variables are matrices, even if simply 1 × 1 matrices. Variable names can be up to 19 characters in length, which is extremely useful in constructing programs with adequate readability. The first character must be a letter, but all remaining characters can be any letter or number; the underscore (_) character may also be used. Variable names in MATLAB are case-sensitive. MATLAB includes several predefined variables. Relevant predefined variables for the material presented in this text include:

eps	The machine accuracy
realmin	The smallest (positive) floating-point number handled by the computer
realmax	The largest floating-point number handled by the computer
inf	Infinity (defined as 1/0)
NaN	Literally, "Not a Number." This includes situations such as 0/0
pi	π (3.14159 . . .)
i, j	Both are initially defined as $\sqrt{-1}$. They may be assigned other values by the user

A complete list of currently defined variables can be obtained with the command *who*. Variables are assigned by using an equal sign (=). If the statement is terminated with a semicolon (;), then another prompt appears. If the statement is simply terminated by a carriage return (i.e., by pressing the Enter key), then the variable is repeated. For example,

```
≫input_voltage = 5;
≫input_current = 1e − 3
input_current =
1.0000e−003
≫
```

Second color has been used to differentiate program-generated text from user-generated text for the convenience of the reader only. We should note that the most recent versions of MATLAB use color to separate different types of text (functions, variables, etc.) and to highlight potential typographical errors.

Complex variables are easy to define in MATLAB; for example,

```
≫s = 9 + j*5;
```

creates a complex variable **s** with value $9 + j5$.

A matrix other than a 1 × 1 matrix is defined using brackets. For example, we would express the matrix $\mathbf{t} = \begin{bmatrix} 2 & -1 \\ 3 & 0 \end{bmatrix}$ in MATLAB as

```
≫t = [2 −1;3 0];
```

Note that the matrix elements are entered a row at a time; row elements are separated by a space, and rows are separated by a semicolon (;). The same arithmetic operations are available for matrices; so, for example, we may find $t + t$ as

```
>>t + t
ans =
  4 -2
  6  0
```

Arithmetic operators include:

^	power	\	left division
*	multiplication	+	addition
/	right (ordinary) division	−	subtraction

The order of operations is important. The order of precedence is power, then multiplication and division, then addition and subtraction.

```
>>x = 1 + 5 ^ 2 * 3
  x =
  76
```

The concept of left division may seem strange at first, but is very useful in matrix algebra. For example,

```
>>1/5
ans =
  0.2000
>>1\5
ans =
  5
>>5\1
ans =
  0.2000
```

And, in the case of the matrix equation $\mathbf{Ax} = \mathbf{B}$, where

$$\mathbf{A} = \begin{bmatrix} 2 & 4 \\ 1 & 6 \end{bmatrix} \quad \text{and} \quad \mathbf{B} = \begin{bmatrix} -1 \\ 2 \end{bmatrix}$$

we find $\mathbf{x}$ with

```
>>A = [2 4; 1 6];
>>B = [-1; 2];
>>x = A\B
x =

  -1.7500
   0.6250
```

Alternatively, we can also write

```
>>x = A^ - 1 * B
x =
  -1.7500
   0.6250
```

or

```
>>inv (A) * B
ans =
  -1.7500
   0.6250
```

When in doubt, parentheses can help a great deal.

Some Useful Functions

Space requirements prevent us from listing every function contained in MATLAB. Some of the more basic ones include:

abs(x)	$\|x\|$	log10(x)	$\log 10x$		
exp(x)	e^x	sin(x)	$\sin x$	asin(x)	$\sin^{-1}x$
sqrt(x)	$\sqrt{x}$	cos(x)	$\cos x$	acos(x)	$\cos^{-1}x$
log(x)	Ln	tan(x)	$\tan x$	atan(x)	$\tan^{-1}x$

Functions useful for manipulating complex variables include:

real(s)	Re{s}
imag(s)	Im{s}
abs(s)	$\sqrt{a^2 + b^2}$, where $\mathbf{s} \equiv a + jb$
angle(s)	$\tan^{-1}(b/a)$, where $\mathbf{s} \equiv a + jb$
conj(s)	complex conjugate of **s**

Another extremely useful command, often forgotten, is simply *help*.

Occasionally we require a vector, such as when we plan to create a plot. The command *linspace* (min, max, number of points) is invaluable in such instances:

```
>>frequency = linspace (0, 10, 5)
frequency =
  0 2.5000 5.0000 7.5000 10.0000
```

A useful cousin is the command *logspace*().

Generating Plots

Plotting with MATLAB is extremely easy. For example, Fig. A6.2 shows the result of executing the following MATLAB program:

```
>>x = linspace (0, 2 * pi,100);
>>y = sin (x);
>>plot (x, y);
>>xlabel ('Angle (radians)')
>>ylabel ('f(x)');
```

Writing Programs

Although the MATLAB examples in this text are presented as lines typed into the Command Window, it is possible (and often prudent, if repetition is an issue) to write a program so that calculations are more convenient.

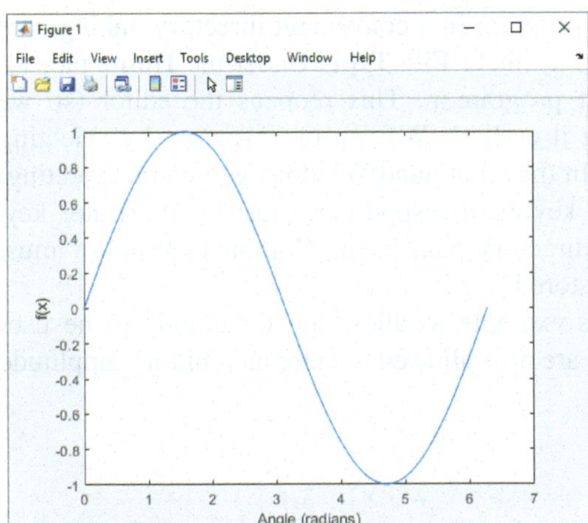

FIGURE A6.2 An example plot of sin(x), $0 < x < 2\pi$, generated using MATLAB. The variable x is a vector comprised of 100 equally spaced elements.
MATLAB

This is accomplished in MATLAB by writing what is termed an m-file, which is simply a text file saved with a ".m" extension (for example, first_program.m). In a nod to Kernighan and Ritchie, we pull down **New Script** under the **Home** tab, which opens up the editor. (Note that you can use another editor, for example, WordPad, if you prefer. In such instances, the file will have to be renamed to have a .m, instead of .txt, extension.)

We type in

```
r = input('Hello, World')
```

as shown in Fig. A6.3.

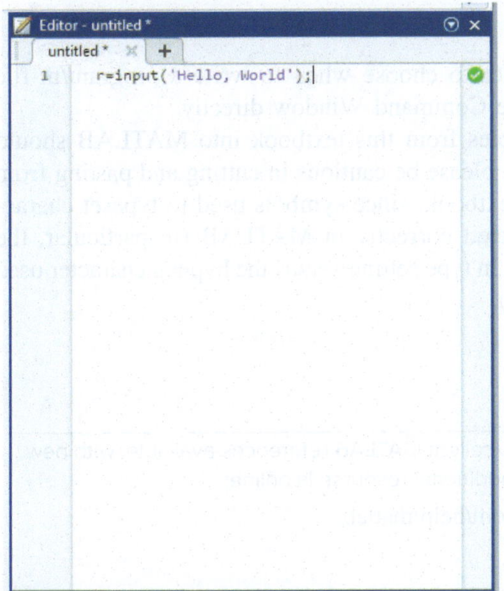

FIGURE A6.3 Example m-file created in m-file editor.
MATLAB

We next save it as first_program in a convenient directory, taking care to select MATLAB Files (*.m) under **File Type.** Under the **Home** tab, we select **Open,** and find first_program.m. This reopens the editor (so we could have skipped closing it earlier). We run our program by selecting **Run** under the **Editor** tab. In the Command Window, we see our greeting; MATLAB is waiting for a keyboard response, so just hit the Enter key. Note for this to work, the directory path in the Command Window must be set to where the file is stored.

Let's expand a previous example to allow the magnitude to be user selected as in Fig. A6.4. We are now allowed to enter an arbitrary amplitude for our plot.

Editor - C:\Users\Steve\Dropbox\HKD10E\Revised Chapters\Appendix 6\unti...

untitled2.m

```
amplitude = input('Enter sine wave amplitude: ');
x = linspace(0,2*pi,100);
y = amplitude*sin(x);
plot(x,y);
xlabel('Angle (radians)');
ylabel('f(x)');
```

FIGURE A6.4 Example m-file for generating sine wave plot.
MATLAB

We leave it to the reader to choose when to write a program/m-file and when to simply use the Command Window directly.

Directly entering examples from this textbook into MATLAB should work as intended. However, please be cautious in cutting and pasting from electronic versions of the textbook, since symbols used to typeset characters are not always interpreted correctly in MATLAB (in particular, the "minus sign" character used in type setting versus the hyphen character used in MATLAB programming).

READING FURTHER

There are a large number of excellent MATLAB references available, with new titles appearing regularly. An additional resource is online:

https://www.mathworks.com/help/matlab/

ADDITIONAL LAPLACE TRANSFORM THEOREMS

In this appendix, we briefly present several Laplace transform theorems typically used in more advanced situations in addition to those described in Chap. 14.

Transforms of Periodic Time Functions

The time-shift theorem is very useful in evaluating the transform of periodic time functions. Suppose that $f(t)$ is periodic with a period T for positive values of t. The behavior of $f(t)$ for $t < 0$ has no effect on the (one-sided) Laplace transform, as we know. Thus, $f(t)$ can be written as

$$f(t) = f(t - nT) \qquad n = 0, 1, 2, \ldots$$

If we now define a new time function which is nonzero only in the first period of $f(t)$,

$$f_1(t) = [u(t) - u(t - T)]\, f(t)$$

then the original $f(t)$ can be represented as the sum of an infinite number of such functions, delayed by integral multiples of T. That is,

$$\begin{aligned} f(t) &= [u(t) - u(t - T)]\, f(t) \\ &\quad + [u(t - T) - u(t - 2T)]\, f(t) \\ &\quad + [u(t - 2T) - u(t - 3T)]\, f(t) + \cdots \\ &= f_1(t) + f_1(t - T) + f_1(t - 2T) + \cdots \end{aligned}$$

or

$$f(t) = \sum_{n=0}^{\infty} f_1(t - nT)$$

The Laplace transform of this sum is just the sum of the transforms,

$$\mathbf{F(s)} = \sum_{n=0}^{\infty} \mathscr{L}\{f_1(t - nT)\}$$

so that the time-shift theorem leads to

$$\mathbf{F(s)} = \sum_{n=0}^{\infty} e^{-nTs}\,\mathbf{F}_1(\mathbf{s})$$

where

$$\mathbf{F}_1(\mathbf{s}) = \mathscr{L}\{f_1(t)\} = \int_{0^-}^{T} e^{-st} f(t)dt$$

Since $\mathbf{F}_1(\mathbf{s})$ is not a function of n, it can be removed from the summation, and $\mathbf{F}(\mathbf{s})$ becomes

$$\mathbf{F}(\mathbf{s}) = \mathbf{F}_1(\mathbf{s})[1 + e^{-Ts} + e^{-2Ts} + \cdots]$$

When we apply the binomial theorem to the bracketed expression, it simplifies to $1/(1 - e^{-Ts})$. Thus, we conclude that the periodic function $f(t)$, with period T, has a Laplace transform expressed by

$$\mathbf{F}(\mathrm{s}) = \frac{\mathbf{F}_1(\mathrm{s})}{1 - e^{-T\mathrm{s}}} \qquad [1]$$

where

$$\mathbf{F}_1(\mathbf{s}) = \mathscr{L}\{[u(t) - u(t - T)]f(t)\} \qquad [2]$$

is the transform of the first period of the time function.

To illustrate the use of this transform theorem for periodic functions, let us apply it to the familiar rectangular pulse train, Fig. A7.1. We may describe this periodic function analytically:

$$v(t) = \sum_{n=0}^{\infty} V_0[u(t - nT) - u(t - nT - \tau)] \qquad t > 0$$

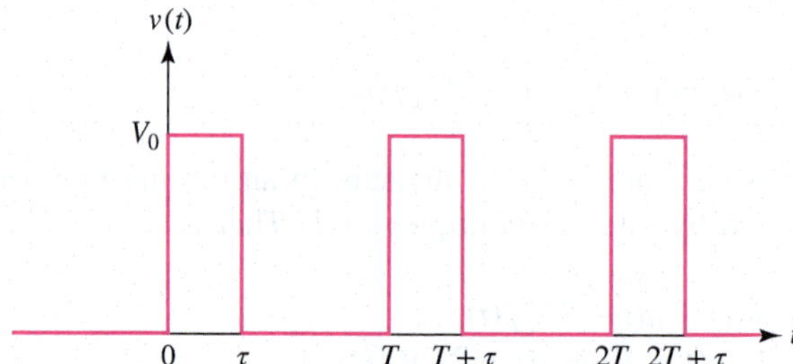

FIGURE A7.1 A periodic train of rectangular pulses for which $\mathbf{F}(\mathbf{s}) = (V_0/\mathbf{s})(1 - e^{-\mathbf{s}\tau})/(1 - e^{-\mathbf{s}T})$.

The function $\mathbf{V}_1(\mathrm{s})$ is simple to calculate:

$$\mathbf{V}_1(\mathrm{s}) = V_0 \int_{0^-}^{\tau} e^{-st} dt = \frac{V_0}{s}(1 - e^{s\tau})$$

Now, to obtain the desired transform, we just divide by $(1 - e^{-sT})$:

$$\mathbf{V}(\mathrm{s}) = \frac{V_0(1 - e^{-s\tau})}{s(1 - e^{-sT})} \qquad [3]$$

We should note how several different theorems show up in the transform in Eq. [3]. The $(1 - e^{-sT})$ factor in the denominator accounts for the periodicity of the function, the $e^{-s\tau}$ term in the numerator arises from the time delay of the negative square wave that turns off the pulse, and the V_0/s factor is, of course, the transform of the step functions involved in $v(t)$.

EXAMPLE **A7.1**

Determine the transform of the periodic function of Fig. A7.2.

We begin by writing an equation which describes $f(t)$, a function composed of alternating positive and negative impulse functions.

$$f(t) = 2\delta(t-1) - 2\delta(t-3) + 2\delta(t-5) - 2\delta(t-7) + \cdots$$

Defining a new function f_1 and recognizing a period $T = 4$ s,

$$f_1(t) = 2[\delta(t-1) - \delta(t-3)]$$

we can make use of the time periodicity operation as listed in Table 14.2 to find $\mathbf{F(s)}$

$$\mathbf{F(s)} = \frac{1}{1 - e^{-T_s}}\mathbf{F}_1\mathbf{(s)} \qquad [4]$$

where

$$\mathbf{F}_1\mathbf{(s)} = \int_{0^-}^{T} f(t)e^{-\mathbf{s}t}\,dt = \int_{0^-}^{4} f_1(t)e^{-\mathbf{s}t}\,dt$$

There are several ways to evaluate this integral. The easiest is to recognize that its value will remain the same if the upper limit is increased to ∞, allowing us to make use of the time-shift theorem. Thus,

$$\mathbf{F}_1\mathbf{(s)} = 1[e^{-\mathbf{s}} - e^{-3\mathbf{s}}] \qquad [5]$$

Our example is completed by multiplying Eq. [5] by the factor indicated in Eq. [4], so that

$$\mathbf{F(s)} = \frac{2}{1 - e^{-4\mathbf{s}}}(e^{-\mathbf{s}} - e^{-3\mathbf{s}}) = \frac{2e^{-\mathbf{s}}}{1 + e^{-2\mathbf{s}}}$$

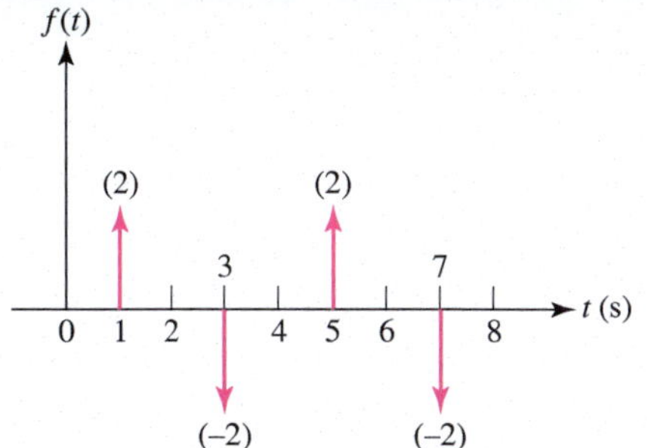

FIGURE A7.2 A periodic function based on unit-impulse functions.

PRACTICE

A7.1 Determine the Laplace transform of the periodic function shown in Fig. A7.3.

Ans: $\left(\frac{8}{\mathbf{s}^2 + \pi^2/4}\right)\frac{\mathbf{s} + (\pi/2)e^{-\mathbf{s}} + (\pi/2)e^{-3\mathbf{s}} - \mathbf{s}e^{-4\mathbf{s}}}{1 - e^{-4\mathbf{s}}}$

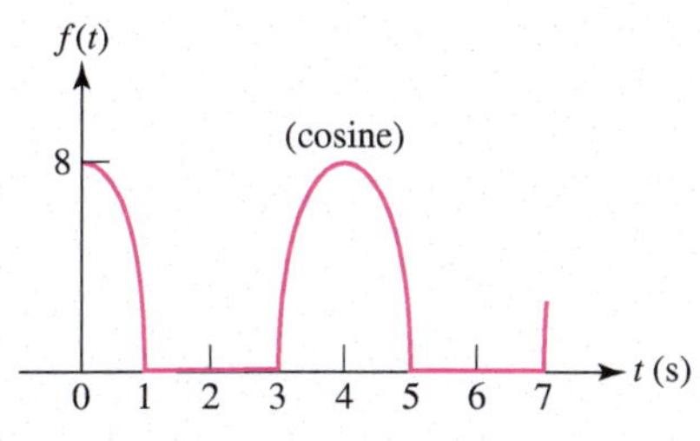

FIGURE A7.3

Frequency Shifting

The next new theorem establishes a relationship between $\mathbf{F(s)} = \mathscr{L}\{f(t)\}$ and $\mathbf{F(s} + a)$. We consider the Laplace transform of $e^{-at}f(t)$,

$$\mathscr{L}\{e^{-at}f(t)\} = \int_{0^-}^{\infty} e^{-\mathbf{s}t}e^{-at}f(t)dt = \int_{0^-}^{\infty} e^{-(\mathbf{s}+a)t}f(t)dt$$

Looking carefully at this result, we note that the integral on the right is identical to that defining $\mathbf{F(s)}$ with one exception: $(\mathbf{s} + a)$ appears in place of $\mathbf{s}$.

Thus,

$$e^{-at}f(t) \Leftrightarrow \mathbf{F(s} + a) \qquad [6]$$

We conclude that replacing $\mathbf{s}$ with $(\mathbf{s} + a)$ in the frequency domain corresponds to multiplication by e^{-at} in the time domain. This is known as the ***frequency-shift*** theorem. It can be put to immediate use in evaluating the transform of the exponentially damped cosine function that we used extensively in previous work. Beginning with the known transform of the cosine function,

$$\mathscr{L}\{\cos \omega_0 t\} = \mathbf{F}(\mathbf{s}) = \frac{\mathbf{s}}{\mathbf{s}^2 + \omega_0^2}$$

then the transform of $e^{-at} \cos \omega_0 t$ must be $\mathbf{F}(\mathbf{s} + a)$:

$$\mathscr{L}\{e^{-at} \cos \omega_0 t\} = \mathbf{F}(\mathbf{s} + a) = \frac{\mathbf{s} + a}{(\mathbf{s} + a)^2 + \omega_0^2} \qquad [7]$$

PRACTICE

A7.2 Find $\mathscr{L}\{e^{-2t} \sin(5t + 0.2\pi)u(t)\}$.

Ans: $(0.588\mathbf{s} + 4.05)/(\mathbf{s}^2 + 4\mathbf{s} + 29)$.

Differentiation in the Frequency Domain

Next let us examine the consequences of differentiating $\mathbf{F}(\mathbf{s})$ with respect to $\mathbf{s}$. The result is

$$\frac{d}{ds}\mathbf{F}(\mathbf{s}) = \frac{d}{ds}\int_{0^-}^{\infty} e^{-\mathbf{s}t} f(t)dt$$

$$= \int_{0^-}^{\infty} -te^{-\mathbf{s}t} f(t)dt = \int_{0^-}^{\infty} e^{-\mathbf{s}t}[-tf(t)]dt$$

which is simply the Laplace transform of $[-tf(t)]$. We therefore conclude that differentiation with respect to $\mathbf{s}$ in the frequency domain results in multiplication by $-t$ in the time domain, or

$$-tf(t) \Leftrightarrow \frac{d}{ds}\mathbf{F}(\mathbf{s}) \qquad [8]$$

Suppose now that $f(t)$ is the unit-ramp function $tu(t)$, whose transform we know is $1/\mathbf{s}^2$. We can use our newly acquired frequency-differentiation theorem to determine the inverse transform of $1/\mathbf{s}^3$ as follows:

$$\frac{d}{ds}\left(\frac{1}{\mathbf{s}^2}\right) = -\frac{2}{\mathbf{s}^3} \Leftrightarrow -t\mathscr{L}^{-1}\left\{\frac{1}{\mathbf{s}^2}\right\} = -t^2 u(t)$$

and

$$\frac{t^2 u(t)}{2} \Leftrightarrow \frac{1}{\mathbf{s}^3} \qquad [9]$$

Continuing with the same procedure, we find

$$\frac{t^3}{3!} u(t) \Leftrightarrow \frac{1}{\mathbf{s}^4} \qquad [10]$$

and in general

$$\frac{t^{(n-1)}}{(n-1)!} u(t) \Leftrightarrow \frac{1}{\mathbf{s}^n} \qquad [11]$$

PRACTICE

A7.3 Find $\mathscr{L}\{t \sin(5t + 0.2\pi)u(t)\}$.

Ans: $(0.588\mathbf{s}^2 + 8.09\mathbf{s} - 14.69)/(\mathbf{s}^2 + 25)^2$.

Integration in the Frequency Domain

The effect on $f(t)$ of integrating $\mathbf{F(s)}$ with respect to $\mathbf{s}$ may be shown by beginning with the definition once more,

$$\mathbf{F(s)} = \int_{0^-}^{\infty} e^{-\mathbf{s}t} f(t)dt$$

performing the frequency integration from $\mathbf{s}$ to ∞,

$$\int_{\mathbf{s}}^{\infty} \mathbf{F(s)}ds = \int_{\mathbf{s}}^{\infty} \left[\int_{0^-}^{\infty} e^{-\mathbf{s}t} f(t)dt\right] ds$$

interchanging the order of integration,

$$\int_{\mathbf{s}}^{\infty} \mathbf{F(s)}ds = \int_{0^-}^{\infty} \left[\int_{\mathbf{s}}^{\infty} e^{-\mathbf{s}t} ds\right] f(t)dt$$

and performing the inner integration,

$$\int_{\mathbf{s}}^{\infty} \mathbf{F(s)}ds = \int_{0^-}^{\infty} \left[-\frac{1}{t} e^{-\mathbf{s}t}\right]_{\mathbf{s}}^{\infty} f(t)dt = \int_{0^-}^{\infty} \frac{f(t)}{t} e^{-\mathbf{s}t} dt$$

Thus,

$$\frac{f(t)}{t} \Leftrightarrow \int_{\mathbf{s}}^{\infty} \mathbf{F(s)}ds \qquad [12]$$

For example, we have already established the transform pair

$$\sin \omega_0 t\, u(t) \Leftrightarrow \frac{\omega_0}{\mathbf{s}^2 + \omega_0^2}$$

Therefore,

$$\mathscr{L}\left\{\frac{\sin \omega_0 t\, u(t)}{t}\right\} = \int_{\mathbf{s}}^{\infty} \frac{\omega_0 d\mathbf{s}}{\mathbf{s}^2 + \omega_0^2} = \tan^{-1} \frac{\mathbf{s}}{\omega_0}\bigg|_{\mathbf{s}}^{\infty}$$

and we have

$$\frac{\sin \omega_0 t\, u(t)}{t} \Leftrightarrow \frac{\pi}{2} - \tan^{-1} \frac{\mathbf{s}}{\omega_0} \qquad [13]$$

PRACTICE

A7.4 Find $\mathscr{L}\{\sin^2 5tu(t)/t\}$.

Ans: $\frac{1}{4}\ln[(\mathbf{s}^2 + 100)/\mathbf{s}^2]$ $\ln[(\mathbf{s}^2 + 100)/\mathbf{s}^2]$.

The Time-Scaling Theorem

We next develop the time-scaling theorem of Laplace transform theory by evaluating the transform of $f(at)$, assuming that $\mathscr{L}\{f(t)\}$ is known. The procedure is very simple:

$$\mathscr{L}\{f(at)\} = \int_{0^-}^{\infty} e^{-\mathbf{s}t} f(at)dt = \frac{1}{a}\int_{0^-}^{\infty} e^{-(\mathbf{s}/a)\lambda} f(\lambda)d\lambda$$

where the change of variable $at = \lambda$ has been employed. The last integral is recognizable as $1/a$ times the Laplace transform of $f(t)$, except that $\mathbf{s}$ is replaced by $\mathbf{s}/a$ in the transform. It follows that

$$f(at) \Leftrightarrow \frac{1}{a}\mathbf{F}\left(\frac{\mathbf{s}}{a}\right) \qquad [14]$$

As an elementary example of the use of this time-scaling theorem, consider the determination of the transform of a 1 kHz cosine wave. Assuming we know the transform of a 1 rad/s cosine wave,

$$\cos t\, u(t) \Leftrightarrow \frac{\mathbf{s}}{\mathbf{s}^2 + 1}$$

the result is

$$\mathscr{L}\{\cos 2000\pi t\, u(t)\} = \frac{1}{2000\pi}\frac{\mathbf{s}/2000\pi}{(\mathbf{s}/2000\pi)^2 + 1} = \frac{\mathbf{s}}{\mathbf{s}^2 + (2000\pi)^2}$$

PRACTICE

A7.5 Determine $\mathscr{L}\{\sin^2 5t\, u(t)\}$.

Ans: $50/[\mathbf{s}(\mathbf{s}^2 + 100)]$.

THE COMPLEX FREQUENCY PLANE

Circuits with even a comparatively small number of elements can lead to rather unwieldy **s**-domain expressions. In such instances, a graphical representation of a particular circuit response or transfer function can provide useful insights. In this appendix, we introduce one such approach, based on the idea of the complex-frequency plane (Fig. A8.1). Complex frequency has two components (σ and ω), so we are naturally drawn to representing our functions using a three-dimensional model.

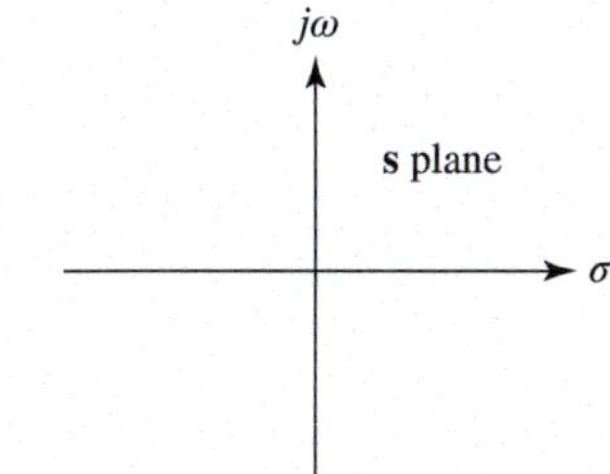

FIGURE A8.1 The complex-frequency plane, also referred to as the **s** plane.

Since ω represents an oscillating function, there is no physical distinction between a positive and negative frequency. In the case of σ, however, which can be identified with an exponential term, positive values are increasing in magnitude, whereas negative values are decaying. The origin of the **s** plane corresponds to dc (no time variation). A pictorial summary of these ideas is presented in Fig. A8.2.

To construct an appropriate three-dimensional representation of some function $\mathbf{F}(\mathbf{s})$, we first note that we have its magnitude in mind, although

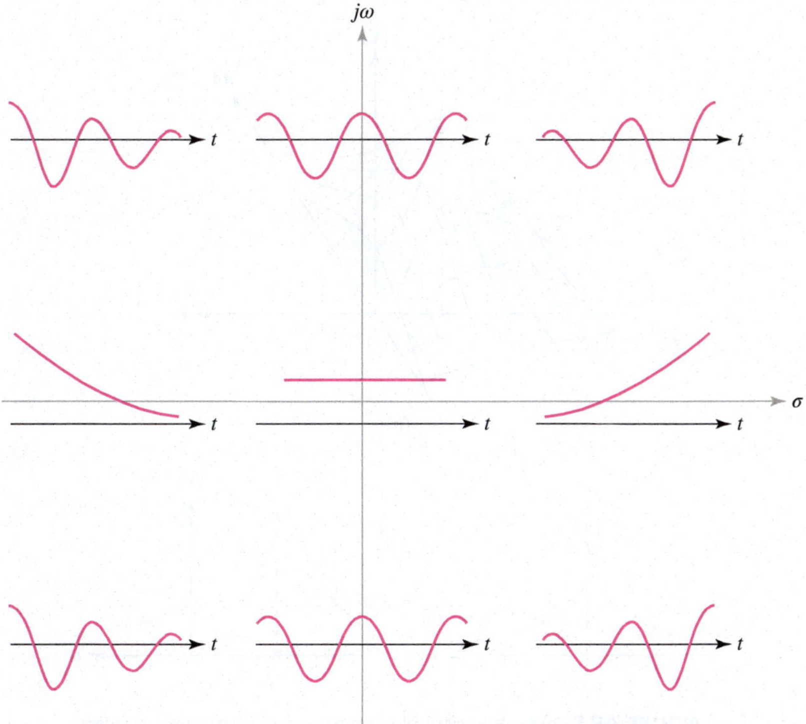

FIGURE A8.2 An illustration of the physical meaning of positive and negative values for σ and ω, as would be represented on the complex-frequency plane. When $\omega = 0$, a function will have no oscillatory component; when $\sigma = 0$, the function is purely sinusoidal except when ω is also zero.

the phase will have a strong complex-frequency dependence as well and may be graphed in a similar fashion. Thus, we will begin by substituting $\sigma + j\omega$ for $\mathbf{s}$ in our function $\mathbf{F}(\mathbf{s})$, then proceed to determine an expression for $|\mathbf{F}(\mathbf{s})|$. We next draw an axis perpendicular to the $\mathbf{s}$ plane, and we use this to plot $|\mathbf{F}(\mathbf{s})|$ for each value of σ and ω. The basic process is illustrated in Example 8.1.

EXAMPLE A8.1

Sketch the admittance of the series combination of a 1 H inductor and a 3 Ω resistor as a function of both $j\omega$ and σ.

The admittance of these two series elements is given by

$$\mathbf{Y}(\mathbf{s}) = \frac{1}{\mathbf{s} + 3}$$

Substituting $\mathbf{s} = \sigma + j\omega$, we find the magnitude of the function is

$$|\mathbf{Y}(\mathbf{s})| = \frac{1}{\sqrt{(\sigma + 3)^2 + \omega^2}}$$

When $\mathbf{s} = -3 + j0$, the response magnitude is infinite; when $\mathbf{s}$ is infinite, the magnitude of $\mathbf{Y}(\mathbf{s})$ is zero. Thus our model must have infinite height over the point $(-3 + j0)$, and it must have zero height at all points infinitely far away from the origin. A cutaway view of such a model is shown in Fig. A8.3*a*.

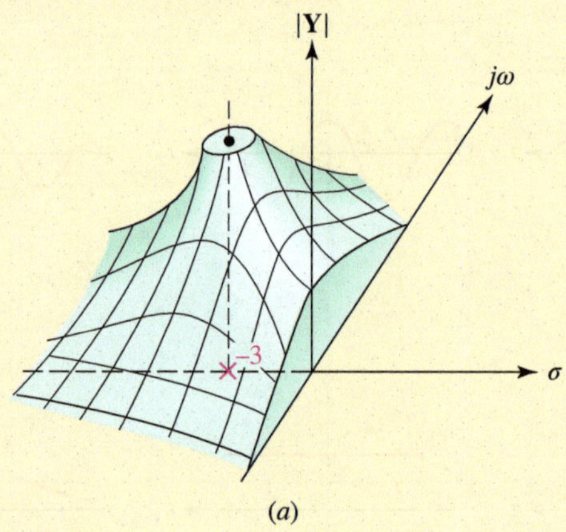

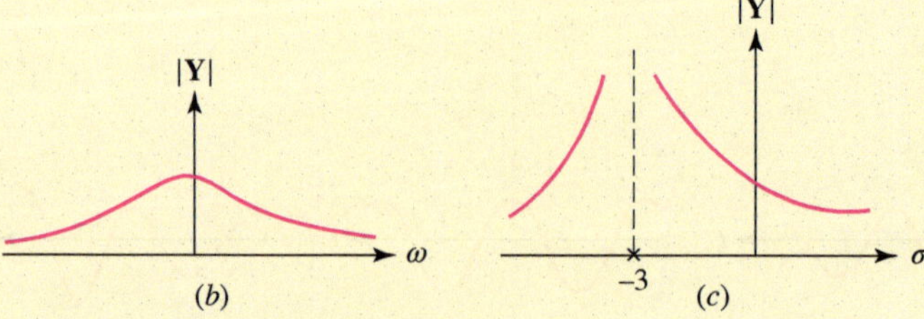

FIGURE A8.3 (*a*) A cutaway view of a clay model whose top surface represents $|\mathbf{Y}(\mathbf{s})|$ for the series combination of a 1 H inductor and a 3 Ω resistor. (*b*) $|\mathbf{Y}(\mathbf{s})|$ as a function of ω. (*c*) $|\mathbf{Y}(\mathbf{s})|$ as a function of σ.

Once the model is constructed, it is possible to visualize the variation of $|\mathbf{Y}|$ as a function of ω (with $\sigma = 0$) by cutting the model with a perpendicular plane containing the $j\omega$ axis. The model shown in Fig. A8.3*a* happens to be cut along this plane, and the desired plot of $|\mathbf{Y}|$ versus ω can be seen; the curve is also drawn in Fig. A8.3*b*. In a similar manner, a vertical plane containing the σ axis enables us to obtain $|\mathbf{Y}|$ versus σ (with $\omega = 0$), as shown in Fig. A8.3*c*.

PRACTICE

A8.1 Sketch the magnitude of the impedance $\mathbf{Z(s)} = 2 + 5\mathbf{s}$ as a function of σ and $j\omega$.

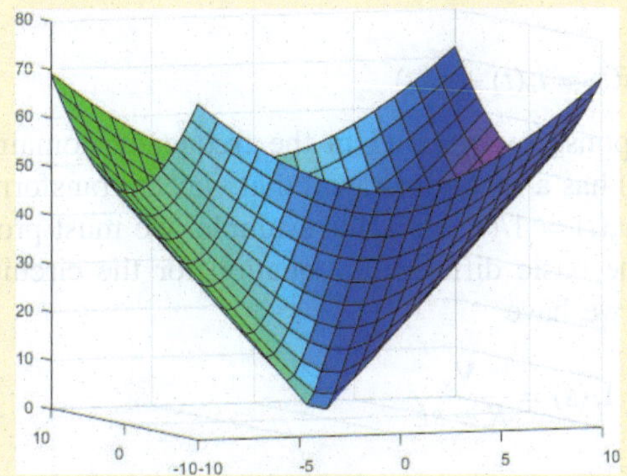

FIGURE A8.4 Solution for Practice Problem A8.1, generated with the following code:

```
>> sigma = linspace( −10, 10, 21);
>> omega = linspace( −10, 10, 21);
>> [X, Y] = meshgrid(sigma,omega);
>> Z = abs(2 + 5*X + j*5*Y);
>> colormap(hsv);
>> s = [−5 3 8];
>> surfl(X,Y,Z,s);
>> view (−20, 5)
```

Ans: See Fig. A8.4.

Building on this concept, there is a *tremendous* amount of information contained in the pole-zero plot of a forced response. We will now consider how such plots can be used to obtain the *complete* response of a circuit—natural plus forced—provided the initial conditions are known. The advantage of such an approach is a more *intuitive* linkage between the location of the critical frequencies, easily visualized through the pole-zero plot, and the desired response.

Let us introduce the method by considering the simplest example, a series *RL* circuit as shown in Fig. A8.5. A general voltage source $v_s(t)$ causes the current $i(t)$ to flow after closure of the switch at $t = 0$. The

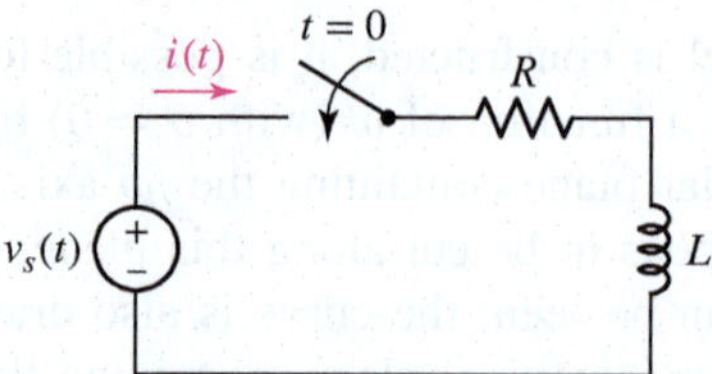

FIGURE A8.5 An example that illustrates the determination of the complete response through a knowledge of the critical frequencies of the impedance faced by the source.

complete response $i(t)$ for $t > 0$ is composed of a natural response and a forced response:

$$i(t) = i_n(t) + i_f(t)$$

We may find the forced response by working in the frequency domain, assuming, of course, that $v_s(t)$ has a functional form that we can transform to the frequency domain; if $v_s(t) = 1/(1 + t^2)$, for example, we must proceed as best we can from the basic differential equation for the circuit. For the circuit of Fig. A8.5, we have

$$\mathbf{I}_f(\mathbf{s}) = \frac{\mathbf{V}_s}{R + sL}$$

or

$$\mathbf{I}_f(\mathbf{s}) = \frac{1}{L}\frac{\mathbf{V}_s}{\mathbf{s} + R/L} \qquad [1]$$

Next we consider the natural response. From previous experience, we know that the form will be a decaying exponential with the time constant L/R, but let's pretend that we are finding it for the first time. The *form* of the natural (source-free) response is, by definition, independent of the forcing function; the forcing function contributes only to the *magnitude* of the natural response. To find the proper form, we turn off all independent sources; here, $v_s(t)$ is replaced by a short circuit. Next, we try to obtain the natural response as a limiting case of the forced response. Returning to the frequency-domain expression of Eq. [1], we obediently set $\mathbf{V}_s = 0$. On the surface, it appears that $\mathbf{I}(\mathbf{s})$ must also be zero, but this is not necessarily true if we are operating at a complex frequency that is a simple pole of $\mathbf{I}(\mathbf{s})$. Specifically, the denominator and the numerator may *both* be zero so that $\mathbf{I}(\mathbf{s})$ need not be zero.

What does it mean to "operate" at a complex frequency? How could we possibly accomplish such a thing in a real laboratory? In this instance, it is important to remember how we invented complex frequency to begin with: It is a means of describing a sinusoidal function of frequency ω multiplied by an exponential function $e^{\sigma t}$. Such types of signals are very easy to generate with real (i.e., nonimaginary) laboratory equipment. Thus, we need only set the value for σ and the value for ω in order to "operate" at $\mathbf{s} = \sigma + j\omega$.

Let us inspect this new idea from a slightly different vantage point. We fix our attention on the ratio of the desired forced response to the forcing function. We designate this ratio $\mathbf{H}(\mathbf{s})$ and define it to be the circuit transfer function. Then,

$$\frac{\mathbf{I}_f(\mathbf{s})}{\mathbf{V}_s} = \mathbf{H}(\mathbf{s}) = \frac{1}{L(s + R/L)}$$

In this example, the transfer function is the input admittance faced by $\mathbf{V}_s$. We seek the natural (source-free) response by setting $\mathbf{V}_s = 0$. However,

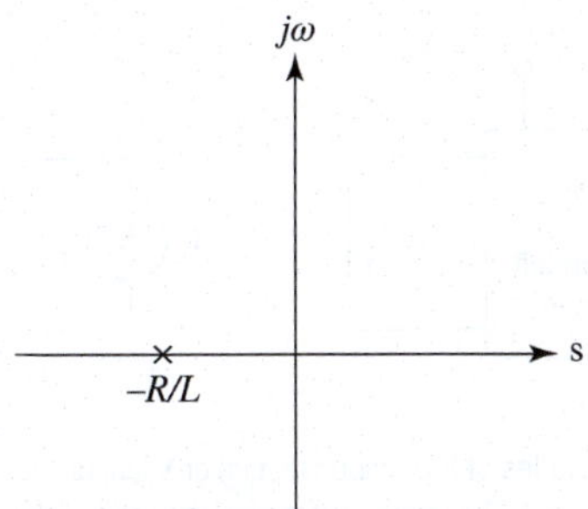

FIGURE A8.6 Pole-zero constellation of the transfer function $\mathbf{H}(\mathbf{s})$ showing the single pole at $\mathbf{s} = -R/L$.

$\mathbf{I}_f(\mathbf{s}) = \mathbf{V}_s\mathbf{H}(\mathbf{s})$, and if $\mathbf{V}_s = 0$, a nonzero value for the current can be obtained only by operating at a pole of $\mathbf{H}(\mathbf{s})$. The poles of the transfer function therefore assume a special significance.

In this example, we see that the pole of the transfer function occurs at $\mathbf{s} = -R/L + j0$, as shown in Fig. A8.6. If we choose to operate at this particular complex frequency, the only *finite* current that could result must be a constant in the $\mathbf{s}$-domain (i.e., frequency-independent). We thus obtain the natural response

$$\mathbf{I}\left(\mathbf{s} = -\frac{R}{L} + j0\right) = A$$

where A is an unknown constant. We next wish to transform this natural response to the time domain. Our knee-jerk reaction might be to try to apply inverse Laplace transform techniques in this situation. However, we have already specified a value of $\mathbf{s}$, so that such an approach is not valid. Instead, we look to the real part of our general function $e^{\mathbf{s}t}$, such that

$$i_n(t) = \text{Re}\{Ae^{\mathbf{s}t}\} = \text{Re}\{Ae^{-Rt/L}\}$$

In this case we find

$$i_n(t) = Ae^{-Rt/L}$$

so that the total response is then

$$i(t) = Ae^{-Rt/L} + i_f(t)$$

and A may be determined once the initial conditions are specified for this circuit. The forced response $i_f(t)$ is obtained by finding the inverse Laplace transform of $\mathbf{I}_f(\mathbf{s})$.

A More General Perspective

Figure A8.7 shows single sources connected to networks containing no independent sources. The desired response, which might be some current $\mathbf{I}_1(\mathbf{s})$ or some voltage $\mathbf{V}_2(\mathbf{s})$, may be expressed by a transfer function that displays all the critical frequencies. To be specific, we select the response $\mathbf{V}_2(\mathbf{s})$ in Fig. A8.7*a*:

$$\frac{\mathbf{V}_2(\mathbf{s})}{\mathbf{V}_s} = \mathbf{H}(\mathbf{s}) = k\frac{(\mathbf{s} - \mathbf{s}_1)(\mathbf{s} - \mathbf{s}_3)\ldots}{(\mathbf{s} - \mathbf{s}_2)(\mathbf{s} - \mathbf{s}_4)\ldots}$$

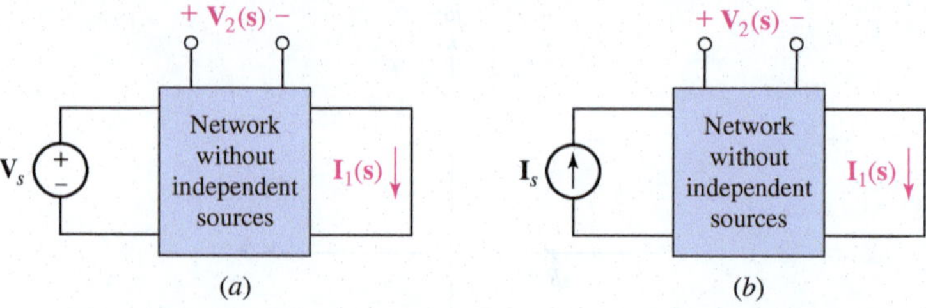

FIGURE A8.7 The poles of the response, $\mathbf{I}_1(\mathbf{s})$ or $\mathbf{V}_2(\mathbf{s})$, produced by (*a*) a voltage source $\mathbf{V}_s$ or (*b*) a current source $\mathbf{I}_s$. The poles determine the form of the natural response, $i_{1n}(t)$ or $v_{2n}(t)$, that occurs when $\mathbf{V}_s$ is replaced by a short circuit or $\mathbf{I}_s$ by an open circuit and some initial energy is available.

The poles of $\mathbf{H}(\mathbf{s})$ occur at $\mathbf{s} = \mathbf{s}_2, \mathbf{s}_4, \ldots$, and so a finite voltage $\mathbf{V}_2(\mathbf{s})$ at each of these frequencies must be a possible functional form for the natural response. Thus, we think of a zero-volt source (which is just a short-circuit) applied to the input terminals; the natural response that occurs when the input terminals are short-circuited must therefore have the form

$$v_{2n}(t) = \mathbf{A}_2 e^{\mathbf{s}_2 t} + \mathbf{A}_4 e^{\mathbf{s}_4 t} + \cdots$$

where each $\mathbf{A}$ must be evaluated in terms of the initial conditions (including the initial value of any voltage source applied at the input terminals).

To find the form of the natural response $i_{1n}(t)$ in Fig. A8.7*a*, we should determine the poles of the transfer function, $\mathbf{H}(\mathbf{s}) = \mathbf{I}_1(\mathbf{s})/\mathbf{V}_s$. The transfer functions applying to the situations depicted in Fig. A8.7*b* would be $\mathbf{I}_1(\mathbf{s})/\mathbf{I}_s$ and $\mathbf{V}_2(\mathbf{s})/\mathbf{I}_s$, and their poles then determine the natural responses $i_{1n}(t)$ and $v_{2n}(t)$, respectively.

If the natural response is desired for a network that does not contain any independent sources, then a source $\mathbf{V}_s$ or $\mathbf{I}_s$ may be inserted at any convenient point, restricted only by the condition that the original network is obtained when the source is set to zero. The corresponding transfer function is then determined and its poles specify the natural frequencies. Note that the same frequencies must be obtained for any of the many source locations possible. If the network already contains a source, that source may be set equal to zero and another source inserted at a more convenient point.

A Special Case

Before we illustrate this method with an example, completeness requires us to acknowledge a special case that might arise. This occurs when the network in Fig. A8.7*a* or *b* contains two or more parts that are isolated from each other. For example, we might have the parallel combination of three networks: R_1 in series with C, R_2 in series with L, and a short circuit. Clearly, a voltage source in series with R_1 and C cannot produce any current in R_2 and L; that transfer function would be zero. To find the form of the natural response of the inductor voltage, for example, the voltage source must be installed in the R_2L network. A case of this type can often be recognized by an inspection of the network before a source is installed; but if it is not, then a transfer function equal to zero will be obtained.

When $\mathbf{H}(\mathbf{s}) = 0$, we obtain no information about the frequencies characterizing the natural response, and a more suitable location for the source must be used.

EXAMPLE **A8.2**

For the source-free circuit of Fig. A8.8, determine expressions for i_1 and i_2 for $t > 0$, given the initial conditions $i_1(0) = i_2(0) = 11$ A.

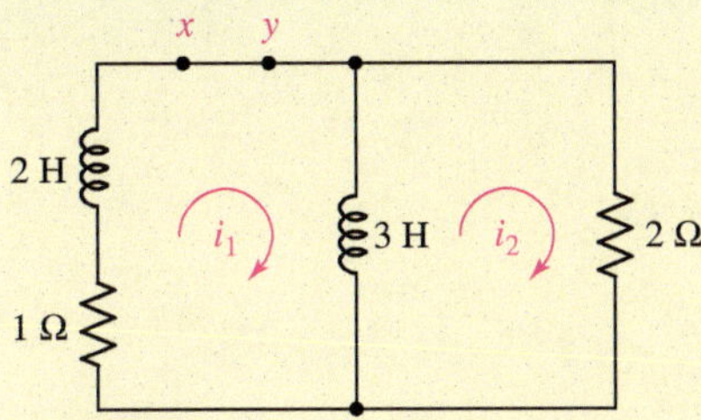

FIGURE A8.8 A circuit for which the natural responses i_1 and i_2 are desired.

Let us install a voltage source $\mathbf{V}_s$ between points x and y and find the transfer function $\mathbf{H}(\mathbf{s}) = \mathbf{I}_1(\mathbf{s})/\mathbf{V}_s$, which also happens to be the input admittance seen by the voltage source. We have

$$\mathbf{I}_1(\mathbf{s}) = \frac{\mathbf{V}_s}{2\mathbf{s} + 1 + 6\mathbf{s}/(3\mathbf{s} + 2)} = \frac{(3\mathbf{s} + 2)\mathbf{V}_s}{6\mathbf{s}^2 + 13\mathbf{s} + 2}$$

or

$$\mathbf{H}(\mathbf{s}) = \frac{\mathbf{I}_1(\mathbf{s})}{\mathbf{V}_s} = \frac{\frac{1}{2}\left(\mathbf{s} + \frac{2}{3}\right)}{(\mathbf{s} + 2)\left(\mathbf{s} + \frac{1}{6}\right)}$$

From recent experience, we know at a glance that i_1 must be of the form

$$i_1(t) = Ae^{-2t} + Be^{-t/6}$$

The solution is completed by using the given initial conditions to establish the values of A and B. Since $i_1(0)$ is given as 11 amperes,

$$11 = A + B$$

The necessary additional equation is obtained by writing the KVL equation around the perimeter of our circuit:

$$1i_1 + 2\frac{di_1}{dt} + 2i_2 = 0$$

and solving for the derivative:

$$\left.\frac{di_1}{dt}\right|_{t=0} = -\frac{1}{2}[2i_2(0) + 1i_1(0)] = -\frac{22 + 11}{2} = -2A - \frac{1}{6}B$$

Thus, $A = 8$ and $B = 3$, and so the desired solution is

$$i_1(t) = 8e^{-2t} + 3e^{-t/6} \quad \text{amperes}$$

(Continued on next page)

The natural frequencies constituting i_2 are the same as those of i_1, and a similar procedure used to evaluate the arbitrary constants leads to

$$i_2(t) = 12e^{-2t} - e^{-t/6} \quad \text{amperes}$$

PRACTICE

A8.2 If a current source $i_1(t) = u(t)$ A is present at a-b in Fig. A8.9 with the arrow entering a, find $\mathbf{H}(\mathbf{s}) = \mathbf{V}_{cd}/\mathbf{I}_1$, and specify the natural frequencies present in $v_{cd}(t)$.

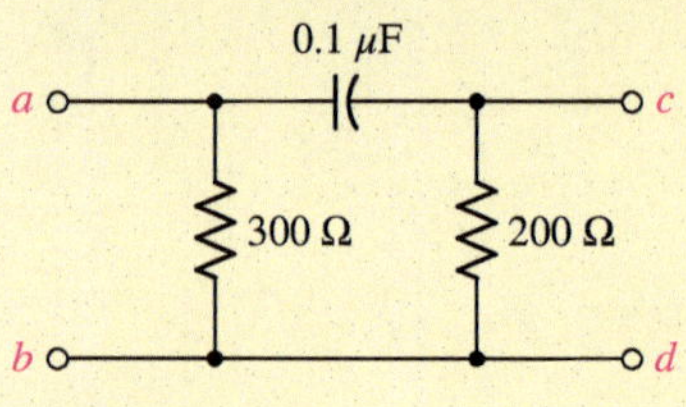

FIGURE A8.9

Ans: 120s/(s + 20,000) Ω, −20,000 s^{-1}.

C

D

E

F

J

K

L

M

N

O

P

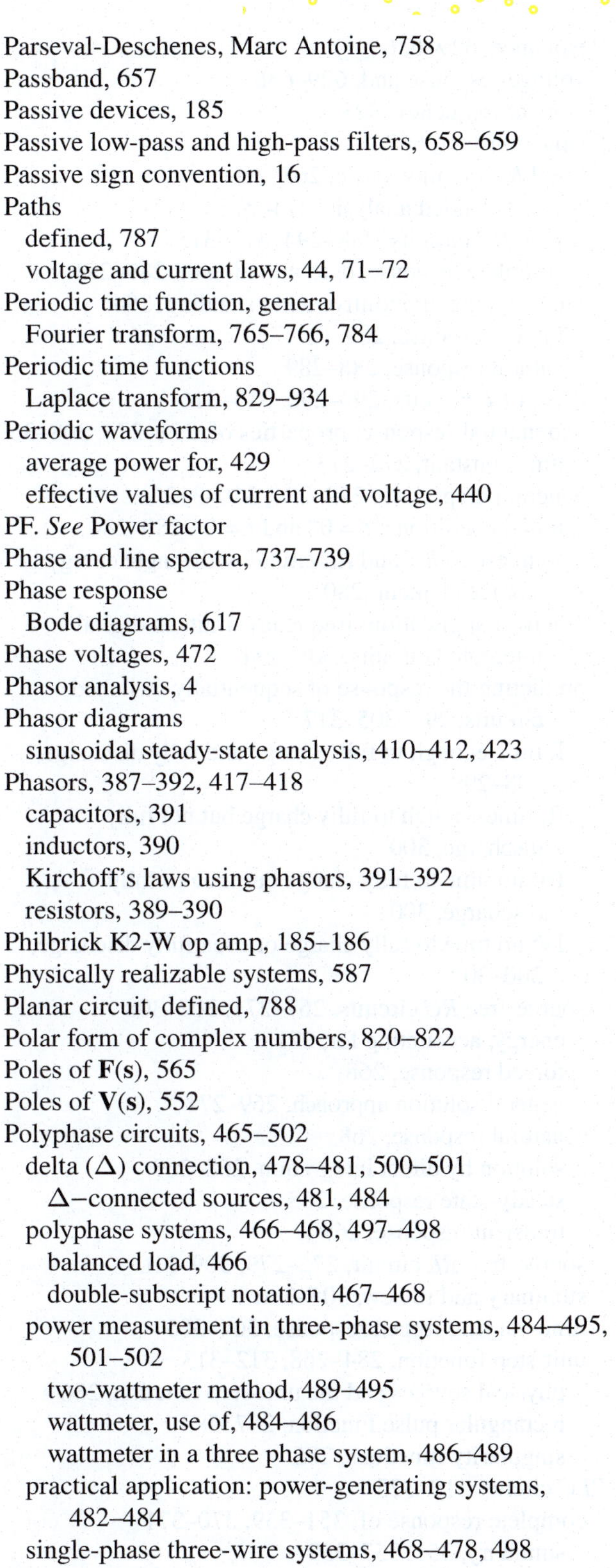

Q

R

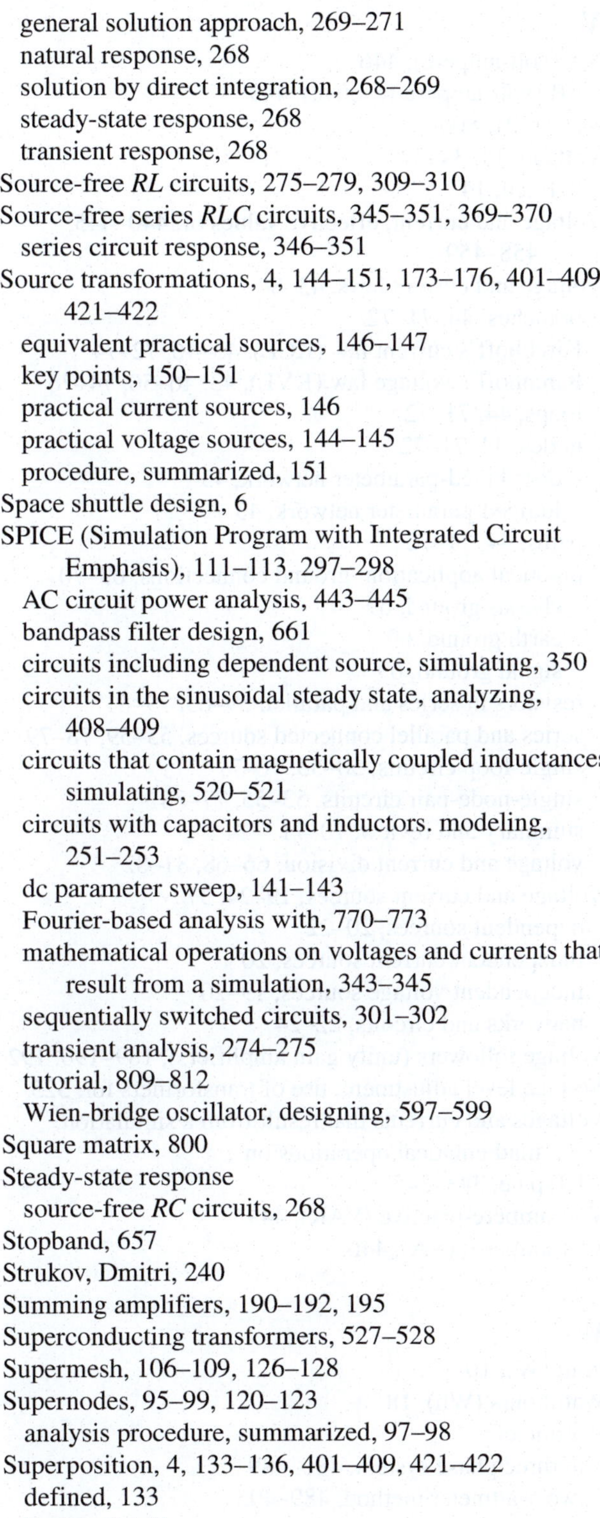

T

U

V

W

Y

Z

The Resistor Color Code

Band color	Black	Brown	Red	Orange	Yellow	Green	Blue	Violet	Gray	White
Numeric value	0	1	2	3	4	5	6	7	8	9

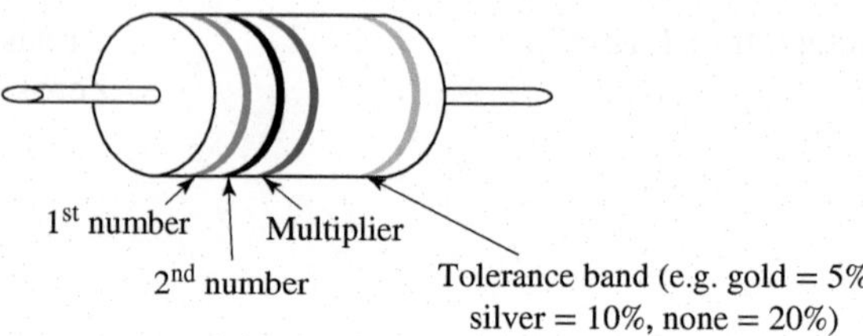

1. Write down the numeric value corresponding to the first band on the left.
2. Write down the numeric value corresponding to the second band from the left.
3. Write down the number of zeros indicated by the multiplier band, which represents a power of 10 (black = no extra zeros, brown = 1 zero, etc.). A gold multiplier band indicates that the decimal is shifted one place to the left; a silver multiplier band indicates that the decimal is shifted two places to the left.
4. The tolerance band represents the precision. So, for example, we would not be surprised to find a 100 Ω 5 percent tolerance resistor that measures anywhere in the range of 95 to 105 Ω.

Example

Red Red Orange Gold	= 22,000	or 22×10^3	= 22 KΩ, 5% tolerance
Blue Grary Gold	= 6.8	or 68×10^{-1}	= 6.8 Ω, 20% tolerance

Standard 5 Percent (E24 Preferred Number Derived) Tolerance Resistor Values

1.0	1.1	1.2	1.3	1.5	1.6	1.8	2.0	2.2	2.4	2.7	3.0	3.3	3.6	3.9	4.3	4.7	5.1	5.6	6.2	6.8	7.5	8.2	9.1	Ω
10.	11.	12.	13.	15.	16.	18.	20.	22.	24.	27.	30.	33.	36.	39.	43.	47.	51.	56.	62.	68.	75.	82.	91.	Ω
100	110	120	130	150	160	180	200	220	240	270	300	330	360	390	430	470	510	560	620	680	750	820	910	Ω
1.0	1.1	1.2	1.3	1.5	1.6	1.8	2.0	2.2	2.4	2.7	3.0	3.3	3.6	3.9	4.3	4.7	5.1	5.6	6.2	6.8	7.5	8.2	9.1	KΩ
10.	11.	12.	13.	15.	16.	18.	20.	22.	24.	27.	30.	33.	36.	39.	43.	47.	51.	56.	62.	68.	75.	82.	91.	KΩ
100	110	120	130	150	160	180	200	220	240	270	300	330	360	390	430	470	510	560	620	680	750	820	910	KΩ
1.0	1.1	1.2	1.3	1.5	1.6	1.8	2.0	2.2	2.4	2.7	3.0	3.3	3.6	3.9	4.3	4.7	5.1	5.6	6.2	6.8	7.5	8.2	9.1	MΩ

TABLE 14.1 Laplace Transform Pairs

$f(t) = \mathscr{L}^{-1}\{F(s)\}$	$F(s) = \mathscr{L}\{f(t)\}$	$f(t) = \mathscr{L}^{-1}\{F(s)\}$	$F(s) = \mathscr{L}\{f(t)\}$
$\delta(t)$	1	$\frac{1}{\beta - \alpha}(e^{-\alpha t} - e^{-\beta t})u(t)$	$\frac{1}{(s + \alpha)(s + \beta)}$
$u(t)$	$\frac{1}{s}$	$\sin \omega t \, u(t)$	$\frac{\omega}{s^2 + \omega^2}$
$tu(t)$	$\frac{1}{s^2}$	$\cos \omega t \, u(t)$	$\frac{s}{s^2 + \omega^2}$
$\frac{t^{n-1}}{(n-1)!} u(t), n = 1, 2, \ldots$	$\frac{1}{s^n}$	$\sin (\omega t + \theta) \, u(t)$	$\frac{s \sin \theta + \omega \cos \theta}{s^2 + \omega^2}$
$e^{-\alpha t}u(t)$	$\frac{1}{s + \alpha}$	$\cos (\omega t + \theta) \, u(t)$	$\frac{s \cos \theta - \omega \sin \theta}{s^2 + \omega^2}$
$te^{-\alpha t}u(t)$	$\frac{1}{(s + \alpha)^2}$	$e^{-\alpha t} \sin \omega t \, u(t)$	$\frac{\omega}{(s + \alpha)^2 + \omega^2}$
$\frac{t^{n-1}}{(n-1)!} e^{-\alpha t}u(t), n = 1, 2, \ldots$	$\frac{1}{(s + \alpha)^n}$	$e^{-\alpha t} \cos \omega t \, u(t)$	$\frac{s + \alpha}{(s + \alpha)^2 + \omega^2}$

TABLE 6.1 Summary of Basic Op Amp Circuits

Name	Circuit Schematic	Input-Output Relation
Inverting Amplifier		$v_{out} = -\frac{R_f}{R_1} v_{in}$
Noninverting Amplifier		$v_{out} = \left(1 + \frac{R_f}{R_1}\right) v_{in}$
Voltage Follower (also known as a Unity Gain Amplifier)		$v_{out} = v_{in}$
Summing Amplifier		$v_{out} = -\frac{R_f}{R}(v_1 + v_2 + v_3)$
Difference Amplifier		$v_{out} = v_2 - v_1$

A Short Table of Integrals

$$\int \sin^2 ax\, dx = \frac{x}{2} - \frac{\sin 2ax}{4a}$$

$$\int \cos^2 ax\, dx = \frac{x}{2} + \frac{\sin 2ax}{4a}$$

$$\int x \sin ax\, dx = \frac{1}{a^2}(\sin ax - ax \cos ax)$$

$$\int x^2 \sin ax\, dx = \frac{1}{a^3}(2ax \sin ax + 2 \cos ax - a^2x^2 \cos ax)$$

$$\int x \cos ax\, dx = \frac{1}{a^2}(\cos ax + ax \sin ax)$$

$$\int x^2 \cos ax\, dx = \frac{1}{a^3}(2ax \cos ax - 2 \sin ax + a^2x^2 \sin ax)$$

$$\int \sin ax \sin bx\, dx = \frac{\sin(a-b)x}{2(a-b)} - \frac{\sin(a+b)x}{2(a+b)};\ a^2 \neq b^2$$

$$\int \sin ax \cos bx\, dx = -\frac{\cos(a-b)x}{2(a-b)} - \frac{\cos(a+b)x}{2(a+b)};\ a^2 \neq b^2$$

$$\int \cos ax \cos bx\, dx = \frac{\sin(a-b)x}{2(a-b)} + \frac{\sin(a+b)x}{2(a+b)};\ a^2 \neq b^2$$

$$\int x\, e^{ax}\, dx = \frac{e^{ax}}{a^2}(ax - 1)$$

$$\int x^2 e^{ax}\, dx = \frac{e^{ax}}{a^3}(a^2x^2 - 2ax + 2)$$

$$\int e^{ax} \sin bx\, dx = \frac{e^{ax}}{a^2 + b^2}(a \sin bx - b \cos bx)$$

$$\int e^{ax} \cos bx\, dx = \frac{e^{ax}}{a^2 + b^2}(a \cos bx + b \sin bx)$$

$$\int \frac{dx}{a^2 + x^2} = \frac{1}{a}\tan^{-1}\frac{x}{a}$$

$$\int_0^\infty \frac{\sin ax}{x}\, dx = \begin{cases} \frac{1}{2}\pi & a > 0 \\ 0 & a = 0 \\ -\frac{1}{2}\pi & a < 0 \end{cases}$$

$$\int_0^\pi \sin^2 x\, dx = \int_0^\pi \cos^2 x\, dx = \frac{\pi}{2}$$

$$\int_0^\pi \sin mx \sin nx\, dx = \int_0^\pi \cos mx \cos nx\, dx = 0;\ m \neq n,\ m \text{ and } n \text{ integers}$$

$$\int_0^\pi \sin mx \cos nx\, dx = \begin{cases} 0 & m - n \text{ even} \\ \frac{2m}{m^2 - n^2} & m - n \text{ odd} \end{cases}$$

A Short Table of Trigonometric Identities

$$\sin(\alpha \pm \beta) = \sin\alpha\cos\beta \pm \cos\alpha\sin\beta$$

$$\cos(\alpha \pm \beta) = \cos\alpha\cos\beta \mp \sin\alpha\sin\beta$$

$$\cos(\alpha \pm 90°) = \mp \sin\alpha$$

$$\sin(\alpha \pm 90°) = \pm \cos\alpha$$

$$\cos\alpha\cos\beta = \frac{1}{2}\cos(\alpha+\beta) + \frac{1}{2}\cos(\alpha-\beta)$$

$$\sin\alpha\sin\beta = \frac{1}{2}\cos(\alpha-\beta) - \frac{1}{2}\cos(\alpha+\beta)$$

$$\sin\alpha\cos\beta = \frac{1}{2}\sin(\alpha+\beta) + \frac{1}{2}\sin(\alpha-\beta)$$

$$\sin 2\alpha = 2\sin\alpha\cos\alpha$$

$$\cos 2\alpha = 2\cos^2\alpha - 1 = 1 - 2\sin^2\alpha = \cos^2\alpha - \sin^2\alpha$$

$$\sin^2\alpha = \frac{1}{2}(1 - \cos 2\alpha)$$

$$\cos^2\alpha = \frac{1}{2}(1 + \cos 2\alpha)$$

$$\sin\alpha = \frac{e^{j\alpha} - e^{-j\alpha}}{j2}$$

$$\cos\alpha = \frac{e^{j\alpha} + e^{-j\alpha}}{2}$$

$$e^{\pm j\alpha} = \cos\alpha \pm j\sin\alpha$$

$$A\cos\alpha + B\sin\alpha = \sqrt{A^2 + B^2}\cos\left(\alpha + \tan^{-1}\frac{-B}{A}\right)$$